Synthesis of Feedback Systems

Synthesis of Feedback Systems

ISAAC M. HOROWITZ

Senior Scientist
Guidance and Controls Division
Hughes Aircraft Company
Culver City, California

Academic Press
New York and London

ACADEMIC PRESS INC.
111 Fifth Avenue, New York 3, New York

United Kingdom Edition published by
ACADEMIC PRESS INC. (LONDON) LTD.
Berkeley Square House, London W.1

LIBRARY OF CONGRESS CATALOG CARD NUMBER: 63-12033

PRINTED IN THE UNITED STATES OF AMERICA

Preface

What are the reasons for deliberately introducing feedback into a system? It is generally agreed that the following are among the principal reasons: inadequate knowledge of system parameter values, parameter variations, disturbances, system nonlinearities, and inherently unstable system elements. The sensitivity of the system to these factors can be drastically reduced by means of feedback. However, the mere assumption of a feedback configuration does not guarantee that the required insensitivity has been attained; or there may be more feedback than is necessary and therefore a greater price is being paid than the specifications inherently require. It is manifest that the amount of feedback that is applied should be related to the extent of parameter variations, etc., and to the desired system insensitivity to the latter.

In a scientific study of the subject, it is essential to find the basic system functions that determine the benefits of feedback. These functions should be related to the important system constraints and should distinguish clearly between the essential and the trivial. In addition, it is important to determine the fundamental capabilities of these functions under realistic constraints, and the optimum values of these functions. Only then is the subject a science rather than an art. For it to be an engineering science, it is furthermore necessary to have practical design procedures which enable the designer to converge as closely as desired to the optimum realizable functions.

The foundation of such an engineering science of feedback was established by H. W. Bode, who related the benefits of feedback to his sensitivity function and showed how the latter was determined by the system transmission, leakage transmission, and the loop transmission functions. Bode intensively investigated zero leakage, single loop systems whose feedback properties are completely determined by a single loop transmission function $L(s)$. Bode definitively established the capabilities and limitations of the minimum phase $L(s)$ function. Furthermore, his work readily permits extension of the results to other kinds of constraints on the $L(s)$ function. This book attempts to follow in the tradition of Bode and to build further on his scientific foundation.

In control systems it is realistic to use the concept of the "plant," which denotes a fixed assembly of equipment with the inherent power capacity for satisfying the specifications. However, one or more of the factors of plant parameter variations, bounded ignorance of the plant, plant disturbances, etc., preclude the attainment of the objectives without feedback. A useful classification of problems is based on the number of independent plants that are available, and the number of plant access points, i.e., the number of distinct plant variables which may be measured and used for feedback purposes. The simplest case is the single plant in which no variables are available, but only the difference between system input and output can be measured. In this book this is called a single-degree-of-freedom system (Chapter 5), because the designer has only one function which he can shape for the achievement of the system specifications. Obviously he may have to compromise between the various desiderata. Much of the linear feedback control literature deals with this class.

Next in the order of complexity is the two-degree-of-freedom system (Chapter 6), wherein the system output and input are separately measurable. Two functions are now available to the designer, and he can use these to independently realize the system filter and feedback properties. However, the resulting system sensitivity to feedback transducer noise is then uncontrollable. Some control over the latter is possible when internal plant variables are measurable, and are used for feedback purposes in multiple-loop systems (Sections 8.2–8.7). There is a remarkable difference in feedback properties and relations when two or more independently controllable plants are available (Sections 8.8 and 8.9). In such systems, conventional taboos of single plant systems may be drastically violated.

Chapter 10 is devoted to systems with a multiplicity of inputs and outputs. The problem is again one of achieving objectives despite plant ignorance, disturbances, etc., and procedures for these purposes are presented. The stability problem is quite complicated because of the existence of many loop transmissions, each with independently varying parameters. However, there is available a rational, straightforward design procedure for shaping these loop transmissions so that the system is stable over the bounds of parameter variation. Chapter 11 is devoted to sampled-data systems with particular attention to the feedback problem (design for insensitivity to parameter variations, disturbances, etc.), and to the limitations in the feedback capabilities due to sampling.

In the various classes of problems described above, a serious attempt is made to present practical engineering design procedures for realizing typical design specifications. These procedures are such as to reveal the price that is being paid for the benefits of feedback and to permit intelligent compromises when such are necessary. These procedures are in all cases illustrated by detailed numerical design examples. It was not possible to check all the numerical calculations, and it will be appreciated if any errors are brought to the author's attention.

In all of the above, the benefits of feedback are determined by loop transmission functions. The number of access points and the number of plants determine the number of functions and how they may be combined. There remains, however, the important problem of the inherent capabilities and limitations of any loop transmission function. Bode's work in this subject is adapted to feedback control systems (Chapter 7) and is further applied to loop transmissions which have one or more of the following properties: conditional stability, nonminimum phase, open-loop instability, minor positive feedback loops. These results are of fundamental importance, for they determine ultimate feedback capabilities under various constraints, and they reveal what is fundamental and what is trivial.

In the matter of parameter variations, it was convenient in all of the above to assume that the plant parameters varied very slowly. The problem of the *rate* of parameter variations is examined in Sections 8.15–8.17, and it is shown how the previous results may be adapted to this problem. The application of linear feedback techniques to systems with nonlinear dynamic plants is treated in Sections 8.19 and 8.20, and design procedures are included.

Throughout the above, primary attention is paid to the feedback problem and to the realization of given specifications. The origin of the specifications is treated in Chapter 9, where three distinct problems are examined:

1. When system design is on the basis of a known typical input signal or disturbance
2. When only the bounds on the magnitude and slope of the inputs are known
3. When there are ensembles of input, disturbance, and feedback transducer noise signals and when design is based on minimizing some mean ensemble error function.

Some approximation problems which continually arise in both practical and theoretical work are considered in Chapter 12. Among these are time domain synthesis, inverse transformation, numerical convolution, and numerical Fourier analysis.

It will be evident to those familiar with the literature that much of this book is based upon the published and unpublished work of the author. It is believed that the following portions unambiguously belong in this category: Sections 4.7, 5.18, 5.19, Chapter 6, Sections 7.12–7.15, Chapter 8 (excluding Section 8.8), Chapter 10, Sections 11.6, 11.14, 11.15, 11.17. A serious attempt was made to acknowledge properly original work of any significance, but undoubtedly there have been omissions. I will be grateful if these are brought to my attention.

The book as a whole is primarily intended for the practicing engineer and the graduate student. However, it is felt that the first six chapters may be used in a first course in feedback theory for senior undergraduate students, providing they are familiar with the Laplace transform. (The instructor will then undoubtedly want to supplement the presentation of the Nyquist criterion in Chapter 4.) The author fully agrees with Professor E. A. Guillemin that treatment of a subject in depth should not be postponed to graduate school.

Much of the research embodied in this book was done during 1959-1961 at the Hughes Research Laboratories. I am indebted to Mr. Al Aronowitz who read and constructively criticized Chapter 2, to Dr. Douglas Anderson for his advice in mathematical niceties, to Dr. Aaron Ksienski for his good counsel and encouragement, and to Dr. Jack Sklansky for reviewing Chapter 11. I am especially grateful to Mr. R. L. Peterson of Hughes Aircraft Company for many provocative and illuminating discussions. I was privileged to present much of Chapters 8 and 10 in a series of lectures excellently organized by Professor C. J. D. M. Verhagen, chairman of the Cooperation Centre for Instrumentation and Control of the Technological Institute of Delft. My wife typed the entire manuscript and the subsequent two or three revisions of most of it, and she also did much of the proof-reading. This book could not have been written without her patient support and encouragement.

Culver City, California
December, 1962

Isaac M. Horowitz

Contents

CHAPTER ONE

Introduction to Feedback Theory

§ 1.1 Formulation of the Feedback Problem

The principal concern of feedback theory is with problems which permit the following formulation:

(*A*) There is an objective or, more generally, a set of objectives. The objectives can be comparatively simple ones, such as to maintain a terminal voltage or a shaft velocity within narrow tolerances, or they can be more complex, such as to operate a vast and complicated process at maximum efficiency. It is assumed that it is possible to express the objectives mathematically by means of a set of variables R consisting of r_1, r_2, ..., r_n in Fig. 1.1-1.

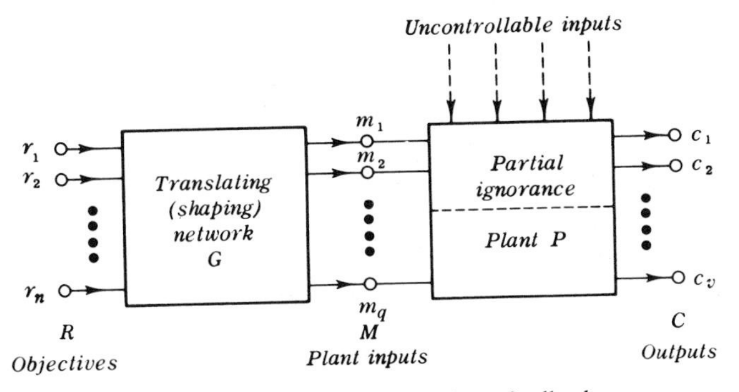

FIG. 1.1-1. Control system without feedback.

(*B*) Equipment is needed to achieve the objectives in (*A*)—chemical equipment if the objectives are chemical products, as in an oil refinery; motors, if the objectives are to direct material to specific positions or at specified velocities. The array of equipment (presumably chosen by the specialists in the particular field) is denoted as the *plant*, with its set of inputs M, consisting of m_1, m_2, ..., m_q (see Fig. 1.1-1). The apparent straightforward solution to the problem would be to insert between R and M a network G which translates (as in Fig. 1.1-1) the set of objectives into the language appropriate to the plant.

(C) The use of G would be satisfactory, but for two problems which very often exist. These problems are:

Partial ignorance of the plant. If there is incomplete knowledge of the plant, then it is not known precisely what inputs to the plant are required in order to achieve the desired outputs. Such incomplete knowledge may be due to a number of reasons. (a) Tolerances in manufacture. There are always tolerances in the values of the parameters of devices contained in the plant, for example, the tolerances in the values of vacuum-tube and transistor parameters. It is theoretically possible to measure the values of the parameters of the specific elements that are used, but this may be inconvenient. If these tolerances are larger than the tolerances permitted in the final results, then input programming, as in Fig. 1.1-1, is ruled out. (b) Parameter variation. Device properties often change with time, due to use, aging, and the accummulated effect of external influences. (c) Ignorance of process. There is often sheer ignorance of some of the laws governing the plant operations, as in many chemical and physical processes, or as in the field of economics where the plant represents an economic system.[1]

Uncontrollable inputs to the plant. In Fig. 1.1-1 the plant has q inputs subject to the designer's control. In addition, there are usually other plant inputs over which the designer has incomplete control or knowledge, for example, the changes in the power demands of consumers in the voltage regulation problem, a meteor striking a satellite in attitude control, the wind torque on a rotating antenna in scanning, the rolling of a ship in platform stabilization, man's inability to control the climate in the problem of planning a nation's food supply. In short, the plant operation is influenced by the environment, and often there is insufficient knowledge of the environment and its effect to permit plant input programming (as in Fig. 1.1-1) to cancel the effect of these uncontrollable inputs.

The two problems listed in (C) can be placed under the single heading of ignorance. For if the uncontrollable inputs and parameter variations are known in time and amount, then their effect on the plant may be canceled by suitably programming the controllable inputs. Such a course is impossible when there is some ignorance of the parameter variations and the uncontrollable inputs.

(D) The problem is to achieve the stated objectives with the given plant, within the permitted tolerances, despite partial ignorance of the system. The specified tolerances may permit inaccuracies of $x\%$, and the "ignorance" may be such that the very best straightforward (no feedback) design of Fig. 1.1-1 will result in maximum inaccuracies of $Mx\%$ with $M > 1$. Nevertheless, despite such ignorance, the desired accuracies may be achievable by means of feedback. The basic idea is to use sensors and comparators to measure the outputs and compare them with the objectives. The knowledge of the differences between

[1] Arnold Tustin, "The Mechanism of Economic Systems" Harvard University Press, Cambridge, Mass. 1953.

the objectives and actual outcomes is used to drive the plant until the differences are satisfactorily small. Feedback theory is concerned with processing this feedback information in such a manner as to secure the desired objectives despite the ignorance.

An example that may be used to illustrate the above is that of a man driving a car to some destination. Without the use of the eyes as sensor and comparator, the result would be disastrous. But the eyes are not really needed to drive the car—only the arms and legs are needed. In order to achieve the objective without feedback, almost perfect knowledge of the plant would be required: of the car, the roads, and the lengths of the blocks to permit the turns at the right points. Almost perfect knowledge would be necessary of the uncontrollable external inputs: the movements of the pedestrians and of other cars, temporary obstacles used in road repairs, the traffic lights, etc. The eyes as comparator and the brain as processor permit precise attainment of the objective despite fantastic ignorance. A little thought suggests that probably all life—both plant and animal—is possible only because of feedback.

§ 1.2 Classification of Feedback Problems

We have emphasized in the above that feedback is a means of coping with ignorance. This does not mean that all problems involving ignorance are the same, and that a single technique is available for any adaptive process. There are, in fact, vast differences in the techniques required for different problems. A convenient way to classify such problems is on the basis of: 1. Nature of ignorance. 2. Nature of objectives. 3. Nature of constraints.

(1) Nature of Ignorance

The ignorance can be of a very simple sort. For example, consider a plant in which it is known that the plant parameters vary very slowly in comparison with the rate at which the signals in the plant vary, as in a system in which the applied signals are step functions, the plant transient response is over after a few seconds, and the plant parameters change at the rate of 1% per hour. Another simple kind of ignorance is when the plant parameters do not vary at all, but it is not known precisely what they are. For example, suppose a plant transfer function[1] is $P = 1/(as + b)$, and it is known that a and b must lie in the region given in Fig. 1.2-1; there is thus a region in two-dimensional space that is the measure

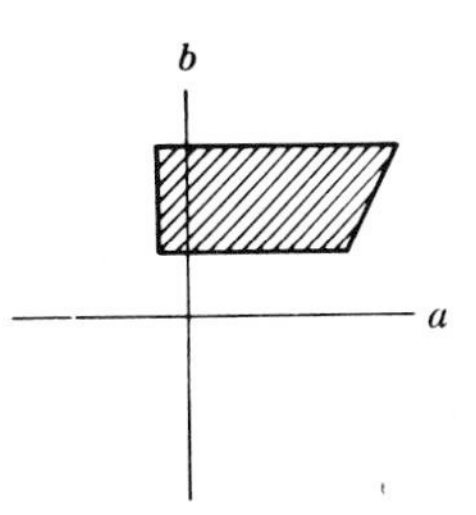

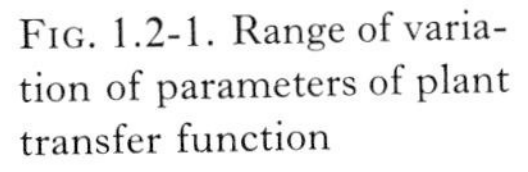

FIG. 1.2-1. Range of variation of parameters of plant transfer function $P(s) = 1/(as + b)$.

[1] The transfer function of a device (which is linear and time invariant) is the Laplace transform of the output due to a unit impulse input; s is the complex variable. The transfer function may also be obtained by finding the forced component of the output

of our ignorance. More generally, a plant such as the above may have a transfer function with n parameters lying in a region in n-dimensional space.

It is possible to formulate more complicated and profound ignorance—for example, the plant saturation levels may be unknown; or the complexity of the equations required to characterize the plant is unknown. It may not be known whether the plant is linear. In a plant with multiple inputs and outputs, there may be substantial ignorance of the relations between the inputs and outputs, or even of the number of inputs and outputs; or the "plant" may include within it an unknown intelligence with a strategy of its own, and a great deal of learning may perhaps be necessary even to deduce this. There may even be ignorance of the "language" that must be used to address the plant.

Another form of ignorance is ignorance of the exact objectives. There may be false input signals mixed with the true input signals and we may have only vague information of the nature of the true signals and none of the false signals. Or there may be some ignorance even of the nature of the objectives, and learning is required (by means of feedback to higher objectives) to deduce the local objectives.

With regard to the uncontrollable inputs, the input points may be known or unknown, the form of the disturbances may be well known except for their time of occurrence, and the converse may be true. The general nature of the inputs may vary in an unknown manner. The external inputs may be directed by an unknown intelligence. In short, the ignorance may be of a relatively simple type, or it may be extremely profound.

(2) Nature of Objectives

The complexity of the required feedback mechanism is strongly influenced by the nature of the objectives. A simple objective is the realization of a relationship between the system input and output that can be described by means of a transfer function, and which is to remain within prescribed tolerances, despite plant parameter variation. Another simple objective is to achieve a prescribed attenuation of uncontrollable external inputs whose form is known. On the other hand, the operation of some plant (whose internal workings are not well known) to produce a given output at minimum cost is a much more sophisticated objective. One can easily formulate very subtle objectives by considering such fields as economics, language learning, games, etc.

(3) Nature of Constraints

When feedback is used to cope with ignorance, the problem is basically one of deciding: (a) what sources of information to use for feedback

due to an input $e^{\sigma t}$. This output component will be in the form $T(\sigma)e^{\sigma t}$. The transfer function is $T(s)$. The Laplace transform is related to the steady-state sinusoidal response as follows: If an input $\sin(\omega t + \theta)$ leads to a steady-state output $A(\omega)\sin[\omega t + \theta + \alpha(\omega)]$, then the function $[A(\omega)\cos\alpha(\omega) + jA(\omega)\sin\alpha(\omega)]$ can always [see Eq. (7.2,2)] be put in the form $R(\omega^2) + j\omega X(\omega^2)$ and the transfer function is obtained by replacing ω^2 by $-s^2$, and $j\omega$ by s.

purposes, (b) how to process this information and apply it to the plant via the controllable plant input variables. By constraints we mean the kinds of elements and devices that may be used for the above tasks. For example, one common constraint is that the elements are such that their variables are related by linear differential equations with constant coefficients. In other cases, difference equations with constant coefficients are also admissible. In more advanced systems, the equations may have time varying coefficients. Finally, elements with nonlinear relations between the variables might be permitted, and it is possible to create hierarchies of various types of nonlinearities. The constraints can be authentic constraints dictated by the problem practicalities. Often they are arbitrarily imposed by the theoretical investigator, in order to intelligently classify feedback theory.

The three axes of Ignorance, Objectives and Constraints suggest a three dimensional region into which the problems of adaptive theory can be fitted. It is reasonable to expect that if the constraints are very severe, then only very simple kinds of ignorance and objectives can be handled, and that the more severe the ignorance and/or the objectives, the less severe must be the constraints, if the objectives are to be achievable.

Let us consider for a moment what ought to constitute feedback theory. Ideally, a genuine theory of feedback is one that describes the fundamental properties of systems that belong in a given region of objectives, ignorance and constraints, no matter whether the specific system is economic, chemical, mechanical, social or any combination of these. All systems, whatever their physical character, that have the same nature of objectives, ignorance, and constraints would be treated in the same manner. It should be possible to formulate plant ignorance and environment ignorance quantitatively. It should be possible to quantitatively formulate the processing of the feedback data. It should be possible to specify quantitatively the cost of the learning or adaptation. The price that must be paid for overcoming ignorance must be in some measure and units that are common to such diverse systems as economic, chemical, and game playing. A genuine feedback theory is one that can definitively describe the kinds of ignorance and objectives that can be handled by different levels of constraints, and the optimum procedures for realizing the objectives.

At the present time, the region defined by the three axes of Ignorance, Objectives, and Constraints is almost completely one vast question mark. For only a very small volume in this region does there exist what may be considered a quantitative feedback theory, although many problems remain unsolved even here. While a great deal of very significant and useful work has been done on problems outside the small volume, the results do not as yet constitute a feedback theory in the sense previously described.

The small region for which a quantitative theory exists is characterized by the following: (1) Nature of Ignorance. The plant is characterized by linear differential or difference equations with parameters which lie within certain bounds, or whose values can be statistically described. To some extent, time varying parameters and nonlinear plant variable relations can also be handled. The saturating levels of the plant are known, so that the plant is not expected to do what it is physically incapable of doing. Similarly, equipment capacities and safety requirements are known, and the applied inputs and system objectives are such that these are not exceeded. The influences of the environment on the plant are known to some extent—at least the bounds are known. (2) Objectives. The objectives must be comparatively simple—relationships between inputs and outputs that can be described by linear differential or difference equations with coefficients that must lie within given bounds; or the response to a specific input or classes of inputs must not deviate more than some prescribed amount from a standard response. The effect of environmental influences on the output must not exceed prescribed values. (3) Constraints. The feedback processing must be done by elements which are characterized by linear, constant coefficient, differential, or difference equations.

We shall see that the formulation of a genuine theory of feedback for the above region has been greatly facilitated by the skillful manipulation of the problems into a form permitting the use of mathematics of linear time-invariant systems. The problems outside this region generally cannot be so manipulated, and despite much brilliant work, the results are far from constituting a genuine theory of feedback, in the sense previously described.

§ 1.3 Coverage of the Book

This book presents the feedback theory which exists for the region outlined in Section 1.2. This is the region for which a quantitative theory of feedback exists, although many problems here too are still unsolved. The balance of Chapter 1 is devoted to the problems of plant identification, representation, and analysis. Clearly, some general, universal method for system representation and analysis is required in order that the theory may be applied to diverse fields. The required technique is presented in Chapter 1. Chapter 2 introduces feedback as a point of view which is mandatory for the synthesis of systems for overcoming ignorance, and is also very useful for analysis. Thus, any system may be examined from the feedback point of view, and such a point of view often considerably simplifies the analysis of fairly complicated systems. The fundamental characterization of linear feedback systems is presented, some general theorems are proven, and they are used in the analysis of complex systems. The main purpose of Chapter 2 is to train the reader in the application of the feedback point of view to any system.

The principal useful properties of feedback are introduced in an elementary manner in Chapter 3. This permits us to consider the design of specific feedback systems. However, it is soon seen that there is always a problem of stability in any practical system, and it is aggravated by plant parameter variation and ignorance of the values of the plant parameters. The emphasis in Chapter 4 is therefore on the problem of stability in the face of plant parameter variations or ignorance. Chapter 5 is devoted to feedback control system synthesis techniques for configurations with only one degree of freedom (in the sense that essentially only one function is subject to the designer's control). The shortcomings inherent in such structures are noted. Chapter 6 presents the class of two-degree-of-freedom feedback configurations (two independent functions available to the designer) and synthesis procedures appropriate for such configurations. The cost of overcoming ignorance becomes apparent, and we then turn to methods of reducing the cost. Chapter 7 is concerned with the optimum cost of overcoming ignorance, for systems satisfying various constraints. Exotic systems that have been suggested as a means of reducing the cost of feedback are examined, and definitive conclusions are reached.

In Chapter 8 we consider a number of advanced topics in linear feedback theory. Some of the constraints applied in previous chapters are lifted, and the resulting economies in the cost of feedback are found. The extent to which linear feedback theory can be applied to time varying and nonlinear plants is also examined.

Up to this point, the characteristics of the desired input-output relation and disturbance rejection have been assumed to be known in advance. In Chapter 9 we pause to examine some of the problems involved in formulating these system objectives. The simplest case is when the nature of the useful system input or disturbance input is well known. The more difficult problem is when these inputs can be described statistically only, and the useful input is contaminated with noise that can only be described statistically.

So far, systems with a single input-output pair have been considered. Chapter 10 is devoted to the overcoming of ignorance in plants with a multiplicity of inputs and outputs. The general stability problem involves a multiplicity of loop transmissions, each with independently varying parameters, and therefore appears to be exceedingly formidable. However, it finally turns out to be amenable to a fairly straightforward and logical design procedure. Chapter 11 is devoted to systems describable by difference equations. We explore the limitations of these sampled-data systems in coping with plant ignorance, parameter variation, and disturbance rejection. Finally, in Chapter 12 we consider a variety of approximation problems that invariably beset the practical designer. Thus, it has been assumed in all the foregoing that the design objectives have been *a priori* available in the language most appropriate for the detailed design techniques. This is rarely so. Chapter 12 considers how to translate specifications from their typical original formulation, to the language appropriate for detailed design.

§ 1.4 The Problem of Plant Identification and Representation

The problem of feedback system synthesis is that of designing a signal-processing network around a plant of which there is some ignorance, operating in an environment of which there is some ignorance, in order to overcome the ignorance, and thereby achieve results with accuracies unobtainable without feedback. The main concern of feedback is therefore with processing of signals. An immediate basic problem is the problem of system representation. The plant may be a very complicated process, such as an oil refinery, or a missile, containing within it a very large number of elements. Among the latter there may be some which are difficult to analyze, and in fact much of the data may be empirical. What information about the plant does the feedback engineer need to enable him to build the required signal processing network around the plant? Must he become an expert in the particular field before he can approach the problem of feedback control of the plant?

The kind of information needed by the feedback engineer is basically the same, no matter what kind of plant is involved. A good first step is to list the plant variables (or functions) under each of the following headings:

The plant input variables. These are controlled by the designer. In a chemical process, these would be the rates of flow of the chemical raw materials and catalysts, the rate of flow of independently applied heat if this is used, the independently applied pressure if this is used, and other factors that can be controlled by the designer. All such independently controllable input variables are listed together with their maximum (and possibly minimum) allowable values, as the set M $(m_1, m_2, \ldots m_m)$.

The plant output variables. These are the desired system outputs which have to be controlled. In the chemical plant they would presumably be the flow rates or production rates of the final chemical compounds. These variables are listed as the set C $(c_1, c_2, \ldots, c_n)$. In a positioning system C would be the positions of the output shafts.

Useful internal variables. There are usually available other variables which serve as good indicators of what is going on in the system, even though they are not output variables. In a chemical process, temperature, pressure, and pH levels are often among such functions. Actually, any internal plant variables which are readily measured and therefore potentially useful for feedback purposes should be listed in this set F $(f_1, f_2, \ldots, f_f)$. It is important at this point to consider the measurement problem, i.e., the instruments (sensors) that are available for the purpose, and the inherent noise level and characteristics of these instruments. The noise data should be included for the sensors of each measurable internal variable, as well as for those of the plant output variables given in the second item.

Variables which are not independent of each other may appear separately. Thus, in a mechanical system, velocity is available directly from displacement by differentiation. Nevertheless, velocity should be listed separately, if it can be measured independently of displacement. The same applies to acceleration. The reason for this apparent duplication is that it may be useful to use for feedback purposes a signal proportional to velocity. If only the displacement signal is available, then the latter would have to be passed through a differentiator to yield a velocity signal. The differentiator in its environment has its own noise level. It is quite possible for the sensor which measures velocity directly to have noise characteristics which are superior to the combination of displacement sensor and differentiator.

Another factor associated with any sensor and which is of fundamental importance is the speed of response of the sensor. Consider a (noise-free) sensor which measures velocity and which, in response to a velocity step input (Fig. 1.4-1), has the output A, and compare it to another whose response for the same input is B. If the responses of the two are equal in all other respects (noise levels, saturation levels, etc.), then the one whose response is A is far superior. This will become apparent from later chapters. The above comments on noise levels and speeds of response apply also to sensors used for measuring the system output variables C listed previously under the second item.

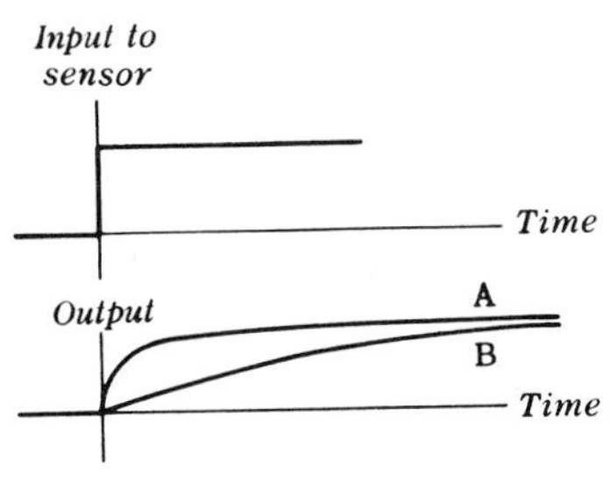

FIG. 1.4-1. Comparison of transducers' step responses.

External disturbance variables. These are the plant variables which are associated with the points of access whereby external influences (uncontrollable by the designer) enter the plant. They should be ascertained, and the nature of the external signals, their maximum values, and maximum derivatives should be ascertained. Actually, it will be seen that the more information available about the disturbances, the more efficient and economical may be the resulting design. These variables are denoted by D (d_1, d_2, ..., d_d).

§ 1.5 Canonical Representation of a Plant

The ideas presented in Section 1.4 are applied to a plant with only one input (m), one output (c), one available (for measurement) internal variable (f), and one disturbance variable (d). These variables are shown as small circles (nodes) in Fig. 1.5-1a. It does not matter if the plant is actually very complicated; the above four variables are sufficient for feedback purposes. The reason is simple. Internal variables are of no use to us if they cannot be measured in some way, so there is really no need to have them clutter up the representation of the plant.

The next step is to find the relations between these four variables. The only two independent inputs are m and d. The other two variables represent depend-

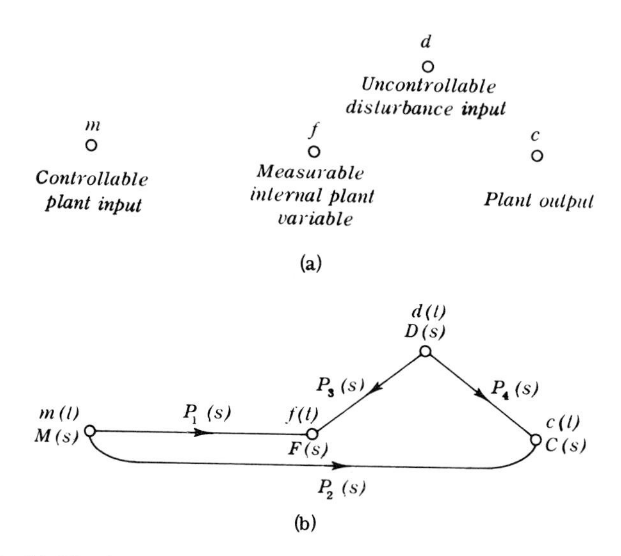

FIG. 1.5-1. (a) The four plant variables essential for feedback design. (b) A sufficiently general plant representation.

ent outputs. Hence, f and c are each influenced by m and d. We therefore have only four independent paths of signal flow: from m to f, from m to c; from d to f, and from d to c (Fig. 1.5-1b). If the plant is linear, the relation between any two variables is independent of the values of any other variables. P_1, P_2, P_3, P_4 are transfer functions, if the relations between the variables do not change with time as well as being linear. If the plant is nonlinear, then P_1 depends on the value of f, and is therefore a function of d and of P_3; similarly, P_3 will depend on m and d. We shall assume henceforth that the plant is linear, and consider the problem of nonlinear plants in Chapter 8. It might well be noted here that even though the plant relations are often nonlinear, the assumption of linearity is very often justifiable and leads to accurate results. This statement will be clarified and amplified in Chapter 8. Another point to note is that the relations need not be exactly known. The whole point of feedback is to permit accurate control with inexact knowledge of the plant and environmental relations.

The really significant point is that only the relations between the signals are required for the purpose of plant representation. There is no need for the details of the mechanisms which produce the signals, and which convert them from one energy form to another. Furthermore, if the fundamental equations describing the plant are linear and time invariant, then P_1 can theoretically be obtained from a single dynamic measurement. Thus, to find P_1 in Fig. 1.5-1b, let $d(t) = 0$ and the independent $m(t)$ can be any convenient time function. The resulting $f(t)$ is recorded and then $P_1(s) = F(s)/M(s)$, where F and M are, respectively, the Laplace transforms of $f(t)$ and $m(t)$. The dimensions of f can be rate of flow of a gas and that of m can be velocity, and the device can be very complicated. Nevertheless, it is only this dynamic relationship as expressed by the transfer function between the variables that is needed. However, it is emphasized that the *dynamic* relationship is needed. Static, final value measurements are insufficient. Similarly, P_4 is obtained by letting $m(t) = 0$, finding $c(t)$ due to $d(t)$ and evaluating $P_4(s) = C(s)/D(s)$. P_2 and P_3 are similarly obtained.

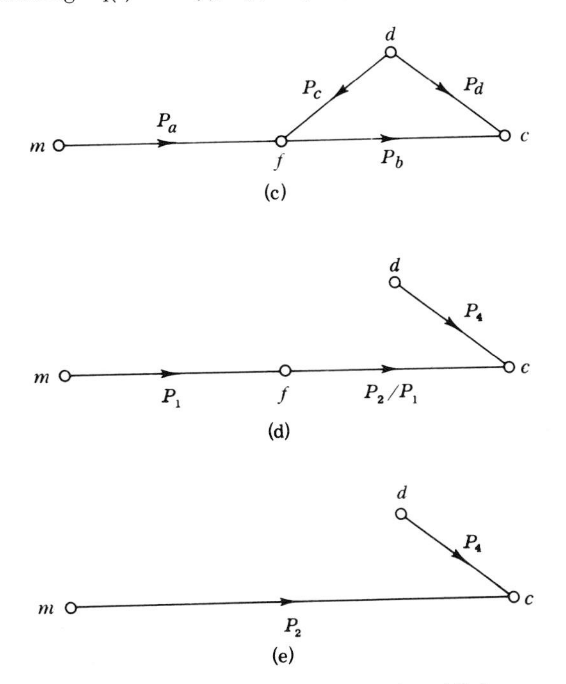

FIG. 1.5-1. (c) An equivalent general plant representation. (d) A representation which is adequate only if f is not used for feedback purposes. (e) A representation equivalent to (d).

The information required to obtain P_1, ..., P_4 in Fig. 1.5-1b can be obtained from specially applied tests, or from normal system operating records.[1] Some of the problems involved in obtaining the values of P_1, ..., P_4 will be discussed

[1] "Plant and Process Dynamic Characteristics," Proceedings of a Conference held at Cambridge, England (Apr. 4-6, 1956), by the Society of Instrument Technology. Academic Press, New York, 1957; S. Lees, Interpreting Dynamic Measurements of Physical

in Chapter 12. It is emphasized that P_1 represents the relation between the independently applied signal m and the resulting dependent signal f, providing that $d(t) = 0$. P_1 is a transfer function (Laplace transform of an impulse response), if this relation is linear and independent of time. We henceforth assume the P's to be transfer functions.

It should be noted that the representation in Fig. 1.5-1b is not unique. Another representation which is equivalent with respect to the effects of m and d on f and c is shown in Fig. 1.5-1c, providing $P_a = P_1$, $P_aP_b = P_2$, $P_c = P_3$, $P_d + P_bP_c = P_4$. On the other hand, the representation shown in Fig. 1.5-1d is equivalent to the previous two, only with respect to the effect of m and d on c and that of m on f. Figure 1.5-1d is not equivalent to Figs. 1.5-1b and c, with respect to the effect of d on f. However, if no feedback is used from f, then Fig. 1.5-1d can be used as the correct plant representation, because there is no need in such a case for a representation which is correct for f; it is better in fact to eliminate f altogether as in Fig. 1.5-1e. Figure 1.5-1d is incorrect if f is used for feedback purposes.

It is therefore important to recognize that the representations in Fig. 1.5-1 are very specialized—they are correct for very special cases only. We have already noted how Fig. 1.5-1d is correct for one type of application and incorrect for another. In the representations of Fig. 1.5-1, the loading effect of the plant on the energy source that supplies m (call the source X) is included in P_1 and P_2. But if a different energy source with a different internal impedance is used, then the values of P_1 and P_2 will be different. In general, in feedback systems, a signal processing network is inserted between the system input point and the plant input point m. The energy source (call it W) supplying the *system* input has its own internal impedance Z_w. The signal delivered by W is $WZ_i/(Z_i + Z_w)$, where Z_i is the load impedance which the system offers to source W. The representations in Fig. 1.5-1 do not contain the value of Z_i. If this information is to be included, then a more complicated representation is needed. Similar comments apply to nodes c and f in Fig. 1.5-1. Thus P_2 in Fig. 1.5-1b gives the value of c due to m for a given load at c. If the load is changed, P_2 (also P_4) must be changed, and the information needed is not available from any of the above figures. The measurements that were used to find the P's must be repeated for the new value of the load. To avoid the repetition of measurements, more general representations of the plant are required. Such models of the plant are described in Section 1.7. The important point that we want to emphasize is that the simpler the representation, the more specialized it is. The simplest class of representations for a single set of m, d, c, and f is that class which has

Systems—Parts I and II, Report R-128, Instrumentation Lab., M.I.T., Cambridge, Massachusetts, Feb., 1957; V. V. Solodovnikov and A. S. Uskov, A Frequency Method for Determining the Dynamic Characteristics of Automatic Control from Data on Their Normal Usage, Automation and Remote Control (English Translation), pp. 1533-1542, 1959.

four nodes (variables) and four independent branches (transmissions). The simplest (and most specialized) plant model that is needed for two independent inputs (m_1, m_2), three available internal variables (f_1, f_2, f_3), two outputs (c_1, c_2), and one disturbance source (d_1) has 15 branches, and one such model is shown in Fig. 1.5-2.

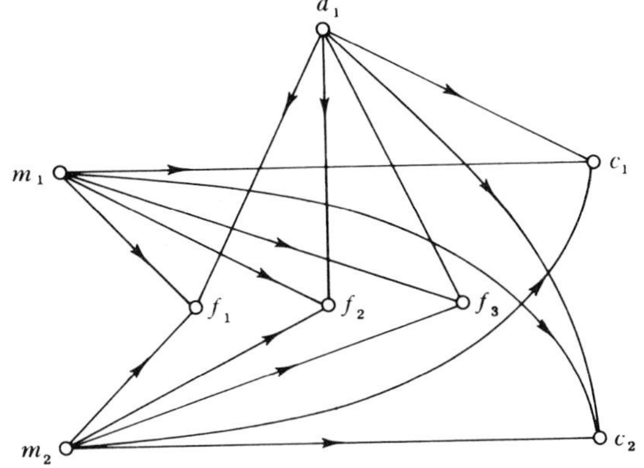

FIG. 1.5-2. A canonical representation of plant with two inputs, three available internal variables, two outputs, and one disturbance source.

§ 1.6 Signal-Flow Graph Representation and Analysis[1]

The plant representations used in Figs. 1.5-1 and 2 are appropriately called signal-flow graphs, because they represent, in the simplest graphical fashion, the flow of signals in a system. Signal-flow graphs are of tremendous usefulness in feedback systems *because feedback theory is primarily concerned with the flow and processing of signals in systems.* The system energy transformations, the chemical, physical, or economic laws, are not needed in the feedback problem—it is the signals representing these physical (or abstract) quantities that are needed. The signal-flow graph presents this information in the simplest possible fashion.

To illustrate how the signal-flow graph of a linear constant coefficient system is obtained, let us begin with a system which is described by differential equations. The following procedure may be used:

(1) Convert all differential equations into transform equations.

(2) Identify and list all variables. Divide them into two groups: (a) input variables (i.e., the forcing functions), (b) the forced or dependent variables.

(3) If there are a total of n forced variables, there must be n independent equations defining these variables. If there are $m > n$ equations, then $m - n$ must be derivable from the remaining n. Select any n independent equations.

[1] S. J. Mason, Feedback theory—Some properties of flow graphs, *Proc. IRE* **41** (9), 1144-1156 (1953).

(4) Use a separate equation to identify each variable. Express each equation by means of a signal-flow graph.

The above rules are illustrated by the following example. Consider a system which is defined by the following three differential equations:

(a) $A\ddot{x} + B\dot{x} + Cx = Dy + E\dot{z} + Fw,$

(b) $G\ddot{y} + H\dot{y} + J\dot{w} + Kx = Lz,$

(c) $M\dot{z} + \ddot{x} = Nw.$

We are told that there is one forcing variable—w. (This information must be supplied.) The differential equations are first converted into algebraic transform equations. Capital letters are used for system parameters, and for transforms of time functions, while lower case letters represent the real time functions, except for s, which represents the complex variable. The transform equations are

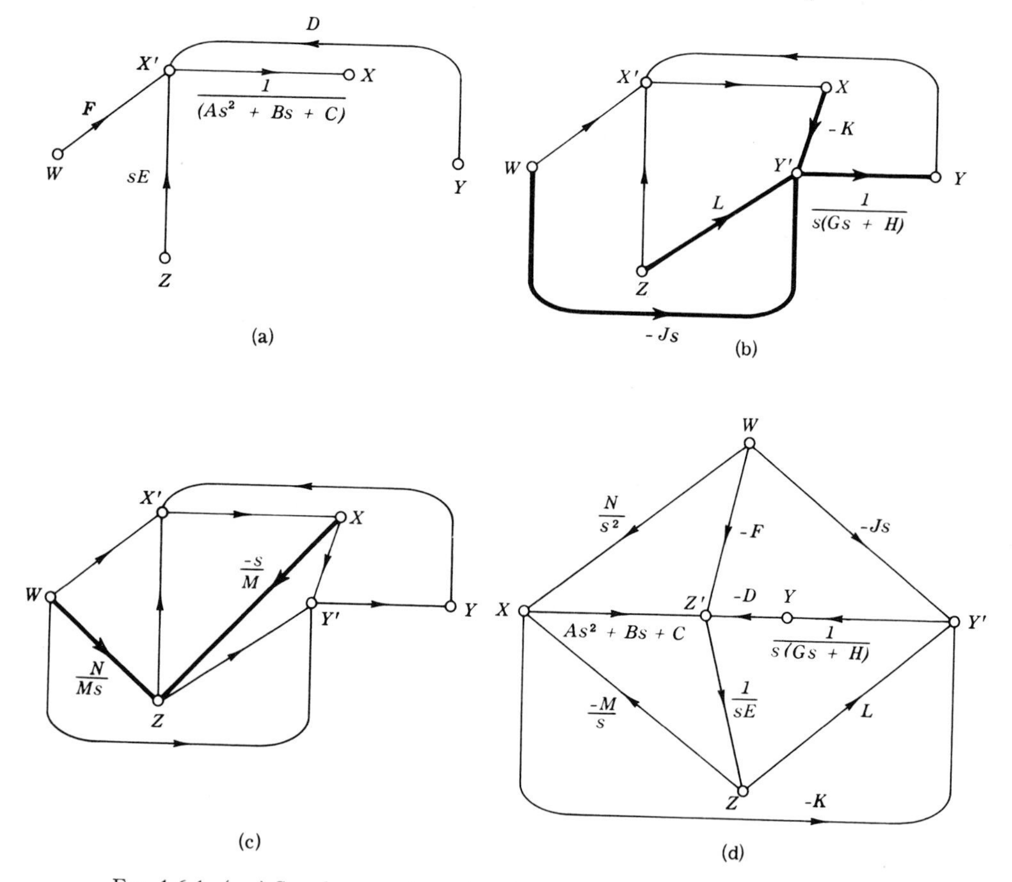

FIG. 1.6-1. (a–c) Step-by-step derivation of system signal-flow graph. (d) Flow graph equivalent of (c).

(a) $(As^2 + Bs + C)X = DY + EsZ + FW$,
(b) $s(Gs + H)Y + JsW + KX = LZ$,
(c) $MsZ + s^2X = NW$.

There are three equations in three unknowns. They are used in the above order (any other order will do) to define X, Y, Z, respectively, i.e.,

(d) $X = (As^2 + Bs + C)^{-1}(DY + EsZ + FW)$,
(e) $Y = (LZ - JsW - KX)/s(Gs + H)$,
(f) $Z = (NW - s^2X)/Ms$.

Equations (d), (e), and (f) are written in signal-flow graph form. The graph of (d) is shown in Fig. 1.6-1a and its meaning is obvious by inspection. In signal-flow graphs, variables are represented by nodes, transmissions by directed branches. Equation (e) is next put in signal-flow graph form, and is shown in darker lines in Fig. 1.6-1b. Finally, Eq. (f) is shown in heavy lines in Fig. 1.6-1c, which represents the complete signal-flow graph of the system. The signal-flow graph shown in Fig. 1.6-1c is not the only one that can be used to represent the system. Equations (a), (b), and (c) could have been used to define Z, Y, X, respectively, in place of X, Y, Z, and the resulting signal-flow graph is shown in Fig. 1.6-1d.

There are two simple checks as to whether obvious errors have been made. There can be no (branches) transmissions *into* notes representing input (forcing) variables; each node representing a forced variable must have at least one transmission entering into it.

The basic rules of the flow graphs are now listed.

1. Nodes represent transforms of variables, real or fictitious.
2. The value of a node (variable) is equal to the sum of incoming signals, thus in Fig. 1.6-1a, $X' = FW + sEZ + DY$.
3. The value of each node variable leaves on all outgoing branches. Thus in Fig. 1.6-1c, W influences X', Y', Z.
4. Transmissions, as indicated by branch arrows, are unidirectional.

Rule for Flow Graph Analysis

Mason[1] has derived a simple and very useful rule for evaluating system transfer functions from signal-flow graphs. The rule is: The transmission from any independent node x to a dependent node y is

$$T_{xy} = \frac{\sum K_{xy}\,\Delta_{xy}}{\Delta} \tag{1.6,1}$$

[1] S. J. Mason, Feedback theory—Further properties of signal-flow graphs. *Proc. IRE* **44**, 920-926 (1956).

where the system determinant

$$\Delta = 1 - \sum_i L_i + \sum_{i,j} L_i'L_j' - \sum_{i,j,k} L_i''L_j''L_k'' + \dots . \tag{1.6,2}$$

In Eq. (1.6,2), L_i represents a loop transmission, i.e., the transmission around any *closed* loop; $\Sigma_i L_i$ is the sum of all loop transmissions. $L_i'L_j'$ is the product of the loop transmissions of any two loops that do not have a node or branch in common; $\Sigma_{i,j} L_i'L_j'$ is therefore the sum of all such products. $L_i''L_j''L_k''$ is the product of the loop transmissions of any three loops that do not have any branches or nodes in common, etc.

Example 1. Find Δ in Fig. 1.6-2. The sum of the individual loop transmissions is $\Sigma L_i = T_2T_3T_{10} + T_3T_4T_5T_{11} + T_3T_4T_5T_6T_{12} + T_4T_5T_6T_7T_9 + T_5T_6T_8$.

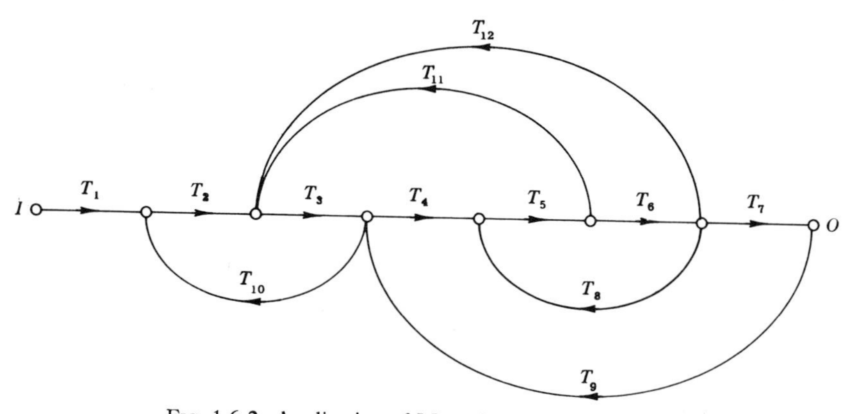

FIG. 1.6-2. Application of Mason's rule. — Example 1.

The above is best done in a systematic manner by starting with T_1 and listing all loops containing T_1, then going on to T_2, and taking all loops containing T_2 that have not already been listed, etc. Next we find $\Sigma_{i,j} L_i'L_j'$ which is the sum of the products of the loop transmissions taken two at a time, omitting all those that have any node or branch in common. To evaluate $\Sigma_{i,j} L_i'L_j'$, the list in ΣL_i is examined. The first term in the latter is $T_2T_3T_{10}$. The other terms are scanned to detect those that do not contain any of T_2, T_3, or T_{10}. The first term satisfying this condition is $T_4T_5T_6T_7T_9$, but Fig. 1.6-2 is examined to check whether these two loops ($T_2T_3T_{10}$ and $T_4T_5T_6T_7T_9$) have any nodes in common. They do, and therefore their product is not included in $\Sigma L_i'L_j'$. However, the last term $T_5T_6T_8$ has no branches or nodes in common with $T_2T_3T_{10}$ and therefore $T_2T_3T_{10}T_5T_6T_8$ appears in $\Sigma L_i'L_j'$. Next we note the second term in ΣL_i, i.e., $T_3T_4T_5T_{11}$, and examine those that follow. In this way it is found that $\Sigma_{i,j} L_i'L_j' = (T_2T_3T_{10})(T_5T_6T_8)$. Since $\Sigma L_i'L_j'$ contains only one term, it is impossible for $\Sigma L_i''L_j''L_k''$ to exist.

The definition of $K_{xy}\Delta_{xy}$ in Eq. (1.6,1) is as follows: K_{xy} is any direct transmission from the input x to the output y. Each Δ_{xy} is associated with a particular K_{xy}. The value of Δ_{xy} is the same as Δ, except for the removal of all terms containing any branches or nodes that appear in K_{xy}. Thus, in the above example, there is only one $K_{io} = T_1T_2T_3T_4T_5T_6T_7$. Since each loop in ΣL_i has at least one branch present in K_{io}, $\Delta_{io} = 1$, and $\Sigma K_{xy}\Delta_{xy} = T_1T_2T_3T_4T_5T_6T_7$.

Example 2. Find T_{xy} in Fig. 1.6-3a. $\Delta = 1 - T_4T_5T_6$, $\Sigma K_{xy}\Delta_{xy} = T_1 + T_3T_6T_5$. Therefore $T_{xy} = (T_1 + T_3T_5T_6)/(1 - T_4T_5T_6)$. In this example, for

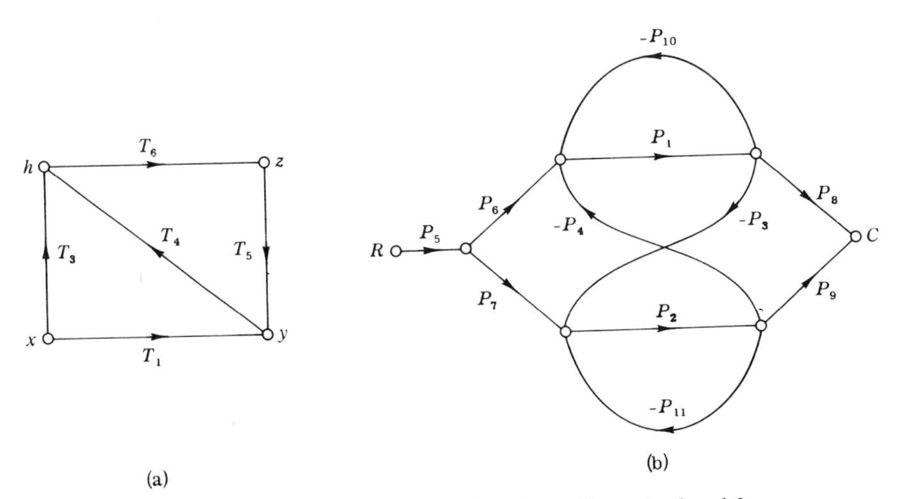

FIG. 1.6-3. Application of Mason's rule. — Examples 2 and 3.

$K_{xy} = T_1$, Δ_{xy} is unity, despite the fact T_1 and $T_4T_6T_5$ do not have any branches in common. There is, however, the node y in common, so this loop transmission is removed from Δ in evaluating Δ_{xy}. [It is noted that in any K_{xy} or L_i in Eqs. (1.6,1) and (1.6,2) *no node or branch may be traversed twice.*]

Example 3 (Fig. 1.6-3b). $\Sigma L_i = -P_1P_{10} - P_2P_{11} + P_1P_3P_2P_4$. $\Sigma L_iL_j = (P_1P_{10})(P_2P_{11})$, $\Sigma L_iL_jL_k = 0$. There are four paths from R to C:

$$\begin{aligned} K_1 &= P_5P_6P_1P_8, & \Delta_1 &= 1 + P_2P_{11}; \\ K_2 &= P_5P_7P_2P_9, & \Delta_2 &= 1 + P_1P_{10}; \\ K_3 &= -P_5P_6P_1P_3P_2P_9, & \Delta_3 &= 1; \\ K_4 &= -P_5P_7P_2P_4P_1P_8, & \Delta_4 &= 1. \end{aligned}$$

Therefore

$$T = \frac{P_5P_6P_1P_8(1 + P_2P_{11}) + P_5P_7P_2P_9(1 + P_1P_{10}) - P_5P_1P_2(P_6P_3P_9 + P_4P_7P_8)}{1 + P_1P_{10} + P_2P_{11} - P_1P_4P_2P_3 + P_1P_{10}P_2P_{11}}.$$

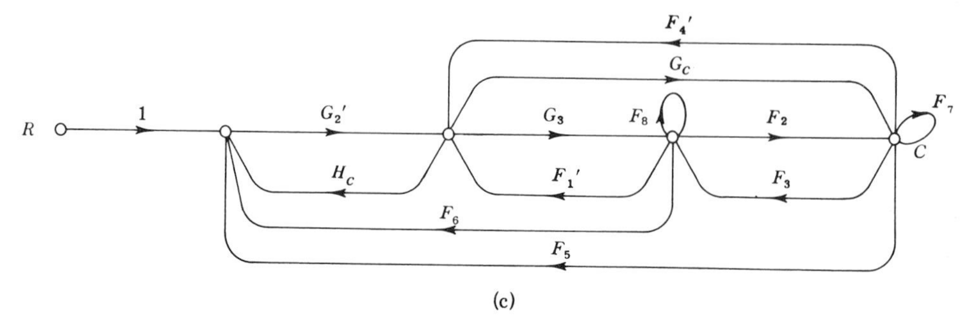

(c)

FIG. 1.6-6. (c) Result of eliminating node N and self-loop H^f.

We next simultaneously remove node N and the self-loop H_f. The result is shown in Fig. 1.6-6c, with

$$G_2' = G_2/(1 - H_f),\quad F_4' = H_5H_a/(1 - H_f),\quad F_1' = (H_b + G_4H_a)/(1 - H_f),$$
$$F_2 = G_4G_a + G_b,\quad F_3 = H_3H_5,\quad F_5 = H_1 + H_eH_5,$$
$$F_6 = H_d + G_4H_e,\quad F_7 = H_5G_a,\quad F_8 = G_4H_3.$$

Finally, Mason's rule is applied, by inspection, to Fig. 1.6-6c, with

$$\Delta = 1 - [G_2'H_c + G_3F_1' + F_8 + F_2F_3 + F_7 + G_2'G_3F_6 + G_2'F_5(G_c + G_3F_2) + G_cF_3F_1' + G_cF_4' + G_3F_2F_4'] + [G_2'H_c(F_7 + F_8 + F_2F_3) + G_3F_1'F_7 + F_8(F_7 + G_2'G_cF_5 + G_cF'_4) + F_7G_2'G_3F_6] - [G_2'H_cF_7F_8].$$

$$\Sigma K_{xy}\Delta_{xy} = G_2'G_c(1 - F_8) + G_2'G_3F_2.$$

Block Diagrams[1]

Block diagrams were used for the same purpose as signal-flow graphs long before signal-flow graphs became popular. However, the tendency in many circles was and remains to use a block diagram to represent the transfer function of an actual physical device. Thus each block diagram would represent a physical element or block of elements. This practice is not adopted here. Block diagrams and signal-flow graphs are used interchangeably in this book.

§ 1.7 Terminal Representation of Systems

The value of the signal-flow graph is in the simple graphical means it provides for the representation of the system in terms of the essential variables. It has been noted in Section 1.5 that the very simplest plant representation is one that con-

[1] T. M. Stout, A Block-diagram approach to network analysis. *Trans AIEE, Part II. Application & Ind.* **71**, 225-260 (1952); Block-diagram solutions for vacuum-tube circuits. *Trans. AIEE, Part I. Communs. & Electronics* **72**, 561-567 (1953).

tains only the input, output, disturbance, and internal variables available for feedback. However, it was also noted that such a representation was accurate only for the specific source impedances (of the independent inputs) and the load impedances (of the dependent outputs) that were used when the various transfer functions were ascertained. The above simplest signal-flow graph cannot be used if any source or load impedances are changed. A more detailed description of the plant is required in order to explicitly show the plant input and output impedances. The advantage of such a more complicated description is that it may be used for any sources driving the plant and any loads applied to the plant.

Consider the simplest case of a single independent input and a single dependent output, and we want a representation that explicitly shows the system input and output impedances, in order that the representation can be used with any source and any load. Suppose the plant input is torque and the output is velocity. Then the plant must be driven by a torque generator with its internal impedance, and the most general load is one with inertia, damping, stiffness, and a velocity source. (Thevenin's theorem may be used to convert it into a torque generator.) The problem is presented in Fig. 1.7-1a. Note that the plant input impedance

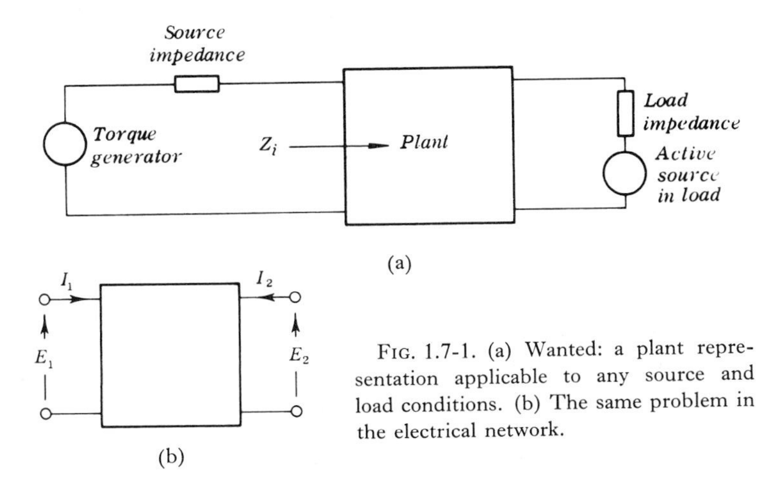

FIG. 1.7-1. (a) Wanted: a plant representation applicable to any source and load conditions. (b) The same problem in the electrical network.

(and therefore the actual torque delivered to the plant) is, in general, a function of the load impedance. In special cases there is no reverse transmission of energy, so the input impedance is independent of the load. Generally, the input and output variables can be any of such diverse variables as voltage, torque, velocity, heat flow, rate of flow of a fluid, displacement from a reference, percentage concentration of an element, pressure, etc. Whatever they may be, we want a representation that can be used for any kinds of sources (with linear source impedances) applied at the inputs and any kinds of (linear) loads applied at the outputs.

In the simplest plant model, correct only for a fixed source and load, only a single transfer function is needed to describe the plant. If the description is to be used for different sources but fixed loads, two functions are needed to characterize the system. If only the load is varied, two *other* functions are needed. If both may vary, then four functions are needed.

To prove the above statements, it is probably easiest to begin with purely electrical systems. The terminal behavior of a two-port network (in lumped electrical networks, a port consists of a pair of terminals) is completely characterized (Fig. 1.7-1b) by the relationship between the terminal E's and the I's. We can regard any two of these as independent sources and the other two as effects. There are many different characterizations possible according as to which two variables are taken as the independent sources.[1] Suppose we take (the h parameters)

$$E_1 = H_{11}I_1 + H_{12}E_2, \qquad I_2 = H_{21}I_1 + H_{22}E_2. \tag{1.7,1}$$

The resulting model is shown in Fig. 1.7-1c. H_{11} is the short-circuit input

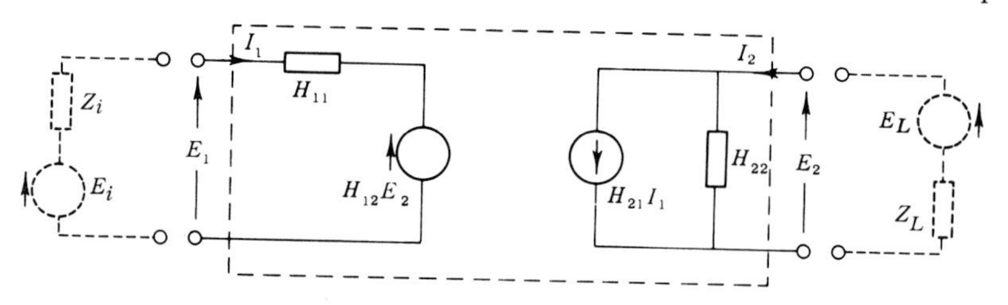

FIG. 1.7-1. (c) A model which is satisfactory for terminal representation of a linear two-port.

impedance, H_{22} is the open-circuit output impedance, H_{21} is the forward short-circuit current gain, and H_{12} is the reverse open-circuit voltage gain. The important point is that the four H's must be known in order that the device description may suffice for any source or load. For example, if the source is E_i with source impedance Z_i and the load is active with E_L and Z_L, then we write

$$E_1 = E_i - Z_iI_1, \quad E_2 = E_L - Z_LI_2 \tag{1.7,2}$$

and use these with Eq. (1.7,1) to obtain $E_i = (Z_i + H_{11})\, I_1 + H_{12}E_2$; $E_L = Z_LH_{21}I_1 + (H_{22}Z_L + 1)\, E_2$, and solve for I_1 and E_2.

The above equations apply to any two-port linear device, if the E's and I's are identified with the proper variables. In any such device, physical or abstract, the drive that actually appears at the input terminals due to an applied active source, depends upon the latter's source impedance and the device input

[1] See, for example, A. E. Guillemin, "Communication Networks," Vol. 2, Chapter 4. Wiley, New York, 1935.

impedance. The same applies at the other end of the device. The only difference is that in place of electrical voltage or current, we may have any kind of quantity—torque, rate of flow, velocity, displacement, etc., and the H's may therefore have any dimensions.

The above terminal representation can of course be displayed in a signal-flow graph. Four nodes and four branches are needed in place of the two nodes and single branch of the simplest representation. As usual, there is no unique flow graph equivalent of Eqs. (1.7,1). If they are used to define E_1 and I_2, the result is shown in Fig. 1.7-2a. If they are used to define E_1 and E_2, respectively, then Fig. 1.7-2b is the result. In the first case (Fig. 1.7-2a), Eqs. (1.7,2) *must* be used to define I_1 and E_2, because in Fig. 1.7-2a, I_1 and E_2 are shown as independent sources; the final graph is then shown in Fig. 1.7-2c. In the second case, they must be used to define I_1 and I_2 (see Fig. 1.7-2d). Note that if Eqs. (1.7,2) are used in the first case to define E_1 and E_2 as in Fig. 1.7-2e, the result is incorrect, because it leads to $E_1 = E_i - Z_i I_1 + H_{11} I_1 + H_{12} E_2$, which is wrong. Figure 1.7-2c does, however, give the correct relations.

Next we consider a plant with several inputs, outputs, and disturbance inputs, as well as internal variables which are available for feedback. If there

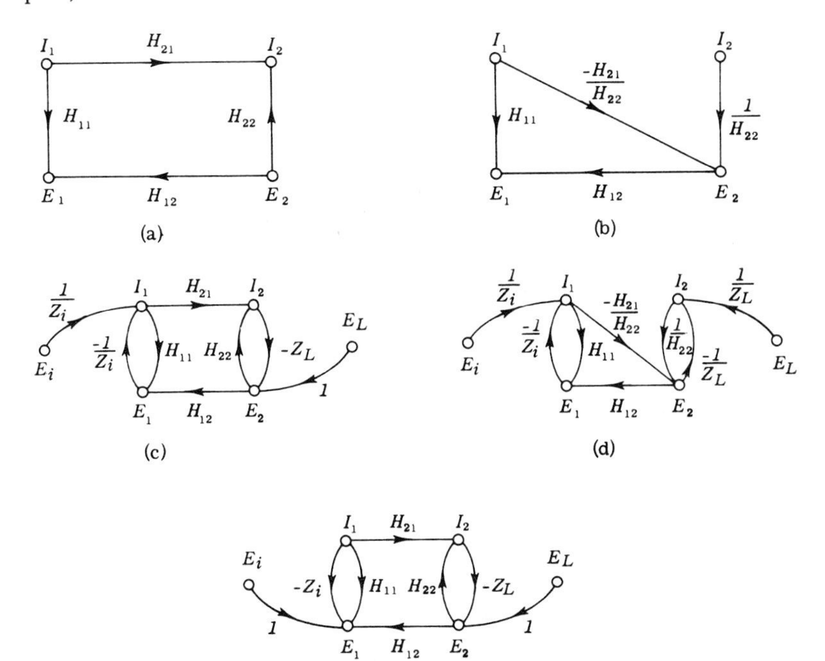

FIG. 1.7-2. (a, b) Two signal-flow graphs of Fig. 1.7-1 (c). (c, d) The corresponding correct flow graphs for systems including source and load. (e) An incorrect signal-flow graph representation.

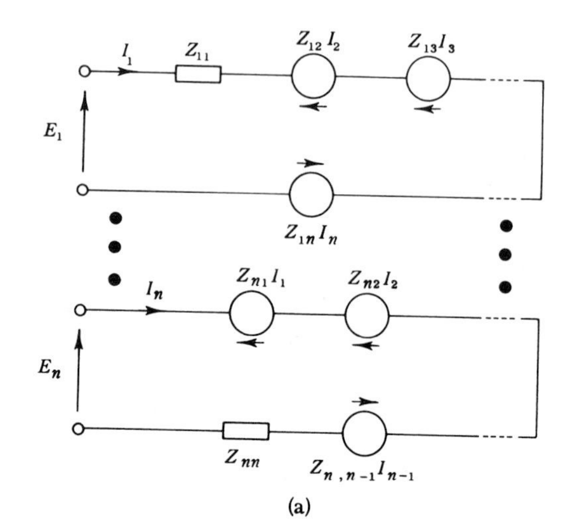

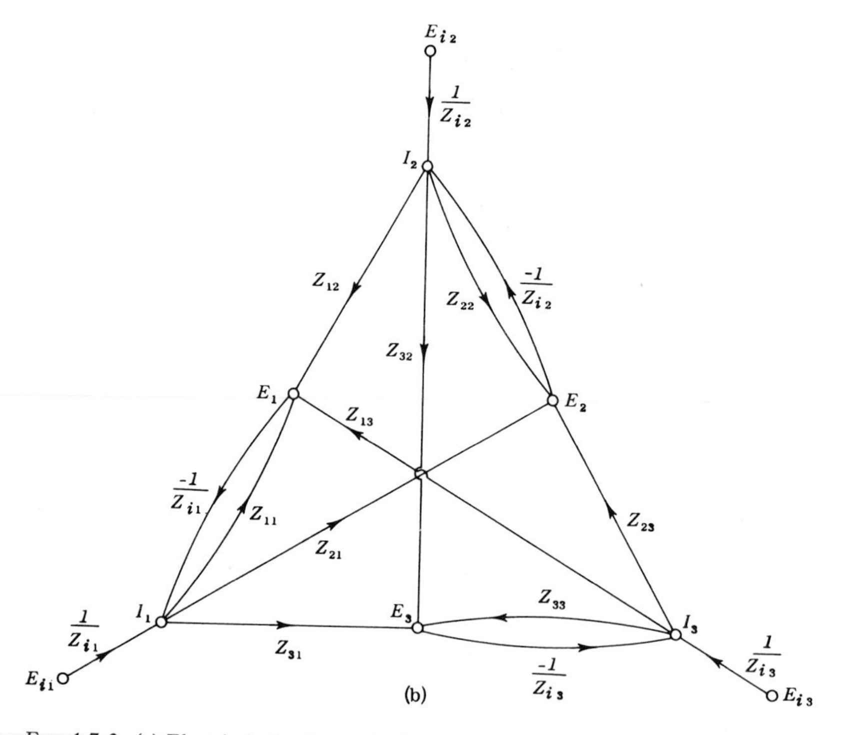

FIG. 1.7-3. (a) Electrical circuit terminal model of an n-port. (b) Flow graph terminal model of a three-port with generalized terminations.

are a total of n ports and we want a representation that can be used for any source or load impedances at each of these ports, we can deduce the required complexity of the representation by considering the purely electrical system. In the latter, we could write, from the principle of superposition: $E_1 = Z_{11}I_1 + Z_{12}I_2 + \dots + Z_{1n}I_n$; $E_2 = Z_{21}I_1 + Z_{22}I_2 + \dots + Z_{2n}I_n$; ...; $E_n = Z_{n1}I_1 + Z_{n2}I_2 + \dots + Z_{nn}I_n$. An electrical circuit model is shown in Fig. 1.7-3a. A signal-flow graph for $n = 3$ is shown in Fig. 1.7-3b, in which the equations are used to define $E_1 \dots E_n$. The following equations must then be used to define $I_1, \dots, I_n$ (not $E_1, \dots, E_n$): $E_1 = E_{i1} - Z_{i1}I_1$, $E_2 = E_{i2} - Z_{i2}I_2$, ..., $E_n = E_{in} - Z_{in}I_n$, where the E_i and Z_i have the same meaning as in Eq. (1.7,2). Again it is noted that the models of Fig. 1.7-3 may be used for any kind of system, except that in place of the E's and I's representing voltages and currents, they can represent any set of variables of any units.

§ 1.8 More Detailed System Representations

The terminal representation used in Section 1.7 permits the designer to handle the interaction between the system and all external influences. It is sufficient so long as there is no desire on the part of the designer to look inside the system. Sometimes he does want to do this in order to modify it or better understand how it works. For this purpose he needs a more detailed representation. One approach to this problem is as follows. Suppose there are x elements inside the system which will be examined. Then consider these elements as additional terminal access points and increase the number of terminal pairs by x. The interaction between these elements and the rest of the system, and the effect of modifying the elements can then be obtained.

An extreme case of the above is when the effect of every element in the system is to be explicitly detailed. In such a case, one obtains the well-known nodal or mesh equations with their less-known signal-flow equivalents, if the system is entirely electrical. Similar equations are obtained for any other kind of system. This type of representation is often used by the device specialist who must concentrate on the specific device in all its details. However, sometimes the feedback engineer must look more carefully into a specific device, especially when its terminal characteristics are undesirable. We shall therefore now consider this detailed type of device representation.

If the device is all electrical (including electronic and solid state devices whose models are known), then the node or mesh equations can usually be written by inspection, by the electrical engineer. This type of analysis is so well known that there is no need to repeat it here. The electrical circuit model (consisting of the symbols of resistance, capacitance, inductance, ideal transformers, and controlled sources, all interconnected in meshes) is so familiar to the engineer, that there is hardly any point in using an equivalent signal-flow graph. In fact, the latter is usually much more complicated than the electrical circuit model, because every bilateral element requires two branches in the flow graph to represent the two-way transmissions. When the device is mechanical, electromechanical, or of other dimensions (and the detailed representation is desired), it is often very convenient to represent the device by an electrical analog. The reason is the greater familiarity of the engineer with electrical circuit models rather than with mechanical and other kinds of models. This is especially true when the original data are in the form of a mechanical or other model. It is often fairly easy to obtain the analogous electrical model. The procedure for doing this is presented in the next few sections.

§ 1.9 Electric Circuit Models of Mechanical Systems

In Section 1.5 we considered the representation of systems in the simplest manner, such that the representation was correct only in a particular context. In Section 1.7 there was discussion of models whose terminal behavior was the same as that of the device. But if we are interested in a more detailed study of a device, then a more detailed representation is desirable, one in which there is a direct relationship between the individual elements in the actual device and those in the electrical model. An electric circuit model in which every electrical element is simply related to a device element is often useful, because of the familiarity of most engineers with electrical networks.

If we are dealing with a mechanical system, then the first step is to obtain a mechanical model of the system. In simple devices the latter may be obvious by inspection, but in complex ones (e.g., that of an airframe), considerable skill and experience is required to make reasonable approximations and obtain a realistic model. However, our present concern is solely with the derivation of the electrical model which may be used in place of the mechanical model.

We are especially interested in finding the conditions under which a passive electrical circuit model of a mechanical device may be obtained, and we shall present methods for the simple derivation (often by inspection) of the electrical circuit model. This problem is sometimes handled by first deriving the differential equations of motion of the mechanical system, and then seeking electrical circuit configurations which have similar equations. Actually, it is rarely necessary to do this. In some classes of problems (wholly rectilinear or wholly rotational motion) the electrical circuit model may be drawn directly from the mechanical model, without any intermediate steps. Even in the more difficult problems (mixed rectilinear and rotational motion), only the scalar energy functions of the mechanical system need be calculated. The procedures are derivable from the Lagrange equations of motion. One of the important advantages of the Lagrange formulation is that the most convenient coordinates for the system characterization may be used. For example, a particle free to move only along a given curve has but one degree of freedom, and the most convenient coordinate is its distance along the curve from a fixed reference point on the curve. The Lagrange method permits us to write the equation of motion in terms of this coordinate. Another advantage of the Lagrange method is that only the scalar energy functions need be calculated. The kinetic, potential, and dissipative energy functions are usually much easier to calculate than the vector functions of torque, angular momentum, etc.

We shall very briefly review the Lagrange formulation.[1] The first step is to identify the minimum number of generalized coordinates ($q_1, q_2, ..., q_n$) needed to completely describe the state of the system. The q's need not be the same dimensionally. In Fig. 1.9-1a, they are $x_1, x_2, ..., x_5$; in Fig. 1.10-1a they are x, θ while in Fig. 1.11-2a they are α, r. We shall confine our attention to systems wherein any constraints between coordinates are such that a corresponding number of coordinates can be entirely eliminated (holonomic systems). Hence the q's represent n independent coordinates.

The second step is to calculate the scalar energy functions, T, V, D. T is the kinetic energy of the system and is available from its fundamental definition $T = 0.5 \sum m_i \dot{x}_i^2$, where m_i, $\dot{x}_i$ are the mass and velocity of the elementary particles of the system. A number of theorems are available which simplify the calculation of T. In Fig. 1.9-1a, $2T = M_1\dot{x}_2^2 + M_2\dot{x}_4^2$; in Fig. 1.9-2a, $2T = J_1\omega_1^2 + J_2\omega_2^2 + J_3\omega_3^2$; in Fig. 1.11-2a, $2T = m(\dot{r}^2 + r^2\omega^2 \sin^2\alpha)$. It is important to note that in general T is a function of the q's and $\dot{q}$'s. V is the system potential energy due to gravity, compression of springs, etc. V is in general only a function of the q's, and represents the stored (conservative) energy which exists due to the values of the coordinates, and not to their rate of change. In Fig. 1.9-1a, $2V = K_1x_1^2 + K_2(x_2 - x_1)^2 + K_3(x_3 - x_2)^2 + K_4(x_4 - x_5)^2 + K_5x_5^2$; in Fig. 1.11-2a, $V = -mgr \cos \alpha$. D represents dissipative power which is proportional to the square of velocity. It is, in general, a function of the q's and $\dot{q}$'s. In Fig. 1.9-2a, $2D = B_1\omega_1^2 + B_2\omega_2^2$; in Fig. 1.9-1a, $2D = B_1(\dot{x}_1 - \dot{x}_2)^2 + B_2(\dot{x}_1 - \dot{x}_5)^2 + B_3(\dot{x}_3 - \dot{x}_4)^2 + B_4\dot{x}_5^2$. It is also necessary to define Q_1, Q_2, ..., Q_n. Q_i represents all external generalized forces operating on the ith coordinate as well as any internal forces which are not derivable from T, V, or D. The dimensions of Q_i are such that $Q_i dq_i$ represents energy.

The n equations of motion of a system are obtained from the n Lagrange equations,

$$\frac{d}{dt}\left(\frac{\partial T}{\partial \dot{q}_i}\right) - \frac{\partial T}{\partial q_i} + \frac{\partial V}{\partial q_i} + \frac{\partial D}{\partial \dot{q}_i} = Q_i \qquad (i = 1, 2, ..., n). \qquad (1.9,1)$$

Equation (1.9,1) may be used as the basis of the analogy between mechanical (and other) systems and electrical systems. Any two systems in which the expres-

[1] For detailed treatments, see H. Goldstein, "Classical Mechanics," Addison-Wesley, Cambridge, Massachusetts, 1950; E. G. Keller, "Mathematics of Modern Engineering," Vol. II, Dover Publ., New York, 1942.

sions for T, V, D and the Q_i are, respectively, similar, are equivalent in their time behavior. We therefore need only examine the character of T, V, D, Q_i for the passive electrical system, in order to determine which mechanical (or other) systems may have passive electrical analogs.

Modified Lagrange Formulation

A modified Lagrange formulation[1] is useful in special cases when it is possible at the very outset to eliminate one or more coordinates. For this purpose, the following conditions must be satisfied:

(1) $\partial T/\partial q_j = \partial V/\partial q_j = \partial D/\partial \dot{q}_j = 0$,

(2) $Q_j = 0$,

(3) $\partial V/\partial q_i + \partial D/\partial \dot{q}_i$, for any $i \neq j$, is independent of q_j, $\dot{q}_j$.

Therefore, from Eq. (1.9,1), $\partial T/\partial \dot{q}_j = c_j$, a constant. In order to eliminate q_j, the equations $c_j = \partial T/\partial \dot{q}_j$ are used to solve for $\dot{q}_j$ in terms of c_j, and the remaining coordinates. The kinetic energy T in Eq. (1.9,1) is replaced by the modified kinetic energy $T^* = T - \Sigma c_j \dot{q}_j$, wherein all the $\dot{q}_j$ are expressed in terms of c_j and the remaining coordinates.

The Force (or Torque)—Voltage Analogy

In an electrical network consisting of resistors, inductors, capacitors (R, L, $C = S^{-1}$), and independent sources, the energies stored in the magnetic and electric fields may be written, respectively, as $T_e = 0.5 \,\Sigma L_{ij}\dot{q}_i\dot{q}_j$, $V_e = 0.5 \,\Sigma S_{ij}q_iq_j$, where i is current, $q = \int i dt$ is electrical charge, and Q is applied emf. The dissipative power function is $D_e = 0.5 \,\Sigma R_{ij}\dot{q}_i\dot{q}_j$. Application of Eq. (1.9,1) then leads to Kirchhoffs first law, that the net voltage drop around a loop is zero. The above expressions for T_e, V_e, D_e, Q_e may be compared to those of mechanical systems. In a mechanical system in which all motion is constrained to be rectilinear, $2T_m = \Sigma M_{ij}\dot{x}_i\dot{x}_j$, $2V_m = \Sigma K_{ij}x_ix_j$, where M, K represent mass and spring constant; $2D_m = \Sigma B_{ij}\dot{x}_i\dot{x}_j$, where B_{ij} represents viscous damping coefficient, and Q_j represents force. In a purely rotational system, angular deflection θ, $\dot{\theta} = \omega$, U (torque), J (moment of inertia) replace x, $\dot{x}$, F (force), M, respectively. Therefore the electrical analog of such mechanical systems is obtained by replacing the rectilinear mechanical M, K, B, F, x, $\dot{x}$ or the rotational mechanical J, K, B, U, θ, $\dot{\theta}$, respectively, by the electrical L, $S = C^{-1}$, R, E (voltage), q, $\dot{q} = i$. Any consistent set of units may be used; the key factor in determining equivalent units is that analogous expressions such as $2T_m = \Sigma M_{ij}\dot{x}_j\dot{x}_i$ and $2T_e = \Sigma L_{ij}\dot{q}_j\dot{q}_i$ must both represent the same units of energy. The analogous quantities are listed in Table 1.9-1.

[1] A. G. Webster, The Dynamics of Particles, 2nd ed., Section 48. Dover Publ., New York, 1959.

TABLE 1.9-1

ELECTRICAL-MECHANICAL ANALOGS

	Analogy 1 (velocity-voltage)		Analogy 2 (force or torque-voltage)	
Electrical	Translational	Rotational	Translational	Rotational
Voltage, $e = \dot{\psi}$	Velocity, v	Angular velocity, ω	Force, f	Torque, u
Current, $i = \dot{q}$	Force, f	Torque, u	Velocity, v	Angular velocity, ω
Capacitance, $C = S^{-1}$	Mass, M	Inertia, J	Reciprocal elastance, $1/K$	Reciprocal compliance, $1/K$
Conductance, G	Damping, B	B	Reciprocal damping, $1/B$	Reciprocal damping, $1/B$
Reciprocal inductance, $\Gamma = L^{-1}$	Elastance, K	Compliance, K	Reciprocal mass, $1/M$	Reciprocal moment of inertia, $1/J$
Impedance, $Z = E/I$	Mechanical impedance, V/F	Ω/U	Mechanical impedance defined as F/V	U/Ω
Charge, q	Momentum, Mv	$J\omega$	Displacement, x	θ
Flux linkage, ψ	Displacement, x	θ	Momentum, Mv	$J\omega$
Electric field energy, $0.5\ Ce^2$, $0.5\ Sq^2$	Kinetic energy, $0.5\ Mv^2$	$0.5\ J\omega^2$	Potential energy, $0.5\ Kx^2$	$0.5\ K\theta^2$
Magnetic field energy, $0.5\ L\dot{q}^2$, $0.5\ \Gamma\psi^2$	Potential energy, $0.5\ Kx^2$	$0.5\ K\theta^2$	Kinetic energy, $0.5\ Mv^2$	$0.5\ J\omega^2$

Example 1. A model of a mechanical system is shown in Fig. 1.9-1a. Exactly enough variables have been assigned to identify the motion of each element. The analogous electrical circuit must have inductors $L_1 = M_1$, $L_2 = M_2$, whose currents are, respectively, $i_2 = \dot{x}_2$, $i_4 = \dot{x}_4$. It must have capacitors $C_1 = K_1^{-1}$, $C_2 = K_2^{-1}$, ..., $C_5 = K_5^{-1}$, whose respective currents are $i_1 = \dot{x}_1$, $i_1 - i_2 = \dot{x}_1 - \dot{x}_2$, ... One way of finding the electrical analog is to list all the electrical elements with their associated currents, as in Fig. 1.9-1b. The final step is to fit together all the elements of Fig. 1.9-1b into an electrical network. This can be done by setting up five loops and assigning the elements to the loops in accordance with their currents. The final result is shown in Fig. 1.9-1c. A positive f in the direction indicated in Fig. 1.9-1a makes x_5 positive; accordingly positive e in Fig. 1.9-1c makes i_5 positive. It is seen that the intermediate step of listing the elements and their currents (Fig. 1.9-1b) can be omitted. Figure 1.9-1c can be obtained directly from Fig. 1.9-1a.

The Velocity-Voltage Analogy

The energy stored in capacitors and inductors can also be expressed as $2T_e = \sum C_{ji}e_je_i = \sum C_{ji}\dot{\psi}_j\dot{\psi}_i$ where[1] $e \triangleq \dot{\psi}$; $2V_e = \sum \Gamma_{ji}\psi_j\psi_i$. It is then neces-

[1] $\triangleq$ means "equal by definition."

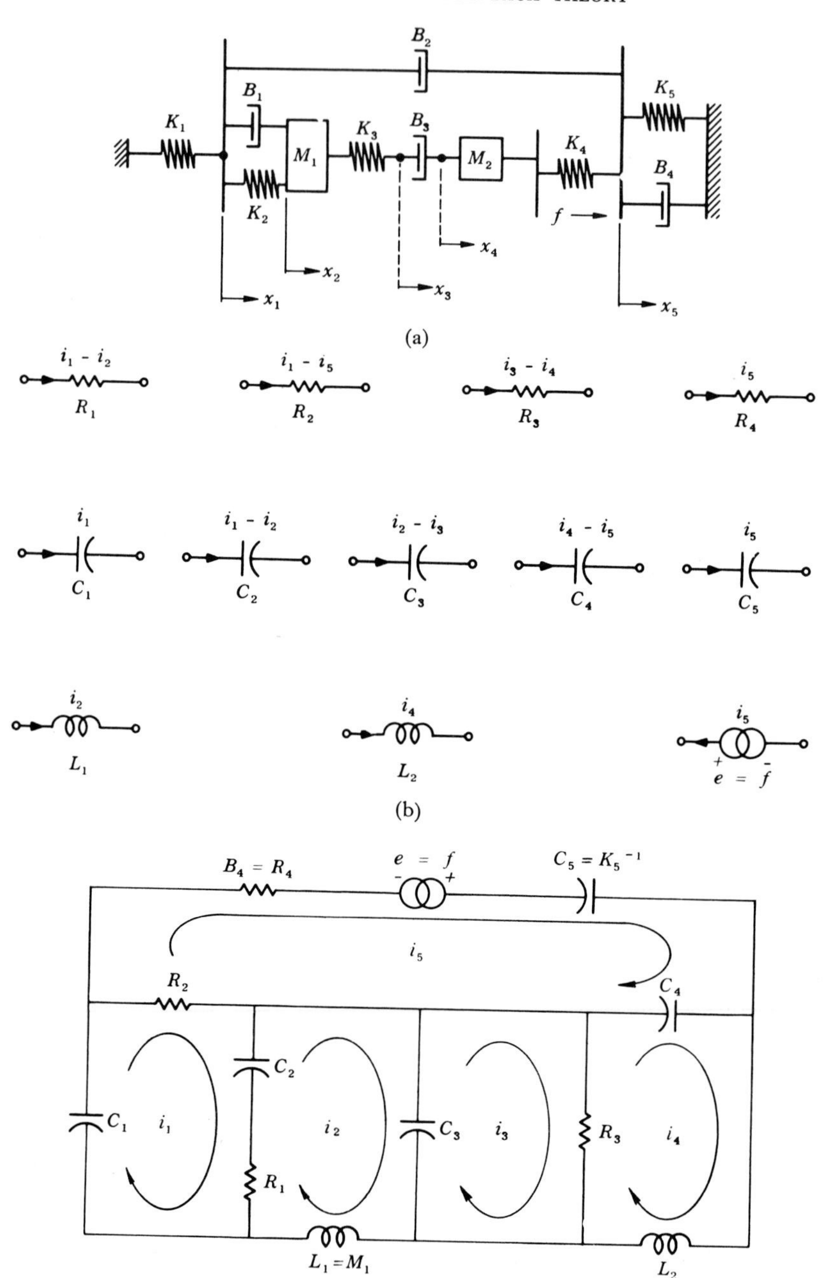

FIG. 1.9-1. (a) Model of rectilinear mechanical system. (b) List of one-to-one electrical analogs and their currents. $C_i = 1/K_i$; $R_i = B_i$; $L_i = M_i$; $\dot{x}_i = i_i$. (c) The electrical elements combined into a consistent circuit.

sary to express the dissipative function D in terms of $\dot{\psi}$, i.e., $2D_e = \sum G_{ij}\dot{\psi}_j\dot{\psi}_i$. The Q_j must therefore represent current sources [from Eq. (1.9,1)]. Comparing with T_m, V_m, D_m of a rectilinear mechanical system, it is seen that the electrical analog is obtained by letting M, K, B, F, x, $\dot{x}$, respectively, equal C, Γ, G, I, ψ, $e = \dot{\psi}$, as in Table 1.9-1. This analogy is applied to Fig. 1.9-1a. Each $\dot{x}$ represents a voltage so the network is derived on a nodal basis, i.e., five nodes and one reference node are set up and the elements assigned by inspection. For example, the velocity of M_1 ($= C_a$) in Fig. 1.9-1a is $\dot{x}_2$, so the voltage of C_a in Fig. 1.9-1d is $e_2 = \dot{x}_2$. The effective velocity of B_3 is $\dot{x}_3 - \dot{x}_4$ (or $\dot{x}_4 - \dot{x}_3$, it does not matter) so $G_c = B_3$ is inserted between nodes 3 and 4, where its effective voltage is $e_3 - e_4 = \dot{x}_3 - \dot{x}_4$. The final result is shown in Fig. 1.9-1d. It is noted that the latter is the dual of Fig. 1.9-1c.

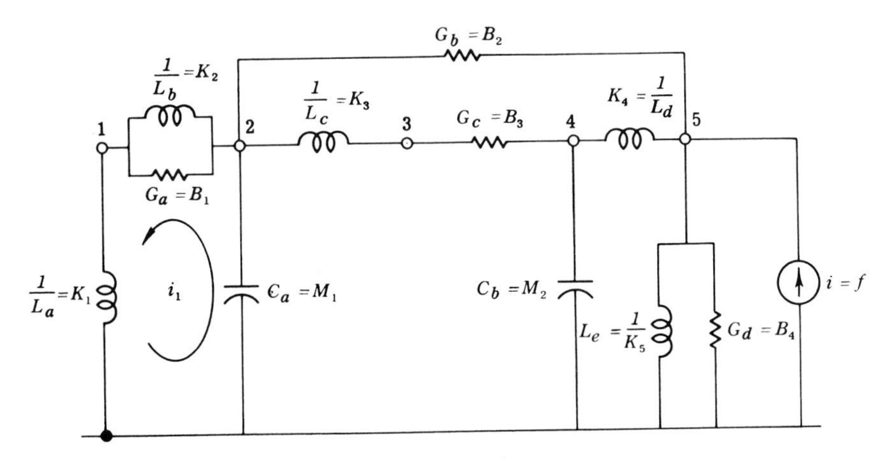

FIG. 1.9-1. (d) Electrical analog of (a) in the velocity-voltage analogy.

In the matter of choice between the two analogies, it should be noted that it is always possible and easy to construct a network on a nodal basis from its matrix of admittance coefficients, when the matrix is reciprocal and dominant, i.e., when $y_{ij} = y_{ji}$ and $y_{ii} \geqslant \sum_{j \neq i} y_{ij} = \sum_{i \neq j} y_{ji}$. Passive reciprocal mechanical circuits lead to such matrices. However, it is not known in general how to construct a network on a mesh basis from a dominant impedance matrix when more than 3 meshes are required, unless ideal transformers are used. Thus it is not known how to construct the dual (without ideal transformers) of Fig. 1.9-1d, if coupling between nodes 1 and 3 is added.

Example 2 (Fig. 1.9-2a). Figure 1.9-2a is a model of a purely rotational system. The four angular velocities ω_1, ..., ω_4 require four node voltages, if the velocity voltage analogy is used. Figure 1.9-2b may be drawn directly by inspection from Fig. 1.9-2a. For example, the effective coordinate of the compliance K_3 is

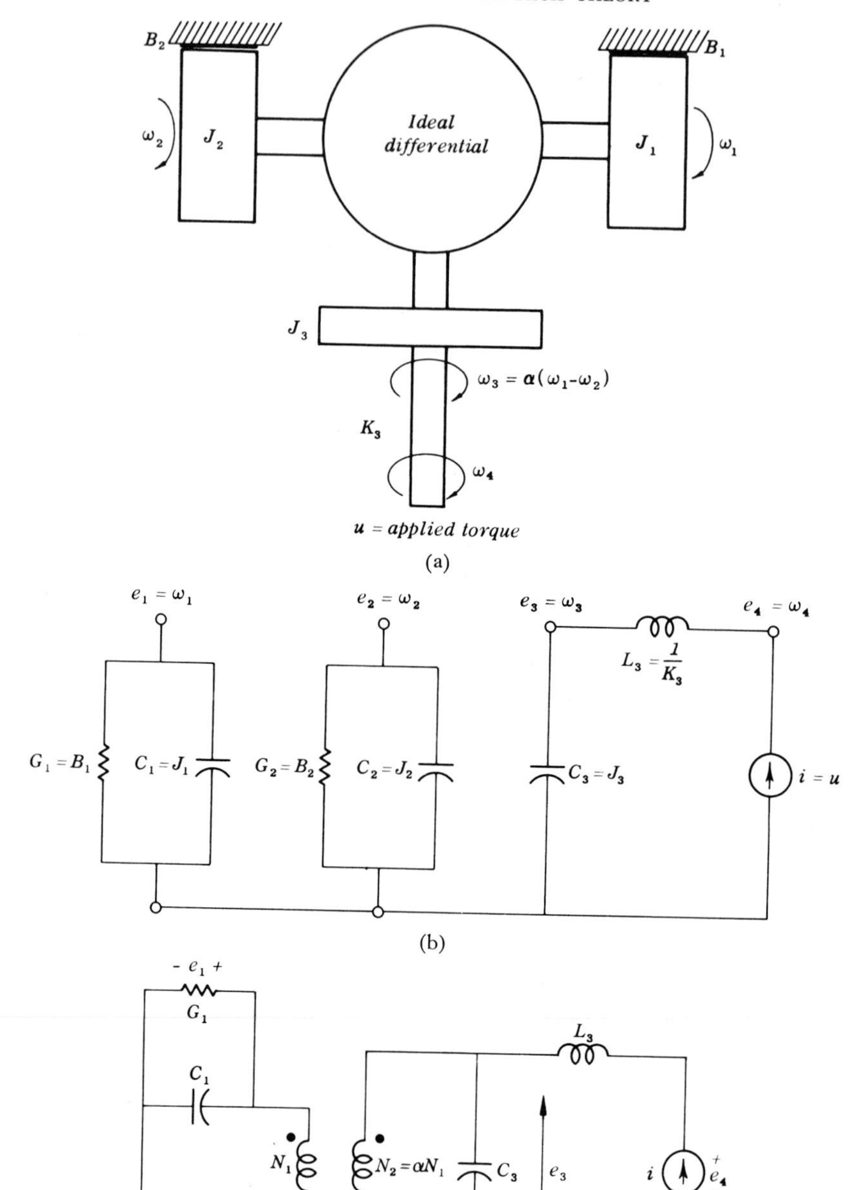

B2
B1
ω2
J2
Ideal differential
J1
ω1
J3
ω3 = α(ω1-ω2)
K3
ω4
u = applied torque
(a)
e1 = ω1
e2 = ω2
e3 = ω3
e4 = ω4
L3 = 1/K3
G1 = B1
C1 = J1
G2 = B2
C2 = J2
C3 = J3
i = u
(b)
- e1 +
G1
C1
N1
N2 = αN1
C3
e3
L3
i
e4
C2
G2
- e2 +

(c)

$\omega_3 - \omega_4$, so $L_3 = K_3^{-1}$ must have the effective voltage $e_3 - e_4 = \omega_3 - \omega_4$. The constraint analogous to $\omega_3 = \alpha(\omega_1 - \omega_2)$ is $e_3 = \alpha(e_1 - e_2)$, and it is implemented by means of an ideal transformer, as in Fig. 1.9-2c. The transformer may be eliminated by reflecting the elements on any one side over to the other. In the velocity-current analogy, Figs. 1.9-2d and e take the place of Figs. 1.9-2b and c, respectively.

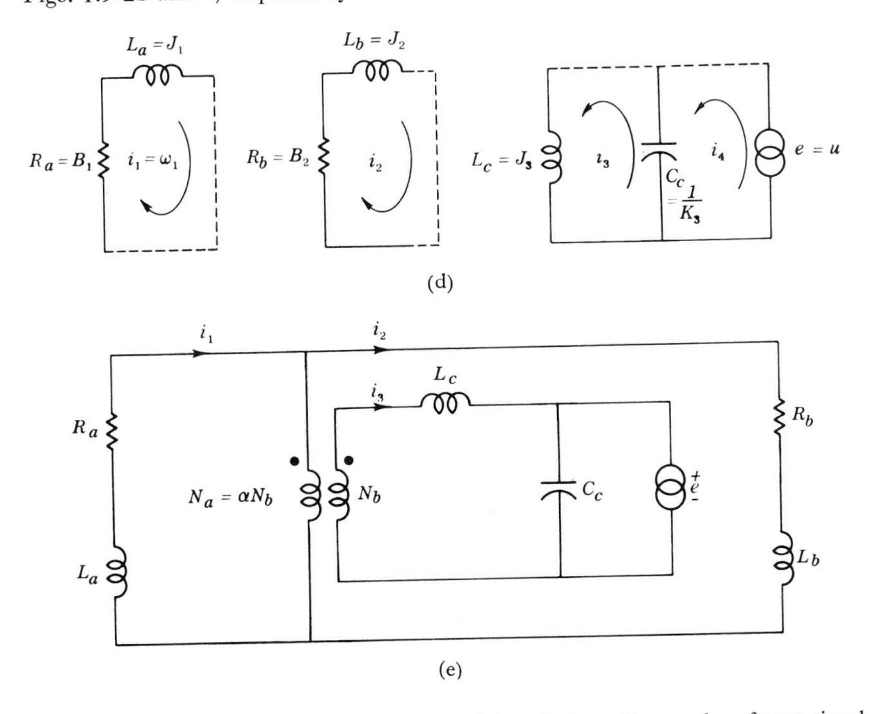

FIG. 1.9-2. (a, b, c) (*opposite*) Development of the velocity-voltage analog of a rotational system. (d, e) Development of the velocity-current analog of (a).

§ 1.10 Combined Translational-Rotational Systems

Combined translational-rotational systems are more difficult to handle than pure translational or pure rotational systems. However, the difficulty is resolved by referring to the scalar energy functions which form the fundamental basis for the analogy. The approach is illustrated by the example of Fig. 1.10-1a, which is a very simple model of an automobile suspension. The kinetic energy of a rigid body of mass M consists of two parts,[1] (a) the kinetic energy of a particle

[1] E. G. Keller, "Mathematics of Modern Engineering," Vol. II, p. 42. Dover Publ., New York, 1942.

of mass M moving with the center of gravity of the body, and (b) the kinetic energy of the body motion relative to that of the center of gravity. In Fig. 1.10-1a, the first part is $0.5\ M\dot{x}^2$, and the second part is $0.5\ Mh^2\dot{\theta}^2$, where h is

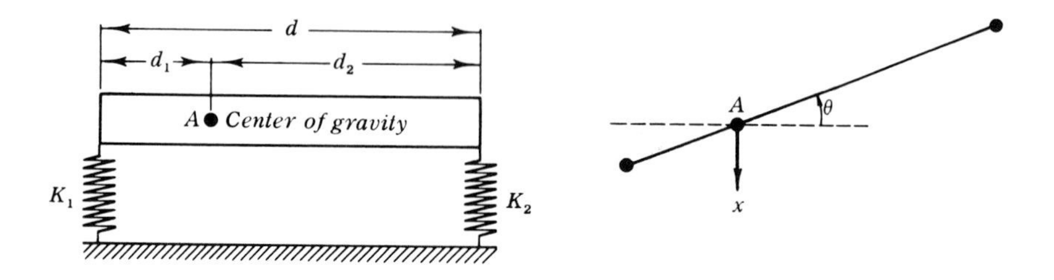

FIG. 1.10-1. (a) A combined translational-rotational system.

the radius of gyration with respect to the center of gravity at A. Therefore $2T_m = M\dot{x}^2 + Mh^2\dot{\theta}^2$, $2V_m \doteq K_1(x + \theta d_1)^2 + K_2(x - \theta d_2)^2 = (K_1 + K_2)x^2 + \theta^2(K_1 d_1^{\,2} + K_2 d_2^{\,2}) - 2x\theta(K_2 d_2 - K_1 d_1) = Ax^2 + B\theta^2 - 2Hx\theta$, if θ is small. (Note that $V_m > 0$ for all finite θ, x. We shall later make use of this.) The coordinate x is measured from the static equilibrium position, so the static spring compression and the gravity potential terms cancel out.

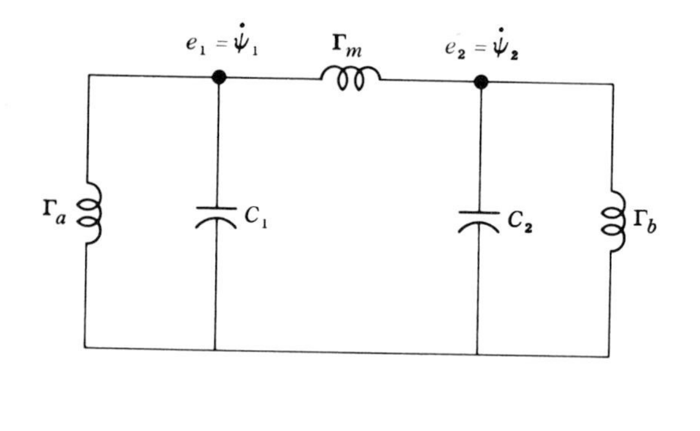

FIG. 1.10-1. (b) The electrical analog of (a).

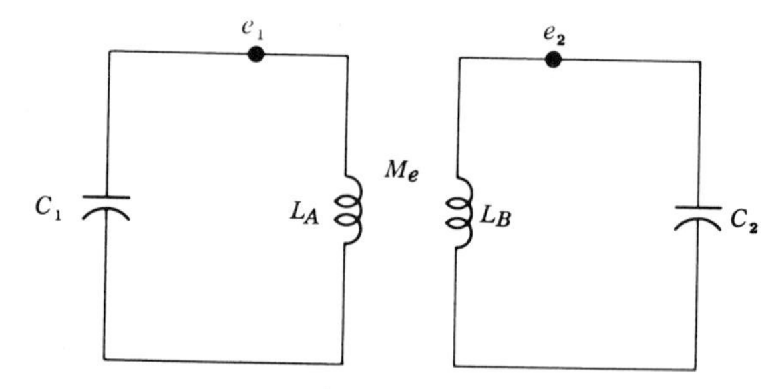

FIG. 1.10-1. (c) Circuit obtained by replacing the three coils by a transformer.

There are two independent coordinates x and θ, so in the voltage-velocity analog, we seek an electrical system in which $2T_m = 2T_e = C_1\dot{\psi}_1^2 + C_2\dot{\psi}_2^2 = C_1e_1^2 + C_2e_2^2$, $2V_m = 2V_e = A\psi_1^2 + B\psi_2^2 - 2H\psi_1\psi_2$, where $C_1 = M, C_2 = Mh^2$, $\dot{x} = e_1 = \dot{\psi}_1$, $\dot{\theta} = e_2 = \dot{\psi}_2$. It is easy to locate C_1 and C_2 in an electrical network (Fig. 1.10-1b) such that the resulting $2T_e = C_1e_1^2 + C_2e_2^2$. The form of V_e suggests the arrangement of inductors shown in Fig. 1.10-1b, where $2V_e = \Gamma_a\psi_1^2 + \Gamma_b\psi_2^2 + \Gamma_m(\psi_1 - \psi_2)^2 = (\Gamma_a + \Gamma_m)\psi_1^2 + (\Gamma_b + \Gamma_m)\psi_2^2 - 2\Gamma_m\psi_1\psi_2$, with $\Gamma_a = A - H$, $\Gamma_b = B - H$, $\Gamma_m = H$. Some of the Γ's may have negative values. We therefore examine the transformer equivalent of the pi network of coils in Fig. 1.10-1b. The terminal equivalence is shown in Fig. 1.10-2. Realization by means of a transformer requires $L_A \geqslant 0$, $L_B \geqslant 0$, which

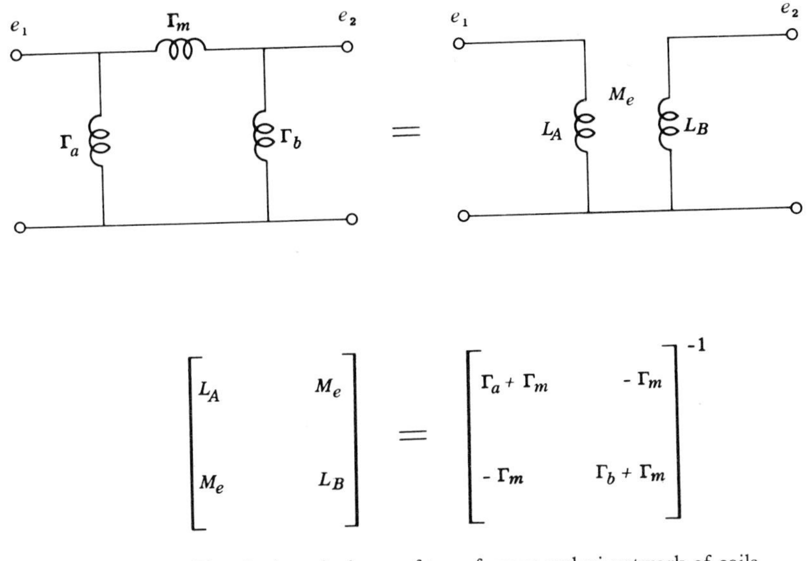

FIG. 1.10-2. Terminal equivalence of transformer and pi network of coils.

requires [with $\Gamma_a\Gamma_b + \Gamma_m(\Gamma_a + \Gamma_b) = \Delta_\Gamma$] that $(\Gamma_b + \Gamma_m)/\Delta_\Gamma = B/(AB - H^2) \geqslant 0$, and $(\Gamma_a + \Gamma_m)/\Delta_\Gamma = A/(AB - H^2) \geqslant 0$. Obviously A, $B > 0$, so it is only necessary to prove $AB - H^2 \geqslant 0$. This inequality follows from the fact that $2V_m = Ax^2 + B\theta^2 - 2Hx\theta$ is a positive definite quadratic function, which is a formal way of saying that the energy $2V_m$ is always positive, no matter what nonzero values x and θ may have. It is known[1] that $AB - H^2 > 0$ follows from the positive-definiteness of V_m. The transformer is therefore realizable; the resulting electrical equivalent of Fig. 1.10-1a is shown in Fig. 1.10-1c.

The above represents only one of several possible ways of finding an equivalent

[1] A. E. Guillemin, "The Mathematics of Circuit Analysis," p. 144 ff. Wiley, New York, 1949.

network in the velocity-voltage analogy. We could have taken the two generalized coordinates as $y_1 = x + \theta d_1$, $y_2 = x - \theta d_2$. We would then have $2V_m = K_1 y_1^2 + K_2 y_2^2$ and $2T_m = [M(\dot{y}_1 d_2 - \dot{y}_2 d_1)^2/(d_1 + d_2)^2] + [Mh^2(\dot{y}_1 + \dot{y}_2)^2/(d_1 + d_2)^2] = E\dot{y}_1^2 + F\dot{y}_2^2 + 2N\dot{y}_1\dot{y}_2$. We should then seek a network of capacitors whose $T_e = T_m$, but the node voltages would now be equivalent to the y_1, y_2 variables.

The Force-Voltage Analogy

A force-voltage analogy of Fig. 1.10-1a is obtainable by equating (using the y's of the last paragraph for the mechanical variables) $2V_m = K_1 y_1^2 + K_2 y_2^2 = 2V_e = S_1 q_1^2 + S_2 q_2^2$ (where $q = \int i dt$); $2V_e = E\dot{y}_1^2 + F\dot{y}_2^2 + 2N\dot{y}_1\dot{y}_2 = L_1 i_1^2 + L_2 i_2^2 + 2Mi_1 i_2$. Again transformer realization of the latter is always possible because $2V_e$ is a positive definite quadratic function.

§ 1.11 Conditions under Which a Passive Linear Electrical Analog Is Possible

It is quite easy to state the conditions under which a passive linear electrical analog of mechanical (or other) systems is possible. The key to this matter is the group of scalar energy functions. In passive, linear, electrical systems, they have the form $T = \sum A_{ij}\dot{q}_i\dot{q}_j$, $V = \sum B_{ij}q_i q_j$, $D = \sum H_{ij}\dot{q}_i\dot{q}_j$ where (1) the A, B, H are constants, (2) T, V, D are positive definite quadratic functions, (3) all the voltage drops (or current flows in the other analogy) are attributable to these three energy functions. Therefore any mechanical (or other) systems whose scalar energy functions satisfy the first two conditions, and in which all the internal forces can be derived from these three energy functions, must have linear passive electrical analogs. When the motion is wholly rectilinear or wholly rotational, and the only elements are masses, springs, and viscous dampers (or their equivalent), then the analogous electrical model can be drawn by inspection.

Systems which satisfy the three conditions of the previous paragraph, but in which the motion is combined linear-rotational, often lead to admittance matrices which are not dominant. The networks cannot then be drawn by inspection. They can, however, be obtained directly from the energy functions. The first step is to derive the Γ, C, G matrices (or the L, S, R matrices in the force, torque-voltage method) from the energy functions. The mechanical kinetic energy has the form $2T_m = \sum_{i=1}^{n} \sum_{j=1}^{n} A_{ij}\dot{y}_i\dot{y}_j$, with $A_{ij} = A_{ji}$. Consider the capacitive electrical network in Fig. 1.11-1a. The electric field energy is $2T_e = (C_1 + C_{12} + C_{13})e_1^2 - 2C_{12}e_1e_2 + \ldots$. If we pick $e_1 = \dot{y}_1$, $e_2 = \dot{y}_2$, etc., then in order that $T_m = T_e$, it is necessary that $C_1 + C_{12} + C_{13} = A_{11}$, $-C_{12} = A_{12}$. Hence in the short-circuit admittance matrix of the electrical network, given by $[Y] = s[C] + [\Gamma]s^{-1} + [G]$, the matrix $[C]$ has the matrix elements $c_{11} = A_{11}$,

$c_{12} = A_{12}$ or, in general, $c_{ij} = A_{ij}$. Similarly, if $2V_m = \Sigma\Sigma\, B_{ij}y_iy_j$, the matrix $[\Gamma]$ has the elements $\gamma_{ij} = B_{ij}$, and if $2D = \Sigma\Sigma\, H_{ij}\dot{y}_i\dot{y}_j$, $[G]$ has the elements $g_{ij} = H_{ij}$. In this way we obtain the short-circuit admittance matrix of a network with n accessible independent nodes plus one reference node (i.e., n node pairs). In the force, torque-voltage analogy, we end up with the open-circuit impedance matrix of a network with n accessible pairs of terminals (n-port network). The problem is then one of formal network synthesis, and it has been treated by several investigators.[1]

The following is an alternative procedure in which the network synthesis part of the problem is simplified. A set of variables is chosen such that the quadratic T function reduces to a sum of squares.[2] Usually there is no problem in realizing the resulting new G matrix. If there is, then instead we simultaneously reduce both the G and C quadratic functions to diagonal forms.[3] The resulting non-diagonal Γ matrix may always be realized (because it is positive definite) by an n coil transformer. We have presented the necessary realization relations in Fig. 1.10-2 for $n=2$. The general relations for any n are obtained as follows: In the mechanical system, let $2V = \Sigma\Sigma\, B_{ij}x_ix_j$ with $B_{ij} = B_{ji}$. Form the Γ matrix, whose elements are $\gamma_{ij} = B_{ij}$. Evaluate the L matrix from the relation $[L] = [\Gamma]^{-1}$. The n-coupled coils are realized as in Fig. 1.11-1b for $n = 3$. The relation $[L] = [\Gamma]^{-1}$ is derived as follows: A typical equation for the ith transformer coil is $e_i = s[l_{i1}i_1 + l_{i2}i_2 + \cdots + l_{in}i_n]$, or, in matrix form, $[E] =$

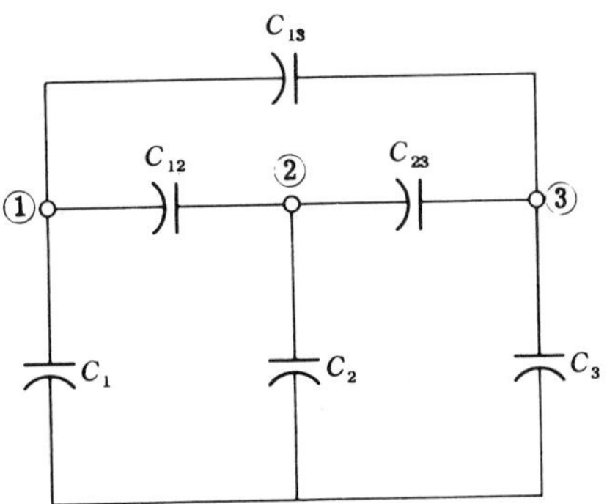

FIG. 1.11-1. (a) Derivation of capacitance network form the kinetic energy quadratic.

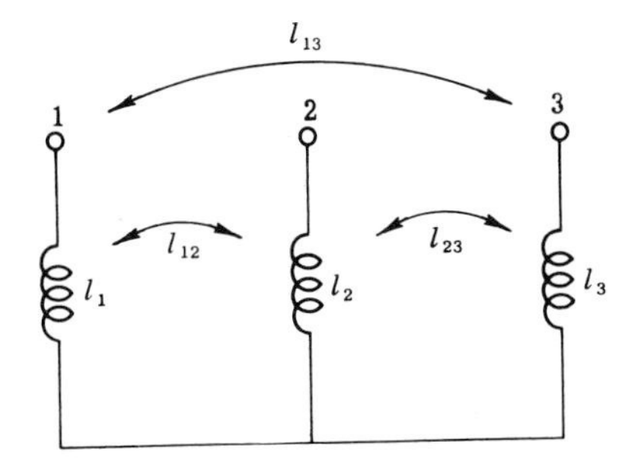

FIG. 1.11-1. (b) Derivation of three-coil transformer from the potential energy quadratic.

---

[1] Y. Oono, Synthesis of a finite 2n-terminal network by a group of networks each of which contains only one ohmic resistance. *J. Math. and Phys.*, April, 1950; B. McMillan, Introduction to formal realizability theory. *Bell System Tech. J.* **31**, 217-279, 541-600 (1952); B. D. H. Tellegen, Synthesis of $2n$-poles by networks containing the minimum no. of elements. *J. Math.* and *Phys.* **32**, 1-18 (1953); V. Belevitch, On the Brune process of n-ports. *Trans. IRE* **CT-7** (3), 280-296 (1960).

[2] A. E. Guillemin, "The Mathematics of Circuit Analysis," p. 137. Wiley, 1949, New York.

[3] See page 156 of A. E. Guillemin.[2]

$[sL]\,[I]$. In the network of uncoupled coils, $[I] \triangleq [s^{-1}\Gamma]\,[E]$. Therefore $[E] = s[\Gamma]^{-1}\,[I] = s[L]\,[I]$, so $[L] = [\Gamma]^{-1}$.

Problems in Which Passive Electrical Analogs Are Not Possible

There are very many mechanical systems which violate the three previously listed requirements for passive electrical equivalents. We may nevertheless find electrical equivalents, except that the latter may have negative or nonlinear elements. Consider, for example, Fig. 1.11-2a, where a particle of mass m is

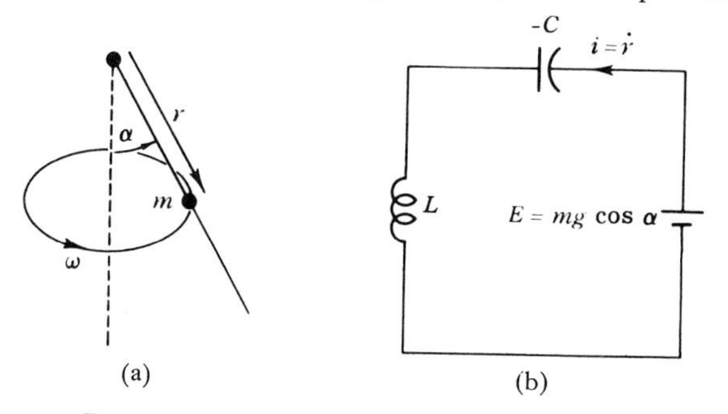

FIG. 1.11-2. System with nonpassive electrical analog.

constrained to move along a track r, which is rotated at constant angular velocity ω, as shown. Here $2T = m(\dot{r}^2 + \omega^2 r^2 \sin^2\alpha)$, which is a function of r, a position coordinate. This violates one of the conditions previously laid down, and a passive electrical analog cannot be obtained. However, the Lagrange equations may be used to obtain the system differential equations and we may then try to find an electrical analog. In such cases, the resulting analog will not be passive; controlled sources, negative elements, and/or nonlinear elements will be present.

We may extend our technique to encompass such an example. Reference is made to the Lagrange equation (1.9,1). If the kinetic energy, T, has a term which is proportional to q^2 (but which is independent of $\dot{q}$, as in the above example), then that particular term should give rise to exactly the same kind of electrical element as $\partial V/\partial q$, but of opposite sign, because of the negative sign associated with $\partial T/\partial q$ in Eq. (1.9,1). Thus, in the above example, using the force-voltage analogy, we have $L = m$ with current $i = \dot{r}$, and a negative capacitor $-C = -(m\omega^2 \sin^2\alpha)^{-1}$ whose current is i. Also $V = -mgr\cos\alpha$; $\partial V/\partial r = -mg\cos\alpha$ is analogous to a constant voltage source, $-mg\cos\alpha$. The resulting circuit is shown in Fig. 1.11-2b. In general, then, if the kinetic energy has terms which have the same form as V (of the form $\sum A_{ij} q_i q_j$ with A_{ij} constant), these terms are represented by the negative of those due to springs in V (depending on the kind of analogy).

What about the stability of a system with a negative capacitance? A positive capacitance may be compared to a bank account with no overdrawing credit, whereas a negative capacitor constitutes a loan acccount. The question is whether the credit source is fictitious, i.e., whether there must be other accounts whose net balance is always more than the amount borrowed. This is when the negative capacitors can be eliminated by transformers, which can always be done if the capacitors are derived from positive definite potential functions (in the force-voltage analogy). Terms in T of the form $T^* = \Sigma A_{ij}q_iq_j$ indicate internal sources (genuine loan accounts) from which energy can be indefinitely extracted, only if the total of $V - T^*$ is not positive definite. In such a case the system is not passive and may be unstable.

Example—*Compound Pendulum* (Fig. 1.11-3a). The velocity of m_1 is $a\dot{\theta}_1$, that of m_2 for very small θ_1, θ_2 is closely $a\dot{\theta}_1 + b\dot{\theta}_2$. (For larger θ_1 or θ_2 the velocity of m_2 is a function of position coordinates and the system is nonlinear.) In the velocity-voltage analogy, a capacitor $C_1 = m_1$ has the voltage $e_1 = a\dot{\theta}_1$ across it, while $C_2 = m_2$ has $e_1 + e_2 = a\dot{\theta}_1 + b\dot{\theta}_2$ across its terminals. Hence they are connected as shown in Fig. 1.11-3b. Gravity forces can be attributed to the

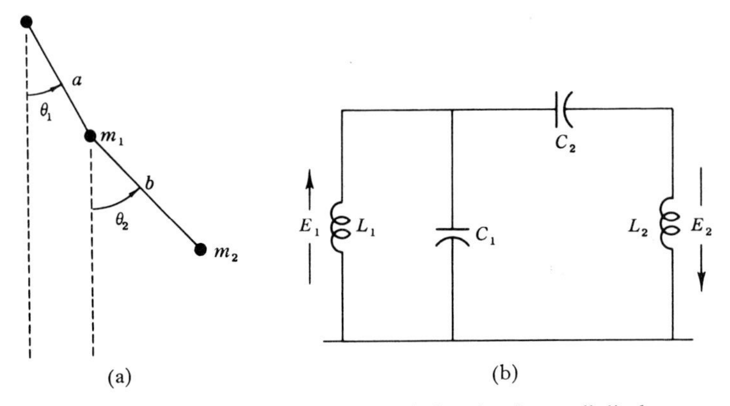

FIG. 1.11-3. Compound pendulum and its electrical analog for small displacement.

potential function $V = -m_1ga\cos\theta_1 - m_2g(a\cos\theta_1 + b\cos\theta_2)$. $\partial V/\partial a\theta_1 = g\sin\theta_1(m_1 + m_2) \doteq g\theta_1(m_1 + m_2) = (ga^{-1})(m_1 + m_2)a\theta$. Therefore an inductor $L_1 = a/g(m_1 + m_2)$ has the voltage $e_1 = a\dot{\theta}_1$ across its terminals. Similarly, $\partial V/\partial b\theta_2 = m_2g\sin\theta_2 \doteq m_2g\theta_2$; therefore, $L_2 = b/gm_2$ is from node 2 to ground. The resulting circuit is shown in Fig. 1.11-3b. Another way to find the electrical elements due to gravity is as follows. The force of gravity on m_1, in the displacement $d\theta_1$, does work $dW = a(m_1g)\sin\theta_1\,d\theta_1 \doteq (m_1g/a)(a\theta_1)d(a\theta_1)$, which may be compared to the work done by a spring of elastance K compressed by amount x, in a displacement dx (signs take care of themselves), $dW = Kxdx$. Therefore the force of gravity is analogous to a spring of elastance m_1g/a with motion coordinate $a\theta_1$. The force of gravity on m_2 does work $dW = m_2ga(\sin\theta_1)d\theta_1 + b\sin\theta_2 d\theta_2 \doteq m_2ga^{-1}a\theta_1 d(a\theta_1) + b^{-1}b\theta_2 d(b\theta_2)$. Therefore gravity on m_2 is analogous to two springs, one of elastance m_2g/a with motion $a\theta_1$, and the second of elastance m_2g/b with motion $b\theta_2$.

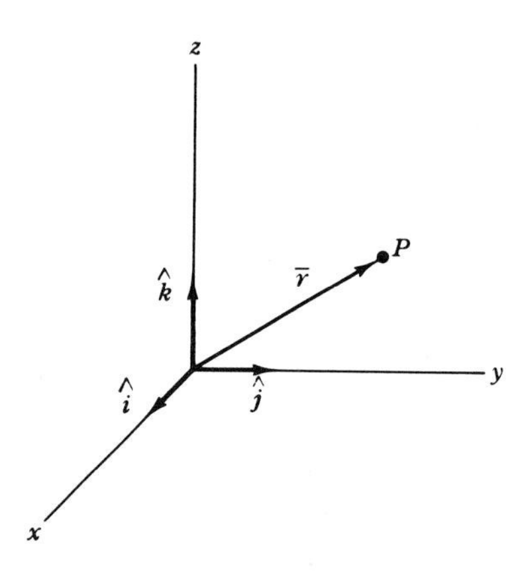

FIG. 1.11-4. (a) Derivation of kinetic energy; xyz-axes fixed in and rotating with earth.

Example — *Coriolis Force.* Newton's laws apply to motion relative to that of a nonaccelerating set of axes. It is desirable to revise them so that they may be applied for motion relative to a set of axes fixed in and moving with the earth. Only the rotation of the earth is considered here. In Fig. 1.11-4a the origin is at the earth's center, and the z axis is the axis of rotation at an angular velocity of $\omega = 2\pi/86{,}164 = (7.29)10^{-5}$ rps. The vectors $\hat{\imath}$, $\hat{\jmath}$, $\hat{k}$ are unit vectors, and the directions of the first two change due to the rotation, i.e., $d\hat{\imath}/dt = \omega\hat{\jmath}$, $d\hat{\jmath}/dt = -\omega\hat{\imath}$. The vector position of a particle of mass m at P, is $\bar{r} = x\hat{\imath} + y\hat{\jmath} + z\hat{k}$. Its velocity is

$$(\dot{\bar{r}})_{\text{true}} = \dot{x}\hat{\imath} + \dot{y}\hat{\jmath} + \dot{z}\hat{k} + x(d\hat{\imath}/dt) + y(d\hat{\jmath}/dt) = \dot{x}\hat{\imath} + \dot{y}\hat{\jmath} + \dot{z}\hat{k} + \omega(x\hat{\jmath} - y\hat{\imath}).$$

Consequently, the true kinetic energy of the particle is

$$0.5m(\dot{\bar{r}})_{\text{true}} \cdot (\dot{\bar{r}})_{\text{true}} = T_{\text{true}} = 0.5m(\dot{x}^2 + \dot{y}^2 + \dot{z}^2) + 0.5m\omega^2(x^2 + y^2) + \omega m(x\dot{y} - y\dot{x}) = T + T_{cp} + T_{\text{cor}}.$$

$T = 0.5m(\dot{x}^2 + \dot{y}^2 + \dot{z}^2)$ is the kinetic energy of the particle as seen by an observer on earth. T_{cp} is a centripetal component, i.e., the force associated with T_{cp} (whose components are $-\hat{\imath}\partial T_{cf}/\partial x = -\hat{\imath}m\omega^2 x$ and $-\hat{\jmath}\partial T_{cf}/\partial y = -\hat{\jmath}m\omega^2 y$) is the centripetal force of rotation, acting in the instantaneous direction of $-\bar{r}$. This force is therefore measured with the gravity force and need not be separately considered. The forces due to $T_{\text{cor}} \triangleq \omega m(x\dot{y} - y\dot{x})$ may be obtained from the Lagrange equation: $(d/dt)(\partial T/\partial\dot{q}) - \partial T/\partial q = 2\omega m(\dot{x}\hat{\jmath} - \dot{y}\hat{\imath}) = \bar{F}_{\text{cor}}$. We would like, in any future problem, to calculate T just as if the earth were motionless, i.e., take $T = 0.5m(\dot{x}^2 + \dot{y}^2 + \dot{z}^2)$. It is therefore necessary to transfer $\bar{F}_{\text{cor}}$ to the right-hand side of Eq. (1.9,1) and add $-\bar{F}_{\text{cor}}$ to any other external forces.

Suppose that in a problem concerning the motion of a particle, T has been calculated as if the earth were motionless, and the resulting electrical analog of the problem (voltage-velocity analogy) is that shown in Fig. 1.11-4b (with the dotted portion omitted). The circuit is in error because of the motion of the earth. The error due to the earth's rotation may be corrected by adding the electrical analog of the Coriolis components, i.e., the controlled current sources

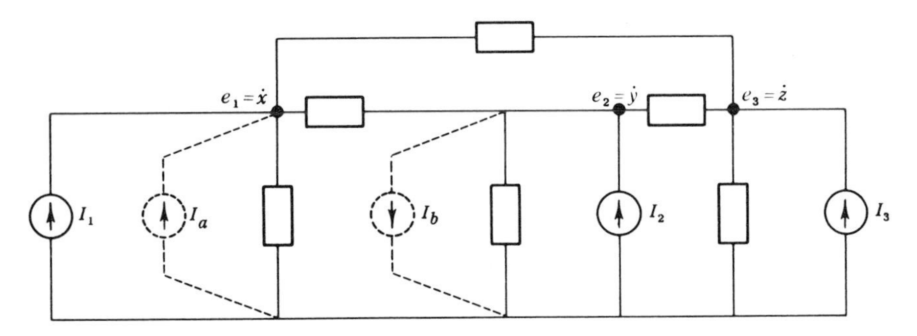

FIG. 1.11-4. (b) Analog for Coriolis force. ($I_a = 2\,\omega m E_2$; $I_b = 2\,\omega m E_1$.)

$I_a = 2\omega m E_2$ into node 1 and $I_b = -2\omega m E_1$ into node 2. A little thought reveals that it is impossible to replace these controlled sources by the passive elements G, Γ, or S. Yet these two controlled sources constitute a passive element because their total power input is $2\omega m e_2 e_1 - 2\omega m e_1 e_2 = 0$. The two sources represent a four terminal passive element known as a gyrator, with the symbolized representation of Fig. 1.11-5.

The equations of the circuit in Fig. 1.11-4b do not have reciprocal elements; thus, $I_1 = y_{11}E_1 + (y_{12} - \alpha)E_2 + y_{13}E_3$, while $I_2 = (y_{12} + \alpha)E_1 + y_{22}E_2 + y_{23}E_3$, where $\alpha = -2\omega m$. It is possible to eliminate the gyrator and obtain a different passive reciprocal circuit, which is equivalent to Fig. 1.11-4b only with respect to the value of e_1. One way to do this is to keep all y'_{ij} of the new circuit equal to the y_{ij} of the old, except for y'_{11}, whose value is set at $y'_{11} = y_{11} + \alpha^2 y_{33}/\Delta_{11}$, where $\Delta_{11} = y_{22}y_{33} - y_{23}^2$. In the new circuit $I_1' = I_1$; also $I_2' = I_2(1 + \alpha y_{33}/\Delta_{21})$, $I_3' = I_3(1 - \alpha y_{23}/\Delta_{31})$. However, the resulting e_2', e_3' are considerably different from e_2, e_3. It would be useful if a new passive reciprocal circuit could be found to replace Fig. 1.11-4b, such that the new e_1', e_2', e_3' were very simply related to e_1, e_2, e_3.

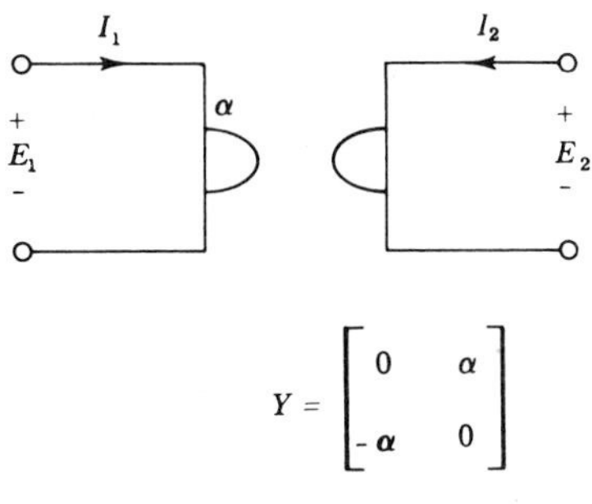

FIG. 1.11-5. Gyrator representation. $I_1 = \alpha E_2$; $I_2 = -\alpha E_1$.

It should be noted that gyrators result whenever T has any terms of the type $q_i\dot{q}_j$. Such terms can be written in the form $0.5(q_i\dot{q}_j - q_j\dot{q}_i) + 0.5(q_i\dot{q}_j + q_j\dot{q}_i)$. The first term leads, as we have seen, to a gyrator. The second term can be deleted, because $d/dt[\partial/\partial\dot{q}_i(q_i\dot{q}_j + q_j\dot{q}_i)] - (\partial/\partial q_i)(q_i\dot{q}_j + q_j\dot{q}_i) = 0$. It is also worth noting that linear systems which contain vacuum tubes, transistors, or equivalent nonelectrical elements (e.g., mechanical or hydraulic amplifiers) are not inherently nonpassive. The circuit representation of these systems usually include controlled sources. But any linear controlled source can be replaced by a combination of gyrator and R, L, C elements (some of the latter may, however, be negative). This follows from the fact that any matrix can always be expressed as the sum of a symmetrical matrix and an antisymmetrical ($q_{ij} = -q_{ji}$ for $i \neq j$) matrix. Therefore the system is always realizable by means of R, L, C (some possibly negative), ideal transformers, and gyrators. The system is nonpassive only if one or more of the appropriate quadratic functions is not positive definite. In that case, some negative elements must appear in the equivalent electrical circuit.

To complete this discussion on testing a system for passiveness, it is only necessary to indicate which quadratic functions must be tested for positive-

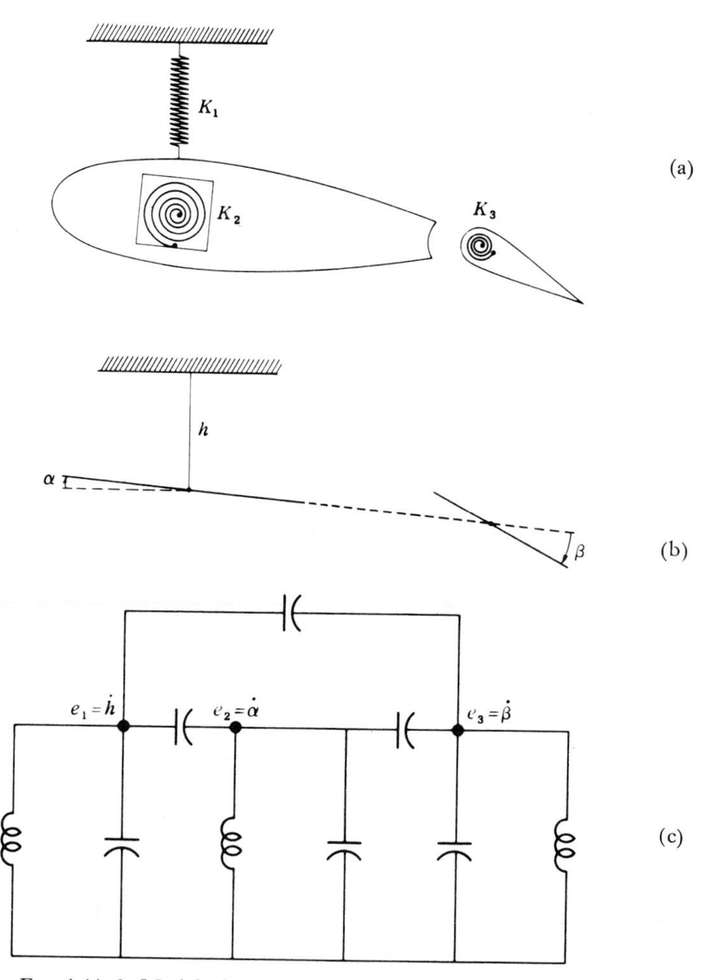

FIG. 1.11-6. Model of an aircraft and its electrical analog.

definiteness. All terms in T of the form $\sum a_{ij}\dot{q}_i\dot{q}_j$ form one quadratic function. In accordance with the discussion around Fig. 1.11-2a, terms in T of the form $\sum a_{ij}q_iq_j$ are subtracted from similar type terms in V to form a second quadratic function. Terms in T of the form $a_{ij}q_i\dot{q}_j$ need not be checked, since they lead to gyrators. Terms in V of the form $a_{ij}(q_i\dot{q}_j - q_j\dot{q}_i)$ also lead to gyrators. However, $a_{ij}(q_i\dot{q}_j + q_j\dot{q}_i)$ terms in V lead to precisely the same elements as $a_{ij}\dot{q}_i\dot{q}_j$ in the dissipation function D, so they may be removed from V, the proper replacements inserted in D, and the latter tested for positive-definiteness. A term $2bq_i\dot{q}_j$ in D leads, in the voltage-velocity analogy, to a controlled source $2b\int e_i dt$ in the jth node ($2b/s$ in transform notation). It can be replaced by a gyrator of value b/s, coupling nodes j and i ($I_j = bE_i/s$, $I_i = -bE_j/s$), and reciprocal inductances b from nodes 1 and 2 to ground, and $-b$ from node 1 to node 2. Therefore, in checking for passiveness, the term $2bq_iq_j$ should be added to V. Finally, it should be noted that in the Lagrange formulation, only q, $\dot{q}$ may appear as variables in T, V, D, and we have in the above listed all possible terms which lead to linear system equations.

Example—*Aircraft Stability*. A simple two-dimensional model of an airplane wing and aileron[1] is shown in Fig. 1.11-6a. In this model, the wing has infinite span and its motion is assumed to be describable by the coordinates h, α, β of Fig. 1.11-6b. The internal forces are taken as the springs, K_1 against vertical bending (h), K_2 against wing torsion about P (α), and K_3 against aileron rotation (β). We rely on the aeronautical engineer that such a model is realistic for some types of airplane motion and are here only interested in finding the general form of the electrical analog. It is not difficult to determine that the kinetic energy per unit length of span has the form $2T = A\dot{h}^2 + B\dot{\alpha}^2 + C\dot{\beta}^2 + 2D\dot{h}\dot{\alpha} + 2E\dot{h}\dot{\beta} + 2F\dot{\alpha}\dot{\beta}$. $2T$ must be a positive definite function of $\dot{h}$, $\dot{\alpha}$, $\dot{\beta}$, because the system is conservative. Therefore the electrical circuit due to T consists of the capacitors shown in Fig. 1.11-6c. The potential energy due to the springs has the form $2V = K_1h^2 + K_2\alpha^2 + K_3\beta^2$, leading to the three coils in Fig. 1.11-6c. It is possible that some of the capacitors are negative, but we are certain that a network with positive elements may be arranged if ideal transformers are added. In any case, the above represents the wing by itself, without any interaction with the surrounding air medium.

In order to find the electrical equivalent of the effect of the surrounding fluid, it is necessary to assume a model of the fluid, derive the resulting local forces, integrate over the surface of the wing, and state the results in terms of the coordinates h, α, β. The electrical analogy may be applied to this problem too, but we will only state the result (for an incompressible fluid), which in electrical equivalent terms and in transform notation ($E_1 = \mathscr{L}e_1$, etc.) have the

[1] R. H. Scanlan and R. Rosenbaum, "Introduction to the Study of Aircraft Vibration arel Flutter," Chapter 8. Macmillan, New York, 1951.

form[1]: $I_1 = (Hs + J)E_1 + (Ks + M + Ns^{-1})E_2 + (Qs + R + Ts^{-1})E_3$. The forms for I_2 and I_3 are similar. It is seen that I_1, I_2, I_3 are not independent sources—such sources would appear in the propulsive components. They are controlled sources, and it is impossible, because of their nonpassive character, to replace all of them by passive elements. In the theory of active networks, there is interest in isolating the "activity" of the system into the minimum number of fundamental active elements.[2] In the present problem, the circuit representation may be completed by the use of controlled sources.

§ 1.12 Electrical Analogs of the Gyroscope[3]

The Lagrange method permits a comparatively simple formulation of gyroscopic problems. We consider a rigid body symmetrical with respect to one axis, and which is constrained to rotate about a point which is fixed both in the body and in space. The usual procedure is to fix a set of axes (x, y, z) in the body, and characterize the motion of the body by means of the motion of this body set of axes (x, y, z) with respect to a set stationary in space (X, Y, Z). The

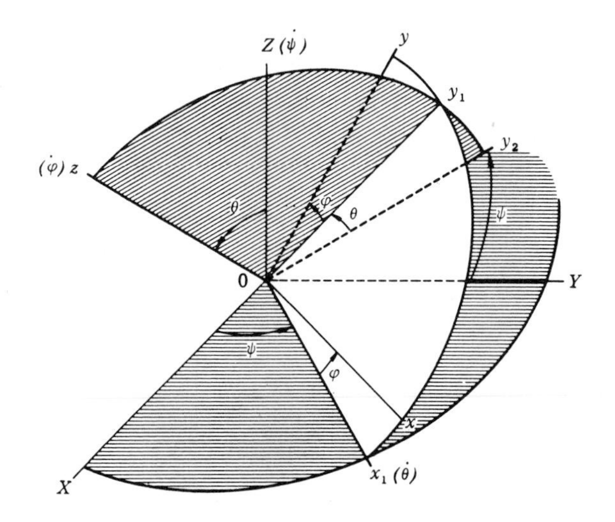

FIG. 1.12-1. XYZ stationary space axis; xyz-body axes; $\theta\varphi\psi$-Euler angles.

[1] R. H. Scanlan and R. Rosenbaum, "Introduction to the Study of Aircraft Vibration and Flutter," Chapter 8. Macmillan, New York, 1951.

[2] H. J. Carlin and D. C. Youla, Network synthesis with negative resistors. *In* "Proceedings of the Symposium on Active Networks and Feedback Systems," pp. 27-67. Interscience, New York, 1961.

[3] The principal references used for this section were: J. B. Scarborough, "The Gyroscope," Interscience, New York, 1958; A. G. Webster, "The Dynamics of Particles and of Rigid, Elastic and Fluid Bodies," 2nd ed., Sections 90-94. Dover Publ., New York, 1959.

two sets of axes are assigned a common origin at the fixed point. The most commonly used coordinates for specifying the motion of the body axes with respect to the space axes, are the Euler angles. These angles are defined by means of a set of three independent angular rotations which move the body set of axes from their instantaneous position into coincidence with the stationary set of axes. Suppose the two sets of axes are as shown in Fig. 1.12-1. The three independent angular rotations are as follows:

(1) Rotate the Oxy plane about the Oz axis through the angle $-\varphi$ until Ox lies in the plane of OXY along the line Ox_1. The Oy axis in the process moves to Oy_1, and Oy_1 lies in the same plane as OzZ.

(2) Rotate the Ozy_1 plane about the Ox_1 axis through the angle $-\theta$ until Oz coincides with OZ. In the process Oy_1 is also rotated until it lies on Oy_2 in the plane OXY.

(3) Rotate the Ox_1y_2 plane through the angle $-\psi$ about the OZ axis until Ox_1 coincides with OX and hence Oy_2 lies on OY.

We choose the z axis along the axis of symmetry of the rigid body and then[1] $2T = A(\omega_x^2 + \omega_y^2) + C\omega_z^2$, where ω_x, ω_y, ω_z are the components of the angular velocity vector $\bar{\omega}$, along Ox, Oy, Oz; A is the moment of inertia of the body with respect to the x or y axis, and C is the same with respect to the z axis. Since θ, ψ, φ are the generalized coordinates, it is necessary to express ω_x, ω_y, ω_z in terms of these angles.[1] It is first noted that the angular velocities $\dot{\psi}$, $\dot{\theta}$, $\dot{\varphi}$, are along OZ, Ox_1, Oz, respectively. It is convenient to replace $\dot{\psi}$ along OZ by the equivalent: $\dot{\psi}\cos\theta$ along Oz and $\dot{\psi}\sin\theta$ along Oy_1. The total angular velocity component along Ox is then $\omega_x = (\dot{\psi}\sin\theta)\sin\varphi + \dot{\theta}\cos\varphi$; also $\omega_y = (\dot{\psi}\sin\theta)\cos\varphi - \dot{\theta}\sin\varphi$, $\omega_z = \dot{\psi}\cos\theta + \dot{\varphi}$. These quantities are substituted into the expression for $2T$, and the result is $2T = A(\dot{\theta}^2 + \dot{\psi}^2\sin^2\theta) + C(\dot{\varphi} + \dot{\psi}\cos\theta)^2$. Also, because of the symmetry, the center of gravity must lie on Oz, say at $z = h$, so that $V = mgh\cos\theta$. Even if we replace $\cos\theta$ by 1 and $\sin\theta$ by θ, the system would be nonlinear because T is a function of the position coordinates. Hence, the derivation of the electrical equivalent by inspection is not possible. We therefore apply the Lagrange equation (1.9,1) for each of ψ, θ, φ and obtain

$$Q_\psi = (d/dt)\,[\dot{\psi}\,(A\sin^2\theta + C\cos^2\theta)] + (Cd/dt)\,[\dot{\varphi}\cos\theta];$$
$$Q_\theta = A\ddot{\theta} + \dot{\psi}^2\sin\theta\cos\theta\,(C - A) + \dot{\varphi}\dot{\psi}\,C\sin\theta - mgh\sin\theta;$$
$$Q_\varphi = (Cd/dt)\,[\dot{\psi}\cos\theta + \dot{\varphi}].$$

It is important to note that the above Q_ψ, Q_θ, Q_φ represent *total* torque vectors in the directions OZ, Ox_1, Oz, respectively. Therefore to find the components of torque along the $Oxyz$ axes, we have immediately $Q_z = Q_\varphi$. To find

[1] E. G. Keller, "Mathematics of Modern Engineering," Vol. II, Sections 1.17, 1.22. Dover Publ., New York, 1942.

the other components, we first find the component along Oy_1, an axis which is normal to Ox_1 and Oz. Since $Q_\psi = Q_\varphi \cos\theta + Q_{y1}\sin\theta$,

$$Q_{y1} = (Q_\psi - Q_\varphi \cos\theta)/\sin\theta = A\ddot{\psi}\sin\theta + (2A - C)\dot{\theta}\dot{\psi}\cos\theta - C\dot{\varphi}\dot{\theta}.$$

Finally, $Q_y = Q_{y1}\cos\varphi - Q_\theta \sin\varphi$, $Q_x = Q_\theta\cos\varphi + Q_{y1}\sin\varphi$.

The analogous electrical circuit (voltage-velocity analog) is shown in Fig. 1.12-2. It consists of nonlinear elements and nonlinear controlled sources. (Note

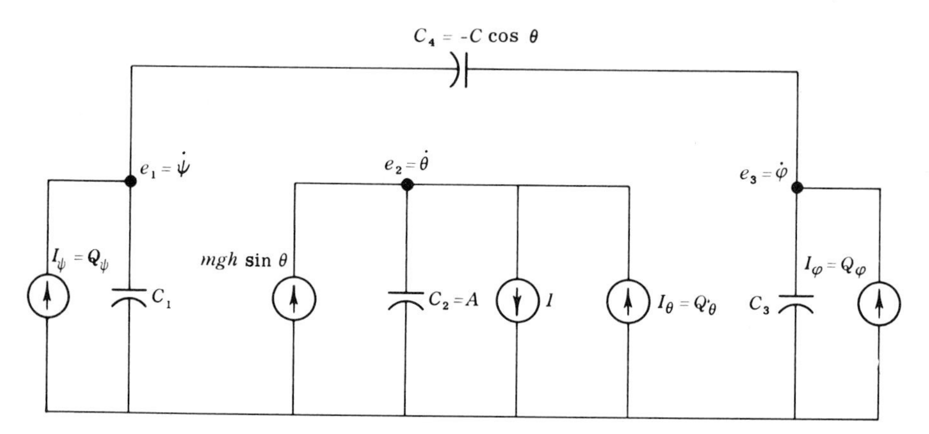

FIG. 1.12-2. Model of gyroscope. $C_3 = C(1 + \cos\theta)$; $I = (C - A)e_1^2 \sin\theta\cos\theta + Ce_3e_1\sin\theta$; $C_1 = A\sin^2\theta + C\cos^2\theta + C\cos\theta$.

in Fig. 1.12-2 that the current in a capacitor is $(d/dt)Ce$ and not $C\dot{e}$. The two are not the same when C is not constant.) The usual procedure in analyzing Fig. 1.12-2 is to assign *a priori* values to a number of variables, such that what remains is solvable. The values of imposed torques (I_ψ, etc.) needed to realize such a motion can then be calculated. Thus, the theory of the gyroscope consists of the study of a large number of special cases. The electrical analogs of some of these simpler cases will be derived.

The gyro precession law. Suppose the motion consists of the following: $\theta = \theta_c$, a constant; $h = 0$, i.e., the center of gravity is the fixed point; the gyro spins around the z axis at $\dot{\varphi} = \omega$, a constant; the spin axis Oz rotates (precesses) about OZ with the constant angular velocity vector $\bar{\Omega}$, whose direction is therefore along the OZ axis, with $\dot{\psi} = \Omega$. Figure 1.12-2 degenerates into the linear circuit of Fig. 1.12-3, with the steady state $i_1 = i_3 = 0$, $i_2 = \Omega\sin\theta\,[C\omega + \Omega(C - A)\cos\theta]$. The latter constitutes an externally applied torque in the vector direction Ox_1, i.e., always perpendicular to the OZz plane. It is therefore a torque which is trying to rotate the spin axis in the θ direction. Instead, the spin axis rotates around the OZ axis with angular velocity Ω, at a constant value of θ. At $\theta = \pi/2$, the applied torque is $Q = \Omega\omega C$, which is the well-known precession law of gyros. It should be noted that after the torque is applied, there is a transient period during which the kinetic energy is increased (i.e., C_4 and C_1 in Fig. 1.12-3 are charged up), and the precession comes up to its steady-state value. Thereafter the constant torque does no work, which indicates that there is an equal and opposite torque reaction from the gyroscope supports.

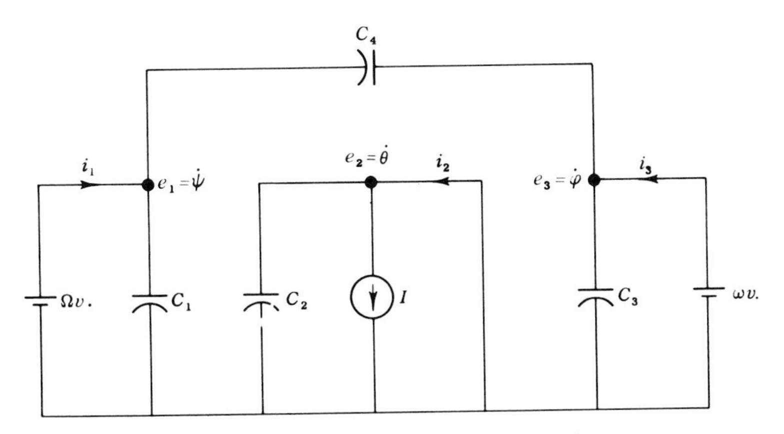

FIG. 1.12-3. Gyroscope model for a special case.

Forced rotation of spin axis. Suppose $\theta = \pi/2$ is fixed, $\dot{\varphi} = \omega$ is constant, and $\dot{\psi}$ originally zero, abruptly changes, i.e., $\dot{\psi}(t) = \Omega u(t)$, where $u(t)$ is the step function. This corresponds approximately to the rotation of the spin axis of the flywheel of a car, of a rotor in the longtitudinal axis of a ship, as the car or ship changes direction. Figure 1.12-2 degenerates into Fig. 1.12-4. When $\dot{\psi} = 0$,

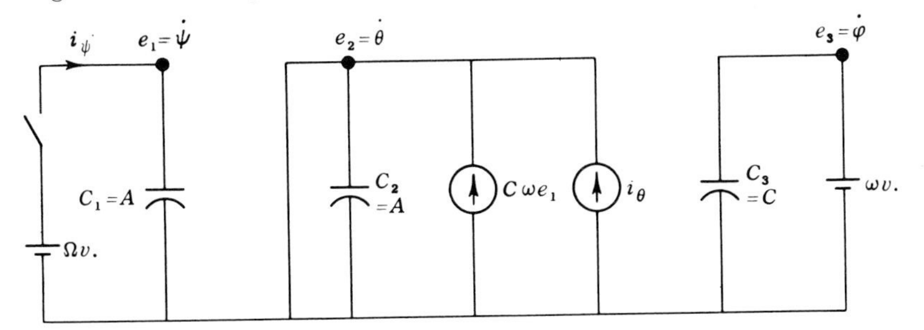

FIG. 1.12-4. Gyroscope model for forced rotation of spin axis.

$i_\psi = i_\theta = i_\varphi = 0$, there are no applied torques (all bearing friction is neglected). When the switch is closed, $i_\psi = \Omega A\delta(t)$ is the impulsive torque input required to accelerate the spin axis; $i_\theta = - C\omega e_1 = - C\omega\Omega u(t)$ is the torque reaction from the gyro bearings which appears during precession, except that there is no opposing input torque. The reaction torque therefore tends to bend the flywheel axle lengthwise.

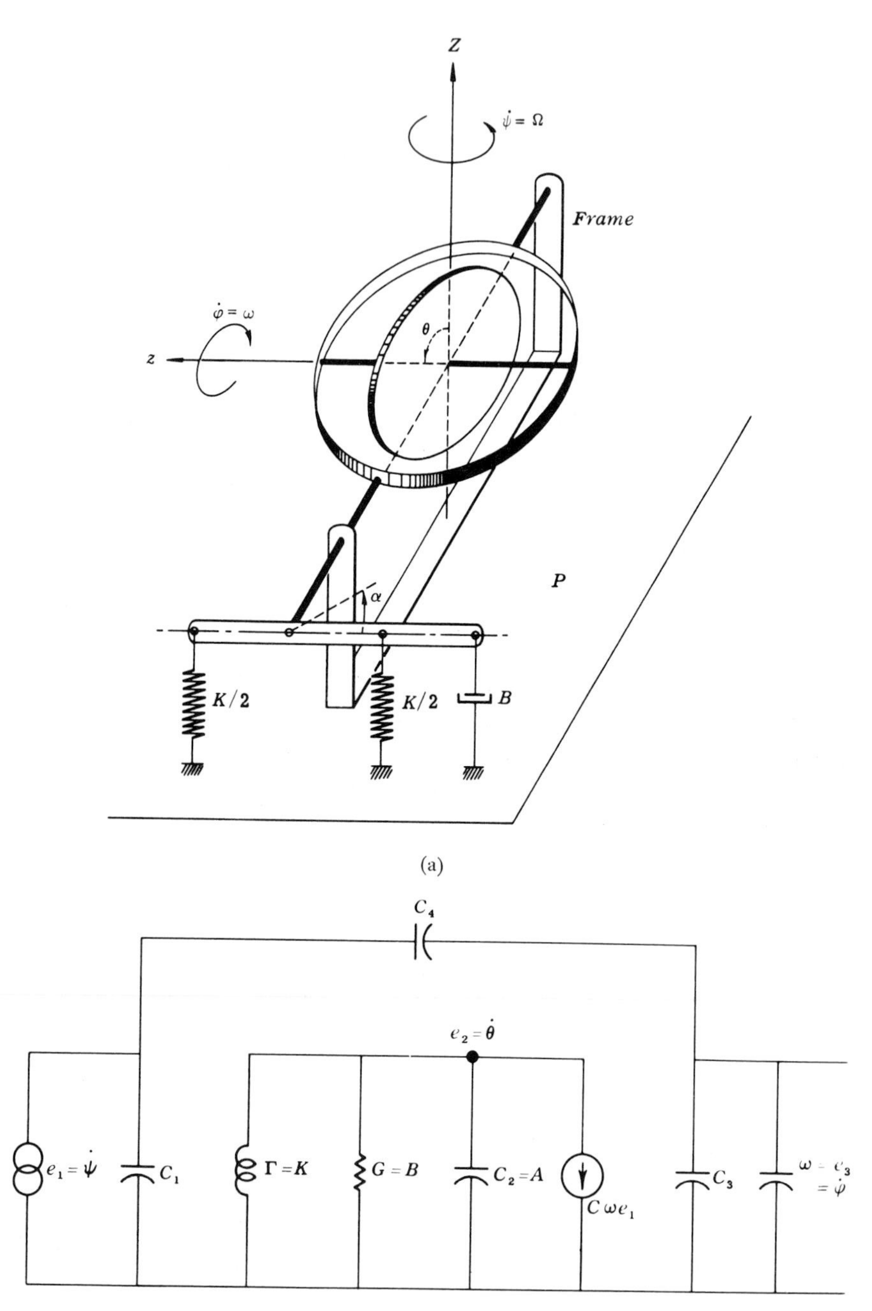

FIG. 1.12-5. Rate gyro and model.

The rate gyro. This is similar to the second case, except that θ is allowed to change slightly against the action of the spring K and damper B in Fig. 1.12-5a, from its quiescent value of $\pi/2$; but $\dot{\varphi} = \omega$, a constant. The frame is rigidly mounted on the platform P, which might represent an airplane or a ship. The object is to measure the rotation ($\dot{\psi}$) of the platform. This is the same as the second case but for the freedom of movement of θ. Figure 1.12-2 degenerates into Fig. 1.12-5b, providing the change in θ, i.e., the angle α, is small. The current at the node $e_2 = \dot{\theta}$ is $C\omega E_1(s) + E_2(s)\,[G + sC_2 + \Gamma s^{-1}] = 0$. Therefore $E_2/E_1 = -\,\omega Cs/(s^2C_2 + Gs + \Gamma)$. Suppose $e_1 = \dot{\psi} = \Omega u(t)$, so that $E_1 = \Omega/s$ and $E_2 = s\mathscr{L}\alpha(t) = -\,\omega C\Omega/(s^2C_2 + Gs + \Gamma)$. By the final value theorem, $\alpha_{s.s.} = -\,\omega\Omega C/\Gamma = -\,\omega\Omega C/K$, i.e., the output deflection angle is proportional to the rate of rotation of the platform; hence the term "rate gyro."

The rate gyro can be used as an integrator if the input is α. Then in Fig. 1.12-5b there is no generator at node 1 and $(d/dt)C_1e_1 = (d/dt)C_4(e_3 - e_1)$. If α is small, the equation becomes $A\dot{e}_1 \doteq \omega\dot{\theta}C$, and therefore $\psi = (\omega C/A)\int\theta dt$.

Gyro reference. In this application, $\theta = \pi$ when there are no disturbances, i.e., the gyro is suspended vertically below its fixed point. It will be seen that for sufficiently large spin speeds, and for disturbances which are not too large, the magnitudes of $\dot{\theta}$ and $\pi - \theta \triangleq \delta$ are small, so that terms which are higher than quadratic in $\dot{\theta}$, δ may be neglected. If friction is neglected, then $Q_\varphi = 0$ because no external torque is needed to maintain the spin velocity. The energy functions then satisfy the conditions for the modified Lagrange formulation, i.e., the coordinate φ may be eliminated because $\partial T/\partial\varphi = 0 = Q_\varphi$, etc. Hence $Q_\varphi = Cd/dt\ \ (\dot{\varphi} + \dot{\psi}\cos\theta) = 0$ and $C(\dot{\varphi} + \dot{\psi}\cos\theta) = H$, a constant. The relation $\dot{\varphi} = (H/C) - \dot{\psi}\cos\theta$ is used to replace $\dot{\varphi}$ in the modified kinetic energy function $T^* = T - H\dot{\varphi}$, which then becomes $T^* = 0.5A(\dot{\theta}^2 + \dot{\psi}^2\sin^2\theta) + H\dot{\psi}\cos\theta - (0.5H^2/C) \doteq 0.5A(\dot{\delta}^2 + \dot{\psi}^2\delta^2) - H\dot{\psi}(1 - 0.5\delta^2) - (0.5H^2/C)$.

In the present application, the disturbance torques are more easily related to the fixed space axes X, Y, Z in Fig. 1.12-1, rather than to the Euler angles. We shall therefore relate the latter to the X, Y, Z coordinates of the gyro. It is also more convenient to use $u = Y$, $v = -\,X$, i.e., to rotate the XY axis 90° about the OZ axis. The u, v coordinates of a point on the gyro z axis located at unit distance from the origin, are $u = \sin\delta\cos\psi$, $v = \sin\delta\sin\psi$. Therefore $\sin\delta = u/\cos\psi = u(1 - \sin^2\psi)^{-0.5}$, etc., which leads to $\cos\delta = [1 - (u^2 + v^2)]^{0.5} \doteq 1 - 0.5(u^2 + v^2)$; $\tan\psi = v/u$, from which $\dot{\psi} = (u\dot{v} - v\dot{u})/(u^2 + v^2)$. The last relations are used to express T^*, V in terms of u, v and the result is $2T^* = A(\dot{u}^2 + \dot{v}^2) - H(2 - u^2 - v^2)\,(u\dot{v} - v\dot{u})\,(u^2 + v^2)^{-1}$; $2V = -\,mgh(2 - u^2 - v^2)$. These expressions for T*, V are used in the Lagrange equation (1.9,1) and lead to: $A\ddot{u} - H\dot{v} + mghu = Q_u$, $A\ddot{v} + H\dot{u} + mghv = Q_v$. The electrical analog is shown in Fig. 1.12-6a, with $m_1 = m_2 = m$.

Suppose the gyro spin axis is initially suspended vertically, and Ou (OY of

Fig. 1.12-1) lies in the longtitudinal axis of the ship. The roll of the ship causes the point of suspension to move along the $Ov = -OX$ axis, and the acceleration of the point of suspension (denoted by $\ddot{r}$) is equivalent to a force $m\ddot{r}$ acting at the center of gravity of the gyro. Therefore, in Fig. 1.12-6, $i_u = mh\ddot{r}$, $i_v = 0$, and the resulting $e_1 = \dot{u}$, $e_2 = \dot{v}$ may be found by straightforward network analysis. The admittance seen looking into the gyrator terminals $\alpha\alpha'$ is $Y_{\alpha\alpha'} =$

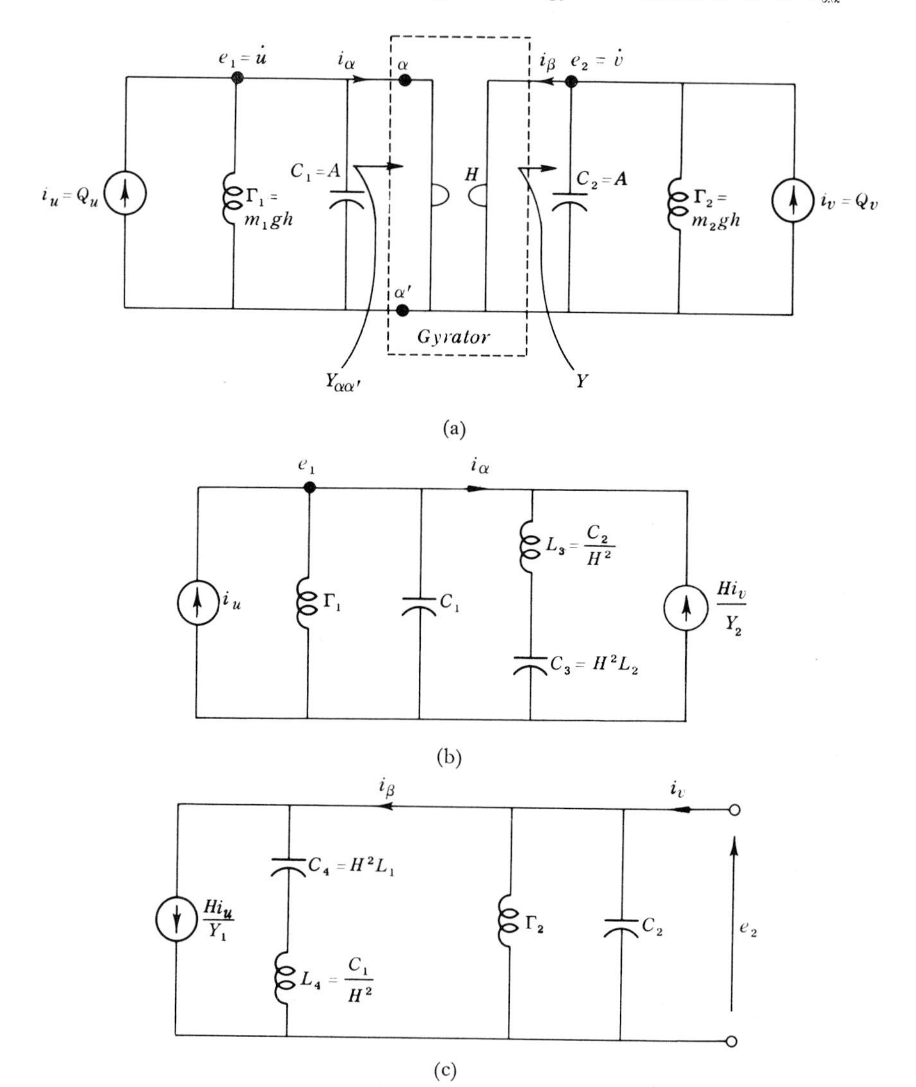

Fig. 1.12-6. (a) Electrical analog of gyro reference. (b) Equivalent of (a) from i_u, e_1 viewpoint; $Y_2 = sC_2 + \Gamma_2 s^{-1}$. (c) Equivalent of (a) from e_2, i_v viewpoint; $Y_1 = sC_1 + \Gamma_1 s^{-1}$.

H^2/Y_2, where $Y_2 = sC_2 + \Gamma_2 S^{-1}$, because $Y_{\alpha\alpha'} = I_\alpha/E_1 = -HE_2/E_1 = -H^2E_2/(I_\beta) = H^2/Y_2$. Thus if a gyrator H is terminated at one end with an admittance Y, the input admittance at the other end is H^2/Y; the gyrator acts as an impedance inverter. From the viewpoint of i_u, e_1, the circuit shown in Fig. 1.12-6b is the same as Fig. 1.12-6a; that in Fig. 1.12-6c is the equivalent for e_2, i_β. The admittance level of the reflected elements is multiplied in each case by the factor $H^2 = C^2(\dot{\varphi} + \dot{\psi}\cos\theta)^2 \doteq C^2\dot{\varphi}^2$, thus significantly reducing $e_1 = \dot{u}$, $e_2 = \dot{v}$, and enabling the gyro to be used as a reference.

Gyro stabilizer. The same circuit applies in the use of the gyro for decreasing the roll of the ship, except that the gyro frame is fixed to the boat so that it cannot rotate about the longtitudinal (Ou) axis. Hence the u coordinate in Fig. 1.12-6 is that of the center of gravity of the entire boat, m_1 is the mass of the ship, and C_1 is the moment of inertia of the ship about the roll axis. Q_u represents the net torque due to the waves. The parameters of the e_2 node are the same as before, since the gyro is allowed to rotate about the transverse axis. Again the admittance values of the appropriate elements are modified by the factor $H^2 = C^2\dot{\varphi}^2$, leading to a remarkable decrease of roll if $\dot{\varphi}$ is sufficiently large.[1]

Summary

Systems in which $T = \sum a_{ij}\dot{q}_i\dot{q}_j$, $V = \sum b_{ij}q_iq_j$, and $D = \sum d_{ij}\dot{q}_i\dot{q}_j$ can have their electrical analogs drawn by inspection. Mechanical systems in which the motion is entirely single-degree rectilinear or rotational have the above property, so there is no need to even formulate T, V, D. Some of the elements may be negative, but if T, V, D are positive-definite (as they must be in passive systems), then it is always possible (if ideal transformers are allowed) to eliminate the negative elements. Terms of the form $q_i\dot{q}_j$ in T, $(q_i\dot{q}_j - q_j\dot{q}_i)$ in V, lead to gyrators. Terms of the form q_iq_j in T can be transferred to V after multiplying by a minus sign. A term $a_{ij}(q_i\dot{q}_j + \dot{q}_iq_j)$ in V may be replaced by $a_{ij}\dot{q}_i\dot{q}_j$ in D. A term $2bq_i\dot{q}_j$ in D requires a gyrator b/s from node j to i, and the term $2bq_iq_j$ should be added to V. Tests for positive-definiteness of the appropriate quadratic functions may be used to check whether the resulting system is passive. Passive electrical circuits can always be realized by positive R, L, C elements, ideal transformers, and gyrators. Terms in T, V other than those cited lead to nonlinear elements.

§ 1.13 Electric Models of Transducers

By transducers we mean devices in which there is a conversion of energy from one form to another, e.g., electrical to mechanical, etc. The first step in

[1] For a numerical example, see Section 78 of J. B. Scarborough, "The Gyroscope." Interscience, New York, 1958.

deriving the transducer model is to choose the form of the model. Figure 1.13-1 illustrates some possibilities for two-port transducers. As usual, e, i may represent electrical, mechanical, or any other variables. The controlled sources represent the basic transduction process, since they reflect the variables of one port into the other port. It is assumed that all the power losses in the transducer are

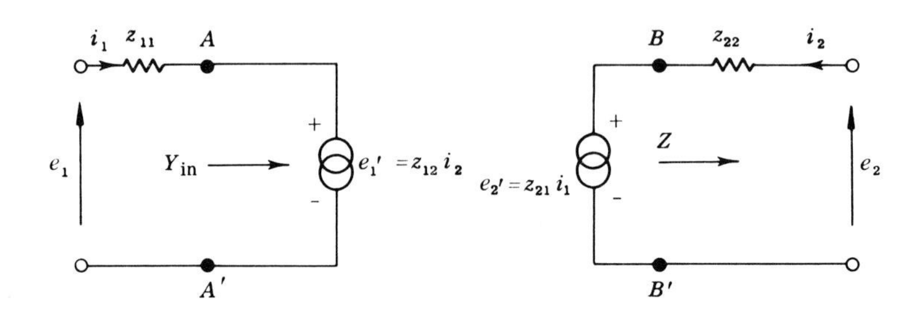

(a)

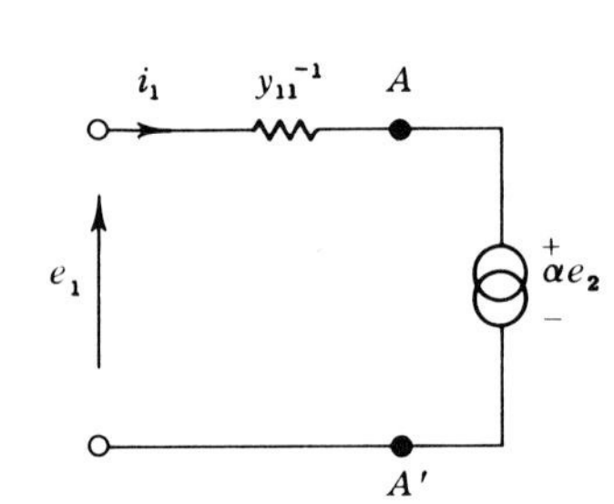

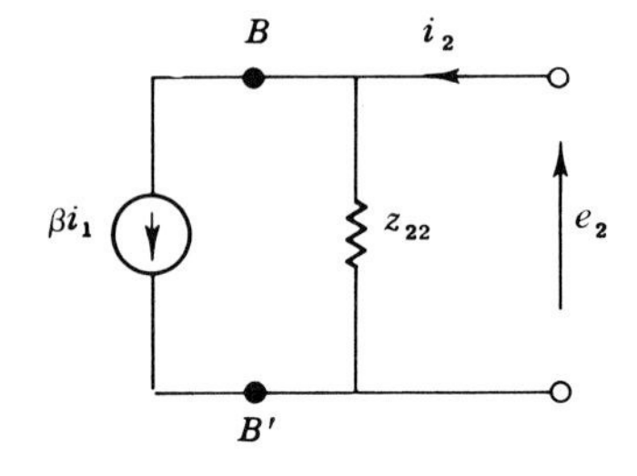

(b)

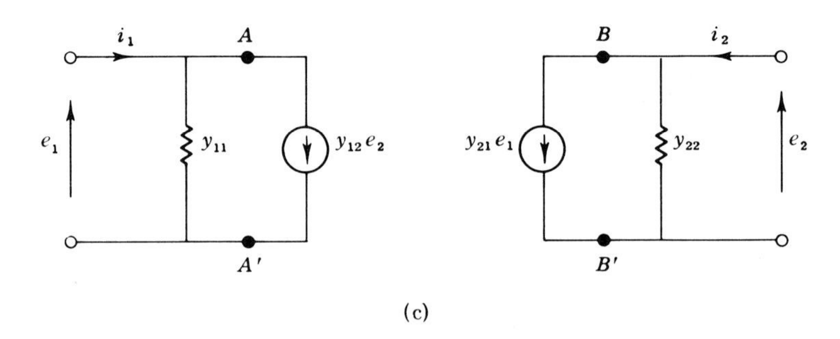

(c)

FIG. 1.13-1. Some two-port electrical models.

represented by resistances included in z_{11}, z_{22}, y_{11}, y_{22}. It is thereby assumed that there are no power losses which are proportional to the product of current (or analogous variable) in one element, and voltage (or analogous variable) across a different element, because such power losses cannot be represented by resistors. If there are no power losses of the latter type, then there are no power losses in the $AA'BB'$ portions of Fig. 1.13-1. In such a case, it is easily shown that z_{12} and z_{21}, α and β, y_{12} and y_{21} are related to each other.

Imagine that an independent source is applied at the e_1, i_1 port of each of the four circuits, and a load is connected at the e_2, i_2 port. All the power delivered at AA' is passed on to BB', because of the assumption that all losses are incorporated into the z's and y's. Hence, in Fig. 1.13-1a, $|E_1'|^2 \mathscr{Re}\, Y_{in} = |z_{12}I_2|^2 \mathscr{Re}\, Y_{in} = |I_2|^2 \mathscr{Re}\, Z$. Also, $Y_{in} = I_1/E_1' = I_1/z_{12}I_2 = E_2'/z_{21}z_{12}I_2 = -Z/z_{21}z_{12}$. Therefore $-\mathscr{Re}\, Z = |z_{12}|^2 \mathscr{Re}\, [Z/z_{12}z_{21}]$ for any Z. This leads to $z_{12} = -\bar{z}_{21}$, i.e., if $z_{12} = a + jb$, then $z_{21} = -a + jb$. Similarly, $\alpha = -\bar{\beta}$, $y_{12} = -\bar{y}_{21}$. These relations apply only if all the power losses have been accounted for in z_{11}, z_{22}, y_{11}, y_{22}. They do not necessarily apply if, for bias purposes, there is a third port, which may supply power to the system.

Example 1. *Model of an Electrostatic Transducer.* The transducer is shown in Fig. 1.13-2a. The attractive force in a parallel-plate capacitor $C = k/x$ farads, whose plates are separated by the distance x, and which has e volts across it,

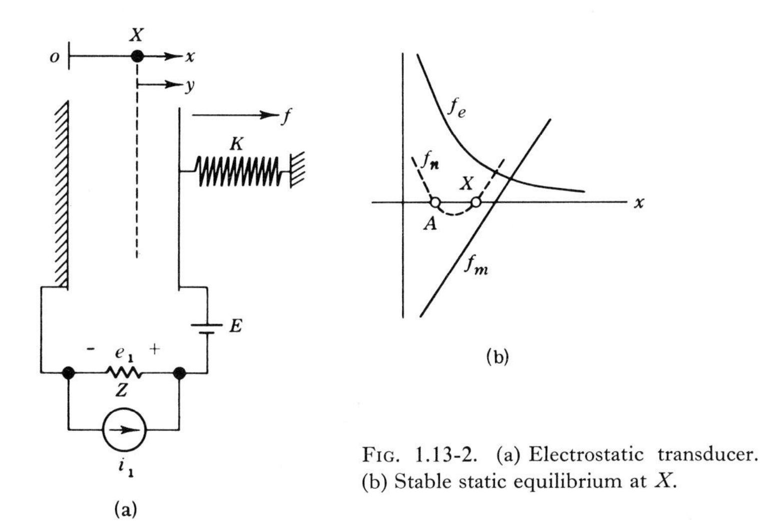

FIG. 1.13-2. (a) Electrostatic transducer. (b) Stable static equilibrium at X.

is $f_e = 0.5ke^2/x^2$, and f_e vs. x is sketched in the figure. The mechanical spring force is $f_m = K(d - x)$, if the spring is quiescent at $d = x$. The net force $f_n = f_e + f_m$ is zero at A, X in Fig. 1.13-2b; however, only point X is one of stable equilibrium. At A, a slight decrease in x results in a net attractive force, which causes x to continue to decrease. It is important to note that the equation $f_e - f_m = 0$, i.e., the equation $0.5ke^2/x^2 - K(d - x) = 0$, must have three real roots for $e = E$; otherwise the two intersections at A and X in Fig. 1.13-2b do not exist. At $x = X$, $0.5kE^2/X = 0.5C_0E^2 = K(d - X)$, where $C_0 = k/X$ is the capacitance at equilibrium, when e_1 is zero.

The total electrical force on the movable plate is

$$f_e = \frac{0.5k(E + e_1)^2}{X + y} \doteq \frac{0.5kE^2}{X^2(1 + 2e_1/E - 2y/X)} = \frac{0.5C_0E^2}{X} + \frac{C_0Ee_1}{X} - \frac{C_0E^2y}{X^2},$$

if e_1/E and y/X are small. This transduction equation suggests that Fig. 1.13-1c (or Fig. 1.13-1b with the port numbers interchanged) should be used for the transducer model. If the former is used, with $i_2 = f$, $e_2 = \dot{y}$, then Fig. 1.13-2c (the por-

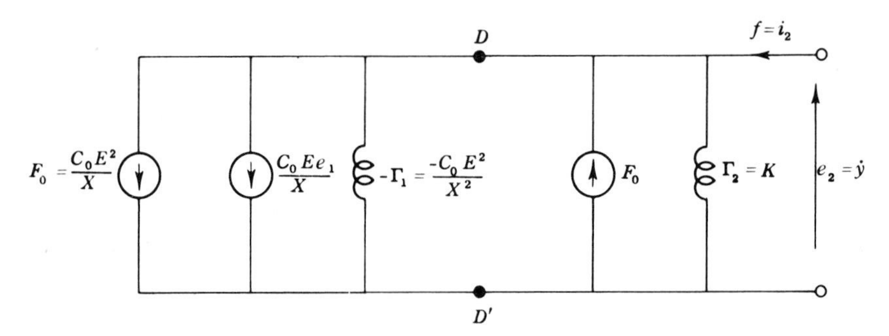

FIG. 1.13-2. (c) Electrical analog of mechanical port.

tion to the left of XX') is obtained from the equation for f_e. The variable f should be considered as an external force generator acting on the movable plate, as shown in Fig. 1.13-2. In Fig. 1.13-2c (if the portion to the right of DD' is ignored), $f = C_0Ee_1X^{-1} - C_0E^2X^{-2}\int e_2dt$ is the external force (in the direction indicated in Fig. 1.13-2a) which would have to be exerted against the attractive electrical force, if there were no spring. The spring force is $f_m = K[d - (X + y)] = K(d - X) - Ky = F_0 - Ky$, and works against the electrical force. It is represented in Fig. 1.13-2c by the circuit to the right of DD'. Note that $\Gamma_2 = K$ is positive because the sign associated with Ky is opposite to that of F_0. Or, in other words, since the spring force decreases as y increases, the external force generator $f = i_2$ must increase to compensate; hence Γ_2 is positive.

System stability demands that $\Gamma = -\Gamma_1 + \Gamma_2 > 0$. Thus, if there is some damping in the system, then the system response has a pole in the right half-plane, if $\Gamma < 0$. It is now shown that the condition $\Gamma_2 > \Gamma_1$ follows from the previously stated condition that $0.5ke^2/x^2 - K(d - x) = 0$ has three real roots for $e = E$. The zeros of the latter are the zeros of $x^3 - x^2d + 0.5kE^2/K = 0 = x^3 + x^2p + r$. Let $a = -p^2/3$, $b = (2p^3 + 27r)/27$. The condition[1] for three real roots is that $(b^2/4) + (a^3/27) < 0$, i.e., $K > 27kE^2/8d^3$. Then $\Gamma_2 > \Gamma_1$

[1] "Mathematical Tables from Handbook of Chemistry & Physics," 8th ed., p. 242. Chemical Rubber Publ. Co., Cleveland, Ohio, 1946.

$(K > C_0E^2/X^2)$ is certainly true, if $27kE^2/8d^3 > C_0E^2/X^2$, i.e., if $X > 2d/3$. We must therefore prove that $X > 2d/3$ follows from $K > 27kE^2/8d^3$. Let $K = 27kE^2/\mu 8d^3$, with $0 < \mu < 1$. X is the point of the second crossing of $f_n = f_e - f_m$ in Fig. 1.13-2b, i.e., of $(0.5kE^2/x^2) - K(d - x)$, whose sign is that of $K(x^3 - x^2d + 0.5kE^2/K) = K(x^3 - x^2d + 4\mu d^3/27)$. Let $x = \lambda d$, and then the sign of $f_n = f_e - f_m$ is the sign of $Kd^3(\lambda^3 - \lambda^2 + 4\mu/27)$. At $\lambda = 2/3$, the sign is negative (unless $\mu > 1$); therefore the crossing at X in Fig. 1.13-2b occurs at $X > 2d/3$. Hence $\Gamma = -\Gamma_1 + \Gamma_2 > 0$, and dynamic stability follows from the existence of static equilibrium.

The mechanical side of the transducer model is basically complete. For complete generality, we could add an external current generator $F = I_2$, with internal impedance Z_m. On the electrical side, the condition $y_{12} = -\bar{y}_{21}$ may be used, because all the losses can be accounted for by resistors. To find y_{11}, let $e_2 = \dot{y} = 0$ (i.e., clamp the movable plate), and then the electrical impedance seen by the current source i_1 in Fig. 1.13-2, is[1] $Z \# (1/sC_0)$. The complete model is shown in Fig. 1.13-2d.

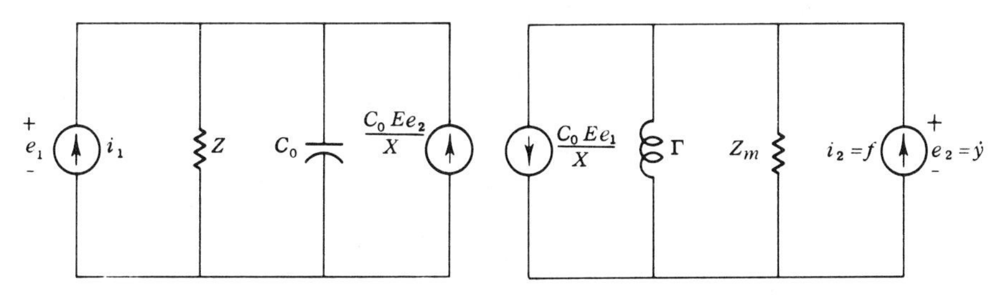

FIG. 1.13-2. (d) Complete electrical analog.

Example 2. *Electromagnetic Transducer*. The loudspeaker of Fig. 1.13-3a is a three-port system, so the two-port transduction relation between the voice

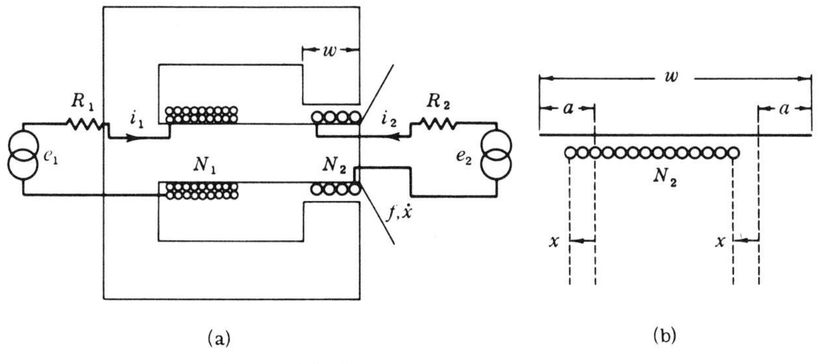

FIG. 1.13-3. Electromagnetic transducer.

[1] $A \# B$ means "A in parallel with B."

coil circuit e_2, i_2, and the mechanical f, $\dot{x}$ port, can be used only if there is no transfer of power to either of these from the field coil circuit e_1, i_1. It is instructive to derive an electrical analog from the Lagrange equations. The magnetic field energy in a coil of current i and flux linkage λ is $0.5\,\lambda i$, and it is kinetic energy.[1] The magnetic flux in the field coil is $\varphi = k(N_1 i_1 + N_2 i_2)$, so the energy associated with the coil is $T_1 = 0.5 k i_1 N_1 \varphi = 0.5 L_1 i_1^2 + 0.5 M i_1 i_2$. The flux linking the voice coil is $\lambda_2 = \varphi N \int_{a+x}^{w-a+x} (\zeta/w) d\zeta$ (see Fig. 1.13-3b) where $N(w - 2a) = N_2$; it is assumed (1) that the coil span is w — 2a; (2) at quiescent conditions the coil is symmetrically located, and (3) $x_{\max} \leqslant a$. If these conditions are not satisfied, then nonlinear terms, in addition to those later noted, will result. Evaluation of λ_2 leads to $\lambda_2 = 0.5 N_2(1 + 2x/w)$, and the energy is $T_2 = 0.5(L_2 i_2^2 + M i_1 i_2)$ $(1 + 2x/w)$. The total kinetic energy is $2T = 2(T_1 + T_2) = L_1 i_1^2 + L_2 i_2^2 + M i_1 i_2 + (x/w)\,[L_2 i_2^2 + M i_1 i_2]$. The term $L_2 i_2^2 x/w$ leads to nonlinear terms, and so does the term $M i_1 i_2 x/w$, unless i_1 is a constant. If a compensation winding with *mmf* $- N_2 i_2$ is located near the voice coil, then the *mmf* would be very closely $N_1 i_1$, instead of $N_1 i_1 + N_2 i_2$, and the term $L_2 i_2^2 x/w$ would not appear in T. Otherwise, applying the Lagrange equation, etc., we get:

$$e_1 = R_1 i_1 + \frac{L_1 di_1}{dt} + \frac{M di_2}{dt} + \frac{M}{w}\left[\frac{x di_2}{dt} + \frac{i_2 dx}{dt}\right];$$

$$e_2 = R_2 i_2 + \frac{L_2 di_2}{dt} + \frac{M di_1}{dt} + \frac{2L_2}{w}\left[\frac{x di_2}{dt} + \frac{i_2 dx}{dt}\right] + \frac{M}{w}\left[\frac{x di_1}{dt} + \frac{i_2 dx}{dt}\right];$$

$$f = -\,L_2 w^{-1} i_2^2 - M w^{-1} i_1 i_2.$$

A linear model is obtained only by assuming i_1 is a constant, and ignoring the other nonlinear terms.

The armature controlled dc motor has basically the same relations. The same assumptions must be made, in order to obtain a linear model. The field controlled dc motor model may also be obtained from the above.

[1] J. H. Jeans, "The Mathematical Theory of Electricity and Magnetism," 5th ed., p. 495. Cambridge Univ. Press, London and New York, 1925.

CHAPTER TWO

Foundations of Linear Feedback Theory

§ 2.1 Introduction

The primary objective of Chapter 2 is to present feedback as a concept which incidentally may be used to analyze linear systems. Techniques are developed which permit the examination of any linear system from the feedback point of view. A mastery of these techniques for seeing systems from the feedback point of view is important for two reasons:

(1) They provide the foundation and framework for the later development of feedback synthesis theory.

(2) They are valuable for their own sake as powerful tools for the analysis of complex active systems.

The usefulness of feedback as an analytical tool provides the justification for presenting the principal results of this chapter in a rather leisurely manner amidst a large number of examples. We feel that this is justified because the assimilation of the feedback point of view is best acquired through practice in specific problems. The important theorems are thus introduced gradually and ostensibly as refinements of the tools provided by the feedback approach. However, it should be emphasized that it is not necessary for the reader to work through all these examples and theorems before going on to Chapter 3. The reader whose principal interest is in synthesis need read only the first five sections before going on to Chapter 3. He will not find it inconvenient to return to Chapter 2 for any other results that he will later need. The reader interested in analysis as well as synthesis will stay longer with Chapter 2 and return to it as the occasion arises. But it is suggested that he too should not attempt to cover all of Chapter 2 before going on to Chapter 3.

Most of the material presented in this chapter was developed originally by Bode.[1] The approach used, however, is different and is one which undoubtedly has been used by many teachers of feedback theory. The networks that are used in the examples are electrical networks; the active elements are vacuum tubes

[1] H. W. Bode, "Network Analysis and Feedback Amplifier Design," Chapters 4-6. Van Nostrand, Princeton, New Jersey, 1945.

or transistors; however, it will be clear that the techniques may be applied to electromechanical or other linear systems.

§ 2.2 The Fundamental Feedback Equation and Signal-Flow Graph

The term "feedback" suggests that there is a variable (call it $\mathscr{S}$), which is determined by other variables, Y_1, Y_2, ..., i.e., $\mathscr{S} = f(Y_1, Y_2, ...)$ and that one (or more) of these latter variables is in turn dependent to some extent on $\mathscr{S}$.

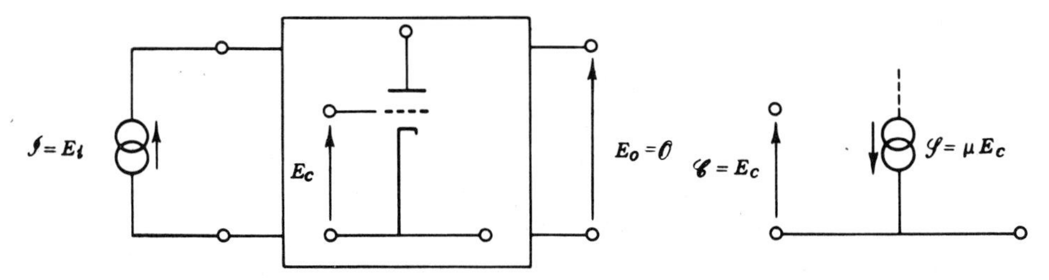

FIG. 2.2-1. (a) Circuit containing an active element.

In this way there is feedback around $\mathscr{S}$, since $\mathscr{S}$ is *feeding* information *back* to its source. In such an arrangement, the input determining $\mathscr{S}$ is not ignorant of the outcome, but is aware of it and adjusts itself to some extent in accordance with the value of the output.

To express the above concept mathematically, consider a circuit (Fig. 2.2-1a)

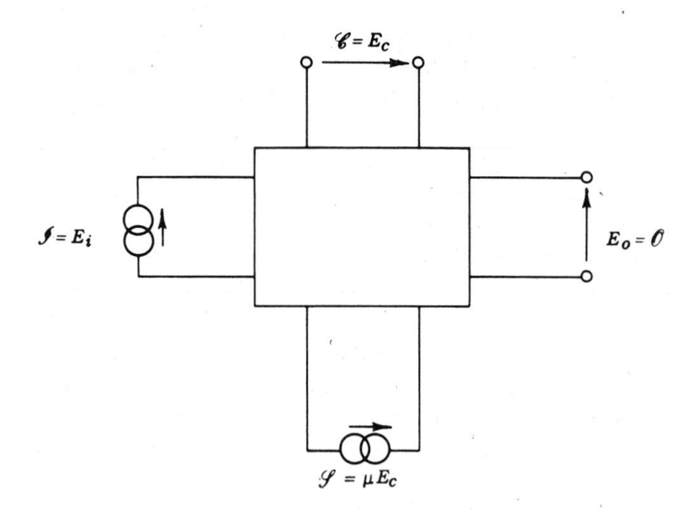

FIG. 2.2-1. (b) Use of controlled source to simplify analysis.

containing within it a vacuum tube. In the model of the vacuum tube we are especially interested in the controlled $\mathscr{S}$ource μE_c denoted by $\mathscr{S}$ and its $\mathscr{C}$ontrolling voltage E_c denoted by $\mathscr{C}$. To emphasize this interest, the controlled $\mathscr{S}$ource and its $\mathscr{C}$ontrolling variable are displayed explicitly in Fig. 2.2-1b. $\mathscr{C} = E_c$ remains a dependent variable of course; two leads have been attached to the appropriate terminals and brought out only for identification purposes. The network in Fig. 2.2-1b is now solved by adding the separate effects of the $\mathscr{I}$nput, $\mathscr{I} = E_i$ and $\mathscr{S} = \mu E_c$ (we are dealing with linear systems, so superposition is permitted). The voltage $\mathscr{O}$utput is

$$\mathscr{O} = E_o = t_{oi}E_i + t_{os}\,\mathscr{S} = t_{oi}E_i + t_{os}(\mu E_c) \tag{2.2,1}$$

where, by definition,

t_{oi} is the transmission to the output $\mathscr{O} = E_o$ from the input $\mathscr{I} = E_i$, with $\mathscr{S} = 0$, when $E_i = 1$ [i.e., t_{oi} is the Laplace transform of the output $e_o(t)$, when the source $\mathscr{S}$ is arbitrarily made zero and the input $e_i(t)$ is the impulse function] (2.2,2)

t_{os} is the transmission to the output $\mathscr{O}$ from the source $\mathscr{S}$ with $E_i = 0$, and $\mathscr{S}$ set at 1. (2.2,3)

E_c in turn is a dependent variable whose value is obtained by adding the effects of E_i and $\mathscr{S}$. It is $E_c = t_{ci}E_i + t_{cs}\,\mathscr{S} = t_{ci}E_i + t_{cs}\mu E_c$, so that

$$E_c = t_{ci}E_i/(1 - \mu t_{cs}) \tag{2.2,4}$$

where

t_{ci} is the transmission to $\mathscr{C} = E_c$ from the input $\mathscr{I} = E_i$ with $\mathscr{S} = 0$ and $\mathscr{I} = 1$, (2.2,5)

t_{cs} is the transmission to $\mathscr{C}$ from $\mathscr{S}$ with $\mathscr{I} = 0$ and $\mathscr{S} = 1$. (2.2,6)

Equation (2.2,4) is used to eliminate E_c in Eq. (2.2,1), resulting in[1] (with $k = \mu$)

$$\boxed{\frac{\mathscr{O}}{\mathscr{I}} \triangleq T(s) = t_{oi} + \frac{kt_{ci}t_{os}}{1 - kt_{cs}}} \tag{2.2,7}$$

Equation (2.2,7) will be called the fundamental feedback equation. It is probably the most important equation in linear feedback theory. All the significant properties of linear feedback systems are derivable from this equation. As an analytic tool it shows how a linear circuit problem can be broken up into

[1] The symbol $\triangleq$ means "equal by definition." It should be noted that $\mathscr{I}$, $\mathscr{O}$, $\mathscr{C}$, $\mathscr{S}$ and the four t_{ij}'s are all functions of the complex variable s.

two simpler circuits. Two transfer functions must be found in each of the simpler circuits. This is the price that must be paid for working with the two simpler circuits in place of the original single, but more complicated, circuit.

The Fundamental Feedback Flow Graph

The signal-flow graph for the circuit of Fig. 2.2-1 is obtained by drawing the flow graphs of Eqs. (2.2,1), etc., in accordance with the rules given in Section 1.6. The result is shown in Fig. 2.2-2. This graph will be called the

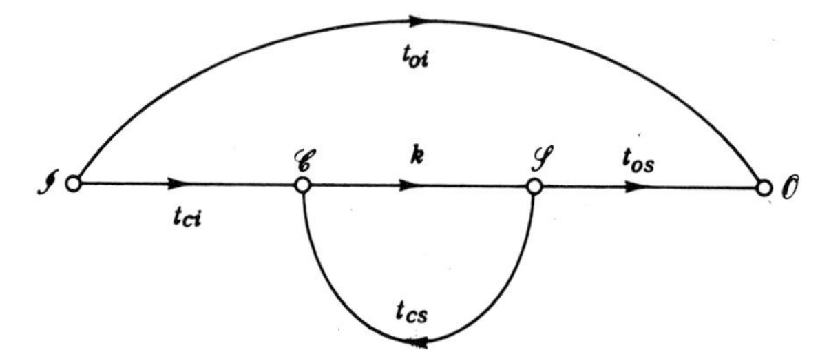

FIG. 2.2-2. The fundamental feedback flow graph.

fundamental feedback signal-flow graph and will be used in conjunction with the fundamental feedback equation (2.2,7) to derive the properties of feedback systems.

In the above, the $\mathscr{C}$ontrolling variable $\mathscr{C}$ is the voltage E_c and the controlled $\mathscr{S}$ource $\mathscr{S}$ is the voltage E_s. But generally the controlling variable and the controlled source can have any units, not necessarily the same. The script letter $\mathscr{C}$ will henceforth be used for the controlling variable, $\mathscr{I}$ for the input, $\mathscr{O}$ for the output, $\mathscr{S}$ for the controlled source, k for the relation between $\mathscr{C}$ and $\mathscr{S}$ ($\mathscr{S} = k\mathscr{C}$), and k is, in general, a transfer function, i.e., $\mathscr{S} = k(s)\mathscr{C}(s)$.

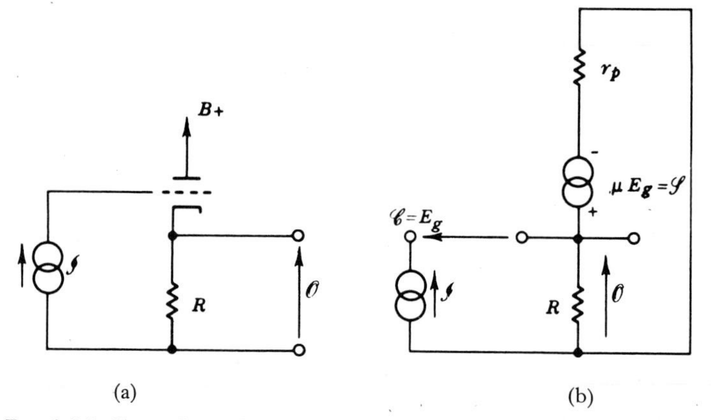

FIG. 2.2-3. Example 1—Cathode follower and its incremental linear model.

The existence of feedback around $\mathscr{S}$ is manifest by the transmission t_{cs} feeding some of the output of the controlled source back to its controlling variable. If t_{cs} is zero, then there is no feedback around $\mathscr{S}$.

The usefulness of the above notions in the analysis of active circuits is illustrated by using them to analyze some specific popular vacuum-tube feedback circuits.

Example 1. The cathode follower circuit (Fig. 2.2-3a) has the electrical circuit model shown in Fig. 2.2-3b. $\mathscr{S} = \mu E_g$ is the controlled source; the controlling variable is $\mathscr{C} = E_g$. To find t_{oi} and t_{ci}, $\mathscr{S}$ is made zero, resulting in the circuit

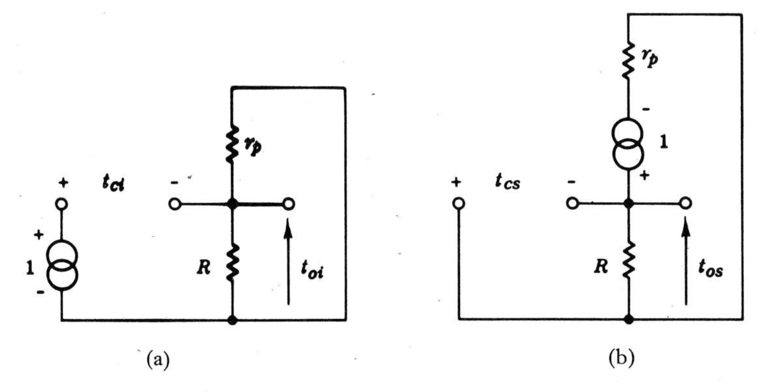

FIG. 2.2-4. (a) Circuit for evaluating t_{oi} and t_{ci}. (b) Circuit for evaluating t_{os} and t_{cs}.

of Fig. 2.2-4a, whose solution is $t_{oi} = 0$ and $t_{ci} = 1$. (When $\mathscr{I}$ is exactly one, and $\mathscr{S} = 0$, the voltages appearing at $\mathscr{C}$ and $\mathscr{O}$ are, respectively, t_{ci} and t_{oi}.)

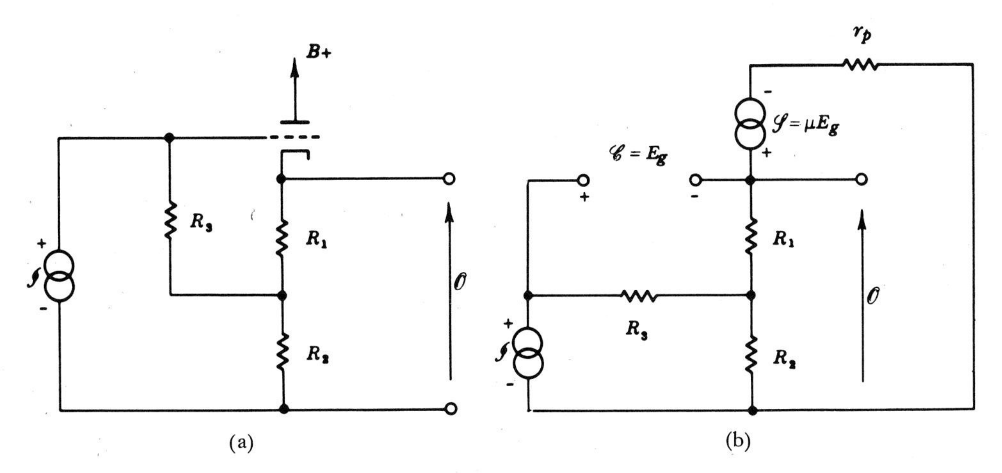

FIG. 2.2-5. Example 2—Self-biased cathode follower and its incremental linear model.

To find t_{os} and t_{cs}, $\mathscr{I}$ is made zero, and $\mathscr{S}$ is made one, resulting in the circuit of Fig. 2.2-4b, with $t_{os} = -t_{cs} = R/(r_p + R)$. Using Eq. (2.2,7), the transmission

$$T = \frac{\mathscr{O}}{\mathscr{I}} = t_{oi} + \frac{kt_{ci}t_{os}}{1 - kt_{cs}} = \left(\frac{\mu R}{r_p + R}\right)\left(1 + \frac{\mu R}{r_p + R}\right)^{-1} = \frac{\mu R}{r_p + R(1 + \mu)}.$$

Example 2. Find $T = \mathscr{O}/\mathscr{I}$ in Fig. 2.2-5a. The vacuum tube is replaced by its incremental linear model in Fig. 2.2-5b. The controlled source is $\mathscr{S} = \mu E_g$, so $E_g = \mathscr{C}$. To evaluate t_{oi} and t_{ci}, $\mathscr{S}$ is short-circuited, and $\mathscr{I}$ made one, giving the circuit of Fig. 2.2-6a. Here[1] $t_{oi} = r_p(R_2 \# R_3)\,[R_3(r_p + R_1 + R_2 \# R_3)]^{-1}$ and $t_{ci} = 1 - t_{oi}$. To find t_{os} and t_{cs}, $\mathscr{I}$ is short-circuited and $\mathscr{S}$ is assigned unit value, resulting in the circuit of Fig. 2.2-6b, with $t_{os} = -t_{cs} = (R_1 + R_2 \# R_3)\,(r_p + R_1 + R_2 \# R_3)^{-1}$. Equation (2.2,7) is then used to find $T = \mathscr{O}/\mathscr{I}$.

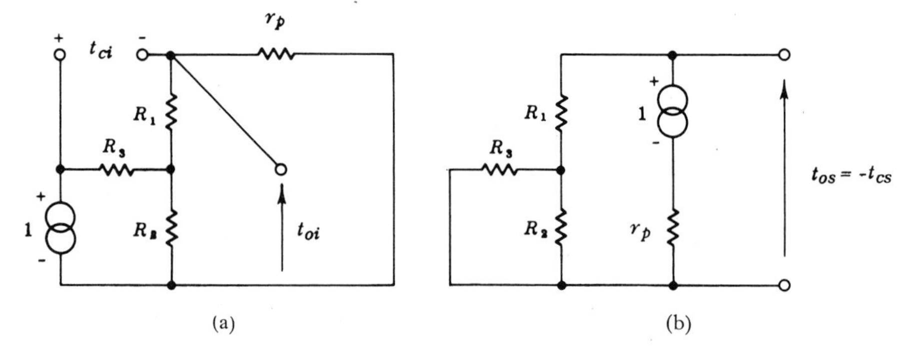

FIG. 2.2-6. (a) Circuit for evaluating t_{oi} and t_{ci}. (b) Circuit for evaluating t_{os} and t_{cs}.

Circuits with several controlled sources. The above approach may be applied to circuits containing many controlled sources. Any one of these may be singled

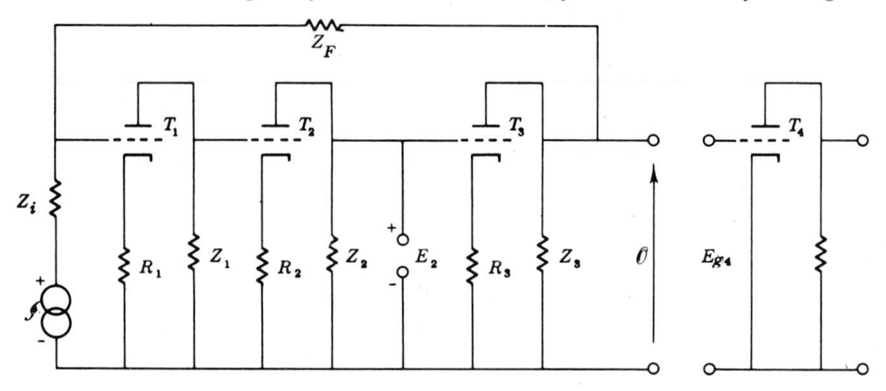

FIG. 2.2-7. Example 3—Feedback amplifier.

[1] The symbol # means "in parallel with."

out and used as $\mathscr{S}$. For analysis purposes, it is sometimes preferable to pick a specific one, while at other times it makes little difference which controlled source is chosen for $\mathscr{S}$. As an example of the latter situation, consider the following example.

Example 3 (Fig. 2.2-7). In this example it is not obvious what to use for the controlled source $\mathscr{S}$. Actually, any choice for $\mathscr{S}$ which leads to two simple circuits, when first $\mathscr{I}$ and then $\mathscr{S}$ is made zero, is satisfactory. With some practice it is not difficult to visualize the two circuits that result. In this example $\mu_1 E_{g1}$, $\mu_2 E_{g2}$, or $\mu_3 E_{g3}$ is satisfactory as $\mathscr{S}$ and the reader may work the problem using one of these as $\mathscr{S}$. By way of variety, however, a grid to ground voltage (E_2 in Fig. 2.2-7) rather than a grid to cathode voltage is used for the controlling variable $\mathscr{C}$. It is then necessary to have as $\mathscr{S}$ a controlled source proportional to E_2. Consequently, the model of the third vacuum tube and its associated

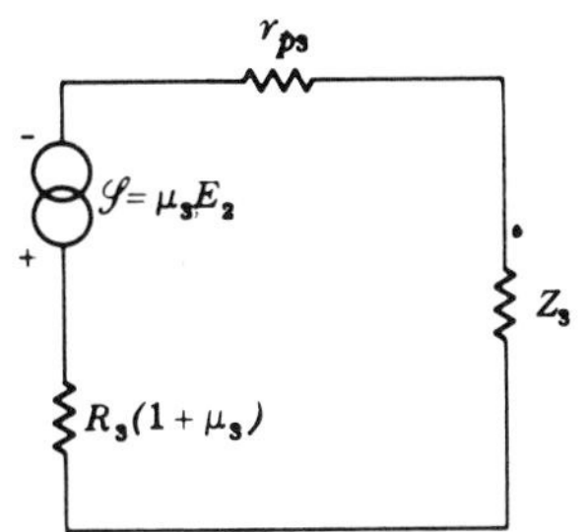

FIG. 2.2-8. Representation of stage 3.

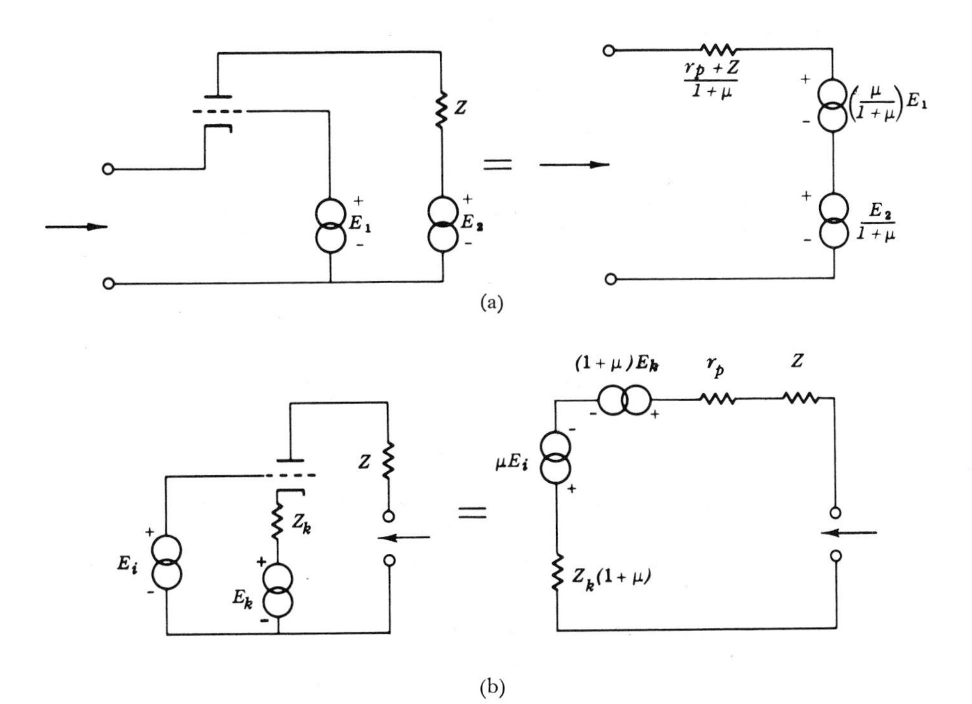

FIG. 2.2-9. Two terminal equivalences that are very useful.

circuitry that must be used is that shown in Fig. 2.2-8[1] (obtained from Fig. 2.2-9b). Here $\mathscr{C} = E_2$ and $\mathscr{S} = \mu_3 E_2$. To find t_{oi} and t_{ci}, the circuit of Fig. 2.2-10a is used. Let $Z_3' = Z_3 \# [r_{p3} + R_3(1 + \mu_3)]$. Then $t_{oi} = Z_3' (Z_i + Z_F + Z_3')^{-1}$; $t_{ci} = A_1 A_2 (Z_F + Z_3') (Z_i + Z_F + Z_3')^{-1}$, where $A_1 = -\mu_1 Z_1 [Z_1 + r_{p1} + R_1(1 + \mu_1)]^{-1}$, and $A_2 = -\mu_2 Z_2 [Z_2 + r_{p2} + R_2(1 + \mu_2)]^{-1}$. To find

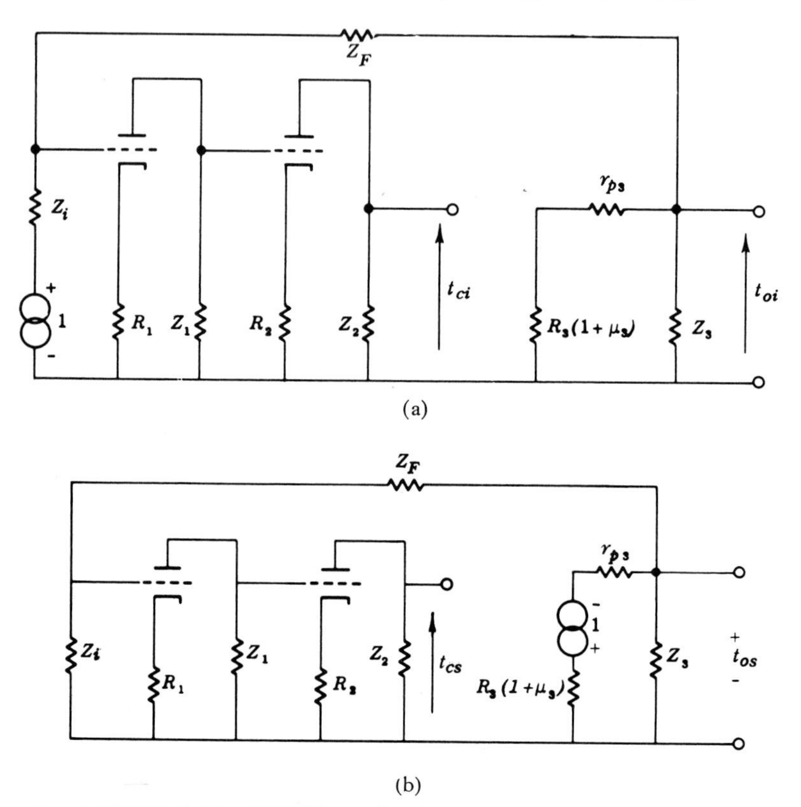

FIG. 2.2-10. Example 3—(a) Circuit for evaluating t_{oi} and t_{ci}. 3—(b) Circuit for evaluating t_{os} and t_{cs}.

t_{os} and t_{cs}, $\mathscr{I}$ is made zero and the appropriate transmissions from $\mathscr{S}$ are found from Fig. 2.2-10b, where $t_{os} = -[Z_3 \# (Z_F + Z_i)] [r_{p3} + R_3(1 + \mu_3) + Z_3 \# (Z_F + Z_i)]^{-1} \triangleq A_3/\mu_3$; $t_{cs} = t_{os} A_1 A_2 Z_i (Z_i + Z_F)^{-1}$. Finally, the transmission from input to output $T = \mathscr{O}/\mathscr{I}$ is found by using Eq. (2.2,7), with $k = \mu_3$. The result is

$$T = \frac{A_1 A_2 A_3 [(Z_F + Z_3')/(Z_i + Z_F + Z_3')]}{1 - A_1 A_2 A_3 [Z_i/(Z_i + Z_F)]} + t_{oi}$$

[1] In the analysis of vacuum-tube feedback circuits by the feedback method, it is very helpful to be familiar with the two simple configurations shown in Fig. 2.2-9.

and t_{oi} may usually be neglected. After a little practice the reader will find that he is able to write a result such as the above directly by inspection of the original circuit, and make reasonable approximations as he goes along.

Example 4 (Fig. 2.2-11a). Find E_o/E_i. In this example some careful thinking must be done in order to pick an $\mathscr{S}$ which will significantly simplify the problem, i.e., one which leads to two simple circuits when each of $\mathscr{I}$ or $\mathscr{S}$ is made zero. Suppose $\mathscr{S} = \mu_1 E_{g1}$ is chosen, then t_{ci} and t_{oi} are easily obtained by inspection as one and zero, respectively, but for t_{os} and t_{cs}, the complicated circuit of Fig. 2.2-11b must be solved. The circuit is complicated not so much by the local feedback in the first stage due to R, as by the fact that the over-all feedback is also via R. It is therefore better to go to the second stage for $\mathscr{S}$. If E_1 is used for $\mathscr{C}$ with $\mathscr{S} = \mu_2 E_1$, then t_{oi} and t_{ci} are evaluated from Fig. 2.2-11c. Here,

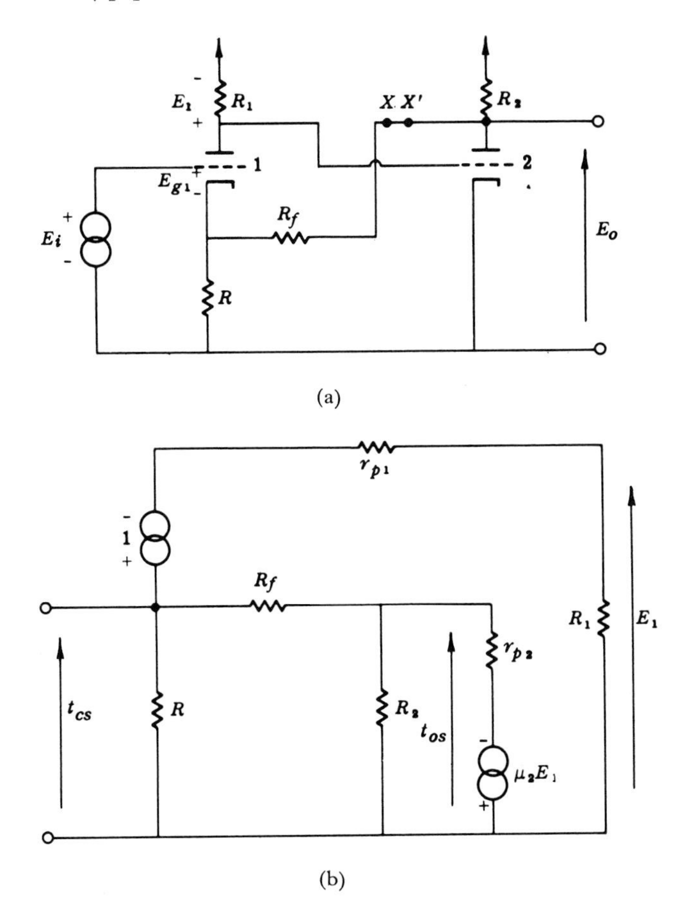

FIG. 2.2-11. (a) Example 4. (b) Circuit for evaluating t_{os} and t_{cs} if $\mathscr{S} = \mu_1 E_{g1}$.

$t_{oi} = \{\mu_1 R'/[r_{p1} + R_1 + R'(1 + \mu_1)]\}[R_2'/(R_f + R_2')]$ is usually so small that it can be neglected, and $t_{ci} = -\mu_1 R_1 [r_{p1} + R_1 + R'(1 + \mu_1)]^{-1}$. For t_{os} and t_{cs}, Fig. 2.2-11d is used. Here (with the help of Fig. 2.2-9a) $t_{cs} = I_{R_1} R_1 = E_x(1 + \mu_1)R_1/(r_{p1} + R_1)$;

$$t_{os} = -R_2''/(r_{p2} + R_2''), \text{ and } R_2'' = R_2 \# [R_f + R \# (r_{p1} + R_1)(1 + \mu_1)^{-1}].$$

Therefore, in Fig. 2.2-11d,

$$E_x = t_{os} [R \# (r_{p1} + R_1)/(1 + \mu_1)] [R_f + R \# (r_{p1} + R_1)/(1 + \mu_1)]^{-1}.$$

Substituting in Eq. (2.2,7), neglecting t_{oi}, and assuming $R_f \gg R, R_2$, $T \doteq A_1 A_2 [1 + A_1 A_2 R/R_f]^{-1}$, where $A_2 = -\mu_2 R_2/(r_{p2} + R_2)$, $A_1 = -\mu_1 R_1/[r_{p1} + R_1 + R(1 + \mu_1)]$.

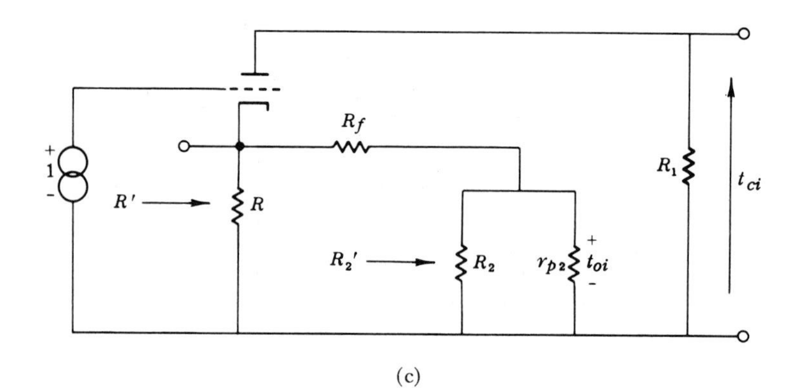

(c)

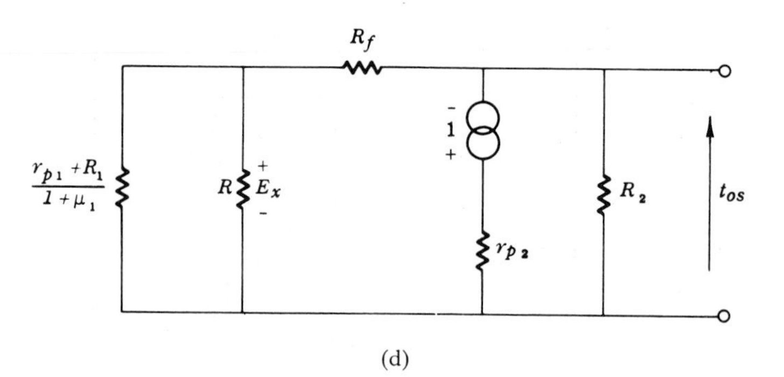

(d)

FIG. 2.2-11. (c) Circuit for finding t_{oi} and t_{ci} when $\mathscr{S} = \mu_2 E_1$. (d) Circuit for finding t_{os} and t_{cs} when $\mathscr{S} = \mu_2 E_1$.

§ 2.3 The Representation of Two-Terminal Elements by Controlled Sources

It may sometimes be convenient to represent a two-terminal element as a controlled source, if by removing the element (opening or shorting it) the resultant circuit becomes much simpler to analyze. It is in any case theoretically important to demonstrate that two terminal elements can be replaced by controlled sources.

There are two ways to replace a two-terminal passive element by a controlled source. One way is to replace Z (Fig. 2.3-1a) by the current-controlled voltage source $\mathscr{S}_1 = ZI$ of Fig. 2.3-1b with $\mathscr{C} = I$; $k = Z$. The other way is to replace Z by the voltage-controlled current source $\mathscr{S}_2 = Z^{-1}E$ of Fig. 2.3-1c, with $\mathscr{C} = E$, $k = Z^{-1} = Y$. The application of this technique is illustrated in the following simple example.

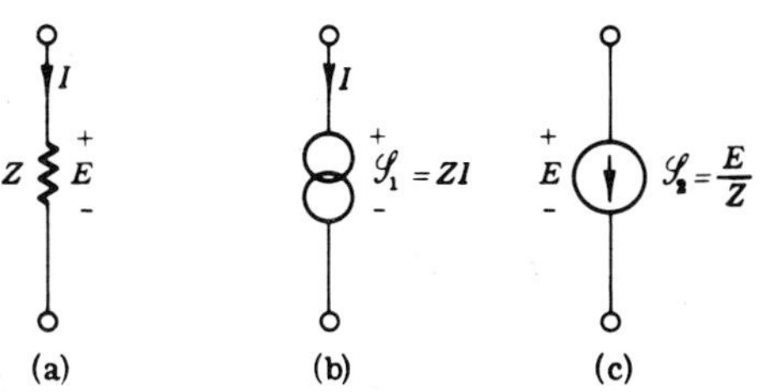

FIG. 2.3-1. Passive element represented by controlled sources.

Example 5 (Fig. 2.3-2a). Replace R_1 by the voltage-controlled current source $I_1 = G_1E_1$, so that $\mathscr{S} = I_1$, $\mathscr{C} = E_1$, $k = G_1$. To find t_{oi}, t_{ci}, set $\mathscr{S} = 0$ (open circuit) and $\mathscr{I} = 1$, resulting in the circuit in Fig. 2.3-2b. The grid-to-cathode voltage is zero because the open circuit prevents the application of any excitation to the grid. Consequently, $t_{oi} = 0$, $t_{ci} = 1$. To find t_{os}, t_{cs}, set $\mathscr{I} = 0$, $\mathscr{S} = 1$, resulting in the circuit of Fig. 2.3-2c where $\mu e_c = \mu(R_2 - IR_3) = (R_3 + R_4 + r_p)I + R_4$, so $I = (\mu R_2 - R_4)\,[r_p + R_4 + (1 + \mu)R_3]^{-1}$,

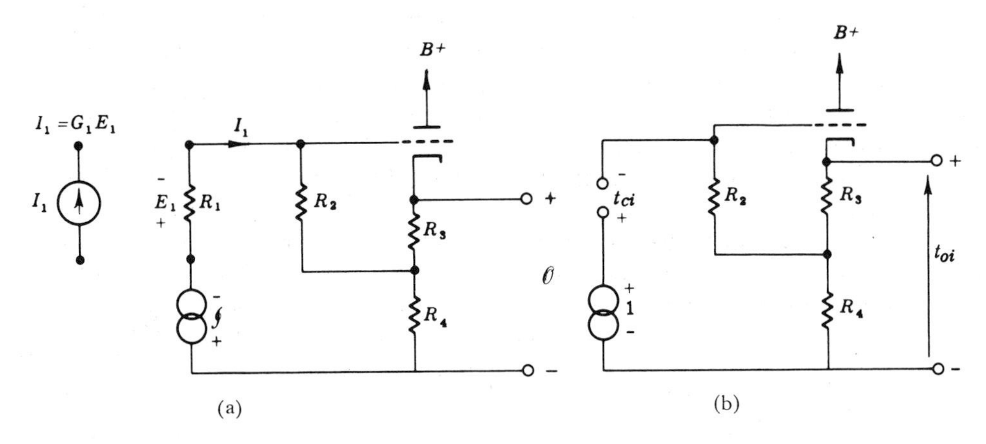

FIG. 2.3-2. (a) Example 5. (b) Circuit for evaluating t_{ci}, t_{oi}; $\mathscr{S} = G_1E_1$.

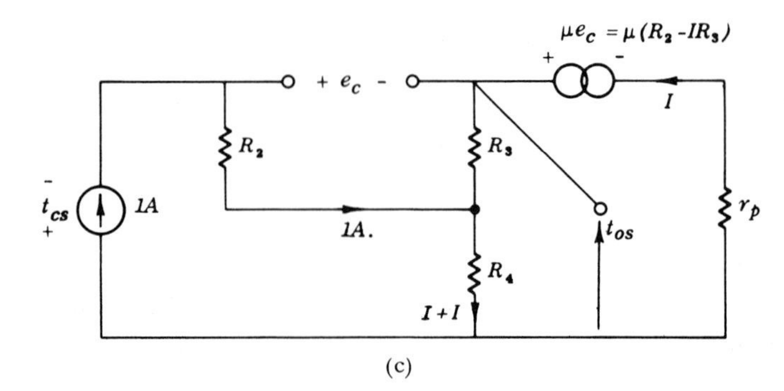

FIG. 2.3-2. (c) Circuit for evaluating t_{os}, t_{cs}.

and

$$t_{os} = IR_3 + (I + 1)R_4 = \frac{\mu[R_2(R_3 + R_4) + R_3R_4] + r_pR_4}{r_p + R_4 + (1 + \mu)R_3};$$

$$t_{cs} = -(1 + I)R_4 - R_2 = -\frac{[r_p + (1 + \mu)R_3](R_2 + R_4) + R_2R_4(1 + \mu)}{r_p + R_4 + (1 + \mu)R_3}.$$

These values are combined [using Eq. (2.2,7)] to give $T = \mathcal{O}/\mathcal{I} = R_1t_{os}/(1 - R_1t_{cs})$.

§ 2.4 The Subjective Nature of Feedback

It is important to note that the term feedback by itself is incomplete. There is feedback around the $\mathcal{S}$ that goes with a specific k in $\mathcal{S} = k\mathcal{C}$. But in the same system there may be sources around which there is no feedback. For example, if in Fig. 2.2-7 the single stage shown at the extreme right is joined to the circuit, and if μ_4E_{g4} is chosen for $\mathcal{S}$, then t_{cs} is zero, so that there is no feedback around μ_4E_{g4}. The feedback around μ_3E_{g3} is, however, the same as it was before. The amount of feedback is generally different for different choices of $\mathcal{S}$. If μ_3E_{g3} is chosen for $\mathcal{S}$ in Fig. 2.2-7, the value of the feedback (i.e., the value of $-kt_{cs}$) is different from the value obtained with μ_3E_2 as $\mathcal{S}$. In Fig. 2.2-11a, the feedback (i.e., the value of $-kt_{cs}$) with respect to $k = \mu_2$ (with $\mathcal{S} = \mu_2E_1$) is not the same as the feedback with respect to $k = \mu_1$ (with $\mathcal{S} = \mu_1E_{g1}$), or the feedback with respect to R_1 (if R_1 is replaced by a current-controlled voltage source).

It is appropriate therefore to note the subjective nature of feedback. When we wish to look upon a system from the feedback point of view (in order to exploit the tools and insight of feedback theory), we manipulate the system into the form of the flow diagram of Fig. 2.2-2, which is the basic system representation which emphasizes the feedback around k. Thus, in the last example in Fig. 2.3-2, we adopted the feedback point of view towards R_1 and found *that* representation of the transfer function in which the feedback around R_1 appeared as a central feature. Any other element might of course have been chosen, leading to different values for the t_{ij} transmissions, but to the same over-all system transfer function. However, given a box with terminals sticking out, it would be impossible by any external measurements to say anything about the feedback properties of the system. Some element inside the box would first have to be chosen, and we could then by internal measurements find the feedback around that element.

§ 2.5 Return Difference: The Bilinear Theorem and Its Exceptions

The previous sections were devoted to the concept of feedback around an element and the exploitation of the concept in the analysis of vacuum-tube circuits. We shall eventually consider the properties of feedback. As a step in that direction and partly for the purpose of extending our analytical tools, consider again the fundamental signal-flow graph of Fig. 2.2-2, reproduced in Fig. 2.5-1. The system transfer function is $T = \mathcal{O}/\mathcal{I} = t_{oi} + kt_{ci}t_{os}/(1 - kt_{cs})$.

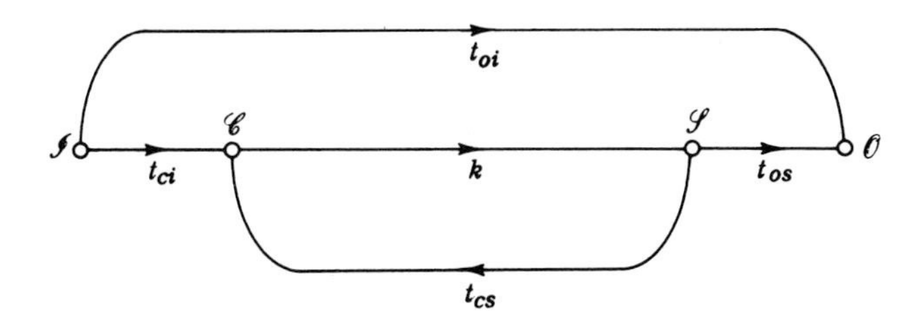

FIG. 2.5-1. The fundamantel feedback flow graph.

Suppose we ask "What is the effect of the feedback around k, on the system function T?" Clearly, the effect is given by the quantity $1 - kt_{cs}$. This quantity, $1 - kt_{cs}$, has a very simple physical interpretation readily seen by opening the closed loop anywhere, for example, at AA' in Fig. 2.5-2. With $\mathcal{I}$ made zero, inject an impulse into the loop at A' and measure the signal returned at A.

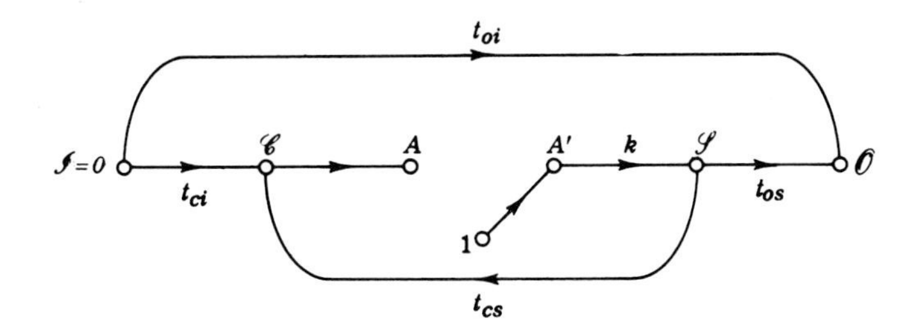

FIG. 2.5-2. Physical interpretation of loop transmission function.

The latter is kt_{cs}. The difference between the injected and the returned signal is precisely $1 - kt_{cs}$. This quantity is accordingly called the *return difference for element k*, with symbol F_k. Thus

$$F_k \triangleq 1 - kt_{cs} \tag{2.5,1}$$

The effect on T of the feedback around k is given by F_k. We may therefore expect that F_k is the quantity that will be significant when we study the properties of feedback. The quantity $-kt_{cs}$ is called the *loop transmission* or *loop gain* for element k, with the symbol L_k, i.e., $L_k \triangleq -kt_{cs}$. When there is no feedback around k, $L_k = t_{cs} = 0$ and $F_k = 1$.

So far as the use of feedback as an analytical tool is concerned, the above definition is sufficient. However, the return difference is intimately associated with, and is the quantitative measure of, the benefits of feedback. In its use for the latter purpose, it is extremely important that the quantity k appear only in the branch from $\mathscr{C}$ to $\mathscr{S}$ in Fig. 2.5-1 and nowhere else. In other words, all the four t's (t_{oi}, t_{ci}, etc.) in the fundamental equation (2.2,7) must be independent of k. For example, in Section 2.2, Example 3, Figs. 2.2-7 and 8, μ_3 is chosen as k, but it also appears in the circuit in $R_3(1 + \mu_3)$. For the purpose of analysis this does not matter. But the $-kt_{cs}$ thereby obtained is not L_k, nor is the resulting $1 - kt_{cs}$ the value of F_k which we want to establish. In order to prevent any ambiguity, we insist henceforth that L_k and F_k be associated with analyses in which k appears only in the branch directed from $\mathscr{C}$ to $\mathscr{S}$ in the fundamental flow graph of Fig. 2.5-1 and absolutely nowhere else. *All four t's must be independent of k, in order that the resulting* $1 - kt_{cs}$ *be indeed* F_k, *and* $-kt_{cs}$ *be* L_k. When this is so, the representation of Fig. 2.5-1 focuses attention on k just as it does on $\mathscr{S}$, in fact k and $\mathscr{S}$ are indivisible.

It is accordingly appropriate to inquire whether it is possible to choose for k *any* system parameter, such that in the resulting fundamental feedback equation (2.2,7) all t_{ij}'s are independent of k. It was seen in Section 2.3 that this could be done for a two-terminal element which is essentially a branch impedance (i.e., its voltage is determined entirely by its own current). It can be done for two-terminal controlled sources that appear in the models of active elements like vacuum tubes and transistors. It cannot be done, however, for coupled coil parameters.[1] It cannot be done whenever the same parameter enters into more than one element or more than one controlled source. In the case of coupled coils, the mutual inductance is $M = k_c(L_1L_2)^{1/2}$ (k_c is the coupling coefficient), so L_1 and L_2 appear in M as well as being self-inductance parameters. It cannot be done for the coefficient of coupling k_c, even though it appears only in M, because M is present in the voltage across L_1 and also in the voltage

[1] L. Tasny-Tschiassny, The return-difference matrix in linear networks, *Proc. IEE (London), Monograph No. 60*, **100**, Part IV, 39-46 (1953). See especially discussion by I. Cederbaum and A. Fuchs, *ibid.* **101**, Part IV, 299 (1954).

across L_2. It is impossible therefore to incorporate k_c into a controlled source $\mathscr{S} = k_c X$ such that k_c is part of $\mathscr{S}$ only and appears nowhere else. The above ideas are incorporated in the following theorem.

Theorem: The transfer function of any two-port linear system can be written in the form of the fundamental feedback equation [Eq. (2.2,7)], such that all the t_{ij} are independent of the system parameter $k(s)$, providing k can be represented as a controlled source $\mathscr{S} = k\mathscr{C}$ and providing k appears only in $\mathscr{S}$ and nowhere else.

The above theorem might be appropriately called the *Bilinear Theorem*[1] because the fundamental feedback equation may be considered as a bilinear transformation of k. This severe constraint on the form in which k appears in the transfer function is the principal reason for the simplicity and elegance of linear feedback theory. Parameters like those in coupled coils can be included in the above by pretending that they are different parameters, and assigning to them different names in the various places they appear in the system function. Another way is to apply the superposition concept to two-port structures instead of to single-port structures (see Section 2.12).

§ 2.6 Null Return Difference

It has previously been noted that a passive element may be represented as a controlled source and used as $\mathscr{S}$ in the feedback method of analysis. It is occasionally convenient to do this. For example, in Fig. 2.6-2a,[2] if Z is replaced by a controlled source $E = IZ$ and used as $\mathscr{S}$, the resulting circuits for evaluating t_{oi}, t_{ci} and t_{os}, t_{cs} are fairly simple. However, the resulting t_{oi} is considerably larger than the net transmission T. Hence in $T = \mathscr{O}/\mathscr{I} = t_{oi} + (kt_{ci}t_{os})/(1 - kt_{cs})$ the two individual parts are large in magnitude and opposite in sign, leading to a net result which is very sensitive to computational errors in each of the two large parts. The only way to avoid this problem and still use Z as the k for the controlled source is if t_{oi} appears as a multiplicative factor rather than as an additive factor in the equation for $T = \mathscr{O}/\mathscr{I}$. In such a case, if we pull out t_{oi} as a factor, the fundamental feedback equation becomes

$$T = t_{oi}\left[\frac{1 - kt_{cs} + kt_{ci}t_{os}/t_{oi}}{1 - kt_{cs}}\right] = t_{oi}\,\frac{Q}{1 - kt_{cs}}. \tag{2.6,1}$$

The above manipulation is useful if and only if the quantity Q in Eq. (2.6,1) is directly available as a single distinct transmission. Otherwise, the individual

[1] H. W. Bode, "Network Analysis and Feedback Amplifier Design," p. 223. Van Nostrand, Princeton, New Jersey, 1945.

[2] This circuit is taken from Bode, Fig. 3.13, p. 41.

elements in Q must be separately calculated, and there is no profit in it. Let us see therefore if we can associate this quantity Q with some physical transmission. Consider Fig. 2.5-2. When a unit signal is injected at A', the signal returned at A is kt_{cs}, and the difference between the injected signal and the returned signal is the return difference $F_k = 1 - kt_{cs}$ for reference k. Now note that Q is the difference between the injected signal and the quantity $kt_{cs} - (kt_{ci}t_{os}/t_{oi})$. In Fig. 2.5-2 if we try to see Q as the difference between the signal at A' and A, we must obtain at A an additional signal $-kt_{ci}t_{os}/t_{oi}$. This additional signal can reach A only via the t_{ci} branch and we can imagine then that it is due to an input $\mathscr{I}$ equal to $-kt_{os}/t_{oi}$. We then have the situation shown in Fig. 2.6-1 where Q is the return difference evaluated under the special

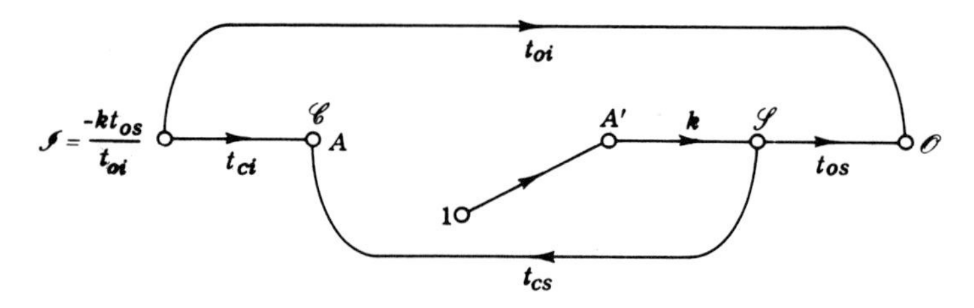

FIG. 2.6-1. Derivation of null return difference.

condition that $\mathscr{I} = -kt_{os}/t_{oi}$. The situation as yet does not appear to have improved at all until we consider the output signal, which from Fig. 2.6-1 is $\mathscr{O} = \mathscr{S}\, t_{os} + \mathscr{I}\, t_{oi} = kt_{os} - t_{oi}(kt_{os}/t_{oi}) = 0$. Hence Q is the return difference evaluated under the special condition that the input is such that the output is null. $Q = 1 - kt_{cs,\,\text{null}} = F_{k,\,\text{null}}$ is called the *Null Return Difference* and Eq. (2.6,1) becomes[1]

$$T = t_{oi}\frac{F_{k,\,\text{null}}}{F_k} \tag{2.6,2}$$

It is never necessary to find the value of the input required to null the output. The fact that the output is zero can be used to simplify the evaluation of F_{null}. This is illustrated by using Eq. (2.6,2) to solve some active circuit problems.

Example 6 (Fig. 2.6-2a). We replace Z by the current-controlled voltage source $\mathscr{S} = IZ$. To find F_{null}, set $\mathscr{O} = 0$ and find I in Fig. 2.6-2b where E_x is such that $E_o = 0$. If $E_o = 0$ then $t_{cs,\,\text{null}} = I = 0$ because I must flow through Z_o. Therefore $F_{\text{null}} = 1 - kt_{cs,\,\text{null}} = 1$. (In an actual problem there

[1] Very often, in the interest of simplicity, the subscript k is omitted and the return difference, loop transmission, null return difference are written, respectively, as F, L, and F_{null}. This will often be done, but it is emphasized that all three quantities have in general different values when different variables are taken for the controlled source $\mathscr{S}$.

would be no need to draw Fig. 2.6-2b—the result is available by inspection.) To find t_{oi}, set $\mathscr{I} = 1$, $Z = 0$ in Fig. 2.6-2a and find the output $\mathscr{O}$. Clearly, t_{oi} is the gain with no feedback and is available by inspection as $t_1 A_1 A_2 A_3 t_2$ where t_1, t_2 are the transmissions through the input and output transformers, respectively, and $A_1 = -\mu_1 Z_2 [r_{p1} + Z_2 + Z_1(1 + \mu_1)]^{-1}$, $A_2 = -\mu_2 Z_4 [r_{p2} + Z_4 + Z_3(1 + \mu_2)]^{-1}$, $A_3 = -\mu_3 Z_o [r_{p3} + Z_o + Z_5(1 + \mu_3)]^{-1}$. To find F, use Fig. 2.6-2b but set $E_x = 0$ and solve for I. The voltage from grid 1 to point B

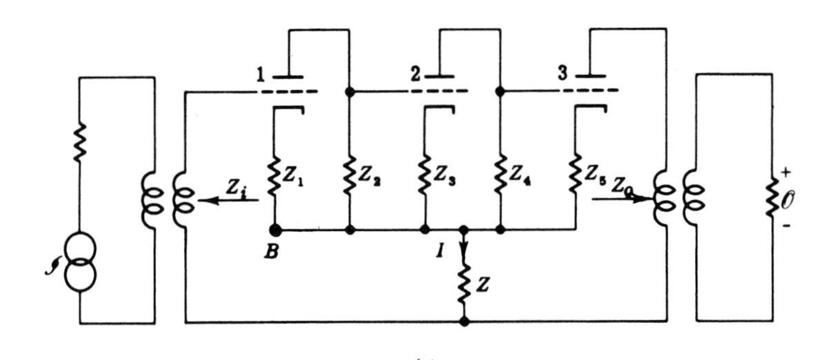

(a)

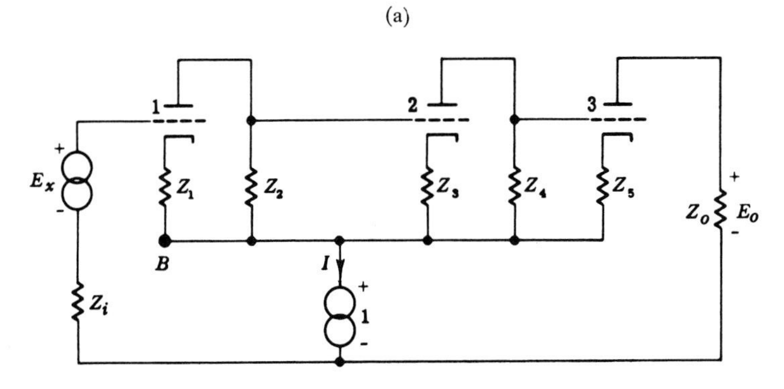

(b)

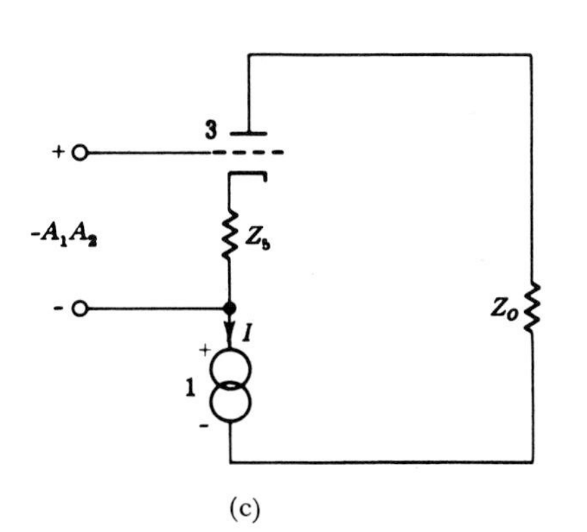

(c)

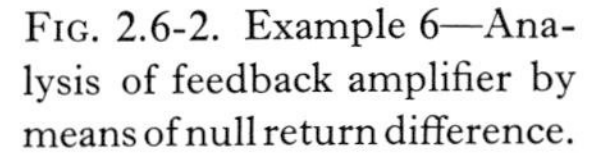

FIG. 2.6-2. Example 6—Analysis of feedback amplifier by means of null return difference.

is then -1, so the voltage from grid 3 to B is $-A_1A_2$, and the last stage is as shown in Fig. 2.6-2c with

$$I = t_{cs} = -(\mu_3 A_1 A_2 + 1)[r_{p3} + Z_o + Z_5(1 + \mu_3)]^{-1} \doteq A_1A_2A_3/Z_o.$$

Consequently, $F \doteq 1 - (A_1A_2A_3Z/Z_o)$ and $T = \mathcal{O}/\mathcal{I} \doteq t_1A_1A_2A_3t_2[1 - (A_1A_2A_3Z/Z_o)]^{-1}$.

§ 2.7 Active Impedances

The expression for T in terms of the null return difference [Eq. (2.6,2)], $T = t_{oi}F_{\text{null}}/F$ is especially convenient for the calculation of active impedances. An active impedance, in contrast to a passive impedance, may be defined simply as the impedance of a circuit containing active elements. In Fig. 2.7-1, Z_{in} is

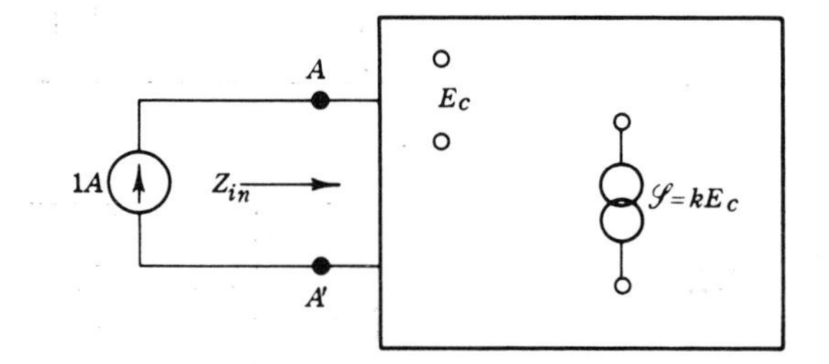

Fig. 2.7-1. Calculation of input impedance by means of null return difference.

to be found. If we assume the input to be a unit current and the output the voltage across AA', then Z_{in} is simply the transmission $T = E_{AA'}$, and Eq. (2.6,2) can be used. Any reference may be selected. To be specific, suppose a voltage-controlled voltage source is used for $\mathcal{S}$; therefore t_{oi} is the input impedance with $\mathcal{S} = 0$. F_{null} is the return difference with respect to k, evaluated when the output is zero, i.e., with a short circuit across AA' (because the output is a voltage in this case). F is the customary return difference with respect to k, evaluated with AA' open.

Example 7. What is the impedance seen by opening the R_f branch at XX' in Fig. 2.2-11a? Let $T = t_{oi}\,F_{\text{null}}/F$ be the voltage across the opened terminals XX' due to a unit current generator applied across XX', with $E_i = 0$ of course. We again choose $\mathcal{S} = \mu_2 E_1$ and then t_{oi} is the impedance seen looking into XX' with $\mathcal{S} = 0$. With the aid of Fig. 2.2-9a,

$$t_{oi} = R_f + (R_2 \# r_{p2}) + R \# [(r_{p1} + R_1)/(1 + \mu_1)] \stackrel{\Delta}{=} Z_A.$$

To find F_{null} note that for zero output XX' is shorted. Therefore here $F_{\text{null}} =$

$F \doteq 1 + (A_1A_2R/R_f)$ of Example 4. To find F, XX' is opened because this corresponds to zero input ($\mathscr{I} = 0$). This obviously eliminates the feedback so that $F = 1$. Hence, $Z_{XX'} \doteq Z_A(1 + A_1A_2R/R_f)$.

Example 8. What is the admittance Y seen by the generator $\mathscr{I}_i$ in Fig. 2.2-7? In that example μ_3E_2 was used as the controlled source $\mathscr{S}$ to find the voltage gain. It is so used here too. To find the admittance Y, let $\mathscr{I}$ in Fig. 2.2-7 equal 1 volt. The current that flows in the generator is then the admittance. To find t_{oi}, $\mathscr{S} = \mu_3E_2$ is shorted and therefore $(t_{oi})^{-1} = Z_i + Z_F + Z_3'$ (see Example 3 for the definition of Z_3'). The return difference F is evaluated with the input $\mathscr{I}$ made zero. It is therefore identical to the value of F used in Example 3. For F_{null} the voltage source is opened in order that the output (which is a current) may be zero. This is equivalent to letting Z_i of Fig. 2.2-7 approach infinity. Therefore

$$F_{\text{null}} = \lim_{Z_i \to \infty} (F) = 1 - A_1A_2A_3', \qquad \text{where } A_3' = \lim_{Z_i \to \infty} A_3.$$

$Y = t_{oi}\, F_{\text{null}}/F$ is then available.

It is seen from the above that if a specific circuit is to be studied intensively, a great deal of information is available directly from F. It appears in the calculation of the gain as well as in the calculation of the impedances. In fact, it will be seen that it appears in every circuit function. Another important property of considerable usefulness in the intensive study of a specific system and of great theoretical importance in feedback theory is the fact that in nontrivial system configurations, the numerator of F is the same, no matter what element is used as reference. This is proved in the next section.

§ 2.8 Invariance of the Numerator of the Return Difference for All References

It has been noted that the return difference has meaning only with reference to some chosen controlled source and has in general different values for different references. Nevertheless in any nontrivial system, there is a common factor in the many return differences that may be defined. This common factor is the numerator of $F_{k_i} = 1 - k_i t_{cs(k_i)}$, where $t_{cs(k_i)}$ is t_{cs} for reference k_i.

Theorem: In all $t_{cs(k_i)}$ which are not identically zero, the numerators of $F_{k_i} = 1 - k_i t_{cs(k_i)}$ are the same no matter which parameter is chosen for the reference k_i. (There are trivial situations in which this theorem is not precisely correct.)

The proof follows. In Fig. 2.8-1 the problem is to evaluate t_{c1s1} which is the t_{cs} when k_1 is used as reference, with $\mathscr{S}_1 = k_1\mathscr{C}_1$. We can use the fundamental feedback equation to evaluate the function t_{c1s1}. Suppose we choose $\mathscr{S}_2 = k_2\mathscr{C}_2$ as the controlled source, for the evaluation of $T = t_{c1s1}$ by means of the fundamental feedback equation. Then

$$t_{c1s1} = (t_{c1s1})_{\mathscr{S}_2=0} + \frac{k_2(t_{c2s1})_{\mathscr{S}_2=0}\,(t_{c1s2})_{\mathscr{S}_1=0}}{1 - k_2(t_{c2s2})_{\mathscr{S}_1=0}}. \tag{2.8,1}$$

In the above $(t_{c1s1})_{\mathscr{S}_2=0}$ represents t_{oi}, $(t_{c2s1})_{\mathscr{S}_2=0}$ is t_{ci}, $(t_{c1s2})_{\mathscr{S}_1=0}$ is t_{os}, and $(t_{c2s2})_{\mathscr{S}_1=0}$ is t_{cs}. Similarly, let t_{c2s2} denote t_{cs} for k_2 as reference, i.e., with $\mathscr{S}_2 = k_2\mathscr{C}_2$. Again the fundamental feedback equation is used to evaluate t_{c2s2} but now k_1 is used as a reference to aid in the evaluation of t_{c2s2}. The result is

$$t_{c2s2} = (t_{c2s2})_{\mathscr{S}_1=0} + \frac{k_1(t_{c1s2})_{\mathscr{S}_1=0}\,(t_{c2s1})_{\mathscr{S}_2=0}}{1 - k_1(t_{c1s1})_{\mathscr{S}_2=0}}. \tag{2.8,2}$$

Next formulate $F_{k_1} \triangleq 1 - k_1 t_{c1s1}$, and $F_{k_2} \triangleq 1 - k_2 t_{s2c2}$, and it is seen that F_{k_1} and F_{k_2} *have precisely the same numerator*, which is

$$N = 1 - k_1(t_{c1s1})_{\mathscr{S}_2} - k_2(t_{c2s2})_{\mathscr{S}_1} + k_1k_2\,[(t_{c1s1})_{\mathscr{S}_2}(t_{c2s2})_{\mathscr{S}_1} - (t_{c2s1})_{\mathscr{S}_2}(t_{c1s2})_{\mathscr{S}_1}].$$

However, their denominators are, respectively, $1 - k_2(t_{c2s2})_{\mathscr{S}_1=0}$, and $1 - k_1(t_{c1s1})_{\mathscr{S}_2=0}$. Therefore

$$\frac{F_{k_1}}{F_{k_2}} = \frac{(1 - k_1 t_{c1s1})_{\mathscr{S}_2=0}}{(1 - k_2 t_{c2s2})_{\mathscr{S}_1=0}} = \frac{(F_{k_1})_{k_2=0}}{(F_{k_2})_{k_1=0}} \tag{2.8,3}$$

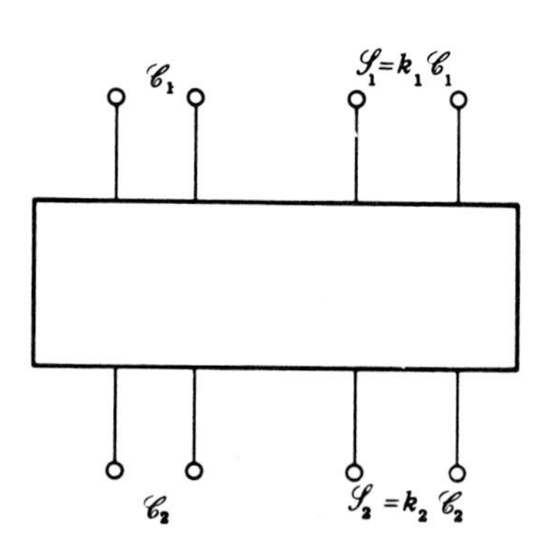

FIG. 2.8-1. Use of two different controlled sources.

The exceptions to the above are:

(1) When for some i, t_{cjsi} is identically zero for all j. For example, in Fig. 2.2-7, if T_4 is joined to the circuit and $\mu_4 E_{g4}$ is taken as $\mathscr{S}_2$, then $t_{cjs2} = 0$ for all j, and $F_{k_2} = 1$.

(2) If either $(t_{c2s1})_{\mathscr{S}_2=0}$ or $(t_{c1s2})_{\mathscr{S}_1=0}$ is zero, then N can be factored into $[1 - k_1(t_{c1s1})_{\mathscr{S}_1}]$ $[1 - k_2(t_{c2s2})_{\mathscr{S}_2}]$ and one of these factors cancels out in F_{k_1}, while the other cancels out in F_{k_2}. Hence F_{k_1} and F_{k_2} do not have the same numerators. An example of this is shown in Fig. 2.8-2, where $t_{c1s2} = 0$.

The theorem can be corrected for the above by stating that it applies only for systems wherein $(t_{c2s1})_{\mathscr{S}_2}(t_{c1s2})_{\mathscr{S}_1}$ is not identically zero.

Equation (2.8,3) may be useful when a particular system is studied intensively. However, it is probably simpler to remember and use the fact that the numerator of F must be the same whatever element is used as reference. Therefore, once the return difference with respect to any element k_1 is found, F_{k_2} may be obtained by manipulating F_{k_1} into the form $1 - k_2 t(s)$, with $t(s)$ independent of k_2. Thus in Example 1 of Section 2.2, Fig. 2.2-3a, it was found that $F_\mu = 1 - \mu t_{cs} = 1 + \mu R/(r_p + R)$. Suppose F_R is desired. The numerator of F_μ is $r_p + R(1 + \mu)$, which is also the numerator of F_R. But F_R must have the form $1 + Rh$ with h independent of R. Therefore $F_R = 1 + R(1 + \mu)/r_p$. Similarly $F_{r_p} = 1 + r_p/[R(1 + \mu)]$. Consider Example 2 of Section 2.2. Here

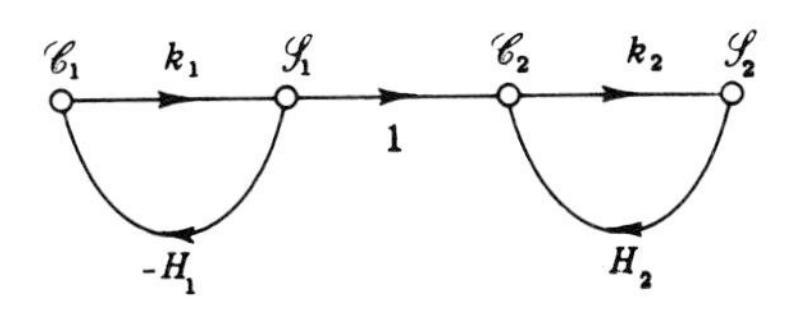

FIG. 2.8-2. A degenerate case.

$$F_\mu = 1 + [\mu(R_1 + R_2 \# R_3)/(r_p + R_1 + R_2 \# R_3)],$$

with numerator $r_p + (R_2 \# R_3)(1 + \mu) + R_1(1 + \mu)$. Therefore,

$$F_{R_1} = 1 + \{[R_1(1 + \mu)]/[r_p + (R_2 \# R_3)(1 + \mu)]\}.$$

To find F_{R_3}, we expand the numerator of F_{R_1}, i.e., write the numerator as

$$[r_p + R_1(1 + \mu)](R_2 + R_3) + R_2 R_3(1 + \mu) = R_3[r_p + (1 + \mu)(R_1 + R_2)] + R_2[r_p + R_1(1 + \mu)].$$

Therefore

$$F_{R_3} = 1 + \{R_3[r_p + (1 + \mu)(R_1 + R_2)]/R_2[r_p + R_1(1 + \mu)]\}.$$

The theorem is also very useful in the stability study of a feedback system, because, from Eq. (2.2,7), the zeros of F_k are the system poles. The theorem tells us that (excluding degenerate cases), any k may be chosen as reference, and the numerator of its return difference examined for the existence of right half-plane system poles.

§ 2.9 The Loaded Transistor Terminal Functions

The feedback analysis technique is applied in the balance of this chapter to transistor feedback circuits. Such circuits are considerably more difficult to analyze than are vacuum-tube circuits. The reason for this is seen by comparing the low frequency incremental models of the two devices in Fig. 2.9-1. The model of the transistor has four elements in it, and this is the largest number that the linear low frequency model of any device need have [this follows from Eq. (1.7,1)]. The nonzero G_A of Fig. 2.9-1b represents a minor difficulty. The greatest trouble is due to G_C. This conductance creates local feedback in each transistor stage. Furthermore, because of G_C, the input (and output) impedance of any transistor stage is a function of the load (or source) impedance. The over-all transmission of a cascade of stages is therefore not equal to the product of the transmissions of the individual isolated stages.

The difficulty is not overcome by using the feedback technique because even after a controlled source is removed, there remains a complicated feedback circuit. Consider, for example, Fig. 2.10-1, and suppose that the transistors are replaced by their low frequency models, and that the controlled source in any one of them is used for $\mathscr{S}$. The circuits which must be solved for t_{oi}, t_{ci},

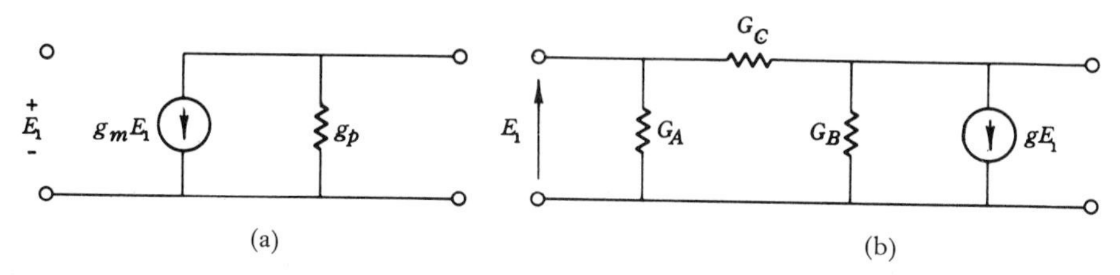

FIG. 2.9-1. (a) Low frequency linear incremental vacuum-tube model. (b) Low frequency linear incremental transistor model.

t_{os}, t_{cs} are still very complicated. It need hardly be noted that nodal or mesh analysis would require hours of computation during which all engineering "feel" for the problem is lost, some numerical errors would probably be made, and the final result would be of little use for suggesting design improvements.

One way to simplify the computation is to effectively eliminate the transistor model from the calculations. This is done by working directly with the loaded transistor terminal functions. It will be seen that such an approach is very well suited for the use of feedback as an analytical tool. There is of course the initial investment of analytical labor that must be made to prepare the list of loaded terminal functions, but it is well worth it if the active device is often used. This approach is now applied to the analysis of transistor feedback circuits.

Four independent terminal functions are needed to characterize a linear device, and probably the most useful are the following (Fig. 2.9-2):

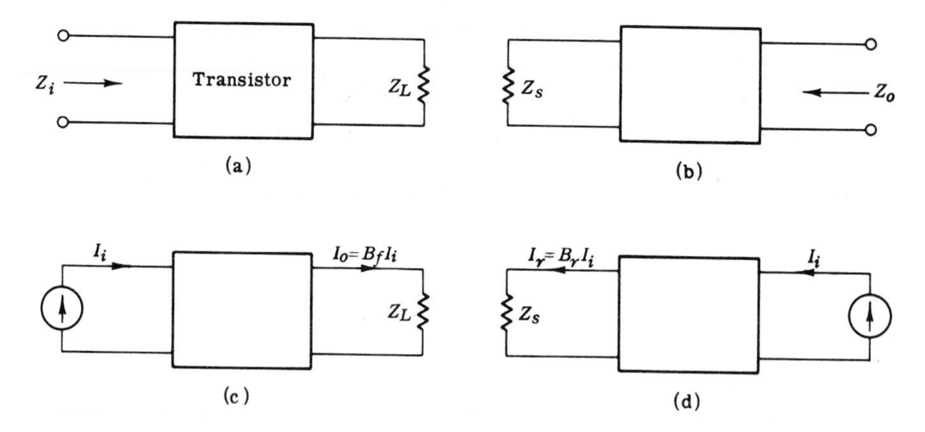

FIG. 2.9-2. Definition of the loaded terminal functions.

(1) the input impedance, Z_i (Fig. 2.9-2a) under load conditions,
(2) the output impedance Z_o (Fig. 2.9-2b) under load conditions,
(3) the forward current gain $B_f = I_o/I_i$ (Fig. 2.9-2c) under load conditions,
(4) the reverse current gain, $B_r = I_r/I_i$ (Fig. 2.9-2d) under load conditions.

The loaded transistor terminal functions for the low frequency range are listed in Table 2.9-1. The terminal functions for three overlapping higher frequency ranges are listed in the reference.[1]

TABLE 2.9-1

LOW FREQUENCY LOADED TERMINAL FUNCTIONS

$$Z_e = z_e + r_e, \quad Z_b = z_b + r_b, \quad Z_{eb} = Z_e + Z_b, \quad Z_{cb} = r_c + Z_b,$$
$$Z_x = r_c(1 - \alpha) + Z_e, \quad \beta = \alpha/(1 - \alpha), \quad Z_n = Z_e + Z_b(1 - \alpha)$$

	C.B.	C.E.	C.C.
Short-circuit forward current transfer ratio	α	$b = -\beta \dfrac{r_c(1 - \alpha)}{Z_x}$ (if $r_c \gg Z_e$)	$\gamma = \dfrac{r_c}{Z_x}$
Loaded forward current transfer ratio	$\alpha' = \alpha \dfrac{Z_{cb}}{Z_{cb} + Z_L}$	$b' = b \dfrac{Z_x}{Z_x + Z_L}$	$\gamma' = \gamma$
Loaded input impedance	$Z_n' = Z_e + Z_b(1 - \alpha')$ (if $r_c \gg z_b$)	$Z_b + Z_e(1 + b')$	$Z_b + \gamma' Z_e$
Loaded output impedance	$\dfrac{Z_{cb} Z_n}{Z_{eb}}$ (if $r_c \gg z_b$)	$\dfrac{Z_{cb} Z_n}{Z_{eb}}$ (if $r_c \gg z_b$)	Z_n (if $r_c \gg z_b$)
Reverse loaded current transfer ratio	$\dfrac{Z_b}{Z_{eb}}$	$\dfrac{Z_e}{Z_{eb}}$	$1 - \alpha$ (if $r_c \gg z_b$)

It is highly desirable to organize the material so as to require memorization of the minimum number of expressions and so as to be able to derive or recall all others that are needed by means of simple, intuitive reasoning. Also one would like to derive expressions:

(1) that can be used with confidence for large ranges of magnitudes without the attendant worry about the validity of the approximations,

[1] I. M. Horowitz, Analysis of Transistor Circuits, Research Report R116. Hughes Research Labs., Sept. 1959.

(2) that can be used with little or no modifications for impedances added in series with the various transistor terminals.

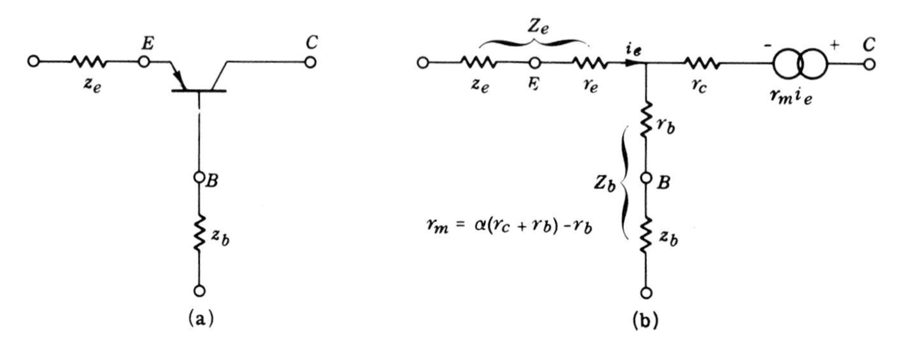

FIG. 2.9-3. Incorporation of external impedances into transistor model.

For the above purposes, the following notation is used in Table 2.9-1 (see Fig. 2.9-3 for the significance of Z_e and Z_b):

$$\left.\begin{array}{l} Z_e = z_e + r_e;\ Z_b = z_b + r_b;\ Z_{cb} = r_c + Z_b;\ Z_{eb} = Z_e + Z_b; \\ r_{eb} = r_e + r_b;\quad Z_x = r_c(1-\alpha) + Z_e;\quad \beta = \alpha/(1-\alpha); \\ Z_n = Z_e + Z_b(1-\alpha); \end{array}\right\} \tag{2.9,1}$$

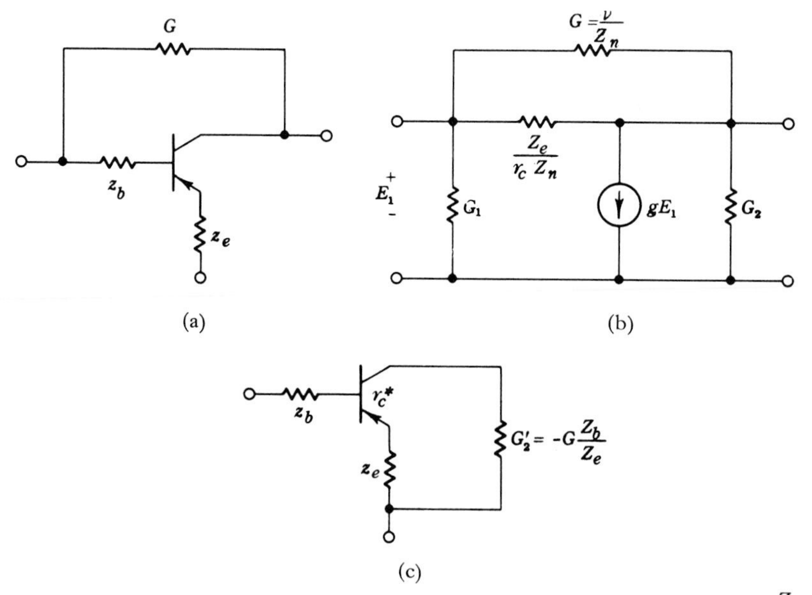

FIG. 2.9-4. (a) Transistor with bridging resistor. (b) Model of (a). $G_2 = \dfrac{Z_b}{r_c Z_n}$, $G_1 = \dfrac{1-\alpha}{Z_n}$, $g = \dfrac{\alpha}{Z_n}$. (c) Transistor circuit equivalent to (a).

where z_e, z_b are external impedances added in series with the emitter and base, respectively, and r_e, r_b, r_c, and α are defined by Fig. 2.9-3. Such approximations as $r_c \gg r_e$, r_b, $r_e/(1-\alpha)$ are used in deriving the loaded terminal functions but it is not assumed that $r_c \gg Z_e/(1-\alpha)$. Others like $r_c \gg Z_b, Z_e, Z_{\text{load}}$ may on rare occasions be incorrect and are avoided whenever possible unless the resulting expressions become very unwieldy, and in such cases the approximations that are made are noted.

By means of a simple modification, the circuit in Fig. 2.9-4a, in which G is often used for bias purposes, may be included in the above scheme. The pi model of Fig. 2.9-4a is shown in Fig. 2.9-4b. Is it possible to find a modified fictitious transistor whose pi model is the same as that of Fig. 2.9-4b? The idea is to use a modified transistor in place of the Fig. 2.9-4a transistor-G combination such that G is absorbed into the modified transistor. Let $r_c{}^*$ be the equivalent of r_c in the modified transistor. It is therefore necessary that $\nu/Z_n + Z_e/r_cZ_n = Z_e/r_c{}^*Z_n$, and hence $r_c{}^*/r_c = Z_e/(Z_e + \nu r_c) = Z_e/(Z_e + r_cZ_nG)$. The values of g^* and $G_1{}^*$ are unaffected by the change in r_c, i.e., $g^* = g$, $G_1{}^* = G_1$, but $G_2{}^*$ is affected by r_c. A conductance G_2' is added in parallel to correct for this, i.e., $(Z_b/r_c{}^*Z_n) + G_2' = Z_b/r_cZ_n$. Hence $G_2' = -\nu Z_b/Z_eZ_n = -GZ_b/Z_e$. The conclusion is that Fig. 2.9-4a may be replaced by Fig. 2.9-4c, providing that in the latter $r_c{}^*$ is used in place of r_c. The only approximation involved in the above is the assumption that $r_c{}^* \gg Z_b$, $r_c \gg Z_{eb}$.

§ 2.10 Analysis of Transistor Feedback Circuits

Example 1. Find I_o/I_i in Fig. 2.10-1a. The dashed portion given by $MM'N'N$ in Fig. 2.10-1a is first replaced by its model. The detailed calculations are given here to help the student see how the loaded transistor terminal function expressions are used in such an analysis. Two forward circuit functions are obtained from Fig. 2.10-1b. (Table 2.9-1 is extensively used in the following.)

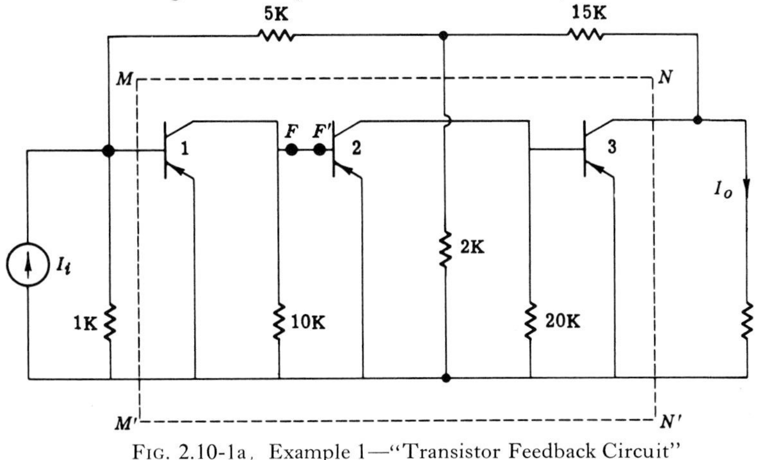

FIG. 2.10-1a. Example 1—"Transistor Feedback Circuit"

Calculations: $Z_{i3} = r_b + r_e(1 + b) = 1400 + 61(30) = 3.23\text{K}$;
$20\text{K} \# 3.23\text{K} = 2.78\text{K}$;
$Z_{x2} = r_c(1 - \alpha) + r_e = 65.3\text{K}$; $b_2' = 60(65.3)/(65.3 + 2.78) = 57.5$;
$Z_{i2} = 1400 + 58.5(30) = 3.16\text{K}$; $10\text{K} \# 3.16\text{K} = 2.40\text{K}$;
$b_1' = 60(65.3)/(65.3 + 2.4) = 57.9$;
$Z_{i1} = 1/y_{11} = 1400 + 58.9(30) = 3.16\text{K}$.

Short-circuit current gain: $= - 57.9(57.5)(60)(2.40/3.16)(2.78/3.23)$
$= - 130{,}000$.

Two backward circuit parameters are found from Fig. 2.10-1c.

Calculations: $Z_{n1} = 30 + 1400(0.98/60) = 52.7$; $r_{eb} = 1430$;
$Z_{o1} = 4(10)^6(52.7/1430) = 148\text{K}$; $10\text{K} \# 148\text{K} = 9.37\text{K}$;
$Z_{n2} = 30 + (1400 + 9370)(0.98/60)$; $Z_{eb2} = 30 + (1400 + 9370)$;
$Z_{o2} = (4)10^6\, Z_{n2}/Z_{eb2} = 77.0\text{K}$; $77\text{K} \# 20\text{K} = 15.8\text{K}$; $Z_{eb3} = 30 + (1400 + 15{,}800)$;
$Z_{n3} = 30 + [\{0.98(1400 + 15{,}800)\}/60]$;
$Z_{o3} = (4)10^6\, Z_{n3}/Z_{eb3} = 72.5\text{K} = 1/y_{22}$.

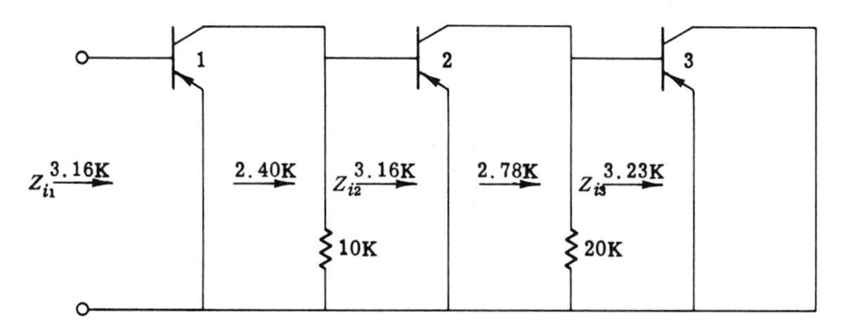

(b)

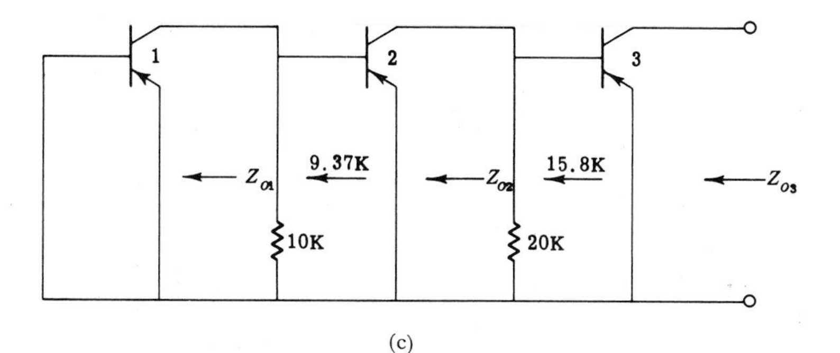

(c)

FIG. 2.10-1(b), (c). Example 1—Calculation of parameters of model of $MNN'M'$ section of FIG. 1a $r_b = 1400\ \Omega$, $r_e = 30\ \Omega$, $r_c = 4$ Meg, $\beta = 60$.

Reverse current gain $=(30/17{,}230)(20/97)(30/10{,}770)(10/158{,}000)(30/1430) \doteq 0$. The model of $MM'NN'$ of Fig. 2.10-1a is therefore the network of Fig. 2.10-2a. Figure 2.10-1a therefore simplifies to the circuit of Fig. 2.10-2b. It is clear that with a bit of practice and a moderate familiarity with the loaded transistor parameters, most of the calculations proceed very quickly and Fig. 2.10-2b

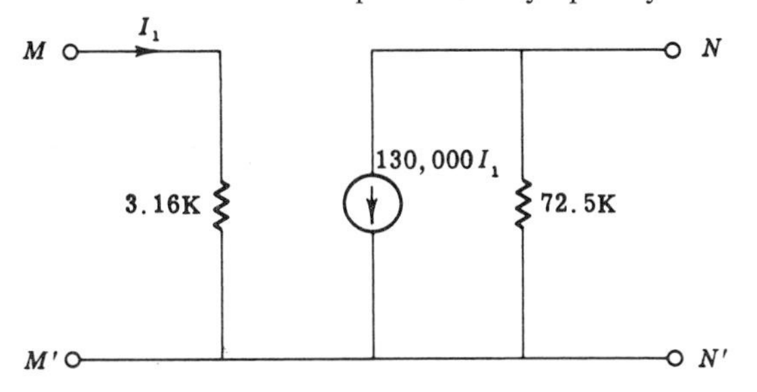

FIG. 2.10-2. (a) Model of $MNN'M'$ section.

may be obtained almost by inspection. Usually one is not interested in three figure accuracy (the transistor parameters are usually not measured with such accuracy and in any case feedback is being used because of their drift or spread

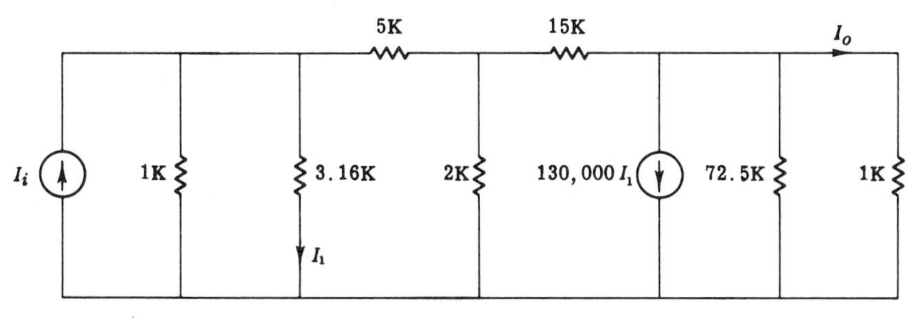

FIG. 2.10-2. (b) Model of Fig. 2.10-1 (a).

in values). In Fig. 2.10-2b if the controlled source is used as reference ($\mathscr{S} = 130{,}000\ I_1$), then t_{oi} and t_{ci} are found from Fig. 2.10-2c with $t_{oi} = 0.11$, $t_{ci} = 0.216$. The values of t_{os} and t_{cs} are found from Fig. 2.10-2d with $t_{os} = -0.943$ and $t_{cs} = -3.55(10)^{-3}$. Therefore $T = t_{oi} + kt_{ci}t_{os}/(1 - kt_{cs}) = -57.5$.

Example 2. The circuit is again the one shown in Fig. 2.10-1. But now the connection FF' (between transistors one and two) is replaced by a current source $GE_{ff'}$ (G is allowed to become infinite eventually) and this source is taken as $\mathscr{S}$. This is one brute force method of choosing a reference whose removal drastically simplifies the remaining circuit. Also approximate calculations will be used. To find t_{oi}, t_{ci}, the current source $\mathscr{S}$ is removed and the result is Fig. 2.10-3a with $t_{oi} \doteq 0$. To find t_{ci}, the energy transferred via the tee network to transistor three and thence to transistor two can obviously be neglected in comparison to that amplified by transistor one. Therefore $t_{ci} \doteq -(60)(65/75)[(0.87)/(2.96+0.87)]\,10^4 = -11{,}800$. For t_{os} and t_{cs}, Fig. 2.10-3b applies.

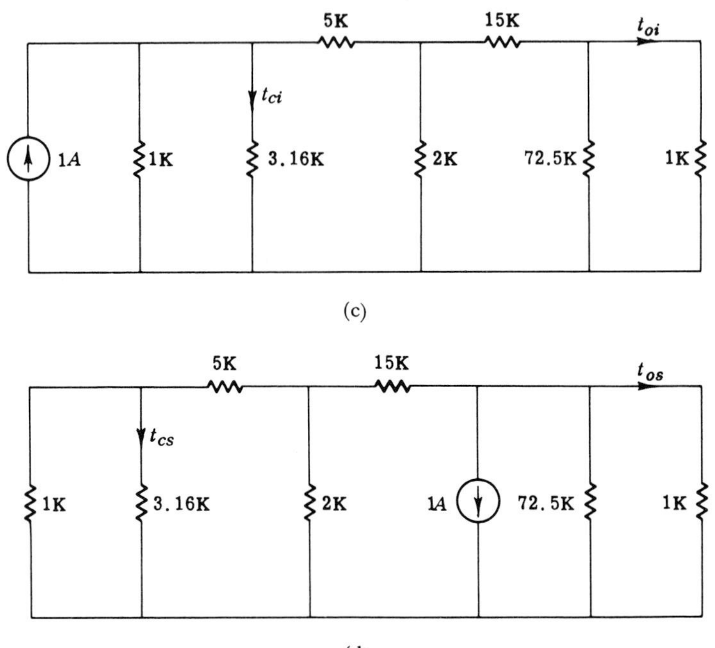

FIG. 2.10-2. (c, d) Calculation of the t_{ij} functions of (b) with $\mathscr{C} = I_1$.

For t_{os}, the transmission through transistor one is ignored and $t_{os} \doteq (60)^2(65/68)(65/66)(20/23)(16.5/17.5) = 2760$. The principal factor determining t_{cs} is the transmission through two and three via the feedback loop and then through transistor one. The contribution of the 1-amp source directly across the output of transistor one is approximately 10^{-4} and may be ignored. Using reasonable approximations, $t_{cs} \doteq -(547)10^4$. The result with all these approximations is -59 for T, as compared to -57.5 obtained by using exact calculations.

Example 3. Find $T = I_o/I_i$ in Fig. 2.10-4a.

It is worthwhile considering some alternative methods of analyzing this circuit for $T = I_o/I_i$. Either mesh or nodal analysis is exceedingly complicated and is ruled out. The feedback method apparently is not too helpful either. Suppose I_{B1} (current into base of transistor 1) is used as $\mathscr{C}$, then the resulting circuit for evaluating t_{oi} and t_{ci} (Fig. 2.10-4b) still has feedback and would require separate application of the superposition technique for its evaluation. If I_{B2} is used as $\mathscr{C}$ the result is the same.

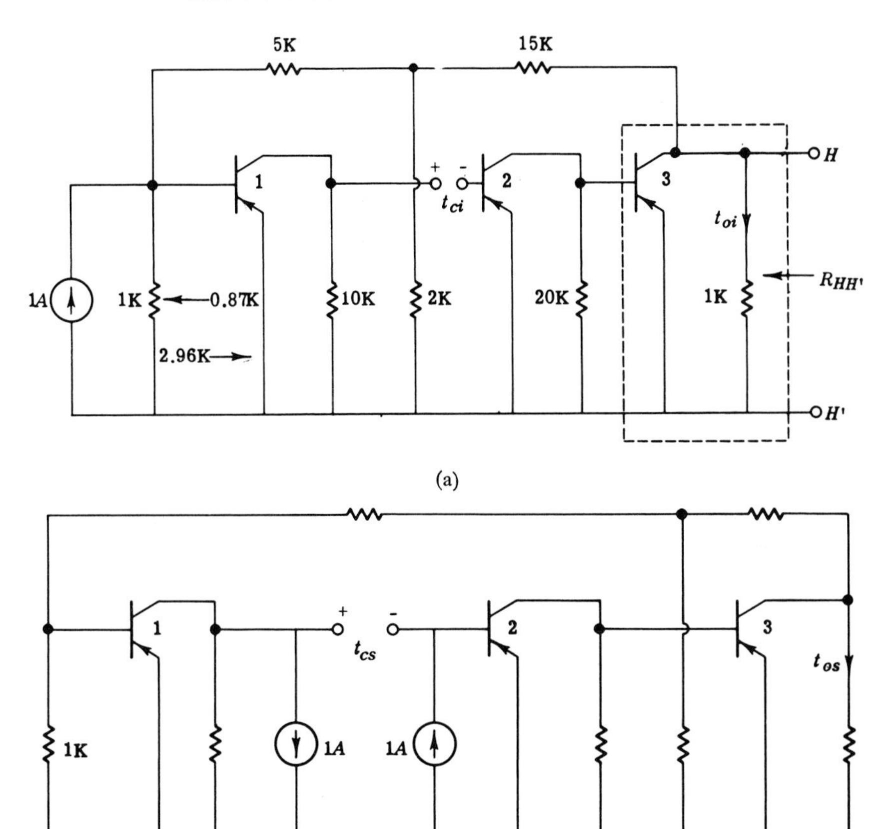

(a)

(b)

FIG. 2.10-3. Example 2—Calculation of t_{ij}'s of Fig. 2.10-1 (a) using $\mathscr{C} = E_{FF'}$.

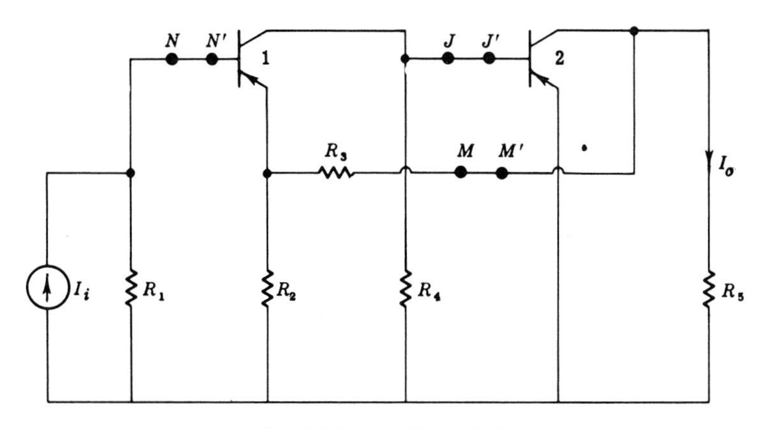

FIG. 2.10-4. (a) Example 3.

What was the original desideratum in the use of the feedback method? It was to obtain simpler circuits in which there was no feedback. One way out of the difficulty is to choose the controlled source in such a manner as to forcibly

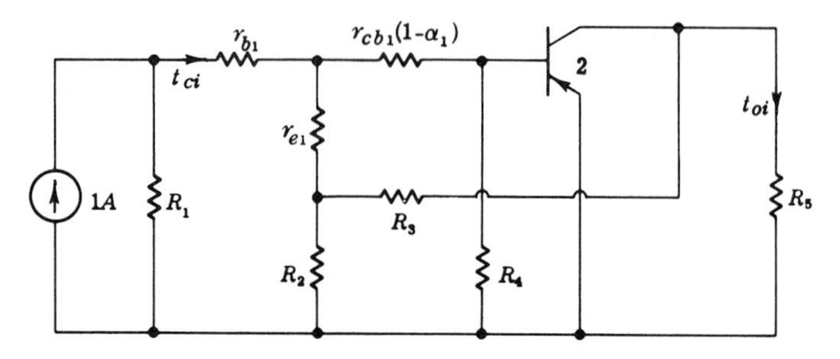

FIG. 2.10-4. (b) Circuit for evaluating t_{oi} and t_{ci} if $\mathscr{C} = I_{B1}$.

kill the feedback. There are two general ways of doing this. One is to choose as $\mathscr{S}$ a current source in series at the input or output side of the transistor. The

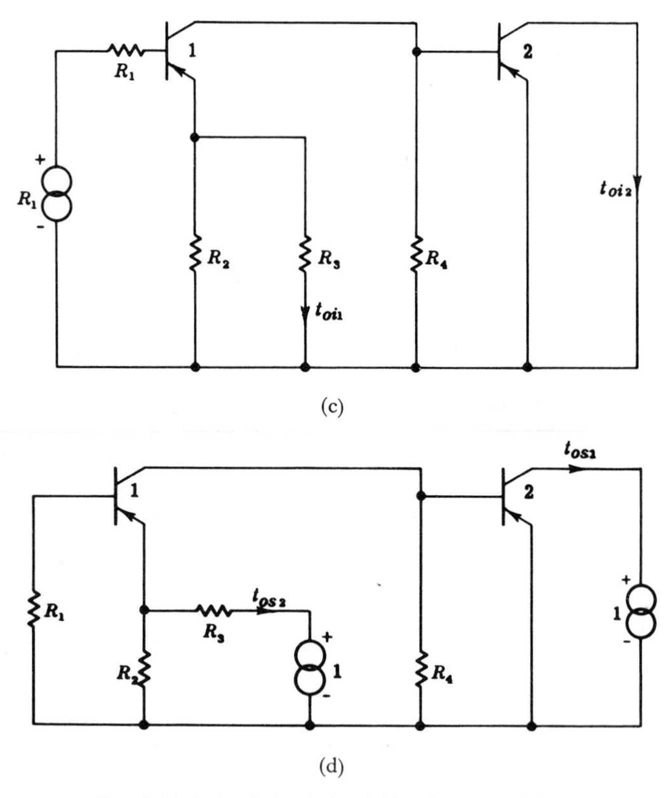

FIG. 2.10-4. (c, d) Analysis of (a) using $\mathscr{S} = I_o R_5$.

other way is to choose as $\mathscr{S}$ a voltage source in shunt at the input or output side of the transistor. In the first case with $\mathscr{S} = 0$, either the input to the transistor or the output becomes an open circuit. In the latter case they become short-circuits. For example, in Fig. 2.10-4a, represent R_5 as a current-controlled voltage source; i.e., $\mathscr{S} = I_o R_5$. Now for t_{oi}, t_{ci} use Fig. 2.10-4c which has no feedback between stages, with $t_{oi} = t_{oi1} + t_{oi2}$, and $t_{ci} = t_{oi}$. For t_{os}, t_{cs}, use Fig. 2.10-4d, with $t_{os} = t_{os1} + t_{os2}$, and $t_{cs} = t_{os}$, and $T = t_{oi} + R_5 t_{ci} t_{os}/(1 - R_5 t_{cs}) = t_{oi}/(1 - R_5 t_{cs})$. Here it is feasible to work with a large t_{oi} only because $\mathscr{C}$ the controlling variable and $\mathscr{O}$ the output are identical, leading to the relation $T = t_{oi}/(1 - kt_{cs})$. This happens whenever the output itself is chosen as the controlled source.

The second way of forcibly obtaining simpler circuits for calculating the t_{ij} functions is illustrated by opening the circuit at JJ' in Fig. 2.10-4a, and inserting a voltage-controlled current source $GE_{JJ'}$ (when the analysis is over G is made infinite). To find t_{oi}, t_{ci}, use Fig. 2.10-5a (the voltage from J' to ground can certainly be ignored). For t_{os}, t_{cs}, use Fig. 2.10-5b.

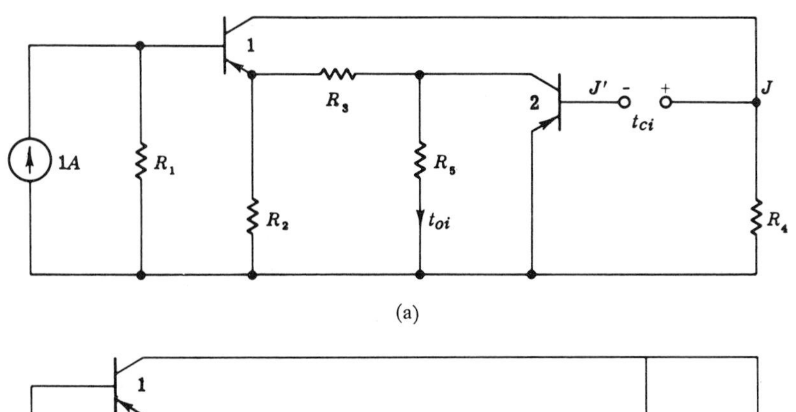

(a)

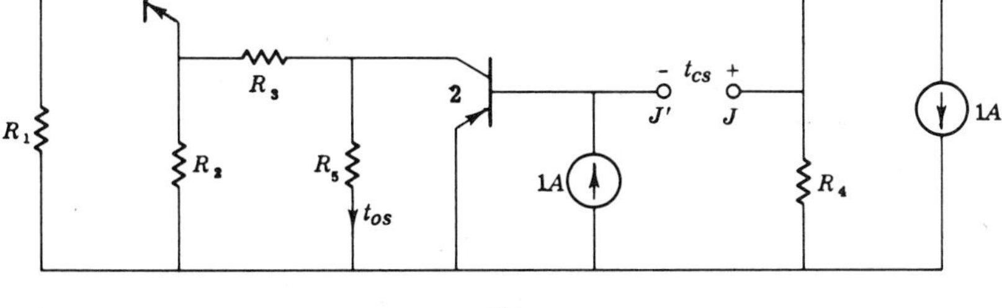

(b)

Fig. 2.10-5. Analysis of Fig. 2.10-4 (a) using $\mathscr{C} = E_{JJ'}$.

There are many alternatives. The circuit may be opened at MM' (Fig. 2.10-4a) or a controlled source may be substituted for R_2. In all such cases, circuits in which there is no feedback between stages must be analyzed. It is

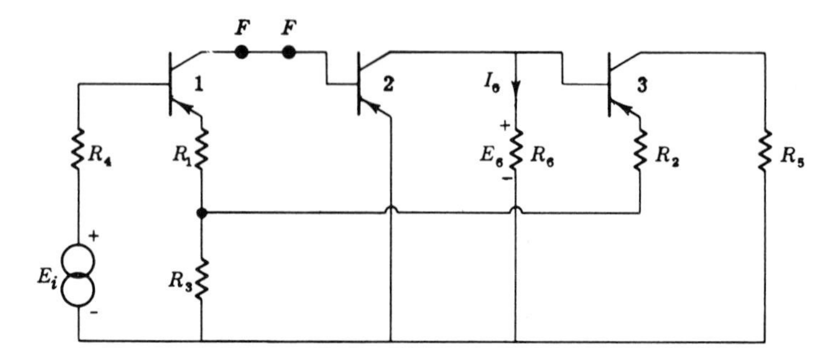

(a)

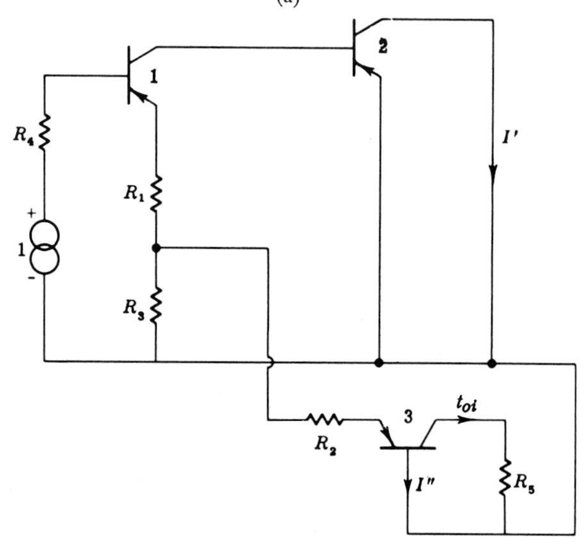

(b)

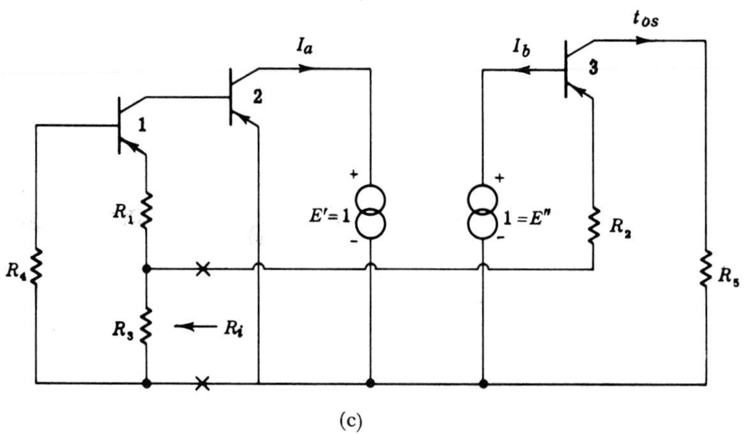

(c)

FIG. 2.10-6. Example 4—Analysis of (a) using $\mathscr{S} = I_6R_6$.

therefore important to be able to quickly analyze the simple cascade type transistor circuit, and for this the loaded terminal functions are helpful.

The above method of forcibly killing the feedback between stages may be used with $T = t_{oi} F_{\text{null}}/F$ [Eq. (2.6,2)].

Example 4. (Fig. 2.10-6a). It is possible to choose $\mathcal{S}$ so that in calculating the t_{ij}'s, no feedback circuits result—i.e., only simple cascade transistor circuits must be solved. This is achieved if R_6 is replaced by the voltage source $\mathcal{S} = I_6 R_6$. To find t_{oi}, t_{ci}, Fig. 2.10-6b is used with $t_{ci} = I' + I'' \doteq I'$. Figure 2.10-6b is fairly easy to solve if the terminal function method is used. To find t_{os}, t_{cs}, Fig. 2.10-6c is used with $t_{cs} = I_a + I_b$. Clearly, the contribution of E' (in assessing it, assume $E'' = 0$) to t_{os} and t_{cs} can be neglected. The contribution of E'' to t_{os} is easily found, since in finding it E' is made zero, and transistors one and two coupled via R_3 merely act as a resistance R_i in series with R_2. R_i is easily found. The contribution of E'' to I_b is similarly easy to find. The contribution of E'' to I_a (with E' set to zero) must not be neglected—it is the major part of t_{cs}, and it is easy to calculate because only a simple cascade of transistors is involved. Thus by the proper choice of $\mathcal{S}$, only simple cascade type circuits need be solved.

§ 2.11 Use of Two Controlled Sources

In complicated feedback circuits where the removal of one controlled source does not suffice to kill the feedback between stages, one may continue to remove controlled sources until there is no longer any feedback. It is probably not practical, however, to remove more than two sources, because of the large number of transmissions that must then be found. The judicious choice of two controlled sources should suffice for most complicated feedback problems. The general signal-flow graph is shown in Fig. 2.11-1, with

$$\begin{aligned}
\mathcal{O} &= t_{oi}\mathcal{I} + t_{os1}\mathcal{S}_1 + t_{os2}\mathcal{S}_2, \\
\mathcal{S}_1 &= k_1\mathcal{C}_1 = k_1\,[t_{c1i}\mathcal{I} + t_{c1s1}\mathcal{S}_1 + t_{c1s2}\mathcal{S}_2], \\
\mathcal{S}_2 &= k_2\mathcal{C}_2 = k_2\,[t_{c2i}\mathcal{I} + t_{c2s1}\mathcal{S}_1 + t_{c2s2}\mathcal{S}_2]\,.
\end{aligned}$$

There are a total of nine functions to be found, some of which are usually zero and the others are often available by inspection. For example, in Fig. 2.10-6a, R_3 could be used for k_1, and FF' could be replaced by a current source $Ge_{FF'} = \mathcal{S}_2$. Then by inspection, $t_{oi} = 0$, $t_{c1i} \stackrel{\Delta}{=} i = (R_4 + r_{eb1} + R_1)^{-1}$, $t_{c2i} = -r_m i + i(r_{eb1} + R_1) = e_{FF'}$. Also by inspection, $t_{os1} = 1/R'_{e2}$, where $R'_{e2} = R_2 + r_{e2} + r'_{b2}(1-\alpha_2)$; $r'_{b2} = r_{b2} + R_6 \# Z_{x2}$; $-t_{c1s1} = (R'_{e2})^{-1} + (R_{eb1})^{-1}$; $R_{eb1} = R_1 + r_{eb1} + R_4$; $t_{c2s1} = (r_{m1}/R_{eb1}) + (r_{b1} + R_4)/R_{eb1}$. Finally, by inspection (see Table 2.9-1 for definitions of Z_x, Z_n, b, b', etc.):

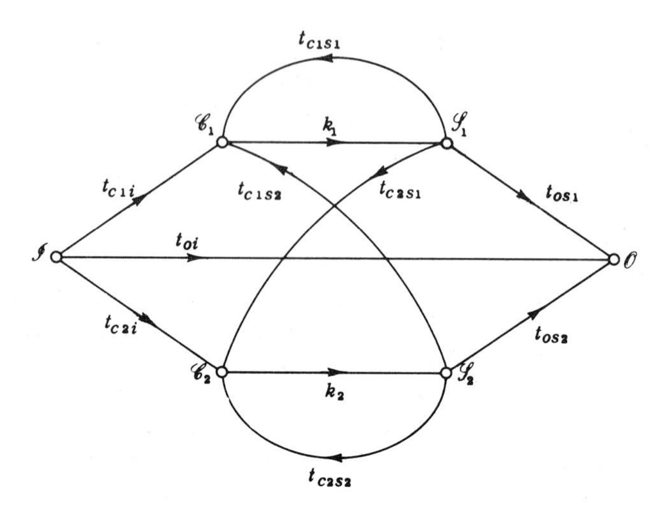

FIG. 2.11-1. Fundamental feedback flow graph for two references.

$$t_{os2} = b_2'\lambda b_3',$$
$$b_2' = b_2 Z_{x2}/\{Z_{x2} + R_6 \# [r_{b1} + (1 + b_3')(R_2 + r_{e2})]\},$$
$$b_3' = b_3 Z_{x3}/(Z_{x3} + R_5), \quad \lambda = R_6/\{R_6 + [r_{b1} + (1 + b_3')(R_2 + r_{e2})]\},$$
$$t_{c1s2} \doteq -\, t_{os2}(1 + b_3')/b_3', \quad t_{c2s2} = -\,(Z_y + r_{c1}Z_{n1}/Z_{eb1}),$$
$$Z_y = r_{b2} + (1 + b_2')r_{e2}.$$

By choosing two appropriate controlled sources, one may have to evaluate fairly simple transmissions. However, there are many more of them; nine instead of four. In general, for the use of n controlled sources $(n + 1)^2$ transmissions must be evaluated.

§ 2.12 The Fundamental Feedback Matrix Equation[1]

Circuits which are separable into sections with two connections only between sections are basically very simple to analyze. The difficulty with feedback circuits can be associated with the fact that there are more than two connections between the various sections of the circuit. We can use a generalization of the superposition approach in conjunction with matrix methods, so that each element in the fundamental feedback equation becomes a matrix of numbers instead of a single number. The method is explained by means of two examples.

Example 1. (Fig. 2.12-1). The cut AA' is made and i_1, i_2 are used as a set of control factors. Let $[\mathscr{C}]$ be a 2×1 matrix (2 rows, 1 column), whose elements

[1] Only elementary matrix theory is needed for this section. See A. E. Guillemin, "The Mathematics of Circuit Analysis," Chaptar 2, Sections 1-6. Wiley, New York, 1949.

are i_1, i_2. The $[\mathscr{S}]$ elements can be any two independent variables completely determined by i_1, i_2; for example, we choose here e_1, e_2, i.e., $[\mathscr{S}]$ is a 2×1

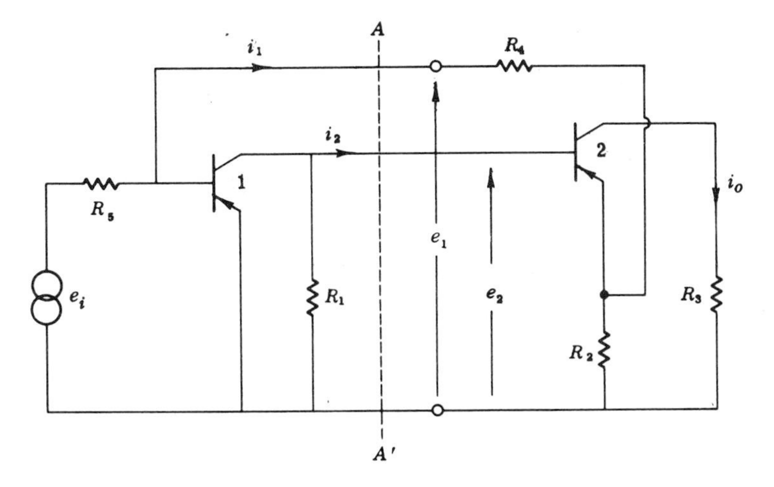

FIG. 2.12-1. Circuit to be solved by the feedback matrix equation.

matrix whose elements are e_1, e_2. The portion to the right of AA' is replaced by the current-controlled voltage sources shown in Fig. 2.12-2a. Figure 2.12-2b

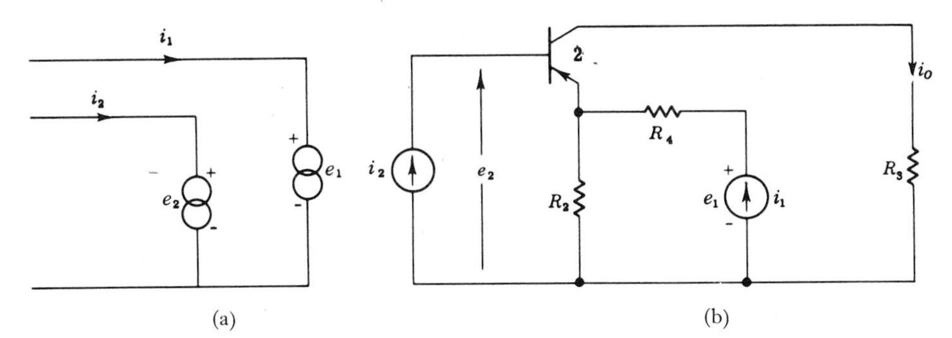

FIG. 2.12-2. (a, b) Evaluation of elements of K matrix. $e_1 = k_{11}i_1 + k_{12}i_2$
$e_2 = k_{21}i_1 + k_{22}i_2$.

is used to find $[K]$ in $[\mathscr{S}] = [K]\,[\mathscr{C}]$. Here

$$e_1 = i_1\{R_4 + R_2 \# [r_{c2}(1 - \alpha_2) + R_3]\} + i_2\,[(1 + b_2')R_2] = i_1 k_{11} + i_2 k_{12};$$

$$b_2' = b_2 Z_{x2}/(Z_{x2} + R_3);$$

$$e_2 = i_1\{R_2 \# [r_{c2}(1 - \alpha_2) + R_3]\} + i_2\,[r_{b2} + (1 + b_2')\,(r_{e2} + R_2)]$$
$$= i_1 k_{21} + i_2 k_{22};$$

$[K]$ is a square matrix whose elements are $k_{11}, k_{12}, k_{21}, k_{22}$. The $[T_{ci}]$ matrix is found by setting $[\mathscr{S}] = 0$, i.e., $e_1 = e_2 = 0$. This leads to Fig. 2.12-2c. Since $i_1 = 1/R_5$, $i_2 = 0$, the elements of the 2×1 $[T_{ci}]$ matrix are G_5 and zero.

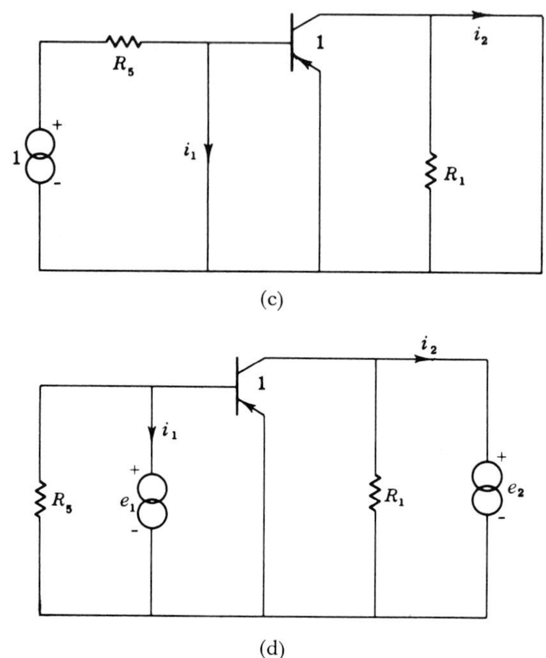

FIG. 2.12-2. (c) Circuit for finding elements of T_{ci} matrix. (d) Evaluation of elements of T_{cs} matrix.

The $[T_{cs}]$ matrix is found by setting $[\mathscr{I}] = 0$ and $[\mathscr{S}] = \left[\begin{smallmatrix}1\\1\end{smallmatrix}\right]$. This leads to Fig. 2.12-2d with

$$i_1 = -\,e_1\,[G_5 + (r_{b1} + b_1 r_{e1})^{-1}] + e_2 r_{e1}/r_{c1} Z_{n1} = b_{11} e_1 + b_{12} e_2;$$
$$i_2 = -\,e_1 b_1/(r_{b1} + b_1 r_{e1}) - e_2(G_1 + r_{eb1}/r_{c1} Z_{n1}) = b_{21} e_1 + b_{22} e_2;$$

$[T_{cs}]$ is a 2×2 matrix whose elements are $b_{11}, b_{12}, b_{21}, b_{22}$. Now, dropping the brackets for convenience, $\mathscr{S} = KC = K\,[T_{ci}\,\mathscr{I} + T_{cs}\,\mathscr{S}]$ where $\mathscr{I}$ is the independent input matrix. Hence $[U - KT_{cs}]\,\mathscr{S} = KT_{ci}\,\mathscr{I}$ where U is the unit diagonal matrix. The solution of this last matrix equation is $\mathscr{S} = [U - KT_{cs}]^{-1}\,KT_{ci}\,\mathscr{I}$. Consequently,

$$\mathscr{O} = T_{oi}\mathscr{I} + T_{os}\mathscr{S} = \{T_{oi} + T_{os}\,[U - KT_{cs}]^{-1}\,KT_{ci}\}\mathscr{I}$$

and

$$T = T_{oi} + T_{os}\,[U - KT_{cs}]^{-1}\,KT_{ci}. \tag{2.12,1}$$

In Eq. (2.12,1) the capital letters represent matrices.

The $\mathscr{C}$ variables need not all have the same dimension. In Fig. 2.12-1, they could be i_1, e_1 and the $\mathscr{S}$ variables e_2, i_2 or i_o, i_2 or i_o alone. But the larger the number of variables that are used, the easier is the step by step analysis, although the matrix calculations are lengthier.

The feedback matrix approach can be applied to any n-port K network imbedded in the system; a four-port K network is shown in Fig. 2.12-3a. The four-port is replaced by controlled sources with the matrix equation $\mathscr{S} = K\mathscr{C}$; for example, the terminal currents are shown as elements of the controlled source matrix $\mathscr{S}$ and the terminal voltages are the elements of the controlling variable matrix $\mathscr{C}$ in Fig. 2.12-3b. (Note that it is not permissible to assume that $I_1 = -I_2$, $I_3 = -I_4$, etc.) Or a set of mixed current and voltage variables could be used. The next step is to apply superposition, $\mathscr{O} = T_{oi}\mathscr{I} + T_{os}\mathscr{S}$, etc., and the final result is Eq. (2.12,1). This approach is used in Chapter 10 which deals with multiple input-output systems.

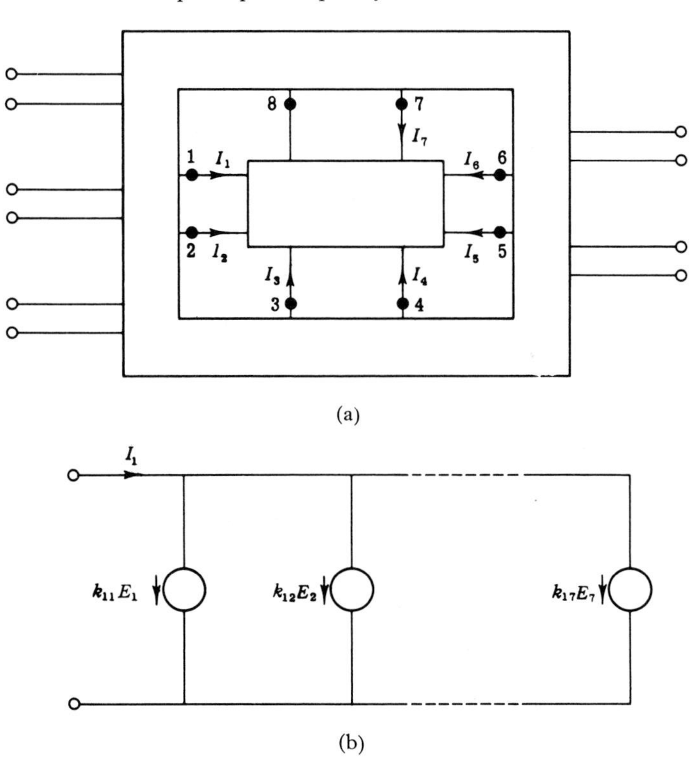

FIG. 2.12-3. Application of feedback matrix method to a four-port.

CHAPTER THREE

Introduction to the Properties of Feedback

§ 3.1 Introduction

The concept of feedback and its analytical formulation were presented in Chapter 2. The decisive result of Chapter 2 is the Bilinear Theorem (Section 2.5) which will be used throughout the book in order to derive the benefits and costs of feedback in a definitive, fundamental manner. This theorem tells us that if we are interested in the effect of a system parameter k upon a linear system function T, then the relation that is needed is the fundamental feedback equation, $T = [kt_{ci}t_{os}/(1 - kt_{cs})] + t_{oi}$, where all the t_{ij}'s are independent of k. The parameter k can be an amplification factor, a passive parameter like a resistance or capacitance, or it can be the transfer function of a complicated system like a missile or a chemical plant. It is only necessary that it be possible to find two system variables $\mathscr{C}$ and $\mathscr{S}$, such that $\mathscr{S} = k\mathscr{C}$, and that k appears in no other relations among any other system variables.

In Chapter 3 the fundamental feedback equation is used in order to introduce the properties of feedback. Benefits of feedback are presented and are quantitatively formulated. Design examples are given to illustrate how the benefits of feedback may be exploited. In the course of working out the designs, some significant problems are noted for intensive treatment in later chapters. There is no attempt at completeness in Chapter 3. It is intended only to break the ice and to serve as an introduction to the reasons and the manner in which feedback is used.

§ 3.2 Reduction in the Sensitivity of a System to Parameter Variation

One of the foremost properties of feedback is its ability to reduce the sensitivity of a system to variations in the system parameters. Suppose there is an element in a system which must be used because of its special properties. However, the element is sensitive to environmental conditions (such as temperature and pressure), or is subject to aging, or has wide manufacturing tolerances such that when it is replaced, the parameters of the new element may be markedly different from those of the old element. Whatever the reason may be, suppose the values of the element parameters vary significantly. It is desired that despite these variations, certain system properties (such as the input-output transfer function or output impedance) should remain substantially constant. The troublesome element could perhaps be replaced by a better one, but this may be very expensive or impossible. In any case we would like at least to have an alternative to replacing the element. In most cases the troublesome elements are the active elements, vacuum tubes, transistors, and energy conversion devices such as motors and generators. But they may be passive elements or transducers. It is possible, by means of feedback around the troublesome element, to achieve the desired reduction in sensitivity. This is seen as follows:

Let the troublesome parameter be k and, in the manner of Chapter 2, we relate k to a controlled source $\mathscr{S}$ and a control variable $\mathscr{C}$, i.e., $\mathscr{S} = k\mathscr{C}$. Also, let the system function of interest be T. Then, using the results of Section 2.5, T can be written in the form of the fundamental feedback equation $T = t_{oi} + kt_{ci}t_{os}/(1 - kt_{cs})$, with the fundamental signal-flow graph of Fig. 3.2-1. Here

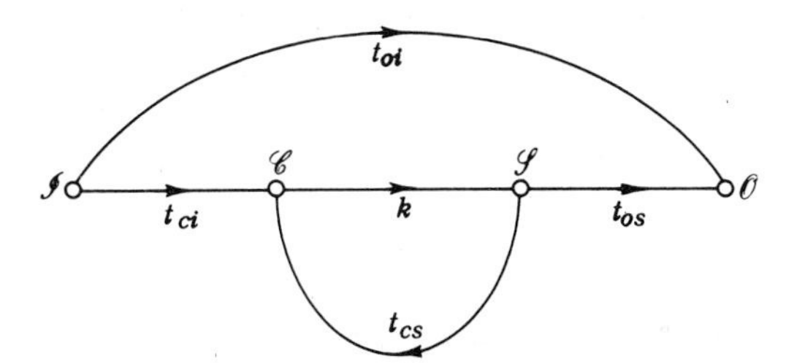

FIG. 3.2-1. The fundamental feedback flow graph.

it is extremely important to note that all the t's in the fundamental feedback equation should be independent of k. (It was noted in Section 2.5 that there may be parameters such as mutual inductance, for which the Bilinear Theorem does not apply. Two ways to handle such elements were noted in Section 2.5.) If k is regarded as a variable, then $T = T(k)$, and the conventional definition[1] of the sensitivity of T to k, symbol S_k^T, is:

$$S_k^T \triangleq \frac{\partial \ln T}{\partial \ln k} = \frac{\partial T/T}{\partial k/k} \tag{3.2,1}$$

S_k^T is thus the relative change in T divided by the relative change in k, for infinitesimally small changes in T and k only. Since T is in general a function of the complex variable s, S_k^T is also a function of s.

[1] Bode's definition is the inverse of the above, but by conventional we mean that the definition is for infinitesimal variations only.

Consider first the case where $t_{oi} \ll T$ (negligible leakage transmission). This is the situation in many feedback systems. The general case is considered later. By straightforward differentiation of the fundamental equation and some manipulation, it is readily found that

$$S_k^T = \frac{1}{1 - kt_{cs}} = \frac{1}{F_k} \qquad (\text{case } t_{oi} = 0 \qquad (3.2,2)$$

where it is recalled (Section 2.5) that $1 - kt_{cs} \triangleq F_k$ has been called the return difference for reference k. It is clear now why we have insisted that all t_{ij}'s in the fundamental feedback equation should be independent of k. For small sensitivity then, $|F_k|$ should be large, i.e., $-kt_{cs}$ (which is usually called the loop gain, symbol L) should be a large number. Since t_{cs} and possibly k itself is a function of frequency, so is S_k^T, and one must consider the frequency range over which a small S_k^T is desired. The phase requirements on the loop transmission $L \triangleq -kt_{cs}$ are considered later in connection with the problem of stability.

A serious shortcoming of Eqs. (3.2,1)-(3.2,2) is that they apply only for infinitesimally small changes of k. There is therefore uncertainty as to their applicability for moderate or large changes in k. We shall therefore find it often useful to use a new definition, defined as follows: Let T_o, k_o represent the nominal or original design value of the system response and of the element under consideration, respectively, and let T_f, k_f be the corresponding (final) values at the new value of k. Thus

$$T_f = T_o + \Delta T, \quad k_f = k_o + \Delta k. \qquad (3.2,3)$$

The new sensitivity function is defined as:

$$\mathcal{S}_k^T \triangleq \frac{\Delta T/T_f}{\Delta k/k_f} \qquad (3.2,4)$$

$\mathcal{S}_k^T$ is first found for $t_{oi} = 0$. In this case it is readily established by means of the fundamental feedback equation and Eq. (3.2,3) that

$$\boxed{\mathcal{S}_k^T = \frac{\Delta T/T_f}{\Delta k/k_f} = \frac{1}{1 - k_o t_{cs}} = \frac{1}{F_{k_0}} = \frac{1}{1 + L_o}} \qquad (\text{case } t_{oi} = 0). \qquad (3.2,5)$$

Thus, when the leakage transmission t_{oi} is zero, the classical and new definition have the same value, although they are defined differently. With the new definition there is no uncertainty as to the effect of large parameter variation. Equation (3.2,5) is exact only if t_{oi} is zero, but it may be used whenever $t_{oi} \ll T_f$ and T_o. Fortunately t_{oi} is zero for a very large class of feedback control systems, and $t_{oi} \ll T_o$ and T_f, for a very large class of feedback amplifiers.

From Eqs. (3.2,3) and (3.2,5), T_f is easily written in terms of the original T_o, F_{k_o} and of Δk. Thus Eq. (3.2,5) is

$$(T_f - T_o)/T_f = 1 - T_o/T_f = (1 - k_o t_{cs})^{-1}(1 - k_o/k_f).$$

Therefore

$$T_o/T_f = 1 - (1 - k_o/k_f)(1 - k_o t_{cs})^{-1} = (k_o/k_f - k_o t_{cs})(1 - k_o t_{cs})^{-1};$$

i.e.,

$$\boxed{\frac{T_o}{T_f} = \frac{(k_o/k_f) + L_o}{1 + L_o}} \qquad \text{(case } t_{oi} = 0) \qquad (3.2,6)$$

where $L_o \triangleq -k_o t_{cs}$ is the nominal loop transmission. The application of these results is illustrated by several examples.

§ 3.3 Design Examples

Example 1.

Specifications: An amplifier is to be designed with an over-all gain of 2500 ± 50. A number of single amplifier stages is available. The gain of any single stage may drift anywhere between 25 and 75.

Design: We attempt to solve the problem by means of feedback and use Fig. 3.2-1 as a guide. To begin with, we assume t_{oi} is zero and take k as a cascade of n amplifier stages, with n unknown. One extreme value of the gain of these n stages is taken as $k_o = (25)^n$. Then the other extreme value of the gain of the n stages is $k_f = (75)^n$. [There is nothing wrong of course in taking $k_o = (75)^n$ and $k_f = (25)^n$.] Then $\Delta k = k_f - k_o = (3^n - 1)25^n$, and from Eq. (3.2,5):

$$\frac{T_f - T_o}{T_f} = \left(\frac{\Delta k}{k_f}\right)\frac{1}{1 - k_o t_{cs}} = \frac{3^n - 1}{3^n}\,\frac{1}{1 - k_o t_{cs}} = \frac{3^n - 1}{3^n}\left(\frac{T_o}{k_o t_{ci} t_{os}}\right).$$

In the last step, use has been made of the fact that $1 - k_o t_{cs} = k_o t_{ci} t_{os}/T_o$ when t_{oi} is zero. Suppose that it is possible to obtain a $t_{ci} t_{os}$ product of 0.90. We take T_o as 2450, T_f as 2550 (since these are the extreme permitted values of T), and then the last equation becomes

$$[(3^n - 1)/3^n]\,[2450/(25)^n(0.90)] = (T_f - T_o)/T_f \leqslant 100/2550.$$

The minimum integral value of n that satisfies the inequality is $n = 4$. With this value of n, $k_o = (25)^4$ and using $T_o = k_o t_{ci} t_{os}/(1 - k_o t_{cs}) = 2450$, it is seen that $t_{cs} = -1/2740$. The system signal-flow graph is sketched in Fig. 3.3-1.

In the above it is implicitly assumed that negative feedback is being used (kt_{cs} negative) rather than positive feedback (kt_{cs} positive). If positive feedback is used, we would write $T_o = 2550$, $T_f = 2450$ and emerge with $k_o t_{cs}$ substan-

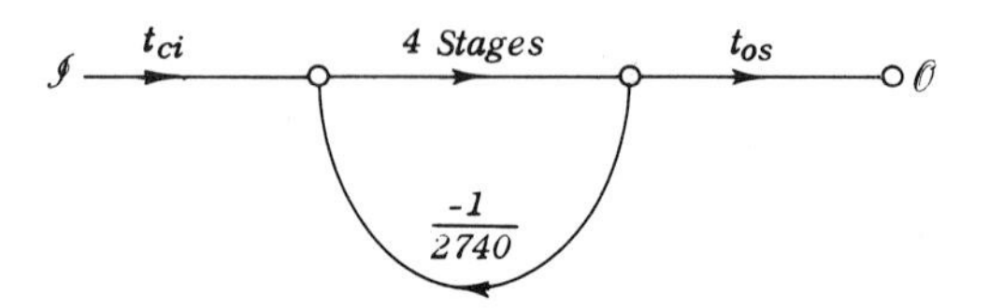

FIG. 3.3-1. Design example 1. ($t_{ci}\, t_{os} = 0.90$).

tially the same in magnitude, but positive. Actually, nothing that has been done so far in the book justifies the prohibition of positive feedback in the above example. It will be shown later why it is impossible to use positive feedback in problems requiring sensitivity reduction. Another point to note in this example is that only the zero frequency values of the various transmissions have been obtained. This is because the system behavior at zero frequency only has been specified. A far more difficult problem is to assign suitable values to the individual transmissions over the balance of the frequency range. This problem is briefly treated in Section 3.5, and is treated in considerable detail in later chapters.

Example 2A.

Specifications: The load R_L at the output of a system may have any value from 10 to 10,000 Ω, and it is desired that the output voltage stay within the limits of 10 ± 0.1 volts for an input of 0.1 volt. The final stage of the system is fixed for various reasons, and its internal resistance $R_i = 1000$. The problem

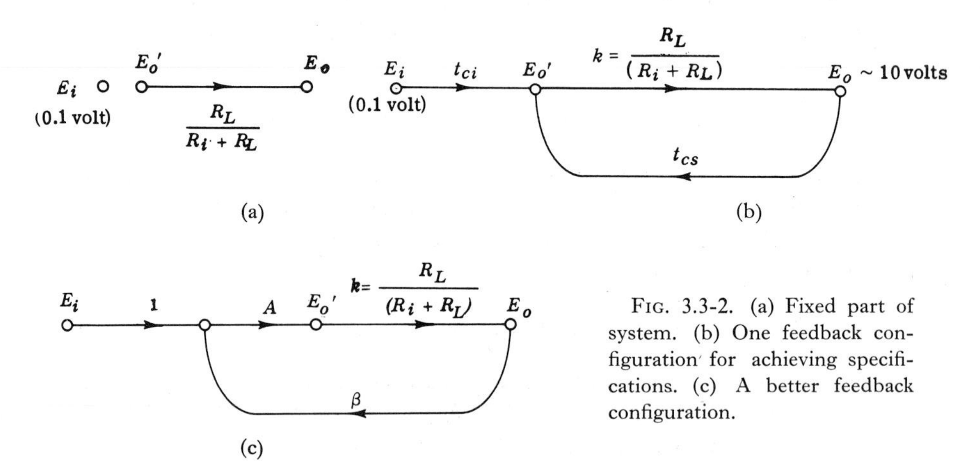

FIG. 3.3-2. (a) Fixed part of system. (b) One feedback configuration for achieving specifications. (c) A better feedback configuration.

is similar to that of a motor driving a load which varies over a wide range and where the load torque must be maintained within narrow tolerances.

Design: For the final stage of the system we write $E_o = E_o' R_L/(R_i + R_L)$ where E_o' is the input to this last stage, and present what we have so far in Fig. 3.3-2a, with E_i representing the system input of 0.1 volt. Using the signal-flow graph of Fig. 3.2-1 as a guide, we could attempt to satisfy the specification by means of the system-flow graph of Fig. 3.3-2b. There are two objections to such a configuration. One is that both t_{ci} and t_{cs} will have to be high-gain amplifiers, the former to achieve the desired over-all system gain and the latter to achieve the required sensitivity reduction. These high-gain amplifiers will have no feedback around them and the system sensitivity to them will therefore not be small. To economize on the total number of amplifiers needed and to obtain small system sensitivity to them as well as to R_L, it is best to include the amplifiers in the loop by means of the configuration of Fig. 3.3-2c.

The representation of the last stage as shown in Fig. 3.3-2c with $k = R_L/(R_i + R_L)$ is a good idea, for if we try to show R_L as k (i.e., manipulate the flow graph so that R_L appears in only one branch transmission as in Fig. 3.3-3[1]), we

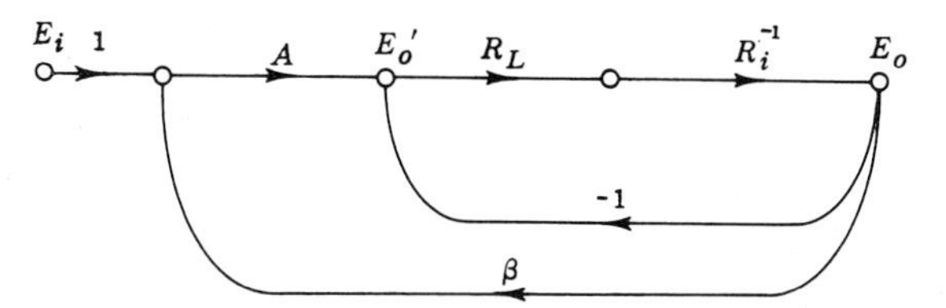

FIG. 3.3-3. A less desirable representation of fixed part of Example 2.

then have a more complicated system to work with. In Fig. 3.3-2c however, not R_L but $R_L/(R_i + R_L)$ is taken as the parameter k of the fundamental feedback equation. This technique is extremely useful and is the basis for the statement that in most feedback control problems it is possible to choose k so that the leakage transmission is zero.

In proceeding with the design we could take into account the maximum variations in the amplifiers to be used in A in Fig. 3.3-2c, but this will not be done here. Suppose the nominal value of R_L is taken as $10\,\Omega$, then $k_o = AR_L/(R_i + R_L) = A/101$, and the extreme $k_f = 10{,}000A/(1000 + 10{,}000) = 10A/11$. The extreme $k_o/k_f \approx 0.01$, so the extreme $\Delta k/k_f = 0.99$. From Eq. (3.2,5), $\Delta T/T_f = S_k{}^T \Delta k/k_f \leqslant (10.1 - 9.9)/10.1$. Therefore $S_k{}^T \leqslant 0.2/10.1$. But $S_k{}^T = 1/(1 - \beta A/101)$ (Fig. 3.3-2c). For negative feedback βA is negative, so we have $[1 + |\beta A|/101] \geqslant 10.1/0.2 = 50.5$, or $|\beta A| \geqslant 5000$. It is also required that E_o/E_i be equal to 9.9 at $R_L = 10$. Therefore $0.01A/$

[1] The minor loop in Fig. 3.3-3 was obtained by writing E_o/E_o' of Fig. 3.3-2 in the form $E_o/E_o' = (R_L/R_i)/(1 + R_L/R_i)$ and noting that this corresponds to the inner loop in the flow graph of Fig. 3.3-3.

$(1 + 0.01 \mid \beta A \mid) = 9.9$. Since $\mid \beta A \mid_{\min} = 5000$, $\mid A \mid_{\min} = 50{,}500$ and $\mid \beta \mid = 1/10.1$.

It is noted that in this example too, only the zero frequency values of the various transmissions have been assigned. While the assumed values of the parameters may be correct to very high frequencies, it is well known how difficult it is to maintain large amplifier gain over a wide range of frequencies. The actual frequency range over which A in the above example must be kept large, is a very important problem in feedback design. It is, however, one which must be postponed for later treatment.

Example 2B. Suppose in the above problem that the load current is to be maintained relatively insensitive to variations in R_L. We would then use the relationship $I_L = E_o'/(R_i + R_L)$ with the signal-flow graph of Fig. 3.3-4,

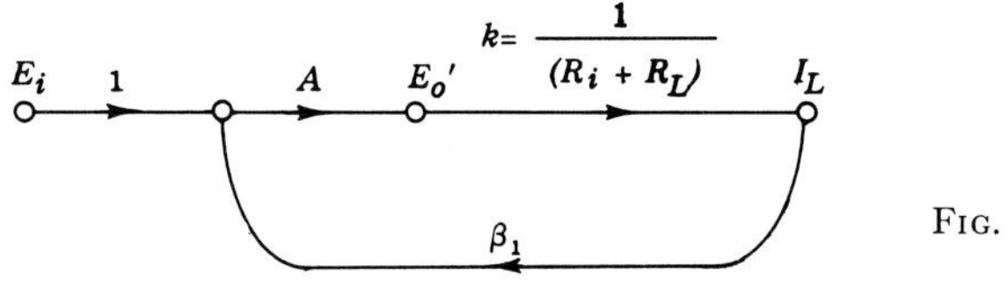

FIG. 3.3-4. Example 2B.

consider the variations in $k = 1/(R_i + R_L)$, and proceed in the same manner. Here the signal fed back must be proportional to the current in R_L, but the procedure is precisely the same.

Example 2C. Suppose in the above that it is the load power which is to be maintained relatively insensitive to variations in R_L. We write $P_L = I_L{}^2 R_L = E_o'^2 R_L/(R_i + R_L)^2$ and use the flow graph shown in Fig. 3.3-5, where the

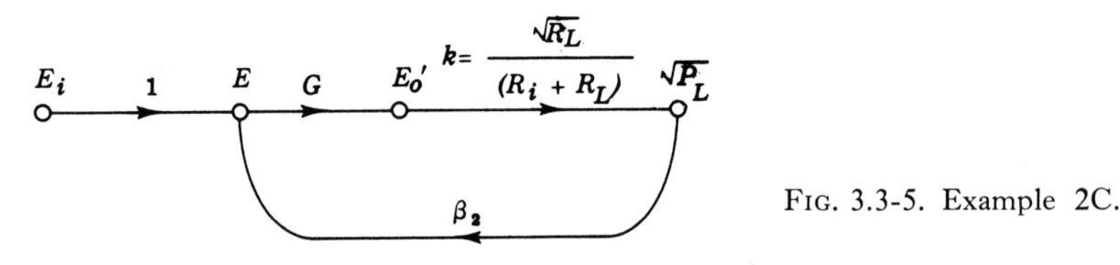

FIG. 3.3-5. Example 2C.

reference $k = (R_L)^{1/2}/(R_i + R_L)$. A device must be found which is sensitive to the square root of load power. The balance of the design procedure is the same as before. There arises the question whether linear feedback theory may be applied to this problem, inasmuch as k is a nonlinear function of R_L, and the output is a nonlinear function of P_L. Let $X = (p_L)^{1/2}$. Then, so long as the integro-differential equations relating X, E_o', k, E_i, and E are linear, the system is linear just as Ohm's law $V = RI$ is linear in V and I, even though R may be a nonlinear function of temperature and of many other variables. If it is possible to instrument the system such that the output of the β element is linearly related to $(p_L)^{1/2}$, then we have a linear system, and the previous theory applies.

By means of this technique, many control problems which at first sight appear to be nonlinear, can be handled by means of linear feedback theory.

§ 3.4 Cost of Feedback

The reduction of S_k^T by means of feedback is achieved by the proper adjustment of the signal applied to the element *k*. As the parameters of *k* vary, the signal fed back to the input to *k* also varies and compensates for the variation in *k*. In Example 2*A* (Fig. 3.3-2c), at $E_o = 1$ volt and $R_L = 1000\ \Omega$, the signal E_o' appearing at the input to the final stage measures 2 volts. If R_L is only $10\ \Omega$, E_o' increases to almost 100 volts, in order that the output may remain nearly the same as before (at 1 volt). In order that $E_o = 10$ volts, E_o' must be 1000 volts when $R_L = 10\ \Omega$.

One wonders whether any of the stages in the forward loop (in the *A* block) do not saturate at such a high signal level. This problem, in fact, is a necessary and extremely important part of the design—the selection of components that are capable of doing what the design specifications call for. The blame for such onerous demands should not be placed on feedback. The large signal level capacities that are required in the system elements have nothing to do with the fact that automatic feedback is used (except in the indirect sense that the properties of feedback may tempt one to attempt extremely difficult tasks such as that in Example 2, Section 3.3). Suppose there is advance knowledge of the variations in R_L, so that feedback is not required. Instead, the preamplifier is programmed to compensate for the variations in R_L. If in Example 2, Section 3.3, the designer wants an output of 10 volts for an input of 0.1 volt, with $R_i = 1000$, $R_L = 10$, then the input of the last stage must be 1000 volts no matter whether feedback is or is not used. And additional amplifier stages of $1000/0.1 = 10{,}000$ total gain are therefore needed. Of course, with feedback, additional amplifier gain is needed over and above what would be required if there were no automatic feedback. In this same example, these *additional* stages have a gain of 5 (since $A = 50{,}500$) with a maximum output level which is ridiculously small (approximately 1/10 volt). There is certainly no problem in securing amplifier stages with such signal level capacity. The extra gain of 5 is a very small price to pay for automatic regulation. Usually, automatic feedback places no specially difficult demands on the signal level capacity requirements of the extra stages that are needed. It will be seen later that the cost of feedback is primarily in the larger bandwidth requirements of the system, rather than in their dynamic range.

§ 3.5 Sensitivity as a Function of Frequency

In the previous examples, only the zero frequency sensitivity specifications and the zero frequency system response were considered. The solutions that were obtained represent therefore only the system design values at zero frequency. The design cannot be completed until the system requirements over the entire frequency range are considered. In this section, the design procedure is extended to cover the "useful" frequency range. By this is meant the system transfer function's significant bandwidth, which in turn is related to the speed of response of the system. The design procedure is best described by means of a numerical example.

Example 3.

Specifications: An instrument includes an indispensable element whose gain and frequency response is sensitive to the environment, such that over the operating conditions the transfer function of the element, $k = m/(\tau s + 1)$, may have m anywhere from 1 to 5, and τ anywhere from 0.5 to 2. It is desired that from zero frequency to 4 radians/sec the system transfer function $T(s)$ should not vary by more than 10%. Also, it is required that $T = 1$ at zero frequency. The output signal must pass through the above element so that $t_{oi} = 0$. It is noted that the specifications recognize the fact that the system function $T(s)$ is actually a function of frequency. Consequently, the statement of system sensitivity includes the frequency range over which the desired insensitivity is to be maintained.

Design: The sensitivity relation best suited for this problem is given by Eq. (3.2,6), and rewritten here:

$$\frac{T_o}{T_f} = \frac{L_o + (k_o/k_f)}{L_o + 1}. \tag{3.5,1}$$

The problem is to select $L_o(j\omega)$ of Eq. (3.5,1) so as to achieve the specified tolerances on T_o/T_f, despite the variation in k_o/k_f. The selection of $L_o(j\omega)$ can be achieved by the following construction. First, choose the nominal value of k;

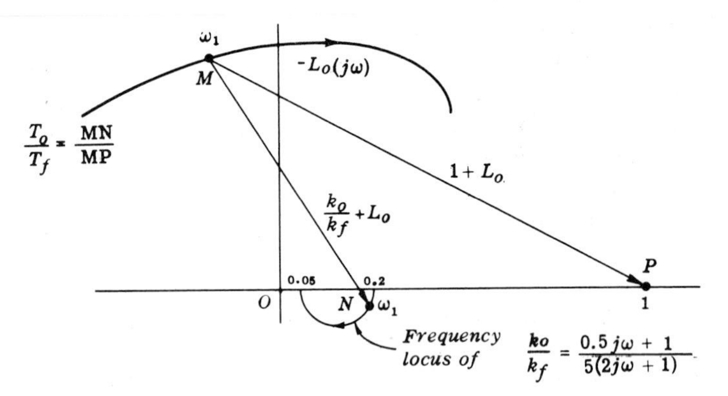

FIG. 3.5-1. Geometrical interpretation of T_o/T_f.

e.g., let $k_o = 1/(2s + 1)$ represent the nominal transfer function of the element. (The choice of k_o is arbitrary.) Next, consider any new value of the element's transfer function within its range of variation, say $k_f = 5/(0.5s + 1)$. Sketch the locus of $k_o/k_f = (0.5s + 1)/5(2s + 1)$ as a function of $s = j\omega$. The resulting circular locus is shown in Fig. 3.5-1. Suppose that the frequency locus of $-L_o(j\omega)$ is as shown in the figure, in particular $-L_o(j\omega_1) = \mathbf{OM}$. Then $\mathbf{MN} = \mathbf{MO} + \mathbf{ON} = (L_o + k_o/k_f)_{s=j\omega_1}$; $\mathbf{MP} = \mathbf{MO} + \mathbf{OP} = L_o(j\omega_1) + 1$. Consequently, from Eq. (3.5,1), $(T_o/T_f)_{s=j\omega_1} = \mathbf{MN}/\mathbf{MP}$, and it is easy to determine the permitted location of M in order that $\mathbf{MN}/\mathbf{MP}$ satisfy the tolerances on $(T_o/T_f)_{s=j\omega_1}$ for all possible values of N. One possible design procedure, therefore, is to sketch the family of k_o/k_f frequency loci for different values of k_f, or at least enough of them to see the extremes of variation of k_o/k_f. Then the frequency locus of $L_o(j\omega)$ may be shaped so as to achieve the required tolerances on T_o/T_f with as reasonable an L_o as is possible. What constitutes "reasonableness" and the means of achieving it is the core of the feedback problem, and constitutes one of the principal topics of the book.

Rather than work at a fixed k_f and sketch k_o/k_f as a function of frequency, it is usually easier to work at a fixed frequency and sketch the locus of k_o/k_f as k_f varies due to the parameter variations. This can be done for a few selected frequencies in the frequency range of interest. In the present problem $k_o/k_f = (\tau s + 1)/m(2s + 1)$. At zero frequency $k_o/k_f = 1/m$, whose locus (as m takes on

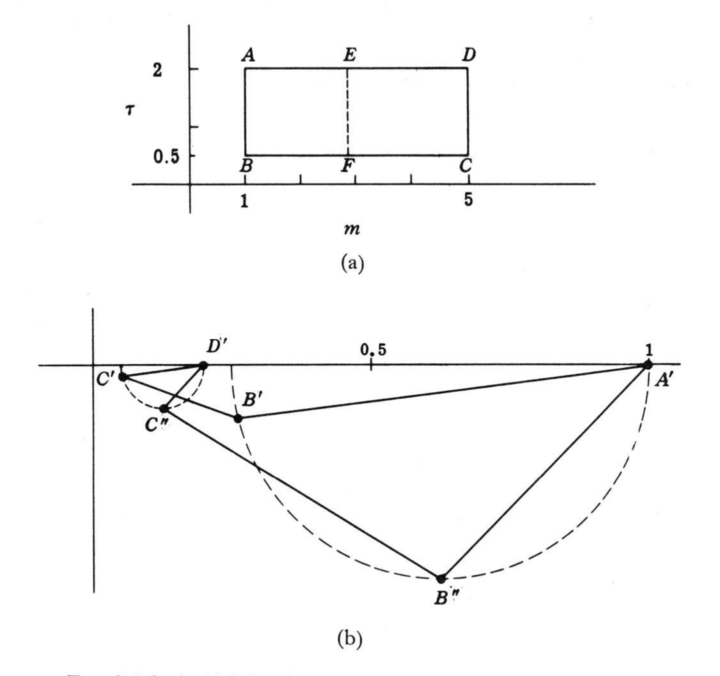

FIG. 3.5-2. (a, b) Mapping from m, τ space into k_o/k_f space.

values from 1 to 5) is the segment of the x axis from 0.2 to 1. At $\omega = 4$, $k_o/k_f = (1 + j4\tau)/m(1 + j8)$ and is a function of the two variables m, τ whose region of variation is the rectangle of Fig. 3.5-2a. The four corners of the rectangle in the m, τ plane map into four points in the k_o/k_f complex plane as indicated in Fig. 3.5-2b by $A'B'C'D'$. Along the line AB, k_o/k_f is a linear function of the single variable τ and therefore the corresponding locus of $k_o/k_f = (1 + j4\tau)/(1 + j8)$ is the straight line $A'B'$ in Fig. 3.5-2b. Each of the four sides of the rectangle $ABCD$ maps as a straight line in the k_o/k_f plane. Along any interior line EF, k_o/k_f is a linear function of τ, so EF maps as a straight line in Fig. 3.5-2b. However, E lies on AD, so that E' lies on $A'D'$. Similarly, F' lies on $B'C'$. Since this is true for any vertical line in $ABCD$, it follows that the interior of the rectangle $ABCD$ maps into the interior of the polygon $A'B'C'D'$ at $\omega = 4$. At $\omega = \frac{1}{2}$ the corresponding polygon is $A'B''C''D'$ in Fig. 3.5-2b. The frequency locus of the primed B's is $(1 + j\omega/2)/(1 + 2j\omega)$, which is a semicircle with diameter extending from $\frac{1}{4}$ to 1. The locus of the primed C's is also a semicircle.

Now that the region covered by k_o/k_f is known, it is a straightforward matter to find a suitable and practical L_o, by means of the construction of Fig. 3.5-1. Considering Eq. (3.5,1) and the construction of Fig. 3.5-1, it becomes apparent that in order to satisfy the specifications with a minimum $| L_o |$, L_o should have a phase angle of approximately $\pm 90°$ from zero frequency to 4 rps. This is impractical because a pole at the origin requires an integrator while a zero at the origin gives $| L | = 0$ there.[1] A feasible L_o is one whose phase is zero at zero frequency and whose phase lag increases with frequency.

The extreme value of $| \mathbf{MN}/\mathbf{MP} |$ (using the notation of Fig. 3.5-1) should, according to the specifications, be less than $0.95/1.05 \approx 0.9$. Therefore, noting

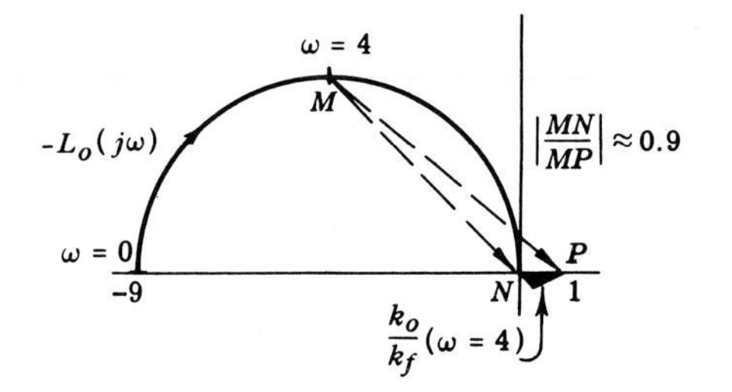

FIG. 3.5-2c. Locus of $-L(j\omega)$ and check of T_o/T_f at $\omega = 4$.

[1] We begin to see here how the relations which exist between the magnitude and the phase of a function of a complex variable, play an important role in feedback theory. Their role is in fact critical, and will be carefully studied in Chapter 7.

the construction of Fig. 3.5-1 and the loci of k_o/k_f of Fig. 3.5-2b, $L_o(0) = 9$ is satisfactory. For the rest, we try $L_o = 9/(1 + \frac{1}{4}s)$ whose negative locus is sketched in Fig. 3.5-2c. At $\omega = 4$, $|\mathbf{MN}/\mathbf{MP}|_{\min} \doteq 0.9$, and therefore this value of L_o is satisfactory. From zero to 4 rps, $|T/T_f| < 0.9$ (but at larger frequencies the ratio is less).

If we use the configuration of Fig. 3.5-3, then $L_o = k_o A\beta = A\beta/(2s + 1) = 9/(1 + \frac{1}{4}s)$ or $A\beta = 9(2s + 1)/(1 + \frac{1}{4}s)$. Also at 0 rps, $|T_o|$ is at the lower extreme of 0.95, i.e., $k_o A/(1 + k_o\beta A)_{s=0} = 0.95 = A(0)/[1 + \beta(0)A(0)] = A(0)/10$. Therefore $A(0) = 9.5$ and consequently $\beta(0) = 9/9.5$. There remains some freedom here because the value of $T_o(j\omega)$ for $\omega > 0$ has not been specified. Once it is specified, then $k_o A/(1 + L_o)$ is set equal to T_o and this determines the unknown A. The equation $L_o = k_o A\beta$ is then used to determine β.

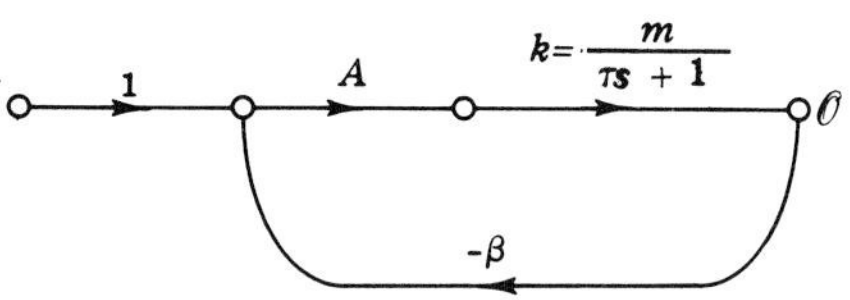

FIG. 3.5-3. Final design—Example 3.

The above example indicates the procedure which may be followed to obtain simultaneously: (1) a desired $T(s)$ function, and (2) a desired insensitivity of $T(s)$, over some frequency range, to variations in a system parameter. However, the design is not as yet complete, because the behavior of T for frequencies beyond 4 rps has not been considered. While it may not matter if there is substantial variation in $T(j\omega)$ for $\omega > 4$ rps, this is only true up to a point. It would be undesirable to have the variation in T in this range so large such that at $\omega = 20$ rps, say, $T(j20) = 10$. The higher frequency range cannot be completely ignored. Some bounds on the permitted variation of T must be set, even in this range. Consequently, the loop gain L_o must be shaped [in accordance with Eq. (3.5,1) and Fig. 3.5-1] to keep the variations in T within the bounds. This part of the design is intimately associated with the stability problem, and is therefore postponed until the latter has been treated.

In the above example, it has been assumed that the area of variation of k is known accurately. When such precise knowledge is lacking, one must estimate or guess the area of variations of k. The vaguer the knowledge, the more liberal one must be in his estimate of the possible variations, and the larger must L_o be. In short, one may be fairly ignorant (with some reservations) of the transfer function of an element, and by means of feedback obtain a very precise system transfer function. (One obvious reservation is that the maximum capabilities of k must not be exceeded. No cascaded compensating network can make up for saturation or pure time delays present in k.) It will be seen however, in Chapters 6 and 7, that the "cost of feedback" is high. It is therefore worthwhile to carefully define the area of element variation, and not attempt to design into the system any more insensitivity than is necessary.

§ 3.6 Sensitivity Function When Leakage Transmission Is Not Zero

In previous sections, it was assumed that in the fundamental feedback equation $T = t_{oi} + kt_{ci}t_{os}/(1 - kt_{cs})$, the leakage transmission t_{oi} was negligibly small in comparison with T, over the range of variation of k. This is not always true. For example, suppose we are constrained to use as the final stage in a system an element whose signal-flow graph representation is that shown in Fig. 3.6-1. Suppose it is the parameter D which varies substantially, and feedback is to be used to reduce the system sensitivity to the variations in D. If we choose $D = k$ of the fundamental feedback equation, then t_{oi} may not be small, and the methods of sections 3.2-3.5 do not apply. However, the problem can be handled by these methods by simply letting k represent the entire element, i.e., $k = A(B + ED)/(1 - AEC)$. The element k is then incorporated into the over-all system feedback configuration. The procedure henceforth is exactly the same as before: one ascertains the loci of k_o/k_f, and, from the construction of Fig. 3.5-1, ascertains the loop gain that is required to obtain the desired sensitivity reduction.

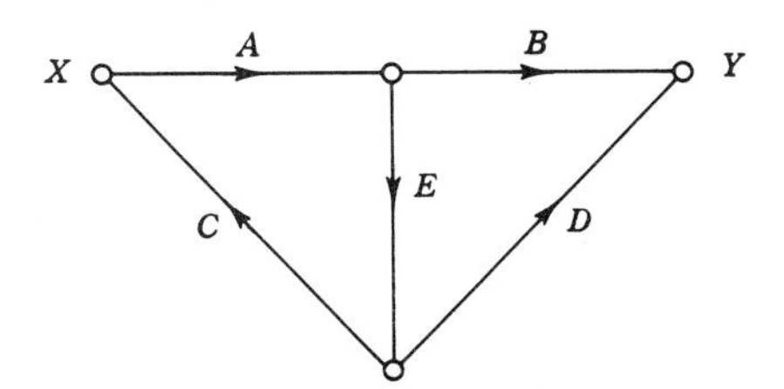

FIG. 3.6-1. Formulation of problem such that $t_{oi} = 0$. $A(B + ED)/(1 - AEC) = Y/X = k$.

In the above manner, the majority of feedback problems can be put into the zero leakage category. However, for those situations *where a parallel path from the over-all system input node to the output node* (from $\mathscr{I}$ to $\mathscr{O}$) exists, some modification of the procedure is necessary, and this is now done. We have, in general, $T = t_{oi} + kt_{ci}t_{os}/(1 - kt_{cs})$. In a straightforward manner [using the notation of Eq. (3.2,3)] it is found that

$$S_k^T \triangleq \frac{\Delta T/T_f}{\Delta k/k_f} = \frac{1}{(1 - k_o t_{cs})}\left(1 - \frac{t_{oi}}{T_f}\right) = \left(\frac{1}{1 + L_o}\right)\left(1 - \frac{t_{oi}}{T_f}\right). \tag{3.6,1}$$

This relation will be useful in Chapter 8 in discussing the properties of systems with parallel plants. Equation (3.6,1) may also be written as $S_k^{(T-t_{oi})} = 1/(1 + L_o)$. Other useful relations are:

$$\frac{T_o - t_{oi}}{T_f - t_{oi}} = \frac{(k_o/k_f) + L_o}{1 + L_o}, \tag{3.6,2a}$$

$$\frac{T_f}{T_o} = \frac{(1 + L_o) - (t_{oi}/T_o)(1 - k_o/k_f)}{L_o + k_o/k_f}, \tag{3.6,2b}$$

$$\frac{\Delta T}{T_o} = \frac{(1 - t_{oi}/T_o)(1 - k_o/k_f)}{L_o + (k_o/k_f)} \tag{3.6,2c}$$

where T_f is the value of T when k has the value k_f; $\Delta T = T_f - T_o$, $\Delta k = k_f - k_o$. Now t_{oi} is a transfer function which is independent of k. If t_{oi} is known and the desired T is known, then one may relate the maximum permissible range of variation of T, over any frequency range, to a corresponding maximum range of variation of $T - t_{oi}$. In other words, one can work graphically with Eq. (3.6,2a) in exactly the same manner as Eq. (3.2,6) was used in Section 3.5. The only difference is that one must first decide what is the permitted range of variations of $T_f - t_{oi}$ rather than that of T_f.

The above is satisfactory if t_{oi} is known. On the other hand, there may be situations when t_{oi} is not precisely known until part of the design has been done. In such a case some cut and try is inevitable.

The following modification of the graphical procedure is an alternative method of handling the nonzero leakage transmission problem. Equation (3.6,2b) is used. Sketch the frequency locus of t_{oi}/T_o (Fig. 3.6-2a) in the complex

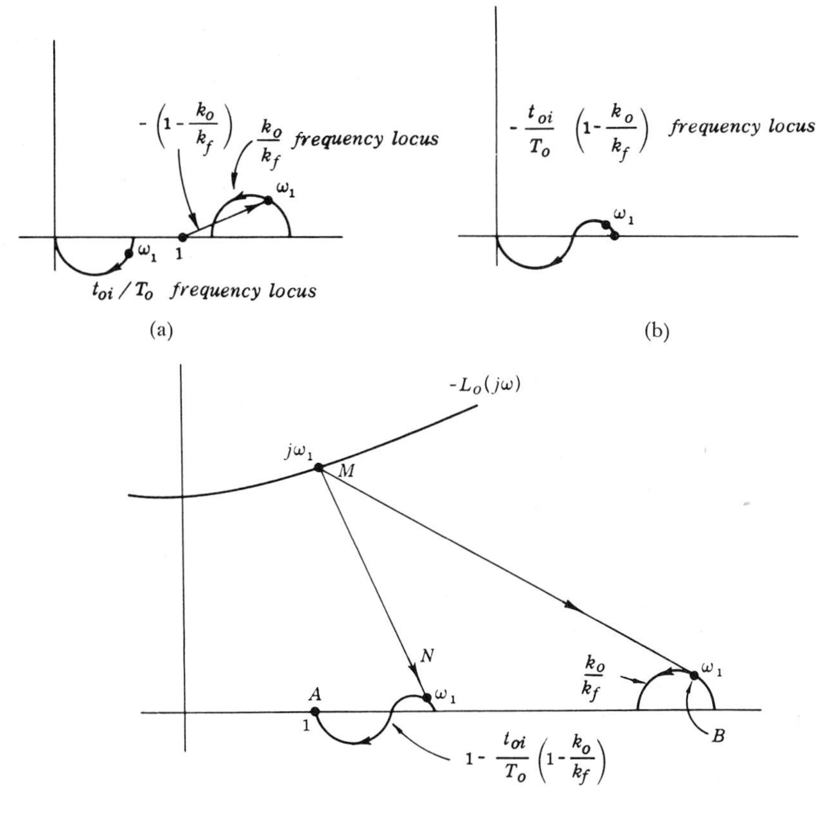

FIG. 3.6-2. Construction for finding T_o/T_f when $t_{oi} \neq 0$. $T_f/T_o = MN/MB$

plane, over the frequency range in which control of variations in T is desired. Also sketch the loci of k_o/k_f (as was done, for example, in Fig. 3.5-2b). From these two figures, obtain the loci of $-(1 - k_o/k_f)(t_{oi}/T_o)$ (Fig. 3.6-2b). Finally, combine the latter and the sketch of k_o/k_f into one figure (Fig. 3.6-2c). In Fig. 3.6-2c $T_f/T_o = \mathbf{MN}/\mathbf{MB}$, in accordance with Eq. (3.6-2b). The design problem is to select the loop transmission L_o so as to satisfy the specifications on T_f/T_o. The procedure henceforth is the same as that followed in Section 3.5.

§ 3.7 Magnitude and Phase Sensitivities

There may be situations when it is important that the magnitude of the system transfer function T should be very insensitive to variations of some system parameters, but there is no concern with the variation in the phase of T (for example, in audio amplifiers or amplifiers of pure sinusoids). Or, the converse may be true. One may want to reduce the sensitivity of the phase of T to parameter variations and not care too much what happens to the magnitude of T. The graphical construction of Section 3.5, based on Eq. (3.5,1), is particularly well suited for this purpose. For example, assume k_o/k_f is independent of frequency and can lie any where on the line segment AB in Fig. 3.7-1, and suppose the interest is primarily in $|T|$. Clearly, the minimum value of L_o that is needed is obtained when L_o lies somewhere along the line XX'. On the other hand, if it is the phase of T that is important, then the minimum required L_o is one which lies near the real axis, excluding the k_o/k_f segment. The general problem of choosing $L_o(j\omega)$ over the entire frequency range such that the magnitude and phase sensitivities are simultaneously satisfied is treated in detail in Chapter 6.

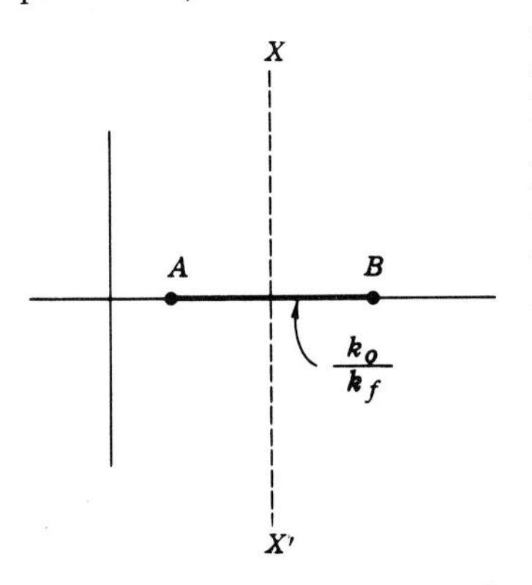

FIG. 3.7-1. Preferred location of $-L(j\omega)$ for small $|S|$.

For analytical purposes, it is convenient to use the following expressions for the magnitude and phase incremental sensitivities: If t_{oi} is zero, then

$$\frac{dT/T}{dk/k} = \frac{d\ln T}{d\ln k} = \frac{1}{1+L_o} = \frac{d\ln|T|}{d\ln k} + \frac{j\,d\angle T}{d\ln k} \triangleq S_k^{|T|} + j(\angle T)S_k^{\angle T}. \tag{3.7,1}$$

If k is real, then

$$\frac{d|T|/T}{dk/k} + \frac{j\,d\angle T}{dk/k} = \mathcal{R}e\,\frac{1}{1+L_o} + j\mathcal{I}m\,\frac{1}{1+L_o} = S_k^{|T|} + j\angle T.S_k^{\angle T} \tag{3.7,2}$$

When the magnitude sensitivity is of paramount interest, one might specify $S_k^{|T|}$ and derive from it S_k^T. The operation is that of obtaining an analytic function from its known real part (or from its imaginary part when $S_k^{\angle T}$ is of interest). The procedure is presented in several texts[1] and is also discussed in Chapter 7. The principal shortcoming of Eqs. (3.7,1) and (3.7,2) is that they apply only to small variations. Their potential usefulness is in theoretical studies, when an analytical formulation of these sensitivity functions is desired.

§ 3.8 Systems with Multiple Inputs

One of the principal reasons for using feedback is because of its ability to reduce the sensitivity of a system to system parameter variations. A second principal reason for using feedback, is that it permits discrimination between signals that enter a system at different points (rejection of disturbances). Thus, consider a system which is designed to achieve a specific objective; for example, to control the rotational velocity of a load. Let P represent the transfer function of the totality of load and equipment needed to achieve the control objective.[2] The plant P has the power capacity to make the output C respond in the desired manner to the input signal R_1 in Fig. 3.8-1. Suppose there are external disturbances which compete with R_1 in the control over C, for example, wind torque on the load. The system with both inputs may be represented by the flow graph of Fig. 3.8-1, with $P = P_1P_2$. In this open-loop system P_2 is fixed, because it represents the given load. The transmission from the input to X is not fixed even if P_1 is fixed, because a prefilter, G, can be inserted ahead of P_1. Let $Q = GP_1$ and Q therefore represents the only degree of freedom that there is in this open-loop system. This freedom can be used only to determine $C/R = QP_2$ because $T_D \triangleq C/D = P_2$ is independent of Q. Of course the objectives could be achieved without feedback, if a filter could be interposed between D and X. But if this is impractical, then the use of feedback is the only means of coping with the problem. (It is also assumed that the designer cannot arrange an independent parallel transmission from D to C.)

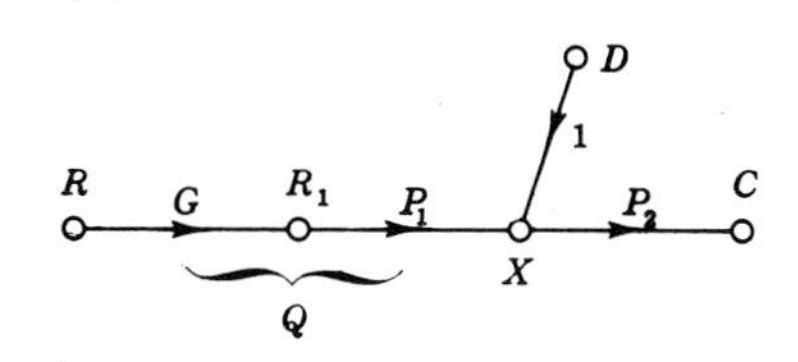

FIG. 3.8-1. Open-loop system for achieving a desired signal to disturbance ratio.

There may be situations when the ratio of signal-to-disturbance transfer function, $(C/R):(C/D) = Q$, rather than the function $T_D = C/D = P_2$, is of

[1] H. W. Bode, "Network Analysis and Feedback Amplifier Design," p. 203. Van Nostrand, Princeton, New Jersey, 1945; E. A. Guillemin, "Synthesis of Passive Networks," Chapter 8. Wiley, New York, 1957.

[2] In accordance with accepted terminology, we henceforth use R and C to represent the system input and output, respectively.

prime interest. Q can then be used to achieve a satisfactory $(C/R):(C/D)$ ratio. However, it is then impossible to use Q to achieve a desired input-output transfer function $C/R = QP_2$. Again feedback is the only means whereby both C/R and the signal-to-disturbance transfer function ratio $(C/R):(C/D)$ can be independently realized.

We are not interested here in finding some specific feedback configurations for coping with the present problem. We are interested in making general definitive statements that apply for all possible structures. The first step is to clearly state the constraints of the problem. In the present case they are as follows: P_1, P_2, X, C, and D form an indivisible fixed unit; no signal from the input can reach the output except through the cascaded P_1P_2; the variable C may be measured and used for feedback purposes (if the variable X may also be measured and so used, then the most general feedback structure has multiple feedback loops, which we postpone to Chapter 8); no signal, subject to our control, is to reach X except through P_1 (this would constitute "plant modification"; furthermore, it would permit feedback from C to X, again leading to a multiple feedback structure).

The second step is to use the simplest but most general structure which satisfies the constraints. To find the structure, begin with the constraints—Fig. 3.8-2a. Consider the variable R. It is the independent input, so it has only

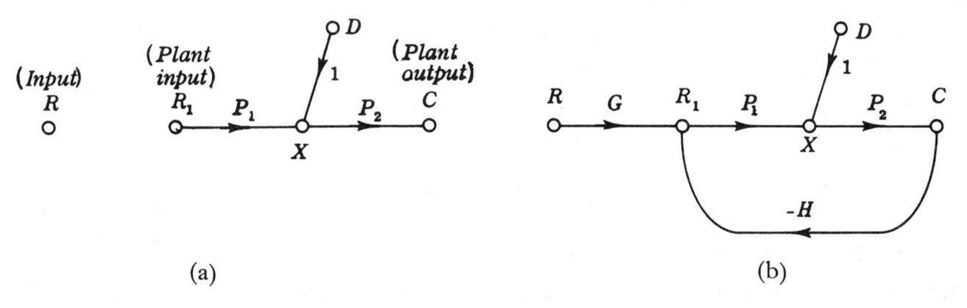

FIG. 3.8-2. Derivation of canonic flow graph for single disturbance.

outgoing branches of which only one, to R_1, is permissible, since R must not affect X and C except through the plant. The designer's control over C is lodged entirely in the value of the variable R_1. The variable R_1 can be influenced only by R and by C; so only two independent transfer functions can be inserted: one from C to R_1, the other from R to R_1, as in Fig. 3.8-2b. We could, of course, add additional feedback from C to R, and to internal points in G. But all of these can be replaced by the simple but general structure of Fig. 3.8-2b. Whatever cannot be achieved with this structure, cannot be achieved with any other structure that satisfies the problem constraints. The fundamental property of this structure is that it possesses two degrees of freedom, G and H. The two degrees of freedom, G and H, can be identified respectively with t_{ci} and $-t_{cs}$ of the fundamental feedback equation (also $t_{oi} = 0$, $t_{os} = 1$, if $k = P_1P_2$).

In fact, the fundamental feedback equation could have been used to derive this structure.

In Fig. 3.8-2b (with $P = P_1P_2$, $L = HP$), $T = C/R = GP/(1 + L)$ and $T_D = P_2/(1 + L)$. The loop transmission L is used to obtain a satisfactory T_D, and G remains as a degree of freedom for the realization of a satisfactory $T(s)$. It should be noted that the disturbance attenuation is achieved by means of a feedback signal $-HC = -HP_2D/(1 + L)$ at the plant input. The resulting $-HPD/(1 + L)$ at the output of P_1 works against D at X. Whatever may be the range of disturbance signals which are to be attenuated in this linear manner, the designer must be certain that P_1 is able to handle such signals (i.e., P_1 will not saturate).

Consider next the case of two disturbance inputs, as shown in Fig. 3.8-3a, and where we cannot interpose filters between D_1 and X_1, or between D_2 and

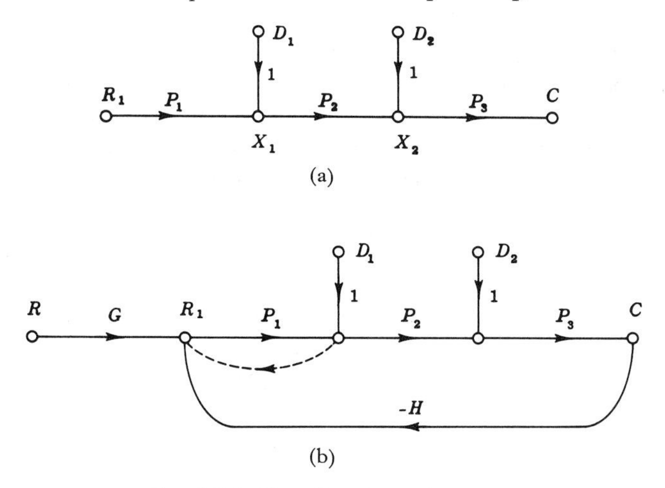

FIG. 3.8-3. Two disturbance inputs.

X_2, cannot change P_2 or P_3, and cannot use X_1 or X_2 for feedback purposes. Here too we can introduce a feedback loop as shown in Fig. 3.8-3b (which is as general a structure as Fig. 3.8-2b). There are only two degrees of freedom, G and H, and we can therefore independently control only two system functions. The problem, generally, is to ensure that C/D_1 and C/D_2 each be kept within some bounds. It may therefore be necessary to overdesign for one of these, in order to satisfy the specifications on the second. However, if the feedback loop shown dotted in Fig. 3.8-3b is added, then there is one more degree of freedom, which permits independent realization of T, C/D_1 and C/D_2.

Design examples for attenuation of disturbances and for simultaneous achievement of a desired T, are detailed in Chapters 6 and 10. The relation between the feedback loops and system degrees of freedom, and when and how to use all the available degrees of freedom is discussed in Chapter 8.

It is important to note that feedback is of no help in discriminating against noise entering at the same point as the useful signal. In such a case the input signal R consists of a useful portion and a noise portion. The conventional filtering approach must be used for such problems, i.e., the system transfer function $T = C/R$ must be chosen so as to discriminate between the useful signal input and the spurious noise inputs.

§ 3.9 Sensitivity to Parameter Variation and Noise in Feedback Return Path

It is important to note that the system sensitivity to noise entering along the return feedback path is not small. This is intuitively obvious, for the feedback loop from the output back to the input may be considered similar to a spy system for observing to what extent the output is responding to the command signals. The input adjusts itself in accordance with the information received from the spy. If, however, the spy conveys information from the wrong sources (noise signals), then the system response is adversely affected. For example, in Fig. 3.9-1, suppose the command signal is of energy form W_1 (electrical) and

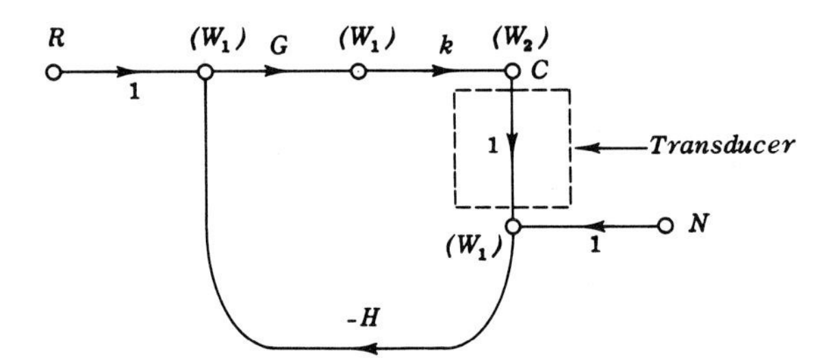

FIG. 3.9-1. Noise in feedback path.

k is an energy conversion device whose output is W_2 (mechanical), then the feedback loop must contain a transducer transforming from W_2 back to W_1. Suppose this transducer (with assumed transfer function 1) has associated with it an inherent noise source N (see Fig. 3.9-1) due to imperfect construction (e.g., gear irregularities). The system response to N is $T_N = C/N = -kGH/(1 + GHk) = -L/(1 + L) \doteq -1$ over the important frequency range, which is the range in which $L > 1$.

There are two ways of coping with this problem (ignoring the obvious one of constructing a better transducer). One way is shown in Fig. 3.9-2a, where $C/N = F - (L)/(1 + L)$. If F is taken as 1, then $C/N = 1/(1 + L)$ which is small in the significant frequency range. Such a solution requires that it should be possible to have the noise separately affect the output. This measure of control over N is seldom available. Figure 3.9-2b shows another solution.

Here it is assumed possible to insert an element M whose output has the same energy form as C (i.e., type W_2). The transducer which transforms W_2 to W_1 follows M. In order that the system feedback properties with respect to the plant P be the same as before, it is necessary that H of Fig. 3.9-1 be equal to

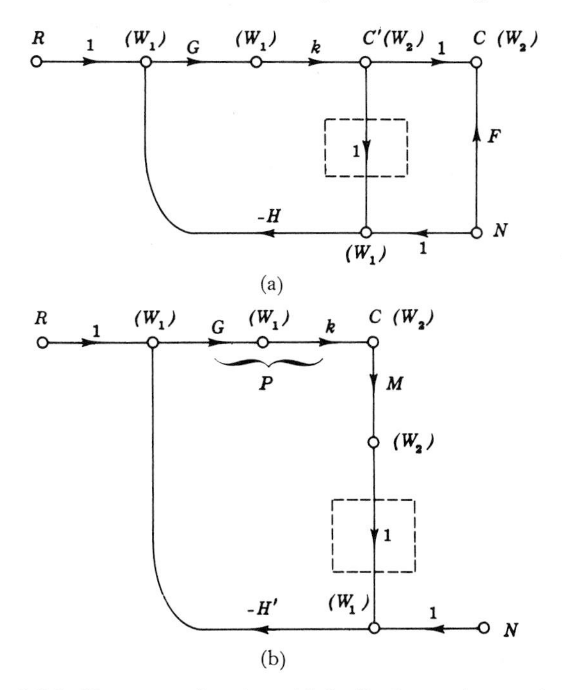

FIG. 3.9-2. Two ways of coping with feedback transducer noise.

MH'. Now $T_N' = -H'kG/(1+L) = T_NH'/H = T_N/M$, where $T_N = C_N/N$ of Fig. 3.9-1. The noise response has been reduced by the amount M. It may, of course, be difficult to instrument either solution without introducing new noise sources or parameter variations, and if so, this problem remains a serious limitation in feedback design. Returning to the spy analogy, it is clear that the usefulness of an intelligence service is seriously limited if one cannot trust his agents.

In many cases the feedback loop (excepting low power control systems such as instrument servos) is at a much lower power level than that of the plant. It is therefore possible to design and construct these lower power level instruments carefully, at considerably less cost than the larger power plant elements in the forward loop. It is similarly important that the feedback element parameters do not vary, for the system sensitivity to their variations is not small. Thus, in Fig. 3.9-1,

$$\frac{dT/T}{dH/H} = \frac{1}{1+L}\left(1 - \frac{kG}{T}\right) = \frac{-L}{1+L} \tag{3.9,1}$$

which is approximately one, in the important frequency range where $|L| > 1$.

The noise sources and parameter variations in the return loop constitute important limiting factors in feedback system design. These factors will be considered in Section 6.14. Attempts at reducing their importance lead to a study of multiple-loop feedback systems in Chapter 8. These same factors are also important in the so-called adaptive systems which are considered in Chapter 8.

§ 3.10 The Effect of Feedback in Nonlinear Systems

It has been emphasized that all the results that have been derived apply only to linear systems. Nevertheless, one of the very useful features of feedback is that it can linearize to some extent a nonlinear system. It will be shown that the linearizing influence of feedback is precisely the same as the sensitivity reduction effect. The treatment in this section is restricted to elements with elementary algebraic nonlinear relations, imbedded in simple frequency independent systems. The more general problem will be considered in Sections 8.19 and 8.20.

In the simple linear system of Fig. 3.9-1, $T = C/R = kG/(1 + kGH) \doteq H^{-1}$ when $kGH \gg 1$. This is true to a certain extent even if there are non-linearities in Gk. In Fig. 3.10-1, N indicates a nonlinear element, i.e., y is related to x by a nonlinear integro-diffe-

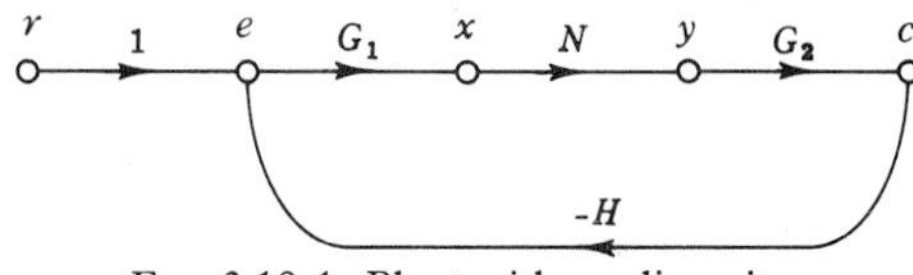

FIG. 3.10-1. Plant with nonlinearity.

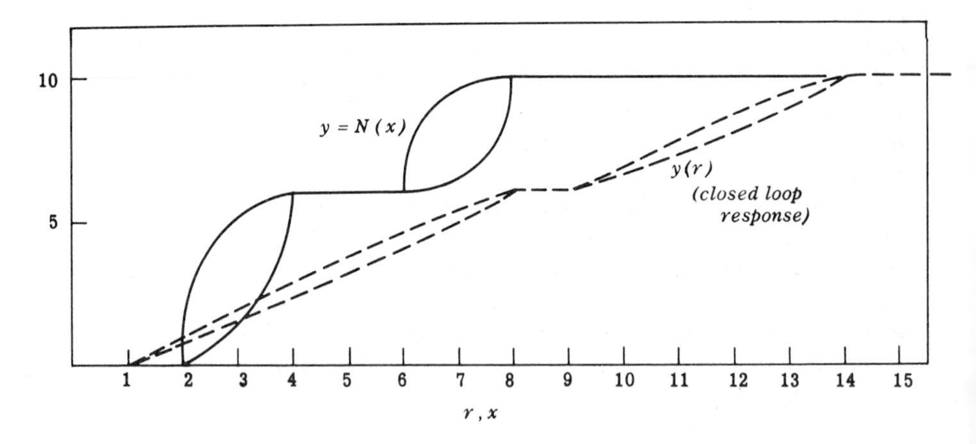

FIG. 3.10-2. Linearizing effect of feedback.

rential equation. Consider the special simple case that all the elements in Fig. 3.10-1 are frequency independent, and the nonlinear relation is a nonlinear algebraic equation, rather than a nonlinear differential equation. Specifically, we assume the nonlinear relation shown in Fig. 3.10-2. There are two relations: $y = N(x)$, and $x = G_1 e = G_1(r - Hc) = G_1 r - HG_1G_2 y$. A graphical solution is possible. The second equation determines a load line on Fig. 3.10-2. The slope of the load line is $dy/dx = -1/HG_1G_2 = -1/L$. The x intercept of the load line is $G_1 r$, which varies as r varies. To be specific, suppose r is a ramp function $r = t$, for $t > 0$, zero for $t < 0$. In Fig. 3.10-2, the load lines are constructed and the resulting $y(t)$ is drawn for $L = 2 = G_1$, $G_2 = H = 1$.

In Fig. 3.10-2, the portions of $y = N(x)$ in which $y'(x) = 0$ constitute one-half of the total range, but in the closed-loop response they constitute only one-seventh of the range. Due to the first dead zone region, an input of two units is required before there is any output. The width of each dead zone region is reduced by half because $G_1 = 2$. Even with such very modest values of G_1 and L, the improvement in linearity is quite startling. Of course, when the final saturating level is reached, feedback is completely ineffective and the closed loop system also saturates.

Quantitative Measure of Linearizing Effect

In a frequency independent linear system, the relation between output and input has the simple form $c = mr$, and $dc/dr = m$ is a constant. This suggests that for the special frequency independent systems, we use as a measure of nonlinearity the quantity $(m - m_o)/m_o$, where m_o is a suitable average ratio of output to input. In Fig. 3.10-1, $dc/dr = (dc/dx)(dx/dr)$. But $x = G_1(r - HG_2 y)$, so that $dx/dr = G_1 - HG_1G_2N'(x)\, dx/dr$, and $dx/dr = G_1/[1 + HG_1G_2N'(x)]$. Therefore,

$$dc/dr = (dc/dx)(dx/dr) = G_2N'(x)dx/dr = G_1G_2N'(x)/[1 + LN'(x)] = m.$$

A measure of nonlinearity is [using $N_o'(x)$ for the reference value of $N'(x)$]:

$$\frac{m - m_o}{m} = \frac{[N'(x) - N_o'(x)]/N'(x)}{1 + LN_o'(x)}. \tag{3.10,1}$$

From Eq. (3.10,1) it is seen that the system nonlinearity which exists due to the fact that $(N' - N_o')/N' \neq 0$, is reduced by the return difference $1 + LN_o'(x)$. Hence the linearizing effect of feedback is precisely the same as the reduction in the sensitivity of the system to parameter variation. The effect of the change in slope of $N(x)$ is exactly the same as that due to a change in gain of an amplifier [compare Eqs. (3.10,1) and (3.2,5)].

The linearization may be better than Eq. (3.10,1) suggests. For example, when $N'(x)$ is zero, m is zero, which suggests that the system is violently nonlinear. However, if G_1 is large, this region may be so narrow that its effect is really a minor one. A good estimate of the system linearity is made by actually sketching the input-output relation as in Fig. 3.10-2.

Another estimate of the nonlinearity that exists in the system output-input ratio is made by finding the harmonic content in the output due to a sinusoidal input. Actually, it is easier to work backwards, assume a sinusoidal output and find the harmonic content in the input. For a given amplitude of a sinusoidal y (Fig. 3.10-1), the required wave shape of x is found and a Fourier analysis is made to find the harmonic content of x. Then $r = e + Hc = G_1^{-1}N^{-1}(c/G_2) + Hc$. For good linearity, it is necessary that $Hc \gg G_1^{-1}N^{-1}(c/G_2)$. Again it is seen how the linearizing effect is due to the loop gain, G_1G_2HN.

§ 3.11 Classification of Reasons for Using Feedback

Several outstanding properties of feedback have been presented in this chapter. We would like to suggest that all the reasons for using feedback can be combined under three general categories:

(1) Use of feedback to contend with ignorance of a plant.

(2) Use of feedback to contend with ignorance of the environment acting on the plant.

(3) Use of feedback to exploit the fact that an element with feedback around it has static and dynamic characteristics which are different from those of the element by itself. This property of feedback is discussed in Section 3.12.

Parameter variations fall under the first heading, for if the plant variations were known precisely in quantity and in time, the input to the plant could be programmed so as to cancel out the effect of the parameter variations. External disturbances acting on the plant belong to the second heading, for similar reasons. The linearizing effect due to feedback could be achieved without feedback by means of cascaded filters with inverse nonlinear characteristics. However, if the nonlinearities are not precisely known, or if they vary, such nonlinear predistortion cannot be used. Another reason sometimes given for using feedback is its ability to reduce or increase the output impedance of a system. Very often the reason for wanting a low (or high) output impedance is to achieve good voltage (or current) regulation of an unpredictable load. This comes under the first heading. Impedance transformation for other reasons, such as efficient power transfer, can be achieved with passive networks.

One of the reasons for using feedback, which has not been mentioned, is the use of feedback around an unstable plant in order to eliminate the instability. If the plant is unstable, with right half-plane poles, then these poles do not appear (Fig. 3.11-1a) in $Y/X = P/(1 + PH)$. It can be argued that this reason too belongs under the first two categories. For if the location of the right half-plane poles were precisely known, the right half-plane poles could be eliminated by prefiltering, e.g., if $P = P_1/(s - a)$, $a > 0$, then in Fig. 3.11-1b, let $G = k(s - a)/(s + b)$ and Y/X has no right half-plane poles. The usual reasons given for the impracticality of such an approach are: (1) it is impossible to

(a) (b)

FIG. 3.11-1. Why cancellation of rhp pole is impractical.

precisely cancel out the pole, and (2) Y/N of Fig. 3.11-1b still has right half-plane poles, and N cannot be eliminated because there is always random noise present. The second reason exists only because of our ignorance of the environment; for if N were precisely known in advance, the input X could be programmed to cancel it. It may be argued that the first reason is valid only because parameter variation *always* exists, and it is this wandering of the pole of P that precludes precise cancellation.

We do not wish to suggest by the above that feedback should be used only if there is ignorance of plant or the environment. Even if future parameter variations or external disturbances are precisely known in advance, it may be easier to use feedback than to suitably program the input. Impedance transformation may be easier to achieve by means of feedback than by passive networks. Nevertheless, the above three categories represent those in which the use of feedback is mandatory because within the constraints of the problem there is no other way to achieve the objectives.

This attempt at a general classification is justified in that it represents the laudable attempt to unite many diverse reasons into as small and as simple a classification as possible. It is furthermore justified in pointing out that there may be situations in which a costly feedback design can be avoided, by exploiting what is known about the parameter variations or disturbances. For example, if a variation in resistance is known to be due to temperature changes, then another matched resistance with opposite characteristics can usually be found and is far easier and cheaper to implement in most cases than is a feedback solution. If one can get to the source (or a similar source) of a disturbance, he can often exploit this to effect a cheaper and easier solution than the feedback method (e.g., 60-cycle hum in an audio amplifier). One should always remember that ultimately it is only ignorance for which the feedback solution is unavoidable, and that reduction of the extent of the ignorance will usually permit an easier and cheaper feedback solution.

§ 3.12 Effect of Feedback on System Response

From the fundamental feedback equation, $T = t_{oi} + kt_{ci}t_{os}/(1 - kt_{cs})$, it is seen that feedback affects the system response. The zeros of the return difference, $1 - kt_{cs}$, are poles of T, unless $t_{ci}t_{os}$ has zeros at these points. But since the zeros of $1 - kt_{cs}$ are not precisely known, or drift,we cannot usually assume such cancellation. If the designer wants a specific system response and is also interested in some specific return difference (in order to achieve the desired insensitivity to parameter variation or reduction of external disturbance, or linearization of a nonlinearity), he cannot achieve both objectives by means of the configuration of Fig. 3.12-1. This configuration has been studied extensively in the literature of feedback control. In this configuration only G is available, and it is therefore impossible to independently realize $T = G/(1 + G)$ and $F = 1 + G$. All that is needed to achieve independent T and F is a structure with two degrees of freedom and any one of many will do. A transfer function — H in the return path would provide the extra degree of freedom. The need for two independent transfer functions in the system, and the corresponding additional design labor, is one of the prices paid for the benefits of feedback.

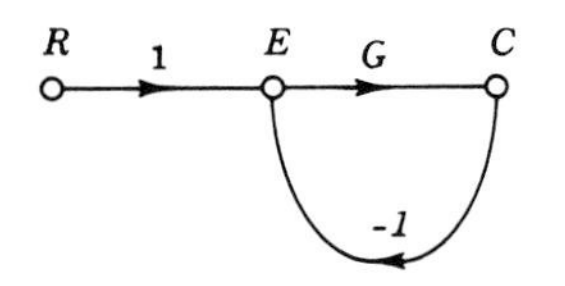

FIG. 3.12-1. Single-degree-of-freedom feedback structure.

The effect of feedback on the system response is not, however, always a liability. This property of feedback is indispensable in the design of oscillators. Another field in which this property of feedback is used is Active *RC* Synthesis. By means of feedback it is possible to eliminate inductors and still obtain in a fairly simple manner, complex function poles. Such poles are unobtainable in networks consisting only of resistors and capacitors.

The beneficial aspects of feedback that belong in the above category will not be studied in this book. Rather we shall be very much concerned with the negative aspects of this property of feedback. One of these negative aspects is the stability problem. The existence of the stability problem is seen by examining the return difference $F_k = 1 - kt_{cs}$. If $F_k = 1$, then there is no feedback around k and no feedback stability problem. The zeros of F_k are poles of T. (It would seem that the zeros of F_k could be canceled out by zeros of $t_{ci}t_{os}$; however, plant parameter variation precludes this.) It is recalled from Section 2.8 that the numerator of F_k in all nontrivial cases is the same, no matter what reference is used for k. Therefore, whenever there is feedback (as evidenced by kt_{cs} not being zero), there must be a stability problem. If F_k has any right half-plane poles, the system is unstable, which in most cases means that it is useless.

We shall demonstrate in Chapter 4 that in almost all feedback systems there is a tendency for the system to be unstable. The design must be sufficiently conservative to ensure stability in the face of the maximum range of parameter variation. The above remarks are very brief and qualitative in nature. Their purpose here is to justify the immediate presentation of a chapter on the problem of stability in feedback systems.

CHAPTER FOUR

The Stability Problem in Feedback Systems with Plant Parameter Variations

§ 4.1 Introduction to Root Locus

The problem of stability is one of fundamental importance in the design of feedback systems. This may be seen from the fundamental feedback equation, $T = t_{oi} + kt_{ci}t_{os}/(1 - kt_{cs})$. The easiest case to consider, from the viewpoint of stability, is when all the t_{ij}'s represent stable transfer functions, with poles in the left half-plane. In such a case, the only possible right half-plane poles of T are the zeros of $1 - kt_{cs}$. These are the poles of T that are due to feedback, and it is the purpose of this chapter to investigate their location and to demonstrate that in almost all practical situations there is a tendency for some of these poles to lie in the right half-plane. The designer must investigate this tendency and curb it sufficiently so that the system remains stable *over the range of the system parameter variations.* Such an investigation and the associated design effort represents a significant portion of the total effort in the design of most feedback systems.

A very useful tool for the above purpose is that of root loci sketching. The system poles that are due to feedback are the zeros of $1 - kt_{cs} = 1 + kB(s)$ [where $-t_{cs}(s) \triangleq B(s)$]. For example, suppose k is a real number independent of s, but varying in magnitude. As k varies, the roots of $1 + kB(s) = 0$ must also vary. The loci of these varying roots are called the root loci of $1 + kB(s) = 0$. In order that the system be stable, it is necessary that the root loci lie in the left half-plane, for all values of k within its region of variation. In addition to checking for system stability, the sketching of root loci is particularly useful for examining the effect of a varying parameter upon a system. The rules for sketching root loci are remarkably simple and provide one with a versatile technique which is extremely useful for a large variety of problems. We therefore begin this chapter on the stability problem by presenting in Section 4.2 a number of rules which permit rapid sketching of root loci. An appreciation of the stability problem in feedback systems is easily obtained from these rules.

Example 1. Suppose $-t_{cs} = B(s) = 1/[(s + 1)(s + 3)]$; then $1 + kB(s) =$

$1 + k[(s + 1)(s + 3)]^{-1} = [(s + 1)(s + 3) + k][(s + 1)(s + 3)]^{-1} = 0$. The roots of $1 + kB(s) = 0$ are at -1, -3, when $k = 0$; -2 ± 0.707 when $k = 0.5$; -2 when $k = 1$, and complex for $k > 1$. The root loci (loci of the roots) are sketched in Fig. 4.1-1a. The arrows indicate the direction the roots move as k increases. The point P is called a breakaway or coalescent point.

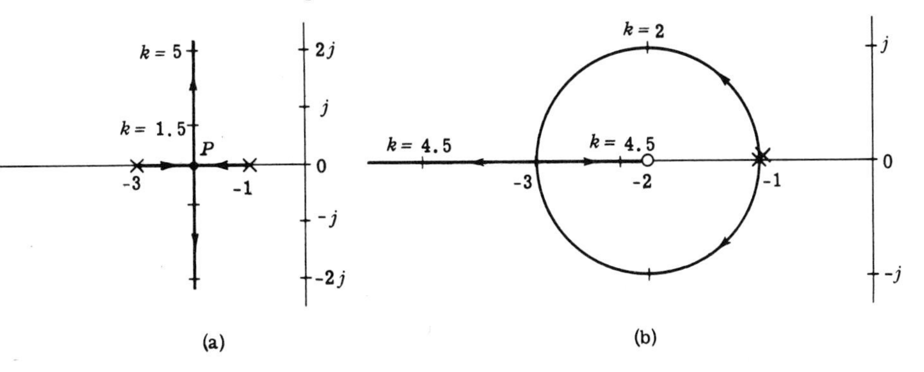

FIG. 4.1-1. Root loci of (a) $1 + k/(s + 1)(s + 3) = 0$ and (b) $1 + k(s + 2)/(s + 1)^2 = 0$.

Example 2. $B(s) = (s + 2)/(s + 1)^2$. The loci of the roots of $1 + k(s + 2)/(s + 1)^2 = 0$ are sketched in Fig. 4.1-1b. The poles and zeros that are shown as such (crosses and circles, respectively) on root locus plots are always those of $B(s)$.

§ 4.2 Rules for Constructing Root Loci[1]

A number of simple rules are now developed which enable one to sketch root loci almost by inspection. The problem is to find the roots of

$$1 - kt_{cs} = 1 + kB(s) = 1 + k\frac{n(s)}{d(s)} = 1 + \frac{k\Pi(s + z_i)}{\Pi(s + p_j)} = 0 \quad (4.2,1)$$

as k varies from $-\infty$ to ∞. The poles and zeros of $B(s)$ are assumed known and any constant multiplier in $B(s)$ has been removed and absorbed in k. The rules are developed for k positive; the changes in the rules for negative k are either obvious or are stated.

Rule 1. Number of root loci. The number of roots of $1 + kB(s) = 0$ is equal to the degree of $n(s)$ or of $d(s)$, whichever is greater. To see this, note that the roots of Eq. (4.2,1) are the roots of $d(s) + kn(s) = 0$, whose degree is that of

[1] These rules are principally due to W. R. Evans, Graphical analysis of control systems, *Trans. AIEE* **67**, 547-551 (1948).

$n(s)$ or $d(s)$, whichever is greater. There are, of course, as many root loci as there are roots.

Rule 2. The root loci are continuous curves. The root loci are continuous curves because the roots of $1 + kB(s) = 0$ are continuous functions of k. This means that a very small change in k leads to a very small displacement of any root. The slopes of the root loci (lines drawn tangent to the root loci at any point) are continuous except for points at which either $B'(s) = 0$, $k = 0$, or $B(s)$ is infinite. This is seen by differentiating Eq. (4.2,1) with respect to s, leading to $ds/dk = -B(s)/kB'(s)$.

Rule 3. Origin and termination of root loci. The loci begin at the poles of $B(s)$ (at $k = 0$) and terminate at the zeros of $B(s)$ (where k is infinite). This is seen by writing Eq. (4.2,1) in the form $d(s) + kn(s) = 0$. When $k = 0$, the roots are at the zeros of $d(s)$, which are the poles of $B(s)$. As k approaches infinity, $n(s)$ must approach zero in order to satisfy the equation. Consequently, the loci terminate at the zeros of $B(s)$, whatever the sign of k.

Rule 4. Number of loci at infinity. When $B(s)$ has p poles and z zeros, $B(s)$ has $p - z = e$ zeros at infinity if $p > z$ (or $z - p$ poles at infinity if $z > p$). In either case it follows from Rule 3 that there are e loci going out to infinity. In Section 4.1, Example 1, $p = 2$, $z = 0$, while in example 2, $p = 2$, $z = 1$.

Rule 5. Slope of the asymptotes to the loci at infinity. We wish to locate the loci at infinity. In order to do so, the equation $1 + kB(s) = 0 = d(s) + kn(s)$ is investigated as k and s approach infinity. It is assumed that $p > z$ and therefore, by Rule 3, very large k corresponds to some roots going out to infinity. If $z > p$, rewrite Eq. (4.2,1) as $1 + d/nk = 0 = 1 + Kd/n$, and consider the root loci with K as the parameter. Equation (4.2,1) is written in the following form:

$$0 = 1 + kn/d = 1 + k[s^z + n_1 s^{z-1} + ...]/[s^{z+e} + d_1 s^{z+e-1} + ...] ,$$

where e is the excess of poles over zeros of $B(s)$. In order to explore the properties of this equation as s approaches infinity, the function $B(s) = n(s)/d(s)$ is expanded in a series at infinity; this can be done by dividing $d(s)$ into $n(s)$. The result is

$$1 + k\left[\frac{1}{s^e} - \frac{(d_1 - n_1)}{s^{e+1}} + ...\right] = 0. \tag{4.2,2}$$

The first approximation of Eq. (4.2,2) as s approaches infinity is: $1 + k/s^e \to 0$, or $s^e \to -k$. When $k > 0$, $s^e \to -k$ leads to $e \angle s \to (2r + 1)180°$, with $r = 0, 1, 2, ..., e - 1$. When $k < 0$, $s^e \to -k$ leads to $e \angle s \to 360°r$. Hence the loci approach infinity along lines whose phase angles are $(2r + 1)180°/e$ for $k > 0$, and $360°r/e$ for $k < 0$. These lines are called the *asymptotes to the loci at infinity*. Thus, in Fig. 4.1-1a, with $k > 0$ and $e = 2$, the phase angles of the two asymptotes are $180°/2$, $540°/2$.

Rule 6. Asymptotes to the loci at infinity meet at the centroid of $B(s)$. Knowing the slope of the asymptotes to the loci at infinity does not suffice to locate the asymptotes, because any number of lines with the same slope may be drawn. A better approximation of Eq. (4.2,2) is needed, viz., $1 + k[s^{-e} - (d_1 - n_1)s^{-e-1}] = 0$, or $s^{e+1} + k(s - E) = 0$, where $E = d_1 - n_1$ is the difference between the sum of the values of the zeros and the sum of the values of the poles of $B(s)$. This equation may be written as $s^{e+1} = -k(s - E)$. Suppose $s = s_o$ with phase θ (point F in Fig. 4.2-1) satisfies the equation $s^{e+1} = -k(s - E)$. If so, $(e + 1)\theta = \angle(-k) + \angle(s_o - E)$, so that (for $k > 0$ and r any integer) $(e + 1)\theta = (2r + 1)180° + \angle(s_o - E)$. To be specific, let $E \triangleq d_1 - n_1 > 0$, then $s_o - E$ is the vector **AF** (see Fig. 4.2-1), with angle $\theta' = \theta + \delta = \angle(s_o - E)$.

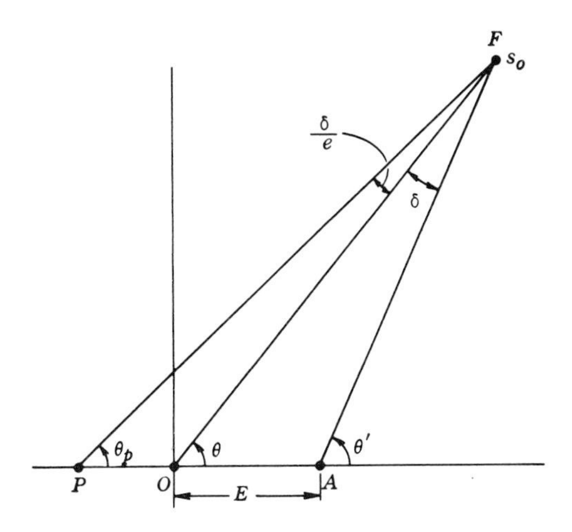

FIG. 4.2-1. Asymptotes intersect at centroid P.

Substituting this into the previous equation, we have $(e + 1)\theta = (2r + 1)180° + \theta + \delta$, or

$$\theta = \frac{(2r + 1)180°}{e} + \frac{\delta}{e}. \tag{4.2,3}$$

Equation (4.2,3), with nonzero δ, indicates that the asymptotes do not meet at the origin with phase angles of $(2r + 1)180°/e$. Let us find the true origin of the asymptotes. From Eq. (4.2,3) it must be the point P in Fig. 4.2-1 such that the angle θ_p of the vector **PF** (from P to s_o) is precisely equal to $(2r + 1)180°/e$. Replacing $(2r + 1)180°/e$ in Eq. (4.2,3) by θ_p, we have $\theta_p = \theta - \delta/e$. This defines the point P, because it is necessary that $OP = E/e$ in order that $\theta - \theta_p = \delta/e$ (recall F is very far away).

Rule 6 therefore is: The asymptotes to the loci at infinity meet at the point P on the real axis, with

$$P = \frac{-(d_1 - n_1)}{e} = \frac{\Sigma \text{ poles of } B(s) - \Sigma \text{ zeros of } B(s)}{\text{no. of finite poles of } B(s) - \text{no. of finite zeros of } B(s)}. \tag{4.2,4}$$

The point P is located at the center of mass of a system consisting of positive unit weight particles at the poles of $B(s)$, and negative unit weight particles at the zeros of $B(s)$. It is easily seen that the rule applies whatever the sign of k.

Rule 7. Portions of real axis that are on root loci. The portions of the real axis that are parts of the root loci are particularly easy to find. The complex equation $1 + kB(s) = 0$ is equivalent to two real equations:

$$|B(s)| = |-1/k|; \tag{4.2,5a}$$

$$\angle B(s) = (2r + 1)\pi, \quad \text{if } k > 0; \tag{4.2,5b}$$

$$\angle B(s) = 2r\pi, \quad \text{if } k < 0 \tag{4.2,5c}$$

where r is any integer. Consider those values of s along the real axis which satisfy Eq. (4.2,5b). The angles of the vectors from any conjugate complex pair of poles or zeros to any point on the real axis are equal in magnitude and opposite in sign. Therefore, only the real poles and zeros of $B(s)$ contribute to $\angle B(s)$, for s real. Since $+180°$ is the same as $-180°$, the angle of a pole vector is the same as that of a zero vector. A pole or zero to the left of a point $s = \sigma$ contributes zero degrees to $B(\sigma)$, while a critical point (pole or zero) to the right of $s = \sigma$ contributes 180° to $B(\sigma)$. Therefore *when* $k > 0$, *those portions of the real axis to the left of an odd number of real axis critical points of* $B(s)$ *constitute segments of the root loci.* (If $k < 0$, substitute even for odd in the above.)

Examples of Application of Rules 1-7

Example 1. Sketch the loci of the roots of

$$1 + kB(s) = 1 + [ks(s + 2)]\,[(s^2 + 2s + 2)(s + 4)(s + 6)]^{-1} = 0$$

for $0 < k < \infty$. The poles and zeros of $B(s)$ are indicated by crosses and circles, respectively, in Fig. 4.2-2a.

(a) Rule 7 is applied first. On the real axis, $0 > \sigma > -2$ and $-4 > \sigma > -6$ are to the left of an odd number of real axis critical points and are therefore parts of the root loci.

(b) There are two more poles than zeros; there are therefore two loci at infinity whose asymptotes are at $\pm 180°/2$. The origin of the asymptotes is at $[-(4 + 6 + 2) + 2]/2 = -5$. The asymptotes are shown dotted in Fig.

4.2-2a. The arrows indicate the direction the roots move along the loci as k is increased. It is known that for sufficiently distant points the loci and the asymptotes almost coincide. To complete the sketch more rules are needed.

Example 2. For the same $B(s)$ as in Example 1, sketch the loci for $0 > k > -\infty$.

(a) From Rule 7, $0 < \sigma$, $-4 < \sigma < -2$, $\sigma < -6$ are to the left of an even number of real axis critical points, and are therefore parts of the root loci—see Fig. 4.2-2b.

(b) The two asymptotes at infinity have angles at 0° and 180°. Again more rules are needed to complete the sketch.

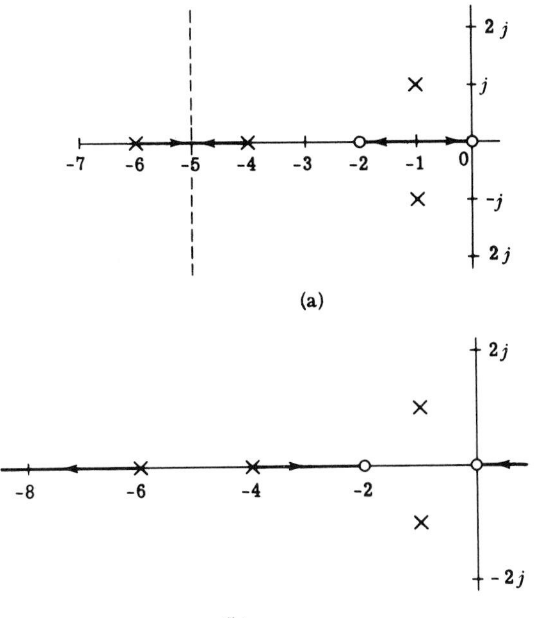

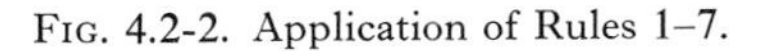

FIG. 4.2-2. Application of Rules 1–7.

Rule 8. Angles at which root loci leave the poles and terminate at the zeros. The rule is best given by means of an example. Figure 4.2-2a is redrawn in Fig. 4.2-3, and the angle of departure of the locus from the pole at $-1 + j$ is sought. Suppose P, a point very close to the pole at $-1 + j$, is a root of $1 + kB(s) = 0$. It must therefore satisfy Eq. (4.2,5b), i.e., $\theta_x + \theta_1 + \theta_2 + \theta_3 - (\alpha_1 + \alpha_2) = 180°$ in Fig. 4.2-3. We take the point P very close to the pole at $-1 + j$, so that in the limit, θ_1 is the angle from -6 to $-1 + j$, θ_2 is the angle from -4 to $-1 + j$, etc. All the θ's and α's in the equation are therefore known, except for θ_x, so the equation may be solved for θ_x. In the above, $\theta_x + 11.3 + 18.4 + 90 - (45 + 135) = 180$. Therefore $\theta_x = 240.3°$. The angles at which the loci terminate at the zeros are obtained in the same manner. When $k < 0$, one writes, similar to the above, $\sum \theta - \sum \alpha = 0$ and solves for θ_x or α_x.

If there is a multiple pole at P in Fig. 4.2-3, of order m, then one gets in exactly the same way $m\theta_x = D° + (2r + 1)180°$, where D is the net angle at P due to the other poles and zeros of $B(s)$. There are m loci leaving the multiple pole at angles $[D°/m] + [(2r + 1)180°/m]$ for $r = 0,1,2$, etc. The difference in angle between two successive loci is $(2)(180°)/m$.

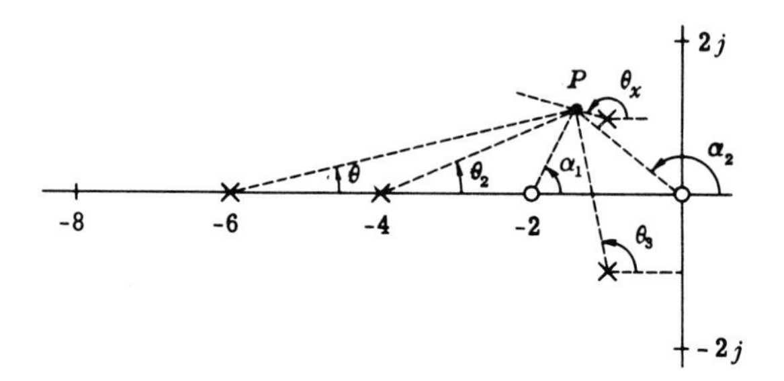

FIG. 4.2-3. Calculation of angle of departure θ_x.

Rule 9. Coalescent points. Coalescent points are points at which two or more loci intersect. If two loci intersect at a point s_o, then $1 + kB(s) = 0$ must have a double root at $s = s_o$, i.e., $1 + kB(s) = (s - s_o)^2H(s) = 0$. Therefore the derivative of $1 + kB(s)$, which is the derivative (with respect to s) of $B(s)$, must be zero at $s = s_o$. Similarly, if three loci meet (coalesce) at a point s_o, then in addition to $s = s_o$ satisfying the equation $1 + kB(s) = 0$, it is necessary that the first and second derivatives of $B(s)$ be zero at $s = s_o$. Coalescent points of great interest are those located on the real axis. They are called breakaway points because they are the points at which the loci leave the real axis and go off into the complex plane. For example, in Fig. 4.2-2a, there must be a breakaway point located somewhere between -4 and -6 and another between 0 and -2. These breakaway points may be located by finding the roots of $B'(s) = 0$. In the example of Fig. 4.2-2a, setting $B'(s) = 0$ leads to the fifth-degree equation:

$$s^5 + 9s^4 + 24s^3 + 12s^2 - 48s - 48 = 0.$$

The solution of the equation is facilitated by the knowledge that a root is located somewhere between -4 and -6 and another between 0 and -2.

Some feeling for the location of breakaway points is obtained by roughly sketching $B(s)$ vs. s for s real, i.e., $B(\sigma)$ vs. σ, where $s = \sigma + j\omega$, and σ, ω are real numbers. This is done in Fig. 4.2-4 for the case $B(s) = [(s + 1)(s + 4)][(s + 0.5)(s + 2)(s + 3)(s^2 + 4s + 8)]^{-1}$. Such qualitative sketches are easily made by inspection. The solution of $B(s) = -1/k$ is now attempted graphically. With $k = 0.5$, the line $-1/k$ intersects $B(s)$ at points near the poles of $B(s)$ at A_1, B_1, C_1 in Fig. 4.2-4. As k increases, the line $-1/k$ moves up and intersects $B(s)$ at points further from the poles, at A_2, B_2, C_2. At some value $k = k_{o1}$, we have the $-1/k_{o1}$ line located as in the figure, tangent to the curve, and corresponding to a breakaway point. The breakaway point is at $s = s_{x1}$, such that $B'(s_{x1}) = 0$. At another value of k ($k = k_{o2}$), there is another breakaway point, at $s = s_{x2}$ (see Fig. 4.2-4), such that $B'(s_{x2}) = 0$.

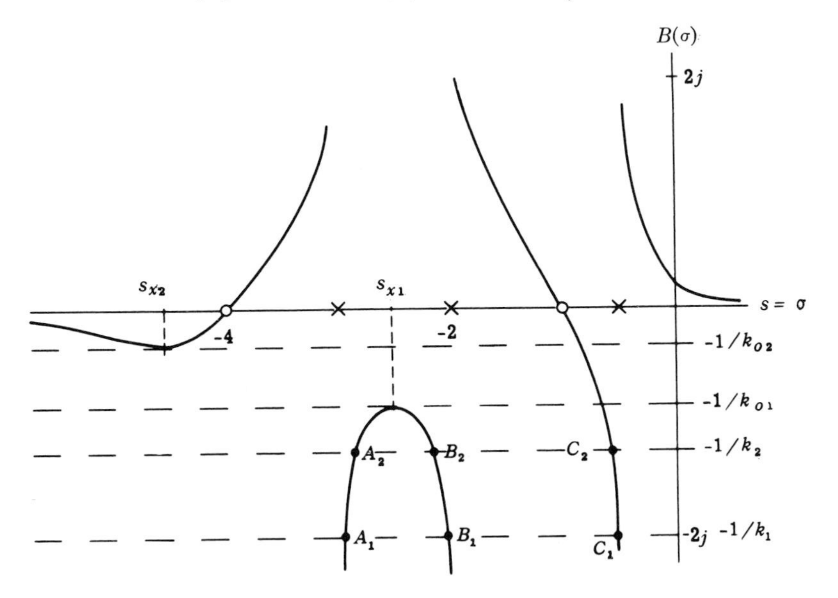

FIG. 4.2-4. Usefulness of sketch of $B(\sigma)$.

Good guesses at coalescent points on the real axis can thus be made by very simple reasoning as to the location of maxima and minima. Exact values of coalescent points are obtained by finding the appropriate roots of $B'(s) = 0$. One knows *a priori* approximately where the roots he seeks are located; it is therefore not necessary to bother finding the other roots. [The other roots of $B'(s) = 0$ correspond to coalescent points for $k < 0$ or k complex.]

In ambiguous cases, a sketch of $B(s)$ vs. σ is often very useful as an aid in root locus plotting. For example, in Fig. 4.2-2b two loci must leave the $B(s)$ poles at $-1 \pm j$. Their angle of departure is found from Rule 8, i.e.,

$$\theta_x + \arctan \tfrac{1}{5} + \arctan \tfrac{1}{3} + \pi/2 - \arctan 1 - (\pi - \arctan 1) = 0$$

leads to $\theta_x = 60.2$. The information so far available leads to the heavy lines in Fig. 4.2-5a. The simple and easily drawn rough sketch of $B(\sigma)$ vs. σ of Fig. 4.2-5b quickly reveals the existence of a breakaway point on the positive real axis and indicates that the root loci sketch must be completed with the dotted portion in Fig. 4.2-5a. In many cases one is not interested in locating the breakaway point exactly, rather he is interested in the general form of the root loci. The above rules, with the aid perhaps of a simple sketch of $B(\sigma)$ vs. σ, permit a rough pattern of the root loci to be sketched very quickly.

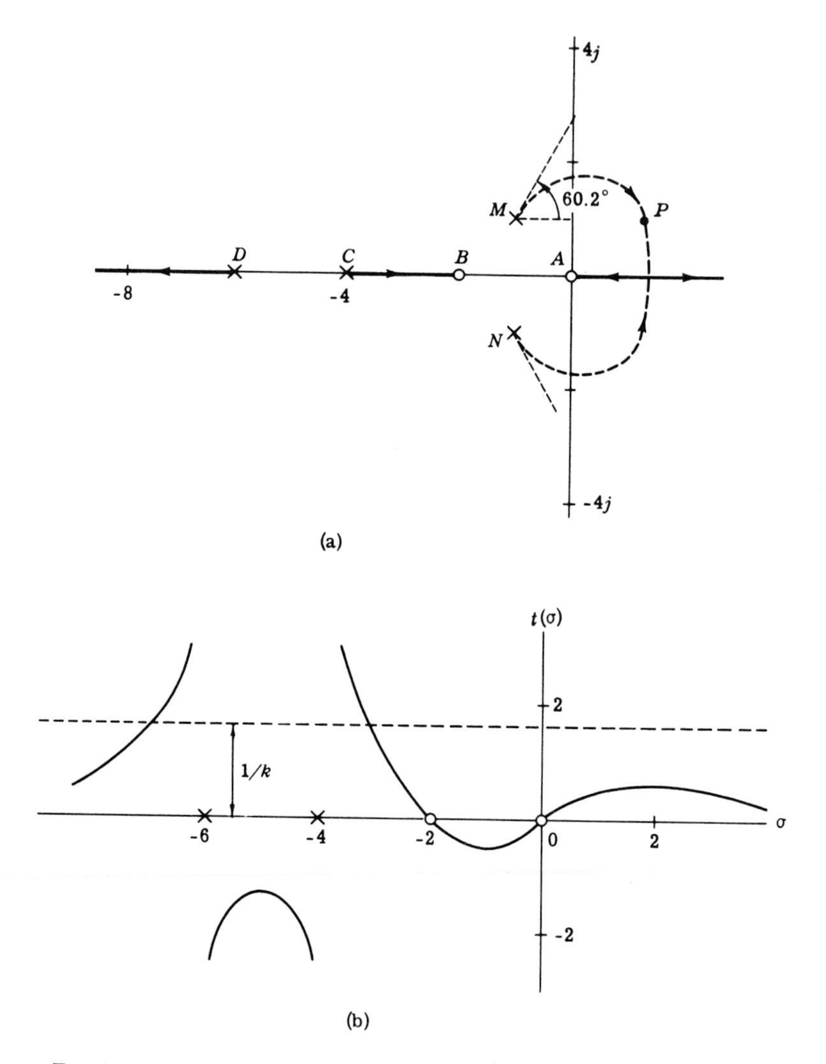

FIG. 4.2-5. Root loci of $1 + ks(s + 2)/(s^2 + 2s + 2)(s + 4)(s + 6) = 0$.

Example. Sketch the loci of the roots of $1 + k/(s + 1)(s + 3)(s^2 + 4s + 5) = 0$ for $0 < k < \infty$. The loci are sketched in Fig. 4.2-6a.

(a) The poles and zeros of $B(s)$ are first marked.

(b) Real axis portions of the loci are marked in heavy lines.

(c) There are four loci at infinity with asymptotes at $\pm 180°/4$, $\pm 540°/4$. The origin of these asymptotes is at $-8/4 = -2$. These asymptotes are shown dotted in Fig. 4.2-6a.

(d) The angle of departure from the pole at $-2 + j$ is found from the following calculation: $\theta_x + 90° + 45° + 135° = 180°$, $\theta_x = -90°$. Because of the conjugate condition, the angle of departure from the pole at $-2 - j$ is $+90°$.

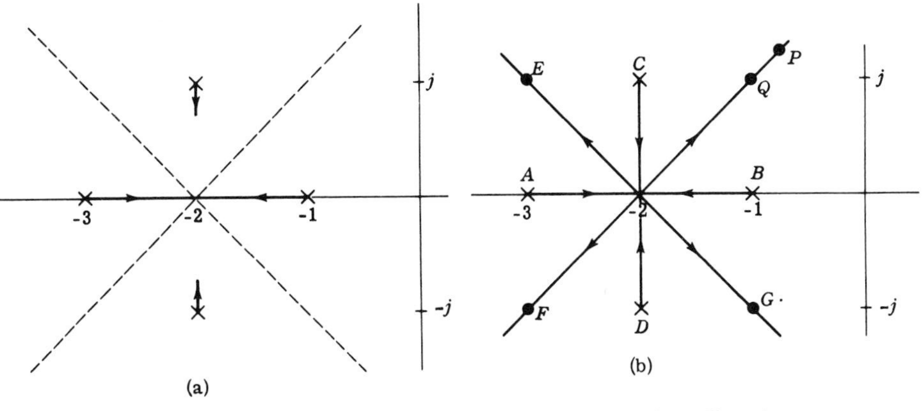

FIG. 4.2-6. Root loci of $1 + k/(s + 1)(s + 3)(s^2 + 4s + 5) = 0$.

(e) There remains the problem of finding the coalescent points. From the symmetry of the pattern, one breakaway point must be at -2. It is not obvious where the others are. To obtain them, solve $B'(s) = 0$, i.e., $4s^3 + 24s^2 + 48s + 32 = 0$. Divide through by $s + 2$ to eliminate the known (or reasonably guessed at) root. The quotient is $s^2 + 4s + 4 = 0 = (s + 2)^2$, i.e., there is a triple root of $B'(s)$ at -2, with $B'(s)$, $B''(s)$, and $B'''(s)$ zero at -2. This indicates that four loci coalesce at -2. The complete sketch is drawn in Fig. 4.2-6b.

Rule 10. Value of k. The value of k corresponding to any point s_o on a locus is found from Eq. (4.2,5a), viz., $k = |1/B(s)|_{s=s_o}$; $B(s_o)$ is the ratio of the product of the vectors from the zeros of $B(s)$ to s_o, to the product of the vectors from the poles of $B(s)$ to s_o. [It is assumed that the leading coefficients of numerator and denominator of $B(s)$ are unity. If not, they are pulled out and absorbed in k.] For example, in Fig. 4.2-6b, the value of k corresponding to the root at P is: $|(CP)(AP)(BP)(DP)|$. In Fig. 4.2-5a, the value of k leading to a zero of $1 + kB(s)$ at P is $|(DP)(CP)(MP)(NP)/(AP)(BP)|$.

Rule 11. Sums and products of the roots on the loci. Additional useful information relating the sum and product of the roots along the loci to the poles and zeros of $B(s)$ is lumped under Rule 11. Consider

$$1 + kB(s) = 1 + k(s^z + n_1 s^{z-1} + \dots + n_z)/(s^p + d_1 s^{p-1} + \dots + d_p) = 0.$$

If $e \triangleq p - z \geqslant 2$ (which is the case in most feedback problems), then the numerator of the above equation becomes $s^p + d_1 s^{p-1} + \dots + (d_p + kn_z) = 0$. Therefore, the sum of the roots along the loci is fixed (independent of k) and equal to the sum of the poles of $B(s)$. For larger values of e, there are more such relations relating the sum of the roots taken two at a time, etc.

The product of the roots along the loci is $d_p + kn_z$ whatever $e = p - z$ may be. If $B(s)$ has a pole at the origin (as is often the case in feedback control problems), then $d_p = 0$ and the product of the roots is proportional to k. On the other hand, if $B(s)$ has a zero at the origin, the product of the roots is fixed and equal to the product of the poles of $B(s)$.

Rule 12. Shifting the origin of the root loci. In the course of developing a sketch of root loci, suppose one has found the loci of the roots for values of k from $k = 0$ to $k = k_o$. To be specific, the roots of $1 + k_o B(s) = 0$ are, say, at s_{1o}, s_{2o}, ..., s_{no}. To complete the sketch for k varying from k_o to infinity, which is equivalent to $k' \triangleq k - k_o$ varying from zero to infinity, the problem is exactly the same as finding the root loci of $1 + k'B^*(s) = 0$. The poles of $B^*(s)$ are at s_{1o}, s_{2o}, ..., s_{no}, while its zeros are those of $B(s)$. In other words, the poles of $B(s)$ may at any time be replaced by a new set of poles picked at a corresponding set of points on the root loci.

To prove this, consider $1 + kB(s) = 0 = 1 + kn/d$, which may be written as

$$0 = d + kn = (d + k_o n) + (k - k_o)n = d^* + k'n,$$

whose zeros are those of $1 + k'n/d^* = 1 + k'B^*(s)$, where the zeros of d^* are the roots of $1 + k_o n/d = 0$. The leading coefficient of d^* in $B^* = n/d^*$ is 1, $1 + k_o$, k_o according to whether $B(s)$ is, respectively, zero, finite, or infinite as $s \to \infty$. It is also possible to retrace the loci using $1 - k_1 B^*(s) = 0$ with $k' = k - k_o = -k_1 < 0$. As k_1 increases from zero value, the root loci are retraced until when $k_1 = k_o$, the roots of $1 - k_1 B^*(s) = 0$ are at the poles of $B(s)$.

The above point of view is useful for estimating the direction a locus is turning when there may be some doubt in the matter. Consider the case of a coalescent point of order m located at any point, i.e., m loci enter the point and then leave. It is known from Rule 8 that the loci which leave the point (or enter) are symmetrically spaced. Can anything be said, however, about the relative orientation of the loci which leave, with respect to those which enter? For the loci which leave the point, one may, from Rule 12, consider the point as a pole of order

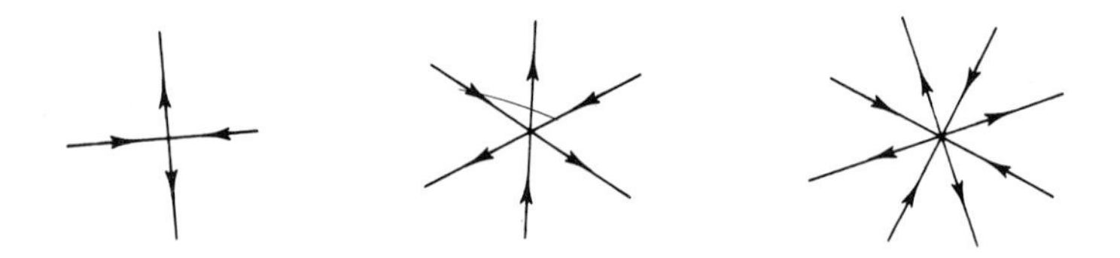

FIG. 4.2-7. Points of coalescence.

m and get the angle of departure (from Rule 8) as $\theta_x = [D^\circ + (2r + 1)180^\circ]/m$, with r any integer. For those which enter, one may again consider the point as a pole of order m, but retrace the loci and find their angles, $\theta_y = (D^\circ + 360r^\circ)/m$. The difference $\theta_y - \theta_x = 180^\circ/m$. Several examples are shown in Fig. 4.2-7.

Example. Consider the root loci of $1 + k/s(s + a)(s + b) = 0$ (Fig. 4.2-8). Asymptotes at infinity are at $\pm 180^\circ/3, \pm 540^\circ/3$, with their origin at $-(a+b)/3$. Breakaway from the real axis is at $(d/ds)[s(s + a)(s + b)] = 0$, i.e., at $-1/3\,[a + b - (a^2 - ab + b^2)^{0.5}]$. The complex loci intersect the imaginary

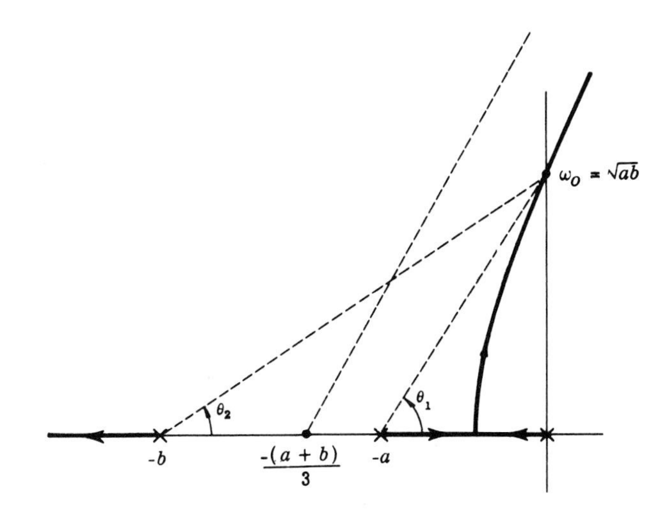

FIG. 4.2-8. Root loci of $1 + k/s(s + a)(s + b) = 0$.

axis at a point such that, in Fig. 4.2-8, $\theta_1 + \theta_2 = 90^\circ$, i.e., $\omega_o^2 = ab$, at which the third root is at $-(a + b)$ and $k = ab(a + b)$ is the $|$ product $|$ of the roots. The root loci of $1 + k/[(s + \alpha)(s + \beta)(s + \gamma)] = 0$ can be obtained from the above, by shifting the $j\omega$ axis to $-\alpha$.

Some root loci which can be drawn by inspection are shown in Fig. 4.2-9. In Fig. 4.2-9e the root loci are on a circle given by

$$[x + (ab - \omega_o^2)/(a + b)]^2 + y^2 = \omega_o^2 + [(ab - \omega_o^2)/(a + b)]^2 .$$

(This can also be used for a complex pole pair at $-m \pm jn$, by writing $a + b = 2m$, $ab = m^2 + n^2$.) The more general case for complex zeros not on the $j\omega$ axis can be handled by suitably shifting the $j\omega$ axis.

The above rules enable one to obtain very quickly an excellent qualitative picture of the root loci of $1 + kB(s) = 0$. If more details are desired, one uses his knowledge of the approximate locations of the roots to test for points that satisfy Eq. (4.2,5). The net angle of $B(s_o)$ is the sum of the vector angles from

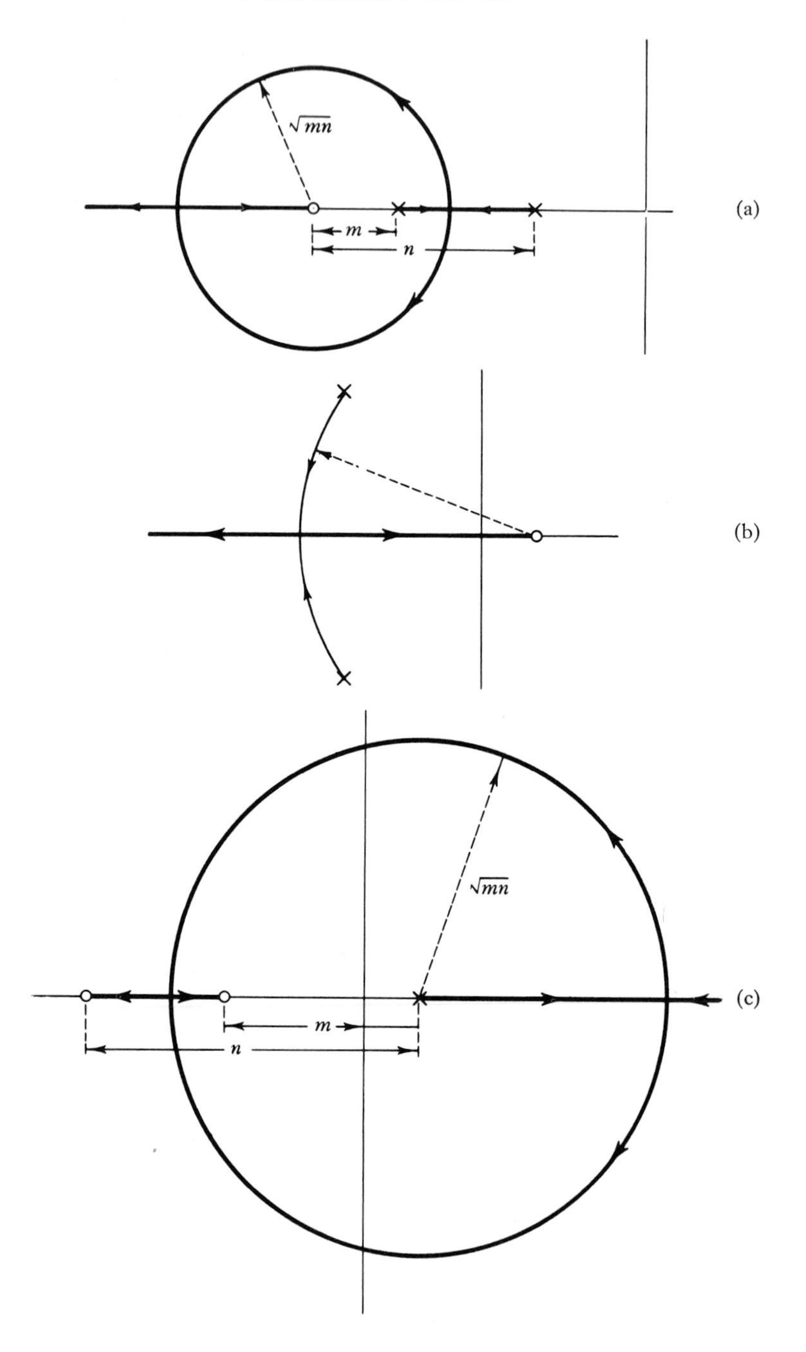

$\sqrt{mn}$
m
n
(a)
(b)
$\sqrt{mn}$
m
n
(c)

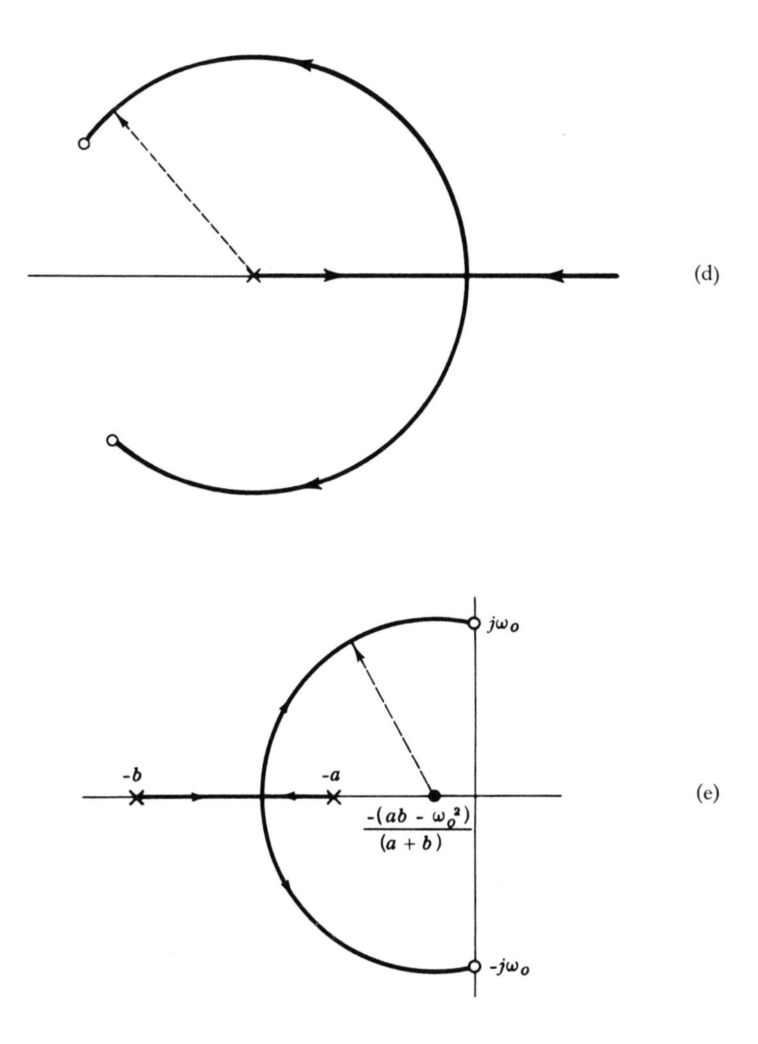

FIG. 4.2-9. Some root loci available by inspection.

the zeros of $B(s)$ to s_o, minus the sum of the vector angles from the poles of $B(s)$ to s_o. The Spirule[1] is useful for summing the vector angles.

---

[1] The Spirule Company, 9728 El Venado, Whittier, California.

§ 4.3 Sensitivity of Roots to Variation in k

It is often useful to know how rapidly the roots of $1 + kB(s) = 0$ move along the loci as k varies[1] (for example, to find how sensitive a system's poles are to parameter variation). The problem may be stated as follows: Given a set of roots of $1 + kn(s)/d(s) = 0$, corresponding to a given k; what is the increment in any root as k is incrementally varied? Let $N(s) = d(s) + kn(s) = K\Pi(s - s_i)$, where $K = 1$, $1 + k$, or k, when $B(s) = n(s)/d(s)$ is, respectively, zero, finite, or infinite at infinite s, and where we have temporarily assumed that all the zeros of $N(s)$ are distinct. Taking the logarithm of $N = d + kn = K\Pi(s - s_i)$, we get $\ln(d + kn) = \ln K + \Sigma \ln(s - s_i)$. We differentiate with respect to k, and obtain

$$(d + kn)^{-1}n = N^{-1}n = K^{-1}\,\partial K/\partial k - \Sigma\,(s - s_i)^{-1}\,\partial s_i/\partial k.$$

Each side of the last equation must have the same residue in any pole. Hence, $n(s_i)/N'(s_i) = -\,\partial s_i/\partial k$. However $d(s_i) + kn(s_i) = 0$, so $n(s_i) = -\,d(s_i)/k$, and we finally have

$$\frac{\partial s_i}{\partial k/k} = \frac{d(s_i)}{N'(s_i)}. \tag{4.3,1}$$

Suppose $N(s)$ has u coincident zeros at s_i, i.e., $N(s) = (s - s_i)^u N_1(s) = d(s) + kn(s)$. If k is replaced by $k + \partial k$, then we must seek the zeros of

$$d(s) + (k + \partial k)n(s) = d(s) + kn(s) + (\partial k)n(s) = N(s) + (\partial k)n(s),$$

whose zeros are the roots of $1 + (\partial k)n(s)/N(s) = 0$. Consider the root loci of the last Equation. From Rule 10 (Section 4.2), if there is a root at $s_i + \partial s_i = s_1$, then $(\partial k) = -\,N(s_1)/n(s_1)$, which in the limit is equal to $-\,(\partial s_i)^u N_1(s_i)/n(s_i)$. But $d(s_i) + kn(s_i) = 0$, so

$$\frac{(\partial s_i)^u}{\partial k/k} = \frac{d(s_i)}{N_1(s_i)} = \frac{d(s_i)}{u!N^{(u)}(s_i)} \triangleq S_k^{s_i} \tag{4.3,2}$$

where $N^{(u)}(s_i)$ is the uth derivative of $N(s)$ evaluated at $s = s_i$. $S_k^{s_i}$ is called the root sensitivity to k. Equation (4.3,2) is more general than Eq. (4.3,1), and may be stated as follows:

Rule for calculating sensitivity of a zero of a polynomial to a parameter: Write the polynomial in the form $N(s) = d(s) + kn(s) = (s - s_i)^u N_1(s)$, where $d(s)$ and $n(s)$ are polynomials which are independent of k, and with unity for their

---

[1] A. Papoulis, Displacement of the zero of the impedance $Z(p)$ due to incremental variations in the network elements. *Proc. IRE* **43**, 79-82 (1955).

leading coefficients. The sensitivity to k, of the zero at s_i, is defined as $(\partial s_i)^u/(\partial k/k) = S_k^{s_i}$, and is equal to $d(s_i)/N_1(s_i) = d(s_i)/u!N^{(u)}(s_i)$. In other words, $S_k^{s_i}$ is the coefficient of $(s - s_i)^{-u}$ in the Laurent expansion of the function $[1 + kB(s)]^{-1}$, where $B(s) = n(s)/d(s)$.

Example. In the system of Fig. 4.3-1a, the sensitivity of the poles of $T = C/R$ to variations in the parameter m are desired.

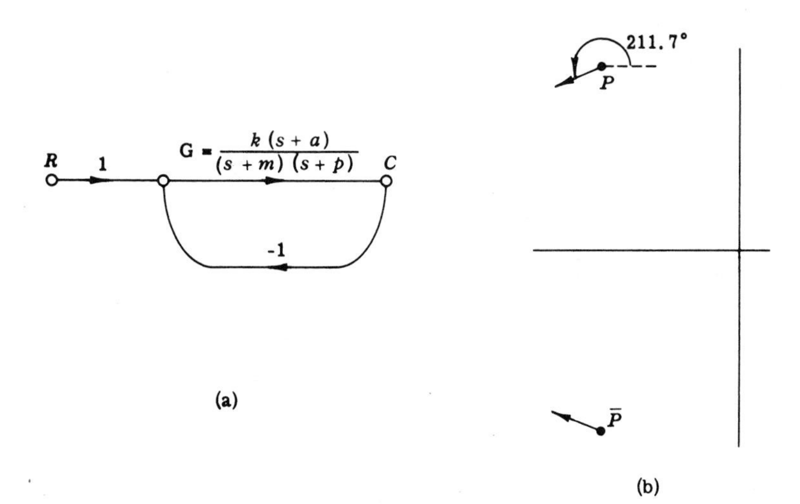

FIG. 4.3-1. Effect of m on the system poles.

Solution: $T = C/R = G/(1 + G)$ and the poles of T are given by the roots of $1 + k(s + a)/(s + m)(s + p) = 0$, which we write in the form of $N(s) = d(s) + kn(s)$, i.e., $(s + m)(s + p) + k(s + a) = [s^2 + s(p + k) + ka] + m(s + p) = N(s)$. Then, using the sensitivity rule,

$$ds_i/(dm/m) = S_m^{s_i} = \text{residue in pole at } s_i \text{ of } [s^2 + s(p + k) + ka]/N(s).$$

Suppose the nominal values are $m = 2, p = 1, a = 5, k = 2$. The nominal poles are then at the zeros of $s^2 + 5s + 12$, i.e., at $-2.5 \pm j2.4$. The sensitivity of the pole at $-2.5 + j2.4$ is therefore the residue in that pole of $(s^2 + 3s + 10)/(s^2 + 5s + 12)$ which is $-(1 + j0.62) = 1.18 \angle 211.7°$. If the relative change in m is 0.2, then the system poles, originally at P, P' in Fig. 4.3-1b, move in the direction indicated approximately by the amount $(0.2)(1.18) = 0.24$. This amount is approximate because the rule applies only for infinitesimally small parameter changes.

It may be noted that the root, whose sensitivity is being found, must lie on a root locus of $1 + kB(s) = 0$. The phase angle of the root sensitivity is simply the angle of the tangent to the root locus at that point.

§ 4.4 Root Sensitivity in Terms of Open-Loop Poles and Zeros

In the above, the sensitivity of the roots of $1 + kB(s) = 0 = 1 + kn/d$ is available in terms of the open-loop poles [zeros of $d(s)$ which are $B(s)$ poles], and the closed-loop roots which are the zeros of $1 + kB(s)$, or of $d(s) + kn(s) = N(s)$. The root sensitivity can also be obtained in terms of the open-loop quantities alone, that is, in terms of the zeros and poles of $B(s)$. Neither k nor the other zeros of $N(s)$ need be known. The derivation follows: In Eq. (4.3,2), $N^{(u)}(s_\alpha) = d^{(u)}(s_\alpha) + kn^{(u)}(s_\alpha)$. We replace k by $-d(s_\alpha)/n(s_\alpha)$, and obtain

$$[(u!)S_k^{s_\alpha}]^{-1} = [d^{(u)}(s_\alpha)/d(s_\alpha)] - [n^{(u)}(s_\alpha)/n(s_\alpha)]$$
$$= (-1)^{u-1}\{\Sigma_i(s_\alpha + p_i)^{-u} - \Sigma_j(s_\alpha + z_j)^{-u}\}. \quad (4.4,1)$$

We do not show the proof of the last step, because of its length. It is a generalization of the following proof for $u = 2$: We have $d(s) = \Pi(s + p_i)$, so $d'(s) = d(s)\,\Sigma\,(s + p_i)^{-1}$ and

$$d''(s) = d'(s)\,\Sigma\,(s + p_i)^{-1} - d(s)\,\Sigma\,(s + p_i)^{-2}$$
$$= d(s)\{[\Sigma\,(s + p_i)^{-1}]^2 - \Sigma\,(s + p_i)^{-2}\}.$$

Consequently,

$$d''/d - n''/n = [\Sigma\,(s + p_i)^{-1}]^2 - \Sigma\,(s + p_i)^{-2} - [\Sigma\,(s + z_j)^{-1}]^2 + \Sigma\,(s + z_j)^{-2}.$$

However at $s = s_\alpha$, $d'/d = n'/n$, which is equivalent to $\Sigma(s + p_i)^{-1} = \Sigma(s + z_j)^{-1}$ at $s = s_\alpha$. This is used to eliminate the first and third terms in the equation for $d''/d - n''/n$ and leaves

$$[d''/d - n''/n]_{s_\alpha} = (-1)\,[\Sigma\,(s_\alpha + p_i)^{-2} - \Sigma\,(s_\alpha + z_j)^{-2}].$$

Equation (4.4,1) suggests the following electrostatic field analogy. Consider a set of infinitely long lines, positively charged lines (2π coulombs per meter) located at the poles of $B(s)$, negatively charged ones located at the zeros of $B(s)$. The electric flux density at s_α, in coulombs per square meter, is exactly the inverse of Eq. (4.4,1) when $u = 1$. Thus when $u = 1$, the root sensitivity at s_α is the inverse of the resulting flux density due to line charges located at the poles and zeros of $B(s)$. When $u > 1$ the analogy must be suitably modified. The root sensitivity which, from Eq. (4.4,1) may be calculated from the poles and zeros of $B(s)$, can be used to aid in the sketching of root loci. One considers the direction of a flux line through the point in question. For small root sensitivity, a large value of flux density is needed.

§ 4.5 Sensitivity of Roots of $1 + kB(s) = 0$ to Variations in the Zeros and Poles of $B(s)$

What is the effect of an incremental change in a zero of $B(s)$, on a root of $1 + kB(s)$ at $s = s_\alpha$, of multiplicity x? We write $B(s) = (s - z_\beta)B_1(s) = n(s)/d(s)$ and seek the roots of $1 + kB_1(s)\,[(s - z_\beta) - \partial z_\beta] = 1 + kB(s)\,[1 - \partial z_\beta/(s - z_\beta)] = 0$. These are the roots of $d^*(s)/d(s) - (\partial z_\beta)kn(s)\,[(s - z_\beta)\,d(s)]^{-1} = 0$ [with $d^*(s) = d + kn = N$, using the notation and technique of Rule 12, Section 4.2]. The displaced roots are therefore on the root loci of $1 - [(\partial z_\beta)kn(s)]\,[(s - z_\beta)N(s)]^{-1} = 0$. From Rule 10, Section 4.2, and using $N(s) = (s - s_\alpha)^x N_1(s)$, and $n(s_\alpha) = -d(s_\alpha)/k$, we get

$$(\partial s_\alpha)^x = \frac{-\partial z_\beta}{(s_\alpha - z_\beta)}\;\frac{d(s_\alpha)}{x!N^{(x)}(s_\alpha)}. \tag{4.5,1}$$

Note that $-\partial z_\beta/(s_\alpha - z_\beta)$ is the relative change in the vector from z_β to s_α and is thus analogous to dk/k. It is not surprising therefore that with this interpretation, Eq. (4.5,1) is identical to Eq. (4.3,2). Note also that if $x = 1$, then the balance of the right-hand side of Eq. (4.5,1) is simply the residue in the pole at s_α of d/N.

In exactly the same manner, the effect on the xth order root of $d + kn = N$ at s_α, due to a pole of $B(s)$ shifting by amount ∂p_γ, is

$$(\partial s_\alpha)^x = \left(\frac{\partial p_\gamma}{s_\alpha - p_\gamma}\right)\frac{d(s_\alpha)}{x!N^{(x)}(s_\alpha)}. \tag{4.5,2}$$

In feedback control system synthesis, one often cancels out the poles and zeros of a fixed part of the control system by means of compensating networks, in order to achieve a more suitable open-loop transfer function. Due to inexact cancellation, dipoles appear in the open-loop transfer function. Equations (4.5,1) and (4.5,2) can be used to find the incremental shift in the closed-loop poles [roots of $1 + kB(s) = 0$] due to the presence of a dipole, $(s - a)/[s - (a + \partial a)]$.

Example. In Fig. 4.2-6b, what is the effect of a shift in the open-loop pole at -1, of amount dp, on the closed-loop pole $s_j = -1 + j$ at Q? Since the other closed-loop poles of the system are easily available, we can use them to calculate the residue in the pole at Q. The residue in the pole at Q of d/N is $\mathscr{R} = [(AQ)(BQ)(CQ)(DQ)]\,[(EQ)(FQ)(GQ)]^{-1} = (5/8(2)^{0.5}) \angle 45°$. Therefore, from Eq. (4.5,2), $ds_j = \mathscr{R}dp/(BQ) = [5dp/16(2)^{0.5}] \angle -45°$.

On the other hand, if only one closed-loop pole is located, and the location of the others is not otherwise needed, it is more convenient to use Eq. (4.4,1), since the latter involves only the open-loop poles and zeros.

§ 4.6 Application of Root Sensitivity to Predistortion for Incidental Dissipation in Filters

Most network synthesis procedures assume that ideal inductors and capacitors are available. When practical lossy elements are used, the resulting functions have poles and zeros which are displaced from the design values. If the Q of the reactive elements is large, the displacements are small. It has been suggested[1] that the poles and zeros of the function to be realized should first be predistorted, such that the dissipation shifts them back to the desired values. This problem can be solved by the root-sensitivity concept.

The coil dissipation is represented by a resistor in series with the inductor L. In general, from the fundamental feedback equation: $T = [a + sLb]/[c + sLd]$, where a, b, c, d are functions of s and independent of L. However, actually $T = [a + (sL + r)b]\,[c + (sL + r)d]^{-1}$, so that the nominal zero of T at s_i is shifted due to $dk = r$ (if we let $k = s_iL$, and then $dk/k = r/s_iL$). From Eq. (4.3,2), and using $a(s_i) + s_iLb(s_i) = 0$,

$$ds_i \doteq \left(\frac{r}{s_iL}\right)\frac{a(s_i)}{\left[\frac{d}{ds}(a + sLb)\right]_{s=s_i}} = \frac{-\,rb(s_i)}{\left[\frac{d}{ds}(a + sLb)\right]_{s=s_i}}. \tag{4.6,1}$$

Similarly, the design pole of T at p_j is shifted by the amount,

$$dp_j \doteq \left(\frac{r}{p_jL}\right)\frac{c(p_j)}{\left[\frac{d}{ds}(c + sLd)\right]_{s=p_j}} = \frac{-\,rd(p_j)}{\left[\frac{d}{ds}(c + sLd)\right]_{s=p_j}}. \tag{4.6,2}$$

For the effect of capacitor losses, one replaces sC by $sC + g$ and the procedure is the same. The total shift in the zero at s_i, due to the coil and capacitor losses, is written as $\Delta s_i = \Sigma ds_i$. Note that it is not necessary that the Q's be identical. Instead of designing for zeros at s_i, one designs for zeros at $s_i' = s_i - \Delta s_i$, poles at $p_j' = p_j - \Delta p_j$. Note, however, that to find the ds_i and dp_j, design on the basis of s_i and p_j must first be made, and $a(s)$ or $b(s)$, and $c(s)$ or $d(s)$ found. The primed poles and zeros are calculated from $s_i' = s_i - \Delta s_i$, etc., and then one synthesizes the network (on the basis of ideal L's and C's) to realize the s_i' and s_j'. The actual poles and zeros obtained are at s_j and s_i.

§ 4.7 Application of Root Sensitivity to Limit the Drift in Feedback System Poles

Suppose the fixed part (plant) of a controlled process has the transfer function $P = k/s(s + 1)(s + 6)$ with k a parameter whose value lies between 2 and 6.

[1] C. A. Desoer, Network design by first-order predistortion technique. *Trans. IRE* **CT-4** (3), 107 (1957).

The system is to have a dominant pole pair centered near $-2 \pm j2$, such that the variations in k do not change the damping factor. Thus in Fig. 4.7-1a, the dominant poles are to move along the 135° line passing through Q, as k varies.

Design: Consider the root loci of $1 + KL(s) = 1 + Kn(s)/d(s) = 0$, where $L(s)$ is the system loop transmission. It is easily seen that it is impossible to achieve the desiderata, if $L(s)$ has no zeros. The simplest alternative is to assign at least one zero to $L(s)$, and for $L(s)$ to have 4 poles and one zero. [We insist on at least 3 more poles than zeros, because $P(s)$ has 3 more poles than zeros.] Qualitatively, we envision the root locus pattern of $1 + KL(s) = 0$ similar to that shown in Fig. 4.7-1b with $\theta_1 + \theta_2 + \theta_3 - \alpha = 45°$, in order that $s_i(-2 + j2)$ may lie on one of the root loci. Also, in order that $\angle S_k^{s_i} = 135°$, we must have in Fig. 4.7-1c (which gives the pole-zero pattern of the sensitivity function): $135° + \theta_1 + \theta_2 + \theta_3 - (90° + \lambda_1 + \lambda_2) = 135°$. This is because the poles of $L(s)$ are the zeros of the sensitivity function. In Fig. 4.7-1c A, B, $\bar{Q}$, Q mark the roots of $1 + KL(s) = 0$.

We could proceed by trial and error, since λ_1, λ_2 in Fig. 4.7-1c are not known until after M, N, R in Fig. 4.7-1b have been chosen. The procedure is as follows: We guess $(\lambda_1 + \lambda_2)$. This leaves two equations in the two unknowns $\sum\theta = \theta_1 + \theta_2 + \theta_3$ and α. The θ's (points M, N, R) are assigned in a reasonable manner (so that A and B in Fig. 4.7-1c are not too close in), and then the remaining two zeros of $1 + KL(s)$ at A and B in Fig. 4.7-1c are found, and it is noted whether the resulting $\lambda_1 + \lambda_2$ is reasonably close to the value that was guessed. If not, a second trial is made. A few trials should suffice. Only the final trial is here detailed.

We guess $\lambda_1 + \lambda_2 = 50°$. Then, from the two equations, we find $\sum\theta = 140°$

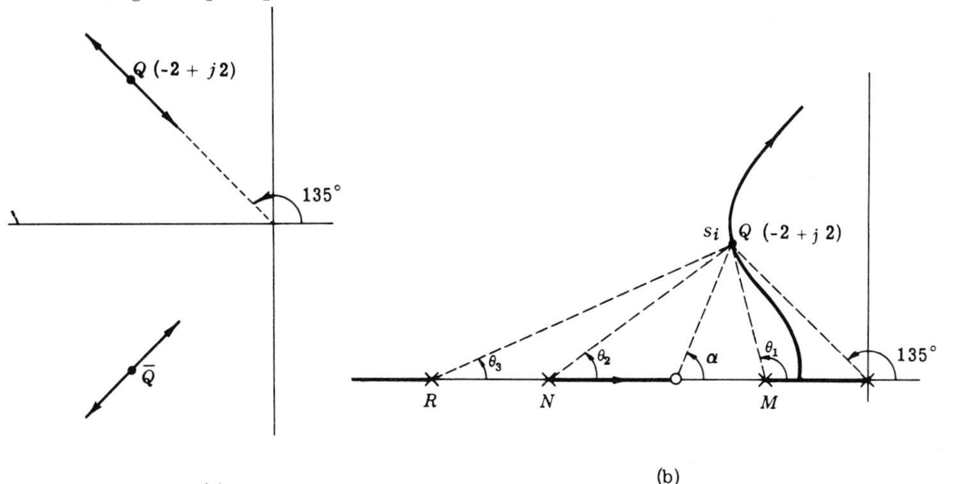

FIG. 4.7-1. (a) Desired motion of dominant poles due to variations in K. (b) Contemplated root loci of $1 + kL(s) = 0$. ×, poles of $L(s)$; ○, zeros of $KL(s)$.

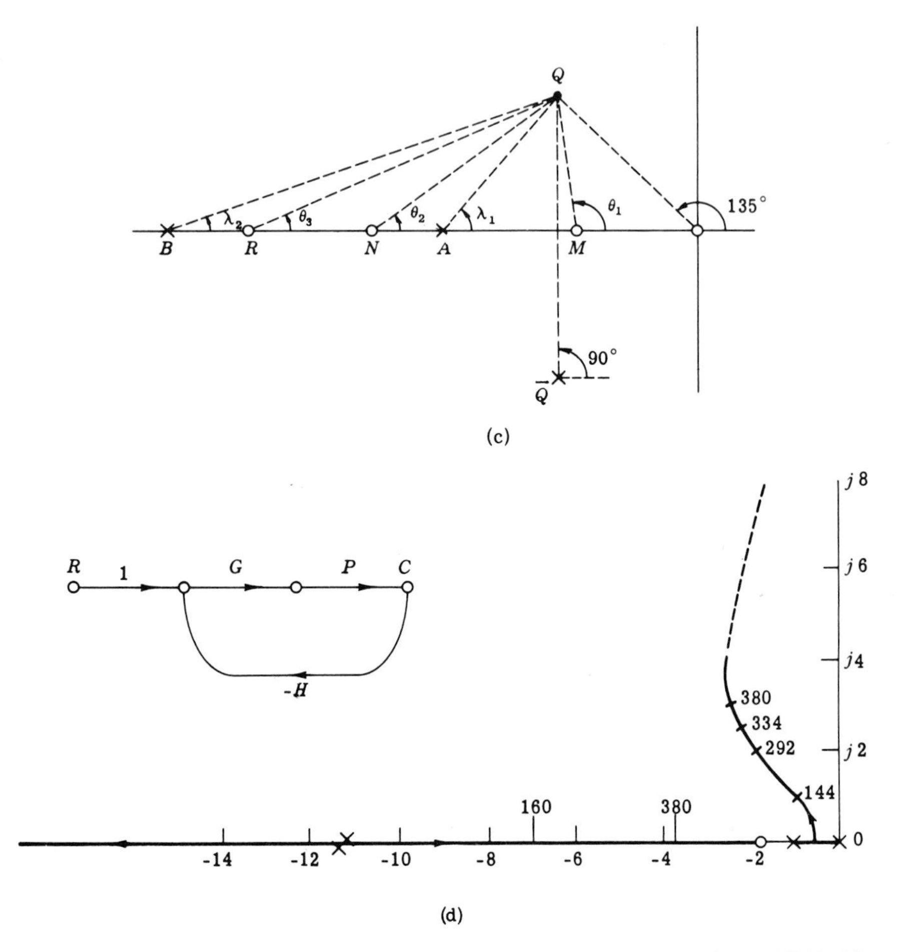

FIG. 4.7-1. (c) Calculation of arg S_k^{si}. ○, poles of $L(s)$; ×, zeros of $1 + KL(s)$. (d) Actual root loci of $1 + KL(s) = 0$.

and $\alpha = 95°$. This fixes the zero of $L(s)$ at -1.8. The poles of $L(s)$ are taken at -1, -11.4, -11.4. The two remaining zeros of $1 + L(s)$ are easily found to be at -4.2 and -15.6, after it is found that $K = 292$. The resulting $\lambda_1 + \lambda_2$ is about 51°, which checks with the assumed value. The actual value of $S_k^{s_i}$ is 0.26. The root loci are sketched in Fig. 4.7-1d and several values of K are marked. From $K = 144$ to 334, the damping factor is between 0.707 and 0.73. At $K = 381$, it decreases to 0.65, so that for a relative increase of K of 250%, the damping factor is sensibly constant. The center of this range is approximately at $-1.7(1 \pm j)$ with $K = 240$. Assuming this is satisfactory, the design can now be completed.

The configuration of Fig. 4.7-1d is assumed, with $GPH = L = 240(s + 1.8)$

$[s(s+1)(s+11.4)^2]^{-1}$ and $P = 4/[s(s+1)(s+6)]$. Therefore, $GH = 60(s+1.8)(s+6)/(s+11.4)^2$. However, $T = GP(1+L)^{-1} = G4(s+11.4)^2[(s+4.7)(s+6)(s^2+3.4s+5.8)(s+15.7)]^{-1}$. If we take $T = (5.8)(16)[(s^2+3.4s+5.8)(s+15.7)]^{-1}$, then $G = 23.2(s+4.7)(s+6)[(s+11.4)^2]^{-1}$ and consequently $H = 4.4(s+1.8)[1.8(s+4.4)]^{-1}$.

Actually, the pole of T at -15.7 will also shift with k, but it is sufficiently far off to have negligible effect on the system response. However, the zero of $1+L$ at -4.7, which is nominally canceled by the zero of G at -4.7, drifts also with k, so that actually there is a dipole in T, consisting of a zero at -4.7, and a pole which moves roughly between -3.8 and -7, as k varies from 380 to 160. The effect of this dipole too is not important in the significant frequency range of T.

In the above example, we did not know *a priori* whether the root sensitivity would be sufficiently small so as to keep the dominant system poles within the permitted range over the entire range of plant parameter variation. Our method, in fact, has been rather primitive. We shall consider this important problem in more detail in Sections 8.11-8.14.

§ 4.8 The Stability Problem from the Root Locus Point of View

The usefulness of the concepts of root loci in all facets of feedback design will become apparent in the development of the material. Extensive use of root loci techniques and concepts will be made throughout the book. In the meantime, we wish to concentrate in the balance of this chapter on the stability aspects of feedback, and demonstrate that in almost all practical feedback design systems there is a real problem in ensuring system stability. This is seen as follows. The system poles due to feedback are the roots of $1 - kt_{cs} = 0$, with $-kt_{cs} = L$, the loop transmission, for reference k. It has been demonstrated in Section 2.8 that the numerator of $1 - kt_{cs} = 1 + L$ is the same whatever reference is chosen.

It is important to note that L is a real physical transmission—it can be measured experimentally. Any real physical transmission must inevitably go to zero as $s = j\omega$ goes to infinity, because of the unavoidable inductances of wires, capacitances between leads and between the various parts of the circuit, inertias and elastances of all physical elements, etc. One or more of these parasitics usually predominate well before the so-called lumped circuit must in reality be considered as a transmission line. In most electronic circuits, it is the stray capacitances that are the dominant parasitics. In any case, L must therefore have more poles than zeros. In a feedback system consisting of x stages, each stage must contribute a minimum of one excess pole over zeros, so that at the very least L has x more poles than zeros. In almost all feedback circuits, $x > 2$. Let the difference between the number of poles and zeros be $e \geqslant x$. From Section 4.2, Rule 4, there are therefore e loci going off to infinity. If $e \geqslant 3$ and $k < 0$ (negative feedback), some of these loci must sooner or later enter and remain in the right half-plane (e.g., if $e = 3$, two loci go off to infinity at $\pm 60°$). Therefore if $e \geqslant 3$, there is always a value of k such that if it is exceeded, the system becomes unstable. For example, in Fig. 4.8-1, crossing of the $j\omega$ axis occurs at $\omega = \omega_o$,

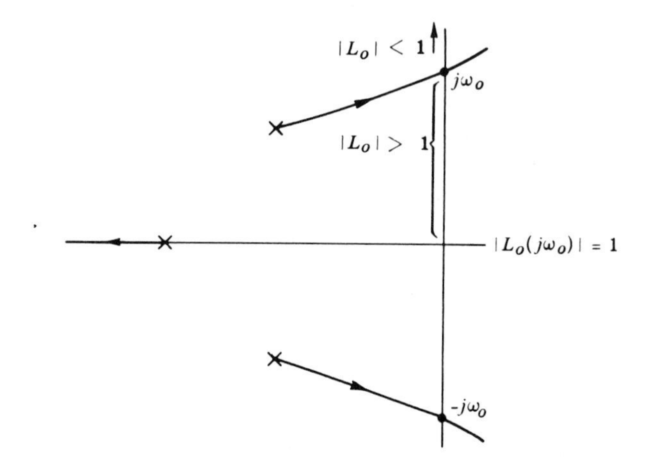

FIG. 4.8-1. Illustration of limited frequency range in which feedback benefits exist.

at which let $k = k_o$, i.e., $| L_o(j\omega_o) | = | k_o t_{cs}(j\omega_o) | = 1$ and for $\omega > \omega_o$ $| L_o(j\omega) | < 1$. Therefore, the benefits of feedback at best extend only for $\omega < \omega_o$.

There are two matters to consider here. Firstly, the amount of negative feedback may be insufficient for our purposes, i.e., the magnitude of kt_{cs} is too small in the range $\omega < \omega_o$. With the existing pattern of poles and zeros of t_{cs}, no improvement is possible, because $k < k_o$ in order that the system be stable. Therefore a rearrangement of the poles and zeros of t_{cs} is necessary. The second point is that ω_o may be too small, i.e., $| -kt_{cs} |$ is large over too small a frequency range. Again the only recourse is to change the pattern of poles and zeros of t_{cs} in order that $| F | = |1 - kt_{cs} |$ be as large as desired over the frequency range of interest.

The above discussion indicates one of the principal problems of feedback design. This is the choice of poles and zeros of t_{cs} so that k may be made sufficiently large to achieve the *amount* of loop transmission (and therefore return difference and consequent sensitivity reduction) over the *desired* frequency range. The problem exists because of the fact that in $t_{cs}(s)$ there is always an excess of poles over zeros, so that increase in k inevitably leads to an unstable system. The problem is further complicated when parameter variation is one of the principal reasons for applying feedback. For in this case, it must be ascertained that the variations in k and in t_{cs} do not suffice to cause instability, even though the system is stable at the nominal values of k and t_{cs}. Root locus techniques are clearly of great help in this design problem of selecting the zeros and poles of t_{cs} to achieve the design objective. They give the designer a physical picture of what is going on, and an idea of which way to turn, even though he may not use root locus methods in the final detailed calculations.

If $e = 2$ only (with k still < 0), the asymptotes to the loci at infinity are at $\pm 90°$, so that the right half-plane may be completely avoided by the root loci. However, the damping factor associated with this pair of complex poles, becomes very small, with consequent large transient overshoot. One usually demands that no system poles be in the shaded region determined by some $\theta < 90°$ (see Fig. 4.8-2); therefore, the design problem is basically the same even for $e = 2$. When $e = 1$ or 0 (and k still < 0), the stability problem does not exist. Unfortunately, $e > 1$ in probably all practical systems.

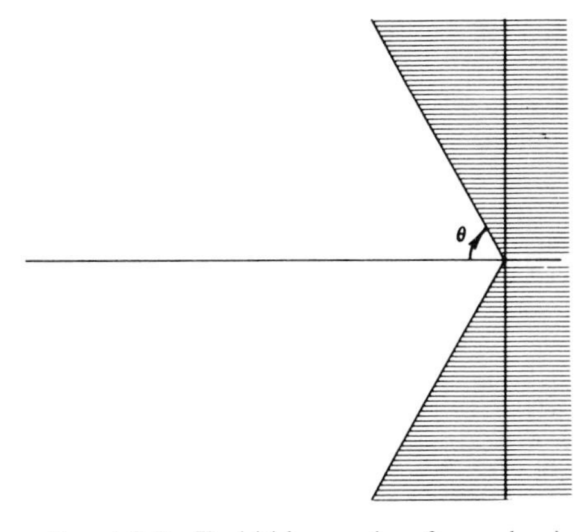

FIG. 4.8-2. Forbidden region for a dominant complex pole pair in order to avoid excessive overshoot.

§ 4.9 The Nyquist Stability Criterion

Another criterion for checking the stability of a system is the Nyquist[1] criterion, which was developed into a very useful and powerful tool well before root locus methods became popular. The Nyquist criterion is rather indirect as compared to root locus, and it does not furnish any record of the zeros of $1 - kt_{cs}$. However it, with its associated techniques have been developed to a very fine point for both analysis and synthesis. Furthermore, the Nyquist criterion may be applied to systems whose frequency response is available only graphically, and to systems with nonrational loop transmissions. The requirement that a feedback system should be stable imposes certain constraints on the loop transmission $L(s)$. It will be seen in Chapter 7 that these constraints drastically increase the price that must be paid for the benefits of feedback. The Nyquist criterion presents the constraints on $L(s)$ in a form which is very convenient for deducing these additional drastic "costs of feedback."

A rigorous derivation of the Nyquist criterion can be made with the aid of

[1] H. Nyquist, Regeneration theory. *Bell System Tech. J.* (January, 1932).

the calculus of residues.[1] Let $w = f(s)$, and consider the following integral around a closed contour, inside which the only singularities of w are poles:

$$\oint \frac{d \ln f(s)}{ds} ds = \oint \left[\frac{d \ln |f(s)|}{ds} + j \frac{d \arg f(s)}{ds}\right] ds = \oint \frac{f'(s)}{f(s)} ds. \quad (4.9,1)$$

The value of $|f(s)|$ is the same at the beginning as at the end of the closed contour, and therefore $\oint d \ln |f(s)| = 0$. The left-hand side of Eq. (4.9,1) is thus equal to $j\Delta \arg f(s)$. From Cauchy's residue theorem, the right-hand side of Eq. (4.9,1) is equal to the product of $2\pi j$ and the sum of the residues in the poles of $f'(s)/f(s)$ that are inside the contour. What are the poles of $f'(s)/f(s)$ and their residues? We may concentrate upon any particular pole of $f'(s)/f(s)$ and then sum over all such poles that are inside the contour. Any pole or zero of $f(s)$ is a pole of $f'(s)/f(s)$. This is seen by considering $f(s) = (s - s_o)^n g(s)$, with $n > 0$ or < 0. Then

$$\frac{f'(s)}{f(s)} = \frac{n(s - s_o)^{n-1}g(s) + (s - s_o)^n g'(s)}{(s - s_o)^n g(s)} = \frac{n}{s - s_o} + \frac{g'(s)}{g(s)}.$$

Thus, a zero of $f(s)$ of multiplicity n leads to a residue n in $f'(s)/f(s)$; while a pole of $f(s)$ of multiplicity n leads to a residue $-n$ in $f'(s)/f(s)$. We may sum over all such poles of $f'(s)/f(s)$, so that, finally

$$\Delta \arg f(s) = 2\pi(Z - P) \quad (4.9,2)$$

where Z is the total number of zeros of $f(s)$ inside the contour and P is the total number of poles inside the contour. The conclusion is: *Given $w = f(s)$, if s describes a closed path in a given sense in the s plane, and thereby encloses Z zeros and P poles of $f(s)$, the corresponding w contour in the w plane encircles the origin $Z - P$ times, in the same sense.*

§ 4.10 Application of the Nyquist Criterion to Open-Loop Stable Feedback Systems

The open-loop stable feedback system is defined as follows: Consider the fundamental feedback equation $T = t_{oi} + kt_{ci}t_{os}/(1 - kt_{cs})$. If it is possible to choose k such that all the four t_{ij} functions represent stable transmissions (left half-plane poles), then the feedback system is open-loop stable. A cascade of vacuum-tube amplifiers with over-all feedback represents such a system. The feedback control system with a total of one loop is also one, if each element in the loop has a stable transfer function. A transistor amplifier with more than one transistor is not necessarily in the same class. This section is devoted to open-

[1] See Section 7.2 for a review of the complex variable theory needed for this derivation.

loop stable systems. Open-loop unstable feedback systems will be studied in Section 7.14.

In open-loop stable systems, the system is unstable only if $1 - kt_{cs} = 1 + L$ has zeros in the right half-plane, because, by definition, all other poles of T are in the left half-plane. The Nyquist criterion may be applied to $1 + L$. To test the system for stability, it is necessary to ascertain whether there are any zeros of $1 + L$ in the right half-plane. The contour enclosing the right half-plane is the imaginary axis from $-jR$ to jR, and the semicircular arc, as R approaches infinity (see Fig. 4.10-1). To apply the theorem, it is necessary to evaluate L along this contour. As s approaches infinity, it is known from physical considerations (see discussion in Section 4.8) that L, a transmission function, must go to zero. Therefore, only values of L along the imaginary axis are required. The number of times the locus of $1 + L$ encircles the origin in a clock-wise sense (which is the same sense as the s contour in Fig. 4.10-1) is equal to the difference between the number of zeros and poles of $1 + L$ in the right half-plane. However, in open-loop stable systems, L has no right half-plane poles. Therefore, there can be only clockwise encirclements, if any, of the origin, and their number is equal to the number of zeros of $1 + L$ (poles of T), in the right half-plane.

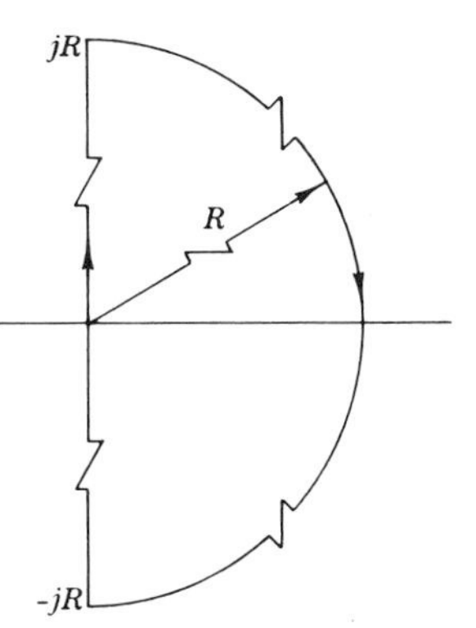

FIG. 4.10-1. Contour used in Nyquist criterion.

An encirclement by $1 + L$ of the origin is equivalent to an encirclement by L of -1 (or by $-L$ of $+1$). It is more convenient to work with L, and therefore the -1 point, in place of the origin, becomes the critical point.

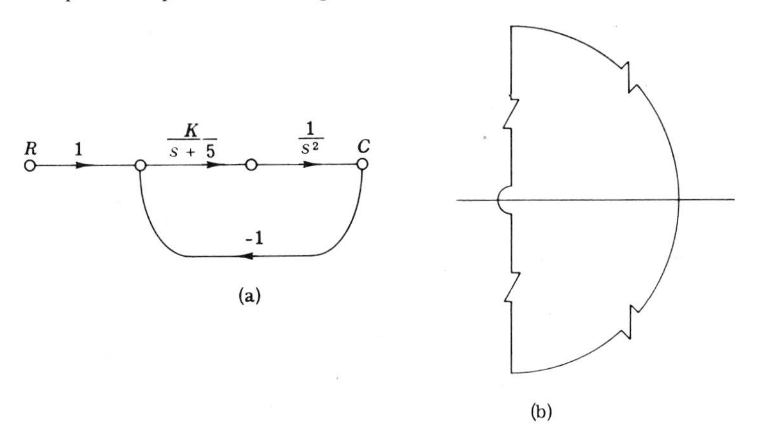

FIG. 4.10-2. Application of Nyquist criterion to a type 2 system. (a) Type 2 system. (b) Rhp boundary for which $Z - P = Z - 2$.

Example. A type 2 feedback control system (defined as one in which the loop transmission has a double pole at the origin) is shown in Fig. 4.10-2a. Here $L = K/s^2(s + 5)$ has a double pole at the origin, which is a point on the contour itself. We can indent the contour to include the origin inside the right half-plane, as in Fig. 4.10-2b and use [Eq. (4.9,2)] $Z - P = Z - 2 =$ number of clockwise encirclements of -1 point. Or we can indent the contour, as in Fig. 4.10-2c, and omit the origin from the region under examination. In the latter case, $Z - P = Z =$ number of clockwise encirclements of -1 point. The latter is usually done.

In order to sketch L as s moves along the contour, only one-half of the contour need be examined, because[1] $L(-j\omega)$ is the conjugate of $L(j\omega)$. Also only a few points usually suffice. The pole-zero pattern of L is shown in Fig. 4.10-2c. At A, the magnitude of L is large and its lag angle is less than 180°; the approximate value of L is given by A' in Fig. 4.10-2d. As s moves along the $j\omega$ axis,

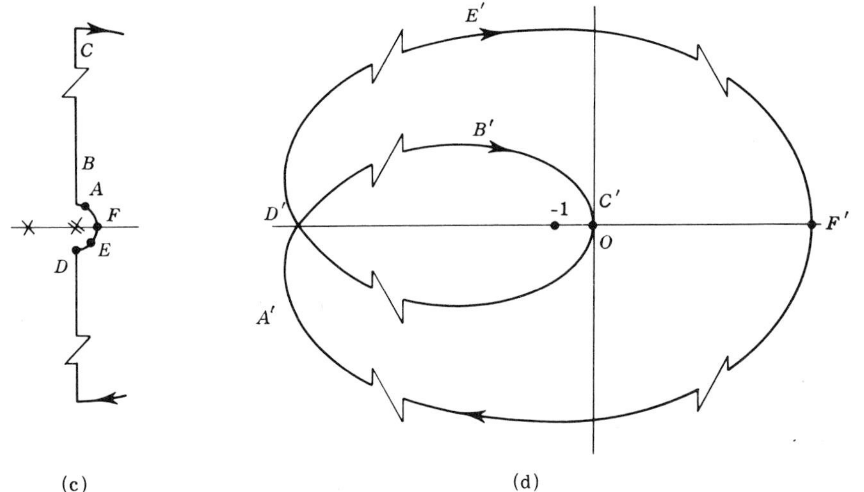

FIG. 4.10-2. (c) Rhp boundary for which $Z - P = Z$. (d) Locus of loop transmission of unstable system.

$|L|$ decreases, approaching zero as s approaches $j\infty$, while its angle approaches $-270°$. As s moves along the semicircular arc at infinity, L remains at zero. As s moves from $-j\infty$ up to E in Fig. 4.10-2c, L describes the mirror image of $A'B'C'$. At E, $s = \epsilon^2 \angle -60°$; as $\epsilon \to 0$, $L \to \infty \angle 120° = E'$. It is clear that the contour encircles the -1 point twice in a clockwise sense. The system is unstable for any finite K, no matter how small K may be. The reader may verify the above by the root locus method.

The Nyquist method is such as to suggest compensation methods for stabilizing a system like the above. (This is one of its outstanding attractive features.)

---

[1] See Chapter 7, Eq. (7.2,2).

In this example, the only portion of the contour we can possibly change (if the double pole at the origin must remain) is the region $OB'A'$ in Fig. 4.10-2d. To prevent -1 encirclement, it must be distorted at least as in Fig. 4.10-2e.

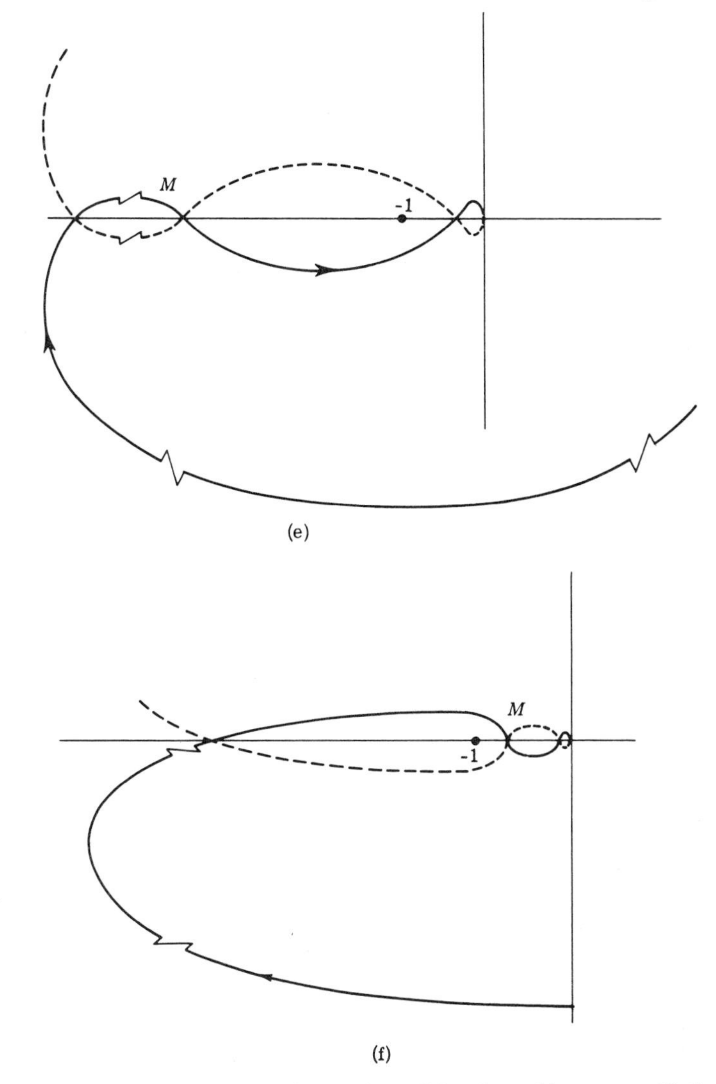

FIG. 4.10-2. (e) Loop transmission locus of conditionally stable system. (f) System is unstable if gain factor decreases sufficiently.

In the latter, there is no encirclement of the -1 point. To perform this compensation, a network is needed with positive phase angle (phase lead) over some finite frequency range. (It is important to note that attention must be paid to

$|L|$ as well, in order to ensure stability.) There arises the question to what extent $|L|$ and arg L can be separately controlled over the same or different frequency ranges. This is a problem of fundamental importance in feedback and network theory, and will be studied in detail in Chapter 7.

For the present, assuming Fig. 4.10-2e has been achieved, it may be noted that if the magnitude only of L decreases sufficiently, the locus in Fig. 4.10-2e contracts into that shown in Fig. 4.10-2f, and the system becomes unstable. Such systems, wherein decrease of gain constant causes instability, are called *conditionally stable systems*. Special precautions must therefore be taken if such systems are contemplated. Conditionally stable feedback systems are very prone to instability when the system is temporarily saturated because the saturation acts like a decrease in gain. It will be seen in Section 7.10 that conditionally stable systems occur when a large loop gain magnitude must be decreased in a small frequency range. A way of avoiding conditional stability by means of parallel plants is presented in Chapter 8.

Stability Margins—Gain and Phase Margins

An open-loop stable system, whose locus of L over the boundary enclosing the right half-plane does not enclose -1, is stable. However, it is not necessarily a useful practical system, if the locus passes very close to the -1 point. In the first place, a slight change of gain or time constant may sufficiently shift the locus so as to lead to an unstable system. In the second place, the closed-loop system response has $1 + L$ for its denominator. At those frequencies for which L is close to -1, $1 + L$ is close to zero, leading to large peaking in the system frequency response. In terms of poles and zeros this is equivalent to one or more lightly damped pole pairs (θ close to 90° in Fig. 4.8-2), with consequent large overshoots in the transient response of the system. For both the above reasons, margins in gain and phase are defined.

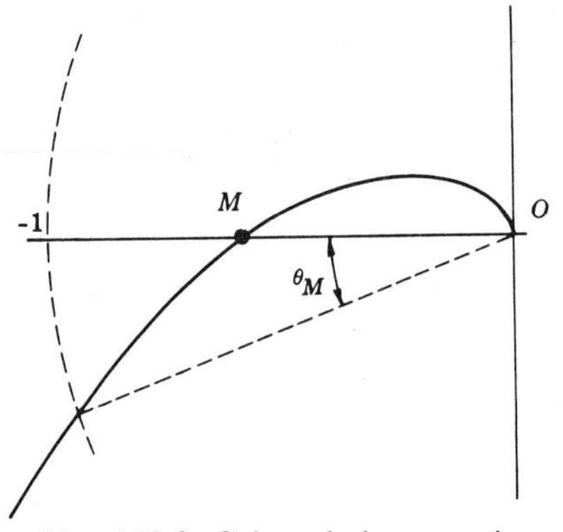

FIG. 4.10-3. Gain and phase margins.

The gain margin is usually specified in decibels (20 log[1] gain magnitude) and is the amount the gain constant must be increased to cause instability. In Fig. 4.10-3, it is 20 log $1/|OM|$. The phase margin is the amount of additional phase lag (with no change in gain constant) required to cause instability. It is the angle θ_M in Fig. 4.10-3.

[1] In this book log is used for the logarithm to the base 10 and ln is used for the logarithm to the base e.

Bode Plots

The application of the Nyquist criterion to a practical system apparently requires a polar sketch of $L(j\omega) \triangleq -kt_{cs}(j\omega)$. The Bode plot is a much more convenient way for doing the same thing. The Bode plot displays the information by means of two separate plots, one of $|L(j\omega)|$ in the logarithmic scale of decibels (db) vs. log ω [defined in Eq. (4.10,1a) below], and the other, of arg $L(j\omega)$ vs. log ω. Although two sketches are required in place of one, each of the two (especially the magnitude sketch) is obtained so easily as to make the Bode plot almost always considerably simpler to prepare than the polar plot of $L(j\omega)$.

In order to obtain the Bode plot, $L(j\omega)$ is first factored, i.e., we write $L(j\omega) = K\prod(s + z_i)/\prod(s + p_j)$. We thereby assume $L(s)$ is a rational function in s. Actually, there is no loss in generality in this assumption, because (as noted in Section 12.6) any practical $L(s)$ can be approximated as accurately as desired by a rational $L(s)$. However, it will be obvious that nonrational $L(s)$ can also be conveniently displayed on a Bode plot; it is merely easier to present the Bode plot by applying it to rational $L(s)$. The magnitude of $L(j\omega)$ in decibels and its phase are, respectively,

$$20 \log |L(j\omega)| = 20 \log |K| + \Sigma 20 \log |j\omega + z_i| - \Sigma 20 \log |j\omega + p_j| \tag{4.10,1a}$$

$$\arg L(j\omega) = \arg K + \Sigma \arctan \omega/z_i - \Sigma \arctan \omega/p_j. \tag{4.10,1b}$$

One of the principal advantages in using the logarithmic measure of $L(j\omega)$ is that the contributions of the individual factors in $L(j\omega)$ add. The reason for using the logarithmic scale for the abscissa is that the contribution of each factor assumes a *standard form* which is furthermore remarkably simple. This is seen as follows.

Consider the typical term, $(s + z_i)$. There are two possibilities: z_i real and z_i complex. The real case is the easiest and is first considered. The amplitude contribution in Eq. (4.10,1a) may be written as $10 \log(\omega^2 + z_i^2)$, which has the asymptotic values $20 \log z_i$ for $\omega^2 \ll z_i^2$, and $20 \log \omega$ for $\omega^2 \gg z_i^2$. Thus, the low frequency asymptote (Fig. 4.10-4a) is a constant $M = 20 \log z_i$, and the high frequency asymptote is a line of slope $+20$ db for a change of unity in log ω, which corresponds to a frequency decade (written as 20 db per decade). Its equivalent in octaves is very closely 6 db per octave. The two asymptotes are equal at $\omega = z_i$ and must therefore meet there.

In order to sketch $10 \log (\omega^2 + z_i^2)$, it is convenient to use the asymptotic sketch as a first approximation, and then add the necessary correction to the asymptotic sketch. The correction as a function of ω/z_i is shown in Fig. 4.10-4b. Similarly, the factor $(s + p_j)$ in the denominator of $L(s)$ has a low frequency magnitude asymptote of $-20 \log p_j$ decibels, and a high frequency magnitude asymptote of $-20 \log \omega$. The slope of the latter is -20 db per decade $\doteq -6$ db per octave. Figure 4.10-4b gives the negative values of the corrections that must be applied to the approximate asymptotic values, in order to obtain the exact $-20 \log | j\omega + p_j |$. It is important to note that the form of the sketch is the same whatever the value of z_i or p_j, providing that they are real numbers. The value of z_i or p_j only dictates the point at which the slope abruptly changes. It is also important to note that when $L(s)$ is a rational function, then the asymptotic sketch of $20 \log | L(j\omega) |$ must consist of straight lines

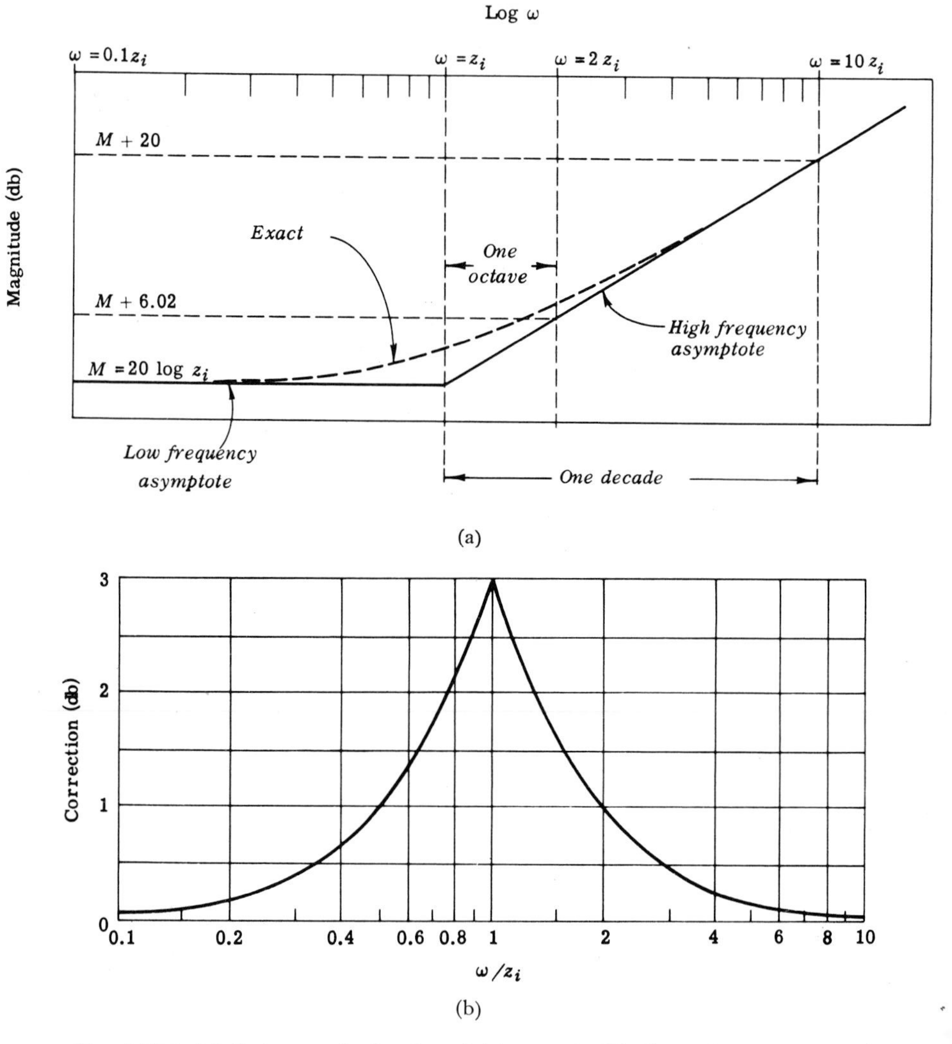

FIG. 4.10-4. (a) Bode magnitude plot of $| j\omega + z_i |$. (b) Correction to asymptotic Bode plot of $| j\omega + z_i |$.

whose slopes are integral multiples of 20 db per decade (i.e., 0, $\pm$ 20, $\pm$ 40 db per decade, etc.). This useful property follows because the terms add in Eq. (4.10,1a), and is the direct result of using logarithmic scales for both the ordinate and abscissa.

With the aid of the above, the magnitude in decibels vs. log ω of any $L(j\omega)$ whose poles and zeros are on the real axis is easily obtained. The steps in the procedure are described by means of an example.

Example. $L(s) = 1000(s + 2)(s + 4)/(s + 0.2)(s + 1)(s + 10)(s + 20)$. The asymptotic sketch is first made. Such a sketch consists of straight lines whose slopes are integral multiples of 20 db per decade. The two asymptotes of any one factor meet at a frequency numerically equal to the magnitude of the real pole or zero. Therefore, at the latter points there is a change of slope by 20 db per decade (for a zero) or $-$ 20 db per decade (for a pole). These critical frequencies at which there is a change of slope are called break or corner frequencies. In this example, there are lead break (or corner) frequencies at 2, and 4 rps; lag corner frequencies at 0.2, 1, 10, and 20 rps.

(1) The first step is to mark the corner frequencies as shown at the top of Fig. 4.10-5. Because the first corner frequency is at 0.2 rps, $|L(j\omega)|_{\text{asy}}$ has zero slope from $\omega = 0$ until $\omega = 0.2$. At the lag corner frequency at $\omega = 0.2$, the slope of $|L(j\omega)|_{\text{asy}}$ abruptly decreases by 20 db per decade. Hence the slope of $|L(j\omega)|_{\text{asy}}$ is $-$ 20 db per decade ($-$ 6 db per octave) for $\omega > 0.2$, up to $\omega = 1$. By the same reasoning, the slope of $|L(j\omega)|_{\text{asy}}$ is $-$ 40 db per decade ($-$ 12 db per octave) for $\omega > 1$ up to $\omega = 2$. The lead corner frequency at $\omega = 2$ causes the slope of $|L(j\omega)|_{\text{asy}}$ to increase abruptly by 20 db per decade at $\omega = 2$; therefore, the slope is $-$ 20 db per decade for $\omega > 2$ up to $\omega = 4$. Similarly, the slope of $|L(j\omega)|_{\text{asy}}$ is zero for $4 < \omega < 10$; $-$ 20 db per decade for $10 < \omega < 20$, $-$ 40 db per decade for $\omega > 20$. As a check, it is noted that $L(s)$ has two more poles than zeros; hence $\lim_{s \to \infty} L(s) = 1000/s^2$, with a slope $-$ 2(20) db per decade at very large frequencies.

(2) It is thus seen that the simple marking of corner frequencies in Fig. 4.10-5 suffices to fix the slope of $|L(j\omega)|_{\text{asy}}$ (in decibels vs. log ω) at all ω. However, this is not sufficient to determine $|L(j\omega)|_{\text{asy}}$, because any constant value can be added to $|L(j\omega)|$ without affecting its slope. The value of $|L(j\omega)|_{\text{asy}}$ at any one value of ω is all that is now needed to completely specify $|L(j\omega)|_{\text{asy}}$. It is noted that along an asymptote $|j\omega + a|$ has the value a for $\omega < a$, and the value ω for all $\omega > a$. Hence, if we take $\omega = 1$, then $|L(j)|_{\text{asy}} = 1000(2)(4)[(1)(1)(10)(20)]^{-1} = 40$, and $20 \log 40 = 32$ db. The sketch of $|L(j\omega)|_{\text{asy}}$ may now easily be made and is shown in Fig. 4.10-5.

For many purposes the asymptotic sketch of $L(j\omega)$ suffices for most of ω (the range of ω for which the accurate sketch is needed will become apparent in Chapters 5 and 6). In order to obtain the exact plot of $|L(j\omega)|$, the correction

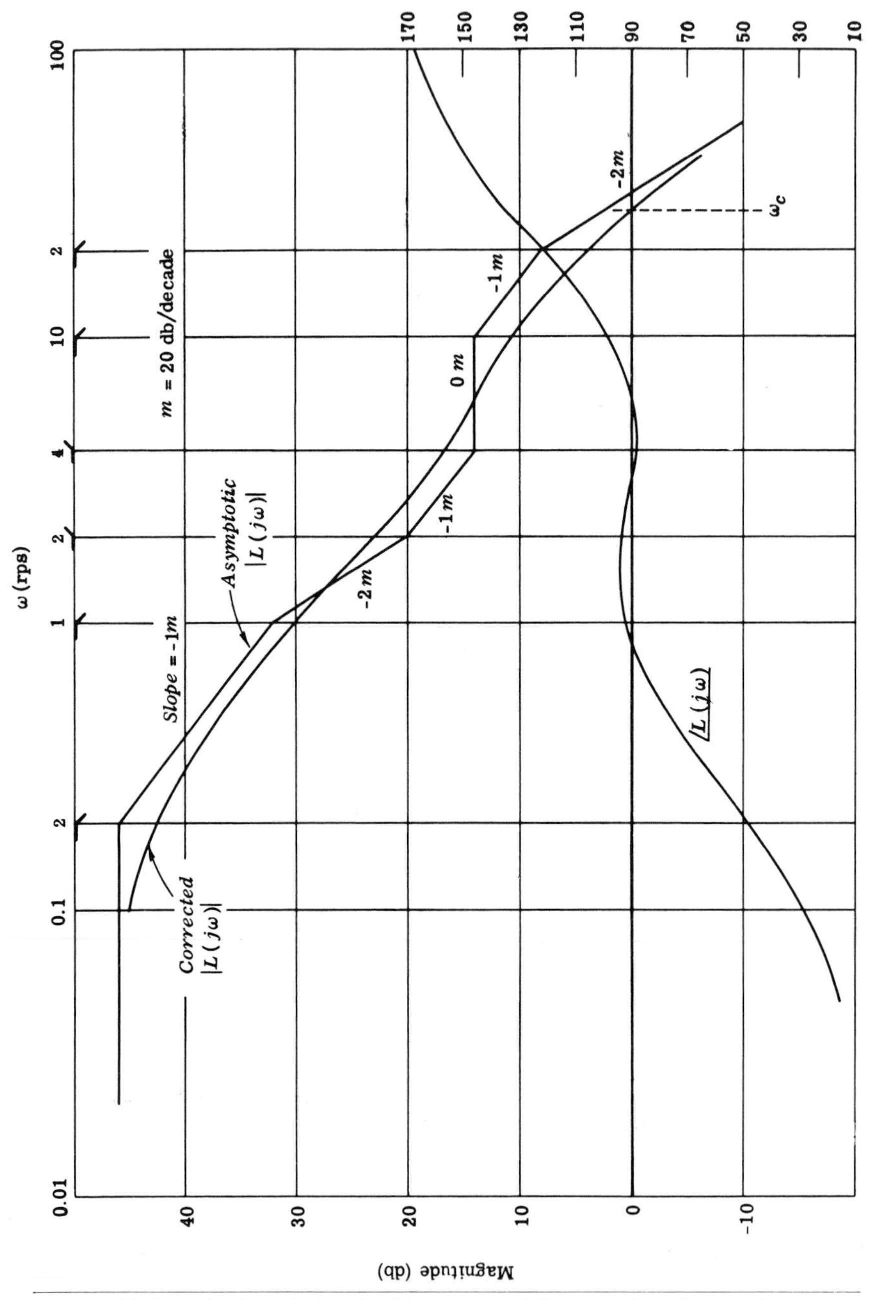

FIG. 4.10-5. Bode plot of $L(s) = 1000(s + 2)(s + 4)/(s + 0.2)(s + 1)(s + 10)(s + 20)$.

of Fig. 4.10-4b must be added for each nonzero finite pole and zero in $L(s)$. The most convenient procedure is to add the corrections at one or more discrete frequencies in the region where accuracy is desired, and not to bother at the other frequency regions. For example, if $\omega = 0.2$ is chosen, then the net correction at $\omega = 0.2$ is calculated as follows: From Fig. 4.10-4b the correction due to the lag corner frequency at $\omega = 0.2$ is -3 db; it is -0.18 due to the lag corner frequency at $\omega = 1$ (see $\omega/z_i = 0.2/1 = 0.2$ in Fig. 4.10-4b) and the contributions of the other factors are even less, so we take the correction as -3.2 db at $\omega = 0.2$. The correction at $\omega = 2$ is 3 db due to the lead corner at 2 rps, -1 db due to the lag corner at 1 rps, 1 db due to the lead at 4 rps, and ignorable for the others, with a resulting net correction of 3 db. The resulting corrected $L(j\omega)$ is drawn in Fig. 4.10-5.

The curve of arg $L(j\omega)$ vs. log ω is obtained by considering a typical term in Eq. (4.10,1b), e.g., arctan ω/z_i, which is 45° at $\omega = z_i$, and closely equal to $57.3(\omega/z_i)$ for $\omega/z_i < \frac{1}{2}$. At $\omega/z_i = \frac{1}{2}$, the approximate value is $\frac{1}{2} \times 57.3 = 28.6°$, while the correct value is 26.6°; at $\omega/z_i = \frac{1}{3}$, the approximate value is $\frac{1}{3} \times 57.3 = 19.1°$, while the correct value is 18.4°; at $\omega/z_i = \frac{1}{4}$, they are, respectively, 14.3° and 14.1°. For $\omega/z_i > 1$, the relation $\tan(90° - \theta) = \cot\theta$ is used. Thus, in the previous example, at $\omega = 1$ the phase angle is: $-\arctan 5 = -[90 - 0.2(57.3)] = -78.6°$, due to the lag corner frequency (c.f.) at 0.2 rps, $-45°$ due to the lag c.f. at 1 rps, 26.6° due to the lead c.f. at 2 rps, 14.0° due to the lead c.f. at 4 rps, $-0.1(57.3) = -5.7°$ due to the lag c.f. at 10 rps, and $-0.05(57.3) = -2.9°$ due to the one at 20 rps, with a net total of $-91.6°$. In this manner a few properly spaced points are evaluated with the result shown in Fig. 4.10-5.

The gain margin is read from the Bode plot by noting the frequency, say ω_x, at which the phase is 180°. The magnitude in decibels at $\omega = \omega_x$ must be less than zero, for the system to be stable. If the magnitude in decibels at $\omega = \omega_x$ is $-A_x$, then the gain margin is A_x db. In the above example the gain margin is infinite. (For conditionally stable systems there are at least two frequencies at which the phase is 180°, and there are two gain margins—see Section 7.10.) The phase margin is read from the Bode plot by noting the frequency (hereafter denoted as the crossover frequency, ω_c) at which the amplitude is 0 db. The phase lag must be less than 180° for a stable feedback system. If θ_c is the phase lag at ω_c, the phase margin is $180° - \theta_c$. In the above example (Fig. 4.10-5), $\theta_c = 134°$, with 46° phase margin.

Complex Pole or Zero Pairs

A complex pole pair factor $1/(s^2 + 2\zeta\omega_n s + \omega_n^2)$ has the horizontal low frequency asymptote $-20 \log \omega_n^2$ and a high frequency asymptote $-20 \log \omega^2$, whose slope is -40 db per decade. The two asymptotes meet at $\omega = \omega_n$.

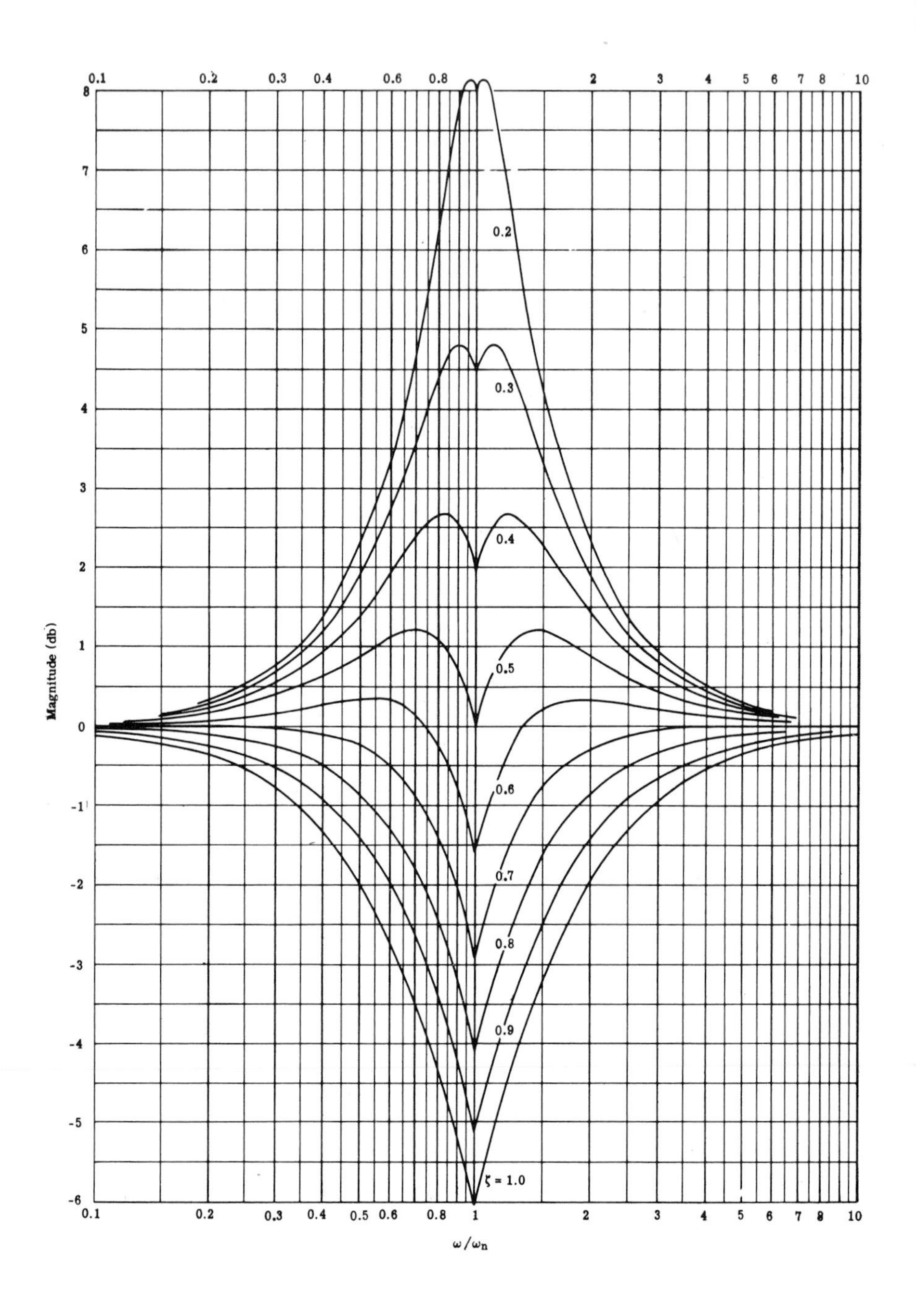

FIG. 4.10-6. (a) Correction curves to Bode asymptote of $1/(s^2 + 2\zeta\omega_n s + \omega_n^2)$.

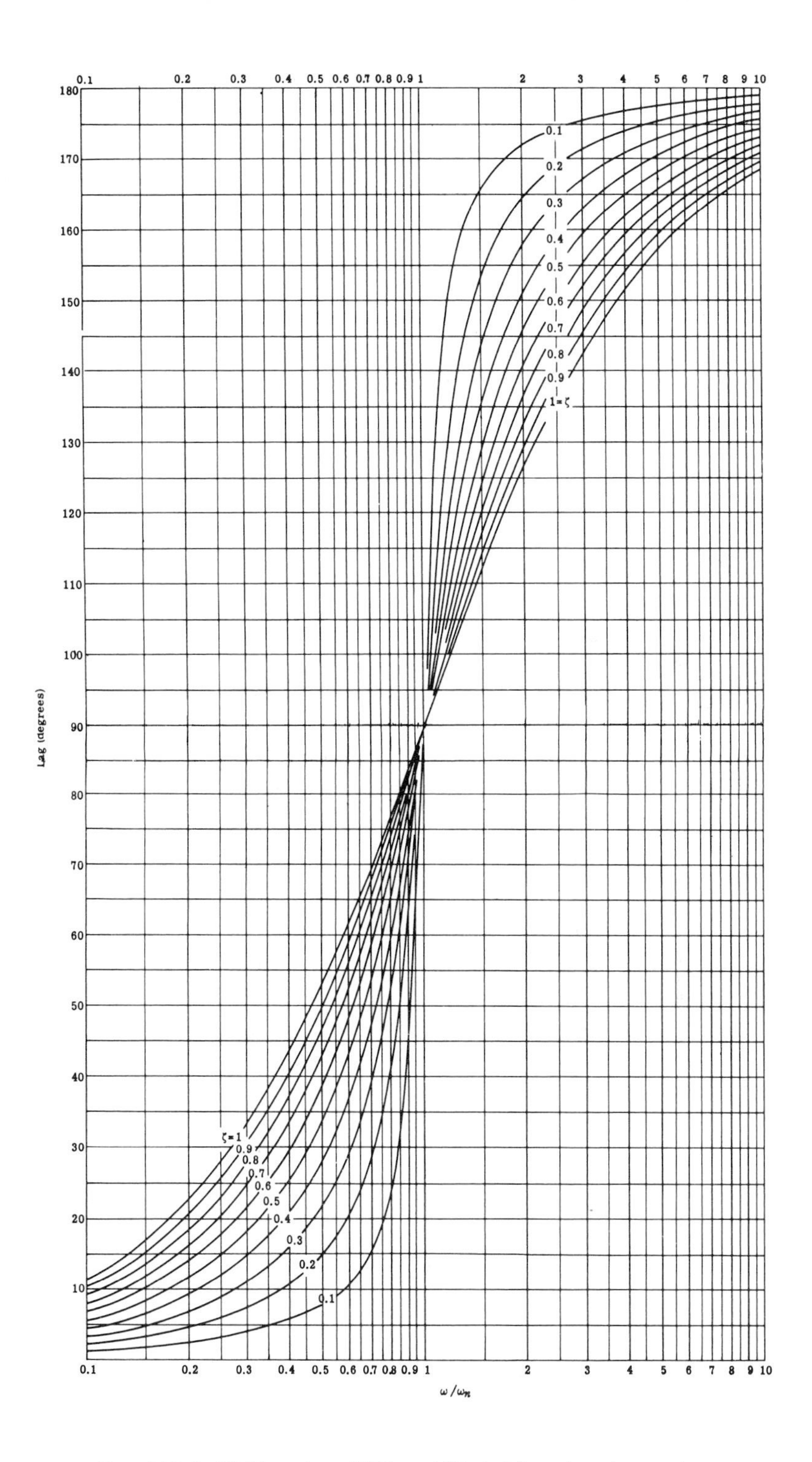

FIG. 4.10-6. (b) Phase lag of $L(s) = 1/(s^2 + 2\zeta\omega_n s + \omega_n^2)$, $s = j\omega$.

Hence $| L(j\omega) |_{\text{asy}}$ still consists of segments of straight lines whose slopes are integral multiples of 20 db per decade. The difference between the actual amplitude and the asymptotic value is now a function of ζ as well as of ω/ω_n. A family of correction curves for different values of ζ is presented in Fig. 4.10-6a. For example, if $\zeta = 0.5$, $\omega_n = 15$, then the correction at $\omega = 45$ is obtained by entering Fig. 4.10-6a at $\omega/\omega_n = 3$, $\zeta = 0.5$ and is 0.45 db. If there is a zero pair given by the factor $(s^2 + 2\zeta\omega_n s + \omega_n^2)$ in the numerator of $L(s)$, then the negative of the values in Figs. 4.10-6a and b must be used. Figure 4.10-6b presents the phase lag due to the complex pole pair at $-\zeta\omega_n \pm j\omega_n(1 - \zeta^2)^{0.5}$, or the phase lead due to a similar complex zero pair.

§ 4.11 Positive Feedback

We have seen that the benefits of feedback are determined by the sensitivity function $(1 - kt_{cs})^{-1} = (1 + L)^{-1}$, so that $| L(j\omega) |$ must be made large in the frequency range for which small sensitivity is wanted. Also, we have been assuming that negative feedback (i.e., $kt_{cs}(0) < 0$) is used for this purpose. Presumably, positive feedback could also be used, so long as $| L(j\omega) |$ were sufficiently large over the frequency range of interest. It is the stability problem that precludes the use of positive feedback. This is seen as follows.

As previously noted, any practical $L(s)$ has more poles than zeros. It is only necessary that this excess of poles over zeros be greater than or equal to one ($e \geqslant 1$) in order to make positive feedback impractical. With $e \geqslant 1$, an $L(j\omega)$ whose magnitude is large at $\omega = 0$ and whose $L(0) < 0$ (i.e., positive feedback) has a polar locus (for positive ω) of the form labeled A in Fig. 4.11-1a. When its mirror image about the real axis (for negative ω) is taken into account, it is easily seen that there is a clockwise encirclement of -1, so that the system is unstable.

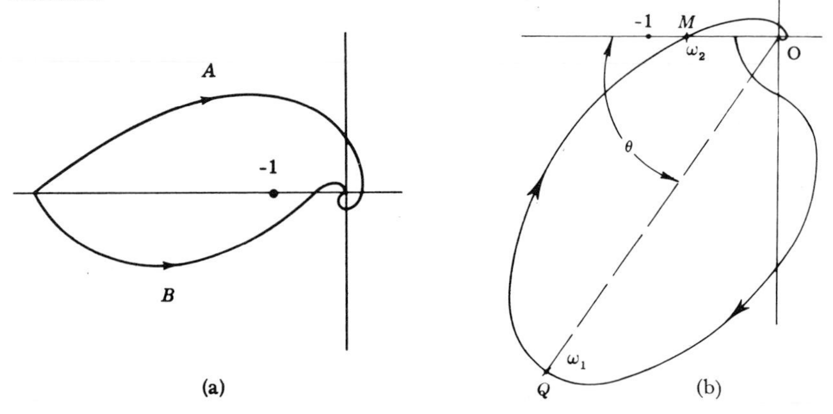

FIG. 4.11-1. Impracticality of positive feedback.

The obvious reaction to this argument is to consider a modified locus, such as the one labeled B in Fig. 4.11-1a. When the latter's mirror image is included, the result is a counter-clockwise encirclement of -1; i.e., in Eq. (4.9,2), $Z - P = -1$, which is impossible, unless $L(s)$ is open-loop unstable with at least one right half-plane pole. Open-loop unstable systems are deferred to Section 7.14. The conclusion is that locus B is impossible in open-loop stable systems. In general, it is impossible to draw a locus which starts at σ_1 (with $\sigma_1 < -1$), and which terminates at the origin without encircling the -1 point at least once (assuming the mirror image is taken into account). Therefore it is impossible to achieve a positive feedback design which is open-loop stable, and which has $|L(0)| > 1$.

In view of the above, any open-loop stable positive feedback design must have $|L(j0)| < 1$, if the system is to be stable. But it would seem that a design such as that shown in Fig. 4.11-1b is feasible, i.e., $|L(j\omega)|$ begins small, but increases and then eventually goes to zero. (It will be shown in Chapter 7 that such a magnitude characteristic must be associated with a phase angle that first increases and then decreases, as in Fig. 4.11-1b.) The difficulty with such a design is that as $OQ = |L(j\omega)|_{\max}$ is increased, OM (see Fig. 4.11-1b) increases and tends to be larger than one, leading to instability. In fact, it is difficult to achieve a design with reasonable stability margins, in which $OQ > 1$ simultaneously with $OM < 1$. For example, consider $L(j\omega) = -As/(s + a)^2$. It is easily found that $(A)_{\max} = 2a$ for a stable design with zero stability margin, and then $|L(j\omega)|_{\max} = 1$ at $\omega = a$. It will be shown in Section 7.7 that the design must be very sophisticated and very wasteful in bandwidth in order to secure a comparatively small frequency range in which $|L(j\omega)| > 1$. It is therefore impractical to use positive feedback for sensitivity reduction.

§ 4.12 Loop Shaping for Stability, with Parameter Variations—Single-Order Systems[1]

Previous sections have shown how the stability problem may be regarded as one of shaping the system loop transmission so that it avoids critical points in the s plane. The problem of ensuring system stability becomes much more difficult when account must be taken of parameter variation, because the loop transmission function changes as its parameters vary. The great desideratum is to be able to work with a fixed function (in the sense that this function is

[1] Techniques similar in some respects to that described in this section have been presented by: R. M. Stewart, A simple graphical method for constructing families of Nyquist diagrams. *J. Aeronaut. Sci.* **18**, 767-768 (1951); F. H. Blecker, Transistor multiple loop feedback amplifiers. *Proc. Natl. Electronics Conf., Chicago, Illinois* **13**, 19-34 (1957); B. R. Myers, A useful extension of the Nyquist criterion to stability analysis of multi-loop feedback amplifiers. *Proc. 4th Midwest Symposium on Circuit Theory, Marquette University, Milwaukee, Wisconsin, December, 1959*, pp. 1-17.

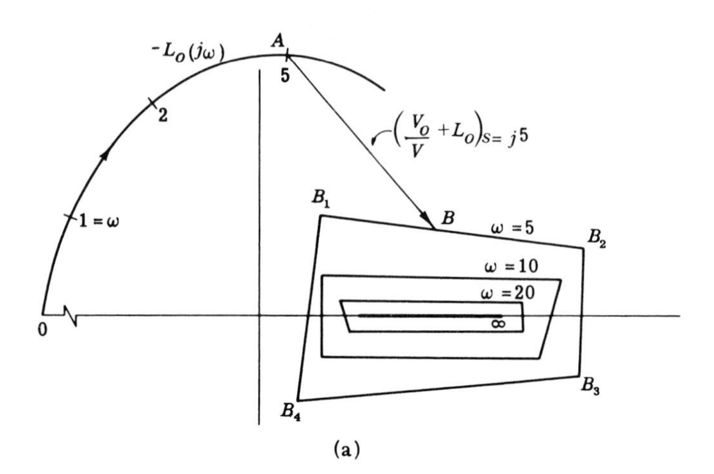

(a)

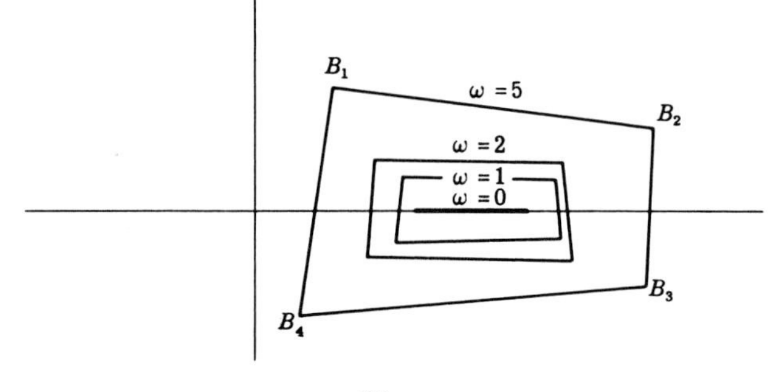

(b)

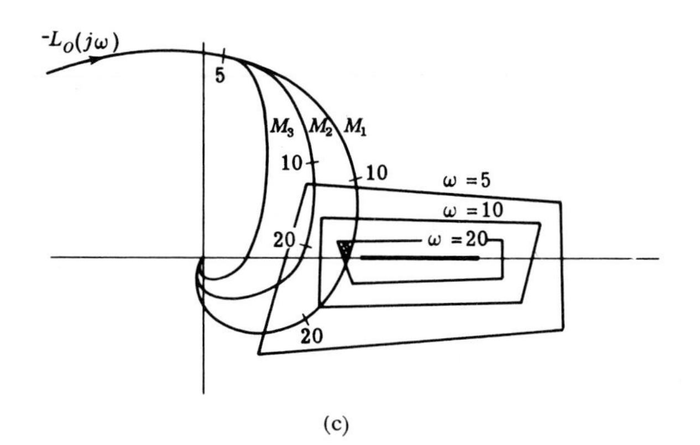

(c)

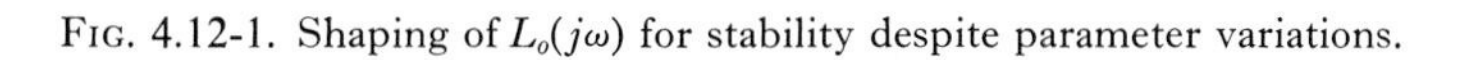
FIG. 4.12-1. Shaping of $L_o(j\omega)$ for stability despite parameter variations.

independent of the parameters which vary). The problem is to make sure that no zeros of $1 - kt_{cs} \triangleq 1 + L$, lie in the right half-plane. The problem is complicated by the fact that L contains within it parts that vary. The simplest type of problem is when it is possible to write $L = FV$, where F is a transfer function with fixed parameters, and V is the transfer function containing all the parameters which vary (slowly enough so that linear time invariant stability theory is applicable). Such systems will be called single order systems. Let V_o be the nominal value of V; it is obtained by picking any set of values of the varying parameters in their given range of variation. The zeros of $1 + L$ are the zeros of $1 + FV = 1 + FV_o(V/V_o)$. The zeros of the latter are the zeros of $(V_o/V) + FV_o = V_o/V + L_o$, where $L_o = FV_o$ is the nominal loop transmission with *fixed parameters.*

The stability problem is now one of shaping the function with nonvarying parameters L_o, such that despite the parameter variation in V_o/V, the zeros of $V_o/V + L_o$ are all in the left half-plane. Let us consider a typical problem. Suppose the loci of V_o/V are as shown in Figs. 4.12-1a and b. For example, at $\omega = 5$, V_o/V may lie anywhere inside or on the boundary marked $B_1BB_2B_3B_4$. Furthermore, it is assumed here that the boundary at $\omega = 5$ represents the extreme boundary, i.e., all other boundaries for different values of ω lie inside it. Generally in feedback problems $L_o(s)$ is assigned a minimum value over some frequency range (in order to obtain the benefits of feedback over this range), and the problem is how to shape it outside this frequency range so as to ensure a stable system. L_o in practice represents a physical transfer function (the nominal loop transmission) with its own parasitics. We shall see that it is desirable that L_o decrease in magnitude as fast as possible with increasing frequency. Suppose L_o is determined by various considerations up to approximately 5 rps, as shown in Figs. 4.12-1a and c (where $-L_o$ is sketched). We must shape L_o for $\omega > 5$ (although in the process L_o for $\omega < 5$ may be slightly modified). In Fig. 4.12-1a the vector **AB** (where B is allowed to range over the entire region whose boundary is $B_1B_2B_3B_4$) represents $[V_o/V + L_o(s)]_{s=j5}$. How should $-L_o(j\omega)$ be continued for $\omega > 5$, to ensure system stability?

Some possibilities are shown in Fig. 4.12-1c. Clearly M_1 represents a system which is unstable for a region of V parameter values. (The student should verify this by making a rough sketch of $V_o/V + L_o$ and noting that its locus encircles the origin for some range of values of V_o/V.) However, M_2, M_3 are stable over the entire range of V variation. In case of doubt it is easy to obtain, from Fig. 4.12-1c, Nyquist sketches of $V_o/V + L_o$. It is seen that for M_1 in Fig. 4.12-1c, the locus of $V_o/V + L_o$ encircles the origin, for at least all parameter combinations of V which cause $V_o/V\ (s = j20)$ to lie inside the shaded area. However, with M_2 or M_3 it is impossible to find a locus of $V_o/V + L_o$ which encircles the origin.

Assuming then that a practical method exists for finding the boundary of V_o/V at a fixed $s = j\omega$, the above procedure may be used to shape the nominal loop transmission L_o to ensure system stability. We next turn our attention to the problem of locating the boundary of V_o/V.

Loci of V_o/V When Gain Factor and Poles Vary

Consider V of the following simple form: $V = k/(s + p)$ in which k and p vary; so that $V_o/V = k_o(s + p)/k(s + p_o)$. At any fixed $s = j\omega$ we want to know the area of variation of V_o/V, as k and p assume all their permitted values. The effect of k is best left to the end. Therefore in the meantime consider only $V_{o1}/V_1 = (p + j\omega)/(p_o + j\omega)$. For a fixed ω, V_{o1}/V_1 describes a straight line as p varies. Suppose $\omega = 1$, $p_o = 1$, and p may assume any values between 1 and 5. Then the two end points of the straight line locus of V_{o1}/V_1 are easily found to be 1 and $3 - j2$ (the line AB in Fig. 4.12-2). Suppose k/k_o can vary from 1 to

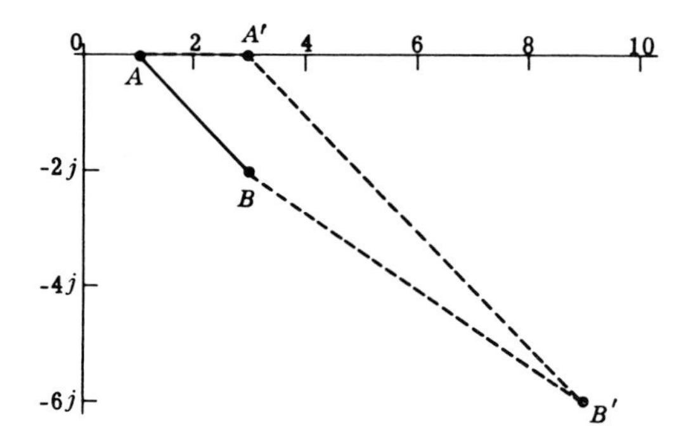

FIG. 4.12-2. Range of $V_o/V = k_o(s + p)/k(s + p_o)$, at fixed $s = j\omega$ due to variations in k and p.

1/3, then the complete region of variation of V_o/V at $s = j$ is the quadrilateral $AA'B'B$ shown in Fig. 4.12-2.

Next consider V of the form $V = k/(s + p_1)(s + p_2)$, and leaving k to the end, $V_{o1}/V_1 = (s + p_1)(s + p_2)/(s + p_{1o})(s + p_{2o})$. What is the region covered by V_{o1}/V_1 as p_1, p_2 assume all possible permitted values? The region of variation of p_1, p_2 covers some area in the two-dimensional space whose axes are p_1, p_2 in Fig. 4.12-3a. To simplify the exposition, numerical values are assigned in Fig. 4.12-3a, and we take $p_{1o} = p_{2o} = 1$. Along HG in Fig. 4.12-3a, $V_1/V_{1o} = (s + p_1)/(s + 1)$, which at any fixed $s = j\omega$, describes a straight line easily defined by its end points. The lines GF, FE, EH, in p_1, p_2 space similarly map on lines in the V_{o1}/V_1 complex space. All that is needed to obtain these four lines is to solve for V_{o1}/V_1 at the four points E, F, G, H. The result is shown in Fig. 4.12-3b for $s = j$. Consider the infinitude of vertical lines extending from the line EF to the line HG in Fig. 4.12-3a. Each of these lines maps as a straight line into the V_{o1}/V_1 space in Fig. 4.12-3b, and must extend from some point on $E'F'$ to some point on the line $H'G'$. At worst this infinitude of lines covers the triangle $H'F'G'$ in Fig. 4.12-3b; actually, the region covered may be the region $H'G'F'T'$. However, to simplify the work, the larger region $H'F'G'$ may

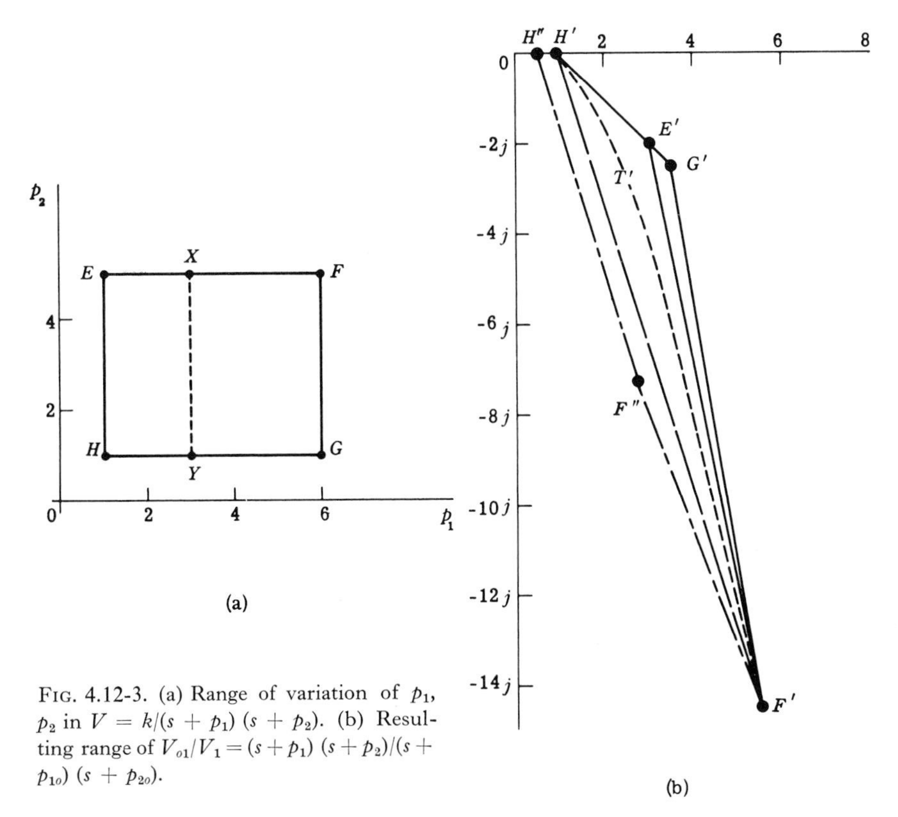

FIG. 4.12-3. (a) Range of variation of p_1, p_2 in $V = k/(s + p_1)(s + p_2)$. (b) Resulting range of $V_{01}/V_1 = (s + p_1)(s + p_2)/(s + p_{1o})(s + p_{2o})$.

be used. The price paid for this simplification is that the final design may perhaps be better than need be. The designer must make his choice between possible economy in the design, versus economy in design effort. If he chooses the former in this problem, he could map a few more points like X, Y in Fig. 4.12-3a. (In many cases, it is possible to tell whether any design economy will result from the additional labor. Consider, for example, Fig. 4.12-4. It does not really matter whether $ABCE$ or the larger area $ABCD$ is used as the region of variation of V_o/V. That part of the V_o/V region which is important for the shaping of L_o is common to both regions.)

To complete the sketch of V_o/V, the variation in k must be taken into account. Suppose k_o/k varies from 1 to 1/2. Then the maximum area of variation of V_o/V is the closed region $H''H'G'F'F''$ in Fig. 4.12-3b.

Next in order of complexity, consider V with three poles so that $V_o/V = k_o(s + p_1)(s + p_2)(s + p_3)/k(s + p_{1o})(s + p_{2o})(s + p_{3o})$, in which the variables (excluding k) are defined in the three-dimensional space shown in Fig. 4.12-5a. From the work previously done, we know that each one of the six faces of the solid, maps at most on a quadrilateral in the complex V_o/V plane. (Some of these quadrilaterals may overlap or degenerate into simpler figures.) Now consider the two faces $ADHG$ and $BCEF$. Suppose these map (at a fixed $s = j\omega$) as shown in Fig. 4.12-5b. The entire region in p space is generated by the infinitude of horizontal lines extending from the face $AGHD$ to the face $BCEF$ in Fig. 4.12-5a. Each one of these straight lines in p space maps as a straight line in V_o/V space, with one end in or on $A'D'H'G'$, and the other end of the line in or on $B'C'E'F'$. Therefore the infinitude of mapped lines must all lie inside the region defined by $A'D'C'E'G'$ in Fig. 4.12-5b (at $k = k_o$).

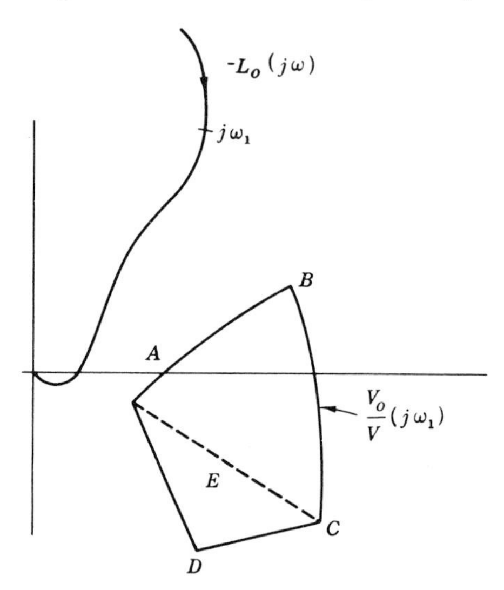

FIG. 4.12-4. Case where exact V_o/V is not critical.

The above reasoning is obviously easy to extend to any number of poles. Therefore, for n poles which vary in V_o/V, one must solve V_o/V for the 2^n extreme values of the poles and draw the biggest possible polygon using these 2^n values as corners. All values of V_o/V (at a constant k) must lie inside or on this largest polygon. The polygon *may* include additional points too, but this is the

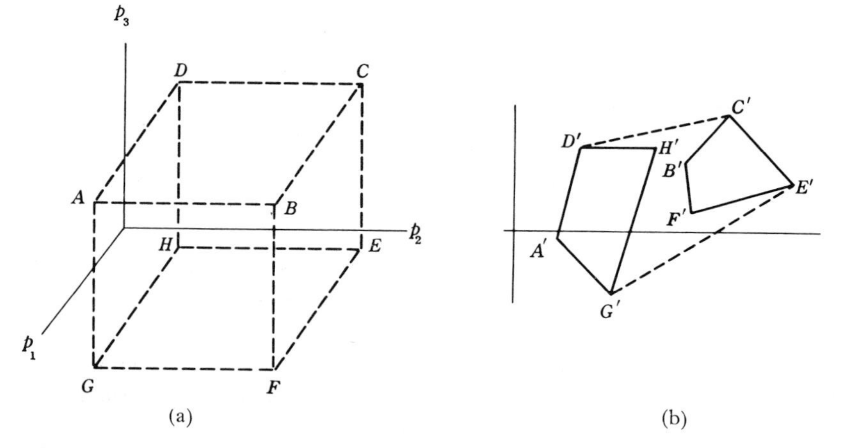

FIG. 4.12-5. (a) Region of variation of three poles. (b) Mapping of $AGHD$, $BFEC$ into V_o/V space.

price paid for using this very simple method. The effect of variations in the gain factor k is then easily taken into account.

Loci of V_0/V When Zeros Vary

Suppose V includes zeros whose positions vary. For example, if V has one pole and one zero, the two-dimensional space (z representing the zero) of Fig. 4.12-6a (solid lines) must be mapped. Along AB (with $k = k_o$ and at a fixed

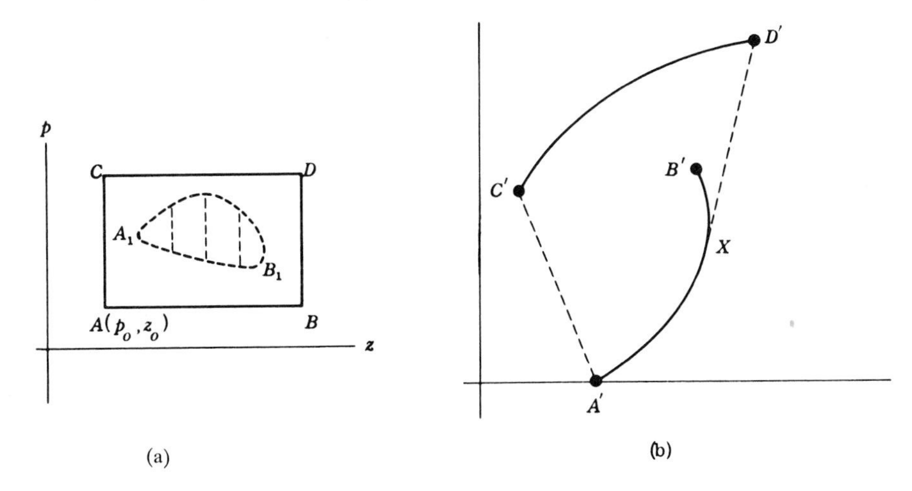

FIG. 4.12-6. Mapping of $ABCD$ from p, z space into space of $V_o/V = (s + z_o)(s + p)/(s + z)(s + p_o)$.

$s = j\omega$), $V_o/V = (z_o + j\omega)/(z + j\omega)$ is an arc of a circle which passes through the origin (corresponding to infinite z). Three points determine a circle. One is at the origin and for the other two the values at A and B can be used. An arc of a circle passing through these three points can be drawn, but only the $A'B'$ portion of it represents the mapping of the line AB on the V_o/V space. Similarly, CD maps on an arc of a circle $C'D'$ passing through the origin. The result may have the form shown in Fig. 4.12-6b. The lines AC, BD map as straight lines. Now consider the infinitude of vertical lines with end points on AB and CD. These map as lines with ends on $A'B'$, $C'D'$, respectively. Therefore the totality of these lines must map in or on the region $C'D'XA'C'$ in Fig. 4.12-6b. Again the latter region may include more points than those contained in $ABCD$.

It is not difficult to see what must be done for the general case where there are n poles and m zeros. A total of $2^{(n+m)}$ points are mapped (at a fixed $s = j\omega$ and a fixed k) on the complex V_0/V plane. These $2^{(n+m)}$ points in the $(n + m)$-dimensional space determine $2^{(n+m)}(m + n)/2$ edges. The number of these edges which correspond to variations in a zero is $[(m/2)2^m]2^n$. [The m zeros determine $(m/2)2^m$ edges for each combination of pole values, and there are 2^n of the latter.] Therefore one must construct $m2^{(m+n-1)}$ arcs of circles. (It is easy to determine the latter because each circle passes through the origin.) Using the $m2^{(m+n-1)}$ arcs and the $2^{(m+n)}$ points, construct the largest required region (e.g., as in Fig. 4.12-5b) in which must surely lie V_{o1}/V_1 for its entire range of variation of poles and zeros. The boundaries of the region will consist of straight lines joining points; arcs, and tangents from points to arcs. Again variations in k are easily taken into account.

So far we have shown how to find V_o/V at a single frequency for the special case when the poles, zeros, and gain constant of V_o/V independently vary. But suppose the poles and zeros do not vary independently. For example, instead of p and z, in $V = (s + z)/(s + p)$, having the range of variation shown by the solid lines in Fig. 4.12-6a, suppose they have that shown by the dotted lines in Fig. 4.12-6a. One way out is to include this region in the smallest possible rectangle and proceed as before. This of course is to a greater or lesser extent uneconomical, by leading to a design which is better than required. One again has a choice here between economy in the final design result and economy in design labor. For example, in Fig. 4.12-6a it should suffice to map A_1, B_1, and the four dotted lines. As the number of variables increases, the additional number of points needed for such irregular regions increases enormously and a computer becomes useful. However, even in such a case, the important point is that the designer knows what he is doing and is not working blindly.

Parameters other than the poles and zeros may be pertinent in many cases. For example, suppose the coefficients of the quadratic denominator polynomial $s^2 + zs + p$, of V, have the region of variation (solid lines) shown in Fig. 4.12-6a. Along AB, $V_o/V = (s^2 + zs + p_o)(s^2 + z_o s + p_o)^{-1}$ and for a fixed $s = j\omega$, the locus is a straight line. The same applies for any horizontal or vertical line in the z, p plane. Clearly then the procedure is identical to that previously discussed for pole variations, viz., the four points $ABCD$ are mapped and the largest polygon determined. The extension to any order denominator polynomial is straightforward. Similarly, if the numerator polynomial of V has coefficients which vary independently, lines such as AB and BD map as arcs of circles (which pass through the origin). The mapping is exactly analogous to that which results when the zeros of V vary independently.

There may be situations when the mapping is not so simple. Suppose, for example, that $V_o/V = [1 + (a^2 + b)F_1(s) + abF_2(s)]\,[b_2F_3(s) + aF_4]^{-1}$. As a moves along a horizontal line, or b along a vertical line, V_o/V does not map as a straight line, and a great deal of work may be necessary to obtain the entire range of variation of V_o/V. Nevertheless, this is an indispensable part of solving the stability problem and must be done.

§ 4.13 Extension to Higher Order Systems

The method outlined in Section 4.12 is satisfactory when the design involves only one loop transmission function, such as in Fig. 4.13-1a. Here P represents

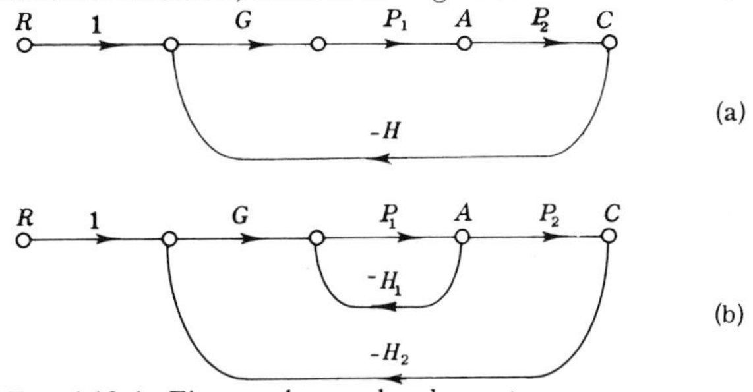

FIG. 4.13-1. First- and second-order systems.

the plant which is V, the varying portion. It should be emphasized that P *itself may be very complicated, with a signal-flow diagram containing many loops.* The essential point is that the varying parameters are contained in P, and that there is only one loop transmission function. Sometimes there are advantages (as will be shown in Chapter 8) for using Fig. 4.13-1b rather than Fig. 4.13-1a to achieve the design objectives. Satisfaction of the design objectives may require certain levels of $L_1 = P_1H_1$, $L_2 = P_2H_2P_1G(1 + P_1H_1)^{-1}$ over the same or different frequency ranges. Here the stability problem is how to shape L_1, L_2 (i.e., H_1 and GH_2, which represent the functions under the control of the designer) beyond these frequency ranges, so as to ensure stability despite parameter variation. Now we have a total of two degrees of freedom, which makes the problem more difficult than that considered in Section 4.12.

The polynomial whose zeros must be examined for stability[1] is $(1 + L_2) = 1 + P_1P_2H_2G(1 + P_1H_1)^{-1}$, whose zeros are those of $1 + P_1H_1 + P_1P_2H_2G$. In the latter form, we see the basic problem. In Section 4.12, the critical polynomial could be written as $1 + V_oF$, with F representing the compensation networks under our control, and V representing the varying part beyond our control. In Fig. 4.13-1b we must write the critical polynomial in the form $1 + V_1F_1 + V_1V_2F_2$, with $V_1 = P_1$, $F_1 = H_1$, $V_2 = P_2$, $F_2 = H_2G$. If in Fig. 4.13-1b there was, in addition, a feedback path (with $- H_3$) around P_2 from C to A, we would have, in place of the previous polynomial, the polynomial $1 + V_1F_1 + V_2F_2 + V_1V_2F_3$, with $V_1 = P_1$, $V_2 = P_2$, $F_1 = H_1$, $F_2 = H_3$, $F_3 = GH_2$.

This point of view appears to be a very reasonable one to use as a basis for classifying feedback systems. Thus, if Section 4.12 can be used when considering

[1] Since the numerator of F is the same for any reference (see Section 2.8), and t_{oi} is zero in Fig. 4.13-1b, any loop may be opened anywhere to find the numerator of the return difference F.

the stability problem (no matter how complicated P may be), we have a stability problem of order 1, while Fig. 4.13-1b represents a system of order 2. The corresponding expression for a system of order 3 is: $1 + V_1F_1 + V_2F_2 + V_3F_3 + V_1V_2F_{12} + V_1V_3F_{13} + V_2V_3F_{23} + V_1V_2V_3F_4$. The order of a system is directly related to the number of input and output points used in a plant. In the case of a single plant with a single input and single output the order is given by the number of feedback points (A, C in Fig. 4.13-1b). It is again emphasized that V_1, V_2, etc., may be very complicated functions, with large numbers of varying parameters.

As far as analysis is concerned, i.e., determining whether a system with V's and F's is stable over the entire range of variation of the V's, the method of Section 4.12 may be used. Consider, for example, the second-order problem. If the varying parameters of V_1 are independent of those of V_2, then at a fixed $s = j\omega$, from the area of variation of V_1F_1 and that of V_2F_2, one may obtain the total area of variation of their sum. The area of variation of V_1V_2 may be likewise obtained. By this means (repeating the process for a sufficient number of discrete frequencies and the use of the Nyquist criterion), it is possible to check the system for stability. Needless to say, this may involve a great deal of computational labor in those cases where the mapping is not in the form of straight lines or circles. But this is the minimum labor, and at least it is intelligent effort, not hapless cut and try. A computer can of course be enlisted to make the computations.

So much for analysis. The synthesis problem is vastly more difficult. Consider the order 2 system. It can be written in the form $1 + L_{1o}(V_1/V_{1o}) + L_{2o}(V_2/V_{2o}) + L_{3o}(V_1/V_{1o})(V_2/V_{2o})$, where $L_{1o} = V_{1o}F_1$, etc., are the various nominal loop transmissions. The most general problem is to shape each of L_{1o}, L_{2o}, and L_{3o} so that the system is stable over the entire range of parameter variations of V_1 and V_2. This general problem is best postponed until later, when we will have a better understanding of the background of such systems. The problem will be considered in detail in Chapter 10.

CHAPTER FIVE

Design of Feedback Control Systems with Single-Degree-of-Freedom Configuration

§ 5.1 Comparison of Feedback Amplifiers with Feedback Control Systems

Previous chapters have dealt with fundamental concepts and properties of feedback which are common to all systems employing feedback. We now turn to the application of feedback theory to a specific class of systems.

The two principal areas in which feedback theory has been applied on a large scale are feedback control systems and feedback amplifiers. The reasons for using feedback in these two areas are basically the same, and are those given in Chapter 3. From the point of view of the feedback specialist, whose concern is with the signal-processing in the system and not with the specific mechanical, aerodynamic, or chemical problems, the differences between feedback control and feedback amplifiers are in the problem constraints and problem specifications. However, these do lead to important differences in the problems which must be solved. What are these differences?

In most feedback control problems it is possible to divide the system into two parts. One part, denoted in this book as the *plant*, consists of the combination of load, equipment, and machinery which have the physical capacity to achieve the system objectives. The feedback engineer can often assume the plant is a fixed piece of equipment which he must perforce work with. Partial ignorance of the plant, plant parameter variations, plant nonlinearities, unstable plants, partial ignorance of the environment, and disturbances make it impossible for the objectives to be achieved by the plant alone, despite the fact that the latter has the muscle that is needed for the job. An intelligence (feedback), data-processing network must be built around the plant, in order that the objectives may be accomplished. There is a significant difference between the power level and the dynamic range of the plant, and those of the data-processing portion of the system. Because of the appreciable difference in power level, it is realistic, and extremely convenient for theoretical work, to assume (at least at the beginning) that the low power level data-processing portion of the system can, at reasonable cost, be built free of parameter variation, nonlinearities, and external disturbances. This assumption tremendously simplifies the feedback problem in control systems.

In control systems we usually deal with motions of masses, flow of fluids, etc.; so the speed of response of the plant is in the order of seconds, not microseconds, and the system bandwidth is therefore in the order of cycles per second, not megacycles per second. Now the fundamental ingredient in feedback is the signal, not the power or physical quantity it represents. Therefore the data-processing portion of the system can be constructed from simple electrical passive elements and electronic or solid state amplifiers, with their very fast responses (large bandwidths). In the feedback control problem, we can therefore usually assume that whatever gain-bandwidth may be assigned to the data-processing portion can be obtained.

There are systems, usually considered as feedback amplifiers, which have some of the properties ascribed above to feedback control systems. Consider, for example, an all-electronic power amplifier whose final output stage consists of a very expensive tube operating at a large power level, and where the major parameter variation, nonlinearities, and disturbances occur in this output stage. This last stage is equivalent to the plant of the control system, and it is reasonable to consider such a system a feedback control system, in the sense previously described.

However, in amplifiers without this marked difference between stages, it is impossible to assume the simplifying assumptions made in control systems. The parameter variations of all stages must be considered, because it is impossible to divide the system into a plant portion and a data-processing portion. Since the system is sensitive to parameter variation in the feedback path (see Section 3.9), this usually excludes any active elements from the feedback path. There is not available for such systems a reservoir of devices with gain-bandwidths much greater than those of the plant portion of the systems. One of the important problems in feedback amplifier design is, therefore, that of extracting the maximum gain-bandwidth capabilities from the electronic or solid state device. In feedback amplifier design we must economize on gain-bandwidth and use relatively complicated passive networks in order to extract the maximum possible gain-bandwidth from the active element. In feedback control we can be more wasteful and use simpler designs and networks. In control systems there is not too much concern about loading effects when stages are connected, because it is assumed that isolating stages can always be added. Such an assumption cannot be made in feedback amplifiers because we must always check the effect of the device parasitics on the high frequency response. Because of all these factors, the theoretical feedback problems in linear feedback amplifier design are often much more sophisticated than those in linear feedback control system design.

Another significant difference between the two types of feedback systems is that nonlinearities are more serious in control systems than in amplifiers, because the nonlinearities encountered in motors, gears, fluid flow, chemical reactions, etc., are of a much more serious nature than those encountered in electronic or solid state devices. This is especially true of the violent nonlinearities such as saturation, dead-zone, Coulomb friction, and hysteresis. The study of nonlinear feedback is therefore of greater importance in control systems. Also, the approximation inherent in a linear design is greater in feedback control systems than it is in feedback amplifiers. The balance of this book is devoted to feedback control systems, although some parts are also applicable to feedback amplifiers.

§ 5.2 Distinction between the Feedback Problem, the Filter Problem, and the Control Problem

There are certain problems present in most systems, whether or not feedback is intentionally introduced into the system. These problems are also present in feedback systems, but they would be there even if there were no feedback. One such problem is that of false (noise) inputs which enter the system at the *same* points as the useful input signals. In a linear design the problem is to find an optimum linear input-output relation (a transfer function) which admits the useful signals but rejects the false signals. The basic approach is to seek some measure in which there is a difference between the two types of signals, and to appropriately choose the optimum system transfer function to discriminate between the two on the basis of this measure. Obviously, this problem, usually called the *filter problem*, has no direct relation to feedback. Indirectly there is a relation, because in a feedback system the optimum transfer function must eventually be realized by a feedback configuration, and so must be of a form which admits such a realization, with whatever constraints there may be imposed by the configuration.

Another problem which exists whenever one builds a system from devices is to understand how the devices work and to decide how to use them most efficiently to obtain the desired outputs. In a system where there is a variety of methods of achieving an output, one must decide which to use, on the basis of efficiency, saturation levels, the system performance indices, etc. This might be called the *control problem* (with its host of optimization problems) but it has nothing directly to do with feedback. There is again the indirect relation, because if feedback is required around the plant, then we have the constraint of a configuration or class of configurations imposed. But this constraint usually imposes no particular hardship.

It seems worthwhile to comment on these matters because many filter and control problems, often classified under the heading of feedback control, are not really *feedback* problems, although they are important problems in their own right. These comments are appropriate in presenting the single-degree feedback control design theory because this theory is very much a mixture of filter theory, control theory, and to a lesser extent, feedback theory. This should not be surprising because the theory was evolved in the course of making systems work, and therefore perforce had to contend with all the system problems. It is therefore also not too surprising that the feedback portion of the theory is able to cope only with the simpler kinds of feedback problems. This will be clarified in the course of the chapter.

§ 5.3 Introduction to Single-Degree-of-Freedom Feedback Control Design

Single-degree-of-freedom feedback control theory accepts at the very outset the constraint of a feedback configuration because it is recognized that parameter variations, disturbances, and nonlinearities preclude the attainment of the design objectives without feedback. (We do not admit that undesirable plant dynamics justify the use of feedback, because such dynamics can be canceled out by means of a prefilter—see Section 3.11.) The structure to which a good portion of classical feedback control theory is devoted, is that shown in Fig. 5.3-1. The input and output time functions are usually denoted by $r(t)$ and $c(t)$

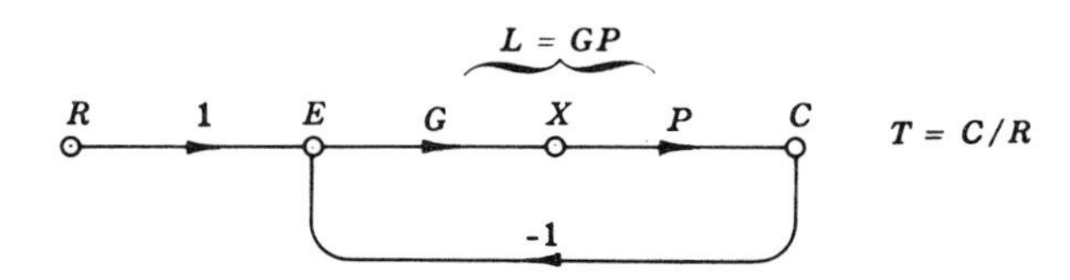

FIG. 5.3-1. Single-degree-of-freedom structure.

and their transforms by $R(s)$ and $C(s)$. P is the transfer function of the *a priori* given plant with which the designer must work. It must be assumed, of course, that the plant has the physical ability to satisfy the design specifications. The transducer which transforms the output form of energy to the input form is taken to have a frequency-independent transfer function equal to one.

An important feature of the structure of Fig. 5.3-1, from the feedback theorist's point of view, is its single degree of freedom $L = GP$. With this one degree of freedom, it is impossible to independently realize the filter specifications, i.e., the desired $T(s) = C(s)/R(s)$, and the feedback specifications [the $L(s)$ which is required for achieving the desired benefits of feedback]. The older classical feedback control theory[1] (frequency response method) does the best it can do under this constraint. It avoids putting all its eggs into the basket of the filter

[1] For a bibliography of the pioneering work in the application of frequency response methods to feedback control, see R. Oldenburger, ed., "Frequency Response". Macmillan, New York, 1956.

problem or into the basket of the feedback problem. Instead, it uses a synthesis technique which enables the designer to keep an eye on both these problems and to compromise when they conflict, according to his inclinations and the circumstances of his specific problem.

In this older method, no attempt is made to formally specify $T(s)$ as a rational function in s. The various design requirements are stated in terms of the demands they make on the loop transmission $L = GP$ of Fig. 5.3-1. The designer therefore has control over the feedback properties of the system, because they are determined by L. Furthermore, he has as good a control of the system response, over a good part of the significant frequency range. This is the frequency range in which $|L(j\omega)| \gg 1$, and therefore $T = L/(1 + L) \doteq L/L = 1$. Now it makes little difference from the point of view of $T(j\omega)$ whether $T(j\omega) = 0.99$ or 0.995 (the difference is about 0.5%). However, E, the error function ($E = R - C$), is in one case $E = 0.01$, and in the other case $E = 0.005$, and such a difference is certainly significant. The classical theory was first developed to handle positional servomechanisms, where the instantaneous error in following or tracking is an important design specification. It was therefore very sensible to state the system response specifications in terms of the error ratio function $E(s)/R(s) \triangleq (R - C)/R = 1/(1 + L)$, which is closely equal to $1/L$ in that part of the significant frequency range in which $|L| \gg 1$. If the error ratio is to be only 0.005 at some frequency ω_1, while it may be 0.01 at ω_2, then $|L(j\omega_1)| \geqslant 200$ while $|L(j\omega_2)| \geqslant 100$; the difference is obvious, and it is clear what has to be done to achieve the difference. If one works with $T(s)$, then the difference between $T(j\omega_1)$ and $T(j\omega_2)$ is about 0.5%, and it is not at all clear how $T(s)$ must be specified in order to achieve this difference, nor why we should bother to achieve it. On the other hand, in the frequency range where $L \approx 1$, the relation between L and $T = L/(1 + L)$ is more tenuous, and the direct specification of $T(s)$ as a rational function in s perhaps affords better control over the system transient response.

To summarize the above, in the classical design theory, all the specifications, both filter type and feedback type, are stated in terms of what they demand of $L(s)$. The design procedure consists of shaping $L(s)$ so as to achieve the specifications. In the older theory one works entirely with $L(j\omega)$ (frequency response), and the natural tools are Bode plots and Nyquist plots. More recently, procedures have been developed wherein one works with the poles and zeros of $L(s)$ and the natural tool is the root locus sketch.[1]

A completely different approach to the problem is afforded by the $T(s)$ pole-zero method.[2] Here the design specifications are used to formally determine

[1] W. R. Evans, "Control-System Dynamics," McGraw-Hill, New York, 1954.

[2] J. G. Truxal, Servomechanism synthesis through pole-zero configurations, *MIT Research Lab. Electronics*, *Tech. Rept.* **162** (1950); M. R. Aaron, Synthesis of feedback control systems by means of pole and zero location of the closed loop function, *Trans. AIEE* **70** (Part 2), 1439-1445 (1951).

$T(s) = C/R$ (Fig. 5.3-1) as a rational function in s. Next, one uses the relation $T = L/(1 + L)$ to solve for L in terms of T, i.e., $L = T/(1 - T)$, and goes ahead with the realization of $L = GP$. A detailed comparison of the relative merits of these various methods will be made at the end of the chapter.

It should be emphasized that this chapter is principally concerned with the implementation of specifications and only incidentally touches upon the important problems of how to choose these, and the evaluation of the various kinds of specifications that are used in feedback control system design. These latter problems are considered in detail in Chapter 9.

§ 5.4 Classical Control System Specifications—the Error Constants

One of the first steps in the development of the classical theory is to relate the requirements on the system transfer function $T(s)$, to the single independent function, which is conveniently taken as the loop transmission, $L = GP$. For this purpose it is first necessary to consider the requirements on $T(s)$. In follow-up systems a classification is used based on the typical inputs to the servomechanism. The following simple typical inputs are used: the step function $u(t)$; the unit ramp ($r = t$ for $t \geqslant 0$); the unit parabola ($r = 0.5\, t^2$ for $t \geqslant 0$), etc. The final steady state values of the position errors (the difference between input and output, $e = r - c$, after the transients have died out) are an important consideration. They are very simply related to the parameters of L and the resulting error constants are therefore convenient for the partial specification of L. (The validity of the error constants for realistic steady state error specification is examined in Section 9.3.)

For the classical configuration of Fig. 5.3-1, $E = R - C = R/(1 + L)$, and by the final-value theorem, $\lim_{t\to\infty} e(t) = \lim_{s\to 0} sR/(1 + L)$. When the input is a step function,

$$R = 1/s \qquad \text{and} \qquad \lim_{t\to\infty} e(t) = \lim_{s\to 0} 1/(1 + L) = 1/[1 + L(0)].$$

$L(0)$ is called the position constant K_p. Zero position error requires that $K_p = L(0)$ must be infinite, so that L must have a pole at the origin. When the input is a unit ramp, $R(s) = 1/s^2$,

$$\lim_{t\to\infty} e(t) = \lim_{s\to 0} sR/(1 + L) = \lim_{s\to 0} 1/s(1 + L) = \lim_{s\to 0} 1/sL(s),$$

which is infinite unless $L(0) = K_p$ is infinite. If K_p is infinite, let $\lim_{s\to 0} L(s) = K_v/s$ and then $\lim_{t\to\infty} e(t) = 1/K_v$. K_v is called the velocity constant. One may continue in this manner with the unit constant acceleration input and define the acceleration constant K_a (only if K_v is infinite): $K_a = \lim_{s\to 0} s^2\, L(s)$. The results are summarized in Table 5.4-1.

With the notions of the error constants and the obvious requirement that the system be stable, one may already consider very elementary system design,

TABLE 5.4-1

THE ERROR CONSTANTS

Input	Definition of constant	Steady-state position error	Steady-state velocity error	Steady-state acceleration error	System type
Step function $r = u(t)$	$K_p = \lim_{s \to 0} L(s)$ (position constant)	$1/(1 + K_p)$	Not applicable	Not applicable	Type 0 if K_p is finite
Ramp $r = tu(t)$	$K_v = \lim_{s \to 0} sL(s)$ (velocity constant)	$1/K_v$ if K_p is infinite, otherwise infinite	$1/(1 + K_p)$	Not applicable	Type 1 if K_p infinite and K_v finite
Parabola $r = 0.5\, t^2u(t)$	$K_a = \lim_{s \to 0} s^2L(s)$ (acceleration constant)	$1/K_a$ if K_v is infinite, otherwise infinite	$1/K_v$ if K_p is infinite, otherwise infinite	$1/(1 + K_p)$	Type 2 if K_v is infinite and K_a finite

using the frequency response characterization. Since T is uniquely determined by L, it is sufficient to consider the latter. A very simple set of specifications consists of a statement of the magnitude of an appropriate error constant and specified stability margins. The design procedure is presented in the following section.

§ 5.5 Design for Simultaneous Achievement of Error Constant and Phase Margin by Means of Lag Compensation

The design procedure can be presented with sufficient generality by means of a numerical example.

Design specifications: The *a priori* specified plant has the transfer function $P = 1/s(s+2)$; the desired $K_v = 20$, the desired phase margin $\theta_M = 45°$.

Design procedure: Begin with a Bode sketch (asymptotes only) of $L_1 = K/s(s+2)$ where K is chosen so that the desired K_v is obtained, i.e., $K = 40$. The resultant Bode sketch is shown in Fig. 5.5-1a. The crossover frequency

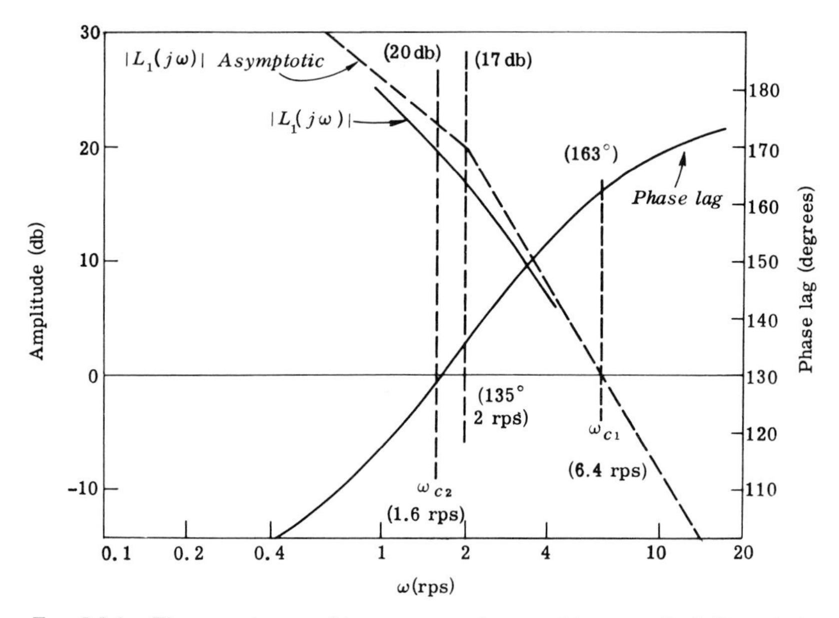

FIG. 5.5-1a. First step in use of lag compensation to achieve specified K_v and phase margin.

ω_{c1} (at which $|L_1| = 0$ db) is approximately 6.4 rps, at which point the phase margin is only $180° - 163° = 17°$. The frequency at which the desired phase margin is achieved is 2 rps, i.e., $\arg L_1 = -135°$ at 2 rps. If the amplitude of

L_1 could be decreased to 0 db at $\omega = 2$ rps, without the phase of L_1 at $\omega = 2$ rps being affected, the phase margin specification would be satisfied. Since $|L_1| = 20 - 3 = 17$ db at 2 rps, $|L_1|$ must be reduced by 17 db at 2 rps. However, $|L_1|$ at very low frequencies must not be changed, in order that the velocity constant K_v remain at the specified value of 20.

The reduction of 17 db at $\omega = 2$ rps, may be achieved by introducing a lag corner frequency at $\omega = \omega_1$ (i.e., a pole at $s = -\omega_1$) well to the left of 2 rps ($\omega_1 \ll 2$). Such a lag corner frequency reduces $|L|$ by 6 db per octave for all $\omega > \omega_1$. After 3 octaves (at $\omega = 2^3\omega_1$), $|L|$ will be reduced by approximately 18 db and therefore at $\omega_2 = 2^3\omega_1$, a leading corner frequency (a zero of L at $-\omega_2$) may be introduced. The amplitude and phase contributions of such a pole and zero combination are shown in Fig. 5.5-1b. Any function which intro-

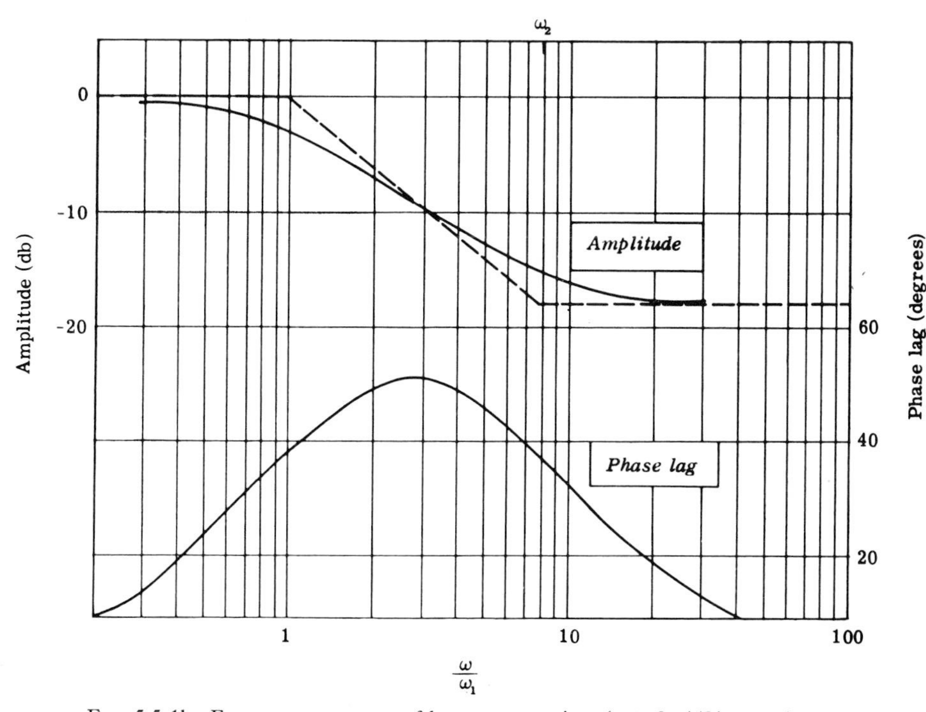

FIG. 5.5-1b. Frequency response of lag compensation, $(s + 8\omega_1)/8(s + \omega_1)$.

duces a change in magnitude must be accompanied by a phase change. Note, however, the very significant fact that while the amplitude change is practically permanent for all $\omega > \omega_2$, the phase change is a temporary one and gradually decreases for $\omega > \omega_2$. We want a change in magnitude of 17 db with very little change of phase at 2 rps. This can be achieved by choosing $\omega_2 \ll 2$ rps, say one decade removed. From Fig. 5.5-1b, (extended) this compensating function at one decade beyond ω_2 has a lag angle of 5°.

To allow for the fact that the compensation will affect the phase slightly, let us at the very outset allow an extra 5° to the phase margin. We therefore examine Fig. 5.5-1a for the frequency at which the phase lag is $180° - 45° - 5° = 130°$. This is at $\omega = \omega_{c2} = 1.6$ rps, at which point the amplitude (exact) must be decreased by 20 db in order that the crossover be at $\omega_{c2} = 1.6$ rps. A 20-db amplitude reduction dictates a separation of one decade between the corner frequencies ($\omega_2 = 10\omega_1$) of the compensating function. In order that the effect of the phase be no more than approximately 5°, we pick $\omega_2 = 0.1\omega_{c2} = 0.16$ rps, and therefore $\omega_1 = 0.1\omega_2 = 0.016$ rps. The new amplitude and phase sketches, as well as the original, are sketched in Fig. 5.5-1c. In this specific problem the gain margin is infinite.

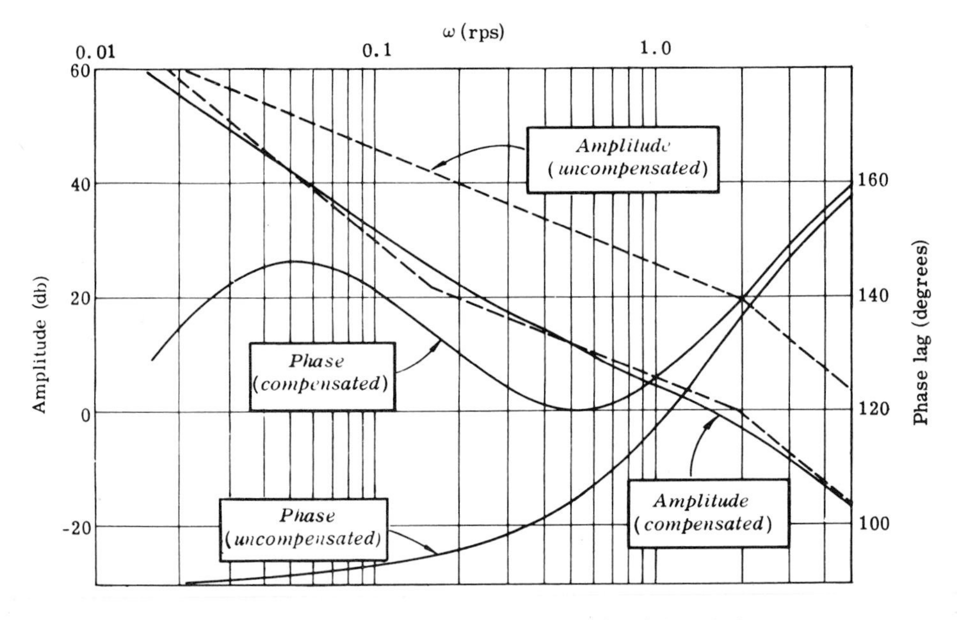

FIG. 5.5-1c. Comparison of uncompensated and final designs.

The compensation used here is called lag compensation because the phase angle due to it is always a lagging (negative) angle. The final value of L is

$$L = \{[(20)(2)(0.016)/(0.16)](s + 0.16)\}/s(s + 0.016)(s + 2) = GP$$

of Fig. 5.3-1. Since $P = [s(s + 2)]^{-1}$, the required $G = 4(s + 0.16)/(s + 0.016)$.

This simple lag compensation method of achieving a specified error coefficient and phase margin can be summarized as follows. Write $L_1 = KP$ and determine K so as to obtain the required error constant. Prepare an asymptotic Bode sketch of L_1. Note the frequency ω_c at which $\angle L_1 = 180° - \theta_M - A$, where θ_M is the desired phase margin and A can be an allowance of about 6°. Note the value of $|L_1|$ at $\omega = \omega_c$. If $|L_1| \leqslant 0$ db, there is no problem. If $|L_1| = Y$ db > 0, then introduce a lag pair (pole at $-\omega_1$, zero at $-\omega_2$) such that $\omega_2 = 2^z\omega_1$, where $Y = 6z$ and $\omega_2 \approx 0.1\,\omega_c$. The value of A can be used to adjust the location of ω_2. For example, if $A = 11.3°$, $\omega_2 \approx 0.2\,\omega_c$.

§ 5.6 Use of Lead Compensation to Achieve Specified Error Constant and Phase Margin

Again the design procedure can be presented with sufficient generality by means of a numerical example. The problem treated is the very same as in Section 5.5. But now instead of adjusting $|L|$ so that the crossover is at the frequency where the phase is approximately $180° - 45°$, we will keep approximately the same crossover frequency as L_1, but adjust the phase at that point to achieve the desired phase margin. Thus, in Fig. 5.5-1a we note that at $\omega =$

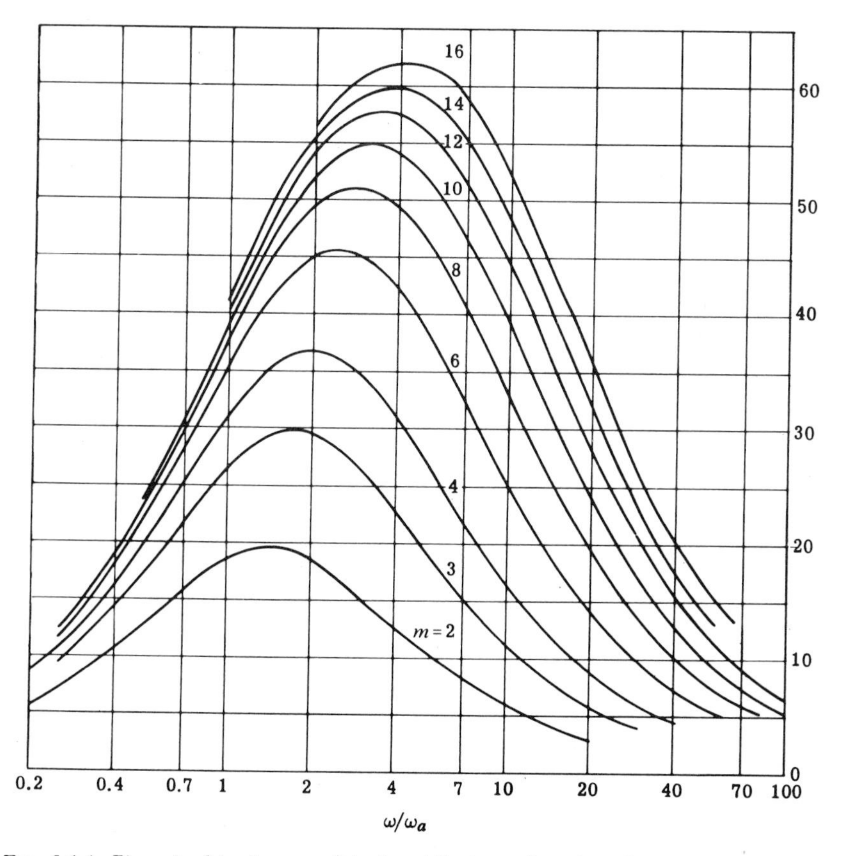

FIG. 5.6-1. Phase lead in degrees of $(s + \omega_a)/(s + m\omega_a)$, or lag of $(s + m\omega_a)/(s + \omega_a)$.

$\omega_{c1} = 6.4$ rps, the phase lag is 163°. If the amplitude there is not changed, then a phase lead of $163° - 135° = 28°$ must be introduced in order to obtain a phase margin of 45°. Consider the combination of a leading corner frequency at $\omega = \omega_a$ and a lagging one at $\omega_b = m\omega_a$. The phase of $m(s + \omega_a)/(s + m\omega_a)$ for different m is shown in Fig. 5.6-1. The magnitude is obvious from Fig. 5.5-1b. It is noted that for $\omega < \omega_a$, the amplitude is hardly affected while there is a phase lead in this region. This zero-pole combination is called lead compensation, because it contributes a leading angle to the loop transmission.

In the present problem, if the amplitude at $\omega_{c1} = 6.4$ rps is unaffected, a phase lead of 28° is needed. From Fig. 5.6-1, a ratio ω_b/ω_a of at least 4 is needed. If ω_a is set at $\omega_{c1} = 6.4$ rps, and ω_b at $(4)(6.4) = 25.6$ rps, the new crossover frequency is approximately at 8 rps, at which point the net phase lag is $166° - 34° = 132°$. This is more than satisfactory. The resulting amplitude and phase of L are sketched in Fig. 5.6-2. In more difficult designs where larger

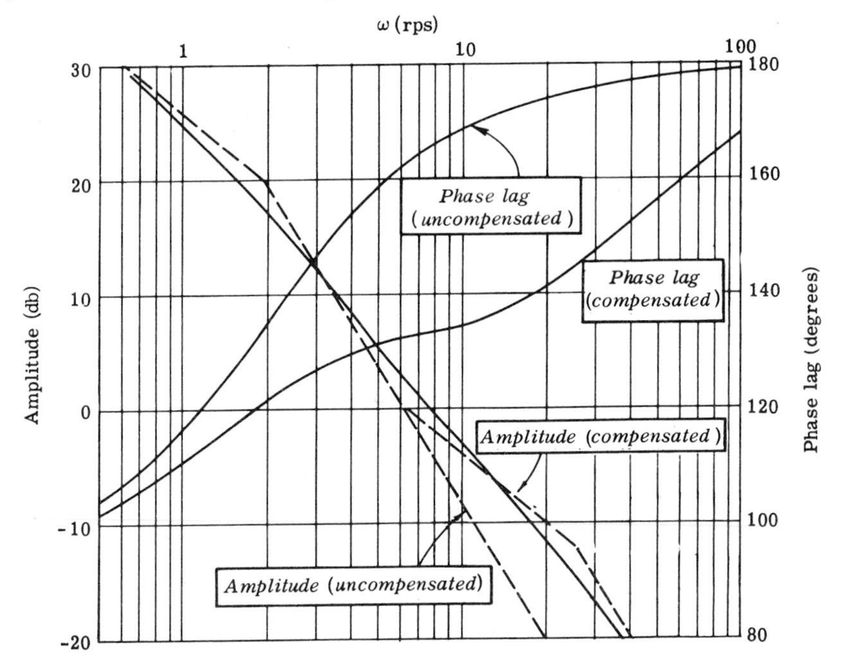

FIG. 5.6-2. Lead compensation design to achieve specified K_v and phase margin.

phase leads are required and it is desired to achieve them with a ratio $m = \omega_b/\omega_a$ that is as small as possible, it is best to position ω_a somewhat to the left of the old crossover frequency. The new crossover will be shifted to the right, but this is more than made up by the larger phase lead thereby obtained. In such more exacting designs, it is best to experiment with a few different positions for ω_a.

From the above examples, it is seen that lag compensation is a means of achieving simultaneously a desired phase margin and error constant, and, in the process, decreasing the crossover frequency (and thereby the bandwidth of T for the system of Fig. 5.3-1). Lead compensation achieves the error constant and phase margin and maintains or increases the value of the crossover frequency (and thereby the system bandwidth). It is not hereby suggested that Figs. 5.5-1c and 5.6-2 display designs which are optimum in any manner whatsoever, for clearly there is no limit to the number of designs which satisfy the few simple specifications that have been so far considered.

In order to proceed with more complex problems, we must introduce additional design specifications. An important specification is the system bandwidth. In a low-pass system this is usually defined as the frequency at which $|T| = -3$ db. The bandwidth is closely related to the speed of response of the system (see Section 5.10). Since we are working with the loop transmission $L(s)$, it is necessary to relate the bandwidth to $L(s)$. From the relation $T = L/(1 + L)$ it can be seen that the bandwidth is in the neighborhood of the crossover frequency [see Eqs. (5.11,1)]. The crossover frequency itself may be used as a measure of the bandwidth, and is therefore usually listed as one of the design specifications. The next section details the procedure for simultaneously realizing specified error constant, stability margins, and crossover frequency.

§ 5.7 Loop Shaping for Simultaneous Achievement of Error Constant, Crossover Frequency, and Stability Margins

The combination of lag and lead compensation permits the simultaneous achievement of all three: error constant, stability margins, and a specific crossover frequency. Once again the design techniques are best presented by means of numerical examples.

Example 1. Consider the example of Fig. 5.5-1a. Suppose the crossover frequency (symbol ω_c) is to be at 0.5 rps and the other specifications are the same as before. At present at 0.5 rps, the gain is 32 db, and the phase lag is 104°. The amplitude is to be decreased by 32 db at 0.5 rps and the phase lag can be increased as much as 31°. This can be achieved by two lag networks properly spaced. There is no limit to the possible combinations, but it is usually desirable to select a combination with corner frequencies as large as possible, in order that the resulting network elements not be too large, since the larger the network time constants, the larger are the element values. Also, the larger the corner frequencies, the greater is the loop gain-bandwidth, which is desirable from the point of view of the benefits of feedback. A bit of cut and try is advisable here. Suppose, for example, that we pick one lag combination to contribute -20 db with say 15° lag at 0.5 rps. From Fig. 5.6-1, for $m = 10$ (because $20 \log 10 = 20$ db), at 15°, the reading is 33, indicating that the crossover frequency of 0.5 rps should be 33 times as great as the pole of the lag network; so the latter corner frequency is set at 1/66 rps. The second lag network must therefore contribute the balance of 12 db and 16° lag. Using $m = 4$, we see that 16° requires ω_c to be 10 times the pole of this lag network; so the latter is set at 1/20 rps. The required L is therefore $L = (s + 0.152)(s + 0.2)[s(s + 0.0152)(s + 0.05)(s + 2)]^{-1}$ and is sketched in Fig. 5.7-1a.

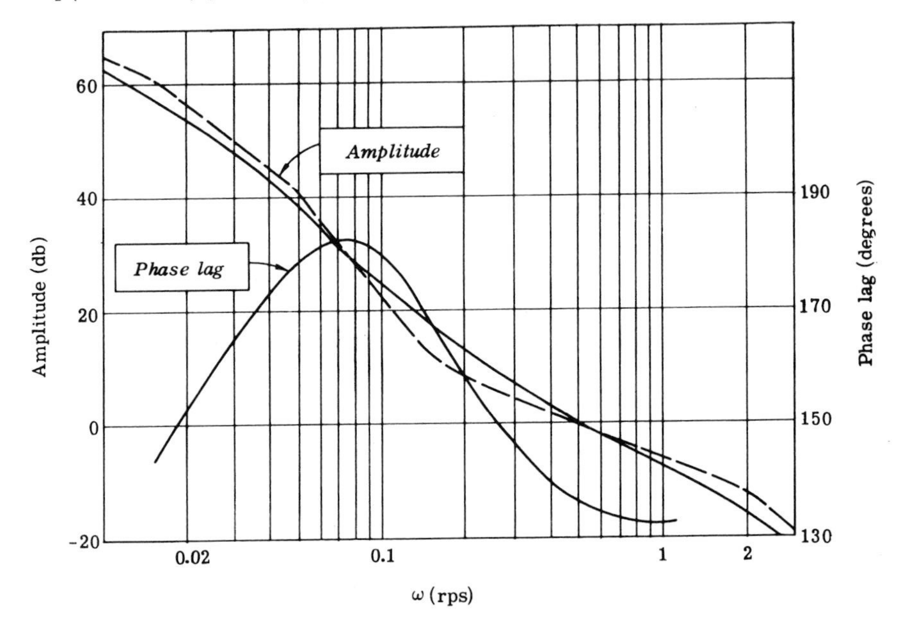

FIG. 5.7-1a. Conditionally stable design which achieves specified K_v, crossover frequency, and phase margin.

While K_v, the bandwidth, and the phase margin have been obtained, the system is conditionally stable because the phase lag is $> 180°$ in a region in which $|L| > 1$. Hence, a decrease in gain of L with no change in phase of L leads to an unstable system. This is usually undesirable. The conditional stability is due to the concentration of lag in one frequency range. It can be avoided by separating the lag pairs. Thus, if the phase contribution at 0.5 rps due to the first pair is reduced to 5°, then the pole of the lag network $\approx 0.01(0.5) = 0.005$ rps (instead of 0.015). Therefore the second lag network must contribute 26° lag at 0.5 rps. In Fig. 5.6-1 draw a horizontal line at 26°, and pick the value of m that supplies approximately 12 db. For example, $m = 4$ at $5.4\,\omega_a$ (ω_a is the lag corner frequency, so $\omega_a = 0.5/5.4 \approx 0.1$) has an attenuation of about 10.7 db, which is probably close enough, although the crossover frequency will be slightly greater than $\frac{1}{2}$ rps. With this choice, $L = (s + 0.05)(s + 0.4)[s(s + 0.005)(s + 0.1)(s + 2)]^{-1}$ and is sketched in Fig. 5.7-1b. The reader should note that over a distinct frequency range (in Fig. 5.7-1b), the phase of L is fairly constant, and that this corresponds to the slope of $|L|$ being reasonably constant over an overlapping range. The reader should compare the phase lag in Fig. 5.7-1a with that in Fig. 5.7-1b and he will note such differences between a poor design and a better design (in a sense later defined) in all ensuing examples.

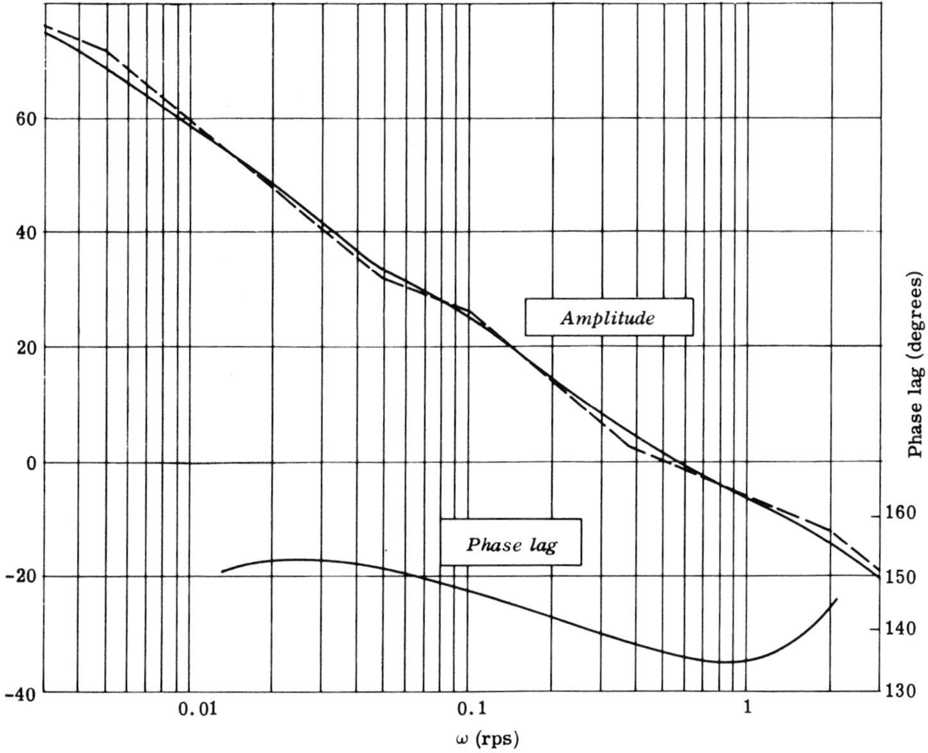

FIG. 5.7-1b. Unconditionally stable design which achieves specified K_v, crossover frequency, and phase margin.

Example 2.

Specifications: $K_v = 40$, $\omega_c = 3$ rps, $\theta_M = 40°$. The plant transfer function is $P = 20[s(s+1)(s+2)]^{-1}$. To guard against conditional stability, the phase lag for all $\omega < \omega_c$ must be less than 150°.

Design: Let the uncompensated L be $L = K/s(s+1)(s+2)$ and choose K such that $K_v = 40$; therefore, $K = 80$. The amplitude and phase of the uncompensated L are sketched in Fig. 5.7-2. At the desired ω_c of 3 rps, the gain is at present 10 db with a phase lag of 217°. It is therefore necessary to reduce the gain at 3 rps by 10 db, and obtain a phase lead of $217° - 140° = 77°$ at 3 rps.

The former requires lag compensation at low frequencies while the latter requires lead compensation in the neighborhood of ω_c. There are an unlimited number of ways in which these two modifications of L can be made. We pick one more or less at random.

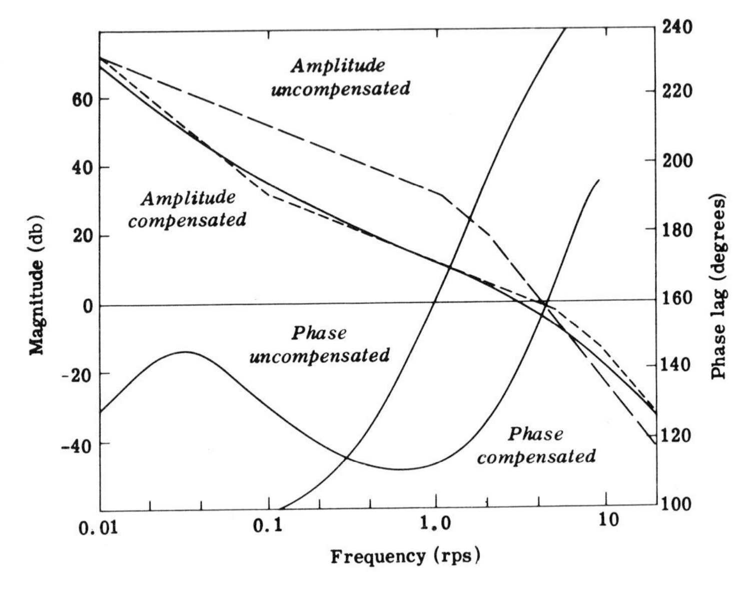

FIG. 5.7-2. Example 2.

In this design we introduce more than 10-db attenuation at low frequencies (which would by itself cause the crossover frequency to be less than 3 rps) so as to permit lead to be inserted below 3 rps, and thereby make it easier to obtain the large phase lead required at 3 rps. Accordingly, lag with $m = 10$ and corner frequencies at 0.01 and 0.10 rps are introduced (with a net decrease in $|L|$ of 20 db), and thereby the lead compensation must provide $+10$ db of gain together with the 77° phase lead required at 3 rps. If we insert two lead corner frequencies, one at 1 rps and the other at 2 rps, the net gain is approximately 0 db at 3 rps, and the total lead angle contributed by the lag compensation and the two lead corner frequencies is 124° at 3 rps. There is an excess of $124° - 77° = 47°$ lead. This permits the placing of poles at -5 and -9. The resulting gain and phase curves are sketched in Fig. 5.7-2. The required compensating network transfer function is $G = 9(s + 0.1)(s + 1)(s + 2)[(s + 0.01)(s + 5)(s + 9)]^{-1}$. Before we make any comments on this specific solution to the design problem, it is advisable to consider the problem of optimization.

§ 5.8 Optimization of Loop Transmission Function

When more than one design is available to achieve a set of design specifications, it is then possible to define a criterion of optimality, and use the available freedom to achieve an optimum design. In view of the fact that we have been experiencing precisely such situations of unlimited design possibilities, it is appropriate to consider optimizing the design in some manner. In general, design optimization is extremely important even in those cases where the absolute optimum requires a design of very great complexity. The reason for its importance, even in such cases, is that it marks the boundary of the ultimate that can be attained. Only when such boundaries of a subject have been thoroughly explored can it be said that the subject is a science rather than a hit or miss affair. In what follows, one special problem in optimization is considered. In Chapter 7 we will develop the theoretical background which will permit the solution of more general optimization problems.

The value of the loop transmission L which was obtained in Fig. 5.7-2 represents one satisfactory L out of an endless number of possibilities. An L function which satisfies all the specifications but which has a larger gain over all or part of the frequency range $\omega < \omega_c = 3$ would be better in terms of desirable feedback properties and also of desirable system transfer function properties. Its superiority in terms of feedback properties is obvious because the benefits of feedback are given by the return difference $1 + L$, so, broadly speaking, the larger $| L |$, the larger is the return difference. It is also superior in terms of properties of $T(s)$ because the error is $E = R - C = R/(1 + L)$. Therefore, for given crossover frequency, stability margins, etc., whatever freedom is left may be usefully used to increase $| L |$ in the region $\omega < \omega_c$. (We emphasize, however, that such a design is not necessarily optimum from other points of view, some of which will be treated in Chapters 8 and 9.)

One important property of the design which maximizes $| L |$ for $\omega < \omega_c$ can be deduced by means of fairly simple reasoning. For this purpose, Figs. 5.8-1a and b are useful. From Fig. 5.8-1a it should be clear that by suitably shaping an evenly symmetrical gain curve, almost any oddly symmetrical phase curve can be obtained. If the gain curve is not symmetrical, neither is the phase curve. (It is proven in Chapter 7 that if the low and high frequency values of gain are the same, the net phase area on a Bode plot is zero.) Figure 5.8-1b shows a gain and phase characteristic which is useful for removing a pocket of phase lead from one region and spreading it out over other regions. Here, too, the precise patterns may be changed by suitably shaping the gain curve. Figures 5.8-1a and b will now be used to derive the condition for optimum $L(s)$.

Consider any L which satisfies the specifications, one of which is that, for $\omega < \omega_c$, the phase lag must not exceed some prescribed θ_e vs. ω, and suppose that in some manner the phase lag for all $\omega > \omega_x < \omega_c$ has been decreased

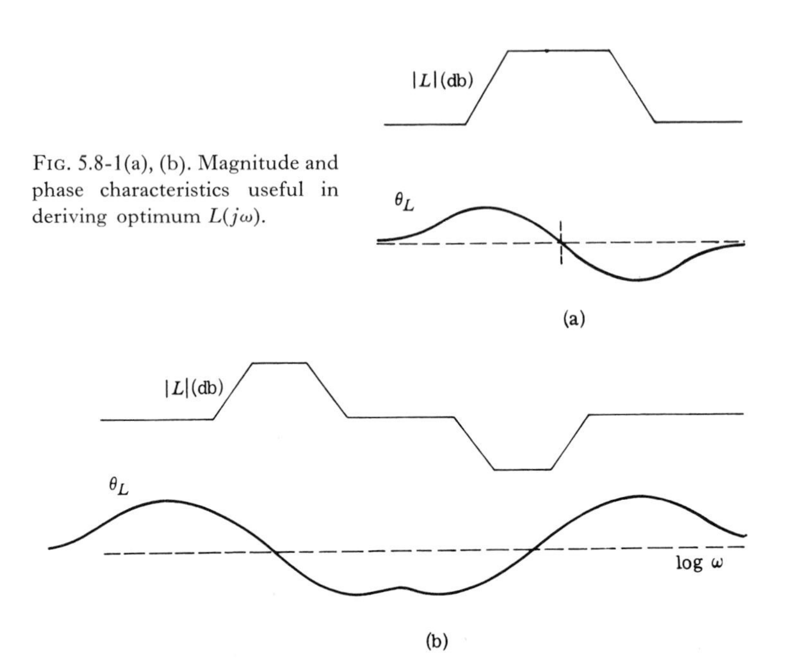

Fig. 5.8-1(a), (b). Magnitude and phase characteristics useful in deriving optimum $L(j\omega)$.

without increasing it anywhere else (ω_c is the crossover frequency). This permits a larger $|L|$ because the phase lead that has been gained for $\omega > \omega_x < \omega_c$ can

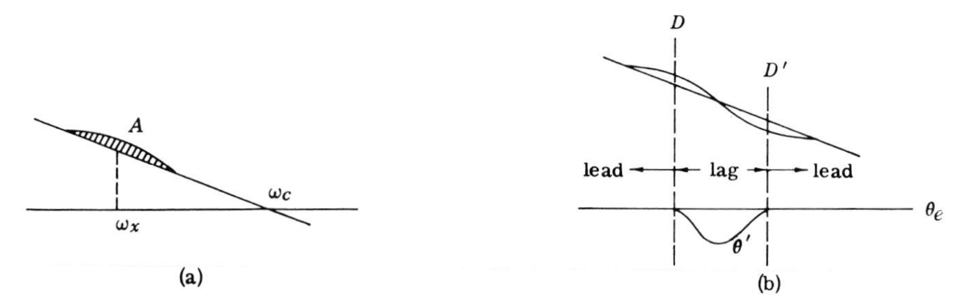

Fig. 5.8-2a, b. Exchange of available phase lag to increase log $|L|$ − log ω area.

be exploited as in Fig. 5.8-2a where a section like that of Fig. 5.8-1a (shown cross-hatched in Fig. 5.8-2a) has been added to the original L. The lag introduced by this section cancels out the surplus lead in the region $\omega > \omega_x$; in the region $\omega < \omega_x$ there is an increase in phase lead. The $|L|$ frequency area (in logarithmic squared units) gained by this operation (the area marked A in Fig. 5.8-2a) may be shaped until the specification of the maximum lag in the $\omega > \omega_x$ interval is barely violated.

It can now be shown that the optimum L is the one whose phase angle has the extreme value of lag angle vs. ω (denoted by θ_e) permitted by the specifications. To prove this, suppose that it is not so, i.e., assume that there is an L' which satisfies the specifications and which over some frequency range has a phase lag $\theta' < \theta_e$. This is shown in Fig. 5.8-2b where $\theta' < \theta_e$ in the frequency range DD'. We employ the technique of Fig. 5.8-1b to remove the pocket of surplus lead in DD' and spread it out to the right of D' and to the left of D. This is done by inserting the pole-zero pattern of Fig. 5.8-1b at DD'. It is seen that the net $|L|$-frequency area (in logarithmic squared units) is thereby unchanged, while the phase lag is increased in region DD' and is decreased everywhere else. We can next use the technique of Fig. 5.8-2a to exchange the excess lead to the right of D' for additional $|L|$-frequency area. The above process may be continued until the phase lag in the range $\omega < \omega_c$ is equal to its maximum permitted value θ_e. Any further increase of the $|L|$-frequency area must result in more phase lag. (The relation between $|L|$ and its phase will be studied in more detail in Chapter 7.)

The above property of the optimum design provides a useful guide in designing $L(j\omega)$. If the specifications are satisfied with a phase lag which is significantly less than the maximum allowable values, the above techniques can be used to improve the design. In this connection, it may be noted that in all the previous examples, the final design has a phase lag curve which, over the important frequency range ($\omega < \omega_c$), is more uniform than the original unsatisfactory design. Thus, note Figs. 5.5-1c and 5.6-2, and compare Fig. 5.7-1a with Fig. 5.7-1b. It is also clear that the design of Fig. 5.7-2 could be substantially improved (i.e., more $|L|$-frequency area for $\omega < \omega_c$).

§ 5.9 Design Example

The loop shaping techniques previously described, and the optimization philosophy of Section 5.8, are applied in this section to a numerical design example.

Specifications: $K_a = 5$; $\omega_c = 5$ rps; $\theta_M = 40°$; gain margin is to be 12 db at 15 rps; system to be unconditionally stable with a margin $\not< 20°$ for $\omega > 0.1$ rps. $P = 10^3 [s^2(s+1)(s+4)]^{-1}$.

Design: In the above it is assumed that the phase and gain margins, crossover frequency, and the 180° lag at 15 rps are wanted fairly precisely, in order to illustrate how one may go about satisfying such specifications. The specified acceleration constant, $K_a = 5$, determines the low frequency behavior of L. At low frequencies L approaches K/s^2 with $K = K_a = 5$. This low frequency asymptote of L is sketched on a Bode graph and marked L_a in Fig. 5.9-1.

We next pick a reasonable shape for L in the crossover region. It is usually easiest to obtain reasonable stability margins with a slope of about 6 db per octave through the crossover frequency (at least asymptotically and assuming we do not want to use complex poles), so such a slope of 6 db per octave through the 0 db line at $\omega_c = 5$ rps is drawn in Fig. 5.9-1 (and labeled XX'). This line at 6 db per octave which crosses $\omega_c = 5$ rps at the 0 db line is extended to meet the $L_a = 5/s^2$ line at X. The result, i.e., L_a, X, X' is used as the skeleton upon which to build up the required L. The skeleton consists of the $-$ 12 db per

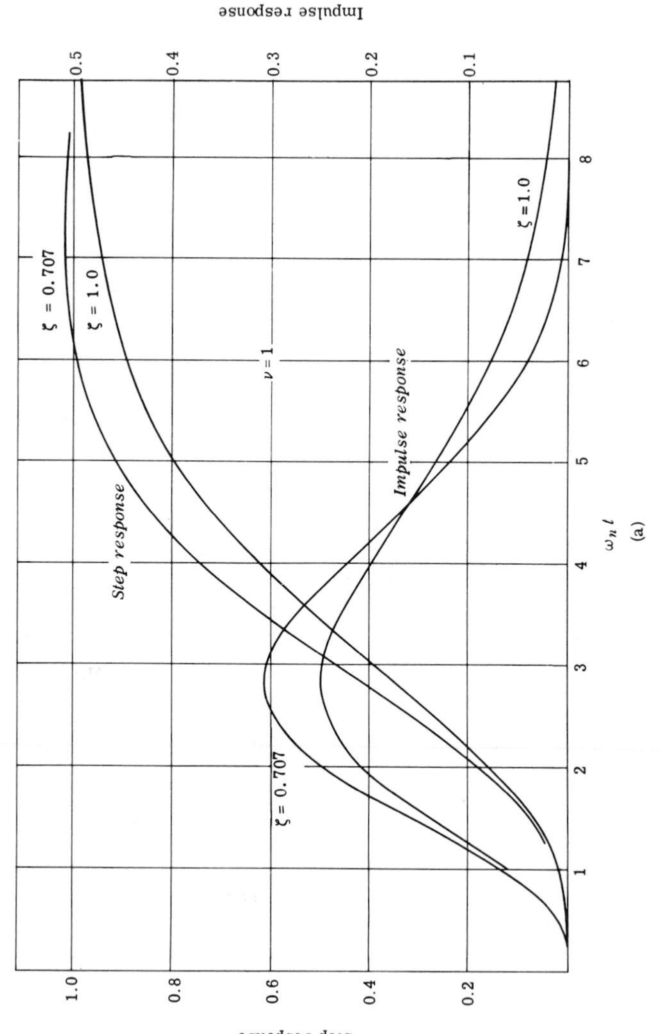

FIG. 5.9-1. Design example.

octave slope from low frequencies up to X in Fig. 5.9-1, followed by the $-$ 6 db per octave slope, line XX'.

It is realized that additional poles must be inserted beyond the crossover frequency, since P has an excess of four poles over zeros, and therefore $L = GP$ should have at least an excess of five poles over zeros (one for G). Because of the requirement that the amplitude should be $-$ 12 db at 15 rps, the additional poles must be beyond 15 rps. One lag corner frequency is tentatively inserted at 20 rps. Next, the phase due to the $MXYY'$ combination of Fig. 5.9-1 is calculated and sketched (labeled θ_a) in the figure. It is seen that already the specification regarding the minimum 20° margin for $\omega > 0.1$ rps has been violated. To correct this situation, a lead network may be introduced in the low frequency range, but this must be followed by a lag network in a higher range (but below 5 rps), in order that the amplitude for $\omega \sim 5$ rps will not be affected. By means of a lead network, the phase lag at low frequencies will be decreased. The net effect on the phase will be to push it in the direction of the approximate ideal θ sketched in Fig. 5.9-1. Where should the lag and lead corner frequencies be placed, and what should be the separation of the corner frequencies? The calculation is made as follows.

At present (curve θ_a), at 0.1 rps, there is an excess phase lag of 15° (175° minus the allowed 160°). The lag network at higher frequencies will contribute some more—guess 3°. The minimum of three additional poles beyond 20 rps will contribute very little more lag at 0.1 rps, so if we arrange the lead network to give about 18° phase lead at 0.1 rps, this should be satisfactory. Referring to Fig. 5.6-1, we see that we can get by with $m = 2$, if the first corner frequency of the lead network is at 0.1 rps. The resulting amplitude correction due to the lead compensation $2(s + 0.1)/(s + 0.2)$ is shown as H_1H_2 in Fig. 5.9-1. In order to achieve the same 5 rps crossover frequency, a lag network $0.5(s + 2)/(s + 1)$ is introduced, so that beyond 2 rps the amplitude (asymptotically) is not affected. The resulting L, so far, is $MH_1H_2H_3YY'$, and its net phase angle is obtained from θ_a and from Fig. 5.6-1, and is labeled θ_b in Fig. 5.9-1. It is noted how in the low frequency range (where the design is basically complete) θ_b is very similar to the approximate ideal phase curve.

Now we must patch up the phase at crossover (5 rps) and at 15 rps, where the phase lag is to be 180° and the amplitude $-$ 12 db. Additional phase lags of 16° at 5 rps and 46° at 15 rps are needed. These are to be achieved by means of additional poles beyond 15 rps. Very little cut and try is required. For example, the combination of three lag corner frequencies at 40, 50, and 100 radians per second provides a total of 46° lag at 15 rps and 15.6° lag at 5 rps. One excess pole over zeros is thereby provided for the compensating network G. The gain margin is now approximately 13 db in place of 12 db, but this is easily taken care of (if deemed necessary) by moving the corner frequency at 20 rps slightly to the right, and slightly readjusting the corner frequencies at 40 and 50 rps. Assuming these minor readjustments are not necessary, $L = (20)\,(10)^6\,(s + 0.1)\,(s + 2)$

$[s^2(s+0.2)(s+20)(s+40)(s+50)(s+100)]^{-1}$, and is sketched in Fig. 5.9-1 (its phase lag is labeled θ_b').

It is noted that θ_b' is not much different from the approximate ideal phase. But since there is some difference, it is possible to achieve more $|L|$-frequency area by means of a more complicated L. Not *all* the permissible phase lag has been used. But in the vast majority of cases, the increased theoretical design effort and labor in synthesizing the resulting more complicated compensating transfer function is not worth the additional small gain in $|L|$-frequency area. (The quantitative relation between $|L|$ and its phase will be derived in Chapter 7.)

In the above, since $P = 10^3[s^2(s+1)(s+4)]^{-1}$, the required cascaded compensating transfer function is $G = L/P = (20)10^3(s+0.1)(s+2)(s+4)(s+1)[(s+0.2)(s+20)(s+40)(s+50)(s+100)]^{-1}$.

This section has presented techniques for shaping the loop transmission function to achieve the typical classical design specifications. Two more tasks must be done in order to make these techniques a part of a reasonably complete design procedure. One is to establish a correlation between the frequency response of a system and its time response. Such a correlation is necessary because our loop shaping technique is done in terms of frequency response. The other task is to correlate the system frequency response characteristics with those of the loop transmission. These two problems are considered in the following two sections.

§ 5.10 Correlation between System Frequency Response and Time Response

The intelligent application of the design techniques presented in the previous sections requires consideration of the following two topics:

(1) the relationship between the frequency response of a system and its time response;

(2) the relationship between the loop gain L and the system response function, $T = C/R = L/(1+L)$.

The relation between the frequency and step response of a system is treated in this section. We want to know the relations between the key features of the frequency response of a system and those of its time response. This knowledge would be very useful in specifying the desired system performance, which is really a time domain performance, in terms of system frequency response. The value of the transform at any value of frequency is obtained by integrating the weighted time function over all of time. Therefore it is unreasonable to expect precise relations between the key features of the time function and its transform. The relations are, in fact, approximate and cannot be used for precise design work. In most design problems the exact specification of the time domain performance is, however, not required. For the unusual case where a time response with fairly close tolerances is desired, the time domain synthesis techniques described in Chapter 12 may be used.

The principal features of the step response and of the system frequency response are defined in Figs. 5.10-1a and b. There are sufficient time domain performance indices defined in Fig. 5.10-1b to enable the desired performance

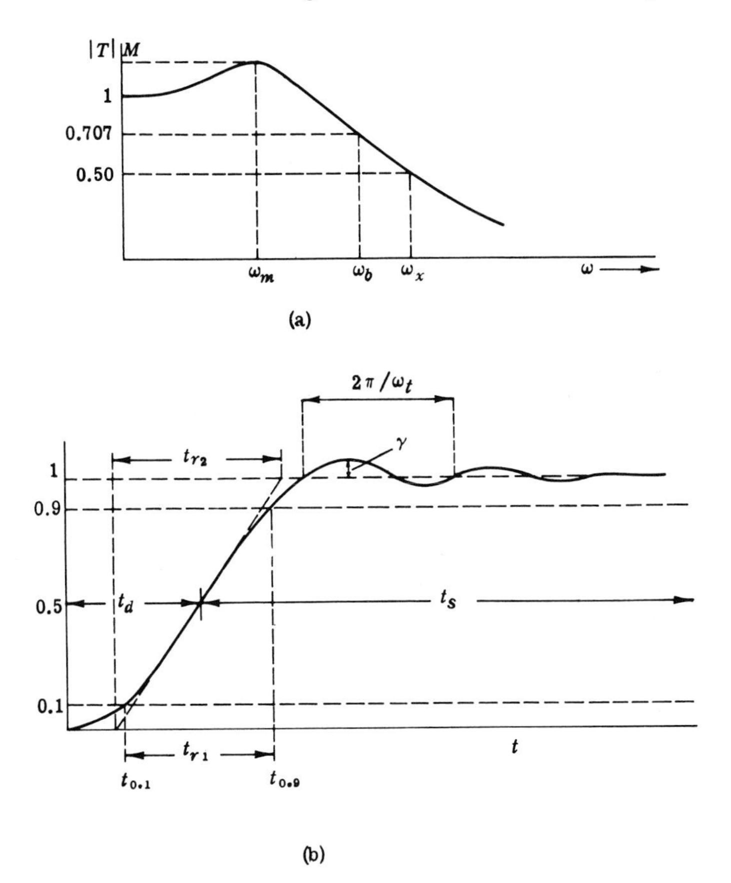

FIG. 5.10-1a, b. Definitions of key features of frequency and time responses. Indices in (b): t_d — time delay; t_r — rise time; γ — peak overshoot; t_s — settling time.

to be adequately described by assigning specific numbers to the various indices. Some of the relations that will be given are based on theory[1]; others were obtained from actual frequency response curves and their corresponding step responses. In view of the approximate nature of the relations, it may be a good idea to check the final design by calculating its time response (see Chapter 12 for techniques).

[1] E. A. Guillemin, "Communication Networks," Vol. 2, p. 477. Wiley, New York, 1947.

It would be useful to know the constraints on $T(j\omega)$ in order that its step response $p(t)$ should have the general form indicated in Fig. 5.10-1b. Some other possibilities are indicated in Fig. 5.10-2. Response C is easily recognized from the fact that its $T(0) = 0$. To recognize A, consider the expansion of $T(s)/s$ at infinity in the form $Ms^{-e} + Ns^{-(e+1)} + \ldots$ where e is the excess of poles over zeros of $T(s)/s$. The time response valid for t near zero is then $[(e - 1)!]^{-1}Mt^{e-1} + (e!)^{-1} Nt^e + \ldots$. Therefore, curve A is excluded if $M > 0$. It is not known what conditions $T(j\omega)$ must satisfy to exclude B or D. The empirical evidence suggests that both are excluded if T has the form shown in Fig. 5.10-1a, the important point in this regard apparently being that the slope of $|T|$ is first positive and then negative, and not positive-negative-positive-negative, or negative-positive-negative, with increasing frequency.

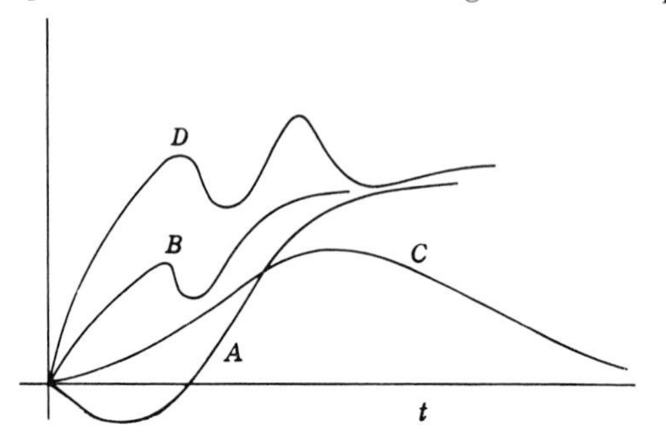

FIG. 5.10-2. Some other types of time response.

Sometimes it is important that the step response be monotonic with no overshoots whatever. A necessary[1] (but not sufficient) condition for monotonicity is that if $T(s)$ has any complex pole pairs, then there be at least one real pole of $T(s)$ to the right of the complex pair nearest the $j\omega$ axis. Another[2] necessary condition is that $d^2T/ds^2 > 0$ for all $s > 0$. This latter condition is satisfied, for example, when $T(s)$ is stable and has no zeros. In frequency response terms, a probably necessary (but not sufficient) condition is that there be no peaking in $|T(j\omega)|$. The scarcity of simple positive results in this subject is noteworthy.[3]

Relations between Time and Frequency Response Indices

Time Delay, t_d. The time delay is often ignored in communication problems where pure time lag may be tolerated, but in control systems the instantaneous difference between input and output is usually of serious concern. The time delay is related to the problem of system saturation. A plant that is complex with many interconnected elements, each with its own significant time constants, will have a large time delay in its response. If a $T(s)$ is specified which does not incorporate within it such a time delay, and if the design is based on working the plant near its maximum capacity, the design will result in such large signals

[1] J. H. Mulligan, Jr., The effect of pole and zero locations on the transient response of linear dynamic systems. *Proc. IRE* **37**, 516-529 (1949).

[2] G. Doetsch, "Laplace Transformation." Dover, New York, 1943.

[3] J. H. Mulligan, Jr., and A. H. Zemanian, Further bounds existing on the transient responses of various types of networks. *Proc. IRE* **43**, 322-326 (1955).

being forced upon the plant as to cause it to saturate. For this not to happen (if we insist on a T that has a small time delay), the plant has to be worked well below its maximum capacity. For example, suppose the plant, in response to a step input, cannot physically respond any faster than as shown in Fig. 5.10-3. If the maximum output for which linear operation is desired is F, then the time delay specified for T could be approximately t_{d2}. However, if t_{d2} is used and the step input is larger than F, then the system will saturate for part of the time.

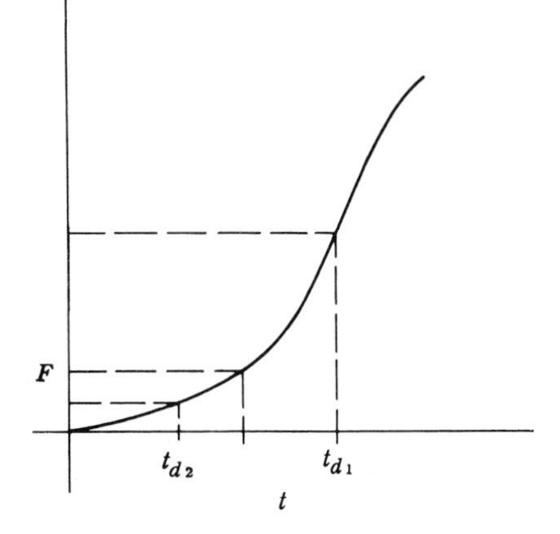

FIG. 5.10-3. Relation between time delay and saturation.

The relation between t_d and frequency response parameters is

$$t_d \approx |\varphi_b|/|\omega_b| \qquad (5.10,1)$$

where φ_b is the phase change of T in radians between ω_b and zero, i.e., $\varphi_b = \arg T(j\omega_b) - \arg T(0)$, and ω_b has been defined in Fig. 5.10-1a as the half-power point. The relation is exact[1] if $\arg T(j\omega)$ is linear in ω. The relation was checked against the results of Figs. 5.19-4 and it was found that for these specific curves, $t_d \approx 0.80 \,|\varphi_b/\omega_b|$ gave far better results.

Equation (5.10,1) reveals that the time delay is important in determining the complexity of T, especially the excess of poles over zeros in the significant frequency range of T. This is due to the fact that ω_b, as later noted, is inversely related to the rise time, i.e., $\omega_b \approx K/t_r$. Therefore, from Eq. (5.10,1),

$$\varphi_b \approx Kt_d/t_r. \qquad (5.10,2)$$

Suppose the sheer physical capabilities of the system dictate, for the specific level of the desired output, a ratio of at least 2 for t_d/t_r (in order that the plant may not saturate). Then, using $K = \pi$ in Eq. (5.10,2), $\varphi_b \approx 2\pi$; which probably requires an excess of at least six poles over zeros in the significant T frequency range. The latter is approximately two octaves beyond ω_b, because a pole or a zero at $-5\omega_b$ contributes only 11.3° to $\arg T(j\omega_b)$. Some writers have suggested that $T(s)$ may very often be characterized by a few dominant poles and zeros. When this is done, φ_b is perforce restricted to a relatively small number and from Eq. (5.10,2), so is t_d (unless one is willing to increase the rise time and obtain very slow response). Therefore, the use of a small number of dominant poles is satisfactory only for systems in which the ratio of time delay to rise time is small.

[1] E. A. Guillemin, "Communication Networks," Vol. 2, p. 477. Wiley, New York, 1947.

For example, Fig. 5.10-4 displays two step-function responses[1] which are quite similar, and one might be tempted to specify $T = T_a = (s^2 + s + 1)^{-1}$ in order to simplify the design. However, for small t, T_a requires the plant output to be approximately twice that of T_b. If the design is based upon the plant working near its maximum capacity, the system may saturate. This can be corrected either by using T_b or, if we persist with the simpler T_a, the rise time of the latter must be increased, for example, by using $T = 0.5/(s^2 + 0.7s + 0.5)$. In the latter case, the rise time is larger, and the over-all system step response is more sluggish.

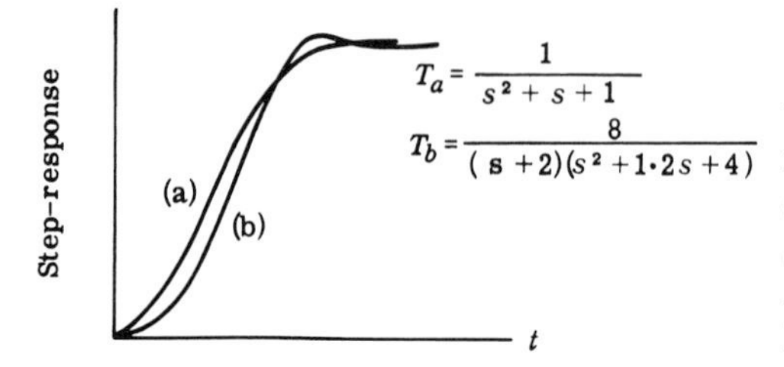

FIG. 5.10-4. Two responses which are superficially similar.

The value of φ_b requires knowledge of the phase of T. If arg T is not available, one may obtain φ_b from $| T |$, if $T(s)$ has no right half-plane zeros. This is proven in Chapter 7, where techniques for finding arg T from $| T |$ are also presented.

The relation between time delay and system complexity may also be seen by noting that if e represents the excess of poles over zeros of T, then the step response $p(t)$, has $p(0) = p'(0) = ... = p^{e-1}(0) = 0$. Obviously, the larger the number of derivatives that are zero at $t = 0$, the larger is the time delay. This might suggest that even the far-off poles are significant. This is not so, as can be seen by considering the effect of an additional far-off pole on $T = H\Pi(s + z_i)/\Pi(s + p_j)$. The effect is noted by finding the difference between the response of T and that of $T_x = H\Pi(s + z_i)q/\Pi(s + p_j)(s + q)$. Now[2] $p_x(t) = p(t)*qe^{-qt}$. If q is large in comparison with the interval over which there are significant changes in $p(t)$, qe^{-qt} is essentially an impulse function and hence $p_x(t) \doteq p(t)$.

Rise time, t_r. Some writers[3] have used the relation

$$t_{r1}f_b = K, \qquad 0.30 \leqslant K \leqslant 0.45, \tag{5.10,3}$$

for systems with overshoots no more than approximately 10%. Values of 0.30 to 0.35 have been recommended for overshoots less than 5%, and the higher figure for the larger overshoots. For larger values of overshoot, the effect is the opposite, i.e., K decreases as the overshoot increases. An empirical relation[4]

[1] See Fig. 1.24 on p. 42 *in* J. G. Truxal, "Control System Synthesis." McGraw-Hill, New York, 1955.

[2] $p(t)*q(t)$ represents the convolution of $p(t)$ and $q(t)$.

[3] G. E. Valley, Jr., and H. Wallman, "Vacuum-Tube Amplifiers," Radiation Lab. Series, Vol. 18, p. 80. McGraw-Hill, New York, 1948; J. G. Truxal, "Control System Synthesis," p. 80. McGraw-Hill, New York, 1955.

[4] Z. E. Jaworski, Empirical transient formulae. *Electronic Eng.* **26**, 319, 396 (1954) (the data is for $|T|$ with a single peak either at $\omega = 0$ or at another frequency).

obtained from data with overshoots mostly in the 8-20% range but with some in the 30's, is

$$t_{r2}f_x = 0.5 \tag{5.10,4}$$

($f_x = \omega_x/2\pi$ is defined in Fig. 5.10-1a). It is reported that the maximum error in about 25 samples is 15%. For the curves of Fig. 5.19-4, the empirical relation $t_{r1}f_b = 0.32 - 0.0011(\gamma - 20)$ gave good results. Out of 27 curves, the error in using the latter formula was $\leqslant 5\%$ for 20 of them, and $\leqslant 8\%$ for all 27. The overshoots varied from 11 to 87% in these 20 curves. (Note in the last equation that the $t_{r1}f_b$ product decreases with overshoot.)

These empirical formulae must be used with caution. They are valid only for the data upon which they are based. For example, note Fig. 5.10-5 for the

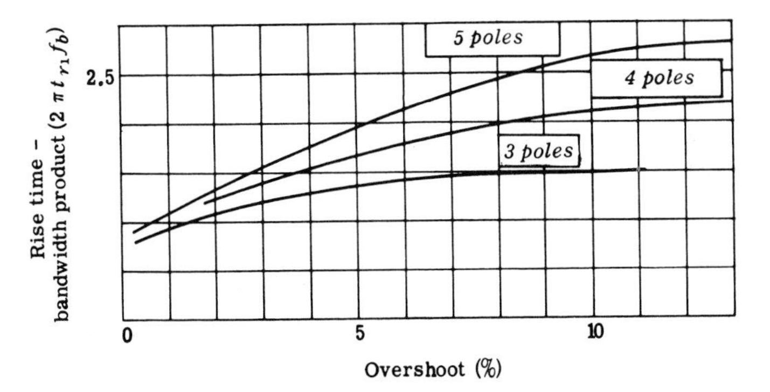

FIG. 5.10-5. Relation between overshoot and $t_{r1}f_b$ product for transitional Butterworth-Thomson filters.

relation between $t_{r1}f_b$ and overshoot for the transitional Butterworth-Thomson filters.[1] It is, however, generally agreed that the bandwidth is the major factor in determining the rise time, the latter being inversely proportional to the bandwidth.

Overshoot, γ. Observations of curves of $|T(j\omega)|$ and the corresponding transient response indicate definitely that the overshoot rises with M (defined in Fig. 5.10-1a as the peak in the frequency response), and with the sharpness of cutoff. Thus, the Butterworth step responses have overshoot which increase with the sharpness of cutoff, despite the fact that their $|T(j\omega)|$ has a maximum at zero frequency.[2] An empirical relation[3] is

$$\gamma = 58(\omega_b/\omega_x)M - 39 \tag{5.10,5}$$

[1] Y. Peless and T. Murakami, Analysis and synthesis of transitional Butterworth-Thomson filters. *RCA Rev.* **18**, 1 (1957).

[2] G. E. Valley, Jr., and H. Wallman, "Vacuum-Tube Amplifiers," Radiation Lab. Series, Vol. 18, p. 283, Fig. 7.7. McGraw-Hill, New York, 1948.

[3] Z. E. Jaworski, *Electronic Eng.* **26**, No. 319 (1954).

for values of ω_b/ω_x ranging from 0.7 to 0.95. The result applies for $T(j\omega)$ with peaks at zero frequency (for which $M = 1$) or at other frequencies (for which $M > 1$). An average error of 14% and a maximum error of 25% (the absolute error increases with overshoot) is reported for the experimental data upon which the formula is based. The relation checks fairly well with the results of Chestnut and Mayer.[1]

The frequency of the damped oscillation, ω_t (see Fig. 5.10-1b) is closely related to ω_m (defined in Fig. 5.10-1a), when the latter exists. If it does not exist, the half-power frequency ω_b should be used. Generally ω_t is somewhat larger than ω_m but approaches ω_m for large overshoot.[2]

Settling time, t_s. Jaworski reports the 1 and 0.1% settling times as

$$t_{s1} = \frac{1}{2f_x}\left(\frac{7\omega_b}{\omega_x} M - 3\right) \tag{5.10,6a}$$

$$t_{s0.1} = \frac{1}{2f_x}\left(\frac{9\omega_b}{\omega_x} M - 3\right) \tag{5.10,6b}$$

with rms errors of 18% and 25%, respectively. These results should be used with caution in view of the relatively small number of samples upon which they are based.

It is clear from the above results that there do not exist precise relations between the key features of the time response and those of the frequency response. The relations that have been given provide the designer with an understanding and circumspection which are very useful in design. We shall use the relations later in comparing the $L(j\omega)$ frequency response method with the $T(s)$ pole-zero method.

The above results may be applied to frequency response curves of the band-pass type which have the same general shape as that in Fig. 5.10-1a, and which are symmetrical with respect to a center frequency ω_o. Figure 5.10-1b then represents the envelope of the response to a step input which modulates a carrier of frequency ω_o. The vertical line at the center frequency ω_o on the band-pass characteristic should be made to correspond to the ordinate axis of Fig. 5.10-1a, e.g., ω_b represents half of what is usually considered to be the bandwidth of the band-pass curve.

§ 5.11 Relation between the Loop Transmission *L* and the System Transfer Function *T*

In the classical single-degree-of-freedom configuration of Fig. 5.11-1, $T = C/R$ is completely determined by $L = GP$, so the latter presumably should be chosen

[1] H. Chestnut and R.W. Mayer, Comparison of steady-state and transient performance of servomechanisms. *Trans. AIEE* **68**, 765-777 (1949).

[2] Z. E. Jaworski, *Electronic Eng.* **26**, No. 319 (1954).

to achieve a satisfactory T. In the low frequency range in which $|L| \gg 1$, $T \doteq 1$; in the high frequency range in which $|L| \ll 1$, $T \doteq L$. The problem of shaping $L(j\omega)$ to achieve a satisfactory $T(j\omega)$ arises near the crossover region.

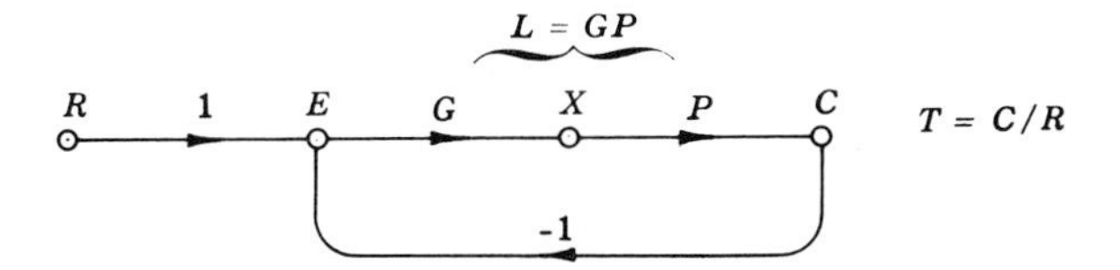

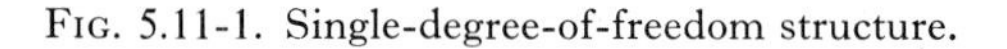

FIG. 5.11-1. Single-degree-of-freedom structure.

However, the need for stability margins in order to ensure stability despite plant parameter variations may dominate the shaping of $L(j\omega)$ in the crossover region. Therefore, the designer is free to shape $L(j\omega)$ in the crossover region only when the shaping [for $T(s)$] leads to stability margins which are even more conservative than those required from stability considerations. This is the inevitable consequence of using a structure with only one degree of freedom.

Because of the simple relation between T and L, it is easy to prepare charts to read one directly from the other. Let $T = |T| e^{j\theta}$. Then on the complex plane of L, loci of constant $|T|$ are circles of radii $|T|/[1 - |T|^2]$, centered at $x = -|T|^2[|T|^2 - 1]^{-1}$. Loci of constant θ are circles centered at $x = -\frac{1}{2}$, $y = (2 \tan \theta)^{-1}$. The radii of the circles are obtained from the fact that they all pass through the origin. Loci of constant $|T|$ and θ superimposed on the polar L plane are shown in Figs. 5.11-2a and b. If the transition from a predetermined desired T to the required L were common, the opposite type of charts (loci of constant $|L|$ and arg L on the T plane) could also be prepared although those of Figs. 5.11-2 may of course be used for the purpose. However, because it is easiest to shape L on a Bode plot (db vs. log frequency), and use of Fig. 5.11-2 requires converting from db to their arithmetic equivalents, charts have been prepared (Nichols[1] charts) in which the superposition of L and T is done on a

[1] H. M. James, N. B. Nichols, and R. S. Phillips, "Theory of Servo-Mechanisms," p. 179. McGraw-Hill, New York, 1947.

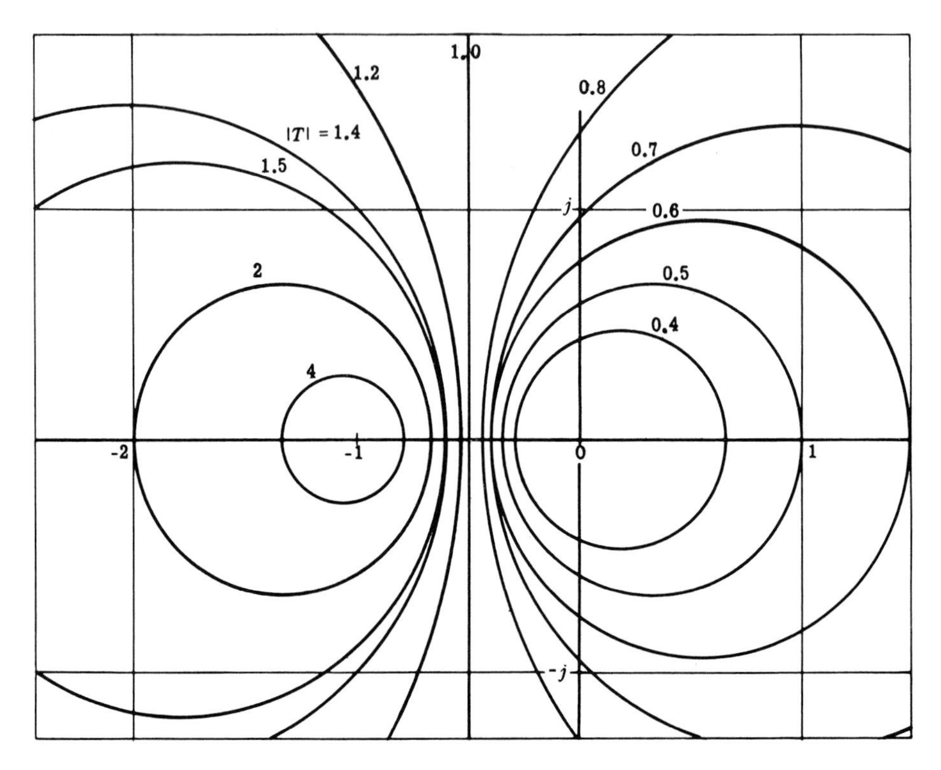

FIG. 5.11-2a. Loci of constant | T | on L plane when $T = L/(1 + L)$.

logarithmic scale (Fig. 5.11-3). The same result could be obtained in Fig. 5.11-2 by using a logarithmic scale for the circles of constant | T | or | L |.

In many design problems the exact specification of T is not considered important. Certain key features of T, such as peak amplitude M, half-power frequency ω_b, phase at ω_b, and sharpness of cutoff, may be chosen from considerations discussed in the previous section. These impose constraints on L, especially around the crossover frequency, ω_c. Whatever the requirements may be on T, the corresponding ones on L are easily ascertained from Figs. 5.11-2 and 3. While this procedure involves some cut and try, and may therefore appear somewhat inelegant, it is nevertheless a very useful practical procedure because very few trials are usually necessary, and because one is constantly working with the L function, which is the actual function that the designer must eventually construct.

The loop transmission functions of many of the simpler servo systems have fairly similar shapes and lend themselves to systematic classification. Key features of the time responses of the various classes have been calculated and are available[1]

[1] H. Chestnut and R. Mayer, Comparison of steady-state and transient performance of servomechanisms. *Trans. AIEE* **68**, 765-777 (1949).

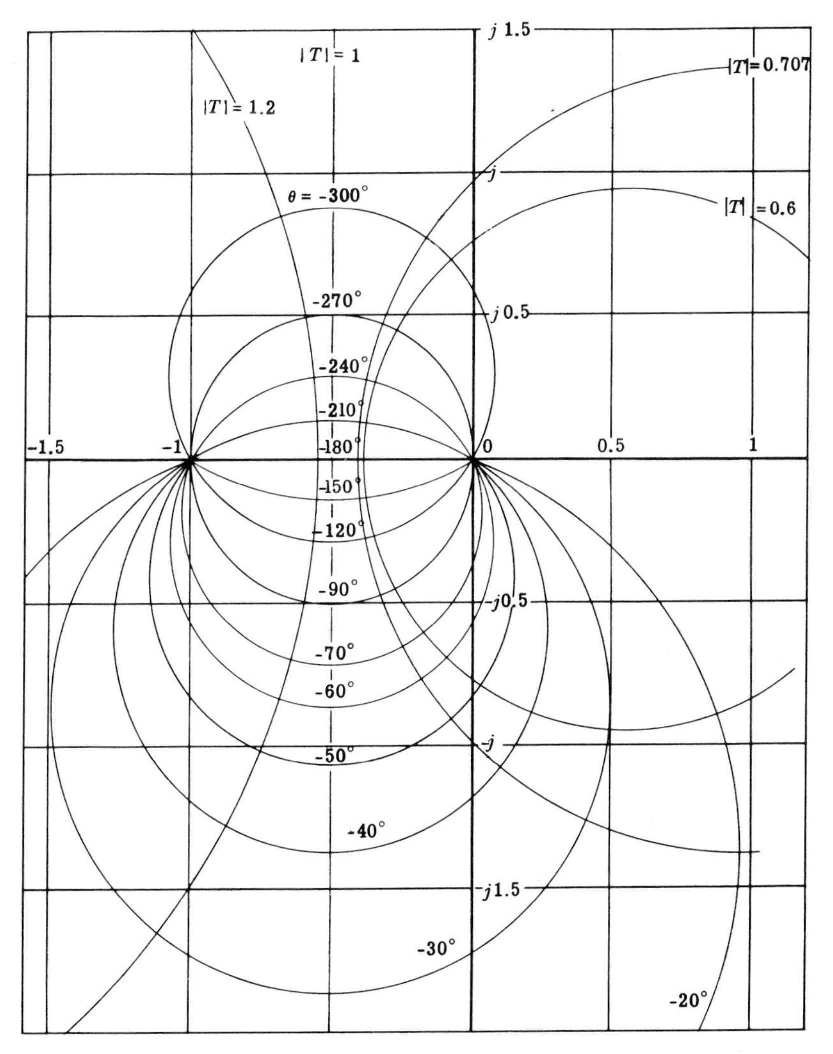

FIG 5.11-2b. Loci of constant $|T|$ and θ = arg. T on L plane when $T = L/(1 + L)$.

in the form of charts which facilitate the choice of L. An approximate relation between the location of the dominant pole pair of $T(s)$ and the gain and phase margins is as follows[1]:

$$\frac{\omega_n^2}{\omega_c^2} \approx \frac{G_M^2 + \theta_M^2(\omega_2^2/\omega_c^2)}{G_M^2 + \theta_M^2} \tag{5.11,1a}$$

$$\zeta \approx \frac{\theta_M G_M}{\theta_M^2 + G_M^2}\left(\frac{\omega_2 - \omega_c}{\omega_n}\right). \tag{5.11,1b}$$

[1] Holt-Smith *in* Discussion on A simple correlation between closed loop transient response and open loop frequency response. *Proc. Inst. Elec. Engrs. (London)*, Part II, **100** p. 210 (1953).

In the above, ω_c is the crossover frequency in rps, ω_2 is the frequency at which $\arg L = -180°$, θ_M is the phase margin in radians, G_M is the gain margin in nepers (1 neper $=$ 8.69 db) and the dominant pole pair of $T(s)$ consists of the

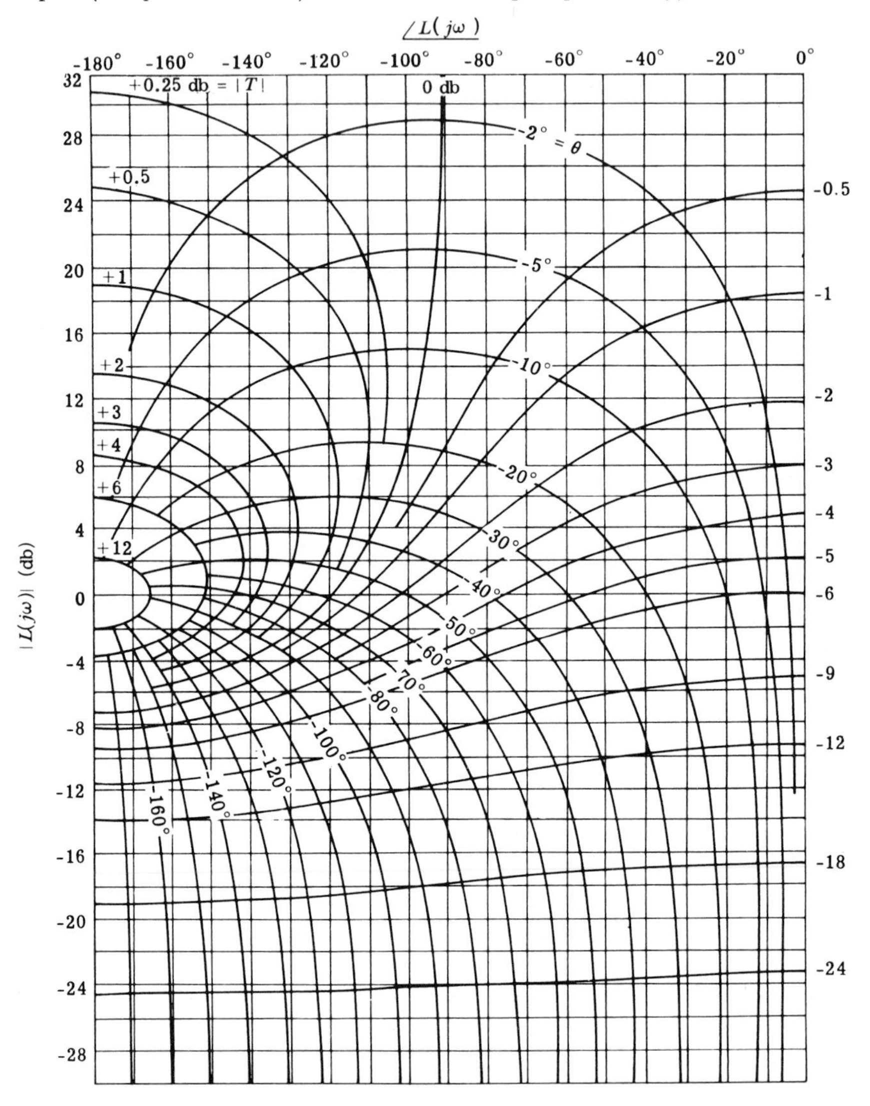

FIG. 5.11-3. Nichols chart—Loci of $|T|$ and $\theta = \arg T$ on logarithmic L plane.

zeros of $s^2 + 2\zeta\omega_n s + \omega_n^2$. For example, if the phase margin is 40°, the gain margin is 8.7 db and $\omega_2/\omega_c = 3$, then Eqs. (5.11,1) give $\zeta \approx 0.5$, $\omega_n/\omega_c \approx 1.9$. Equations (5.11,1) are derived by using the data at ω_c and ω_2 to obtain an approximate dL/ds [the mapping from s to $L(s)$ is conformal]; this value of dL/ds is used to find the value of $s = -\zeta\omega_n + j\omega_n(1 - \zeta^2)^{1/2}$ at which $L = -1$.

§ 5.12 Synthesis from Pole-Zero Specifications of *T*(s)

The synthesis procedure described in this section is based upon the characterization of the system transfer function $T(s)$ by means of its poles and zeros.[1] At the very outset, one specifies analytically the $T(s)$ that is desired, and then goes about realizing it. The procedure is strongly influenced by the philosophy of modern network synthesis. Instead of working with the loop transmission function, and shaping it until the specifications are achieved, one states the specifications in terms of $T(s)$, obtains an analytical expression for the latter, and solves for $L(s)$. If there are any constraints on $L(s)$, it is necessary to find the corresponding constraints on $T(s)$ before specifying the latter. We will present the procedure, as it has been developed and refined, in Sections 5.12-5.16, and will compare it in some detail in Sections 5.17-5.18, with the open-loop frequency response method.

The first step in the development of the $T(s)$ pole-zero method is to relate the customary control system specifications into their equivalent in terms of the closed-loop system response function $T(s)$. The error constants defined in Section 5.4 are very popular design specifications. Their relations with $T(s)$ are easily obtained as follows. Let

$$T = K\Pi(s + z_i)/\Pi(s + p_j). \tag{5.12,1}$$

When the input R represents a step function, the final value theorem gives (Fig. 5.11-1)

$$e(\infty) = \lim_{s\to 0} sE(s) = \lim_{s\to 0} sR(1 - T) = \lim_{s\to 0} (1 - T) = 1 - T(0).$$

However, it was shown in Section 5.4 that for a step input $e(\infty) = 1/(1 + K_p)$. Consequently, $1 - T(0) = 1/(1 + K_p)$, and

$$T(0) = K_p/(1 + K_p) = K\Pi z_i/\Pi p_j. \tag{5.12,2}$$

When the input is a ramp, the final value of the error is finite only if K_p is infinite. Assuming then that K_p is infinite [*i.e.*, $T(0) = 1$], the relation between the error constant K_v and $T(s)$ is found as follows. Expand $T(s)$ in the Taylor

[1] J. G. Truxal, Servomechanism synthesis through pole-zero configurations. *MIT Research Lab. Electronics, Tech. Rept.* **162** (1950); M. R. Aaron, Synthesis of feedback control systems by means of pole and zero location of the closed loop function. *Trans. AIEE* **70** (Part 2), 1439-1445 (1951).

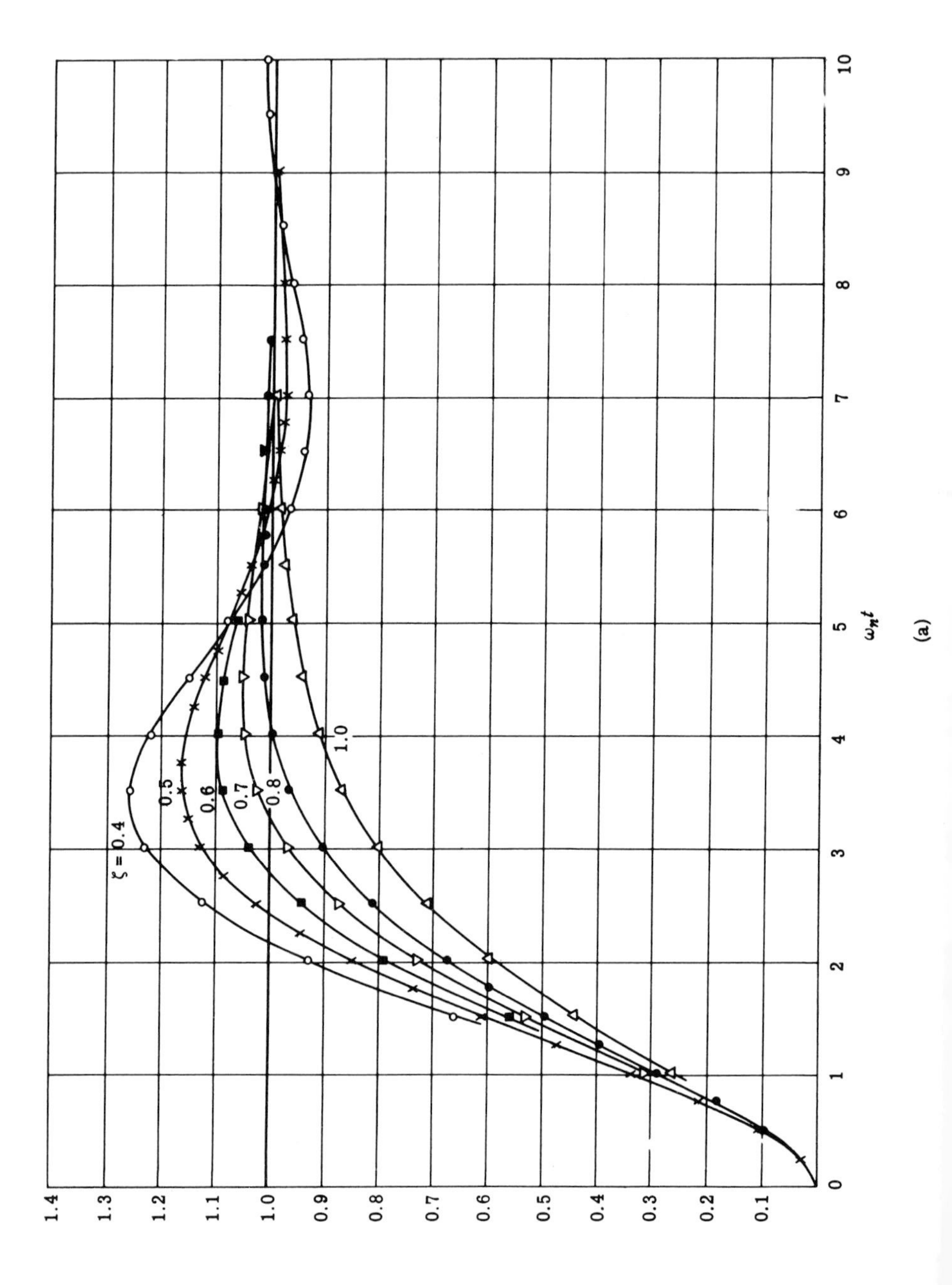

ζ = 0.4
0.5
0.6
0.7
0.8
1.0
0
0.1
0.2
0.3
0.4
0.5
0.6
0.7
0.8
0.9
1.0
1.1
1.2
1.3
1.4
0
1
2
3
4
5
6
7
8
9
10
$\omega_n t$

(a)

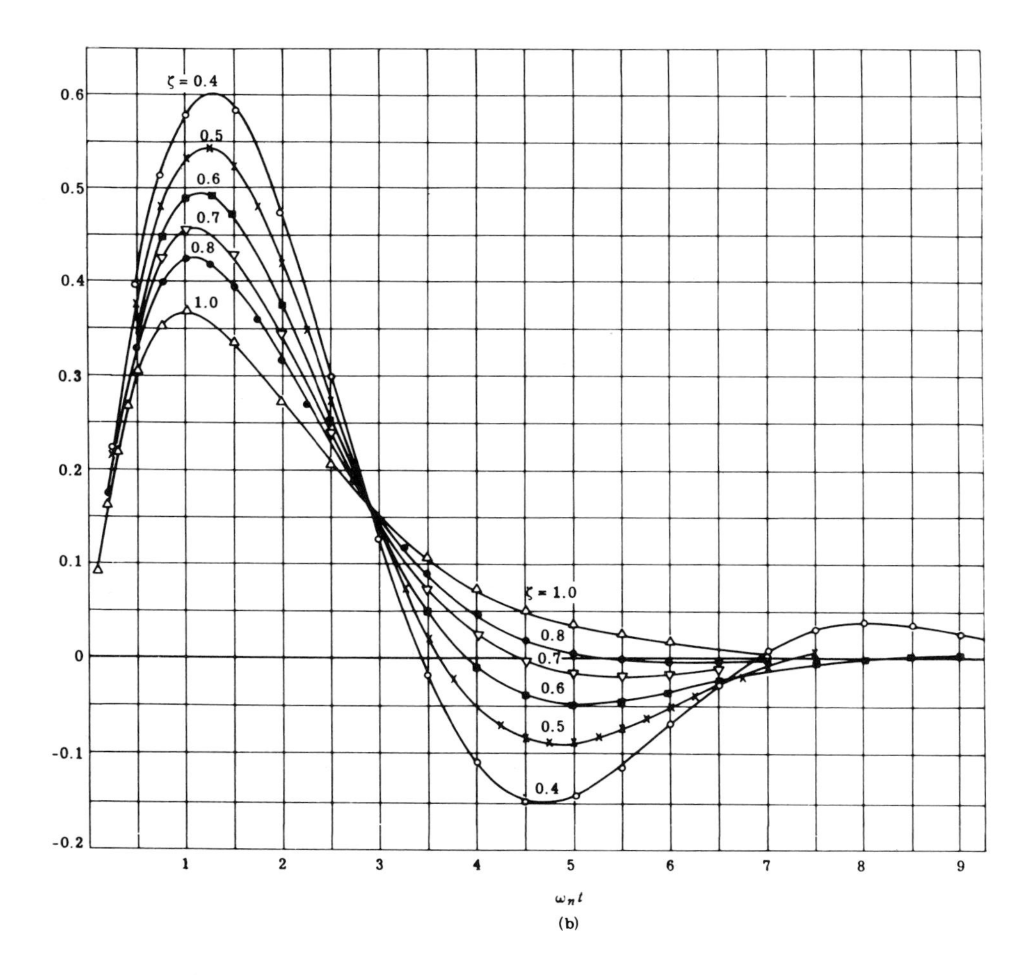

FIG. 5.12-1(a), (opposite), Step and (b) impulse response,

$$T(s) = \frac{\omega_n^2}{s^2 + 2\zeta\omega_n s + \omega_n^2}.$$

series $T(s) = 1 + sT'(0) + 0.5\, s^2T''(0) + \ldots$. The steady state error for a ramp input ($R = 1/s^2$) is

$$e(\infty) = \lim_{s\to 0} s(R - C) = \lim_{s\to 0} sR(1 - T) = \lim_{s\to 0} \frac{1}{s} \{1 - [1 + sT'(0) + 0.5s^2T''(0) + \ldots]\} = - T'(0).$$

Since $T(0) = 1$ here, we may write $e(\infty) = - T'(0)/T(0)$, which in turn is identical to $[- d/ds \ln T(s)]_{s=0}$. Also in Section 5.4 we defined the velocity constant as $K_v^{-1} = e(\infty)$ when $R = 1/s^2$ and K_p is infinite. Therefore

$$e(\infty) = \frac{1}{K_v} = - \left[\frac{d}{ds} \ln T(s)\right]_{s=0} = - \left[\frac{d}{ds} \ln K \frac{\Pi(s + z_i)}{\Pi(s + p_j)}\right]_{s=0}$$

$$= - \left[\sum \frac{1}{s + z_i} - \sum \frac{1}{s + p_j}\right]_{s=0} = \sum \frac{1}{p_j} - \sum \frac{1}{z_i}. \tag{5.12,3}$$

Similarly, if K_v is infinite, $T'(0) = 0$ and the acceleration constant K_a is obtained from

$$e(\infty) = \frac{1}{K_a} = - 0.5T''(0) = - 0.5 \left[\frac{d^2}{ds^2} \ln T(s)\right]_{s=0}$$

$$= \frac{1}{2} \left[\sum \frac{1}{z_i^2} - \sum \frac{1}{p_j^2}\right]. \tag{5.12,4}$$

Some of the other specifications such as gain margin, phase margin, and crossover frequency, to the extent that they are specifications on $T(s)$, are not as significant as they were in Sections 5.5-5.9, because we are now working directly with $T(s)$. Instead, we seek to establish a direct correlation between the poles and zeros of $T(s)$ and the system step response. When one considers a choice for $T(s)$, he is interested in the quantities defined in Fig. 5.10-1b, i.e., overshoot, rise time, settling time, etc., and also in such things as the error constants and system bandwidth. Workers in this field have tabulated a limited amount of such information for relatively simple pole-zero patterns of $T(s)$.

The simplest $T(s)$ is one consisting of a single pole on the negative real axis. Its properties are obvious and will not be considered any further. Next is $T(s)$ with a single pair of poles given by $T(s) = \omega_n^2/(s^2 + 2\omega_n\zeta s + \omega_n^2)$. The corresponding step response for a number of values of ζ are shown in Fig. 5.12-1a. The corresponding impulse response curves are sketched in Fig. 5.12-1b. The bandwidth (frequency at which $|T| = 0.707$) is found to be

$$\omega_b = BW = \omega_n[1 - 2\zeta^2 + (2 - 4\zeta^2 + 4\zeta^4)^{0.5}]^{0.5} \tag{5.12,5}$$

and from Eq. (5.12,3) it is found that

$$K_v = \omega_n/2\zeta. \tag{5.12,6}$$

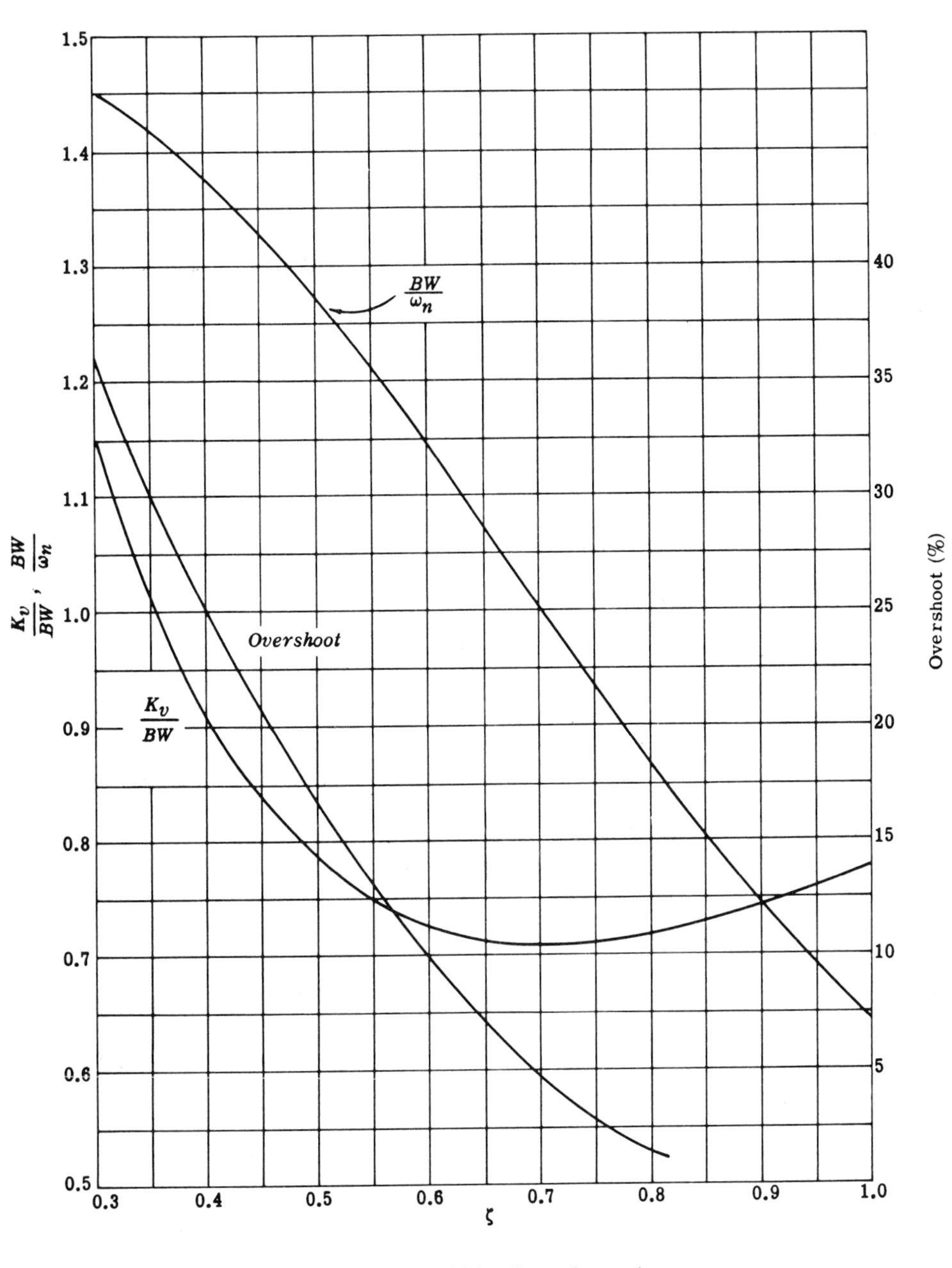

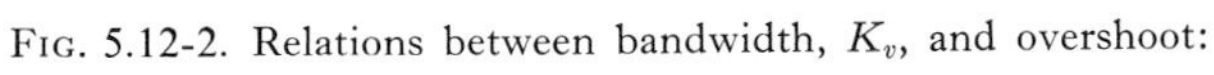

FIG. 5.12-2. Relations between bandwidth, K_v, and overshoot:

$$T(s) = \frac{\omega_n^2}{s^2 + 2\zeta\omega_n s + \omega_n^2} .$$

In Fig. 5.12-2 the relations between overshoot, K_v, BW, and ζ are displayed graphically. These curves tell at a glance whether a particular set of design specifications can be handled by means of a $T(s)$ consisting of a single pair of poles. For example, if the maximum overshoot and the bandwidth are, respectively, 20% and 20 rps, then the minimum ζ is 0.46 at which $K_v = 0.83BW = 16.6$. If a velocity constant larger than 16.6 is wanted, then a more complicated $T(s)$ must be used. If a smaller value is desired, the difference could be used to obtain a larger K_v, a smaller overshoot, a smaller bandwidth, or some combination of these.

§ 5.13 Realization of Any Combination of K_v, Bandwidth, and Overshoot with a Pair of Poles and One Zero

We next consider $T(s) = (\omega_n/\lambda\zeta)(s + \lambda\zeta\omega_n)/(s^2 + 2\zeta\omega_n s + \omega_n^2)$ (Fig. 5.13-1) whose step response is

$$C(s) = \frac{T(s)}{s} = \frac{\omega_n(s + \lambda\zeta\omega_n)}{\lambda\zeta s D} = \frac{\omega_n^2}{sD} + \frac{\omega_n^2}{\lambda\zeta\omega_n D},$$

where $D = s^2 + 2\zeta\omega_n s + \omega_n^2$. Hence

$$c(t) = p(t) + (\lambda\zeta\omega_n)^{-1}h(t) \tag{5.13,1}$$

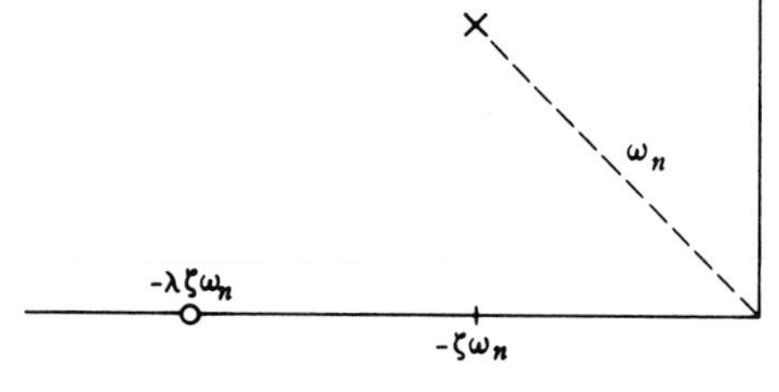

FIG. 5.13-1. $T(s)$ with one zero and complex pole pair.

where $p(t)$ is the step response due to the pole pair alone, and the second term is $(\lambda\zeta\omega_n)^{-1}$ times the impulse response of the pole pair alone. The summing can be achieved by suitably combining appropriate curves from Figs. 5.12-1a and b. For example, if $\zeta = 0.8$, $\lambda = 2$ (and ω_n is as usual normalized to unit value), we add the step response for $\zeta = 0.8$, and $1/1.6 = 0.625$ times the impulse response for $\zeta = 0.8$. The sum of these two is the step response of the two pole-one zero system. The data for $p(t)$ and $h(t)$ are tabulated in Table 5.13-1 to enable the reader to prepare his own accurate curves.

It is readily seen that the added zero at $-\lambda\zeta\omega_n$ always leads to a faster response and a larger overshoot than that due to the pole pair alone. It is also seen, from Eq. (5.12,3), that the zero leads to a larger value of K_v; in fact, by locating the zero at $z = -\omega_n/2\zeta$, infinite K_v [requiring a double pole of $L(s)$ at $s = 0$] is

TABLE 5.13-1

STEP AND IMPULSE RESPONSE DATA FOR $T(s) = 1/(s^2 + 2\zeta s + 1)$

	$\zeta = 1.0$		$\zeta = 0.8$		$\zeta = 0.7$		$\zeta = 0.6$		$\zeta = 0.5$		$\zeta = 0.4$	
t	$h(t)$	$p(t)$	$h(t)$	$p(t)$	$h(t)$	$p(t)$	$h(t)$	$p(t)$	$h(t)$	$p(t)$	$h(t)$	$p(t)$
0.5	0.303	0.090	0.330	0.095	0.345	0.101	0.360	0.102	0.377	0.105	0.395	0.109
1	0.368	0.264	0.423	0.291	0.456	0.305	0.492	0.322	0.534	0.339	0.580	0.360
1.5	0.335	0.442	0.392	0.500	0.430	0.531	0.474	0.568	0.524	0.611	0.588	0.658
2	0.271	0.594	0.313	0.676	0.342	0.725	0.376	0.783	0.419	0.849	0.473	0.927
2.5	0.205	0.713	0.244	0.811	0.238	0.870	0.254	0.941	0.274	1.020	0.302	1.122
3	0.150	0.800	0.147	0.903	0.144	0.965	0.139	1.038	0.133	1.124	0.126	1.228
3.5	0.106	0.864	0.088	0.961	0.072	1.018	0.051	1.084	0.022	1.162	−0.018	1.253
4	0.073	0.908	0.046	0.933	0.024	1.041	−0.007	1.094	−0.049	1.153	−0.110	1.219
4.5	0.050	0.939	0.019	1.009	−0.004	1.046	−0.037	1.083	−0.083	1.118	−0.150	1.152
5	0.034	0.960	0.004	1.015	−0.018	1.040	−0.047	1.061	−0.087	1.074	−0.146	1.076
5.5	0.0225	0.974	−0.003	1.015	−0.021	1.030	−0.044	1.038	−0.074	1.034	−0.115	1.010
6	0.015	0.983	−0.006	1.012	−0.019	1.020	−0.034	1.018	−0.051	1.002	−0.070	0.964
6.5	0.010	0.989	−0.006	1.009	−0.015	1.011	−0.022	1.004	−0.027	0.983	−0.026	0.940
7	0.0064	0.993	−0.005	1.006	−0.010	1.005	−0.012	0.996	−0.008	0.975	0.009	0.936
7.5					−0.006	1.001	−0.004	0.992	0.006	0.974	0.030	0.947
8					−0.003	0.999	0.001	0.991			0.039	0.964
8.5							0.004	0.992			0.036	0.983
9							0.004	0.995			0.027	0.999

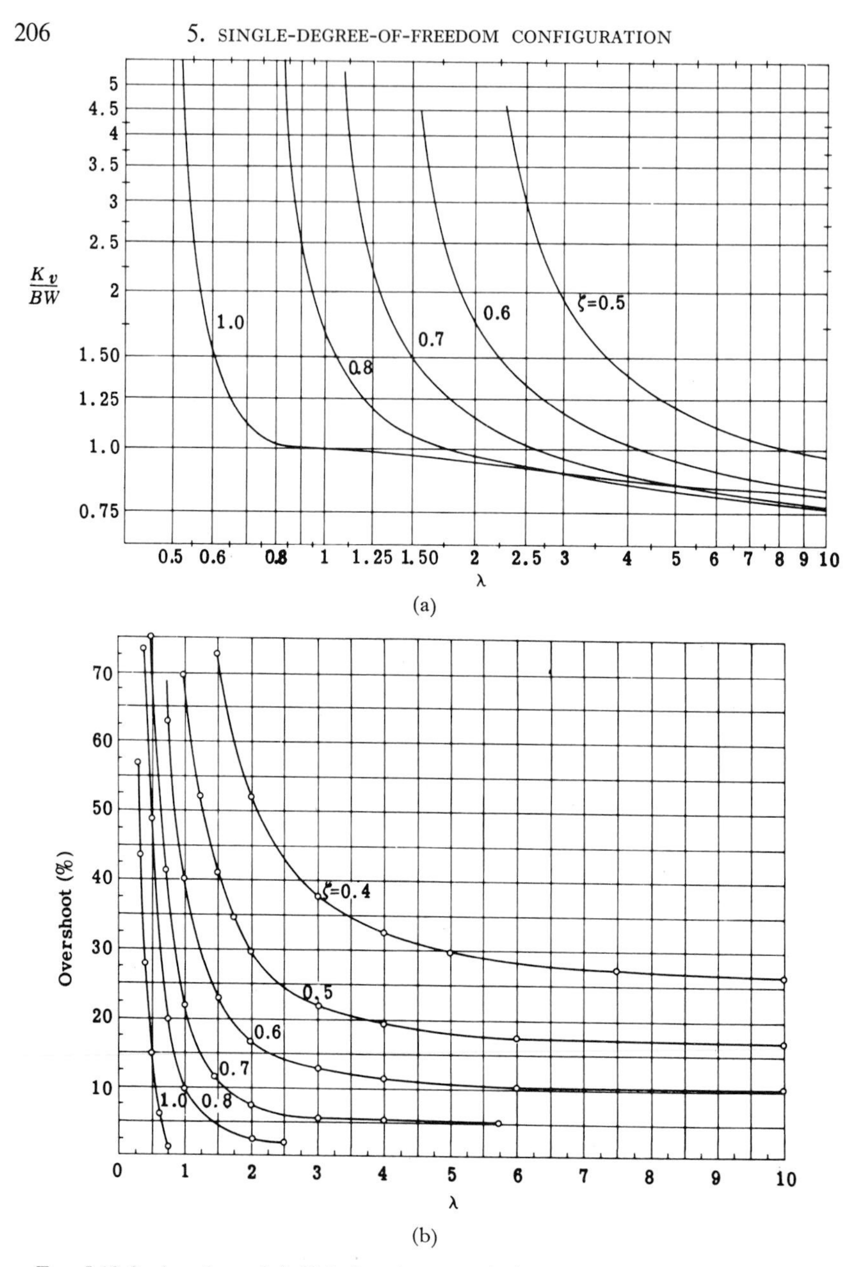

FIG. 5.13-2a, b, c (*opposite*). Relations between design parameters:

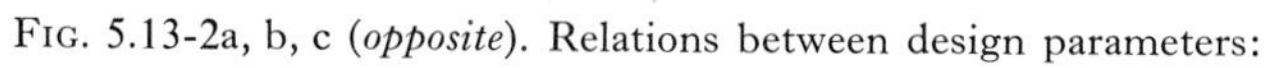

$$T(s) = \frac{\omega_n(s + \lambda\zeta\omega_n)}{\lambda\zeta(s^2 + 2\zeta\omega_n s + \omega_n^2)}$$

obtained. A $T(s)$ consisting of two poles and one zero is therefore useful when a single pole pair cannot deliver the required K_v at the permitted overshoot and

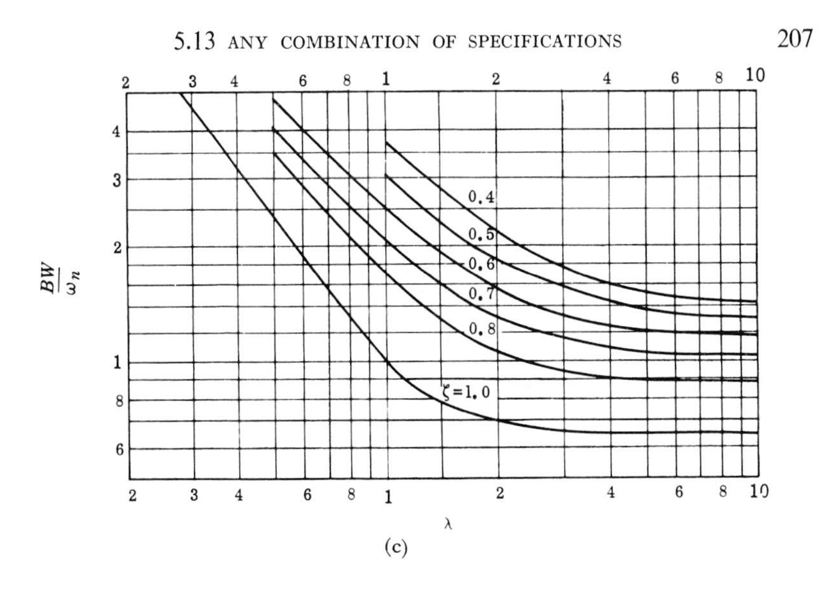

(c)

bandwidth. In fact, any combination of K_v, bandwidth, and peak overshoot can be realized by means of a $T(s)$ with two poles and one zero.

Figures 5.13-2a, b, and c have been prepared to facilitate the selection of ω_n, ζ, and λ to achieve any combination of K_v, bandwidth, and peak overshoot.[1] The bandwidth ω_b was found by writing

$$| T(j\omega_b) |^2 = 0.5 = \frac{(\lambda^2\zeta^2)^{-1} \, (\lambda^2\zeta^2 + \omega_b{}^2)}{(1 - \omega_b{}^2)^2 + 4\zeta^2\omega_b{}^2},$$

and solving for ω_b. K_v was obtained from $\omega_n(K_v)^{-1} = 2\zeta - (\lambda\zeta)^{-1}$, and the overshoot was found by combining the curves of Fig. 5.12-1a with the appropriate multiples of those of Fig. 5.12-1b, in accordance with Eq. (5.13,1). The application of these curves in design is illustrated by the following example.

Example.

Specifications: The bandwidth is to be 20 rps, the maximum overshoot is 15%, and $K_v = 40$.

Design: $T(s)$ with a single pole pair is attempted, so Fig. 5.12-2 is examined. At 15% overshoot, $\zeta = 0.52$ and $K_v/BW = 0.77$, compared with the required ratio of $40/20 = 2$, so a complex pole pair design will not do. We try $T(s)$ with two poles and one zero and turn to Figs. 5.13-2.

In Fig. 5.13-2a the horizontal line corresponding to $(K_v/BW) = 2$ cuts the curves at the points listed in Table 5.13-2. The entries in the first two columns of Table 5.13-2 are obtained from Fig. 5.13-2a. The entries in the third column

[1] Curves similar to these were first prepared by J. G. Truxal, "Control System Synthesis," pp. 40, 294. McGraw-Hill, New York, 1955.

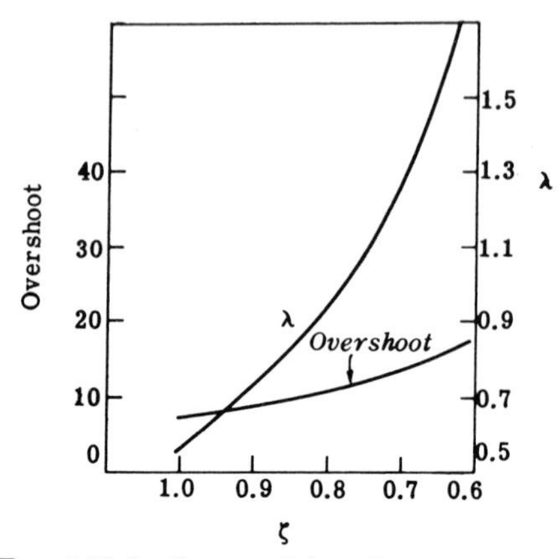

FIG. 5.13-3. Curves of λ and overshoot for which $K_v/BW = 2$.

are taken from Fig. 5.13-2b. For example, $\lambda = 0.57$ on the $\zeta = 1.0$ curve in Fig. 5.13-2b corresponds to 7.5% overshoot.

The data in Table 5.13-2 are next plotted in Fig. 5.13-3 as two curves: overshoot vs. ζ, and λ vs. ζ. It is seen that the specifications are all realizable with $\zeta \geqslant 0.67$. At $\zeta = 0.67$, the overshoot is 15%, but at larger ζ, the overshoot is less than 15%. At one extreme we could use $\zeta > 1$, and have even smaller overshoot; at the other extreme we could retain the 15% overshoot and obtain a larger K_v. However, a large K_v is not necessarily desirable. In fact, we shall see in Section 9.3 that the specification of a number for K_v is really an incomplete specification. For the time being, it suffices to state that for a fixed value of K_v, generally the larger the overshoot (providing it is not more than approximately 18%), the smaller is the rise time and the settling time. (The reason for this will be seen in Section 9.3.) For example, this is seen to be the case in Fig. 5.13-4 where $\zeta = 0.7$, $\lambda = 1.29$, overshoot $= 13.5\%$ is compared with $\zeta = 1.0$, $\lambda = 0.57$, overshoot $= 7.5\%$. Both designs have the same value of $K_v = 40$. These two curves were quickly and easily obtained by using Eq. (5.13,1) and the data of Table 5.13-1.

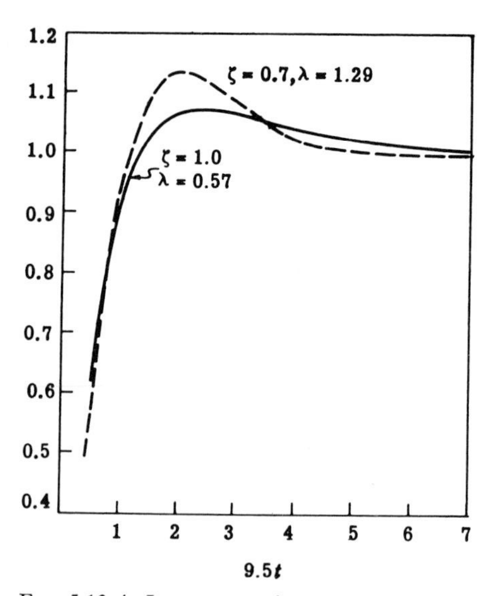

FIG. 5.13-4. Larger overshoot permits smaller settling time at same K_v.

In this design problem we will therefore choose the design with the maximum permitted 15% overshoot. From Fig. 5.13-3, this corresponds to $\zeta = 0.67$, $\lambda = 1.4$. To find ω_n, we write [using Eq. (5.12,3)] $K_v = \omega_n \lambda \zeta / (2\lambda\zeta^2 - 1)$, and solve for ω_n. The result is $\omega_n = 11$ rps. Figure 5.13-2c can be used as an approximate check on the answer (it gives $BW/\omega_n = 1.7$ cf. $20/11 = 1.8$).

The above technique can be used to realize any combination of K_v, band-

TABLE 5.13-2

($K_v/BW = 2$)

ζ	λ	Overshoot
1.0	0.57	7.5
0.8	0.95	11
0.7	1.29	13.5
0.6	1.88	17.5

width, and overshoot (the latter may, however, have to be finite). Suppose that $(K_v/BW) \gg 1$ is wanted with small overshoot. Consider $\zeta = 1$; at $\lambda = 0.5$, K_v/BW is infinite and the overshoot is only 14.5%. If the curves for $\zeta = 1.2$ were drawn, we would obtain infinite K_v at smaller overshoot. However, as will be seen in Section 9.3, large K_v and small overshoot would require very large settling time.

§ 5.14 Increase of Velocity Constant by Lag Compensation

The technique described in the above used a zero on the negative real axis to increase K_v, and it therefore usually corresponds to an extreme case of the lead compensation method described in Section 5.6 (extreme because only a zero is used, not a zero-pole pair as in Section 5.6). If larger K_v/BW ratios are required, it begins to look more like lag compensation, because the open-loop pole that results must inevitably be fairly close to the origin in order to achieve a large K_v/BW ratio. In the extreme case of infinite K_v, two poles of L must be at the origin.

There is a more direct method of using lag compensation in the $T(s)$ pole-zero method. The open-loop lag compensation method described in Section 5.5 used a pole-zero pair located near the origin. This pole-zero pair, $(s + z)/(s + p)$ of the loop transmission L, results in a zero-pole pair of T, because $T = L/(1 + L)$. Therefore T has a zero at $-z$ and a pole very close to z. [Consider the typical root loci of $1 + L = 0$ in Fig. 5.14-1 and note that the pole of $T(s)$, at $-mz$, must be very close to $-z$.] The resulting K_v for the lag compensation system with transfer function $T(s)$ is, from Eq. (5.12,3) (with K_{vo} representing the velocity constant of the original system transfer function T_o),

$$\frac{1}{K_v} = \frac{1}{K_{vo}} + \frac{1}{mz} - \frac{1}{z} = \frac{1}{K_{vo}} - \frac{m-1}{mz}. \tag{5.14,1}$$

In order that $T(0) = 1$ (recall K_p is infinite), $T = T_o m(s + z)/(s + mz)$.

By choosing z appropriately, any value of K_v is achievable. However, the pole of $T(s)$ at $-mz$ gives rise to a component in the step output equal to Ae^{-mzt}, with the large time constant $1/mz$. This tail in the step response substantially increases the settling time. Of course, A can be made very small. From the equation $T = T_0 m(s+z)/(s+mz)$, the residue of $T(s)/s$ in the pole at $-mz$ is $A \doteq m - 1$, because $A = mz(1-m)T_0(-mz)/(-mz) = (m-1)T_0(-mz)$, and the pole at $-mz$ is very close to the origin, so $T_0(-mz) \doteq T_0(0) = 1$. The value of m may be chosen so that A is as small as desired. However, the closer m is to 1, the smaller z must be [see Eq. (5.14,1)], and therefore the longer the tail in the step response. This is also seen from the fact that for a specific increase in K_v, the quantity $(m-1)/mz$ in Eq. (5.14,1) is fixed. This quantity is precisely the area $\int_0^\infty Ae^{-mzt}\,dt$; hence, the smaller A, the longer must be the duration of the tail, in order that the area remain the same. Also, the overshoot is increased by the amount $A \doteq m - 1$.

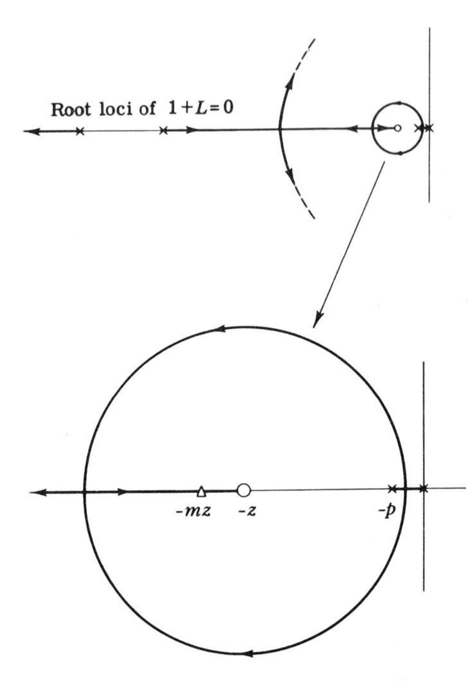

FIG. 5.14-1. $T(s)$ dipole (zero at $-z$, pole at $-mz$) due to lag compensation of $L(s)$. Root loci of $1 + L = 0$.

§ 5.15 Comparison of Lead and Lag Compensation Having the Same K_v, Bandwidth, and Overshoot

The comparison between the two methods for realizing the specifications is made by working out a design problem.

Example.

Specifications: Overshoot $\leqslant 13\%$, $BW = 40$ rps, $K_v = 60$.

Design: A complex pole pair for $T(s)$ cannot realize the K_v/BW ratio of 1.5. Thus, in Fig. 5.12-2, at 13% overshoot, $\zeta = 0.55$ and $K_v/BW = 0.75$. A dipole of T near the origin (lag compensation) can be used, or a zero can be inserted (lead compensation).

Lag compensation. The effect of the dipole on the overshoot must be taken into account. If we choose $m - 1 = 0.03 \doteq A$ (see Section 5.14) so as to have a small amplitude for the term with the large time constant, then this term contributes 3% to the overshoot. The complex pole pair must therefore contribute no more than 10%. From Fig. 5.12-2, this corresponds to $\zeta = 0.60$, at which $BW = 1.15\ \omega_n$. Consequently, $\omega_n = 40/(1.15) = 35$ and $T(s) = (35)^2 m (s + z)/(s + mz)\,[s^2 + (1.2)\,(35)s + (35)^2]$. At $\zeta = 0.60$, $K_{vo} = 0.72BW = 28$, so, from Eq. (5.14,1), $1/60 = (1/28) - (m - 1)/mz$, and since $m = 1.03$, $z = 1.5$. The step response is obtained by adding Ae^{-mzt} to the step response of the single pair of poles. The result is sketched in Fig. 5.15-1b.

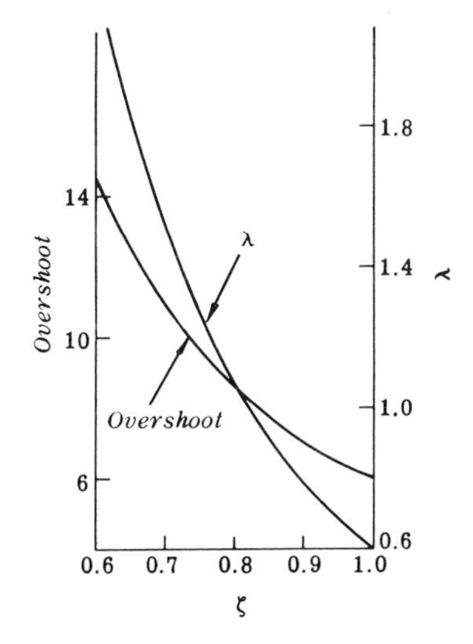

Fig. 5.15-1a. Curves of λ and overshoot for which $K_v/BW = 1.5$.

Lead compensation. The design procedure is identical to that used in the example in Section 5.13. Figure 5.13-2a is first examined along the horizontal line $K_v/BW = 1.5$. The first two columns of Table 5.15-1 are thereby obtained.

TABLE 5.15-1

$(K_v/BW = 1.5)$

ζ	λ	Overshoot
1.0	0.605	6.0
0.8	1.06	8.5
0.7	1.5	10.8
0.6	2.25	14.7

The third-column data are obtained from Fig. 5.13-2b and are plotted in Fig. 5.15-1a. An overshoot of 13% is obtained at $\zeta \approx 0.65$, $\lambda \approx 1.9$. From Fig. 5.13-2c, $BW/\omega_n \approx 1.47$ at these values of ζ and λ, so we pick $\omega_n = BW/1.47 = 26$ rps. [If ω_n is found from $K_v = \omega_n \lambda\zeta/(2\lambda\zeta^2 - 1)$, the result is $\omega_n = 28$.] The step response is obtained by applying Eq. (5.13,1) to the data of Table 5.13-1 and interpolating between the entries for $\zeta = 0.60$ and $\zeta = 0.70$. The result is shown in Fig. 5.15-1b.

It is seen that the lead compensation design results, in this example, to smaller rise time and settling time. But the comparison is not fair with respect to the rise time, because in the lead design, $|\,T(i\omega)\,|$ is decreasing near the band edge at only 6 db per octave, as compared to 12 db per octave for the lag design.

Therefore, despite the nominal equality of bandwidth, the amplitude vs. frequency areas are unequal, and are responsible for the appreciable difference in the rise times. The comparison with respect to the settling time is therefore also unfair, as will be evident from the discussion in Section 9.3.

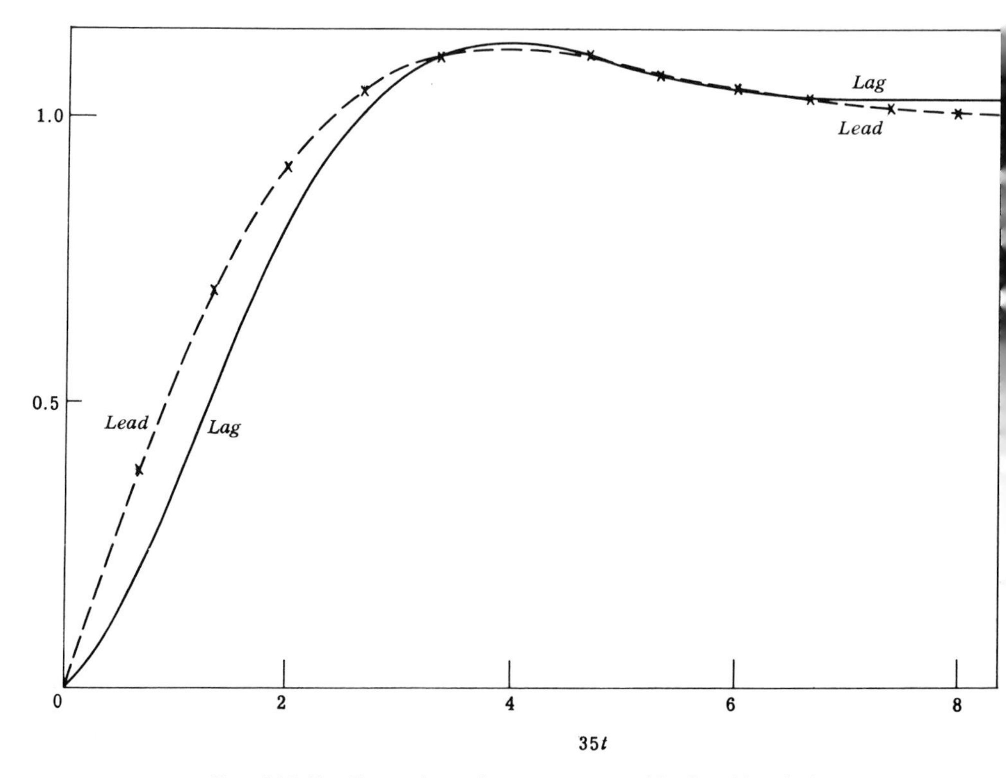

FIG. 5.15-1b. Comparison of step responses of lead and lag designs.

Another important consideration that is often overlooked in the pole-zero $T(s)$ synthesis method is the matter of the loop transmission. The value of $L(s)$ may be obtained from $T = L/(1 + L)$ giving $L = T/(1 - T)$. The value of L_1 (lag compensation) is found by using $T_o = (35)^2 [s^2 + (1.2)(35)s + (35)^2]^{-1}$ and solving for L_o. The result is $L_o = (35)^2/s(s + 42)$. Now we know that $L_1 = L_o \, m(s + z)/(s + \alpha z)$ with $m = 1.03$, $z = 1.5$, and α chosen so that $\lim_{s \to 0} ms \, L_o/\alpha = 60$. Hence $\alpha = 0.50$. The value of L_2 is obtained by solving directly $L = T/(1 - T)$, resulting in $L_2 = 24.5(s + 33)[s(s + 13.5)]^{-1}$. (In a later discussion on the problem of the far-off poles, it is concluded that additional poles of L may be introduced at about three times the bandwidth of T, with negligible effect on the design. Therefore a pole may be assigned to L_2 at $s = -3(40) = -120$, in order that L_1 and L_2 have the same excess of poles over zeros.) The results are compared in Fig. 5.15-1c. $|L_2| > |L_1|$ by 6 db for about four octaves in the important frequency range $\omega < \omega_c$ (ω_c is the cross-over frequency).

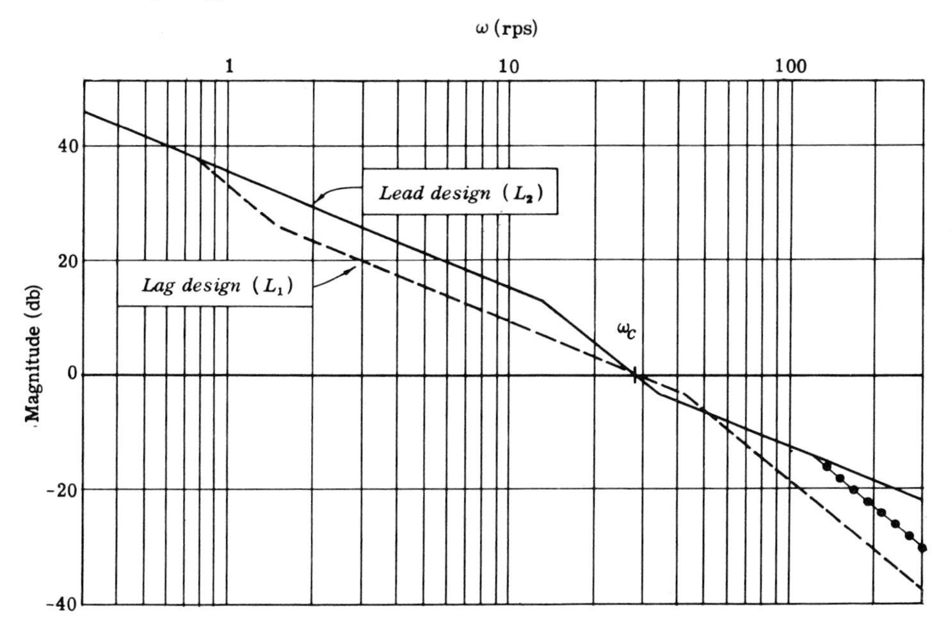

FIG. 5.15-1c. Comparison of loop transmissions of lead and lag designs.

Now the benefits of feedback are given by $1/(1 + L)$, so that over a significant frequency range one design has better feedback properties than the other, despite the fact that there is nothing in the pole-zero $T(s)$ synthesis procedure to indicate this. Also, the error-input ratio $E(s)/R(s) = 1/(1 + L)$, so that the same comments apply to the error function. The reason that appreciable difference in L is obscured when looking at $T(s)$ is that when L is large, $|T| \doteq 1$. Thus, if $L_1 = 50$ at $\omega = \omega_x$, $|T_1(\omega_x)| \doteq 50/51 = 0.98$; if $L_2(\omega_x) = 100$, $|T_2(\omega_x)| \doteq 100/101 = 0.99$ which is hardly distinguishable from 0.98. Yet the second design is twice as good as the first at ω_x, in terms of error, sensitivity, and disturbance rejection properties. The point here is that in the important range where $L \gg 1$, it is difficult to keep track of L by working with T. Yet the magnitude of L is important from the point of view of the benefits of feedback and the error response. This is one weakness of the $T(s)$ pole-zero method.

§ 5.16 Determination of the Loop Transmission $L(s)$

Once $T(s)$ has been chosen, it is formally easy to solve for $L(s)$ from the relation $T = L/(1 + L)$. Let $T = KN(s)/D(s)$ and $L = kn(s)/d(s)$, and it is readily found that $K = k$, $N = n$, and $d = D - KN$. A polynomial must be factored in order to determine the poles of L. When $T(s)$ contains only a few poles and zeros, the factoring of $D - KN$ is easy. Furthermore, the approximate locations of the zeros of $D - KN$ may first be ascertained by means of a simple sketch of the root loci of $1 - (KN/D) = 0$. However, Truxal[1] has noted the desirability of ensuring that the poles of $L(s)$ lie on the negative-real axis, so that the compensating networks can be RC networks. He has chosen a master pole-zero pattern for $T(s)$, which is such as to ensure that the poles of $L(s)$ are always restricted to the negative real axis. This is done as follows.

The general form of the master pole-zero pattern of $T(s)$ that is suggested

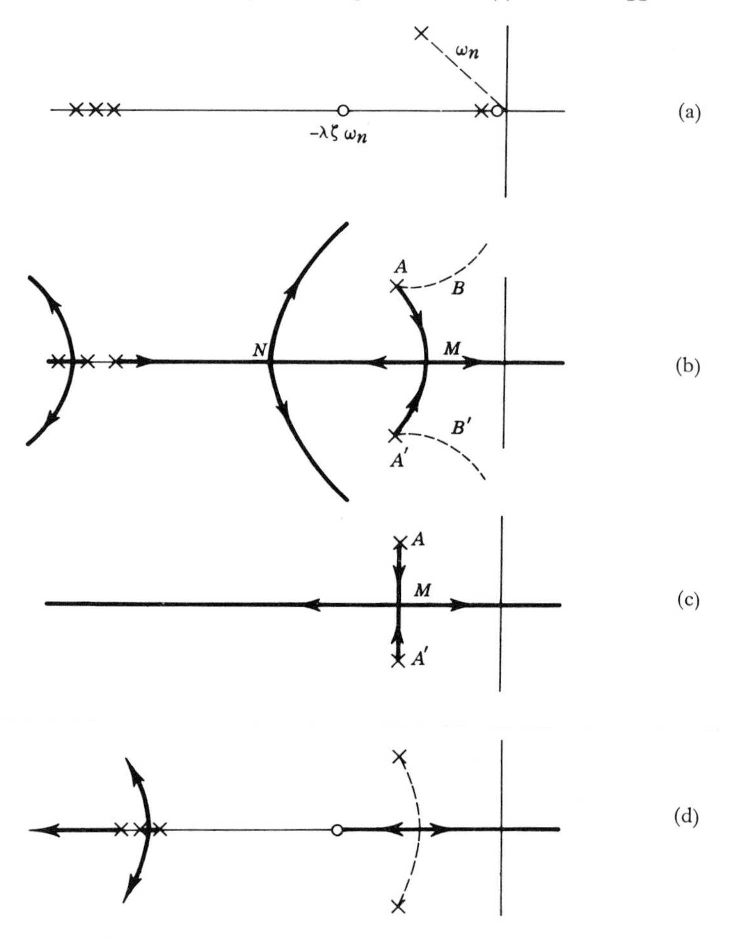

FIG. 5.16-1. (a) Suggested $T(s)$ with one dominant group of poles and zero and one far-off group of poles. (b) Root loci of $1 - KN(s)/D(s) = 0$ when $T = KN/D$ has no dominant zero. (c) Limiting case of (b) when far-off poles are at infinity. (d) Root loci of $1 - KN/D = 0$ when $T(s)$ has a dominant zero at $-\lambda\xi\omega_n$.

[1] J. G. Truxal, "Control System Synthesis," Section 5.3, p. 296. McGraw-Hill, New York, 1955.

is shown in Fig. 5.16-1a (not all the poles and zeros are always needed). The dipole near the origin corresponds to lag compensation, the zero at $-\lambda\zeta\omega_n$ to lead compensation, and we know from our work with the frequency response method that both lag and lead compensation are sometimes needed. We have in previous sections assumed that two or three dominant poles for $T(s)$ were sufficient. More poles are needed, because at large frequencies, where $|L(s)| \ll 1$, $T = L/(1 + L)$ approaches $L(s)$, and therefore T must have the same excess of poles over zeros as L. In the $T(s)$ pole-zero method, these additional poles must be placed "far off," in order that the design conducted on a few dominant poles basis should be negligibly affected. Far-off poles of $T(s)$ are therefore also shown, and the minimum distance of these far-off poles for RC realizability of $L(s)$ will now be determined.

Consider the poles of L. These are on the root loci of $1 - (KN/D) = 0$. The dipole has negligible effect on the root loci in the region of the s plane not close to the origin, so for such roots the dipole is neglected. Let us examine the root loci for the case when the zero at $-\lambda\zeta\omega_n$ does not exist. The root loci for such a case are sketched in Fig. 5.16-1b. In order that the dominant poles of $L(s)$ lie on the negative real axis, it is necessary

(1) that M in Fig. 5.16-1b exist, i.e., that the loci departing from A, A' go to the real axis, and not off into space as shown by the dotted lines BB';

(2) that M, if it exists, must be on the negative real axis, and not on the positive real axis.

(3) Since K must be such that one of the roots of $1 - (KN/D) = 0$ is at the origin [in order that $L(s)$ may have a pole at the origin], this root must reach the origin before its mate reaches the breakaway point at N.

All three requirements are satisfied by locating the distant poles of $T(s)$ sufficiently far off. For in the limiting case when they are at infinity, the point N too goes off to infinity, and the point M is then directly below the point A (as in Fig. 5.16-1c).

It is worthwhile finding the minimum distance of the far-off poles of $T(s)$, such that $L(s)$ has only negative real poles. Suppose there are three far-off poles, and for the present purpose they can be assumed coincident at $-\gamma\omega_n$. If there is a breakaway point at M on the real axis, there must also be one at N in Fig. 5.16-1b. At M and N, $T'(s)$ is zero. Let $1/T(s) = (s + \gamma)^3 (s^2 + 2\zeta s + 1)$, differentiate, etc., and we find that $T'(s)$ has two zeros at $-\gamma$, and two at the zeros of $B(s) = 5s^2 + (8\zeta + 2\gamma)s + (3 + 2\zeta\gamma)$. For these to be real, it is necessary that the discriminant be positive, which is always possible if γ is chosen sufficiently large. If this is satisfied, it automatically follows that M is on the negative real axis. To ensure that one root reaches the origin before the other reaches N, consider the limiting case when they reach these points simultaneously. This limiting situation occurs when $T(-N) = T(0)$. The point $-N$ is the largest magnitude zero of $B(s)$. Since $T(-N) = T(0)$, we must have $\gamma^3 = (\gamma - N)^3(N^2 - 2\zeta N + 1)$, with $5N = (4\zeta + \lambda) + [(4\zeta + \lambda)^2 -$

$5(3 + 2\lambda)]^{1/2}$. These equations are best solved by cut and try. The results are $\gamma \geqslant 6.2$ for $\zeta = 0.5$, and $\gamma \geqslant 7.1$ for $\zeta = 0.7$. In this way we establish minimum distances of the far-off poles for RC realizability of $L(s)$.

Next consider the case where $T(s)$ does have a zero. The root loci then have the form shown in Fig. 5.16-1d. Here it is only necessary to ensure that the breakaway point (which must exist on account of the zero) is to the left of the origin. In the latter case, the sensitivity of the root of $1 - (KN/D) = 0$ at the origin $[S_K^{s_i} = ds_i/(dK/K)$, see Section 4.3] has zero angle, while if the breakaway point is on the positive real axis, its angle is 180°. It is recalled from Section 4.4 that the root sensitivity can be regarded as the electric field intensity due to line charges at the poles and zeros of $T = KN/D$. With this approach and for $T = K(s + \lambda\zeta\omega_n)\,[(s^2 + 2\zeta\omega_n s + \omega_n{}^2)\,(s + \gamma_1\omega_n)\,(s + \gamma_2\omega_n) \ldots]^{-1}$, it is seen that it is necessary that $2\zeta + \Sigma\,(1/\gamma_i) > 1/\lambda\zeta$. In practice there is no difficulty satisfying this inequality because usually $2\zeta > 1/\lambda\zeta$.

We have ignored in the above the requirements for ensuring that the far-off poles of $L(s)$ lie on the negative real axis. This can be done formally. However, the simplest way to do this in practice is not to worry about it, but to use as a first approximation a $T(s)$ that has no far-off poles at all, and go ahead and find the resulting $L(s)$. The factoring of $1 - T$ is then simple. Next we assign to $L(s)$ as many far-off poles as are practically necessary (the required number is considered later), and assign them sufficiently far away.

What constitutes "sufficiently far away?" Our principal concern is that the assumed system time response (based on no far-off poles) should not be seriously affected by the far-off poles of $T(s)$. In general, such poles slow up the response and decrease the overshoot. The minimum distance of a single far-off pole so that its effect is very small can be deduced from the equation $p_x(t) = p(t) * qe^{-qt}$ of Section 5.10, where the effect of an additional pole was considered. In $p(t)$ we take the rise time $t_r \approx 0.5/f_b = \pi/BW$ [Eq. (5.10,3)], and approximate the first part of the step response $p(t)$ by a straight line of slope $1/t_r$. Performing the convolution $p_x(t) = p(t) * qe^{-qt}$, we get $p_x(t) = (1/t_r)\,[t - (1/q) + e^{-qt}/q]$. Let $q = \psi(BW) = \psi\pi/t_r$. Also, suppose we allow the crossing of the line $p_x(t) = 1$ to be at $t_r(1 + \alpha)$, i.e., $p_x(t) = 1$ at $t = t_r(1 + \alpha)$. This means that the rise time is allowed to increase approximately $100\alpha\%$. We then get $\alpha = (1/\psi\pi)$ $[1 - e^{-\psi\pi(1+\alpha)}] \doteq 1/\psi\pi$. If $\alpha = 0.1$, $\psi = 10/\pi$, which means the far-off pole must be about 3.2 times the bandwidth. Similar approximate calculations may be made for two or more poles spaced at certain ratios apart from each other. Clearly, the larger the number of far-off poles, the further off they must be for their effect to be small.

The above gives the minimum permissible distance of the far-off poles of T.[1]

[1] Since the approximate locations of the far-off poles of $T(s)$ are known in advance, it is possible to allow for their effect on K_v in using Eq. 5.12,3 and Figures 5.13-2 to find the dominant poles and zeros of $T(s)$.

They must be related to equivalent restrictions on the far-off poles of $L(s)$. At large frequencies where $|L| \ll 1$, $T = L/(1+L) \doteq L$, and therefore $L \doteq T$ in this frequency range. Therefore we assign the far-off poles of L precisely where the far-off poles of T are to be. Three times the bandwidth of T should correspond to about five to eight times the crossover frequency of L. However, it is a good idea to check their effect on the stability margins and if necessary locate them further off. Thus the far-off poles of $T(s)$ can be directly assigned to $L(s)$ in all cases where the poles of $T(s)$ may be divided into two distinct regions—the dominant region and the far-off region, providing only that in the far-off region $|T| \ll 1$. In this manner we are spared the trouble of including the far-off poles in the factoring of $1 - T$. Finally, if $T(s)$ has a dipole near the origin, it is best to ignore it in calculating $L(s)$, and then insert the corresponding lag compensation dipole directly into $L(s)$, as was done in calculating L_1 in the example of Section 5.15.

The design problem in Section 5.15 can now be completed. The values of $L(s)$ without the far-off poles were previously obtained. With our present appreciation of how far off the distant poles must be, a reasonable choice would be $L_1 = (35)^2(1.03)\,(s+1.5)\,(200)/s(s+42)\,(s+0.75)\,(s+200)$; $L_2 = (24.5)\,(s+33)\,(120)\,(200)/s(s+13.5)\,(s+120)\,(s+200)$. Suppose the plant transfer function is $P = (2)10^4/[s(s+20)\,(s+30)]$. Since we are using the single-degree-of-freedom structure of Fig. 5.16-2 where $L = GP$, we must

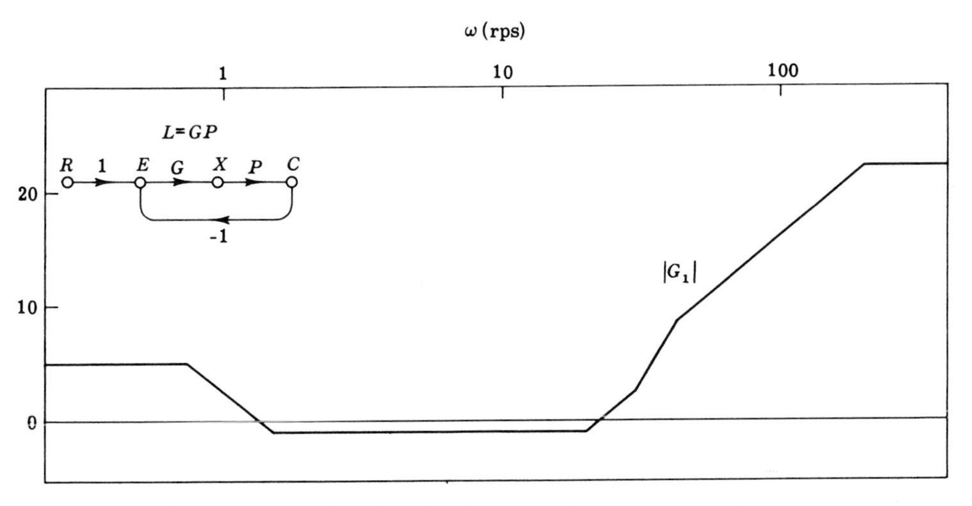

FIG. 5.16-2. Undesirable high frequency character of $G_1 = L_1/P$.

have $G = L/P$ with $G_1 = L_1/P = 12.5(s+1.5)\,(s+20)\,(s+30)\,[(s+0.75)\,(s+42)\,(s+200)]^{-1}$. $|G_1|$ is sketched in Fig. 5.16-2, and its high frequency behavior requires comment. The high frequency value of $|G_1|$ is 22 db, but for how long must this value be maintained? While there may be no difficulty in ensuring that G_1 maintains this value for very large frequencies, it is nevertheless true that eventually G_1 must go to zero value as $1/s^x$ with $x \geqslant 1$. In any efficient design, at least the corner frequency at which G begins its journey to $-\infty$ db should be ascertained. If it is not so easy to secure a large gain-bandwidth product for G, then several more poles should be assigned in order that $|G|$ may be decreased as fast as possible. This is done by noting that at large frequencies, $T \doteq L = GP$. Therefore enough far-off poles should be assigned to T, and consequently to L, to handle the high frequency behavior of both G and P.

We have been ignoring one important matter in the above discussion. We took $P = (2)10^4/[s(s + 20)(s + 30)]$ as strictly correct. Actually, it is possible that there are also additional far-off poles of the plant. It is usually difficult to measure the plant transfer function at high frequencies where its response is exceedingly small. The statement that $P = (2)10^4/s(s + 20)(s + 30)$ might really mean that this is a good estimate up to possibly 150 rps. There may be more far-off poles of P. These should be ascertained, if necessary, from a theoretical analysis. Once the far-off frequency region has been located, the nature of the plant at these frequencies should be considered, and $L(s)$ should be assigned enough far-off poles to take care of the high frequency needs of $P(s)$ and $G(s)$. Fortunately (and necessarily for a reasonable design), in most practical designs the far-off plant poles are well beyond the crossover frequency, so that they and the inevitable far-off poles of G have negligible effect on L and on T. This far-off pole problem does, however, arise in exacting designs and will be treated in Section 6.10.

Constraints Imposed by the Plant

In the design procedure of this section, little attention was paid to the plant transfer function $P(s)$. The design procedure led to a specific loop transmission function $L(s) = P(s)G(s)$, and we assumed that there was no difficulty in realizing $G(s) = L(s)/P(s)$. The network realization of a transfer function $G(s)$, from its poles and zeros, is treated in several texts.[1] In general, $G(s) = L(s)/P(s)$ must cancel out those poles and zeros of $P(s)$ which are not present in $L(s)$, and also supply those poles and zeros of $L(s)$ which are not present in $P(s)$. There are the following possibilities:

(1) All the poles and zeros of $G(s)$ are on the negative real axis. The synthesis problem is then simple.

(2) $G(s)$ has only negative real axis poles, but has some complex zeros in the left half-plane. Design procedures for this problem are available.[1]

(3) $G(s)$ has some complex poles in the left half-plane. Such poles are inconvenient to realize by means of R, L, C elements, if they are located comparatively

[1] E. A. Guillemin, "Synthesis of Passive Networks." Wiley, New York, 1957; J. E. Storer, "Passive Network Synthesis," McGraw-Hill, New York, 1957.

near the origin (which is often the case in control systems with their relatively small bandwidths). Active *RC* synthesis procedures are, however, available, and one such simple procedure is described in the appendix.

(4) If $G(s)$ has any right half-plane zeros, they are presumably due to poles of $P(s)$ and not to zeros of $L(s)$, because it is rarely advisable to deliberately introduce right half-plane zeros into $L(s)$ (see Section 7.9 for the reasons, but note Section 8.14 for an exceptional case where a right half-plane zero is deliberately introduced into the loop transmission). A right half-plane pole of $P(s)$ should not be canceled by a zero of $G(s)$, because inevitably the inexact cancellation or drift leads to a right half-plane pole of $T(s)$. There is a zero located near this pole so its residue is small. If the system is used in a relatively short operating cycle, then the right half-plane dipole may be tolerable if the amplitude of the oscillatory term does not grow sufficiently large before the system is shut off. Therefore $G(s)$ with right half-plane zeros rarely occurs. If it does, then such a $G(s)$ may be realized by the methods discussed in the references.[1]

(5) If $G(s)$ has any right half-plane poles, then they must be due to the attempt to cancel right half-plane zeros of $P(s)$. Only in very exceptional cases is such an attempt justified, because of the inevitable right half-plane dipoles that appear in the system response function. Therefore such poles of $P(s)$ appear in $L(s)$, and as zeros of $T(s)$. The problem of shaping unstable loop transmissions for closed loop stability is discussed in Section 7.14.

Validity of Canceling Plant Dynamics

Sometimes objections are made as to the feasibility and validity of canceling the plant's poles or zeros. It has been suggested that the plant's poles or zeros cannot really be canceled, because they are usually not precisely known, and/or they drift. This is a significant objection, and can be answered only by referring to the feedback (sensitivity) properties of the system. Our answer to this question is that we shall develop (in Chapter 6) feedback design procedures which permit control over the drift in the system poles and zeros due to plant parameter variation. However, the objection that has been raised is just as valid for any design of $G(s)$, whether or not the plant poles and zeros are canceled (despite the implication in the criticism that the objection exists only if the design involves cancellation of the nominal poles and zeros of the plant).

Consider a design (such as the frequency response designs of Sections 5.5 to 5.9, or any s-plane designs) in which the plant poles and zeros are *not* canceled. Nevertheless, the design is based upon an *assumed knowledge* of $P(s)$, and if $P(s)$ is not known accurately, or its parameters drift, then the design will also be in error, in precisely the same manner as one in which the poles and zeros of the plant are assumed canceled. It is incorrect to refer this question of plant

[1] See also I. Horowitz, *R-C* Transistor network synthesis. *Proc. Natl. Electronics Conf.* **12**, 818-829 (1956).

ignorance or parameter variation only to those designs in which $G(s)$ happens to have some poles and zeros at the nominal values of the plant zeros and poles. The problem exists in all designs. Thus, from Eq. (3.2,5) $\Delta T/T$ is determined by the loop transmission $L(s)$ and by $\Delta P/P$. Obviously $\Delta P/P$ is independent of G. As for $L(s)$, it is incorrect to think that $L(s)$ somehow assumes desirable sensitivity properties from the imposed constraint that it has all the poles and zeros of $P(s)$. A much more deliberate effort at choosing $L(s)$ must be made and this will be done in Chapter 6. There is therefore no particular merit in insisting that $L(s)$ contain the poles and zeros of $P(s)$. On the other hand, we have previously noted that $L(s)$ should contain all the right half-plane poles and zeros of $P(s)$ (excluding short operating cycle conditions where right half-plane dipoles may be tolerable).

§ 5.17 Relative Merits of the Open-Loop Frequency Response Method and the $T(s)$ Pole-Zero Method

Much has been made of the argument that the $T(s)$ pole-zero method is a true synthesis procedure, whereas classical open-loop frequency shaping is one of cut and try. The argument is valid if and only if the closed-loop system response is really the logical representation of the system specifications. If $T(s)$ is indeed the ideal (or at least a better) repository of the design intentions, then its analytical formulation and direct realization represents a synthesis procedure considerably superior to $L(j\omega)$ shaping. If not, then we are indeed synthesizing, but not what ought to be synthesized. Therefore, the critical question is whether $T(s)$ really is a better formulation of the design specifications.

We have already noted in Section 5.15 that the benefits of feedback are given by $1/(1 + L)$, with $| L | > 1$ in the significant frequency range. We have seen that it is difficult to control the magnitude of L from the specification of T. At least the techniques that have been worked out for the $T(s)$ pole-zero method have neglected this. The same remarks apply, of course, to the error function, since $E(s) = R/(1 + L)$. In accurate positioning servos and in tracking servos, a very small tracking error is essential. Now $E/R = (1 + L)^{-1}$ and is closely equal to L^{-1} in the important frequency range in which $| E/R | < 1$. Therefore it might well be argued that the direct specification of $L(j\omega)$ is the logical means of expressing this portion of the design specifications.

Another important design factor is the stability margins. They provide a means of taking plant parameter variation into account and ensuring system stability. There is no apparent means of incorporating stability margins into the pole-zero specification of $T(s)$.

The above discussion may be summarized as follows. The $T(s)$ pole-zero method is a true synthesis procedure only to the extent that $T(s)$ is the most convenient formulation of the design specifications. While $T(s)$ is certainly a convenient representation for the *filter* part of the problem, it is definitely not uitable representation for the *feedback* part of the design. In the single-degree-of-freedom structure, there is only one quantity, the loop transmission $L = GP$, which is at the designer's disposal. The design specifications fall under a variety of headings: error response, the benefits of feedback, stability margins on account of parameter variations, and system transient response. Only for the last of these is $T(s)$ perhaps the most convenient formulation. For the others, $L(j\omega)$ is decidedly a more convenient design vehicle. In two-degree-of-freedom configurations (discussed in Chapter 6) it is possible to separate the feedback properties of the design and the filter properties. It is then convenient to use $T(s)$ for the filter specifications and $L(s)$ for the feedback specifications, because these two quantities are independently realizable.

It is also worth noting that it is $L(s)$ which must be physically constructed. If the designer works directly with $L(s)$, it is easy for him to take into account any constraints on $L(s)$ and practical considerations in the realization of $L(s)$. It is not so easy for him to take such matters into account when he is working with $T(s)$. Finally, it should be noted that it is usually easier to obtain the frequency response of a plant than its pole-zero representation. It is often possible to apply sinusoidal signals to the plant and so obtain its frequency response directly. If transient measurements are used, it is still usually easier to evaluate the frequency response from the transient response than it is to find a pole-zero representation (see Chapter 12).

The distinctive superior feature of the $T(s)$ pole-zero method is in its better correlation with the system transient response. However, such better correlation exists only for relatively simple pole-zero patterns of $T(s)$.[1] (For complicated pole-zero patterns, the time-frequency correlation of Section 5.10 is probably as good as the time—s-plane correlation). In order to preserve such superior correlation for complicated systems, it has been necessary to invoke the concept of dominant poles and zeros. By this is meant that only a few poles and zeros of $T(s)$ dominate the system response. The other poles and zeros of $T(s)$ are placed far away from the dominant group (as noted in Section 5.16, at about three times the bandwidth). The dominance concept permits the easy "time-to-s-plane" correlation of the simple $T(s)$ pole-zero patterns to be applied to high-order systems. However, it is often overlooked that a price is paid for enforcing dominance in complex systems. The price that must be paid will now be clarified by considering the problem in both the frequency domain and the time domain.

§ 5.18 The Price That Is Paid for a Dominant Type $T(s)$

The frequency response of a $T(s)$ which consists of a dominant group of poles and zero and a far-off group of poles and zeros has three very distinct regions.

[1] The step responses of some higher order $T(s)$ functions have been presented by O. I. Elgerd and W. C. Stephens, Effect of closed loop pole and zero locations on the transient response. *Trans. AIEE* **78**, (Part 2), 121-127 (1959).

In the low frequency region, $\omega < \omega_b/n$ (where ω_b is the bandwidth of T and $n \approx 2$), $| T(j\omega) | \doteq 1$. In the intermediate region, at best $| T(j\omega) |$ decreases at about 12 db per octave [where by "best" is meant the greatest rate of decrease of $| T(j\omega) |$]. This second region extends from approximately ω_b/n to about three or four times ω_b. Finally, in the last region consisting of $\omega \gtrsim 3.5\ \omega_b$, $| T(j\omega) |$ is allowed to decrease as rapidly as the natural higher frequency characteristics of $L = GP$ require. Accordingly, there is a definite gross shape for any $| T(j\omega) |$ which has a dominant set of poles and zero. This shape (straight line approximations) is sketched in Fig. 5.18-1, where far-off poles of T are assigned at $-3.5\omega_b$, $-7\omega_b$, and $-14\ \omega_b$.

Next consider a plant transfer function $P(j\omega)$. One with only a few poles in the region $\omega < 3.5\ \omega_b$ is sketched in Fig. 5.18-1. $P(j\omega)$ and $T(j\omega)$ of Fig. 5.18-1

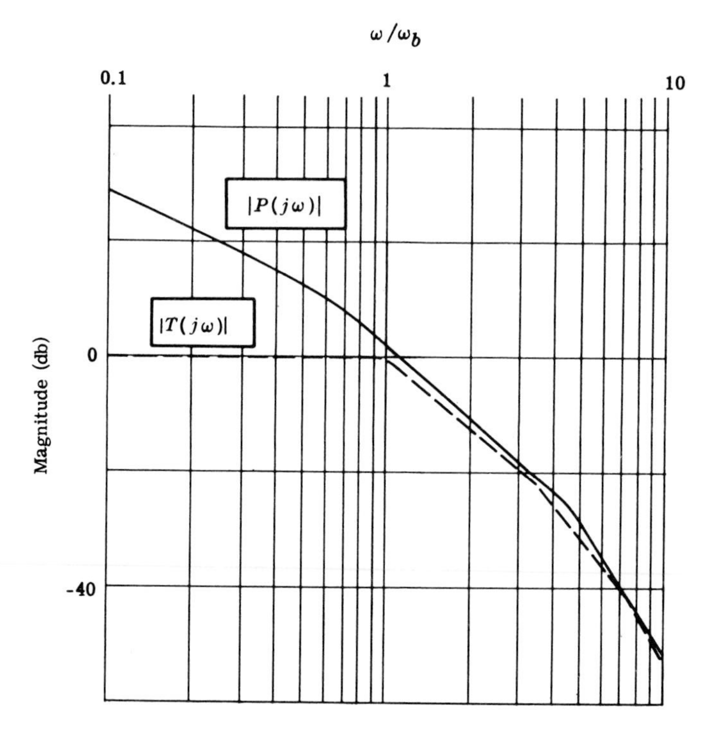

FIG. 5.18-1. A $T(j\omega)$ which is well mated to $P(j\omega)$.

are very similar for $\omega > \omega_b$ and the two are naturally mated—the specified $T(j\omega)$ is natural for this plant. Now consider a more complex plant, one whose magnitude decreases faster in the region $\omega > \omega_b$. Such a $P(j\omega)$ is sketched in Fig. 5.18-2a. Two possible ways that the standard $T(j\omega)$ of Fig. 5.18-1 can be fitted to this P function are shown in Figs. 5.18-2a and b. In the design of

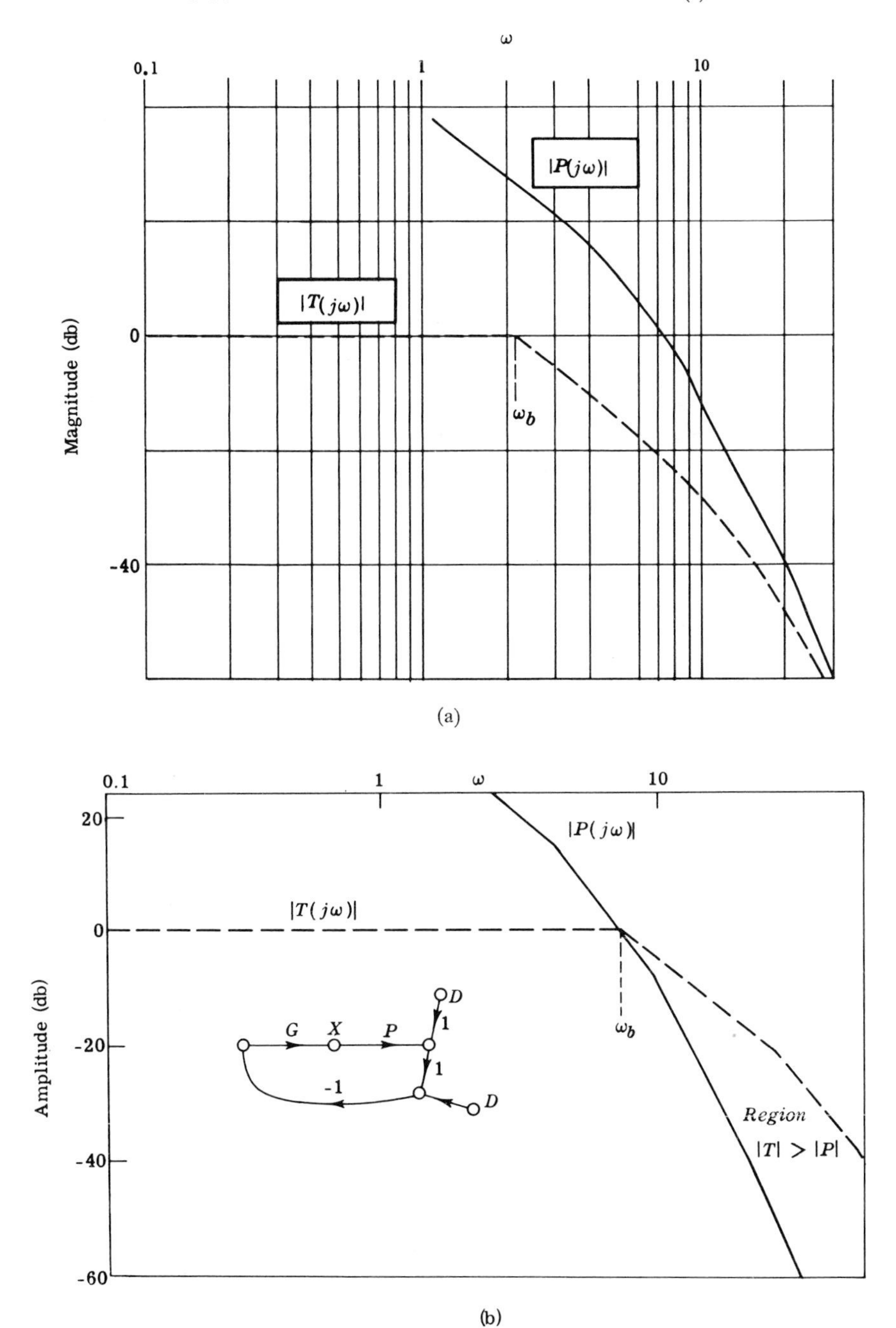

FIG. 5.18-2. (a) Waste in plant capabilities in a dominant $T(s)$ design. (b) Alternate dominant design increases likelihood of saturation.

Fig. 5.18-2a there is a waste of about 1.5 octaves of the plant capabilities. The plant has the physical capacity to respond faster than it is called upon by $T(s)$. The dominance strait-jacket does not permit us to use a $T(s)$ which is "natural" for the given plant. On the other hand, in the design of Fig. 5.18-2b, $T(s)$ imposes demands upon the plant which the latter cannot meet. The plant will saturate at a lower system input level than for a $T(s)$ which is mated to the given plant. There is another difficulty. In Fig. 5.18-2b, in the region $\omega > \omega_b$, $T \doteq L = GP$; since $|T| > |P|$ for $\omega > \omega_b$, $|G| > 1$ by the amount that $|T| > |P|$. Any noise or disturbance D at the load or in the feedback path leads to $X = -GD/(1 + L)$ at the plant input (see Fig. 5.18-2b). Thus, in the region $\omega > \omega_b$, $X \doteq -GD$, so D is amplified at X by the amount that $|T|$ is greater than $|P|$. This amplified intermediate and high frequency noise may saturate the plant. It should also be noted that if the dominant $T(s)$ pattern included a zero, then the situation would be worse by an additional octave or so. Furthermore, no attention was paid to RC realizability of $L(s)$. The latter might require the "far-off" poles to be even further away.

It will be seen in Section 6.14 that similar difficulties result when we design for more of the beneficial properties of feedback than the plant can inherently supply. However, in the latter case, we at least obtain for the system, properties which are beneficial in some respect. There are actually some system benefits realized in return for the liabilities. In the design of Fig. 5.18-2a, the only benefit is to the designer, in that it permits him to use the dominant-type $T(s)$ pole-zero synthesis procedure in a situation which really requires more design effort. However, it would be better for the system if $T(j\omega)$ was designed to have the same shape as $P(j\omega)$ in the region $\omega > \omega_b$. The frequency response method is not as constrained to a fixed pattern, and therefore offers more scope for fitting $T(j\omega)$ to the existing plant. However, even the frequency response design is not completely unconstrained, because of stability considerations; the resulting limitations will be discussed in Section 5.22.

The above disadvantage of the dominant-type $T(s)$ pole-zero synthesis will now be explained from the time domain point of view. In a $T(s)$ with a small number of dominant poles and zeros, the nature of the system time response is

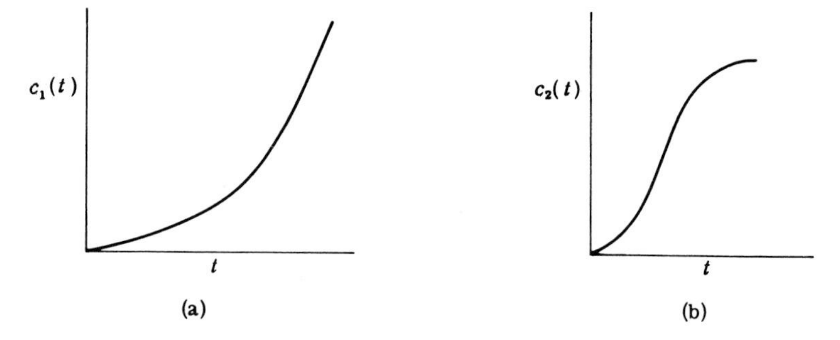

FIG. 5.18-3. Limitation of dominant $T(s)$ seen from time domain point of view.

severely restricted. In particular, there is little flexibility in the relation between the system time delay and the system rise time; the relation cannot be varied over a wide range. A plant with more than an excess of two or three poles over zeros can have a much different relation between its rise time and time delay. Figure 5.10-3 is redrawn here as Fig. 5.18-3a. It represents the fastest step response (without saturating) of which the plant is capable. Now $T(s)$ of the dominant type has a relation between delay and rise time akin to that shown in Fig. 5.18-3b. $T(s)$ must be chosen so that $c_2(t) \leqslant c_1(t)$ for all t, or the plant will saturate. The designer is therefore compelled to assign a very small bandwidth to $T(s)$ in order to achieve the larger delay of $c_1(t)$, and he must take with it a much larger rise time than in $c_1(t)$; this corresponds to the design of Fig. 5.18-2a. The trouble is that the dominant type $T(s)$ time response is constrained and cannot be adapted to any general plant time response.

In order to avoid the above inefficient use of the plant's capabilities, it is necessary to forego the use of a small number of dominant poles and zeros. In its place the $T(s)$ pole-zero method must be broadened to include more complicated dominant pole-zero patterns. One way in which this may be done is presented in the next section.

§ 5.19 $T(s)$ Pole-Zero Method for More Complicated Dominant Pole-Zero Patterns

The $T(s)$ pole-zero method is extended in this section to a more complicated dominant pole-zero pattern than those in previous sections. The pattern chosen is shown in Fig. 5.19-1. Calculations have been performed for $\nu = 1$, 2 and for $\zeta = 1.0$ and 0.707 with λ as parameter. The step and impulse responses [$p(t)$ and $h(t)$, respectively] of

$$T_1(s) = 8\nu^3\omega_n{}^5/(s^2 + 2\zeta\omega_n s + \omega_n{}^2)\,(s + \nu\omega_n)\,(s + 2\nu\omega_n)\,(s + 4\nu\omega_n)$$

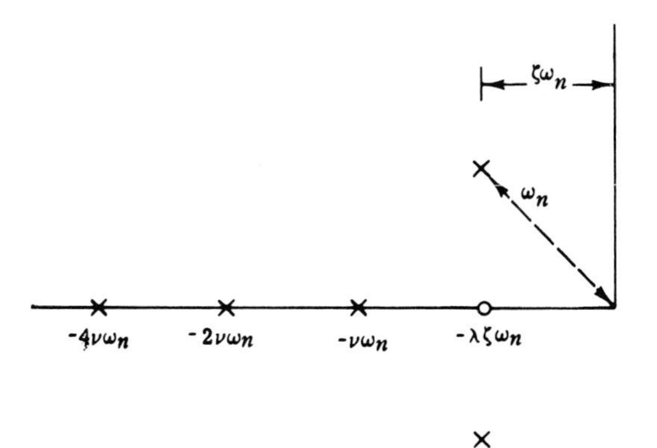

FIG. 5.19-1. Pole-zero pattern of a more complicated $T(s)$.

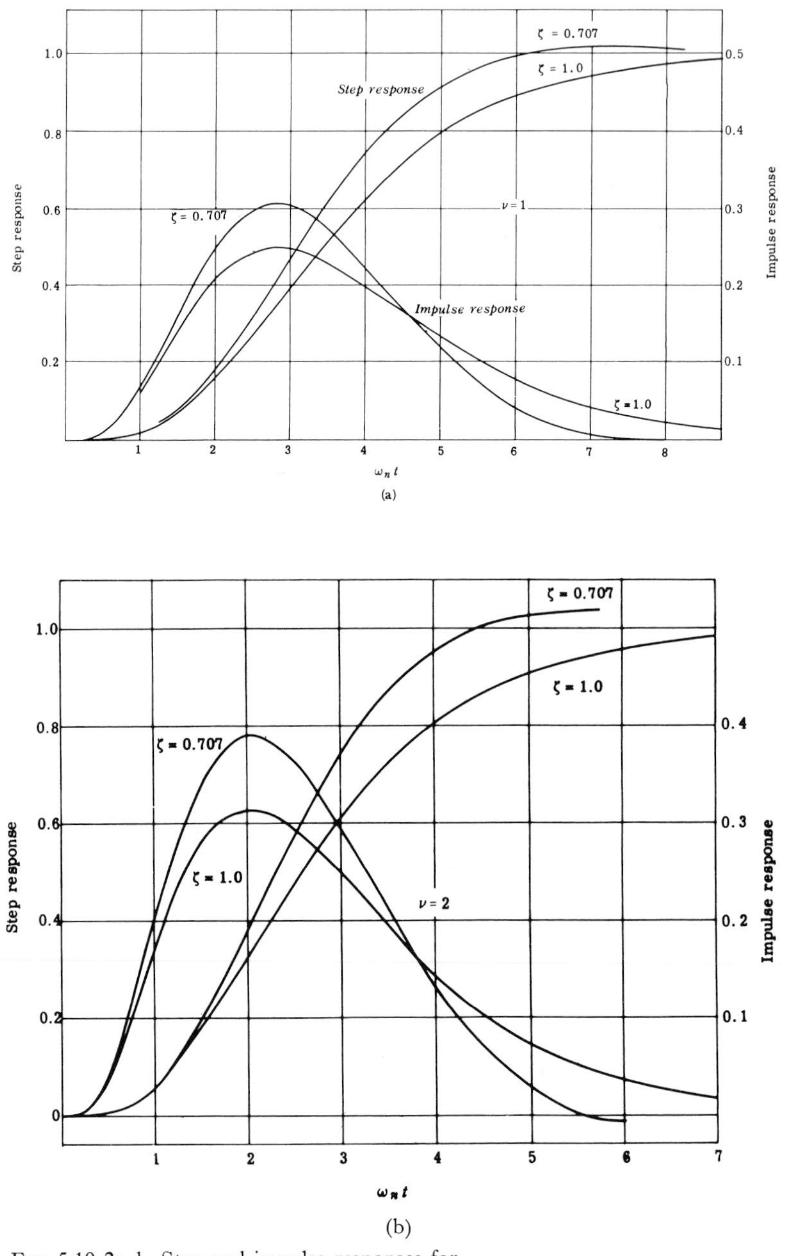

FIG. 5.19-2a, b. Step and impulse responses for

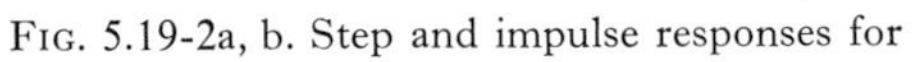

$$\text{(a)}\quad T(s) = \frac{8\omega_n^5}{s(s^2 + 2\zeta\omega_n s + \omega_n^2)(s + \omega_n)(s + 2\omega_n)(s + 4\omega_n)}.$$

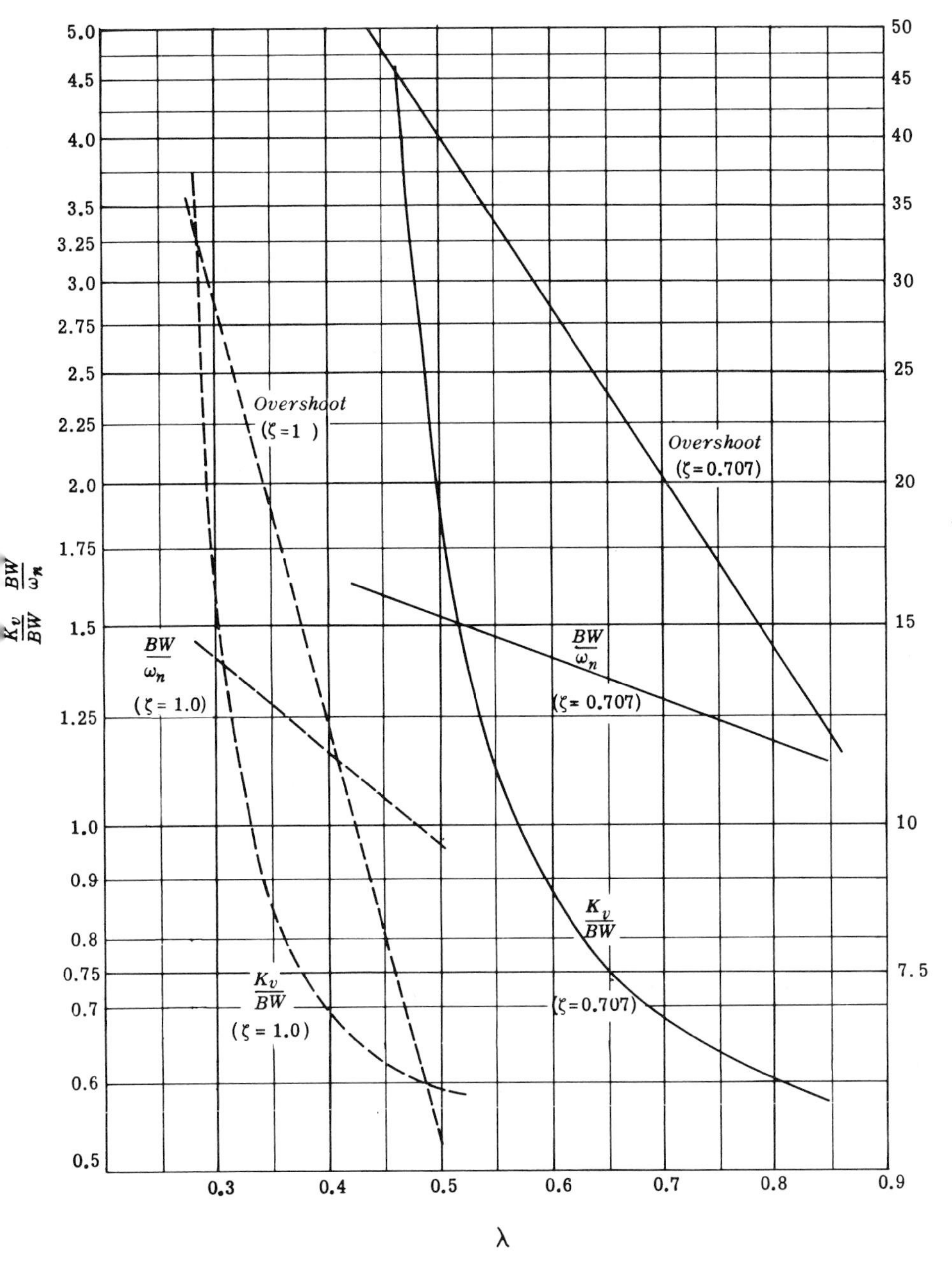

FIG. 5.19-3a. Relations between design parameters, $\nu = 1$.

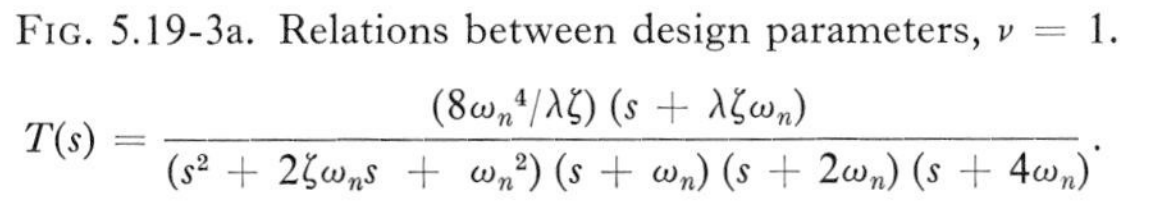

$$T(s) = \frac{(8\omega_n^4/\lambda\zeta)(s + \lambda\zeta\omega_n)}{(s^2 + 2\zeta\omega_n s + \omega_n^2)(s + \omega_n)(s + 2\omega_n)(s + 4\omega_n)}.$$

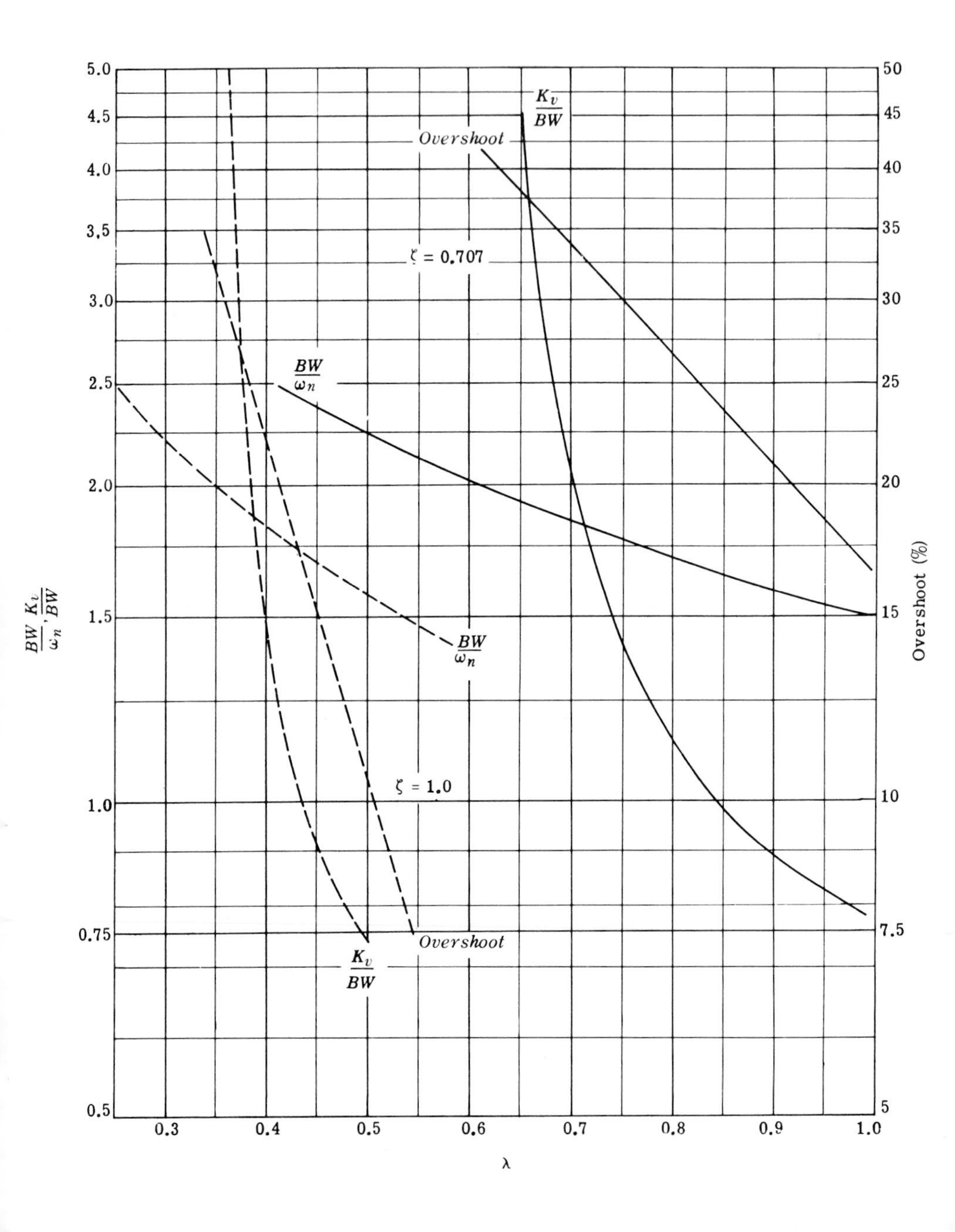

FIG. 5.19-3b. Relations between design parameters, $\nu = 2$.

TABLE 5.19-1

	$\nu = 1$				$\nu = 2$			
	$\zeta = 1.0$		$\zeta = 0.707$		$\zeta = 1.0$		$\zeta = 0.707$	
$\omega_n t$	$p(t)$	$h(t)$	$p(t)$	$h(t)$	$p(t)$	$h(t)$	$p(t)$	$h(t)$
0	0	0	0	0	0	0	0	0
0.5	0.001	0.009	0.002	0.009	0.005	0.038	0.005	0.041
1.0	0.017	0.061	0.018	0.068	0.055	0.172	0.062	0.196
1.5	0.066	0.140	0.076	0.164	0.173	0.279	0.198	0.335
2.0	0.154	0.208	0.180	0.250	0.323	0.313	0.385	0.389
2.5	0.267	0.243	0.319	0.299	0.476	0.292	0.575	0.364
3.0	0.390	0.249	0.470	0.304	0.612	0.244	0.740	0.295
3.5	0.513	0.230	0.615	0.274	0.719	0.191	0.865	0.211
4.0	0.620	0.200	0.745	0.224	0.803	0.143	0.951	0.134
4.5	0.711	0.165	0.840	0.169	0.863	0.103	1.002	0.072
5.0	0.787	0.131	0.915	0.117	0.907	0.073	1.027	0.029
5.5	0.847	0.100	0.960	0.074	0.938	0.050	1.034	0.003
6.0	0.887	0.075	0.990	0.041	0.959	0.034		
6.5	0.921	0.055	1.005	0.020	0.972	0.023		
7.0	0.947	0.041	1.01	0.006	0.981	0.015		
7.5	0.960	0.029			0.989	0.010		
8.0	0.974	0.020			0.992	0.007		
8.5	0.983	0.014						
9.0	0.990	0.010						

are sketched in Figs. 5.19-2a and b for $\nu = 1, 2$, respectively. The step response of $T(s) = (\lambda\zeta\omega_n)^{-1}(s + \lambda\zeta\omega_n)T_1(s)$ is $(\lambda\zeta\omega_n)^{-1}h(t) + p(t)$, and may be obtained from Figs. 5.19-2a and b (for $\zeta = 1.0$ and 0.707) for any value of λ. The data are also tabulated in Table 5.19-1 for increments of 0.5 in $\omega_n t$.

Figures 5.19-3a and b are the equivalent of Figs. 5.13-2a, b, and c. They are intended to permit design for specified overshoot, K_v, and bandwidth similar to the procedure outlined in Section 5.13. However, coarser interpolation may be necessary because data for only two values of ζ are available in each case. The designer must decide beforehand whether $\nu = 1$ (Fig. 5.19-3a), $\nu = 2$ (Fig. 5.19-3b), or the simpler pattern of Section 5.13 is to be used. The choice is made using the reasoning of Section 5.18. To aid him in his choice, the step

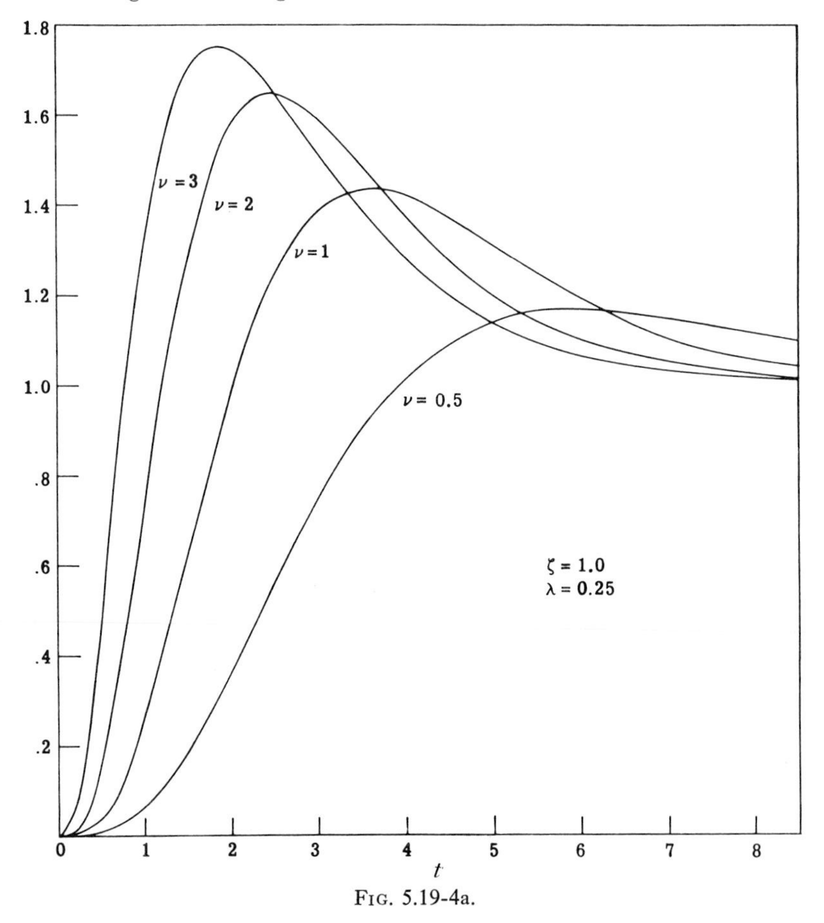

FIG. 5.19-4a.

FIG. 5.19-4a to 4e. Step response curves for

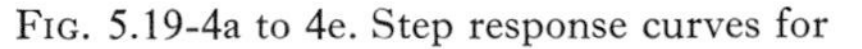

$$T(s) = \frac{8\nu^3(s + \lambda\zeta)}{\lambda\zeta(s^2 + 2\zeta s + 1)(s + \nu)(s + 2\nu)(s + 4\nu)}.$$

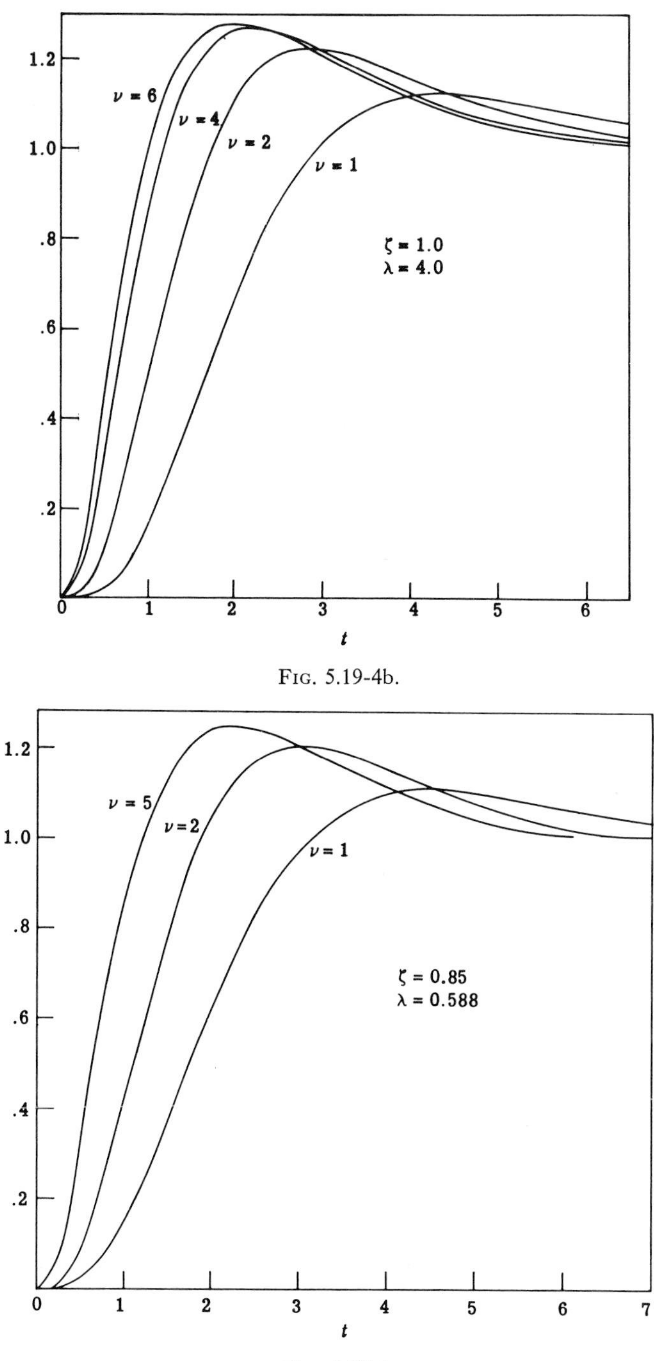

FIG. 5.19-4b.

FIG. 5.19-4c.

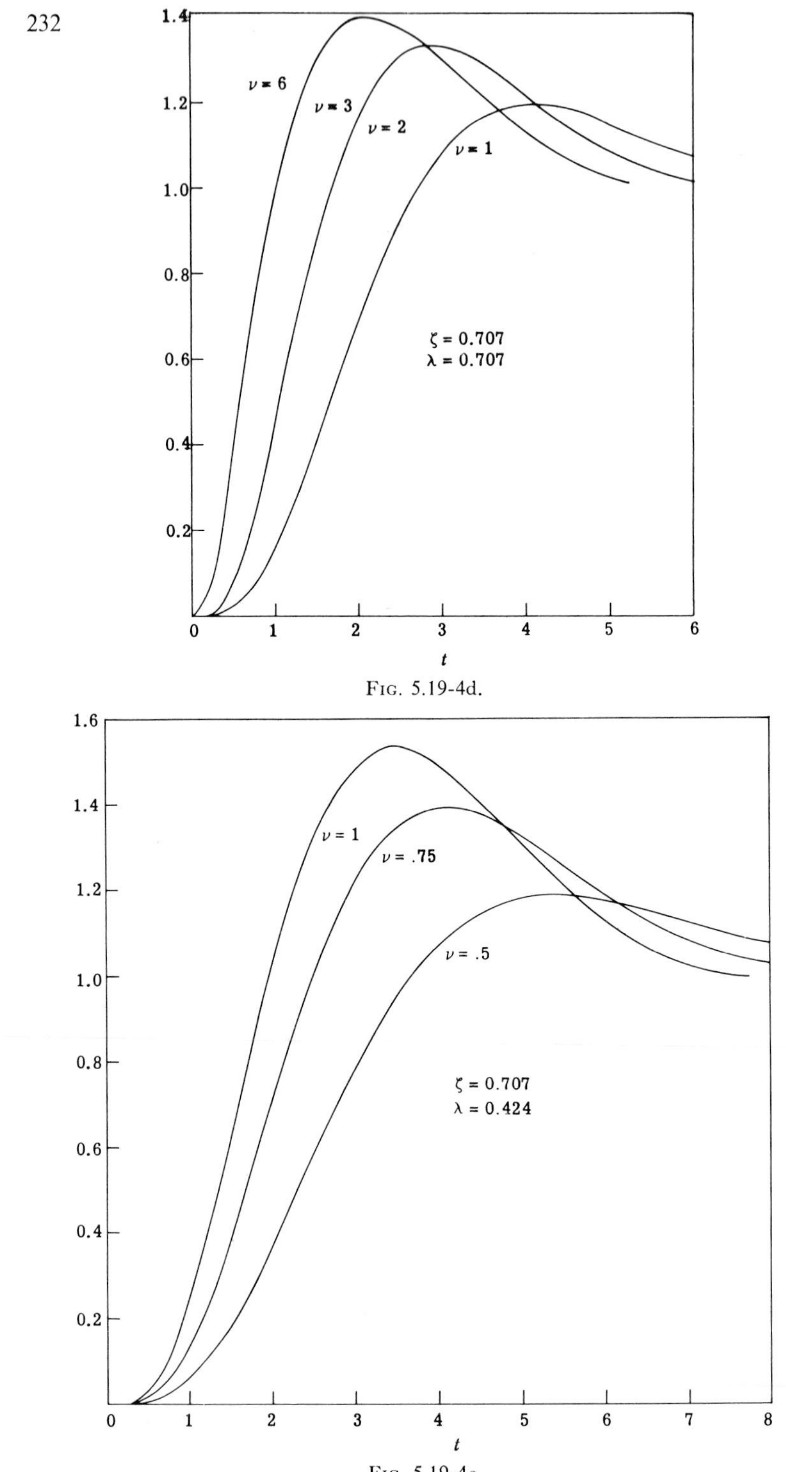

FIG. 5.19-4d.

FIG. 5.19-4e.

responses of some of these (and other) more complicated pole-zero patterns are sketched in Figs. 5.19-4a to e.

It is interesting to note the nature of the loop transmission function $L(s)$ that emerges when one of the above more complicated $T(s)$ is used. The general form of pole-zero pattern of the $T(s)$ here considered is shown in Fig. 5.19-5a. $L(s) = T/(1 - T)$ will have a zero at $-\lambda\zeta\omega_n$ and poles on the root loci of $1 - T = 0$. The root loci of $1 - T = 0$ are shown qualitatively in Fig. 5.19-5b. The pole-zero pattern of L will therefore be generally as shown in Fig. 5.19-5c and may include one pair of complex poles. An active RC technique for realizing complex poles is presented in the appendix.

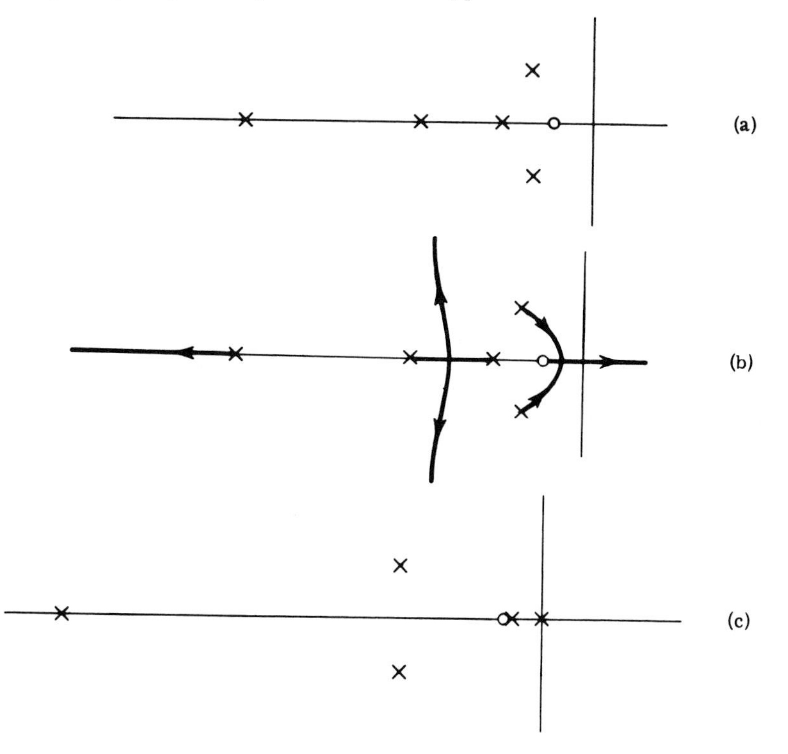

FIG. 5.19-5. (a) Pole-zero pattern of $T(s)$. (b) Root loci of $1 - T(s) = 0$. (c) Pole-zero pattern of $L(s)$.

§ 5.20 Root Locus Method

The methods presented in this chapter are useful as supplements to each other. They should not be considered as competitors, but as different ways of looking at the design problem. The competent control engineer will be familiar with all of them and use each for the properties for which it is best suited. These remarks apply especially to the root locus design approach. The root locus approach begins as one which is very similar to the frequency response design in that one begins with the constraint of a given plant and patches it up until a satisfactory $L(s)$, and therefore $T(s)$, is attained. The difference between the two is that in one case the system is characterized by its poles and zeros and in the other case it is described by its behavior along the $j\omega$ axis. But in proceeding with the root locus technique, the $T(s)$ pole-zero specifications must of necessity be considered and the root locus method becomes similar to the $T(s)$ pole-zero method. One advantage of root locus is that the designer never loses sight of what he actually must construct (the loop transmission), but at the same time he can observe directly the effect of $L(s)$ on $T(s)$. Because of the similarity between root locus design and the $T(s)$ pole-zero method, the presentation here is very brief.

Consider a simple plant with the pole-zero pattern of Fig. 5.20-1a, $L(s) = k/s(s + a)$. To avoid excessive overshoot and inadequate stability margins, no poles of $T(s)$ are allowed in the shaded region. The root loci of $1 + L(s) = 0$ are sketched (Fig. 5.20-1a); $k_{\max} = |OA|^2$ and therefore $(K_v)_{\max} = k_{\max}/a$.

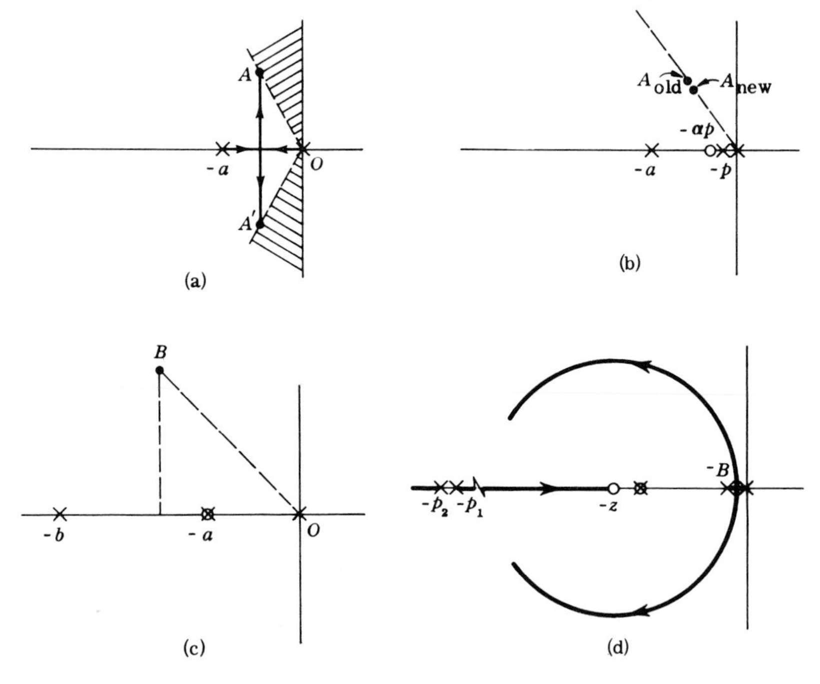

FIG. 5.20-1. Root locus method of design. (a) Root loci of $1 + L(s) = 0$, $L(s) = k/s(s + a)$, $K_v = |OA|^2/a$. (b) Use of lag compensation to increase K_v by factor α. (c) Use of lead compensation to increase K_v. (d) Another form of compensation to increase K_v.

Suppose this value of K_v is not sufficient. There are the two simple methods of increasing K_v. In Fig. 5.20-1b, lag compensation is used. Here k_{max} is practically the same as before, but $(K_v)_{max} = k_{max}\alpha/a$. The important poles of $T(s)$ are unaffected.

In Fig. 5.20-1c, lead compensation is used. The pole at $-a$ is canceled (or its effect decreased by a nearby zero). The system bandwidth is increased and now $k_{max} = |OB|^2$ and $(K_v)_{max} = k_{max}/b$. The increase in k_{max} is proportional to b^2/a^2, so the increase in K_v is proportional to b/a. Another form of lead compensation is shown in Fig. 5.20-1d. If the poles at $-p_1$, $-p_2$ are ignored, the loci are on a circle centered at $-z$. By suitably adjusting the latter, any desired closed-loop pair is attainable. K_v can also be made as large as desired by letting B lie close to the origin. This type of compensation was discussed in Section 5.13. It has the advantage that any combination of K_v and dominant closed-loop pole pair is achievable. In Fig. 5.20-1c increase in K_v is attainable only by increasing ω_n.

The root locus picture gives the designer some feeling for the kind of compensation needed to achieve the design objectives. It is therefore a very useful supplement to the other design methods. Consider, for example, Fig. 5.20-2a with the plant poles at O, A, A'. Without compensation, the closed-loop poles are

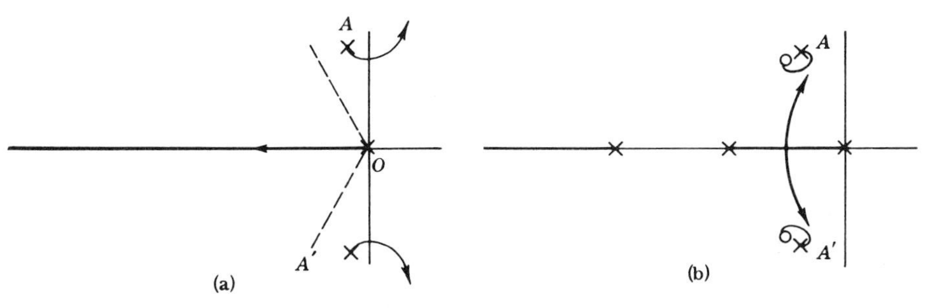

FIG. 5.20-2. Root locus design for plant with low damped poles. (a) Undesirable root loci of $1 + L(s) = 0$. (b) Satisfactory root loci of $1 + L(s) = 0$.

found to be in the forbidden region accompanied by a very low value of K_v. To force the loci to bend back into the left half-plane, it is clear that zeros of $L(s)$ must be located near AA' to effectively cancel these poles. [In the $T(s)$ pole-zero method, $L(s)$ would emerge with no poles at AA', therefore $G(s)$ would have zeros at AA'.] Then poles are assigned to a more desirable location, as in Fig. 5.20-2b.

If a plant has many dominant poles and zeros, as in Fig. 5.20-3, and only a single dominant pole pair is desired, then the four complex poles must all be canceled out by zeros of $G(s)$; one or more poles are placed at suitable locations in the dominant region and the other poles of $G(s)$ assigned far away. This is basically what is done when the dominance concept is imposed in a system with

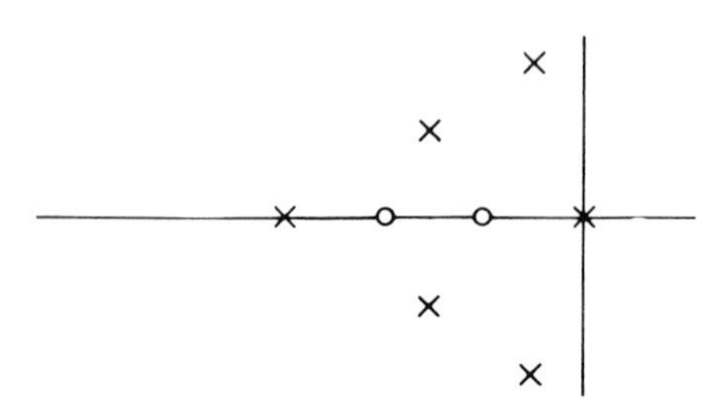

FIG. 5.20-3. Poles and zeros of a high-order plant.

a complex plant. If we attempt a $T(s)$ with four or five poles in the dominant region, then careful placing of zeros and poles of $G(s)$ are needed to ensure that the resulting root loci pass simultaneously through the desired points. This is easier done analytically, but a root locus sketch is a useful supplement, for example, to check against conditional stability.

§ 5.21 Design of High-Order System

The concepts and tools that have been hereto introduced are applied to the design of a system containing a high-order plant, on the assumption that the elementary configuration of Fig. 5.3-1 must be used. The plant poles and zeros are shown in Fig. 5.21-1.[1] Let $L(s) = -L_1(s)(s^2 + \omega_o^2)/(s^2 + \omega_p^2)$. (The minus sign is necessary because of the positive real zero, in order to have negative feedback.) The distance $\delta = \omega_o - \omega_p$ is so small that $-L_1(j\omega)$ is sensibly constant near ω_o, ω_p, and the effect of the dipoles at other frequencies is negligible. The problem is conveniently divided into two parts. The region near the dipole is examined by root locus, and the balance of the design can be handled by the frequency response method. The root locus near the dipole is first considered.

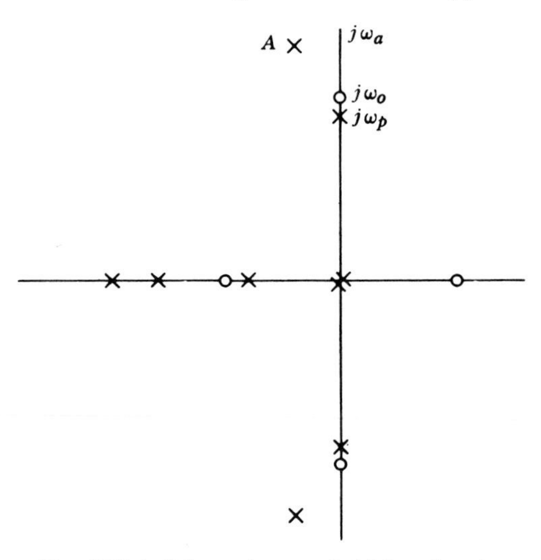

FIG. 5.21-1. Poles and zeros of a high-order plant.

At the outset we discard the idea of using compensation zeros or poles near the dipole in order to control the effect of the dipole on the system response. The reason is that the location of such zeros or poles would be predicated on precise knowledge of ω_o, ω_p. Such precise knowledge is lacking in any realistic

[1] The pole-zero pattern is somewhat similar to the one considered by J. A. Aseltine. The methods of feedback synthesis-A survey, *in* "Proceedings of the Symposium on Active Networks and Feedback Systems," Vol. 10. Interscience, New York, 1961. Our approach to the problem differs considerably from that of the author.

design. A decision must also be made as to whether the closed-loop pole (denoted by s_d) near the dipole must be in the left half-plane (lhp). If the system has an operating cycle of not too long duration (after which it is switched off, until the next cycle), then s_d may be in the rhp, providing the residue of $T(s)$ in s_d is sufficiently small. If the system must be on for a long time, then s_d must be in the lhp, and its residue must also be small.

The shape of the root locus near the dipole is easy to ascertain. From the equation

$$1 + L(s) = 1 - (s^2 + \omega_o^2)L_1(s)/(s^2 + \omega_p^2) = 0,$$

the angle of departure θ_p of the root locus from $j\omega_p$ is given by $\theta_1(j\omega_p) - 90° - \theta_p = -180°$, where $\theta_1(j\omega) = \arg[-L_1(j\omega)]$. Hence $\theta_p = \theta_1(j\omega_p) + 90°$. If s_d is to be in the lhp, then $270° > \theta_p > 90°$, so $180° > \theta_1(j\omega_p) > 0°$, i.e., $-L_1(j\omega_p)$ must lie in the first two quadrants. Let $\theta_1(j\omega_p) = \alpha > 0$. The angle of departure from the pole is then $90° + \alpha$ (see Fig. 5.21-2). If ω_p and ω_o are much closer to each other than to any other poles and zeros, then the angle of arrival at $j\omega_o$ is $270° - \alpha$, and the root locus joining ω_p, ω_o is an arc of a circle which is easily drawn from the known value of α. If α is much more than 90°, then the circle bulges far out into the lhp, and only the small portion near the dipole is part of the root locus.

The seriousness of the condition that $-L_1(j\omega_p)$ lie in the first two quadrants depends on the ratio of the crossover frequency ω_c to ω_p. Suppose the desired system bandwidth (which in the one-degree-of-freedom configuration is closely related to ω_c) is such as to require ω_c near ω_p. The two requirements are then incompatible. The crossover frequency must be sufficiently larger than ω_p to permit $-L_1(j\omega)$ to decrease from its positive value of $\alpha°$ at ω_p to about $-150°$ (for 30° phase margin) at ω_c. Also, in such a case the compensation must contribute a very large lead angle, because the plant (without the dipole) contributes a very large lag angle at ω_p. The problem is much easier if ω_c may be less than ω_p by a comfortable margin. In such a case $-360° < \theta_1(j\omega_p) < -180°$, or $-720° < \theta_1(j\omega_p) < -540°$, etc., are satisfactory and the required compensation is much more reasonable. Of course, there is no difficulty if s_d is allowed to be in the rhp. The lag angle of $-L_1(j\omega_p)$ is then determined by other considerations.

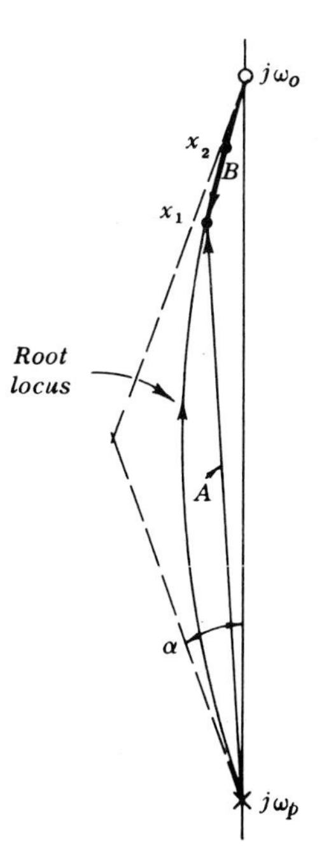

FIG. 5.21-2. Loci of roots near the dipoles.

We next consider the requirement of small residue of $T(s)$ in the pole at s_d. Since Fig. 5.3-1 has been prescribed as mandatory, $T = L/(1 + L) = - L_1(s^2 + \omega_o^2) [(s^2 + \omega_p^2) - (s^2 + \omega_o^2)L_1]^{-1}$. Since s_d is presumably very close to ω_o, ω_p, its residue is $\mathscr{R} \doteq - BL_1(j\omega_o)/[1 - L_1(j\omega_o)]$ (see Fig. 5.21-2 for B). Since s_d is a pole of $T(s)$, it must be a zero of $1 + L(s)$, i.e., $1 + L(s_d) \doteq 1 - L_1(j\omega_o)(B/A) = 0$; hence $L_1(j\omega_o) = A/B$. If $\omega_c \ll \omega_o$, then $|L_1(j\omega_o)| \ll 1$ and $\mathscr{R} \doteq - BL_1(j\omega_o) = - A$. For example, suppose $\omega_p = 20$, $\omega_o = 20.5$, $\theta_1(j\omega_o) = - 340°$, and $|L_1(j\omega)| = 0.05$. Then $\alpha = 20°$ in Fig. 5.21-2, and the arc of the circle is easily obtained. The location of s_d is found from the relation $A/B = L_1(j\omega_o)$. The result is $|B| \doteq 0.5$, $|\mathscr{R}| \doteq 0.025$. Clearly, $\mathscr{R} \doteq -j(\omega_o - \omega_p)L_1(j\omega_o)$, if $|L_1(j\omega_o)| \ll 1$.

If the crossover frequency $\omega_c > \omega_o$, then $\mathscr{R} \doteq - BL_1(j\omega_o)/[1 - L_1(j\omega_o)]$ must be used. For example, suppose $\omega_p = 20$, $\omega_o = 20.5$; it is required that $|\mathscr{R}| \leqslant 0.05$ and $\alpha = 20°$ is used (to avoid conditional stability). The arc of the circle is drawn (Fig. 5.21-2) and some points on the arc are tried. For example, at x_1 in Fig. 5.21-2, $B/A = 0.25$. At x_1, $1 + L(x_1) = 1 - (x_1^2 + \omega_o^2)L_1(x_1)/(x_1^2 + \omega_p^2) \doteq 1 - BL_1(j\omega_o)/A = 0$; hence $- L_1(j\omega_o) = |A/B| \angle 20° = 4 \angle 20°$ and $|\mathscr{R}_1| = (0.10)4/4.95 = 0.08$. We try x_2 closer to $j\omega_o$, and the result is $- L_1(j\omega_o) = 9 \angle 20°$, $|\mathscr{R}_2| = 0.45$. Hence $- L_1(j\omega_o) \approx 8 \angle 20°$ by extrapolation. Clearly, it is necessary that $\omega_c > \omega_o$ by a reasonable margin, so that $- L_1(j\omega)$ may decrease from $8 \angle 20°$ at ω_o, to about $1 \angle - 150°$ at ω_c. A fast change in phase would require a pole very near ω_o and make the design very sensitive to the dipole location, i.e., error or drift in ω_o, ω_p may lead to s_d in the rhp. In the design procedure which has been described, the response is only slightly affected if the dipole drifts as a unit vertically, or horizontally to the left, or if ω_o, ω_p each move vertically so long as $\omega_o > \omega_p$. If the drift is such that $\omega_o < \omega_p$, then $- L_1(j\omega_o)$ in the first and second quadrants leads to a rhp s_d. Let $T = T_d + T_1$, with $T_d = \mathscr{R}(s - s_d)^{-1} + \bar{\mathscr{R}}(s - \bar{s}_d)^{-1}$ and $T_1 = - L_1/(1 - L_1)$. $\mathscr{R}$ is chosen so that T_d is satisfactorily small and L_1 must be chosen on the basis of a reasonable compromise between the system filter and feedback properties. The balance of the design consists of shaping $- L_1(j\omega)$ on the Bode plot. This can be done in the usual manner with the dipole completely ignored (unless the crossover frequency is to be very close to the dipole—but this would be unwise). The high order of the plant does not in itself make the loop shaping on the Bode plot any more difficult. One proceeds in the usual manner to achieve the very low frequency specifications (the appropriate error constant), the crossover frequency ω_c, and the stability margins. However, the resulting compensation (G of Fig. 5.3-1) will be high order if the two plant poles at the origin are retained in $- L_1(s)$, if conditional stability is prohibited, and if ω_c is to be larger than ω_o. But if infinite K_v is unnecessary, then one plant pole can be canceled out, which eases the problem. The more poles at the origin that can be canceled out, the easier is the loop shaping problem. The problem is also considerably simplified if ω_c may be less than ω_p by a comfortable margin.

If these simplifications are prohibited, i.e., if infinite K_v and $\omega_c > \omega_o$ are specified, then the loop shaping problem is more difficult, but not because of the high order of the plant. The latter only affects the order of the compensating network $G(s)$. The difficulty is due to the fact that $-L_1(s)$ must have a zero in the rhp, and because of this zero it may be *impossible* to achieve $|-L_1(j\omega)| > 1$, up to the desired ω_c. It will be shown in Sections 7.9 and 7.12 (also Section 11.11) that rhp zeros impose very significant limitations on the achievable loop gain and crossover frequency. The specifications must be compatible with these limitations, as well as with that due to the specified maximum residue in the pole at s_d.

§ 5.22 Inadequacies of the Single-Degree-of-Freedom Configuration

The inadequacies of the structure are due to its dependence on the loop transmission to achieve both the filter-type and feedback-type system specifications. It is impossible to achieve both in general, with only one degree of freedom. In the pole-zero $T(s)$ design method, there is complete neglect of the system feedback properties. In the frequency response method, there is greater awareness and ability to cope with the feedback problem, but it is of no avail in many cases. The principal inadequacies will now be discussed.

Frequency Range of Large Sensitivity

The one-degree-of-freedom system is inherently very sensitive to parameter variation in an important frequency range. This is seen as follows. The system has only one degree of freedom, hence the system sensitivity which is given by the inverse of the return difference is [from Eq. (3.2,5)]

$$S_p{}^T \triangleq \frac{\Delta T/T_f}{\Delta P/P_f} = \frac{1}{1 + L_o} = 1 - T_o. \tag{5.22,1}$$

Consider, for example, $T_o = \omega_n{}^2/(s^2 + 2\zeta\omega_n s + \omega_n{}^2)$ whose polar locus is sketched for $s = j\omega$ in Fig. 5.22-1 for $\zeta = 0.5$, and for $\zeta = 0.707$. At $\omega = \omega_1$, $T_o(j\omega_1)$ is the vector from the origin to the corresponding point on the locus. Now $S_P{}^T = 1 - T_o$, so the sensitivity function is directly available from $T_o(j\omega)$ by using the point $+1$ as its origin. With $+1$ as center, a circle of unit radius is drawn. For all frequencies outside this unit circle, $|S_P{}^T| > 1$, and *for these frequencies the system sensitivity to plant variations is worse than if no feedback were used at all.*

One cannot argue that the region in which $|S_P{}^T| > 1$ is one of negligible importance. At $\zeta = 0.707$, this region extends for all $\omega > 0.7\omega_n$, and since $|T(j0.7\omega_n)| = 0.87$, it is not of negligible importance. The system bandwidth extends to ω_n rps, so that in arithmetic units the feedback is positive for 30% of the system bandwidth. And $\zeta = 0.707$ is a conservative design. At $\zeta = 0.5$

the feedback is positive (in the sense of greater sensitivity than a zero feedback system) for slightly more than 50% of the system response bandwidth. The system overshoot is very sensitive to the behavior of $T(j\omega)$ at the edge of the pass band. It is obvious from Fig. 5.22-1 that one requirement for avoiding this positive feedback region, at least over the pass band of T, is that the phase lag of $T(j\omega)$ must be less than 70° at its half-power point. This condition means that $T(s)$ must have only one dominant pole or, in other words, that the system impulse response can be fairly well approximated by the single term Ae^{-kt}. Clearly, this is unrealistic for any but the most elementary control systems.

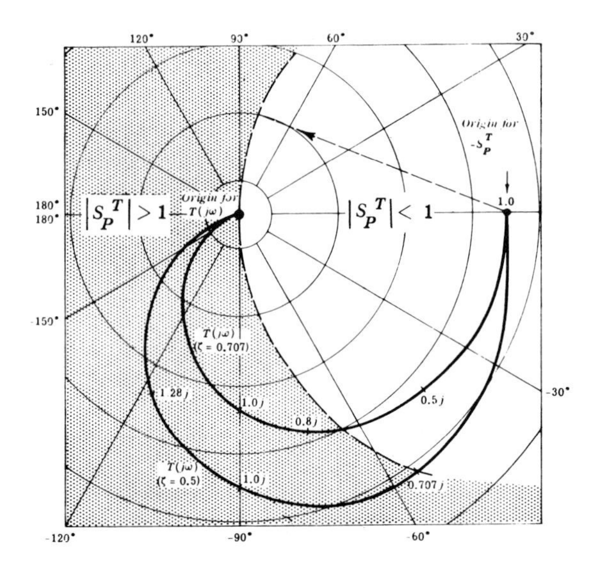

FIG. 5.22-1. Inherently poor sensitivity in the single-degree-of-freedom structure: $|S_P^T| > 1$ in shaded region.

This limitation in the feedback capabilities of the one-degree-of-freedom structure becomes more severe in higher order systems. Higher order T functions are required for plants which contain many elements in cascade, and which therefore have larger time delays. It is noted that this type of time delay is not necessarily associated with a large inertia. A large inertia also involves a large value of rise time and a small bandwidth, while the excess of poles over zeros in the significant frequency range can be small. The time delay considered here can be accompanied by small rise time (large bandwidth). It is primarily due to a large increase in phase lag over the significant frequency range [recall Eq. (5.10,1)], and therefore involves a large excess of poles over zeros in the important frequency range of T. [Such a larger ratio of time delay to rise time can also be achieved with a smaller excess of poles over zeros, if $T(s)$ has right half-plane zeros (see Section 7.9). However, such zeros must be zeros of $L(s)$ and consequently this requires that additional time delay, over and above that due to the plant, be deliberately inserted into the system. This is of course extremely undesirable.]

The difficulty in achieving the above kind of T function is seen by considering the loci of constant $|T|$ and constant $\theta = \angle T$ in Fig. 5.11-2b. Suppose $|L|$ and its phase angle decrease monotonically after $L(j\omega)$ has crossed the 180° line. The locus $|T| = 0.707$ cuts the x axis at -0.414, which corresponds to 7.6 db gain margin, at which $\theta = \angle T = -180°$. If T is to have a larger lag than 180° at its half-power point, then L must cross the locus of $|T| = 0.707$ higher in the second quadrant, and this will be accompanied by a smaller gain margin and larger peaking of $|T|$. The only way to secure a larger lag angle for T without sacrificing gain margin or peaking of T is to have $|L|$ temporarily increase after it has crossed the 180° line (e.g., if $\theta = -300°$ at $|T| = 0.707$, then $L = 0.78\angle -257°$). However, this temporary increase of $|L|$ must be accompanied by increase of phase lag, and this combination is secured by means of a sufficiently underdamped complex pole pair. Obviously, $T(s)$ is very sensitive to small changes in $P(j\omega)$ in this frequency range. This can also be seen from Fig. 5.22-1; the larger the phase lag of $T(j\omega)$ at its half-power point, the larger is the significant range of frequencies for which $|S_P{}^T| > 1$.

Poor Sensitivity of the Time Response

It is interesting to find the system sensitivity of the time response. From Eq. (5.22,1), for incremental variations, $dT = T_o(1 - T_o)dP/P$. Let $h(t) = \mathscr{L}^{-1}T_o(s)$ denote the system impulse response, and $dh(t)$ its incremental change due to plant parameter variations. Then, for the case dP/P independent of frequency,[1]

$$dh(t) = h(t)*[1 - h(t)]dP/P. \tag{5.22,2}$$

It is seen that $dh(t)$ is available directly from $h(t)$ and dP/P, without any need to refer to the system. A typical $h(t)$ is sketched in Fig. 5.22-2a and $dh(t)$ is obtained by convolution using Eq. (5.22,2). [The technique used to obtain $dh(t)$ is described in Chapter 12.] If no feedback at all is used, $dh(t)/dP/P = h(t)$ exactly. The difference in the sensitivity of the impulse response between a feedback design and no feedback is far from impressive. One would imagine that a good feedback design would result in an impulse response sensitivity similar to the fictional dotted curve included in Fig. 5.22-2a. However, the single-degree-of-freedom configuration is inherently unable to produce such an impulse response sensitivity.

The sensitivity of the step response is shown in Fig. 5.22-2b. Curve A is derived from Eq. (5.22,2), which is strictly correct for incremental variations

[1] $a*b$ means "the convolution of a and b."

only. For large variations, from Eq. (5.22,1), and using $C_o = T_o R$, $C_f = T_f R$,

$$\Delta C = \left(\frac{\Delta P}{P_f}\right)\left(\frac{C_f}{1 + L_o}\right) = \left(\frac{\Delta P}{P_f}\right) T_f(R - C_o). \tag{5.22,3}$$

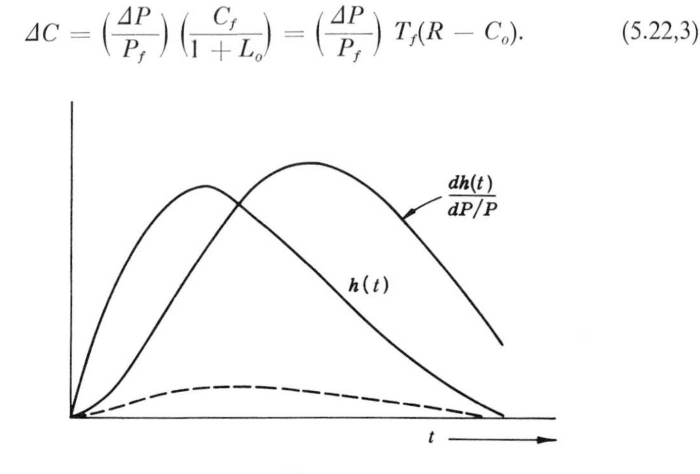

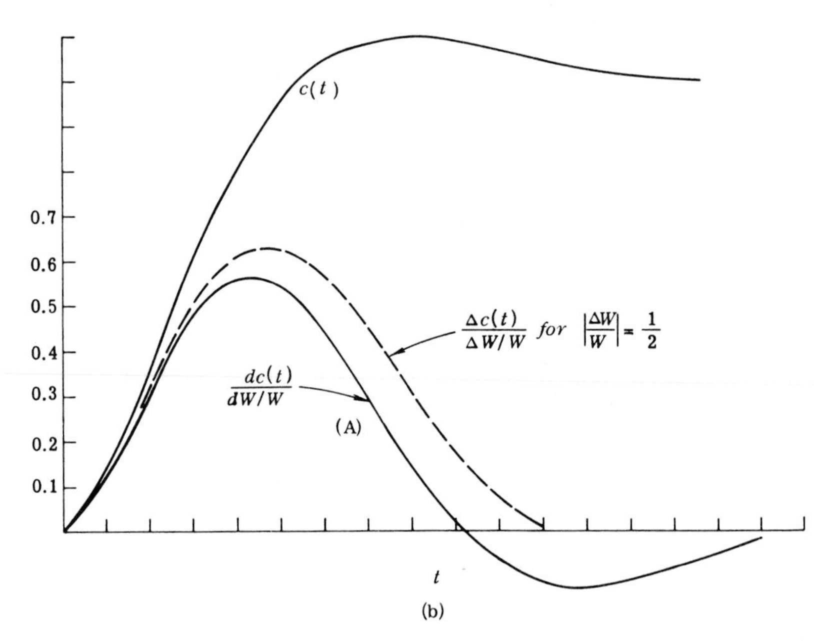

FIG. 5.22-2. (a) Sensitivity of the impulse response in the single-degree-of-freedom system. (b) Sensitivity of the step response in the single-degree-of-freedom system.

When $\Delta P/P_f$ represents a constant, K, $\Delta c(t) = Kh_f(t) * [1 - c_o(t)]$. There is an unknown on the right-hand side of the equation. The method used fo. its solution is presented in Chapter 12. The result for $K = 1/2$ is included in Fig. 5.22-2b. Clearly, the system overshoot is very sensitive to variations in P, while the final steady state value is unaffected. The reason for the latter is that $S_P{}^T$ is zero at zero frequency (because L is infinite there) and $T(0)$ determines the final value of the step response in the time domain.

Rejection of Disturbances

The single-degree-of-freedom configuration also has an inherent weakness in the matter of rejection of external disturbances or corrupting signals. In Fig. 5.22-3,

$$\frac{C_{D_1}}{D_1} = T_{D_1} = \frac{P}{1+L} = P(1 - T) \tag{5.22,4a}$$

$$\frac{C_{D_2}}{D_2} = T_{D_2} = \frac{1}{1+L} = 1 - T. \tag{5.22,4b}$$

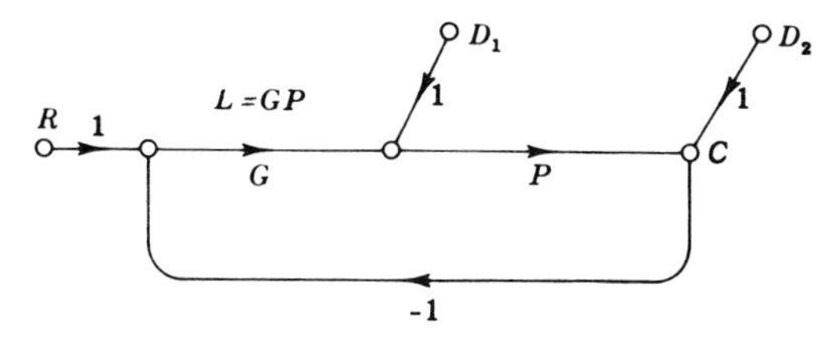

FIG. 5.22-3. System subject to plant disturbances.

Since L is completely determined by T, there is no freedom available in specifying T_{D_1} or T_{D_2}. Therefore, in general, there is not sufficient flexibility in this configuration to achieve T independently of T_D.

Consider, for example, $C_{D_2} = (1 - T)D_2$. At any frequency ω_1 for which $|S_P{}^T| = |1 - T| > 1$, the system response to D_2 is worse than if no feedback at all were used. And we have seen that $|S_P{}^T| > 1$ for a significant portion of the system bandwidth. If we make $1 - T$ small for the important frequency range of D_2, then $|T| \doteq 1$ for all these frequencies. Thus, only when the bandwidth of T is considerably larger than that of D_2 does the system behave satisfactorily with respect to the disturbance D_2. It may sometimes be possible to provide such a bandwidth for T. However, it is shown in Chapter 9 that the choice of bandwidth of T is strongly influenced by the noise that enters the system directly at the input point R. This may prohibit increasing the bandwidth of T. There simply is not enough flexibility in the system to handle any but very special kinds of D_2.

The same criticism applies to $T_{D_1} = P(1 - T) = PS_P{}^T$. In most cases, P is small when $S_P{}^T$ is large (because when P is small, L is usually small) and $S_P{}^T$ is small when P is large. However, there are situations where P is large when $S_P{}^T$ is not small. A typical case is when P has a resonant characteristic due to highly underdamped complex poles (as, for example, in some hydraulic systems). Let $P = k_p/(s^2 + 2\zeta_p\omega_p s + \omega_p{}^2)$. If ζ_p is small, it will be necessary in any reasonable design to have this underdamped plant pole pair canceled out by a corresponding underdamped zero pair of G (as in Figs. 5.22-4a and b). Thus,

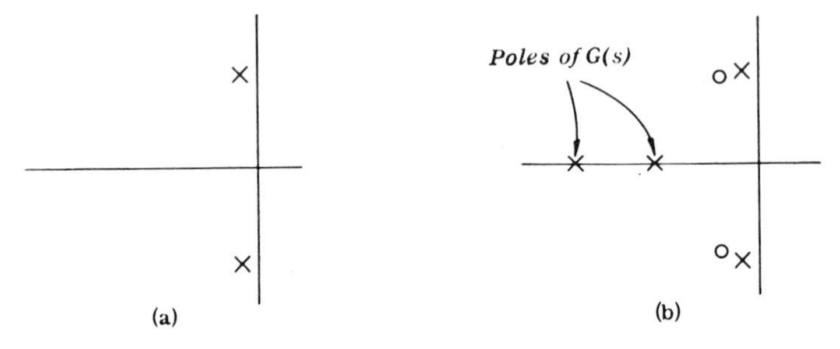

FIG. 5.22-4. Problem of low damped plant poles. (a) Highly underdamped plant poles. (b) ○ — zeros of $G(s)$.

let $G = (s^2 + 2\zeta_p\omega_p s + \omega_p{}^2)G_1$, and G_1 supplies a more reasonable set of poles for the loop transmission, in order to realize a satisfactory $T(s)$. However, the response to D_1 is $C_{D_1} = PD_1/(1 + L)$, so the highly underdamped poles of P are excited by D_1 (they are not canceled out by similar poles in L because $L = GP$ does not have these poles), and they lead to a relatively large disturbance output. This problem is aggravated when there are several such poles, and even more so when their locations drift due to plant parameter variations.

Conflict between Stability and System Transient Response

It is not difficult to imagine a situation where either stability or system response must be sacrificed in the classical configuration. Suppose there is substantial ignorance or variation of the plant parameter values. It is then necessary to have large gain and phase margins. These are associated with overdamped dominant poles (see Section 5.11) and hence with a more sluggish response (for a range of parameter values) than the system bandwidth is capable of delivering. It is impossible to decrease the damping without decreasing the stability margins and running the risk of instability at some values of the plant parameters. Only with the two-degree-of-freedom structure (Chapter 6) it is possible to avoid such a conflict.

In view of the above criticisms it is appropriate to inquire whether the single-degree-of-freedom configuration is at all useful. The answer is that it certainly is not sufficiently flexible to handle any combination of filter, sensitivity, and disturbance rejection specifications for any plant. However, it can be used for the many cases where these specifications are reasonably compatible. Consider

$T_{D_1} = P/(1 + L)$ in Fig. 5.22-3. Suppose D_1 is strong in the frequency range in which $P/(1 + L) \ll D_1$. The inequality may be strong enough to satisfy the noise rejection specifications. If the inequality is not sufficiently strong, increase in $|L|$ may be contemplated, but this will affect T of course. The same considerations apply to D_2.

Consider the system sensitivity to parameter variations. If the parameter variation is such as to decrease the gain, the effect is to slow up the response and decrease the overshoot. In many cases, this is not serious. In fact, when the system is working near its maximum capacity, it is advantageous to have the speed of response decrease as the plant's performance capacity decreases. For if it were not so, the system might saturate. Finally, there are control problems in which final steady state accuracy to step inputs is extremely important and transient response of secondary importance. So long as the motor can turn, L has a pole at the origin and $S_P{}^T$ is therefore zero at zero frequency. The steady state step response is therefore completely insensitive to plant changes. The extent to which the single-degree-of-freedom structure can cope with parameter variations is explored in more detail in Section 8.11.

The final conclusion is that each design problem should be separately considered. It may be that the single-degree-of-freedom configuration is able to satisfy all the design specifications. However, the most general feedback problem requires a structure with at least two degrees of freedom and these are discussed in Chapter 6.

CHAPTER SIX

Design of Feedback Control Systems for Independent Control of Transmission and Sensitivity Functions

§ 6.1 Configuration with Two Degrees of Freedom

It has been seen in Chapter 5 that the one-degree-of-freedom configuration is, in general, unable to cope simultaneously with the two problems of (1) attaining a desired system response as specified by a system transfer function $T(s)$, and (2) achieving sufficient feedback (loop gain) to handle plant parameter variations and/or disturbances acting on the plant. To simultaneously attain these two system objectives, a configuration with at least two degrees of freedom is necessary. Several such configurations are shown in Fig. 6.1-1. The significant feature of any two-degree-of-freedom configuration is that the system sensitivity (to the plant) function $S_P{}^T$ and the system transmission function T can be independently realized. These two functions, T and $S_P{}^T$, fix the values of the G and H functions in the various configurations of Fig. 6.1-1. Furthermore, if the plant has only two access points (the plant input and output points), and if access to the output is only via the plant (zero leakage transmission), then the plant is a potential two-degree-of-freedom system in the sense that it is impossible to create more degrees of freedom, no matter how complicated the configuration that is built around the plant. Thus, none of the structures of Fig. 6.1-1 can achieve any more than any of the others in Fig. 6.1-1 in terms of plant sensitivity and/or plant disturbance reduction and in attaining a desired system response.

Some of the structures in Fig. 6.1-1 have been presented as fundamentally different from the others. It has been suggested that they have virtues not possessed by the others, and have been given special names. For example, that in Fig. 6.1-1d has been called conditional feedback, because if P has its nominal value P_o, and $D_1 = D_2 = 0$, then $Y = \alpha T_o R - \alpha P_o G_4 R = \alpha T_o R - \alpha T_o R = 0$ and there is no feedback. In Fig. 6.1-1d feedback signals are applied at X via Y only when there is a reason for the existence of feedback, i.e., only when plant parameters change ($P \neq P_o$) or when there are disturbances. The structure of Fig. 6.1-1f has been called model feedback, because αP_o is a model of the plant.

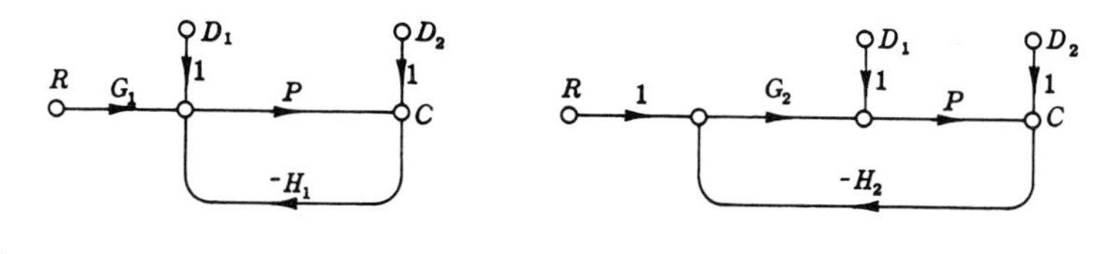

(a)

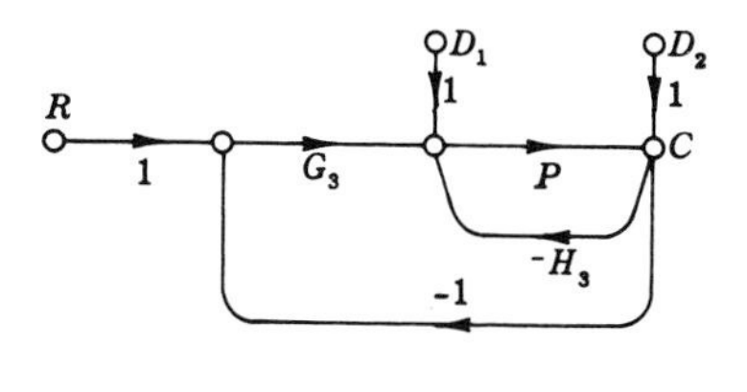

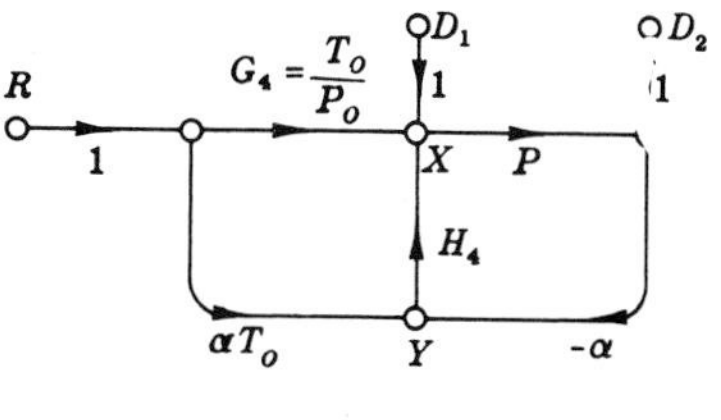

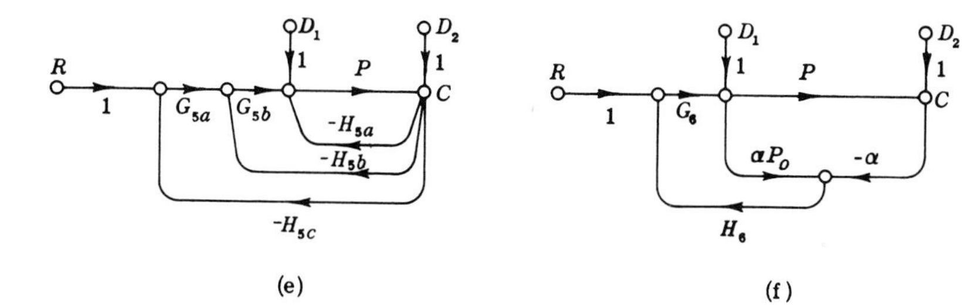

(e) (f)

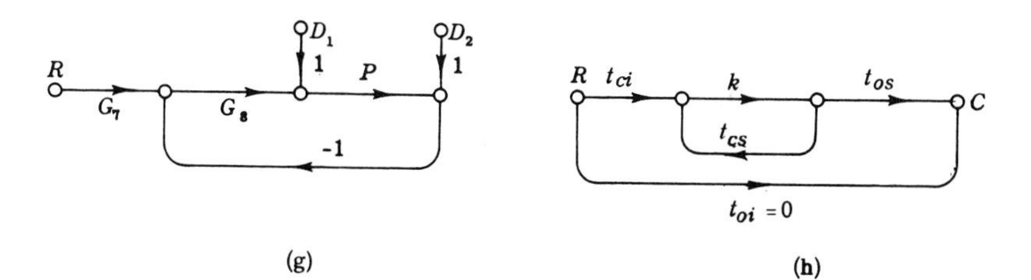

(g) (h)

FIG. 6.1-1. Some-two-degree-of-freedom feedback structures.

$$T = \frac{GP}{1 + L}, \qquad S_P^T = \frac{1}{1 + L},$$

$$G = G_1 = G_2 = G_3 = G_4 + \alpha T_o H_4 = G_{5a} G_{5b} = \frac{G_6}{1 - \alpha P_o H_6 G_6} = G_7 G_8 = t_{ci}$$

$$L = PH_1 = G_2 P H_2 = P(H_3 + G_3) = \alpha H_4 P$$

$$= P[H_{5a} + G_{5b}(H_{5b} + H_{5c} G_{5a})] = \frac{\alpha P H_6 G_6}{1 - \alpha G_6 P_o H_6} = P G_8 = -k t_{cs}$$

Here, too, there is no feedback when $P = P_o$ and $D_2 = 0$. It has been suggested that design for simultaneous realization of T and $S_P{}^T$ is easier with these configurations than with the others of Fig. 6.1-1.[1] In this subject, as in any other, it is often possible to make persuasive qualitative arguments for some position. The only way to truly appraise the situation is by a *quantitative* formulation of design objectives and the quantitative comparison of how the different structures satisfy these objectives. When this is done, it is found that with respect to the enumerated design objectives of the first paragraph, all two-degree-of-freedom configurations have basically the same properties and potentialities, and the price paid for the benefits of feedback is the same for all of them. The differences between the various configurations of Fig. 6.1-1 are in such matters as system sensitivity to the compensation networks, and the effect of disturbances acting on the compensating networks; even in these matters there is no particular advantage in using Fig. 6.1-1d or f.

Proof of the above statements is obtained by applying the fundamental feedback equation (2.2,7), using the plant P as the parameter k. Consider the fundamental feedback graph of Fig. 2.2-2, redrawn here as Fig. 6.1-1h with the fundamental feedback equation $T = t_{oi} + (kt_{ci}t_{os})/(1 - kt_{cs})$. It is seen that $t_{oi} = 0$, because zero leakage transmission has been postulated. Also, k is the plant $P(s)$, and $t_{os} = 1$ because the system output is the plant output. Therefore, the fundamental feedback equation becomes $T = Pt_{ci}/(1 - kt_{cs})$. Let t_{ci} be represented by G, and the loop transmission $-kt_{cs}$ by L, and then the general expression for the two-degree-of-freedom system is

$$T(s) = \frac{G(s)P(s)}{1 + L(s)} \tag{6.1,1}$$

It is emphasized that Eq. (6.1,1) is the general expression providing (1) that the plant has only two access points—an input and an output, and (2) that there is no control of the output (by the designer) except through the plant, i.e., $t_{oi} = 0$.

It is easily seen that $S_P{}^T \triangleq (\Delta T/T_f)/(\Delta P/P_f)$ [as defined by Eq. (3.2,4)] and the disturbance transmissions of Fig. 6.1-1, are given by $S_P{}^T = 1/(1 + L)$, $C/D_1 = T_{D_1} = P/(1 + L)$, $C/D_2 = T_{D_2} = 1/(1 + L)$. Therefore control of T, $S_P{}^T$, and the T_D's is lodged completely in the two functions L and G. Of course, each of these two functions can consist of many others, as in Fig. 6.1-1e, where $G = G_{5A}G_{5B}$, $L = (H_{5A} + H_{5B}G_{5B} + H_{5C}G_{5A}G_{5B})P$. However, there is no point at all in postulating *a priori* any such complicated configurations in a fundamental study of the two-degree-of-freedom system, or even in a practical design. The proper approach is to realize that there are two fundamental functions that must be

[1] For references see I. Horowitz, Fundamental theory of automatic linear feedback control systems, *IRE Trans.* **AC-4**, 5-19 (1959); Plant adaptive systems vs ordinary feedback systems, *ibid* **AC-7**, 48-56 (1962).

specified. These are (1) T, and (2) $S_P{}^T$ (or $L = S^{-1} - 1$). In any numerical design problem, they can be specified with hardly any reference to the actual configuration that will eventually be used. After T and S have been chosen, one can choose the specific configuration on the basis of ease of realizing T and $S_P{}^T$, constraints of the particular problem, etc. Any one of the structures in Fig. 6.1-1, or of the infinitude of other two-degree-of-freedom structures, can be chosen to implement the design of specified T and $S_P{}^T$, for a given plant.

This point is of substantial importance, for it clearly separates the problem of basic system capabilities from the problem of structure. The basic system capabilities (as far as plant sensitivity and plant disturbances are concerned) are completely determined by the number of access points of the plant. The plant access points determine the potential degrees of freedom of the system. If these are n in number and are all to be used, then any one of the infinite structures with n degrees of freedom may be used. It should be emphasized that all the advantages and disadvantages of feedback are precisely the same, whatever structure is used. These advantages and disadvantages are completely determined by $S_P{}^T$ (by the $S_{P_i}{}^T$'s in the case of n degrees of freedom with $n > 2$). Therefore all structures with the same $S_{P_i}{}^T$ will have the same advantages (sensitivity reduction, etc.) and disadavantages (stability problem, need for large loop gain-bandwidth, etc.).

This separation of basic system capabilities from the problem of structure is actually helpful to the designer, for it destroys the mystique of structure which seems to some to be of great importance in feedback theory. The designer need not fear that, if he were only clever enough, he could find some exotic structure with new and wonderful properties. The important factors determining the feedback capabilities and the cost of feedback are the system constraints: whether t_{oi} is or is not zero, the number of access points in the plant, the kind of stability that is permitted. The configuration that is thereafter used is not of major importance.

This chapter is devoted exclusively to the two-degree-of-freedom system, and to general design procedures for achieving the two independent quantities achievable: (1) the system transfer function $T = C/R$; (2a) insensitivity of T to plant parameter variations, (2b) attenuation of disturbances acting on the plant. Sections 6.2-6.6 are devoted to root locus s-plane techniques, and Sections 6.7-6.12 present frequency response methods. The two methods are compared in Section 6.12. Plants with more than two degrees of freedom are treated in Chapter 8.

§ 6.2 Root Locus Synthesis to Control System Sensitivity to Variations in Plant Gain Factor

In systems wherein the over-all system transfer function $T(s)$ may be characterized by relatively few dominant poles and zeros, it is feasible to use root locus methods for realizing a desired insensitivity of T to plant parameter variations. The specifications for the sensitivity of the system transfer function $T(s)$ to plant parameter variation may be formulated in terms of the maximum permissible drift in the dominant system poles and zeros. Thus, let $T(s) = K\Pi(s + z_j)/\Pi(s + p_i)$, $h(t) = \mathscr{L}^{-1}T(s) = \Sigma_i A_i e^{-p_i t}$, and only the few dominant terms need be considered.

The permitted variation of $h(t)$ in the time domain may be related to a corresponding area of variation of the dominant poles and their residues. If the variation in the dominant poles and zeros is small, their residues, too, will vary only slightly, although the far-off poles and zeros may change considerably. This is because $T(0) = K\Pi z_j/\Pi p_i = K\Pi z_d \Pi z_f/\Pi p_d \Pi p_f$ is either completely insensitive (if L has a pole at the origin, as is often the case) or is highly insensitive to plant variations, because invariably $L(0)$ is large. The z_d are the dominant zeros of T, the z_f are the far-off zeros of T, and similar notation is used for the poles. If, as is postulated, there is little variation in the dominant z_d, p_d, then $K\Pi z_f/\Pi p_f$ also varies little, because the variation in $T(0)$ is zero or very small. The residue in a dominant pole of T is determined by the vectors from all the other zeros and poles of T to the pole in question. The contributions from the remaining dominant poles and zeros are relatively small and, in any case, it is assumed that there is small variation of the latter. The contributions from the far-off poles and zeros of T is closely equal to $K\Pi z_f/\Pi p_f$, whose net variation, as has just been seen, is also small; and certainly in any case readily estimated, and is approximately the same for all the dominant poles. Thus the variation in the dominant poles of T is fairly simply related to the variation in $h(t) = \mathscr{L}^{-1}T(s)$.

The procedure for root locus s-plane synthesis to achieve a specified T and a desired insensitivity of T to variations in the plant gain factor is presented with sufficient generality by means of a specific numerical example.

Design Example.

Specifications: The plant transfer function is $P = K/s(s + 2)$, in which K may vary from 8 to 20. The system transfer function $T(s)$ is to have a dominant pole pair at $-4 \pm j4$. It is desired that the variations in K should not cause the dominant pole pair of $T(s)$, to move outside a circle of radius 1 centered at $-4 \pm j4$.

Design: It is fairly easy to see that the desired insensitivity of the dominant poles of T cannot be achieved with the single-degree-of-freedom structure of Chapter 5. This is seen as follows. If the latter is used, then $T = L/(1 + L)$. Using root locus, it is obvious that the simplest pole-zero pattern of L is that shown in Fig. 6.2-1, with $L = 32/s(s + 8)$. Since the nominal $P_o = K_o/s(s + 2)$, $G = L/P_o = 32(s + 2)/K_o(s + 8)$. Suppose $K_o = 14$. Then, as K varies from 8 to 20, it is found, by factoring $1 + L$, or by root locus rules, that the dominant poles move vertically from $-4 \pm j1.51$ to $-4 \pm j5.44$. It is possible to make this region of variation symmetrical about $-4 \pm j4$, by writing $s(s+8) + K_o' = s^2 + 8s + 16 + (4-a)^2$, and $s(s+8) + 2.5K_o' = s^2 + 8s + 16 + (4+a)^2$ and solving for K_o' and a. The results are $K_o' = 20.4$ and $a = 1.91$, i.e., the nominal loop transmission is $L_o = 20.4/s(s+8)$, $P_o = 8/s(s+2)$; so that the required $G = 2.55(s+2)/(s+8)$, and the dominant poles of T vary along the vertical from $-4 \pm j2.09$ to $-4 \pm j5.91$.

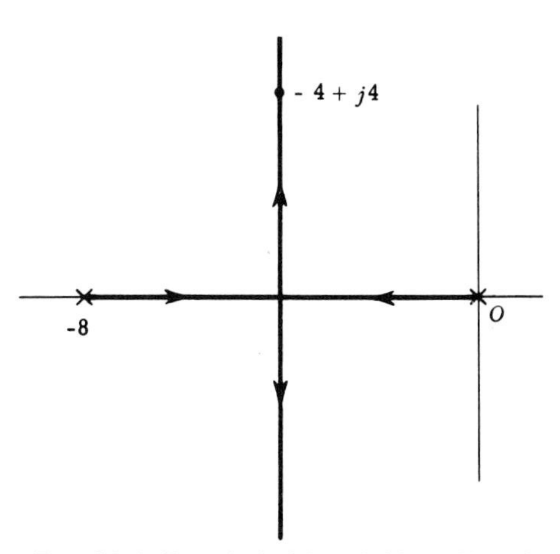

FIG. 6.2-1. Root loci of $1 + K/s(s+8) = 0$.

Is it possible to decrease the region of variation of the dominant poles by assigning more poles and zeros to L? There is certainly no improvement possible without adding at least one zero to L. The trouble is that this zero cannot be very far away from $-4 \pm j4$, if it is to have any significant effect. This zero (in the single-degree-of-freedom structure) must also appear as a zero of T, and therefore significantly changes T as well as L. There is insufficient freedom to independently control T and the sensitivity of the poles of T.

The designer is compelled to turn to a two-degree-of-freedom structure, which permits him to assign zeros (and, to a lesser extent, poles) to L without having to worry about their effect on T. Thus L is chosen primarily to attain the desired insensitivity of T, and there still remains one degree of freedom to achieve the desired T.

What is the fundamental principle by means of which the proper choice of L restricts the movement of the dominant zeros of $1 + L$? An illustration of the principle is shown in Fig. 6.2-2. Zeros of L are assigned quite close to $-4 \pm j4$. (In a two-degree-of-freedom system, zeros of L need not be zeros of T.) The poles of the device or network supplying these zeros can be placed far away (at $-a$, $-b$, $-c$ in Fig. 6.2-2), in order that their effect at $-4 \pm j4$ is small. It is possible to choose the gain factor such that at the lowest value of K, the dominant roots of $1 + L = 0$ are near $-4 \pm j4$, and at the largest value of K they are still fairly close to $-4 \pm j4$. The variation in their position can, in fact, be made arbitrarily small by assigning the zeros of L sufficiently close to $-4 \pm j4$. However, the closer the zeros are placed to $-4 \pm j4$, the larger must be the gain factor in order that $-4 \pm j4$ may be roots of $1 + L = 0$ (this follows from Rule 10 of Section 4.2). Also, the root loci which emerge from the far-off poles of the device supplying the zeros, must be considered. Thus, their root loci are qualitatively as shown in Fig. 6.2-2. The larger the gain

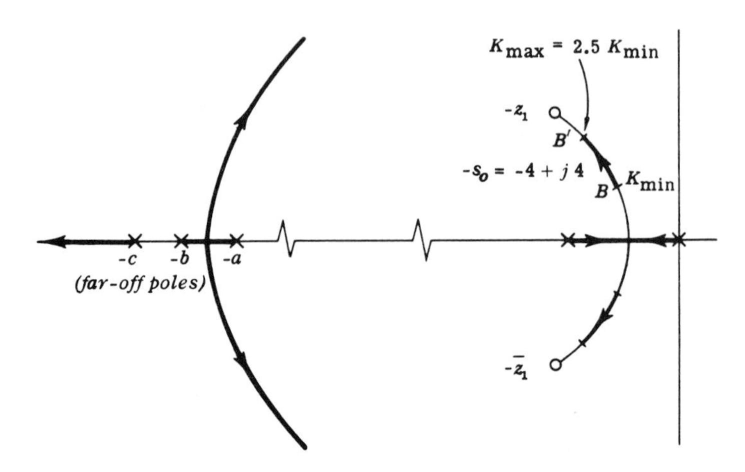

FIG. 6.2-2. Philosophy of root locus design for control of sensitivity; ○ — zeros of $L(s)$, × — poles of $L(s)$.

factor (for fixed a, b, c), the closer do some of the far-off roots of $1 + L = 0$ approach the $j\omega$ axis (consider the loci at infinity), and the greater is the danger of high frequency oscillations. This danger is avoided by moving the far-off poles even further to the left (increasing the values of a, b, c). The smaller the permitted movement of the dominant roots, the larger must be the gain factor, and the further must be the far-off poles. Consequently, the device or network, with transfer function $k(s + z_1)(s + \bar{z}_1)[(s + a)(s + b)(s + c)]^{-1}$, must have very large gain and bandwidth.

The above is the situation in a design which insists upon the dominant poles concept, and small movement of the dominant poles. As the latter requirement is eased, it is possible to get by with zeros further away from the dominant poles of T, or even with a single zero on the negative real axis (see Section 8.12). In the present numerical example, it would appear at first guess that a reasonable qualitative pole-zero pattern for $L(s)$ and root loci of $1 + L(s) = 0$ would be similar to that shown in Fig. 6.2-2. The first problem is to find the required location of the zeros of $L(s)$. This is done as follows.

Suppose B in Fig. 6.2-2 is a root of $1 + L = 0$ when the plant gain factor is K. From the rules of root loci construction (Section 4.2, Rule 10), $K = \Pi(p_iB)/\Pi(z_jB)$, where p_iB is the vector from a pole of L at $-p_i$ to the root at B, and similarly for the zero at $-z_j$. When K varies in value and becomes K', the root of $1 + L = 0$ originally at B moves to a new point B', such that $K' = \Pi(p_iB')/\Pi(z_jB')$. If the root at B is not allowed to drift much, then each $p_iB' \approx p_iB$, and each $z_jB' \approx z_jB$, except for the zero z_1 near B, B' (see Fig. 6.2-2). Obviously, such an approximation is valid only when z_1 is significantly closer to B and B' than any other poles and zeros of L. If there is such a marked difference, then $K' \approx \Pi(p_i B)/(z_1 B')\Pi_{j \neq 1}(z_j B)$. Combining the above equations, we have

$$z_1 B'/z_1 B \approx K/K'. \tag{6.2,1}$$

It also follows from this approximation that z_1, B, and B' lie approximately on a straight line. In the present example, if K corresponds to the minimum value of the gain factor, then $K/K' = 0.4$ and, from Eq. (6.2,1), $| z_1 B' | \approx 0.4 | z_1 B |$ (see Fig. 6.2-2). However, the design specifications dictate that $| BB' | \leqslant 2$. If $| z_1 B | = x$, then $x - 2 \approx 0.4x$ with $x \approx 3$.

The above does not precisely locate the zero, because its phase angle is not as yet specified. This permits the designer to pick the direction in which the poles of T will move as the gain factor varies. A reasonable direction might be at 135° with the origin, so that the damping factor of the dominant poles of T does not change. This choice locates $-z_1$ at $-5.4(1+j)$.

The next step is to fix the other poles of L. It is necessary that $\arg L(s_o) = 180°$, where $s_o = -4 \pm j4$. The vectors to $-4 + j4$ from the plant poles at 0, -2, and from the zeros at $-5.4(1 \pm j)$, contribute a total of $-214°$. There-

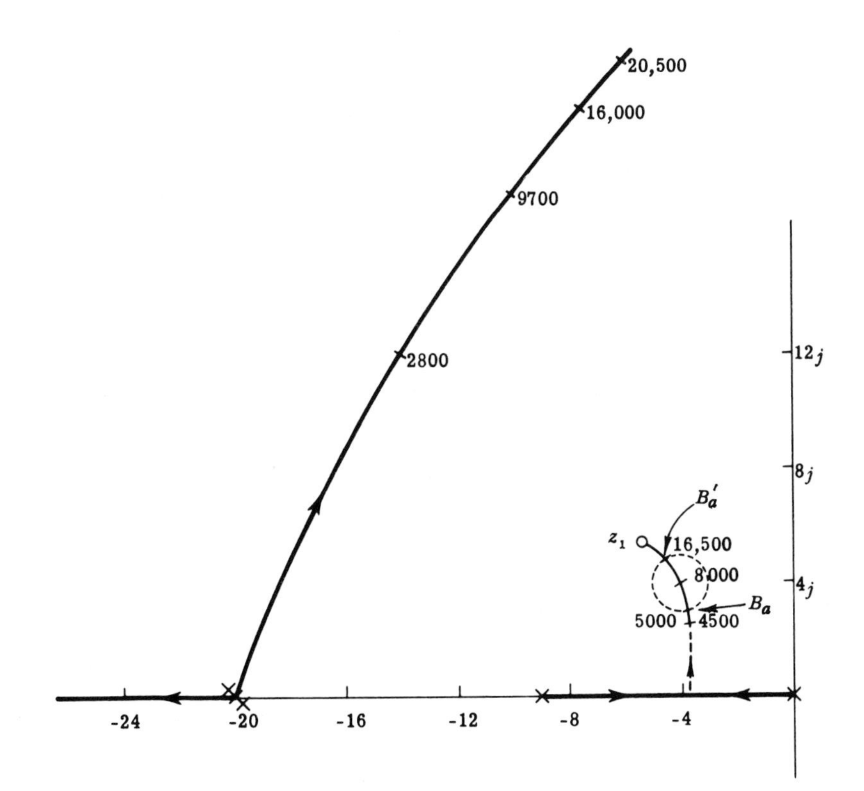

FIG. 6.2-3. Root loci of $1 + L = 0$.

fore the pole at -2 is moved further out, say to -9, and the new angle is $-138°$. This leaves $42°$ to be assigned to the far-off poles; three coincident poles are therefore located at -20.

We are now ready to check the design. The root loci of $1 + L = 0$ are sketched in Fig. 6.2-3. It is seen that the variation in the dominant roots is less than the maximum permitted. The gain may vary by a factor of 3 without violating the specifications. The designer can therefore, if he likes, use Fig. 6.2-3 to obtain a better approximation. Thus,

$$\frac{K'}{K} = \frac{\Pi(p_i B')\Pi(z_j B)}{\Pi(p_i B)\Pi(z_j B')} \approx \frac{\Pi(p_i B_a')\Pi_{j\neq 1}(z_j B_a) z_1 B_a}{\Pi(p_i B_a)\Pi_{j\neq 1}(z_j B_a') z_1 B_a''}. \tag{6.2,2}$$

In the above, B_a and B_a' are the first approximations to the unknown B and B'. They are obtained from Fig. 6.2-3, with $B_a = -3.8 + j3$, $B_a' = -4.6 + j4.8$. B_a'' is the second approximation to the unknown B'. This leads to: $2.5 = K'/K \approx 1.1[(z_1 B_a)/(z_1 B_a'')] \angle 23°$ or $z_1 B_a / z_1 B_a'' \approx 2.27 \angle -23°$; z_1 is chosen to satisfy this equation, and is found to be $-5.7 + j5.4$, which is slightly further from the closed loop roots than the original value.

Since there is not very much difference between the two locations of the zeros, the original value will be used to complete the design. From Fig. 6.2-3, the extreme value of L is $L' = 16{,}500(s^2 + 10.8s + 58.3)\,[s(s + 9)\,(s + 20)^3]^{-1}$, at the extreme plant transfer function value of $P' = 20/s(s + 2)$. Suppose the structure of Fig. 6.1-1a is used, with $L = PH_1$. Then $H_1 = L/P = L'/P' = 825(s+2)$ $(s^2 + 10.8s + 58.3)\,[(s + 9)\,(s + 20)^3]^{-1}$. Finally, G is chosen to achieve the desired system transfer function

$$T = \frac{C}{R} = \frac{GP}{1 + L} = \frac{G(s + 9)(s + 20)^3}{(s + 2)(s^2 + 9.2s + 44.2)(s^2 + 14.7s + 480)(s + 45)}.$$

To simplify G as much as possible, T is chosen as $T = 119(s + 20)^2\,[(s^2 + 9.2s + 44.2)\,(s^2 + 14.7s + 480)\,(s + 45)]^{-1}$. The net effect of these far-off poles and zeros of T, on the system response, is negligible. If Fig. 6.1-1b is used, then $G_2 = T(1 + L)/P$ and

$$H_2 = \frac{L}{PG_2} = \frac{L}{T(1 + L)} = \frac{16{,}500(s^2 + 10.8s + 58.3)}{T(s^2 + 9.2s + 44.2)(s^2 + 14.7s + 480)(s + 45)}.$$

In this case, it is convenient to let $T = 53.04(s + 20)^2\,[(s^2 + 9.2s + 44.2)$ $(s^2 + 14.7s + 480)]^{-1}$, and then $G_2 = [53.04(s + 2)\,(s + 45)]\,[20(s + 9)\,(s + 20)]^{-1}$, $H_2 = [312(s^2 + 10.8s + 58.3)]\,[(s + 45)\,(s + 20)^2]^{-1}$. Note that G_2 has emerged finite at infinity, which is, of course, impossible. If the designer knew beforehand that the structure of Fig. 6.1-1b would be used, with $L = G_2PH_2$, then L should have been chosen with at least four more poles than zeros, in order to provide P with its excess of two poles, and at least one for G_2 and for H_2.

If the conditional feedback structure of Fig. 6.2-1d is used, then $\alpha H_4 = L/P = 825(s^2 + 10.8s + 58.3)(s + 2)[(s + 9)(s + 20)^3]^{-1}$, which is precisely the same as H_1 of Fig. 6.1-1a. It may be noted that the G and H networks in the above require active elements, and hence have their own sensitivity problems. As noted in Section 5.1, in feedback control systems, it is generally assumed that there is no difficulty in obtaining active elements such as vacuum tubes or transistors, which have sufficient gain-bandwidths to provide both the gain needed and the local loop gain needed for their own sensitivities (see Appendix).

It is important to make sure in Fig. 6.2-3 that the far-off roots of $1 + L = 0$ do not approach too closely to the imaginary axis over the range of variation of the gain factor. The design is satisfactory in this respect. It is, however, only a matter of luck that this is the case here, because nothing was done in the course of the design to assure this. Suppose that it had turned out that the far-off roots of $1 + L = 0$ did come too close to the $j\omega$ axis, or even entered the right half-plane, for some range of K. In such a case, it would have been necessary to move the far-off poles of L further out to the left, and modify somewhat the other poles of L. However, this matter need not be left to chance. In Section 6.4 it will be shown how the proper precautions with respect to the far-off poles may be incorporated into the design procedure.

§ 6.3 Root Locus s-Plane Synthesis for General Plant Parameter Variations

In the preceding section it was assumed that only the plant gain factor varied. In this section the design technique is extended for the more general case where both the gain factor and the plant poles vary. Again it is convenient to present the procedure by means of a numerical example.

Design Example.

Specifications: The plant transfer function is $P = 1250K[s(s^2 + 2\zeta_p\omega_p s + \omega_p^2)]^{-1}$, with K varying from 1 to 4 and ζ_p, ω_p are such that the complex pole pair of P may be anywhere in the rectangles $ABCD$, $\bar{A}\bar{B}\bar{C}\bar{D}$ in Fig. 6.3-1. T is desired with a dominant pole pair inside a circle of radius 1.2 centered at $-10 \pm j10$ (R, $\bar{R}$ in Fig. 6.3-1). It is realized that T will undoubtedly have more (far-off) poles, and it is specified that they must always be to the left of the line $UVV'U'$ in Fig. 6.3-1 (given by $x = -30$). One may argue over the specific location and shape of $UVV'U'$, but the important point here is that such a choice must be made if one is to work with dominant poles and zeros. Such a choice is also necessary to assure stability and prevent ringing at the higher frequencies at any of the possible plant parameter values. The example demonstrates how to proceed after the choice has been made.

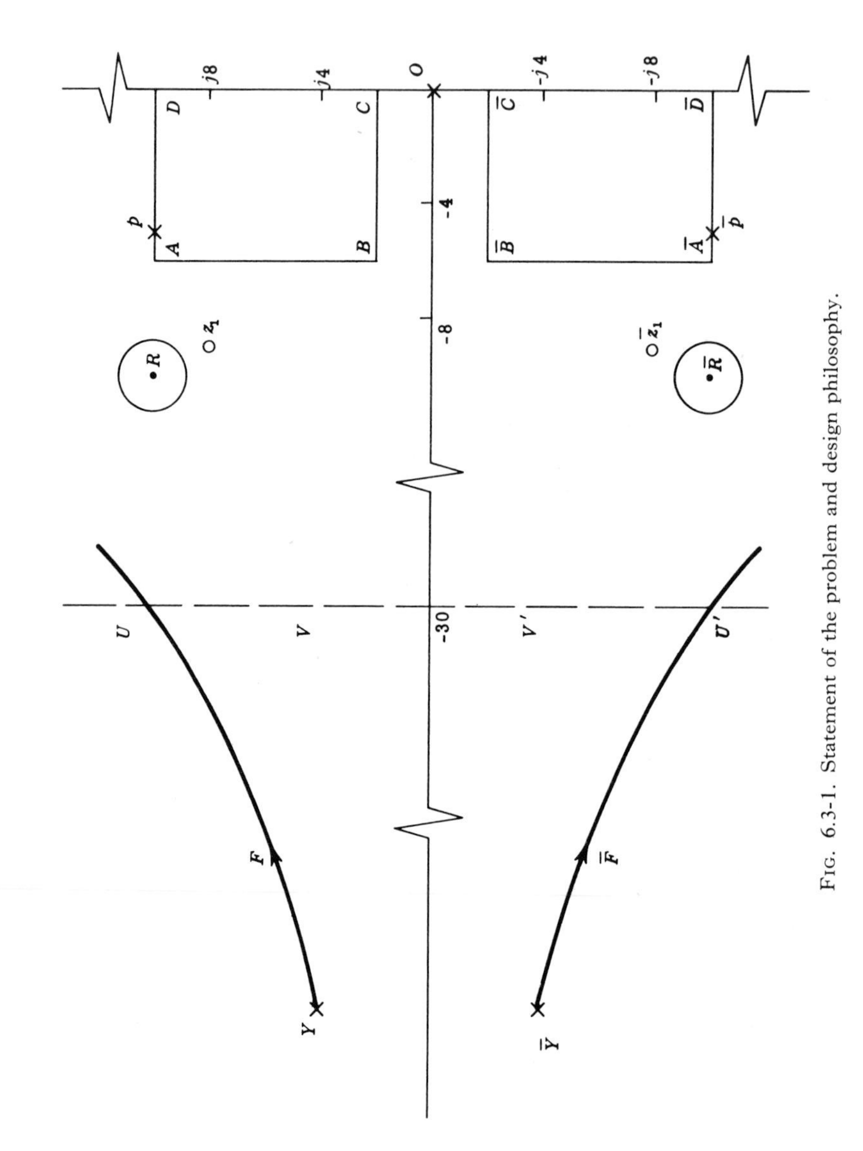

FIG. 6.3-1. Statement of the problem and design philosophy.

Design philosophy: The following is the design philosophy. It is basically the same as that used in Section 6.2. Loop transmission zeros (say at z_1, $\bar{z}_1$ in Fig. 6.3-1) near the desired dominant T poles (at R, $\bar{R}$), in conjunction with large loop gain, ensure closed-loop roots of $1 + L(s) = 0$ near the zeros at z_1, $\bar{z}_1$, irrespective of the plant-pole locations. The device or network which supplies the zeros at z_1, $\bar{z}_1$ must have at least two finite poles. These must be located sufficiently far away to ensure that the specifications with respect to the far-off poles of $T(s)$ are not violated. Suppose the poles of the device are located at Y, $\bar{Y}$ in Fig. 6.3-1. The danger is that the large loop gains needed to keep the dominant roots relatively stationary will cause the roots along the loci F, $\bar{F}$ to cross the $UVV'U'$ line. To prevent this, the device poles at Y, $\bar{Y}$ must be sufficiently far off to the left. Here is seen the price that is paid for keeping T so insensitive to such large plant variation over the significant frequency range. First, a large loop gain is needed, so that the dominant roots stay near the zeros at z_1, $\bar{z}_1$. Second, the large loop gain causes the need for compensating networks with large bandwidths, in order to preserve relative stability in the high-frequency range.

Design procedure: Determination of the approximate shape and orientation of region of dominant pole movement. A design procedure is now developed for finding the required locations of the zeros and poles of the loop transmission $L(s)$. In the present problem, there is drift in the roots of $1 + L(s) = 0$, because the gain factor of $L(s)$ varies by a factor of 4, and because one pair of complex poles of $L(s)$ moves over the large area $ABCD$, $\bar{A}\bar{B}\bar{C}\bar{D}$ in Fig. 6.3-1. Due to these two independent variations in the parameters of $L(s)$, the dominant roots of $1 + L = 0$ will vary in position over some area. The first step in the design is to find the general shape and orientation of this area. This is done by letting the gain factor K vary, and then letting the poles of L vary, and noting the effect on the dominant roots of $1 + L = 0$. The procedure for finding the shape of this area of variation is as follows.

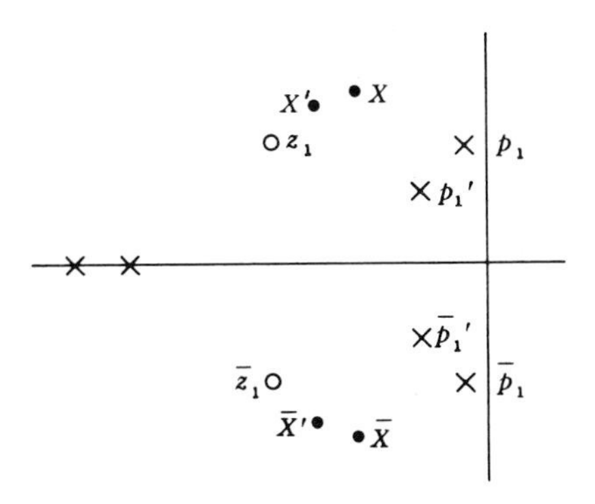

FIG. 6.3-2. Diagram for studying the effect of drift in plant poles from p_1, $\bar{p}_1$ to p_1', $\bar{p}_1'$.

The effect of variations in the gain factor has been noted in Section 6.2, Eq. (6.2,1) and is therefore not repeated here.

To find the change in the dominant root of $1 + L = 0$ due to the change in position of a complex pole pair of L, let p_1, $\bar{p}_1$, and p_1', $\bar{p}_1'$ respectively, denote the old and new complex pole pair positions of L; and X, X' denote the old and new positions of the dominant root of $1 + L = 0$ (see Fig. 6.3-2). Then

$$K = \frac{(p_1X)(\bar{p}_1X)\Pi_{i\neq 1}(p_iX)}{(z_1X)\Pi(z_jX)} = \frac{(p_1'X')(\bar{p}_1'X')\Pi(p_iX')}{(z_1X')\Pi(z_jX')}$$

$$\approx \frac{(p_1'X')(\bar{p}_1'X')\Pi(p_iX)}{(z_1X')\Pi(z_jX)}$$

and therefore

$$\frac{z_1X'}{z_1X} \approx \frac{(p_1'X')(\bar{p}_1'X')}{(p_1X)(\bar{p}_1X)}. \tag{6.3,1}$$

The approximation is valid if $|p_iX| \approx |p_iX'|$, except possibly for $i = 1$, and if $|z_jX| \approx |z_jX'|$, except for z_1. When the maximum variation of the dominant root must be kept small, then it is permissible to also assume that $p_1'X' \approx p_1'X$, $\bar{p}_1'X' \approx \bar{p}_1'X$. Equations (6.2,1) and (6.3,1) are used to find the approximate shape and orientation of the region of variation of the dominant roots of $1 + L = 0$, due to the variations in the gain factor and the poles of L. This is done in Fig. 6.3-3 as follows.

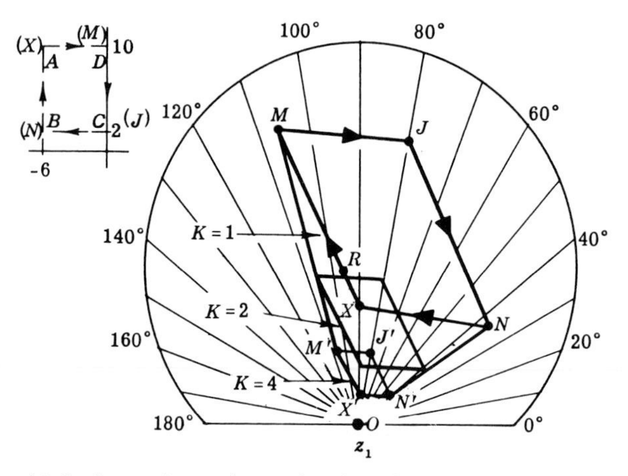

FIG. 6.3-3. Approximate shape of region of variation of dominant poles.

In Fig. 6.3-3, the points O and X correspond, respectively, to the location of the zero near the dominant root of $1 + L = 0$ (i.e., z_1 of Fig. 6.3-2), and to the nominal location of this dominant root. These two points, O and X, are selected at random in Fig. 6.3-3. Their relative location determines only the scale of Fig. 6.3-3 and its orientation. When K changes from its nominal value of unity to its maximum value of 4, the new position of the dominant root (denoted by X'), is given by Eq. (6.2,1), i.e., $z_1X'/z_1X = 0.25$. This locates the point X' in Fig. 6.3-3 on the straight line $OX'X$ (with $OX' = 0.25\, OX$). Thus, XX' is the approximate locus of the dominant root, as K changes from 1 to 4.

The effect of the plant poles moving from A to D (and $\bar{A}$ to $\bar{D}$) in Fig. 6.3-1 is found from Eq. (6.3,1), with $p_1X \approx -4$, $\bar{p}_1X \approx -4 + j20$, $p_1'X \approx -10$, $\bar{p}_1'X \approx -10 + j20$. Letting M correspond to the new root position (with $K = 1$), $z_1M/z_1X \approx (p_1'X)(\bar{p}_1'X)/(p_1X)(\bar{p}_1X) = 2.7 \angle 15.2°$, which locates M in Fig. 6.3-3. At $K = 4$, with the plant poles still at D, $\bar{D}$, Eq. (6.2,1) is used, with $z_1M'/z_1M \approx 0.25$. This locates M' in Fig. 6.3-3. The location of the dominant root when the plant poles are at C, $\bar{C}$ in Fig. 6.3-1 (at $K = 1$) is likewise found from Eq. (6.3,1), with $z_1J/z_1X \approx (-10 + j8)(-10 + j12)/(-4)(-4 + j20) = 2.45 \angle -10°$. This determines J in Fig. 6.3-3. When $K = 4$, Eq. (6.2,1) is used with $z_1J' \approx 0.25(z_1J)$. Similarly, the points N, N' in Fig. 6.3-3 correspond to the plant poles at B, $\bar{B}$ for $K = 1$ and 4, respectively. In Fig. 6.3-3, $X'M'MJNN'X'$ marks the approximate boundary of the drift in the dominant roots of $1 + L = 0$, due to the specified drift in the plant parameters.

Now that the general shape of the region of variation of the dominant root of $1 + L = 0$ is available in Fig. 6.3-3, it is used to determine the required scale of Fig. 6.3-3 and its orientation in the s plane. The dominant poles of T (roots of $1 + L = 0$) lie inside the region $X'M'MJNN'X'$ in Fig. 6.3-3. By trial and error, the circle of minimum radius which contains the region $X'M'...N'X'$ is found. Its center is at R in Fig. 6.3-3. The specifications dictate that the dominant roots of $1 + L = 0$ must lie inside a circle of radius 1.2 centered at $-10 + j10$. The magnitude of RM is therefore 1.2 and R must correspond to the point $-10 + j10$. It is found that the distance OR, which is the distance of z_1 from $-10 + j10$, is also 1.2. It is clear from Fig. 6.3-3, that the point R is a root of $1 + L = 0$ when $K = 1$ and when the plant poles are at approximately $-5 \pm 10j$. The latter is obtained by noting that R lies on the line XM in Fig. 6.3-3, which corresponds to the line AD in Fig. 6.3-1. By measuring on Fig. 6.3-3, it is found that $XR \approx (0.18)XM$. Since $(0.18)(AD) \approx 1$, the point $-5 + j10$ is thus determined.

With the above information, the required location of the zeros z_1, $\bar{z}_1$ of Fig. 6.3-1 can be obtained fairly closely. It is known that the net angle of the vectors from the poles and zeros of L to $-10 + j10$ must be 180°. The far-off poles of L are not known as yet, so a few degrees may be assigned for their contribution, or, as a first approximation, they may be neglected altogether. Therefore (see Fig. 6.3-1): $180° = \angle OR + \angle pR + \angle \bar{p}R - \angle z_1R - \angle \bar{z}_1R$. Since $\angle \bar{z}_1R \approx 90°$, this leads to $\angle z_1R \approx 146°$. It has previously been determined that $|z_1R| = 1.2$. Since R is at $-10 + j10$, this locates z_1 at $-9 + j9.3$. With this information, it is possible to turn to the problem of locating the far-off poles of L.

§ 6.4 Location of the Far-Off Poles of L

Let $L = PM$. Then M must supply the two zeros of L at z_1, $\bar{z}_1$ (Figs. 6.3-1 and 6.3-2). How many poles should be assigned to M? If Fig. 6.1-1a is used, $M = H_1$, and should have at least three poles in order that H_1 realistically be zero at infinity. If Fig. 6.1-1b is used, $M = G_2H_2$, and should therefore have at least four poles to allow each of G_2 and H_2 to be zero at infinity. There are two conflicting factors here. The more poles are assigned, the more complicated is a root locus s-plane type of synthesis. On the other hand, the less poles are assigned, the greater must be the bandwidth of the G and H compensating networks. In this problem, four additional poles will be assigned to L, but the procedure which is described may be applied to any even number of far-off poles, providing they are assumed to be in coincident pairs. Thus, in order that a solution may be obtained with a reasonable amount of labor, the four far-off poles of L are assumed as two coincident pairs, if complex.

With four far-off poles of L, the root loci of $1 + L = 0$ appear in Fig. 6.4-1. As far as the far-off loci are concerned, the three near poles and two

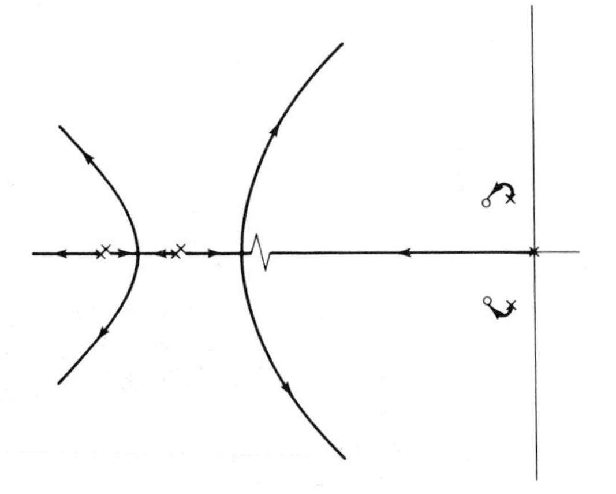

FIG. 6.4-1. Root loci of $1 + L = 0$.

zeros may be replaced by one equivalent pole at the origin. It is desirable to choose the far-off poles of L as close in as possible to minimize the gain-bandwidth requirements of L. At the same time, the dominance concept is not to be violated, and in addition the far-off roots of $1 + L = 0$ must not become so highly underdamped as to lead to high frequency ringing (as the plant parameters vary). In fact, the specifications demand that no far-off roots of $1 + L = 0$ lie to the right of the line $x = -30$. Clearly then, the optimum design is the one in which the specifications are barely satisfied, i.e., at $K = 4$, a pair of roots of $1 + L = 0$ lies on the vertical line $x = -30$.

Suppose we guess that this pair of far-off roots is at $-30 \pm j85$ (Q, $\bar{Q}$ in Fig. 6.4-2). The pole from the origin to Q (this pole at the origin is the equivalent of the two near zeros and three near poles of L) has the angle 110°. The four far-off poles must therefore contribute the balance of $180° - 110° = 70°$. The locus of the pole positions (if they are in coincident complex pairs) whose

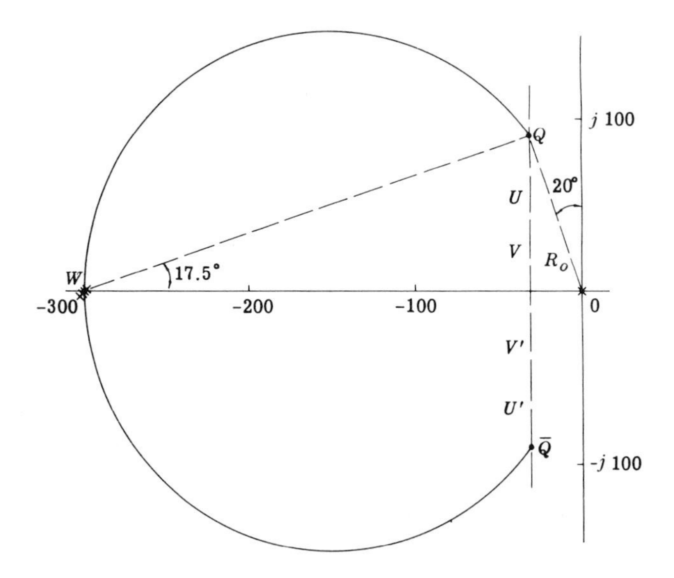

FIG. 6.4-2. Construction for locating far-off poles of $L(s)$.

vectors to Q contribute a constant angle is a segment of a circle passing through $Q, \bar{Q}$. The third point on the circle is easily located. This is the point on the real axis, at which there are four coincident roots, with each contributing $70/4 = 17.5°$. This locates the third point of the circle at $W = -300$, and the circle $QW\bar{Q}$ is drawn.

Suppose that the four far-off poles of L are located at $W = -300$. K, the gain factor of L, is then obtainable as follows: A root of $1 + L = 0$ is at $-10 + j10$. Therefore, from Figs. 6.3-1 and 6.4-2, $K = (pR)(\bar{p}R)(OR)(WR)^4/(z_1R)(\bar{z}_1R) = (64)(290)^4 = (4.5)10^{11}$. Now consider the value of the gain factor when a root of $1 + L = 0$ is at $Q = -30 + j85$. It is easily found (Fig. 6.4-2) that it is $(5.7)10^{11}$, which is considerably less than $(4)(4.5)10^{11}$. The maximum tolerable change in the gain factor is therefore only $5.7/4.5 = 1.27$. Consequently, the specifications are violated if the four far-off poles of L are located at $W = -300$.

If we continue to consider a pair of far-off roots at $Q, \bar{Q}$ in Fig. 6.4-2, then the largest tolerable ratio K'/K is achieved when the far-off poles of L are as far away as possible. Their extreme locations consist of two coincident poles at

-150 (with their vectors to Q contributing the required angle of 70°), and the other two poles at $-\infty$. (This is the extension on the real axis of the locus of the four poles in two pairs, which contribute a total of 70° at Q. Figure 6.4-1 shows an intermediate location of the poles.) The ratio K'/K is then $(146)^2(90)/(64)(140)^2 = 1.53$.

Clearly, points much further off on $UVV'U'$ must be attempted. If $Q, \bar{Q}$ at $-30 \pm j200$ is attempted, then $\angle OQ = 98.5°$, and the largest tolerable gain factor ratio K'/K is obtained by placing two poles of L at -264 (their vectors to Q then contribute $180° - 98.5° = 81.5°$) and the other two at $-\infty$. The maximum $K'/K = (309)^2(202)/(252)^2(64) = 4.75$. These values for $Q, \bar{Q}$ are therefore satisfactory. By cut and try, it is found that $K'/K = 4$ is achieved with one pole pair at -330, and another at -1720. The effective compensating network $M_1 = L_1/P$ has the frequency response shown in Fig. 6.4-3 (where it has been

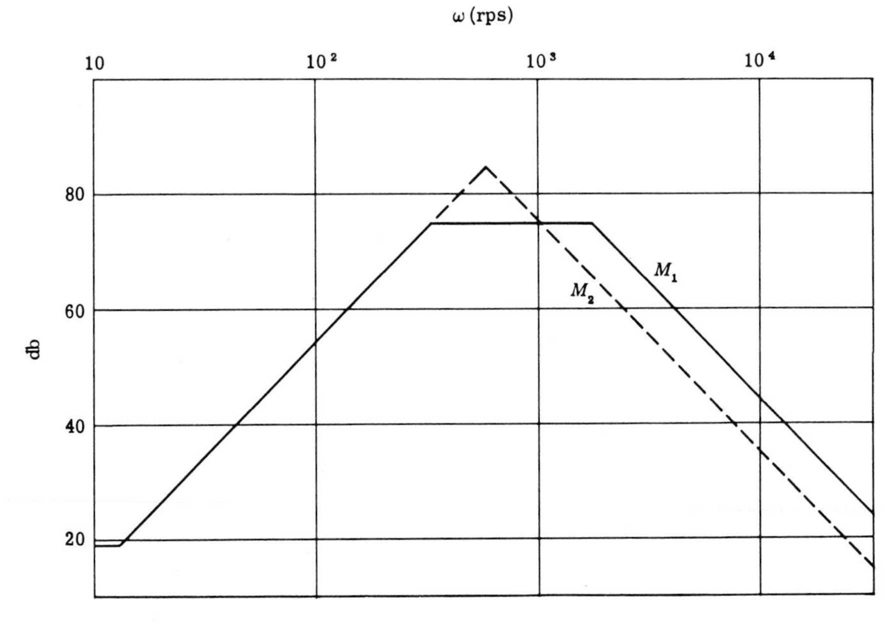

FIG. 6.4-3. Frequency response of $M = L/P$.

assumed that the plant behaves as $10/s$ at zero frequency). The tremendous gain-bandwidth of $M_1 = L_1/P$ is the price paid for the extremely small system sensitivity to the exceedingly large plant parameter variations.

Can the gain-bandwidth of $M = L/P$ be significantly decreased? One way is to experiment with other positions of $Q, \bar{Q}$ of Fig. 6.4-2. Suppose, for example, that one tries $Q = -30 + j240$ with $\angle OQ = 97.1°$. By cut and try it is found that $K'/K = 4$ is achieved with two complex pole pairs at $-520 \pm j280$. The resulting (asymptotic) frequency response of $M_2 = L_2/P$ is shown dotted in Fig. 6.4-3. This is nearly the best that can be done for the given design specifications, with M consisting of two zeros and two coincident complex pole pairs. The loci of the far-off roots of $1 + L_2 = 0$ are sketched in Fig. 6.4-4.

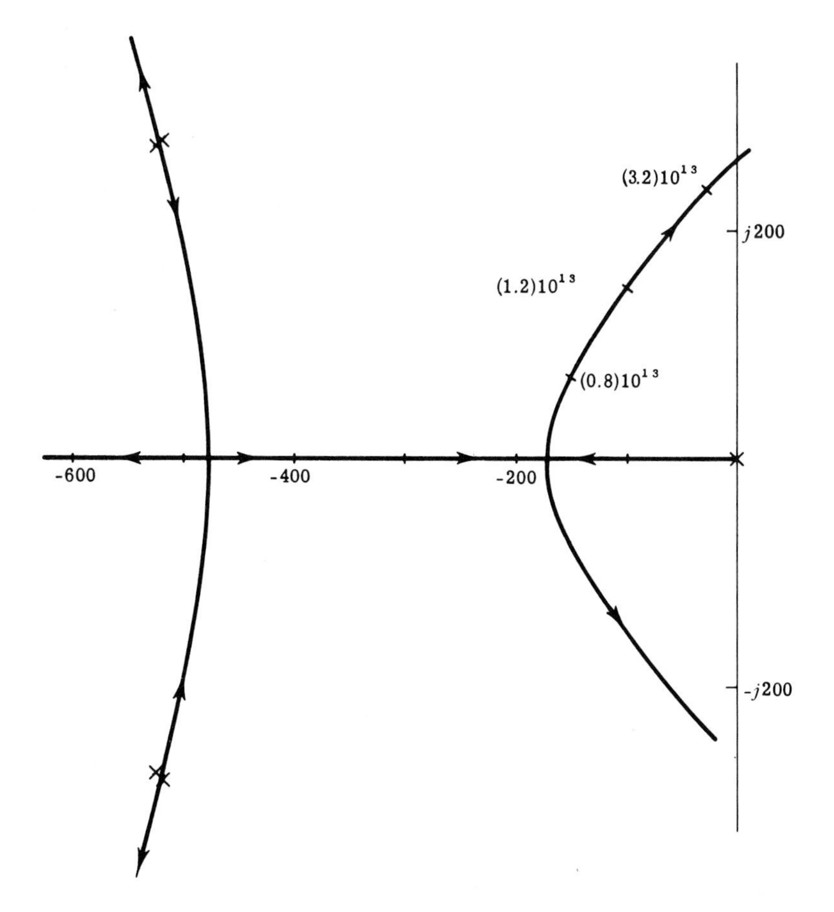

FIG. 6.4-4. Loci of far-off roots of $1 + L_2 = 0$.

Additional economy in the gain-bandwidth of $M = L/P$ is achievable only if a more complicated pole-zero pattern of M is permitted. However, root locus s-plane design is no longer convenient when complicated pole-zero patterns of L are used. It will be seen in Chapter 7 that invariably the best L (in terms of minimum gain-bandwidth) is one with an infinitude of poles and zeros. A reasonable approximation to the optimum is achievable only by frequency response methods (shaping of L on a Bode plot). Thus, one pays a price in gain-bandwidth for the simpler analytical root locus s-plane design procedure.

The root locus s-plane design is now completed. It has been found that a satisfactory $M = L/P$ is $M = (6.2)10^9(s^2 + 18s + 167.5)\,[s^2 + 1040s + (590)^2]^{-2}$, with the nominal plant transfer function $P = 1250/s(s^2 + 10s + 125)$. If the structure of Fig. 6.1-1a is used, then $H_1 = M = L/P$ and $G_1 = T(1 + L)/P$. It is necessary to factor $1 + L$. The dominant roots are known to be very nearly at $-10 \pm j10$. The others are obtained with sufficient accuracy from the root loci of Fig. 6.4-4. The result is $G_1 = T(s^2 + 200s + 200)\,(s^2 + 300s + 28{,}900)\,(s + 690)\,(s^2 + 1080s + 350{,}000)/[s^2 + 1040s + (590)^2]\,1250$. The far-off poles of T can be chosen to simplify G, since they hardly affect the response of $T(s)$ as expressed by its dominant poles and zeros. Thus, if $T(s)$ is to be characterized by a pair of dominant poles at $-10 \pm j10$, we choose $T = [(140)10^{13}]\,[(s^2 + 20s + 200)\,(s^2 + 300s + 28{,}900)\,(s + 690)\,(s^2 + 1080s + 350{,}000)]^{-1}$, and then $G_1 = [(112)10^{10}]\,[s^2 + 1040s + (590)^2]^{-2}$. If $|T|$ is allowed to decrease faster than in the above, then G_1 too can decrease faster.

The above procedure, although detailed for a specific numerical design example, is clearly able to handle any problem in which the plant and T have a small number of dominant poles. The drift in the dominant T poles due to the variations in the plant gain factor and to the plant's dominant poles can be made as small as desired. This small sensitivity is attained by inserting zeros of L near the desired positions of the dominant poles of T. The other poles of L are placed sufficiently far away to ensure that the far-off roots of T due to them do not too closely approach the $j\omega$ axis. The price paid for the small sensitivity of T is in the large gain-bandwidth required of $M = L/P$.

In the next section, it is shown how the procedure may be extended to handle more general types of plant transfer functions.

§ 6.5 Sensitivity of the Dominant Zeros of $T(s)$

In Section 6.4, neither $P(s)$ nor $T(s)$ had any zeros. When a $T(s)$ is specified which has one or more dominant zeros, and if $P(s)$ has no dominant zeros, then the dominant zeros of $T(s)$ must be contributed by a compensating network. This is seen from the fact that in the two-degree-of-freedom system, the general expression for T is given by Eq. (6.1,1), $T = GP/(1 + L)$. Zeros of T not due to P must therefore be zeros of G or poles of L that are not poles of GP. In any case, these zeros of T are due to zeros or poles of a compensating network; there is therefore no problem of drift in these zeros, because it is assumed that the parameters of the compensating networks have negligible drift.

There is a problem in the drift of zeros of T only when P has zeros which drift. Any drifting zeros of P must appear as drifting zeros of T, because it is impossible to precisely cancel such zeros. Suppose the plant has poles at M, M', F and a zero at B in Fig. 6.5-1a, and the zero drifts.

To minimize the effect on T of a large drift in the zero of the plant, a pole of L

is placed near the zero—the pole at A in Fig. 6.5-1a. (Zeros at N, N' are also shown to handle the drift in the plant poles at M, M'.) If the gain factor of L is sufficiently large, then one pole of T is near B, and stays close to B even as it moves about. The resulting dipole has negligible effect on the system response despite the drift in B. This action need be taken only for dominant zeros of P. The drift in far-off zeros of P can be ignored. The procedure for locating the pole of L at A and the required gain is illustrated by considering a specific example.

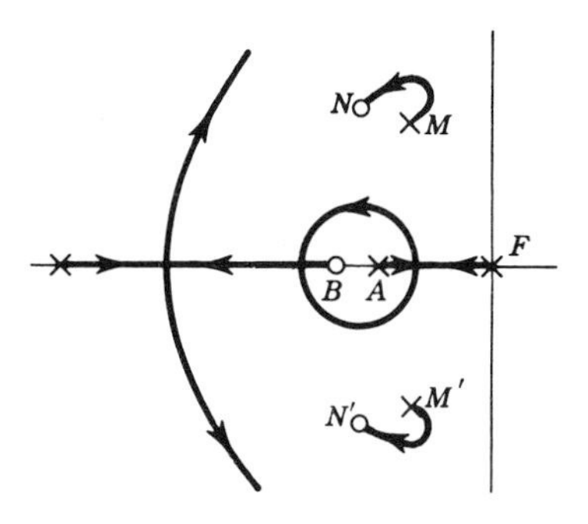

FIG. 6.5-1a. Use of dipole to reduce effect of drift in plant zero. Root loci of $1 + L = 0$.

In the example of Section 6.3, suppose P has a zero whose position drifts between -4 and -10, while the variations in the gain factor and the complex poles of P are the same as before. When the zero of P is at -4, one root of $1 + L = 0$ is between 0 and -4 (Q in Fig. 6.5-1b), and it is furthest

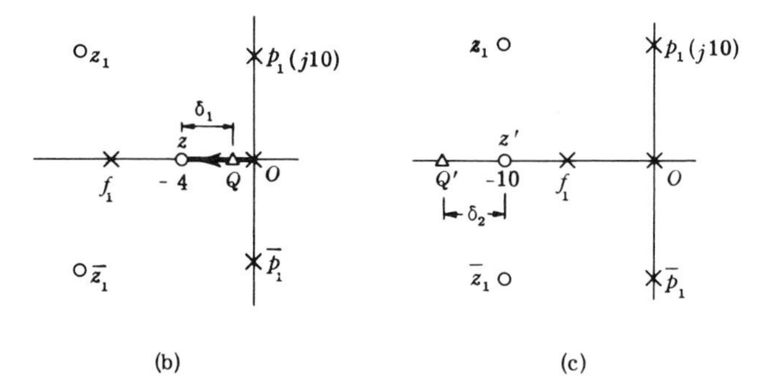

FIG. 6.5-1. (b) One extreme position of plant zero: $|f_1| > |z|$. (c) Other extreme position of plant zero: $|f_1| < |z'|$.

from the zero at -4 when K is at its lowest value and when the complex poles of L are furthest away at $\pm j10$. At this point, $K = |(OQ)(p_1Q)^2(f_1Q)H/(z_1Q)^2(zQ)|$. H refers to the contribution of the far-off poles and zeros of $L(s)$; z_1, $\bar{z}_1$ may be taken approximately at $-10 \pm j10$, and f_1 refers to the pole of $L(s)$ intentionally inserted near the varying plant zero. The question is whether the previous design of sections 6.3 and 6.4, which handled the variations in the plant poles and gain factor, is satisfactory if suitably amended by the additional pole at f_1. If the maximum dipole separation when $z = -4$ is to be δ_1 then we write $|zQ| \leqslant \delta_1$, which imposes a restriction on f_1.

The other extreme location of the plant zero is at -10. Let Q' in Fig. 6.5-1c be the resulting new position of the corresponding root of $1 + L = 0$. The maximum separation between Q' and the zero (now at -10) again occurs when the complex poles of P are at $\pm j10$ and when K has its lowest value. Therefore (Fig. 6.5-1c) $K \doteq |(OQ')(p_1Q')^2(f_1Q')H/(z_1Q')^2(z'Q')|$. The approximate relation is due to the fact that the value of H is really slightly different from its previous value. If the maximum dipole separation when $z = -10$ is to be δ_2, then $|z'Q'| \leqslant \delta_2$, which imposes another restriction on f_1. One way to check whether (for given K, z, etc.) it is possible to find f_1 which satisfies these inequalities is to set the two equations for K equal to each other, leading to

$$\frac{(OQ)(p_1Q)^2(f_1Q)}{(z_1Q)^2(zQ)} \doteq \frac{(OQ')(p_1Q')^2(f_1Q')}{(z_1Q')^2(z'Q')}. \qquad (6.5,1)$$

Let $|zQ| = \delta_1$, $|z'Q'| = \delta_2$ exactly. Assuming that f_1 lies between -4 and -10, the quantities in Eq. (6.5,1) are: $f_1Q = f_1 - 4 + \delta_1$; $f_1Q' = \delta_2 + 10 - f_1$; $OQ = (4 - \delta_1)$; $OQ' = 10 + \delta_2$; $(p_1Q)^2 = 100 + (4 - \delta_1)^2$; $(p_1Q')^2 = 100 + (10 + \delta_2)^2$; $(z_1Q)^2 \approx 100 + (6 + \delta_1)^2$; $(z_1Q')^2 \approx 100 + \delta_2{}^2$. The only unknown that remains in Eq. (6.5,1) is f_1, so the equation may be solved for f_1. Using this value of f_1 it is possible to solve for K. If this calculated value of K is less than the actual minimum K, then the design is satisfactory, because the larger value of K will lead to $|zQ| < \delta_1$, $|z'Q'| < \delta_2$. If the opposite occurs, then the specifications on the drifting plant zero require a greater level of loop gain that those on the drifting plant poles.

§ 6.6 Feasibility of Root Locus Sensitivity Design in High-Order Systems

The basic philosophy of the root locus approach to the sensitivity problem is to assign loop transmission zeros in the neighborhood of the desired system poles (roots of $1 + L = 0$), and loop transmission poles near the drifting plant zeros. In conjunction with large loop gain, this assures that the system poles cannot drift very far away from these zeros of L, and that the drifting zeros of P are effectively canceled by nearby roots of $1 + L = 0$.

Formally, therefore, the root locus approach to the sensitivity problem can be applied to higher order systems with many poles and zeros. It is very questionable, however, whether this s-domain approach is justified in higher order systems. One reason is that it is not really necessary for the pole-zero pattern of $T(s)$ to remain substantially the same in order that the time response should not change significantly. It is quite possible to vary the $T(s)$ pole-zero pattern significantly and still obtain substantially the same time response (see Section 12.6). It appears, therefore, that the root locus approach to the sensitivity problem is justified only in systems where there are few dominant poles and zeros. A design which insists that each dominant pole and zero of $T(s)$ stays near its original position is really unnecessary, and is overly expensive in gain-bandwidth demands on $L(s)$. It was also noted in Section 5.18 that in high-order plants, a design with only a few dominant poles and zeros can be very wasteful of plant capabilities. Thus, in both cases [in realizing $T(s)$, and small sensitivity of $T(s)$], the designer may be paying considerably for the privilege of having to concern himself with only a few dominant poles and zeros.

The pole-zero root-locus approach to the sensitivity problem is overly costly in gain-bandwidth demands, even when $T(s)$ consists of just a very few dominant poles and zeros. This has previously been noted in Section 6.4 in discussing the location of the far-off poles of $L(s)$. As noted in Section 6.4, and as will be shown later in Chapter 7 and 8, a more complicated $L(s)$ can result in a more economical $L(s)$.

The root-locus s-plane approach to realizing a desired sensitivity of $T(s)$ to plant parameter variations is useful when the plant is sufficiently simple to justify the concept of dominance and when the sensitivity demands are not very severe. For example, it may suffice that despite plant parameter variations, the system complex pole pair should have a damping factor not less than 0.5, but otherwise its variation may be very large. This type of problem was briefly considered in Section 4.7 and will be treated in more detail in Sections 8.11-8.13.

§ 6.7 Philosophy of the Frequency Response Approach to the Sensitivity Problem

The second principal method of dealing with the sensitivity problem is by considering sensitivity as a function of frequency. The distinguishing feature of the frequency response approach is that the sensitivity specifications are given in terms of the system's behavior on the $j\omega$ axis. This is in contrast to the

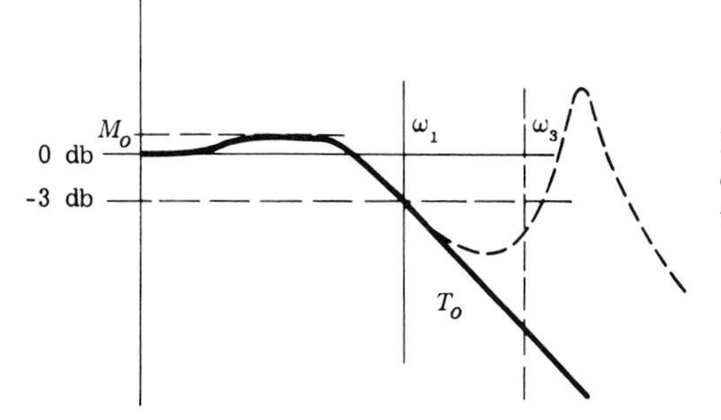

FIG. 6.7-1. Undesirable high frequency peaking in $T(j\omega)$, due to plant parameter variations.

s-plane approach, where the specifications are in terms of the system poles and zeros. It is possible that the allowed variation in $T(j\omega)$ may be exceedingly small over one frequency range and fairly large in other frequency ranges. The specifications may obviously be in a wide variety of forms. For example, the nominal T_o may be as shown in Fig. 6.7-1, with bandwidth ω_1, amplitude peaking M_o, and the specifications may dictate that despite parameter variations, the magnitude peaking should never exceed 30%, the bandwidth should be in the range $0.8\omega_1$ to $1.2\omega_1$, and that from zero frequency to $0.5\omega_1$, $|T|$ should never be less than 0.95. Finally, some thought should be given to the intermediate and the far-off frequency range. While very large variations in T in the far-off range are tolerable, this is true only up to a point. For example, the behavior shown by the dotted line in Fig. 6.7-1 is usually intolerable, as it results in high frequency ringing. A reasonable statement might be that for $1.2\omega_1 < \omega < \omega_3$, $|T(j\omega)| < 3$ db and for $\omega > \omega_3$, $|T(j\omega)|$ must not exceed $-K$ db, with K some suitable value. In some designs, the sensitivity requirements may be stringent only at very low frequencies, and quite lenient at other frequencies.

In this section, we do not consider the problem of how to draw up sensitivity specifications in terms of frequency response. We consider only the problem of realizing the frequency response sensitivity specifications, whatever they may happen to be. Furthermore, it is clearly not feasible to consider the variety of specifications that may occur in practice. Instead, one rather exacting design problem is treated numerically in detail. The procedure that is followed is sufficiently general to enable the reader to treat a large variety of design problems that may arise in practice.

The basic philosophy of the frequency response approach to sensitivity reduction, is deduced from Eq. (3.2,6), rewritten here:

$$\frac{T_o}{T} = \frac{(P_o/P) + L_o}{1 + L_o} \tag{6.7,1}$$

where T, P are used in place of T_f, k_f. This equation applies whenever the leakage transmission, t_{oi}, is zero, and therefore is correct for each of the configurations in Fig. 6.1-1. Suppose that at some specific frequency $s = j\omega_x$, plant parameter variations are such that $(P_o/P)(j\omega_x)$ may lie anywhere inside the region indicated in Fig. 6.7-2a. Suppose also that $-L_o(j\omega_x)$ is given by the complex number $A + jB$ located at Q. Then Eq. (6.7,1) states that $T_o/T = \mathbf{QV}/\mathbf{QN}$. The range of variation of the vector **QV** fixes the range of variation of T_o/T at $s = j\omega_x$.

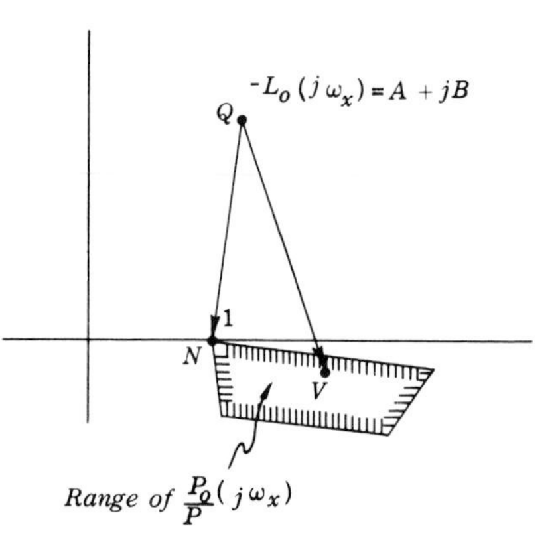

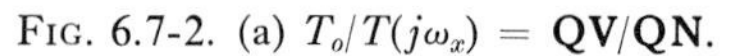

FIG. 6.7-2. (a) $T_o/T(j\omega_x) = \mathbf{QV}/\mathbf{QN}$.

Suppose, for example, that it is required that $1.20 > |T_o/T| > 0.80$ at $s = j\omega_x$.

The problem is to find the locus of $-L_o(j\omega_x)$ which barely satisfies this inequality.

A suitable procedure is as follows. Suppose we seek the locus of the boundary of $|T_o/T| \geqslant 0.80$. With $N = 1$ as center, draw a circle C_1 of any radius R (Fig. 6.7-2b). Next draw a circle C_2 of radius $0.8R$, using as center the point

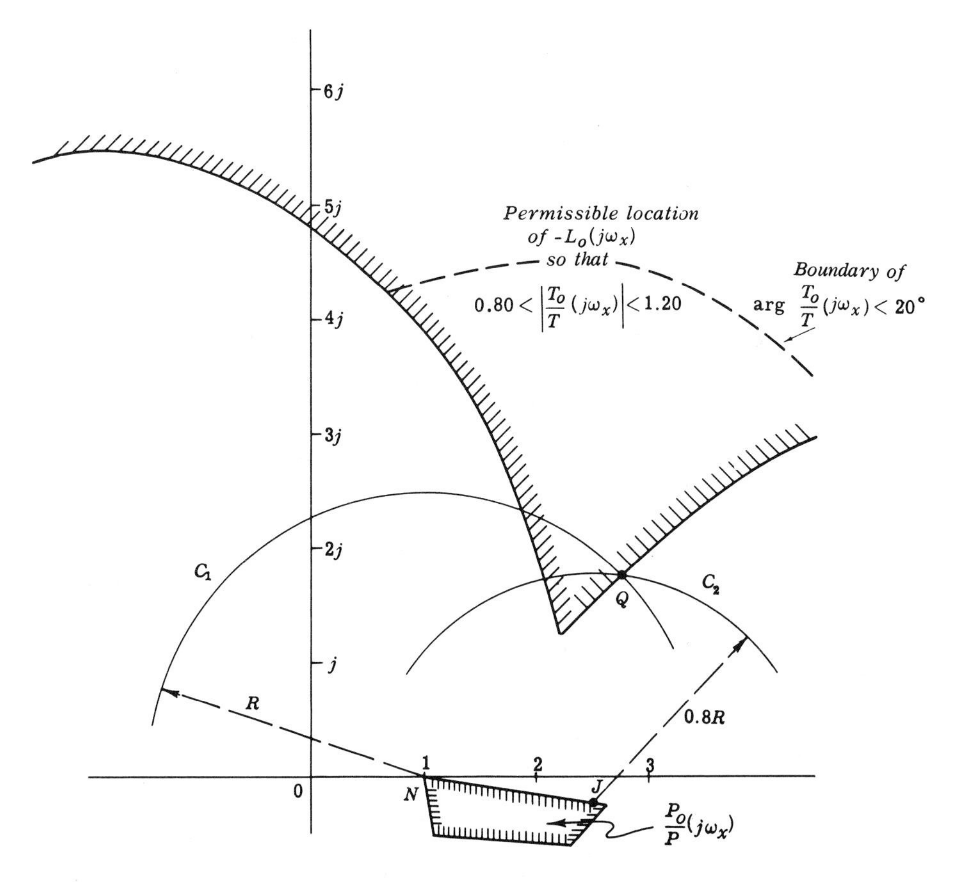

FIG. 6.7-2. (b) Construction for locating boundary of $-L_o(j\omega_x)$ which maintains $T_o/T(j\omega_x)$ within specified bounds.

(on the boundary of P_o/P) which appears to be the closest to the first circle—this appears to be point J in Fig. 6.7-2b. The intersection of the two circles determines a point Q. As a final check, use Q as center, and check whether a circle of radius $0.8R$ cuts $P_o/P(j\omega_x)$ only at J. If so, Q is on the boundary of the locus of $|T_o/T| \geqslant 0.80$ at $s = j\omega_x$. (Alternatively, seek that point on C_1 which, when used as the center of a circle of radius $0.8R$, grazes the boundary of P_o/P.) The above construction is repeated, using different values for the radius

R, until the locus of Q is obtained. Such a locus for $1.20 > |T_o/T| > 0.80$ is shown in Fig. 6.7-2b. Clearly, $-L_o(j\omega_x)$ must be located inside the indicated region in Fig. 6.7-2b. On the other hand, if there is the additional requirement that the phase of $T_o/T(j\omega_x)$ must not change more than 20°, then the permissible range of $-L_o(j\omega_x)$ is reduced to the smaller region shown in Fig. 6.7-2b.

In this manner, one may deduce the boundaries of permissible $-L_o(j\omega)$ for a number of frequencies. The requirements of the loop transmission then become apparent. The same procedure is followed in the high frequency range as well, where it has previously been noted (Fig. 6.7-1) that it is also necessary to make some restrictions on the permissible variation of $T(j\omega)$. These statements, in fact, replace the usual equivalent but vaguer stability margin requirements. Thus, designing for desired sensitivity automatically takes care of the stability problem. It is, of course, important that the boundary of $(P_o/P)(j\omega_x)$ truly enclose all possible values of P_o/P at $s = j\omega_x$. The procedure for obtaining the boundary of $(P_o/P)(j\omega)$ has been treated in considerable detail in Sections 4.12 and 4.13 and is therefore not repeated here.

It is extremely important to note that due to (Eq. 6.7,1) *the sensitivity requirements essentially determine the gain and bandwidth of $L_o(s)$ with hardly any need to refer to the specific feedback configuration that is used.* Hence, the cost of sensitivity reduction, in terms of gain and bandwidth of $L_o(s)$, is independent of configuration. At best, one configuration may be more efficient than another in the practical realization of the required $L_o(s)$. The latter is actually a problem in network synthesis. In this book the design is terminated as soon as the compensating transfer functions have been specified. The realization of a transfer function in the form of a network consisting of active and passive elements is, in general, not treated in this book. At the low frequencies in which control systems operate, it is usually preferable to omit inductors and obtain the required functions from networks consisting of resistors, capacitors, and active elements. Networks consisting of resistors and capacitors alone have transfer functions whose poles are restricted to the negative real axis. There are fairly straightforward procedures for designing unbalanced RC networks (i.e., with common ground between input and output) to obtain transfer functions with negative real axis poles, and zeros in the left half-plane, including the $j\omega$ axis. Right half-plane zeros excluding the positive real axis are also obtainable, but with greater difficulty. Such zeros are, however, less often required. It will be shown in Chapters 7 and 8 that the assignment of complex poles and zeros to $L_o(s)$ can reduce the gain bandwidth of $L_o(s)$. Such complex poles cannot be realized by means of networks with resistors and capacitors alone. But they can be realized by means of networks consisting of resistors, capacitors, and active elements such as vacuum tubes or transistors. A simple active RC circuit for realizing complex poles is described in the Appendix.

§ 6.8 Realization of Sensitivity Specifications—Frequency Response Method

The design procedure is best presented by means of a numerical example. Consider the numerical example previously treated by s-plane root locus methods in Sections 6.3-6.6. The plant transfer function is $1250\ K/s(s^2 + 2\zeta_p\omega_p s + \omega_p^2)$ with K varying from 1 to 4, and the complex pole pair lying anywhere in the rectangles $ABCD$, $\bar{A}\bar{B}\bar{C}\bar{D}$ in Fig. 6.3-1. Sensitivity requirements (not the same as those in Section 6.3) are given here in terms of frequency response. They are as follows: $|\ T(j\omega)\ |$ must always be exactly unity at zero frequency, must not vary more than 5% at 5 rps, 17.5% at 15 rps, and 33% at 25 rps. At 50 rps $|\ T\ |$ must not be more than -17 db; at 100 rps $|\ T\ |$ must never be more than -25 db. Beyond 200 rps $|\ T\ |$ must never exceed -40 db. (The sensitivity requirements for $\omega \geqslant 50$ are in accordance with the discussion centered around Fig. 6.7-1.)

The sensitivity specifications in the higher frequency range are such that we must state the $T(s)$ specifications in order to define the permissible relative change in $T(j\omega)$. A reasonable set of specifications might be as follows: $T(s)$ is to be dominated by a complex pole pair at $-10 \pm j10$. In order not to have too fast a cutoff characteristic (since it leads to larger overshoot—see Section 5.10), it is prescribed that at 50 rps $|\ T\ | \not< -35$ db, and for this purpose we tentatively assume an additional $T(s)$ pole at -50. The resulting $T(s)$ and maximum per-

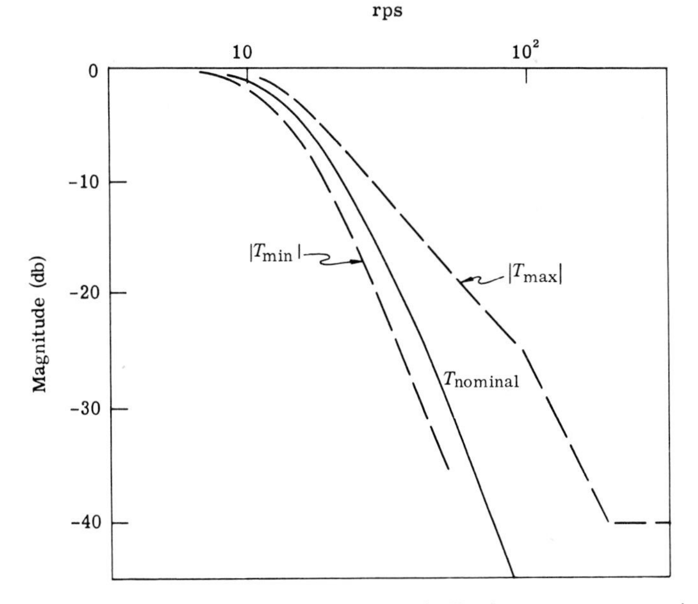

FIG. 6.8-1. Specified maximum tolerable range of $|\ T\ |$ due to parameter variations.

mitted extremes in $| T(j\omega) |$ are therefore as shown in Fig. 6.8-1. At 50 rps, $| T(j50) | = -28$ db, and since $| T(j50) | = -17$ db, $| T(j50) |$ is allowed to increase by 11 db or by 35%. Also $| T(j100) | = -48$ db, and since it is allowed to increase to -25 db, $| T(j100) |$ may increase by 23 db or a factor of 14.

If it should turn out that it is difficult to achieve the high frequency specifications, we could let $| T_o(j\omega) |$ decrease faster in this range, and thus permit a larger per cent increase in $| T(j100) |$, etc. In view of the small allowed changes in $| T(j\omega) |$ over its significant frequency range (0 to 20 rps), we can reasonably expect that the system time response will be only slightly affected, despite the fantastically large plant parameter variations. These sensitivity specifications may well be unduly stringent for most practical design problems, but we deliberately want to consider an extreme case, in order to see the price that must be paid, and the difficulties that arise.

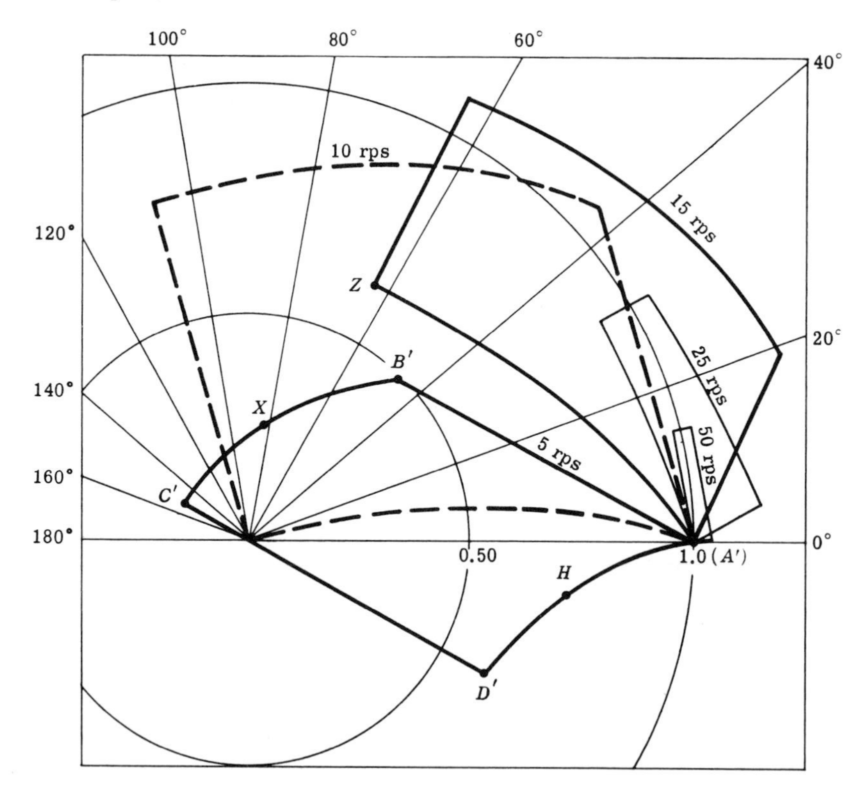

FIG. 6.8-2. (a) Range of P_o/P at selected frequencies.

Design procedure:

Step 1. Boundaries of $(P_o/P)(j\omega)$. The first step is to prepare polar sketches of the boundary of $(P_o/P)(j\omega)$ for a number of frequencies in the low and high

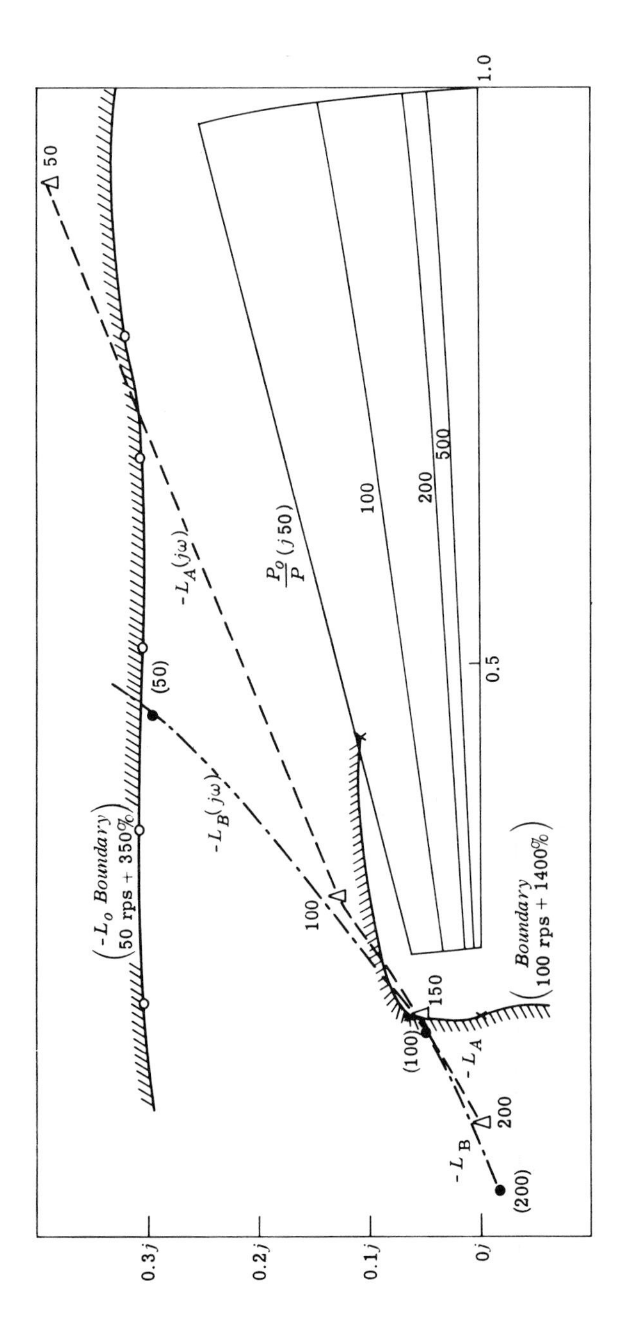

FIG. 6.8-2. (b) Range of P_o/P and boundaries of required $-L_o(j50)$, $-L_o(j100)$.

frequency ranges. Normally, the information on plant variations would be more likely available in terms of frequency response than in terms of poles and zeros. In this particular academic problem, the opposite is true. The procedure for finding the boundary of $(P_o/P)(j\omega)$ has been treated in Sections 4.12 and 4.13. It will be reviewed here by describing how $(P_o/P)(j5)$ is obtained. In this problem

$$\frac{P_o}{P}(j\omega) = \frac{(s^2 + 2xs + x^2 + y^2)}{K(s^2 + 12s + 136)} \tag{6.8,1}$$

where we have taken $P_o = 1250[s(s^2 + 12s + 136)]^{-1}$. The reason for this choice is that it is, in general, a good idea to take for P_o that combination of parameter values which results in the lowest gain over all frequencies. With such a choice of P_o, the region of variation of P_o/P will occupy a small area in the complex plane, and so facilitate the graphical design procedure. With $|P_o| < |P|$, $|P_o/P|$ tends to be in the neighborhood of the origin and the point 1 in the s plane. Thus, it may be noted how, with the above choice for P_o, the boundaries of P_o/P occupy a fairly small area in Figs. 6.8-2a and b.

Consider (P_o/P) (j5) along the lines CD, $\bar{C}\bar{D}$ in Fig. 6.3-1, at $K = 1$. Along these lines, $x = 0$ in Eq. (6.8,1), and $[P_o/P(j5)]_{CD} = (y^2 - 25)/(111 + j60) = 0.008(y^2 - 25) \angle -28.4°$, which is a straight line passing through the origin at an angle of $-28.4°$. The end points on the line are easily obtained: D (of Fig. 6.3-1) with $y = 10$ leads to D' in Fig. 6.8-2a. C with $y = 2$ leads to C'. Along the lines BA, $\bar{B}\bar{A}$, $[P_o/P(j5)]_{BA} = 0.008(11 + y^2 + j60) \angle -28.4°$. This is the equation of a straight line (if it is rotated 28.4°, its ordinate is a constant), so the locus is easily obtained from its end points: at B, $y = 2$ leads to B' in Fig. 6.8-2a; at A, $y = 10$ gives $A' = 1$. Along AD, $[P_o/P(j5)]_{AD} = 0.008(75 + x^2 + j10x) \angle -28.4°$, which describes a parabola. However, it is just as easy to obtain the locus by solving the above at several values of x. This leads to the curve $D'HA'$ in Fig. 6.8-2a. Similarly, BC of Fig. 6.3-1 maps into the parabolic segment $C'XB'$. It is easily seen that as P ranges in the interior of $ABCD$, $\bar{A}\bar{B}\bar{C}\bar{D}$ in Fig. 6.3-1, $(P_o/P)(j5)$ maps into the interior of $A'HD'C'XB'$ in Fig. 6.8-2a. The boundaries at the other frequencies are obtained in the same manner. They all apply for $K = 1$. The effect of K varying from 1 to 4 is obvious, but it is not shown in the figure in order not to clutter up the diagram. The loci of $(P_o/P)(j\omega)$ for the high frequency range are similarly obtained, and are sketched in Fig. 6.8-2b. Here the boundaries include the effect of the variations in K.

Step 2. Boundaries of required $L_o(j\omega)$. The next step is to find the required magnitude of $L_o(j\omega)$ in order to satisfy the specifications. This is done in the manner of Fig. 6.7-2b and the results are shown in Fig. 6.8-3 for the low and intermediate frequency regions; for example, the curve labeled "15 rps + 17.5%" is the boundary of $-L_o(j15)$, for which $|T| < 1.175 |T_o|$. Similar boundaries for the high frequency region are sketched in Fig. 6.8-2b. The loci of two choices of loop transmission (labeled L_A, L_B) are also shown in these figures.

Step 3. Choice of $L_o(s)$. Step 2 has facilitated the choice of $L_o(s)$, but has not uniquely fixed it. Some cut and try is inevitable because both the phase and amplitude of $L_o(j\omega)$ must be properly shaped. The approximate minimum $|L_o|$ levels at various frequencies have been fixed by Step 2. One tentatively chooses an $L_o(j\omega)$ that more or less follows these levels, finds the phase angle for this choice of $L_o(j\omega)$, plots a few values of $L_o(j\omega)$, notes what corrections

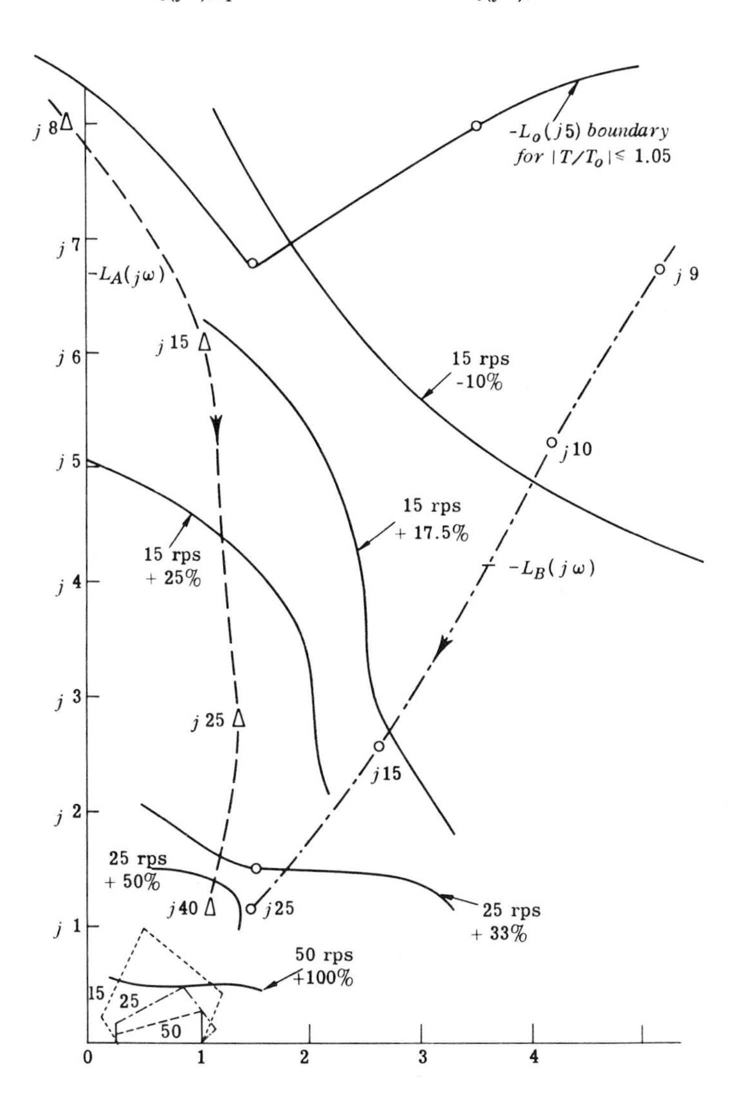

FIG. 6.8-3. Boundaries of $-L_o(j\omega)$ which satisfy various sensitivity specifications (for $\omega = 5, 15, 25, 50$ rps).

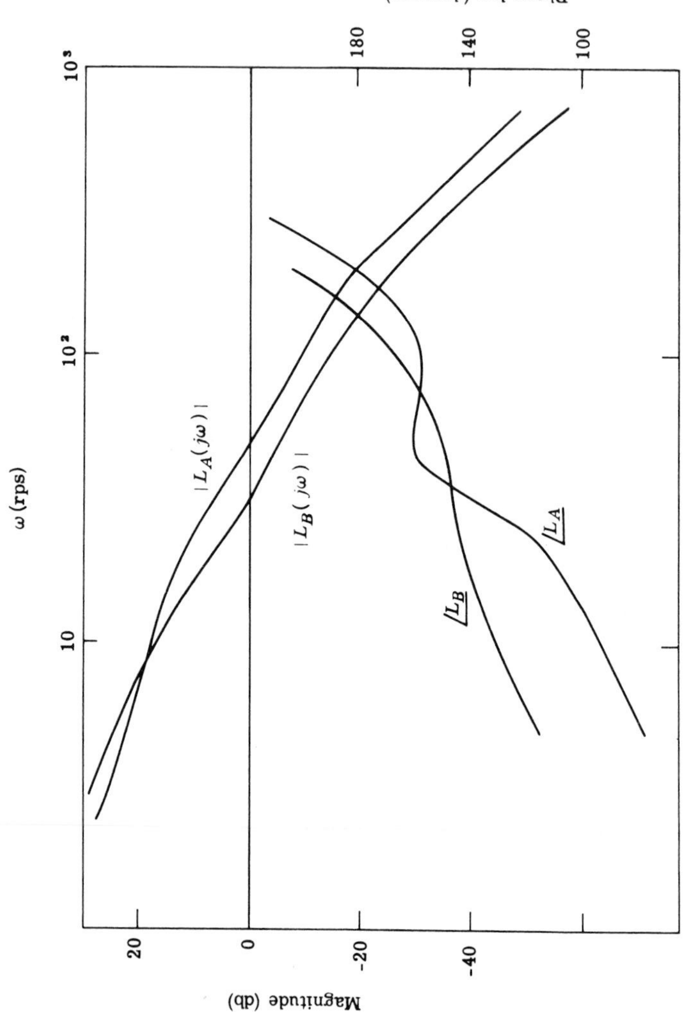

FIG. 6.8-4. Bode plots of L_A, L_B.

are necessary, readjusts $L_o(j\omega)$, etc., until a satisfactory $L_o(j\omega)$ is obtained. [A systematic way of doing this so as to economize on the gain-bandwidth of $L_o(s)$ by means of a strategic choice of complex poles and zeros is described in Sections 8.2 and 8.3.] In this problem the gain margin of $L_o(j\omega)$ must be quite large because of the shaping that must be done around 50 to 100 rps. A satisfactory $L_o(j\omega)$ is sketched in Fig. 6.8-4. Its polar locus is shown in Figs. 6.8-2b and 3. It is $L_A(s) = [(2.1)10^6(s+5)(s+45)^2][s(s+15)(s+25)^2(s+150)(s+300)]^{-1}$.

If the specifications in the intermediate frequency range are slightly relaxed, viz., a maximum of 25% at 15 rps and 50% at 25 rps, the loop transmission can have a smaller crossover frequency and a smaller gain-bandwidth. A suitable loop transmission is then given by $L_B(s) = (5.9)10^5(s+22)^2[s(s+12)^2(s+150)^2]^{-1}$, whose Bode plot is also sketched in Fig. 6.8-4. Polar sketches of L_B are also included in Figs. 6.8-2b and 3.

Step 4. Considerations at high frequencies. It is necessary to specify the values of the G and H compensating networks (Fig. 6.1-1) in order to complete the design. One degree of freedom has been used up in the specification of the loop transmission function $L_o(s)$. Let $L_o = P_oM$, where $M = H_1$ for the structure of Fig. 6.1-1a, G_2H_2 for Fig. 6.1-1b, $G_3 + H_3$ for Fig. 6.1-1c, etc. The value of $M = L_o/P_o$ is, of course, independent of configuration. M is sketched in Fig. 6.8-5 for the two choices of L_o. They are: $L_A/P_o = M_A = [(1680(s+5)$

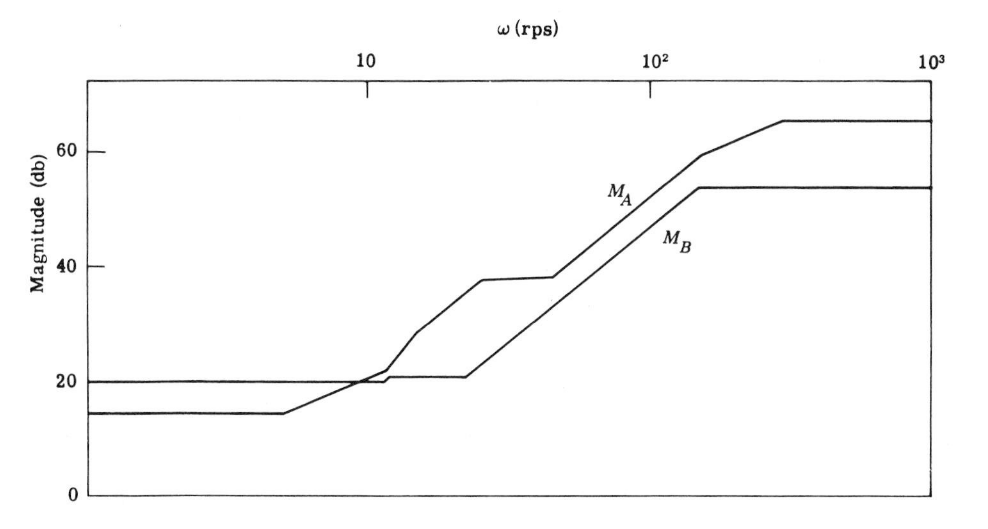

FIG. 6.8-5. Undesirable high frequency character of compensation network $M = L/P$.

$(s+45)^2(s^2+12s+136)][(s+15)(s+25)^2(s+150)(s+300)]^{-1}$, $L_B/P_o = M_B = [472(s+22)^2(s^2+12s+136)][(s+12)^2(s+150)^2]^{-1}$.

There are two startling features in M_A, M_B. One is their gain levels at zero frequency; the other is their behavior at high frequencies, where it appears that a high level of gain must be maintained indefinitely. The latter feature is first examined. Consider the asymptotic behavior of $M = L_o/P_o$ as $s \to \infty$. If the structure of Fig. 6.1-1a is used, $M = H_1$ and our concern is therefore with the high frequency behavior of H_1. We know that at a sufficiently high frequency, $|H_1|$ will decrease in amplitude and approach zero as $1/\omega^x$ with $x \geqslant 1$. What effect will this have on $L(s)$ and consequently on the system sensitivity? Consider Figs. 6.8-2b and 3. If the inevitable additional poles of H_1 occur far enough away (roughly beyond 150 rps) such that $L_o(j\omega)$ for $\omega \leqslant 100$ rps is only mildly affected, then the original design is as good as ever. So we check whether in actual practice H_1 can readily be built so as to maintain the desired level and phase at least up to 150 rps. If there is no difficulty in building H_1 to do this, then the design is satisfactory.

The above is one convenient design approach, which is summarized as follows: One performs the design with a loop transmission which has an excess of poles over zeros equal to that of the plant. This results in $M = L_o/P_o$ finite at infinity. From the design data and the specifications, it is ascertained that $L_o(j\omega)$ may change significantly only for $\omega > \omega_x$ say, without violating the specifications. One checks whether $M(s)$ can in practice readily be built so as to maintain the desired design value up to at least ω_x. If it can, the simple design of $L_o(s)$ is satisfactory. The above remarks apply whatever configuration is used. In Fig. 6.1-1b, $M = G_2H_2$, and the question is whether G_2H_2 as a product can be readily built to have the desired value up to at least the frequency ω_x. In Fig. 6.1-1c the question is addressed to $(G_3 + H_3)$, etc.

It is necessary to add one important proviso here. It has been assumed that the plant transfer function is a true representation of the plant behavior up to ω_x. Often the plant is readily measured and characterized only over the low frequency range, since it is usually difficult to measure the plant response at the higher frequencies. It is, therefore, necessary to find the additional plant poles up to ω_x—if not by measurement, then at least from a theoretical analysis, and to check whether $L_o = P_oM$ as a whole (and not M alone) can maintain the desired design value up to ω_x. If the answer is in the affirmative, the above design, with a relatively simple $L_o(s)$, is satisfactory. We shall assume, in the meantime that this is the case, and examine the alternative later in Section 6.10.

We now examine the first startling feature of the $M(j\omega) = L_o/P_o$ of Fig. 6.8-5, namely, its level of gain at 0 rps. Is this gain level to be considered as part of the cost of feedback? The answer is no! A prescribed value of $T(j\omega)$, in conjunction with a known input, determines the output level at all frequencies. The plant must be able to deliver this output, irrespective of whether feedback is or is not used. Thus, if we insist, as we did in Fig. 6.8-1, that $|T(j\omega)| \geqslant |T_{\min}(j\omega)|$, we must make sure that $|P(j\omega)| \geqslant |T_{\min}(j\omega)|$. The latter applies even if no feedback is used: it is a *control*, not a *feedback*, requirement. It is seen that the present level of $P(j\omega)$ must be raised, in order to satisfy the requirement that $|P(j\omega)| \geqslant |T_{\min}(j\omega)|$, at any combination of plant parameter values. If sketches of $|T_{\min}(j\omega)|$ and $|P_o(j\omega)|$ are compared, then it is readily found that the gain level of the plant must be raised by about 10 db, in order that $|P(j\omega)| \geqslant |T_{\min}(j\omega)|$. When the gain level of the plant is so adjusted, the zero frequency value of M is only a few decibels.

Step 5. Choice of compensating networks. The final step in the design is to fix the values of G and H, by using the second degree of freedom to realize the desired T_o. From Eq. (6.1,1), $G = T_o(1 + L_o)/P_o$. In Fig. 6.1-1, $G = G_1 = G_2 = G_3 = (T_o/P_o) + \alpha T_o H_4 = G_{5A}G_{5B}$, etc., for the respective configurations shown in the figure. The procedure is therefore clear: $1 + L_o$ is factored, and G is obtained. (As an aid in the factoring, a rough sketch of the root loci of $1 + L_o = 0$ may be used to obtain the approximate location of the roots.) Assuming $L_A(s)$ is used, we have here[1] $1 + L_A(s) \doteq [(s + 4.72)(s + 37.5)(s + 89.8)(s + 331.5)(s^2 + 51.5s + 4000)][s(s + 15)(s + 25)^2(s + 150)(s + 300)]^{-1}$. The far-off zeros of $1 + L_A(s)$ need not be located very accurately, because the previously referred to inevitable additional poles of the compensating networks will affect their position.

Let $T_o(s) = [8000F_oF(s)]/[(s^2 + 20s + 200)(s + 40)]$, where the dominant part of $T_o(s)$ is specified in accordance with the original specifications and $F(s)$ represents the far-off portion. The far-off portion $F_oF(s)$ may be specified so as to simplify the design of G. Thus

$$G = T_o(1 + L_o)/P_o = \frac{8000F_oF(s)(s+4.72)(s+37.5)(s+89.8)(s+331.5)(s^2+51.5s+4000)(s^2+12s+136)}{(s^2+20s+200)(s+40)(s+15)(s+25)^2(s+150)(s+300)}.$$

In order to simplify G, we pick $F(s) = [(s + 40)(s + 300)(s + 150)][(s + 37.5)(s + 331.5)(s + 89.8)]^{-1}$, all of which has a negligible effect on $T_o(j\omega)$ in its significant range. The resulting $G(s)$ is now $G(s) = 8000F_o(s + 4.72)(s^2 + 12s + 136)(s^2 + 51.5s + 4000)[(s^2 + 20s + 200)(s + 15)(s + 25)^2]^{-1}$. The value of F_o is chosen so that $T(0) = 1$. With $L_o(s)$ and $T_o(s)$ thus finally specified, the specific values of the compensating networks for the various configurations of Fig. 6.1-1 are uniquely determined (except for Fig. 6.1-1e).

Suppose Fig. 6.1-1a is used, then $G_1 = G$ and $H_1 = L_o/P_o = [(2.1)10^6(s + 5)(s + 45)^2(s^2 + 12s + 136)][(s + 15)(s + 25)^2(s + 150)(s + 300)]^{-1}$. We have already observed that this expression for H_1 is finite (approximately 55 db) at infinity. It has been previously noted that H_1 can have additional poles, providing that these do not substantially affect $L_A(s) = PH_1(s)$ for $\omega > 100$ rps. The additional inevitable poles of G_1 will only affect $T(s)$ (if Fig. 6.1-1a is used),

[1] If Eqs. (5.11, 1a, b) are used to obtain approximately the dominant roots of $1 + L_A(s) = 0$, the result is $\omega_n = 62$, $\zeta_n = 0.45$, which agrees closely with the correct $\omega_n = 63$, $\zeta_n = 0.41$.

and one may decide how strict he wants to be on this matter. If Fig. 6.1-1b is used, the additional poles of $G(s)$ will also affect $L_o(s)$, with its sensitivity and stability properties, so more care must be taken.

Assuming that a configuration has been picked, the design is now complete. If the configuration has not been preordained, however, it would be useful to study the design data in order to decide if there is any advantage in one configuration over another. This will be done in the next section, where we will also gain some insight on the choice of $T(s)$ in the far-off frequency region.

§ 6.9 Cost of Feedback, and Comparison of Two-Degree-of-Freedom Structures

The highlights of the frequency response method for achieving sensitivity reduction were presented in the last section. In this section the various two-degree-of-freedom configurations of Fig. 6.1-1 are compared. The comparison is done by considering a hypothetical design.

The nominal (hypothetical) T_o and P_o are displayed on a Bode plot in Fig. 6.9-1. The sketch of T_o is only over its significant frequency range, for the time being, for reasons which will soon be clear. It has been noted that the physical output that is achieved is only via the exertion of the plant. In order to make a valid study of the cost of feedback, we insist that $|P| \geqslant |T|$ at all frequencies, and that this relationship hold for the smallest $|P|$, over its range of variation. This requirement is correctly chargeable to the plant, because obviously the plant must have the physical capacity that is being demanded of the system. Therefore, in the figure we use for P_o the expression which gives the smallest $|P|$ over the entire frequency range, and raise its level, such that $|P_o| \geqslant |T|$ for all ω. (If P_o has a zero at $s = j\omega_1$, then T_o should be assigned a zero there.)

Next consider the required value of L_o. Suppose the plant parameter variations are large, and the specifications demand that the variation in $|T|$ be kept fairly small, down to its half-power point. Consequently, $|L_o|$ may have to be moderately large up to 10 rps. Stability considerations (as accentuated by the plant parameter variations) may require that L_o be as shown in the figure. The precise shape of L_o is unimportant here; what is important is that in this example it so happens that $|L_o| > |P_o|$ over a fairly large range in the higher frequency region.

Let $L_o = MP_o$, and therefore M must make up the logarithmic difference between $|L_o|$ and $|P_o|$. Accordingly, in Fig. 6.9-1 $|M| > 1$ for $2 < \omega < \omega_3$ (the cross-hatched region in Fig. 6.9-1). This frequency range extends over a region considerably larger than the dominant frequency range, and over this large frequency range $|M| > 1$ in order to supply the loop gain that P_o is unable to supply. M is the true measure of the cost of feedback. It is needed only because of feedback, and it is the price that must be paid for the benefits of feedback.

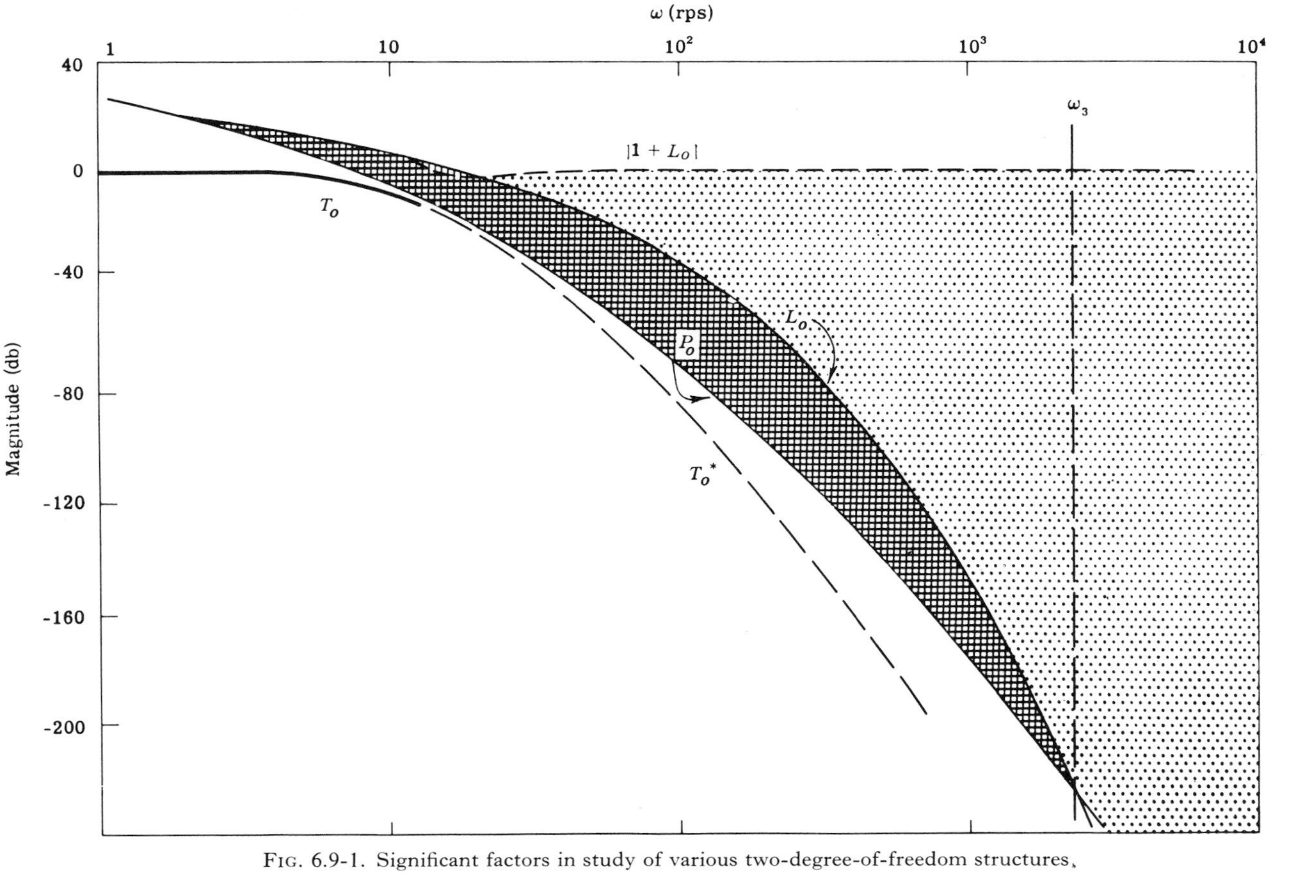

FIG. 6.9-1. Significant factors in study of various two-degree-of-freedom structures.

It is obvious now why $|L_o|$ should be decreased as fast as possible. The smaller the frequency ω_3 beyond which $|L_o| < |P_o|$, the sooner is $|M|$ allowed to be less than one. However, as may already be apparent, and as will be discussed in detail in chapter 7, there is a best L_o (in the above sense) that cannot be improved upon without sacrificing sensitivity requirements.

It is desirable to have the level of M small and its bandwidth (which we shall take approximately as the value of ω_3) small for several important reasons. One reason which we shall later examine in detail (Section 6.14) is that the susceptibility of the system to feedback transducer noise (or generally to noise and parameter variation in the return portion of the feedback loop) depends strongly on M. Another reason is that devices with gain-bandwidth areas corresponding to the value of M must be added to the system. A third reason is that we must know the plant transfer function at least up to ω_3, and this knowledge may be difficult to obtain. In exacting designs it is, therefore, desirable to minimize the gain-bandwidth area of M. For this purpose some effort may be required for the optimum shaping of $L_o(j\omega)$. A good part of Chapter 7 is devoted to the basic properties of a function such as $L(s)$ and its optimum shaping under various conditions.

It will be found that some economy in the gain bandwidth of $L(s)$ is usually achievable if $L(s)$ is allowed to have some complex poles and zeros in addition to real poles and zeros. In fact, for the usual kind of sensitivity specifications which were described in the last section (very small sensitivity at low frequencies which gradually increase with increasing frequency), it is possible to formulate a fairly systematic loop shaping technique. It is felt that this chapter has its fair share of new topics, so this technique is postponed to Chapter 8, after the study, in Chapter 7, of the fundamental restrictions on $L(s)$.

Next consider the role of G in $T_o = GP_o/(1 + L_o)$ with

$$G = T_o(1 + L_o)/P_o. \tag{6.9,1}$$

A sketch of $1 + L_o$ is included in Fig. 6.9-1. It is easy to see the value of $\lambda \triangleq |(1 + L_o)/P_o|$, since in decibels it consists of the subtraction of P_o from $1 + L_o$. This difference has been shaded in Fig. 6.9-1. From Eq. (6.9,1) the value of $|G|$ is given by the difference between λ and $|T_o|$. If we let T_o simply follow P_o, then it is necessary that $|G| > 1$ from about 2 rps to about 12 rps. But before we commit ourselves, consider again $M = L_o/P_o$. At this point, we can no longer treat this generally, but must turn to specific structures. Consider Fig. 6.1-1a. Here $M = H_1$ so the choice of G does not at all affect H_1; H_1 and G_1 are, respectively, equal to M and G. If we indeed let T_o follow P_o up to ω_3, then $|G| = 1$ approximately from 12 rps up to ω_3. The requirement on G_1 is not as severe as on H_1, but it is still serious to have to maintain this value of unity up to so high a frequency. We could make things easier for G_1 by letting $|T_o|$ fall off faster, as shown by $T_o{}^*$ in Fig. 6.9-1. In the latter case, we can relax control of G_1 at about 20 rps. In this configuration, therefore, it is desirable to let $|T_o|$ decrease as rapidly as the specifications on T_o may permit.

Consider Fig. 6.1-1b. Here $M = G_2H_2$, and since M is prescribed by $L(s)$, it is best to choose T_o so that the burden of M is shared by G_2 and H_2. (Note that the cross-hatched region represents $|M|_{db} = |L|_{db} - |P|_{db}$ in Fig. 6.9-1.) Thus, if at some frequency ω_x, $|M(j\omega_x)| = 30$ db, it would be wasteful to have $|H_2(j\omega_x)| = 40$ db, and $|G_2(j\omega_x)| = -10$ db. Such a situation should be avoided, unless there are good and compelling reasons for it. Therefore, $|T_o|$ should not be less than $|P_o|$ (i.e., $|G_2| \nless 1$) up to ω_3 (where $|M| = 1$). For if it is, $|G_2| < 1$ over some frequency range in which $|M| = |G_2H_2| > 1$.

Consider the structure of Fig. 6.1-1c. Here $G_3 = G_2 = G$, so the demands on G_3 in the high frequency range depend, as in the others, on the logarithmic difference between $|P_o|$ and $|T_o|$. But now $M = L/P = G_3 + H_3$. The contributions of G_3 and H_3 add vectorially and not logarithmically (as they did in Fig. 6.1-1b) to achieve M. Therefore G_3 and H_3 cannot share the burden of M as efficiently as G_2 and H_2. For example, suppose $|M(j\omega_x)| = 40$ db. Even if the phase angles of G_3 and H_3 happen to be the same and we split the value of M equally between them, each must be 50, i.e., 34 db. When $G_2H_2 = M$, then G_2 and H_2 need only be each 20 db (magnitude of 10), no matter what their phase angles are. However, if H_3 (or much less likely G_3) is to bear the burden of M practically alone, ($|T_o| \leqslant |P_o|$ over $12 < \omega < \omega_3$), then, as far as the gain-bandwidth demands on G and H are concerned, it doesn't make any difference whether Fig. 6.1-1a, b, or c is used.

Next consider the conditional feedback structure of Fig. 6.1-1d. Is the price of sensitivity reduction any less here than in the other structures? Here $M = \alpha H_4$, so that αH_4 must be precisely the same as H_1 of Fig. 6.1-1a. Two more transmissions are needed. One is αT_o, which offers no special problems. The other is T_o/P_o, which in the range $\omega > 12$ is practically identical to $G_1 = G_2 = G_3 = T_o(1 + L_o)/P_o$. So the difference between the gain-bandwidth costs for this structure and the others is negligible. Finally, consider the model feedback structure of Fig. 6.1-1f. Here $M = \alpha G_6H_6/(1 - \alpha G_6H_6P_o)$. One might be tempted to consider making the denominator close to zero, i.e., using positive feedback to achieve $M(s)$. The use of positive feedback as a tool in active network synthesis is discussed in Section 7.15. It is there noted that it has no spectacular virtues, although it may sometimes be convenient. In the present case it seems to be an overly complicated way of synthesizing the transfer function $M(s)$. We thus see that for the situation depicted in Fog. 6.9-1 there is no profit in any of the more exotic structures of Fig. 6.1-1. It is perhaps conceivable that there may be situations in which there is some advantage in one configuration over another. It is not difficult to check if there is any such advantage. The given plant design specifications, parameter variations, etc., determine P_o, L_o, $1 + L_o$, and T_o, which may be displayed on a Bode chart just as we did in Fig. 6.9-1. It is then easy to see the requirements on the G and H functions, and decide which configuration is the best to use.

§ 6.10 The Problem of the Far-Off Poles

We have seen how important it is to secure the fastest possible decrease of the loop transmission L. In the numerical example of Section 6.8, where $L \to K/s^3$ and $P \to K/s^3$ as $s \to \infty$, it is theoretically necessary to maintain control of G and M over an infinite frequency range (see Fig. 6.8-5). Of course, more far-off poles of L_o may be introduced, but they should be introduced at as low a frequency as possible in order to minimize the gain-bandwidth demands on the compensating networks. Must we then really design a much more complicated L_o function? Thus, if the plant, at the frequencies corresponding to ω_3 in Fig. 6.9-1, has a slope of $-$ 36 db per octave, and if we want M to have a slope of $-$ 6 db per octave there, we need an L_o with a slope of $-$ 42 db per octave in that frequency range. There are several alternatives:

(1) A relatively simple $L_o(s)$ which satisfies the design specifications may be chosen. At the very least, however, $L_o(s)$ has the same excess of poles over zeros as the plant. One is left with G and H functions which are finite at infinity. From the Bode sketch of $L_o(s)$ and from the design specifications, parameter variations, etc., one ascertains the frequency range in which $L_o(j\omega)$ may materially decrease in amplitude and increase in lag angle without the design specifications being violated. If there is no difficulty ensuring that the practical $L_o(s)$ that is constructed satisfies these requirements, then the design is satisfactory. Here we are able to use a simple design only because our devices have gain-bandwidth to spare. We are economizing in design effort at the price of gain-bandwidth of M. This is the method discussed in Section 6.8.

(2) Suppose there is some difficulty in ensuring that the practical $L_o(s)$ follows its nominal design value right up to the required frequency range. We should then do our best to determine the theoretical optimum loop transmission function for our specific problem—go to some trouble and work with a more complicated theoretical loop transmission function, introduce more poles, etc., and emerge with an $L_o(s)$ which decreases faster and sooner than the original simple (but more costly in gain-bandwidth) design. Techniques for this purpose are given in Chapters 7 and 8.

At this point there are several possibilities:

(A) It may be impossible to build $L_o(s)$. In such a case, either the specifications are eased, or a multiloop design is attempted (if the plant has more access points). Multiloop design is treated in Chapter 8.

(B) Suppose it is possible to construct a system whose loop transmission does not decrease any faster than the theoretical optimum design. It was perhaps impossible to build a system when we used a relatively simple design for $L_o(s)$ with its larger gain-bandwidth cost, but suppose it does prove possible after we have gone to some trouble and designed a complicated $L_0(s)$ carefully and emerged with a more economical design. There is, however, no actual need to physically implement the design. For example, suppose we originally thought that $| L_0(j\omega) |$ must be $\nless -20$ db and could have a slope of only -24 db per octave at 500 rps, but we now find that thanks to the more painstaking design, it can be -30 db and could have a slope of -30 db per octave at 500 rps. And we find that if we build *simple* G and H functions, their natural parasitics will not cause the real $L_0(j\omega)$ to violate the requirements on the theoretical optimum $L_o(s)$. In this case, it is necessary to use a complicated theoretical design, but it is not necessary to fully implement it.

(C) In this final possibility, we visualize that things are so close that we must actually implement the complicated design. Not only is it necessary to prepare an economical theoretical design of $L_o(s)$, but it is also necessary to implement the design. There is no margin of gain-bandwidth available. Both the design and the implementation require much work and care.

It is clear that one cannot be dogmatic in these matters. It is certainly desirable to use simple designs and simple compensating networks. However, the simple designs are the most wasteful in gain-bandwidth requirements of the plant and the compensating $M = L_o/P_o$ networks. Therefore simple designs are feasible only when there is gain-bandwidth to spare. Then there is the case where there is very little gain-bandwidth that can be wasted. We must at least *know* the least that is required of $L_0(s)$ so that we can check whether the simple construction will do. If it is satisfactory, we have here the case of a theoretical design which is economical in gain-bandwidth, complicated in theoretical design effort, and accompanied by a relatively simple design implementation. In the final case, we must use the design costly in effort but most economical in gain-bandwidth and also implement this design. This last case has occurred more often in feedback amplifiers than in feedback control systems. However, in modern technology, more and more control problems are arising where there are large parameter variations, more severe disturbances, and small sensitivity is required.

The above completes, for the time being, the problem of designing a two-degree-of-freedom system to achieve a desired insensitivity to parameter variations. Another principal reason for using feedback is to control the effect of disturbances acting on the plant. The next few sections are devoted to this problem.

§ 6.11 Design for Multiple Inputs

Our attention here is confined to the case where the designer has only two access points to the plant (C and V in Fig. 6.11-1). However, for the sake of generality, we assume disturbances can also occur at internal nodes. It has been noted in Section 6.1 that this flow graph is a perfectly general representation of the system when $t_{oi} = 0$ and the plant output is the system output. All the disturbances can be collected together as one equivalent disturbance at node Y, with $D_e = P_1D_1 + D_2 + (D_3/P_2)$. Then $T_D = C/D_e = P_2/(1 + L)$ and there is no need to consider the specific structure that is being used.

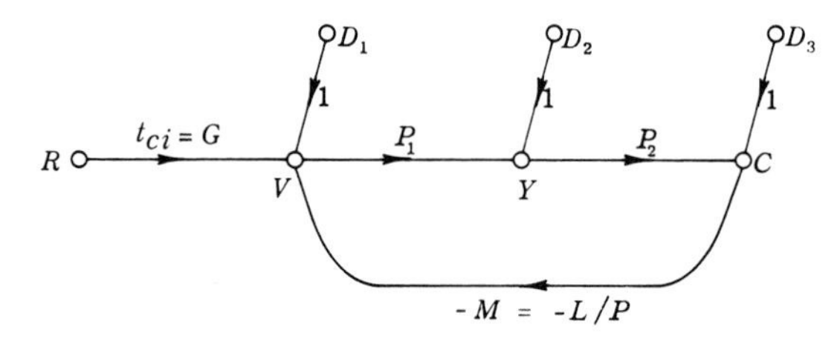

FIG. 6.11-1. System with plant disturbances.

If the desired system transfer function and the desired disturbance transfer function are analytically available, then one formally solves for the two independent functions G and $M = L/P$ and the design is complete. But this is not realistic. What we really want is a design procedure which gives us some feeling for the price that is paid for the disturbance attenuation, and which therefore enables us to understand specifications that are reasonably attained and others for which a heavy price must be paid. If we have such a design procedure, then if we do run into difficulties, we know to what extent the design specifications must be relaxed in order to permit a reasonable design. Such an insight is readily obtained by the frequency response methods. However, such an insight is not as readily obtained by s-plane pole-zero methods (see Section 6.13). Consequently, only the frequency response method is presented here.

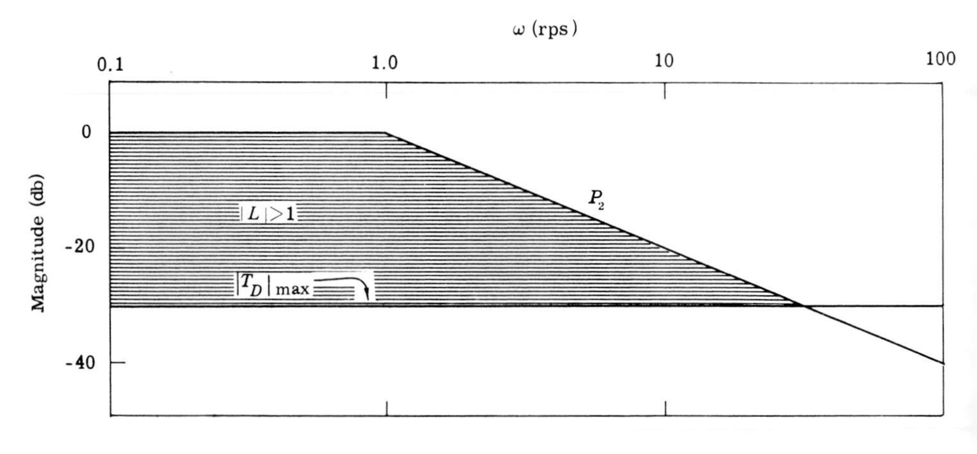

FIG. 6.11-2. Bode plots of $P_2(j\omega)$ and $|TD(j\omega)|_{max}$.

Consider $T_D = P_2/(1 + L)$. We shall assume at first that the plant parameter variations are negligible. The frequency spectrum of the disturbance input and its desired attenuation determine the required T_D. It is assumed that the designer knows how small he wants $|T_D|$ to be, as a function of frequency. The problem of translating time domain specifications into frequency domain specifications is treated in detail in Sections 9.4-9.6. Our principal interest at this point is the implementation of specifications, not in their enunciation. Assuming then that the maximum value of $|T_D|$ is known as a function of frequency, this completely specifies L. In decibels, $|T_D|_{\text{db}} = |P_2|_{\text{db}} - |1 + L|_{\text{db}}$ and in the significant frequency range where $|L| \gg 1$, $|T_D|_{\text{db}} \doteq |P_2|_{\text{db}} - |L|_{\text{db}}$.

For example, suppose it is required that $|T_D| \leqslant -30$ db for all ω, and P_2 is as shown in Fig. 6.11-2. If $L = 0$, there is no feedback and $(T_D)_{\text{n.f.}} = P_2$, which is more than the permitted maximum of -30 db for $0 < \omega < 32$ rps. The loop gain L must provide the necessary reduction from 0 rps up to 32 rps. Beyond 32 rps, P_2 needs no help from L. Therefore, the minimum $|L|$ is given by the shaded region in Fig. 6.11-2, e.g., at $\omega = 2$, $|P_2| = -6$ db, and $|T_D|_{\max} = -30$ db, so $|L|_{\min} = 24$ db.

The next step is to complete the design of L. The procedure is as follows. In the region in which $|T_D| < |P_2|$, $|L| > 1$ and we take $|1 + L| \doteq |L|$. This determines the minimum $|L|$, up to about 32 rps. This $|L|_{\min}$ (the shaded region of Fig. 6.11-2) is sketched in Fig. 6.11-3. We also sketch P, which is

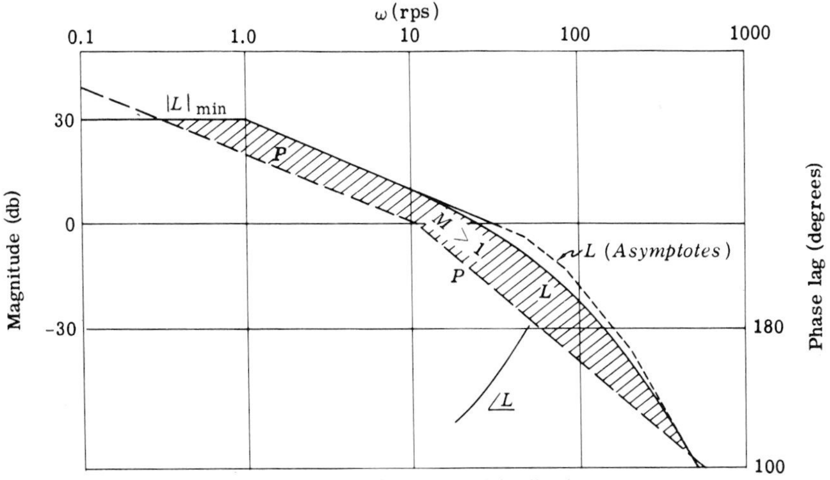

FIG. 6.11-3. The cost of feedback.

known of course, in the same figure. For $\omega < 0.3$, $P > |L_{\min}|$, but it would be foolish to throw away the infinite loop gain at 0 rps which we have anyhow. Therefore, we let $|L| = |P|$ wherever $|P| > |L_{\min}|$ and choose $|L| = |L_{\min}|$ where $|P| < |L_{\min}|$. In this way $|L|$ is specified up to $\omega \approx 30$ rps. Beyond 30 rps, it is desirable to decrease $|L|$ as fast as possible, consistent with reasonable stability margins. We assume in this section that plant parameter variation is not a serious problem; therefore, a simple statement of gain and phase margins suffices. So we shape L beyond 30 rps, bring it down reasonably fast, and obtain the desired gain and phase margins. This is done in Fig. 6.11-3. Suppose the stability margins are $\theta_M = 40°$ and $G_M = 10$ db. We shape L accordingly, using the techniques of Chapter 5. A suitable value is $L = (25)10^6 (s + 0.3) [s(s + 50) (s + 80) (s + 200) (s + 1)]^{-1}$, which is sketched in Fig. 6.11-3.

Let $L = PM$ so M must make up the difference between L and P. With P as shown in Fig. 6.11-3 the value of M is the shaded area and represents that portion of the cost of feedback which the plant by itself is unable to bear. This area is the additional gain-bandwidth area that must be supplied by M. Here $| M | = 1$ for $0 \leqslant \omega < 0.3$, and $| M | > 1$ for $0.3 < \omega < 500$. The faster $| L |$ is decreased, the smaller the required gain-bandwidth area of M. The smaller T_D the larger L must be, and consequently the larger M must be. Design in terms of frequency response displays the cost of feedback simply and clearly.

The above determines L and uses up one degree of freedom. The second degree of freedom is used to fix the desired system response $T = C/R$. If we use Fig. 6.1-1a, $H_1 = M = L/P$, and $G = T(1 + L)/P$. The faster $| T |$ is allowed to decrease, the faster G may decrease. If we use Fig. 6.1-1b, $H_2 = L/PG_2$ and the discussion of Section 6.10 applies. The design procedure henceforth parallels that in Sections 6.8 and 6.9 and is therefore not treated any further.

§ 6.12 Design for Disturbance Attenuation Accompanied by Plant Parameter Variation

In Section 6.11 it was assumed that the plant parameter variation was negligible. If this is not so, then in designing for disturbance attenuation, account must be taken of the plant parameter variation. Note that here we are not concerned with the sensitivity of $T(s)$ to changes in P. The procedure presented in previous sections of this chapter is used for this purpose. We are concerned that the desired disturbance attenuation is secured for the entire range of plant parameter values. The design procedure is as follows.

Consider the disturbance at Y in Fig. 6.11-1, with $T_{D_2} = P_2/(1 + L) = P_2/(1 + PM)$. Let P_{2o}, $L_o = P_oM$ be the appropriate values when the plant function has its nominal value P_o, and let P_2, $L = PM$ be the corresponding values when the plant transfer function has the value P. Then

$$T_{D_2} = \frac{P_2}{1 + PM} = \frac{P_2}{1 + PM(P_o/P_o)} = \frac{P_2}{1 + L_o(P/P_o)} = \frac{(P_o/P)P_2}{(P_o/P) + L_o}$$

$$= \frac{P_o/P_1}{(P_o/P) + L_o} = \frac{\mathbf{OB}}{\mathbf{WQ}} \quad \text{(in Fig. 6.12-1).} \qquad (6.12,1)$$

The procedure is straightforward. At any frequency $j\omega_x$, the boundaries of all possible values of P_o/P_1 and of P_o/P are obtained. In Fig. 6.12-1 [using Eq. (6.12,1)], $T_{D_2} = \mathbf{OB}/\mathbf{WQ}$. Suppose it is required that $|T_{D_2}(j\omega_x)| \leqslant \delta$. The permissible location for W is then easily obtained. One can easily sketch a locus of satisfactory minimum $|L_o(j\omega_x)|$ in the manner of Fig. 6.7-2b. In deriving the boundary of minimum $|L_o(j\omega)|$, one can simplify the numerical work by taking the worst possible combination of P_o/P_1 and $(P_o/P) + L_o$, i.e., that combination which gives the largest T_{D_2}. Or one can note the segments on the boundary of P_o/P_1 which correspond to segments of the boundary of P_o/P, and in general obtain a more economical design.

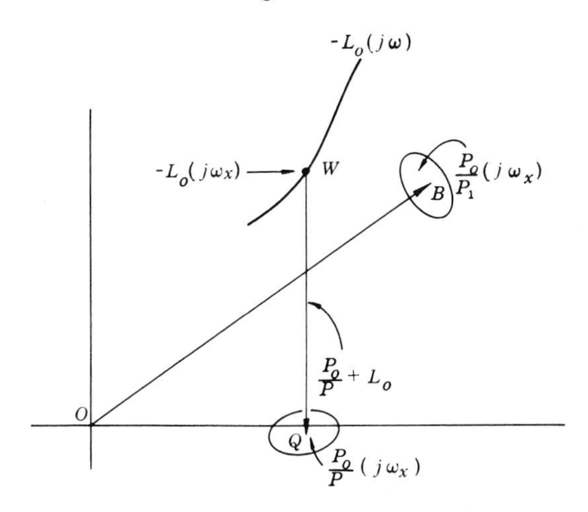

FIG. 6.12-1. $T_D(j\omega_x) = \mathbf{OB}/\mathbf{WQ}$.

The above is repeated for several discrete frequencies in the significant frequency range, after which the requirements on L_o are obvious.

This procedure may be extended to the case of simultaneous disturbances at V, Y, C in Fig. 6.11-1. Thus,

$$C_D = C_{D_1} + C_{D_2} + C_{D_3} = \frac{P_o[D_1 + (D_2/P_1) + (D_3/P)]}{(P_o/P) + L_o} \qquad (6.12,2)$$

A satisfactory procedure is to state the specifications in terms of C_D. At $s = j\omega_x$, sketch the boundary of $(P_o/P)(j\omega_x)$ and the boundary of $D_1 + D_2/P_1 + D_3/P$ and proceed as before.

It is important to note that throughout this entire chapter we have been assuming that the designer has no access to any internal points in the plant. The design, both for sensitivity reduction and disturbance rejection, may sometimes be considerably more economical (in gain-bandwidth) when there is access to internal plant variables. This more complex problem is treated in Chapter 8.

Desing Example. Consider the plant described in Section 6.3, whose parameter variations were so large (Fig. 6.3-1). Suppose that the plant is subjected to white noise at its input. This particular plant is exceedingly sensitive to any noise, when the single-degree-of-freedom configuration of Chapter 5 is used. The reason is as follows: When the plant poles are on the $j\omega$ axis, it is obviously essential that no such poles appear in the system transfer function $T = C/R$. In the single-degree-of-freedom system design, the usual procedure is to have zeros of $M = L/P$ near these $j\omega$ poles of P (Fig. 6.12-2b). The root loci of $1 + L = 0$ and the pole-zero pattern of T are then as shown in Fig. 6.12-2. The resulting pole-zero pattern of $T_D = P/(1 + L)$ is shown in Fig. 6.12-2e.

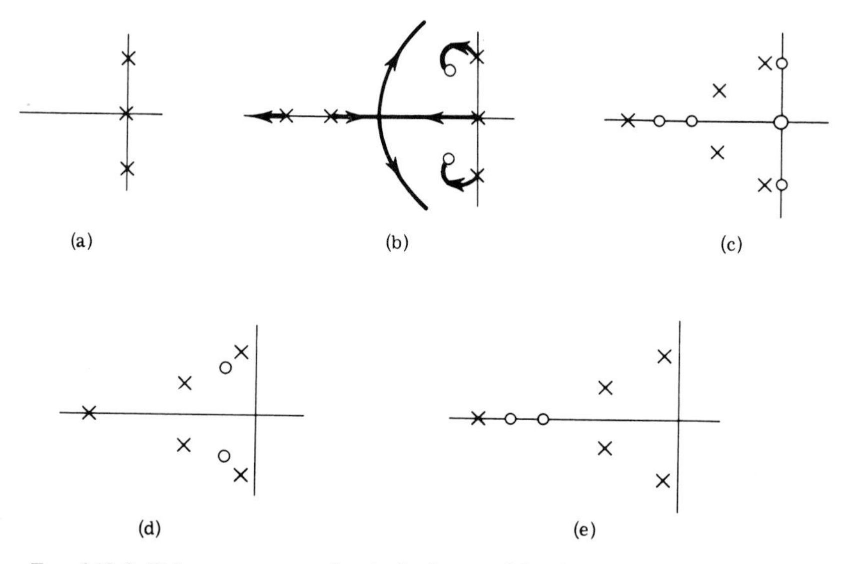

FIG. 6.12-2. Pole-zero patterns in single-degree-of-freedom structure for plant with low-damped poles. (a) Pole-zero pattern of $P(s)$. (b) Root loci of $1 + L = 0$. (c) Pole-zero pattern of $1/(1 + L)$. (d) Pole-zero pattern of $T = L/(1 + L)$. (e) Pole-zero pattern of $T_D = P/(1 + L)$.

It is seen that the highly underdamped roots of $1 + L = 0$ appear as poles of T_D and there are no zeros close by to make their residues small. A slight noise input leads to highly underdamped oscillations.

The only way to overcome this problem with the single-degree-of-freedom configuration is by means of a pole-zero configuration of L similar to that shown in Fig. 6.12-3, accompanied by very large loop gain. The latter assures zeros of $1 + L = 0$ near A, $\bar{A}$ in Fig. 6.12-3, despite plant parameter variations. The roots of $1 + L = 0$ at A, $\bar{A}$ which appear as poles of T_D are then no longer highly underdamped. However, this large loop gain leads to an exceedingly large system bandwidth. The dominant poles of $T(s)$ are at B, $\bar{B}$, because the poles of T at A, $\bar{A}$ are effectively canceled by the nearby zeros. The poles at F,

N, Q must be far away in order that the system be stable. Such a large system bandwidth is nearly always undesirable, because of its susceptibility to high frequency noise entering with the useful input signal (see Chapter 9). This inability to satisfy in general the specifications for T and T_D simultaneously is,

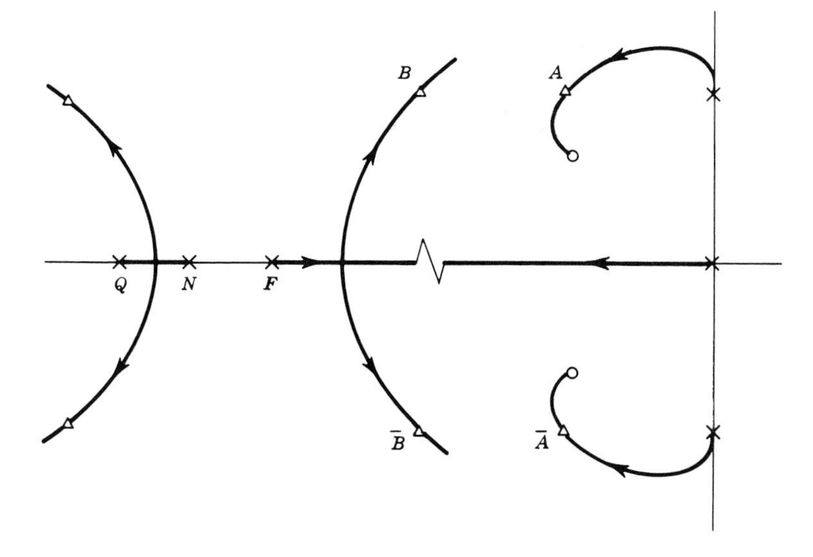

FIG. 6.12-3. Solution of the low-damped plant pole problem with a two-degree-of-freedom structure; ○ — zeros of $L(s)$, × — poles of $L(s)$, △ — zeros of $1 + L(s)$.

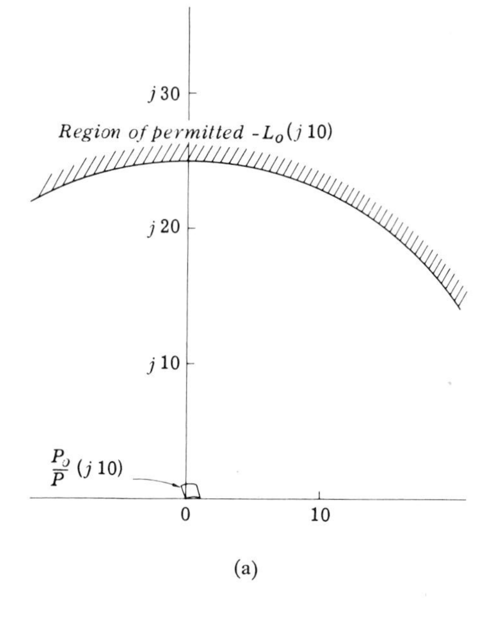

FIG. 6.12-4. (a) Permissible range of $-L_o(j10)$ which satisfies disturbance specifications.

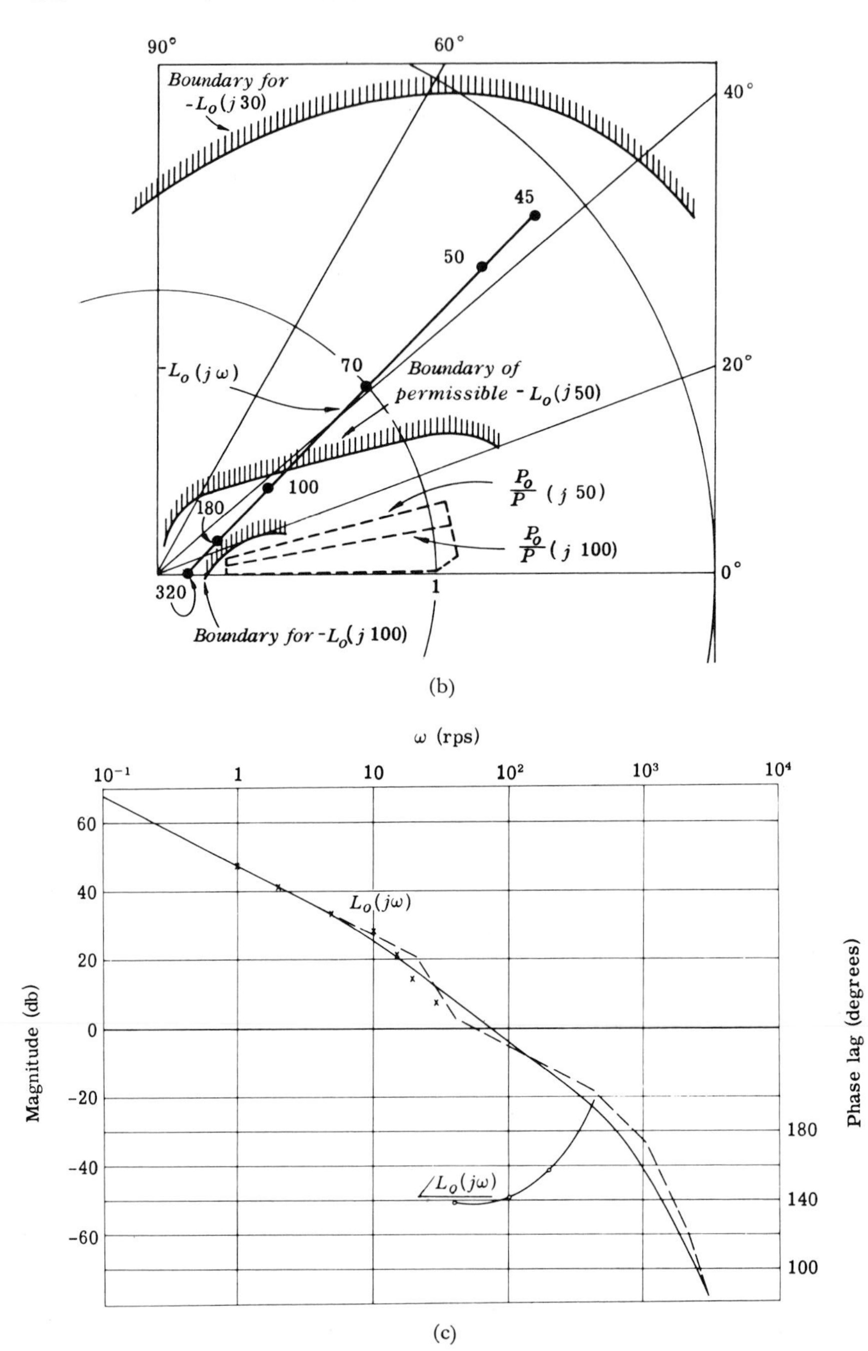

FIG. 6.12-4. (b) Permissible range of $-L_o(j\omega)$ at $\omega = 30, 50, 100$ rps. (c) Bode plot shaping of $L_o(j\omega)$. $\times$ indicates values of minimum $|L_o|$ to satisfy specifications.

of course, a characteristic feature of the single-degree-of-freedom configuration. We shall now obtain a satisfactory design for this problem with a two-degree-of-freedom configuration.

Suppose that a study of the noise spectrum and the desired minimum noise output leads us to the conclusion that T_D must be less than -20 db for all ω (how such conclusions are reached is considered in Sections 9.4-9.6). In this example, since the noise enters at the plant input, Eq. (6.12,1) becomes $T_D = P_o/[L_o + (P_o/P)]$.

Design procedure: The first step is to obtain sketches of the boundaries of P_o/P at a number of values of $s = j\omega$ in the low, intermediate, and high frequency regions. These have previously been calculated and sketched in Figs. 6.8-2a and b. We now use the technique described in Fig. 6.12-1. For example, at $s = j10$, the boundary of $(P_o/P)(j10)$ is as shown in Fig. 6.12-4a. Also $P_o = 3000/s(s^2 + 12s + 136)$ which at $s = j10$ is $2.4 \angle -163.3°$. In order that $|T_D| \leqslant 0.1$, $|(P_o/P) + L_o| \geqslant 24$, and the boundary of permissible $-L_o(j10)$ is easily obtained. In the low frequency range, $|L_o(j\omega)|$ must be so large that its permissible boundary can be taken as a circle. Thus, it is found that the requirements at low frequencies are as follows: $K_v \geqslant 30{,}000/136 = 220$. $|L_o(j1)| \geqslant 220$, $|L_o(j2)| \geqslant 110$, $|L_o(j5)| \geqslant 50$, $|L_o(j10)| \geqslant 25$, $|L_o(j15)| \geqslant 11$, $|L_o(j20)| \geqslant 5$. In the intermediate frequency range, where $|L_o| \sim 1$, the boundary of permissible L_o should be found more carefully. This is done in Fig. 6.12-4b.

From an examination of the above requirements of $L_o(j\omega)$ and their comparison with Figs. 6.8-2b and 3, it is seen that the disturbance rejection demands on L_o in the low frequencies are even greater than those due to the sensitivity requirements of the design of Section 6.8. The balance of the design procedure is straightforward. One prepares a Bode sketch of $L_o(j\omega)$ which satisfies its low frequency requirements (Fig. 6.12-4c), and then works fairly carefully in shaping it in the crossover region. Here one must watch the stability and sensitivity specifications, as well as the specifications on T_D. The sensitivity specifications in the intermediate frequency range require a more conservative design for L_o in this region, than that due to the T_D specifications. A suitable value of $L_o(s)$ is: $L_o(s) = (s + 40)^2(44)(10)^{12}[s(s + 20)^2(s + 400)(s + 1000)^2(s + 2000)]^{-1}$ which is sketched in Figs. 6.12-4b and c. The procedure thereafter for finding $M = L_o/P_o$ and the G and H functions is straightforward.

Discussion: There are several attractive features of the frequency response approach in the design for plant disturbance rejection. One is in its ease and simplicity. Another is in the insight it provides the designer in the relation between the benefits of feedback and the cost of feedback. If the cost is too high in some specific aspect, the designer knows precisely how he must relax the specifications (in terms of frequency response). Also in the matter of economy in the gain-bandwidth requirements of the compensating networks, one can carefully shape $L_o(s)$ to obtain the most economical $L_o(s)$. One knows precisely how soon the far-off poles may be introduced. In all respects except one, the frequency response method is superior to any other method. The one drawback is in relating it to time domain behavior. However, it should be noted that in the pole-zero approach, it is also not easy to relate the time response to the poles and zeros, unless the system is a dominant one. In fact, in nondominant systems, the relations between frequency and time response (Section 5.10) are much simpler than those between pole-zeros and time response. This problem has been discussed in Section 5.10. It will be considered again in Chapters 9 and 12.

§ 6.13 Analytical Specification of the Sensitivity Function

In previous sections, we have emphasized graphical methods of design for the attainment of the benefits of feedback. The graphical method of shaping $L(j\omega)$ has some very attractive features. The required benefits of feedback can be specified graphically in terms of the required magnitude of $L(j\omega)$ as a function of frequency. Then the shaping of $L(j\omega)$ to achieve these levels, and simultaneously to decrease $L(j\omega)$ as fast as possible, is readily done graphically. On the other hand, the analytical (pole-zero) specification of the sensitivity function $S_P{}^T = (1 + L)^{-1}$ appears to be singularly unattractive for numerical design work. This was briefly noted in Section 6.11. We shall now examine the reasons for this. Let $L = K\Pi_m(s + z_i)/\Pi_n(s + p_j)$, with $n > m$ because all physical transfer functions must go to zero at infinite frequency. Then the sensitivity function $S = (1 + L)^{-1} = \Pi_n(s + p_j)\,[\Pi_n(s + p_j) + K\,\Pi_m(s + z_i)]^{-1}$. Let $n - m = e$, an integer. Then

$$S = \frac{1}{1 + L}$$

$$= \frac{s^n + N_1 s^{n-1} + \dots + N_{e-1} s^{n-e+1} + N_e s^{n-e} + \dots + N_n}{s^n + N_1 s^{n-1} + \dots + N_{e-1} s^{n-e+1} + (N_e + K)s^{n-e} + \dots + (N_n + K\Pi\, z_i)}. \tag{6.13,1}$$

It is seen that the first e coefficients of the numerator of S must be equal, respectively, to the corresponding first e coefficients of the denominator of S. In addition, S must be chosen to have the desired properties (presumably small residues in its dominant poles or small magnitude over some frequency range), in order to achieve the desired feedback properties for the system. If e is only 2, it is not too difficult to choose an S function which provides the sensitivity reduction, which tends to unity at infinite s, and the sum of whose zeros is equal to the sum of its poles. But the task becomes very difficult even for $e = 3$, and well nigh impossible for larger e.

In practice then, one must take $e = 1$ or 2, and after factoring, end up with an $L = S^{-1} - 1$ which goes to infinity as Ks^{-1} or Ks^{-2}. The inevitable additional poles must then be placed "far away" in order not to spoil the assumed S. This of course constitutes an unnecessary waste of gain-bandwidth. If they are not placed "far away," then their effect on S should be checked, and we no longer have the pure synthesis which was presumably our objective. If we do not use additional poles, we end up with transfer functions which are infinite as $s \to \infty$ (unless we are constrained to plants with only two more poles than zeros). Then we are confronted with the question of how soon may these functions stop increasing with frequency. We may trust to luck, i.e., construct the functions to have differentiating characteristics (of a high order for even moderately complex plants), and hope that they have such characteristics for as large a frequency range as necessary. But this is hardly an attractive procedure; and a somewhat strange price to pay for having a "pure synthesis" procedure.

In addition to the above difficulties, there is the difficulty of handling the stability problem with varying parameters. There is also the fact that one loses sight of the price being paid for the benefits of feedback. For these reasons, there seems to be no advantage in trying to use pole-zero methods of design for the attainment of the benefits of feedback. On the other hand, the pole-zero specification of $L(s)$ may be feasible. We have considered this approach in Sections 6.2-6.6 for those cases in which very small sensitivity is desired. The design procedure where only moderate sensitivity is desired is presented in Sections 8.11-8.13.

§ 6.14 Achievable Benefits of Feedback in Two-Degree-of-Freedom Structure

This chapter has presented design procedures for realizing specified insensitivity of a system response function to plant parameter variations, and for attenuating the effect of disturbances acting on the plant. These beneficial properties of feedback are attained by assigning to the nominal loop transmission $L_o(j\omega)$, a sufficiently large magnitude over a sufficiently large bandwidth. The loop transmission function $L_o(j\omega)$ in general consists of $L_o = P_oM$, where P_o is the nominal plant transfer function, and M represents the added compensation, data-processing networks. The function M must therefore make up the logarithmic difference between the gain-bandwidth of L_o and that made available by the plant.

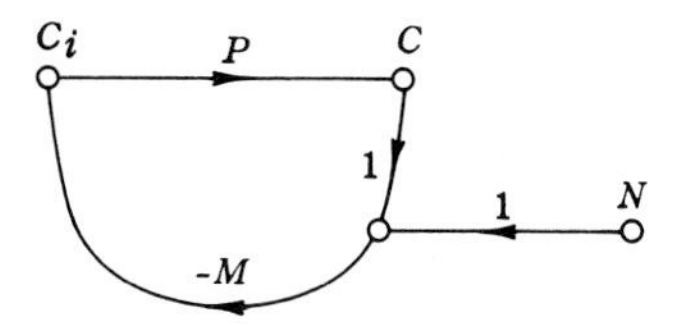

FIG. 6.14-1. The feedback transducer noise problem.

There are several limitations on the achievable benefits of feedback. One of these limitations is due to noise (or parameter variation) in the return path of the feedback loop. Consider the effect of the noise source N in Fig. 6.14-1. The output $C_N = -PMN/(1 + PM) = -LN/(1+L) \approx -N$ in the frequency range in which $| L | > 1$. The means whereby $T_N = C/N$ can be reduced in this significant frequency range, have been discussed in Section 3.9. We are here concerned with the noise level that exists at C_i (the plant input) because

$$\frac{C_i}{N} = \frac{-M}{1+L} = \frac{-L/P}{1+L}. \tag{6.14,1}$$

The example of Fig. 6.11-3 is useful here. In Fig. 6.11-3, $| L | < 1$ for all $\omega > 25$ and we may therefore write $C_i/N \approx -M = -L/P$ for $\omega > 25$. This quantity L/P is more than 1 for $25 < \omega < 450$ rps. Its maximum value is almost 20 db at $\omega = 100$ rps. The high level of the higher frequency components of the noise input to the plant may seriously affect the plant's performance and even cause it to saturate. To cope with this problem, the techniques suggested in Section 3.9 may be helpful if they are implementable.

It is seen that this problem exists when $| L | > |P|$ in the frequency range in which $| L | < 1$. For then $C_i/N = -M/(1 + L) \approx -M = -L/P$. It will also exist in any region in which $| L | > 1$ and $| P | < 1$, because then $C_i/N \approx -M/L = -1/P$. The problem is not so serious when $| L | > 1$ and $| P | > 1$, because then $C_i/N = -M/(1 + L) \approx -M/L = -1/P$. It is important to recognize that this problem can occur only when the demand for the benefits of feedback is greater than what the plant by itself is capable of supplying. In such a case, $M = L/P$ must make up the difference. The trouble is aggravated by the fact that stability considerations apparently require $| L |$ to decrease comparatively slowly near the crossover region.

The above represents one serious limitation on the achievable benefits of feedback. To reduce this limitation, it is important to determine the maximum rate at which $| L |$ may be decreased. Only then is it possible to define the absolute limits in the realizable benefits of feedback under the present constraint. This is treated in Chapter 7. In Chapter 7 we shall consider this problem for the easiest case of plants that have only left half-plane poles and zeros, and for the more difficult cases of plants with right half-plane zeros, or with right half-plane poles, or with pure time lags. The only other means of reducing the limitations is to remove one or more constraints, for example, by assuming that one or more internal plant variables are available for feedback purposes. This leads to multiple loop feedback systems, and is treated in Chapter 8.

Another limitation on the achievable benefits of feedback is due to finite gain-bandwidths available from active elements. Thus even if $N = 0$ in Fig. 6.14-1, so that $| M | \gg 1$ is feasible, there remains the problem of finding active elements to supply the required gain-bandwidth. Here it should be recalled that we are assuming small parameter variation in M. Therefore, part of the gain-bandwidth capabilities of the active elements must be used to reduce their own parameter variations. Another significant problem is, therefore, to find the means of extracting the maximum gain-bandwidth capabilities from any specific active element.

This problem is more serious in feedback amplifiers, with their large bandwidths, than it is in control systems with their small bandwidths.

A third limitation is due to a factor which we have not at all considered as yet. It will be seen in Chapter 7 that if a plant has more than one rhp zero, then there is an absolute limit on the loop gain-bandwidth achievable, even if an unlimited number of active elements with unlimited gain bandwidths are available. In a specific problem, it may therefore be impossible to sécure for such plants all the desired benefits of feedback.

Another limitation is the limit on the loop transmission crossover frequency that exists in any practical design. The model that is used to represent the plant is often reasonably accurate only over the low and medium frequency ranges. The plant higher order modes are usually neglected in the plant model. The practical designer is extremely reluctant to attempt design in those frequencies where the higher order plant modes become significant. The reasons are the greater complexity that is required for the plant model, the large plant phase lag in this frequency range and the greater importance of plant nonlinearities at these frequencies. In order that the designer may be reasonably sure that his paper design is realistic and stable, he must have the loop transmission crossover frequency ω_c at a frequency where the plant parasitics have little effect, i.e., several octaves below the higher order modes. Since the crossover frequency is limited in this way, it is important that the designer be able to pack in as much loop transmission magnitude as possible, in the limited range that is available. Chapter 7 presents the maximum $L(j\omega)$ that is realizable for a given ω_c under various constraints.

Some of the above difficulties may be alleviated if "plant modification" at high frequencies is permitted. Suppose it is possible to insert a smaller capacity but faster responding second plant (P_2), in parallel with the original plant (P_1). A crossover filter network at the plant inputs is included to ensure that the low frequency (large energy) components of the input signals are fed only to the original plant. Thus at low frequencies $P = P_1 + P_2 \doteq P_1$. However at high frequencies $|P_2| \gg |P_1|$ and then $P \doteq P_2$. It is necessary that the second plant has the capacity to respond to the high frequency components of the input (for which the linear design is supposed to apply) without saturating. Obviously it would be very helpful if $|P_2| > |L|_{\min}$ for $\omega > \omega_c$, where $|L|_{\min}$ is the minimum $|L|$ required for purposes of feedback benefits and stability requirements. The added network in the feedback loop, $M = L/P$, will then not have differentiating characteristics, and high frequency feedback transducer noise is then not amplified by the M network.

Another limitation in the benefits of feedback is due to system nonlinearities. Nonviolent nonlinearities are equivalent to external disturbances, or parameter variations which are functions of the input signal, and their effect is reduced by feedback (see Sections 8.19 and 8.20). The violent nonlinearities may, however, lead to stability problems. For example, saturation will usually cause a conditionally stable system to oscillate at a frequency in the range in which the phase lag is more than 180°. Nonlinear compensation can be used to prevent instability. Hysteresis (backlash) under certain conditions has a destabilizing effect, which is aggravated by a large loop gain and bandwidth. The nonlinear factor is mentioned here, but its treatment is outside the scope of this book.

Our concern now is to try to contend with the linear factors which limit the attainable benefits of feedback. Chapter 7 is devoted to this problem.

CHAPTER SEVEN

Fundamental Properties and Limitations of the Loop Transmission Function

§ 7.1 Introduction

The previous chapter dealt with the problem of designing a system for prescribed insensitivity to parameter variation and/or rejection of disturbances. It was noted that one price paid for these benefits of feedback was the need to control the loop transfer function $L(s)$ over a specific frequency range. However, it did appear that any amount of sensitivity reduction and disturbance rejection could be attained. In this chapter we seek clear, definitive answers to related questions, such as the following: Given a prescribed level of loop gain $L(j\omega)$ up to a given frequency (for sensitivity reduction, etc.), what is the minimum additional frequency range over which $L(j\omega)$ must be controlled before it may assume some specified small value, or before it is allowed to decrease at some specified rate? Given a prescribed crossover frequency for $L(j\omega)$, what is the maximum loop-gain frequency area achievable? What improvement is there in the above if the system is allowed to be conditionally stable? What are the answers to these questions when the system includes a pure time lag or right half-plane (rhp) zeros? What are the answers when $L(s)$ must have one or more rhp poles, i.e., $L(s)$ is open-loop unstable? What definitive statements can be made about systems with minor positive feedback loops? Are "zero-sensitivity" systems achievable, and to what extent, if any, are they better than conventional feedback designs? What are the basic limitations on the gain-bandwidth products achievable from an active device?

The answers to many of these questions were found by Bode[1] who, in the process, transformed feedback theory from an art to a science. The scientific foundation established by Bode will be used in this chapter to answer those of the above questions which he did not formally solve. Bode's work is applied here to the problem formulated in Chapter 6. It is important to understand the constraints imposed on the problem. They are as follows: There is a given

[1] H. W. Bode, "Network Analysis and Feedback Amplifier Design," Chapters 13-19. Van Nostrand, Princeton, New Jersey, 1945.

unalterable piece of equipment called the plant, whose transfer function is $P(s)$; the system input signal reaches the output only via the plant. We are interested in the benefits of feedback only as they apply to the plant, i.e., to the plant parameter variations, and to disturbances acting anywhere in the plant. Such benefits are completely determined by the loop transmission function $L(s) = P(s)M(s)$. Thus $P(s)$ must appear as part of $L(s)$. The compensation function $M(s)$ can be relatively complicated with its own feedback loops. The problem is to find the limitations on $L(s)$, and thereby to conclusively establish the maximum benefits of feedback achievable under the above constraints.

§ 7.2 Mathematical Background

The mathematical background that is needed for this chapter is remarkably simple in proportion to the imposing results that become available. Linear feedback theory deals with transforms of time functions and with transfer functions (which are transforms of the impulse response of linear systems). These are all functions of a complex variable. Hence the powerful tools of complex variable theory may be applied. This section consists of a review of the complex variable theory that is pertinent for our work.

A single-valued function of a complex variable $F(s)$ is analytic (also called regular and holomorphic) in any region in which it satisfies the Cauchy-Riemann equations $\partial A/\partial\sigma = \partial B/\partial\omega$, $\partial B/\partial\sigma = -\partial A/\partial\omega$, where the notation used is $F(s) = F(\sigma + j\omega) = A(\sigma, \omega) + jB(\sigma, \omega)$ and A and B are real functions of the real variables σ and ω. Equivalently, it is analytic if it can be expanded in a Taylor series about any point of the region. Those points at which $F(s)$ is not analytic are called singularities. One kind of singularity of importance here is the *pole*, defined as follows: If $F(s)$ has a singularity at $s = s_1$ such that as $s \to s_1$, $(s - s_1)^m F(s) \to K$, with m an integer and K finite, then $F(s)$ has a pole at s_1 of order (or multiplicity) m. A simple test for such a pole is that $(s - s_1)^m F(s)$ must be analytic at s_1. Thus $F(s) = 1/(e^{sT} - 1)$ has simple (order 1) poles at $s = j2n\pi$ (n is any integer) because $(s - j2n\pi)/(e^{sT} - 1)$ is analytic at $s = j2n\pi$.

If $F(s)$ has a pole of order m at $s = s_1$, then it is always possible to find some circle centered at s_1, in whose interior the following expansion of $F(s)$ is valid:

$$F(s) = A_o + A_1(s - s_1) + A_2(s - s_1)^2 + \dots + B_1/(s - s_1) + B_2/(s - s_1)^2 + \dots + B_m/(s - s_1)^m.$$

In this expression, at least B_m must be nonzero. B_1 is defined as the *residue* of $F(s)$ at the pole s_1.

Cauchy's Residue Theorem

Consider any region defined by a closed boundary in which and on whose boundary $F(s)$ is single-valued and analytic, except for poles of any finite number and multiplicity. Then the line integral of $F(s)$ around this boundary is related to the residues of those poles of $F(s)$ which are located inside the region as follows:

$$\oint F(s)\, ds = 2\pi j \sum \text{residues} \tag{7.2,1}$$

For example, if $F(s) = (s + 1)/(s - 1)(s - 2) = [-2/(s - 1)] + [3/(s - 2)]$, then $\int_{C_1} F(s)\, ds = -2(2\pi j)$, and $\int_{C_2} F(s)\, ds = (3 - 2)(2\pi j)$ (see Fig. 7.2-1a for definitions of C_1 and C_2).

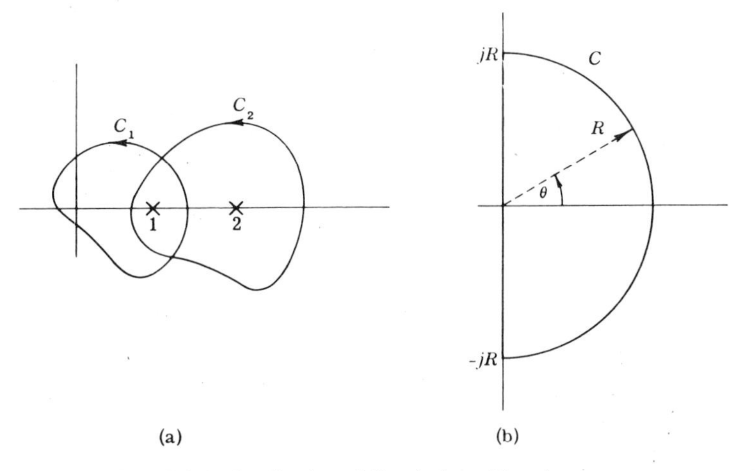

FIG. 7.2-1. Application of Cauchy's residue theorem.

As noted in the introduction to this chapter, we are interested in the value of loop gain achievable over finite frequency regions. This will require evaluation of integrals such as $\int_{j\omega_1}^{j\omega_2} F(s)\, ds$, and generally $F(s)$ is too complicated to evaluate the integral directly. It is often possible to let the limits go to infinity and we can then evaluate such integrals by applying Cauchy's residue theorem to the function or some suitable modification of it around the contour shown in Fig. 7.2-1b, letting $R \to \infty$. $\oint F(s)\, ds = 0$ if $F(s)$ has no singularities in the right half-plane and on the boundary. Also,

$$\oint F(s)\, ds = \lim_{R\to\infty} \int_{-jR}^{jR} F(s)\, j d\omega + \lim_{R\to\infty} \int_{\pi/2}^{-\pi/2} F(R, \theta) d(Re^{j\theta}) = I_1 + I_2 = 0.$$

If $s^x F(s) \to B_\infty'$, a constant, as $s \to \infty$ and $x = 1$, then I_2 is equal to $-j\pi B_\infty'$ and I_1 is then known. If $x > 1$, $I_2 = 0$; if $x < 1$, I_2 is infinite and I_1 is indeterminate. This technique can be modified to handle simple poles on the $j\omega$ axis and some singularities in the right half-plane. The basic idea is to find a contour which includes the $j\omega$ axis, such that Cauchy's residue theorem can be applied.

Application of Cauchy's theorem to multivalued functions. Consider a multivalued function such as $F(s) = (s - a)^{0.5}$. Write s-a in the form $s-a = Me^{j\theta}$ with M a positive number. Ordinarily, one may take $F(s)$ either as $|M^{0.5}| e^{j0.5\theta}$ or as $|M^{0.5}| e^{j0.5(\theta+2\pi)}$. One might attempt to make $F(s)$ single-valued (in order to be able to apply Cauchy's residue theorem to such a function), by arbitrarily defining $F(s)$ to have consistently the first of the above values. This would be satisfactory, if we were concerned only with such operations as addition, multiplication, or integration along an open curve (one that does not cross on itself and does not close on itself). However, it does not suffice for use in Cauchy's residue theorem, because the multivaluedness of $F(s)$ persists in contour integration.

Consider $\oint_{C_1}(s - b)^{0.5}\, ds$ for the contour of Fig. 7.2-2a. As s advances along

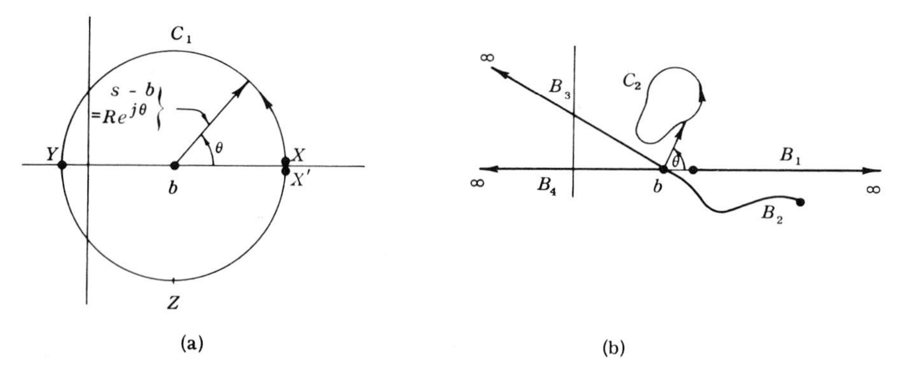

FIG. 7.2-2. (a) Contour for which Cauchy's theorem does not apply if $F(s) = (s - b)^{0.5}$. (b) Contour for which Cauchy's theorem applies, $F(s) = (s - b)^{0.5}$.

the contour, $\theta = \arg(s - b)$ increases from zero at X to π at Y, 1.5π at Z, and almost 2π at X'. Consequently $(s - b)^{0.5}$ approaches $R^{0.5}e^{j\pi}$ as s passes through X' and on to X. This is different from the original value of $(s - b)^{0.5} = R^{0.5}e^{j0}$ at the point X. Thus, when contour integration is being performed, $(s - b)^{0.5}$ remains multivalued. However, consider $\oint_{C_2}(s - b)^{0.5}\, ds$ for the contour of Fig. 7.2-2b. As s moves around the contour, $\arg(s - b)$ returns to its original value, and hence $(s - b)^{0.5}$ too returns to its original value. For such a contour (which does not enclose b), $F(s)$ is single-valued.

The conclusion is that Cauchy's residue theorem may be applied to $F(s) = (s - b)^{0.5}$ only for those contours which do not enclose the point b. To ensure that we do not forget this injunction, we create a barrier (called a branch cut) which must not be crossed by any contour. This branch cut must be chosen in such a manner that it prevents encirclement of the point b (called a branch point). Obviously, the barriers B_1, B_2 in Fig. 7.2-2b are unsatisfactory, because one can draw closed contours around b without crossing either barrier. However, the branch cuts B_3,B_4 in Fig. 7.2-2b are satisfactory.

Example. $F(s) = [(s - b)(s - c)]^{0.5}$. Consider the contour C_1 in Fig. 7.2-2c and note that as s moves around C_1 once, arg $(s - b)(s - c)$ increases by 4π. Hence, if $F(X) = Me^{j\theta}$, then as X' approaches X, $F(X')$ approaches $Me^{j(\theta+2\pi)}$, which is precisely the same as $F(X)$, since $e^{j2\pi} = 1$. Contour C_1 may therefore be used in Cauchy's residue theorem. Now consider C_2 (or C_3). Here arg $(s - b)(s - c)$ increases by 2π, so if $F(Y) = Ne^{j\alpha}$, after one encirclement $F(Y')$ approaches $Ne^{j(\alpha+\pi)} \neq F(Y)$. Hence contours C_2, C_3 are forbidden. Thus a suitable branch cut is B_1 in Fig. 7.2-2c with branch points at b and c.

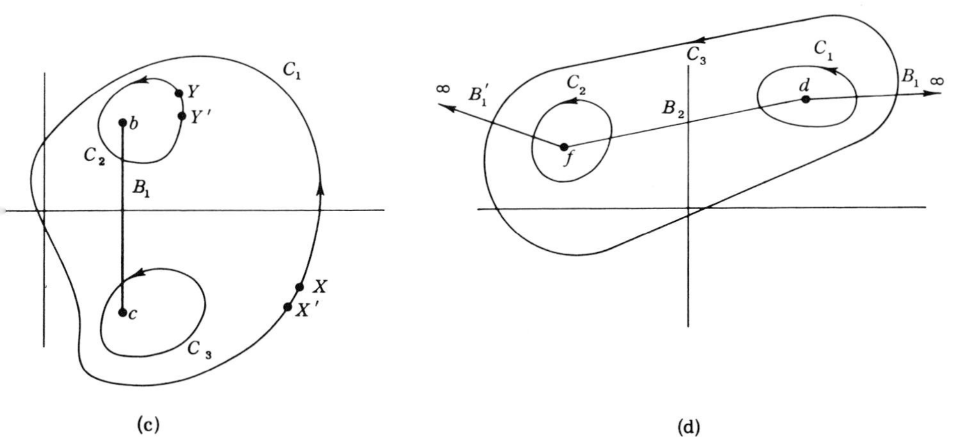

FIG. 7.2-2. (c) C_1 is satisfactory; C_2, C_3 unsatisfactory, $F(s) = [(s - b)(s - c)]^{0.5}$. (d) C_1, C_2, C_3 are forbidden, $F(s) = \ln (s - d)(s - f)$.

Example. $F(s) = \ln [(s - d)(s - f)] = \ln |(s - d)(s - f)| + j \arg (s - d)(s - f)$. Consider contours C_1, C_2 in Fig. 7.2-2d. As s moves once around either contour, arg $(s - d)(s - f)$ increases by 2π. Therefore the imaginary part of $F(s)$ increases by 2π. Also, in contour C_3 the increase is one of 4π in the imaginary part of $F(s)$. Hence none of these contours is permissible. Accordingly, a suitable branch cut among a double infinitude of possibilities is B_1B_1' in Fig. 7.2-2d.

Example. $F(s) = \ln [(s - d)/(s - f)]$. It is readily seen that contours C_1 and C_2 in Fig. 7.2-2d are not permissible, because arg $(s - d)(s - f)^{-1}$ increases by 2π, -2π, respectively. However, C_3 is permitted. Hence the branch cut B_2 in Fig. 7.2-2d is satisfactory, whereas it was unsatisfactory for $\ln (s - d)(s - f)$.

Functions of interest. In this chapter we are interested in loop transmission functions or their modifications. One property of such functions is that they can be regarded as transforms of impulse responses of real systems. The transform of a time functions is $F(s) = \int_{-\infty}^{\infty} f(t)e^{-st}\,dt$, so that $F(j\omega) = \int_{-\infty}^{\infty} f(t)\,[\cos \omega t - j \sin \omega t]\,dt = A(\omega) + jB(\omega)$. Hence $A(j\omega) = \int_{-\infty}^{\infty} f(t) \cos \omega t\,dt$ is clearly an even function of ω, while $-B(\omega) = \int_{-\infty}^{\infty} f(t) \sin \omega t\,dt$ is an odd function of ω. Thus

$$F(j\omega) = A(\omega) + jB(\omega) = \overline{F(-j\omega)} = A(-\omega) - jB(-\omega). \qquad (7.2,2)$$

It is easily seen that products and ratios of such functions have this same property, and that this is also true for $\ln F(s)$. In other words, any function considered in this chapter has the property that its real part is an even function of frequency and its imaginary part is an odd function of frequency.

It will often be necessary to focus attention on $F(s)$ for s near zero, and for s near infinity. The notation that will be used for the expansion of $F(s)$ about the origin is as follows:

$$F(s) = A_o + B_o' s + A_o'' s^2 + B_o''' s^3 + \dots \tag{7.2,3}$$

where all the A_o's and B_o's are real, in view of Eq. (7.2,2). Any poles of $F(s)$ at the origin will be removed and explicitly shown. The notation for the expansion of $F(s)$ at infinity is

$$F(s) = A_\infty + (B_\infty'/s) + (A_\infty''/s^2) + \dots \tag{7.2,4}$$

with all A_∞'s and B_∞'s real in view of Eq. (7.2,2). Any poles of $F(s)$ at infinity will be removed and explicitly shown.

All relations which will be derived in the balance of this chapter apply to what we call the *type* X function which is defined as follows: $F(s)$ is type X if it satisfies Eq. (7.2,2), is finite (or zero) at infinity, and is analytic in the right half-plane. If $F(s)$ has a singularity at $s = j\beta$ (β real), $F(s)$ is type X only if $\lim_{s\to j\beta}(s - j\beta)F(s) = 0$. All rational functions with no poles in the right half-plane or at ∞ or on the $j\omega$ axis, are type X. $F(s) = e^{-sT}f(s)$ with $f(s)$ type X is type X, and so is $F(s) = \ln g(s)$, providing $g(s)$ in neither zero nor infinite at infinity, and has no poles or zeros in the right half-plane. However, $g(s)$ may have poles and zeros of any multiplicity on the $j\omega$ axis.

Only some of the relations hold for the Y *type* function which has the form $F(s) = \ln h(s)$, with $h(s)$ zero or infinite at infinity, but analytic in the right half-plane. The restriction for its $j\omega$-axis behavior is the same as for the type X function.

§ 7.3 Resistance Integral Theorem and the Equality of Positive and Negative Feedback Areas

Consider $F(s)$ type X and apply Cauchy's theorem to $F(s) - F(\infty)$ over the right half-plane whose boundary is the $j\omega$ axis and the right half-infinite-semicircle (Fig. 7.2-1b). Recall that in expanding $F(s)$ at infinity we use the notation of Eq. (7.2,4). For $F(s)$ type X,

$$\int_C [F(s) - A_\infty]\, ds = 0$$

$$= \lim_{R\to\infty} \int_{-R}^{R} [F(j\omega) - A_\infty]\, j d\omega + \lim_{R\to\infty} \int_{+\pi/2}^{-\pi/2} [F(s) - A_\infty]\, d(Re^{j\theta})$$

$$= I_1 + I_2.$$

Let $F(j\omega) = A(\omega) + jB(\omega)$ and note that $A(\omega) = A(-\omega)$, and $B(\omega) = -B(-\omega)$. Consequently I_1 becomes $j2\int_0^\infty [A(\omega) - A_\infty]d\omega$. As s approaches infinity, the only term of $F(s) - A_\infty$ that contributes to I_2 is B_∞'/s, and with it I_2 becomes $\lim_{R\to\infty} \int_{\pi/2}^{-\pi/2} B_\infty' j d\theta = -j\pi B_\infty'$. [If we had used $F(s)$ rather than $F(s) - A_\infty$ as the integrand, I_2 would be infinite unless $A_\infty = 0$.] The result is

$$\int_0^\infty [A(\omega) - A_\infty]\, d\omega = 0.5\pi B_\infty'. \tag{7.3,1}$$

Equation (7.3,1) relates the real part of the frequency area of $F(s)$ (type X only) to the behavior of $F(s)$ at infinity. It is useful in problems involving maximization of power transfer, and for obtaining general gross properties of feedback systems.

Maximization of Power Transfer

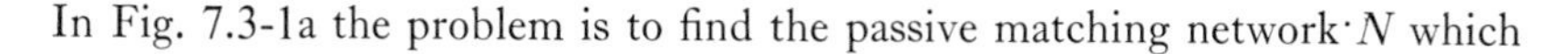

In Fig. 7.3-1a the problem is to find the passive matching network N which

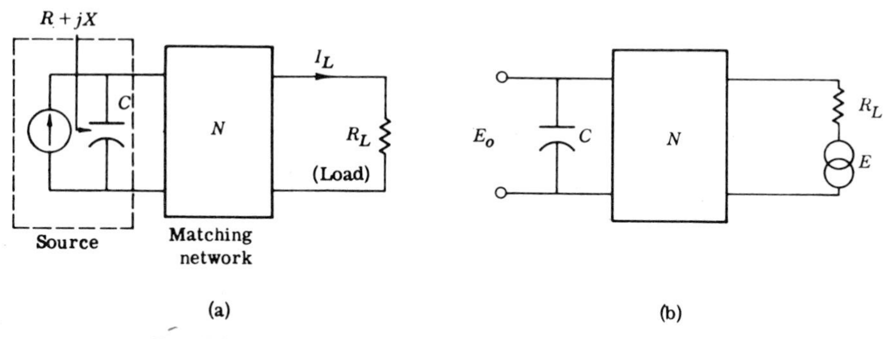

FIG. 7.3-1. Problem: find N which maximizes (a) I_L, (b) E_o.

maximizes the power transferred to R_L over some frequency band $\omega_2 - \omega_1$. The source has an unavoidable capacitance C.[1]

Assuming that the desired characteristics of N (whatever they may turn out to be) are realizable with N purely reactive, the power transferred to $R + jX$ is all delivered to R_L. We must therefore maximize $\int_{\omega_1}^{\omega_2} R(\omega)d\omega$, so we apply Eq. (7.3,1) to $R(j\omega)$. Here $A_\infty = R(\infty) = 0$, $(B_\infty')_{\max} = 1/C$, and Eq. (7.3,1) becomes

$$\left[\int_{\omega_1}^{\omega_2} R(\omega)d\omega\right]_{\max} = \frac{\pi}{2C} \tag{7.3,2}$$

[1] The reciprocity theorem permits us to interchange source and load. The latter may be taken as the current I_L. Therefore the results which follow also apply to the maximization of the voltage output E_o in Fig. 7.3-1b.

The maximum value is achievable only if R in Fig. 7.3-1a can be made zero for all frequencies outside the $\omega_2 - \omega_1$ band. This, in fact, is the function of N. Suppose also that P_d, the desired power transfer in this band, is some given function of frequency, $P_d = Kf(\omega)$, and K is to be maximized. Then the ideal $R(\omega)$ is $kf(\omega)$, and the scale factor k is determined by Eq. (7.3,2), since $\int_{\omega_1}^{\omega_2} f(\omega) d\omega = \pi/2kC$.

The problem is now entirely one of formal network synthesis. The input impedance $R + jX$ in Fig. 7.3-1a is completely determined by its real part (this is proven in Section 7.4), and the usual approximation[1] techniques may be used to approximate the ideal $R(\omega)$. Darlington's[2] method of synthesis can then be used to realize $Z = R + jX$ in the required form of a pure reactance network terminated by a resistance. It is guaranteed that after the shunt capacitance C is removed, the remainder is realizable, because the constraint imposed by the presence of C has been taken into account.

The Equality of Positive and Negative Feedback Areas

It is not possible to apply Eq. (7.3,1) directly to ln $L(s)$, with $L(s)$ zero at infinity, because ln $L(s)$ is then not type X. However, $F(s) = \ln [1 + L(s)]$ is type X, providing $L(s)$ has no poles in the right half-plane, and $1 + L(s)$ has no zeros there. However, $L(s)$ may include pure time lags, and it may have poles and zeros on the $j\omega$ axis, and zeros in the right half-plane. If $L(s)$ approaches zero as s approaches infinity, then $F(s) = \ln [1 + L(s)]$ approaches $L(s)$ as s approaches infinity. Almost every practical loop transmission function has at least two more poles than zeros (goes to zero as $1/s^e$ with $e \geqslant 2$). Hence, for such functions [using the notation of Eq. (7.2,4)] $A_\infty = B_\infty' = 0$. Therefore, applying Eq. (7.3,1),

$$\int_0^\infty A(\omega)\, d\omega = \int_0^\infty \ln | 1 + L(\omega) |\, d\omega = 0. \tag{7.3,3}$$

This means that in almost all feedback systems the net feedback area (in decibel-radians per second units) is zero. This is true for both nominal positive and negative feedback systems. The real difference between the two is that in the latter it is possible to concentrate the sensitivity reduction area (where $| 1 + L | > 1$) in a relatively small region where $| L |$ is large, and to spread the equal negative area (where $| 1 + L | < 1$) over a much larger arithmetic frequency range. The opposite is true of positive feedback. In the ideal (but non-existent) case when the loop transmission is finite at infinity, $\int_0^\infty A(\omega)\, d\omega$ is infinite. Some of the feedback systems that have been suggested as having

[1] E. A. Guillemin, "Synthesis of Passive Networks," Chapter 14. Wiley, New York, 1957.

[2] E. A. Guillemin, Chapter 9, Sections 6-10.

unusually wonderful sensitivity reduction properties are systems in which it is implicitly assumed that the loop transmission is finite at infinite frequency. In such systems, there is an infinite fund of loop gain-frequency area available, but such systems are unfortunately impossible to realize. If $L(s)$ approaches B_∞'/s as s approaches infinity, then $\int_0^\infty A(\omega)\,d\omega = \int_0^\infty \ln|1 + L(\omega)|\,d\omega = \pi B_\infty'/2$, and a finite nonzero net sensitivity reduction area is obtained. However, it is doubtful whether it is possible to build any useful feedback system in which $L(s)$ behaves in this manner at infinite frequency.

Unstable open-loop systems. Suppose the loop transmission $L(s)$ has poles in the right half-plane. Such poles constitute rhp singularities (branch points) of $F(s) = \ln[1 + f(s)]$. Cauchy's residue theorem can still be applied, providing we distort the boundary so as to exclude such singularities from the region of integration. In Fig. 7.3-2, where $L(s)$ has one pole on the positive real axis at $s = a$, the branch cut for that pole is chosen as shown in the figure (branch cuts for the lhp poles and zeros are chosen completely in the left half-plane, so they are of no concern). The contour used does not cross the branch cut. The only differences between this contour and that in Fig. 7.2-1b are due to the terms

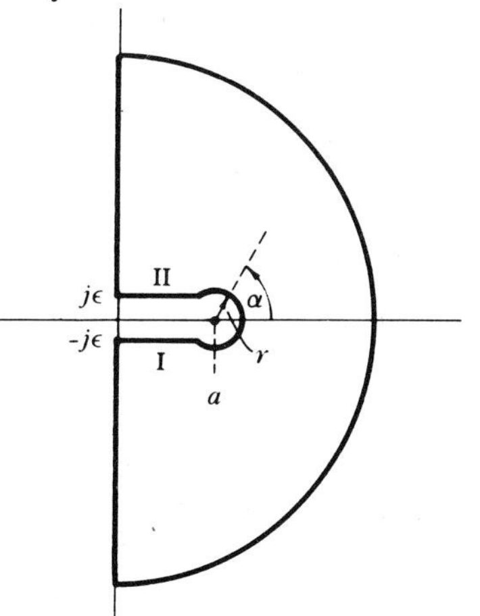

FIG. 7.3-2. Contour when $L(s)$ has rhp pole.

$$\lim_{\epsilon, r\to 0}\left\{\int_{\mathrm{I}} F(s)d\sigma + \int_{-\pi}^{\pi} F(s)d(re^{j\alpha}) + \int_{\mathrm{II}} F(s)d\sigma\right\} \triangleq I_1 + I_2 + I_3.$$

It is seen that $I_2 = 0$, because $\lim_{r\to 0} r \ln r = 0$, and $I_1 = -I_3$.[1] Hence, for this case (and in general for any number of rhp poles), the net logarithmic feedback-frequency area is zero.

§ 7.4 Real or Imaginary Part Sufficiency

In this section, it is shown that if the real or imaginary part of $F(s)$ is known over the complete frequency range, this suffices to determine $F(s)$. Suppose $F(j\omega) = A(\omega) + jB(\omega)$, and $A(\omega)$ is known. Cauchy's residue theorem is applied to a suitable modification of $F(s)$ around the usual contour. In order that the

[1] Along I in Fig. 7.3-2, $\ln[1 + L(s)] = \ln|1 + L(s)| + j\arg[1 + L(s)] = A_1(\sigma) + j\arg[1 + L_1(\sigma)/|\sigma - a|e^{-j\pi}] = A_1 + j\theta_1$ [where $L(s) = L_1(s)/(s - a)$]. Along II, $\ln[1 + L(s)] = A_1 + j\arg[1 + L_1(\sigma)/|\sigma - a|e^{j\pi}] = A_1 + j\theta_1$ [because $L_1(s)$ is a real number]. Therefore I_1 and I_3 cancel each other out.

value of $F(s)$ at a particular frequency, $\omega = \omega_x$, may appear in the final result, it is necessary to create a pole of the integrand at that frequency. Thus, let the integrand be $F(s)/(s - j\omega_x)$, and the contour be that shown in Fig. 7.4-1. Applying the residue theorem,

$$\oint \frac{F(s)}{s - j\omega_x} ds = \lim_{\substack{r\to 0 \\ R\to\infty}} \left\{ \int_{-jR}^{j(\omega_x - r)} \frac{F(s)}{s - j\omega_x} ds + F(j\omega_x) \int_{-\pi/2}^{\pi/2} \frac{d(re^{j\alpha})}{re^{j\alpha}} \right.$$

$$\left. + \int_{j(\omega_x + r)}^{jR} \frac{F(s)}{s - j\omega_x} ds + \int_{\pi/2}^{-\pi/2} \frac{A_\infty}{R'e^{j\theta'}} d(Re^{j\theta}) \right\} = \sum_1^4 I_i = 0.$$

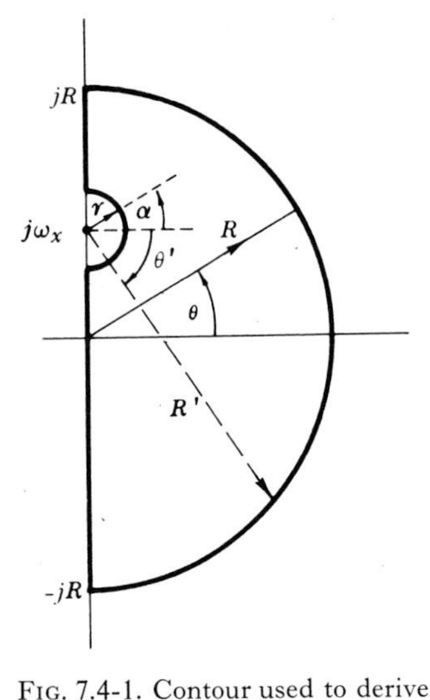

FIG. 7.4-1. Contour used to derive relation between $A(\omega)$ and $B(\omega)$.

It is readily found that $I_2 = j\pi[A(\omega_x) + jB(\omega_x)]$, $I_4 = -j\pi A_\infty$. Hence the above becomes $\int_{-\infty}^{\infty} (A + jB) d\omega/(\omega - \omega_x) + j\pi[A(\omega_x) + jB(\omega_x) - A_\infty] = 0$. Equating the reals and imaginaries, replacing ω_x by ω, and using ζ as the dummy variable, there results

$$\pi[A(\omega) - A_\infty] = \int_{-\infty}^{\infty} \frac{B(\zeta)d\zeta}{\omega - \zeta}; \quad (7.4,1a)$$

$$-\pi B(\omega) = \int_{-\infty}^{\infty} \frac{A(\zeta)d\zeta}{\omega - \zeta}. \quad (7.4,1b)$$

Since $\int_{-\infty}^{\infty} d\zeta/(\omega - \zeta) = 0$, an arbitrary constant can be added to the above integrands, without affecting the final results.

The above may be repeated with $F(s)/(s + j\omega_x)$ as the integrand, resulting in $\pi[A(-\omega) - A_\infty] = \int_{-\infty}^{\infty} B(\zeta)d\zeta/(\omega + \zeta)$; $\pi B(-\omega) = \int_{-\infty}^{\infty} A(\zeta)d\zeta/(\omega + \zeta)$. The latter may be combined with Eq. (7.4,1), using the fact that $A(\omega) = A(-\omega)$, and $B(\omega) = -B(-\omega)$. This leads to

$$B(\omega) = \frac{-2\omega}{\pi} \int_0^{\infty} \frac{A(\zeta)d\zeta}{\omega^2 - \zeta^2}; \quad (7.4,2a)$$

$$A(\omega) - A_\infty = \frac{2}{\pi} \int_0^{\infty} \frac{\zeta B(\zeta)d\zeta}{\omega^2 - \zeta^2}. \quad (7.4,2b)$$

Since $\int_0^{\infty} d\zeta/(\omega^2 - \zeta^2) = 0$, any arbitrary constant can be added to $A(\zeta)$ or $\zeta B(\zeta)$ without affecting the resulting $B(\omega)$ or $A(\omega)$, respectively.

Equation (7.4,2a) also applies to type Y functions [i.e., $F(s) = \ln f(s)$], providing $f(s)$ has no singularities in the interior of the right half-plane, but poles and zeros on the $j\omega$ axis and at infinity are permitted. This can be verified by deriving Eq. (7.4,2a), by applying the residue theorem to the integrand $F(s)/(s^2 + \omega_x^2) = [\ln f(s)]/(s^2 + \omega_x^2)$ around the usual contour indented at both $j\omega_x$ and $-j\omega_x$. Equation (7.4,2b) applies to functions with logarithmic singularities on the $j\omega$ axis, but not at infinity. This is seen from the fact that it does not converge at infinity if B is finite [and therefore $A(s)$ is infinite] there. If $F(s)/s$ is used in place of $F(s)$, and the contour suitably modified at $s = 0$, the result is

$$\pi[A(\omega) - A_o] = 2\omega^2 \int_0^\infty \frac{B\zeta^{-1}d\zeta}{\omega^2 - \zeta^2}. \tag{7.4,2c}$$

This last relation is useful for functions with logarithmic singularities at infinity, but not at the origin.

These equations may be regarded as constraints on the integrals of $A(\omega)$ and $B(\omega)$ over frequency. If either one is specified at any one frequency, there is a corresponding constraint on the integral of the other. They are useful for obtaining $B(\omega)$ or $A(\omega)$ when the appropriate integrands are known either analytically or graphically. For graphical work, Guillemin's[1] method is useful because it readily lends itself to very high order accuracy.

Guillemin's Method

The method for evaluating $B(\omega)$ graphically from a known $A(\omega)$ will be developed. Only the results will be stated for the opposite case.

In Eq. (7.4,1b), $-\pi B(\omega) = \int_{-\infty}^{\infty} A(\zeta)d\zeta/(\omega - \zeta)$. Let $\eta = \omega - \zeta$, and it is found that $-\pi B(\omega) = \int_{-\infty}^{\infty} A(\omega - \eta)d\eta/\eta$. Note that $B(\omega)$ is simply the convolution of $A(\omega)$ with $1/\omega$; hence the variables in A and $1/\omega$ may be interchanged. Also, $-\pi B'(\omega) = \int_{-\infty}^{\infty} A'(\omega - \eta)d\eta/\eta = \int_{-\infty}^{\infty} A'(\zeta)d\zeta/(\omega - \zeta)$. Integrating the latter with respect to ω, we get $-2\pi B(\omega) = \int_{-\infty}^{\infty} A'(\zeta) \ln (\omega - \zeta)^2 d\zeta = \int_{-\infty}^{\infty} A'(\omega - \zeta) \ln \zeta^2 d\zeta$. By continuing as in the above, i.e., by differentiation with respect to ω, interchange of the variables followed by integration with respect to ω, there results

$$-2\pi B(\omega) = \int_{-\infty}^{\infty} A''(\zeta)[(\omega - \zeta) \ln (\omega - \zeta)^2 - 2(\omega - \zeta)]d\zeta. \tag{7.4,3}$$

For example, consider $A(\omega)$ in Fig. 7.4-2a. $A(\omega)$ may be approximated by the straight line segments shown, and this linear approximation is then differen-

[1] E. A. Guillemin, "Synthesis of Passive Networks," Chapter 8. Wiley, New York, 1957.

tiated twice until the final result consists of the series of impulses of Fig. 7.4-2c. Equation (7.4,3) applies and is easily integrated, leading to

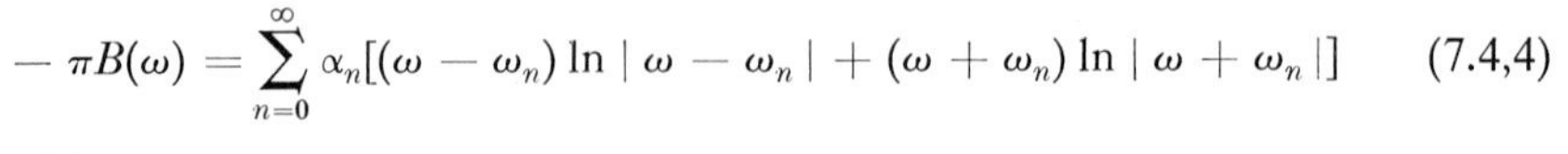

$$-\pi B(\omega) = \sum_{n=0}^{\infty} \alpha_n[(\omega - \omega_n) \ln | \omega - \omega_n | + (\omega + \omega_n) \ln | \omega + \omega_n |] \qquad (7.4,4)$$

with

$$\alpha_n = m_n - m_{n-1}, \qquad m_n = V_{n+1} - V_n, \qquad V_n = A(\omega_n). \qquad (7.4,5)$$

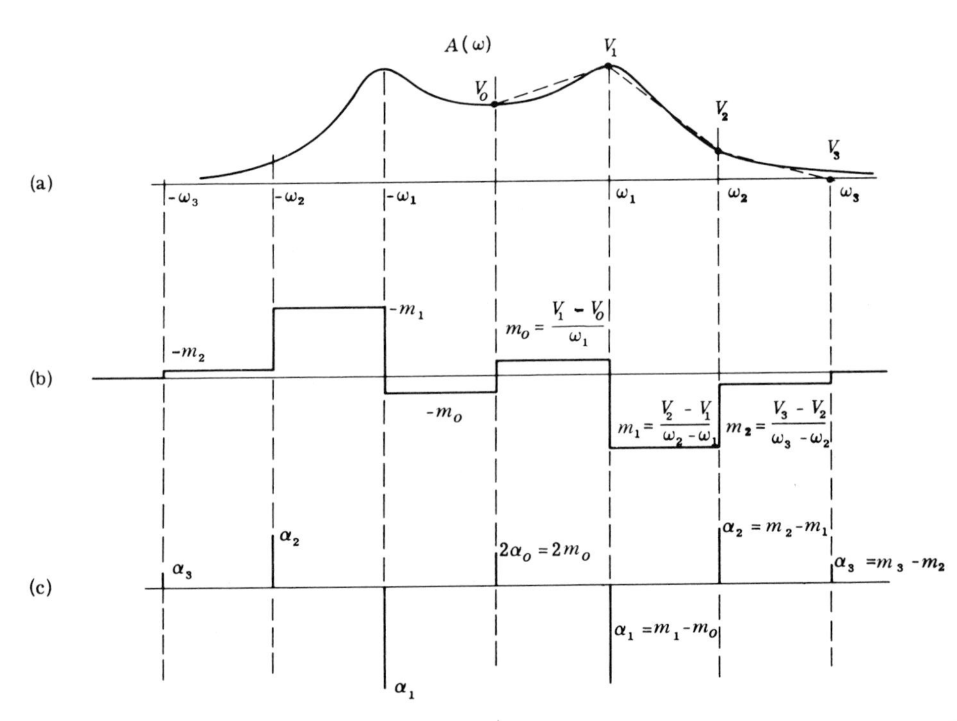

FIG. 7.4-2. Successive differentiation in Guillemin's method.

The $\omega \Sigma \alpha_n$ term drops out because $\alpha_n = m_n - m_{n-1}$, so that in such a summation only the last m remains, and it can always be taken as zero because, as has previously been noted, any arbitrary constant can be added to $A(\omega)$ without affecting the $B(\omega)$ that is thereby obtained.

One can continue in this manner with higher order approximations, which are better only if the original function has the characteristic of a higher order curve (see Section 12.8). After the segments have been chosen, there is no need to actually find the values of the various derivatives. One works directly with the V's, as in Eq. (7.4,5), where the α's are given in terms of the V's. The latter are read directly from the graph of $A(\omega)$. However, it is necessary to evaluate the $x \ln x$ quantities accurately, because the final result at any frequency may be due to the subtraction of large numbers fairly close to each other.

To find $A(\omega)$ from a known $B(\omega)$, one may, in a similar manner, obtain the following results: For a linear approximation of $B(\omega)$,

$$\pi[A(\omega) - A_\infty] = \sum_{n=1}^{\infty} \alpha_n \left[(\omega - \omega_n) \ln |\omega - \omega_n| - (\omega + \omega_n) \ln |\omega + \omega_n|\right] \tag{7.4,6}$$

with

$$\alpha_n = m_n - m_{n-1}, \qquad m_n = [V_{n+1} - V_n]/(\omega_{n+1} - \omega_n). \tag{7.4,7}$$

In Eq. (7.4,6) it is assumed that $B(0)$ and $B(\infty)$ are zero.

Example 1. The phase of $f(j\omega)$ in radians, which is the imaginary part of $F(j\omega) = \ln f(j\omega)$, is given in Fig. 7.4-3a. The real part of $F(j\omega)$, which is

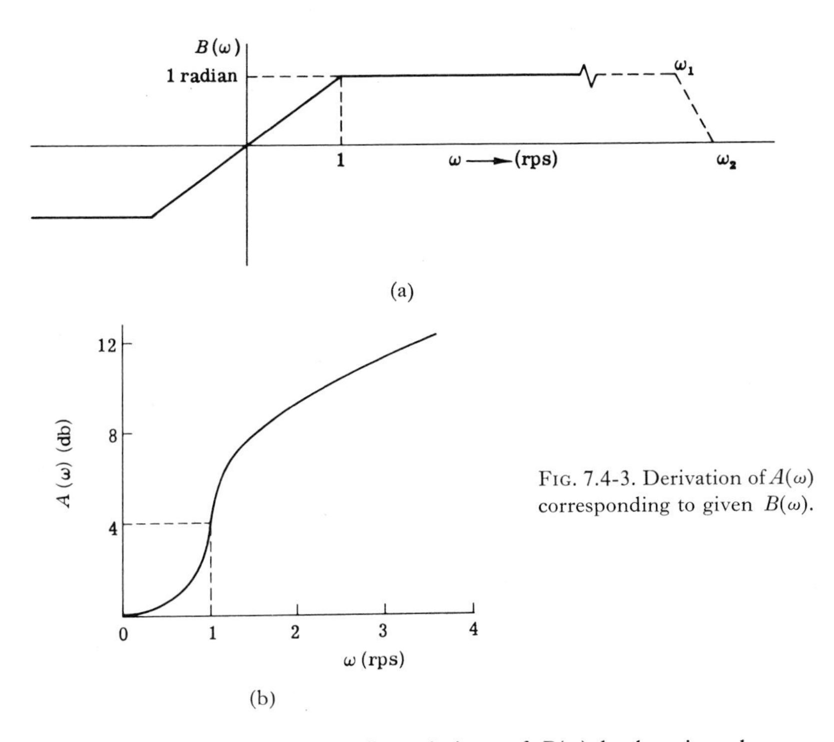

FIG. 7.4-3. Derivation of $A(\omega)$ corresponding to given $B(\omega)$.

$\ln |f(j\omega)|$, is to be found. Two differentiations of $B(\omega)$ lead to impulses, so Eq. (7.4,6) is used, with $m_0 = 1$, $m_1 = 0$, $\alpha_1 = -1$, $-\pi[A(\omega) - A_\infty] = [(\omega - 1) \ln |\omega - 1| - (\omega + 1) \ln |\omega + 1|]$ nepers. To convert to decibels, multiply by 20/2.303. The result is sketched in Fig. 7.4-3b, and is correct despite the fact that $B(\infty) \neq 0$. This can be explained as follows: In Fig. 7.4-3a, imagine that at very high frequencies we let $B(\omega)$ go down to zero at ω_1, ω_2 as shown by the dotted lines in the figure. It is readily seen that the contributions of the resulting impulses at $\pm\,\omega_1$, $\pm\,\omega_2$ tend toward a constant for $\omega_1, \omega_2 \gg \omega$, and this inequality can be made as strong as we please. Since the real part is in any case specified within an arbitrary constant by the phase, this constant quantity due to the impulses at $\pm\,\omega_1$, $\pm\,\omega_2$ may be ignored.

The above example can be generalized to some extent. If the slope at the origin in Fig. 7.4-3a was k in place of 1, the real part scale should also be multiplied by k. Also, the frequency scale in the figure can be considered normalized, i.e., ω/ω_o instead of ω and then the frequency scale of Fig. 7.4-3b is also ω/ω_o in place of ω.

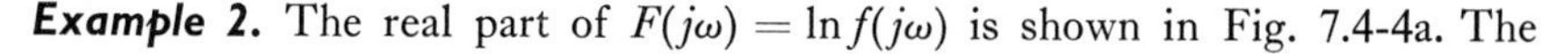

Example 2. The real part of $F(j\omega) = \ln f(j\omega)$ is shown in Fig. 7.4-4a. The

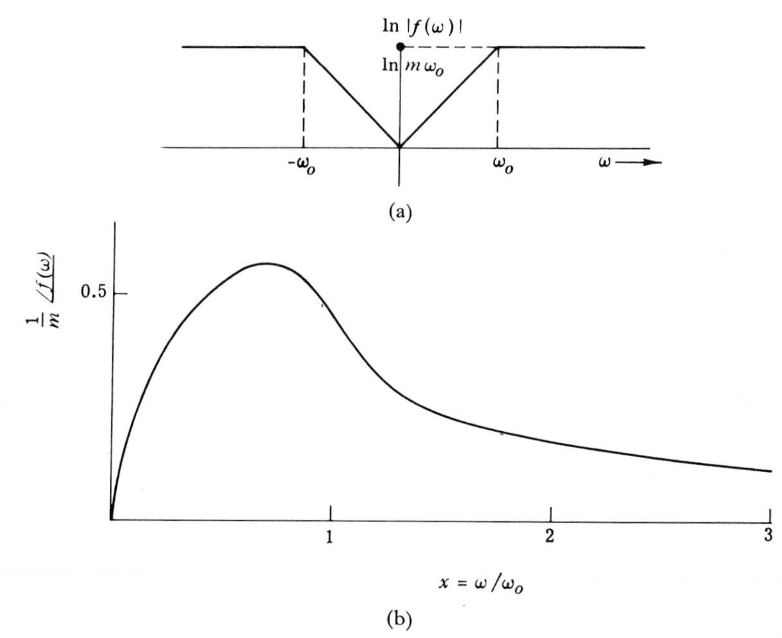

FIG. 7.4-4. Derivation of $B(\omega)$ corresponding to given $A(\omega)$.

corresponding $\mathscr{I}\mathrm{m}F(j\omega)$ [i.e., $\arg f(j\omega)$] is to be found. This characteristic is not zero at infinite frequency, but we may subtract from it the constant quantity $m\omega_o$, so that the resulting function is zero at infinity. However, this step need not be carried out because there is no difference between the derivatives of the two characteristics. Using Eq. (7.4,4), $\pi B(\omega) = -\,m\,\{2\omega \ln \omega - [(\omega - \omega_o) \ln |\,\omega - \omega_o\,| + (\omega + \omega_o) \ln |\,\omega + \omega_o\,|]\}$ which, if we let $\omega = x\omega_o$, can also be written as $\pi B(\omega) = -\,m\omega_o\,\{2x \ln x - [(x - 1) \ln |\,x - 1\,| + (x + 1) \ln (x + 1)]\}$. The latter is sketched in Fig. 7.4-4b. (If A is in nepers, B is in radians.)

Bode's Method

Guillemin's graphical procedure is not convenient for the calculation of $\mathscr{I}m\, F(s) = \mathscr{I}m \ln f(s)$ from $\mathscr{R}e\, F(s) = \ln |f(s)|$, when $f(s)$ is zero at infinity. In Guillemin's procedure, $\mathscr{R}e\, F(s)$ must be finite at infinity. Bode has suggested a procedure which is convenient both for $F(s)$ a logarithmic function (type Y), and for $F(s)$ type X.

We can use Eqs. (7.4,2a, b) as starting points (they apply for both type X and type Y functions) and manipulate[1] them into the form

$$\pi B(\omega_x) = \int_{-\infty}^{\infty} \frac{dA}{du} \ln \coth 0.5\,|u|\,du \tag{7.4,8a}$$

where

$$u = \ln \omega/\omega_x; \tag{7.4,8b}$$

$$\pi[A(\omega_x) - A_\infty] = -\frac{1}{\omega_x} \int_{-\infty}^{\infty} \frac{d(\omega B)}{du} \ln \coth 0.5\,|u|\,du \tag{7.4,9a}$$

$$= -\,\omega_x \int_{-\infty}^{\infty} \frac{d(B/\omega)}{du} \ln \coth 0.5\,|u|\,du. \tag{7.4,9b}$$

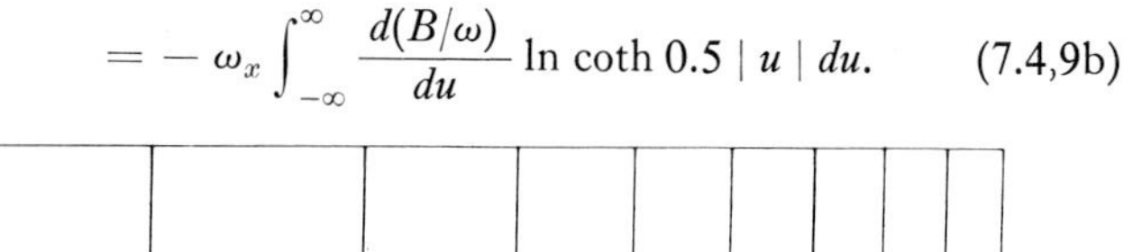

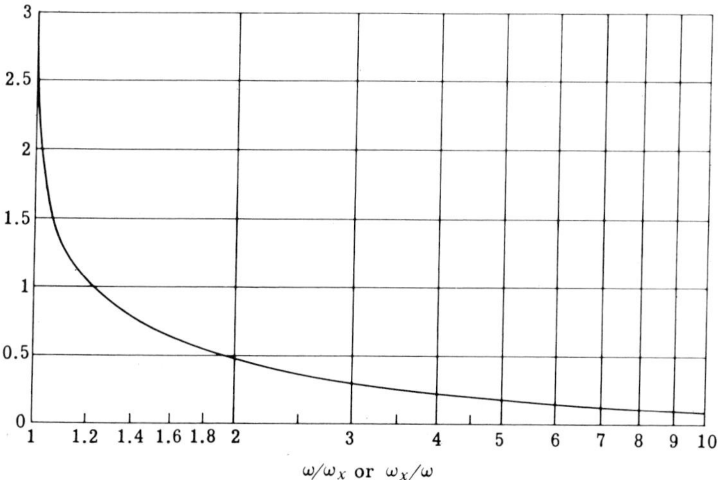

FIG. 7.4-5. Phase shift in radians due to an abrupt change in gain of 0.434π nepers (11.84 db) at $\omega = \omega_x$. $\log \coth |u|/2 = 0.434 \ln \coth |u|/2$; $u = \ln \omega/\omega_x$.

The above equations are interesting for their physical interpretation, and for their usefulness in revealing how one set of charts may apply for finding A from B and the reverse. As regards physical interpretation, note in Eq. (7.4,8a)

[1] H. W. Bode, "Network Analysis and Feedback Amplifier Design," pp. 313 and 320. Van Nostrand, Princeton, New Jersey, 1945.

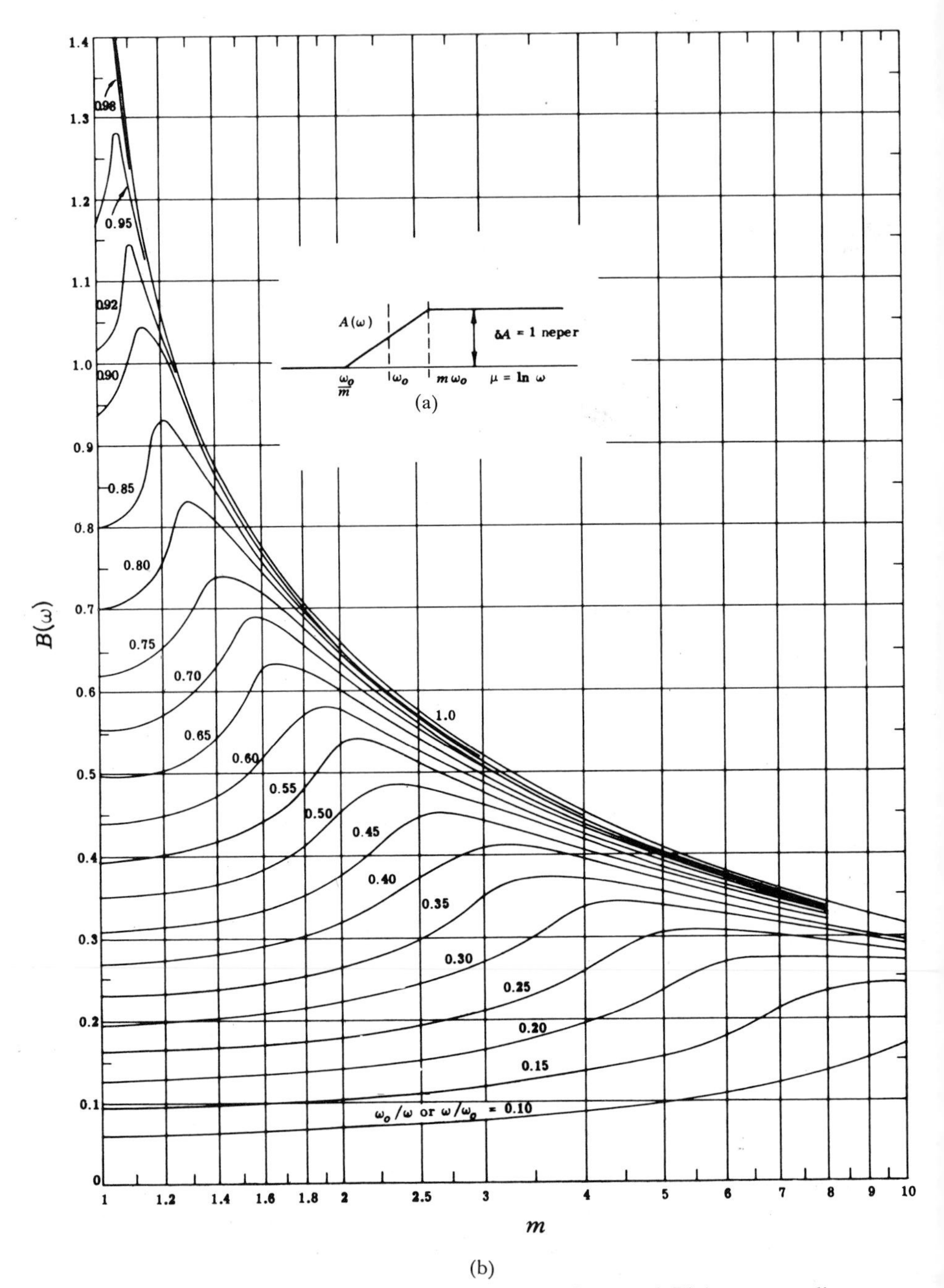

(b)

FIG. 7.4-6. (a) The finite line $A(\omega)$ segment. (b) Curves of $B(\omega)$ corresponding to finite line $A(\omega)$ segment as functions of m with ω_o/ω (or ω/ω_o) as parameter.

that it is the rate of change (on a logarithmic frequency scale) of the real part of $F(s)$ [or of the logarithmic magnitude of $f(s)$ when $F(s) = \ln f(s)$] which determines the imaginary part of $F(s)$ [or the phase of $f(s)$]. Also, $\ln \coth |u|/2$, which is sketched in Fig. 7.4-5, acts as a weighting function. While dA/du is independent of ω_x, the weighting function is a function of ω_x. It has, in fact, its infinite peak centered at $\omega = \omega_x$, so that the frequencies closest to the point in question are weighted the most, while those some distance away have little effect. In short, the imaginary part of $F(j\omega)$ at any point is essentially determined by the "rate of change" of the real part in the neighborhood (a few octaves on either side) of the point. If $dA/du = 0$ everywhere except at ω_1, where A changes abruptly by the amount ΔA, then, from Eq. (7.4,8a), $B(\omega_x) = (\Delta A/\pi) \ln \coth (0.5\omega_1/\omega_x)$ and is directly available from Fig. 7.4-5.

Suppose charts exist which enable us to find B from A. Then [from Eqs. (7.4,8) and (7.4,9)] to use these charts to find A from B, one must first calculate ωB (or B/ω), use the charts in exactly the same manner substituting ωB (or B/ω) for A, and then multiply the results by $-1/\omega$ (or $-\omega$ if $1/\omega$ is used originally as the multiplier). The final result is $A(\omega)$ within an arbitrary constant. The curves in Fig. 7.4-6b[1] give $B(\omega)$ that goes with an $A(\omega)$ characteristic, which

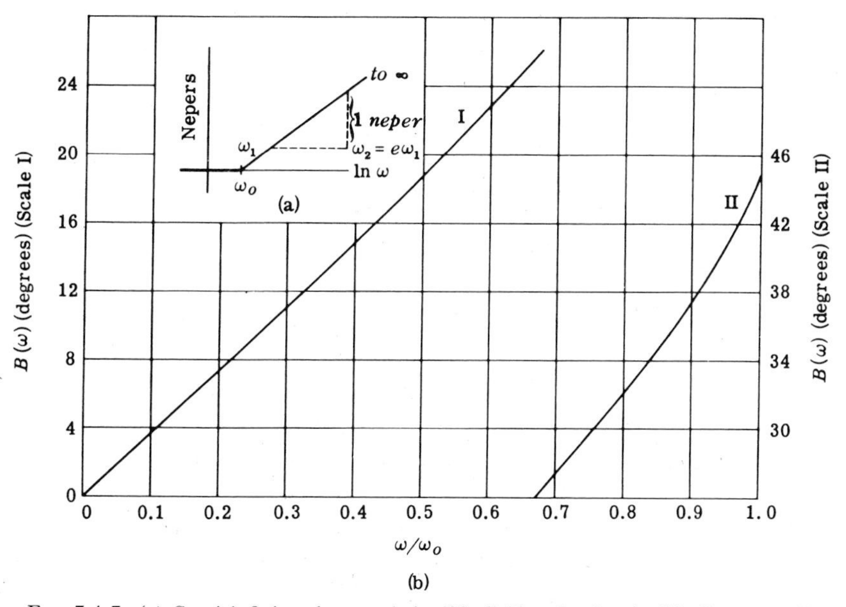

FIG. 7.4-7. (a) Semi-infinite characteristic (20 db/decade slope). (b) Corresponding $B(\omega)$ characteristic.

[1] The curves were prepared from data originally due to H. W. Bode, "Network Analysis and Feedback Amplifier Design," Chapter 15, Van Nostrand, Princeton, New Jersey, 1945; and to D. E. Thomas, Tables of phase associated with a semi-infinite unit slope attenuation. *Bell System Tech. J.* **26**, 870-899 (1947).

Bode calls the *finite line segment*, shown in Fig. 7.4-6a. This segment is completely described by the parameters ω_0, m, δA, but the first of these is immaterial because it only indicates where the resulting curve belongs on the u axis. Also, the $B(\omega)$ that results is directly proportional to δA, so the charts have been prepared for $\delta A = 1$. However, it is necessary to have a set of curves for different values of m. Figure 7.4-6b presents the data as a function of m, with ω/ω_o as a parameter. When the charts are applied to logarithmic functions, then the ordinate represents radians (1 radian $= 57.3°$). The unit value of δA is 1 neper, which corresponds to 8.686 db. In applying the charts to rational functions, $B(\omega)$ is obtained in the same units as $A(\omega)$. The curves are given for $\omega/\omega_o < 1$; the data for $\omega/\omega_o > 1$ are available from the fact that $B(\omega/\omega_o) = B(\omega_o/\omega)$.

Another $A(\omega)$ characteristic which is useful is the *semi-infinite characteristic* of Fig. 7.4-7a. The unit slope here is 1 neper for $\omega_2/\omega_1 = e$. This corresponds to a slope of 8.686 db for a frequency ratio of $e = 2.718$, which is equivalent to a slope of 20 db per decade (or 6 db per octave). The resulting $B(\omega)$ characteristic (in degrees) for unit slope[1] is given in Fig. 7.4-7b for $\omega/\omega_o < 1$. The values for $\omega/\omega_o > 1$ are obtained from the relation $B(\omega/\omega_o) = 90° - B(\omega_o/\omega)$.

Example. As an exercise in the application of these charts, we shall find the imaginary part corresponding to the real part of $\ln [k/(s + 0.4)(s + 2.5)]$. Here $A(\omega) = \ln k - \ln |(j\omega + 0.4)(j\omega + 2.5)|$, and is plotted in the usual decibel scale in Fig. 7.4-8. The value of k is immaterial, so $A(0)$ is taken as zero.

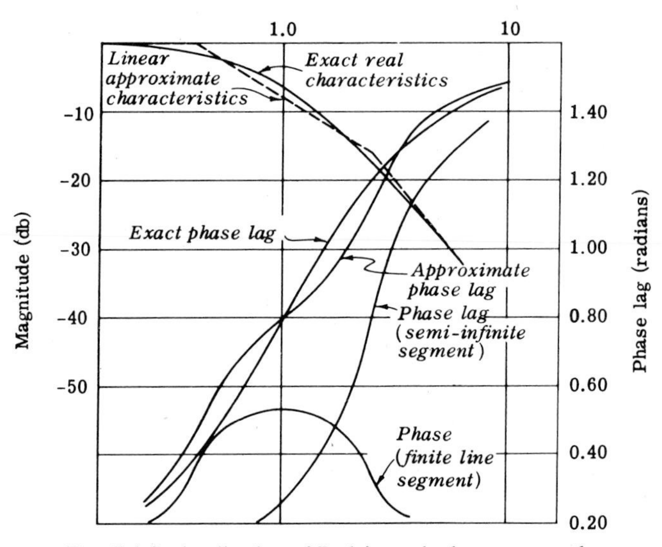

FIG. 7.4-8. Application of Bode's method to an example.

[1] For a very accurate table of values, see D. E. Thomas, Tables of phase associated with semi-infinite unit slope attenuation. *Bell System Tech. J.* **26**, 870-899 (1947).

$A(\omega)$ is approximated by three straight lines with break points at 0.4 and 2.5 rps, so we have one finite line segment (Fig. 7.4-6), with $\omega_o = [(0.4)\,(2.5)]^{1/2} = 1.0$, $m = -[(2.5)/(0.4)]^{1/2} = -2.5$, and one semi-infinite characteristic ($\omega_o = 2.5$) with 6 db per octave, or one unit negative slope. The individual components and their sum are shown in Fig. 7.4-8. The maximum error is approximately 6°.

§ 7.5 Relation between Loop Transmission Lag Angle and Optimum Loop Transmission Function

We have seen how the benefits of feedback in two-degree-of-freedom systems are determined by $(1 + L)$, or by the loop transmission L when $|L| \gg 1$. In a practical feedback control problem, the level of $|L|$ and its functional dependence on frequency may therefore be dictated by the required benefits of feedback, from zero frequency up to a specific frequency ω_o. We have also seen in Section 6.14 that in the frequency range $\omega > \omega_o$, it is extremely important to reduce $|L|$ as fast as possible. In this section we shall deduce an important property of such an optimum $L(j\omega)$.

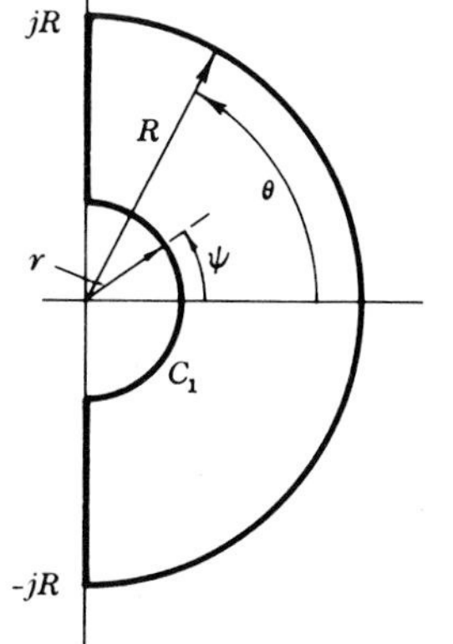

FIG. 7.5-1. Contour for $F(s)/s$.

We apply Cauchy's theorem to $F(s)/s$ over the contour of Fig. 7.5-1. In the notation of Eqs. (7.2,3) and (7.2,4),

$$\oint \frac{F(s)}{s} ds = \lim_{\substack{r \to 0 \\ R \to \infty}} \int_r^R \frac{F(j\omega) d\omega}{\omega} + \int_{\pi/2}^{-\pi/2} \frac{A_\infty}{Re^{j\theta}} d(Re^{j\theta}) + \int_{-R}^{-r} \frac{F(j\omega)}{\omega} d\omega$$

$$+ \int_{-\pi/2}^{\pi/2} \frac{A_o}{re^{j\psi}} d(re^{j\psi}). \qquad (7.5,1)$$

When the first and third terms of Eq. (7.5,1) are combined, the odd parts cancel and the even parts add, so that, writing $F(j\omega) = A(\omega) + jB(\omega)$, Eq. (7.5,1) becomes

$$-2 \int_{-\infty}^{\infty} B(\omega) d \ln \omega = \pi(A_o - A_\infty). \qquad (7.5,2)$$

The application of Eq. (7.5,2) to $F(s) = e^{-sT}$, where $A_o = 1$, $A_\infty = 0$, and $B(\omega) = -\sin \omega T$, leads to $\int_0^\infty (\sin \omega T/\omega) d\omega = \pi/2$. Another application is to Figs. 5.8-1a and b where $\int_{-\infty}^{\infty} \theta \, d \ln \omega = 0$, because $A_\infty = A_o$.

In order to apply Eq. (7.5,2) to the problem postulated at the beginning of the section, it is necessary that $A_\infty \triangleq \lim_{\omega \to \infty} \ln |L(j\omega)|$, and $A_o \triangleq \lim_{\omega \to 0} \ln$

$|L(j\omega)|$ are both finite. Therefore, at very large and (if necessary) at very small frequencies, we modify $L(j\omega)$, so that A_∞ and A_o are finite. This is shown by the dashed lines in Fig. 7.5-2. We know from experience (and from Fig.

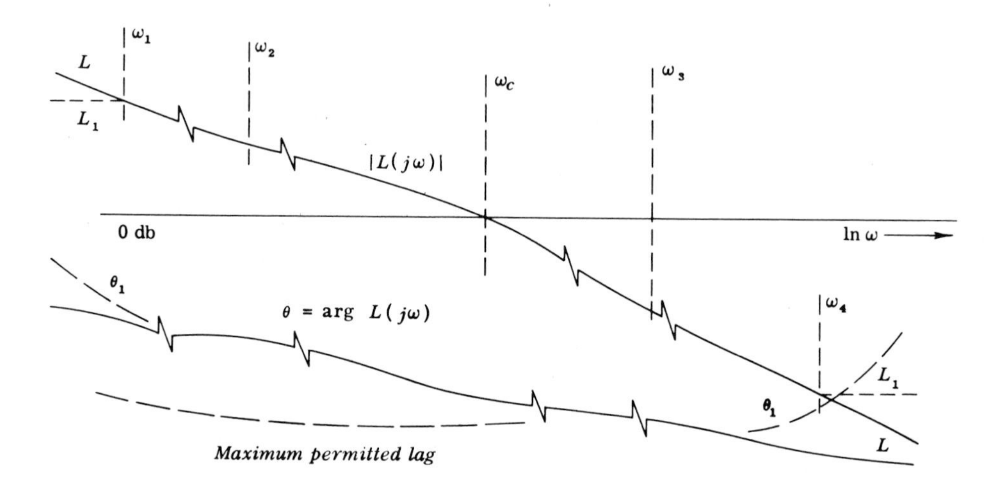

FIG. 7.5-2. Derivation of optimum phase lag curve.

7.4-7) that the effect of these modifications on the phase of $|L(j\omega)|$ is concentrated near ω_1 and ω_4 and on $\omega < \omega_1$, $\omega > \omega_4$. Therefore, by choosing ω_1 and ω_4 sufficiently small and large, respectively, we are sure that the phase angle in the region $\omega_2 < \omega < \omega_3$ is unaffected by the modifications. Thus, if $\lim_{s\to\infty} L(s) = k_1 s^{-n}$, and $\lim_{s\to 0} L(s) = k_2 s^{-m}$, we could pick $L_1(s) = L(s)s^m(s + \omega_4)^n / \omega_4^n(s + \omega_1)^m$. Equation (7.5,2) is applied to $L_1(s)$, whose $A_\infty = \ln k_1\omega_4^{-n}$, $A_o = \ln k_2\omega_1^{-m}$, $B(\omega) = \theta(\omega) + 0.5m\pi - [m \tan^{-1} \omega/\omega_1] + n \tan^{-1} \omega/\omega_4$, where $\theta(\omega) = \arg L(j\omega)$. Equation (7.5,2) becomes

$$-2\int_{-\infty}^{\infty} \left[\theta(\omega) + 0.5m\pi - m \tan^{-1}\frac{\omega}{\omega_1} + n \tan^{-1}\frac{\omega}{\omega_4}\right] d \ln \omega$$
$$= \pi[\ln k_2\omega_1^{-m} - \ln k_1\omega_4^{-n}].$$

In this equation it has been postulated that all terms are fixed, except $\theta(\omega)$ and k_1. The faster $|L(j\omega)|$ decreases, the smaller is k_1, and it is readily seen that a minimum value of k_1 is secured by minimizing $\theta(\omega)$. Therefore, whatever freedom is available should be used to make the phase lag of $L(j\omega)$ as large as possible. The design cannot be completed as yet, because we do not know just what freedom is available to the designer in this problem where $|L(j\omega)|$ has been specified for $0 < \omega < \omega_o$. Section 7.6 is devoted to determining what this freedom is.

§ 7.6 Specification of F(s) from Its Real and Imaginary Parts in Different Frequency Ranges

In this section, we shall show that if $A(\omega)$ is specified in one frequency range, and $B(\omega)$ for all other ω, then $F(s)$ is completely determined. To do this, let $F(s) = G(s)/(s^2 + \omega_x{}^2)^{0.5}$, with branch points at $\pm j\omega_x$. The branch cuts may be located entirely in the left half-plane, so they are of no concern. It is readily determined that, for $s = j\omega$ and $|\omega| < \omega_x$, $F(j\omega) = [G_1(\omega) + jG_2(\omega)]/(\omega_x{}^2 - \omega^2)^{0.5} = A + jB$. For $\omega > \omega_x$, $(s^2 - \omega^2) = (s + j\omega_x)(s - j\omega_x) = j(\omega + \omega_x)j(\omega - \omega_x)$; so $(s^2 - \omega^2)^{-0.5} = -j(\omega^2 - \omega_x{}^2)^{-0.5}$, and hence $F(j\omega) = (G_2 - jG_1)/(\omega^2 - \omega_x{}^2)^{0.5}$. Therefore, for $0 < \omega < \omega_x$,

$$A(\omega) = G_1(\omega)/(\omega_x{}^2 - \omega^2)^{0.5}, \qquad B(\omega) = G_2(\omega)/(\omega_x{}^2 - \omega^2)^{0.5}; \quad (7.6,1a)$$

for $\omega > \omega_x$,

$$A(\omega) = G_2(\omega)/(\omega^2 - \omega_x{}^2)^{0.5}, \qquad B(\omega) = -G_1(\omega)/(\omega^2 - \omega_x{}^2)^{0.5}. \quad (7.6,1b)$$

Equation (7.4,2a) (which is appropriate for type X and Y functions) is applied to $A(\omega)$ defined by Eq. (7.6,1). The result is

$$\int_0^{\omega_x} \frac{G_1 d\zeta}{(\omega^2 - \zeta^2)(\omega_x{}^2 - \zeta^2)^{0.5}} + \int_{\omega_x}^{\infty} \frac{G_2 d\zeta}{(\omega^2 - \zeta^2)(\zeta^2 - \omega_x{}^2)^{0.5}}$$

$$= \frac{-0.5\pi G_2}{\omega(\omega_x{}^2 - \omega^2)^{0.5}} \qquad \text{for } \omega < \omega_x \quad (7.6,2a)$$

$$= \frac{0.5\pi G_1}{\omega(\omega^2 - \omega_x{}^2)^{0.5}} \qquad \text{for } \omega > \omega_x. \quad (7.6,2b)$$

On the other hand, if we apply Eq. (7.4,2b) to the $B(\omega)$ defined in Eqs. (7.6,1a, b), the result is

$$\int_0^{\omega_x} \frac{\zeta G_2(\zeta) d\zeta}{(\omega^2 - \zeta^2)(\omega_x{}^2 - \zeta^2)^{0.5}} - \int_{\omega_x}^{\infty} \frac{\zeta G_1(\zeta) d\zeta}{(\omega^2 - \zeta^2)(\zeta^2 - \omega_x{}^2)^{0.5}}$$

$$= \frac{0.5\pi G_1(\omega)}{(\omega_x{}^2 - \omega^2)^{0.5}} \qquad \text{for } \omega < \omega_x \quad (7.6,3a)$$

$$= \frac{0.5\pi G_2(\omega)}{(\omega^2 - \omega_x{}^2)^{0.5}} \qquad \text{for } \omega > \omega_x. \quad (7.6,3b)$$

Equations (7.6,2) and (7.6,3) show that if the real (imaginary) part of a function is specified for $0 < \omega < \omega_o$, and the imaginary (real) part is specified for $\omega > \omega_o$, then the function is completely specified. This result will be applied in the next section to the optimum loop transmission problem formulated in Section 7.5.

§ 7.7 The Ideal Bode Characteristic

We are finally able to complete the problem formulated in Section 7.5, where $| L(j\omega) |$ is *a priori* known for $0 < \omega < \omega_o$, and the object is to decrease $| L(j\omega) |$ as rapidly as possible for $\omega > \omega_o$. In Section 7.5 it was shown that the latter objective requires that $\theta(\omega) = \arg L(j\omega)$ should be negative and as large as possible in magnitude at all ω. However, Section 7.6 demonstrated that its lag angle for only $\omega > \omega_o$ could be independently controlled, because $| L(j\omega) |$ is fixed for $\omega < \omega_o$.

There is one factor which conflicts with our desire for large negative $\theta(\omega)$ for $\omega > \omega_o$, and this is the stability problem. If unconditional stability is desired, then the lag angle must be less than 180° for all $\omega < \omega_c$, where ω_c is the crossover frequency. (We are assuming that the nature of $| L(j\omega) |$ for $\omega < \omega_o$ is such that the lag angle of the $\theta(\omega)$ which finally results, is less than 180° for $\omega < \omega_o$.) The lag angle of $L(j\omega)$, for $\omega > \omega_o$, may therefore be assigned as in Fig. 7.7-1, and this choice is explained as follows.

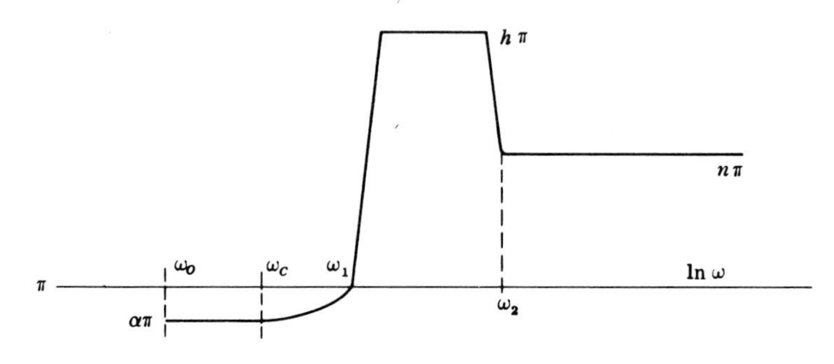

FIG. 7.7-1. Tentative phase lag characteristic.

Unconditional stability requires that $\theta(\omega) = -\alpha\pi$ for $\omega_o < \omega < \omega_c$, with the phase margin $\theta_M = \pi(1 - \alpha)$. Some gain margin G_M is required, so that for $\omega_c < \omega < \omega_1$, the phase lag cannot be made too large [i.e., $| L(j\omega_1) | = - G_M$ db]. The lag angle can presumably be made extremely large for all $\omega > \omega_1$. However, we must assume some finite excess of poles over zeros for $L(s)$, and if this excess is the integer $2n$, then $\lim_{\omega\to\infty}\theta(\omega) = - n\pi$. Nevertheless, $- \theta(\omega)$ can temporarily be made larger than $n\pi$, as shown in Fig. 7.7-1. There are two difficulties involved in the above choice of $\theta(\omega)$; one is theoretical, the other is practical. The theoretical difficulty is that if we assigned $\theta(\omega)$ as shown, with specific numbers for ω_c, ω_1, ω_2, there is no guarantee that the $| L(j\omega) |$ that would emerge [from Eq. (7.6,2b)] would be 0 db at precisely $\omega = \omega_c$, $- G_M$ db at $\omega = \omega_1$. In other words, we cannot find ω_c, ω_1 (and hence ω_2) until the computations have been completed. Therefore it is necessary to carry them as unknowns until the very end. The transcendental equations in ω_c, ω_1, ω_2 that thereby result are very difficult to solve, and have discouraged work in this direction. The second difficulty with Fig. 7.7-1 is that involved in the practical synthesis and network design of such a $L(j\omega)$. The reason is that the design specifications are often such that the plant parasitics (higher modes) become significant at $\omega \approx \omega_c$ to ω_1 (certainly this is true in systems with high performance requirements). It is therefore unrealistic to assume that it is possible to exert the kind of design control over $L(s)$, for $\omega > \omega_c$, that is envisioned in Fig. 7.7-1.

For the above two reasons, we abandon the previous approach, and turn to a practical approximation of the ideal design, due to Bode, which has become known as the "ideal Bode Characteristic." This characteristic is for the special case where the specified $|L(j\omega)|$ is constant (say equal to M), for $0 < \omega < \omega_o$. The approximation to the ideal choice of $\theta = \arg L(j\omega)$ is made by first trying

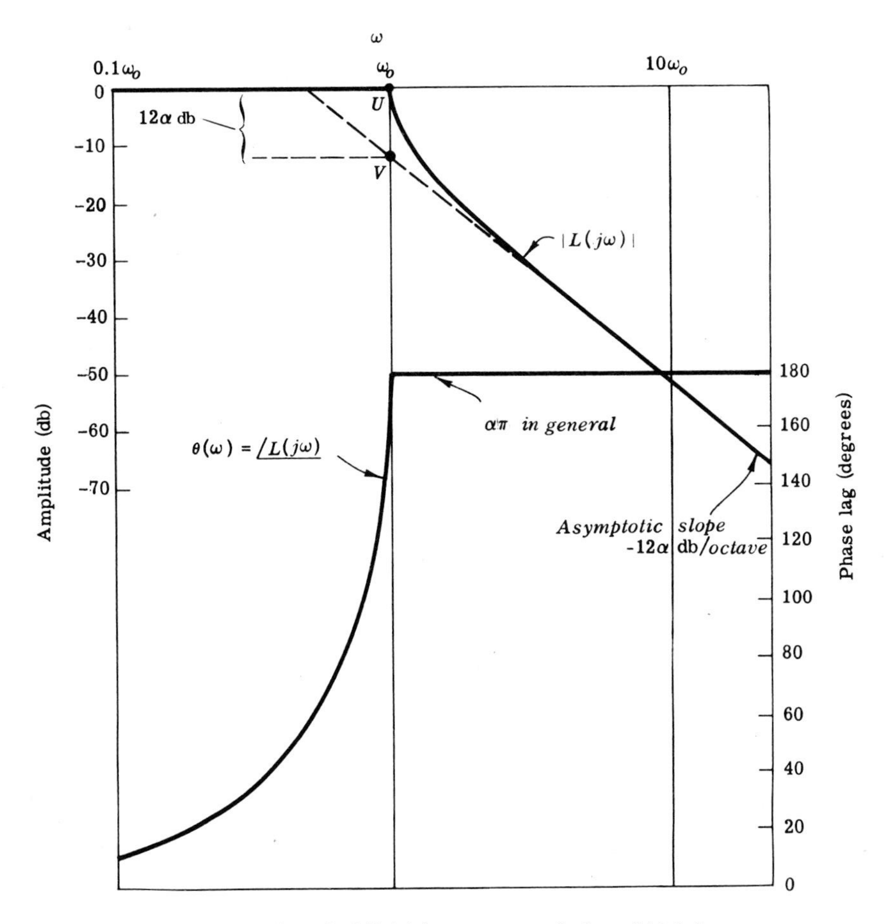

FIG. 7.7-2. Derivation of $|L(j\omega)|$ for $\omega > \omega_o$ and of $\arg L(j\omega)$ for $\omega < \omega_o$.

a very simple tentative $\theta = \arg L(j\omega)$, obtaining from the latter the approximate location of ω_c and ω_1, and then completing the design.

The first step is to choose the simple $\theta = -\alpha\pi$ for all $\omega > \omega_o$. Equations (7.6,2) are then used to derive the resulting $|L(j\omega)|$ for $\omega > \omega_o$, and $\theta(\omega)$ for $\omega < \omega_o$. In Eqs. (7.6,2) $G_1 = \ln M$ for $\omega < \omega_x = \omega_o$ and $G_2 = -\alpha\pi$ for $\omega > \omega_x = \omega_o$. An arbitrary constant can always be subtracted from G_1. It is convenient to subtract $\ln M$ from G_1, so that the first integral on the left-hand side of Eqs. (7.6,2) is zero. Solving Eqs. (7.6,2),

$$\left.\begin{aligned} G_2 &= \theta(\omega) = -2\alpha \sin^{-1} \omega/\omega_o && \text{for } \omega < \omega_o \\ G_1 &= \ln M - 2\alpha \ln \{(\omega/\omega_o) + [(\omega^2/\omega_o^2) - 1]^{0.5}\} && \text{for } \omega > \omega_o. \end{aligned}\right\} \quad (7.7,1)$$

The resulting complete $G = \ln L(j\omega) = G_1 + jG_2$ is sketched in Fig. 7.7-2 for $M = \alpha = 1$. In the general case, the gain curve decreases asymptotically at 12α db per octave. If this asymptote is extended to the left, it cuts the vertical line $\omega = \omega_o$ at a point 12α db below the horizontal $\ln M$ line; i.e., the point V in Fig. 7.7-2 is in general 12α db below the point U. Thus, at y octaves beyond ω_o, where $\omega = 2^y\omega_o$, the decrease in gain from its original constant value is

$$12\alpha(1 + y) \text{ db} = \text{decrease in gain at } \omega/\omega_o = 2^y. \quad (7.7,2)$$

For an idea of the results, let the desired level of gain be 40 db. The -40 db

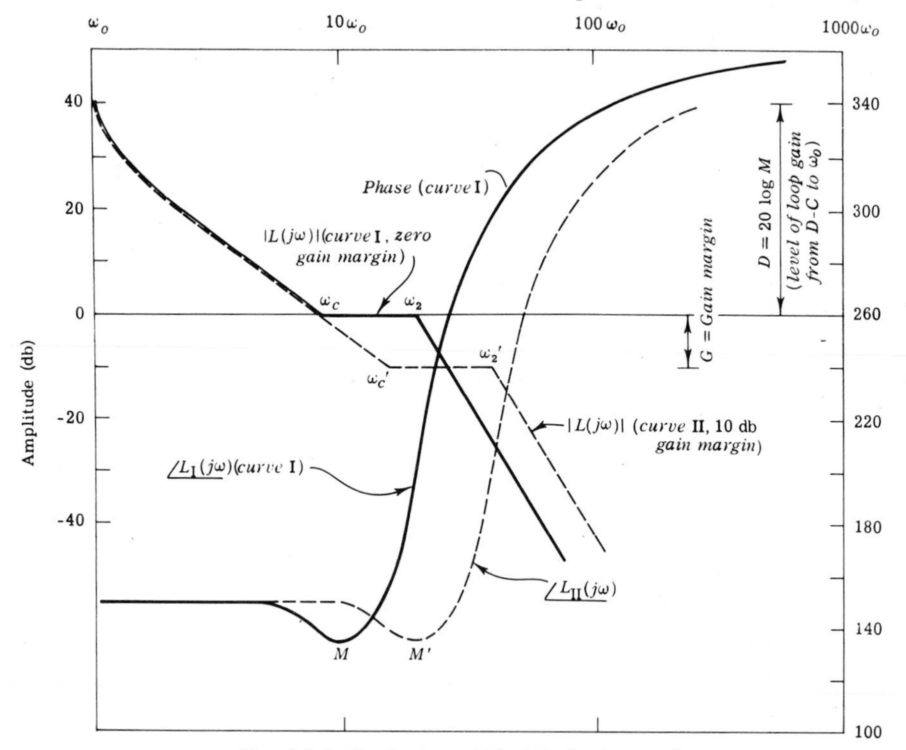

FIG. 7.7-3. Derivation of ideal Bode characteristic.

point in Fig. 7.7-2 is at $\omega_c = 5\omega_o$, which is therefore the crossover frequency. For $\alpha = 5/6$ ($\theta_M = 30°$), if the crossover point is to be at $5\omega_o$, the level of gain is no longer 40 db, but (5/6)40 = 33.3 db. Or, for the same 40 db, we find the point at which the gain is $(-6/5)40 = -48$, corresponding to $\omega_c = 8\omega_o$.

Actually, we must let the phase lag approach $-n\pi$ at large frequencies, if $2n$ is equal to the excess of poles over zeros of $L(j\omega)$. This corresponds to the loop gain decreasing at high frequencies, at the rate of $12n$ db per octave, instead of at 12α decibels per octave. If we introduce this faster falloff at some $\omega_2 > \omega_c$, more phase lag will result at the lower frequencies, thereby violating the prescribed maximum phase lag there. To offset this extra lag, a modification is made of $|L(j\omega)|$ in the region between ω_c and ω_2. The modification is shown in Fig. 7.7-3 (curve I). It consists of Fig. 7.7-2 (as modified for $\alpha \neq 1$) plus two semi-infinite characteristics, one with positive 40α decibels per decade slope at ω_c and the other with negative $40n$ db per decade slope at ω_2. We want ω_2 to be as small as possible without violating the maximum phase lag specifications at low frequencies [i.e., $\pi(1-\alpha)$ phase margin against conditional stability from ω_o to ω_c]. The value of ω_2 is calculated as follows.

The phase due to a semi-infinite characteristic, at frequencies considerably less than its break point, is almost linearly related to the slope of the characteristic and to the distance from the break point (see Fig. 7.4-7). Let us assume ω_o is sufficiently less than ω_c for this approximation to hold (the error leads to more phase lead than necessary). This gives the required ratio as $\omega_2/\omega_c = n/\alpha$, in order that the phase change at ω_o, due to the two semi-infinite segments at ω_c and ω_2, total zero. For example, with a level of 40 db loop gain, 30° phase margin, and asymptotic falloff as $1/s^4$ at high frequencies, $\omega_2 = (2\omega_c)/(5/6) = 2.4\,\omega_c$. The resulting $L(j\omega)$ is curve I in Fig. 7.7-3. Note that if ω_2 is the frequency at which the final asymptote begins, then the ratio ω_2/ω_o is $19 = 2^{4.3}$, so that 4.3 octaves must be allowed from the useful band edge (ω_o) to the frequency at which the loop gain is allowed to drop at its final rate.

In the above, no provision has been made for gain margin and the latter is in fact zero. To accommodate a gain margin, the horizontal step is introduced, not on the 0 db line, but on the line of the desired gain margin. For example, in Fig. 7.7-3, curve II has been drawn for 10 db gain margin. The ratio ω_2'/ω_c' for curve II is equal to that of ω_2/ω_c of curve I, and the number of octaves between ω_2' and ω_o is not 4.3 but 5.2. It is in fact possible to find general relations between M, α, and ω_2'/ω_o as follows.

In Fig. 7.7-3, curve II, the characteristic of Fig. 7.7-2 extends from ω_o to ω_c', in which interval the amplitude decreases by $(D+G)$ db, where $D = 20\log M$, and G is the gain margin in decibels. The phase margin is $\pi(1-\alpha)$ radians. Suppose $\omega_c'/\omega_o = 2^y$. Then, from Eqs. (7.7,2), $12\alpha(1+y) = D+G$. Next we have $\omega_2'/\omega_c' = n/\alpha$ so that

$$\frac{\omega_2'}{\omega_o} = \frac{n}{\alpha}\frac{\omega_c'}{\omega_o} = \frac{n}{\alpha}2^y; \qquad y = \frac{D+G}{12\alpha} - 1. \tag{7.7,3}$$

For example, when $D = 40$ db, $G = 10$ db, $n = 2$, $\alpha = 5/6$, then $\omega_2'/\omega_o = 38.4$, which checks with curve II, Fig. 7.7-3.

The above, which is often called the *Bode ideal cutoff characteristic*, provides a very useful, simple, and good approximation to the fastest cut-off possible. However, as Bode has pointed out,[1] and as we have earlier noted (Fig. 7.7-1, etc.), it is not really the best. This can be seen by studying Fig. 7.7-3. There is some waste of phase area in the pocket at M'—the phase lag there is less than the permitted maximum value. This phase area is not much, however. Of greater importance is the fact that more phase lag area could be obtained by having the phase lag increase faster from approximately ω_c' and at higher frequencies. For example, the phase lag could be 180° at ω_c', without changing the phase margin. To improve matters, we could assign a phase lag characteristic which, for $\omega > \omega_c$, increases faster than that in Fig. 7.7-3. As previously noted, in connection with Fig. 7.7-1, it is difficult to obtain the truly ideal characteristic analytically, because, for a given phase curve, one does not know *a priori* what gain margin will result. It is, of course, possible to work out some examples, find the resulting gain margins, and catalog the results. No one has bothered to carry out these calculations. However, for the simplified case of zero gain and phase margins, and for a phase lag curve that rises abruptly from π to $n\pi$, Bode has found that if ω_2 is kept fixed, then the value of M may be increased by 4 db for $n = 1.5$, and 8 db when $n = 3$.[2] We have previously noted the practical difficulties in exerting control over $L(j\omega)$ at these higher frequencies, and therefore the "ideal Bode characteristic" is generally considered the best that can be done, despite the fact that it is not really so.

The practical importance of Eq. (7.7,3) deserves emphasis. If 40 db of feedback is wanted from 0 to 10 rps, and if the system is to be unconditionally stable, then one must exert control over the loop gain up to about 400 rps. In fact, if $| L(j400) | > | P(j400) |$, we must go beyond 400, up to the frequency at which $| P(j\omega) | > | L(j\omega) |$. Only then can the gain of the additional devices introduced into the loop be less than one. This is one of the important costs of feedback that must be paid in return for its benefits. Thus, although the bandwidth for which a large level of loop gain is desired is only 10 radians per second, $| L(j\omega) | > 1$ at least up to 160, and a final asymptotic slope of 24 db per octave can be initiated no earlier than 400 rps (if 30° phase margin and 10 db gain margin are desired. This problem has been noted in Chapter 6 in specific numerical design problems.

We do not consider here the actual network realization of $L(s)$. This is a problem in network synthesis. In control work the ideal characteristic is hardly ever attempted. Its importance lies in that it establishes the minimum price that

[1] H. W. Bode, "Network Analysis and Feedback Amplifier Design," pp. 471-476. Van Nostrand, Princeton, New Jersey, 1945.

[2] H. W. Bode, "Network Analysis and Feedback Amplifier Design," pp. 474-476. Van Nostrand, Princeton, New Jersey, 1945.

must be paid for a specified level of loop gain over some frequency band. The same technique [Eqs. (7.6,2)] may be used to determine the "ideal" characteristic for any other shape of $|L(j\omega)|$ vs. ω, for $\omega < \omega_o$.

Positive Feedback

We digress to prove the statement made in Section 4.11 regarding the impracticality of positive feedback for attaining large $|L(j\omega)|$ over a respectable frequency range. (Section 4.11 should be re-read.) It was stated that in Fig. 4.11-1b (redrawn as Fig. 7.7-4a) it was impossible to obtain much of a frequency

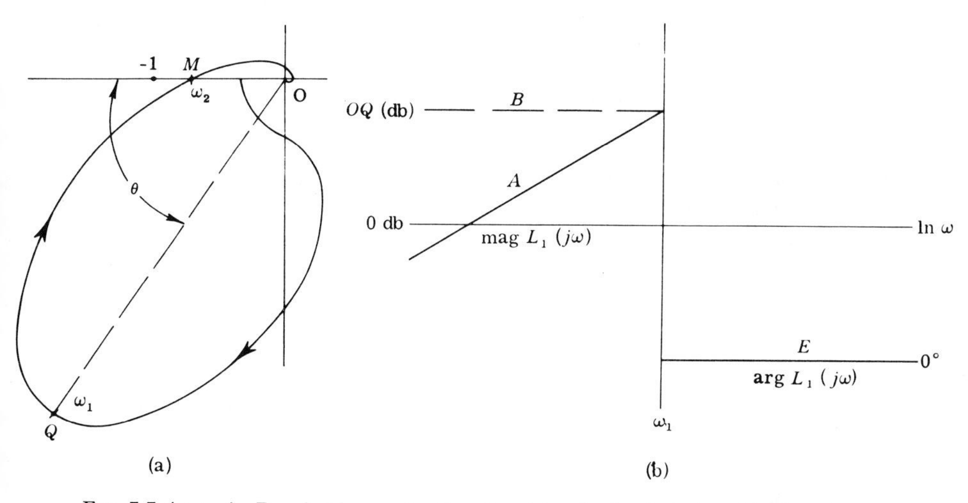

FIG. 7.7-4 a to 4e. Proof of impracticality of positive feedback design. (a) Hypothetical $-L_1(j\omega)$ locus. (b) A, E are known parts of optimum $L_1(j\omega)$.

range with large OQ [OQ is $|L(j\omega)|_{max}$], and with $OM < 1$. The difficulty is due to the fact that the decrease of $|L(j\omega)|$ from its maximum value of OQ to $OM < 1$ is accompanied by phase lag. Before $|L(j\omega)|$ has decreased by much, the phase lag is already 180°. What is the best that can be done under the circumstances?

There are two conditions which maximize the ratio OQ/OM. (1) For a given OQ, the lag of $L(j\omega_1)$ should be made as small as possible. This maximizes θ in Fig. 7.7-4a, and makes more phase lag available for the reduction of $|L(j\omega)|$. Let $L_1(s) = -L(s)$, and then the above means that the lead of $L_1(j\omega_1)$ should be as large as possible. (2) The available angle θ should be used to secure the largest ratio OQ/OM.

The first condition is achieved by letting $L_1(s) = Ks^x$ (x being a positive integer) for $0 < \omega < \omega_1$. Thus, A in Fig. 7.7-4b has a larger phase lead for all ω than does B (assuming their magnitudes are identical for $\omega > \omega_1$). Of course, x can be made very large, but if the maximum slope of $|L_1(j\omega)|$ for $\omega < \omega_1$

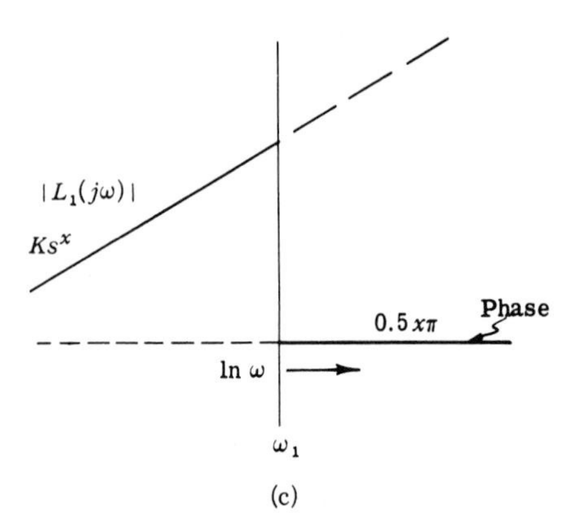

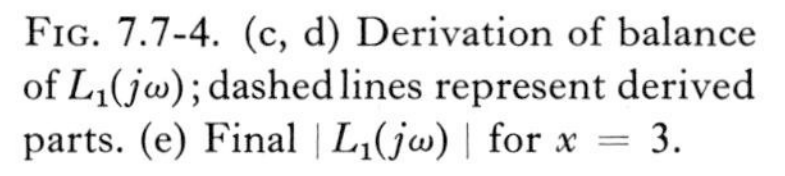

FIG. 7.7-4. (c, d) Derivation of balance of $L_1(j\omega)$; dashed lines represent derived parts. (e) Final $|L_1(j\omega)|$ for $x = 3$.

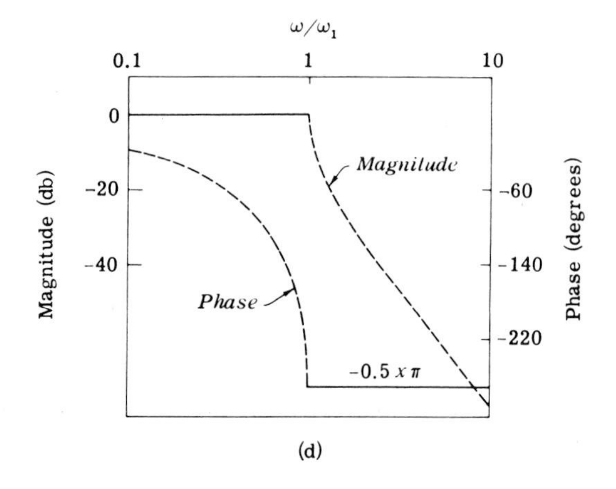

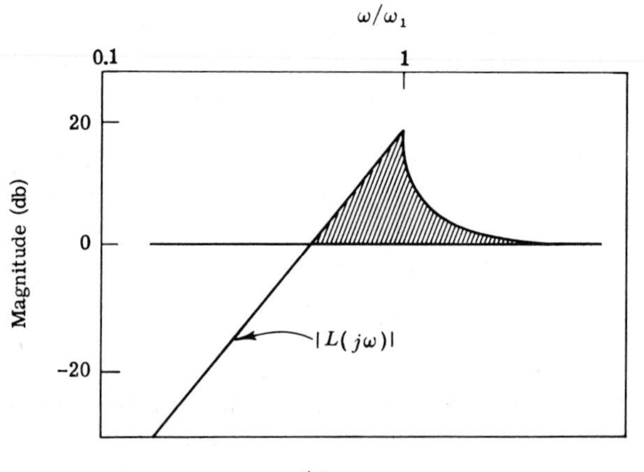

is prescribed, then A (with maximum permitted slope) is the best. The second condition is achieved by assigning the maximum permissible phase lag to $L(j\omega)$ for $\omega > \omega_1$. Tentatively (as in the derivation of the ideal Bode characteristic), it is set at 180°, i.e., zero for $L_1(j\omega)$. From Eq. (7.6,2), this combination of A and E in Fig. 7.7-4b determines $L_1(j\omega)$ for all ω. It is easier to solve for $L_1(j\omega)$ by expressing the A, E combination as the sum of Figs. 7.7-4c and d (solid lines). The solution for Fig. 7.7-4c is obvious, and is indicated by its dotted extension. The solution of Fig. 7.7-4d is given in Eq. (7.7,1), with $x = 2\alpha$. The resulting $|L(j\omega)|$ (i.e., the sum of Figs. 7.7-4c and d) is sketched in Fig. 7.7-4e for $x = 3$; the 0 db line must be located as shown in the figure in order that $|L(j\omega)| = 1$ when $\arg L(j\omega) = 180°$. The shaded area in Fig. 7.7-4e represents the region in which $|L(j\omega)| > 1$ in this idealized design which has no stability margins and in which the final asymptotic decrease of $|L(j\omega)|$ has yet to be inserted. There is no need to continue; the positive feedback design suggested in Fig. 4.11-1b is obviously useless for achieving the usual benefits of feedback.

§ 7.8 A Different Kind of Optimum $L(j\omega)$

Sections 7.5-7.7 were concerned with achieving the fastest decrease of $|L(j\omega)|$ for $\omega > \omega_o$, when $|L(j\omega)|$ has been *a priori* specified for $\omega < \omega_o$. Another problem in optimization is as follows: Suppose the crossover frequency is fixed at ω_c, and we want to maximize the area under the curve of $|L(j\omega)|$ vs. ω, for $\omega < \omega_c$, subject to unconditional stability; i.e., $|L(j\omega)|$ is not *a priori* fixed. We may expect that the solution will involve exploitation of the unused phase lag area in the region $\omega < \omega_o$ in Figs. 7.7-2 and 3 where $\theta(\omega) > -\pi\alpha$, for $\omega < \omega_o$. Actually, we must add, in our statement of desiderata that, for $\omega > \omega_c$, $|L(j\omega)|$ is to be decreased as rapidly as possible, for the usual reasons. If this statement is not added, i.e., if we are only interested in securing the maximum area under the curve of $|L(j\omega)|$ vs. ω for $\omega < \omega_c$, such an objective is secured by setting $\theta = -\pi\alpha$ for $\omega < \omega_c$ and $|L(j\omega)| = 1$ for $\omega > \omega_c$. This is, of course, unrealistic, hence the need for the additional statement.

From the results of Sections 7.5-7.7 (compare Section 5.8), it should be clear that the objectives are secured by assigning to $\theta(\omega) = \arg L(j\omega)$ the maximum permissible phase lag for all ω. The relations or graphs of Section 7.4 may then be used to determine the resulting $|L(j\omega)|$.

However, as noted in connection with Fig. 7.7-1, there is some difficulty in specifying the maximum permissible phase lag. For illustrative purposes, we choose the phase lag curve of Fig. 7.8-1, with $\pi(1 - \alpha)$ the prescribed phase margin against conditional stability. Some gain margin G_M is required, so the 180° lag value is set at $m_1\omega_c$, and thereafter the phase lag is increased rapidly

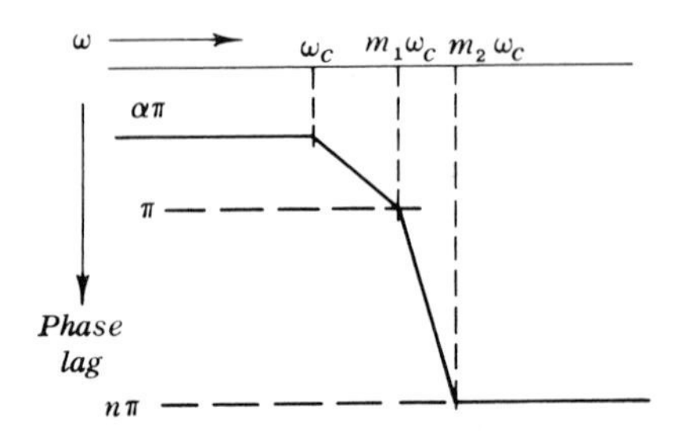

FIG. 7.8-1. Assumed arg $L(j\omega)$.

to its final value of $n\pi$, where it is thus implicitly assumed that as $\omega \to \infty$, $L \to 0$ at $12n$ db per octave. Of course, one could always choose a larger n, and obtain a faster falloff of $|L|$, but presumably the system is to be of finite complexity, so that once n is chosen, the above gives a reasonably fast cutoff for that value of n.

The problem then is to find the magnitude curve that corresponds to the phase function of Fig. 7.8-1. The first step is to subtract out $-\alpha\pi$ from

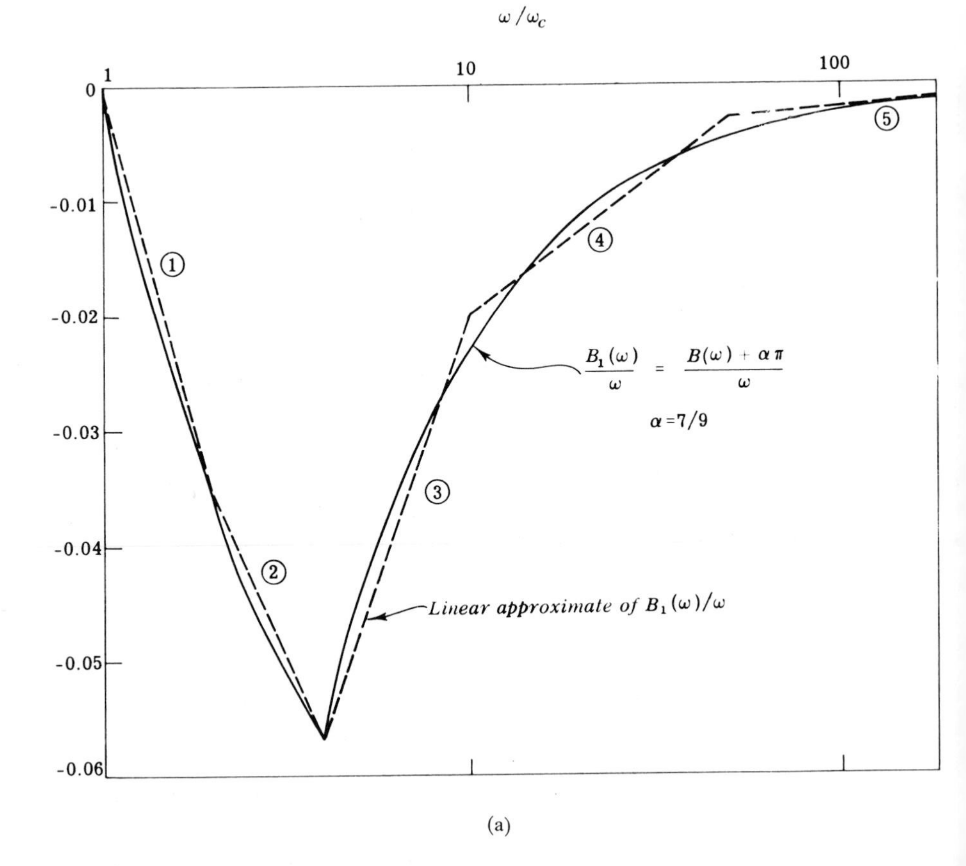

(a)

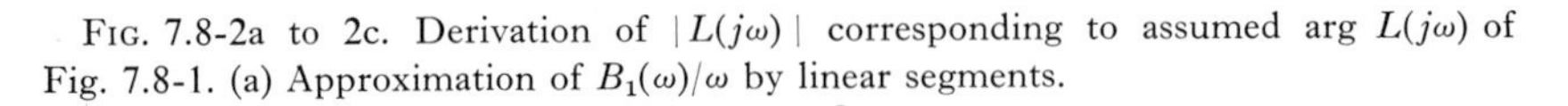

FIG. 7.8-2a to 2c. Derivation of $|L(j\omega)|$ corresponding to assumed arg $L(j\omega)$ of Fig. 7.8-1. (a) Approximation of $B_1(\omega)/\omega$ by linear segments.

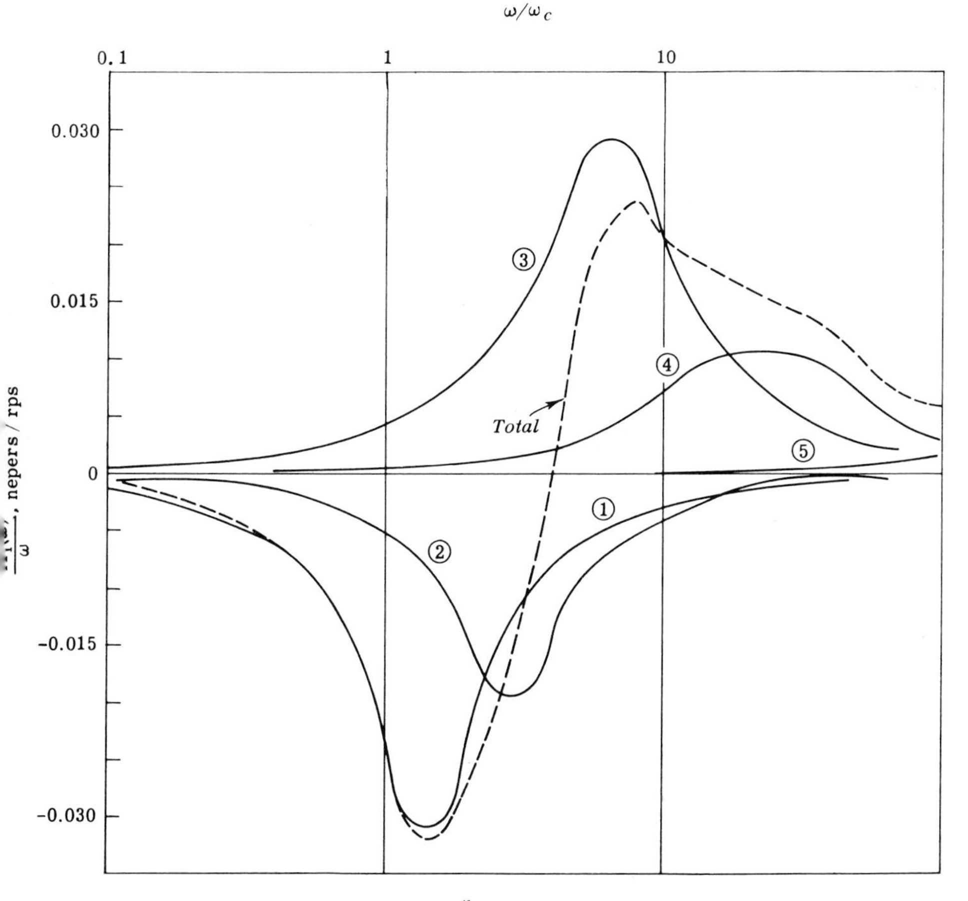

(b)

FIG. 7.8-2b. Resulting curves of $A_1(\omega)/\omega$.

$B(\omega)$ and obtain the contribution to $A(\omega)$ due to it.[1] Next we divide the resulting $B(\omega) + \alpha\pi = B_1(\omega)$ by ω and approximate $B_1(\omega)/\omega$ by linear segments. This is done in Fig. 7.8-2a, where we have taken (in the notation of Fig. 7.8-1) $\omega_c = 10$, $m_1 = 2$, $m_2 = 4$, $\alpha = 7/9$, and the ordinate is in radians. As a result of the approximation, there are five finite line segments, with the data (in the notation of Fig. 7.4-6a), given in Table 7.8-1. Fig. 7.4-6b is used, and the results

[1] Consider the case $A = \ln \omega^n = n \ln (\omega_x e^u)$ in the notation of Eqs. (7.4,8a, b). Then $dA/du = n$, and the value of $B(\omega_x)$ that results from such an A is, from Eq. (7.4,8a), $(n/\pi) \int_{-\infty}^{\infty} \ln \coth |u/2| \, du = n\pi/2$. Thus, a magnitude characteristic with a constant slope of $6n$ db per octave is accompanied by constant phase angle of $n\pi/2$ radians. In this problem, the lag angle is $\alpha\pi$ radians, so the accompanying magnitude characteristic has a negative slope of $(2\alpha)(6) = 12\alpha$ db per octave.

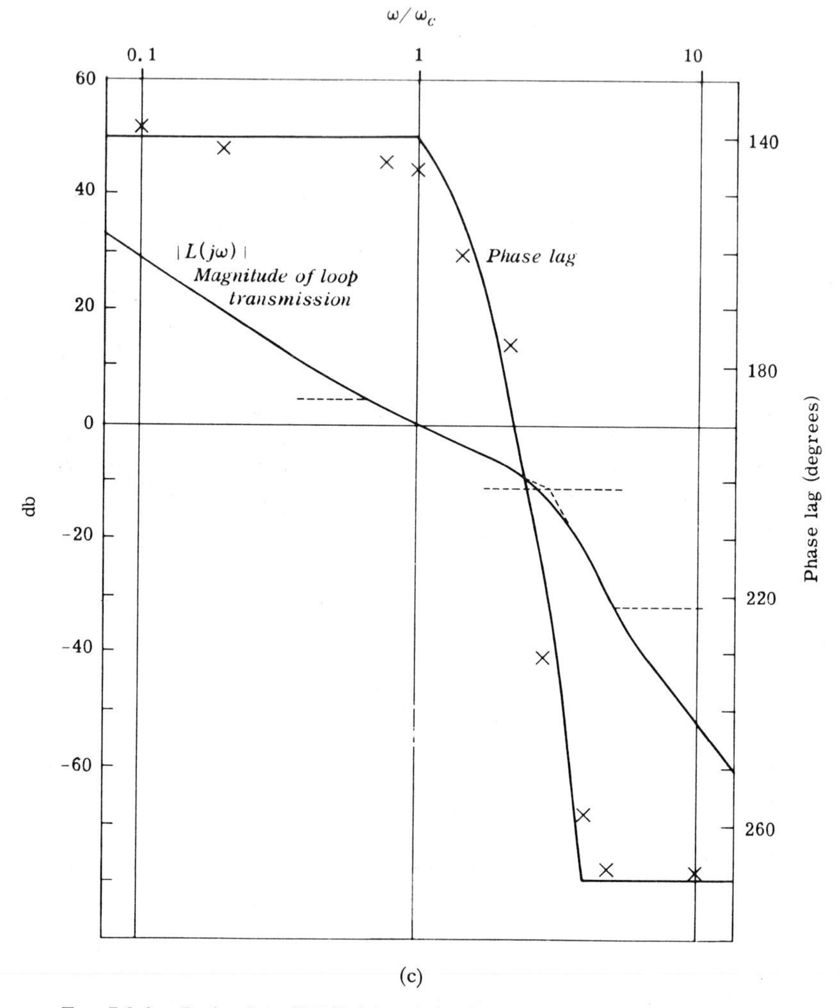

(c)

FIG. 7.8-2c. Bode plot of $|L(j\omega)|$ and check on accuracy of approximation.

are plotted in Fig. 7.8-2b. The final total is multiplied by ω, in accordance with Eq. (7.4,9). The contribution of $-\alpha\pi$ is now added. The latter is a characteristic with a constant slope of -12α db per octave, and its value at any one frequency may be chosen at random, since we can always add a constant to $A(\omega)$ without affecting $B(\omega)$. All these characteristics added together give the $|L(j\omega)|$ which is shown in Fig. 7.8-2c. The level of $A(\omega)$ is determined by having it cross the 0 db line at $\omega = 10$. The gain margin is almost 7 db. It can be increased by letting the phase lag increase more slowly for $\omega > \omega_c$.

As a final check on the accuracy of the work, the magnitude characteristic in Fig. 7.8-2c was approximated by two finite line segments and two semi-infinite

TABLE 7.8-1

PARAMETERS OF FINITE LINE SEGMENTS

Segment	ω_0	m	δA
1	$14.14 = \sqrt{(10)(20)}$	1.414	− 0.035
2	$28.28 = \sqrt{(20)(40)}$	1.414	− 0.022
3	$63.2 = \sqrt{(40)(100)}$	1.58	+ 0.037
4	$223 = \sqrt{(100)(500)}$	2.24	0.017
5	$1580 = \sqrt{(500)(5000)}$	3.16	0.003

characteristics, as shown by the dotted lines in Fig. 7.8-2c. Figure 7.4-6b was used to calculate the phase of $L(s)$. The sum of the contributions of these four items is given by the crosses in Fig. 7.8-2c. The resulting phase curve is a good approximation of the original $B(\omega)$.

It should be noted that $L(j\omega)$ of Fig. 7.8-2c is unrealistic at very low frequencies, because, in any practical design, $\lim_{s\to 0} L(s) = ks^x$, with x equal to zero

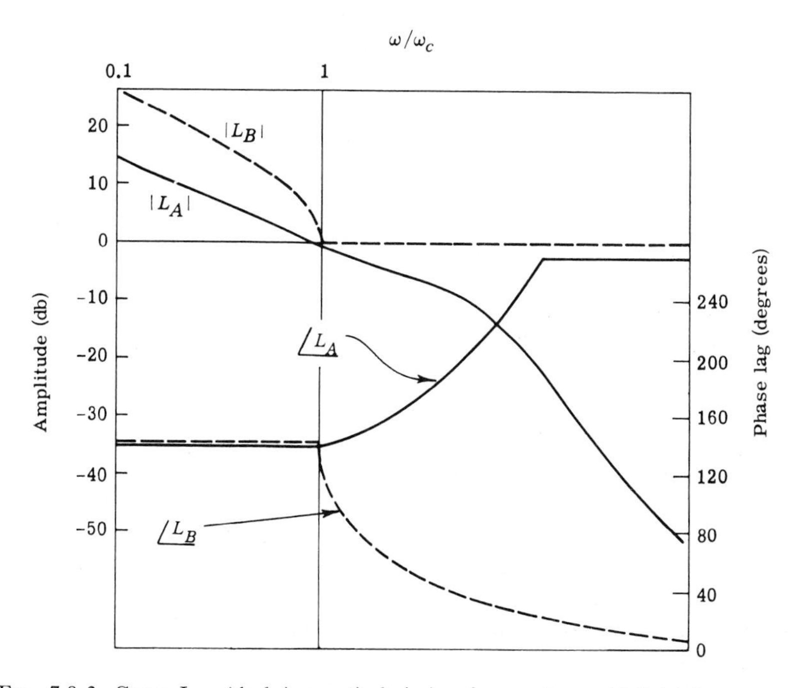

FIG. 7.8-3. Curve L_B—ideal impractical design for maximum $|L(j\omega)|$ for $\omega < \omega_c$; curve L_A—compromise between $|L|_{\max}$ for $\omega < \omega_c$ and $|L|_{\min}$ for $\omega > \omega_c$.

or an integer. In practice, then, the loop transmission must be modified at low frequencies.

We have previously noted that if maximization of the area under $|L(j\omega)|$ vs. ω, for $\omega < \omega_c$, was the sole objective, then this would be secured by setting $\theta = -\pi\alpha$ for $\omega < \omega_c$ and $|L(j\omega)| = 1$ for $\omega > \omega_c$. It is interesting to compare the resulting ideal but wholly unrealistic design with that of Fig. 7.8-2c. Equations (7.6,3a, b) (with $\omega_x = \omega_c$) may be used to solve for the impractical ideal design. In the notation of these equations, $G_1 = \ln|L| = 0$, for $\omega > \omega_x$, $G_2 = -\alpha\pi$ for $\omega < \omega_x$. The integrals may be solved by letting $x^2 = \omega^2 - \zeta^2$. The result is $\ln|L(j\omega)| = 2\alpha \ln [\omega_c + (\omega_c^2 - \omega^2)^{0.5}]/\omega$ for $\omega < \omega_c$ and $\theta = -2\alpha \sin^{-1} \omega_c/\omega$ for $\omega > \omega_c$. The two designs are compared in Fig. 7.8-3. It is seen that for most of $\omega < \omega_c$, the level of the impractical ideal L_B is larger by (140/180)12 db than that of Fig. 7.8-2c (labeled L_A in Fig. 7.8-3).

§ 7.9 Minimum Phase Functions

It has been assumed in the above that $\ln L(s)$ has no singularities in the right half-plane. Consider $L(s)$ with zeros in the right half-plane. These zeros are explicitly noted by writing $L(s) = L_1(s)(s-a)(s-b)\dots(s-g)$, in which $L_1(s)$ has only lhp zeros and poles, and a, b, etc., have positive real parts and are in conjugate pairs, if complex. It is convenient to write

$$L(s) = [L_1(s)(s+a)(s+b)\dots(s+g)]\left[\frac{(s-a)(s-b)\dots(s-g)}{(s+a)(s+b)\dots(s+g)}\right] \triangleq [L_M(s)][A(s)]. \tag{7.9,1}$$

In the above, $A(s) \triangleq [(s-a)(s-b)\dots(s-g)]/[(s+a)(s+b)\dots(s+g)]$, and it is easily seen that $|A(j\omega)| = 1$ for all ω. $A(s)$ is therefore appropriately called an *all-pass* function. Hence $L(j\omega)$ and $L_M(j\omega)$ have the same magnitudes, but differ in phase by the phase angle of the all-pass function $A(j\omega)$. The pole-zero pattern of a typical all-pass network is shown in Fig. 7.9-1.

Consider arg $A(j\omega)$, for example, for $A(s) = (s-a)/(s+a)$; $a > 0$. The angle decreases from 180° at zero frequency to zero degrees at infinite frequency (or from $-180°$ to $-360°$). It is the change in angle (encirclements of the -1 point) that is decisive in stability considerations, not the original value at any one frequency. Any constant can be added to the phase of a function at all frequencies, without affecting the stability of the system, because the constant does not affect the encirclement of the -1 point—it merely rotates the frequency locus. The all-pass function has an angle which always decreases with frequency. If it has n zeros, the total amount of increase of phase lag from zero to infinite frequency is $180n°$.

Thus any function $L(s)$ with rhp zeros and lhp poles is equivalent to a function $L_M(s)A(s)$, where $A(s)$ is an all-pass function, $L_M(s)$ has all its zeros and poles in the left half-plane, and $|L_M(j\omega)| = |L(j\omega)|$ for all real ω. The phase lag (or really the increase in phase lag from zero to infinite frequency) of $L(j\omega)$ is greater than that of $L_M(j\omega)$ by that of $A(j\omega)$, i.e., by an amount which is 180° times the number of zeros of $A(s)$. $L_M(s)$ is called a *minimum phase* function, because for its magnitude characteristic (as a function of frequency), it has the minimum phase lag possible. No other stable function (i.e., which has no rhp poles) with the same magnitude vs. frequency characteristic can have any less phase lag.

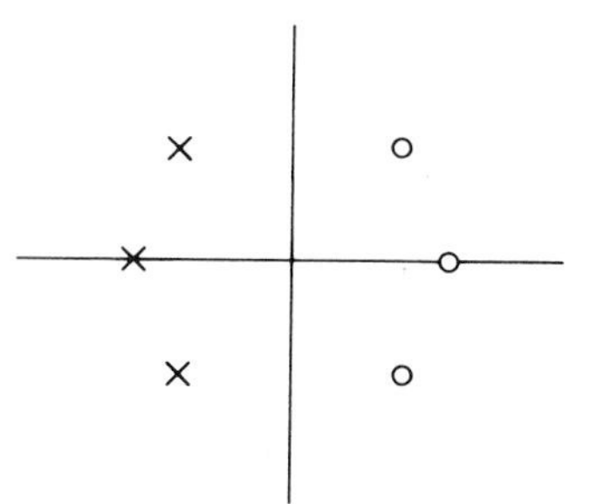

FIG. 7.9-1. An all-pass pole-zero pattern.

The harmful effect of rhp zeros is seen from Eq. (7.4,8a) and its physical interpretation. The faster the loop transmission decreases, the larger is its phase lag. $|L(j\omega)|$ cannot be allowed to decrease too fast, or the phase lag will be so great as to cause conditional or complete instability. If $L(s) = L_M(s)A(s)$, then the effective phase lag of $L(j\omega)$ increases by $180n°$, even if $|L_M(j\omega)|$ is constant. Hence the rate of decrease of $|L_M(j\omega)|$ must be less than before, for the same stability margins. In fact, if $L(s)$ has more than one rhp zero, there is an upper bound (see Section 7.12) on the loop-gain level that is achievable for any range $0 < \omega < \omega_o$.

The above may be restated as follows. Stability considerations determine a certain amount of available phase lag area. In the minimum phase loop transmission, none of the available phase lag area is wasted. However, in the nonminimum phase function, $L_{NM}(j\omega) = L_M(j\omega)A(j\omega)$, and the all-pass network $A(j\omega)$ contributes phase lag while its magnitude is constant. It therefore eats up part of the available phase lag area, leaving exactly that much less for $L_M(j\omega)$. Any pure delay characteristic e^{-sT} can be approximated as closely as desired by an all-pass network, so that the above conclusions also apply to systems with pure time delays.

Are the above conclusions applicable to loop transmission functions with rhp poles? If the same procedure as in the above is used, the result is that the nonminimum phase $L_{NM'}(s)$, with rhp poles, is written as $L_{NM'} = L_M A'$, where L_M is minimum phase, and A' is an all-pass network, in which the poles are in the right half-plane, and the zeros are in the left half-plane. Such a function has a phase angle which increases from zero to infinite frequency by $180n°$, where n is the number of poles in A'. It would seem, therefore, that such a loop transmission is better than a minimum phase loop transmission. The trouble is that the phase lag requirements (for closed-loop stability) of unstable loop transmissions are radically different from those of stable loop transmissions. The problem is considered in Section 7.14. It is there shown that unstable loop transmissions are inferior to minimum phase loop transmissions.

§ 7.10 Conditionally Stable Systems

Two optimization problems were studied in Sections 7.5-7.9. Both were treated under the constraints of (1) open-loop stability, (2) minimum phase $L(s)$, and (3) unconditionally stable systems. In Sections 7.10-7.14 we shall determine the profit or loss in relaxing these three constraints. In this section we consider the profit in permitting the system to be conditionally stable for the optimization problem treated in Section 7.8.

The problem is one of maximizing the area under $| L(j\omega) |$ vs. ω for $\omega < \omega_c$, subject to achieving maximum falloff beyond the crossover frequency, with the constraint of unconditional stability lifted. The optimum design is the one with the maximum phase lag for all ω. In contrast with Fig. 7.8-1, the phase lag is not necessarily $\alpha\pi$ ($\alpha < 1$) up to the crossover frequency ω_c. We may, in fact, make it much more than π, as in Fig. 7.10-1a. The difficulty in *a priori* choosing the "maximum" phase lag was previously noted in Section 7.7. In Fig. 7.10-1a we arbitrarily choose what seems to be reasonable for a conditionally stable design, with no advance knowledge of the resulting gain margins. The corresponding Nyquist sketch (for $n = 2$, $\alpha = 5/6$, $\lambda = 1$) is shown in Fig. 7.10-1b for the contour indicated in Fig. 7.10-1b. The indentation of the contour at the origin is around a pole of order $2(\lambda + 1) = 4$. A portion of the locus of Fig. 7.10-1b is shown magnified in Fig. 7.10-1c. Note that there are now two gain margins, and possibly two phase margins. One of the latter does not exist if $L(j\omega)$ avoids the unit circle, as does the dotted characteristic in the figure.

The gain characteristic corresponding to Fig. 7.10-1a must be found. For this purpose we need the gain characteristic which goes with the phase characteristic of Fig. 7.10-2a (note the linear ω scale). Equation (7.4,2c) is used to obtain the magnitude corresponding to the phase characteristic of Fig. 7.10-2a. [Equation (7.4,2b) cannot be used, because $F(s)$ of Fig. 7.10-2 has a logarithmic singularity at infinity.] The result is (for $\nu = \omega/\omega_o$)

$$A(\omega) = \frac{\Delta\theta}{\pi}\left[\ln | 1 - \nu^2 | + \frac{n_1}{1 - n_1} \ln \left| \frac{1 - \nu^2}{n_1{}^2 - \nu^2} \right| - \frac{\nu}{1 - n_1} \ln \left| \frac{(1 - \nu)(n_1 + \nu)}{(1 + \nu)(n_1 - \nu)} \right| \right]. \tag{7.10,1}$$

This is plotted in Fig. 7.10-2b (on logarithmic scales) for several values of n_1, for $\Delta\theta = \pi$. The levels have been adjusted so that they are all 0 db at zero frequency. [$A(\omega)$ can be shifted by a constant without affecting $B(\omega)$.]

Example. — *Design of Conditionally Stable Loop Transmission.* We can now solve the problem formulated earlier in the section. A numerical example is treated in a sufficiently general manner to permit the solution of other such problems. Our approach is somewhat similar to that used in deriving the "ideal

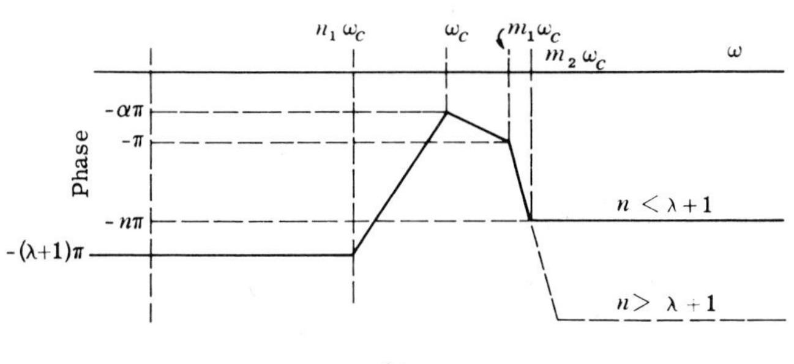

(a)

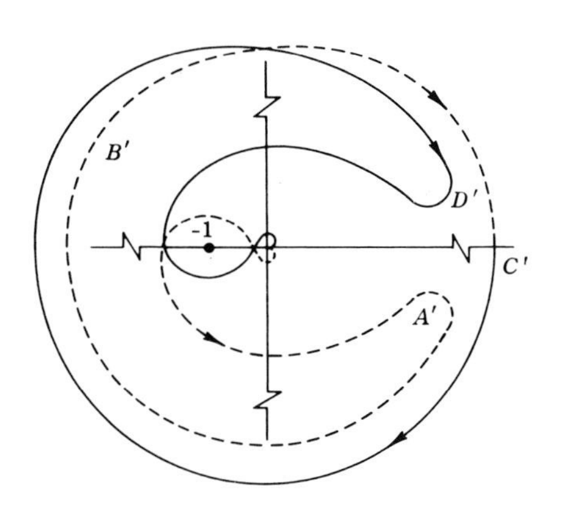

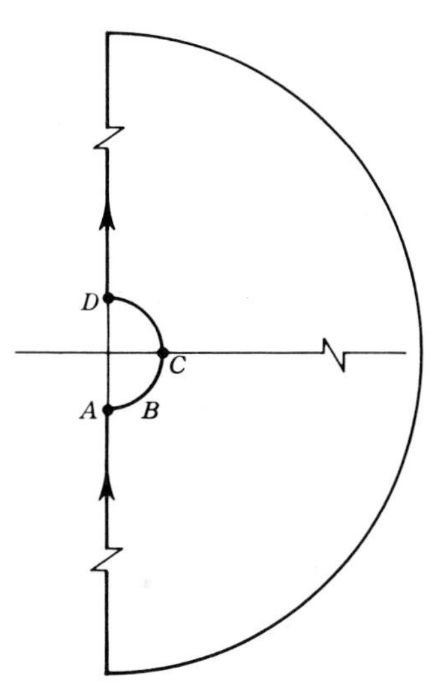

(b)

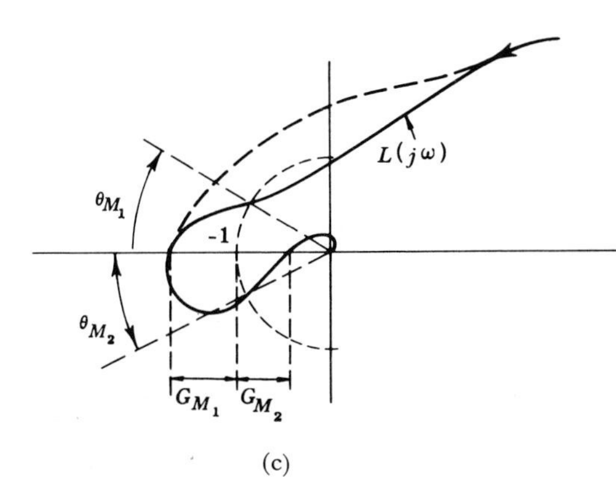

(c)

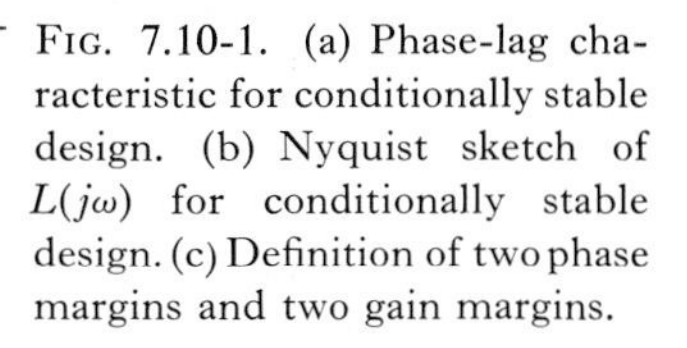
FIG. 7.10-1. (a) Phase-lag characteristic for conditionally stable design. (b) Nyquist sketch of $L(j\omega)$ for conditionally stable design. (c) Definition of two phase margins and two gain margins.

Bode characteristic" of Section 7.7, in order to be able to assign the phase and gain margins. We begin by taking the lag as 2π up to $\omega_c/2$, and decreasing it arithmetically with frequency until it is π at the frequency ω_c. (These choices are obviously arbitrary, but they will illustrate what can be achieved by a conditionally stable design.) Beyond ω_c, we temporarily take it to be constant at π for all $\omega > \omega_c$. This corresponds to a constant lag of 2π and a characteristic similar to that in Fig. 7.10-2a with $\Delta\theta = \pi$, $n_1 = 0.5$. The former contributes

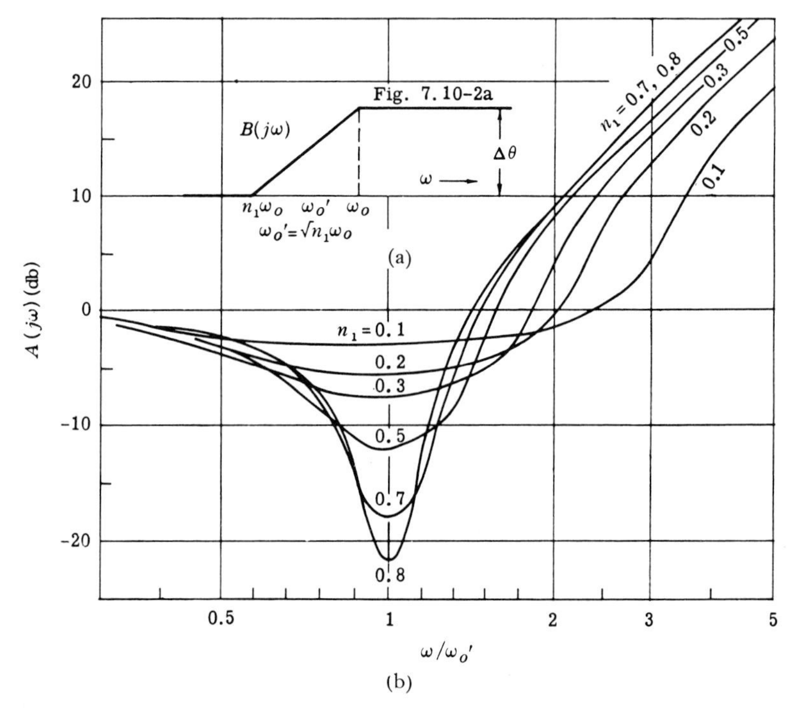

FIG. 7.10-2. (a) The finite line phase segment. $F(j\omega) = A(\omega) + jB(\omega)$. (b) The corresponding magnitude characteristics.

an $|L(j\omega)|$ with a negative slope of 24 db per octave. The magnitude corresponding to the latter is available in Fig. 7.10-2b. Their sum is plotted as curve A in Fig. 7.10-3. Its high frequency slope is -12 db per octave, corresponding to the high frequency lag of 180°.

Suppose gain margin No. 1 (G_{M1} in Fig. 7.10-1c) is to be 7 db. We take the 0 db line 7 db below the peak X at ω_m in Fig. 7.10-3, noting that G_{M1} corresponds to a gain greater than 1 at 180° phase. We really should take the 0 db line 7 db below point Y, but a slight adjustment of the phase characteristic would make the 180° point correspond to X. We see that phase margin No. 1 (θ_{M1} in Fig. 7.10-1c) is not relevant here, because $|L(j\omega)|$ does not decrease to less than unity before 180° lag is reached.

So far, only G_{M1} has been attained. Next we want to achieve θ_{M2} and G_{M2} (Fig. 7.10-1c) which are similar to the stability margins for unconditionally stable systems. We note that the present crossover frequency is at $\omega_3 = 4.6$. If $\theta_{M2} = 30°$, then 30° lead must be introduced at 4.6 rps, with the gain at X and at 4.6 rps only negligibly affected. Suppose we add the finite phase line segment of Fig. 7.10-2a to $L(j\omega)$, with $n_1\omega_o = 2.3$, $\omega_o = 4.6$, and $\Delta\theta = \pi/6$ radian $= 30°$. The resulting over-all $L(j\omega)$ is curve B in Fig. 7.10-3. The contribution to

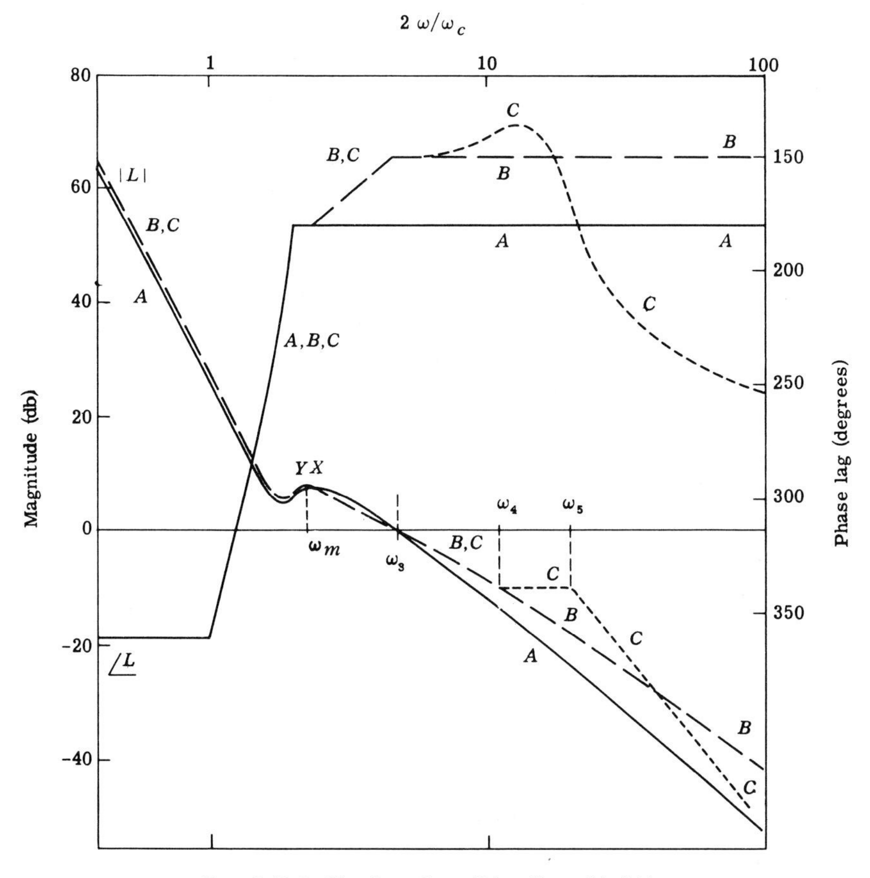

FIG. 7.10-3. Shaping of conditionally stable $L(j\omega)$.

$|L(j\omega)|$ was obtained from Fig. 7.10-2b (after multiplying by 1/6). We are free to adjust the level of the magnitude contribution, and adjust its level to be 0 db at $\omega = \omega_m = 2.3$ rps, in order not to affect the first gain margin, G_{M1}. The gain level at 4.6 rps is then raised by 0.5 db. But this does not matter. The crossover frequency is now slightly larger than 4.6, say at 4.8, but the phase lag there is also 150°, so we have achieved the desired $\theta_{M2} = 30°$. The high frequency slope of curve B is $-$ 10 db per octave, corresponding to the high frequency lag of 150°.

We have yet to obtain the desired second gain margin G_{M2}, and to introduce the final asymptotic slope as soon as possible. Suppose the desired $G_{M2} = 10$ db (which corresponds on curve B to $\omega_4 \approx 11$ rps), and that the final asymptote is $-$ 18 db per octave. A simple way to do this is to introduce a horizontal step in $|L(j\omega)|$ at 11 rps on curve B, and introduce the final $-$ 18 db per octave slope at ω_5, such that the phase at ω_m is not affected. The horizontal step beginning at $\omega_4 = 11$ rps corresponds to a semi-infinite characteristic with positive slope of $(5/6)(12) = 10$ db per octave. The phase corresponding to such a slope is available from Fig. 7.4-7b. The location of the final negative slope of $-$ 18 db per octave is obtained as follows: The positive break at $\omega_4 = 11$ rps contributes $+$ 13° at 2.3 rps. From Fig. 7.4-7b it is seen that a semi-infinite characteristic contributes 4.2° at a frequency 0.115 times its break point. Hence the break point is $2.3/0.115 = 20$ rps. The net effect of the positive and negative slopes is zero for $\omega \leqslant 2.3$, slightly leading for $2.3 < \omega < 18$, and thereafter lagging. The value of θ_{M2} is consequently very slightly increased. In Fig. 7.10-3 the phase lag at high frequencies could be made to reach its final value of $-$ 270° somewhat faster, and so have a faster falloff of the gain curve.

Figure 7.10-3 should be compared with Fig. 7.8-2c in order to see the increase

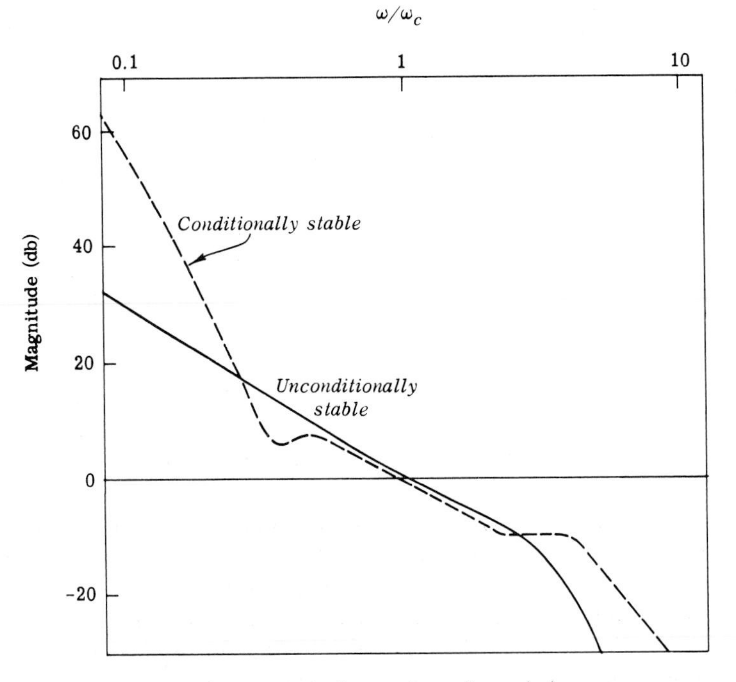

FIG. 7.10-4. Comparison of two designs.

in gain area when conditional stability is allowed. This is done in Fig. 7.10-4. Because of the larger number of stability margins in the conditional design, the final break point is at a higher frequency than in the Fig. 7.8-2c design. Actually, the difference of about 16 db at the higher frequencies in Fig. 7.10-4 can be decreased to about 10 db, if a phase characteristic of the type of Fig. 7.10-2a is used (instead of the two semi-infinite characteristics), with $n_1 = 0.5$, $\omega_o = 30$, and $\Delta\theta = 2\pi/3$.

It will be useful to have the ratio ω_5/ω_m (of Fig. 7.10-3) in terms of G_{M1}, G_{M2}, etc. The calculation is approximate but conservative. From ω_m to ω_3 we take the slope as 6 db per octave, so that if $\omega_3/\omega_m = 2^x$, then $G_{M1} = 6x$. From ω_3 to ω_4 the slope is $12(1 - \theta_{M2}/\pi)$ db per octave, where θ_{M2} is the phase margin. Therefore, if $\omega_4/\omega_3 = 2^y$, $G_{M2} = 12y(1 - \theta_{M2}/\pi)$. The value of ω_5/ω_4 is [as in the Bode ideal cutoff characteristic] $\omega_5/\omega_4 = n/(1 - \theta_{M2}/\pi)$. Combining these,

$$\frac{\omega_5}{\omega_m} = \left[\frac{n}{1 - \theta_{M2}/\pi}\right] 2^{(G_{M1}/6)}\, 2^y; \quad y = \frac{G_{M2}}{12(1 - \theta_{M2}/\pi)} \tag{7.10,2}$$

§ 7.11 Maximum Rate at Which | $L(j\omega)$ | May Be Decreased for Conditionally Stable Systems

This section[1] is the counterpart of Section 7.7 (where $|L|$ is specified for $\omega < \omega_o$), except that conditional stability is permitted here. The problem is to decrease $|L|$ for $\omega > \omega_o$, as fast as possible, subject to attaining specified stability margins. This is a problem where the real part of $G(j\omega) = G_1 + jG_2 = \ln L(j\omega)$ is known over one frequency range, and the imaginary part over the balance. As in Section 7.7, we shall take G_1 constant from zero to ω_1 (which is used in place of ω_o of Section 7.7), but the phase lag of $L(j\omega)$ is $r\pi$ from ω_1 to ω_2 (with $r > 1$, for conditional stability), and $\lambda\pi$ (with $\lambda \leqslant 1$ to ensure stability) for $\omega > \omega_2$ (see Fig. 7.11-1). The value of ω_2 is as yet undetermined. Equations (7.6,2) are used to find the unknown parts of $G(j\omega)$. As usual, an arbitrary constant can be added to G_1, such that the first integral on the left-hand side of Eq. (7.6,2) is zero. The result is, for $\omega > \omega_1$,

$$G_1 = \alpha_o - \lambda \ln \left[\left(1 - \frac{\omega_1^2}{\omega^2}\right)^{0.5} + 1\right] \Big/ \left[\left(1 - \frac{\omega_1^2}{\omega^2}\right)^{0.5} - 1\right]$$
$$- (r - \lambda) \ln \left[\left(1 - \frac{\omega_1^2}{\omega^2}\right)^{0.5} + \left(1 - \frac{\omega_1^2}{\omega_2^2}\right)\right] \Big/ \left[\left(1 - \frac{\omega_1^2}{\omega^2}\right)^{0.5} - \left(1 - \frac{\omega_1^2}{\omega_2^2}\right)^{0.5}\right]; \tag{7.11,1}$$

[1] This section is based to some extent on the work of J. Oizumi and M. Kimura, Design of conditionally stable feedback systems. *Trans. IRE* **CT-4**, 157-166 (1957).

for $\omega < \omega_1$,

$$-0.5G_2 = (r - \lambda) \arctan \left[\left(1 - \frac{\omega_1^2}{\omega_2^2}\right)^{0.5} \Big/ \left(\frac{\omega_1^2}{\omega^2} - 1\right)^{0.5} \right] + \lambda \arcsin \frac{\omega}{\omega_1}. \tag{7.11,2}$$

The complete ln $L(j\omega)$ is sketched in Fig. 7.11-1 for the case $r = 2$, $\lambda = 1$, $\omega_2 = 2.87\omega_1$.

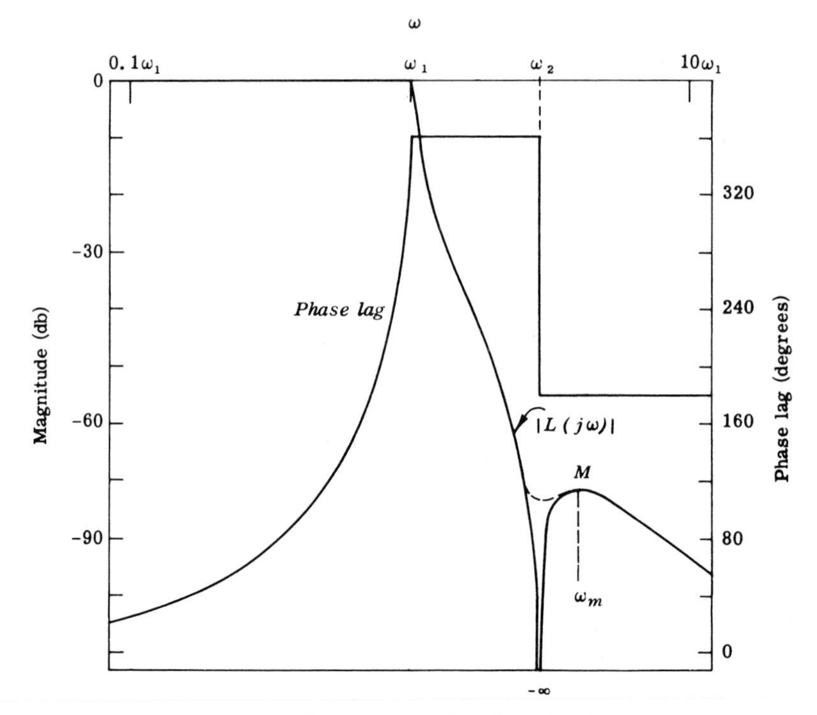

FIG. 7.11-1. Tentative conditionally stable design.

If the 0 db line is chosen tangent to $|L(j\omega)|$ at its peak at M in Fig. 7.11-1, then the system is barely stable. The corresponding Nyquist sketch of $L(j\omega)$ is drawn in Fig. 7.11-2a. It can be seen that the -1 point is encircled in the manner shown, by considering what happens if a very small linear phase segment, like that shown in Fig. 7.11-2b, is added to $L(j\omega)$, and the 0 db line is slightly adjusted. This type of characteristic (Fig. 7.11-2) cannot be used for $r > 3$, for then it is readily seen that the system is unstable.

The characteristic of Fig. 7.11-1 is useful only as a means of arriving at a reasonable approximation to the optimum design. It would be better if we could adjust the phase so as to obtain the dotted modification shown in Fig.

7.11-1 in the neighborhood of M. For then the 0 db line could be picked G_{M1} db below M, and there would be no need for θ_{M1} (see Fig. 7.10-1c for definitions of G_{M1} and θ_{M1}). This can indeed be done as is seen from the following.

Suppose we tackled the problem at the very outset by choosing $\ln|L(j\omega)|$ constant from zero to ω_1 and $\angle(j\omega L) = -r\pi$ from ω_1 to infinity. The resulting

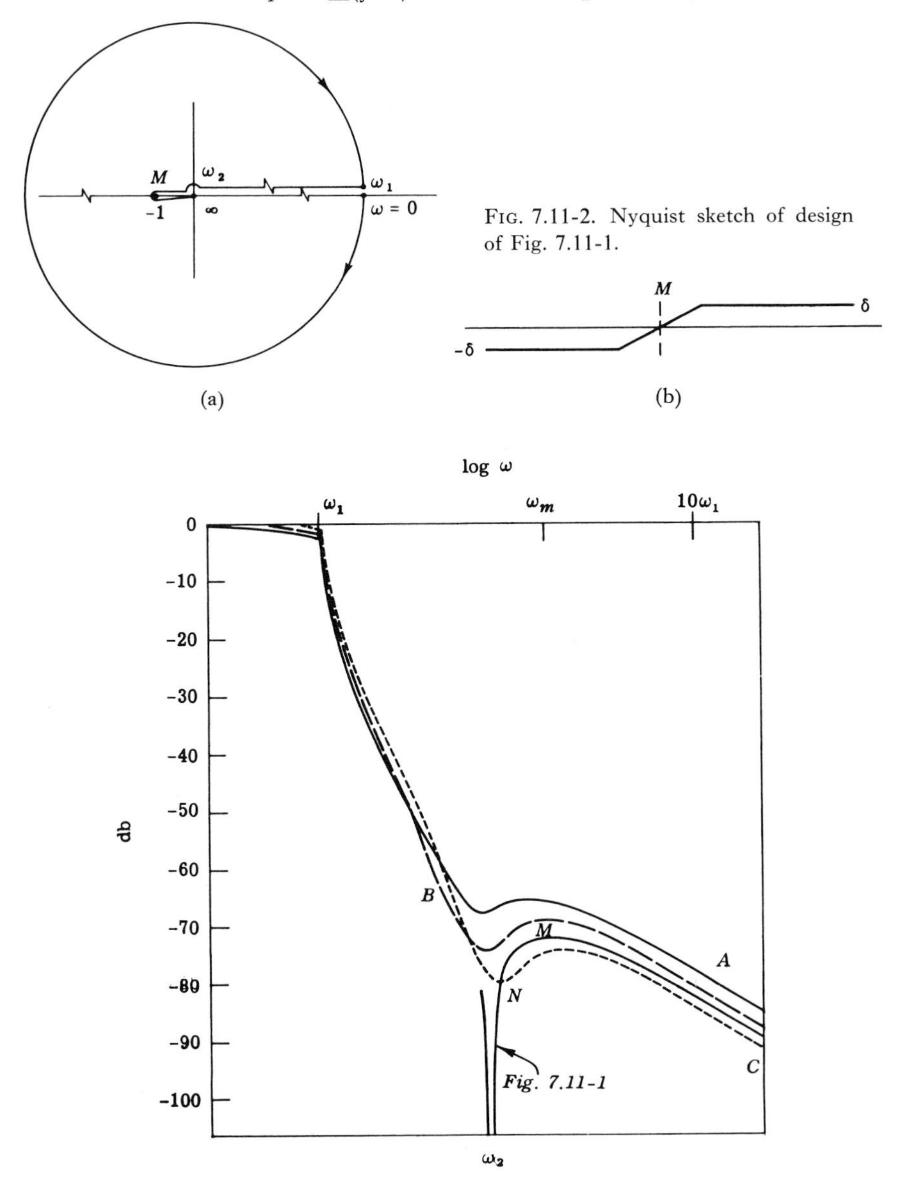

FIG. 7.11-2. Nyquist sketch of design of Fig. 7.11-1.

FIG. 7.11-3. Steps in modifying $L(j\omega)$ of Fig. 7.11-1 at ω_2.

$|L(j\omega)|$ for $\omega > \omega_1$, and $\angle L(j\omega)$ for $\omega < \omega_1$, are immediately available from our work in Section 7.7—in fact, by simply changing the scale of Fig. 7.7-2 by the factor r ($r > 1$). Next, to reduce the high frequency phase lag, we add a linear phase segment such as the one in Fig. 7.10-2a, and in the notation of this figure, $\Delta\theta = \pi$, $n_1 = 0.5$, $\omega_o = 3$. The result is curve A in Fig. 7.11-3. Suppose we used $n_1 = 0.7$, $\omega_o = 3$; the result is curve B in Fig. 7.11-3. If we use $n_1 = 0.7$, $\omega_o = 3.4$, then the result is curve C. These curves should be compared to that of Fig. 7.11-1, which is included in Fig. 7.11-3 for the sake of comparison. We would also like the 180° lag point to coincide with the peak at M, just as we argued that it could be done (Section 7.10). To achieve this, the phase can be adjusted by means of segments of the type of Fig. 7.10-2a, and simultaneously the dip in the region N in Fig. 7.11-3 is partly eliminated. The final conclusion is that the dotted gain characteristic of Fig. 7.11-1, with the 180° lag point at ω_m is achievable, and that in the process the phase is affected in the region ω_1 to ω_m only.

Next consider the region $\omega > \omega_m$ in Fig. 7.11-1. The very close similarity of the phase and gain characteristics with those of curve A for $\omega > \omega_m$ in Fig. 7.10-3 should be obvious by inspection, and, if necessary, by using Fig. 7.11-3 as an intermediate step. Therefore, from here on, we use precisely the same procedure as in Section 7.10 to secure the desired three stability margins. Equation (7.10,2) gives the frequency at which the final asymptote may be introduced, in terms of the gain margins and ω_m.

All that is needed now to finish our problem is to find ω_m in terms of ω_1, and the gain level in Fig. 7.11-1. To locate ω_m in terms of ω_1, note that ω_m is the frequency at which $dG_1/d\omega = 0$. Differentiating Eq. (7.11,1) and setting the derivative equal to zero leads to

$$\zeta^2 \triangleq 1 - \frac{\omega_1^2}{\omega_m^2} = \frac{[m + (r - \lambda)/\lambda]m}{1 + m(r - \lambda)/\lambda} \tag{7.11,3}$$

where

$$m \triangleq [1 - (\omega_1^2/\omega_2^2)]^{0.5}. \tag{7.11,4}$$

Solving Eq. (7.11,1) at $\omega = \omega_m$ results in

$$G_{1(\omega_m)} = G_{M2} = \alpha_o - \lambda \ln \left| \frac{\zeta + 1}{\zeta - 1} \right| - (r - \lambda) \ln \left| \frac{\zeta + m}{\zeta - m} \right|. \tag{7.11,5}$$

Here we always use $\lambda = 1$. In the special case when $r = 2$ and $\zeta^2 = m$,

$$G_{M2} = \alpha_o - 2 \ln |(\sqrt{m} + 1)/(\sqrt{m} - 1)|, \text{ or, in decibels,}$$

$$(G_{M2})_{db} = (\alpha_o)_{db} - 40 \log (1 - m)/(m^{0.5} - 1)^2. \tag{7.11,6}$$

For example, suppose $\alpha_o = 40$ db, both gain margins are 10 db, $\theta_M = 30°$, and $n = 2 = r$ is used. We solve Eq. (7.11,6) for m, i.e., $10 = 40 - 40 \log$

$[(1 - m)/(m^{0.5} - 1)^2]$, giving $m = 0.49$, and finally $\omega_2/\omega_1 = 1.15$, $\omega_m/\omega_1 = 1.40$. Equation (7.10,2) is used to find $\omega_5/\omega_m = 15.3$, so that $\omega_5/\omega_1 = 17.6$. If an unconditionally stable system is used, the result is 38.4, from Eq. (7.7,3). The improvement is by 1.12 octaves.

If $r = 3/2$ is used, them $m = 0.685$, $\zeta = 0.776$, $\omega_m/\omega_1 = 1.60$. However, ω_5/ω_m is unchanged, so that $\omega_5/\omega_1 = 24.5$. Now the improvement is 0.65 octaves. There is hardly any point in taking $r > 2$, since most of the frequency spread from ω_1 to ω_5 is due to the stability margins and is independent of r.

§ 7.12 Loop Transmissions for Systems with Time Delay (Unconditional Stability)

We now consider a system whose loop transmission contains an unavoidable pure time delay T, with transfer function e^{-sT}. We wish to determine (1) the limit of loop gain (for $\omega < \omega_c$) that is achievable for a given crossover frequency ω_c, and (2) the fastest cutoff characteristic for $\omega > \omega_c$.

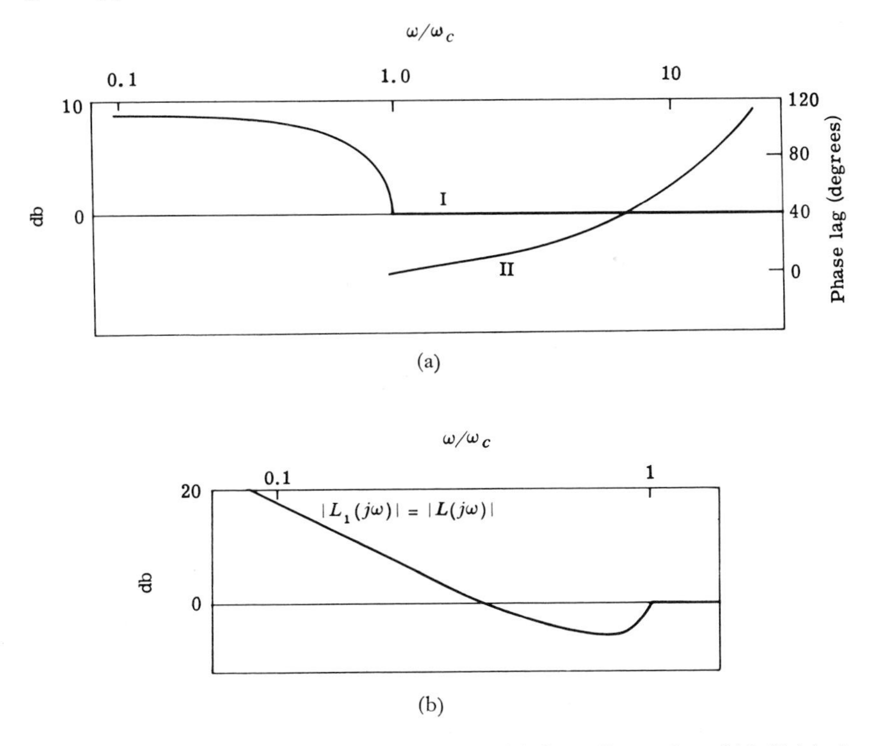

Fig. 7.12-1. Idealized $L(s)$ with pure time delay. (a) Curve I—portion of $|L_1(j\omega)|$ of Eq. 7.12,1 due to time delay; II—part of $\angle L_1(j\omega)$ for $\omega > \omega_c$. (b) Idealized $L_1(j\omega)$ for $\omega_c T = 3$ and 30° phase margin.

To answer the first question, we assume that everything is subordinated to obtaining the maximum loop gain for $\omega < \omega_c$. This problem was discussed in Section 7.8, and it was concluded that L_B in Fig. 7.8-3 presented the best that could be done. In the present case, let $L(s) = L_1(s)e^{-sT}$, and if the maximum permissible phase lag of $L(j\omega)$ for $\omega < \omega_c$ is $\alpha\pi$, then the maximum permissible phase lag of $L_1(j\omega)$ for $\omega < \omega_c$ is $\alpha\pi - \omega T$. In the region $\omega > \omega_c$, $| L_1(j\omega) | = 1$. Equations (7.6,3) are used to solve the problem. The solution for the $\alpha\pi$ portion of the lag has been presented at the end of Section 7.8. For the ωT portion, we must solve the integral $T\int_0^{\omega_c} \zeta^2 d\zeta/(\zeta^2 - \omega^2)(\omega_c^2 - \zeta^2)^{0.5}$. Partial fraction expansion leads to integrands of the form $A + B/(C + \sin\theta)$, where $\zeta = \omega_c \sin\theta$. The final result[1] is

$$| L_1(j\omega) | = [\omega_c + (\omega_c^2 - \omega^2)^{0.5}/\omega]^{2\alpha} - \omega_c T(1 - \omega^2/\omega_c^2)^{0.5} \qquad (7.12,1a)$$

for $\omega < \omega_c$; and for $\omega > \omega_c$,

$$\arg L_1(j\omega) = -2\alpha \arcsin(\omega_c/\omega) + \omega_c T\{\omega/\omega_c - [(\omega^2/\omega_c^2) - 1]^{0.5}\}. \qquad (7.12,1b)$$

If $T = 0$, so that there is no time delay, we get the characteristic plotted as L_B in Fig. 7.8-3. The portions in Eqs. (7.12,1a, b), which are due to the time lag are plotted in Fig. 7.12-1a. For the total $L_1(j\omega)$, these must be subtracted from L_B of Fig. 7.8-3. For example, if $\omega_c T = 3$ and $\alpha = 5/6$, the net result is that shown in Fig. 7.12-1b. It is seen that it is actually not possible to obtain $| L_1(j\omega) | > 1$ for all $\omega < \omega_c$. The larger the value of $\omega_c T$, the smaller is the

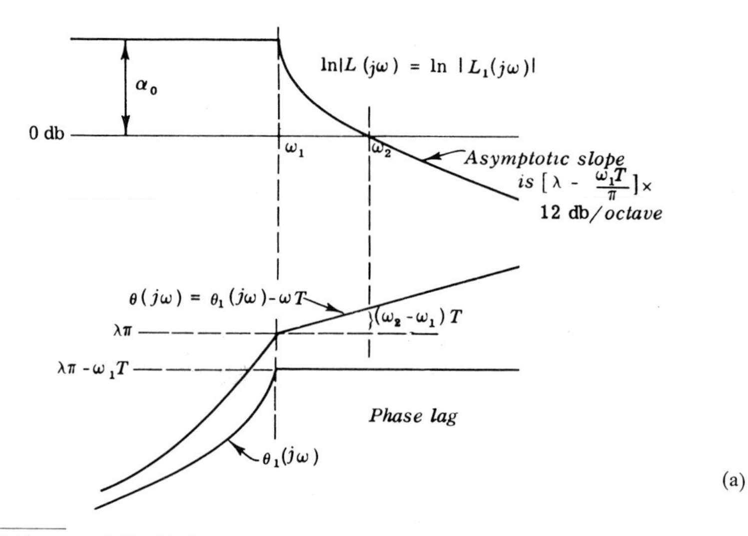

[1] W. Gröbner and N. Hofreiter, "Integraltafel," Part 1, 331.16c,d. Springer, Vienna, 1957.

frequency range over which it is possible to obtain $|L_1(j\omega)| > 1$, in this idealized situation.

Let us now consider the more realistic problem when a constant level of loop gain is wanted from zero to ω_1 and thereafter the loop gain is to be decreased as

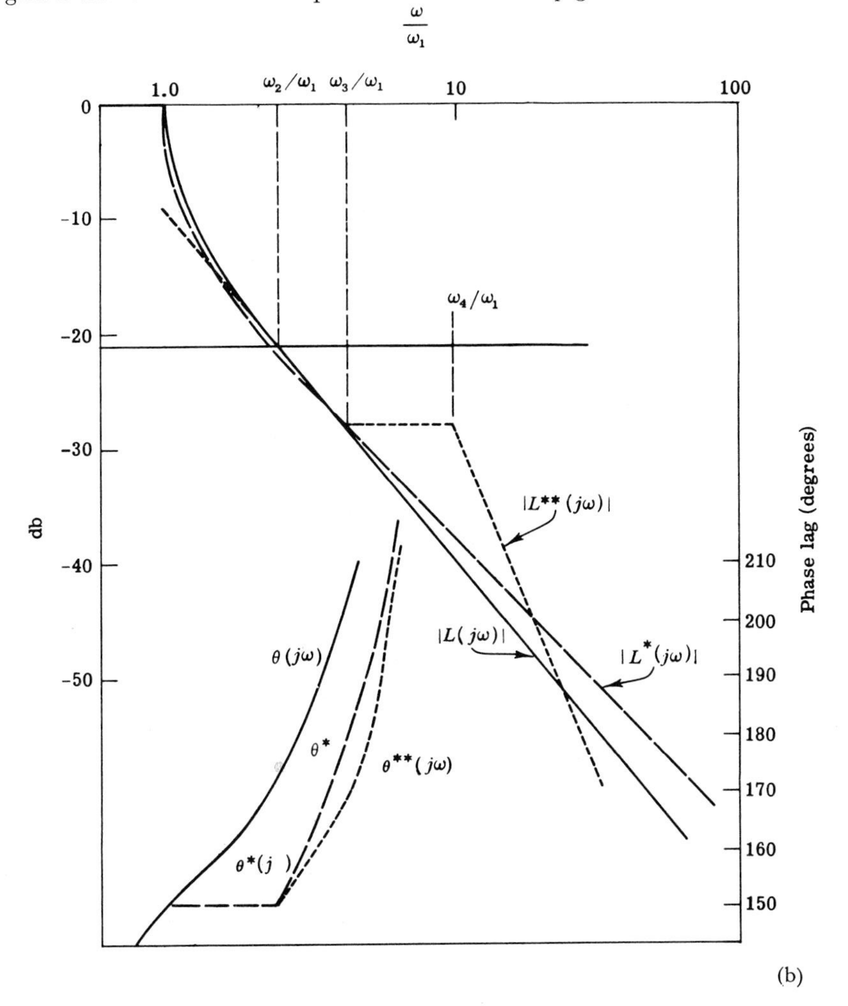

(b)

(c)

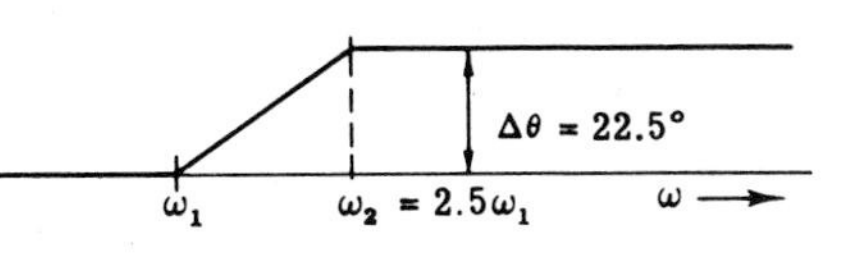

FIG. 7.12-2. Practical design of $L(s)$ with pure time delay $T = \pi/12\omega_1$. (a) (*opposite*) Skeleton of design derived from $L(s)$ with no time delay. (b) Shaping of loop transmission. (c) Linear phase lead segment to cancel part of phase lag due to pure time delay.

fast as possible, subject to attaining given stability margins. For the moment, unconditional stability is also assumed. We approach the problem by choosing $\ln | L_1(j\omega) | = \alpha_o$ for $\omega < \omega_1$, and $\theta_1(\omega) = \angle L_1(j\omega) = -(\lambda\pi - \omega_1 T)$ for $\omega > \omega_1$. As before, $L(s) = L_1(s)e^{-sT}$. This is the problem treated in Section 7.6, whose solution (if $\alpha\pi$ is replaced by $\lambda\pi - \omega_1 T$) is given in Eq. (7.7,1), and is sketched in Fig. 7.7-2; i.e., for $\omega > \omega_1$, $\ln | L_1(j\omega) | = \alpha_o - (2/\pi)(\lambda\pi - \omega_1 T)$ $\ln \{(\omega/\omega_1) + [(\omega^2/\omega_1^2) - 1]^{0.5}\}$ and for $\omega < \omega_1$, $\theta_1(\omega) = -2(\lambda - \omega_1 T/\pi)$ $\sin^{-1}\omega/\omega_1$. Since $L(s) = L_1(s)e^{-sT}$, $\theta(\omega) = -\omega T - 2(\lambda - \omega_1 T/\pi)\sin^{-1}\omega/\omega_1$ and $\theta(\omega) = -\lambda\pi + \omega_1 T - \omega T$ for $\omega > \omega_1$ (Fig. 7.12-2a). The procedure from here on is perhaps best explained by means of a numerical example.

Example. Suppose $\omega_1 T = \pi/12$, $\lambda = 5/6$. Then $\theta(\omega_1) = -\lambda\pi + \omega_1 T = -135°$, and the asymptotic slope of the amplitude characteristic is $-(\lambda - \omega_1 T/\pi)12 = -9$ db per octave (Figs. 7.12-2a and b). Suppose 21 db gain for $\omega < \omega_1$ is desired. We note that the -21 db line in Fig. 7.12-2b intersects the magnitude characteristic at $\omega = 2.5\,\omega_1$. However, the phase of $L(j\omega)$ at this frequency is $-150° + 15° - (2.5)15° = -172.5°$. We therefore introduce the linear phase lead segment sketched in Fig. 7.12-2c. This superposed on the original $\theta(\omega)$ gives a phase lag constant at 150° from ω_1 to ω_2. This is labeled $\theta^*(\omega)$ in Fig. 7.12-2b. Figure 7.10-2b is used to obtain the contribution (due to the linear phase segment) to the gain curve. The latter is now labeled $| L^*(j\omega) |$ in Fig. 7.12-2b. The asymptotic negative slope of $L^*(j\omega)$ is $9 - (12)(225)/180 = 7.5$ db per octave. $| L^*(j\omega) |$ is not strictly constant for $\omega < \omega_1$, but the discrepancy is negligibly small. On $\theta^*(\omega)$, 180° lag is at $\omega/\omega_1 = 4.6$, and corresponds to 7 db gain margin. If this is satisfactory, we can turn to the last part of the problem, which is to introduce the final asymptote as soon as possible. Suppose its slope is -18 db per octave. A simple procedure is to introduce a horizontal step of sufficient duration (as in Section 7.7), so that the phase at low frequencies is not affected, i.e., in Fig. 7.12-2b, $\omega_4/\omega_3 = 11/4.6 = 2.4$. The final design is labeled $L^{**}(\omega)$ in Fig. 7.12-2b. Note that $\omega_4/\omega_1 = 11$ for 21db gain, whereas if the system had no time delay, then, from Eq. (7.7,3), $\omega_4/\omega_1 = 6.3$ only.

It is obvious, in the above, that we intentionally selected a problem which was solvable in a fairly simple manner. Let us make the problem more difficult. Suppose 30 db loop gain is wanted. The -30 db line intersects the amplitude characteristic at $\omega = 5\omega_1$ in Fig. 7.12-2b, at which point the phase lag is 210°. The next step is to introduce a linear phase segment, similar to that in Fig. 7.12-2c, but with $\Delta\theta = 210° - 150° = 60°$. The effect on the amplitude is obtained from Fig. 7.10-2b, with $n_1 = 0.2$. The slope of the asymptote is $9 - (60)(12)/180 = 5$ db per octave. Clearly, the resulting $L^*(j\omega)$ is not satisfactory because the gain margin is only 1.2 db (see Fig. 7.12-3).

In the above, we must allow for the gain margin at the very outset. Suppose it is to be 6 db. The intersection of $| L(j\omega) |$ with the -36 db line is at $\omega = 8\omega_1$, at which point the lag is 255°. A linear phase segment, with $\Delta\theta = 255° - 150° =$

105°, must extend from ω_1 to $8\omega_1$, in order that the phase lag should stay constant at 150° up to $8\omega_1$. The resulting amplitude characteristic has a final asymptote of only $9 - (105)/(180)(12) = 2$ db per octave. The amplitude characteristic,

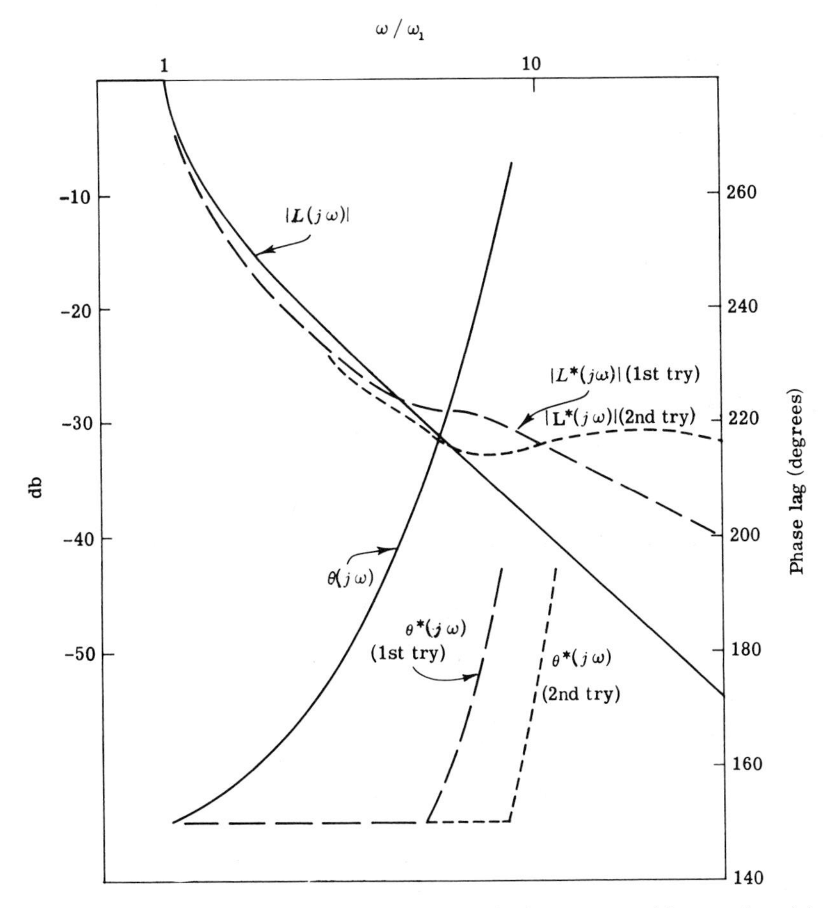

FIG. 7.12-3. Difficulty in design for 30 db loop gain for $\omega \leqslant \omega_1$, with pure time delay $T = \pi/12\omega_1$.

as seen in Fig. 7.12-3, is still not satisfactory, because the gain margin has not been secured. If we continue in the same direction, it is obvious that the difficulty would only be accentuated. Also, it appears that nothing can be done about it, because we are compelled to use linear phase compensation if we are to satisfy the unconditional stability specification. In short, there is an absolute limit on the attainable loop gain and crossover frequency when a time delay is included in the loop transmission.

One obvious upper bound is obtained as follows: Suppose α_o, G_M are, respectively, the desired loop gain for $\omega < \omega_1$ and the gain margin in decibels. The frequency ω_2 is the one at which the horizontal $\alpha_o + G_M$ db line intersects the magnitude characteristic (as in Fig. 7.12-2b). If $\omega_2/\omega_1 = 2^x$, then $(x + 1)12(\lambda - \omega_1 T/\pi) = \alpha_o + G_M$. The effect of the linear phase segment on the asymptote of $| L^*(j\omega) |$ is to reduce its slope by $(\Delta\theta/\pi)12$ db per octave. The linear phase segment must have its $\Delta\theta = (\omega_2 - \omega_1)T$, in order that $\theta(\omega_2) = -\lambda\pi$. The net slope of $L^*(j\omega)$ is then $12(\lambda - \omega_1 T/\pi) - (\omega_2 - \omega_1)T12/\pi$. When this is zero, one can insert the final asymptote ($12n$ db per octave) only at very large frequency. Therefore it is necessary that $(\lambda - \omega_1 T/\pi) > (\omega_2 - \omega_1)T/\pi$, or

$$\lambda > \omega_2 T/\pi = 2^x \omega_1 T/\pi. \tag{7.12,2}$$

For example, when $\lambda = 5/6$, and $\omega_1 T = \pi/12$, we have $\omega_2 T/\pi = (\omega_2/\omega_1)(\omega_1 T/\pi) < 5/6$, $\omega_2/\omega_1 < 10$, $x < 3.3$, and $(\alpha_o + G_M) < 38.7$. We have seen that we were not able to secure the smaller amount of 36 db for $\alpha_o + G_M$, due to the peaking effect of the linear phase segment on the amplitude characteristic, which was ignored in this derivation. The exact maximum for $\alpha_o + G_M$ is difficult to obtain, but Eq. (7.12,2) may be used as an approximate upper bound.

§ 7.13 Conditionally Stable Systems with Pure Time Delay

What improvement is possible in the realizable loop gain level from 0 rps to ω_1, if conditional stability is permitted in the system which includes a pure time lag? Conditional stability definitely permits a significant improvement over the unconditionally stable system. This is seen as follows: In the unconditionally stable system (refer to Fig. 7.12-3), for 40 db of feedback, the crossover point is about a decade away from ω_1, and in that decade the phase lag increases by $9\omega_1 T$. A positive linear phase segment with $\Delta\theta = 9\omega_1 T$ is needed, and this immediately decreases the magnitude of the asymptotic high frequency by $12(9\omega_1 T/\pi)$ db per octave. On the other hand, in the conditionally stable system (refer to Fig. 7.10-3), the analogous point is ω_3, whose distance from ω_1 is determined by $(\omega_3/\omega_m)(\omega_m/\omega_1)$. Now ω_m/ω_1 can be made not much more than one by letting the lag angle be very large and $\omega_3/\omega_m \approx 2^{G_{M1}/6}$ [see discussion preceding Eq. (7.10,2)]. Therefore, if, for example, $G_{M1} = 6$ db, and $\omega_m/\omega_1 = 2$, then $(\omega_3/\omega_1) \approx 4$, which is a definite improvement over the value of 10 for the unconditionally stable system.

We shall now find an upper bound for the realizable gain level, in relation to $\omega_1 T$, stability margins, etc. We commence with the characteristics of Fig. 7.13-1, which are those of $L_1(j\omega)$, where $L(s) = L_1(s)e^{-sT}$. The dotted phase curve is that of $\angle L(j\omega)$ and equals $\angle L_1(j\omega) - \omega T$. So far, we have secured exactly 180° lag at ω_m. Next refer to Fig. 7.10-3. Curve A is precisely $| L_1(j\omega) |$ so far, for $\omega > \omega_m$. At the crossover frequency $\angle L(j\omega) = -\pi - (\omega_3 - \omega_m)T$.

We therefore need a positive linear phase segment extending from ω_m to ω_3, with $\Delta\theta = (\omega_3 - \omega_m)T + \theta_M$, where θ_M is the phase margin. We need another positive linear phase segment from ω_3 to ω_4, with $\Delta\theta = (\omega_4 - \omega_3)T$. The total

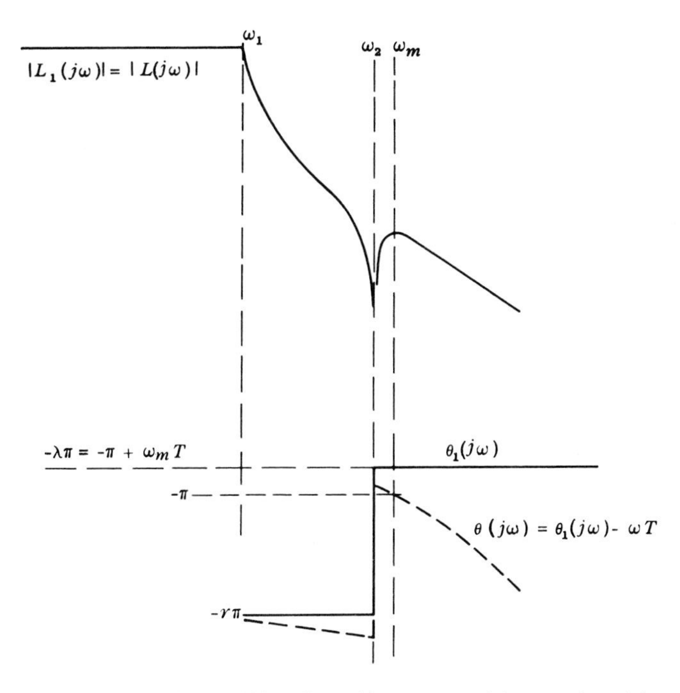

FIG. 7.13-1. $L(j\omega)$ for conditionally stable systems with pure time delay.

value of $\Delta\theta$ is $\Delta\theta_T = \theta_M + (\omega_4 - \omega_m)T$, and it reduces the magnitude of the asymptotic slope of $L_1(j\omega)$ by $12[\theta_M + T(\omega_4 - \omega_m)]/\pi$ db per octave. It is possible to complete the design only if the asymptotic slope is still downward, i.e., if

$$12[\theta_M + T(\omega_4 - \omega_m)]/\pi < 12. \tag{7.13,1}$$

We next relate the above to α_o, which is the level of loop gain for $\omega < \omega_1$, and to G_{M1}, G_{M2}, etc. The desired relations are given in the discussion preceeding Eq. (7.10,2). We have $\omega_4/\omega_m = (\omega_4/\omega_3)(\omega_3/\omega_m) = 2^{x+y}$ with $x = G_{M1}/6$ and $G_{M2} = 12y(1 - \theta_M/\pi)$. Hence inequality (7.13,1) becomes

$$\theta_M + (\omega_1 T)\frac{\omega_m}{\omega_1}\left(\frac{\omega_4}{\omega_m} - 1\right) < \pi. \tag{7.13,2}$$

We need ω_m/ω_1. This is obtained from Eqs. (7.11,3)-(7.11,6). For example, suppose $\omega_1 T = \pi/12$, $r = 2$, and then $\lambda = 1 - \omega_m/12\omega_1$. Try $\alpha_o = 40$ db, $G_{M2} = 10$ db. Equation (7.11,6) is solved, giving $m = 0.49$, so that $\omega_2/\omega_1 = 1.15$.

This value of m is used in Eq. (7.11,3) whose only unknown is ω_m/ω_1. The solution of Eq. (7.11,3) is $\omega_m/\omega_1 = 1.18$, in contrast to 1.15 for the same problem with no time lag. This value substituted into Eq. (7.13,2) with ($\theta_M = \pi/6$, $G_{M1} = 10$ db) gives $0.69\ \pi < \pi$, so that these values are less than the upper bound by a very comfortable margin. It is recalled that when unconditional stability was prescribed, the upper bound was approximately 29 db for α_o, when $G_M = 10$ db. Clearly there is a substantial improvement when conditional stability is allowed.

The actual value of the upper bound can be obtained from Eq. (7.13,2) for given values of $\omega_1 T$, θ_M, etc. For the previous case, Eq. (7.13,2) becomes $(\omega_m/\omega_1) < 1.85$. This makes $\lambda = 0.85$ and, from Eq. (7.11,3), $m = 0.84$. Consequently, from Eq. (7.11,6), $\alpha_o = 64$ db. This number may be increased by using a larger value of r.

These last two sections have dealt with systems whose loop transmission includes a pure time delay. It is clear from the results obtained that pure time delays in the loop transmission should be avoided if at all possible. This work can be extended also to an $L(s)$ which contains rhp zeros. For example, suppose there is only one such zero. Write $L(s) = L_x(s)\,(s - a) = L_x(s)\,(s + a)\,[(s - a)/(s + a)] = L_1(s)\,(s - a)/(s + a) = L_1(s)A(s)$. $A(s)$ is an all-pass function because $|\,A(j\omega)\,| = 1$ for all ω. However, the phase of $A(j\omega)$ changes from 180° at $\omega = 0$ in a clockwise direction to zero as $\omega \to \infty$. As we have previously noted in Section 7.9, it is the change in phase that is important in stability studies. $A(j\omega)$ contributes a total of 180° lag without any contribution to the gain function. $A(s)$ can, in fact, be considered as a first approximation to e^{-sT}, with $aT = \pi/2$. This approximation is exact for the gain, but is poor for the phase except near $\omega = a$. The important point is that $A(s)$ has the same effect as a pure time delay. It decreases the rate at which $|\,L(j\omega)\,|$ may be decreased, and forces $L(j\omega)$ to be controlled over a larger frequency range before it can assume its final asymptotic value. Nevertheless, if $A(s)$ consists of a single pole-zero pair, it is still possible to theoretically obtain any level of loop gain over a given frequency region. However, in extreme cases (when $a \ll \omega_c$), $|\,L(j\omega)\,|$ must decrease at such an extremely slow rate that the design is impractical, so that either the level of loop gain or its bandwidth must be decreased.

When the all-pass network $A(s)$ has more than one pole-zero pair, it is not even theoretically possible to realize any combination of loop gain level and bandwidth. The precise limitations can be obtained by means of the techniques used in Sections 7.12 and 7.13.

Comments on Application of Conditionally Stable Systems

We have developed the theory for linear conditionally stable systems in order to have a reasonably complete treatment of the fundamental capabilities and limitations of single loop linear feedback systems. It is, however, very much in order to comment on difficulties that result when a conditionally stable system is designed. The basic problem is that many nonlinearities affect a linear conditionally stable design much more seriously than they do an unconditionally stable design. Any nonlinearity which has the effect of gain reduction (e.g., saturation, dead zone) tends to destabilize the conditionally stable design. If the effective gain reduction is sufficient, the system becomes unstable. Many practical systems are often driven into saturation because of the magnitude of the input, and/or because of disturbances. It is possible to overcome this problem and retain the benefits of the conditionally stable design by means of a nonlinear stability design, in which effectively the phase of the loop transmission changes with the gain reduction. However, such a nonlinear design is outside the scope of this book.

It should not be concluded that conditionally stable systems are very rare. When the loop transmission crossover frequency ω_c is fixed (see Section 6.14 for the reasons), and when it is important to pack in as much loop transmission magnitude as possible for $\omega < \omega_c$, then the conditionally stable design may provide one possible solution and is indeed used. The linear theory presented here is then useful for estimating the additional benefits to be gained by such a design. It is, however, necessary to consider the possibilities of saturation and if necessary, add the nonlinear stabilization.

§ 7.14 Systems with Unstable Loop Transmissions

Suppose the loop transmission $L(s)$ is constrained to be unstable due to the plant transfer function having one or more rhp poles. [Such poles should not be canceled out (even if they could), because the system response to noise at the plant input, i.e., $P/(1 + L)$, will then have rhp poles.] What is the effect of these poles on the loop gain achievable in the range $\omega < \omega_c$ for a given fixed crossover frequency ω_c, and for $\omega > \omega_c$? What is the effect on the maximum rate at which $L(j\omega)$ may be allowed to decrease? It is readily seen that the effect is precisely the same harmful effect as that due to rhp zeros.

For example, let $L(s) = KL_x(s)/(s - m) = KL_x(s)\,(s + m)/(s + m)\,(s - m) = KL_1(s)A(s)$, with m positive, where $A(s)$ is an unstable all-pass transfer function, and $L_1(s)$ is minimum phase with unity for its leading coefficient. The phase of $A(s) \triangleq (s + m)/(s - m)$ changes from 180° at $\omega = 0$, in a counterclockwise (leading) direction, to zero degrees at infinite frequency. Suppose too that K is positive. If $L_1(s)$ is a constant, then the locus of $L(j\omega)$ is the semicircular locus shown in Fig. 7.14-1a. There is no limit on the loop gain-bandwidth achievable, but in any practical case $L_1(s)$ has its own poles and zeros with more poles than zeros. Therefore the locus of $L(j\omega)$, for positive ω, may be like either one of those shown in Fig. 7.14-1b. Recalling the Nyquist criterion (Section 4.9), $L_A(j\omega)$ of Fig. 7.14-1b, has $Z - P = -1$ but $P = 1$, so $Z = 0$, where Z and P are,

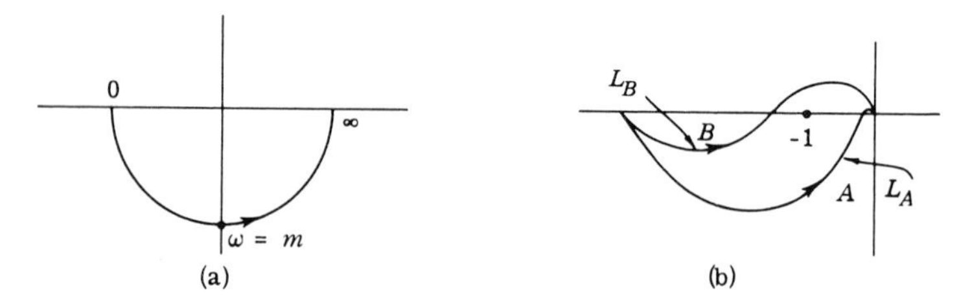

FIG. 7.14-1a to 1h. Nyquist sketches for $L(s)$ with rhp poles. (a) Case $L(s) = A(s) = K(s+m)/(s-m)$, $K > 0$. (b) $L(s) = A(s)L_1(s)$; $L_1(s)$ minimum phase and zero at infinity.

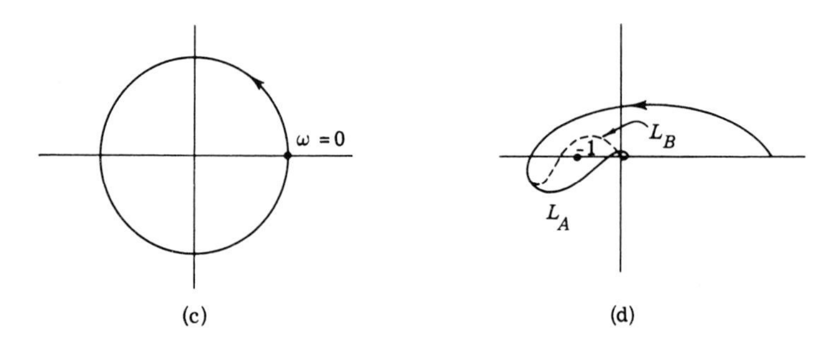

FIG. 7.14-1. (c) Case $L(s) = A(s) = K(s+m)(s+n)/(s-m)(s-n)$, $K > 0$. (d) $L(s) = A(s)L_1(s)$.

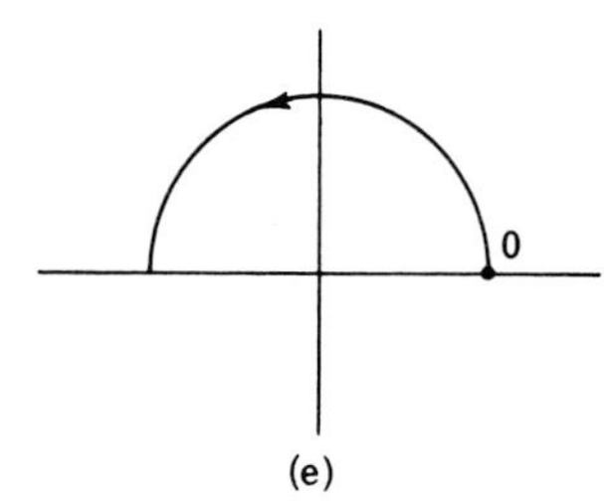

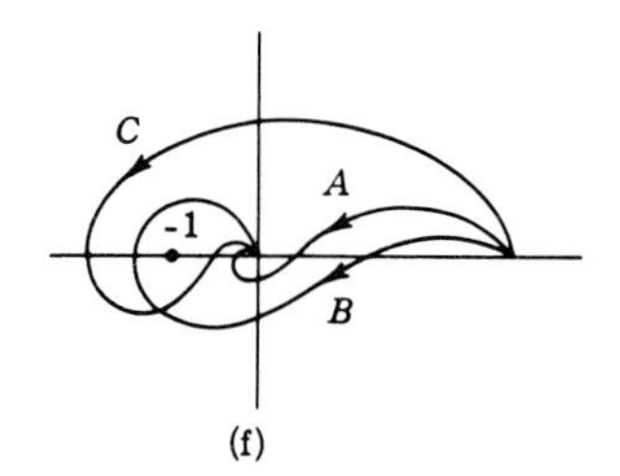

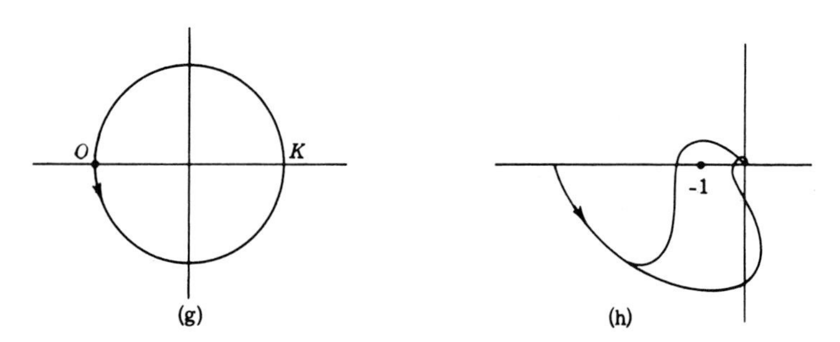

FIG. 7.14-1. (e) $L(s) = A(s) = K(s+m)/(s-m)$, $K < 0$. (f) $L(s) = A(s)L_1(s)$. (g) $L(s) = A(s) = K(s+m)(s+n)/(s-m)(s-n)$, $K < 0$. (h) $L(s) = A(s)L_1(s)$.

respectively, the number of zeros and poles of $L(s)$ in the right half-plane, L_A therefore represents a stable system. On the other hand, for $L_B(j\omega)$. $Z - P = 1$, but $P = 1$, so $Z = 2$ and the system is unstable.

The conclusion is that if $|L(j\omega_c)| = 1$, then $\arg L(j\omega_c) = \theta_L(\omega_c) > -\pi$, in order that $L(j\omega)$ cross the axis to the right of the -1 point. But $\theta_L(\omega_c) = \theta_A(\omega_c) + \theta_{L1}(\omega_c)$. Therefore, $\theta_{L1}(\omega_c) > -\pi - \theta_A(\omega_c)$ [θ_A is $\arg A(s)$]. However, $-\pi < \theta_A(\omega_c) = -\psi < 0$, so

$$\theta_{L1}(\omega_c) > -\pi + \psi, \qquad \text{with } \pi > \psi > 0. \tag{7.14,1}$$

If $L(s)$ did not have a rhp pole, the requirement would be that $\theta_{L1}(\omega_c) > -\pi$; but on account of the rhp pole, the lag angle of $L_1(j\omega_c)$ must be less than 180° by the amount ψ. The fund of phase lag available for reducing $|L_1(j\omega)|$ is reduced by the magnitude of ψ. Hence $|L_1(j\omega)|$ for $\omega < \omega_c$ cannot be as large as when $\psi = 0$. The magnitude of ψ is a good measure of the degradation in the loop transmission capabilities. Let $\omega_c = \nu m$ and then $|\psi| = \pi - 2 \arctan \nu$. The larger the crossover frequency ω_c, the smaller is $|\psi|$ and the smaller is the degradation. This is in sharp contrast to the case when $L(s)$ has a rhp zero, i.e., where $L(s) = L_2(s)\, A_2(s)$ with L_2 minimum-phase and $A_2(s) = (m - s)/(m + s)$. At $\omega_c = \nu m$, $\psi_2 = \angle A_2 = -2 \arctan \nu$. Here too the lag angle of $L_2(j\omega_c)$ must be less than 180° by the amount $|\psi_2|$, exactly as for L_1 in the above. However, $|\psi_2|$ and the degradation *increases* as $\nu = \omega_c/m$ is increased.

This argument is extended to $L(s)$ with two poles in the right half-plane, and it will then be clear what the results are for more rhp poles. Again we have $L(s) = KL_1(s)A(s)$; $A(s)$ is an unstable all-pass function, with two rhp plane poles, and K is again taken as positive. The locus of $A(j\omega)$ is the circle shown in Fig. 7.14-1c. The locus of $L(j\omega)$ for $\omega > 0$ must be like one of those sketched in Fig. 7.14-1d. L_A represents a stable system, because $Z - P = -2$, with $P = 2$, while L_B represents an unstable system, because $Z - P = 0$. The criterion for stability is that if $|L(j\omega_c)| = 1$, $\theta_L(\omega_c) = \theta_{L1}(\omega_c) + \theta_A(\omega_c) > \pi$. However, $2\pi > \theta_A(\omega_c) = 2\pi - \psi > 0$, and therefore $\theta_{L1}(\omega_c) > -\pi + \psi$, which is exactly the same as (7.14,1). In this case $|\psi| = 2\pi$ at $\omega = 0$ and decreases to zero at infinite frequencies. Hence there is a lower bound on the achievable crossover frequency. It is certainly essential that $\omega_c > \omega_x$, where $|\psi(j\omega_x)| = \pi$. In the analogous stable nonminimum phase problem, there is an upper bound on the achievable crossover frequency, i.e., $\omega_c < \omega_x$. The conclusion is that if $L(s)$ has rhp poles, then the system is perforce conditionally stable and it may be necessary to use a larger value of ω_c than the desired benefits of feedback inherently require. It is usually undesirable to do this (see Section 6.14).

Finally, consider $K < 0$. If $A(s)$ has one pole in the right half-plane, the locus of $KA(j\omega)$ is that shown in Fig. 7.14-1e, and that of $L(j\omega) = KL_1(j\omega)A(j\omega)$ must be similar to either A or B in Fig. 7.14-1f. In both cases the system is unstable.

C is impossible if $P = 1$, because $Z - P = -2$ and then $Z = -1$. If $A(s)$ has two poles in the right half-plane, the locus of $KA(j\omega)$ is the circle shown in Fig. 7.14-1g, while that of $L(j\omega)$ must be similar to either one of those shown in Fig. 7.14-1h. In both cases, the system is unstable. In all cases when $L(s)$ has more poles than zeros and $A(s)$ has rhp poles, the system is unstable for K negative.

§ 7.15 Systems with Combined Positive and Negative Feedback; Zero-Sensitivity Systems

The literature on feedback systems is replete with claims for the extraordinary benefits available from systems with minor positive feedback loops. The fallacy in these claims is usually due to the lack of consideration of sensitivity (or loop transmission) as a function of frequency. The infinite loop gain that is achievable by an $L(s)$ which contains a minor positive feedback loop on the verge of instability appears at first sight to be very attractive. For when $L(s)$ is infinite, the sensitivity is zero, and this, of course, is the ultimate in the use of feedback. However, it must be noted that zero sensitivity is thereby achieved at only a single frequency, and the benefits are consequently not as great as they first appear. Furthermore, when the functional dependence on frequency is ignored, it is impossible to ascertain the price that must be paid for the benefits.

The scientific theory of feedback, due to Bode, to which this chapter has been devoted, can be readily used to investigate zero sensitivity systems, and this will now be done.

In order to compare a zero-sensitivity system with a more conventional system, we shall begin with a conventional system, and add to it a zero-sensitivity portion. For our conventional system, we choose a loop transmission which has constant magnitude for $0 < \omega < \omega_1$, and a lag of $\alpha\pi$ radians for $\omega > \omega_1$. The latter form of $L(j\omega)$ was studied intensively in Section 7.7. To this conventional loop transmission denoted by $L_c(j\omega)$, we add the zero-sensitivity portion $L_z(j\omega)$, i.e., $L = L_c L_z$. L_z is defined as follows: $| L_z(j\omega) | = |(s + \omega_o)^2/(s^2 + \omega_o^2)|$ for $0 < \omega < \omega_1$ (where $\omega_o < \omega_1$); and the phase of $L_z(j\omega)$ is zero for $\omega > \omega_1$. From Section 7.6, we know that this characterization of L_z suffices to fix L_z over the entire frequency range.

L_z was chosen with poles at $\pm j\omega_o$ in order to secure infinite loop gain, and consequently zero sensitivity, at $\pm j\omega_o$. Two zeros were assigned at $-\omega_o$ in order that the low and high frequency properties of $L = L_c L_z$ should be unaffected by L_z. Zero phase lag was assigned to L_z for $\omega > \omega_1$, because this is consistent with our desire that L_z does not affect the high frequency properties of L.

The magnitude of L_z for $\omega < \omega_1$ (and assuming $\omega_o \ll \omega_1$) is shown on a Bode plot in Fig. 7.15-1. The area under the curve in Fig. 7.15-1 therefore represents the added feedback benefits in decibels—log frequency measure, due to the zero sensitivity portion $L_z(j\omega)$. It is seen in Fig. 7.15-1 that the additional feedback area due to L_z is not so very impressive, because of the steepness of the curve near ω_o.

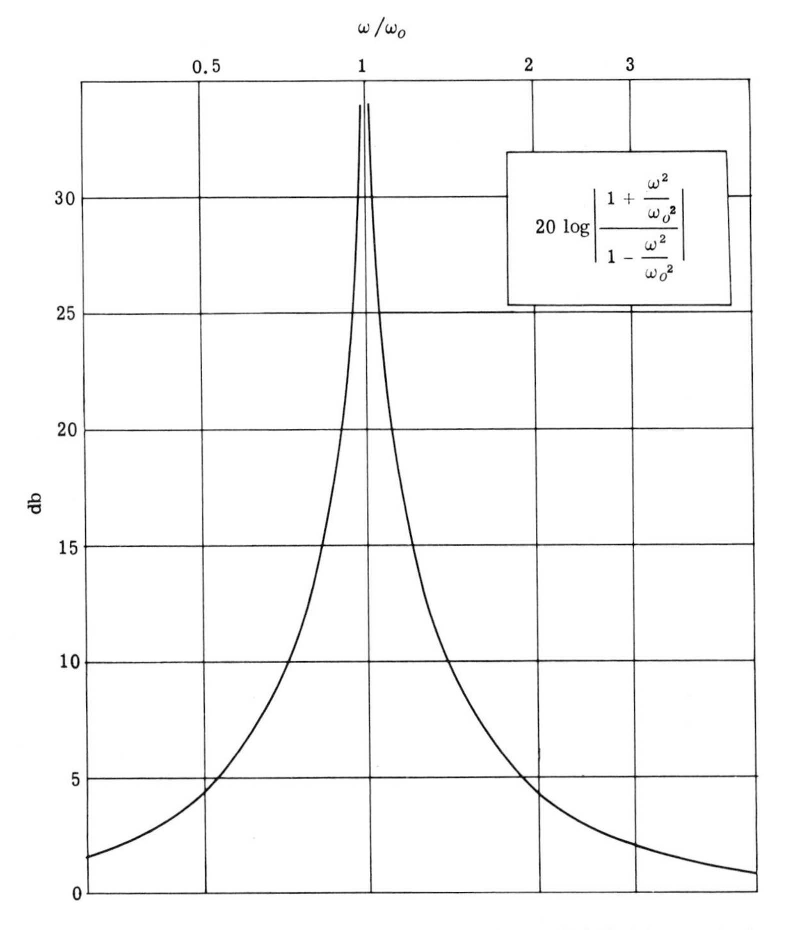

FIG. 7.15-1. Frequency response of $(s + \omega_o)^2/(s^2 + \omega_o^2)$ ($|L_z|$ for $\omega < \omega_1$).

Equations (7.6,2) are next used to find the balance of L_z, i.e., $|L_z|$ for $\omega > \omega_1$, and the phase of L_z for $\omega < \omega_1$. These equations become

$$\int_0^{\omega_1} d\zeta \frac{\ln [(\omega_o^2 + \zeta^2)(\omega_o^2 - \zeta^2)^{-1}]}{(\zeta^2 - \omega^2)(\omega_1^2 - \zeta^2)^{0.5}} = \frac{0.5\pi G_2}{\omega(\omega_1^2 - \omega^2)^{0.5}}, \qquad \text{for } \omega < \omega_1$$

$$= \frac{-0.5\pi G_1}{\omega(\omega^2 - \omega_1^2)^{0.5}} \qquad \text{for } \omega > \omega_1 \qquad (7.15,1)$$

where $G_2 = \arg L_z(j\omega)$, and $G_1 = \ln | L_z(j\omega) |$. To perform the integration, it is convenient to let $\zeta/\omega_1 = y(1 + y^2)^{-0.5}$, and then we must evaluate integrals of the form $I_1 \triangleq \int_0^\infty \ln(y^2 - a^2)dy/(y^2 - b^2)$, and $I_2 \triangleq \int_0^\infty \ln(y^2 + d^2)dy/(y^2 - b^2)$, where $a^2 = \omega_o{}^2/(\omega_1{}^2 - \omega_o{}^2)$, $b^2 = \omega^2/(\omega_1{}^2 - \omega^2) > 0$, when $\omega < \omega_1$, and $d^2 = \omega_o{}^2/(\omega_o{}^2 + \omega_1{}^2) < 1$. In order to find I_1, consider $J \triangleq \int_{C_1} \ln(s^2 + a^2)ds/(s^2 + b^2)$, with $s \triangleq x + jy$, and C_1 defined in Fig. 7.15-2 for the case $a < b$, which corresponds to $\omega > \omega_o$. For the case $\omega < \omega_o$, $a > b$, but otherwise the contour is the same. The branch cuts at $\pm ja$ lie entirely in the left half-plane.

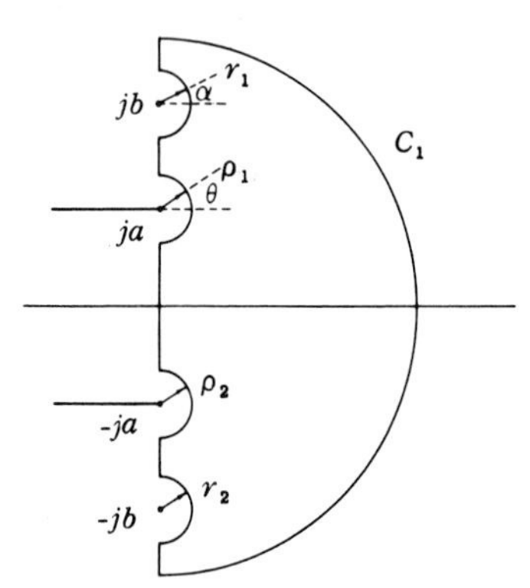

FIG. 7.15-2 Contour for evaluating integrals I_1, I_2.

From Cauchy's residue theorem, $J = 0$. The integrals associated with ρ_1, ρ_2 in Fig. 7.15-2 are zero because the integrands are of the form $\rho \ln \rho$. The integral over the infinite semi-circular contour is zero, because the integrand behaves at infinity as $s^{-2} \ln s$. In the arc centered at jb, $(s^2 + a^2) = | a^2 - b^2 | e^{j\pi}$, but in the one centered at $-jb$, $(s^2 + a^2) = | a^2 - b^2 | e^{-j\pi}$. Consequently, the former contributes $(\pi/2b) \ln | a^2 - b^2 | + j\pi^2/2b$, while the latter contributes $-(\pi/2b) \ln | a^2 - b^2 | + j\pi^2/2b$, which add to $j\pi^2/b$. The real parts of the integrals along the $j\omega$ axis cancel out, and the final result is $\int_0^\infty \ln | a^2 - y^2 | dy/ (b^2 - y^2) = -\pi^2/2b$ for $a < b$. $I_2 = \int_0^\infty \ln(y^2 + d^2)dy/(y^2 - b^2)$ is available in tables,[1] and is $I_2 = \pi b^{-1} \arctan b/d$.

It is now possible to find G_2 in Eq. (7.15,1) for $\omega_1 > \omega > \omega_o$. The result is (for $\omega_1 > \omega > \omega_o$)

$$G_2 = \angle L_z = -\pi + 2 \arctan \frac{\omega}{\omega_o} \left[\frac{1 + \omega_o{}^2/\omega_1{}^2}{1 - \omega_o{}^2/\omega_1{}^2}\right]^{0.5}. \qquad (7.15,2)$$

For $\omega < \omega_o < \omega_1$, the contour C_1 of Fig. 7.15-2 is used, except that $a > b$. But now, along the arc at r_1, $(s^2 + a^2) = | a^2 - b^2 |$ with zero angle, and similarly along the arc at r_2, $(s^2 + a^2)$ has zero angle. The result (for $\omega < \omega_o < \omega_1$) is

$$\arg L_z = G_2 = 2 \arctan \frac{\omega}{\omega_o} \left[\frac{1 + \omega_o{}^2/\omega_1{}^2}{1 - \omega^2/\omega_1{}^2}\right]^{0.5}. \qquad (7.15,3)$$

Thus $\arg L_z$ has a discontinuity of π radians at $\omega = \omega_o$, which is as expected, in view of the poles of L_z at $\pm j\omega_o$.

[1] D. Bierens de Haan, "Nouvelles Tables d'Intégrales Définies," No. 15, T136. Hafner Publ., New York, 1957.

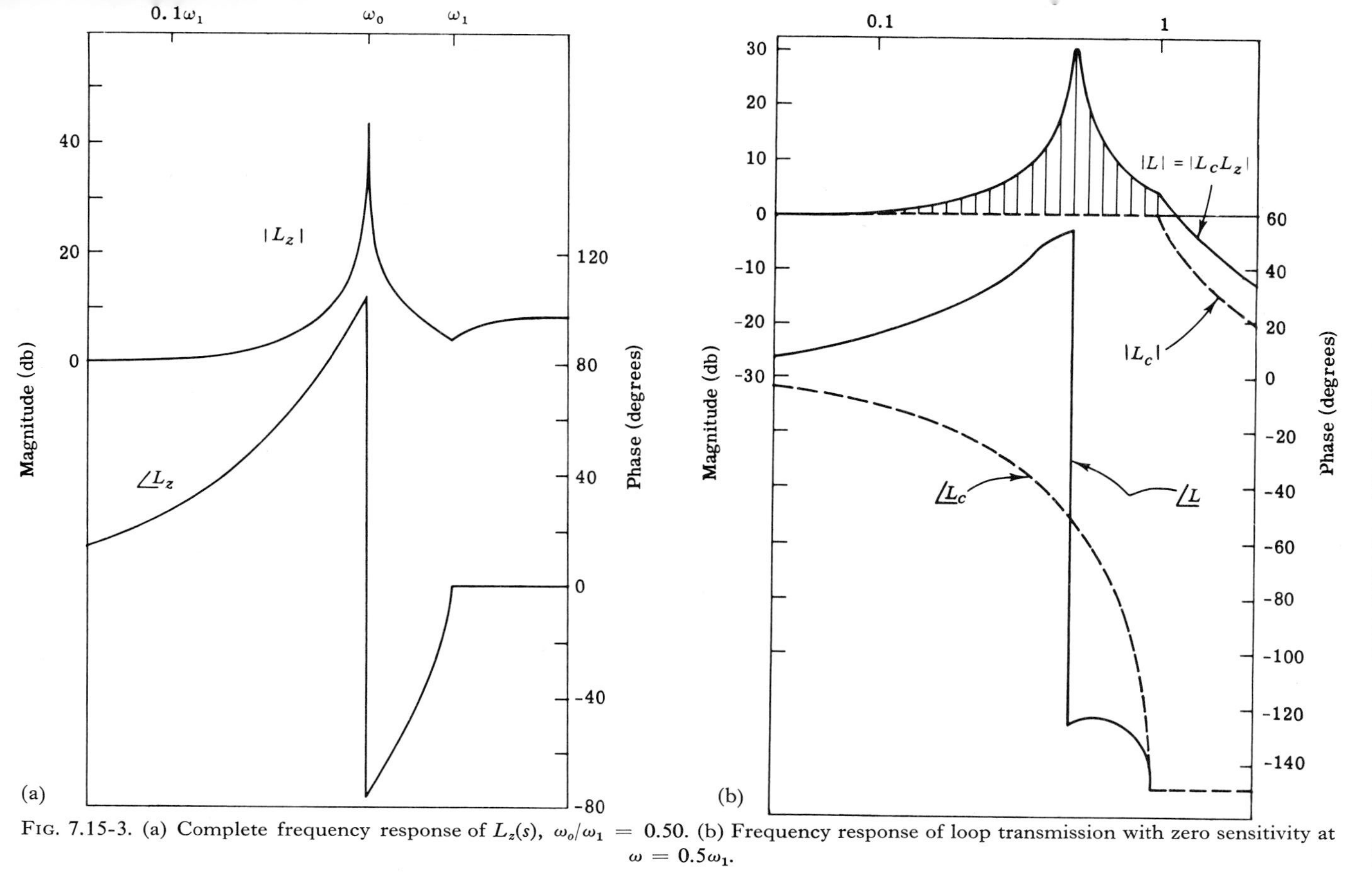

FIG. 7.15-3. (a) Complete frequency response of $L_z(s)$, $\omega_o/\omega_1 = 0.50$. (b) Frequency response of loop transmission with zero sensitivity at $\omega = 0.5\omega_1$.

For $\omega > \omega_1$, we must evaluate integrals of the form

$$I_3 \triangleq \int_0^\infty \frac{\ln | y^2 - a^2 | \, dy}{y^2 + g^2}, \qquad I_4 \triangleq \int_0^\infty \frac{\ln (y^2 + h^2) dy}{y^2 + g^2}$$

where $g^2 = \omega^2/(\omega^2 - \omega_1{}^2)$, $h^2 = \omega_o{}^2/(\omega_o{}^2 + \omega_1{}^2)$. The solution[1] for I_3 is $I_3 = (\pi/2g) \ln (a^2 + g^2)$, and[2] for I_4, $I_4 = (\pi/g) \ln (g + h)$. Putting these together, the result is, for $\omega > \omega_1$,

$$-\ln | L_z | = \ln \frac{1 - \omega_o{}^2/\omega_1{}^2}{1 + \omega_o{}^2/\omega_1{}^2} - 2 \ln \left[\frac{\omega/\omega_1}{(\omega^2/\omega_1{}^2 - 1)^{0.5}} + \frac{\omega_o/\omega_1}{(1 + \omega_o{}^2/\omega_1{}^2)^{0.5}} \right]$$

$$+ \ln \left[\frac{\omega_o{}^2/\omega_1{}^2}{1 - \omega_o{}^2/\omega_1{}^2} + \frac{\omega^2/\omega_1{}^2}{\omega^2/\omega_1{}^2 - 1} \right]. \tag{7.15,4}$$

Equations (7.15,2)-(7.15,4) have been evaluated for $\omega_o/\omega_1 = 0.50$. The resulting $L_z(j\omega)$ is sketched in Fig. 7.15-3a. Now we add L_z (logarithmically) to a more conventional design, which is denoted as L_c. The latter is taken from Fig. 7.7-2, with $\alpha = 5/6$. The two designs, L_c and $L = L_c L_z$, are compared in Fig. 7.15-3b.

Suppose the feedback level is to be 40 db. With L_c alone, in Fig. 7.15-3b, the intersection with the 0 db line (the present $-$ 40 db line) is at $\omega/\omega_1 = 8$, whereas with $L = L_c L_z$, the intersection is at $\omega/\omega_1 = 13.8$. This ratio of $13.8/8 = 1.73$ (0.79 octaves) is one part of the price that must be paid for securing the additional gain-bandwidth area indicated by the shaded region in Fig. 7.15-3b. Thus L must be controlled over 0.79 octave more than L_c before it can be allowed to merge with its high frequency asymptote. However, the smaller the ratio of ω_o/ω_1, the less is this price, because the smaller is the infinite frequency level of L_z at $\omega \gg \omega_1$. (If we use an L_c design with the same high frequency asymptote as for $L = L_c L_z$, then 8 more db for *all* $\omega < \omega_1$ is obtainable in place of the shaded area due to $| L_z |$.)

In order to obtain improvements more impressive than that shown in Fig. 7.15-3b, it is possible to use a larger number of zero-sensitivity sections, each tuned to a different frequency. Such a design is sketched in Fig. 7.15-4. It has resonances at $\omega/\omega_1 = 0.01$, 0.02, 0.05, 0.10, 0.20. When superimposed on L_c (with $\alpha = 1$), the resulting phase lag exceeds 180° for $0.20 < \omega/\omega_1 < 0.25$, so the resonance at 0.20 should probably be omitted. Again the results are not overly impressive, especially when one considers that the design labeled L_x is so much easier to realize (L_x requires negative real poles and a complex zero pair), and can be designed so that it is unconditionally stable. The loop gain areas for $\omega < \omega_c$ (ω_c is the crossover frequency), available from the design of the kind shown in Fig. 7.8-2c, are much more impressive than those available

[1] D. Bierens de Haan, "Nouvelles Tables d'Intégrales Définies," No. 16, T136. Hafner Publ., New York, 1957.

[2] W. Gröbner and N. Hofreiter, "Integraltafel," Part 1, Table 325, No. 18. Springer, Vienna; de Haan, T136, No. 13.

from using large numbers of zero-sensitivity sections. The reason is that the design of Fig. 7.8-2c uses all the permissible phase lag, whereas it is impossible to do this with a zero sensitivity design. Figure 7.8-2c is also easier to implement.

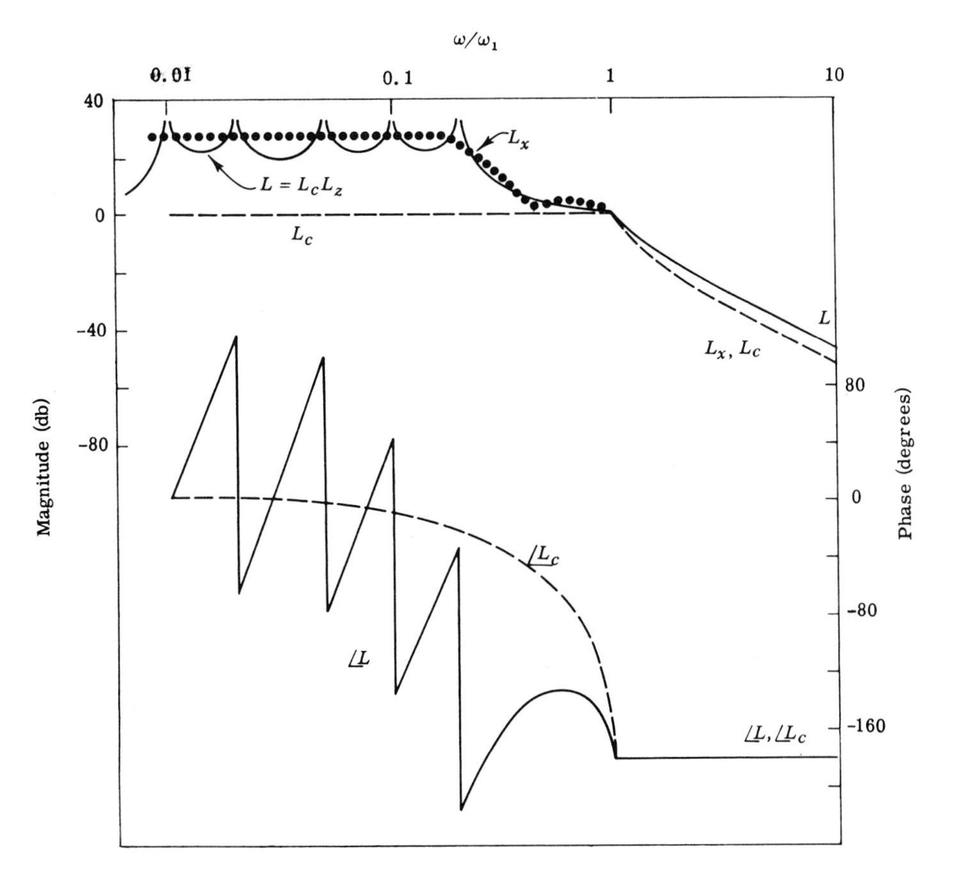

FIG. 7.15-4. $L(j\omega) = L_c L_z$ with 5 zero-sensitivity sections, compared with other loop transmissions.

Of course, sections tuned to resonance are mandatory if one insists on zero sensitivity at a specific frequency or frequencies. The significant question is whether there is any value in a design in which the sensitivity is zero at a specific frequency. The problem of choosing sensitivity specifications is discussed in Chapter 9. We therefore postpone discussion on this point.

It should be apparent from the above that no startling improvements in loop gain-frequency areas, for given crossover frequencies, are available from zero-sensitivity sections. It is apparent that the sensible approach is to decide on the feedback benefits needed for the design, from which the loop transmission as a function of frequency becomes available. Thereafter the realization of the desired loop transmission as a function of frequency is a problem of active network synthesis. Positive feedback around active elements is a legitimate tool of active network synthesis, and it should be regarded as such and no more. For example, if very small sensitivity at a specific frequency should be desired, then positive feedback is one convenient means of realizing a pole of $L(j\omega)$ very close to the $j\omega$ axis. A transformer may be used to obtain a very large amplifier gain (larger than the vacuum-tube amplification factor), over a very small frequency range, providing that the gain-bandwidth product is within the capabilities of the vacuum-tube. But it may be more convenient to achieve this specific relation between gain and frequency, by means of positive feedback. However it has never been shown that the gain-bandwidth available from an active device can be increased by using positive feedback.

§ 7.16 Summary

It is extremely important that the area of applicability of the contents of this chapter be thoroughly understood. The theory applies to one or two-degree-of-freedom systems, wherein there is one, and only one, loop transmission around the plant. We are concerned only with the feedback properties relating to the plant, i.e., plant parameter variations, disturbances at any point in the plant, and those plant nonlinearities which are equivalent to parameter variations or disturbances (Section 8.19). Also, there is the constraint that signals from the command point can reach the output only via the plant [i.e., t_{oi} in the fundamental feedback equation (2.2,7) is zero]. In such systems, the feedback properties are entirely determined by the single loop transmission function L. L must be of the form $L = PM$, with M representing the compensation added in the loop.

In this chapter we have found that for minimum-phase plants there are no theoretical limitations on the realizable gain-bandwidth of L. This statement is correct only if it is assumed that there is available (for use in M, in $L = PM$) a reservoir of active devices with infinite gain-bandwidth capacity. This assumption is realistic in most control systems because of their relatively small bandwidths. In nonminimum phase plants, there are theoretical limits on the achievable gain-bandwidth of L.

However, even in minimum-phase plants, noise and plant nonlinearities make it generally desirable to reduce $|L|$ as quickly as possible, and to let it assume its natural value at as low a frequency as possible. By the latter is meant that $|M|$ in $L = PM$ should be allowed to be less than one at as low a frequency as possible. The optimum means of achieving this objective have been derived in this chapter. The optimum means are different for conditionally stable systems and for nonminimum phase plants.

There are two costs that must be borne for the benefits of feedback. One is the gain (over the required bandwidth) of L. Here we refer to signal gain, and not to power gain. The gain needed for L is not available as over-all system output/input gain. Thus a positioning system does not inherently require a motor with its infinite gain at zero frequency, since a simple mechanical linkage is sufficient for the desired input-output relations. The motor is needed because of the desired zero-sensitivity properties at zero frequency (which corresponds to zero steady state error to step inputs despite any parameter variations, so long as the motor can turn). The above is one of the costs.

The second cost arises out of the stability problem, and is related to the fact that a physically realizable $L(s)$ is a function of a complex variable. Therefore, independent realization of its real and imaginary parts is impossible. This cost manifests itself in that $|L(j\omega)|$ cannot decrease beyond a certain rate, if the system is to be stable. Consequently, $|L(j\omega)|$ must be controlled well beyond the frequency range in which the feedback is of benefit to the system.

The above costs of feedback cannot be reduced by resorting to systems with combined positive and negative feedback. The latter should be regarded as a tool in active network synthesis.

The only way to change any of the above conclusions is by changing the problem constraints. If internal plant variables are available for feedback purposes, then there can be several plant loop transmissions instead of only one. If the constraint of zero leakage is dropped (i.e., $t_{oi} \neq 0$), then parallel plants are possible. Such systems are studied in Chapter 8. If time-varying, modulated, or nonlinear compensation networks are permitted, then the constraints on $L(s)$ and the relations between its real and imaginary parts are no longer the same as before.

CHAPTER EIGHT

Advanced Topics in Linear Feedback Control Theory

§ 8.1 Introduction

A variety of advanced topics are treated in this chapter. In order to establish their relationship to the previous material, let us first briefly review Chapters 5-7.

Chapter 5 dealt with one-degree-of-freedom systems, which permitted the realization of the desired system transfer function $T(s)$. However, the sensitivity function, which determines the effect of plant variations and the system response to plant disturbances, could not in general be independently realized unless the properties of $T(s)$ were sacrificed. Two-degree-of-freedom systems were presented in Chapter 6. Independent realization of $T(s)$ and the sensitivity function $S(s)$ was then possible. It was found that the price paid for the benefits of feedback was in the gain and bandwidth of the loop transmission function $L(s)$. It was noted how noise in the feedback loop is amplified by the amount that $|L(j\omega)| > |P(j\omega)|$, in the range in which $|L(j\omega)| < 1$. For this and other reasons (see Section 6.14) it was considered desirable to realize the benefits of feedback with an $L(s)$ which has the minimum possible gain and bandwidth. Chapter 7 was therefore devoted to a detailed study of the optimum loop transmission function for minimum phase, nonminimum phase, and open-loop unstable systems. The role of minor positive feedback loops, zero-sensitivity systems, and conditionally stable systems was also explored.

It is evident, from the results of Chapter 7, that the best possible two-degree-of-freedom design may require a loop transmission function which is so large in the high frequency range that the design is unsatisfactory, because of the feedback transducer noise problem. In addition, if the plant is nonminimum phase, with at least two zeros in the right half-plane (rhp), then there is an absolute limit on the realizable loop transmission crossover frequency, even when there is no feedback transducer noise problem. In the face of these difficulties, either the design specifications or the system constraints must be modified. The latter course is followed in this chapter.

(*A*) In Sections 8.2-8.7, it is assumed that internal plant variables are available for feedback purposes. A feedback structure with several independent loop transmission functions may be built around such a plant. It may be possible to apportion the feedback burden between the various loop transmission functions, in such a manner as to significantly ameliorate the feedback transducer noise problem. Sections 8.2 and 8.3 are devoted to multiple-loop design for attenuating disturbances and Sections 8.4-8.7 concentrate on the parameter variation problem.

(*B*) A different kind of constraint is dropped in Sections 8.8 and 8.9. In the multiple-loop feedback systems of Sections 8.2-8.7, the sensitivity reduction is at best proportional to the sum of the various loop transmissions; therefore the loop transmission with the largest magnitude dominates. Different terms are allowed to dominate over different frequency ranges. However, a radical departure from this situation is obtained when the constraint of zero leakage transmission is dropped. This leads to parallel plants and to sensitivity reduction which is proportional to the product of loop transmissions, rather than to their sum. One important result of this property is that the total effective loop transmission can be decreased at a very fast rate, without the system becoming conditionally stable. The restrictions on the effective feedback achievable in nonminimum phase plants and in open-loop unstable plants, which apply to two-degree-of-freedom plants, are considerably eased in parallel plants. In short, the limitations deduced in Chapter 7 are radically altered in the parallel plant feedback system.

(*C*) When the parameter variation problem was considered in previous chapters, it was usually assumed that the specifications permitted only small variations in the system response. Such specifications were assumed in order that the resulting severe demands on the feedback system would reveal the design difficulties and would compel consideration of more advanced systems. This is a reasonable way of exploring the boundaries of a subject—by making exorbitant demands, seeing what gives way, and then considering how to remedy the situation. In Sections 8.11-8.14, the sensitivity specifications are made much more moderate, and resemble what may reasonably be expected in practical systems. Systematic design procedures for the design of systems with more liberal sensitivity specifications are presented. Root locus methods are studied in Sections 8.11-8.13. The frequency response approach is considered in Section 8.14.

(*D*) Up to this point, parameter variations were assumed to be so very slow that the techniques applicable to linear differential equations with constant coefficients were sufficiently accurate. In Sections 8.15-8.17 there are no restrictions set on the rate of parameter variations. Nevertheless, it is found that most of the older concepts can be retained and used in the design of feedback systems around fast varying plants.

(*E*) Sections 8.19 and 8.20 are devoted to the application of linear feedback techniques to systems with nonlinear plants. In Section 8.19 the emphasis is on the use of feedback to improve the system linearity. The nonlinearity may be represented as a parameter variation of a linear plant, but the variation is a function of the signal being processed. Linear feedback theory can be used in the usual manner to reduce the system sensitivity to parameter variations. However, the results are correct only for the specific signals considered. In Section 8.20 the objective is not system linearization per se, but the direct achievement of system specifications by means of feedback. The solution of the resulting nonlinear integral equations is greatly facilitated by the "engineering realities" of feedback control systems.

(*F*) In Sections 8.21 and 8.22 we consider some of the fundamental limitations in the adaptive capabilities of linear time invariant feedback systems and how some of these limitations may possibly be overcome by the so-called "active adaptive" systems.

§ 8.2 Design of Multiple-Loop Systems for Disturbance Attenuation (Cascade Plants)

This section is devoted to the disturbance attenuation problem in the three-degree-of-freedom plant of Fig. 8.2-1a. It is so called because the designer has

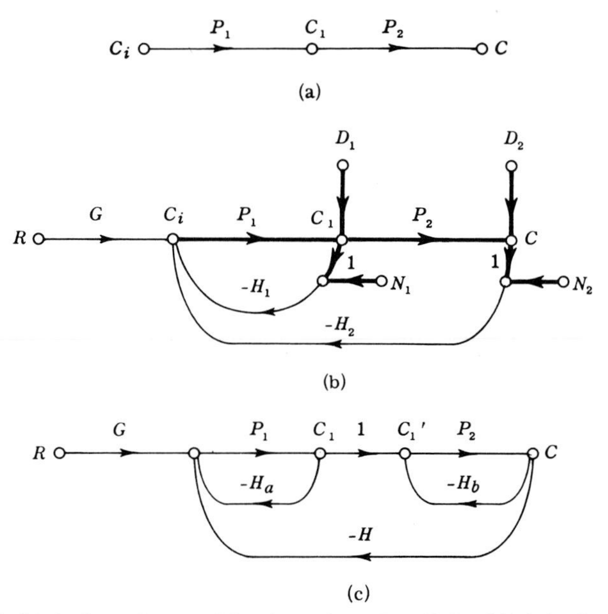

FIG. 8.2-1. (a) A three-degree-of-freedom plant. $P = P_1P_2$. (b) A feedback structure which utilizes the plant's three degrees of freedom

$$P = P_1P_2; \qquad L_1 = P_1H_1 + PH_2; \qquad L_2 = PH_2/(1 + P_1H_1).$$

(c) Forbidden four-degree-of-freedom feedback structure, because of plant modification.

only three essential independent functions which he may specify if (1) the plant output can be influenced only via the given plant, (2) plant modification is not allowed. A structure in which all three degrees of freedom are utilized is shown in Fig. 8.2-1b. The second constraint of "no plant modification" prohibits the application of feedback from C to C_1', shown in Fig. 8.2-1c. The reason is that if the same output C is to be achieved in both structures, then in Fig. 8.2-1b, $C_1 = C/P_2$, whereas in Fig. 8.2-1c, $C_1' = C/P_2$ but $C_1 = C(1 + P_2H_b)/P_2$. Hence the output of P_1 is different in the two systems, and the plant has in effect been modified.

In Fig. 8.2-1b all the disturbances acting on the plant are lumped in D_1 and D_2. There is no resulting loss in generality, as there would be if they were all lumped under D_2 (because H_1 is effective in attenuating any disturbances acting on P_1, whereas it is useless for disturbances acting anywhere on P_2, except at C_1). Noise sources N_1, N_2 are assumed to be an inherent part of the transducers required for measuring C_1, C. These noise sources may also be used to represent variation of parameters of H_1, H_2 (see Section 8.15). It is assumed that the effect of D_1, D_2 on the final output C must be controlled and that there is no concern with their effect on the internal plant values. Thus we are concerned with keeping C/D_1 satisfactorily small, not with C_1/D_1. The disturbance transfer functions are

$$\frac{C}{D_1} = \frac{P_2}{1 + P_1H_1 + PH_2} \triangleq \frac{P_2}{1 + L_1}; \qquad \frac{C}{D_2} = \frac{1 + P_1H_1}{1 + L_1} \triangleq \frac{1}{1 + L_2}$$

Attenuation of D_1, D_2 is achieved in those frequencies in which $|1 + L_1| > 1$, $|1 + L_2| > 1$, respectively. As a first approximation, it is convenient to let $1 + L_1 \doteq L_1$ when $|L_1| > 1$ and $1 + L_1 \doteq 1$, when $|L_1| < 1$, and similarly for L_2. The time specifications on the effects of the disturbances will determine the minimum magnitudes of L_1 and L_2 vs. frequency for the frequency ranges in which $|L_1| > 1$, $|L_2| > 1$, respectively (recall Section 6.11 and see Sections 9.6 and 10.15, how such minimum levels are ascertained from time specifications). In the vast majority of problems, the minimum values of L_1, L_2 (denoted henceforth as L_{1m}, L_{2m}) are thereby determined from zero frequency up to some specific frequencies, hereafter denoted as the minimum crossover frequencies ω_{1m}, ω_{2m}. Beyond ω_{1m}, ω_{2m} it is of course desirable to decrease $|L_1|$, $|L_2|$ as rapidly as possible subject to avoidance of undue peaking in $|1 + L_1|^{-1}$, $|1 + L_2|^{-1}$ and subject to stability considerations. Therefore L_{1m}, L_{2m} are also known for $\omega > \omega_{1m}$, ω_{2m}.

Whatever freedom is available in the design may be usefully used to minimize the input to the plant due to the noise sources N_1, N_2 in Fig. 8.2-1b, i.e., to minimize $C_i/N_1 = -H_1/(1 + L_1)$, $C_i/N_2 = -H_2/(1 + L_1)$. Let us first consider the component due to N_2: $C_i/N_2 = -H_2/(1 + L_1)$. It is readily seen that this can be written as $C_i/N_2 = -L_2/P(1 + L_2)$. For $\omega < \omega_{2m}$, $|L_2| \geqslant |L_{2m}| > 1$ and $C_i/N_2 \doteq -1/P$ independent of the specific values of

L_1, L_2. Therefore, in the range $\omega < \omega_{2m}$, ideally $H_1 = 0$ so that at least $C_i/N_1 = 0$ because in the range $\omega < \omega_{2m}$, C_i/N_2 is unaffected by the value of H_1. Hence, in the range $\omega < \omega_{2m}$, all the disturbance attenuation is ideally achieved by H_2, and H_2 must therefore be the larger of L_{2m}/P, L_{1m}/P in order to satisfy the specifications on both C/D_1 and C/D_2. What about the range $\omega > \omega_{2m}$? Is there any value in using H_1 in the latter range? To answer this question it is necessary to distinguish between two cases.

Case $\omega_{2m} > \omega_{1m}$

In this case, H_2 ideally achieves the entire attenuation of D_1 and D_2, and therefore ideally $H_1 = 0$ for $\omega < \omega_{2m}$. No attenuation is required for $\omega > \omega_{2m}$, but $L_1 = P_1(H_1 + P_2H_2)$ cannot be decreased too rapidly in order that the zeros of $1 + L_1$ (poles of C/D_1, C/D_2) may not be in the rhp or insufficiently damped. It is necessary that $|L_1|$ be more than some minimum magnitude denoted by L_{1m}. Of course $L_{1m} < 1$ for $\omega > \omega_{2m}$. There was no value in using H_1 in the range $\omega < \omega_{2m}$, but this is no longer true in the range $\omega > \omega_{2m}$ because it is no longer necessary to maintain $L_2 > 1$. If the entire burden of having $L_1 \geqslant L_{1m}$ is thrown on H_1, then $H_2 = 0$ and $|H_1| = |L_{1m}/P_1|$ and $C_{iN} = -H_1N_1/(1 + H_1P_1) \doteq -H_1N_1 = L_{1m}N_1/P_1$. If it is thrown on H_2, then $H_1 = 0$, $|H_2| = |L_{1m}/P|$ and $C_{iN} = -H_2N_2/(1 + H_2P) \doteq -H_2N_2 = L_{1m}N_2/P$. Hence, ideally, let $H_1 = 0$ and let H_2 bear the burden when $|L_{1m}N_2/P| < |L_{1m}N_1/P_1|$, i.e., when $|N_1P_2| > |N_2|$ and vice versa. It is actually impossible to switch instantanteously from one to the other because of phase considerations, as will be seen in a later example (Figs. 8.2-2 to 5). Of course if the ideal transition frequency is less than ω_{2m}, it cannot be used and it must actually occur as soon as feasible after ω_{2m}.

Case $\omega_{1m} > \omega_{2m}$

The situation is the same as before for the range $\omega < \omega_{2m}$, i.e., ideally $H_1 = 0$ and H_2 is the larger of L_{1m}/P, L_{2m}/P. However, there is a difference for $\omega > \omega_{2m}$. In the range $\omega > \omega_{2m}$, we set $|L_1| = |L_{1m}|$ and then $C_{iN} \doteq -(H_1N_1 + H_2N_2)/L_{1m}$ for $\omega < \omega_{1m}$ and $-(H_1N_1 + H_2N_2)$ for $\omega > \omega_{1m}$. If $H_1 = 0$, then $H_2 = L_{1m}/P$ and $C_{iN} \doteq -N_2/P$; if $H_2 = 0$, $H_1 = L_{1m}/P_1$ and $C_{iN} \doteq -N_1/P_1$. Consequently, let $H_1 = 0$, $H_2 = L_{1m}/P$ in the range where $|N_2/P| < |N_1/P_1|$; $H_2 = 0$, $H_1 = L_{1m}/P_1$ in the range $|N_2| > |N_1P_2|$. The conclusion is the same for $\omega > \omega_{1m}$. In practice, the switch from H_2 to H_1 cannot be made instantaneously. In fact, PH_2 and P_1H_1 are often opposite in phase at the transition point, so that in practice a few octaves must be allowed for the transition. This is illustrated in the following example.

Example. L_{1m}, L_{2m}, which are the minimum required loop gain levels for D_1, D_2 attenuation, are shown in Fig. 8.2-2. The respective minimum crossover frequencies are $\omega_{1m} = 20 > \omega_{2m} = 0.5$. Also, it is known that $N_1 = N_2$. Hence, ideally $H_1 = 0$, $H_2 = L_{1m}/P$ from 0 rps up to the frequency at which $N_2P_1 =$

N_1P, i.e., up to 1 rps (note that $\omega_{P_1=P} = 1 > \omega_{2m}$; in the case $\omega_{2m} > \omega_{P_1=P}$, we would ideally take $H_1 = 0$ up to ω_{2m}). For $\omega > 1$, H_1 ideally assumes the

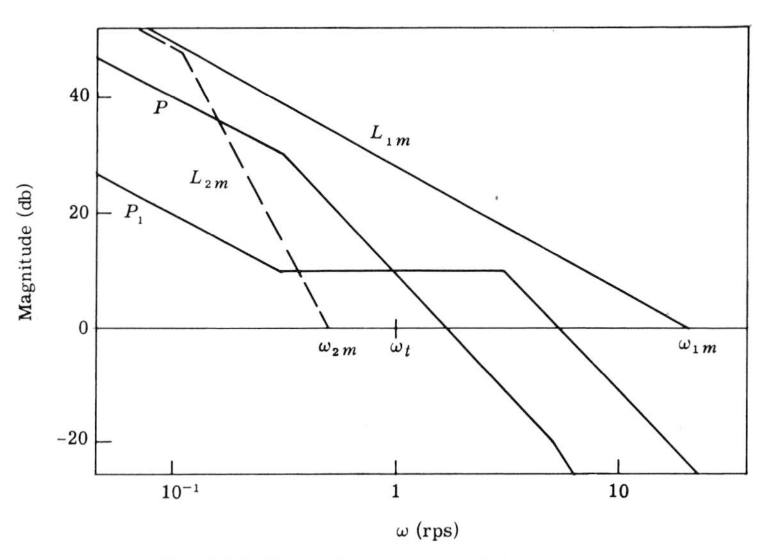

FIG. 8.2-2. Example—statement of the problem.

burden and PH_2 may be decreased very rapidly. Thus it is not necessary that $1 + PH_2$ should be free of rhp zeros. For example, suppose the loci of PH_2

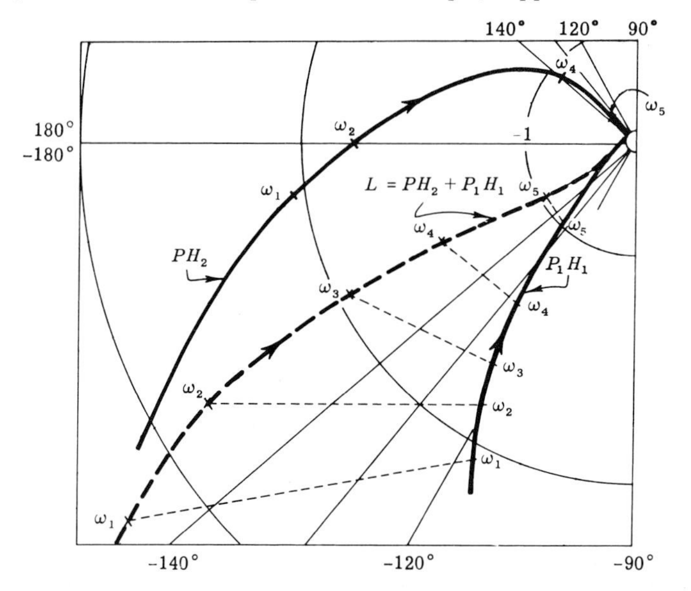

FIG. 8.2-3. A hypothetical stable design with outer loop PH_2 (of Fig. 8.2-1b) encircling -1.

and P_1H_1 are as shown in Fig. 8.2-3. Their sum $L_1 = P_1H_1 + PH_2$ does not encircle the -1 point, although PH_2 does. It is conceivable that such a design may sometimes be forbidden, because an open circuit in H_1 will lead to an unstable system. If it is permitted, then rapid reduction of H_2 and of C_i/N_2 is possible.

First trial design: From 0 rps to 1 rps, let $PH_2 = L_{1m}$ (region AB in Fig. 8.2-4) and let $P_1H_1 = L_{1m}$ (region BC) for $\omega > 1$. A simple, self-explanatory transition between the two is shown in Fig. 8.2-4. Beyond $\omega_{1m} = 20$, P_1H_1 is decreased rapidly, subject to stability considerations (45° phase margin). Four excess poles over zeros are assigned to P_1H_1, two for P_1 and two for H_1, so that the latter may be rapidly decreased. PH_2 is allowed to decrease even more rapidly, in accordance with our previous comments regarding Fig. 8.2-3. The resulting phase lags of P_1H_1, PH_2 are sketched in Fig. 8.2-4. It is seen that the two functions are almost equal in magnitude, and 180° out of phase in the transition region 1-3 rps, and therefore almost cancel each other out in this frequency range. Hence the transfer of the feedback burden from PH_2 to P_1H_1 must be more gradual.

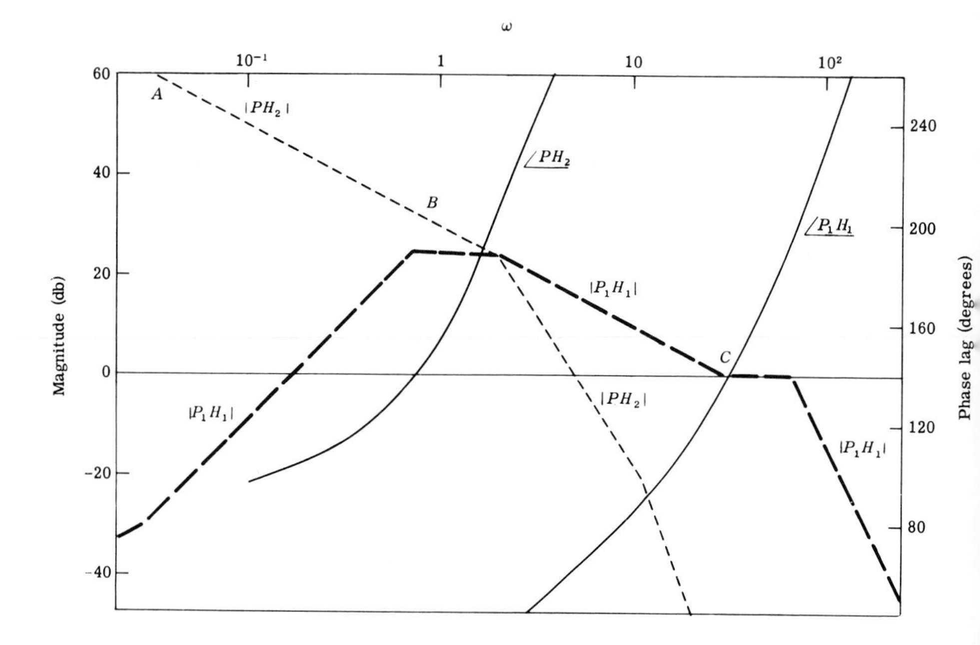

FIG. 8.2-4. First trial design, in which feedback burden is transferred too abruptly from PH_2 loop to P_1H_1 loop.

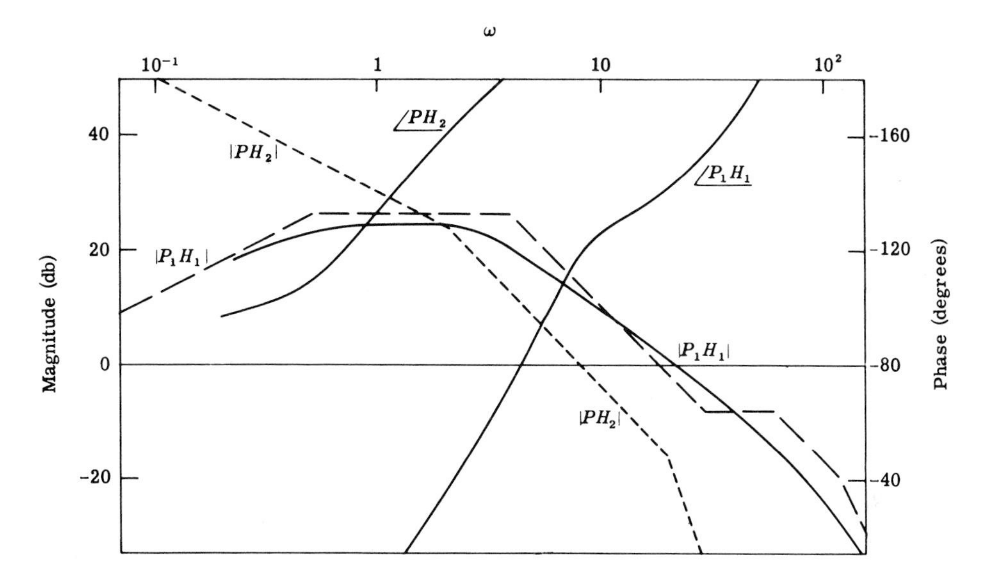

FIG. 8.2-5. (a) Satisfactory design with a more gradual transition from PH_2 loop to P_1H_1 loop.

Second design: It is necessary to increase the phase lag of P_1H_1, and to decrease that of PH_2, in the transition range. The former requires the low frequency level of P_1H_1 to be increased, and the latter leads to a larger bandwidth for PH_2. Both tend to increase C_{iN}, but this is unavoidable. Such a more conservative design is shown in Fig. 8.2-5a, with $P_1H_1 = (18.4)10^6(s + 30)^2(s + 0.01)[(s + 0.5)(s + 4)^2(s + 60)^2(s + 120)^2]^{-1}$, $PH_2 = 10^7/s(s + 2)[s^2 + 2(0.6)(20)s + 400]^2$. Far-off complex poles are assigned to PH_2, so that the lag angle due to them should be small at low frequencies. Polar plots of $P_1H_1 + PH_2$ are sketched in Figs. 8.2-5b and c. The design is like that shown in Fig. 8.2-3, i.e., the PH_2 loop by itself is unstable, while the over-all system is stable.

Comparison with single-loop design: If the P_1H_1 loop is not used, then PH_2 must follow ABC in Fig. 8.2-4, and thereafter is decreased as rapidly as possible subject to stability requirements. The resulting noise transfer ratio C_i/N_2 is sketched in Fig. 8.2-6, and compares very poorly with the noise transfer ratios of the two-loop design. In the single-loop design, the large high frequency noise input to the plant may cause saturation of the early stages of the plant. Even if the plant has a very high saturation level, the high frequency noise may cause excessive heating and wear of parts. Furthermore, the interaction between the

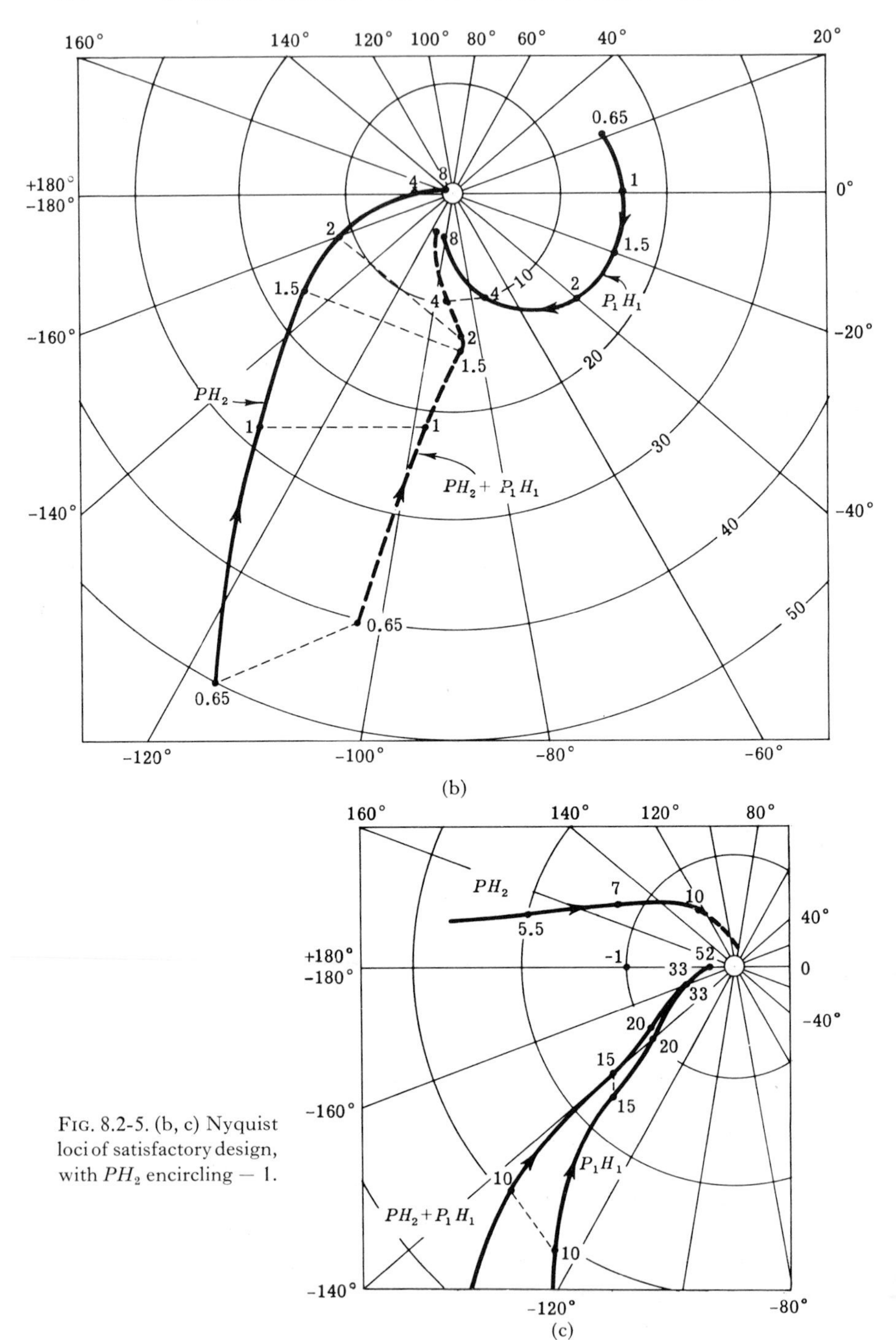

FIG. 8.2-5. (b, c) Nyquist loci of satisfactory design, with PH_2 encircling -1.

high frequency noise and plant nonlinearities may cause lower frequency components to appear in the output.[1]

It should be clear from Fig. 8.2-6 that C_i/N_2 can be decreased in the range DE, by means of the following procedure. In the range $\omega > 40$, PH_2 (and with it C_i/N_2) may be substantially increased with very small effect on the total C_{iN}. For example, if C_i/N_2 is increased by 19 db from its present value of 0 db at 100 rps, then the rms value of $C_{iN} = C_{iN1} + C_{iN2}$ increases by only $20 \log \sqrt{1.25} \approx 2$ db, if N_1 and N_2 are independent. A substantial increase in $|PH_2|$ in the range $\omega > 40$ provides a fund of phase lead for all ω. This fund of phase lead permits some reduction in $|PH_2|$ for $\omega < 40$, with very minor effect on arg PH_2 for $\omega < 40$. Thus, let

$$(PH_2)_{\text{new}} = (3.1)\, 10^6[s(s+2)(s+4)(s+40)(s+300)]^{-1}.$$

Note that $|PH_2|_{\text{new}} < |PH_2|_{\text{old}}$ for approximately $4 < \omega < 40$ (asymptotic values), and so has smaller C_{iN2} in this range; while for $\omega > 40$,

$$|PH_2|_{\text{new}} > |PH_2|_{\text{old}}$$

with larger C_{iN2}. The resulting $(C_i/N_2)_{\text{new}}$ is shown as dashed lines in Fig. 8.2-6, and clearly the total $C_{iN,\text{new}}$ is considerably less than the old. The new PH_2 has only a slightly larger phase lag than the old for $\omega < 7$ rps, and therefore negligibly affects the sum $P_1H_1 + PH_2$ in the transition range.

Third design: A difficulty with the second design (Fig. 8.2-5a) is that in the region $0.025 < \omega < 0.25$, the attenuation of D_2 is not as large as is required by the specifications. Thus $C/D_2 = (1 + L_2)^{-1} = (1 + P_1H_1)/(1 + P_1H_1 + PH_2)$, and at $\omega = 0.1$, for example, $|P_1H_1| = 12$ db, $|P_1H_1 + PH_2| \approx |PH_2| = 50$ db, so that $|L_2| \approx 38$ db, instead of the required value of 50 db indicated by L_{2m} in Fig. 8.2-2. This situation begins at $\omega = 0.025$ because $|P_1H_1| < 1$ for $\omega < 0.025$. It persists up to $\omega = 0.25$ because, in Fig. 8.2-5a, $|L_2(j0.25)|_{\text{db}} \approx |PH_2(j0.25)|_{\text{db}} - |P_1H_1(j0.25)|_{\text{db}} = 42 - 20 = 22$ db, and in Fig. 8.2-2, $|L_{2m}(j0.25)| = 22$ db. This particular trouble did not occur in the first trial design (Fig. 8.2-4), because in the latter $|P_1H_1| < 1$ for $\omega < 0.16$, and L_{2m} decreases rapidly for $\omega > 0.1$.

The above design shortcoming may be overcome either by increasing $|PH_2|$ or by decreasing $|P_1H_1|$ in the frequency range involved, or by some combination of both. For example, one way is to introduce a horizontal step in $|PH_2|$ at $\omega = 0.025$ up to about $\omega = 0.1$, followed by two lag corner frequencies at $\omega = 0.1$, until the new PH_2 merges with the old at $\omega = 0.4$, i.e., let $(H_2)_{\text{new}} = (s + 0.25)(s + 0.1)^{-2}(s + 0.4)(H_2)_{\text{old}}$. Or P_1H_1 may be modified in the opposite way, with $(H_1)_{\text{new}} = (s + 0.25)^{-1}(s + 0.1)^2(s + 0.4)^{-1}(H_1)_{\text{old}}$. It should be recognized, however, that any such modification constitutes in effect a

[1] John C. West, "Analytical Techniques for Non-Linear Control Systems," p. 148. The English Universities Press, London, 1960.

more rapid transition in shifting the disturbance attenuation burden from PH_2 to P_1H_1. The magnitude of $PH_2 + P_1H_1$ in the transition range (for example, at $\omega = 1$) tends to be decreased by either modification. The decrease is not intolerable in the present example.

Practical Shaping of PH_2 and P_1H_1 to Minimize Plant Noise Input

It has previously been shown that the idealized optimum design is one in which $H_1 = 0$ for $\omega < \omega_{2m}$, so that at least $C_i/N_1 = 0$ there; and $H_2 = 0$ for $\omega > \omega_t$,[1] so that $C_i/N_2 = 0$ in the latter range. In practice, the transition from PH_2 to P_1H_1 must cover several octaves. The results are then (for the case $\omega_{1m} > \omega_{2m}$) of the form shown in Fig. 8.2-6, where it is particularly noted that C_{iN1} dominates in the higher frequency range EF, and C_{iN2} dominates in the lower frequency range DE. We now ask to what extent can the sum $(C_{iN1} + C_{iN2})$ be reduced, and seek a general practical means of securing such minimum $(C_{iN1} + C_{iN2})$.

C_i/N_1 is first considered. A fundamental question is whether its values (>1) for $\omega > 5$ are due to faulty design, or are inherent and unavoidably follow from the system specifications. It is readily seen from Fig. 8.2-2 that the latter is the case. Thus, for $5 < \omega < 20$, (assuming $L_1 \approx P_1H_1$ for $\omega > 5$), $C_i/N_1 \approx -H_1/P_1H_1 = -1/P_1$, which checks with Fig. 8.2-6, so that no improvement is possible in this range (unless PH_2 is made larger than P_1H_1, which would be ridiculous). For $\omega > 20$, $C_i/N_1 \approx -H_1 = -L_1/P_1$. Improvement in this range is possible only if L_1 can be decreased faster. Of course there is a "best" L_1, in this sense, which cannot be decreased any further without affecting the system stability. Assuming a reasonable effort has been made to secure an L_1 which is no larger than absolutely necessary for $\omega < \omega_{1m}$ and decreases as fast as possible thereafter, the resulting C_i/N_1 necessarily follows from the specifications, and cannot be improved. There remains the possibility of deliberately increasing C_i/N_1 in the range DE if this should permit a sufficient decrease in the dominant C_i/N_2 there. This possibility will be considered later.

Let us now consider C_i/N_2. The range (EF) in which it is less than C_i/N_1 merits careful consideration. Suppose, in Fig. 8.2-6, that H_2 is increased by 25 db at $\omega = 100$, such that $|C_i/N_2| = |C_i/N_1| = 25$ db at this frequency. Nevertheless, $C_{iN,\mathrm{rms}} = (C_{iN1} + C_{iN2})_{\mathrm{rms}}$ only increases by 3 db (if N_1 and N_2 are independent, otherwise the maximum increase is 6 db). It has previously been noted how the phase lead thereby made available may be used to decrease C_i/N_2 in the region where it dominates, and so decrease $C_{iN,\mathrm{rms}}$ in the latter range by much more than 3 db. It is even conceivable that, in some cases, it may be worthwhile to continue with the process so far as to let $|C_{iN2}| = |C_{iN1}|$ for a good portion of the range corresponding to EF, if in going so far, more is gained in the reduction of C_i/N_2 in the analogous DE range.

---

[1] ω_t is the ideal transition frequency.

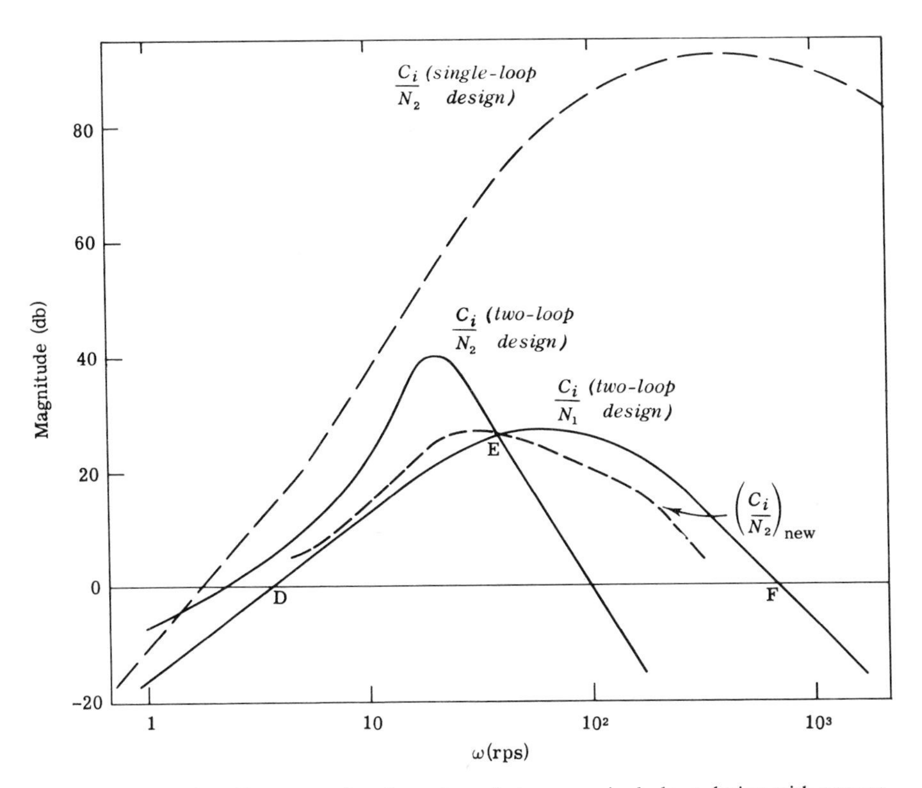

FIG. 8.2-6. Superior properties of two-loop design over single-loop design with respect to feedback transducer noise.

The only range left to consider is the lower one, DE, in which $|C_{iN2}| > |C_{iN1}|$ (bearing in mind that our last adjustments have reduced the difference between the two). It has already been noted that C_{iN1} cannot be reduced, so there only remain two problems. One is to what extent C_{iN2} can be decreased in this range with the given minimum C_{iN1}. The second is whether it is possible and worthwhile to increase C_{iN1} in the range DE, if by so doing C_{iN2} may be decreased, such that the total C_{iN} is reduced. The former possibility is examined first. There is again the fundamental question of whether the existing C_i/N_2 level is definitely unavoidable, or is due to faulty design. Certainly, for $\omega > \omega_{2m}$ (0.5 in the present example), H_2 is not needed for attenuating disturbances, so the basic problem is really one of reducing PH_2 as fast as possible for $\omega > \omega_{2m}$ (C_i/N is so small in the range $\omega \sim \omega_t$ that fast decrease of PH_2 for $\omega > \omega_{2m}$ rather than for $\omega > \omega_t > \omega_{2m}$ is the real desideratum). The important factor which limits faster decrease of PH_2 is the tendency for P_1H_1 and PH_2 to cancel each other in the transition range, so we shall next examine the fundamental limitations in the transition range.

Discussion of the Transition Problem

It is seen that P_1H_1 tends to have a small phase angle in the transition range, while PH_2 tends to have a large lag angle there, especially if advantage is taken of the previously noted fact (Fig. 8.2-3) that the PH_2 loop by itself can be unstable. Therefore the two tend to cancel each other out (e.g., as in Fig. 8.2-4). More precisely, consider the situation at the ideal transition frequency (ω_t) determined by $|N_1P_2| = |N_2|$ (1 rps in the previous example), at which ideally we choose $|PH_2| = |P_1H_1|$. Their sum,

$$|L_1| = |PH_2 + P_1H_1| < |PH_2| = |P_1H_1|,$$

if the difference in their angles is more than 120°, and vice versa. Suppose that the former occurs in the first trial design, and therefore $|L_1| < |L_{1m}|$ in the transition frequency range. The situation must be improved either by (1) decreasing the phase (increasing the lag) of P_1H_1 in the transition range, or (2) by increasing the phase (decreasing the lag) of PH_2 in the same range, or by a combination of the two. How may the lag of P_1H_1 be increased? This is done by letting P_1H_1 be as large as possible in the lower frequency range (i.e., at $\omega < \omega_t$), because less lead corner frequencies are then needed to bring $|P_1H_1|$ up to equal $|PH_2|$ at the transition frequency ω_t. However, there is an upper limit for $|P_1H_1|$, which is a function of frequency. What is this upper limit?

If $\omega_{1m} > \omega_{2m}$, then $L_{1m} > L_{2m}$ for some portion of $\omega < \omega_{2m}$ (see, for example, Fig. 8.2-2—the reader is urged to follow the argument by making sketches similar to those in Fig. 8.2-2). In this range, PH_2 must handle the larger of L_{1m}, L_{2m} so $|PH_2 + P_1H_1| = |L_1| \geqslant L_{1m}$. On the other hand, it is necessary that $|L_2| = |PH_2/(1 + P_1H_1)| \geqslant L_{2m}$, and if $|P_1H_1| > 1$ by a reasonable margin, we may write this as $|P_1H_1| \leqslant |PH_2/L_{2m}|$. Thus PH_2 must be larger than P_1H_1 at least by the magnitude of L_{2m}. Hence, for that portion of the range in which L_{2m} is not small, we may write $|PH_2| \approx |PH_2 + P_1H_1| \geqslant L_{1m}$. The relation $|P_1H_1| \leqslant |PH_2/L_{2m}|$ may be written as $|P_1H_1|_{\max} = |PH_2|_{\max}/L_{2m}$. Now $|PH_2| \geqslant L_{1m}$, but there is no point in obtaining a larger P_1H_1 by taking $|PH_2| > L_{1m}$, because this larger PH_2 must then be decreased faster at higher frequencies, if it is to have the same level there as the old (and small PH_2 at higher frequencies is the principal objective of the multiple-loop design). This faster decrease of PH_2 increases its lag angle and cancels the benefit gained in the increase in lag angle of P_1H_1 (as in the "second design" in the previous example). The conclusion therefore is that in the range under consideration (i.e., where $L_{1m} > L_{2m}$), $(P_1H_1)_{\text{db, max}} \approx (L_{1m})_{\text{db}} - (L_{2m})_{\text{db}}$.

Applying the above to Fig. 8.2-2 indicates that for $\omega \leqslant 0.1$, $|P_1H_1|_{\max} \approx 1$; for $0.1 \leqslant \omega \leqslant \omega_{2m}$, $|P_1H_1|_{\max}$ is given by the straight line joining 0 db at $\omega = 0.1$, to 34 db at $\omega = \omega_{2m} = 0.5$. However, it is not necessarily worthwile to let P_1H_1, in the range $\omega < \omega_t$, be larger (i.e., 34 db at 0.5 rps) than the value of 28 db which is required at the ideal transition frequency $\omega_t = 1$ rps in

Fig. 8.2-2. It may be better to let P_1H_1 increase only up to the value required at ω_t, i.e., up to 28 db at $\omega = 0.37$, and then for $0.37 \leqslant \omega \leqslant \omega_t$, let $|P_1H_1|$ be constant at 28 db. The former P_1H_1 has more phase lead than the latter for $\omega < 0.7$ approximately, and more lag for $\omega > 0.7$ (the difference between the two is an asymmetrical version of Fig. 5.8-1a). Which is better depends on where there exists the greater difficulty in securing $[\arg (PH_2) - \arg (P_1H_1)] < 120°$. In the present example the latter is therefore better. The above represents the best that can be done to increase the lag of P_1H_1. Faster decrease of P_1H_1 at higher frequencies would help, but cannot be considered, if, as postulated, P_1H_1 has already been decreased there as much as possible subject to stability considerations.

The second way to improve the situation is to decrease the lag of PH_2, by letting $|PH_2|$ decrease more slowly in the range DE. This is undesirable in that it increases C_i/N_2. Hence the "best" PH_2 in the transition range (up to about the frequency at which $|PH_2| = 0.25 |P_1H_1|$) is the one which results in $[\arg P_1H_1 - \arg PH_2] = 120°$.

The foregoing completely specifies the best that can be done to decrease C_i/N_2 in the range DE, while keeping C_i/N_1 at its minimum possible value. The limit is reached when PH_2 has been decreased in the transition to the point where $[\arg P_1H_1 - \arg PH_2] = 120°$. The only remaining possibility is whether increase of P_1H_1 (and consequent increase of C_{iN1}) may be used to sufficiently decrease PH_2, such that the total $C_{iN1} + C_{iN2}$ is decreased. This postulated decrease of PH_2 (by means of asymmetrical versions of Fig. 5.8-1a) leads to $[\arg P_1H_1 - \arg PH_2] > 120°$ in the range DE, which requires that P_1H_1 must be increased (and with it C_{iN1}) in order that $|P_1H_1 + PH_2| \nless L_{1m}$ in the transition range. This increase of P_1H_1 tends to magnify the difference in the angles of P_1H_1 and PH_2, at least over part of the frequency range (for it involves the addition to P_1H_1 of an unsymmetrical version of Fig. 5.8-1a). If the process converges, then it is certainly worthwhile if $|C_{iN2}| \gg |C_{iN1}|$ in the range analogous to DE of Fig. 8.2-6. It will be necessary to experiment a bit to find where to stop. Certainly there is no point continuing when there is little difference between $|C_{iN2}|$ and $|C_{iN1}|$.

Summary of Procedure for Shaping PH_2, P_1H_1 (Case $\omega_{1m} > \omega_{2m}$)

1. For $\omega < \omega_{2m}$, $(P_1H_1)_{\text{db}} \approx (L_{1m})_{\text{db}} - (L_{2m})_{\text{db}}$, in the range where $L_{1m} > L_{2m}$. Otherwise take P_1H_1 as approximately 0 db.

2. For $\omega_{2m} < \omega < \omega_t$ any P_1H_1 between the following two extremes may be used: (a) same as No. 1 above, (b) $|P_1H_1| = L_{1m}(j\omega_t)$. The difference between these two is in general an asymmetrical version of Fig. 5.8-1a. If the greater difficulty in securing $[\arg P_1H_1 - \arg PH_2] \leqslant 120°$ is at the lower frequencies, then (b) is preferable.

3. $|P_1H_1| = L_{1m}$ for $\omega_{1m} > \omega > \omega_t$.

4. For $\omega > \omega_{1m}$, P_1H_1 is decreased as rapidly as possible subject to stability requirements.

5. Prepare a Bode sketch of $|C_i/N_1|$ in the range in which it is more than one. $C_i/N_1 \approx 1/P_1$ for $\omega_x < \omega < \omega_{1m}$, where $|P_1(j\omega_x)| \triangleq 1$. For $\omega > \omega_{1m}$, $-C_i/N_1 \approx H_1 = (P_1H_1)/P_1$.

6. For $\omega < \omega_{2m}$, $|PH_2| \approx L_{1m}$ if $L_{1m} > L_{2m}$; otherwise $|PH_2| = L_{2m}$.

7. For $\omega > \omega_{2m}$, decrease PH_2 as fast as possible subject to [arg PH_2 − arg P_1H_1] $\leqslant 120°$ in the transition range, which extends approximately up to the frequency at which $|PH_2/P_1H_1| = 0.25$. When $|PH_2/P_1H_1| = 0.25$, then at worst $(PH_2 + P_1H_1)$ is less than P_1H_1 by only 2.5 db. The difference in phase can be greater beyond the transition range.

8. However, there is no value in having C_{iN2} much less than C_{iN1}. Therefore increase (if necessary) PH_2 in the range in which presently $C_{iN1} > C_{iN2}$. This will permit a smaller PH_2 in a lower frequency range (in which presently $C_{iN2} > C_{iN1}$). Continue to increase PH_2 in the higher range so long as the resulting permissible decrease of PH_2 (in the lower range) results in a net improvement (from whatever point of view is considered realistic for the given problem) in the net $C_{iN} = C_{iN1} + C_{iN2}$. Conceivably a point may even be reached where $C_{iN2} < C_{iN1}$ for most of the frequency range.

9. In the above, C_{iN1} has been kept at its minimum possible value for all $\omega > \omega_t$. If the result is a C_{iN2} which is considerably larger than C_{iN1} over part of the frequency range (analogous to region DE in Fig. 8.2-6), then there is one more step. It is to decrease PH_2 faster (and therefore decrease C_{iN2}) in this range, which leads to [arg PH_2 − arg P_1H_1] $> 120°$. P_1H_1 must therefore be increased (with increase of C_{iN1}) to maintain $|P_1H_1 + PH_2| = L_{1m}$. One may continue with this process up to the point where the total $C_{iN2} + C_{iN1}$ no longer decreases.

Nonminimum Phase Plants

Nonminimum phase properties in P_2 and/or P_1 will obviously affect the optimum shaping of PH_2 and P_1H_1. A rhp zero in P_1 will be a factor in determining the fastest decreasing P_1H_1 in the range $\omega > \omega_{1m}$, and so influence the minimum level of C_{iN1}. Rhp zero will make themselves felt through the requirement that [arg PH_2 − arg P_1H_1] $= 120°$, in that stage of the design procedure where C_{iN1} is kept at its minimum level, etc. The presence of rhp zeros in P_2 and/or P_1 does not invalidate the design procedure that has been outlined. Precisely the same procedure may be followed, but the specific results will of course be affected by the rhp zeros.

Extension to Higher Order Cascade Plants

The design procedure for higher order cascade plants is very similar to the above, and is briefly indicated for the three-loop system of Fig. 8.2-7.

Here
$$\frac{C}{D_1} = \frac{P_2P_3}{1 + P_1H_1 + P_1P_2H_2 + PH_3} \triangleq \frac{P_2P_3}{1 + L_1},$$

$$\frac{C}{D_2} = \frac{P_3(1 + P_1H_1)}{1 + L_1} \triangleq \frac{P_3}{1 + L_2}, \qquad \frac{C}{D_3} = \frac{1 + P_1H_1 + P_1P_2H_2}{1 + L_1} \triangleq \frac{1}{1 + L_3}.$$

Let L_{1m}, L_{2m}, L_{3m} be the minimum values required for the attenuation of C/D_1, C/D_2, C/D_3, and let ω_{1m}, ω_{2m}, ω_{3m} be their respective minimum crossover frequencies. Then $C_i/N_3 = - L_3/P(1 + L_3)$ is closely independent of the choice of H_1, H_2, H_3 because for $\omega < \omega_{3m}$, $C_i/N_3 \doteq - 1/P$, and for $\omega > \omega_{3m}$, there is

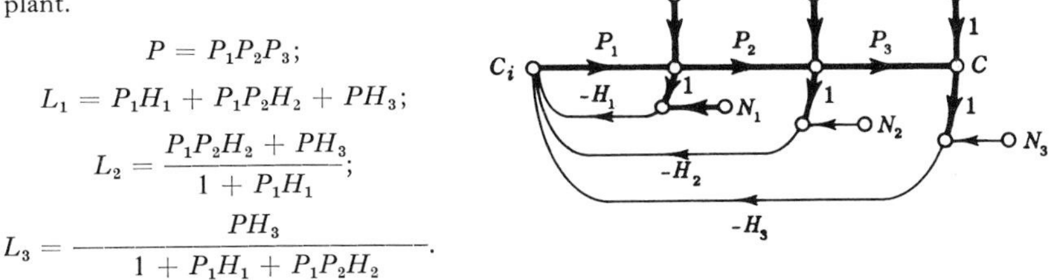

FIG. 8.2-7. A canonic feedback structure for four-degree-of-freedom plant.

$$P = P_1P_2P_3;$$
$$L_1 = P_1H_1 + P_1P_2H_2 + PH_3;$$
$$L_2 = \frac{P_1P_2H_2 + PH_3}{1 + P_1H_1};$$
$$L_3 = \frac{PH_3}{1 + P_1H_1 + P_1P_2H_2}.$$

a minimum L_3 for peaking and for stability. Hence ideally, for $\omega < \omega_{3m}$, let $H_1 = H_2 = 0$ and let H_3 be the larger of L_{1m}/P, L_{2m}/P, L_{3m}/P. If $\omega_{3m} > \omega_{1m}$, ω_{2m}, then (just as in the corresponding two-loop case) for $\omega > \omega_{3m}$ the burden of maintaining $|L_1| > |L_{1m}|$ is ideally assigned to one of H_1, H_2, H_3 according as to which of N_1/P_1, N_2/P_1P_2, $N_3/P_1P_2P_3$ has the smallest value.

If $\omega_{2m} > \omega_{3m}$, then $L_2 \triangleq (P_1P_2H_2 + PH_3)/(1 + P_1H_1)$ must be larger or equal to L_{2m} in the range $\omega > \omega_{3m}$. H_2 or H_3 must bear the burden because it is impossible for the P_1H_1 loop to attenuate D_2. In fact, we might as well let (ideally) $H_1 = 0$ and have $C_i/N_1 = 0$ in this range. The transition between H_3 and H_2 may be made so as to minimize the noise input to the plant. Obviously, the transition between the two should be at the frequency at which $|C_{i(N_2)}| = |C_{i(N_3)}|$, i.e., at $|P_3N_2| = |N_3|$. In the range $|P_3N_2| < |N_3|$, ideally let $H_3 = 0$ and H_2 is the larger of L_{1m}/P_1P_2, L_{2m}/P_1P_2. In the range $|P_3N_2| > |N_3|$, ideally let $H_2 = 0$ and H_3 is the larger of L_{1m}/P, L_{2m}/P. If $\omega_{2m} > \omega_{1m}$, then beyond ω_{2m} the assignment is made exactly as in the range $\omega > \omega_{3m}$ in the previous paragraph, i.e., to H_1, H_2, or H_3 according as to which one of N_1/P_1, N_2/P_1P_2, N_3/P has the smallest magnitude.

If $\omega_{1m} > \omega_{2m} > \omega_{3m}$, and if H_2 is bearing the burden for the frequencies immediately less than ω_{2m}, then there should be a transition from H_2 to H_1 at that frequency at which $|P_2N_1| = |N_2|$. In the range $|P_2N_1| < |N_2|$, ideally let $H_2 = 0$, $H_1 = L_{1m}/P_1$. In the range $|P_2N_1| > |N_2|$, ideally let $H_1 = 0$, $H_2 = L_{1m}/P_1P_2$. If H_3 is bearing the burden for the frequencies immediately less than ω_{2m}, this means that at ω_{2m}, $|P_3N_2| > |N_3|$. For $\omega > \omega_{2m}$, we assume that only one of H_1, H_2, H_3 assumes the burden of L_{1m}. The question then is: Which of the resulting C_i/N_1, C_i/N_2, C_i/N_3 is smallest, i.e., which of N_1/P_1, N_2/P_1P_2, N_3/P is the smallest. If $N_i/P_{ij...}$ is the smallest in any frequency range, then ideally $H_i = L_{1m}/P_{ij...}$ is used in that range and the other H's are made zero. As previously noted, the transition from any one H to another cannot be too fast, in order that the terms may not cancel each other in the transition range.

Multiple-Loop Design with Limited Plant Modification

In the following, we assume that plant modification in the form of feedback from C to C_1 in Fig. 8.2-1a is permissible. A canonic feedback structure around the plant of Fig. 8.2-1a is now that shown in Fig. 8.2-8, with

$$\frac{C}{D_1} = \frac{P_2}{1 + P_1H_1 + P_2H_3 + PH_2} \triangleq \frac{P_2}{1 + L_1};$$

$$\frac{C}{D_2} = \frac{1 + P_1H_1}{1 + L_1} \triangleq \frac{1}{1 + L_2}.$$

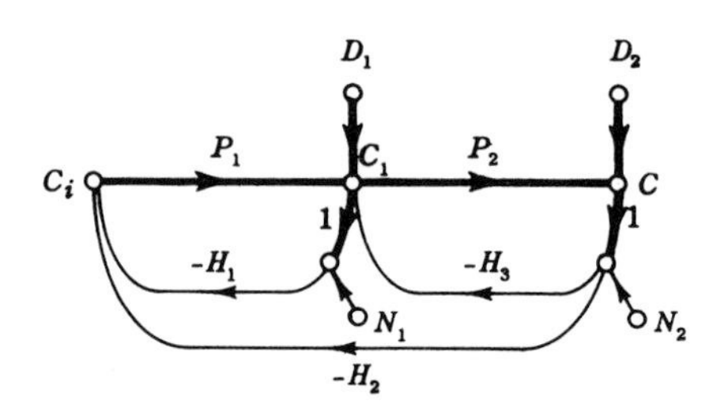

FIG. 8.2-8. A canonic structure for two-stage plant which has four degrees of freedom due to permissible plant modification.

$L_{1m}, L_{2m}, \omega_{1m}, \omega_{2m}$ are defined exactly as before. Again, for $\omega < \omega_{2m}$, we ideally let $H_1 = 0$ in this range because $C_1/N_2 = -(H_3 + H_2P_1)/(1 + L_1) = -L_2/P_2\ (1 + L_2)$ is independent of H_1. And again, if $\omega_{2m} > \omega_{1m}$, then H_1 may as well be zero for $\omega < \omega_{2m}$. In the latter case, $L_2 = P_2H_3 + PH_2$ must be the larger of L_{1m}, L_{2m}. C_1/N_2 is also independent of the division of L_2 between H_2, H_3, at least up to ω_{2m}. The transfer ratio C_i/N_2 does, however, depend on the division between H_2, H_3. The possibility of saturating P_1 and of high frequency wear of P_1 should be taken into account in making the assignments between H_2 and H_3. In the range that $|P_1| < 1$, the required $|H_3|$ (for $H_2 = 0$, $L_2 = P_2H_3$) is smaller than the required $|H_2|$ (for $H_3 = 0$, $L_2 = PH_2$) by the amount $|1/P_1|$. It certainly then appears worthwhile to transfer the burden to H_3, at the very least in the range in which $|P_1| \ll 1$.

If $\omega_{1m} > \omega_{2m}$, then it may be worthwhile using H_1 to assume the burden of L_{1m} in the range $\omega > \omega_{2m}$. If the noise at C_1 is used as the criterion, then the transition is ideally at that frequency at which $|C_{1(N_2)}| = |C_{1(N_1)}|$, when each of these values is calculated as if either H_1 alone or $H_3 + H_2P_1$ alone assumes the entire burden of L_{1m}. The transition is therefore (just as in the case of no plant modification) at the frequency at which $|N_1P_2| = |N_2|$. The result may be different if the noise at C_i is used as the sole criterion, for then ideally

$H_1 = H_2 = 0$, $|H_3| = |L_{1m}/P_2|$ and $C_{iN} = 0$; however, C_1/N_2 may then be larger than with the previous criterion. However, plant modification obviously becomes more significant, because the attenuation of D_1 by H_3 alone is obtained by reducing the transmission from C_1 to C. A compromise must therefore be made among the factors of plant modification, effect of noise at C_i and at C_1.

Supplement to Design Procedure

The material presented in this section may be criticized on the basis of the design objectives. It is not always realistic to assume that the design *must* satisfy the disturbance specifications and only thereafter minimize the effect of noise (and/or parameter variations—see Section 8.15) in the feedback return paths. It is possible to have a situation where the feedback transducer noise sources are so strong that the disturbance specifications must be relaxed (this problem is treated by statistical methods, for a special criterion of goodness, in Section 9.15). At the other extreme, the feedback transducer noise sources

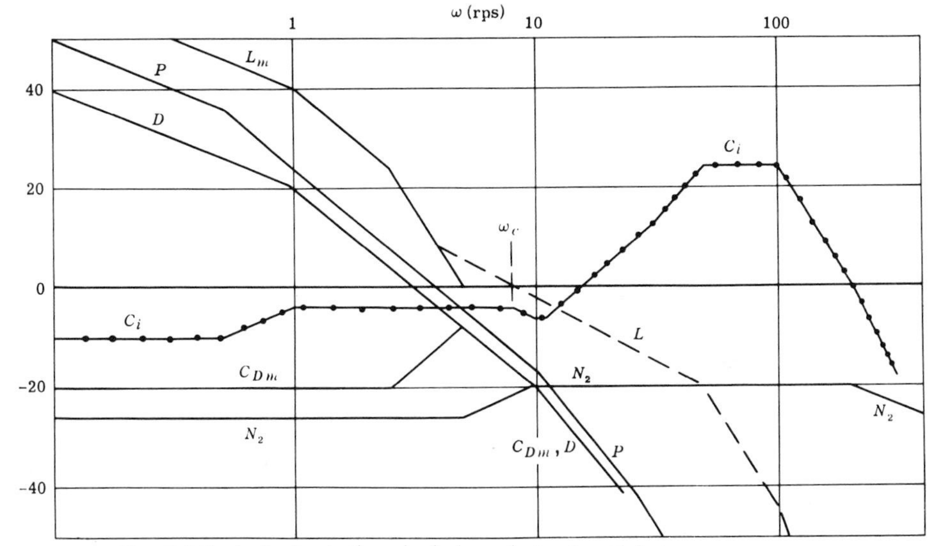

FIG. 8.2-9. A check whether single-loop design is satisfactory

may be so weak that a single loop design is satisfactory—i.e., a two-loop design would certainly decrease the noise output, but it would then be, say, 0.5ϵ in place of ϵ, and ϵ is anyway so small that it is not worth the bother and added cost of a two-loop design to achieve this small improvement. Therefore it is advisable to first attempt a single-loop design. Some of the important features of the single-loop design are indicated in Fig. 8.2-9.

Figure 8.2-9 presents the results of a single-loop design ($H_1 = 0$ in Fig. 8.2-1b) for a hypothetical problem. The curves are obtained in the following step-by-step procedure. The plant transfer function $P = P_1P_2$, the transducer noise output N_2, and the net disturbance $D = D_2 + P_2D_1$ acting at C are presumed known. C_{Dm} indicates the specified minimum system output due to the disturbance D. While C_{Dm} may be assumed to be *a priori* given or deducible from the specifications (see Section 9.6), it is nevertheless worth noting that the value of having $|C_{Dm}(j\omega)| < |N_2(j\omega)|$ for much of the significant (disturbance attenuation) frequency range, is questionable. The reason is that in the latter range the system output due to N_2 is $C_{N2} = -LN_2/(1 + L) \approx -N_2$. The total system output (when $R = 0$) is therefore $C_D - N_2$, and when these are independent, the addition must be rms addition. In any case, there is hardly any point in making C_D much smaller than N_2, which is fixed for a given feedback transducer. On the other hand, N_2 may be sufficiently small in the significant range that there is no need to have $|C_{Dm}| = |N_2|$, i.e., $|C_{Dm}| > |N_2|$ by a fair amount may be considered satisfactory. It is also noted that in Fig. 8.2-9 the bandwidth of N_2 is assumed to be much larger than that of D. This is a reasonable assumption in view of the assumed relative bandwidths of the plant and the feedback transducer.

With D and C_{Dm} known, the next step is to determine the loop transmission $L = PH_2$, from the relation $|C_{Dm}| \geqslant |C_D| = |D/(1 + L)| \approx |D/L|$, in the range where $|L| > 1$. This relation determines the minimum magnitude of L, denoted by L_m, i.e., $(L_m)_{\text{db}} = (D)_{\text{db}} - (C_{Dm})_{\text{db}}$, only for those frequencies in which $|L| > 1$. When $|L| < 1$, $C_D \approx D$, and there is no attenuation of D. L_m determines L over the significant frequency range (assuming that L_m is also large enough to handle the plant parameter variation problem). In the higher frequencies, L is decreased as fast as possible, subject to stability despite plant parameter variations. Let us suppose that the result is the dashed curve labeled L in Fig. 8.2-9.

The next step is to find the total noise and disturbance input to the plant, i.e., $C_i = -H_2(N_2 + D)/(1 + H_2P)$, which, for $\omega < \omega_c$ may be approximated by $C_i \approx -(N_2 + D)/P$, and in the range $\omega > \omega_c$ by $C_i \approx -H_2(N_2 + D) = -(N_2 + D)L/P$. Furthermore, the larger of N_2 and D dominates in $(N_2 + D)$, especially if they are independent, so we shall use only the larger of the two terms. The result is the C_i curve shown in Fig. 8.2-9. At this point the designer must consider whether his plant is capable of processing this input without excessive heating or other harmful effects. The actual system noise output $C_{N2} = -N_2L/(1 + L)$ should also be considered. However, this is usually no problem, because in the range $\omega < \omega_c$, $C_{N2} \approx -N_2$, which, as previously noted, must be less than or of the same order of magnitude as C_{Dm}, or else a better feedback transducer is absolutely mandatory; and in the range $\omega > \omega_c$, $C_{N2} \approx -N_2L$, which is usually small, but which of course should be examined whether it is satisfactorily small.

If the resulting single-loop design is found unsatisfactory, then one may consider a two-loop design, providing of course that an internal plant variable is available for feedback purposes. The design procedure described in this section gives closely the best that can be done by a multiple-loop design. If it too is unsatisfactory, then either the specifications must be relaxed, or less noisy feedback transducers are essential, for a design based on linear feedback techniques.

Generality of Structure of Fig. 8.2-1b

We wish to point out the generality of the structure of Fig. 8.2-1b, even though the transducer models appear to be rather specialized—i.e., with transfer functions which are unity. Sufficiently general models are shown in Fig. 8.2-10, where only J_1 and J_2 are available to the designer. If we let $H_1 = F_1J_1$, $H_2 = F_2J_2$,

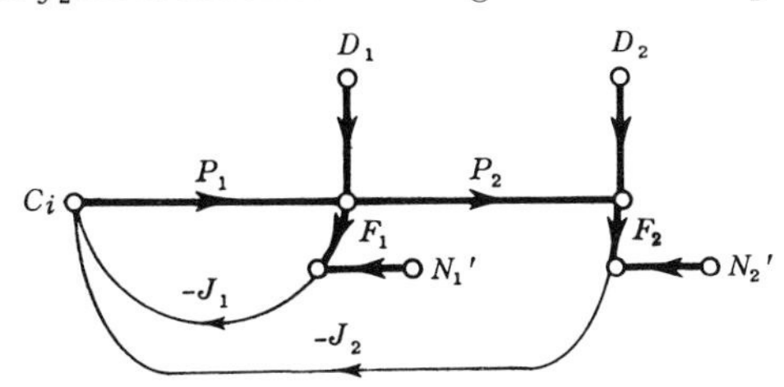

FIG. 8.2-10. Two-loop structure reducible to Fig. 8.2-1b.

then the disturbance attenuation expressions and their realization are precisely the same as before (if F_1, F_2 are minimum phase). Also, $-C_{iN} = (J_1N_1' + J_2N_2')/(1 + L_1)$. Let $J_1N_1' = H_1N_1$, $J_2N_2' = H_2N_2$, and then C_{iN} is precisely the same as for Fig. 8.2-1b. Hence the latter structure is sufficiently general.

§ 8.3 Multiple-Loop Design for Noncascade Plants

In cascade plants, the multiple-loop design problem, with or without plant modification, is relatively simple because there is only one path of signal flow from plant input C_i to plant output C. The problem is more difficult whenever there is more than one path from any node to the output. Such a plant is shown in Fig. 8.3-1 which is embedded in a perfectly general feedback structure for the case of no plant modification. As far as the disturbance acting at C is concerned, there is no difference between this structure and that of Fig. 8.2-7, i.e., if $C/D_2 = 1/(1 + L_2)$, then $C_i/N_2 = -L_2/P(1 + L_2) \doteq -1/P$ for $|L_2| > 1$ and $-L_2/P$ for $|L_2| < 1$, and there is a definite minimum value of L_2 because of peaking and stability. Hence C_i/N_2 is independent of the division of values between H_1 and H_2, so we may as well let $H_1 = 0$ and make sure that H_2 is large enough to attenuate both D_1 and D_2 in the range $\omega < \omega_{2m}$. This result is appli-

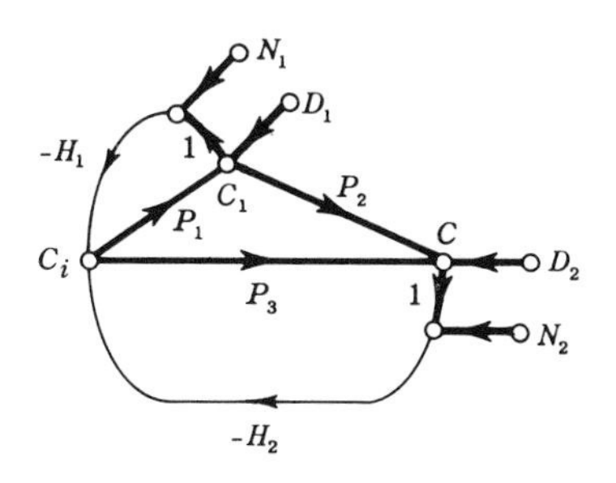

FIG. 8.3-1. A canonic feedback structure for plant with parallel paths from input to output.

cable to *all plants* in which there is only a single input (C_i) to the plant and only a single plant output (C). In the range $\omega > \omega_{2m}$ (when $\omega_{2m} > \omega_{1m}$), the transition from H_2 to H_1 is ideally made when $| P_1 N_2 | = | N_1(P_3 + P_1 P_2) |$ or at ω_{2m}, whichever is larger.

In noncascade plants the difficulty arises with internal nodes through which flow only a part of the signal which eventually reaches the output. Consider D_1 in Fig. 8.3-1. Feedback via H_1 comes back to C_1 through P_1, and if $P_1 H_1$ is large, the signal returned to C_1 is almost equal and opposite to D_1 and almost cancels it out, so that H_1 can be used to make C_1/D_1 very small. However, there is in addition a signal sent directly to C via $H_1 P_3$ (i.e., which does not pass through C_1), so C/D_1 can be large even though C_1/D_1 is small. Formally $C/D_1 = (P_2 - H_1 P_3)/[1 + H_1 P_1 + (P_1 P_2 + P_3) H_2]$. If $H_1 P_1$ dominates, then $C/D_1 \doteq (P_2/H_1 P_1) - (P_3/P_1)$. H_1 can satisfactorily attenuate D_1 only if P_3/P_1 is sufficiently small.

Is it possible to have the two terms in the numerator of C/D_1 cancel out, i.e., let $H_1 = P_2/P_3$ and thereby attenuate D_1 without a large value of the denominator? This is highly desirable in the frequency range in which P_2/P_3 is not large. But it is feasible only if the variations in P_2/P_3 are not too large. For example, suppose that at a specific frequency, P_2 ranges from 9 to 11, P_3 from 18 to 22. If $H_2 = 0$, $H_1 = 0.5$ are used, then C/D_1 ranges from $-4/P_1$ to $4/P_1$ as compared with $C/D_1 = P_2$ for no feedback. Because of parameter variations, it is possible that some but not all the necessary attenuation is achievable. The minimum amount of attenuation achievable is deducible from the bounds of variation of P_2/P_3. The balance of the required attenuation must be obtained from H_2. This is one alternative. The other is to let $H_1 = 0$ and let H_2 alone achieve the required attenuation of D_1. The choice between these two methods may be made so as to minimize the noise input to C_i.

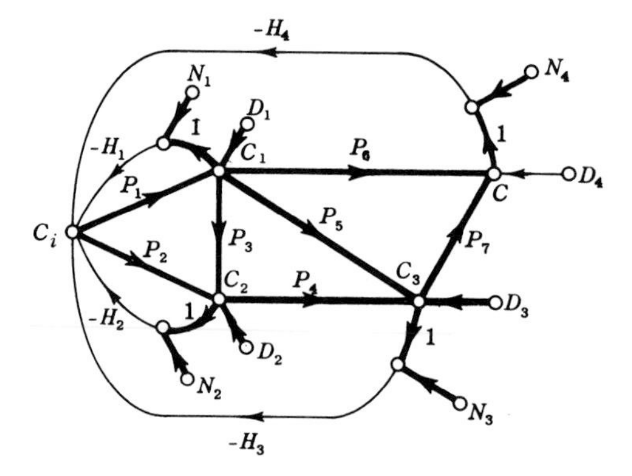

FIG. 8.3-2. Feedback system with complex noncascade plant.

The above discussion indicates the major difficulty in the multiple-loop design of a noncascade plant. Consider Fig. 8.3-2. The same difficulty arises in the attenuation of D_1, D_2, D_3. D_4 is no problem. Ideally H_4 only is used in the range $\omega < \omega_{4m}$, and if $\omega_{4m} > \omega_{1m}, \omega_{2m}, \omega_{3m}$, then design is straightforward. But if any one of these exceeds ω_{4m}, then in the range $\omega > \omega_{4m}$, it is not at all easy to decide how to divide up the feedback burden. For example, if $\omega_{3m} > \omega_{4m}$, then in the cascade plant we would let $H_1 = H_2 = 0$ and would decide between H_3 and H_4, knowing that either H_3 or H_4

can also handle D_1, D_2. In this noncascade plant, H_3 by itself cannot necessarily handle D_1 and D_2 or even D_3. In such plants it is generally not possible to relieve H_4 from the feedback burden as much as in cascade plants.

The above difficulties indicate that in noncascade plants it is important to consider the feedback problem when designing the individual portions of the plant, i.e., the idealized situation we have postulated of the feedback engineer being confronted with a complete plant, should be revised. Any node from which parallel transmissions branch out should be noted and one should study the advisability of modifying these parallel transmissions so as to permit individual control of the branches. For example, the plant of Fig. 8.3-1 could be altered as in Fig. 8.3-3. It is now possible for H_1' to control D_1. It is necessary that $P_1'/(1 + P_1'H_1')$ of Fig. 8.3-3 should be equal to P_1 of Fig. 8.3-1 if the value of C_1' of the former is to be the same as that of C_1 in Fig. 8.3-1. One must decide beforehand how much of the attenuation of D_1 should be assigned to H_2' and how much to H_1', and P_1' suitably modified in order to obtain the proper level of C_1'.

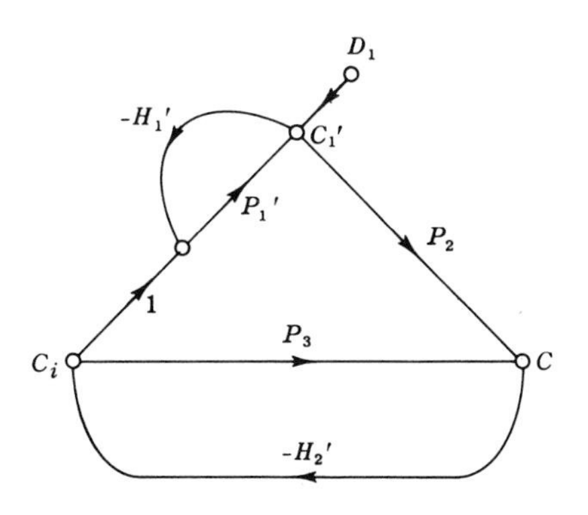

FIG. 8.3-3. Modification of plant in Fig. 8.3-1 with same plant signal levels but allowing more design flexibility.

§ 8.4 Comparison of Single-Loop and Multiple-Loop Systems for Insensitivity to Parameter Variation

Sections 8.2 and 8.3 have been devoted to multiple-loop system design for the attenuation of disturbances. Sections 8.4-8.7 are devoted to the parameter variation problem in multiple-loop systems. We shall first consider the simple cascade three-degree-of-freedom plant of Fig. 8.4-1a, and compare the two-loop

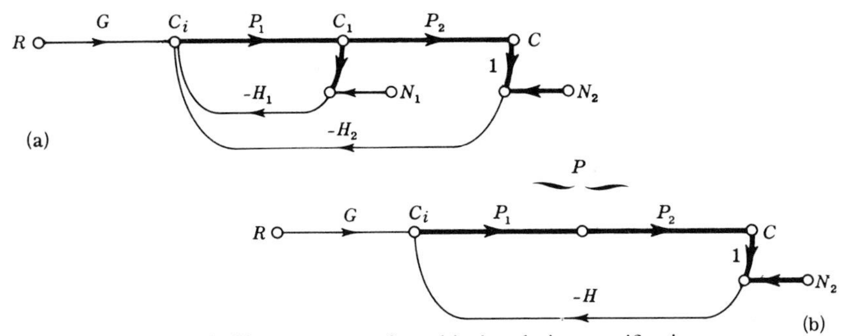

FIG. 8.4-1. Two structures for achieving design specifications.

design of that figure with the single-loop design of Fig. 8.4-1b. If the comparison is to be a fair one, it is necessary that both designs accomplish the same amount of system insensitivity to parameter variation. We shall consider a specific numerical design problem. The generality of the results will become evident, and is discussed in the latter part of Section 8.5. The following problem is considered.

Example—*Use of Complex Poles and Zeros in Loop Transmission.*

Specifications: Suppose nominal $P_1 = P_2$, and the gain factors of P_1 and P_2 may each independently vary by a factor of 2. It is required that the system

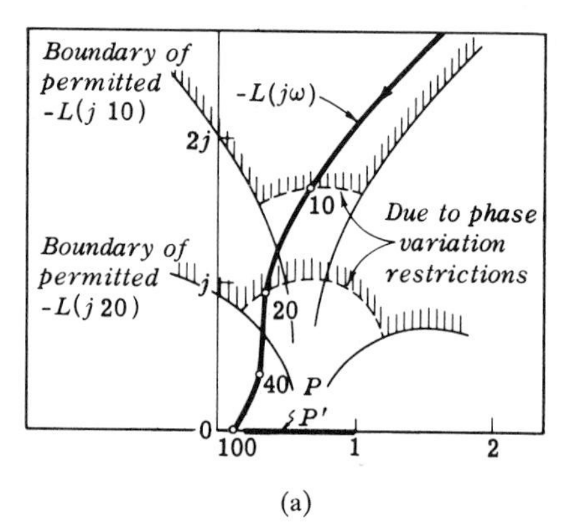

FIG. 8.4-2. a and b. Design details for single-loop system of Fig. 8.4-1b. (a) Boundaries of $-L(j10)$ and $-L(j20)$ which satisfy sensitivity specifications. (b) Bode plot of $L(j\omega)$ with complex zeros and poles, which satisfies sensitivity specifications.

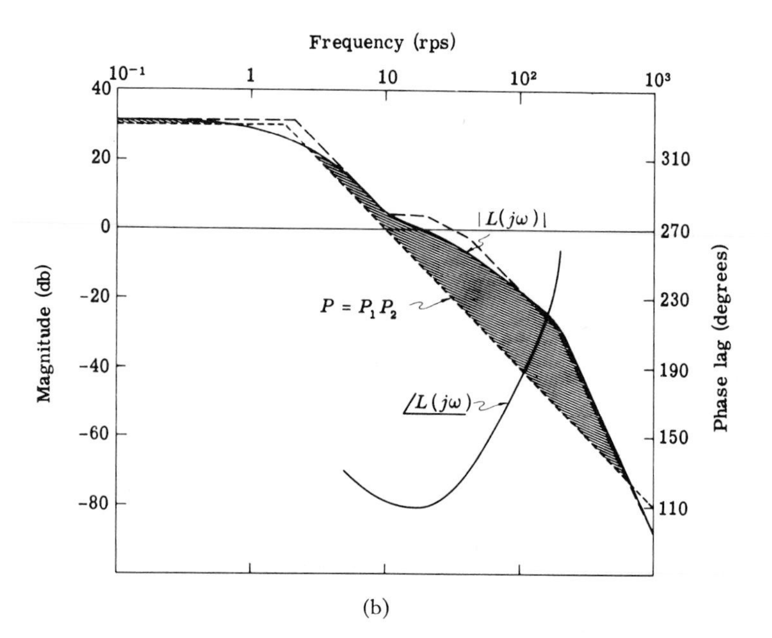

response must not vary by more than 2% in magnitude at zero frequency, 10% in magnitude and 25° in phase at 10 rps, and 40% and 40° at 20 rps.

The design procedure described in Sections 6.7 and 6.8 is applied to find the loop transmission $L = P_1P_2H$ of Fig. 8.4-1b that is required to achieve the specifications. The boundaries of permissible L at 10 and 20 rps are shown in Fig. 8.4-2a. The required zero frequency value of L is easily found to be 37. With the above information, it is possible to find in a fairly straightforward manner an $L(j\omega)$ that decreases in magnitude just about as fast as the specifications permit. It was mentioned in Section 6.9 that an optimum $L(j\omega)$ (whose magnitude decreases as fast as possible) generally will have complex poles and zeros. It is recognized that in low frequency feedback control applications, complex poles and zeros of $L(s)$ are usually undesirable. Nevertheless, very economical loop transmission designs are sometimes mandatory in high performance systems. This numerical example is therefore used to illustrate how to proceed to find a reasonably simple $L(j\omega)$, which is fairly close to optimum. The procedure is as follows.

At zero frequency, a gain of 37, i.e., 31.5 db, is needed. From Fig. 8.4-2a it is seen that the crossover frequency (where $|L|$ is 0 db) can be near 20 rps, providing the lag angle of L is approximately 100°-120°. This suggests a phase margin for L of about 70°, with crossover near 20 rps. At 10 rps, about 4 to 6 db is needed at the same phase lag. Also, from Fig. 8.4-2a, when $L(j\omega)$ has 180° phase lag, its magnitude must be less than 0.25. A conservative gain margin is therefore 18 db. These values suffice as guides for selecting $L(s)$.

The Bode sketch of $L(j\omega)$ is initiated with 31.5 db at low frequencies (see Fig. 8.4-2b). In order to maintain this high level as long as possible, a slope of − 12 db per octave (a somewhat larger slope is possible, but is hardly worthwhile) is used to bring $|L|$ down to the minimum level of 4 db needed at 10 rps. Working backwards at the rate of − 12 db per octave from a value of 4 db at 10 rps, it is found that the 31.5 db level is reached at 2 rps. Consequently, a lag pair must be located at 2 rps. A pair of complex zeros, at 10-rps corner frequency, is then inserted in order to decrease the phase lag and to maintain $|L|$ at the required level from 10 to 20 rps. A complex pair of zeros (with $\zeta = 0.6$) is used, in order that a large lead angle may thereby quickly (as a function of frequency) be obtained—the smaller the value of ζ, the faster is the lead angle obtained. This action permits the inevitable later lags to be introduced at a lower frequency than would otherwise be possible.

Simple lags are thereafter introduced as soon as possible, subject to satisfying the sensitivity requirements at 20 rps. Complex poles are reserved for the higher frequency range, because of the practical difficulties involved in realizing complex poles at low frequencies. Obviously, some cut and try is necessary—break points at 20 and 30 rps appear to be reasonable. So far, the design leads to a high frequency asymptote of − 12 db per octave. Since 18 db gain margin is needed, the frequency is noted at which this present asymptote gives $|L| = -18$ db.

This is at 100 rps. The phase lag at 100 rps due to the present tentative design is found to be 153°. Only 27° more lag is therefore permissible at 100 rps. The desired high frequency asymptotic slope of $L(j\omega)$ is taken to be -24 db per octave, in order to permit an excess of two poles over zeros for the plant, and two more for the compensating networks. Since the design of $L(j\omega)$ so far has a high frequency slope of only -12 db per octave, two more lags are needed. It is advisable to use a complex pole pair with low damping for these two lags, in order that this complex pair may be inserted as soon as possible. Thus, if real poles are used, they must be approximately at 400 rps, in order that their phase lag at 100 rps is only 27°. However, if a complex pole pair with $\zeta = 0.40$ is used, their break point can be at 200 rps. Accordingly, we choose $L(s) = (47.2)10^6(s^2 + 12s + 100)\,[(s + 2)^2(s + 20)\,(s + 40)\,(s^2 + 160s + 40{,}000)]^{-1}$.

Suppose now that $|P(j\omega)|$ has the value shown in Fig. 8.4-2b. Since $L = PH$ (Fig. 8.4-1b), $|H|$ must make up the logarithmic difference between $|L|$ and $|P|$. The gain-bandwidth of H is truly chargeable to feedback, and it serves as an excellent measure for comparing various designs. H is easily obtained by subtracting (logarithmically) $|P|$ from $|L|$ in Fig. 8.4-2b. (The shaded area in Fig. 8.4-2b represents H in that region where $|H| > 1$.) The peaked high frequency character of H represents a major difficulty of the design. The system is sensitive to high frequency feedback transducer noise (see Fig. 8.5-5) because the input to the plant due to such noise is $C_i = -N_2H/(1 + L)$, and at high frequencies where $|L| \ll 1$, $C_i \doteq -N_2H$. For example, at 200 rps, where $|H| = 24$ db, the noise is amplified by 24 db and may saturate the early stages of the plant. This difficulty with the single-loop design (when $|L| \gg |P|$) has previously been noted in Section 6.14. It is shown in the next section how this high frequency noise problem may be considerably eased if the two-loop structure of Fig. 8.4-1a is used.

§ 8.5 Design for Parameter Variations in the Two-Loop System

This section is devoted to a two-loop feedback design in which there is no plant modification. Some three-degree-of-freedom structures are shown in Fig. 8.5-1. It should be noted that there are many other equivalent configurations (just as there are many equivalent two-degree-of-freedom configurations, some of which were shown in Fig. 6.1-1). The design procedure that is developed is based on Fig. 8.5-1a. However, the results may be used for any other configuration in Fig. 8.5-1 by using the following relations:

(1) The transmission from R to C_i (t_{ci} of the fundamental feedback equation), must be the same for each. Therefore $G = G_B = G_{C1}G_{C2} = G_D$.

(2) The minor loop transmission L_1 around P_1 (with $P_2 = 0$) is the same for all. Therefore $H_1 = H_{1B} = G_{C2} = H_{1D}$.

(3) The loop transmission around P_2 is the same for all, so that after canceling out $P/(1 + L_1)$, there results $H_2 = G_BH_B = G_{C1}G_{C2}H_C = H_{2D} + G_D$.

The basic problem is to find the advantages in using the three-degree-of-freedom configurations of Fig. 8.5-1 in place of the simpler single-loop, two-degree-of-freedom configuration (Fig. 8.4-1b), and how to implement these advantages. The problem will be studied by referring to the numerical design problem introduced in Section 8.4. (The generality of the conclusions will be discussed later.) It was there solved with the single-loop structure of Fig.

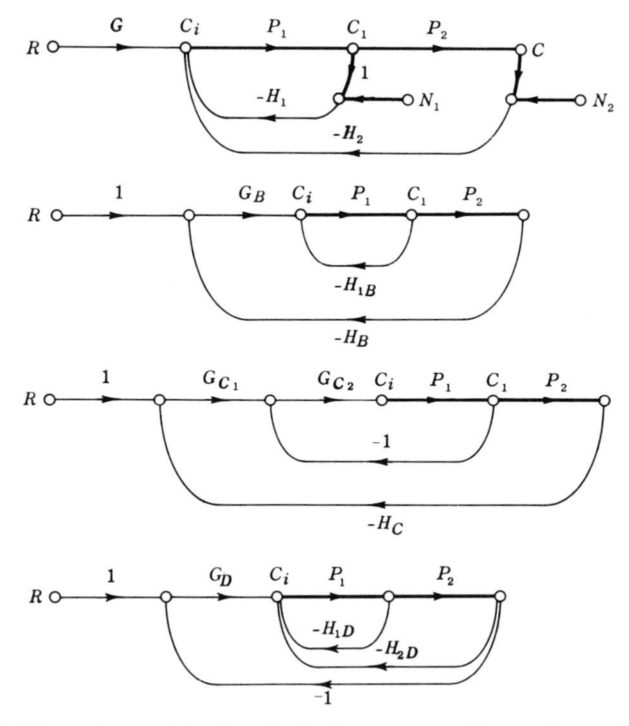

FIG. 8.5-1. Three-degree-of-freedom feedback structures with no plant modification. $P = P_1P_2$; $L_1 = P_1H_1$; $P_{1e} = P_1/(1+L_1)$; $P_e = P_{1e}P_2$; $L_2 = P_eH_2$.

8.4-1b. The difficulty with the design was the peaked high frequency character of H. Can the three-degree-of-freedom structure of Fig. 8.5-1 be of any help? The answer is that it can significantly improve matters, and the means whereby it does this is now presented.

As is evident from our work in Sections 8.2 and 8.3, the optimum division of the feedback burden between H_1 and H_2 in Fig. 8.5-1a depends very much on the relative strengths of the noise sources N_1 and N_2 in Fig. 8.5-1a. The design procedure for the parameter variation problem is more complicated than that of the disturbance attenuation problem because of the need in the former to design for stability over the bounds of parameter variation. Because of this greater complexity, we should like in this section to ignore the factor of the relative magnitudes of N_1 and N_2, in order to focus attention on the stability problem in the two-loop design. We do this by assuming that $N_1 = 0$. If $N_1 = 0$, it does not matter how large H_1 may be. Our only concern then is to make H_2 as small as possible, and this is done by having H_2 cope only with the parameter variation of P_2 (because obviously H_1 can do nothing for the variations in P_2). After we have developed the design procedure for this simplified situation, it will be possible to see the modifications necessary for nonzero N_1.

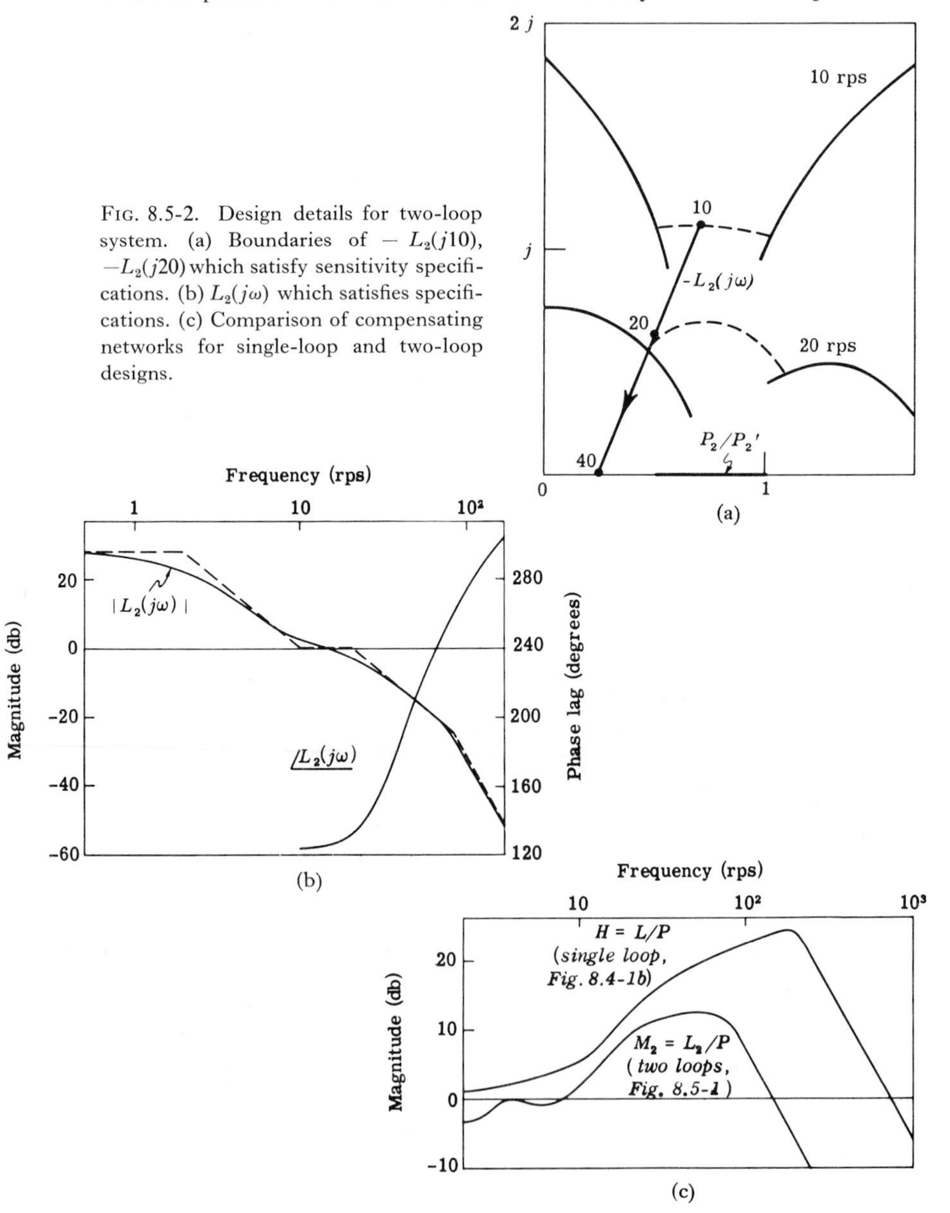

FIG. 8.5-2. Design details for two-loop system. (a) Boundaries of $-L_2(j10)$, $-L_2(j20)$ which satisfy sensitivity specifications. (b) $L_2(j\omega)$ which satisfies specifications. (c) Comparison of compensating networks for single-loop and two-loop designs.

Consider the minor loop $L_1 = P_1H_1$ of Fig. 8.5-1a. Let it be assumed, for the time being, that L_1 has sufficiently large gain and bandwidth, such that the variations in $P_{1e} \triangleq P_1/(1 + L_1)$ are extremely small over a very large bandwidth. Therefore $L_2 \triangleq P_{1e}P_2H_2$ must be designed so as to handle the parameter variations of P_2 only. L_2 is therefore less than L (of the single-loop design). In the high frequency range where $L_1 \ll 1$, $P_{1e} \doteq P_1$ so $H_2 = L_2/P_{1e}P_2 \doteq L_2/P_1P_2$ is less than $H = L/P$ of the single-loop design by the amount L_2 is less than L. The effect of the high frequency feedback transducer noise (N_2 of Figs. 8.4-1 or of Fig. 8.5-1a) on C_i is therefore reduced by precisely this difference between L and L_2. It should be noted that the effect is precisely the same whatever structure in Fig. 8.5-1 is used. In each of the latter structures, $-C_i/N_2 = L_2/P(1 + L_2)$. The absolute best that can be done in this three-degree-of-freedom system, with respect to the high frequency value of C_i/N_2, is to design L_2 to handle the P_2 variations only. The above ideas are implemented in the following example.

Example. The problem is that presented in Section 8.4. The notation is defined in Fig. 8.5-1a. P'_{1e}, P_e', etc., are as usual the values assumed as a result of parameter variations. The design is initiated by assuming that P_{1e} suffers no variations whatsoever, so L_2 is designed to handle P_2 variations only. It is recalled that the sensitivity specifications are 2% at zero frequency, 10% magnitude and 25° phase at 10 rps, 40% and 40° at 20 rps. The boundaries of permitted L_2 at several frequencies are obtained in the usual manner and are sketched in Fig. 8.5-2a. The corresponding L_2, sketched in Fig. 8.5-2b, is chosen in a manner similar to that presented in Section 8.4. It is $L_2 = (2.56)10^6(s^2 + 12s + 100)\,[(s + 2)^2(s^2 + 28.28s + 400)\,(s^2 + 96s + 6400)]^{-1}$. At high frequencies, where $L_2 \ll 1$, C_i/N_2 (of Fig. 8.5-1) $\doteq L_2/P$. This quantity L_2/P, labeled M_2, is sketched in Fig. 8.5-2c where it compares very favorably with the analogous $H = L/P$ of the single-loop design of Fig. 8.4-1b.

In the design of $L_2 = H_2P_2P_{1e}$, it was assumed that P_{1e} had no parameter variations. The second part of the procedure is to find the requirements on $L_1 = P_1H_1$ so that the above choice of L_2 is satisfactory, or is only slightly affected. For this purpose, it is convenient to divide the frequency spectrum into two parts. The first part consists of all ω for which the phase lag of $L_2(j\omega)$ is less than 180°. In the above example, it is seen from Figs. 8.5-2a and b that it consists of $\omega < 40$. A careful study of the problem indicates that in this frequency range the variations in P_{1e} can be fairly large (so L_1 need not be large) without any adverse effect on the variations in T.

The reason for this result is seen by noting the character of P_{1e}/P'_{1e} due to variations in P_1. In Fig. 8.5-3a, a polar sketch is made of a typical $-L_1$ (of moderate value). At 0 rps, $P_1/P_1' = 0.5$ and $L_1 = 4$. Now $P_{1e}/P'_{1e} = [(P_1/P_1') + L_1]/(1 + L_1)$, so $P_{1e}/P'_{1e} = (0.5 + 4)/(1 + 4) = 0.9$. Consequently, since $L_2(0) = 25$, and $P_e = P_{1e}P_2$, $T/T' = [(P_e/P_e') + L_2]/(1 + L_2) = 25.45/26$.

The maximum change in T is 2.16% in place of the desired 2%. If the very low value of $L_1(0) = 1$ is used, the result is a maximum change in T of 2.45% at 0 rps.

Next consider the situation at ω_1, where, in Fig. 8.5-3a, $-L_1(j\omega_1) = j2$. It is easily found that the locus of P_{1e}/P'_{1e} is the curve QT shown in Fig. 8.5-3b. The region covered by $P_e/P_e' = (P_{1e}/P'_{1e})(P_2/P_2')$ is therefore that enclosed by $QTNM$ in Fig. 8.5-3b. Now consider $T/T' = (P_e/P_e' + L_2)/(1 + L_2)$ and

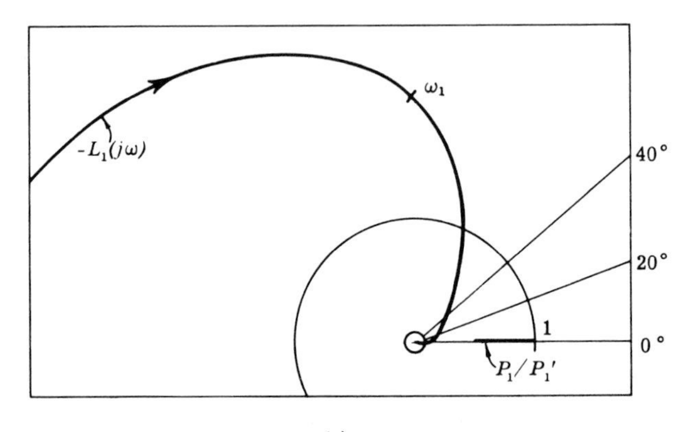

(a)

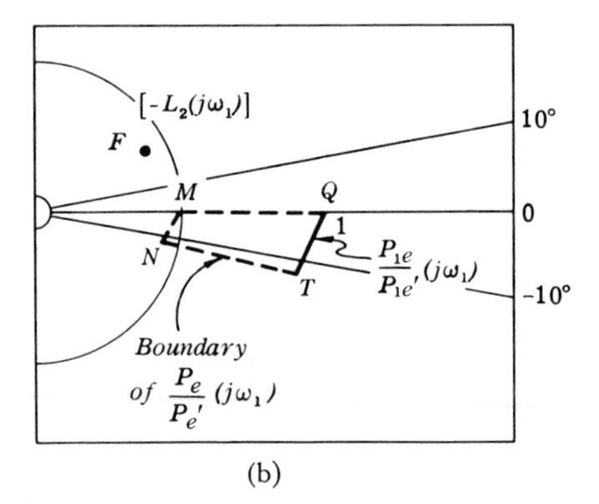

(b)

FIG. 8.5-3a to 3c. Hypothetical example illustrating why minor loop transmission can be small. (a) Hypothetical L_1 and P_1/P_1'. (b) Resulting boundary of $P_e/P_e'(j\omega_1)$.

suppose $\omega_1 = 30$, so that $L_2(j\omega_1) = 0.44\angle -152°$ (F in Fig. 8.5-3b). It is easily seen the maximum change in $|T|$ is no more than if $P_{1e}/P'_{1e} = 1$, but the maximum phase change of T is 60° in place of the previous 45°. The significant point in all problems in which only gain factors change is as follows: For all $L_1(j\omega)$ with phase lags less than 180°, the loci of P_{1e}/P'_{1e} are such as to lead to boundaries of P_e/P_e' of the form shown in Fig. 8.5-3b. Now consider $-L_2(j\omega)$ given by the point F in Fig. 8.5-3b. If $|FM| < |FN|$, then the maximum change in $|T|$ is the same as if P_e/P_e' consisted only of the line MQ, i.e., just as if P_{1e} had no variations. The maximum phase change of T, however, is determined by N and not by M. For the above kind of behavior, two conditions are necessary: (a) $L_1(j\omega)$ must have a phase lag less than 180°, (b) $|FM| < |FN|$, which generally requires that the phase lag of $L_2(j\omega)$ be somewhat less than 180°.

Next consider the situation in which the phase lag of $L_2(j\omega_1)$ in Fig. 8.5-3b is more than 180°. Clearly, the point N now determines the maximum change in T. Finally, suppose the phase lag of $L_1(j\omega)$ is more than 180°. From the usual construction it is seen that the locus of P_{1e}/P'_{1e} has the form of the curve VW in Fig. 8.5-3c. Consequently, in this example where the P_2 gain constant can change by a factor of 2, the region occupied by P_e/P_e' is $VWJR$ in Fig. 8.5-3c. If the phase lag of L_2 is less than 180°, the effect of P_{1e}/P'_{1e} can be substantial. The effect is very slight, however, if the lag angle of L_2 is more than 180°. Therefore, the 180° crossing by L_1 and L_2 should be approximately the same.

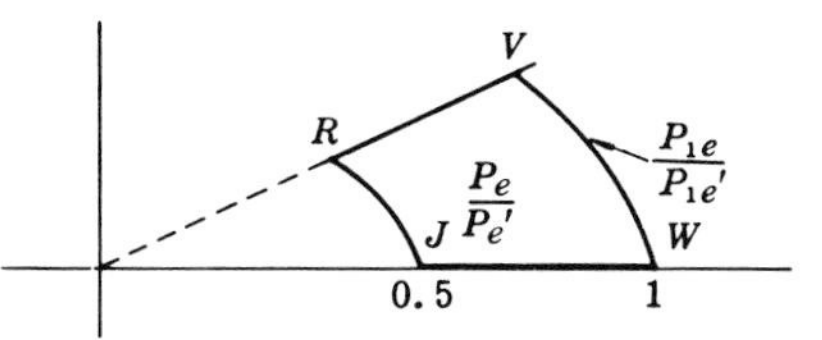

FIG. 8.5-3c. Locus of P'_{1e}/P_{1e} and range of P_e/P_e' when $\arg L_1 < 180°$.

The purpose of this discussion has been to establish some feeling for the problem, and to indicate why a design based on no variation in P_{1e} can be satisfactory despite substantial variation in P_{1e}. The actual design procedure is now presented in detail. The objective is to find the permitted values of $L_1(j\omega)$ in Fig. 8.5-1a, in order that the previous design of $L_2(j\omega)$, which was based on a P_{1e} with no parameter variations, should be satisfactory. The procedure is as follows:

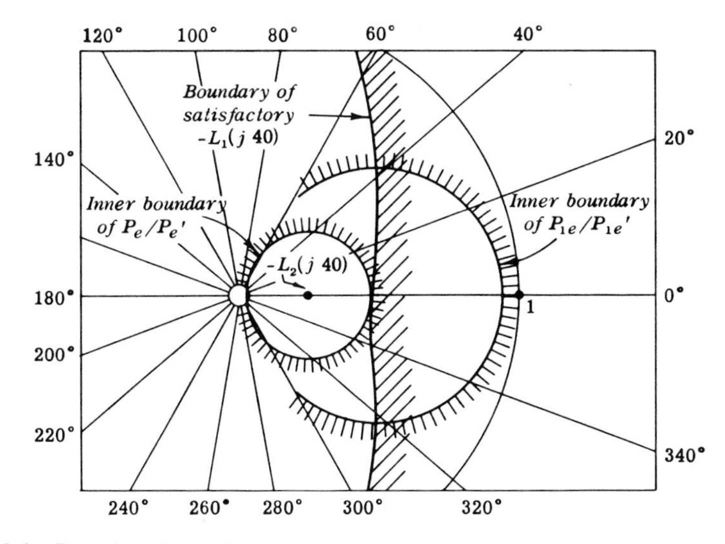

FIG. 8.5-4a. Procedure for finding boundary of $-L_1(j40)$ which satisfies specifications.

Design details: Consider the situation, for example, at the specific value $\omega = 40$, where, from Fig. 8.5-2b, $-L_2(j40) = 0.25$.

(1) Suppose that at 40 rps it is permissible to have $3 \geqslant |T/T'| \geqslant 0.3$. Then one boundary of the region in which P_e/P_e' may lie must be the circle centered at 0.25 with radius (0.3) (0.75) = 0.225 (see Fig. 8.5-4a). Thus, at any point on this circle and its exterior, $|T/T'| = [(P_e/P_e') + L_2]/(1 + L_2) \geqslant 0.3$. If there are any restrictions on the maximum change in phase of $T(j40)$, then

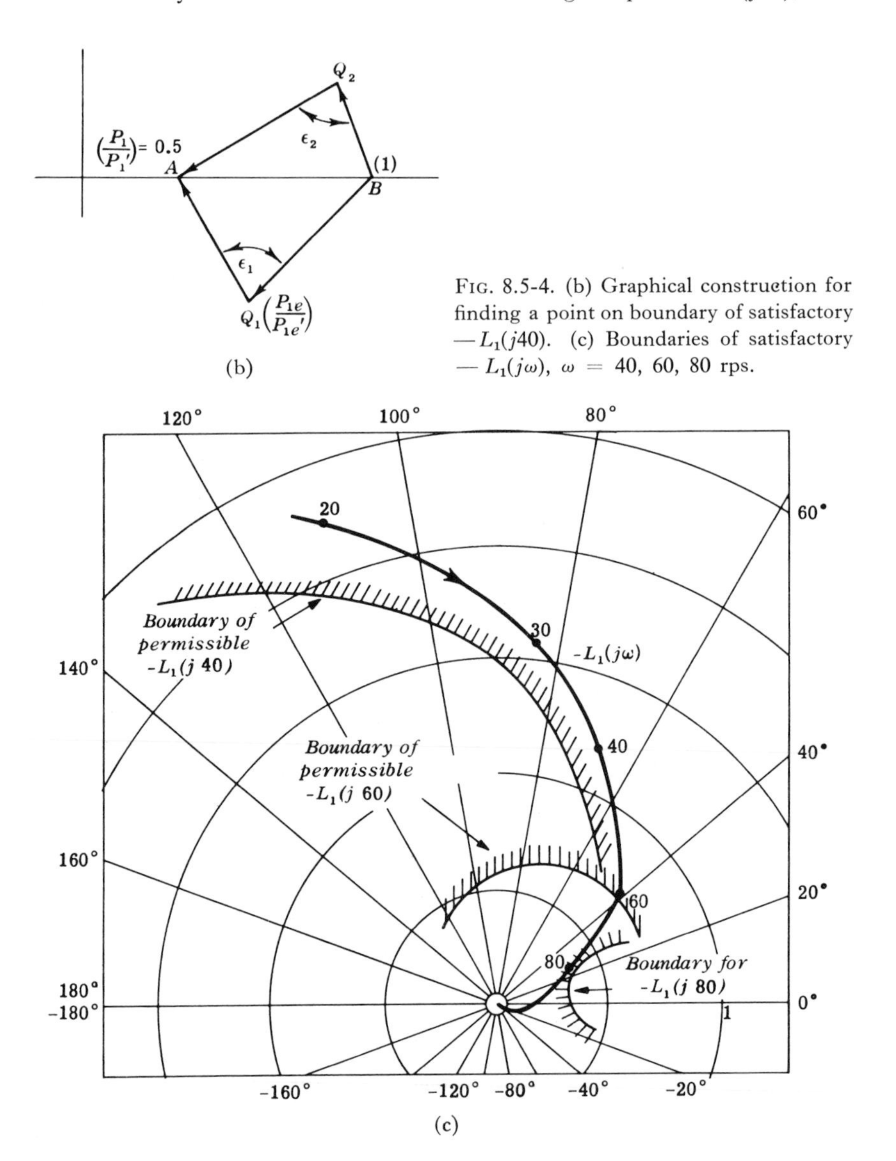

FIG. 8.5-4. (b) Graphical construction for finding a point on boundary of satisfactory $-L_1(j40)$. (c) Boundaries of satisfactory $-L_1(j\omega)$, $\omega = 40, 60, 80$ rps.

the part of the above region which is inadmissible is easily ascertained. The outer boundary of P_e/P_e' is the circle centered at 0.25 with radius (3) (0.75) = 2.25. It is not needed here as there is no danger of trespassing beyond this boundary.

(2) The above boundary is that of $P_e/P_e' = (P_{1e}/P'_{1e})(P_2/P_2')$. From the known boundaries of P_e/P_e' and of P_2/P_2', it is easy to get the boundary of permissible P_{1e}/P'_{1e}. In this problem, it is simply twice that of P_e/P_e' and is included in Fig. 8.5-4a.

(3) Finally, from the known boundary of P_{1e}/P'_{1e}, it is possible to find the boundary of permitted L_1 because $P_{1e}/P'_{1e} = [(P_1/P_1') + L_1]/(1 + L_1)$. Therefore $L_1 = [(P_1/P_1') - (P_{1e}/P'_{1e})]/[(P_{1e}/P'_{1e}) - 1]$. The procedure is quite straightforward. It is assumed that $P_1/P_1' = 0.5$ is responsible for the maximum variation in P_{1e}/P'_{1e}. For every point on the locus of P_{1e}/P'_{1e}, there is a corresponding point on L_1. The points on L_1 can be determined arithmetically or graphically; the latter is easier. Thus it is noted that from the above equation, $L_1 = \mathbf{QA}/\mathbf{BQ}$ in Fig. 8.5-4b and the angle of L_1 is $+(180° + \epsilon_1)$ for $Q = Q_1$ and $-(180° + \epsilon_2)$ for $Q = Q_2$. In this way, the locus of the boundary of $-L_1(j40)$ is obtained, and is sketched in Fig. 8.5-4a. It is easily ascertained that the permissible region for $-L_1(j40)$ is on the right-hand side of the boundary.

The above process is repeated for several discrete frequencies (60 and 80 rps).

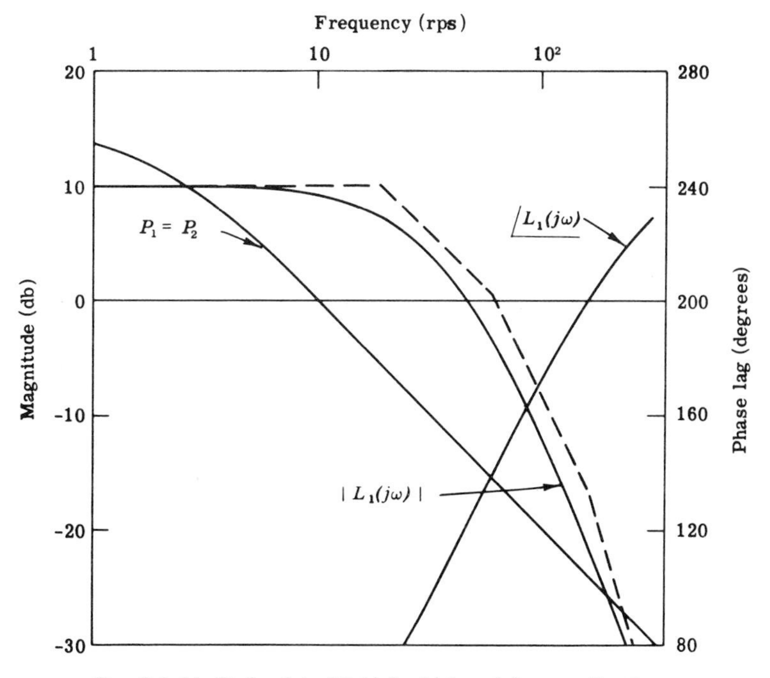

FIG. 8.5-4d. Bode plot of $L_1(j\omega)$ which satisfies specifications.

The resulting boundaries and the permitted (shaded) regions are shown in Fig. 8.5-4c. At higher frequencies, the boundaries permit the lag of L_1 to be more than 180°. Thus $L_1(j60)$ must have a lag angle less than 180°, but $L_1(j80)$ may have a lag angle larger than 180°. In obtaining Fig. 8.5-4c, it was assumed that the maximum permitted changes in T were given by $|T/T'| = 0.20$ at $\omega = 60$ and 0.15 at $\omega = 80$ rps. The smaller these values, the less the gain-bandwidth demand on L_1. Similar boundaries may be sketched for as many frequencies as may be deemed necessary, but the above procedure is best used for the higher frequency range only. It is found in this example that the specifications regarding the permitted change in the phase of T must be slightly relaxed, if the original L_2 is to be satisfactory. This is because L_2 (see Fig. 8.5-2a) barely satisfies the phase specification at 10 rps. It was previously noted (Fig. 8.5-3b) that there is no difficulty with the low frequency magnitude specifications, when L_2 is chosen as if P_{1e} has no parameter variation, but that there is an effect on the phase changes. Therefore, in choosing $L_2(j\omega)$, one should satisfy the phase requirements with a few degrees to spare. The actual amounts involved in this specific example are so small that they do not justify repeating the design.

There are sufficient data in Figs. 8.5-4a and c to permit the design of L_1, which is sketched in Figs. 8.5-4c and d. It is $L_1 = (57)10^4[(s + 20)(s + 60)(s + 150)]^{-1}$. The effect of N_1 on C_i in Fig. 8.5-1a is $C_i/N_1 = (L_1/P_1)[(1 + L_1)(1 + L_2)]^{-1} \doteq L_1/P_1 = H_1$ at high frequencies. (In designing L_1, it was assumed that there was little noise in the L_1 loop; therefore, no attempt was made to economize on L_1 by using complex poles and zeros.) The design is

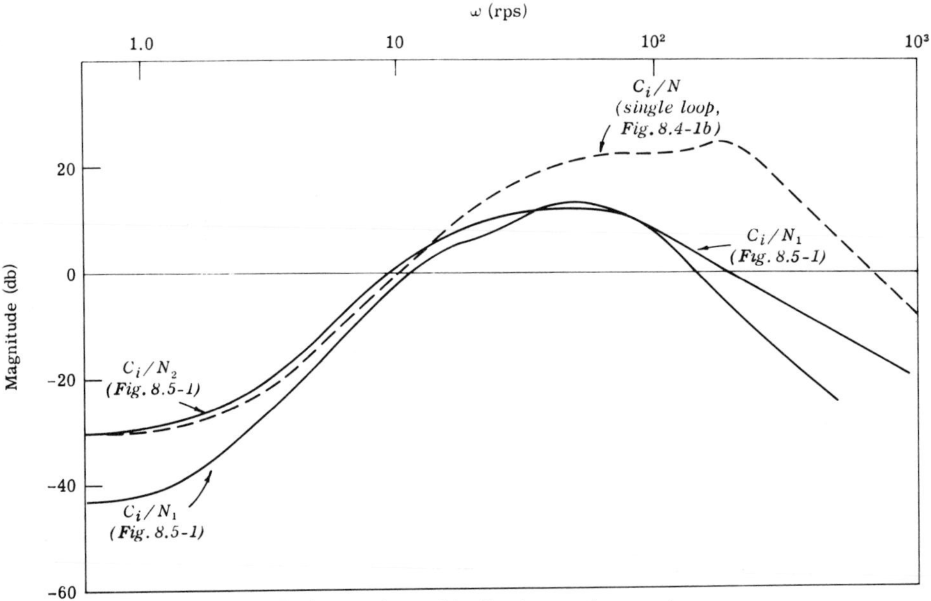

FIG. 8.5-5. Comparison of feedback transducer noise responses.

now complete. Since $L_1 = P_1H_1$ and $P_1 = 10.1/(s + 1.8)$, $H_1 = [(5.7)10^4(s + 1.8)]\,[(s + 20)\,(s + 60)\,(s + 150)]^{-1}$. Also, $P_{1e} = P_1/(1 + L_1) = 10.1(s + 20)(s + 60)(s + 150)\,[(s + 1.8)\,(s + 180)\,(s^2 + 50s + 4200)]^{-1}$. Since $L_2 = P_2H_2P_{1e}$,

$$H_2 = \frac{L_2}{P_2P_{1e}} = \frac{(2.5)10^4(s + 1.8)^2(s^2 + 12s + 100)\,(s^2 + 50s + 4200)\,(s + 180)}{(s+2)^2(s^2+28.28s+400)(s+20)(s+60)(s^2+96s+6400)(s+150)}.$$

Several obvious cancellations can be made, with very minor effects on the total result, i.e., cancellation of $[4/(1.8)^2]\,(s + 1.8)^2$ by $(s + 2)^2$, and $(5/6)\,(s + 180)$ by $(s + 150)$.

The feedback noise responses of the designs of Figs. 8.4-1b and 8.5-1 are compared in Fig. 8.5-5. The difference would be even more significant if $|P_2|$ decreased faster.

General Plant Variations

The same procedure is followed in the more general case, when plant poles and zeros as well as the gain factor vary. For example, suppose P_2/P_2' (of Fig. 8.5-1) occupies at some particular frequency the region K in Fig. 8.5-6, and L_2 has the

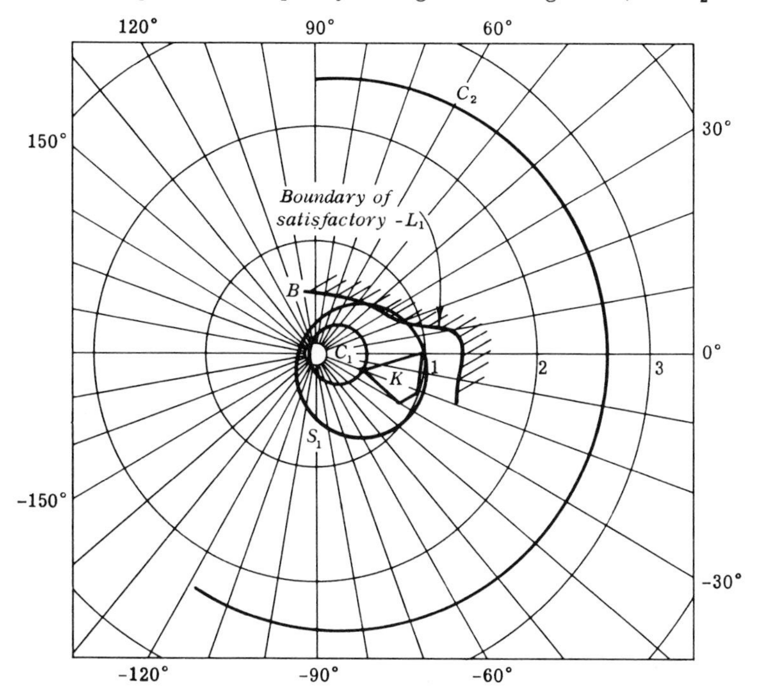

FIG. 8.5-6. Case in which plant poles and zeros also vary.

value -0.2, while it is required that $3 \geqslant |T/T'| \geqslant 0.3$, at this same frequency. Following the previous procedure, with 0.2 as center, draw a circle C_1 of radius

(0.3) (1 − 0.2) = 0.24, and a circle C_2 of radius 3(0.8) = 2.4. Outside the small circle $|T/T'| \geqslant 0.3$, and inside the larger circle $|T/T'| \leqslant 3$. Therefore the annulus is the region of permitted variation of $P_e/P_e' = (P_{1e}/P_{1e}')(P_2/P_2')$. From the known variation of P_2/P_2', the permitted variations of P_{1e}/P_{1e}' can be deduced. The point $0.42 \angle -20°$ on the boundary of the region K clearly dominates here. Thus one boundary of P_{1e}/P_{1e}' is obtained by dividing the points on C_1 by $0.42 \angle -20°$. The resulting locus is labeled S_1, i.e., P_{1e}/P_{1e}' cannot occupy any points interior to S_1, because then P_e/P_e' would occupy points interior to C_1. The circle C_2 is the other boundary of P_{1e}/P_{1e}', i.e., if P_{1e}/P_{1e}' occupies any points outside C_2, P_e/P_e' would also occupy points outside C_2.

Now that the boundaries of $P_{1e}/P_{1e}' = [(P_1/P_1') + L_1]/(1 + L_1)$ are known, the permitted location of L_1 can be determined. The region occupied by P_1/P_1' is first determined from the original data. In order not to clutter up the diagram, it is supposed here that it is the same as that occupied by P_2/P_2', i.e., the region K. It is straightforward if somewhat tedious to find the permitted location of L_1. The curve marked B is part of the boundary of permitted $-L_1$. Thus the procedure for the more general kind of plant parameter variation is basically the same as before. Thus the conclusions regarding the improvement in the system response to feedback transducer noise also apply for general plant parameter variation. There may be more effort involved in obtaining the boundaries. Actually, in the higher frequency range, the loci of P_2/P_2', P_1/P_1' tend to be almost straight lines (narrow regions), because the important plant poles and zeros are in the low frequency range, so that the plant phase change is pretty well complete in the high frequencies. For example, if a plant pole does change from its original position at −1 to a new position at −5, the phase at 40 rps changes only by 7°.

§ 8.6 Extension to More Than Two Loops

The approach used in Section 8.5 can be directly extended to configurations with more than two loops. In Fig. 8.6-1, let $P_{Ae} = P_A/(1 + P_AH_A)$, $P_{Be} = P_{Ae}P_B/(1 + P_{Ae}P_BH_B)$, and $L_A = P_AH_A$, $L_B = P_{Ae}P_BH_B$, $P_e = P_CP_{Be}$,

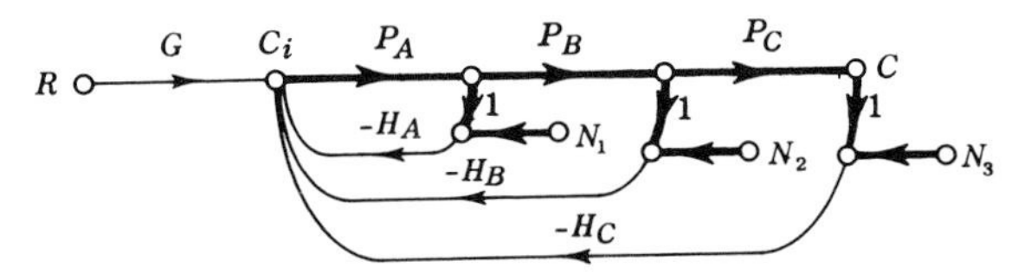

FIG. 8.6-1. Four-degree-of-freedom feedback structure with no plant modification.

$L_C = P_eH_C$. Just as in Section 8.5, $L_C = P_CH_CP_{Be}$ is chosen almost as if the only variations were in P_C and none in P_{Be}. (The qualification of almost is made because the phase variations in the low frequencies should be somewhat less than the permitted values.) This results in the minimum transmission of noise from N_3 to C_i. Next, L_B is deduced (as in the previous example), almost as if the only variations in P_{Be} were caused by P_B. This results in the minimum transmission of noise from N_2 to C_i, subject to the minimum transmission from N_3 to C_i. Finally, L_A is deduced in precisely the same manner, on the basis of the variations in P_A. H_A is found from $H_A = L_A/P_A$, H_B from $L_B = P_B H_B P_{Ae} = P_B H_B P_A/(1 + L_A)$, and H_C from $L_C = P_C H_C P_{Be}$. The procedure is illustrated by a numerical example.

Example.

Specifications: In Fig. 8.6-1 the gain factors of each of P_A, P_B, P_C independently vary by a factor of 2. The sensitivity specifications are those of the previous example—a maximum of 2% at 0 rps, 10% at 10 rps, 40% at 20 rps, 333% at 40 rps, 500% at 60 rps, 670% at 80 rps, and maximum phase changes of 25° at 10 rps and 40° at 20 rps.

Design: The value of L_C used here is precisely that of L_2 of the previous example in Section 8.5 (Fig. 8.5-2b). Also, the value of L_B here can be the same as that of L_1 in the previous example (Fig. 8.5-4d), because L_B is chosen as if the only variations are in P_B. Thus only L_A need be deduced. The procedure is exactly the same as that used in finding L_1 in the example in Section 8.5.

It is necessary to know the permitted region of P_{Be}/P'_{Be}, which is P_{1e}/P'_{1e} of the previous example (Section 8.5). The permitted region of P_{Be}/P'_{Be} for various frequencies is obtained as an integral part of designing L_B (the previous L_1). Thus Fig. 8.5-4a has the boundary of permitted $P_{Be}/P'_{Be} = P_{1e}/P'_{1e}$ for $\omega = 40$. But $P_{Be}/P'_{Be} = [(P_B P_{Ae}/P'_B P'_{Ae}) + L_B]/(1 + L_B)$. Since $L_B(j40) = 1.19 \angle -112°$ (Q in Fig. 8.6-2a), and the boundary of P_{Be}/P'_{Be} is known (it is that of P_{1e}/P'_{1e} in Fig. 8.5-4a), the boundary of $P_B P_{Ae}/P'_B/P'_{Ae}$ can be deduced. For example, consider the point $0.95 \angle -10°$ on the boundary of $P_{Be}/P'_{Be} = P_{1e}/P'_{1e}$. Draw the line QF (Fig. 8.6-2a), where F is the point 1.0. Draw the line QM at an angle of 10° with QF. Measure QN along QM such that $QN = 0.95$ (QF). Therefore, N is that value of $P_B P_{Ae}/P'_B P'_{Ae}$ which corresponds to $P_{Be}/P'_{Be} = 0.95 \angle -10°$. In this way the boundary of $P_B P_{Ae}/P'_B P'_{Ae}$ is obtained. Since the range of P_B/P'_B is known to be from 1 to 0.5, the boundary of P_{Ae}/P'_{Ae} is obtainable by multiplying every point on the boundary of $P_B P_{Ae}/P'_B P'_{Ae}$ by 2. This is only the inner boundary. The outer boundary is similarly obtained from that of P_{Be}/P'_{Be}. In the higher frequency range, the outer boundary can usually be disregarded, because the high frequency region is the positive feedback region in which $|T'| > |T|$.

The next step is to deduce the permitted $L_A(j40)$ from the fact that $P_{Ae}/P'_{Ae} = [(P_A/P'_A) + L_A]/(1 + L_A)$ and P_A/P'_A is known. The result is shown in Fig. 8.6-2a. Similar permitted regions of $-L_A(j80)$ and $-L_A(j150)$ are obtained step-by-step in Figs. 8.6-2b and c, respectively.

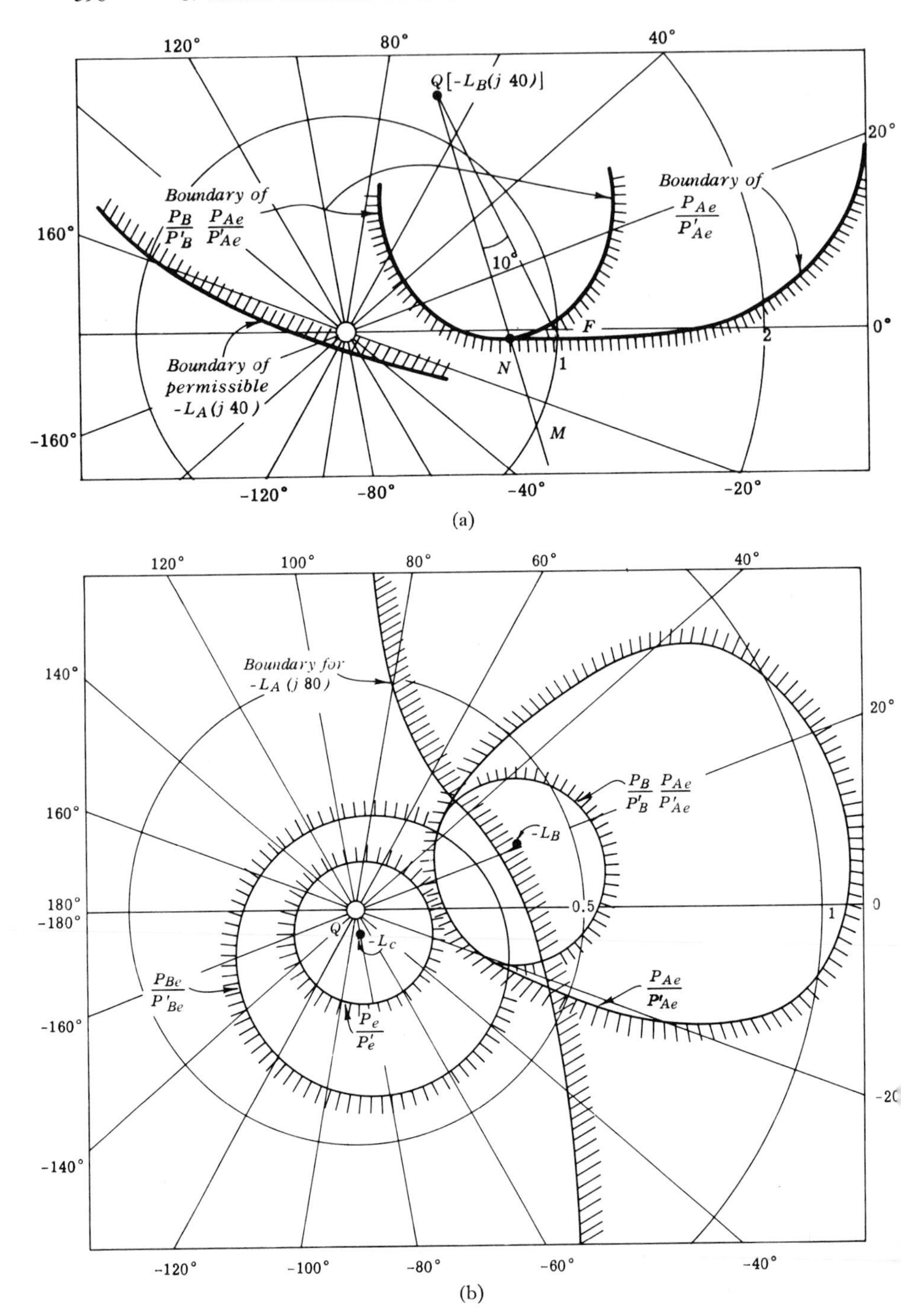

120°
80°
40°
$Q[-L_B(j\,40)]$
20°
Boundary of
$\frac{P_B}{P'_B}\ \frac{P_{Ae}}{P'_{Ae}}$
Boundary of
$\frac{P_{Ae}}{P'_{Ae}}$
160°
10°
F
0°
N
1
2
Boundary of
permissible
$-L_A(j\,40)$
M
-160°
-120°
-80°
-40°
-20°
120°
100°
80°
60°
40°
Boundary for
$-L_A(j\,80)$
140°
20°
$\frac{P_B}{P'_B}\ \frac{P_{Ae}}{P'_{Ae}}$
160°
$-L_B$
180°
-180°
0.5
1
0
Q
$-L_c$
$\frac{P_{Be}}{P'_{Be}}$
$\frac{P_{Ae}}{P'_{Ae}}$
-160°
$\frac{P_e}{P'_e}$
-140°
-120°
-100°
-80°
-60°
-40°

(a)

(b)

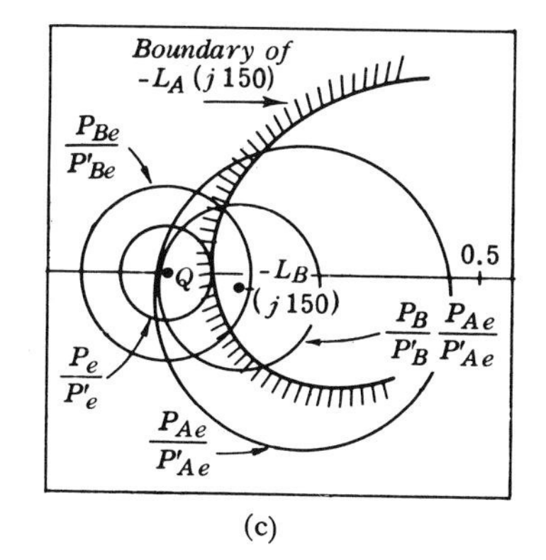

(c)

FIG. 8.6-2a to 2c (a *and* b *opposite*). Steps in determining boundary of satisfactory $-L_A(j\omega)$ at $\omega = 40$, 80, 150.

It is found that the allowed phase variation at 10 rps must be increased, in order that any finite $L_A(j10)$ should be satisfactory. This difficulty is a result of (see Fig. 8.5-2a) having $L_2(j10)$ barely satisfy the phase specification. Note, however, that $L_2(j20)$ in Fig. 8.5-2a satisfies the phase specification with only 3° to spare, and it is found that this suffices. If the maximum phase variation at 10 rps is increased from 25° to 30°, then it is found that $L_A(j10)$ too may have a very small value. At 0 rps, if 2.25% variation can be tolerated, then $|L_A(0)|$ may be only one. Thus it is clear from the above and from Figs. 8.6-2 that the gain-bandwidth demands on L_A are very moderate indeed, and the step-by-step design, on the basis of no parameter variations in the earlier loops, is justified.

§ 8.7 Effect of Finite N_1

In Section 8.5 it was assumed that N_1 (in Fig. 8.5-1a) was zero, and therefore the relative strengths of N_1 and N_2 was of no consequence in the design. No attention was paid to economizing on H_1. We shall now consider the effect of finite N_1. It was noted in Section 8.2 that in the frequency range in which L_2 (of Fig. 8.5-1a) was greater than 1, $C_i/N_2 \doteq -1/P$ irrespective of the actual values of H_1 and H_2. Ideally, therefore, H_1 is made zero in this range and L_2 is allowed to handle the parameter variations of both P_1 and P_2 in this range. From Figs. 8.5-2a and b, it is seen that this range extends from 0 rps to approximately 5 rps. (We should not try to extend the range to 10 rps because in modifying L_2 to handle both P_1 and P_2 variations right up to 10 rps, we would be compelled to use a higher value of L_2.) Therefore up to approximately 5 rps we could let L_2 handle both P_1 and P_2 variations. This would make C_i/N_1 lower than its value in Fig. 8.5-5, for $\omega < 5$ rps.

For higher frequencies we inquire whether we prefer to use H_2 or H_1 to handle the parameter variations of P_1. It is much more difficult in the parameter variation problem (than in the disturbance attenuation problem) to compare the two alternatives, because of the need to obtain diagrams like those Figs. 8.5-4a and c when H_1 is assigned the entire burden. In the present case, the result is given in Fig. 8.5-4d (actually, as noted in the preceding paragraph, L_1 could be much less for $\omega < 5$). C_i/N_1 is available from Fig. 8.5-4d and one may judge whether it is preferable to assign the burden of P_1 parameter variations to H_2 for $\omega > 5$ rps. Such would be the case, for example, if $N_1 \gg N_2$. It is obviously not possible in the parameter variation problem to make the decision as easily as in the disturbance attenuation problem.

If plant modification is allowed (Fig. 8.2-8), then it may be worthwhile using the H_3 loop around P_2. As noted in Section 8.2, for a given L_2 this loop has no effect on the noise level at C_1, but it can be used to significantly decrease the noise level at the plant input C_i. This could be useful if, for example, $|P_1| \ll 1$ in the range $\omega < \omega_{2m}$. It is possible to postpone the decision to use H_3 until after L_2 has been found. This is because H_3 by itself is ineffective for reducing the system sensitivity to variations in P_1. Since $L_2 = H_3P_2 + H_2P_e$, one may then decide how to divide L_2 up between H_2 and H_3.

§ 8.8 Multiple-Loop Systems with Parallel Plants

Sections 8.8 and 8.9 are devoted to feedback systems around parallel-type plant structures such as those in Fig. 8.8-1. It is essential that the designer

FIG. 8.8-1. Parallel plant structures.

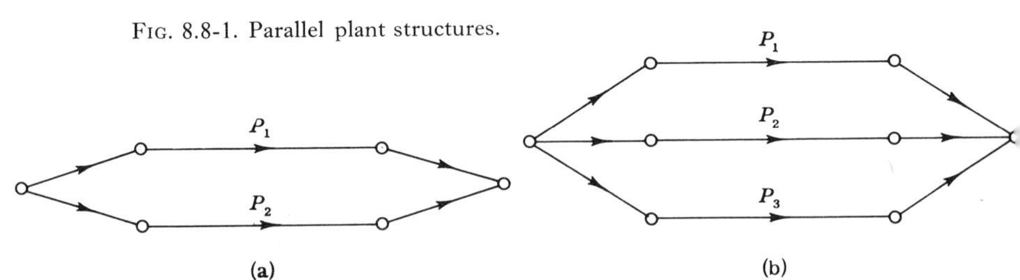

have access to the individual plant inputs and outputs; otherwise the parallel plant degenerates into the single plant with $P = P_1 + P_2$, etc. Some of the significant properties of parallel plants in feedback amplifiers were apparently independently discovered by J. J. Zaalberg van Zelst[1] and by B. McMillan.[2] The approach is modified here, in order to demonstrate how the invention and its novel properties are directly derivable from a fundamental approach to feedback theory.

---

[1] J. J Zaalberg van Zelst, Dual Channel Amplifying Circuit, U. S. Patent 2,775,657, Dec. 25, 1956; Amplificateurs à Amplification Constante, *Rev. techn. Philips* **9**, No. 1, 24-31 (1947).

[2] B. McMillan, Multiple-Feedback Systems, U.S. Patent 2,748,201, May 29, 1956.

Throughout the book, up to this point, it has been assumed that there is a single plant and that the output can be reached only through this plant. The sensitivity properties and limitations of single-loop feedback around such plants were presented in Chapters 5-7. The pioneering work in this field is due to Bode. From time to time there have appeared in the literature claims for new feedback structures with wonderful properties not possessed by the old. In many cases, these claims only revealed that the authors were ignorant of, or did not understand, Bode's work. In other cases, the advantages were due to the fact that conditionally stable feedback systems were being permitted, or, in this same class, that designs with different stability margins, or different values of phase lags vs. frequency, were being (unfairly) compared. The point is that the fundamental properties (in the frequency domain) of zero leakage, single plant, single-loop feedback systems, with linear time invariant compensation, have been thoroughly explored. If any new inventions are to be forthcoming, they must involve lifting one or more of the constraints. The removal of the single-loop constraint in cascade-type plants led to the multiple-loop systems of earlier sections. Removing the constraint of zero leakage transmission leads to the parallel plant structure and its novel properties.

Consider a system in which the leakage transmission (denoted here by t_{o1}) is not zero. From Eq. (3.6,2c),

$$\frac{\Delta T/T}{\Delta k_1/k_1'} = \frac{1 - (t_{o1}/T)}{L_1 + k_1/k_1'} \tag{8.8,1}$$

in which T, T', k_1, k_1', L_1, t_{o1}, respectively, replace T_o, T_f, k_o, k_f, L_o, t_{oi}; $\Delta T = T' - T$, $\Delta k_1 = k_1' - k_1$, and k_1 represents one of the two parallel branches of the plant.

Clearly, from Eq. (8.8,1), the sensitivity can be greatly decreased (without increasing L_1) by letting t_{o1} be nearly equal to T, providing that in the process L_1 is not decreased. This appears to be absurd, if it means letting most of the output pass through the second branch k_2, for in such a case, what about the system sensitivity to the leakage path, k_2? Obviously, by analogy with Eq. (8.8,1),

$$[\Delta T/T]/[\Delta k_2/k_2'] = [1 - (t_{o2}/T)]/[L_2 + (k_2/k_2')],$$

and can also be made very small (without increasing L_2), if $t_{o2} \doteq T$.

The above is feasible only if a structure can be found in which both t_{o1} and t_{o2} are very nearly equal to T. (At first thought, such a condition appears ridiculous, but it is remarkable how often important discoveries are made, by stubbornly thinking through apparently ridiculous situations.) The search for such a structure is logically begun by considering the simplest but most general feedback structure built around the plant of Fig. 8.8-1a. The fundamental feedback structure built around two references (in place of only one in Fig.

2.2-2) is derived from the fundamental feedback equation, which permits us to write

$$T = (t_{o1}) + [k_1 t_{c1i} t_{os1}/(1 - k_1 t_{c1s1})],$$

in which t_{o1} is the transmission from $\mathscr{I}$ to $\mathscr{O}$ when $k_1 = 0$, and all the t_{xy}'s are independent of k_1, but are functions of k_2, e.g., t_{c1i} is the transmission from the input to $\mathscr{C}_1$ with $k_1 = 0$.

We wish now to explicitly display the dependence of t_{o1}, t_{c1i}, etc., on k_2. For this purpose, the fundamental feedback equation is applied to each of these four transmissions, but now the reference is k_2. When this is done, t_{o1}, t_{c1i} are represented by Fig. 8.8-2a [i.e., the independent input is $\mathscr{I} = 1$ and there are

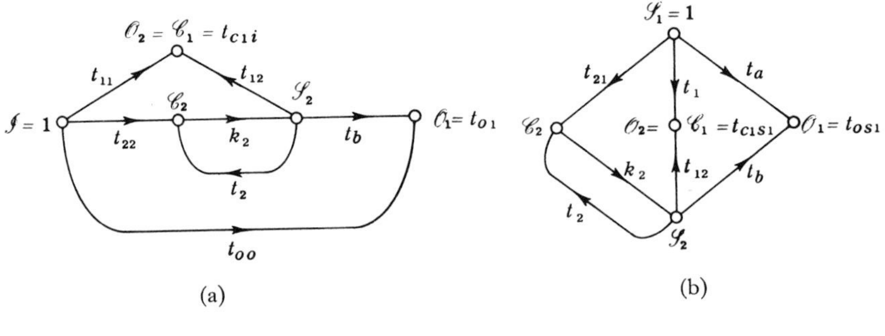

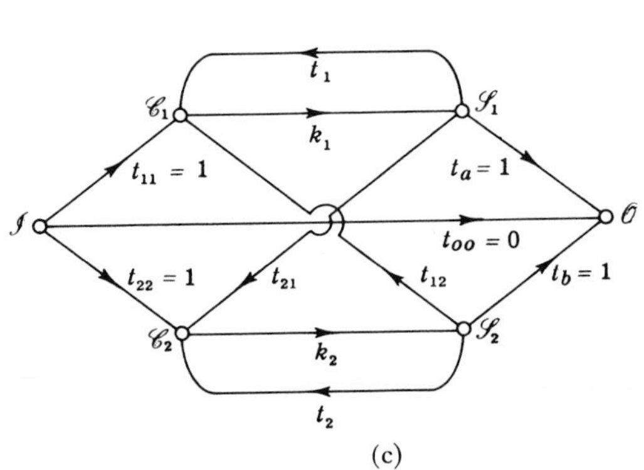

FIG. 8.8-2. Derivation of fundamental feedback flow graph for two controlled sources. (a) Use of fundamental feedback equation (with k_2 as reference) to evaluate t_{o1} and t_{c1i}. (b) Evaluation of t_{os1} and t_{c1s1} (with k_2 as reference). (c) Fundamental feedback flow graph for two references.

two outputs, $\mathscr{O}_1 = t_{o1}$ and $\mathscr{O}_2 = \mathscr{C}_1 = t_{c1i}$. Let $t_{11} = t_{12} = 0$; then Fig. 8.8-2a consists of the fundamental feedback flow graph (Fig. 2.2-2) with $\mathscr{O}_1 = t_{o1}$ as the output. Let $t_b = t_{oo} = 0$; then the figure is the fundamental feedback graph with $\mathscr{O}_2 = t_{c1i}$ as the output. Most of Fig. 8.8-2a is common to both and that is why the two fundamental graphs are combined into one flow graph.] The transmissions t_{os1}, t_{c1s1} are represented by Fig. 8.8-2b. (Here the independent input is taken as $\mathscr{S}_1 = 1$, and there are two outputs, $\mathscr{O}_1 = t_{os1}$, and $\mathscr{O}_2 = \mathscr{C}_1 = t_{c1s1}$.) Figures 8.8-2a and b may be combined into Fig. 8.8-2c, in which the relation $\mathscr{S}_1 = k_1 \mathscr{C}_1$ has also been added. Figure 8.8-2c therefore represents the fundamental feedback flow graph for two references. Some simplifications can be made. There are only two parallel plants, so $t_{oo} = 0$. Also, t_{11}, t_{22}, t_a, t_b act only as prefilters or postfilters, and are not involved in the system sensitivity (feedback) properties, so that they can all be made equal to one.

Next, the conditions for making $t_{o1} \doteq T$ are sought. If $t_{o1} \doteq T$, then, from the last equation for T, $t_{o1} \gg (k_1 t_{c1i} t_{os1})/(1 - k_1 t_{c1s1})$, which, from Fig. 8.8-2c, is the condition that

$$\frac{k_2}{1 - k_2 t_2} \gg \frac{k_1[1 + (k_2 t_{12})/(1 - k_2 t_2)]\,[1 + (k_2 t_{21})/(1 - k_2 t_2)]}{1 - k_1[t_1 + (k_2 t_{12} t_{21})/(1 - k_2 t_2)]}.$$

Cross-multiplying, etc., leads to inequality (8.8,2a), and by analogy, due to the complete symmetry of Fig. 8.8-2c, the requirement that $t_{o2} \doteq T$ leads to inequality (8.8,2b):

$$k_2[1 - k_1 t_1 - k_2 t_2 + k_1 k_2 (t_1 t_2 - t_{12} t_{21})] \gg k_1[1 + k_2(t_{12} - t_2)]\,[1 + k_2(t_{21} - t_2)] \qquad (8.8,2a)$$

$$k_1[1 - k_2 t_2 - k_1 t_1 + k_2 k_1 (t_2 t_1 - t_{21} t_{12})] \gg k_2[1 + k_1(t_{21} - t_1)]\,[1 + k_1(t_{12} - t_1)]. \qquad (8.8,2b)$$

The critical question is whether it is possible to satisfy (8.8,2a) and (8.8,2b) simultaneously? In the useful frequency range, the k's will be large, so it is the terms with the most k's that dominate. One might try $t_1 = t_2 = t_{12} = t_{21} = t$. Then (8.8,2) become $k_2[1 - t(k_1 + k_2)] \gg k_1$; $k_1[1 - t(k_1 + k_2)] \gg k_2$. If we let $k_1 = k_2 = k$, this leads to the requirement that $2kt \gg 1$, which is satisfactory. However, in Eq. (8.8,1) and its analog, making $t_{o1}, t_{o2} \doteq T$ is only one part of having small sensitivity. There is L_i in the denominator which should not become small in the process. It is easily found that $k_1 = k_2 = k$ leads to $L_1 = -k_1[t_1 + (k_2 t_{12} t_{21})/(1 - k_2 t_2)] = k_1 t_1/(1 - k_2 t_2) \doteq 1$, when $t_1 = t_2 = t_{12} = t_{21} = t$ and $kt \gg 1$. Consequently, the choice of $k_1 = k_2 = k$ is not satisfactory.

Another way to satisfy (8.8,2) is to let either $t_1 = t_2 = t_{12}$, or $t_1 = t_2 = t_{21}$. One is the dual of the other, so we take the first, i.e., $t_1 = t_2 = t_{12} = t$. The desired inequality (8.8,2a) becomes $k_2[1 - t(k_1 + k_2) + k_1 k_2 t(t - t_{21})] \gg k_1[1 + k_2(t_{21} - t)]$, and when the dominant terms only are taken, it becomes $k_2 t(t - t_{21}) \gg (t_{21} - t)$, or $|(1 + k_2 t)(t - t_{21})| \gg 1$. Similarly, (8.8,2b) becomes $|(1 + k_1 t)(t - t_{21})| \gg 1$. Next check if L_1 and L_2 are thereby affected. If $t_1 = t_2 = t_{12} = t$, then $L_1 = -k_1 t[1 + k_2(t_{21} - t)]/(1 - k_2 t) \doteq k_1(t_{21} - t)$, and $L_2 \doteq k_2(t_{21} - t)$. Therefore L_1, L_2, $(1 - t_{o1}/T)^{-1}$ and $(1 - t_{o2}/T)^{-1}$ can each have the order of magnitude of $l = kt$; from Eq. (8.8,1) and its analog, the sensitivities are then in the order of $1/l^2$ instead of the usual $1/l$.

The above is rigorously justified by finding the exact values of the sensitivities. But before doing so, it should be noted that in the above argument, the effect of variations in k_1 only, or k_2 only, are being considered. It is, in fact, possible to make $\Delta T/\Delta k_1 = 0$ for any change in k_1 only, and $\Delta T/\Delta k_2 = 0$ for any change in k_2 only. This is achieved if $k_1 = k_2 = k$, $t_1 = t_2 = t_{12} = t$, and $t_{21} = t - (1/k)$. [To see this, substitute in Eq. (8.8,1).] However, the effect is not any better than the single-degree-of-freedom system, if both k_1 and k_2 simultaneously vary. In practice it is impossible to have $t_{21} = t - k^{-1}$ over the entire frequency range because k^{-1} becomes infinite at infinite frequencies. A structure which is totally insensitive to changes in the parameters of only one plant is shown in Fig. 8.8-3, where $C = P_2E + C_1$. However, $E = GR - HC_1 = GR - (C_1/P_2)$. Hence $C = P_2[GR - (C_1/P_2)] + C_1 = P_2GR$ independent of any changes in P_1. If P_2 is a small fraction of the nominal P_1, then only a small fraction of the output passes through P_2. It seems, therefore, that P_2 could be a very low level plant, carefully constructed so that it had no parameter variations. However, when P_1 changes by any amount, P_2 must deliver the amount missing from P_1, if the output is to be completely insensitive to P_1. Hence the capacity of P_2 must be large if the system is to behave as postulated for large changes in the transmission of P_1. It should be noted that complete insensitivity to P_1 exists even if P_1 is violently nonlinear, providing of course that P_2 is capable of delivering the output missing from P_1. It is, of course, impossible to have $H = 1/P_2$ for all ω, but it could be done over the frequency range of interest. Also, $C/D = 0$ for disturbances at C_1. Of course P_2 can have its own feedback from C to E.

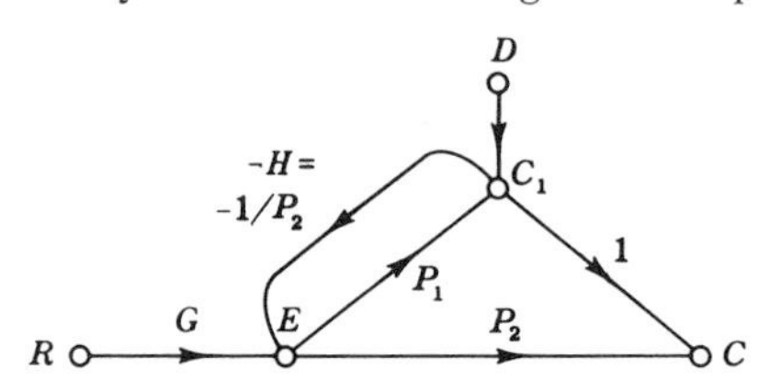

FIG. 8.8-3. Structure with zero sensitivity to variations in P_1 only.

These zero sensitivity systems are interesting but rarely useful, because of the special conditions they must satisfy. In a practical parallel plant system it is necessary to consider simultaneous and independent variations in both plants. Referring to Fig. 8.8-2c, it is therefore insufficient to have small $S_{k_1}^T$ (written as S_1) and $S_{k_2}^T$ (written as S_2). It is also necessary that $S_{k_1}^{S_1}$, $S_{k_2}^{S_1}$ (i.e., the sensitivity of the sensitivity functions), etc., should not be large. This is next checked, using Eq. (3.6,1) for small variations in k_i, and results in

$$S_{k_i}{}^T \triangleq \frac{dT/T}{dk_i/k_i} = \frac{1}{1 + L_i}\left[1 - \frac{t_{oi}}{T}\right]. \tag{8.8,3}$$

A small $S_{k_i}{}^T$ ensures small sensitivity even for large variations in k_i, if L_i stays large and $1 - (t_{oi}/T)$ stays small over the range of variation of k_i.

Using Eq. (8.8,3), it is readily found that, when $t_1 = t_2 = t_{12} = t$,

$$\begin{aligned} S_1 \triangleq S_{k_1}{}^T &= \frac{k_1[1 + k_2(t_{21} - t)]}{[1 - (k_1 + k_2)t + k_1k_2t(t - t_{21})]\,[k_1k_2(t_{21} - t) + k_1 + k_2]} \\ &\doteq \frac{1}{k_1k_2t(t - t_{21})}, \end{aligned}$$

if $k_1t,\ k_2t \gg 1$. Similarly, $S_2 \triangleq S^T_{k_2} \doteq 1/k_1k_2t(t - t_{21})$. Thus the sensitivities are proportional to $1/l^2$ (with $l = kt$) in place of $1/l$. It is also readily found that $S^{S_1}_{k_1} \doteq S^{S_1}_{k_2} \doteq S^{S_2}_{k_1} \doteq S^{S_2}_{k_2} \doteq -1$. Therefore, the system sensitivity to simultaneous changes in k_1 and k_2 are also as $1/l^2$.

Implementation

So far t_{21} has not been specified. It can conveniently be taken as zero. The result is shown in Fig. 8.8-4a. The structure can be simplified in several ways.

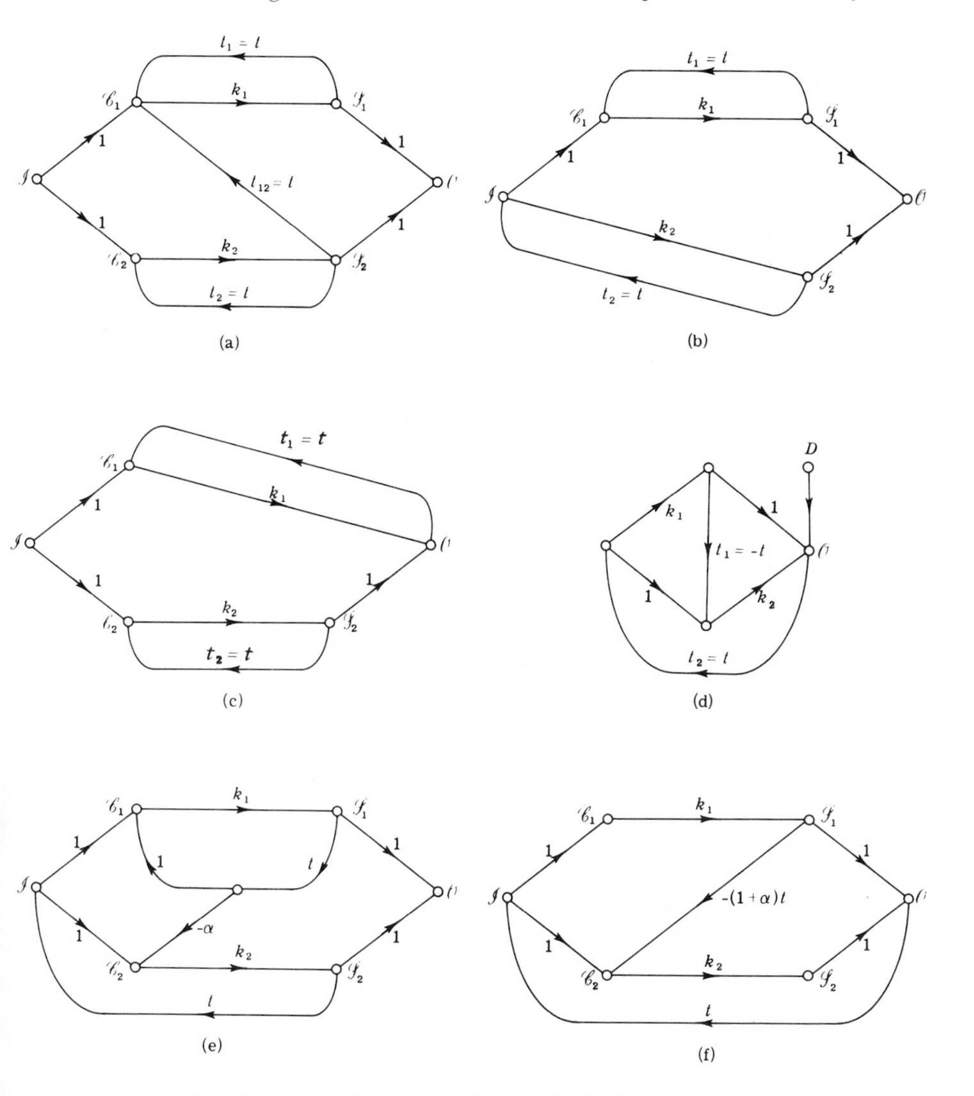

FIG. 8.8-4. A variety of parallel plant feedback structures.

For example, if the rules of node elimination (Section 1.6) are used to eliminate node C_2, the result is the simpler structure of Fig. 8.8-4b. Similarly, node S_1 in Fig. 8.8-4a can be eliminated, resulting in Fig. 8.8-4c. In the first two structures, disturbances at $\mathcal{O}$ are not attenuated, while in the third it is attenuated by $1/(1 - k_1 t)$. An equivalent structure can be found in which a disturbance at $\mathcal{O}$ is attenuated by $1/l_1 l_2$. This is the structure shown in Fig. 8.8-4d, in which disturbances at $\mathcal{O}$ are attenuated by $1/\Delta$ (i.e., $\mathcal{O}/D = 1/\Delta$) with $\Delta = 1 - (k_1 + k_2)t + k_1 k_2 t^2 = (1 - k_1 t)(1 - k_2 t) \doteq 1/k_1 k_2 t^2 = 1/l_1 l_2$. The structures are not sensitive to adjustments of t_1 and t_2. Thus in each of the last three structures of Fig. 8.8-4, it is readily found that $S_{t_1}^T = k_1^2 t_1/(1 - k_1 t_1)[k_1 + k_2(1 - k_1 t_1)] \doteq 1/k_2 t_1$, and $S_{t_2}^T = k_2 t_2/(1 - k_2 t_2) \doteq -1$. Therefore t_1 can be considerably misaligned, while t_2 can be moderately misaligned, with small effect on the system sensitivities to k_1 and k_2.

The sensitivity can be decreased even more by choosing $t_{21} = -\alpha t$ with α positive and finite in place of zero. For example, if $t_{21} = -t$, all sensitivities are halved. Two equivalent structures are shown in Figs. 8.8-4e and f. However, one advantage of the others in Fig. 8.8-4 is that if $-t_1$ and t_2 are very similar, then the stability problem is the same as in two distinct single-loop systems, because, as we have seen, $\Delta = (1 - k_1 t)(1 - k_2 t)$, if $-t_1 = t_2 = t$.

Signal Levels in the Parallel Plants

In Figs. 8.8-4e and f it is readily found that $C_2/\mathcal{I} = [1 - (1 + \alpha)k_1 t]C_1/\mathcal{I}$. Consequently, the signal level in k_2 is greater than that in k_1 by approximately $(1 + \alpha)k_1 t$. If $\alpha = 0$, for example, and a sensitivity of 0.01 is desired, then $-kt = 10$ (if $k_1 = k_2$), and the signal level in k_2 is 11 times that in k_1. Thus the k_1 portion of the plant is not fully utilized. Should, however, either one of k_1 or k_2 break down, the other will supply almost the same output as both, if they both have the same saturation levels. It is, however, possible to build k_1 with much less capacity (much smaller saturation values, etc.) than k_2, and so eliminate the expense of two plants, each with the same capacity, doing the work of only one plant. As the plant parameters vary, the ratio of plant signal levels remains large so long $(1 + \alpha) k_1 t$ is large. The above is also true for the other structures in Fig. 8.8-4.

§ 8.9 Application of Parallel Plants in Design (to Conditionally Stable Systems and Nonminimum Phase Plants)

In single-loop feedback amplifier design, there is an optimum number of stages for realizing the maximum possible feedback.[1] If a larger number is used, the amount of achievable loop gain-bandwidth area is decreased. Parallel plants

[1] H. W. Bode, "Network Analysis and Feedback Amplifier Design," p. 478. Van Nostrand, Princeton, New Jersey, 1945.

offer some possibility of removing this limitation. This property of parallel plants is not of interest in feedback control design (see Section 5.1). There are, however, very significant problems in feedback control design in which parallel plants have great potential usefulness.

A difficult design problem is often encountered in advanced control systems in which there are severe demands on sensitivity reduction and/or disturbance attenuation. The problem is presented in Fig. 8.9-1. A large level of feedback is

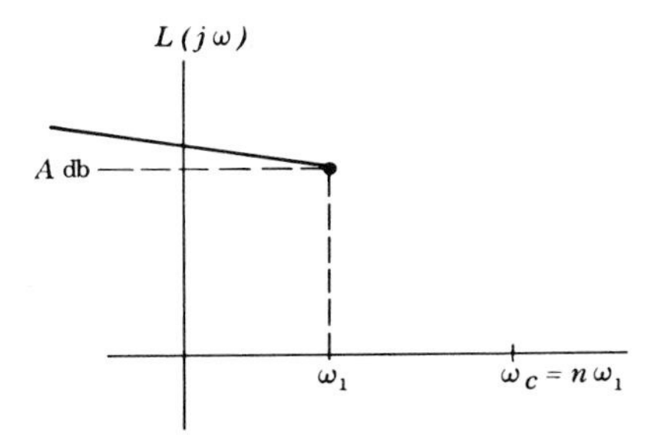

FIG. 8.9-1. Design specifications which cannot be satisfied or which lead to conditionally stable design in a single-loop system.

required up to $\omega = \omega_1$, in order to achieve the essential benefits of feedback for the system. However, various considerations dictate that the crossover frequency ω_c is limited. (See Section 6.14 for a discussion of these considerations.) The situation may be alleviated somewhat by using the multiple-loop design of Sections 8.2-8.7. Let us assume that the very best has been done in that respect, and the problem is that A in Fig. 8.9-1 is too large to permit an unconditionally stable design for L for the constrained value of $\omega_c = n\omega_1$. For example, suppose $A = 50$ db and $\omega_c = n\omega_1 = 10\omega_1$. It is impossible to effect a reduction of 50 db in one decade with an unconditionally stable $L(s)$ function. This should be clear to the reader from our work in Chapter 7. A conditionally stable $L(s)$ (e.g., L in Fig. 8.9-3) is therefore used. The system will become unstable if the gain decreases by the factor α, such that $\alpha\epsilon < 1$. Unfortunately, saturation acts like a decrease in gain, and it may be well-nigh impossible to prevent saturation. Such systems are consequently very susceptible to low frequency instability.

The usual way of achieving a stable system, despite conditional stability and saturation, is to use nonlinear compensation. Detailed treatment of the latter is outside the scope of this book, but the basic idea is to use a minor feedback loop around the saturating element,[1] or a suitable additional cascaded nonlinearity,[2] such that the loop dynamics [i.e., the poles and zeros as well as the gain factor of $L(s)$] are affected by saturation. It is arranged that when saturation occurs, the "equivalent" system loop transmission is no longer a conditionally

[1] R. J. Kochenburger, Limiting in feedback control systems, *Trans. AIEE Part II, Applications & Ind.* **72**, 180-194 (1953).

[2] E. S. Sherrard, Stabilization of a servomechanism subject to large amplitude oscillation. *Trans. AIEE Part II, Applications & Ind.* **71**, 312-324, Nov. (1952).

stable one. This is the "describing function" approach,[1] and it is often successful.

The parallel plant approach offers an alternative linear method of achieving a satisfactory stable system design. It does this by removing entirely the existence of conditional stability. In Fig. 8.9-2 (which is Fig. 8.8-4d with the signs changed for convenience),

$$\frac{\mathcal{O}}{D} = \frac{1}{1 + k_2 t_2 + k_1 t_2 + k_1 k_2 t_1 t_2} = \frac{1}{(1 + k_1 t)(1 + k_2 t)} = \frac{1}{(1 + l_1)(1 + l_2)}, \tag{8.9,1}$$

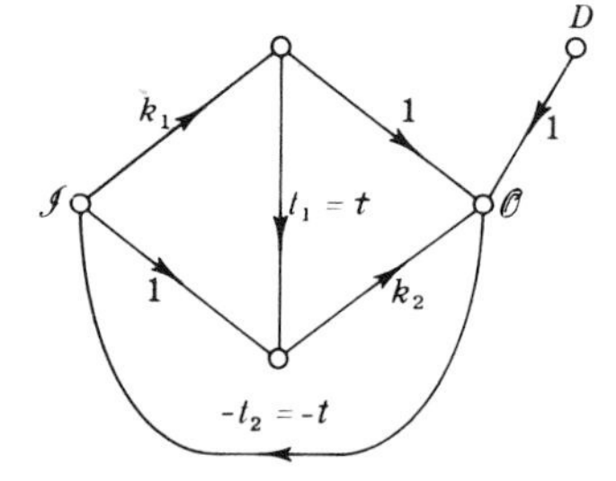

FIG. 8.9-2. Parallel plant feedback structure.

if $t_2 = t_1 = t$, $l_1 \triangleq k_1 t$, and $l_2 \triangleq k_2 t$. In the problem posed in Fig. 8.9-1, $A = 50$ db, so 25 db may be assigned to each of l_1 and l_2. There is no difficulty in shaping l_1 and l_2 such that the phase lag of each of these two loop transmissions is less than 180° for $\omega < \omega_c$. The effect of saturation depends on the element which saturates. If k_2 saturates (and almost all the output and feedback effect passes through k_2, so that it is the most important factor here), then only $1 + k_2 t = 1 + l_2$ is affected, and the system is stable. Similarly, if k_1 saturates, $1 + k_1 t = 1 + l_1$ only is affected, and the system is stable.

If t_2 saturates, the system will be unstable (if the ordinary single plant conditionally stable design leads to instability). This is seen as follows: We write Eq. (8.9,1) in the form $\mathcal{O}/D = 1/(1 + L)$. If a single plant is used, the expression for $\mathcal{O}/D$ will have exactly the same form, and we know that L must be conditionally stable. Consequently, in Eq. (8.9,1) the quantity $L = t_2(k_1 k_2 t_1 + k_1 + k_2)$ represents a conditionally stable loop transmission. If t_2 saturates, the system will be susceptible to instability, precisely to the same extent as the single plant design which achieves the same benefits of feedback. The reason why saturation of k_2 or k_1 does not cause instability is because the return difference with respect to k_2 is $1 + k_2 t_2$, and $l_2 = k_2 t_2$ is an unconditionally stable loop function.

For the effect of saturation in t_1, the return difference with respect to t_1 must be studied. It is obtained by manipulating the denominator of Eq. (8.9,1) (recall the theorem of Section 2.8) into the form $1 + t_1 f$. $L_{t1} \triangleq t_1 f$ represents the loop transmission which must be examined for conditional stability. From Eq. (8.9,1) it is found that $L_{t1} = k_1 k_2 t_1 t_2/[1 + (k_1 + k_2)t] = l_1 l_2/(1 + l_1 + l_2)$. In order to simplify the determination of the nature of L_{t1}, let $l_1 = l_2 = l$, and then $L_{t1} = l^2/(1 + 2l)$. Now, in Eq. (8.9,1) the denominator is $(1 + l)^2 = 1 + L$,

[1] A. Tustin, The effects of backlash and of speed-dependent friction on the stability of closed-cycle control systems, *J. Inst. Elec. Engrs. (London)* **94**(1), Part 2A (1947).

with L conditionally stable, for example, as shown in Fig. 8.9-3. Given L, it is easy to find $1 + l = (1 + L)^{0.5}$, and finally $L_{t1} = l^2/(1 + 2l)$. This has been done in Fig. 8.9-3, and it is seen that L_{t1} is an absolutely stable loop transmission.

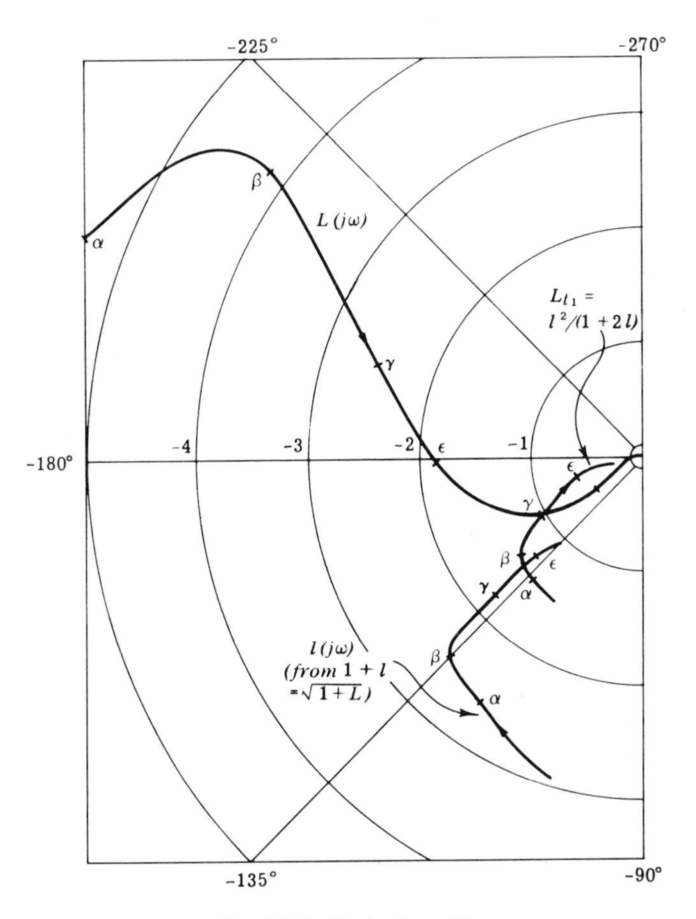

FIG. 8.9-3. Derivation of L_{t_1}.

It is therefore seen that for the difficult design constraints of Fig. 8.9-1, the parallel plant design permits the attainment of an unconditionally stable design with respect to all elements, except those in t_2 of Fig. 8.9-2. (This claim is, perhaps, somewhat exaggerated with respect to t_1, because we have only done one example; but it is readily seen that L_{t1} in general has at least far less lag than L.) The price paid for this significant improvement is the need for an additional plant. It has, however, previously been noted that this added plant need have far smaller capacity and saturation levels than the original plant. Furthermore, its dynamics may be considerably different from the original. If it is so desired, it may provide the parallel path only over a critical frequency range, and be negligibly small over other frequency ranges. Furthermore, it is possible that the original plant has, in fact, parallel paths, so that the above presents a method for the useful exploitation of such a situation.

Only two parallel plants have been considered in the above. However, the idea may be extended to any number of parallel plants. A structure with three parallel plants may be obtained by replacing k_1 of Fig. 8.9-2 by the forward portion of a parallel structure, as in Fig. 8.9-4a. Or k_2 of Fig. 8.9-2 may be replaced by the forward portion of a parallel structure, as in Fig. 8.9-4b. In

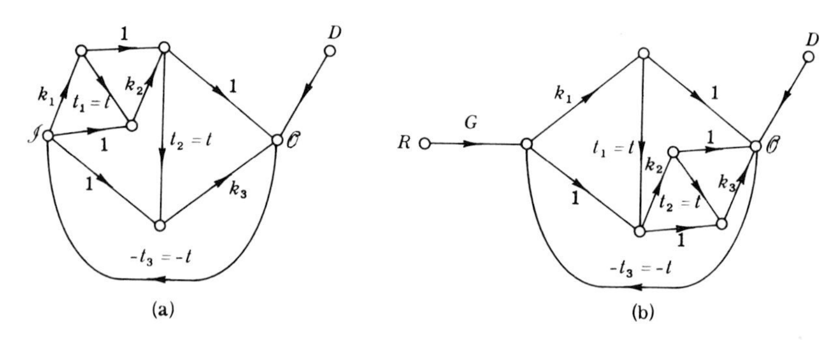

FIG. 8.9-4. Three-parallel-plant feedback structures.

both of these structures, it is readily found that $\mathscr{O}/D = [(1 + k_1t)(1 + k_2t)(1 + k_3t)]^{-1} = \Delta^{-1}$. A fourth-order parallel plant may be similarly obtained, i.e., by replacing any one of k_1, k_2, or k_3 in Figs. 8.9-4a and b by the forward portion of a second-order parallel plant; or by replacing k_1 or k_2 in Figs. 8.9-2 by the forward portion of a third-order plant. Again, it is noted how simple the stability problem is if t_1, t_2, t_3, etc., are made closely equal to each other. The sensitivity of the system transfer function $T = \mathscr{O}/\mathscr{I}$ to each of k_1, k_2, k_3 is also proportional to $1/\Delta$. The signal levels at the inputs to k_1, k_2, k_3 are in the ratios 1, $(1 + l_1)$, $(1 + l_1)(1 + l_2)$, respectively. Therefore the main burden is borne by k_3, much less by k_2, and far less by k_1. The structures in Fig. 8.9-4 could be used to achieve a decrease of 90 db in one decade, with a loop transmission which is absolutely stable with respect to k_1, k_2, k_3, and probably also with respect to t_1 and t_2. However, the loop transmission with respect to t_3 would be conditionally stable.

Application of Parallel Plant Design in Nonminimum Phase Plants

It was seen in Chapter 7 that the available benefits of feedback are seriously limited when the plant transfer function has either poles or zeros in the right half-plane. The significant reason for this is the relatively slow rate at which the loop transmission may be decreased in unconditionally stable designs. The slow rate of decrease gives the lags due to the rhp zeros a large frequency range in which to grow and smother the design. On the other hand, the conditionally stable design (Section 7.13) permits much faster decrease of loop gain, and affords a much smaller range for the angle due to the rhp zeros to accumulate. Accordingly, it was noted in Section 7.13 that much more feedback can be secured with a conditionally stable design.

The parallel plant technique permits the total effective loop gain to be decreased even faster than in a conditionally stable design. It does this by dividing the feedback burden among the different parallel loop transmissions. The limitations of Chapter 7 apply to each individual loop transmission. For example, if, in a nonminimum phase plant the maximum rate of decrease is 6 db per octave, and three parallel plants are used, then the total effective feedback may be reduced at the rate of 18 db per octave. For a given crossover frequency, a much larger total level of feedback may be obtained; or the bandwidth over which there is sensitivity reduction and disturbance rejection may be increased. Furthermore, this can be done with an absolutely stable design. Hence the limitations on the available feedback in nonminimum phase plants are radically modified when the parallel plant design is used. Should a parallel plant conditionally stable design be used, then even more feedback is available in nonminimum phase plants.

Design in the Higher Frequency Range

While parallel plants increase the sensitivity reduction achievable in the frequency range where $l > 1$, it offers no appreciable advantage in the frequency range where $l \sim 1$ or $l < 1$. This is because the effective sensitivity reduction is by the factor $(1 + l)^2$, and in the crossover region where $|l| \sim 1$ and its phase is $180° - \theta_M$ (θ_M is the phase margin), $(1 + l)^2$ is at best also approximately unity. In this range the kt terms become more important than the k^2t^2 terms. There is the advantage that for the same feedback benefits, $|l|$ is smaller in the low frequency region, and can therefore be decreased somewhat faster than in the nonparallel system. It may also be possible to build the smaller capacity plant to have a better high frequency response than the main plant. In that case, the former becomes the effective plant in the high frequency range.

Relations Suitable for Design

Relations suitable for actual numerical design work are readily obtained. The structure of Fig. 8.9-2 is used, in which $T = (k_1 + k_2 + k_1k_2t)/(1 + k_1t)(1 + k_2t)$. It is readily found that $T/T' = [(l_1 + l_2 + l_1l_2)(\alpha_1 + l_1)(\alpha_2 + l_2)][(\alpha_2 l_1 + \alpha_1 l_2 + l_1 l_2)(1 + l_1)(1 + l_2)]^{-1}$, where $\alpha_1 = k_1/k_1'$, $\alpha_2 = k_2/k_2'$, $l_1 = k_1t$, $l_2 = k_2t$. We replace $(l_1 + l_2 + l_1l_2)$ by $[(1 + l_1)(1 + l_2) - 1]$, $(\alpha_2 l_1 + \alpha_1 l_2 + l_1 l_2)$ by $[(\alpha_1 + l_1)(\alpha_2 + l_2) - \alpha_1\alpha_2]$, and then obtain

$$\frac{T}{T'} = \frac{1 - [(1 + l_1)(1 + l_2)]^{-1}}{1 - (\alpha_1\alpha_2)[(\alpha_1 + l_1)(\alpha_2 + l_2)]^{-1}} \tag{8.9,2}$$

which is convenient for exact calculations. In the low and medium frequency ranges, where $(\alpha_1\alpha_2)/(\alpha_1 + l_1)(\alpha_2 + l_2) \ll 1$ (this inequality is facilitated by using the lowest values of k_1, k_2 for nominal values, and therefore $|\alpha_1|$, $|\alpha_2| < 1$), the above may be approximated as $T/T' \doteq 1 - [(1 + l_1)(1 + l_2)]^{-1} + \alpha_1\alpha_2 [(\alpha_1 + l_1)(\alpha_2 + l_2)]^{-1}$. In the higher frequency range, where $l_1 l_2 \ll 1$, while $l_1, l_2 \not\ll 1$, Eq. (8.9,2) becomes

$$\frac{T}{T'} \doteq \frac{(l_1 + l_2)(\alpha_1\alpha_2 + \alpha_2 l_1 + \alpha_1 l_2)}{(1 + l_1 + l_2)(\alpha_1 l_2 + \alpha_2 l_1)} = \frac{1 + \alpha_1\alpha_2(\alpha_1 l_2 + \alpha_2 l_1)^{-1}}{1 + (l_1 + l_2)^{-1}}$$

$$= \frac{1 + (\alpha_1 \# \alpha_2)(2l)^{-1}}{(1 + 0.5l)^{-1}},$$

if $l_1 = l_2 = l$.

Greater effort is required to shape l_1 and l_2 in the crossover region, where $l_1, l_2 \sim 1$. However, the stability problem is hardly any more difficult than in the single-loop design, because the denominator of T' is $(1 + k_1't)(1 + k_2't) = (\alpha_1\alpha_2)^{-1}(\alpha_1 + l_1)(\alpha_2 + l_2)$. Therefore l_1 and l_2 must be each shaped in the usual fashion (just as in the single-loop design), to avoid the appropriate regions occupied in the complex plane by α_1 and α_2, respectively.

Finally, it is noted that in the above discussions, only the sensitivity and disturbance attenuation properties have been discussed. From these considerations, l_1 and l_2 are chosen. In order that these be independent of the over-all system transfer function, it is necessary to add a prefilter, as in Fig. 8.9-4b. G is chosen to obtain the desired T, after l_1 and l_2 have been chosen. It is almost needless to add that the prefilter can be incorporated in many ways into the feedback loops.

§ 8.10 Synthetic Multiple-Loop Feedback Systems

It is possible, and sometimes useful, to have a multiple-loop design around a plant which at first sight appears to be inherently a two-degree-of-freedom plant, i.e., no internal plant variables are available for feedback purposes. This situation can occur in the following manner. Suppose that the output variable only is available, and the design is such that H, in Fig. 8.10-1a has rising fre-

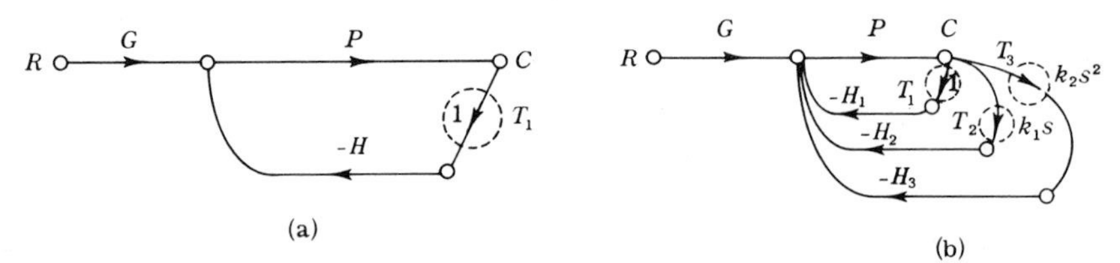

FIG. 8.10-1. Single-loop and synthetic multiple-loop feedback structure for the same plant.

quency characteristics (e.g., H in Fig. 8.5-2c). H may consist of a transducer (T_1) whose output is proportional to C, followed by filters with essentially differentiation characteristics (of first, second, or higher order). Every transducer has its inherent noise output. Unless the noise output of T_1 is exceedingly low in the high frequency region, the noise output of H will be very large in the high frequency region, because of the high-order differentiation characteristics of H.

On the other hand, suppose a transducer, T_2, is available, which measures directly the first derivative of c, and whose inherent high frequency noise output is less than that of the T_1H combination. The comparison must be made on the basis of the same useful signal transmission, i.e., the superior technique is the one which leads to the higher signal to noise ratio, for the same magnitude of signal output. It is then advantageous to use T_2 to supply the feedback for this frequency range. The same conclusions apply to transducers which measure second and higher order derivatives of output. The final design may therefore be a synthetic multiple-loop feedback design, like the one shown in Fig. 8.10-1b with the feedback appropriately divided among H_1, H_2, and H_3, in accordance with the above considerations.

§ 8.11 Control of Effect of Parameter Variations by Shaping of Root Loci

In previous chapters, when the problem of parameter variation was discussed, it was nearly always assumed that only small variations were permitted over the significant frequency range of $T(s)$. When the parameter variations are very large, this leads to the need for large loop gains over large bandwidths, and to the feedback transducer noise problem. As previously noted, the latter may be alleviated by a multiple-loop feedback design. Often, however, the sensitivity specifications are not so severe. For example, in the case $T(s)$ is dominated by a single complex pole pair, it may be required that $0.5 < \zeta < 0.8$; $K_v \geqslant 10$ and $15 > \omega_n > 10$ rps. In a more complicated system, it may suffice if the amplitude peaking is kept below some quantity, say 1.2, and the bandwidth between limits such as 10 and 30 rps. It ought to be possible to realize such specifications with a smaller loop gain and bandwidth, than would be necessary for narrower tolerances on $T(s)$. Furthermore, if a change in the plant parameters means physically that the plant capacity for work has increased, then it is desirable to have the system bandwidth increase, in order to take advantage of the plant's increased capabilities. The following sections are devoted to various forms of this design problem.

The material is developed in the following order. At first, simple plants and forms of $T(s)$ are assumed. The system response variations that result with the simplest type of $L(s)$, in the single-degree-of-freedom configuration (Fig.

8.11-1) are found. Then, step by step, more complicated forms of $L(s)$ are considered, and the resulting improvement noted. So far, the work involves analysis, rather than synthesis. Eventually, a form for $L(s)$ is assumed, which is sufficiently general to handle any amount of parameter variation (for relatively simple plants), and for this case, a synthesis procedure is developed (Section 8.13). More complicated plant functions are treated in the same manner. Finally, the plant function, and consequently $T(s)$, is so complicated that root locus design is considered impractical, and the frequency response approach is used (Section 8.14).

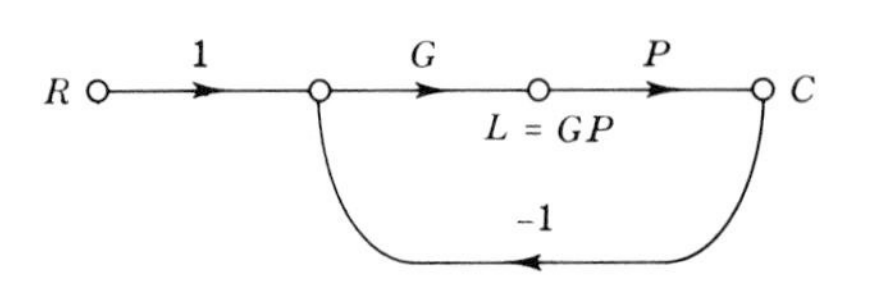

FIG. 8.11-1. One-degree-of-freedom configuration.

Gain Variation in Single-Degree-of-Freedom System

The first problem is constrained to be of the following very simple form: The system is single degree in the sense of Chapters 5 and 6, i.e., $T(s) = L/(1 + L)$ (Fig. 8.11-1). The loop transmission is $L(s) = G(s)P(s)$, in which the plant function has the form $P(s) = k_1 n(s)/sd_1(s)$; k_1 is the only element which varies and $d_1(s)$ is a polynomial with unity for its leading coefficient. The objective is a $T(s)$ with a dominant complex pole pair and other far-off poles (hence G cancels out the zeros of P). With this simple type of $L(s)$, it is implicitly assumed that if the resulting velocity constant K_v is insufficient, lag compensation will suffice to increase it to the desired value. Consequently (ignoring any possible lag compensation, which has negligible effect on the dominant and far-off system poles), the simplest loop transmission has the form $L(s) = k/sd(s)$. The problem is to choose k and the zeros of $d(s)$. A number of conflicting desiderata are involved here. These are clarified by considering the simple case of $L(s) = k/s(s + \alpha)(s + \beta)$, so that the problem is how to choose k, α, β. Here it is assumed that $P(s)$ has two more poles than zeros and one excess pole is assigned to the compensating network $G(s)$. It is also assumed that the poles of $L(s)$ are to be real.

Suppose that the nominal dominant poles of $T(s)$ [zeros of $1 + L(s)$] are specified at $-\zeta\omega_n \pm j\omega_n\sqrt{(1 - \zeta^2)}$ in Fig. 8.11-2a, so $\theta_1 + \theta_2 = 180° - \theta_o$ is fixed. It is desirable to have the high frequency loop gain as small as possible, and since L approaches ks^{-3} as s becomes large, this means that k should be minimized. Also, it might be considered desirable to have the dominant root sensitivity $S_k^{s_i} = ds_i/(dk/k)$, as small as possible, and it is also probably desirable to maximize K_v. In this specific problem (α and β are constrained to be real positive numbers), it is readily found that the smallest value of k is achieved when $\theta_1 = \theta_2$ in Fig. 8.11-2a, and that this corresponds to the largest value of $S_k^{s_i}$ and the smallest value of K_v, in fact, to the smallest $|L(j\omega)|$ at every single value of ω. These results are readily ascertained by formulating $k = \omega_n V_1 V_2$

(Fig. 8.11-2a), $K_v = k/\alpha\beta$, and $S_k^{s_i} = \omega_n V_1 V_2/[2\omega_n(1 - \zeta^2)^{0.5} W]$, as functions of θ_1, θ_2, using the fact that $\theta_1 + \theta_2$ is a constant, and differentiating, etc.

It might seem best to choose θ_1, θ_2 so as to minimize $S_k^{s_i}$, which also leads to the maximum K_v (under the constraint of $L(s) = k/[s(s + \alpha)(s + \beta)]$). However,

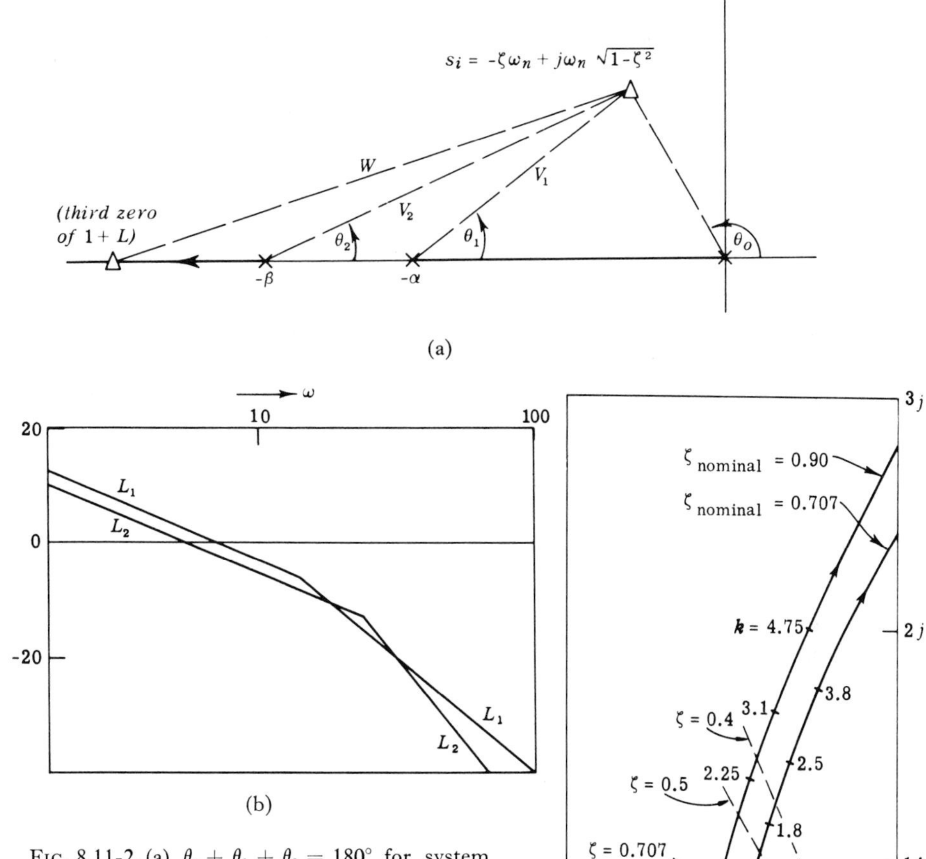

FIG. 8.11-2. (a) $\theta_o + \theta_1 + \theta_2 = 180°$ for system pole at s_i. (b) Justification for assigning coincident real poles to $L(s)$. (c) Loci of dominant roots due to gain variations.

this condition corresponds to $\theta_1 = 180° - \theta_o$ and $\theta_2 = 0$, which is the worst design in terms of the high frequency response of L_1, since it results in only two poles of L_1. For example, if $\omega_n = 10$, $\zeta = 0.707$, then $L_1 = 100/[s(s + 14.14)]$ (Fig. 8.11-2b). L_1 represents one extreme design. In the other extreme, $\theta_1 = \theta_2$ leads to the maximum value of $S_k^{s_i}$, the minimum value of K_v, but to the smallest value of k. The latter corresponds to the best design in terms of the high frequency response of L. For $\omega_n = 10$, $\zeta = 0.707$, the resulting L is $L_2 = 3420/[s(s + 24.1)^2]$ of Fig. 8.11-2b.

The difference in K_v between L_1 and L_2 is only 1.5 db, and the difference in $S_k^{s_i}$ is only by a factor of 1.22. To attain these small improvements in K_v and in $S_k^{s_i}$, $G(s)$ in Fig. 8.11-1 must be finite at infinite frequency. The price paid, in bandwidth of $G(s)$, is much too high for such small benefits. It is therefore assumed that L_2 is the most desirable, since it permits considerable reduction in the bandwidth of $G(s)$, with little loss in K_v and $S_k^{s_i}$. The resulting variations in the dominant poles of T, as k varies, i.e., the root loci of $1 + kL(s)$, are sketched in Fig. 8.11-2c for $\omega_n = 1$, $\zeta = 0.707$, and $\zeta = 0.9$. In any problem satisfying the above constraints, it is readily ascertained whether the above very simple design is sufficient to keep the dominant roots of $1 + L(s) = 0$ within a satisfactory range. If the design is not satisfactory, a more complicated $L(s)$ must be chosen.

The above procedure may be applied to all problems involving a $T(s)$ with only one dominant pair of poles in which Fig. 8.11-1 is used, and $L(s)$ is constrained to have only negative real poles, and no zeros. If $L(s) = k/[s(s + \alpha_1)(s + \alpha_2)(s + \alpha_3) \ldots (s + \alpha_n)]$, due to the nature of the plant and the desired excess of poles of $G(s)$, then the choice of $\alpha_1 = \alpha_2 = \ldots \alpha_n$ leads to the smallest high frequency value of $L(s)$, but not to the smallest $S_k^{s_i}$ or largest K_v (under the constraint of positive α's). However, again, small improvements in the latter two quantities require very much larger high frequency values of $L(s)$, so that it is reasonable to consider as optimum that design which leads to the smallest k. This optimum design requires $\alpha_1 = \alpha_2 = \ldots \alpha_n$.

In the above it was assumed $P(s)$ had a pole at the origin. If this is not the case, then one pole of $L(s)$ is placed as close as necessary to the origin to obtain the desired position constant K_p. This pole, in place of the pole at the origin, is taken as fixed. The latter, together with the conditions $\alpha_1 = \alpha_2 \ldots \alpha_n$, and the known position of the dominant pole pair of $T(s)$, completely determine the design.

Variation in a Pole of $P(s)$[1]

The effect of large variations in a pole of $P(s)$ can be relatively easily analyzed. The procedure is clarified by means of a numerical example where $P =$

[1] For several practical problems where such variations occur, see Kan Chen, Analysis and design of feedback systems with gain and time constant variations. *Trans. IRE* **AC-6** (1), 73-79 (1961).

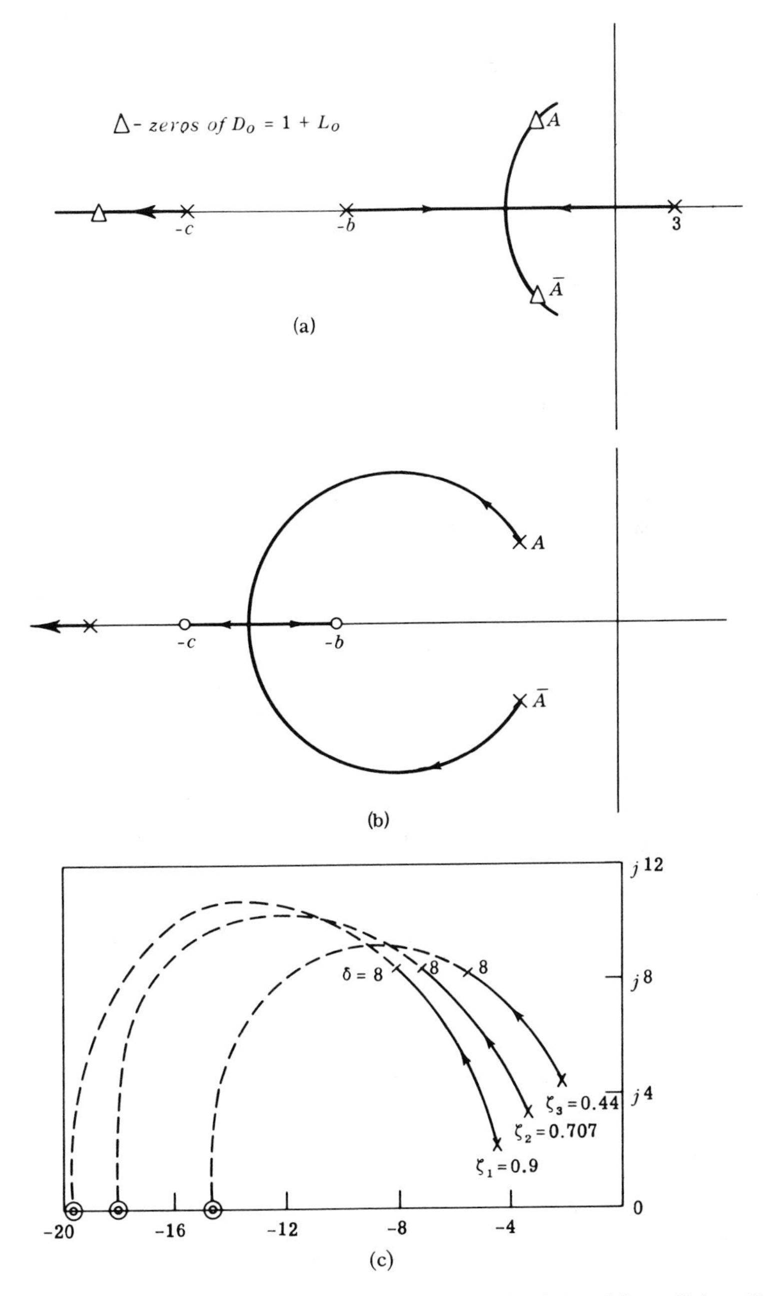

FIG. 8.11-3. (a) Nominal dominant roots from root loci of $1 + k/(s - 3)(s + b)(s + c) = 0$ (not to scale). (b) Root loci of $1 + \delta(s + b)(s + c)/D_o(s) = 0$ (not to scale. (c) Effect of variations in plant pole for three different nominal ζ.

$k_1/(s + a)(s + b)$ and a varies from -3 to 5. We allow one excess pole for G and therefore have $1 + L = 1 + k/(s + a)(s + b)(s + c) = 1 + k/(s - 3 + \delta)(s + b)(s + c)$, where δ varies from zero to $+8$. [Note that again the simplest form of $L(s)$ is being considered.] The numerator of $1 + L$ is $(s - 3 + \delta)(s + b)(s + c) + k = (s + c)(s - 3)(s + b) + k + \delta(s + b)(s + c) = D_o(s) + \delta(s + c)(s + b)$, where $D_o(s)$ is the denominator of the nominal $T_o = L_o/(1 + L_o)$, with $L = L_o$ at $a = -3$. The roots of $1 + L = 0$ are therefore on the root loci of $1 + \delta(s + b)(s + c)/D_o(s) = 0$, which are sketched in Fig. 8.11-3b. The zeros of $D_o(s)$, which are the zeros of $1 + L_o$, are shown (as poles) in Fig. 8.11-3b.

The only freedom that exists in this problem is in the values of b and c. Again, for fixed A, $\bar{A}$ (and positive b, c) in Fig. 8.11-3, it is found that to minimize $|L_{\text{hf}}|$ (the high frequency magnitude of L), it is best to choose $b = c$. More generally, if $L(s)$ is assigned n negative real poles in addition to the one which varies, these n poles should be coincident, in order to minimize $|L_{\text{hf}}|$. If any of the poles are fixed, the remaining ones are taken coincident. Therefore, in Fig. 8.11-3a, we choose $b = c$, and locate them so that the dominant roots of $1 + L_o = 0$ are at the desired points A, $\bar{A}$. Then the root loci of $1 + \delta(s + b)^2/D_o(s)$ are sketched to find the ffect of the variation in "a." The results are shown in Fig. 8.11-3c for three different values of A, $\bar{A}$, corresponding to $\omega_n = 5$, $\zeta_1 = 0.9$, $\zeta_2 = 0.707$, and $\zeta_3 = 0.44$. For example, at $\zeta = 0.707$, it is easily found that $b = c = 18$; the third zero of $D_o(s)$ is found from the fact that the sum of the poles of L_o is equal to the sum of the zeros of $D_o(s)$. Therefore the third zero of $D_o(s)$ is at $-36 + 3 + 7.07 = -25.9$. Hence the root loci of $1 + \delta(s + 18)^2/(s^2 + 7.07s + 25)(s + 25.9)$ give the effect of variations of $a = -3 + \delta$. It is seen that the movement in the pole $P(s)$ affects primarily the bandwidth of $T(s)$, with a smaller effect on the damping factor.

The above procedure may be used in any problem where a single pole of the plant varies, and the loop transmission has the form $L(s) = k/(s + a)(s + b)(s + c)\ldots$, with some poles (say c, d...) unassigned. We set $c = d = \ldots$ and proceed as above. It is quickly ascertained whether the above simple design of $L(s)$ suffices to keep the dominant poles of $T(s)$ within a satisfactory range. Figures 8.11-2c and 3c can be used for all problems with the same number of poles, etc., by suitably normalizing. If the variations in the dominant poles of $T(s)$ are too large, then a more complicated $L(s)$ must be chosen.

Independent Variations in a Plant Pole and in Plant Gain Factor

It is again assumed that $T(s)$ is dominated by a single complex pole pair, and that the simplest type of $L(s)$ is being used. Hence $L(s) = k_o(1 - \delta_1)[(s - a + \delta_2)(s + b)(s + c)\ldots]^{-1}$, where δ_1 and δ_2 are used to denote the independent variations in the gain factor and in the plant pole. Then the numerator of $1 + L(s)$ is $[(s - a)(s + b)(s + c) + k_o] + [\delta_2(s + b)(s + c)\ldots] - \delta_1 k_o = D_o(s) + \delta_2(s + b)(s + c)\ldots - \delta_1 k_o$. Some feeling for the effect of δ_1

and δ_2 is obtained by working out a numerical example. Suppose $a = 3$, the maximum value of δ_2 is 15 (i.e., the pole changes from $+3$ to -12), and it is assumed that the smallest permitted $T(s)$ dominant pole damping factor is 0.44 at $\omega_n = 9$. Decrease in gain factor (i.e., increase of δ_1) tends to increase the damping factor (see Fig. 8.11-2c). From Fig. 8.11-3c, it is seen that at $\zeta = 0.44$, increase of δ_2 also tends to increase the damping factor; therefore $D_o(s)$ is taken to correspond to $\omega_n = 9$, $\zeta = 0.44$ (A in Fig. 8.11-4). It is assumed here that

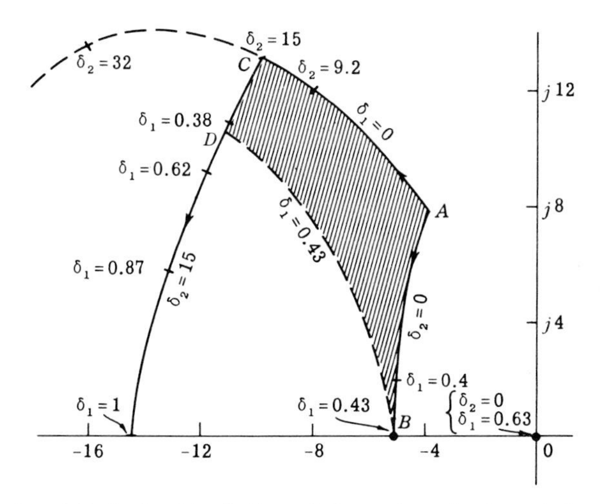

FIG. 8.11-4. Combined effect of plant gain and pole variations.

$L(s)$ has a total of three poles. One pole of $L_o(s)$ is at $+3$, and the other two (taken coincident) must be at -22 in order that $1 + L_o(s)$ may have a zero at $A = -4 + j8$. Also $k_o = 4150$. Hence $D_o(s) = (s^2 + 8s + 80)(s + 33)$, and we must consider the root loci of

$$1 + \frac{\delta_2(s + 22)^2}{(s^2 + 8s + 80)(s + 33)} - \frac{4150\delta_1}{(s^2 + 8s + 80)(s + 33)}$$

$$= \frac{D_1(s, \delta_2) - 4150\delta_1}{(s^2 + 8s + 80)(s + 33)}.$$

The following procedure is used to find the effect of δ_1 and δ_2. First the root loci of $1 + \delta_2(s + 22)^2/(s + 33)(s^2 + 8s + 80) = 0$ are sketched for the range $\delta_2 = 0$ to $\delta_2 = 15$ (AC in Fig. 8.11-4). This determines the zeros of $D_1(s, \delta_2)$ as a function of δ_2. Then the root loci of $1 - (4150\,\delta_1)/D_1(s, \delta_2) = 0$ give the zeros of $1 + L$ for any values of δ_1, δ_2. It is a good idea to first take the two extremes of $D_1(s, \delta_2 = 0) = (s^2 + 8s + 80)(s + 33)$ and of $D_1(s, \delta_2 = 15) = (s^2 + 20s + 274)(s + 36)$. In the first case, the root loci of $1 - 4150\delta_1/(s^2 + 8s + 80)(s + 33) = 0$ are sketched—AB in Fig. 8.11-4, and then for the second case, the root loci of $1 - 4150\delta_1/(s^2 + 20s + 274)(s + 36) = 0$ are sketched—CD in Fig. 8.11-4. Therefore it is clear that for δ_1 varying from 0 to 0.43, and δ_2 from 0 to 15, the dominant pole of $T(s)$ lies within the shaded region $ACBD$ in Fig. 8.11-4. It is noted that at $\delta_2 = 0$, $\delta_1 = 0.63$, one zero of $1 + L$ is crossing into the right half-plane, and the system is unstable.

The above technique can be used to ascertain if this simplest kind of design of $L(s)$ is satisfactory. If it is not, a more complicated $L(s)$ must be considered. A similar procedure may be used if a zero of $P(s)$ is varying.

§ 8.12 Design of $L(s)$ with a Single Zero

If the very simple $L(s)$ of Section 8.11 does not satisfactorily restrict the variation in the dominant poles of $T(s)$, then one may add a single zero to $L(s)$, in order to decrease the root sensitivity. The effect of this single zero may be seen from Figs. 8.12-1a and b. In Fig. 8.12-1b, because of the added pole-

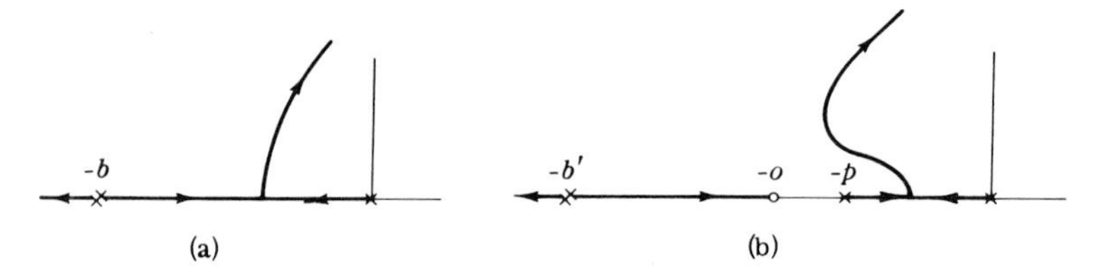

FIG. 8.12-1. Effect of a loop transmission zero on loci of dominant roots due to gain variations (not to scale). (a) Root loci of $1 + k/s(s + b)^2 = 0$. (b) Root loci of $1 + k(s + o)/(s + p)(s + b')^2 = 0$.

zero pair at $-o$ and $-p$ the far-off poles (originally at $-b$ in Fig. 8.12-1a), must be placed further away (to $-b'$); therefore, the center of the asymptotes is further to the left than in Fig. 8.12-1a. The zero also forces the loci to curve to the left. Because of these two effects, the relative change in gain factor needed to cause instability is greater in Fig. 8.12-1b than it is in Fig. 8.12-1a. A design procedure is now presented for choosing the poles and zero of $L(s)$ for this case, wherein it is assumed, for the present, that only the plant gain factor is varying.

It is assumed that two extreme end-points of the root loci have been picked, and it is desired that a root locus pass through these two points. It is assumed too that this is to be done with an $L(s)$ of the type of Fig. 8.12-1b, i.e., one zero and a fixed number of poles. Whatever freedom is available will be used to minimize $|L_{\text{hf}}|$, in order to satisfy the specifications with the most economical $L(s)$. This means the far-off poles of $L(s)$ are chosen coincident.

Suppose A and B in Fig. 8.12-2a are the two points through which the locus must pass. Consequently, $\angle L_A = \angle L_B = 180°$, i.e., $\theta_1 + \theta_A + n\theta_\alpha = 180°$,

$\theta_2 + \theta_B + n\theta_\beta = 180°$. Therefore $\theta_1 - \theta_2 + \theta_A - \theta_B = n(\theta_\beta - \theta_\alpha)$, which is written as $\theta_1 - \theta_2 + \Delta = \Delta' \triangleq n(\theta_\beta - \theta_\alpha)$. It is clear, from Fig. 8.12-2a that $\theta_\beta > \theta_\alpha$; therefore Δ' is positive, and the smaller the value of w in Fig. 8.12-2a (so as to have the loop gain decrease quickly), the larger is Δ'. The value of

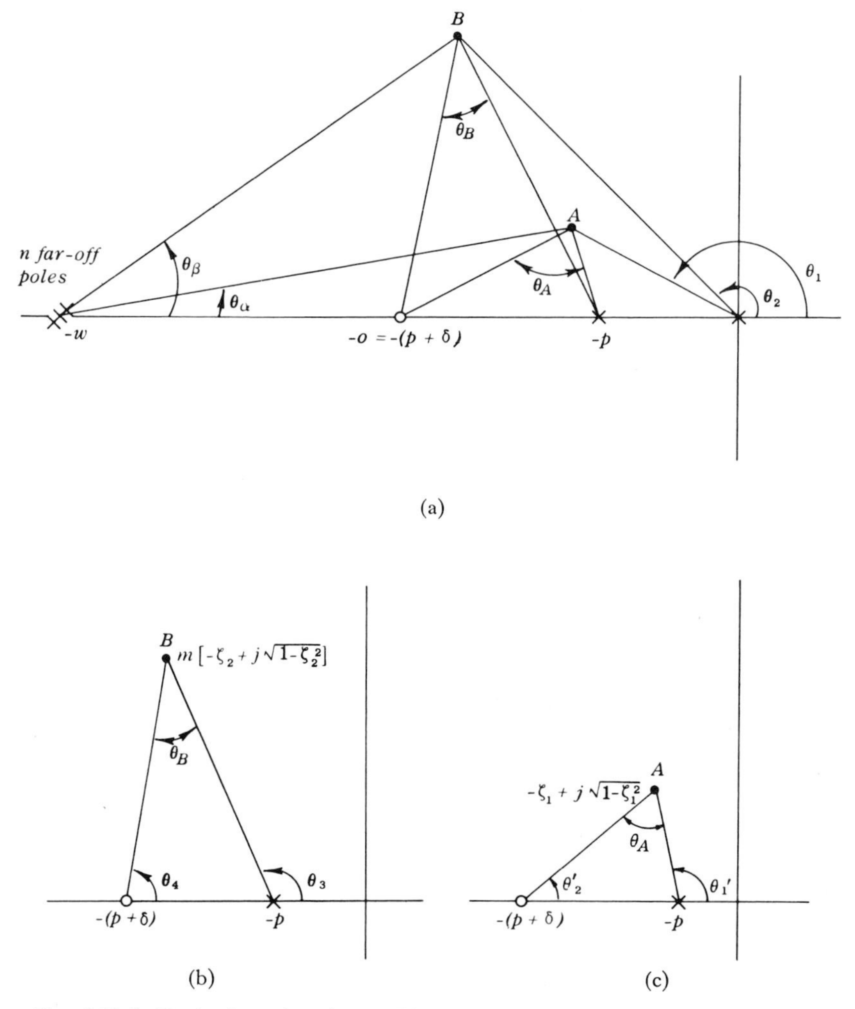

FIG. 8.12-2. Derivation of optimum $L(s)$ such that A and B are on root loci. (a) $\arg L_A = \arg L_B = 180°$. (b) $\tan\theta_B = \tan(\theta_3 - \theta_4)$. (c) $\tan\theta_A = \tan(\theta_1' - \theta_2')$.

$\theta_1 - \theta_2$ is of course fixed, so that the problem is to have Δ as large as possible in order to permit a large Δ', i.e., small w. To do this, choose p and $o = p + \delta$, such that $\Delta = \theta_A - \theta_B$ is maximized for a given θ_A. This is the best that we can do.

The relation which maximizes Δ for a given θ_A is now derived. In Fig. 8.12-2b, $\tan \theta_B = \tan(\theta_3 - \theta_4) = [\tan \theta_3 - \tan \theta_4][1 + \tan \theta_3 \tan \theta_4)]^{-1}$. Since $\tan \theta_4 = m(1 - \zeta_2^2)^{0.5}/(p + \delta - m\zeta_2)$, and $\tan \theta_3 = m(1 - \zeta_2^2)^{0.5}/(p - m\zeta_2)$, $\tan \theta_B = [\delta m(1 - \zeta_2^2)^{0.5}][(p - m\zeta_2)(p + \delta - m\zeta_2) + m^2(1 - \zeta_2^2)]^{-1}$. Therefore

$$\tan \Delta = \frac{\tan \theta_A - \tan \theta_B}{1 + \tan \theta_A \tan \theta_B} = \frac{F_1(p, \delta) \tan \theta_A - F_2(\delta)}{F_1(p, \delta) + F_2(\delta) \tan \theta_A} \tag{8.12,1}$$

where $F_1(p, \delta) = (p - m\zeta_2)(p + \delta - m\zeta_2) + m^2(1 - \zeta_2^2)$, and $F_2(\delta) = m\delta(1 - \zeta_2)^{0.5}$.

The objective is to choose p and $o = p + \delta$ so as to maximize Δ for a given θ_A. It is noted that since $\tan \theta_A$ is a fixed number, p and δ are related (Fig. 8.12-2c) by $\tan \theta_A = \tan(\theta_1' - \theta_2') = \delta(1 - \zeta_1^2)^{0.5}[(p - \zeta_1)(p + \delta - \zeta_1) + (1 - \zeta_1^2)]^{-1}$. Hence $\delta = (p^2 - 2\zeta_1 p + 1)/(E + \zeta_1 - p)$, and $d\delta/dp = (2p + \delta - 2\zeta_1)/(E + \zeta_1 - p)$, where $E = (1 - \zeta_1^2)^{0.5}/\tan \theta_A$. In order to maximize Δ, we set the derivative of $\tan \Delta$ [in Eq. (8.12,1)] equal to zero, i.e.,

$$\left\{\left[\frac{\partial F_1}{\partial p} + \left(\frac{\partial F_1}{\partial \delta}\right)\left(\frac{d\delta}{dp}\right)\right] \tan \theta_A - \left(\frac{dF_2}{d\delta}\right)\left(\frac{d\delta}{dp}\right)\right\}\left\{F_1 + F_2 \tan \theta_A\right\}$$
$$= (F_1 \tan \theta_A - F_2)\left[\frac{\partial F_1}{\partial p} + \left(\frac{\partial F_1}{\partial \delta}\right)\left(\frac{d\delta}{dp}\right) + \left(\frac{dF_2}{d\delta}\right)\left(\frac{d\delta}{dp}\right) \tan \theta_A\right].$$

This simplifies to $(d\delta/dp)[F_1(dF_2/d\delta) - F_2(\partial F_1/\partial\delta)] = F_2(\partial F_1/\partial p)$. The expression for $d\delta/dp$ is known; $\partial F_1/\partial p$, $\partial F_1/\partial \delta$, $dF_2/d\delta$ are available from the expressions for F_1 and F_2. Thus $\partial F_1/\partial p = 2p + \delta - am\zeta_2$, $\partial F_1/\partial \delta = p - m\zeta_2$, $dF_2/d\delta = m(1 - \zeta_2^2)^{0.5}$. When these are used, the result is $(p^2 - 2mp\zeta_2 + m^2)(d\delta/dp) = \delta(2p + \delta - 2m\zeta_2)$, and after grinding through, it finally becomes

$$p^2[2(m\zeta_2 - \zeta_1)(\zeta_1 - E) + 1 - m^2] + 2p[E(m^2 - 1) + \zeta_1(m^2 - 1) - 2m\zeta_2]$$
$$+ [2m(\zeta_2 - m\zeta_1)(\zeta_1 + E) + m^2 - 1] = 0. \tag{8.12,2}$$

Equation (8.12,2) determines the optimum p for a given $E = (1 - \zeta_1^2)^{0.5}/\tan \theta_A$. Then $\delta = (p^2 - 2\zeta_1 p + 1)/(E + \zeta_1 - p)$ may be obtained, and finally Eq. (8.12,1) determines Δ.

Example. The detailed design procedure can now be described by means of a numerical example. The specifications are: $\zeta_1 = 0.707$, $m = 2.7$, $\zeta_2 = 0.65$, and $n = 2$ (see Figs. 8.12-2 for the meaning of ζ_1, ζ_2, n, etc.). These parameters determine A, B (in Fig. 8.12-3b) through which the locus is to pass. Equations (8.12,1) and (8.12,2) are used to find the optimum values of p, δ, and Δ, which correspond to a number of assumed values of $\tan \theta_A$. The value of θ_A vs. Δ is plotted as curve I in Fig. 8.12-3a. It is also convenient to sketch p vs. Δ on the same figure. It is necessary (see Fig. 8.12-2) that $\theta_1 + \theta_A + n\theta_\alpha = 180° = \theta_2 +$

$\theta_B + n\theta_\beta$, which may be written as $\theta_A + n\theta_\alpha = 180° - \theta_1$, $\theta_1 - \theta_2 + \Delta = \Delta' \stackrel{\Delta}{=} n(\theta_\beta - \theta_\alpha)$. In this example, the latter become $\theta_A + n\theta_\alpha = 45°$, $15° + \Delta = \Delta'$. The next step is to sketch $n\theta_\alpha$ as a function of $\Delta = \Delta' - 15°$, where $\Delta' = n(\theta_\beta - \theta_\alpha)$ (see Fig. 8.12-2 for definitions of w, n, θ_α, θ_β). To do this, we pick any value for w, and compute θ_α, θ_β, $\Delta' = n(\theta_\beta - \theta_\alpha)$. For example, if $w = 7$,

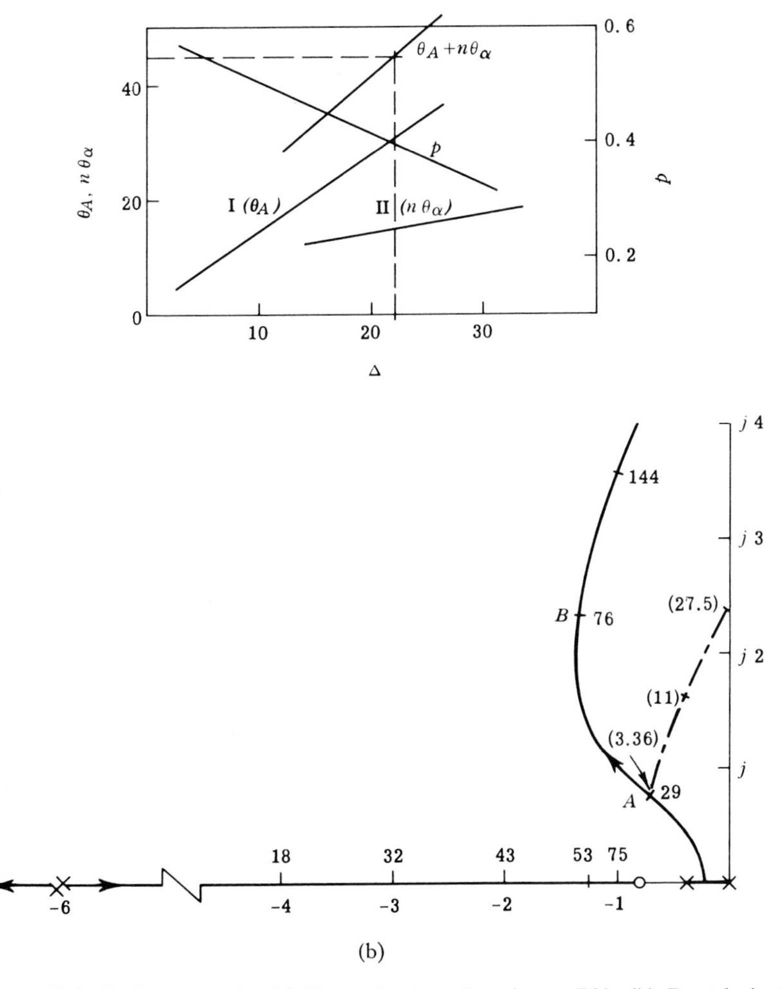

FIG. 8.12-3. Design example. (a) Determination of optimum $L(s)$. (b) Root loci of $1 + L = 0$. Dashed line—$L(s)$ with no zero.

then $\theta_\beta = 22.5°$, $\theta_\alpha = 6.5°$, so that $\Delta' = 32°$ and $n\theta_\alpha = 13°$. This determines the point $n\theta_\alpha = 13°$, $\Delta = 32° - 15° = 17°$ on curve II in Fig. 8.12-3a. Other points are similarly obtained, and are used to sketch curve II in Fig. 8.12-3a. Next we sketch $(\theta_A + n\theta_\alpha)$ vs. Δ, i.e., curves I and II are added together, and the point at which $\theta_A + n\theta_\alpha = 45°$, is found. This determines $\Delta = 22°$, $p = 0.4$, $\delta = 0.4$, and $m = 6$ (see Fig. 8.12-3a). The design is complete. $L = k(s + 0.8)/s(s + 0.4)(s + 6)^2$, and the root loci of $1 + L = 0$ are sketched in Fig. 8.12-3b.

The superiority of the above design over one without a zero is seen by choosing $L_2 = 3.36/s(s + 2.4)^2$. Its locus is shown dotted in Fig. 8.12-3b. An increase in k by a factor of 13 is needed to cause instability in the better design, while a factor of 8.2 suffices in the latter design. Actually, the design with the zero gives a satisfactory response (reasonable damping factor), for a much larger range of gain factor, than the design with poles only. The price paid for this improvement is shown in Fig. 8.12-4, where it is seen that the better design involves

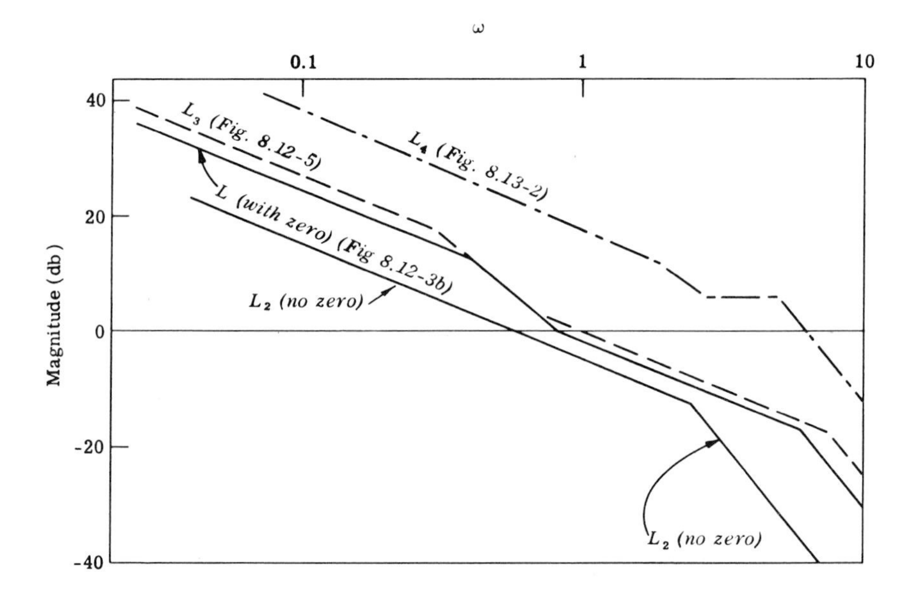

FIG. 8.12-4. Comparison of loop transmissions for various designs.

a loop transmission with larger gain and bandwidth. However, we have obtained this improved design with a loop transmission which, for the given constraints, decreases most rapidly at high frequency.

Completion of the Design

When the above more complicated L function is used, it may be necessary to use a two-degree-of-freedom configuration in order to attain the desired T. In the above example, if Fig. 8.11-1 is used, $T = L/(1 + L)$; therefore T has a zero at -0.8, and has a negative real pole which varies in position from approximately -3 to -1, as the dominant complex pole of T moves from A to B in Fig. 8.12-3b. This may be satisfactory, because the movement of this negative real pole is such as to partially cancel the tendency for larger overshoot due to the movement of the dominant complex pole pair. The zero at -0.8 increases the overshoot, but it can be offset by picking A with a larger ζ. On the other hand, if for any reason T is not to have a zero at -0.8, then a two-degree-of-freedom structure such as the one in Fig. 6.1-1b can be used, with $G_2 = T(1 + L)/P$ and $H = L/PG_2$. T unavoidably has a pole which moves from -3 to -1, as the gain varies by a factor of 76/29. If this is satisfactory, then the nominal T_o is found as follows.

It is readily found that the numerator of $1 + L_o$ is $(s^2 + 1.4s + 1)(s + 3)(s + 8)$. Suppose $T_o = 8/(s^2 + 1.4s + 1)(s + 8)$, and $P = 5/s(s+2)$, then $G_2 = 1.6(s + 2)(s + 3)/(s + 0.4)(s + 6)^2$, and $H = L/PG_2 = (29/8)(s + 0.8)/(s + 3)$. As the gain varies, $T = GP/(1 + L) = [8(s + 3)]/[(\text{zeros of } 1 + L)]$. The zero at -3 tends to cancel the effect of the $1 + L$ zero which varies from -3 to -1. If better cancellation is desired, T can be assigned a zero midway between -3 and -1. In such a case, we take $T_o = 8(s + 2)[(s^2 + 1.4s + 1)(s + 3)(s + 8)]^{-1}$, and proceed as before. It is noted in the above that H is finite at infinity. In view of the fact that $L = G_2PH_2$ in Fig. 6.1-1b, L should have been chosen with four more poles than zeros, to allow two excess poles for P, and one each for G_2 and H_2. If Fig. 6.1-1a is used, the present design is satisfactory. G_1 is as before, and $H_1 = L/P = 5.8(s + 0.8)(s + 2)/(s + 0.4)(s + 6)^2$ is zero at infinity. The design is also satisfactory if Fig. 6.1-1c is used, because in the latter

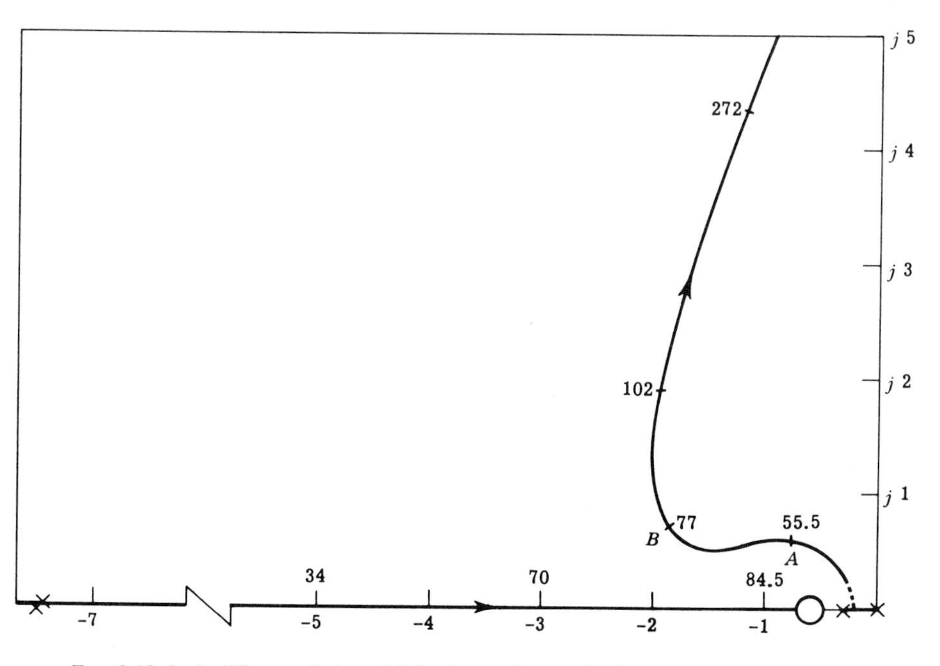

FIG. 8.12-5. A different choice of B leads to a locus which curves more to the left.

$L = P(G_3 + H_3)$; $G_3 = G_2 = G_1$, and $H_3 = (L/P) - G_1 = 4.2s(s+2)/(s+0.4)(s+6)^2$. Thus, in deciding on the number of excess poles over zeros, one should consider the structure that is to be used.

It is possible to force the dominant root loci to curve more to the left than in Fig. 8.12-3b. Thus, if B is chosen as in Fig. 8.12-5, and the above design procedure is repeated, the result is that shown in Fig. 8.12-5. Here we took $m = 2$, $\zeta_2 = 0.94$, $\zeta_1 = 0.76$ and obtained $L(s) = 55(s+0.65)/s(s+0.3)(s+7.5)^2$. This is labeled L_3 in Fig. 8.12-4. Its gain-bandwidth is slightly greater than that of the previous design, and in exchange, the allowable gain variation which leads to system instability is slightly larger (a factor of 14 instead of 13).

Basic Capabilities of L with a Single Zero

An extreme L of the above type (i.e. with a single zero) is capable of providing a design which remains stable for extremely large gain variation of L. However, T may have at most two dominant poles. This is seen in Fig. 8.12-6. To secure

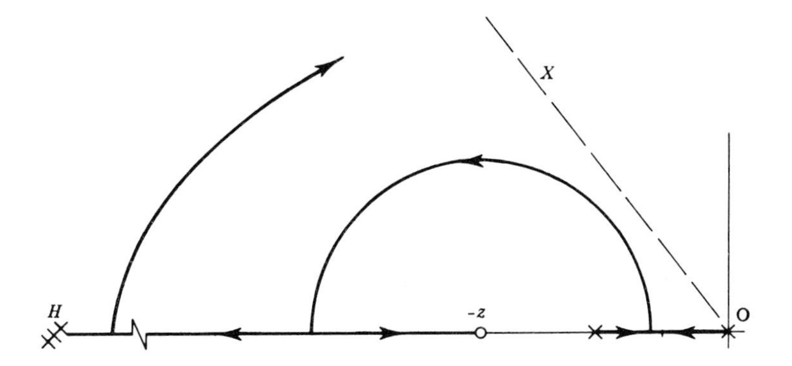

FIG. 8.12-6. Basic capabilities of $L(s)$ with a single zero.

such a design, it is necessary to place the far-off poles at H very far away, in order to ensure that loci emanating from them remain to the left of some preassigned XOX' boundary. The required positions of these far-off poles can be found by the methods of Section 6.4. Such a design may not be satisfactory, because at very large gain the system response is dominated by a single pole near $-z$. The logical step, therefore, for handling even larger parameter variations, is to use a more complicated $L(s)$, one with two zeros. However, before going into this, the effect of variation in a pole or a zero of $L(s)$ is briefly considered.

Does the above $L(s)$ with a zero lead to decreased sensitivity of the dominant poles of $T(s)$, to the variation in a pole of $L(s)$? Since such an $L(s)$ has a larger loop gain-bandwidth, it must perforce lead to smaller sensitivity. The procedure for finding the effect of a change in a plant pole is the same as before. Thus let the pole be at $a - \delta$, so that $1 + L(s) = 1 + k(s + o)/(s - a + \delta)d(s)$. The zeros of $1 + L(s)$ are those of $(s - a)d(s) + k(s + o) + \delta d(s) = D_o(s) + \delta d(s)$. Consider a pole-zero pattern of $L(s)$ of the type shown in Fig. 8.12-3b with $D_o(s) = (s^2 + 1.414s + 1)(s + 3.2)(s + 7)$. If the pole at the origin varies, then $d(s) = (s + 0.4)(s + 6)^2$, and the effect of δ is shown in Fig. 8.12-7.

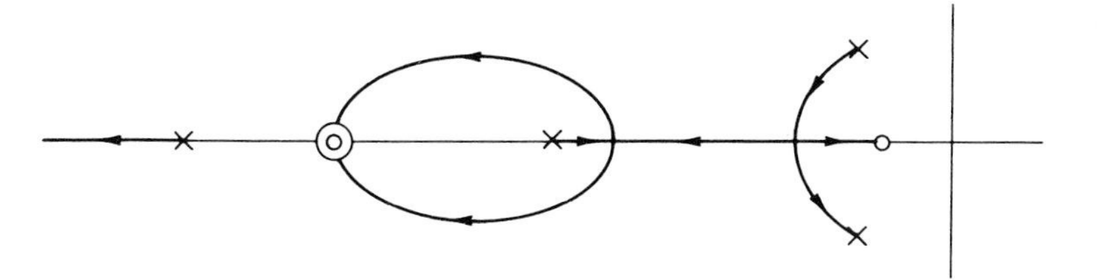

FIG. 8.12-7. Effect of variation of plant pole—root loci of $1 + \delta d(s)/D_o(s) = 0$ (not to scale).

There is, therefore, no difficulty in securing a satisfactory design, even for large δ. If it is a zero of $P(s)$ which is varying, then $1 + L = 1 + k(s + o + \delta)/d(s)$, whose zeros are those of $d(s) + k(s + o) + k\delta = D_o(s) + k\delta$, and the procedure is the same as before. Simultaneous variations in gain and a pole or zero are handled in exactly the same way as in Section 8.11, and no further comments are necessary.

§ 8.13 Design of $L(s)$ with Two Zeros

It has been noted how an $L(s)$ with a single zero may not be satisfactory for very large parameter variations and/or small dominant pole movement, even when there are only two dominant poles. A minimum of two zeros are needed in the latter case. In fact, in general, if the parameter variation is very large, and/or small permitted pole variation is desired, than as many zeros are needed as there are dominant poles. If the variation in the dominant roots is to be very small, then the methods of Section 6.2 and 6.3 are applicable. In this section, we treat the more realistic case where moderate excursions of the roots are permissible, so that the approximations used in Sections 6.2 and 6.3 are not valid. The problem is first treated for the case where the gain factor only is changing.

The following is the synthesis philosophy. Suppose that the gain factor changes by a factor $m = k_2/k_1$, and that at the two extreme values, k_1, k_2, the dominant poles are to be at A, B (Fig. 8.13-1a) respectively. A zero of $L(s)$ will have to be near B. So we tentatively have an $L(s)$ of the qualitative form shown in Fig. 8.13-1a. [If $L(s)$ does not have a pole at the origin, then we assume, as before, that a pole of $L(s)$ is placed as close to the origin as necessary to obtain the desired K_p]. It is also necessary to assign some boundary for the far-off roots

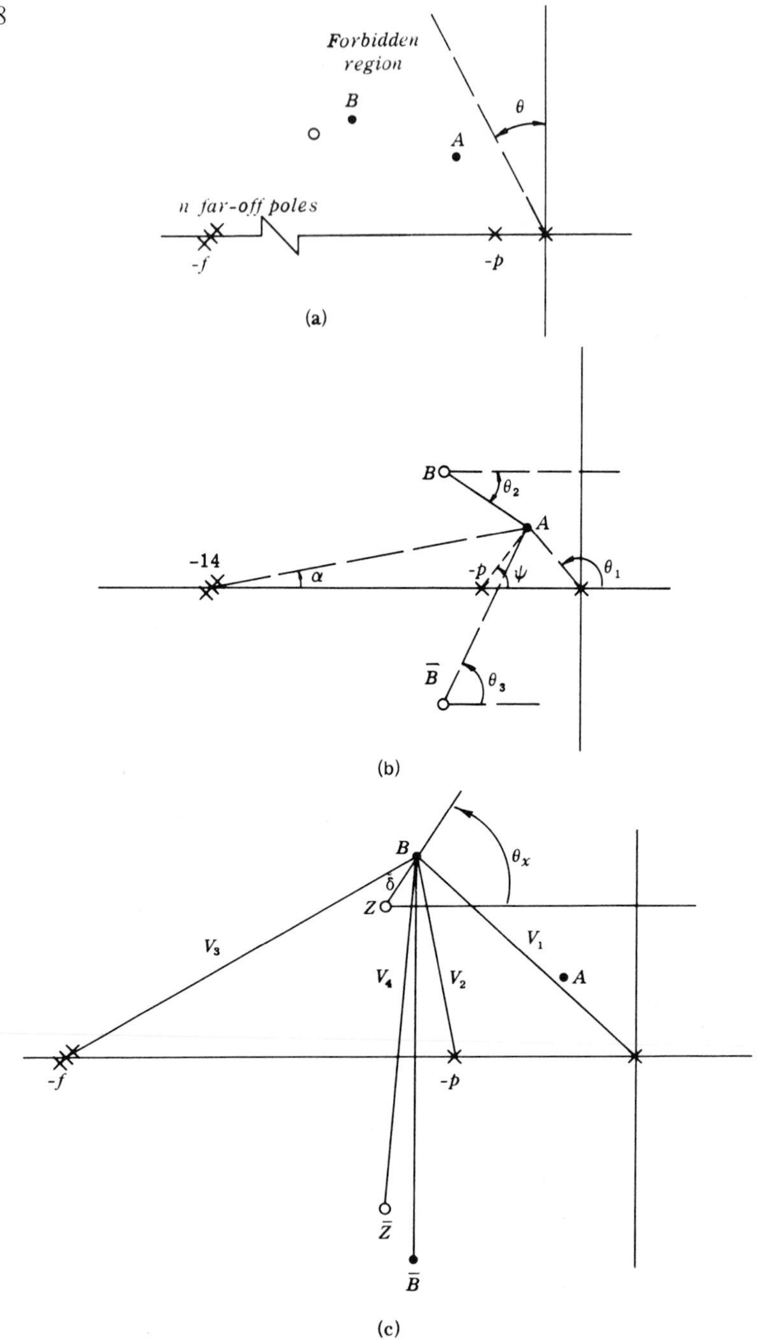

FIG. 8.13-1. Use of zero pair to achieve extreme dominant root variation (from given A to B) for specified gain variation. (a) Tentative qualitative form of $L(s)$. (b) Approximate determination of p. (c) Determination of $k_2 = | V_1 V_2 V_3^3 / V_4 \delta |$.

of $1 + L = 0$, beyond which they must not lie. Suppose this boundary is determined by the angle θ in Fig. 8.13-1a.

To a specific, let $m = k_2/k_1 = 10$, $\theta = 15°$, and $A = -2 + j2$, $B = -5 + j6$ (Fig. 8.13-1b). Try $f = 14$. Next assume that the zeros of L are precisely at B, $\bar{B}$, and locate p so that $\angle L_A = 180°$. In this numerical problem (Fig. 8.13-1b), $3\alpha + \theta_1 + \theta_2 - \theta_3 + \psi = 180°$ leads to $p = 5$. The next step is to check whether the above satisfies the constraint on the far-off roots of $1 + L = 0$. For this purpose, first find k_1 (the gain factor of L) from Fig. 8.13-1b (still assuming that the zeros of L are at B, $\bar{B}$). The result is $k_1 = 425$. Next consider the far-off loci of $1 + L = 0$. There is no need to sketch the loci. All that need be done is to check along the line $\theta = 15°$ with a spirule, until the point at which $\angle L = 180°$ is located. The value of k corresponding to this point is found. In this numerical example, the point is $-3.5 + j12.7$, and $k_2 = 5500 = 13k_1$ at this point. Therefore $f = 14$ is better than necessary, i.e., f can be made smaller, with consequent smaller loop gain-bandwidth for $L(s)$. However, if this is done, the extreme far-off pole, along the line $\theta = 15°$, would have a lower value and probably would not really be sufficiently "far-off." Therefore the present value of $f = 14$ is retained. If it had turned out that $k_2 < 10k_1$, then a larger value of f would have been required.

Now that f is fixed, it is possible to find the precise value of p (Fig. 8.13-1b), and the precise positions for the zeros of L, by an iterative procedure. In Fig.

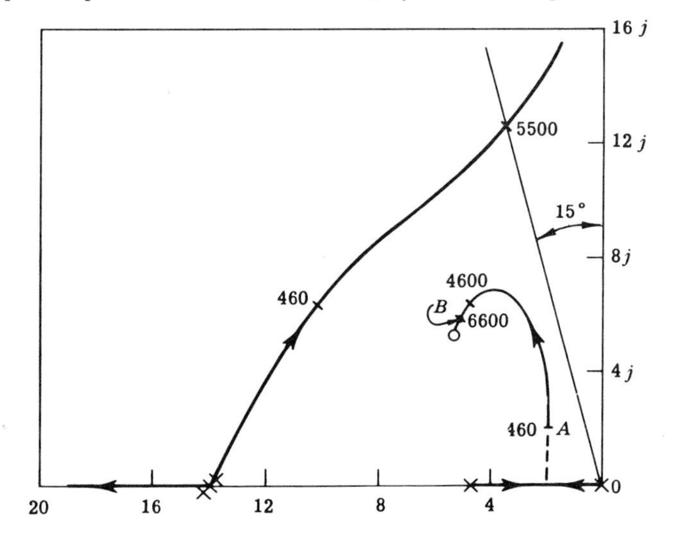

FIG. 8.13-2. Root loci for design example.

8.13-1c $k_2 = | V_1 V_2 V_3^3 / \delta V_4 |$. Since Z is not precisely known, we use $\bar{B}B$ in place of V_4, and the equation determines δ. Next find θ_x by demanding that $\angle L_B = 180°$. The next step is to recalculate p so that $\angle L_A = 180°$, using Z,

$\bar{Z}$ for the zeros in place of the previous B, $\bar{B}$. Next Z, $\bar{Z}$ can be slightly readjusted to make up for the slight change in p, and δ should be recalculated. One can continue in this way to any desired accuracy. In this numerical example, we stop at $Z, \bar{Z} = -5.3 \pm j5.3$, $p = -4.7$. The resulting root loci are shown in Figs. 8.13-2. It is seen that the design is slightly conservative. This can be corrected by increasing the value of δ, but it is not done here. We ought to expect that in this design L must have a larger gain-bandwidth than in the simpler, less demanding designs of Section 8.12. This is verified by normalizing our L, by dividing the frequencies by 2.828 (so that $\omega_n = 1$). The result is shown as L_4 in Fig. 8.12-4. It is seen that it is substantially more than the other loop transmissions. In each case, the most economical L has been chosen, subject to the constraints of the problem.

The above is a fairly complete description of the design procedure. Now we ask whether any amount of variation in k can be handled by the above procedure. To answer this question, consider the extreme case when f (Fig. 8.13-1c) is very far off. When f is very far off, then the far-off loci are nearly as shown

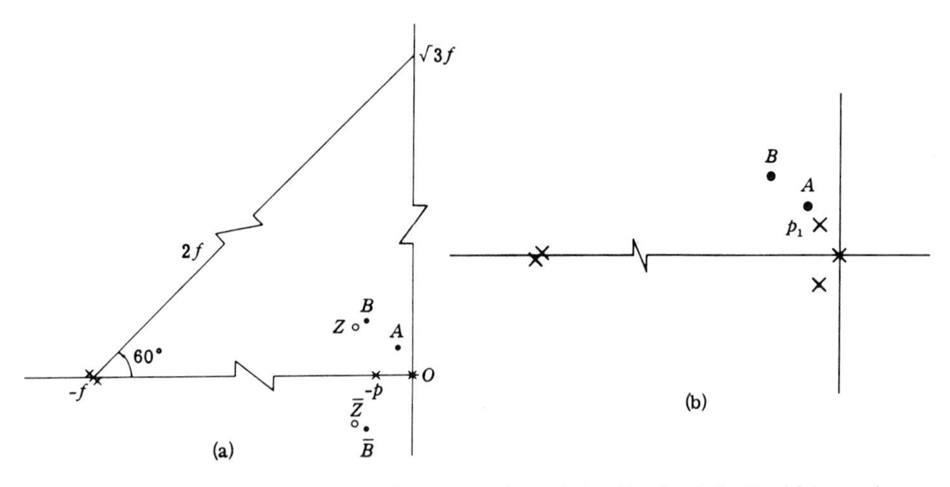

FIG. 8.13-3. (a) Determination of extreme gain variation (for fixed A, B) which may be handled by two zeros. (b) Use of complex poles to increase maximum range of gain variation.

in Fig. 8.13-3a (for the case $n \doteq 3$); and, for simplicity, taking zero for θ (of Fig. 8.13-1a), $k_2 \doteq 8f^3$. On the other hand, $k_1 \doteq |f^3(AO)(Ap)/(AB)(A\bar{B})|$. Therefore

$$\left(\frac{k_2}{k_1}\right)_{\max} \doteq \left|\frac{8(AB)(A\bar{B})}{(Ap)\ (AO)}\right|. \tag{8.13,1}$$

This is the best that can be done. But suppose that this is not enough. In that case, complex poles may be assigned to L, as in Fig. 8.13-3b. In the latter, the equivalent of (Ap) in Eq. (8.13,1) becomes (Ap_1) of Fig. 8.13-3b, and this quantity can be made as small as desired. Since it appears in the denominator of the expression equivalent to Eq. (8.13,1), any ratio of k_2/k_1 can be handled.

It is also important to note the shape of the locus between A and B in Fig. 8.13-2. Although the locus passes through A and B with adequate ratio of k_2/k_1, it may be unsatisfactory, because, for some intermediate values of k, the damping factor of the root becomes as small as 0.39 (cos 67° = 0.39). This problem is handled by noting that generally, in practice, the points A and B would not be *a priori* given. More likely some acceptable *area* for the location of the dominant roots would be assigned. For example, suppose it is the area $A_1A_2B_2B_1$ shown in Fig. 8.13-4a. Then one can use the above procedure after suitable A and B in or on $A_1A_2B_2B_1$ have been chosen.

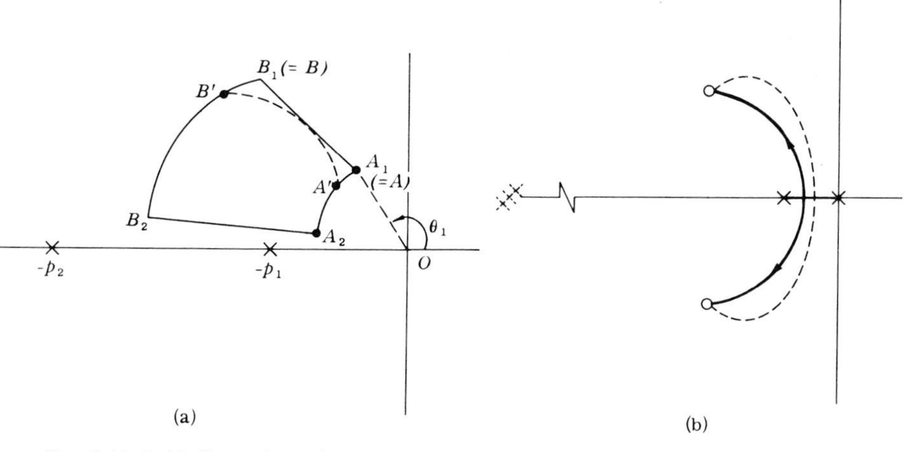

FIG. 8.13-4. (a) Case where dominant poles must lie in specified area. (b) Distortion of locus from circle, due to far-off poles of $L(s)$.

How are points A and B of our design procedure chosen? It is desirable to choose A and B so as to minimize the gain-bandwidth area of L. It is reasoned that this result is obtained by choosing A and B so as to obtain the maximum possible $(k_2/k_1)_{\max}$ of Eq. (8.13,1). It is seen that this is done by choosing AB so as to maximize the distances $(AB)\,(A\bar{B})/(AO)$. A is chosen as close as possible to the origin, but if there is a choice, as in Fig. 8.13-4a (where $OA_1 = OA_2$), we take $A = A_1$, and not A_2 because this gives a smaller θ_1; therefore, $p = p_1$ is closer to the origin, which minimizes Ap in the denominator of Eq. (8.13,1). In the case of Fig. 8.13-4a this is achieved with $A = A_1$, $B = B_1$. However, what of the shape of the locus in between A and B? If it is like that in Fig. 8.13-2, then B should be shifted. It is easily seen that, in general, the locus does have the shape shown in Fig. 8.13-2. Thus, if the far-off poles of $L(s)$ are ignored, one has the pole-zero pattern of $L(s)$ shown in Fig. 8.13-4b, and the loci of $1 + L(s) = 0$ are segments of a circle. When far-off poles are added, then the circle is distorted, as shown by the dotted loci in Fig. 8.13-4b. Therefore, in Fig. 8.13-4a, B and A should be suitably shifted (to A', B'), to allow for this effect. This, too, is best a matter of cut and try.

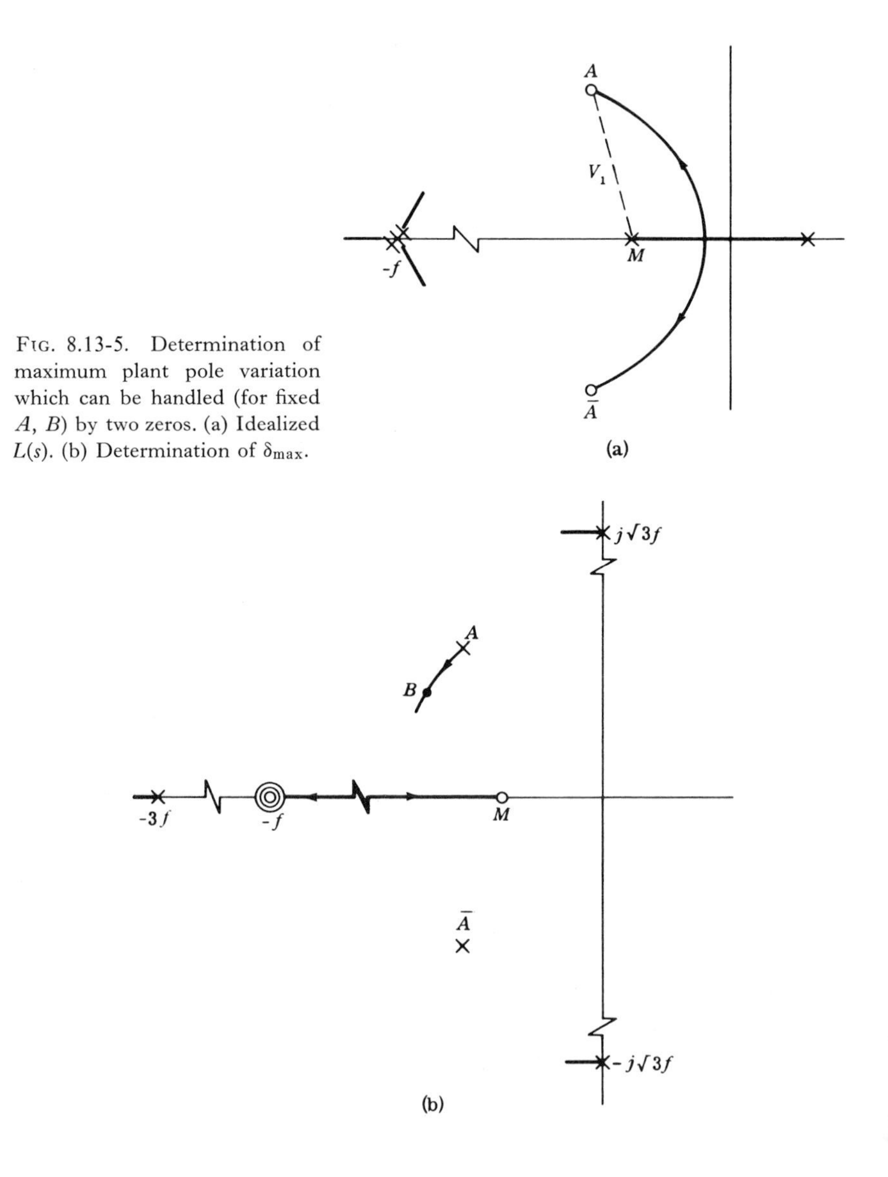

FIG. 8.13-5. Determination of maximum plant pole variation which can be handled (for fixed A, B) by two zeros. (a) Idealized $L(s)$. (b) Determination of δ_{max}.

Variation in a Plant Pole

An $L(s)$ with two zeros can handle much larger variations in a plant pole than one with only one zero. An upper bound for the best that can be done is obtained by means of the argument used in Fig. 8.13-3a. Consider the zeros of $1 + L(s) = 1 + kn(s)/d_1(s)(s + h + \delta)$. These are the zeros of $d_1(s)(s + h) + kn(s) + \delta d_1(s) = D_o(s) + \delta d_1(s)$. In Fig. 8.13-5a, the extreme case of very far off poles is assumed. In addition, it is assumed that in the limit, the far off roots of $1 + L = 0$ are on the $j\omega$ axis at $\pm j1.732f$. Also, the dominant roots are very near A, $\bar{A}$. The effect of δ is given by the root loci of $1 + \delta d_1(s)/D_o(s) = 0$, which are shown in Fig. 8.13-5b. Suppose that the extreme magnitude of the variation in the dominant pole is given by the magnitude of AB in Fig. 8.13-5b. Then

$$(\delta)_{\max} \doteq \frac{(AB)(3f)(1.732f)^2(B\bar{A})}{f^3(AM)} = \frac{9(AB)(B\bar{A})}{(AM)}. \tag{8.13,2}$$

If the actual maximum value of δ is more than the above, then the pole of $L(s)$ at M must be removed, and replaced by a complex pole pair, as close to A, $\bar{A}$ (as in Fig. 8.13-3b) as may be necessary.

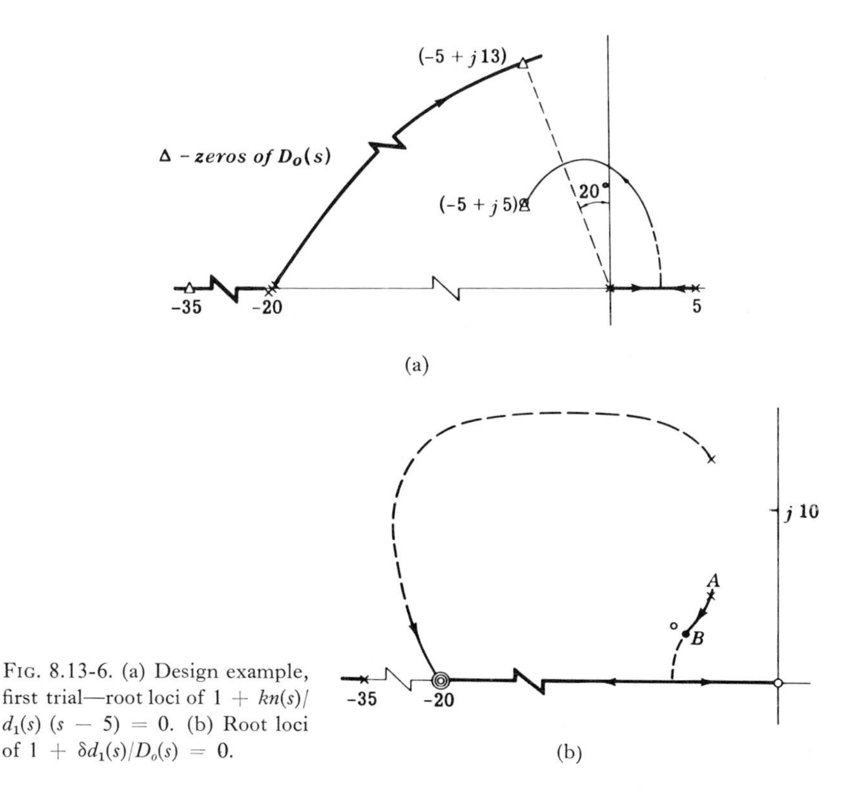

FIG. 8.13-6. (a) Design example, first trial—root loci of $1 + kn(s)/d_1(s)(s - 5) = 0$. (b) Root loci of $1 + \delta d_1(s)/D_o(s) = 0$.

Assuming that the actual δ in the problem is less than $(\delta)_{\max}$ of Eq. (8.13,2), what is a practical design procedure to ensure that the dominant root does not vary by more than some prescribed amount? The procedure is described by means of a numerical example.

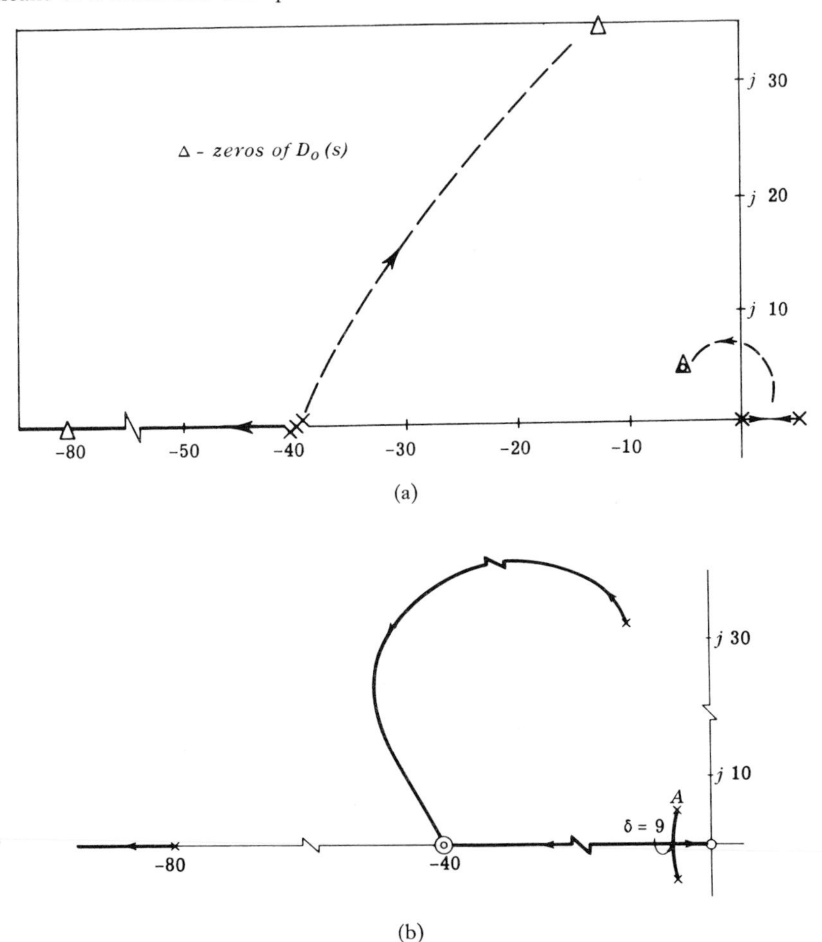

FIG. 8.13-7. (a) Second trial—root loci of $1 + kn(s)/d_1(s)\,(s - 5) = 0$. (b) Root loci of $1 + \delta d_1(s)/D_o(s) = 0$.

Example. Suppose a pole of the plant P varies from 5 to -4, and that the dominant root is to be within a circle of radius 3, centered at $A = -5 + j5$. The first step is to check Eq. (8.13,2), with $AB = 3$, $(B\bar{A})_{\min} = 8$, $(AM) \approx 7$ (on the assumption that M is near the origin, in order to obtain a large K_p). This leads to $(\delta)_{\max} > 30$, so that the specifications should be realizable with dominant poles of L on the real axis.

First trial: Suppose the desired $K_p = 100$. Assume, in the meantime, that $M = 0$ in Fig. 8.13-5a. Now try some value of f, say $f = 20$. This gives the tentative $L(s)$ pole-zero pattern shown in Fig. 8.13-6a. Suppose $\theta = 20°$ marks the boundary for the far-off roots of $1 + L(s) = 0$. In Fig. 8.13-6a we find where the far-off loci pass the $\theta = 20°$ line. It is easily found that a far-off locus crosses $\theta = 20°$ at $-5 + j13$. (The root loci are sketched in Fig. 8.13-6a, but this is not necessary.) Assuming that the dominant roots are very near the zeros, the fifth root must be at $-(3)(20) + 5 + (4)(5) = -35$. The root loci of $1 + \delta d_1(s)/D_o(s) = 0$ are next considered—see Fig. 8.13-6b (the sketch of the root loci is unnecessary), with $\delta = (AB)(8)(18)(35.4)(10)/(7)(15.8)^3$. If $AB = 4$, then δ is appreciably less than 9, so that f is not sufficiently large.

Second trial: In the second trial, we try $f = 40$, resulting in the far-off locus intersecting the $\theta = 20°$ line at $-13 + j35$. The far-off root on the negative real axis is at $-(3)(40) + 5 + 26 + 10 = -79$. The angle of departure from the dominant nominal pole at A is approximately $-120°$, and $\delta = 9$ corresponds to roots coincident approximately at -6 (Figs. 8.13-7a and b). This is a larger variation than specified, but we will assume that it is satisfactory; otherwise a larger value of f must be used.

Using $f = 40$, the design can be completed. First k is found from the fact that one far-off root of $1 + L_o(s)$ is at $-13 + j35$. The result is $k = 10^5$. Consequently $L_o(s) \approx 10^5(s^2 + 10s + 50)[(s - 5)(s + \alpha)(s + 40)^3]^{-1}$. Suppose $(K_p)_{\min} = 100 = 10^5(50)/5\alpha(40)^3$, so $\alpha \leqslant 10/64$; we therefore take $\alpha = 0.16$. Next, the precise positions of the zeros are found, in order to place one extreme value of the dominant root (when the pole of P is at $+5$) wherever it is wanted—say at $-5 + j5$, at $k = 10^5$. It is found that the zero is sufficiently far away (about 4 units) from A as to require readjustment, i.e., the position of the zeros must be recalculated. A second adjustment leads to the zeros at $-7 \pm j$, which is judged satisfactory, with $L(s) = 10^5(s^2 + 14s + 50)[(s + 0.16)(s - 5)(s + 40)^3]^{-1}$. The root loci of $1 + [\delta(s + 0.16)(s + 40)^3/D_o(s)]$ are sketched in Fig. 8.13-7b.

Simultaneous Variations in Plant Gain Factor and Pole

The previous methods for separate gain and pole variations can be combined to handle simultaneous variations in these two parameters. Suppose, for example, that despite the variations in k and δ of $P = k(1 - \delta_1)/(s + a + \delta_2)x(s)$, the dominant plant poles are to lie within the area indicated in Fig. 8.13-4a. The points $A'B'$ might be reasonably selected to be the extreme points on the loci at $\delta_2 = 0$, as δ_1 varies from zero to its maximum value. The dotted line represents a likely path. A minimum distance for the far-off poles, and tentative positions of the zeros, is determined in the usual manner. Than a check is made whether the variation in δ_2 does not cause variations greater than AA_2 and A_2B_2. If it does cause greater variation, then the far-off poles must be located further away, etc.

§ 8.14 Control of Effect of Parameter Variations by the Frequency Response Method

In this section we attempt to control the effect of parameter variation by appropriately shaping the loop transmission in the frequency domain. The justification is the same as that noted in the early part of Section 8.11. The frequency response approach is appropriate for systems which cannot be characterized by only a small number of poles and zeros. The design procedure, its difficulties and limitations, will become apparent by working some examples.

Example 1. Suppose the plant gain factor may vary by a factor of 2, and that the latter means that the actual capacity for work of the plant increases by a factor of 2. In order to exploit this increased capacity for work, we would like the system bandwidth to be proportional to its gain, i.e., its time delay and its rise time halved, but its overshoot unaffected.

Suppose $T_o(j\omega)$ at $k = k_o$ is as shown in Fig. 8.14-1a. The desired $T(j\omega)$ at

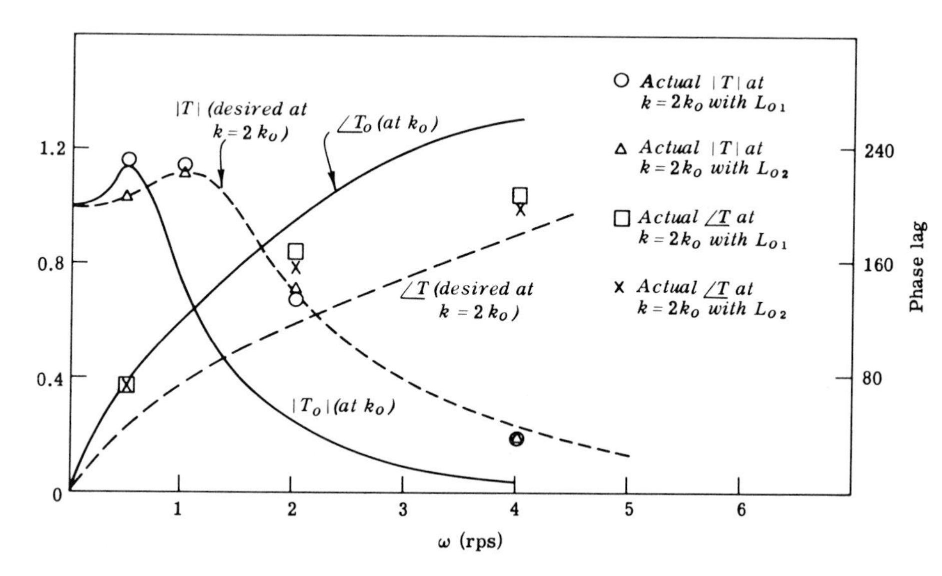

FIG. 8.14-1a. Comparison of desired and actual effect of gain variations on $T(j\omega)$.

$k = 2k_o$ is also shown. The problem is to choose $L_o(j\omega)$ in $T/T_o = [1 + L_o(j\omega)]/[P_o/P + L_o(j\omega)]$, so as to achieve this. For example, at $s = j$, from Fig. 8.14-1a, we want $T/T_o = 1.15 \angle -80°/0.7 \angle -120° = 1.65 \angle 40°$; at $s = 2j$, we want $T/T_o = 0.7 \angle -120°/0.25 \angle -170° = 2.8 \angle 50°$, etc.

The first step is to plot loci of $L_o(j\omega)$ which achieve constant $|T/T_o|$ and constant $\angle(T/T_o)$ at $k = 2k_o$. The construction of Section 6.7 is used, and the result is shown in Fig. 8.14-1b. For example, the curve AB is a locus of points for which $|T/T_o| = 1.7$ at $k = 2k_o$. The curve CD is a locus of points for which $\angle(T/T_o) = 30°$ at $k = 2k_o$. From Fig. 8.14-1a and the loci of Fig.

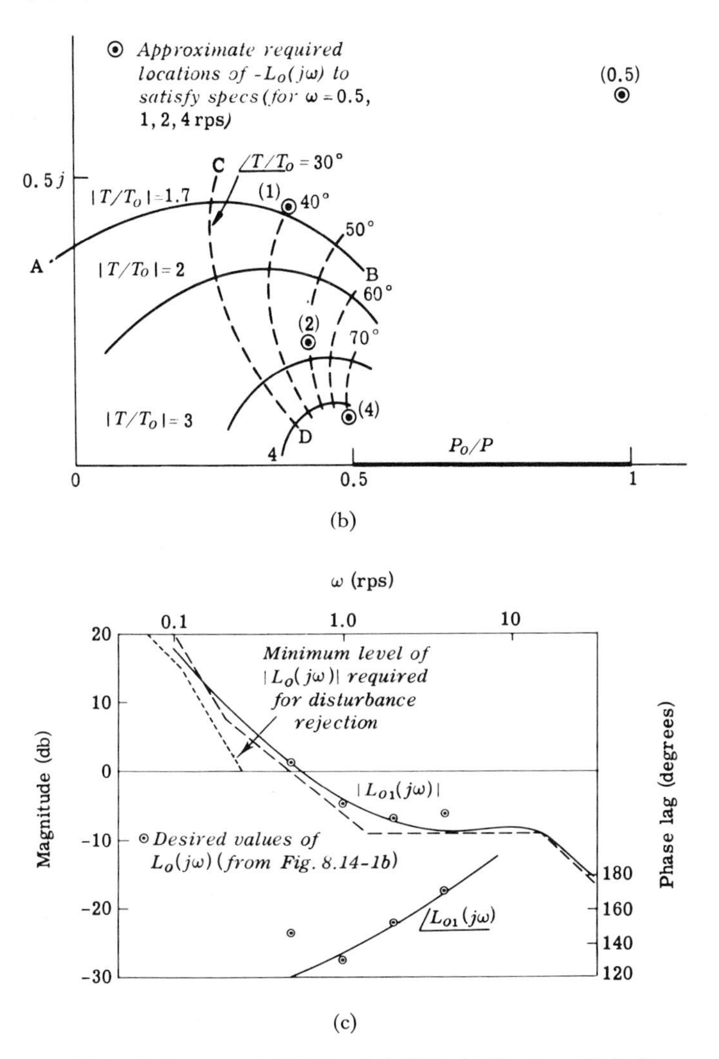

FIG. 8.14-1. (b) Loci of constant $|T/T_o|$ and $\angle T/T_o$. (c) Shaping of $L_o(j\omega)$ to pass near required points—first trial.

8.14-1b, the approximate required values of $L_o(j\omega)$ become available, and are indicated in Figs. 8.14-1b and c.

The next step is to try to select $L_o(j\omega)$ to pass through the indicated points in Fig. 8.14-1c. Nothing has as yet been said about the low frequency requirements of $L_o(j\omega)$. Suppose that disturbance rejection specifications require the $L_o(j\omega)$ level magnitude indicated in Fig. 8.14-1c. The sensitivity and disturbance rejection specifications then appear to be compatible, for realization with a two-degree-of-freedom system.

How do we proceed to shape $L_o(j\omega)$ to pass through the desired points in Fig. 8.14-1c? Analytical procedures for complex curve fitting have been developed[1] However, it is very rarely advisable for the control engineer to use any of these methods. One reason is the large amount of analytical labor that is required, and which increases very rapidly with the number of points of fit. Another reason is that these procedures are so remote from the customary loop shaping techniques of the feedback engineer, that all engineering "feel" for the problem is lost in the process of computation.

In view of the feedback engineer's experience with loop shaping on the Bode plot, he is better off with cut and try on the Bode plot. One important advantage of this method is that the designer has firm control over $L_o(j\omega)$ for all ω. He is aware of the conflicting factors and can compromise, as he nearly always must, between the various desiderata. After a design has been achieved which approximates the acceptable, Linvill's method[2] may be used to improve the design.

In the present problem, it soon becomes evident that if $L_o(s)$ is minimum phase, then the magnitude plateau in the frequency range $1 < \omega < 4$ (Fig. 8.14-1c) is incompatible with the desired phase characteristic. A nonminimum phase $L_o(s)$ must be used. After some cut and try, it is found that a reasonable choice is

$$L_o(s) = \frac{-K_1(s+0.2)(s-1.4)(s+1.5)}{s(s+0.1)[s^2+(2)(0.7)(1.5)s+2.25]} \triangleq L_{o1}(s)$$

(plotted in Figs. 8.14-1c and 8.14-2a).

The above choice of $L_{o1}(s)$ was made as follows: At low frequencies, the level needed for disturbance rejection is approximately followed, with a slope of -6 db per octave. This same slope could be retained until about 1 rps, and then a zero introduced, in order to obtain the required plateau for $1 < \omega < 4$. The zero would have to be in the right half-plane, in order to achieve the increase (not decrease) in phase lag which is required in Fig. 8.14-1c for $0.5 < \omega < 4$. (A rhp zero has the same magnitude characteristic as a lhp zero, but its phase angle may be taken as exactly the same as that of a lhp pole.) The above is almost satisfactory, but would involve a smaller $|L_0(j\omega)|$ than desired at low frequencies. Therefore, instead, a pole is introduced at $s = -0.1$, followed

[1] J. L. Walsh, Interpolation and approximation, *Am. Math. Soc. Colloquium Publs.* **20**, Chapter 8 (1956); R. Steinberg and R. M. Redheffer, Simultaneous trigonometric approximation, *J. Math. Phys.* **31** (4), 260-266 (1953); S. Darlington, Network synthesis using Tchebycheff polynomial series, *Bell System Tech. J.* **31**, 613-665 (1952).

[2] J. G. Linvill, The approximation with rational functions of prescribed magnitude and phase characteristics, *Proc. IRE* **40**, 711-721 (1952).

by a zero at $s = -0.2$, and a zero at $s = 1.4$. The desired magnitude plateau is thereby secured. $|L_o(j\omega)|$ must be decreased at a frequency $\omega > 4$. This has to be done with very little effect on the phase lag for $0.5 < \omega < 4$. To accomplish

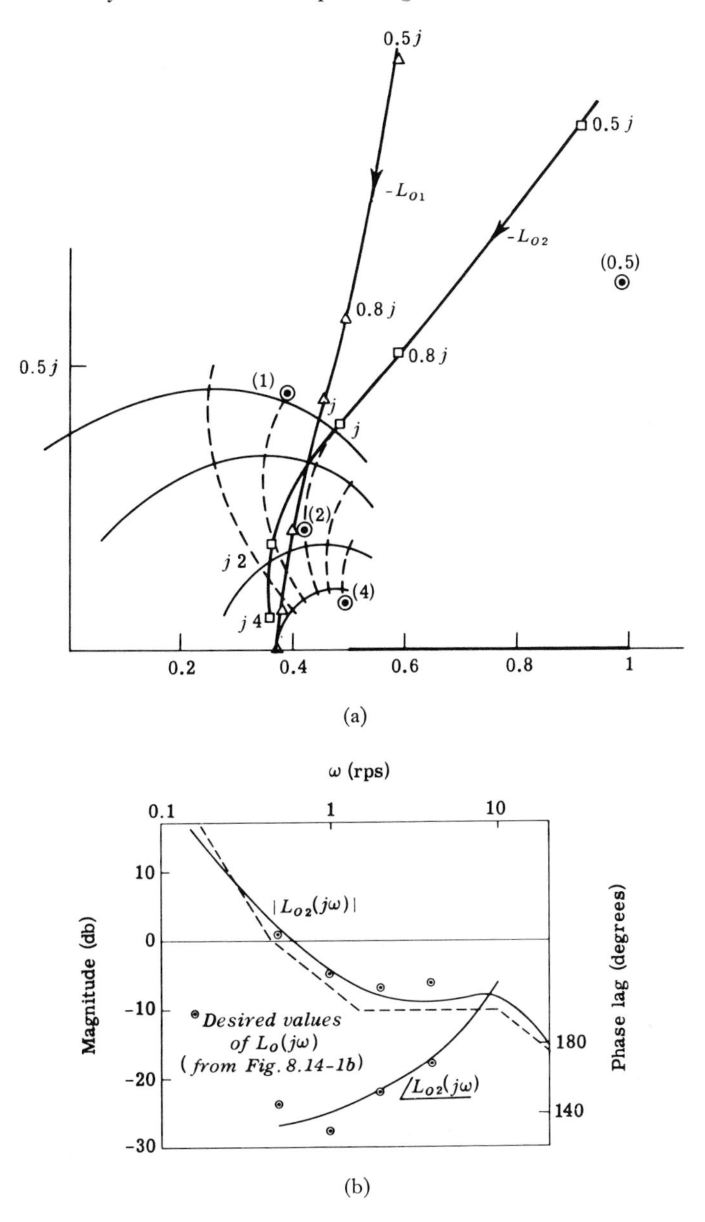

FIG. 8.14-2. (a) Comparison of actual $L_o(j\omega)$ and desired values. (b) Shaping of $L_o(j\omega)$ —second trial.

this, a single lead and a complex lag pair (with $\zeta = 0.7$) are assigned at 15 rps. The phase angle of $L_o(j\omega)$ for $0.5 < \omega < 4$ is only slightly affected by this combination of single lead and double lag.

If a smaller ζ is used, $|L_o|$ would be larger and closer to the desired value at $\omega = 4$, but the system would be unstable, i.e., $|L_o(j\omega)| > 0.5$ at $\angle - L_o(j\omega) = 180°$. In order to improve matters at $\omega = 0.5$, we could move one zero from 0.2 to about 0.50 and change the low frequency level.

A second design is sketched in Figs. 8.14-2a and b. It is

$$L_{o2} = \frac{-K_2(s + 0.45)(s - 1.5)(s + 10)}{s(s + 0.15)[s^2 + 2(0.60)(10)s + 100]}.$$

The $T(j\omega)$ that result at $k = 2k_o$ for L_{o1} and L_{o2} are indicated in Fig. 8.14-1a. The design specifications have been fairly closely achieved. It is furthermore clear from Fig. 8.14-2a that when $k = \alpha k_o$, with $1 < \alpha < 2$, then $T(j\omega)$ lies approximately proportionately in between $T_o(k = 1)$ and $T(k = 2)$. It is also evident from Figs. 8.14-1c and 8.14-2b that if we had designed $L_o(j\omega) = L_A$ so as to secure small changes in $T_o(j\omega)$ over its original bandwidth of 1 rps, that $|L_A| > |L_{o1}|$ or $|L_{o2}|$ for $0.2 < \omega < 10$; but there would probably not be very much difference between the two for $\omega > 10$.

Example 2. In this second example, $P = P_1(s)k/(s + b)$. $P_1(s)$ does not vary much, but k and b each independently vary from 1 to 4. Suppose study of the

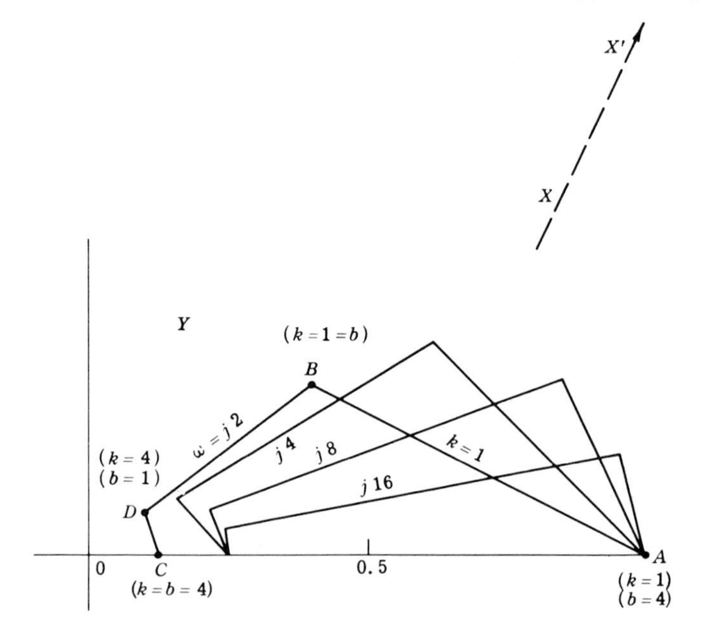

FIG. 8.14-3. Case in which plant gain and pole vary independently.

plant indicates that the plant capacity for work increases primarily with k, but is only secondarily inversely proportional to b. It is therefore desired that, as k increases from 1 to 4, the system bandwidth should proportionately double, as in the first problem. However, the system response is to be relatively insensitive to b. The bandwidth at $k = 1$, $a = 4$, is to be 4 rps.

Design procedure: We choose $P_o = P_1(s)/(s + 4)$ and therefore have $P_o/P = (b + j\omega)/k(4 + j\omega)$. Since the nominal bandwidth is 4 rps, we are especially interested in frequencies approximately in the region $2 < \omega < 16$. At $\omega = 2$, $P_o/P = (b + j2)/k(4 + j2)$. The locus of P_o/P due to the variation in b, at $k = 1$, is the line AB in Fig. 8.14-3. At $k = 4$, the corresponding locus is CD. These two extreme loci, together with the desired $T/T_o(j2)$, must be used to determine an acceptable region of location for $L_o(j2)$. It is clearly impossible (nor is it desired) to locate $L_o(j2)$ so that $T/T_o = (1 + L_o)/(P_o/P + L_o)$ is completely insensitive to b. However, a reasonably good choice for $- L_o(j2)$ is evidently further off, in the direction XX' in Fig. 8.14-3. Loci of P_o/P at $s = j4$, $j8$, $j16$ are also shown in Fig. 8.14-3. Again, it is evident that it is possible to locate $- L_o(j\omega)$ (in the region marked Y in Fig. 8.14-3), so that there is a good chance of approximately satisfying the design specifications at these frequencies. The design details are henceforth similar to those of Example 1.

The above examples indicate some of the highlights and difficulties associated with the attempt at more deliberate control of the effect of parameter variations. Further research is needed in order to determine the kinds of specifications that are achievable, and those that cannot be achieved with the above design approach. It would also be desirable to extend the above to multiple loop systems where presumably the effects of variations in different plant parameters could be individually controlled. The additional flexibility of a multiple loop design might also permit the attainment of disturbance attenuation and parameter variation effects which are incompatible in a single loop design.

§ 8.15 The Rate of Parameter Variations

In all our work involving the sensitivity of feedback systems to parameter variations, nothing was said about the rate of parameter variations. Our previous results apply exactly only to systems in which the parameters do not vary while a signal is being processed; the parameter variations may occur only in between periods of signal processing. The results also apply exacıly to systems in which feedback is used because of ignorance of parameter values, but in which there is no parameter variation. However, the problem must be re-examined for systems in which the parameters vary while a signal is being processed.

Systems in which parameters vary during the processing of a signal, are characterized by differential equations with time-varying coefficients. Such equations are exactly solvable only in special cases. It is therefore unrealistic to expect that the more difficult synthesis problem can be solved in an easy straightforward manner. It is possible, however, to formulate the problem in a manner which is very familiar to the feedback engineer. It is, in fact, shown that the instantaneous effect of the parameter variations is identical to the effect of disturbances acting on the system. The familiar feedback techniques for reducing disturbances can be used. The designer can then intelligently make the inevitable trade-off between design complexity, the cost of feedback, and its benefits.

The problem is presented in Fig. 8.15-1a, which is sufficiently general for

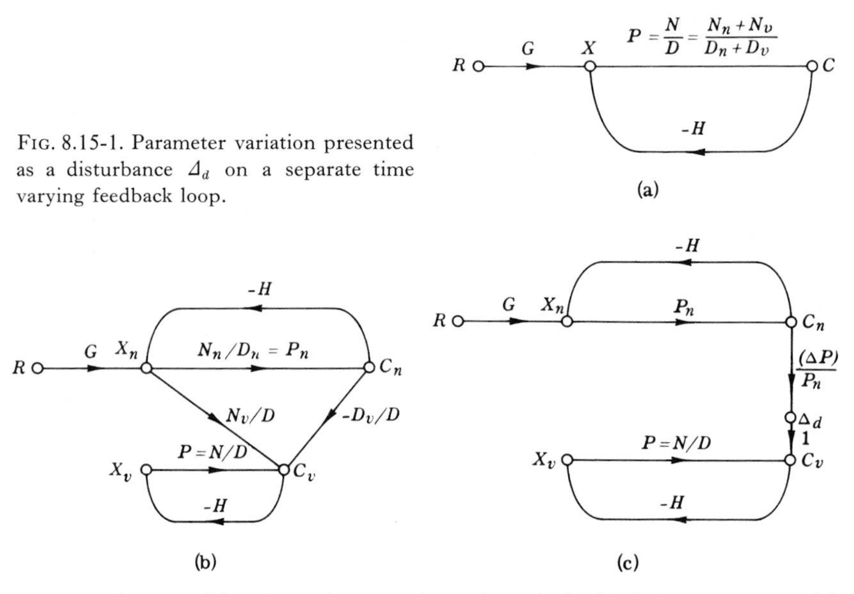

FIG. 8.15-1. Parameter variation presented as a disturbance Δ_d on a separate time varying feedback loop.

any two-degree-of-freedom time-varying plant, imbedded in a system with zero leakage transmission. However, P is not the usual transfer function plant representation; P is symbolic only of the operation $Dc = Nx$, where D and N may have time-varying coefficients (the lower case letters refer to the time functions). H is a transfer function. There is no point in letting H be time-varying, because the exact time at which the plant parameters begin their pattern of variation is, in general, not known. (If it is known, and the pattern of variation is also known, then there is no ignorance of the system, and feedback is not mandatory.)

It is convenient to divide each of X and C into two parts: $C = C_n + C_v$, $X = X_n + X_v$. C_n and X_n represent the transforms of the nominal values of the variables (if there were no parameter variation), while C_v and X_v are the transforms of the portions due to the variation of the plant parameters. Also in $N = N_n + N_v$, $D = D_n + D_v$, N_n and D_n represent terms with fixed coefficients, while N_v and D_v contain all the time-varying coefficients. Hence $P_n \triangleq N_n/D_n$ is a transfer function representing the nominal value of the plant.

In Fig. 8.15-1a, $X = GR - HC = X_n + X_v = GR - H(C_n + C_v)$ and $D(c_n + c_v) = N(x_n + x_v)$. We arbitrarily define $X_n \triangleq GR - HC_n$; $D_n C_n \triangleq N_n X_n$, i.e., $C_n = P_n X_n$. These last two defined relations are diagrammed in the upper part of Fig. 8.15-1b.

Since $Dc = Nx$, i.e., $(D_n + D_v)(c_n + c_v) = (N_n + N_v)(x_n + x_v)$, and $D_n c_n = N_n x_n$, it follows that $Dc_v = Nx_v + N_v x_n - D_v c_n$, which is also shown in Fig. 8.15-1b. Also from $X = GR - HC$, and from $X_n = GR - HC_n$, it follows that $X_v = -HC_v$, which is also shown in Fig. 8.15-1b. If we use $x_v = -h{*}c_v$ to eliminate x_v in the above equation, the result is

$$Dc_v + Nh{*}c_v = N_v x_n - D_v c_n. \tag{8.15,1}$$

The two "disturbances" which act to create C_v can be combined symbolically into the single equivalent "disturbance"

$$\Delta_d = (N_v X_n - D_v C_n)/D = [(N_v D_n/N_n) - D_v]C_n/D = C_n(\Delta P)/P_n,$$

with $\Delta P = P - P_n$; the result is shown in Fig. 8.15-1c.

Figures 8.15-1b and c present the problem in a form which is very familiar to the feedback engineer. The parameter variations are equivalent to disturbances acting on the time-varying system. The feedback must be chosen so that c_v remains within the desired tolerances. The only freedom available for this purpose is H. It is evident that usually (excluding pure time delays in P, or the equivalent of nonminimum-phase plants), it is possible to choose H so as to satisfy the specifications on c_v. The real challenge is to satisfy the specifications with an H whose gain-bandwidth is as small as possible. Design procedures are presented in the following sections.

Equation (8.15,1) clearly reveals how the "equivalent" loop transmission achieves the insensitivity to the instantaneous parameter variation. In Equation (8.15,1) if $Nh{*}c_v \gg Dc_v$ (corresponding symbolically to $Ph{*}c_v \gg c_v$, i.e., $L \gg 1$), then $Nh{*}c_v \doteq N_v x_n - D_v c_n = f(t)$ with known $f(t)$. Also, if h is much faster than c_v, then effectively $h \approx h_o\delta(t)$ and $Nh{*}c_v \approx h_o N c_v = f(t)$, and c_v is reduced to the desired value by choosing h_o sufficiently large.

Experimental Determination of Equivalent Disturbance

The equivalent disturbance $\Delta_d(t)$ can be experimentally obtained as follows. Let the plant have its nominal value P_n. Apply the nominal input x_n and thereby obtain the nominal output $C_n = P_n X_n$. Next let the plant assume its actual time-varying behavior and simultaneously let the plant input be again x_n. Denote the resulting plant output as c_o, i.e., $Dc_o = Nx_n$. The equivalent disturbance $\Delta_d(t)$, is $\Delta_d(t) = c_o - c_n$. To prove this we must show that Eq. (8.15,1) results when $c_o - c_n$ is used in place of Δ_d in the lower part of Fig. 8.15-1c. Using the latter figure we therefore write $c_v = c_o - c_n -$ (plant output). If the plant output in the lower part of Fig. 8.15-1c is denoted by y, then $Dy = Nx_v$. Therefore $Dc_v = Dc_o - Dc_n - Dy = Nx_n - Dc_n - Nx_v$. But $x_v = -h * c_v$, so

$$Dc_v + N[h * c_v] = Nx_n - Dc_n = (N_n + N_v)\, x_n - (D_n + D_v)\, c_n. \qquad (8.15, 2)$$

However, we have previously defined $N_n x_n \triangleq D_n c_n$, so the right-hand side of Eq. (8.15,2) becomes $N_v x_n - D_v c_n$. Eq. (8.15,2) is therefore identical to Eq. (8.15,1). This proves that the equivalent disturbance $\Delta_d(t)$ is precisely $c_o - c_n$, and may be experimentally obtained in the manner indicated above.

§ 8.16 Abrupt Parameter Variations

In this section, we consider the case of abrupt parameter changes. It is useful to study this case for two reasons. One reason is that such changes sometimes occur in a specific system and it is usually necessary to limit their effect. The second reason is that it may be useful, for design purposes, to approximate a nonabrupt variation by an abrupt one. Such an approximation is convenient because in the abrupt case the system is piecewise linear, so that exact analysis is possible. In nonabrupt variations, there result differential equations with time-varying coefficients, and such equations are exactly solvable only in special cases. It is intuitively felt that an abrupt change in any parameter ought to have a greater effect than a slow change (if the net change is the same for both). Consequently, it is conjectured that if the system is designed so that the effect of abrupt changes is tolerable, then the design should be more than satisfactory for the nonabrupt change. Of course, this vague reasoning cannot be relied upon. The designer must check the design. It is often possible to determine, from theoretical considerations, whether the conjecture is correct. In any case, such an assumption permits a relatively simple design procedure to be used. The resulting design can then be checked numerically (see Section 12.10) in order to ascertain whether the design is satisfactory.

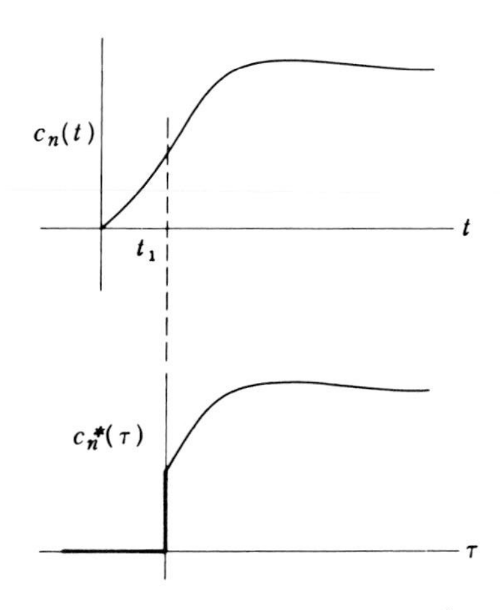

FIG. 8.16-1. Abrupt parameter variation at t_1; determination of c_n^* in $\Delta_d = C_n^* \Delta P/P_n$.

When the plant transfer function changes abruptly, N_v, D_v, and therefore ΔP in Fig. 8.15-1 are zero until some time $t = t_1$, after which N_v and D_v are operators with constant coefficients. The equivalent disturbance $\Delta_d(t)$ is therefore zero for $t < t_1$. Suppose the nominal output $c_n(t)$ is that shown in Fig. 8.16-1. Let $c_n^*(t)$ be defined as follows: $c_n^* = 0$ for $t < t_1$, $c_n^* = c_n$ for $t > t_1$.

For the purpose of calculating c_v, it is convenient to shift the time axis by t_1, i.e., let $\tau = t - t_1$, and the Laplace transform from this point on is with respect to τ. Hence $C_v(s) = \Delta_d(s)/(1 + HP)$, where $\Delta_d(s) = C_n{}^*(s)\,(P - P_n)/P_n$. The problem is identical to that of a disturbance acting on a feedback system whose parameters are fixed. The actual output is $c_n(t)$ for $t < t_1$; $c_n(t) + c_v(\tau)$ for $t > t_1$. Analysis is straightforward. In synthesis, $c_v(\tau)$ must be kept within prescribed bounds. The problem is then identical to the disturbance attenuation problem, of which one phase has been treated in Section 6.11. A second phase of this problem is treated in Section 9.6.

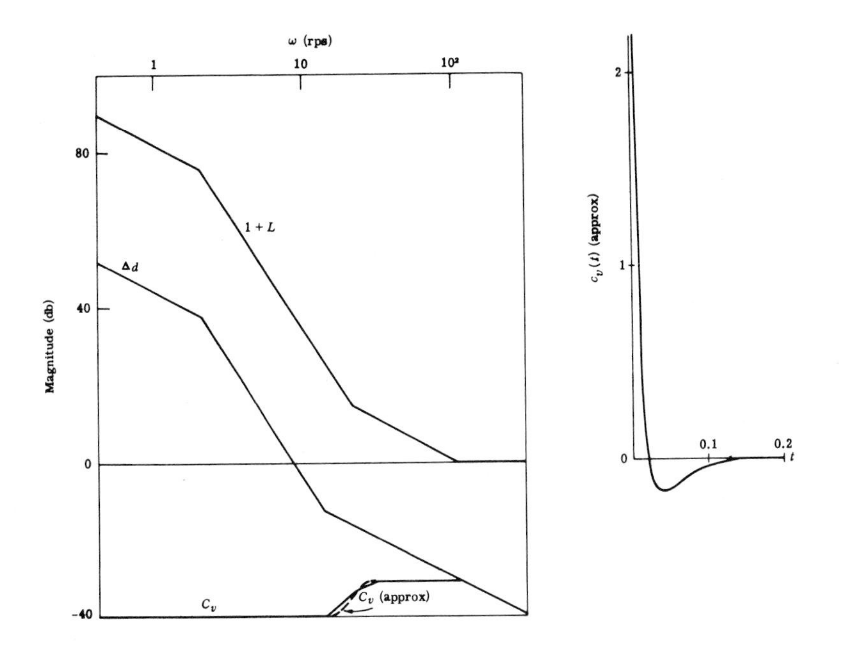

FIG. 8.16-2. Calculation of effect of abrupt parameter variations in problem of Fig. 6.3-1.

The above is illustrated numerically by considering the most extreme parameter variation in the problem of Fig. 6.3-1, i.e., the plant poles change abruptly from $-6 \pm j10$ to $\pm j2$ and simultaneously the plant gain factor increases abruptly by a factor of 4. Also, suppose this occurs after the system has settled down to a constant value, so that $c_n{}^*$ consists of a step function, which is also an extreme case. The resulting equivalent disturbance Δ_d (asymptotic) is sketched on a Bode plot in Fig. 8.16-2a. The disturbance is attenuated by the factor $1 + L$, whose asymptotic sketch is also shown. The effect on the output is given by C_v which is fairly well approximated by the function

$C_{v(\text{app})} = 25(s+15)^2/9(s+25)^2\,(s+100)$ in Fig. 8.16-2a. The time function $c_{v(\text{app})}$ is easily calculated and is sketched in Fig. 8.16-2b. It is seen that the system adjusts itself within about 0.02 second to this fantastically large and abrupt parameter variation.

Effectiveness of Usual Feedback Design for Fast Parameter Variations

The above results enable us to evaluate the effect of the usual feedback design (e.g., the methods of Chapter 6) on fast parameter variations. For example, consider the severe parameter variation design problem of Chapter 6 (see Fig. 6.3-1). Nothing was said at the time about the rate of parameter variation and presumably the design is exact only for plant ignorance with no parameter variation, or if the variations occur in between periods of signal processing. It is often assumed that the design is very closely correct when the rate of parameter variation is much less than 1/15 second. [The latter is the order of magnitude of the time of response for $T(s)$ with a bandwidth of approximately 15 rps.] This is not true. The feedback may be quite effective even for fast parameter variation. This is seen as follows.

Let the nominal $P_n(s)$ have any value within its specified range, and to be specific, let the input be a step function, and the output $c_n(t)$ of Fig. 8.16-1. It is recalled that the design was such that $T(s)$ was quite insensitive to parameter variation. So the output $c_n(t)$ is basically the same whatever the value of $P_n(s)$. If there is an abrupt change in the plant function *precisely at* $t = 0$, then the equivalent disturbance is $\Delta_d = C_n(s)\,(\Delta P/P_n)$, i.e., $\mathscr{L}^{-1}(\Delta P/P_n)$ must be convolved with $c_n(t)$ of Fig. 8.16-1 in order to find the equivalent disturbance in time. We know that the feedback is very effective in attenuating *this* disturbance. If the plant changes abruptly at $t = t_1$, then the equivalent disturbance is obtained by convolving $\mathscr{L}^{-1}(\Delta P/P_n)$ with $c_n{}^*(\tau)$ of Fig. 8.16-1, etc. Is the feedback also very effective for this new disturbance? The question is how much more severe is the second disturbance $C_n{}^*\Delta P/P_n$ than the first $C_n \Delta P/P_n$. While it is certainly more severe because of the abrupt discontinuity in $c_n{}^*$, the difference is not enormous. Also the loop transmission L_B of Section 6.8, with its minimum crossover frequency of 33 rps (Fig. 6.8-4), will reasonably attenuate the disturbance within or less than the time interval 1/33 second. The nominal ω_c is the minimum crossover frequency because the nominal plant transfer function $P_o(s)$ was deliberately chosen such that nearly always $|P_o(j\omega)| < |P(j\omega)|$ (see Fig. 6.8-2). The relation between the bandwidth of $L(s)$ and the speed of disturbance attenuation is discussed in Sections 9.6 and 10.15. Furthermore, we are here considering the extreme case of abrupt parameter variation.

The conclusion is that the design of Chapter 6 with its lack of concern over rate of parameter variation nevertheless leads to a feedback system which is quite insensitive to fast variation of parameters. The precise insensitivity depends on the signal being processed, since the equivalent disturbance is $\Delta_d(s) =$

$C_n \Delta P/P_n$. Hence the greater the magnitude and bandwidth of C_n, the greater are those of $\Delta_d(s)$. The speed with which the equivalent disturbance is attenuated is closely related to ω_c, the crossover frequency of $L(s)$, because the magnitude of the dominant roots of $1 + L(s) = 0$ is not far removed from the magnitude of ω_c [recall Eqs. (5.11,1)].

This property of linear feedback theory deserves emphasis. A design which is made on the basis of parameter ignorance with no variation, or of variation which is very slow compared to the speed of the signals being processed, nevertheless leads to a system which is fairly insensitive to fast parameter variation. The feedback is effective for disturbances whose bandwidths are in the order of ω_c, and the disturbances are related to the speed of parameter variation by the relation $\Delta_d = C_n \Delta P/P_n$.

It is noteworthy that the equation $C_v = \Delta_d/(1 + HP)$ can be manipulated into the form:

$$(C_v + C_n^*)/C_n^* = (1 + L_n)/[(P_n/P) + L_n], \tag{8.16,1}$$

and into $C_v/(C_n^* + C_v) = (\Delta P/P)/(1 + L_n)$. These equations are exactly the same as Eqs. (3.2,6) and (3.2,5), if C_n^*, $C_n^* + C_v$, L_n are used in place of T_o, T_f, L_o, respectively. In other words, the ordinary sensitivity relations can be used for finding the exact time domain effect of abrupt parameter variations. If the abrupt variation occurs at $t = t_1$, then the normal system output for $t \geqslant t_1$ takes the place of T_o and Eq. (3.2,6) solved for T_f gives the new output $c_n^* + c_v$ for $t \geqslant t_1$.

Approximate Calculation of $c_v(t)$

The frequency response of the right-hand side of Eq. (8.16,1), denoted by $F(s)$, usually has the form shown in Fig. 8.16-3, i.e., at very low frequencies it is unity or very close to unity, and at high frequencies it is P_n/P. If the smallest value of the plant transfer function is taken for P_n, as has been consistently recommended, then P_n/P will usually tend at high frequencies as shown in Fig. 8.16-3. $F(s)$ can be written in the form $F = 1 + J(s)$, and then, from

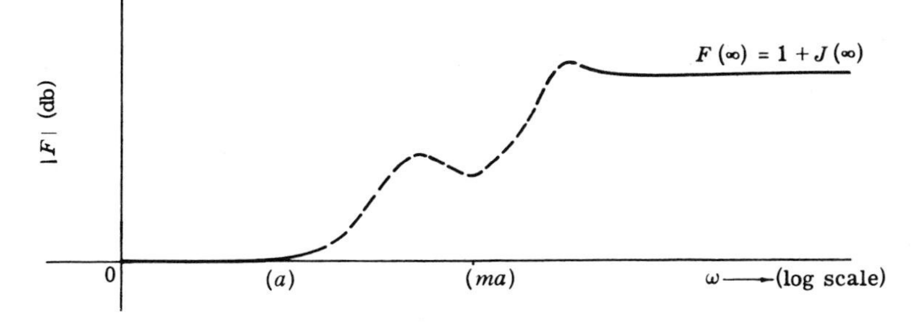

FIG. 8.16-3. Usual form of $(C_v + C_n^*)/C_n^* = F = \dfrac{1 + L_n}{(P_n/P) + L_n}$

Eq. (8.16,1), $c_v + c_n^* = c_n^* * [\delta(t) + j(t)]$, or $c_v = c_n^* * j(t)$. For example, a very simple F is $F = m(s + a)/(s + ma)$ and then $J = F - 1 = s(m - 1)/(s + ma)$. Suppose we consider the extreme case of c_n^* consisting of a step function. Then $c_v = (m - 1)\, e^{-mat}$. It is important to note the broad generality of this result. The low and high frequency character of F shown in Fig. 8.16-3 is certainly very often correct. Therefore (for the extreme case of step function c_n^*), c_v has the value $J(\infty)$ at the instant the abrupt parameter variation occurs, and thereafter decreases. The nature and rate of decrease of c_v depend on the character of F in the region $a < \omega < ma$. If F is reasonably well approximated by $m(s + a)/(s + ma)$ then the decrease in c_v is exponential, with time constant $1/ma$. If a reasonable approximation for F is $m^2(s + a)^2/(s^2 + 2\zeta mas + m^2a^2)$, then $c_v(0) = m^2 - 1$ and $c_v(t)$ is precisely the impulse response of a function with a pole pair at $ma[-\zeta \pm (\zeta^2 - 1)^{0.5}]$, and a zero at $-2am(m - \zeta)/(m^2 - 1)$. Hence a good practical way of finding $c_{v(\text{app})}(t)$ is to sketch F on a Bode plot, and if F is minimum-phase, then a reasonable magnitude approximation of F is satisfactory. Then $C_{v(\text{app})} = C_n^* J_{(\text{app})}$ where $J_{(\text{app})} = F_{(\text{app})} - 1$, and $c_{v(\text{app})}(t)$ is calculated.

§ 8.17 Design Example—Time-Varying Plant

The design procedure for the more general case of time-varying parameters is described by means of the following example. The example concerns a regulation problem in which the output variable must stay within 100 ± 5. In Fig. 8.15-1b, $N = 1$ and $D = [J(t)p + J'(t) + B(t)]$, where p represents the differentiation operation. Also $J = 2 + 10te^{-t}$, $B = 2 - e^{-2t}$. The changes in J and B

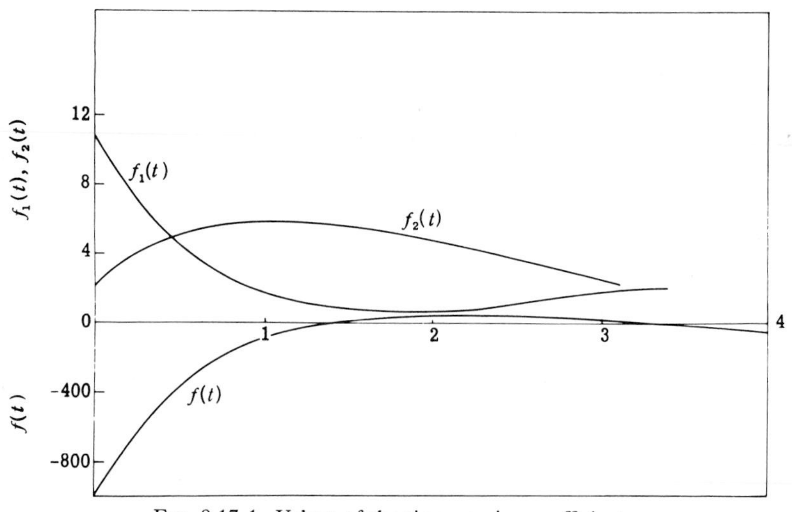

FIG. 8.17-1. Values of the time varying coefficients.

begin at $t = 0$, but prior to that, the output c_n was 100. We take $N_n = 1$, $D_n = 2s + 1$. Also $x_n = 100$, $c_n = 100$, so $N_v = 0$, $D_v = 10te^{-t}p + 10e^{-t}(1 - t) + 1 - e^{-2t}$.

We may now use the design equation (8.15,1) which becomes $(2 + 10te^{-t})\dot{c}_v + [10e^{-t}(1 - t) + 2 - e^{-2t}]c_v + h{*}c_v = -100[1 - e^{-2t} + 10e^{-t}(1 - t)]$. This is written as

$$\dot{c}_v f_2(t) + c_v f_1(t) + h{*}c_v = f(t) \tag{8.17,1}$$

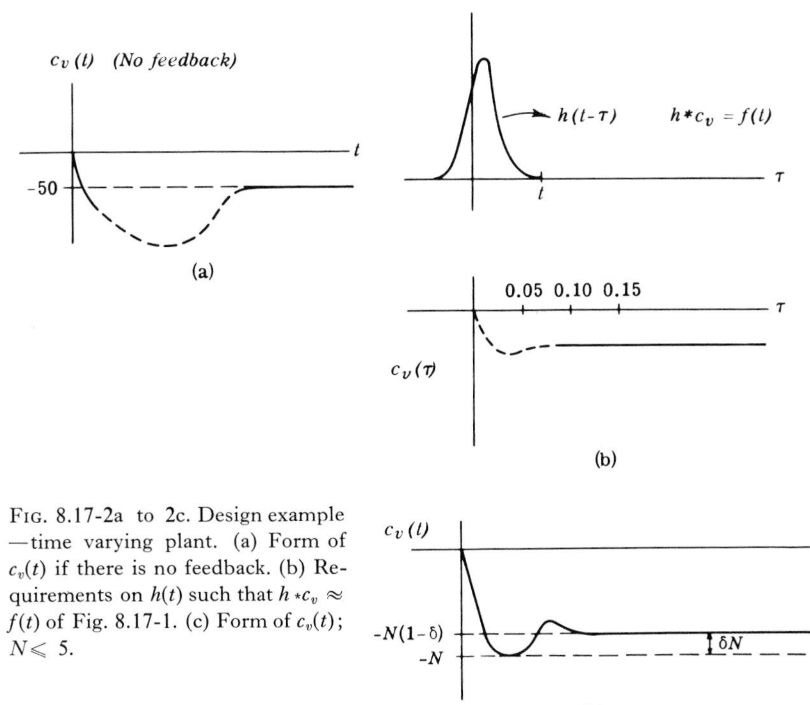

FIG. 8.17-2a to 2c. Design example—time varying plant. (a) Form of $c_v(t)$ if there is no feedback. (b) Requirements on $h(t)$ such that $h * c_v \approx f(t)$ of Fig. 8.17-1. (c) Form of $c_v(t)$; $N \leqslant 5$.

where $f_2(t)$, $f_1(t)$, and $f(t)$ are known and are sketched in Fig. 8.17-1. In Eq. (8.17,1) the object is to choose h so that c_v is kept within the desired bounds. It must be recognized that since H cannot have infinite bandwidth, $h(t)$ cannot contain an impulse at $t = 0$, and therefore $h{*}c_v$ is zero at $t = 0$ (unless c_v has an impulse at $t = 0$, but this is impossible because $\dot{c}_v$ would then have a higher order singularity at $t = 0$). It is easy to see, from Eq. (8.17,1), that $c_v(0) = 0$, because otherwise c_v would have an impulse at $t = 0$, and $f(t)$ on the right-hand side of Eq. (8.17,1) has no impulses. At $t = 0$, Eq. (8.17,1) becomes $f_2(0)\dot{c}_v(0) = f(0)$, from which $\dot{c}_v(0) = -500$. With no feedback, $h = 0$ and $c_v(t)$ would appear as in Fig. 8.17-2a. It is necessary to choose h sufficiently large and fast so that $h{*}c_v$ in Eq. (8.17,1) is large enough to keep c_v less than 5.

We seek now the time interval in which it is most difficult to keep c_v small, and which therefore imposes the greatest burden on h. For this purpose, we carefully examine Eq. (8.17,1). Since $|c_v| \leqslant 5$, and $(f_1)_{max} = 11$ [whereas $f(t) \approx 1000$ in this interval], it is clear that the term $c_v f_1(t)$ is of very little help and can be ignored. This is not the case for $\dot{c}_v f_2(t)$, for it assumes the entire burden at $t = 0$. However, because $-\dot{c}_v(0)$ is so large, the term $\dot{c}_v f_2(t)$ is useful only for a short time interval. After this interval (which, we guess, ends no later than $t = 0.05$), Eq. (8.17,1) becomes effectively $h * c_v \approx f(t)$. The situation is as shown in Fig. 8.17-2b. Since $|c_v|_{max} = 5$, it is necessary that $\int h\, dt > 0.2 f(t)$. Therefore $\int h\, dt > 160$, approximately. It is also clear from the shape of $f(t)$ in Fig. 8.17-1 that the sharpness of $h(t)$ is dictated by the situation at very small t, because for larger t it is not necessary for $h(t)$ to be very narrow [i.e., $H(s)$ could then have a small bandwidth].

We therefore focus our attention on Eq. (8.17,1) for very small t ($t < 0.1$). For this time interval only, Eq. (8.17,1) is approximately

$$2.5\, \dot{c}_v + h * c_v = -900. \tag{8.17,2}$$

The simplest H, which is zero at infinity, is $H = Ab/(s + b)$. If this is used in Eq. (8.17,2), and the latter is transformed, the result is

$$-C_v = 360(s + b)/s(s^2 + bs + 0.4Ab). \tag{8.17,3}$$

The value of $c_v(t)$ is therefore identical to the step response of a system with one zero and two poles, with $\omega_n^2 = 0.4Ab$, $2\zeta\omega_n = b$, so $\zeta^2 = 0.62b/A$; $c_v(t)$ therefore has the form shown in Fig. 8.17-2c.

In Fig. 8.17-2c, $N(1 - \delta) = -\lim_{s \to 0} sC_v = 900/A$, and since $N \leqslant 5$, $A \geqslant 900/(1 - \delta)5$. It is desirable to minimize the bandwidth of H, by using a small value of b. However, this decreases the damping factor of the complex pole pair of $-C_v$, which increases the overshoot δ and therefore the zero frequency level of H, i.e., $A \geqslant 180/(1 - \delta)$ is increased. Suppose the first design trial leads to curve I for H in Fig. 8.17-3a. Also suppose that halving

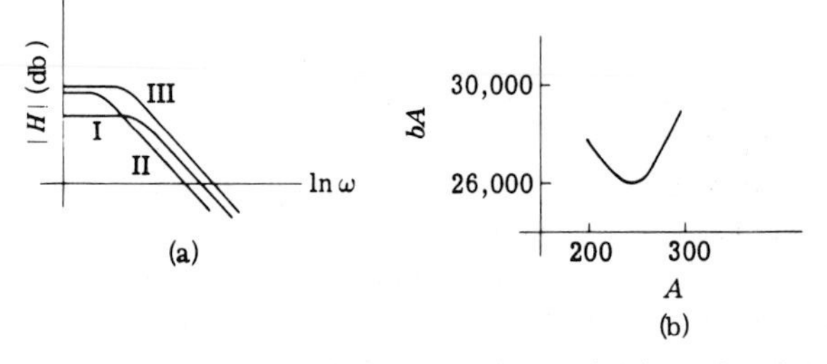

FIG. 8.17-3. (a) L_{II} is considered preferable. (b) Determination of optimum $H(s) = Ab/(s + b)$.

of b forces A to increase by 50%. This leads to curve II. Which design is preferable? We assume curve II is better, bacause of its smaller high frequency response. However, if halving of b forces A to more than double, then the result is curve III, and curve I is better. We therefore consider as optimum that design which minimizes $sH(\infty) = Ab = 180b/(1 - \delta)$. Since an analytical relation between b and δ is not available, the optimum design is found by cut and try. The procedure is as follows.

We guess a value for A, for example, $A = 200$. This corresponds to a minimum $1 - \delta = 900/(5)(200) = 0.90$; i.e., a maximum overshoot of 10%. From Fig. 5.13-2b, 10% overshoot at $\lambda = 2$ corresponds to $\zeta = 0.66$. But $\zeta^2 = 0.62b/A$, so $b = 1.6\zeta^2 A = 139$ and $bA = 27{,}800$. In this way, bA is found for several trial values of A and the results are sketched in Fig. 8.17-3b. The optimum choice is $A = 250$, $b = 104$, with $\delta = 0.28$, $\zeta = 0.51$. This results in $\omega_n = 102$, so that the transient condition is over in about 0.05 second, and c_v is then $5(1 - \delta) = 3.6$.

It should be clear that if temporary $|c_v| > 5$ is permitted, then a better $H(s)$ can be used (in the sense of Fig. 8.17-3a). For example, suppose $c_v(t)$ may temporarily be as much as 7 units, providing this excess of 2 units does not endure for more than 0.05 second. We may therefore write (see Fig. 8.17-2c) $N \leqslant 7$, and therefore $A \geqslant 900/(1 - \delta)7 = 128/(1 - \delta)$, etc., and proceed as before. However, we must be certain that $A \geqslant 180$ and that $c_v(t)$ does not exceed 5 units for more than 0.05 second.

It is possible to achieve greater economy in the gain-bandwidth of $H(s)$ by using a more complex form for $H(s)$, i.e., one with a larger number of parameters. However, the resulting computational effort is greatly increased, especially since curves of the kind shown in Fig. 5.13-2b are not available for more complex system transfer functions.

Our design is almost complete. We have $H = 26{,}000/(s + 104)$. From Eq. (8.17,1) and Fig. 8.17-1, it should be clear that the specifications are satisfied for all time, because $f(t)$ sharply decreases with time. At large t, the equation becomes $2\dot{c}_v + 2c_v + h*c_v = -100$, and with the above $H(s)$, $c_v(\infty) < 0.4$. It is also clear from this relation that, at large t, the system is stable.

To complete the design, G in Fig. 8.15-1a is needed. G is chosen in accordance with the desired system response, at the normal value of P. The procedure therefore is to choose $G = T(1 + L_n)/P_n$, with $L_n = HP_n$. Any one of the many possible two-degree-of-freedom configurations may of course be used.

Discussion

It is apparent from Fig. 8.15-1c that the system must be designed on the basis of the worst combination of Δ_d and $P = N/D$. Presumably a variety of input signals and of variations in the plant parameters are possible. In order to ensure that the sensitivity specifications are not exceeded, it is necessary to pick the most extreme Δ_d and $P = N/D$. This may require some experimentation. The procedure previously outlined indicates how this experimentation may be conducted. The basic design equation is Eq. (8.15,1) and the procedure outlined in the above example is one which may generally be used.

Equation (8.15,1) may also be used for purpose of analysis; for example, as a check as to whether a design based upon the techniques of Chapter 6 leads to acceptable results when the parameter variation is not very slow. A technique for approximately solving Eq. (8.15,1) is required, because such an equation cannot always be exactly solved. An approximate method is described in Section 12.10.

§ 8.18 Sensitivity of the Transient Response

The procedure described in Sections 8.15-8.17 may be used to directly control the sensitivity of the system time response to slow plant parameter variations, in place of the indirect control offered by the frequency response method of Chapter 6. The parameter variations are assumed to be very slow. Figure 8.15-1 is used, and c_v is the difference between the time response when the plant transfer function is $P = P_n + \Delta P$, and the response when it is P_n. It is best to choose for P_n that combination of plant parameters which gives the smallest gain vs. frequency for the plant transfer function, and for P, those parameters leading to the largest gain vs. frequency values. In this way, the maximum equivalent disturbance is found, and the maximum demand on H is ascertained. This approach to the parameter variation problem may permit greater economy in the gain and bandwidth of the loop transmission. The reason is that the designer is working directly with the system time response, and sees precisely how much is sacrificed by using smaller loop transmission gain and bandwidth.

The design procedure is precisely that used in Section 8.16, and requires no further elaboration. The major shortcoming of this approach is in ensuring that the two (supposedly extreme) values of the plant transfer function lead to the extreme values of the system time response. It is conceivable that intermediate values of the plant transfer function might lead to greater departures in the system time response. There appears to be no simple way (as there is in the frequency response method) to achieve definite control of the extremes of the system time response, due to all the specified plant parameter variations.

§ 8.19 Application of Linear Techniques to Nonlinear Plants

Probably the simplest method of applying linear techniques to nonlinear plants is to approximate the latter by a linear plant, for a restricted region of operation. A well-known example of this is in the small signal analysis that is commonly used in vacuum-tube and transistor circuits. The theoretical basis for such an approximation is indicated in the following.

Small Signal Linearization of Nonlinear Differential Equations

Consider nonlinear differential equations which may be written in the form $\Sigma_i f_i(y) g_i(y^{(n)}, y^{(n-1)}, \ldots) = x(t)$, where $y^{(n)} \triangleq d^n y/dt^n$, and f_i, g_i are nonlinear algebraic equations which are piecewise continuous. Suppose that at $t = t_o$ (and the time axis may be shifted so that $t_o = 0$) $y = y_o$, $\dot{y} = \dot{y}_o$, etc. Let $y = y_o + \delta$. If there is no abrupt change in y at t_o (and this can be checked from the differential equation), then for some time interval, δ will be very small. In small signal operation, it is assumed that $x(t)$ is such that δ is very small for the entire operation.

The small signal linearization permits the approximate replacement of the above by a linear differential equation. This is done by expanding each nonlinear function into its power series and neglecting all but the constant and linear terms. For example, consider $y\ddot{y} + y^2\dot{y} + y^{0.5} = x(t)$, with $y_o\ddot{y}_o + y_o^2\dot{y}_o + y_o^{0.5} = x(t_o)$. Let $\tau = t - t_o$, $\delta = y - y_o$, $x(t) = x(t_o) + \eta(\tau)$. Then $y\ddot{y} = (y_o + \delta)(\ddot{y}_o + \ddot{\delta}) = y_o\ddot{y}_o + \delta\ddot{y}_o + y_o\ddot{\delta} + \delta\ddot{\delta}$. The last term may be neglected if it is very small compared to the others. However, it is incorrect to assume that $\ddot{\delta}$ or $\dot{\delta}$ is small, even though δ may be very small. Therefore, neglect of $\delta\ddot{\delta}$ must be based on the assumption that $\delta \ll y_o$. Similarly,

$$y^2\dot{y} = (y_o + \delta)^2(\dot{y}_o + \dot{\delta}) = [y_o^2\dot{y}_o + 2\delta y_o\dot{y}_o + y_o^2\dot{\delta}] + [\delta^2(\dot{y}_o + \dot{\delta}) + 2\delta\dot{\delta}y_o].$$

Again neglect of the second group of (nonlinear) terms is based on $\delta \ll y_o$, rather than on $\dot{\delta} < \dot{y}_o$, etc. Also $y^{0.5} = (y_o + \delta)^{0.5} \doteq y_o^{0.5} + 0.5\delta y_o^{-0.5}$. Hence the small signal linear differential equation approximant is $y_o\ddot{\delta} + y_o^2\dot{\delta} + \delta(\ddot{y}_o + 2y_o\dot{y}_o + 0.5y_o^{-0.5}) = \eta(\tau)$.

The above representation is useful and can be made as accurate as desired, by suitably restricting the magnitude of δ. It is often a reasonable approach, especially in regulator systems, where a specified operating condition is to be maintained, and excursions are kept small by means of feedback. Linear analysis may also be used to determine stability, providing of course that the assumptions of small excursions about an operating point are not violated. A better approximation is obtained by solving for δ in the approximate linear equation, and using the resulting δ, $\dot{\delta}$, etc., in the nonlinear terms of the expansion. The nonlinear terms in the expansion are thus approximated by known time functions which modify the forcing function $\eta(\tau)$.

The Effect of Feedback on Plant Nonlinearities

It was shown in Section 3.10 how nonlinearities are linearized by the factor $1 + L$ in idealized infinite bandwidth systems (which may be characterized by algebraic equations). This result is still valid, in a certain sense, for realistic finite bandwidth systems (which must be characterized by differential equations). We shall first consider nonlinearities which are free of time derivatives (e.g., $y = \Sigma a_i x^i$, $\sinh x$, etc.) imbedded in realistic feedback systems.

The problem is presented in a fairly general manner in the two-degree-of-freedom configuration of Fig. 8.19-1, where the nonlinearity is part of the known plant. In Fig. 8.19-1, $X = GR - GHPY = GR - L'Y$, where $L' \triangleq GHP$. If

$l'{*}y \gg x$, then $GR \doteq L'Y = L'C/P$, and $C \doteq GPR/L'$, just as in linear systems. Note that if the loop is opened at node x, and the signal x is injected, then the returned signal is $l'{*}y$. The condition $l'{*}y \gg x$ means that the returned signal must swamp the injected signal for all t.

The effect of feedback may be quantitatively formulated by *assuming a given output* and working backwards to find the required input. In Fig. 8.19-1, if C

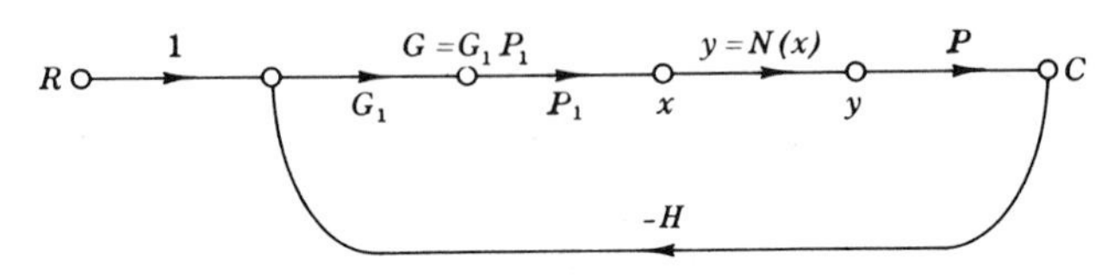

FIG. 8.19-1. Nonlinearity embedded in plant.

is known, then $Y = C/P$, and X can be found (if necessary by numerical methods). Also $R = XG^{-1} + L'YG^{-1}$. We write symbolically $x = Ay\cdot + n^{-1}y$, and its transform is $X = AY + (N^{-1}Y)$, where the nonlinearities are absorbed in $n^{-1}y$, i.e., A is a constant. Therefore $R = G^{-1}\ [AY + N^{-1}Y + L'Y] = G^{-1}(A + L')Y + G^{-1}(\mathrm{N}^{-1}Y)$. Let $R_l \triangleq G^{-1}(A + L')Y$, i.e., R_l is the transform of the input when there is no nonlinearity. Then it is readily found (replacing R by R_{nl} and letting the loop transmission $L = L'/A$) that

$$\frac{R_{nl}}{R_l} = \frac{L + [(AY + N^{-1}Y)/YA]}{L + 1} = \frac{L + (X_{nl}/X_l)}{L + 1} = \frac{T_l}{T_{nl}}. \quad (8.19,1)$$

This is identical in form to Eq. (6.7,1), $T_o/T = [(P_o/P) + L_o]/(1 + L_o)$, which describes the effect of feedback on parameter variations; $1 + (N^{-1}Y)/YA$ is analogous to $1 + (\Delta P)/P = P_o/P$. In Eq. (8.19,1), R_{nl}/R_l is set equal to T_l/T_{nl} by letting $T_l = C/R_l$, $T_{nl} = C/R_{nl}$. T_l is a genuine transfer function, while T_{nl} is not. T_{nl} is obtained by arbitrarily taking the ratio of the transforms of the output and input.

Equation (8.19,1) enables us to express the nonlinearity as equivalent to a parameter variation. The *amount of parameter variation is, however, a function of the signal passing through the nonlinear element*. For a known desired output $c(t)$, one may deduce $y(t)$ and then $x(t)$. The frequency locus of X_{nl}/X_l may be sketched in exactly the same manner as P/P_o (e.g., as in Fig. 8.19-2). The process may be repeated for different desired outputs of various amplitudes, giving a family of loci, possibly as in Fig. 8.19-2 (at infinite frequency presumably X_{nl} and X_l is each zero). If $-L(j\omega_1) = \mathbf{OM}$, then for the output c_1, $T_l/T_{nl}(j\omega_1) = \mathbf{MN}/\mathbf{MV}$, and the boundary of $L(j\omega_1)$ may be found such that $T_l/T_{nl}(j\omega_1)$ is within specified tolerances. If T_l/T_{nl} is close to one over the significant frequency range of the input, then the system is sensibly linear for this class of inputs. It may, of course, be strongly nonlinear for other signals.

The above approach and Eq. (8.19,1) can be used for any kind of nonlinearity imbedded in the plant. It is usually easier to solve for the input to the nonlinear element, for a known output, than for the opposite. In any case, numerical methods may be used to find the input, and then Eq. (8.19,1) is used to find the

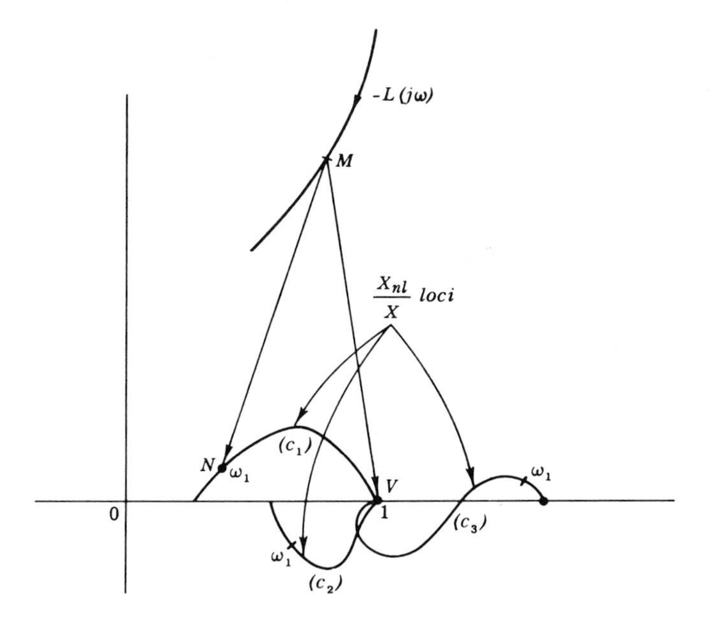

FIG. 8.19-2. Nonlinearity treated as parameter variation for specified range of output signals.

requirements on the linear function $L = GPH/A$, such that $T_l/T_{nl}(j\omega)$ is maintained within specified tolerances, providing of course that the plant can deliver the desired output. However, it should be noted that feedback is not always mandatory for achieving a linear response function. A nonlinear element with a characteristic which is the inverse of the one imbedded in the plant can be inserted (in some cases inverse characteristics are not realizable, e.g., saturation). If $P_1 = 1$ in Fig. 8.19-1, then the added inverse nonlinear element may be adjacent to N and will compensate for it, *for all* inputs. If the two cannot be adjacent, then the linearization may be possible only for some classes of inputs. But a compensating nonlinear element may at least reduce the overall nonlinearity for other inputs. This approach is not feasible, however, if there is some ignorance of the nonlinearity, or if the nonlinear parameters change in value. However, the approach of Fig. 8.19-2, etc., is still valid, by finding and sketching X_{nl}/X_l of Eq. (8.19,1), over the range of nonlinear parameter values.

§ 8.20 Practical Design Techniques for Feedback Control Systems with Nonlinear Plants

Equation (8.19,1) is probably more of philosophical interest than of practical use in the design of feedback systems with nonlinear plants. The reason is that linearity per se is not necessarily always important or desirable in a feedback control system. Linear techniques are used so extensively because they provide simple and reliable means (when the plant is linear) of achieving design objectives. If the plant is nonlinear, then application of Eq. (8.19,1) to achieve a sensibly linear response function (for a given class of output signals) may lead to a larger loop gain and bandwidth than the feedback specifications inherently require. It is more sensible to concentrate directly on the system specifications, and, for achieving them, to seek techniques which are suitable for nonlinear plants.

It would seem that such an approach must contend with the formidable analytical problem of solving nonlinear integral equations. Fortunately, the feedback control specifications are very often of such a nature as to radically simplify this problem. This will be illustrated by considering the following two feedback control problems: (1) attenuation of typical disturbances with a plant which is nonlinear and whose parameters vary, and (2) the achievement of a specified minimum time domain error response for typical inputs, with a plant which is nonlinear and whose parameters vary.

(1) Design Procedure for Attenuation of Disturbances

In Fig. 8.20-1, the nonlinearity is symbolically represented by the equation $y = n(x)$ and for generality "n" is imbedded in the linear plant elements P_1, P_2. Given a class of disturbance signals $d(t)$, the problem is to find $H(s)$ so that the corresponding values of the output $c(t)$ are not more than their specified maximum permitted values. It is furthermore desirable to accomplish this with an H whose gain and bandwidth are as small as possible.

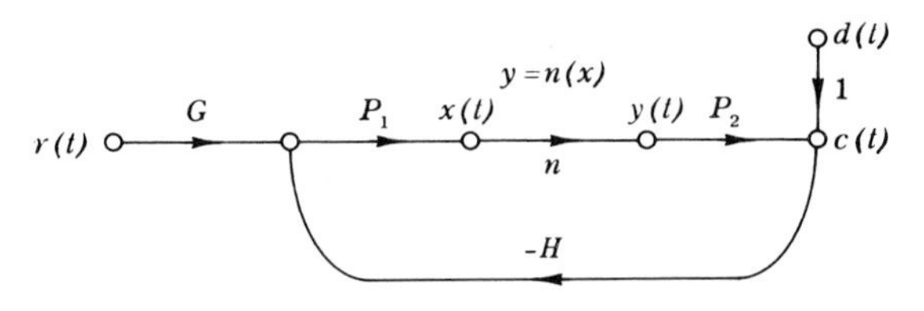

FIG. 8.20-1. System with nonlinearity imbedded in plant.

We shall first consider the case when $r(t)$, in Fig. 8.20-1, is zero. This corresponds (if necessary, by shifting the axes of the variables) to all regulation problems wherein a variable must be maintained at a given value within close tolerances. Although $y = n(x)$ represents a nonlinear relation between y and x, we may, for a specific known $x(t)$ and the resulting (assumed known) $y(t)$, write $Y(s) = N(s)X(s)$. $N(s)$ is a transfer function which is valid only for that specific $x(t)$, $y(t)$ pair from which it was derived. In linear time-invariant systems, $N(s)$ is independent of $Y(s)$ and of time. In the nonlinear system, $N(s)$ is different for different $x(t)$. If the nonlinear function also slowly varies or if there is some ignorance of the precise nonlinear relation (if the variation is rapid, then it is incorporated into the nonlinear equation), then $N(s)$ has a range of variation for a specific $x(t)$. In any case, we may write $Y(s) = N(s)X(s)$ with the understanding that $N(s)$ is different for different $X(s)$, $Y(s)$. The variation in $N(s)$ can be found only by solving the nonlinear relation $y = n(x)$ for the assumed $x(t)$ or $y(t)$.

Fortunately, the solution of the nonlinear integral equation (for $r = 0$) $x = -p_1*h*[d + p_2*n(x)]$ is facilitated by the realities of a feedback control system whose *raison d'être* is the attenuation of disturbances. This is seen as follows.

In Fig. 8.20-1, $c = d + p_2*y = d + p_2*n(x)$. Also $X = -(P_1H)C = -H_1C$ (where $H_1 \triangleq P_1H$), because it has been postulated that $r = 0$. The attenuation achieved by feedback is evidenced by writing the above in the form $c + p_2*n(h_1*c) = d$. Good disturbance attenuation presumably requires $|c(t)| \ll |d(t)|$ and therefore it seems that $p_2*n(h_1*c) \doteq d$. One important qualification must, however, be made. It may be inherently impossible for $c(t)$ to be much less than $d(t)$ for all t. For example, if $d(t)$ is a step function, then at $t = 0+$, $c(t) = d(t)$ because in any practical system, at least P_1, P_2, H do not have infinite bandwidth. This problem is avoided by writing $c = c_1 + c_2$,

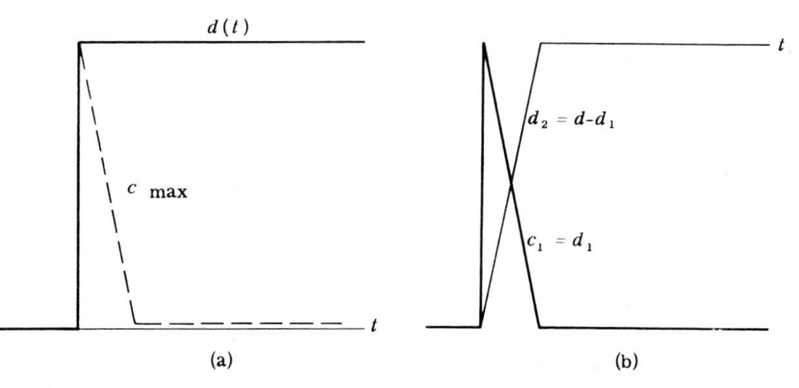

FIG. 8.20-2. Splitting of c, d into $c_1 + c_2$, $d_1 + d_2$ with $c_1 = d_1$.

$d = d_1 + d_2$ with $c_1 = d_1$ and intentionally choosing c_1 so that $|c_2| \ll |d_2|$. For example, consider the step function $d(t)$ of Fig. 8.20-2a. A realistic specification for $|c|_{max}$ is also shown, and $c_1(t) = d_1(t)$ is deliberately chosen in such a manner that the remaining $c_2 = c - c_1$ must (if it is to satisfy the specifications) be much smaller than $d_2 = d - d_1$. A satisfactory and rather obvious choice for $c_1 = d_1$ is shown in Fig. 8.20-2b. Clearly, $|c_2| \ll |d_2|$ if $|c| \leqslant |c|_{\max}$.

We may therefore write

$c_1 + c_2 + p_2{*}n(h_1{*}c) = d_1 + d_2$, $c_2 + p_2{*}n(h_1{*}c) = d_2$, and since $|c_2| \ll |d_2|$, it follows that

$$p_2{*}n(x) = p_2{*}n(h_1{*}c) \doteq d_2, \text{ or } P_2Y \doteq D_2 \tag{8.20,1}$$

Let us for the moment assume that p_2 and the nonlinear relation $y = n(x)$ are precisely known. Then $y = \mathscr{L}^{-1}Y \doteq \mathscr{L}^{-1}D_2/P_2$ is known and x must be found in order to deduce $N = Y/X$. It is usually much easier to solve a nonlinear equation backwards, i.e., find the input required for a known output. In any case, numerical methods can always be used, and thereby $x = -h_1{*}c$ is known. All that remains to be done is to find an $H_1(s) = P_1H$ with as economical a gain and bandwidth as possible and which satisfies the specifications on $|c(t)|_{\max}$. This problem is discussed and solutions presented in Section 9.4. As indicated there, $H_1(s) \approx X(s)/C_x(s)$ provides a good initial trial solution, where $|c_x(t)| \approx |c|_{\max}$ and $c_x(t)$ is deliberately chosen so that its transform $C_x(s)$ is easily calculated.

Let us now consider the problem for the case when P_1, P_2, and the nonlinear relation vary, or there is some bounded ignorance of their parameter values. We may formulate the problem in the s domain by writing (for Fig. 8.20-1)

$$C = D/(1 + PH) = (DP_o/P)/[(P_o/P) + L_o] \tag{8.20,2}$$

where $P = P_1NP_2$, $P_o = P_{1o}N_oP_{2o}$, $L_o = P_oH$. It is again emphasized that $N(s)$ is defined by the relation $Y(s) = N(s)X(s)$ and is different for different $x(t)$. Equation (8.20,2) is the basic design equation for determining $L_o = P_oH$. It may be used in exactly the same manner as the analogous linear Eq. (6.12,1) (see Section 6.12). The requirements on $C(j\omega)$ may be obtained by the methods of Section 9.4. The important difference is in the work involved in finding the range of $P_o/P(j\omega)$. In Section 6.12, P_o/P is independent of $d(t)$, $c(t)$, etc. Here, P_o/P can be determined only after examining $c(t)$, because of the dependence of $N(s)$ on x, y. The method previously described (Fig. 8.20-2, etc.) is used to find the range of variation of $N_o/N(j\omega)$. Equation (8.20,1) is very convenient for this purpose, i.e., y or Y is available from Eq. (8.20,1), from which x is obtained and then $N = Y/X$ deduced.

In a specific design problem, in order to get a good first approximation to the gain and bandwidth demands on $H(s)$, it is easier to pick for $P_o = P_{1o}N_oP_{2o}$ that particular set of parameters for which $P = P_1NP_2$ has the smallest gain and bandwidth. $L_o = P_oH$ is then chosen so as to achieve a satisfactory $c(t)$. It is reasonable to assume that the specifications will be more than satisfied at the other plant parameter values at which $|P(j\omega)| > |P_o(j\omega)|$. (However, see Fig. 9.4-1 for a case where this is violated.) When there is a "typical" disturbance signal, the tuned response approach of Section 9.4 is definitely applicable, i.e., $C/D = 1/(1 + PH)$ is assigned zeros at the poles of $D(s)$. The cancellation of the poles of $D(s)$ is independent of $P = P_1NP_2$ because the cancellation is achieved by the poles of H. In such a case, the objection (of Fig. 9.4-1) to this approach of designing for the smallest gain vs. frequency $P_o(j\omega)$ is at least weakened.

If there is a variety of disturbance signals for which prescribed maximum tolerable outputs have been specified, then a certain amount of repetition may be necessary. It is worth trying to circumvent this by picking the worst combination of plant parameter, disturbance input and tolerable output values. It is easy to see which is the worst when one disturbance input is obviously much "faster" and larger than the others. The frequency responses of $d(t)$ may also be examined for this purpose. If there are not too many poles of the various $D(s)$, then $H(s)$ may be assigned poles at all of them, in order to effectively eliminate the forced components of all the disturbance signals.

In the above, it was assumed that in Fig. 8.20-1, $r(t)$ was zero. The general case is postponed for later discussion.

(2) Design Procedure for Achieving Specified Maximum Time Domain Error Response

It is assumed that the system is driven by a typical input signal $r(t)$, and the problem is to choose G in Fig. 8.20-3 so that the error $e(t)$ is kept within specified

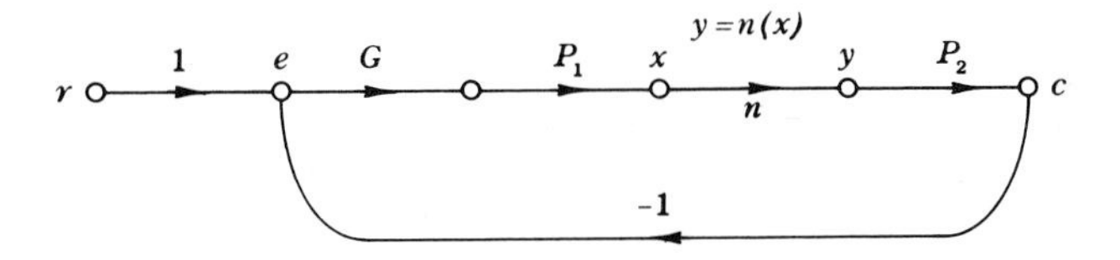

FIG. 8.20-3. Single-degree-of-freedom system with nonlinearity.

bounds. We write $E = R/(1 + L)$, $L = GP_1NP_2$ with the understanding that N is defined by the relation $Y = NX$ and differs for different y and x. The range of variation of N can be deduced only after the ranges of y and x are known. Knowledge of the latter is again facilitated by the design realities in a follow-up feedback control system, in which $|e(t)| \ll |r(t)|$, except possibly for certain time intervals which are easily identified. We write (by analogy with the first problem treated, and using similar notation) $e(t) = e_1 + e_2 = r_1 + r_2 - p_2 * n(g_1 * e)$, with $e_1 = r_1$, $G_1 = GP_1$, and $|e_2(t)| \ll |r_2(t)|$. This leads to $r_2 = p_2 * n(g_1 * e)$. N is thereby made available. If R represents a "typical" input, then G may be assigned poles at the poles of R in accordance with the "tuned response" technique of Section 9.4. The discussion and procedure for this problem are precisely the same as for the previous problem.

When there is substantial parameter variation in $P = P_1NP_2$ of Fig. 8.20-3, the demands on the loop transmission [in order that $e(t)$ may be small for the whole range of plant parameters] may be such as to lead to an unacceptably large system response bandwidth over some range of plant parameter values. In such a case, a two-degree-of-freedom structure must be used. The sensitivity requirements on $T_E = E/R = 1 - T$ are easily related to those of T, because $\Delta T_E = -\Delta T$. The design procedure for finding $L_o = P_1N_oP_2$ is the same as that described in Sections 6.7-6.9. The difference is in the larger amount of labor involved in finding the range of variation in $P = P_1NP_2$, because N is a function of the signal passing through the nonlinear portion of the plant. After L_o has been found, $G = T_o(1 + L_o)/P_o$ (of Section 6.1) is chosen to achieve a satisfactory T_o.

Disturbances While the Input Signal Is Being Processed

We finally consider the case when $r(t)$ and $d(t)$ in Fig. 8.20-1 exist simultaneously. A two-degree-of-freedom structure is of course essential in order to independently control $T = C/R$ and the disturbance response. The problem is more complicated than the previous, because the signals in the nonlinear element are dependent both on $r(t)$ and $d(t)$. Hence there is more work involved in calculating $N(s)$. Nevertheless, if the restriction on $N(s)$ is borne in mind, we may write for Fig. 8.20-1

$$C = \frac{GP_1NP_2R + D}{1 + HP_1NP_2} = C_R + C_D.$$

We may separately consider C_R and C_D remembering, however, that in each case the c_r, c_d combination must be considered in order to find N. The procedure as described in Section 6.12 is used to determine the nominal loop transmission and finally $G(s)$. The novelty here is to deduce the variations in $N(s)$ due to the combined effect of $r(t)$ and $d(t)$.

In order to find the range of variations of $N(s)$, it is necessary to consider the different possible combinations of $r(t)$ and $c(t)$ and solve a nonlinear integral equation. In this case, as in the previous one, the solution is facilitated by the engineering realities of a feedback control design. In Fig. 8.20-1, $c = d + p_2*y$, $y = n(x)$, and $X = P_1(GR - HC)$, which we write as $X = AR - BC$, i.e., $A = P_1G, B = P_1H$. Hence $c = d + p_2*n(a*r - b*c)$. Consider $e = r - c$, and let $r = r_1 + r_2$, $d = d_1 + d_2$, and $e = (e_{r1} + e_{r2}) - (e_{d1} + e_{d2})$, where (as in Fig. 8.20-2, etc.) the division is purposely made so that $r_1 = e_{r1}$, $d_1 = e_{d1}$, and $|e_{r2}| \ll |r_2|$, $|e_{d2}| \ll |d_2|$. The latter inequalities constitute the engineering realities previously mentioned. Their realization despite parameter variation is the reason for using feedback. The equation $e = r - c$ becomes $e_{r2} - e_{d2} = r_2 - d_2 - p_2*n(a*r - b*c)$. The inequalities permit the equation to be approximated by $r_2 - d_2 \doteq p_2*n(a*r - b*c)$, which is, of course, $r_2 - d_2 \doteq p_2*y$ or

$R_2 - D_2 \doteq P_2 Y$. (There is, of course, the unlikely possibility that $| r_2 - d_2 | \sim | e_{r2} - e_{d2} |$ even though $| e_{r2} | \ll | r_2 |$, $| e_{d2} | \ll | d_2 |$. In such a case, $p_2 * y$ may be very small and a loop transmission with small gain and bandwidth suffices for this specific case. However, the loop transmission must also handle the more difficult case when $| r_2 - d_2 |$ is large, so we may ignore the case of small $| r_2 - d_2 |$.) Since r_2, d_2, and p_2 are known, the equation $r_2 - d_2 \doteq p_2 * y$ may be solved for y. The nonlinear equation $y = n(x)$ is solved backwards for x, from which $N = Y/X$ is available. The process is repeated for the different possible combinations of r and d and over the range of variation of the nonlinear parameters and of P_2, etc. Of course, it is a good idea to endeavor to economize on the design labor by seeking the extreme combinations of $r(t)$, $d(t)$, and the plant parameter values.

Summary

The nonlinear relation $y = n(x)$ may be replaced by the transform equation $Y = NX$, providing a different value of $N(s)$ is used for different x, y combinations. In this way the nonlinearity is replaced by a transfer function which is different for different signals passing through it. Hence our design techniques for satisfying specifications on system response, disturbance attenuation, etc., in the face of parameter variations, may be used without any change. An important feature of this approach is that the problem of achieving realistic system specifications is directly attacked. The specifications are achieved with compensating networks whose gain and bandwidth are as economical as they possibly can be.

A vital part of the problem is to find the range of $N(s)$ due to the known range of input signals, disturbance signals, and plant parameter values. To do this exactly would require the solution of a nonlinear integral equation of the form $e = r - p*n(g_1*e)$ and would be extremely difficult, making the entire approach impractical. The problem is tremendously simplified by the "engineering realities" which demanded the use of feedback in the first place. The result is that only the nonlinear relation $y = n(x)$ need by solved. Furthermore, it must be solved backwards by finding the input required to achieve a known output. This is usually much easier to do than to find y for a given x.

Nevertheless, a great deal of work may be necessary to cover the range of input signals, disturbance signals, plant parameter values, and to go back and forth between the time domain and transform domain. The amount of labor can be drastically reduced if the extreme combinations of these factors can be deduced. It is conjectured that one extreme consists of that combination consisting of (a) nonlinear and other plant parameters with the smallest gain and slowest response, accompanied by (b) the largest and fastest changing input and disturbance signals which must be faithfully reproduced and drastically attenuated, respectively. Sometimes artificial composite signals may be used for this purpose. It is very reasonable to expect even better system performance for the slower and smaller signals at plant parameter values corresponding to larger plant gain and faster plant response.

It is also evident that the methods described in Sections 8.19 and 8.20 are effective only if the plant is capable of achieving the specified desired plant output. For example, if the plant saturates, then the breakdown in the design procedure occurs in the fact that X_{nl}/X and P_o/P [of Fig. 8.19-2 and Eq. (8.20,2)] are infinite (if the saturation is ignored in formulating the output). It is also difficult to use this procedure even if the saturation is taken into account. For the designer no longer knows in advance what constitutes a reasonable output, and the "engineering realities" are no longer valid. A necessary condition for the practicality of the procedures described in Sections 8.19 and 8.20 is that the equivalent plant transfer function is always nonzero, and this condition excludes such nonlinearities as saturation and dead zone.

§ 8.21 Fundamental Limitations in Adaptive Capabilities of Linear Time-Invariant Feedback Systems

Our work to date has been devoted to feedback systems wherein all the data processing elements around the plant are linear and with impulse responses which are independent of the time at which the impulse is applied (time invariant linear elements). We have been engaged in deriving the benefits and costs of feedback available from such systems, under various plant constraints. There are, however, inherent limitations in the ultimate adaptive capabilities of such feedback systems. Some of these limitations are given below.

(1) Parameter Variations Which Severely Limit Loop Transmission

We have already seen (Chapter 7) the fundamental limitations in the realizable loop transmission around nonminimum phase plants. There are also parameter variations which severely limit the realizable loop transmissions. For example, consider an element whose gain factor k may be negative or positive, say from 0.5 to 1, and from -0.5 to -1. The loop transmission around this element must have the form $L = kM(s)$, and the stability of the system (recall Theorem of Section 2.8) is determined by the zeros of $1 + kM(s)$. These are the zeros of $(k_o/k) + L_o(s)$, where $L_o(s) = k_o M$ is the nominal loop transmission around the element k. Following the construction of Fig. 4.12-1, the loci of k_o/k consist of AB, CD in Fig. 8.21-1 (taking $k_o = 0.5$). It is easily seen that L_{o1}, L_{o2} are unsatisfactory choices for the nominal loop transmission because they result in instability, while L_{o3} leads to a stable system. However, the benefits of feedback that L_{o3} can provide are negligible. The techniques of Chapter 7 (e.g., see Section 7.7) may be used to prove that $-L_{o4}$ is impossible. The conclusion is that the loop transmission (and hence the benefits of feedback) around an element whose gain factor changes sign is severely limited. More generally, it is readily seen that this is true for any plant whose $P_o/P(s)$ at infinite frequency has portions of its loci on both the positive and negative real axis.

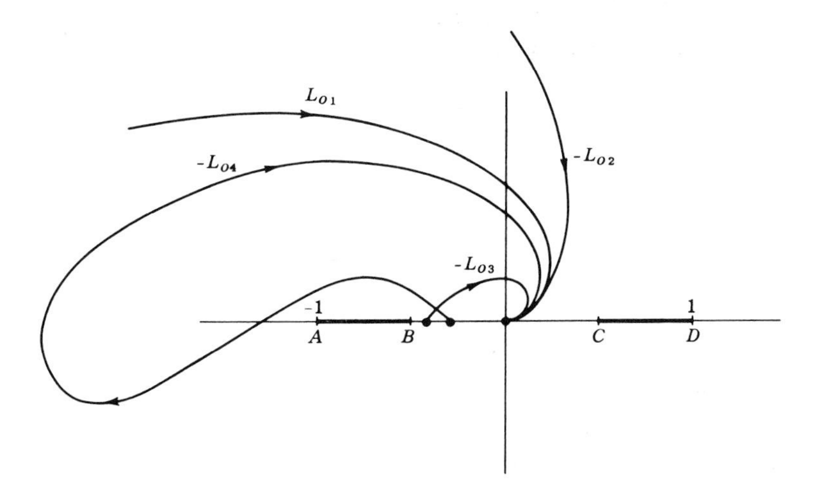

FIG. 8.21-1. Range of k_o/k which severely limits loop transmission function.

(2) Insufficient Flexibility for Optimum Control

The loop shaping techniques that have been described are very well suited for securing small sensitivity to parameter variations and for attenuating plant disturbances. However, in systems in which the plant itself or the plant operation is quite expensive, it is very desirable to secure optimum plant operation. If there is good knowledge of the plant operation, then the optimum settings of the various plant inputs, the biases, etc., as functions of the process as it proceeds in time, can be predicted and the adjustments implemented with little need for feedback. But if such knowledge is lacking, then the outputs must be monitored and the data used in appropriate (and usually nonlinear) equations, in order to determine the optimum control functions. Sections 8.11-8.14 have described procedures whereby the plant operation can to some extent be made to change in a desired manner as plant parameters vary. These procedures could (and ought to) be extended to multiloop systems for more detailed control over the plant operation as a function of its parameter values. However, it is obvious that time-invariant feedback of this type is inherently limited in its abilities to achieve optimum plant operation under a variety of conditions, and that more sophisticated nonlinear feedback systems are required for this purpose. For example, optimum operation often requires the plant to be operating at its maximum (saturated) levels for *any* input. The system is then nonlinear and linear design theory is inadequate.

(3) Input Signal Adaptation

As noted in Chapter 9, the optimum $T(s)$ is very much influenced by the nature of the useful message input signals and the false noise input signals. If either the noise or message statistics change, then the optimum $T(s)$ should also be changed, i.e., it is desirable for the system to adapt itself to variations in the nature of the input signals. The feedback theory and techniques that have been described appear to be inadequate for this purpose.

(4) Inflexible Relationship between Sensitivity over System Response Bandwidth and Sensitivity to Rate of Parameter Variation

Suppose $T(s)$ has a bandwidth of 20 rps and small sensitivity is desired over this bandwidth. It follows from the design procedure of Sections 6.7-6.9, and assuming even moderate parameter variations, that the crossover frequency ω_c of $L(j\omega)$ will be beyond 20 rps. The rate of parameter variations was not noted and often it is assumed that they must be slow in comparison with the system response time. However, from Sections 8.15 and 8.16, it is known that parameter variations can be represented as equivalent disturbances, and it will be seen in Section 9.6 that a loop transmission with $\omega_c = 20$ can cope with disturbances whose time constants are in the order of 1/20 second. Hence our system with small sensitivity over some bandwidth automatically can cope with parameter variations whose equivalent disturbance representation has the same order of bandwidth. But suppose that in practice the parameter variations are very slow. It therefore seems that the design is wasteful in its ability to cope with faster parameter variations than actually occur. It would be extremely desirable to be able to exchange this unrequired benefit of feedback for something else, specifically for reduced system sensitivity to feedback transducer noise.

As an extreme case of the above, suppose that the system considered in Chapter 6 (Fig. 6.3-1) has exceedingly slow parameter variations, such that a year may elapse before the plant poles move from $\pm j2$ to $-6 \pm j10$. The final design is very sensitive to high frequency feedback transducer noise (see Fig. 6.8-5), but it seems ridiculous that it should be so, in view of the extremely slow parameter variations. Common sense tells us the feedback data may be evaluated much more slowly than $1/\omega_c$, such that the high frequency noise has negligible effect. However, Chapter 6 definitely and irrevocably establishes that slower evaluation by means of linear time-invariant networks cannot secure the desired insensitivity. Some other kind of feedback data-processing is therefore required.

In this same connection, it is noted that if a sufficiently large pure time delay were present in the plant, it would be impossible (see Section 7.12) to achieve the desired crossover frequency ω_c. We are then unable to secure for $T(s)$ the desired insensitivity to the very slow parameter variations. But it is obvious that when the parameter variations are very much slower than the time delay, then it should certainly be possible somehow to have $T(s)$ very insensitive to these parameter variations. In short, our work has indicated that feedback transducer noise, nonminimum phase characteristics, and plant parasitics (which tend to put a limit on ω_c—see Section 6.14) impose a limit on the attainable benefits of feedback. This is certainly true if the feedback data processing is constrained to consist of linear time-invariant elements. However, it is intuitively obvious that the situation can be radically altered for *slow* parameter variations, if the feedback data is processed in a nonlinear manner.

The reason for this inflexibility of linear time-invariant feedback is seen from the discussion in Section 8.16. Suppose that at the end of a year's operation, the plant function has changed from its initial value of P_n by the amount $\Delta P = P - P_n$, and that a step function is then applied (letting $t = 0$ at this instant) at the system input. The difference in the system output (as compared to the output $c_n(t)$ of a year ago) is precisely the same as if the plant function instantaneously changed at $t = 0$ from P_n to P. The equivalent "disturbance" is exactly $C_n \Delta P/P_n$, no matter whether the plant abruptly changed or whether it takes a year to make the change.

§ 8.22 The Role of Plant Identification in Adaptive Systems[1]

We have repeatedly emphasized that feedback is a means of achieving objectives despite ignorance. It is important to note that in the feedback theory and techniques that have been presented, there is no attempt made to remove the ignorance. Nowhere in the system are there any variables which represent the true plant transfer function. The philosophy has been one of coping with the ignorance, not of removing it; of directly examining the objectives and seeking to achieve them despite ignorance. Clearly this is not the only way to do the job. A more obvious (and, it may be argued, more primitive) way is to first remove the ignorance, by means of plant identification. In this approach the plant transfer function is measured by means of automatized instrumentation and the resulting information is used to appropriately modify the input to the plant. We shall see that ideally plant identification and the accompanying system modification does not necessarily involve feedback, and ideally, therefore, there is no stability problem. At first this appears to be a significant advantage of the identification-modification (or, as it is sometimes called, the "active adaptive") approach. However, when the theoretical difficulties and practical complexities of the identification problem are considered, it is then seen how elegant and simple is the feedback method.

Nevertheless, there are problems, as outlined in Section 8.21, which cannot

[1] For references see I. M. Horowitz, Plant adaptive vs. ordinary feedback systems, *Trans. IRE* **AC-7** (1), 48-56 (1962).

be satisfactorily handled by linear time invariant processing of the feedback data. One alternative is to turn to nonlinear feedback and to explore the fundamental capabilities of nonlinear feedback. Another is to directly remove the ignorance by automatized plant measurement. One possible configuration is shown in Fig. 8.22-1. The special input for plant identification $f(t)$ is chosen so that its

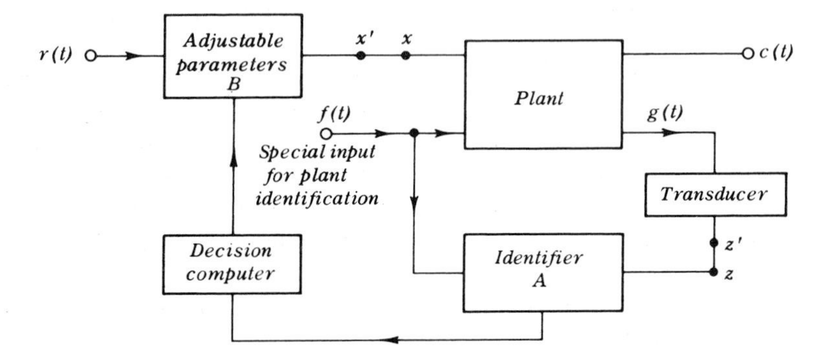

FIG. 8.22-1. A plant identification structure for coping with plant ignorance by removing it.

effect on $c(t)$ is as unobjectionable as possible from the point of view of the desired output. (The ideal measuring instrument has no effect on what it measures and while this cannot be exactly achieved, it can at least theoretically be closely realized.) Identifier A operates on $f(t)$ and $g(t)$ to achieve the desired identification. If the plant parameters change very slowly, then A can take its time, and be quite insensitive to noise in the $gz'z$ branch. The resulting identification data can be used to modify parameters in B, by adjusting potentiometers, biases, etc.

The system works by removing the ignorance, and it may be argued that in such a system feedback plays a minor role. Suppose the system is opened at xx' and a signal is injected into x. Will a returned signal appear at x'? If $r(t) = 0$ and there is no energy stored in B, then no signal will appear at x', even though the plant parameters are meanwhile changing. When $r(t)$ exists, or there is energy stored in B, then there is true feedback only if x' is different from what it would have been for zero value of x. The Decision Computer output is determined by the Identifier output. If the plant is linear, then its parameter values are not affected by the input at x; the output of A is then independent of x, and there is no feedback. If the plant is nonlinear, then its parameter values may be changed by x; the identifier output is changed, and with it, the settings in B and the value of x'; there is then feedback. However, it is clear that this feedback is not the heart of the adaptive process, as it is in the systems of Chapter 6, etc. Furthermore, the special plant identification input might be coded and the identifier matched to this code such that the component of g due to the non-coded input at x is largely ignored by the identifier.

Suppose the system is opened at zz' and a signal injected at z. There certainly is feedback if $r(t) \neq 0$, or if there is energy stored in B. The system is obviously very sensitive to noise in the gzz' branch. The identifier must operate slowly on the data in order to reduce such sensitivity, or, as previously noted, if $f(t)$ is coded, then the identifier may be made very insensitive to noise in the $gz'z$ branch. Accordingly, the signal returned at z' due to an input at z ought to be negligible. In any case, as previously noted, this closed loop of signal flow is not the essential adaptive ingredient in this system. The essential ingredient is the direct open-loop removal of ignorance by automatized measurement of the plant.

How may one choose between ordinary feedback and the method of Fig. 8.22-1? If there are m independently varying plant parameters with known ranges and maximum rates of parameter variation, then the techniques that have been presented may be used to find the required loop transmissions for ordinary feedback. The design may be unattainable because of the plant parasitics or because the plant is nonminimum phase, or because of feedback transducer noise. The plant identification method may then be attempted. Ideally, a definitive theory of identification would make available the optimum $f(t)$ code and identification technique for discriminating against gzz' noise and satisfying the specifications, for the bounds and maximum rate of variation of plant parameter values. It would then be possible to divide up the adaptive burden in an optimum manner between the identification method and the ordinary feedback method. A great deal of effort is currently being made towards establishing such a theory of identification.

While final conclusions are as yet unwarranted, the following speculations appear to be reasonable. There are parameter variations which can be identified very quickly, as, for example, the change in the magnitude or sign of the gain factor of a transfer function. If the plant poles and zeros are known to be fixed, then it is possible to identify the variations and adapt the system to them very quickly. However, consider a plant whose impulse response is mostly over after t_0 seconds (this is approximately equal to the time delay plus half the rise time, if the overshoot is small, see Fig. 5.10-1b). If the identification process does not include extrapolation into the future, then the identification process would require at least t_0 seconds. This is probably too low a figure, for it is based on the special input $f(t)$ of Fig. 8.22-1 consisting of impulses into a quiescent plant. A coded input into a non quiescent plant would probably require considerably more time. We shall assume that the computer operations for deducing the required modification and the implementation of these modifications take zero time (although they may require fair complexity in equipment). Hence a non-extrapolating identifier in a noise-free system appears to be inherently unable to cope with parameters changing faster than t_0 seconds. The minimum-phase ordinary feedback system is not so limited. Therefore the ordinary feedback system compares favorably with the nonextrapolating active adaptive system, with respect to fast parameter changes.

Extrapolation in the identifier requires differentiation of the data and makes the system exceedingly sensitive to noise. A noise free system with no pure time delays could theoretically be identified, by means of extrapolation, in infinitesimal time. A noise-free minimum-phase ordinary feedback system can also cope with very fast parameter variations, because its crossover frequency is unlimited. But if the plant has a pure time delay t_d, then the loop transmission crossover frequency cannot be more than $\omega_c = \pi/2t_d$, if the slope is -6 db per octave for several octaves on either side of crossover. The dominant roots of $1 + L = 0$ are then approximately at $2\omega_c$ [recall Eq. (5.11-1)]. If their dampling factor ζ is 0.7, then the time constant of the feedback loop is $1/1.4\omega_c \approx 0.5t_d$. A noise-free plant with pure time delay t_d requires t_d seconds to be identified, even with extrapolation, so the active adaptive system is more sensitive to noise than ordinary feedback in coping with fast parameter variations. It may also be noted that there is no alternative to feedback around an unstable plant.

The identification-modification approach does, however, have superior capabilities when the parameter variations are slow. The fundamental limitation in ordinary feedback systems was discussed under item (4) in Section 8.21. It was there noted that small sensitivity of a function over a bandwidth ω_o required the system to be able to cope with rates of parameter variation in the order of ω_o. A nonminimum-phase system might easily not have the latter capability because of the limitation on its crossover frequency. Also, the design might be impractical if the system had much feedback transducer noise in the frequency range where $| L(j\omega) | < 1$ and $| P(j\omega) | < | L(j\omega) |$. There is no such inflexible relationship in the identification-modification method. In fact, the latter appears to be ideally suited for slow parameter variations. The identification can proceed very leisurely and be extremely insensitive to noise (but not to slow drift), and result in a system response function (which can have a large bandwidth) which is very insensitive to slow parameter variations. On the other hand, the active adaptive approach appears to be inferior in its ability to cope with external disturbances acting on the plant. Furthermore, such disturbances interfere with the identification process and increase the need for coding $f(t)$ in Fig. 8.22-1.

CHAPTER NINE

Problems in the Specification of System Functions

§ 9.1 Introduction

Chapters 5-8 have been devoted to procedures for attaining given design specifications by means of structures with various constraints. The specifications were always assumed to be available in a form suitable for the specific design technique. For example, the system filter properties, as denoted by the system response function $T(s) = C(s)/R(s)$, was presumed to be available analytically, or at least its bandwidth and amplitude peaking were known. It was concluded in Section 6.11 that design for disturbance attenuation is conveniently done in terms of frequency response; such specifications were thereafter stated in terms of frequency response.

Unfortunately, the above is rarely the case in practice. There are, in fact, at least two stages which precede the detailed design procedures of the previous chapters. The first stage consists of deciding what is wanted. This may seem obvious, but it is surprising how often designs of complex systems are initiated without a clear enunciation of what the system is supposed to achieve. We are not concerned with this first stage, except to re-emphasize that if it is a feedback system which is being designed, then the desired achievements of the system can be divided into two classes. One class consists of the system filter and control properties, namely the nature of its response to the input command signals. This class of problems exists whether feedback is or is not used around the plant. The other is the class of system feedback properties. The latter design achievements may be stated in terms of maximum tolerable effects of slow and fast parameter variation, the linearization of nonlinearities, the precision of the output despite estimated bounds of ignorance, the nature of the disturbances, their maximum tolerable effect, etc.

The second stage consists of translating the above desiderata into a language suitable for detailed design. This stage should ideally end with $T(s)$ in analytic or graphical form, with the desired disturbance attenuation as a function of frequency, etc. The techniques of Chapters 5-8 may then be applied. Compromise between conflicting desiderata may be made in this stage.

This chapter is devoted to problems belonging in the second stage of the system design. There are many factors which enter into choosing $T(s)$ and the sensitivity functions. Some of the principal factors are: the plant capabilities, its safe working limits and saturation levels, the nature of the command signals and of the noise contaminating the latter, and the characteristics of the disturbances entering the plant. These problems are treated individually in the earlier sections of the chapter. Various combinations of these problems are treated in the later sections.

§ 9.2 The Physical Capacity of the Plant

In any type of design, it is unreasonable to demand performance which the plant is physically incapable of achieving. If the plant saturates, the linear design theory is not valid. How may the plant saturation levels be used to determine the system transfer function $T(s)$? If the system output must pass through the plant, then the signal levels in the plant are completely determined by the system input signals and the desired system performance. For example, let us compare two designs, $T_1(s) = 1/(s^2 + s + 1)$ and $T_2(s) = 4/(s^2 + 2s + 4)$, for the same plant, $P(s) = 1/s(s + 1)$. The time responses of T_1 and T_2 have precisely the same form, but the latter is exactly twice as fast as the former. No matter what the system structure is (providing there is no plant modification), the input to the plant is $X(s) = TR/P$. Let $R = 1/s$. Then $X_1 = (s + 1)/(s^2 + s + 1)$, $X_2 = 4(s + 1)/(s^2 + 2s + 4)$, and the corresponding $x_1(t)$, $x_2(t)$ are displayed in Fig. 9.2-1. The T_2 design is more liable to cause plant saturation.

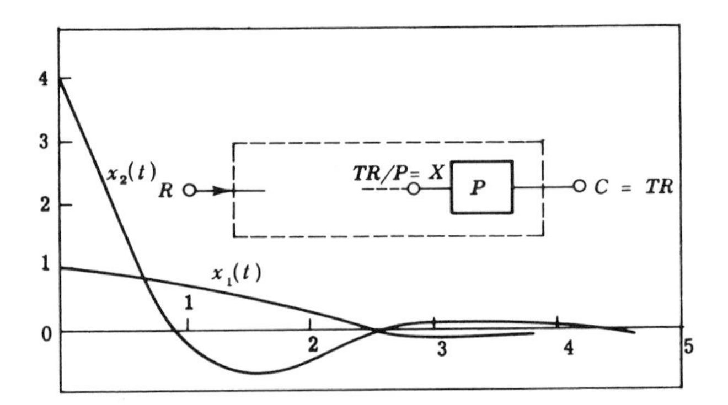

FIG. 9.2-1. Comparison of inputs to plant; $T_1 = (s^2 + s + 1)^{-1}$, $T_2 = 4(s^2 + 2s + 4)^{-1}$.

It is a fairly straightforward matter to check whether any contemplated $T(s)$ will lead to plant saturation. The extreme inputs are considered, i.e., those with combinations of largest magnitudes, derivatives, etc. The output is $C(s) = T(s)\,R(s)$, so it is simply a matter of determining whether the plant is capable of supplying such an output. For example, in a positioning system, the output follows the input. Let the plant consist of a power amplifier with constant gain, whose maximum output is 5 units, in cascade with a motor whose transfer function is $1/s(s+1)$. The motor velocity saturates at 10 units per second, and its acceleration saturates at 50 units per sec^2. The motor velocity and acceleration values are obtained by successively differentiating $c(t)$. It is easily ascertained whether they exceed the plant saturation levels. As for the power amplifier, its required output is $x(t) = \mathscr{L}^{-1}C/P$. If $C(s)$ is available analytically, $\mathscr{L}^{-1}X(s)$ can be obtained formally. If $c(t)$ only is available, and is of such a form that its transform is not readily deduced, then $x(t)$ may be obtained by methods presented in Chapter 12. When $x(t)$ is known, it is possible to check whether the saturation level of the power amplifier has been exceeded.

If the above analysis reveals that one or more saturation levels are exceeded, then there are a number of alternatives. (1) The plant can perhaps be modified

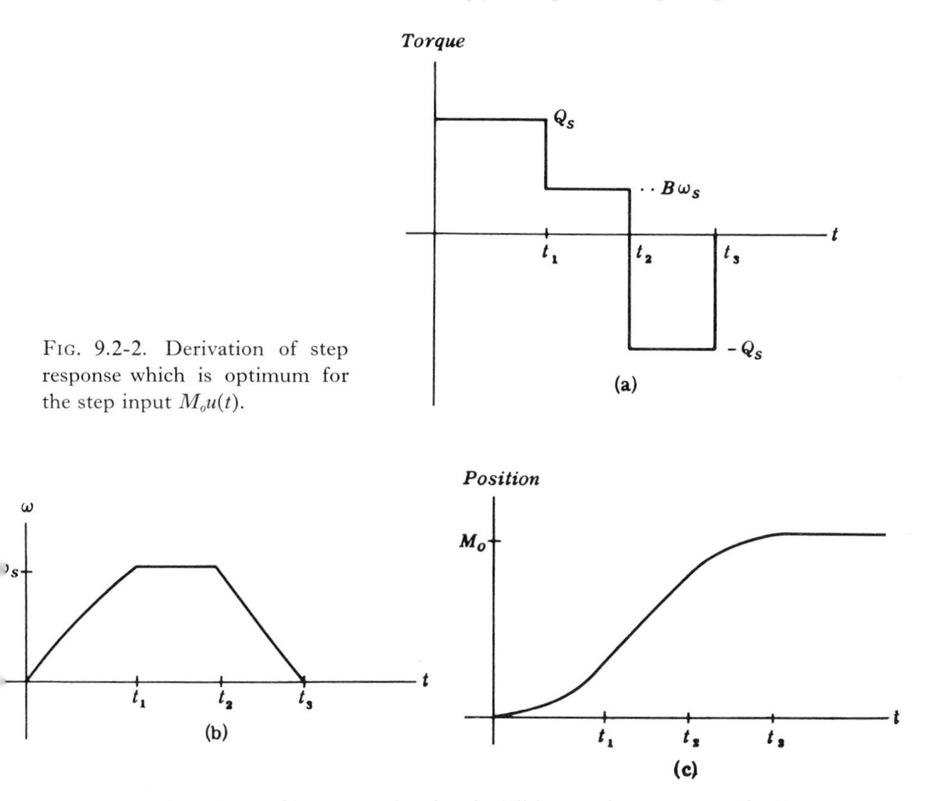

FIG. 9.2-2. Derivation of step response which is optimum for the step input $M_0u(t)$.

to increase the values of its saturation levels. This may be a matter of adjustment of settings (e.g., bias in amplifiers), rather than that of plant replacement. (2) $T(s)$ can be modified so that even with the most extreme inputs the saturation levels are not reached. (3) It is accepted that for some inputs the system opera-

tion will be partly nonlinear. The nonlinear modes are examined. (4) A completely nonlinear system design is attempted.
The last two alternatives are outside the scope of this book.

The following is a simple example of choosing $T(s)$ so that saturation levels are not exceeded. We consider a position servo where the typical inputs are step functions. It is desired that the output reach its final value as soon as possible and remain there. The plant has two saturation levels—a maximum torque Q_s, and a maximum angular velocity ω_s. The best response for a step function input of magnitude M_o consists of the following pieceswise linear segments: *Interval No. 1*: $0 < t < t_1$ in Fig. 9.2-2, in which maximum torque is applied until the velocity reaches its maximum value ω_s, at which point the torque drops to the value needed to maintain the velocity ω_s. If the plant has the transfer function $P = 1/(Js^2 + Bs)$, then in interval No. 1, $\Omega = \mathscr{L}\omega(t) = sC = Q_s/Js(s + B/J)$, and $\omega = \omega_x(1 - e^{-t/\tau})$, $\tau = J/B$, $\omega_x = Q_s/B$. The above applies until t_1, when $\omega = \omega_s$, i.e., $(Q_s/B)(1 - e^{-t_1/\tau}) = \omega_s$. *Interval No. 2*: $t_1 < t < t_2$ in which the torque is $B\omega_s$. Finally, at the appropriate instant, t_2 (determined later), the maximum reverse torque $-Q_s$ is applied to deccelerate the motor. When the velocity is zero (at t_3), the torque is shut off, and the motor stops—the final steady state value has been reached. From t_2 to t_3, letting $t' = t - t_2$, $\omega = \omega_s e^{-t'/\tau} - (Q_s/B)(1 - e^{-t'/\tau})$. The value of t_3 is determined by the fact that $\omega = 0$ at t_3, viz., $0 = \omega_s e^{-t_3'/\tau} - (Q_s/B)(1 - e^{-t_3'/\tau})$ with $t_3' = t_3 - t_2$. For $t > t_3$, the torque is zero. The value of t_2 is determined by the fact that $\int_0^{t_3} \omega dt = M_o$.

Therefore the over-all system transfer function is the sum of the transforms of the segments of Fig. 9.2-2c, divided by s. Exact realization of $T(s)$ requires pure time delay elements and is usually impractical. There are two alternatives. One (called time domain synthesis) is to approximate the time function of Fig. 9.2-2c by another whose transform is available as a rational function in s. The approximation errors are then available in the time domain. The other is to approximate the transcendental $T(s)$ function by means of a rational function in s. The disadvantage of the latter approach is that the approximation errors in the time domain are not directly available. Techniques for time domain synthesis are presented in Chapter 12.

It is noteworthy that the above $T(s)$ is optimum only for a step input of magnitude M_o. If the step input has smaller magnitude, the final value is not reached until t_3 (in Fig. 9.2-2c), even though the plant has the ability to achieve it in less time. In general, a linear time-invariant system response function can be optimum (in the sense of working the plant at its maximum capacity) only for a specific input signal.

§ 9.3 Limitations of Design Based on Step or Ramp Response

An important consideration in choosing the system transfer function $T(s)$ is the typical signal which the system will have to handle, if indeed it is possible to select a specific time function as being representative of the class of input signals that the system must process. If the typical signal is one whose value changes slowly, then there is no need for a fast responding $T(s)$. It is desirable to choose $T(s)$ with as small a bandwidth as possible (time response as slow as possible) and which still manages to satisfy the design specifications. The system is then less sensitive to high frequency noise which enters with the useful input signals; also, the bandwidth over which the sensitivity to parameter variations must be controlled (loop gain more than 1) is thereby reduced. Elementary design theory is based upon very simple typical input signals, such as the step function and the ramp. The steady state (large t) system response is then given by the appropriate error constants, K_p and K_v. These were defined in Section 5.4.

A great deal of work has been done in the preparation of optimum response functions when the typical input is a step function. Different definitions of optimum lead to different optimum designs. Among those defined have been the minimization of $\int_0^\infty |e|\,dt$, $\int_0^\infty e^2dt$, $\int_0^\infty e^2t\,dt$, and $\int_0^\infty |e|\,t\,dt$, where $e(t)$ represents the error between the desired step response and the actual output. In all cases, a constraint on the bandwidth available (or one similar to this) must be imposed, or else the optimum $T(s)$ (for the case where the desired output is the input) is obviously $T(s) = 1$. Values of optimum $T(s)$ for assigned degrees of complexity of $T(s)$ have been calculated, and are available in the literature.[1] These results are useful, but only for those cases where the typical input is indeed a step function. If they are used when the typical input is not a step function, they may result in extremely inefficient designs. A vivid example illustrating this point is presented later in the section.

It is important to be aware of the limitations of designs based on step (or ramp, etc.) inputs. One limitation is related to the question whether the step, or ramp, is of sufficient time duration for the error constants to be effective. For example, it is true that for a unit ramp input, the steady state error is $1/K_v$, but what if the ramp exists for only a finite amount of time? How much time must the ramp exist before the error constant concept has any meaning? To be specific, suppose the step response of the system is that shown in Fig. 9.3-1a. The ramp response is the integral of the step response and is therefore, as a function of time, the shaded area marked A in the figure. Hence, instantaneous

[1] D. Graham and R. C. Lathrop, The synthesis of optimum transient responses: Criteria and standard forms, *Trans. AIEE Part II, Applications & Ind.*, **72**, 273-286 (1953); W. C. Schultz and V. C. Rideout, Control system performance measures: Past, present and future, *Trans. IRE* **AC-6** (1), 22-35 (1961). (This paper includes a large bibliography.)

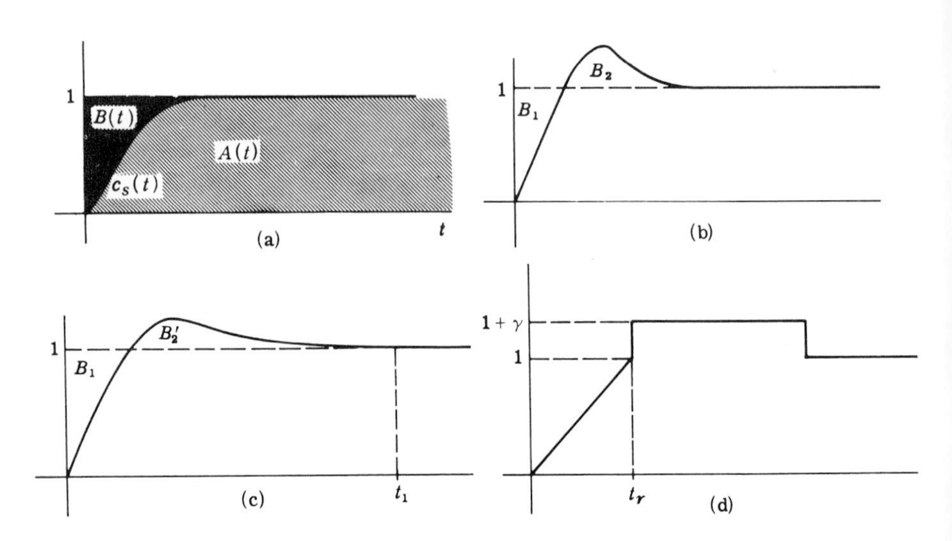

FIG. 9.3-1. (a) $B(t) = t - \int_0^t c_s(\tau)d\tau$ = instantaneous position error for ramp input. $K_v = [\int_0^\infty B(t)dt]^{-1}$. (b) Decrease of error by means of overshoot in step response, $K_v = [B_1 - B_2]^{-1}$. (c) Large settling time required to achieve large $K_v = [B_1 - B_2']^{-1}$ if rise time and peak overshoot are fixed. (d) The fastest response curve for fixed t_r, K_v, and peak overshoot.

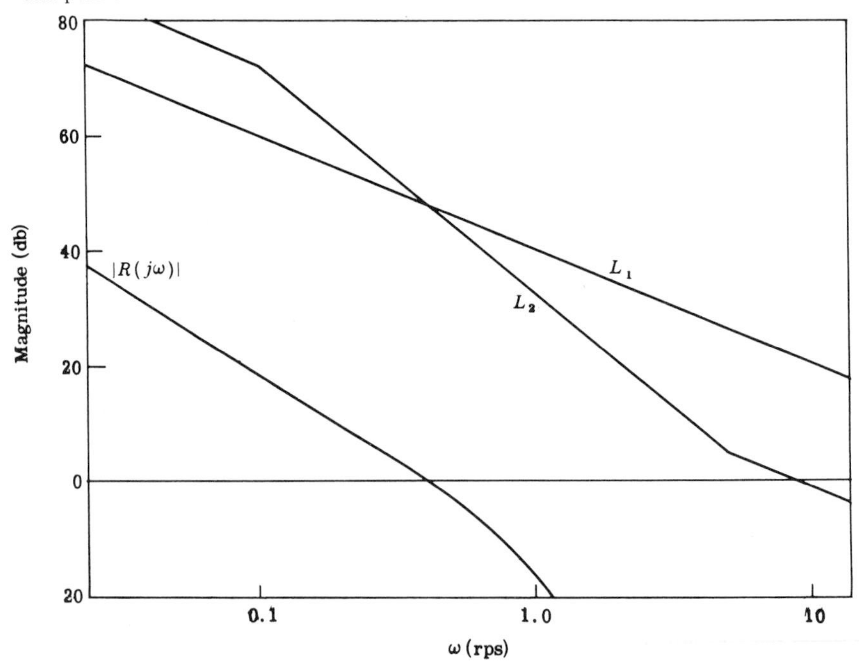

FIG. 9.3-2. Importance of tailoring $T(s)$ to typical input. (a) Two significantly different choices of $L(s)$.

position error *vs.* time must be the cross-hatched area B in Fig. 9.3-1a. How can the final steady state value of this error be decreased? One way is by decreasing the system rise time (increasing its bandwidth). However, if the bandwidth is fixed, the only recourse is to increase its overshoot, as in Fig. 9.3-1b, in which K_v is equal to [Area B_1 − Area $B_2]^{-1}$. If the maximum overshoot is limited, further increase of K_v (increase of B_2 to offset B_1) requires a long tail in the step response, as in Fig. 9.3-1c. If the overshoot is limited to $\gamma\%$, and the "rise time" to t_r, then the ideal (in terms of fastest settling time) step response is approximately that shown in Fig. 9.3-1d. In Fig. 9.3-1c, it takes approximately t_1 seconds before the error coefficient K_v is really effective. Therefore the ramp must exist for at least t_1 seconds if the velocity constant is to be related to the position error (for $t > t_1$) in the accepted manner. Thus the specification of K_v by itself is incomplete—the time interval in which the error is to reach its final value should be included. Better still, the maximum instantaneous error as a function of time ought to be specified. The above considerations also reveal the futility of searching for optimum pole-zero patterns to achieve large K_v with small bandwidth and small overshoot. The price must then invariably be paid in large settling time, and no manipulation of poles and zeros can help.

We have previously noted that a design based on the system step response can be quite poor if the actual input is not a step function. A better design procedure is to tailor the system response to the characteristics of the typical input signal. This is done as follows. Let the desired output be $C_d(s) = R$, and the actual output be $C = RT$, where R is the system input. The transform of the error is $E = R - C = R(1 - T) \triangleq RT_E$. It seems reasonable to make T_E small wherever R is large, and not to worry too much about the frequency range in which R is small. The ideal T_E is identically zero (with $T = 1$), but this would require an infinite bandwidth for T. In the classical one-degree-of-freedom configuration of Chapter 5 (Fig. 5.3-1), $C(s) = RL/(1 + L)$ and $E(s) = R/(1 + L)$. It is undesirable to maintain a large $|L|$ over a frequency range larger than is absolutely essential. At sufficiently high frequencies, $|L|$ approaches zero, and therefore E approaches R. The equation $E = R/(1 + L)$ suggests that L be made large in the frequency range where R is large, and allowed to become small in the range where R is small.

An impressive example of the advantage of tailoring $T_E = 1/(1 + L)$ to the frequency spectrum of R is given by Graham,[1] who considered a system where the typical input signal is $r(t) = \tan^{-1} 0.524t$ (the azimuth angle of a target over a constant linear-velocity cource). Its Fourier transform is $R(j\omega) = \pi e^{-1.9|\omega|}/j\omega$, whose magnitude is sketched in Fig. 9.3-2a. It is assumed that at the beginning of the run the output coincides with the input, and that the maximum azimuth rate is 30°/second. Two different choices of L are shown

[1] R. E. Graham, Linear servo theory, *Bell System Tech. J.* **25**, 616-651 (1946).

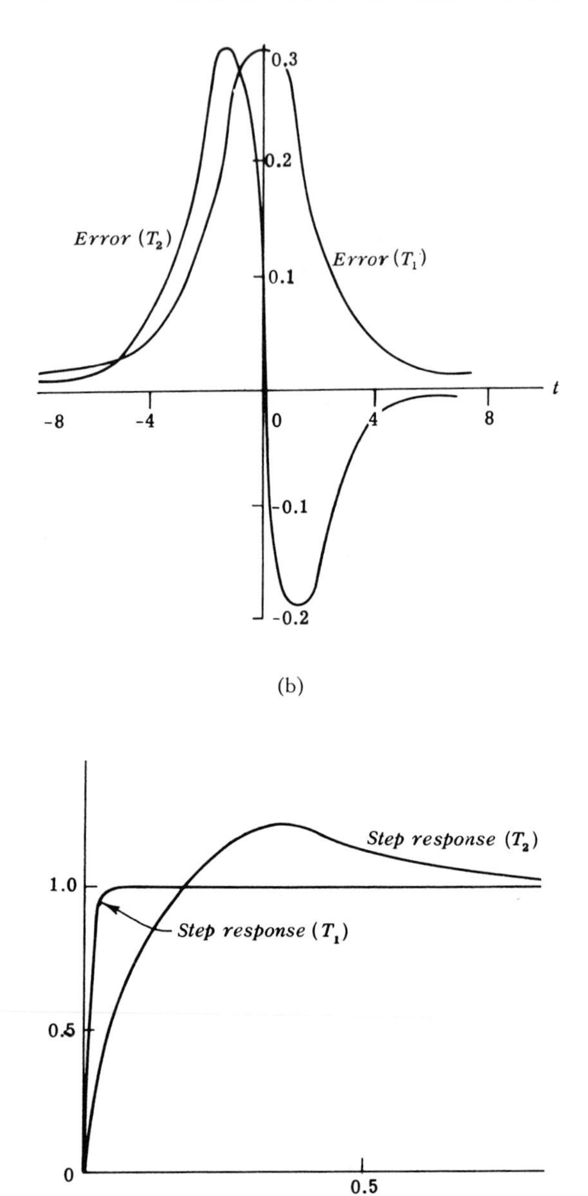

FIG. 9.3-2. Importance of tailoring $T(s)$ to typical input. (b) Peak position error is nevertheless same for both choices. (c) The radical difference in the two step responses.

in the same figure. The tracking errors for the two choices are shown in Fig. 9.3-2b (with the time axis shifted), and the step function responses of the two are sketched in Fig. 9.3-2c. Even though T_2 (corresponding to L_2) represents a system considerably more sluggish than T_1 (corresponding to L_1), there is not much difference in the magnitude of the instantaneous error, $e = r - c$. (In this example, the system transient response is negligible over most of the run, because the system time delay and rise time total a fraction of a second, which is very small in comparison with the duration of the signal.)

System T_1 is considerably more sensitive to noise than T_2, because the bandwidth of the former is approximately 10 times that of the latter. And yet despite the much snappier step response of T_1, T_2 is at least as good in following the signal. This is due to the fact that $E(s) = R/(1 + L)$. Although $|L_1| > |L_2|$ (Fig. 9.3-2a) in the higher frequency range, this is more than compensated by the fact that $|L_2| > |L_1|$ in the low frequency range, where R is large. This example serves as a warning that the system step response is not a good performance criterion when the typical input signal is not a step function.

§ 9.4 S-Plane Design Based on the Typical Input Signal

We have seen that a design based on the system step response may be quite inefficient when the input is not a step. This section describes a design procedure[1] which is based on the actual typical input to the system.

It is noted in Section 12.6 that any real input signal may be approximated as closely as desired by a finite sum of exponentials. Hence the transform of any practical input signal is a rational function in s. The system error response $E = RT_E$ has two parts. One part consists of the forced components, due to the poles of R. The other part consists of the natural components, due to the poles of T_E. The poles and zeros of R are known, while those of T_E are to be chosen. A realistic design procedure is to choose the zeros of T_E so as to cancel out the poles of R. For example, if R has a pole of multiplicity m at $s = s_i$, then T_E is assigned a factor $(s - s_i)^m$ in its numerator. In this way, the forced components of the error are all zero. Only the natural error components exist. The final step is to choose the poles of T_E, and if necessary additional zeros, so that in conjunction with the zeros of R, they provide an acceptable transient error. Basically, this approach is similar to that previously suggested by Graham. The difference is that Graham works with the frequency response of the input signal, while here the pole-zero representation of the signal is used. A possible advantage of the pole-zero method is that if R is not too complex, there is

[1] This procedure is partly based on J. G. Truxal and I. M. Horowitz, Sensitivity considerations in active network synthesis, *Proc. 2nd Midwest Symposium on Circuit Theory, Kellogg Center, Michigan State University, East Lansing, December*, 1956, pp. 6-1 to 6-11.

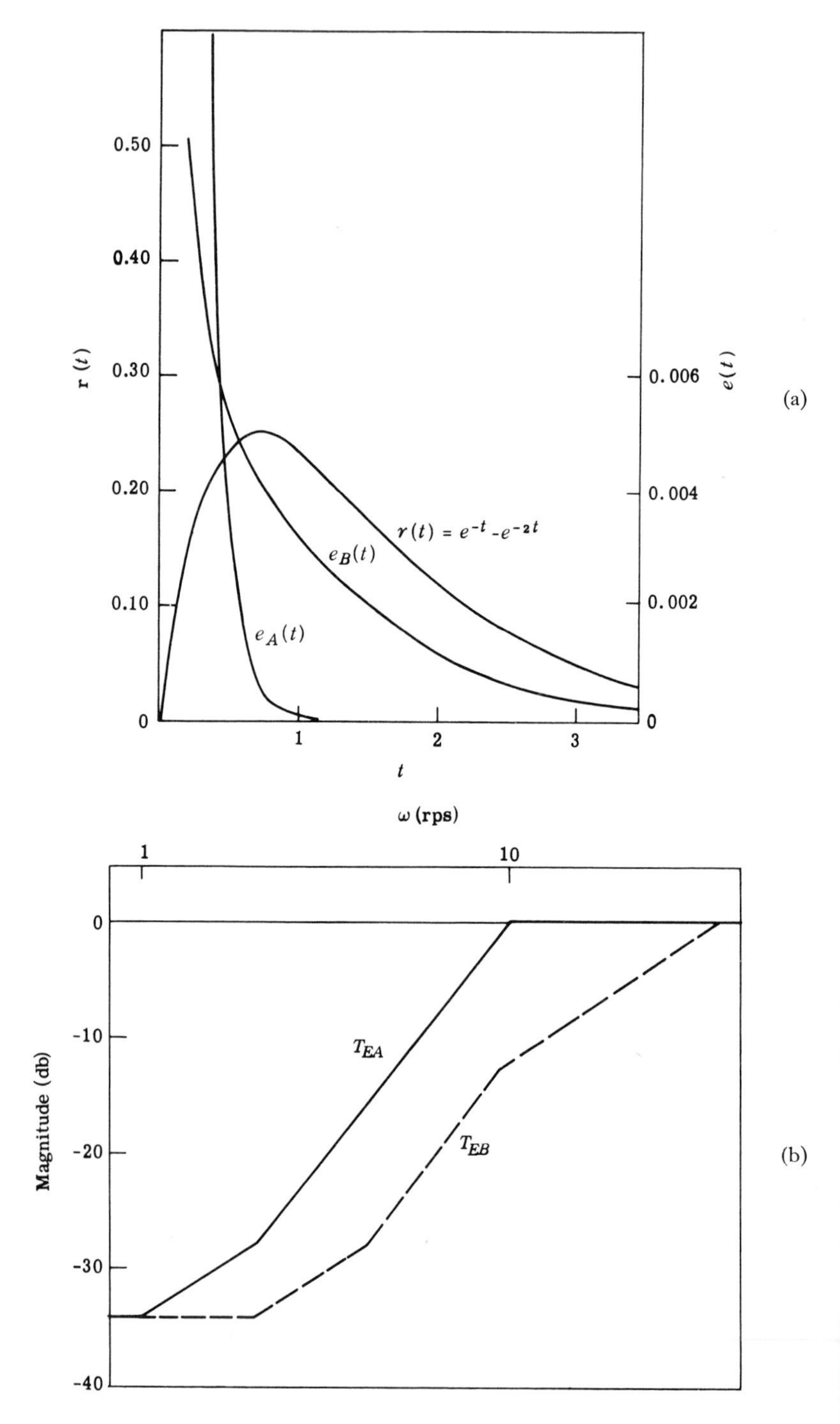

r (t)
0.50
0.40
0.30
0.20
0.10
0
e(t)
0.006
0.004
0.002
0
r(t) = e^-t - e^-2t
e_B(t)
e_A(t)
1
2
3
t
(a)
ω (rps)
1
10
Magnitude (db)
0
-10
-20
-30
-40
T_EA
T_EB
(b)

better control of the time domain error. An advantage of the frequency response representation is that the cost in system gain and bandwidth is more readily available.

To illustrate the value of the above approach, suppose $r(t) = e^{-t} - e^{-2t}$, which is sketched in Fig. 9.4-1. Then $R(s) = [(s + 1)(s + 2)]^{-1}$. In accordance with the above, we let $T_E(s)$ have zeros at the poles of $R(s)$. For illustrative purposes we arbitrarily assign T_E poles at -10. (We shall later present a rational procedure for choosing the poles.) If $T_E = 1 - T$, then T_E is 1 at infinite s. Hence this first choice of T_E is $T_{EA} = (s + 1)(s + 2)/(s + 10)^2$, leading to $E_A = RT_{EA} = (s + 10)^{-2}$, and $e_A(t) = te^{-10t}$, which is sketched in Fig. 9.4-1a. The above design is compared with $T_{EB} = (s + 2)(s + 4)/(s + 10)(s + 40)$. The frequency responses of T_{EA} and T_{EB} are shown in

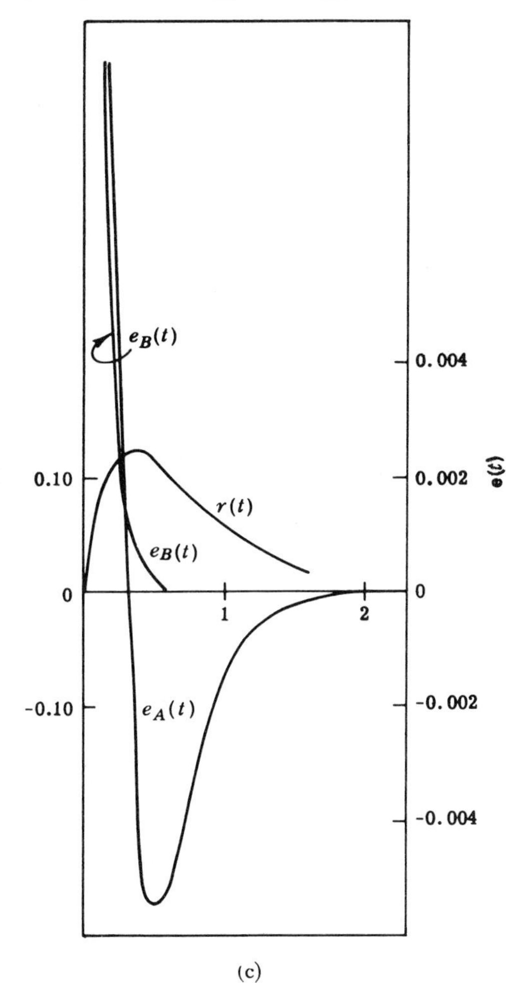

FIG. 9.4-1. S-plane design based on typical input. (a) Comparison of error time functions for two designs. A is better for $t > 0.4$. (b) Comparison of error frequency responses designs, $|T_{EA}| > |T_{EB}|$ for all ω. (c) Input for which design B is better for nearly all t.

Fig. 9.4-1b. It is noted that T_{EB} is considerably smaller than T_{EA} over a large frequency range, so that it is expected that the second design is better than the first. But this is not the case. The error $e_B(t) = 0.00854e^{-t} + 0.0222e^{-10t} - 0.03074e^{-40t}$ is sketched in Fig. 9.4-1a. It is less than $e_A(t)$ for $t < 0.4$ second. However, for $t > 0.4$, e_A is appreciably smaller than e_B, in spite of the fact that $| T_{EB} | < | T_{EA} |$ for all frequencies. If the error magnitude in the interval $t < 0.4$ is of no consequence, then the T_{EA} design is better, in addition to being more economical (in terms of bandwidth—see Fig. 9.4-1b) than the T_{EB} design. However, the situation is reversed (see Fig. 9.4-1c) if the input is $R = 1/(s + 2)(s + 4)$. In such a case, T_{EB}, with its zeros at the poles of R, leads to a smaller error for almost all t.

It is known (Section 12.6) that the approximation of a time function $r(t)$ by one with a rational transform is not unique. It is possible, for example, to satisfactorily approximate $r(t)$ with an approximant $r_a(t)$ such that the poles of $\mathscr{L}r_a(t) = R_a(s)$ are all coincident on the negative-real axis (the Laguerre expansion). Or a satisfactory approximation in time may be secured by an $r_a(t)$ whose transform has distinct poles, complex and real. What does this mean in view of the previous conclusion that the zeros of $T_E(s)$ should cancel out the poles of $R(s)$? The answer is as follows. If different functions are used to approximate $r(t)$, and if there are only minor differences among the approximants in the time domain, then there are only minor differences between their transforms in the significant region of the frequency domain. Therefore any pole-zero representation of the input may be used with confidence, providing its time function is a satisfactory approximation of the actual input. (An exception to this is when there any zeros close to the $j\omega$ axis—see Section 12.6.) The real question is that of finding the simplest way of characterizing R and hence T_E. This is related to the question of finding the "natural" poles of a time function, i.e., those which, for a given permissible approximation error, lead to the simplest expression for the transform. This problem is considered in Section 12.6.

To summarize this section, a procedure is described for specifying the system error response T_E (and hence T) to respond satisfactorily to a typical input signal. The rational transform $R(s)$ of the signal (or of a satisfactory approximation of the signal) is found by any of the methods of Chapter 12. T_E is assigned zeros at the poles of R. The poles of T_E (and more zeros, if they are needed to realize the error response) are assigned so that in conjunction with the zeros of R, the error response is within the specified bounds. Such a design procedure leads to an economical (in terms of bandwidth) design—other designs with larger bandwidths, which are not based on the typical input, are likely to have larger error outputs. All that remains is to show how to choose the poles of T_E so as to maintain the instantaneous error within the allowed limits, by means of a design which is as economical as possible in terms of gain-bandwidth. The procedure is presented by means of a design example.

Example. The design procedure is applied to the previous problem, where $R = 1/(s+1)(s+2)$. We choose $T_E = s(s+1)(s+2)\,T_{E1}$. The zero at the origin is required when the plant includes a motor, for then $T(0) = 1$ and $T_E(0) = 0$. The problem is to choose T_{E1}, such that $e(t) = \mathscr{L}^{-1}RT_E$ is within the specified bounds. T_{E1} must have three more poles than zeros, in order that $T_E(\infty) = 1$. The latter is necessary because $T_E = 1 - T$, and any practical T is zero at infinite s. Suppose the error bound vs. time is that shown in Fig. 9.4-2a. The simplest $E(s)$ has the form $E = RT_E = [(s+1)(s+2)]^{-1}\,[s(s+1)(s+2)]\,[(s^2 + 2\zeta\omega_n s + \omega_n^2)(s+\alpha)]^{-1} = s/[(s^2 + 2\zeta\omega_n s + \omega_n^2)(s+\alpha)]$.

Our first approach is to choose α so that $s + \alpha$ effectively cancels out the s in the numerator. Such a small α leads to a term $Ae^{-\alpha t}$ with a very small damping factor. The value of $|A| \doteq |-\alpha/\omega_n^2|$ must therefore be less than the maximum permitted error of 10^{-4} for $t > 0.6$; i.e., $\alpha/\omega_n^2 \leqslant 10^{-4}$, providing of course that the other components are much less than 10^{-4} at $t > 0.6$. If α is so chosen, then $e(t) \doteq -10^{-4}\,e^{-\alpha t} + \mathscr{L}^{-1}[1/(s^2 + 2\zeta\omega_n s + \omega_n^2)]$ because $s/(s+\alpha)$ is very closely equal to 1 at $s = -\omega_n$ (obviously $\omega_n \gg \alpha$). We are therefore left with $\mathscr{L}^{-1}\,1/(s^2 + 2\zeta\omega_n s + \omega_n^2)$, which is graphically available (Fig. 5.12-1). It is desirable to satisfy the specifications with a T_E (which must be very small) as large as possible over as large a bandwidth as possible. The reason is that $T_E = 1/(1+L)$ exactly in single-degree-of-freedom systems; hence L must be large over the frequency range in which T_E is small. In two-degree-of-freedom systems, $T_E = 1 - T = 1 - [GP/(1+L)] = [1 + L - (L/H)]/(1+L)$, in the notation of Section 6.1 and Fig. 6.1-1. Over the significant bandwidth of T, $H \doteq 1$, because in this frequency range $T = GP/(1+L) = (L/H)/(1+L) \doteq 1$, and L is large. In the present problem, we therefore try to pick ω_n as small as possible, in order to increase T_E. Study of Fig. 5.12-1 suggests that $\zeta = 1.0$ and $\omega_n = 20$ should be tried (with $\alpha = 10^{-4}\omega_n^2 = 0.04$). The resulting $e_{\text{I}}(t)$ and $T_{E\text{I}}(j\omega)$ are shown in Figs. 9.4-2a and b, respectively. There is a slight violation of the error bound at $t \approx 0.05$, but it may probably be overlooked.

Examination of Fig. 9.4-2a indicates that $e_{\text{I}}(t)$ is quite a bit less than $e_{\max}(t)$ for an appreciable value of t. There is an instinctive feeling that the larger $e(t)$ is, the larger may T_E be, and consequently the smaller the value of L. This is verified grossly by the Parseval relation,[1] viz., $\int_0^\infty e^2(t)\,dt = \int_{-\infty}^{\infty} |E(j\omega)|^2\,df = \int_{-\infty}^{\infty} |1 + L|^{-2}\,df$. We therefore attempt to improve the design by letting $e(t)$ come closer to its maximum value. The curves in Fig. 5.12-1 suggest that this is possible by choosing a larger value of ζ. We try $E_{\text{II}}(s) = s/(s+10)(s+30)(s+0.03)$; the resulting $e_{\text{II}}(t)$ and $T_{E\text{II}}(j\omega)$ are shown in Figs. 9.4-2a and b. In the latter figure, the exact $T_{E\text{I}}$ and $T_{E\text{II}}$ are almost the same for $\omega > 8$, but $T_{E\text{II}}$ is larger for all $\omega < 8$. The improvement in decibels is not as impressive as it is in arithmetic units. In order to economize to the utmost on $L(j\omega)$, it is

[1] E. C. Titchmarsh, "Introduction to the Theory of Fourier Integrals," 2nd ed., p. 50. Oxford Univ. Press, London and New York, 1948.

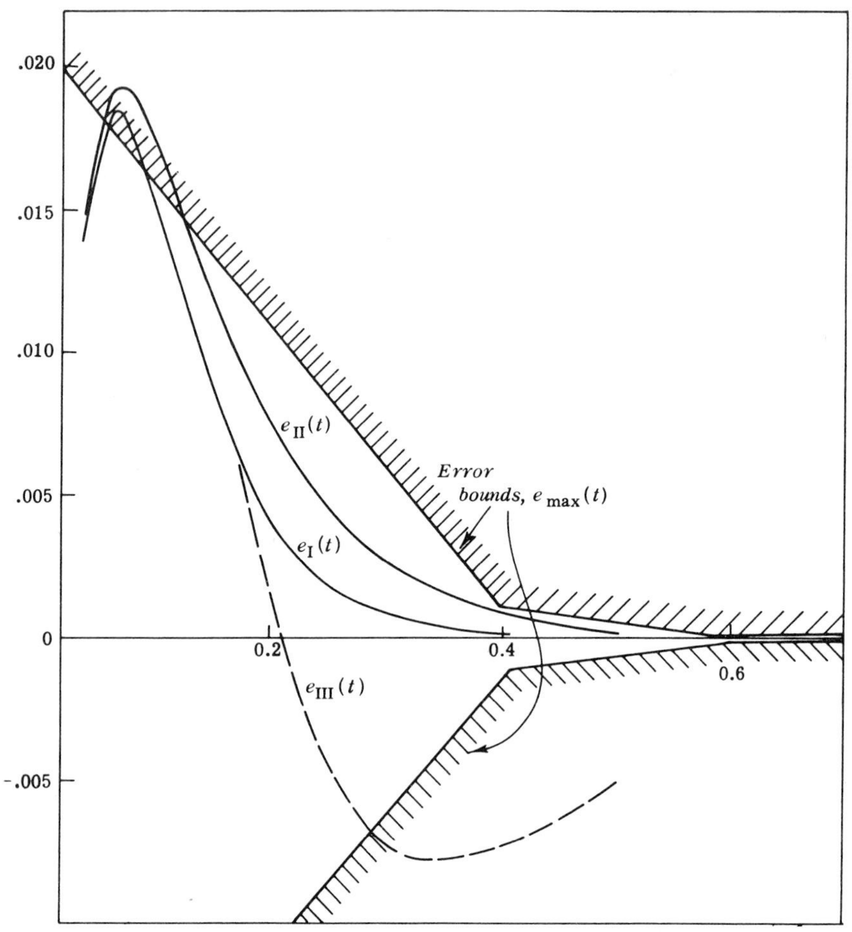

FIG. 9.4-2. (a) Design for specified maximum $|e(t)|$ - time domain curves.

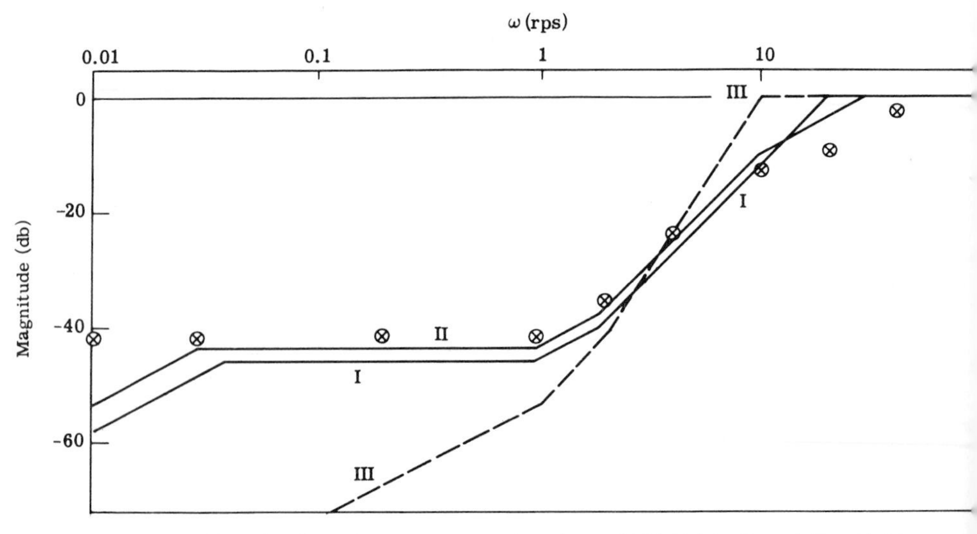

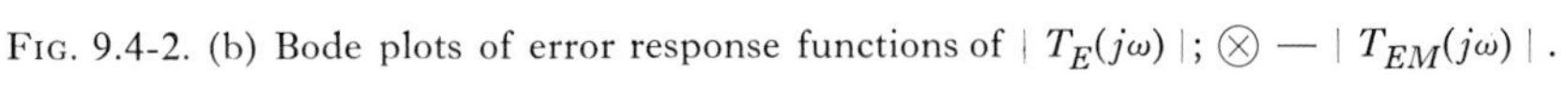
FIG. 9.4-2. (b) Bode plots of error response functions of $|T_E(j\omega)|$; ⊗ — $|T_{EM}(j\omega)|$.

necessary to shape $e(t)$ to follow closely the bounded value. Such shaping (time domain synthesis) requires a more complicated $T_E(s)$ and techniques for this purpose are presented in Chapter 12. However, comparison of designs I and II in Figs. 9.4-2a and b indicate that the maximum improvement cannot be very large.[2]

A third trial was attempted in order to see whether increase of T_E in the higher frequency range ($\omega \sim 10$) could be compensated by decrease of T_E at smaller ω. In this trial $E_{\text{III}} = s/(s+10)^3$ with $e_{\text{III}}(t)$ and $T_{E\text{III}}$ of Figs. 9.4-2a and b. It is seen that the error bound is greatly exceeded.

It should be noted that the above procedure can be used to specify $L(s)$ only over its significant frequency range, i.e., up to about an octave beyond its crossover frequency ω_c. This is because a $T_E(s)$, which is unity at infinite s, leads to an $L = T_E^{-1} - 1$ which has only one more pole than zeros. If L is to have x more poles than zeros, then it is necessary that the first x numerator coefficients of T_E should be equal, respectively, to the first x denominator coefficients of T_E (see Section 6.13). This is difficult to do when $x > 2$. Instead, the above procedure is used to specify $L(j\omega)$ up to about $2\omega_c$. For $\omega > 2\omega_c$, $1 + L(j\omega) \doteq 1$, so that additional corner frequencies may be assigned to $L(j\omega)$ beyond $2\omega_c$, with minor effect on $1 + L(s)$. Of course these poles should be assigned to achieve adequate stability margins and/or the desired damping for the dominant roots of $1 + L(s) = 0$.

The above procedure is more satisfactory and complete than a design based on error constants. Design for step and ramp inputs is then only a special case of the above general procedure. Furthermore, the designer is able to control the error as a function of time, and not just the steady state value. The latter is meaningless if the step or ramp does not exist for sufficient time. However, the design philosophy is appropriate only when it is possible to single out a specific signal as the typical signal input. The system response function is then tuned to this signal. The system response to a significantly different signal may be unsatisfactory. When the system must respond to an ensemble of signals

[2] The benefit that is realized by increasing $e(t)$, until it is everywhere equal to its maximum permitted value, may be obtained as follows. Let $e(t)$ have its maximum permitted value $e_M(t)$. In this calculation, $e_M(t)$ is approximately taken to be the straight line (see Fig. 9.4-2a) from 0.02 at $t = 0$ to zero at $t = 0.4$; i.e., $e_M(t) = 0.02 - 0.05t + 0.05(t - 0.4)u(t - 0.4)$, and $E_M(s) = (2)10^{-2}s^{-2}(s - 2.5 + 2.5e^{-0.4s})$. Since $E_M = RT_E$ and $R = 1/(s+1)(s+2)$, $T_{EM} = 0.02s^{-2}(s+1)(s+2)(s - 2.5 + 2.5e^{-0.4s})$. $|T_{EM}(j\omega)|$ is evaluated, and its values are indicated by the circled crosses in Fig. 9.4-2b. It is conjectured (but not proven) that in order for $|e(t)| \leqslant |e_M(t)|$, it is necessary, more or less, that $|T_E(j\omega)| \leqslant |T_{EM}(j\omega)|$. (By "more or less" we mean that the inequalities can probably be only slightly and locally violated, so that for practical purposes they are satisfactory.) For example, it is noted that $|T_E| > |T_{EM}|$ near $\omega = 20$ rps in Fig. 9.4-2b and this is related to the fact that $e(t) > e_M(t)$ near $t = 0.05$ in Fig. 9.4-2a. It is clear from Fig. 9.4-2b that our initial simple designs for T_E are very close to the optimum. This technique may generally be used to find the worth of a simple choice for T_E. If the difference between the design and $|T_{EM}(j\omega)|$ is large, then a better choice for T_E shoud be sought.

which are not individually very similar, then it is better to design on the basis of the statistical properties of the ensemble of signals. This is done in Sections 9.8-9.17.

Time Domain Specification of System Response

The following is a simple method for acquiring some feeling for the system response function that is required to minimize the error. When the output is to follow the input, the error output $E = (1 - T)R = (1/s - T/s)\,sR$. The corresponding relation in the time domain is

$$e(t) = [u(t) - c_s(t)] * r'(t) \triangleq f(t) * r'(t) \tag{9.4,1}$$

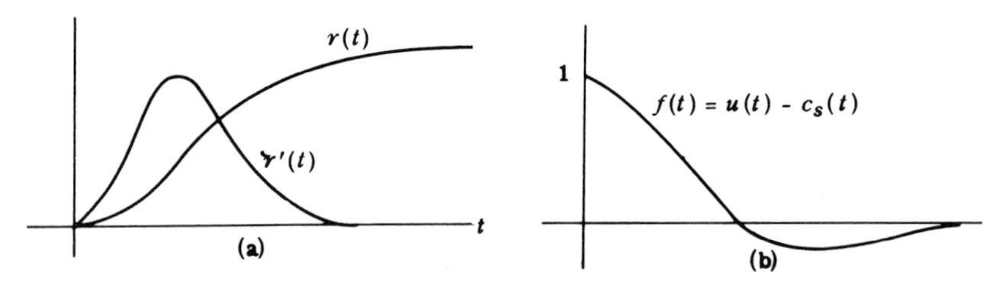

FIG. 9.4-3. Curves of input $r(t)$, its derivative $r'(t)$, and typical $f(t)$.

where $u(t)$ is the unit step function, and $c_s(t)$ is the system step response. For example, let $r(t)$ and $r'(t)$ be as shown in Fig. 9.4-3a. The function $f(t) \triangleq u(t) - c_s(t)$ has the general form shown in Fig. 9.4-3b. To find $e(t)$, $f(t)$ is convolved with $r'(t)$ as per Eq. (9.4,1). The objective is, for a given maximum $e(t)$, to get by with as slow a response as possible (small bandwidth). Let us compare $c_s(t)$ with and without overshoot. The case of no overshoot is shown in Fig. 9.4-4a. The convolution is performed by multiplying the two curves in the figure and integrating the product. Clearly, the maximum error is approximately at $t = \tau_0$. Next we consider a $c_s(t)$ with overshoot. (Fig. 9.4-4b). The negative overshoot area can be adjusted to cancel out a significant part of the positive error area. Hence, for all $r(t)$ whose derivative has the same sign, the optimum system step response is one with a large overshoot. Increase in overshoot results in a

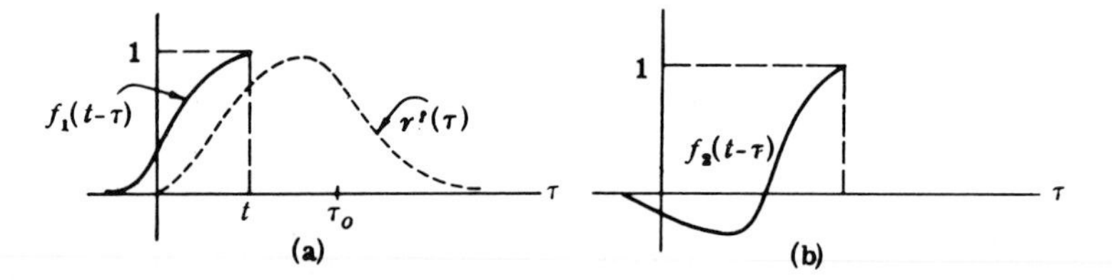

FIG. 9.4-4. Use of convolution to obtain feeling for optimum $f(t)$.

decrease in rise time *for the same bandwidth* (this should be clear from Fig. 5.12-1). Furthermore, the overshoot results in negative values of $f(t)$ over some range. The negative values of $f(t)$ act to decrease the value of $e(t) = f(t) * r'(t)$ especially near $t = \tau_0$. It is not always possible to use the optimum so deduced, in the one-degree-of-freedom configuration, because the resulting low damped poles are also the poles of the response to disturbance signals. However, this objection does not exist for the configurations of Chapter 6.

The above is exact—the transient components are included. However, it does not readily lend itself to mathematical formulation. But in those cases where there is a definite "typical" signal input, the above is a useful aid in obtaining some feeling for what constitutes an optimum response. Also, it is possible to see the requirements on $c_s(t)$ for weighting the error, for those systems where the error over some specific time interval is to be minimized, and the error in other time intervals is of lesser consequence.

§ 9.5 Effect of Plant Parameter and Input Signal Variations

Do the forced components of the error remain zero as the plant parameters vary? In order that the forced error components be zero, it is necessary that $T_E(s_i) = 0$, where $R = \sum A_i/(s - s_i)$. In the classical one-degree-of-freedom configuration of Chapter 5, $T_E = 1/(1 + L)$, and $T_E(s_i) = 0$ is obtained by letting L have a pole at s_i. So long as the pole at s_i is unaffected by plant parameter variations, this forced component of the error remains zero. The poles of L at s_i should therefore be obtained from the compensation networks, and they are then insensitive to plant parameter variation. In the two-degree-of-freedom configuration of Fig. 9.5-1a, $T_E = 1 - T = (1 + PH)/[1 + P(H + G)]$.

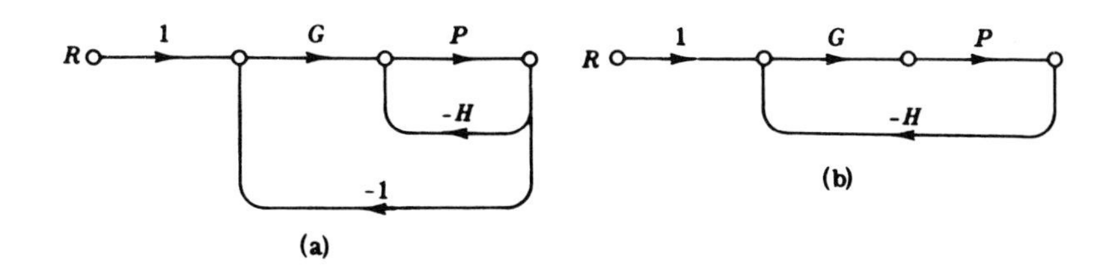

FIG. 9.5-1. Two-degree-of-freedom structures.

Therefore, poles of G are zeros of T_E, and are independent of the plant parameter values. The forced components of the error may therefore be zero, despite plant parameter variations. It is necessary to make sure that the natural components of the error remain small over the range of plant parameter variations. In the structure of Fig. 9.5-1b, $T_E = [1 + GP(H - 1)]/(1 + GPH)$. The poles of the forced components at $s = s_i$ are zero, independently of the plant parameter values, if G is assigned poles at s_i and if $H(s_i) = 1$. In a positioning system $H(s_i) \doteq 1$ in any case, because $T = GP/(1 + GPH) \doteq 1$ over the significant frequency range; the latter must surely include the region of the significant poles of the input to the system.

The effect of variations in the input signal is obtained from the relations: $E_o = R_o T_E$, $E = RT_E$; therefore, $E = E_o(R/R_o)$, $e(t) = e_o(t) * \mathscr{L}^{-1}(R/R_o)$. A simple way to find $e(t)$ is to evaluate $\mathscr{L}^{-1}(R/R_o)$ and perform a numerical convolution with $e_o(t)$, using the method of Section 12.8.

§ 9.6 Specification of Disturbance Attenuation

In previous chapters, when disturbance attenuation was considered, it was assumed that the required amount of attenuation was known as a function of frequency. In this section, it is assumed (1) that there is a known time function which represents the typical disturbance, and (2) that the maximum allowable instantaneous effect of the disturbance is specified as a function of time. We shall refer all disturbances to one equivalent disturbance at the plant output (Fig. 9.6-1). The disturbance output is then always $C_D = D/(1 + L) = DT_D$.

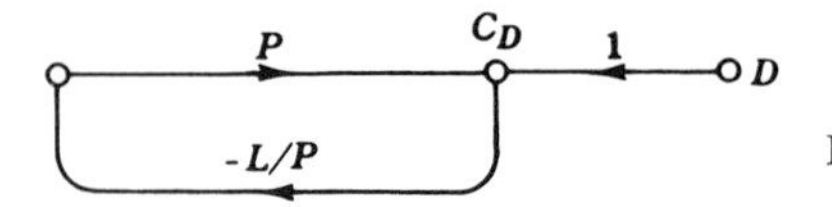

FIG. 9.6-1. System with disturbance input.

The problem is to find the requirements on $T_D = 1/(1 + L)$ in order that the time domain specifications are achieved.

The above problem is exactly the same as that treated in Section 9.4 for $E = RT_E = R/(1 + L)$. Exactly the same procedure may be used for specifying T_D as for T_E, i.e., poles of L are chosen at the poles of D, and the poles of T_E are chosen so that the natural disturbance output components satisfy the time domain specifications. The design example (Figs. 9.4-2a and b) of Section 9.4 could have been one of disturbance attenuation. A Bode plot of frequency response (Fig. 9.4-2b) is used, together with the time response (Fig. 9.4-2a), in order to satisfy the time domain specifications, by means of a design which is economical in gain-bandwidth demands on L. It was also shown in Section 8.16 that this same procedure may be used to design for abrupt parameter variations.

The above procedure is suitable for a specific typical disturbance signal. When there does not exist such a unique disturbance signal, statistical methods (Sections 9.8-9.17) can be used.

§ 9.7 Inputs Bounded in Slope and/or Magnitude

Sections 9.4-9.6 were devoted to the design of systems driven by a unique command signal or disturbed by a unique disturbance signal. Sections 9.8-9.17 are devoted to systems whose command, disturbance, and noise inputs each consist of families or ensemble of signals. This section is concerned with systems in which only the bounds on the inputs are known.[1] In one case, only their magnitude is bounded, while the derivatives are unbounded. In the second case under consideration, there are bounds on both the magnitude and first derivative of the input signals. The analysis problem is one of finding, for a system with a known response, the specific input which leads to the maximum value of the output. The synthesis problem is one of finding the system response function which satisfies the design specifications and is optimum in some sense. For example, if the input signals are disturbances acting on a feedback system, their effect must be kept less than some amount, preferably by a loop transmission function which has as small a gain and bandwidth as possible.

We first consider a system whose input signal $x(t)$ of Fig. 9.7-1 may lie anywhere within the bounds $\pm M$ and may instantaneously switch from one

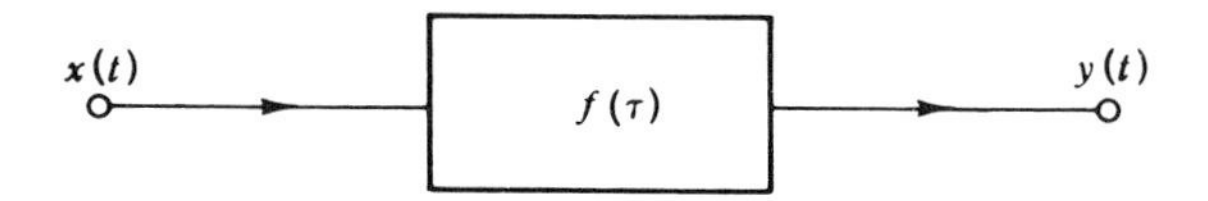

FIG. 9.7-1. The analysis problem—only the bounds of $x(t)$ or of $x(t)$, $x'(t)$ are known.

value to another within these bounds. The output $y(t) = \int_0^t f(\tau)\, x(t - \tau)\, d\tau$. Since $x(t)$ may be reversed in time and since we are interested in $|y(t)|_{\max}$, the above may be replaced by $y = \int_0^\infty f(t)\, x(t)\, dt$. Obviously, y is a maximum if $x(t) = M$ when $f(t) > 0$, $x(t) = -M$ when $f(t) < 0$. The maximum value of y is then $y_m = M \int_0^\infty |f(t)|\, dt$.

A synthesis problem is shown in Fig. 9.7-2a. The input $x(t)$ consists of the bounded disturbances, whose effect is to be attenuated by the feedback network, i.e., $F \triangleq Y/X = 1/(1 + L)$ if the disturbance is at node 2, and $F = P/(1 + L)$ if it is at node 1. The loop transmission $L = GPH$ must be chosen to attenuate the disturbances, and it is desirable to choose L with as small a gain and bandwidth as possible. The synthesis problem is trivial for the node 2 case, because

[1] This section is based on the work of: P. E. Pfeiffer, The maximum response ratio of linear systems, *Trans. AIEE, Part II, Applications & Ind.* **74**, 480-484 (1955); B.J. Birch and R. Jackson, The behavior of linear systems with inputs satisfying certain bounding conditions, *J. Electronics and Control* **6**, 366-375 (1959); I. Horowitz, Analysis and synthesis of linear systems with inputs satisfying certain bounding conditions, *J. Electronics and Control* **12** (1), 195-208 (1962).

Y can be written in the form $Y = X/(1 + L) = X - XQ$, which defines $Q = L/(1 + L)$. $\mathscr{L}^{-1}XQ \triangleq \psi(t)$ is zero at $t = 0$ when $X = 1/s$, because L must go to zero at infinite s, and so must Q. Hence, if $x(t)$ is as shown in Fig. 9.7-2b, $\psi(t)$ will be as shown (if the step response of Q has $100\gamma\ \%$ overshoot).

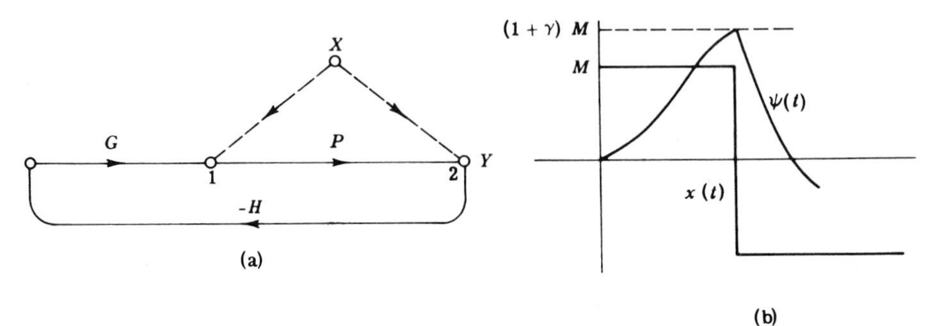

FIG. 9.7-2. (a) A synthesis problem. $L = GPH$. (b) Maximum error is $M(2 + \gamma)$, so choose $\gamma = 0$.

The maximum magnitude of y is that of $x - \psi$, which is $|-M - (1 + \gamma)M| = M(2 + \gamma)$. Hence $L/(1 + L)$ should be so chosen that its step response has no overshoot, and then $y_m = 2M$.

The synthesis problem is only slightly more complicated if the input is at node 1 in Fig. 9.7-2a. Then $Y = FX = PX/(1 + L)$ does not instantaneously follow a step, because any practical P has more poles than zeros. The maximum

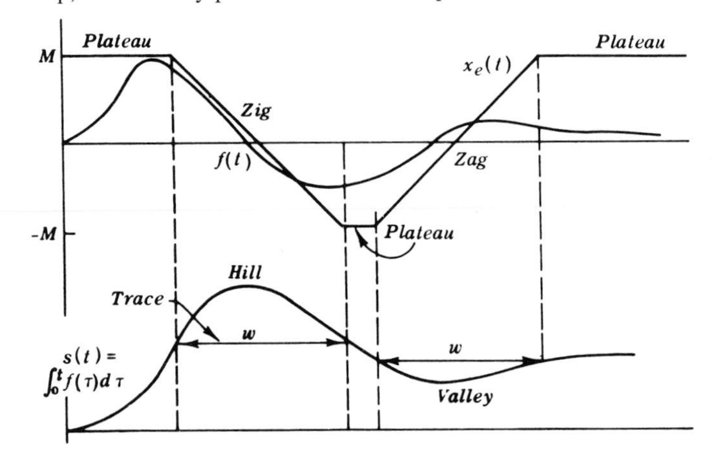

FIG. 9.7-3. (a) Definition of terms. $wm = 2M$.

Figures 9.7-3, 4. 5 are taken from I. Horowitz, Analysis and synthesis of linear systems with inputs satisfying certain bounding conditions. *J. Electronics and Control* **12** (1), 195-208 (1962).

y is $y_m = \int_0^\infty |f(t)\, x(t)|\, dt = M \int_0^\infty |f(t)|\, dt$. It will be seen that $\int_0^\infty f(t)\, dt$ is independent of the overshoot in the step response of F. Therefore $\int_0^\infty |f(t)|\, dt > \int_0^\infty f(t)\, dt$ when there is overshoot; so y_m is minimized by choosing $f(t)$ with no overshoot. To prove that $\int_0^\infty f(t)\, dt$ is independent of overshoot, it is noted that $\int_0^\infty f(t)\, dt = \lim_{s\to 0} F(s) = \lim_{s\to 0} P(s)/[1 + L(s)]$. The overshoot is determined by the shaping of $L(s)$ in its crossover region, while $\int_0^\infty f(t)\, dt$ is determined by its zero frequency behavior. Hence, the two are independent, and $y_m = M \lim_{s\to 0} P(s)/[1 + L(s)]$. The zero frequency value of $L(s)$ must be assigned so as to keep y_m within the desired bounds.

Inputs Bounded in Magnitude and Slope

In Fig. 9.7-1, $f(t)$ is known, $|x(t)| \leqslant M$, $|dx/dt| \leqslant m$. The analysis problem is to find the extreme input $x_e(t)$, for which $y = \int_0^\infty x(t) f(t)\, dt$ is a maximum.

Figure 9.7-3a consists of a hypothetical $f(t)$ and the corresponding extreme $x_e(t)$ function. The function $s(t) \triangleq \int_0^t f(\tau)\, d\tau$ is also included. The terms zig, zag, plateau, hill, and valley are defined in the figure. It is noted that the first three are determined by $x(t)$, the latter two by $s(t)$. A complete zig or zag is one whose end points are at M, $-M$ (note $wm \triangleq 2M$). It is useful to project the end points of any zig or zag on $s(t)$ and join each such pair of end points by a line called a "trace" (see Fig. 9.7-3a). It will later be noted that in $x_e(t)$ all traces must be horizontal. From the definition of complete zigs and zags, it follows that there must be a plateau between a complete zig (zag) and any zag (zig). In degenerate cases, the plateau can have zero width. An incomplete zig or zag has a trace whose width is less than w.

The following restrictions are made on $f(t)$:

(1) $s(t) = \int_0^t f(\tau)\, d\tau$ is finite for all t, otherwise the output y will be unbounded.

(2) It is always possible to find a t_1 such that $|M \int_{t_1}^\infty f(\tau)\, d\tau| < \epsilon$, with ϵ any small nonzero positive number. It is also necessary to amend the statement of the problem to the effect that the extreme $x_e(t)$ may be found such that the resulting $y = \int_0^\infty f(\tau)\, x(\tau)\, d\tau$ has the maximum value within an accuracy of ϵ. This amendment permits an $f(t)$ with one or more damped oscillatory terms to be made zero for all $t > t_1$. It is henceforth assumed that this operation has been performed on an $f(t)$ of the latter type, and therefore in all $f(t)$ under consideration, t_1 can always be found such that $f(t)$ is either $\geqslant 0$, or $\leqslant 0$, for $t > t_1$.

The following rules[1] enable $x_e(t)$ to be found.

Rule 1. The derivative of $x_e(t)$ has the maximum permissible magnitude m, at all its zigs and zags.

Rule 2. At very large t, $x_e(t) = M$ if $f(t)$ approaches zero positively as t approaches infinity, and vice versa.

---

[1] B. J. Birch and R. Jackson, The behavior of linear systems with inputs satisfying

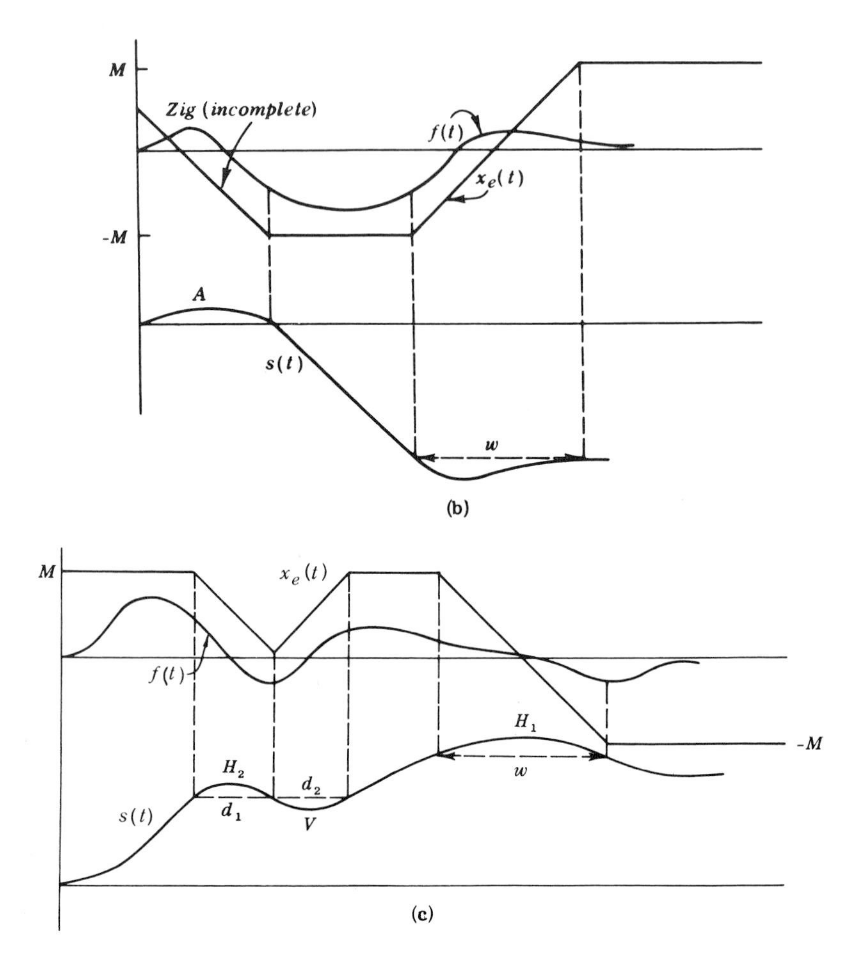

FIG. 9.7-3. (b, c) Derivation of extreme input $x_e(t)$ for given system response $f(t)$.

Rule 3. Any trace of $x_e(t)$ is a horizontal line which extends across a hill or valley.

Rule 4. In $x_e(t)$ a zig (or zag) of width w (i.e., a complete zig) exists at every hill (or valley) wherein there is no possibility of overlapping zigs and zags.

Rule 5. At any plateau, $x_e(t) = \pm M$.

Rule 6. In a hill-valley pair in which a complete zig and complete zag cannot both exist, if $x_e(t)$ does have a trace there, and if it is impossible that $x_e(t) = \pm M$

---

certain bounding conditions. *J. Electronics and Control* **6**, 366-375 (1959); I. Horowitz, Analysis and synthesis of linear systems with inputs satisfying certain bounding conditions. *J. Electronics and Control* **12** (1), 195-208 (1962).

in between the zig and zag, then the trace consists of a single horizontal line which cuts across the hill and valley.

Rule 7. Given a hill-valley (valley-hill) pair in which the lack of a zig-zag (zag-zig) pair results in a constant value for $x_e(t)$ on both sides of the pair, then $x_e(t)$ must have zig-zag (zag-zig) pair associated with the hill-valley (valley-hill).

The derivation of $x_e(t)$ is illustrated by applying the above rules to several examples:

Example 1 (Fig. 9.7-3a). $f(t)$ is given. The first step is to graph $s(t) \triangleq \int_0^t f(t)\, dt$. Rule 2 determines $x_e(t)$ at large t. Rule 4 is used to find the two unique complete traces, and $x_e(t)$ is then uniquely determined.

Example 2 (Fig. 9.7-3b). A complete trace cannot be assigned to the hill at A. However, it may be imagined that $f(t)$ has infinitesimally small positive

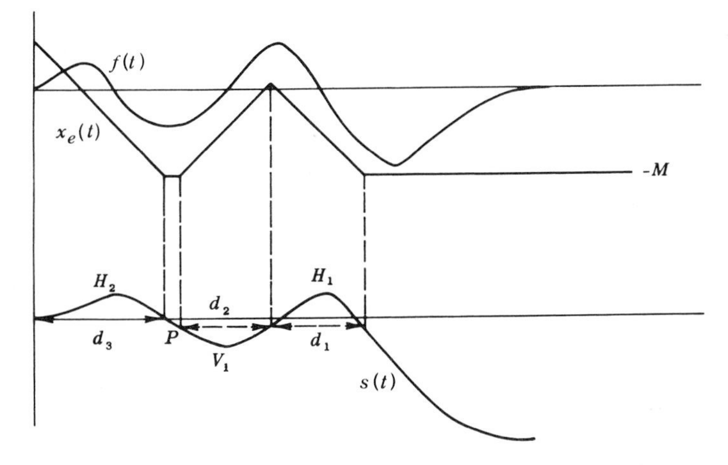

FIG. 9.7-4. (a-d) Examples for application of rules. (a) $d_1 = d_2$.

value for some range of $t < 0$, and then the hill at A may accommodate a degenerate complete trace, which must be located as shown in the figure.

Example 3 (Fig. 9.7-3c). From Rule 1, $x_e(\infty) = -M$. A complete zig, (with no possibility of overlap with other zags) is possible at H_1, and it is therefore inserted there in accordance with Rule 4. The above is impossible at H_2, V; but there must be a zig-zag pair at H_2, V, otherwise Rule 7 is violated. From Rule 6 the trace at H_2V must be a single horizontal line. Such a trace must have $d_1 = d_2$, otherwise there will be a plateau at $x_e \neq M$, and Rule 5 is thereby violated. The resulting $x_e(t)$ is shown in the figure. It is seen that it is unique and that the theorems listed have been sufficient for deriving it.

Example 4 (Fig. 9.7-4a). Clearly, $x_e(\infty) = -M$. Rule 7 is violated if H_1, V_1 has no zig-zag pair. Therefore d_2, d_1 exist and $d_2 \leqslant d_1$, because it is necessary

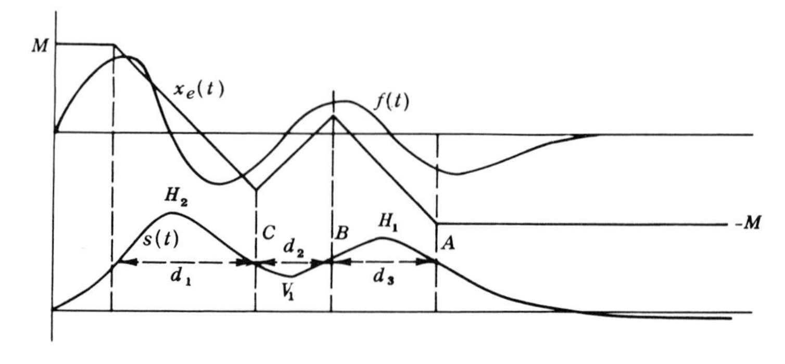

FIG. 9.7-4b. $m(d_1 + d_3 - d_2) = 2M$.

that $|x_e| \leqslant M$. If $d_2 < d_1$, then the trace must be a single horizontal line across H_1, V_1, H_2. Since $s(t) > 0$ at H_2, such a trace across H_1, V_1, H_2 would have to lie above the abscissa. But this would require $d_2 > d_1$, which is forbidden. Therefore, the alternative $d_2 = d_1$ must be used. There is only one trace for which $d_2 = d_1$, and it results in a plateau at $-M$ between V_1 and H_2. The trace at H_2 must then be of the "degenerate complete" type, similar to that at A in Fig. 9.7-3b. Again, $x_e(t)$ is unique and the theorems have been sufficient for finding it.

Example 5 (Fig. 9.7-4b). Obviously, $x_e(\infty) = -M$, and for the same reason as in Fig. 9.7-4a, there must be a trace at H_1V_1. It is necessary that $d_2 \leqslant d_3$. The equality sign would require the d_2, d_3 trace to be at A, B, C in the figure, and would lead to a plateau (with $x_e = -M$) between H_2 and V_1. It would then be necessary to have a complete zig at H_2, or else Rule 5 would be violated. However, the resulting complete trace in H_2 would overlap with that at A, B, C. Therefore, $d_2 = d_3$ is impossible. It is therefore necessary that $d_2 < d_3$. The resulting trace must then be a single horizontal line cutting across $H_2V_1H_1$, and it is necessary that $d_1 + d_3 - d_2 \leqslant w$. The inequality sign leads to a violation of Rule 5. Therefore, $d_1 + d_3 - d_2 = w$. It may be noted that uniqueness of $x_e(t)$ is due to the fact that the theorems cannot be satisfied by both $d_2 < d_3$ and $d_2 = d_3$.

Example 6 (Fig. 9.7-4c). In this case, $x_e(\infty) = M$ and from Rule 7, d_3, d_4 exist and $d_3 \leqslant d_4$. If $d_3 = d_4$, then by the same reasoning, $d_1 \leqslant d_2$, whose trace then overlaps that due to d_3, d_4. Hence, $d_3 < d_4$, and d_2, d_3, d_4 must provide a single trace with $d_2 + d_4 - d_3 \leqslant w$. If $d_2 + d_4 - d_3 = w$, then $d_5 \leqslant d_1$ and the trace of d_5, d_1 overlaps that of d_2, d_3, d_4; therefore, $d_2 + d_4 - d_3 < w$, and d_1, d_2, d_3, d_4 must consist of a single trace with $d_1 + d_3 \leqslant d_2 + d_4$. If $d_1 + d_3 = d_2 + d_4$, the resulting trace does not interfere with that of d_5, d_6 (where it is established by the same kind of argument that $d_5 = d_6$). If it could be proven that $x_e(t)$ is unique, then the alternative $d_1 + d_3 < d_2 + d_4$ need not be examined. Due to the absence of such a proof, the latter is considered. It would require that d_5 be included in the trace, and therefore $d_5 + d_2 + d_4 - d_1 - d_3 \leqslant w$. The equality sign would require $d_6 = w$, which is impossible.

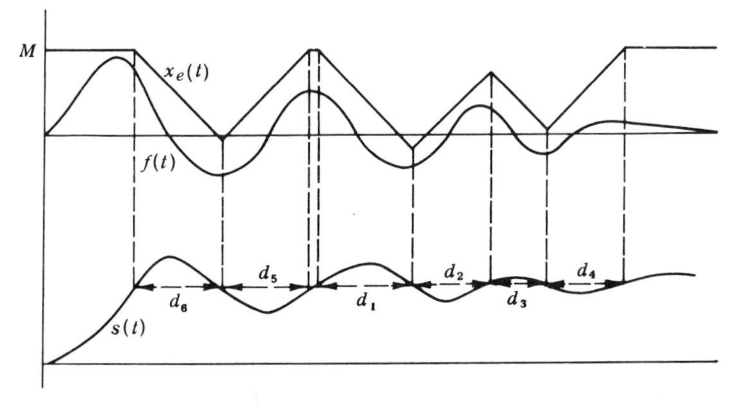

FIG. 9.7-4c. $d_1 + d_3 = d_2 + d_4$; $d_5 = d_6$.

The inequality sign would require that d_6 also be included in the single trace, and that $d_6 + d_1 + d_3 = d_5 + d_2 + d_4$. Consider the $x_e(t)$ shown in the figure (based on $d_1 + d_3 = d_2 + d_4$) in which $d_6 + d_1 + d_3 = d_5 + d_2 + d_4$, and in which there are two traces in contrast to the single trace which would exist if $d_1 + d_3 < d_2 + d_4$. The present $d_1 \ldots d_4$ line would have to be raised in order that $d_1 + d_3 < d_2 + d_4$; and the present $d_5 d_6$ line would also have to be raised in order that there be only one trace, which would result in $d_5 > d_6$ and make it impossible for $d_6 + d_1 + d_3 = d_5 + d_2 + d_4$. Hence, the alternative $d_1 + d_3 < d_2 + d_4$ is impossible.

Example 7 (Fig. 9.7-4d). $x_e(\infty) = M$ and a complete trace (denoted by d_1') is assigned to V_1. From the usual arguments, $d_3 \leqslant d_2$. If $d_3 = d_2$, then $d_4 = w$,

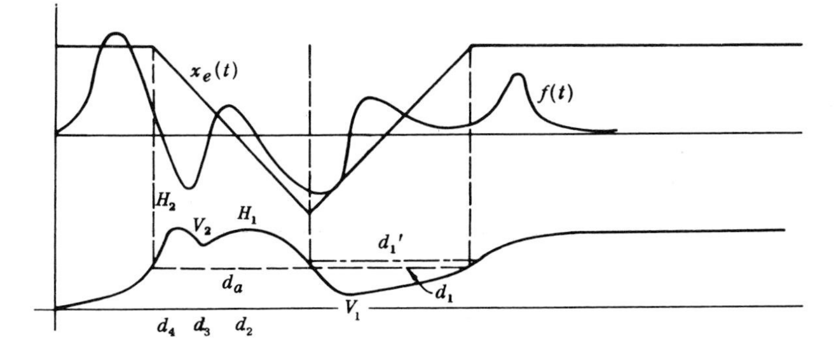

FIG. 9.7-4d. $d_1 = d_a$.

which is impossible. If $d_3 < d_2$, then $d_2 + d_4 - d_3 \leqslant w$. The inequality sign violates Rule 5. The equality sign is impossible for nonzero d_3. Therefore, V_2 is ignored and $H_1V_2H_2$ is considered as a single hill H_a. Nevertheless, a complete trace at H_a overlaps the one at V_1. Therefore, there is a single trace for H_aV_1 with $d_1 = d_a$. Such a trace is unique.

Discussion of Sufficiency of the Rules and Uniqueness of $x_e(t)$

Uniqueness and sufficiency may be proven by showing that there is only one $x_e(t)$ which satisfies the previous rules. There is no need for any proof when all hills and valleys of $s(t)$ are wide enough to admit complete zigs and zags without any overlapping. Similarly, there is no problem for those specific hills and valleys of any $s(t)$ which have the above property. The problem exists only for a group of adjacent hills and valleys in which it is absolutely impossible to assign complete zigs and zags, or to do so without overlapping. In such a case, it is known that the first hill-valley pair (working backwards from large t) are included in one trace. The derivation of $x_e(t)$ is initiated by writing (as in the examples of Figs. 9.7-3 and 4) $d_{n-1} \leqslant d_n$; trying one of these conditions to see if it is compatible with the rules, etc. The key to proving uniqueness lies in proving that only one possibility exists, i.e., that if $d_{n-1} = d_n$ is compatible with the theorems, then $d_{n-1} < d_n$ is incompatible with them. The difficulty in proving the exclusiveness of the two conditions is in the large variety of possibilities which require examination, although, as in the previous examples, no difficulty has been encountered in all the specific cases that have been examined.

Synthesis

The solution to the synthesis problem for inputs with bounded magnitudes and slopes is the same as that for inputs with only bounded magnitudes, i.e., $F(s)$ should be chosen such that there is no overshoot in its step response. It has previously been noted that the quantity $A \triangleq \int_0^\infty f(t)\,dt$ is independent of the overshoot in the system step response. It is therefore necessary to prove that if A is fixed, then the minimum value of $y_e = \int_0^\infty f(\tau)\,x_e(\tau)\,d\tau$ is secured if $f(t)$ has the same sign for all t.

Proof: If A is fixed, then it is contended that the minimum value of y_e, denoted by $y_{em} = M \int_0^\infty f(\tau)\,d\tau = MA$. The resulting $s(t) = \int_0^t f(\tau)\,d\tau$ is monotonic as, for example, $s_1(t)$ in Fig. 9.7-5. An $f(t)$ with overshoot has a step response like that of $s_2(t)$ or $s_3(t)$ in Fig. 9.7-5. Consider s_2: the zig trace must be above or at A, i.e., $\eta \geqslant 0$. Let $y_{ei} = \int_0^\infty x_{ei}(\tau) f_i(\tau)\,d\tau$. Then, $y_{e2} = \int_0^{t_o} + \int_{t_o}^{t_1} + \int_{t_1}^{t_2} + \int_{t_2}^{\infty}$. The first of these is $y_{e1} = AM$. The second and fourth are each $\eta M > 0$; therefore, it is certain that $y_{e2} > AM = y_{e1}$, if

$\int_{t_1}^{t_2} > 0$. To prove the inequality, let $\delta = t - t_1$, and then $\int_{t_1}^{t_2} = \int_{t_1}^{t_2} f_2(t)x_e(t)dt = \int_0^{t_2-t_1} g(\delta)(-m\delta + M)\,d\delta = -m\int_0^{t_2-t_1} \delta g(\delta)\,d\delta + M\int_0^{t_2-t_1} g(\delta)\,d\delta$, where $g(\delta) = f_2(\delta + t_1)$ for $0 \leqslant \delta \leqslant t_2 - t_1$, and is zero otherwise. The second term is zero because $s_2(t_2) = s_2(t_1)$. The first term can be written as $-m\,[\delta g^{(-1)}]_0^{t_2-t_1} + m\int_0^{t_2-t_1} g^{(-1)}(\delta)\,d\delta$. Since $\int_0^{\delta} g(\delta)\,d\delta \triangleq g^{(-1)}(\delta)$ is zero at $\delta = 0$, and at $\delta = t_2 - t_1$, $\int_{t_1}^{t_2} = m\int_0^{t_2-t_1} g^{(-1)}(\delta)\,d\delta > 0$. Therefore, $y_{e2} > y_{e1}$. Similarly, it can be proven that the y_{ei} for any $f_i(t)$ with overshoot is greater than that for $f_1(t)$, providing of course that $s_i(\infty)$ is the same for all the $f_i(t)$.

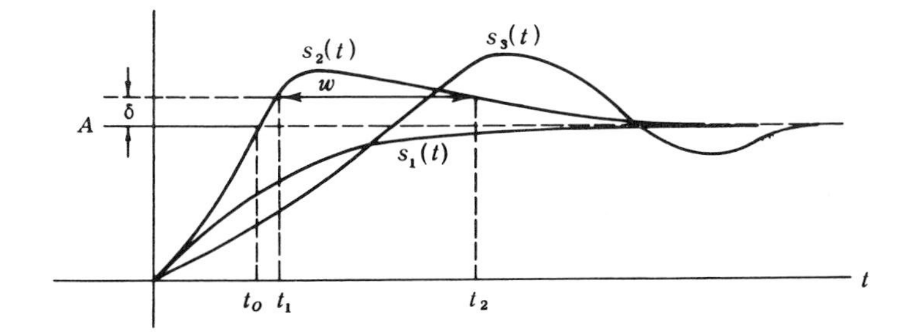

FIG. 9.7-5. Synthesis problem—the optimum step response has no overshoot.

§ 9.8 Choice of *T(s)* from the Statistical Properties of the Input

The preceding sections have emphasized the importance of tailoring $T(s)$ to the typical signal input. This is satisfactory if there is an unmistakable unique typical input, or if the system is to be optimized for one particular input signal, and there is minor concern with its response to other inputs. However, there is often no unique typical input signal, but rather a family or ensemble of typical input signals to the system. In addition, there is almost always an ensemble of false signals (noise) which contaminate the useful signal (message) inputs. The problem is to find that system response function $T(s)$ which is optimum in some manner in responding to the useful message components, and in attenuating the corrupting noise components of the input signals. The problem may be treated with no reference to practical constraints, such as the saturation levels or safe limits which exist in any practical system, or these may be taken into account. In the following sections, the solutions of a representative number of these problems are presented. In all cases, it is assumed that a linear time-invariant $T(s)$ is wanted.

The first problem under consideration is formulated as follows: The control system is subjected to message signals wherein it is possible to define the instant at which any message signal begins. If necessary, it may be imagined that there is a detection process which decides when there is a message, and turns the system on. The system is turned off when the message component falls below some ratio to the noise, as, for example, in tracking an aircraft. (In short, the detection problem and the filtering problem are separated. It is also presumed that abrupt discontinuities at the on and off switching instants may be smoothed out with insignificant effect on the signals.) It is furthermore assumed that after one message signal is over, there is sufficient time or there exists an automatic action which ensures that all the stored energy in the system is discharged before another message arrives. Let us imagine that there is no noise contamination, and the resulting message signals entering the system are recorded and appear as shown in Fig. 9.8-1. The information may have been gathered over

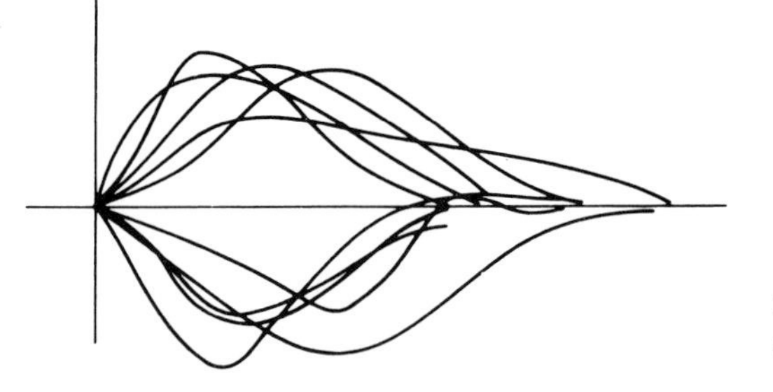

FIG. 9.8-1. Ensemble of message signals.

hours, days, or months. The important point is that a record has been made of the various kinds of messages to which the system is supposed to respond, and the relative number of times a particular message appears is equal to the probability of its occurrence in the typical operation of the system. If among the recorded messages there are some that are not of interest (i.e., we do not care how the system responds to them), then such samples are deleted.

A similar record of the noise signals is available. Actually, it may be difficult to separate the two, and the character of the noise signal may depend upon the character of the message signal. However, we assume that a record of the ensemble of noise signals is available, or that the essential information that must be extracted from such a record (we shall see what it is) may be obtained. Let the total number of sample message signals be β and the total number of sample noise signals be γ. If the noise and message signals are unrelated, then for each sample message there are γ possible noise signals, leading to a total of $\beta\gamma = \mu$ combinations of net input signals. If the two are related, there will be a different total number of possible combinations.

The problem is graphically formulated in Fig. 9.8-2. The subscript α refers to a sample input. $T(s) \triangleq \mathscr{L}f(t)$ is the transfer function of the practical system. The output of the latter consists of the message component $m_{\alpha o}$ and the noise component $n_{\alpha o}$. The ideal output is $c_{I\alpha}(t)$. In a positioning system, $c_{I\alpha}(t) = m_\alpha(t)$; in a system which predicts the message T seconds in the future, $c_{I\alpha}(t) = m_\alpha(t+T)$. The ideal response function $T_I(s)$ is one which realizes the ideal output $c_{I\alpha}(t)$, when the input is $m_\alpha(t)$. Hence $T_I(s) = 1$ in a positioning servo, and $T_I(s) = e^{sT}$ in the predicting system. The sample error $e_\alpha(t)$ is defined as the difference between the ideal output $c_{I\alpha}(t)$ and the actual output $c_\alpha(t) = m_{\alpha o}(t) + n_{\alpha o}(t)$.

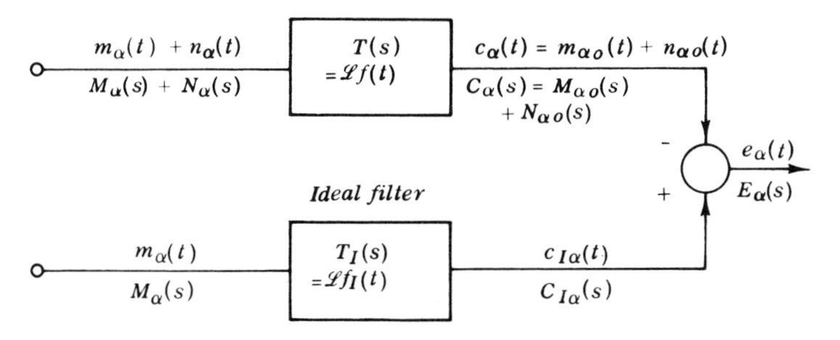

FIG. 9.8-2. Formulation of the filter problem.

The objective is to find the transfer function, $T(s)$ (i.e., of a system which is characterized by a linear differential equation with constant coefficients) such that some measure of the collective error is minimized. It is conceivable that a linear time-varying or a nonlinear response function would be better, but the arbitrary restriction is here made that the filter must be linear, time invariant. In addition, the resulting system is not necessarily the best for each individual combination of message-noise signals. It is the best that can be done for the ensemble of message and noise signals. This problem is often called the *filter problem*, because $T(j\omega)$ acts as a filter to remove the noise components from the useful message components.

The error measure that is used is the ensemble average of the square of the error. The reasons for using it are because with it the problem is solvable and because it is very often a reasonable measure.[1] It is therefore necessary to find $T(s)$ which minimizes the average over the ensemble of inputs of $\int_0^\infty [e_\alpha(t)]^2\, dt$. This is written as $\epsilon^2 = \mathsf{E} \int_0^\infty [e_\alpha(t)]^2\, dt$. The averaging is done over the ensemble of all combinations of sample message-noise signals. Also, the errors at all instants of time are equally weighted. The above formulation is satisfactory only if the time integrals of the message and noise signals are finite. It is possible to work with signals of infinite area whose time averages are finite by using as the error measure the ensemble average of the time average of each squared error function. The procedure and final results are very similar.

The key mathematical tool that is used in the derivation is the Parseval relation

$$\int_0^\infty f_1(t) f_2(t)dt = \int_{-\infty}^\infty F_1(s)F_2(-s)df \tag{9.8,1}$$

[1] See H. James, N. Nichols, and R. Phillips, "Theory of Servomechanisms," pp. 309-10. McGraw-Hill, New York, 1947.

where $F_1(s) = \mathscr{L}f_1(t), F_2(s) = \mathscr{L}f_2(t)$, and $s = \sigma + j\omega = \sigma + j2\pi f$. The relation may be derived from the complex convolution integral[1] $\mathscr{L}f_1(t)f_2(t) = \int_0^\infty e^{-st}f_1(t)f_2(t)\,dt = (2\pi j)^{-1}\int_{c-j\infty}^{c+j\infty} F_1(w)F_2(s-w)\,dw$. We are dealing with signals which are zero at very large t, and may therefore take $c = 0$ in the above. At $s = 0$ the latter equation becomes $\int_0^\infty f_1(t)f_2(t)\,dt = (2\pi j)^{-1}\int_{-j\infty}^{j\infty} F_1(w)$ $F_2(-w)\,dw$. The dummy variable w may be replaced by s, leading to Eq. (9.8,1).

In order to make use of Eq. (9.8,1), we write, from Fig. 9.8-2, $\mathscr{L}e_\alpha(t) = \mathscr{L}[c_{I\alpha}(t) - c_\alpha(t)] = \mathscr{L}[c_{I\alpha} - m_{\alpha o} - n_{\alpha o}] = T_I(s)\,M_\alpha(s) - T(s)\,[M_\alpha(s) + N_\alpha(s)]$. This is used in the Parseval equation, $\epsilon^2 = \mathsf{E}\int_0^\infty e_\alpha{}^2(t)\,dt = \mathsf{E}\int_{-\infty}^\infty E_\alpha(s)\,E_\alpha(-s)\,df$ and results in $\epsilon^2 = \mathsf{E}\int_{-\infty}^\infty \{T_I(s)\,M_\alpha(s) - T(s)\,[M_\alpha(s) + N_\alpha(s)]\}\,\{T_I(-s)\,M_\alpha(-s)$ $- T(-s)\,[M_\alpha(-s) + N_\alpha(-s)]\}\,df$. The averaging operation is next brought under the integral sign, and the following notation is used:

$$\begin{aligned} \Phi_m &\triangleq \mathsf{E}\,M_\alpha(s)\,M_\alpha(-s), & \Phi_{mn} &\triangleq \mathsf{E}\,M_\alpha(s)\,N_\alpha(-s), \\ \Phi_n &\triangleq \mathsf{E}\,N_\alpha(s)\,N_\alpha(-s), & \Phi_{nm} &\triangleq \mathsf{E}\,N_\alpha(s)\,M_\alpha(-s). \end{aligned} \tag{9.8,2}$$

The result is

$$\begin{aligned} \epsilon^2 = \int_{-\infty}^{\infty} &\{[T_I(s) - T(s)]\,[T_I(-s) - T(-s)]\,\Phi_m + T(s)T(-s)\Phi_n \\ &+ T(-s)\,[T(s) - T_I(s)]\Phi_{mn} + T(s)\,[T(-s) - T_I(-s)]\Phi_{nm}\}df. \end{aligned} \tag{9.8,3}$$

The functions Φ_m is the average energy density spectrum of the m_α ensemble. This term arises by considering the energy dissipated by a current equal to $m_\alpha(t)$ in a 1-ohm resistor. The (ensemble) average energy is $\mathsf{E}\int_0^\infty m_\alpha{}^2(t)\,dt = \mathsf{E}\int_{-\infty}^\infty M_\alpha(s)\,M_\alpha(-s)\,df = \int_{-\infty}^\infty \Phi_m\,df$. Therefore, Φ_m indicates the density of the signal energy as a function of frequency, and may be appropriately called the average energy density spectrum of the ensemble of message signals.

§ 9.9 Energy Density Spectra and Correlation Functions

It is worthwhile to pause at this point in order to examine the functions Φ_m, Φ_n, etc., of Eq. (9.8,2). It is evident from Eq. (9.8,3) that the optimum $T(s)$ will perforce be a function of the energy density spectra Φ_m, Φ_n, etc. These functions will therefore have to be evaluated from computations on the message and noise signal samples. Since $\Phi_m = \mathsf{E}\,M_\alpha(s)\,M_\alpha(-s)$, it seems that the

[1] M. Gardner and J. Barnes, "Transients in Linear Systems," p. 275. Wiley, New York, 1942.

transform of *each* message signal must be found, multiplied by its image, and the average of all the products calculated. The first part alone, that of finding the transform of each sample message signal, would involve so much labor (see Chapter 12), that the operation would, in fact, be impractical. Is it possible, instead, to do most of the work on the *ensemble of sample time functions*, and emerge with some sort of average time function? Hopefully, then the Laplace transform of the latter function would suffice for finding Φ_m. Fortunately, this is in fact the case, and is seen as follows.

Consider any sample $M_\alpha(s)\,M_\alpha(-s) = \int_0^\infty m_\alpha(t_1)\,e^{-st_1}\,dt_1 \int_0^\infty m_\alpha(t_2)\,e^{st_2}\,dt_2 = \int_0^\infty \int_0^\infty m_\alpha(t_1)\,m_\alpha(t_2)\exp[-s(t_1 - t_2)]\,dt_1\,dt_2$. [It is possible to find an $s = \sigma + j\omega$, with a σ for which each integral converges, because $m(t)$ is zero at infinite t.] The region of integration in the above is the first quadrant of the t_1, t_2 plane (see Fig. 9.9-1). The double integral can therefore be written as the sum of two integrals, one over the region A, and the other over the region B in Fig. 9.9-1. Thus,

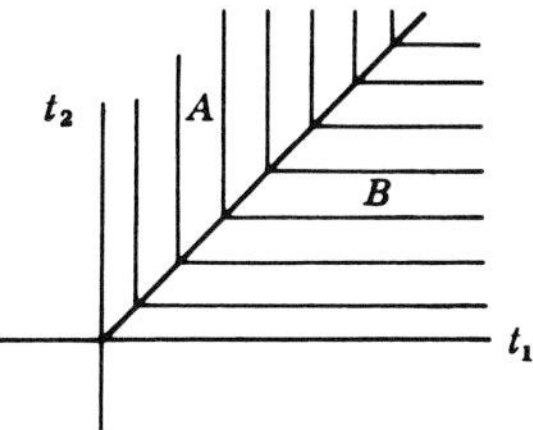

FIG. 9.9-1. Region of integration.

$$M_\alpha(s)\,M_\alpha(-s) = \int_0^\infty m_\alpha(t_1)\,dt_1 \int_{t_1}^\infty m_\alpha(t_2)\exp[-s(t_1 - t_2)]\,dt_2$$

$$+ \int_0^\infty m_\alpha(t_1)\,dt_1 \int_0^{t_1} m_\alpha(t_2)\exp[-s(t_1 - t_2)]\,dt_2 \triangleq I_1 + I_2.$$

Let $\tau = t_1 - t_2$, and in $\int_{t_1}^\infty m_\alpha(t_2)\exp[-s(t_1 - t_2)]\,dt_2$, t_1 is fixed so $dt_2 = -d\tau$. Hence $I_1 = \int_0^\infty m_\alpha(t_1)\,dt_1 \int_{-\infty}^0 m_\alpha(t_1 - \tau)\,e^{-s\tau}\,d\tau = \int_{-\infty}^0 e^{-s\tau}\,d\tau \int_0^\infty m_\alpha(t_1)\,m_\alpha(t_1 - \tau)\,dt_1$.

The function

$$\phi_{m_\alpha}(\tau) \triangleq \int_0^\infty m_\alpha(t_1)\,m_\alpha(t_1 - \tau)\,dt_1 \tag{9.9,1}$$

is called the *autocorrelation function* of $m_\alpha(t)$. If $m(t_1)$ is as shown in Fig. 9.9-2a, then $m(t_1 - \tau)$ for $\tau > 0$ is shown in Fig. 9.9-2b, and $m(t_1 - \tau)$ for $\tau < 0$ is shown in Fig. 9.9-2c. Obviously, $\phi_m(\tau) = \phi_m(-\tau)$. In I_1, we are interested in $\phi_m(\tau)$ for $\tau < 0$, because the integration limits are from $-\infty$ to zero. By analogy with the definition of the usual Laplace transform with limits from zero to infinity, we use the following definition for the transform of the negative time portion of a function:

$$\mathscr{L}^- f(t) \triangleq \int_{-\infty}^0 e^{-st} f(t)\,dt. \tag{9.9,2}$$

Consequently, $I_1 = \mathscr{L}^- \phi_m(\tau)$.

It is easy to prove that if $G(s) = \mathscr{L}g(t)$, with $g(t) = 0$ for $t < 0$, then $\mathscr{L}^{-} g(-t) = G(-s)$. The proof is as follows: In $G(s) = \int_0^{\infty} g(t)\, e^{-st}\, dt$, let $x = -t$, and then $G(s) = -\int_0^{-\infty} g(-x)\, e^{sx}\, dx = \int_{-\infty}^{0} g(-x)\, e^{sx}\, dx$. But $\mathscr{L}^{-} g(-t) = \int_{-\infty}^{0} g(-t)\, e^{-st}\, dt$. Therefore, $\mathscr{L}^{-} g(-t) = G(-s)$. Hence,

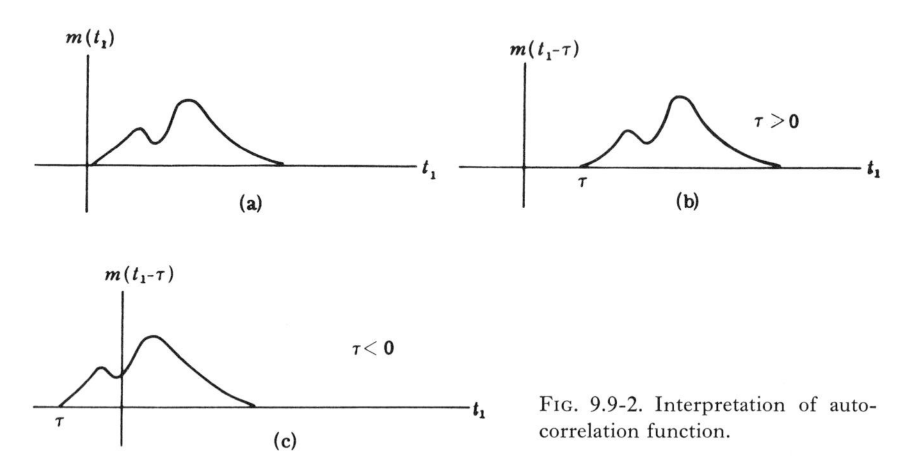

FIG. 9.9-2. Interpretation of autocorrelation function.

transforms with right half-plane (rhp) poles correspond to time functions which are defined for negative time. Any possible conflict with their interpretation as transforms of exponentially increasing time functions for $t > 0$ may be resolved by considering the physics of the problem. If the negative time interpretation applies, then the inverse transform may be found as follows: Let $H(s)$ be the function all of whose poles are in the right half-plane. Its inverse may be found by replacing each s by $-s$, and finding the corresponding inverse $h(t)$ in the usual manner; the desired time function is $h(-t)$. For example, to invert $F(s) = 1/(s - a)$, find the inverse of $-1/(s + a)$, which is $-e^{-at}$ for $t \geqslant 0$. The required function is $-e^{at}$ for $t \leqslant 0$ only. Also, the Parseval relation may be generalized for time functions which exist for both positive and negative time, i.e., $\int_{-\infty}^{\infty} h_1(t)\, h_2(t)\, dt = \int_{-\infty}^{\infty} H_1(s)\, H_2(-s)\, df$, where $H_1(s)$, $H_2(s)$ may have both left half-plane (lhp) poles and rhp poles.

We have found that $I_1 = \mathscr{L}^{-} \phi_{m_\alpha}(\tau)$ of Eq. (9.9,1). Next, we consider $I_2 \triangleq \int_0^{\infty} m_\alpha(t_1)\, dt_1 \int_0^{t_1} m_\alpha(t_1 - \tau)\, e^{-s\tau}\, d\tau$. Since $m_\alpha(t_1 - \tau) = 0$ for $\tau > t_1$, the upper limit of t_1 may be changed to infinity. We may then write $I_2 = \int_0^{\infty} e^{-s\tau}\, d\tau$ $\int_0^{\infty} m_\alpha(t_1)\, m_\alpha(t_1 - \tau)\, dt_1 = \int_0^{\infty} \phi_{m_\alpha}(\tau)\, e^{-s\tau}\, d\tau = \mathscr{L}\phi_{m_\alpha}(\tau)$. Hence $M_\alpha(s)\, M_\alpha(-s)$ $= (I_1 + I_2) = (\mathscr{L}^{-} + \mathscr{L})\, \phi_{m_\alpha}(\tau)$, and

$$\Phi_m \triangleq \mathsf{E}\, M_\alpha(s)\, M_\alpha(-s) = (\mathscr{L} + \mathscr{L}^{-})\, \mathsf{E}\, \phi_{m_\alpha}(\tau) = (\mathscr{L} + \mathscr{L}^{-})\, \phi_m(\tau) \quad (9.9,3)$$

where $\phi_m(\tau) \triangleq \mathsf{E}\, \phi_{m_\alpha}(\tau)$. Since $\mathscr{L}\phi_m(\tau)$ has only lhp poles, and $\mathscr{L}^{-} \phi_m(\tau)$ is its mirror image with only rhp poles, it follows that the poles (and the zeros) of Φ_m are symmetrical with respect to the $j\omega$ axis. This also follows from the fact that $M_\alpha(j\omega)\,M_\alpha(-j\omega)$ is positive for all ω. $M_\alpha(s)\,M_\alpha(-s)$ must therefore have all its poles and zeros symmetrical with respect to the $j\omega$ axis.

The function $\phi_m(\tau)$ may be obtained as follows. The individual samples are put on a tape, as shown in Fig. 9.9-3. It is noted that the T_{odd} in Fig. 9.9-3 are

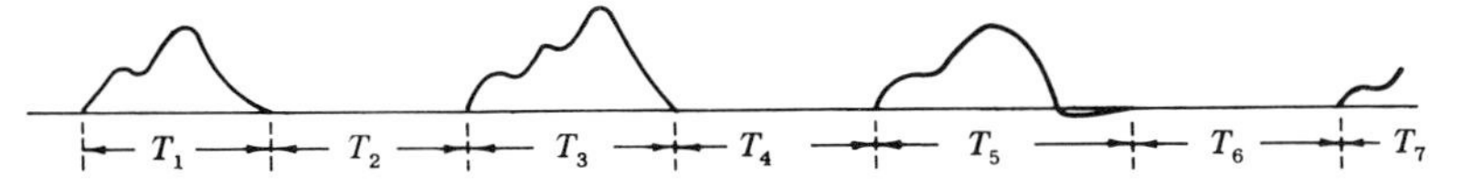

FIG. 9.9-3. A procedure for evaluating autocorrelation function.

in general unequal. T_{even} may or may not be equal, but it is necessary that $T_2, T_4 > T_3$; $T_4, T_6 > T_5$, etc. Two playback heads are used, one displaced from the other by the spatial equivalent of the time interval τ. The signals from the two pick-up heads are multiplied, and the product is integrated and divided by the number of samples, to give $\phi_m(\tau) = \mathsf{E}\,\phi_{m_\alpha}(\tau)$. The process is repeated for different values of τ. The resulting $\phi_m(\tau)$ is the essential information that is needed to characterize the ensemble of message signals. To find $\Phi_m \triangleq \mathsf{E}\,M_\alpha(s)\,M_\alpha(-s)$ in Eq. (9.9,3), we first find $\mathscr{L}\phi_m(\tau) \triangleq \psi(s)$ and then $\Phi_m = \psi(s) + \psi(-s)$. Methods of deriving the transform of a function which is available graphically are described in Chapter 12. Conversely, if $\Phi_m(s)$ is available, $\phi_m(\tau)$ may be found by performing a partial fraction expansion of $\Phi_m(s)$ and writing $\Phi_m(s) = \psi(s) + \psi(-s)$. In the latter, $\psi(s)$ consists of all the terms with lhp poles, and $\psi(-s)$ must consist of those with rhp poles. Then $\phi_m(\tau) = \mathscr{L}^{-1}\psi(s)$. For example, suppose $\Phi_m(\omega) = 5/(4+\omega^2)$; then, replacing ω^2 by $-s^2$, $\Phi_m(s) = 5/(4-s^2) = 5/(2-s)(2+s) = 1.25[(s+2)^{-1} - (s-2)^{-1}]$. Hence, $\psi(s) = 1.25/(s+2)$ and $\phi_m(\tau) = 1.25e^{-2\tau}$. In exactly the same way, the autocorrelation function of $n_\alpha(t)$ is $\phi_{n_\alpha}(\tau) \triangleq \int_0^\infty n_\alpha(t)\,n_\alpha(t-\tau)\,dt$;

$$\phi_n(\tau) \triangleq \mathsf{E}\,\phi_{n_\alpha}(\tau) \text{ and } \Phi_n(s) = \mathsf{E}\,N(s)\,N(-s) = (\mathscr{L} + \mathscr{L}^-)\,\phi_n(\tau).$$

Finally, we need $\Phi_{mn} = \mathsf{E}\,M_\alpha(s)\,N_\alpha(-s)$ and $\Phi_{nm} = \mathsf{E}\,M_\alpha(-s)\,N_\alpha(s)$ in Eq. (9.8,3). $M_\alpha(s)\,N_\alpha(-s) = \int_0^\infty m_\alpha(t_1)\,e^{-st_1}\,dt_1 \int_{-\infty}^0 n(-t_2)\,e^{-st_2}\,dt_2 = \int_0^\infty m_\alpha(t_1)$ $e^{-st_1}\,dt_1 \int_0^\infty n(t_2)\,e^{st_2}\,dt_2$. The procedure used for finding $\Phi_m(s)$ (Fig. 9.9-1, etc.) is exactly repeated. The result is $M_\alpha(s)\,N_\alpha(-s) = \int_{-\infty}^0 e^{-s\tau}\,d\tau \int_0^\infty m_\alpha(t_1)$ $n_\alpha(t_1-\tau)\,dt_1 + \int_0^\infty e^{-s\tau}\,d\tau \int_0^\infty m_\alpha(t_1)\,n_\alpha(t_1-\tau)\,dt_1$. Let

$$\phi_{ab}(\tau) \triangleq \int_0^\infty a(t)b(t-\tau)dt \tag{9.9,4}$$

be known as the cross-correlation function of the signals $a(t)$, $b(t)$. Therefore

$$\mathsf{E}\,M_\alpha(s)\,N_\alpha(-s) = (\mathscr{L} + \mathscr{L}^-)\,\phi_{mn}(\tau) = \Phi_{mn}(s) \tag{9.9,5a}$$

where

$$\phi_{mn}(\tau) \triangleq \mathsf{E} \int_0^\infty m_\alpha(t) n_\alpha(t-\tau) dt. \tag{9.9,5b}$$

It is noteworthy that the rhp poles of $\Phi_{mn}(s)$ are not the mirror image of its lhp poles. However, the pole-zero pattern of $\Phi_{nm}(s) \triangleq \mathsf{E}\, N_\alpha(s)\, M_\alpha(-s)$, must be the mirror image of that of $\Phi_{mn}(s)$. Similarly, $\Phi_{nm}(s) \triangleq \mathsf{E}\, N_\alpha(s)\, M_\alpha(-s) = (\mathscr{L} + \mathscr{L}^-)\, \phi_{nm}(\tau)$, where $\phi_{nm}(\tau) = \mathsf{E} \int_0^\infty n_\alpha(t)\, m_\alpha(t-\tau)\, dt$.

It should be noted that the cross-correlation functions $\phi_{nm}(\tau)$, $\phi_{mn}(\tau)$, in common with the autocorrelation functions $\phi_m(\tau)$, $\phi_n(\tau)$, exist for both negative and positive τ. However, unlike the latter, the cross-correlation functions are not even functions of τ, i.e., if $\psi_{nm}(s) = \mathscr{L}\phi_{nm}(\tau)$, then $\mathscr{L}^-\phi_{nm}(\tau) \neq \psi_{nm}(-s)$. However, $\Phi_{nm}(s) = \Phi_{mn}(-s)$, and $\phi_{nm}(\tau) = \phi_{mn}(-\tau)$ for all τ. This follows from the definition of these functions in Eq. (9.8,2), or from their relation to the cross-correlation functions. Hence, if $\Phi_{mn}(j\omega) = \Phi_1 + j\Phi_2$, then $\Phi_{nm}(j\omega) = \Phi_1 - j\Phi_2$, where Φ_1, Φ_2 are real functions of the real variable ω.

§ 9.10 Optimum but Unrealizable Response Function

We return to the task of finding the optimum $T(s)$. The starting point is Eq. (9.8,3). We shall first take the case when the message and noise are uncorrelated, i.e., Φ_{mn} and $\Phi_{nm} = 0$. Some obvious necessary conditions for this to be true are that the message and noise signals are totally unrelated, that neither ensemble has an average component, and that there is a sufficiently large number of samples to permit chance discrepancies to cancel out. Also, let $T(j\omega) = Ae^{j\theta}$, $T_I(j\omega) = A_I \exp(j\theta_I)$, where the A's and θ's are real functions of ω. Then Eq. (9.8,3) becomes

$$\epsilon^2 = \int_{-\infty}^\infty \{[A_I^2 + A^2 - 2AA_I \cos(\theta_I - \theta)]\, \Phi_m + A^2\Phi_n\}\, df. \tag{9.10,1}$$

The coefficient of Φ_m is minimized by choosing $\theta_I = \theta$, and this choice does not restrict our freedom[1] in choosing the coefficient of Φ_n. With $\theta_I = \theta$, $\epsilon^2 = \int_{-\infty}^\infty [(A_I - A)^2\, \Phi_m + A^2\Phi_n]\, df$. Differentiation with respect to A leads to opt $A = A_I\Phi_m/(\Phi_m + \Phi_n)$. Combining this with $\theta_I = \theta$, we get

$$T_{\text{opt}}{}^* = T_I \frac{\Phi_m}{\Phi_m + \Phi_n}, \tag{9.10,2a}$$

$$[\epsilon^2{}_{\text{opt}}]^* = \int_{-\infty}^\infty \frac{\Phi_m\Phi_n}{\Phi_m + \Phi_n} A_I^2\, df \tag{9.10,2b}$$

[1] Note that we do not apply the constraint of the relationship between magnitude and phase for transfer functions with lhp poles (Section 7.4). Since this constraint is not applied at this point, it should not be surprising if the resulting optimum $T(s)$ turns out to be unrealizable.

The notation T_{opt}^* is used to distinguish the above theoretical but, as we shall see, unrealizable optimum, from the realizable optimum. The latter will be denoted by T_{opt}.

For example, suppose $T_I = 1$, $\Phi_m(\omega) = 5/(4 + \omega^2)$, and $\Phi_n = 1$. Then $\Phi_m(s)$ is obtained by writing $\omega^2 = -(j\omega)^2 = -s^2$. Hence $T_{\text{opt}}^* = [5/(4 - s^2)][1 + 5/(4 - s^2)]^{-1} = 5/(9 - s^2) = -5/(s - 3)(s + 3)$. In general, T_{opt}^* has rhp poles, because the poles of T_{opt}^* are the zeros of $\Phi_i \triangleq \Phi_m + \Phi_n$, and from the nature of Φ_m and of Φ_n [Eq. (9.9,3), etc.], the zeros of Φ_i are symmetrically distributed with respect to the $j\omega$ axis. The rhp poles of Φ_m/Φ_i are associated with a time function which exists for negative time. This can be seen by writing $\mathscr{L}^{-1} T_{\text{opt}}^* = [\mathscr{L}^{-1} T_I] * \mathscr{L}^{-1}[\Phi_m/\Phi_i]$. In accordance with Eq. (9.9,3), $\Phi_m/\Phi_i = (\mathscr{L} + \mathscr{L}^-) g(t)$. Hence $g(t)$ must exist for negative time, and $\mathscr{L}^{-1} T_{\text{opt}}^*$ exists for negative time. Consequently, the optimum theoretical system is one which responds before it is excited. It is therefore not realizable. It would be ridiculous to interpret the rhp poles of Φ_m/Φ_i as corresponding to exponentially increasing positive-time functions, because ϵ^2 would then be infinite, while ϵ^2 in Eq. (9.10,2) is finite.

The unrealizability of T_{opt}^* should not be surprising, because nowhere in the above development was the constraint of physical realizability imposed. We can, in fact, pinpoint where it was violated. This was in assuming that the phase vs. $j\omega$ characteristic of T_{opt}^* could be chosen without any reference to its magnitude frequency characteristic. Thus it is recalled that we chose $\angle (T_{\text{opt}}^*(j\omega)) = \angle (T_I(j\omega))$ and then assumed that we could choose $| T_{\text{opt}}^*(j\omega) | = \Phi_m/(\Phi_m + \Phi_n)$, without the phase of T_{opt}^* being in any way affected. The only way to construct a function with a constant value of phase on the $j\omega$ axis is by having all the poles (and the zeros) of that function symmetrically distributed with respect to the $j\omega$ axis. Each symmetrical pole (or zero) pair contributes the constant angle $\theta_1 + \theta_2 = 180°$ (see Fig. 9.10-1). The only way that T_{opt}^* can be realizable is if the phase characteristic of $T_I(j\omega)$ should accidentally turn out compatible with the magnitude characteristic of $| T_{\text{opt}}^* | = | T_I | \Phi_m/(\Phi_m + \Phi_n)$; or, in other words, if T_I has zeros at the rhp zeros of $\Phi_m + \Phi_n$.

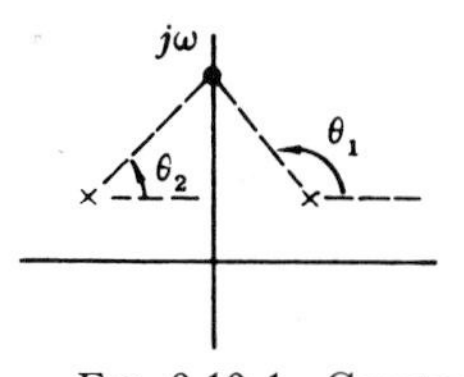

FIG. 9.10-1. Constant phase due to symmetrical pole pair.

§ 9.11 Optimum Realizable Response Function

The constraint of physical realizability must be imposed in order to determine the realizable optimum $T(s)$, denoted henceforth as $T_{\text{opt}}(s)$. The constraint can be imposed in the time domain, or in the frequency domain. In this section, we use a time domain argument in order to obtain T_{opt}. The argument leads to the correct answer, but it is not at all rigorous. It has the merit of being easy to follow, and it leads to an approach which is easy to apply in specific problems. This approach can be used with confidence only because the correct answer has previously been found by other rigorous methods. Nevertheless, we feel justified in presenting it, if only because of its usefulness as a mnemonic.

The argument is based on setting up a fictitious model analogous to that of Fig. 9.8-2. In the fictitious model, a single signal, which is "equivalent" to the ensemble of $m_\alpha(t) + n_\alpha(t)$, is used as the input to the upper T in Fig. 9.8-2. A single "equivalent" signal is also used as the input to T_I in Fig. 9.8-2, in place of the ensemble of $m_\alpha(t)$. These equivalent signals are deliberately chosen so that the resulting expression for $\int_0^\infty e^2(t)\,dt$ is identical to that obtained [Eq. (9.8,3)] in the real problem. Since the two expressions are mathematically identical, the value of T_{opt}^* (i.e., disregarding physical realizability) is the same for the fictitious problem as it is for the real problem. However, because the fictitious problem involves specific signals (not ensembles), it is easy to use physical reasoning to derive the realizable T_{opt}. The derivation is not rigorous, because we cannot assume that the physical reasoning appropriate for the two specific signals is valid for the ensemble of signals.

To find the two "equivalent" signals, let $\Phi_i \triangleq \Phi_m + \Phi_n = \Phi_i^+\Phi_i^-$, where Φ_i^+ contains all the lhp poles and zeros of Φ_i. The constant factor in Φ_i is split up so that $|\Phi_i^+(j\omega)| = |\Phi_i^-(j\omega)|$, i.e., if it is k, then $(k)^{0.5}$ is assigned to each. The signal which is the equivalent of the $m_\alpha + n_\alpha$ ensemble of Fig. 9.8-2, is taken as $\mathscr{L}^{-1}\Phi_i^+ \triangleq \psi_i(t)$. Similarly, the equivalent of the m_α ensemble is taken as $\mathscr{L}^{-1}\Phi_m^+ \triangleq \psi_m(t)$, where $\Phi_m = \Phi_m^+\Phi_m^-$. The real justification for these choices is that they lead to the correct answer. But it can be made more palatable by arguing that the filtering is done on the basis of the energy density spectra of the ensembles. Therefore any signals which have the same energy density spectra as in the real problem may be used. There is an infinitude of signals which have the same energy density spectra, but there is only one whose transform is minimum phase. It is therefore chosen because it is unique and because, as it will be seen, any other choice leads to an unstable T_{opt}.

With the above choice of "equivalents," and the additional assumption that the cross-correlation terms are zero, the fictitious model leads to Eq. (9.10,1) for ϵ^2. Therefore the optimum $T(s)$, with no regard to physical realizability, is T_{opt}^* of Eq. (9.10,2a). At this point, the following physical argument is used: T_{opt}^*, if it could be constructed, would give an output for negative time as well as positive time. It is the part that exists for negative time that makes the filter unrealizable. The best that one can hope to achieve is to find a realizable filter T_{opt} whose output for positive time only is exactly the same as that of the unrealizable optimum. Hence the output of T_{opt}, which is $\mathscr{L}^{-1}[\Phi_i^+T_{\text{opt}}]$, is set equal to $\mathscr{L}^{-1}[\Phi_i^+T_{\text{opt}}^*]$ for $t \geqslant 0$ only. In view of the relation $\Phi_i = \Phi_i^+\Phi_i^-$ and Eq. (9.10,2a),

$$\mathscr{L}^{-1}[\Phi_i^+T_{\text{opt}}] = \mathscr{L}^{-1}[\Phi_i^+T_{\text{opt}}^*] = \mathscr{L}^{-1}\left[\Phi_i^+\frac{T_I\Phi_m}{\Phi_i}\right] = \mathscr{L}^{-1}\left[\frac{T_I\Phi_m}{\Phi_i^-}\right]$$

for $t \geqslant 0$ only. It is necessary to find that portion of $\mathscr{L}^{-1}[T_I\Phi_m/\Phi_i^-]$ which exists for $t \geqslant 0$. This is done by expanding $T_I\Phi_m/\Phi_i^-$ in partial fractions and picking the terms with the lhp poles. The terms with rhp poles correspond to negative time functions. Let the transform of the positive time portion of $\mathscr{L}^{-1}[T_I\Phi_m/\Phi_i^-]$ be denoted by $[T_I\Phi_m/\Phi_i^-]^+$. Then

$$T_{\text{opt}} = \frac{1}{\Phi_i^+}\left[\frac{T_I\Phi_m}{\Phi_i^-}\right]^+. \tag{9.11,1}$$

***Example 1**—Optimum Positioning Filter.* In this case, $T_I = 1$. Suppose $\Phi_n = 1$, and $\Phi_m(\omega) = 5/(4+\omega^2)$, so that, replacing ω^2 by $-s^2$, $\Phi_m(s) = 5/(4-s^2)$. $\Phi_i = \Phi_m + \Phi_n = (9-s^2)/(4-s^2) = [(s+3)/(s+2)]\,[(s-3)/(s-2)] = \Phi_i^+\Phi_i^-$. Consequently, $[T_I\Phi_m/\Phi_i^-]^+ = \{[5/(2-s)\,(2+s)]\,(s-2)/(s-3)\}^+ = [(s+2)^{-1} - (s-3)^{-1}]^+ = 1/(s+2)$. Therefore $T_{\text{opt}} = (1/\Phi_i^+)\,[1/(s+2)] = 1/(s+3)$.

***Example 2**—Optimum Predictor.* In this problem, $T_I = e^{\alpha s}$, if the filter is to predict the value of the message component α seconds in the future. $\mathscr{L}^{-1}[T_I\Phi_m/\Phi_i^-] = \mathscr{L}^{-1}e^{\alpha s}[(s+2)^{-1} - (s-3)^{-1}] = e^{-2(t+\alpha)}u(t+\alpha) + u^*(t+\alpha)\,e^{3(t+\alpha)}$, where $u^*(t+\alpha) \overset{\Delta}{=} 1$ for $t < -\alpha$ and is zero otherwise. The portion of the above which exists for $t > 0$ is $u(t)\,e^{-2(t+\alpha)}$, whose transform is $e^{-2\alpha}/(s+2)$. Hence, $[T_I\Phi_m/\Phi_i^-]^+ = e^{-2\alpha}/(s+2)$ and $T_{\text{opt}} = e^{-2\alpha}/(s+3)$.

***Example 3**—Optimum Estimation of the Past.* At $t > \beta$, the ideal output is the message input at $t-\beta$, i.e., $T_I = e^{-\beta s}$. $\mathscr{L}^{-1}[T_I\Phi_m/\Phi_i^-] = \mathscr{L}^{-1}e^{-\beta s}[(s+2)^{-1} - (s-3)^{-1}] = e^{-2(t-\beta)}u(t-\beta) + e^{3(t-\beta)}u^*(t-\beta)$, where $u^*(t-\beta) = 1$ for $t < \beta$ and is zero otherwise. To find $\mathscr{L}^{-1}[-e^{-\beta s}/(s-3)]$, we replace s by $-s$ and find $\mathscr{L}^{-1}[e^{\beta s}/(s-3)] = e^{-3(t+\beta)}u(t+\beta)$, which, when reflected around the ordinate, gives $e^{3(t-\beta)}u^*(t-\beta)$. Therefore, $[T_I\Phi_m/\Phi_i^-]^+ = e^{-\beta s}/(s+2) + \int_0^\beta e^{-st}\,e^{3(t-\beta)}\,dt = e^{-\beta s}/(s+2) + (e^{-3\beta} - e^{-s\beta})/(s-3)$, and $T_{\text{opt}} = 1/\Phi_i^+[T_I\Phi_m/\Phi_i^-]^+ = [(s+2)/(s+3)]\,[e^{-\beta s}/(s+2) + (e^{-3\beta} - e^{-\beta s})/(s-3)]$. It appears superficially that T_{opt} has a rhp pole at $s = 3$. However, it is canceled out by the zero of $e^{-3\beta} - e^{-\beta s}$ at $s = 3$.

§ 9.12 Derivation of the Realizable Optimum Filter Function

In this section a rigorous solution of the optimum filter problem is presented. The starting point is Eq. (9.8,3), which, for the zero cross-correlation case ($\Phi_{mn} = \Phi_{nm} = 0$) is

$$\epsilon^2 = \int_{-\infty}^{\infty} \{[T_I(s) - T(s)]\,[T_I(-s) - T(-s)]\Phi_m + T(s)T(-s)\Phi_n\}df \tag{9.12,1}$$

The calculus of variations[1] is used to find the unknown function $T = T_{\text{opt}}(s)$,

[1] F. B. Hildebrand, "Methods of Applied Mathematics," Chapter 2. Prentice-Hall, New York, 1952.

which minimizes ϵ^2. The procedure is to replace T in Eq. (9.12,1) by $T = T_{\text{opt}} + k\Delta(s)$, with k real and $\Delta(s)$ an arbitrarily chosen function, except for the condition that $T_{\text{opt}} + k\Delta(s)$ is in the class of suitable response functions. A suitable response function is one which is physically realizable, and which satisfies the specifications of the problem. For example, if the problem constraints demand that T_{opt} have a specific zero, then $\Delta(s)$ must have such a zero. With this substitution for T, Eq. (9.12,1) becomes

$$\epsilon^2 = \int_{-\infty}^{\infty} \{[T_I(s) - T_{\text{opt}}(s)]\,[T_I(-s) - T_{\text{opt}}(-s)]\Phi_m + T_{\text{opt}}(s)T_{\text{opt}}(-s)\Phi_n\}\,df$$

$$+ k^2 \int_{-\infty}^{\infty} \Delta(s)\Delta(-s)\Phi_i\,df + k \int_{-\infty}^{\infty} [T_{\text{opt}}(s)\Phi_i - T_I(s)\Phi_m]\Delta(-s)\,df$$

$$+ k \int_{-\infty}^{\infty} [T_{\text{opt}}(-s)\Phi_i - T_I(-s)\Phi_m]\Delta(s)\,df \overset{\Delta}{=} P + k^2Q + k(W_1 + W_2). \tag{9.12,2}$$

In order that ϵ^2 be a minimum, it is necessary that $\partial\epsilon^2/\partial k = 2kQ + W_1 + W_2 = 0$, for all k. But the minimum is postulated at $k = 0$. Therefore, a necessary condition for T_{opt} to result in a minimum value of ϵ^2 is that $W_1 + W_2 = 0$. The integration is along $s = j\omega$; hence $W_1 = W_2$. Accordingly, we may take the necessary condition as

$$\int_{-\infty}^{\infty} [T_{\text{opt}}(s)\Phi_i - T_I(s)\Phi_m]\Delta(-s)\,df = 0, \tag{9.12,3}$$

for all permissible $\Delta(-s)$.

An obvious choice for T_{opt}, which satisfies Eq. (9.12,3), is $T_{\text{opt}}\Phi_i = T_I\Phi_m$. This is T_{opt}^{+} of Eq. (9.10,2); it is inadmissible if we impose the constraint that $T = T_{\text{opt}} + k\Delta$ must have no rhp poles. The following derivation[1] is used to find a realizable optimum $T(s)$. Let $A(s) \overset{\Delta}{=} T_{\text{opt}}\Phi_i^{+} - T_I(\Phi_m/\Phi_i^{-})$, and divide $A(s)$ into $A(s) = A^+ + A^-$, where the impulse response of A^+ $[a^+(t) = \mathscr{L}^{-1}A^+]$ exists for $t \geqslant 0$ only, and that of A^- [denoted by $a^-(t) = \mathscr{L}^{-1}A^-$] exists for $t \leqslant 0$ only. Equation (9.12,3) may then be written as

$$\int_{-\infty}^{\infty} A(s)\Phi_i^{-}\Delta(-s)df = \int_{-\infty}^{\infty} A^+\Phi_i^{-}\,\Delta(-s)df + \int_{-\infty}^{\infty} A^-\Phi_i^{-}\,\Delta(-s)df = 0. \tag{9.12,4}$$

From the generalized Parseval relation, $\int_{-\infty}^{\infty} A^-\Phi_i^-\Delta(-s)\,df = \int_{-\infty}^{\infty} a^-(t)x(t)dt$, where $x(t) \overset{\Delta}{=} \mathscr{L}^{-1}\Phi_i^{+}\Delta(s)$ exists for positive time only, while $a^-(t)$ of course

[1] This derivation is due to Otto J. M. Smith, Separating information from noise. *Trans. IRE* **PGCT-1**, 81-100 (1952).

exists for negative time only. Hence the last integral in Eq. (9.12,4) is zero, and (9.12,4) becomes $\int_{-\infty}^{\infty} A^+\Phi_i^-\Delta(-s)\,df = \int_0^{\infty} a^+(t)\,x(t)\,dt = 0$. Since $\Delta(-s)$ is arbitrary, so is $x(t)$; therefore $a^+(t) = 0$ and its transform $A^+ = 0$; i.e., $[T_{\text{opt}}\Phi_i^+]^+ = [T_I\Phi_m/\Phi_i^-]^+$. But $[T_{\text{opt}}\Phi_i^+]^+ = T_{\text{opt}}\Phi_i^+$. Equation (9.11,1) follows. Finally, it must be shown that $d^2t^2/dk^2 > 0$, in order that T_{opt} give a minimum ϵ^2 rather than a maximum. This is obviously so, because Q [of Eq. (9.12,2)] > 0.

Constraints on Φ_m, Φ_n

The above results are valid only if ϵ^2 in Eq. (9.12,1) is finite, for only then does a minimum exist. For this purpose it is necessary that as s approaches infinity, the integrand in Eq. (9.12,1) must go to zero as s^{-x} (denoted as "Integrand $= Os^{-x}$"), with $x > 1$. It is assumed that all the variables in Eq. (9.12,1) behave at infinity as s^y with y an integer. Hence $x \geqslant 2$. Inspection of Eq. (9.12,1) reveals that each of the following must be Os^{-x}: $T_I^2\Phi_m$, $T^2\Phi_i$, $TT_I\Phi_m$. Let $T_I = Os^{-I}$, $T = Os^{-p}$, $\Phi_m = Os^{-2m}$, $\Phi_n = Os^{-2n}$, $\Phi_i = Os^{-2i}$ with $i = n$ if $m > n$, $i = m$ if $n > m$. Therefore $x \geqslant 2$ requires that $I + m \geqslant 1$, $p + m \geqslant 1$, $p + n \geqslant 1$. The last condition $p + I + 2m \geqslant 2$ (from $TT_I\Phi_m = Os^{-x}$) then follows.

The only inequality the designer must check in advance is

$$I + m \geqslant 1. \tag{9.12,5}$$

If (9.12,5) holds, then the others also are true, i.e., Eq. (9.11,1) must then result in a value of p which satisfies the other inequalities. This should be clear from the fact that in Eq. (9.12,1) $T(s)$ can always be chosen with a p [$T(s) = Os^{-p}$] sufficiently large so that ϵ^2 is finite [if Eq. (9.12,5) holds]. Hence $T(s)$ which minimizes ϵ^2 certainly must have a satisfactory value of p. If (9.12,5) does not hold, then T_I (or Φ_m) can be modified, viz., let $T_{I(\text{new})} = T_{I(\text{old})}\, ab \ldots/(s + a)(s + b)\ldots$, such that (9.12,5) holds. Equation (9.12,1) is used to find T_{opt} as a function of $a, b, \ldots$. Finally, $a, b, \ldots$ are allowed to approach infinity.

§ 9.13 Optimum Filter When Message and Noise Signals Are Related

In this section, we do not assume that $\Phi_{mn} = 0 = \Phi_{nm}$; i.e., the message and noise signals may be related. The starting point is Eq. (9.8,3). The calculus of variations may be used, just as in Section 9.12. If $T_{\text{opt}}(s)$ is the optimum admissible response function, then $T(s) = T_{\text{opt}}(s) + k\Delta(s)$ is the class of all admissible response functions. This expression for $T(s)$ is used in Eq. (9.8,3), and it results in $\epsilon^2 = \epsilon_o^2 + k^2H + k(F_1 + F_2)$, where ϵ_o^2 is identical to Eq. (9.8,3), if T is replaced by T_{opt}, $H = \int_{-\infty}^{\infty} \Delta(s)\,\Delta(-s)\,(\Phi_m + \Phi_n + \Phi_{mn} + \Phi_{nm})\,df$. F_2 is the conjugate of F_1, and $F_1 = \int_{-\infty}^{\infty} \{T_{\text{opt}}(s)\,(\Phi_m + \Phi_n + \Phi_{mn} + \Phi_{nm}) - T_I(s)\,(\Phi_m + \Phi_{mn})\}\, \Delta(-s)\, df$. A necessary condition for an optimum is that $F_2 = F_1 = 0$. A T_{opt} that satisfies this condition must be shown to be unique, and it must be shown that $\partial^2\epsilon/\partial k^2 > 0$, in order that ϵ^2 be a minimum for this value of T_{opt}. The last condition is satisfied, because H is always positive. Let $\Phi_i' \triangleq \Phi_m + \Phi_n + \Phi_{mn} + \Phi_{nm}$, $\Phi_m' \triangleq \Phi_m + \Phi_{mn}$ and $A(s) = A^+ + A^- \triangleq T_{\text{opt}}\Phi_i'^+ - T_I(\Phi_m'/\Phi_i'^-)$. The procedure is henceforth exactly the same as in Section 9.12, and leads to

$$T_{\text{opt}} = \frac{1}{\Phi_i'^+}\left[\frac{T_I\Phi_m'}{\Phi_i'^-}\right]^+. \qquad (9.13,1)$$

The bracketed factor on the right-hand side of Eq. (9.13,1) is the transform of the positive time portion of what is inside the bracket, exactly as in Eq. (9.11,1). There is a very simple physical interpretation of Eq. (9.13,1), similar to that of Eq. (9.11,1). The concept of the single signal equivalent of the entire ensemble is used. In this case, the "equivalent" signal is $\mathscr{L}^{-1}\Phi_i'^+$. Also, the unrealizable optimum is obviously, from the expression for F_1, $T_{\text{opt}}^* = T_I\Phi_m'/\Phi_i'$. The "equivalent" signal passing through T_{opt}^* produces the best "equivalent" output for $t > 0$, but also gives an output for $t < 0$. The realizable optimum is the one that gives the same positive time output as T_{opt}^*. The transform of the latter is $[T_{\text{opt}}^*\Phi_i'^+]^+$. Hence $T_{\text{opt}}\Phi_i'^+ = [T_{\text{opt}}^*\Phi_i'^+]^+$, which is Eq. (9.13,1).

Example. $T_I = 1$, $\Phi_{mn} = -1/(s-1)(s+2)$, $\Phi_{nm} = -1/(s+1)(s-2)$, $\Phi_m = -7/(s^2-9)$, $\Phi_n = 1$. Therefore $\Phi_i = (s^2-16)/(s^2-9)$, $\Phi_i' = [(s-4.14)(s-2.13)(s-1.13)(s+4.14)(s+2.13)(s+1.13)]\ [(s^2-9)(s^2-1)(s^2-4)]^{-1}$, $\Phi_m' = (-8s^2-7s+23)[(s-1)(s+2)(s^2-9)]^{-1}$. $[T_I\Phi_m'/\Phi_i'^-]^+ = 0.252(s+2)^{-1} + 0.965(s+3)^{-1} = 1.217(s+2.21)/(s+2)(s+3)$, and $T_{\text{opt}} = 1.217(s+2.21)(s+1)[(s+4.14)(s+2.13)(s+1.13)]^{-1}$.

§ 9.14 Conditions under Which the Optimum Linear Time-Varying Filter Degenerates into a Time-Invariant Filter

Under what conditions is the optimum linear time-invariant filter of Sections 9.8-9.13 as good as the optimum linear time varying filter? This requires qualification, as it may be argued that the optimal filter of Sections 9.8-9.13 is time varying, because it has been postulated that the filter is switched off after the signal-to-noise ratio has fallen below some value. To answer this argument, we may assume that all the input signals appear at $t = 0$ and simply disappear at the appropriate time. Or we may rephrase the question to: Would a filter with unrestricted time-varying properties give a smaller ensemble-averaged squared error, than the semi-time-invariant filter that was arbitrarily postulated in Section 9.8?

In order to answer the above question, it is necessary to return to the original formulation of the problem, remove the restriction of a linear time-invariant filter, and impose the lesser restriction of a linear time-varying filter (the least restrictive would be a nonlinear time-varying filter). The time-invariant restriction was imposed soon after Eq. (9.8,1) in the substitutions $c_{I\alpha}(t) = \mathscr{L}^{-1}T_I(s)M_\alpha(s)$ and $c_\alpha(t) = \mathscr{L}^{-1}T(s)\,[M_\alpha(s) + N_\alpha(s)]$, and the use of the Parseval relation. The resulting formulation in the frequency domain permitted a tremendous simplification of the optimum filter problem. In the time-varying case, this cannot be done. In addition, in the time-varying problem, it is permissible to consider the optimization problem as one of minimizing the ensemble average, $\mathsf{E}[e_\alpha(t)]^2$ at every instant of time, rather than one of minimizing $\mathsf{E}\int_0^\infty [e_\alpha(t)]^2\,dt$. Certainly the latter is a minimum when the former is a minimum. The same time-varying filter results whichever measure is used. We shall formulate the problem, but present only a few steps in its solution.[1] The purpose is to clarify what conditions must be satisfied in order that the general linear optimum filter, which is a time-vaying filter, degenerates into a time-invariant filter.

The problem is as follows: Given an ensemble of message signals $m_\alpha(t)$, and one of noise signals $n_\alpha(t)$; find the time-varying linear filter $f(t, \tau)$, which minimizes $\mathsf{E}\,[e_\alpha(t)]^2$ at every t.

The problem is formulated in Fig. 9.14-1, with $\mathsf{E}\,e_\alpha{}^2(t) = \mathsf{E}\,[m_{\alpha I}(t) - s_{\alpha o}(t)]^2$

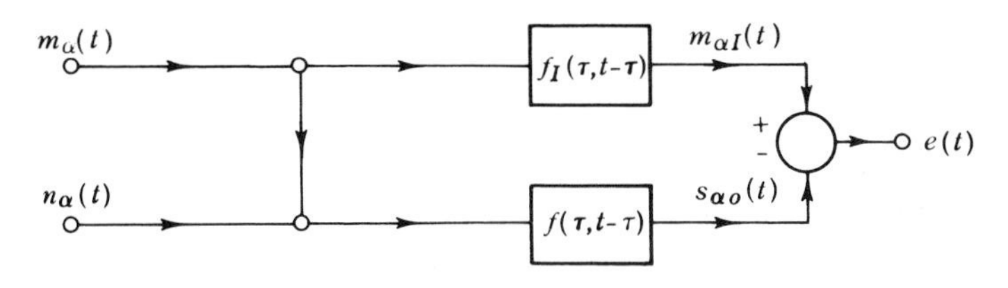

FIG. 9.14-1. The time varying filter problem.

where $s_{\alpha o}(t) \triangleq m_{\alpha o}(t) + n_{\alpha o}(t)$. The time-varying impulse response function $f(\tau, t - \tau)$ is defined as follows: If an impulse is applied at time τ, the output at time t (with $t > \tau$) is $f(\tau, t - \tau)$. In the time-invariant filter the output is a function of $t - \tau$ only, but in the time-varying filter it is also a function of the instant at which the impulse is applied. Therefore, if the input is $s_\alpha(t) = m_\alpha(t) + n_\alpha(t)$, the output is $s_{\alpha o}(t) = \int_0^t s_\alpha(\tau) f(\tau, t - \tau)\,d\tau$, and

$$\epsilon^2(t) \triangleq \mathsf{E}[e(t)]^2 = \mathsf{E}\left[m_{\alpha I}(t) - \int_0^t s_\alpha(\tau) f(\tau, t - \tau)d\tau\right]^2 = \mathsf{E}\Big\{m_{\alpha I}{}^2(t)$$

$$- 2\int_0^t m_{I\alpha}(t) s_\alpha(\tau) f(\tau, t - \tau)d\tau + \int_0^t\int_0^t s_\alpha(\tau)s_\alpha(\zeta) f(\tau, t - \tau) f(\zeta, t - \zeta)d\tau d\zeta\Big\}.$$

[1] R. C. Booton, Jr., An optimization theory for time-varying filter systems with non-stationary inputs, *Proc. IRE* **40** (8), 977-981 (1952).

The following correlation functions are defined:

$$\phi_{Is}(t_1, t_2) = \mathsf{E}\, m_{\alpha I}(t_1)\, s_\alpha(t_2), \qquad \phi_{ss}(t_1, t_2) = \mathsf{E}\, s_\alpha(t_1)\, s_\alpha(t_2),$$

$$\phi_{II}(t_1, t_2) = \mathsf{E}\, m_{\alpha I}(t_1)\, m_{\alpha I}(t_2)$$

and with these,

$$\epsilon^2(t) = \phi_{II}(t, t) - 2\int_0^t \phi_{Is}(t,\tau) f(\tau, t-\tau)\, d\tau + \int_0^t \int_0^t \phi_{ss}(\tau, \zeta) f(\tau, t-\tau) f(\zeta, t-\zeta)\, d\tau\, d\zeta. \tag{9.14,1}$$

It is not necessary to develop the time-varying filter problem much further, because we are not seeking the solution to the problem, but only wish to find the conditions under which the time-invariant filter is as good as the time-varying filter. It is noted that the time-varying filter can minimize the ensemble error $\epsilon^2(t)$ of Eq. (9.14,1) at every single instant t. If the time invariant filter is to do as well, it too must minimize the error at every single instant. But how can it aspire to do this when it is unable to adjust itself by changing its characteristics? Obviously, it can be as good only if the time-variation property is superfluous, i.e., if there is no need for the filter to have this additional flexibility. The filter problem is then a stationary one—whatever is the optimum filtering at any one instant, it remains the optimum filtering for all time. In such a case, there must be the same ensemble average squared error at all times, and $\epsilon^2(t)$ is a constant. This condition requires that time be indistinguishable in the filtering process. We certainly cannot have a semi-infinite interval of zero error, and then instantaneously the error changes to a constant value for all later time. A time-varying filter would be better if there was such a violent change in the input. We therefore are forced to the conclusion that the time invariant filter can be as good as the time varying only if the input was applied an infinitely long time ago. Hence the lower limits of integration must be changed to $-\infty$ in Eq. (9.14,1).

The above is a necessary condition that no instant in time be distinguishable in the optimum time invariant filter. But we shall see that it is not sufficient. It is also necessary that the characteristics of the noise ensemble and of the filter ensemble remain the same at all time. In such a case, the optimum at any one time remains optimum for all time. In short, in this contest between the mixture of noise and message signals and the filter, the nature of the ideal filter is time invariant only if (1) no instant of time is distinguishable from any other—hence the origin of time is at $-\infty$, (2) the problem confronting the filter is invariant with time—hence the noise and message ensemble characteristics which are relevant to the problem, must be independent of time.

The above is verified by Eq. (9.14,1). First we change the lower limits of integration to $-\infty$. It is also convenient to replace τ by $t+\beta$ and ζ by $t+\lambda$. Equation (9.14,1) becomes

$$\epsilon^2(t) = \phi_{II}(t, t) - 2\int_{-\infty}^{0} \phi_{Is}(t, t+\beta) f(t+\beta, -\beta) d\beta$$

$$+ \int_{-\infty}^{0}\int_{-\infty}^{0} \phi_{ss}(t+\beta, t+\lambda) f(t+\beta, -\beta) f(t+\lambda, -\lambda) d\beta d\lambda.$$

It is important to note that the limits of the integrals are independent of t, thanks to the lower limits in Eq. (9.14,1) being changed to $-\infty$.

Under what conditions is ϵ^2 independent of t? If we use a time-invariant filter, then $f(t+\beta, -\beta)$ is only a function of β, but ϵ^2 is still a function of t. It is also necessary that $\phi_{II}, \phi_{Is}, \phi_{ss}$ be independent of t, i.e., $\phi_{Is}(t_1, t_2)$, $\phi_{ss}(t_1, t_2)$, and $\phi_{II}(t_1, t_2)$ must be functions of $t_1 - t_2$ only. In such a case, $\epsilon^2(t) = \phi_{II}(0) - 2\int_{-\infty}^{0} \phi_{Is}(\beta) f(t+\beta, -\beta)\, d\beta + \int_{-\infty}^{0}\int_{-\infty}^{0} \phi_{ss}(\beta - \lambda) f(t+\beta, -\beta) f(t+\lambda, -\lambda) d\beta d\lambda$. There is then no need for $f(t+\beta, -\beta)$ to be a function of t; ϵ^2 is a constant, and the time-varying filter degenerates into a time invariant filter.

It is important to understand the characteristics of this degenerate ensemble, and how it compares with the ensemble postulated in Section 9.8 (Fig. 9.8-1). Consider the two message ensemble members sketched in Fig. 9.14-2a, and

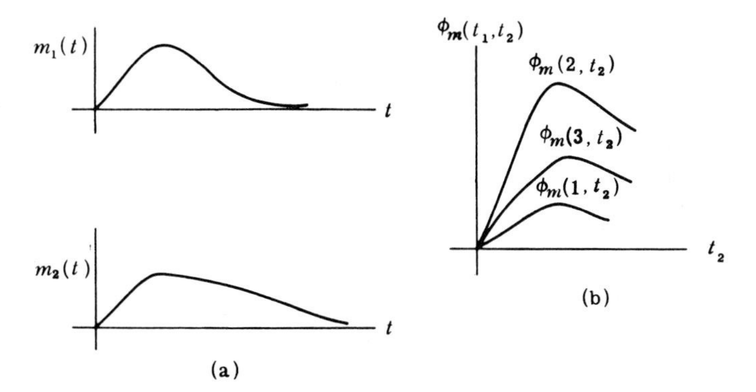

FIG. 9.14-2. Signals whose correlation function is a function of two variables.

suppose they are sufficiently representative for studying the gross properties of the correlation function $\phi_{mm}(t_1, t_2) = \mathsf{E}\, m_\alpha(t_1)\, m_\alpha(t_2)$. In Fig. 9.14-2b some sketches of $\phi_{mm}(t_1, t_2)$ are made as functions of t_2 for fixed t_1. Thus $\phi_{mm}(0, t_2)$ is zero for all t_2; $\phi_{mm}(2, t_2)$ is exactly twice $\phi_{mm}(1, t_2)$, etc. Clearly, $\phi_{mm}(t_1, t_2)$ is not a function of $t_1 - t_2$ alone, since it is zero for all $t_1 - t_2$ when t_1 (or t_2) is zero, and it has finite values otherwise. In this case, ϕ_{mm} is a function of two independent variables—it is impossible to express it as a function of $t_1 - t_2$ alone.

On the other hand, consider an ensemble composed of a very large number of samples, with two typical members sketched in Fig. 9.14-3 (the signals exist and have the same general nature for all time). In this case, $\mathsf{E}\, m_\alpha(t_1)\, m_\alpha(t_2)$ is

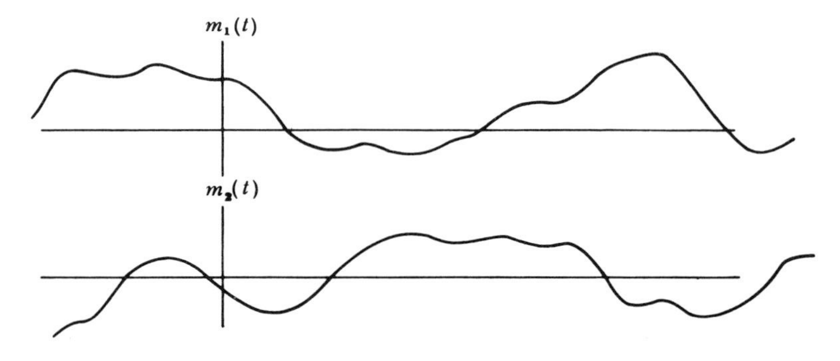

FIG. 9.14-3. Signals whose correlation function is a function of one variable.

a function of $t_1 - t_2$ only. Thus suppose $t_1 = 1$, $t_2 = 0$, and the measurements are made over the ensemble, and averaged. Then let $t_1 = 0$, $t_2 = 1$ or $t_1 = 1$, $t_2 = 2$. The ensemble averages will be the same. Imagine that a set of measurements has been made, and that while one's back is turned, the time axis has been shifted in different amounts for the different members of the ensemble, and then the measurements are repeated. The values of the correlation functions will be the same as before. The ensemble members have basically the same statistical properties at all times—the statistics of the ensemble are stationary, and the probability functions associated with the ensemble are stationary. In Fig. 9.14-2a the probability that any ensemble member has a specific value x is clearly a function of x and of time. However, in Fig. 9.14-3, such a probability function is a function of x only.

Finally, an even greater degeneracy is possible. Suppose that all members of the ensemble are, in a statistical sense, indistinguishable from each other. The latter does not require that they all be exactly alike. It does mean that any statistical measurements made over a long period of time on any one member of the ensemble would be precisely duplicated on any other member, as, for example, measurements of the thermal noise of an ensemble of resistors with the same value of resistance and maintained at the same temperature. In such a case, only one member of the ensemble need be examined in order to obtain any desired statistical information. The value of an ensemble average at any instant of time is obtainable by averaging over time on any one member of the ensemble. This property is called the ergodic property.

It is seen from the above that only when the correlation functions are independent of the time at which measurements are taken, does the best (time-varying) linear filter degenerate into a time-invariant linear filter.[1] This is a very reasonable result. The sample input signal is fed to a box whose function it is to make the best estimate of the message component. The estimate is made on the basis of the *a priori* knowledge it has of the general characteristics of the message and noise. If these characteristics are functions of time, then surely the processing of the signal input should also change with time—the optimum filter is then a time-varying filter. The crucial question is: What characteristics of the ensemble of message and noise signals determine the estimate? We have seen that it is the ensemble-averaged correlation functions. Therefore, when these do not vary in time, the filter characteristics need not vary in time.

Nevertheless, due to the greater complexity of the time-varying filter, and the present state of the art, one very often uses the time-invariant filter, even when the signal statistics are nonstationary. The general approach used in Sections 9.8-9.13 for finding the optimum linear time-invariant filter is, therefore, very useful, even for nonstationary problems. In more exacting design problems (aircraft automatic landing systems, missile guidance, etc.) where the considerations of price and design labor are not as great, the time-varying system is used, and increasing attention is being given to it. However, it is outside the scope of this book.

It should be noted that with infinite duration signals of the stationary statistics type shown in Fig. 9.14-3, it is no longer possible to formulate the problem as one of minimizing $\mathsf{E} \int_{-\infty}^{\infty} [e_\alpha(t)]^2 \, dt$. The quantity to be minimized is $\mathsf{E} \lim_{T\to\infty}(1/2T) \int_{-T}^{T} [e_\alpha(t)]^2 \, dt$. The procedure and results are, however, basically the same. Since all measurements are in practice finite anyhow, one can, in fact, use the same formulation as in Sections 9.8 and 9.9 with finite integral squared error, but use a sufficiently long time interval to ensure that the correlation functions, energy or power spectra, have been correctly evaluated. The practical problems of how to evaluate the power spectra and correlation functions, how to decide whether the ensemble statistics are stationary, etc., are discussed in many references.[2]

---

[1] The solution of the optimum filter problem was done independently by N. Wiener and by A. Kolmogoroff, for stationary random processes: N. Wiener, ,,The Extrapolation, Interpolation & Smoothing of Stationary Time Series.'' Wiley, New York, 1948; A. Kolmogoroff, *Bull. Acad. Sci.* (*U.S.S.R.*) **5**, 3 (1941). It has also been shown that the optimum filter designed on a minimum mean-square error basis for Gaussian inputs remains optimum for any other weight function $\phi(\epsilon)$ which satisfies the following conditions: $\phi(\epsilon)$ is nonnegative, nondecreasing for $\epsilon \geqslant 0$, and nonincreasing for $\epsilon \leqslant 0$. See J. L. Brown, Assymetric non-mean-square error criteria, *Trans. IRE* **AC-7** (1), 64-66 (1962).

[2] W. Davenport and W. Root , ,,An Introduction to the Theory of Random Signals and Noise,'' McGraw-Hill, New York, 1958; R. B. Blackman and J. W. Tukey, "The Measurement of Power Spectra from the Point of View of Communications Engineering.'' Dover, New York, 1959.

Discussion of the Optimization Technique

In the original formulation of the filter problem in Section 9.8, nothing was said about the relationship of the individual message (and noise) samples to each other. The results apply even if they radically differ from each other. But if they do so differ, the resulting optimum is just a compromise in a difficult situation, and the system behaves poorly to all or most of the individual message signals. The point is that if the characteristics of the individual samples are very much dispersed about the average, then the results are of very little benefit in practice. The design is being asked to handle too wide a variety of message and noise signals with the simple linear time invariant filter.

It is obvious that from this point of view the ideal ensemble is one in which the correlation functions of all the individual sample signals are the same. This, in fact, is the situation for ensembles in which the ergodic property applies. For other ensembles, one may, of course, compare the correlation functions of the individual sample signals with the ensemble-averaged correlation functions, and note their dispersion about the latter.

§ 9.15 Applications to Joint Filter and Feedback Problems

The optimization technique that has been developed is very useful for a large variety of problems. In Fig. 9.15-1a $m(t)$ and $n(t)$ are, respectively, the useful message and false noise components of the signal input. In addition, there is a disturbance $d(t)$ acting on the system. The problem is to choose a

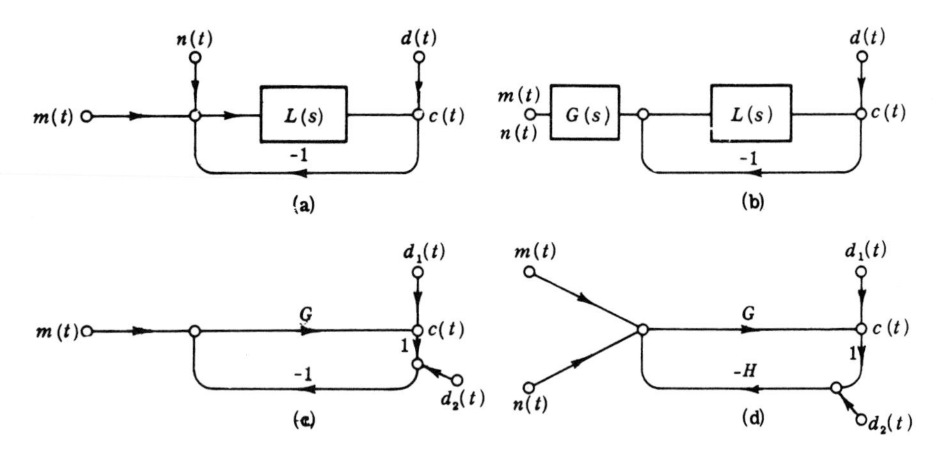

FIG. 9.15-1. (a) One-degree-of-freedom system subject to ensemble of message, noise and disturbance signals. (b) The problem is trivial in the two-degree-of-freedom system. (c) Message and two disturbance signal ensembles in one-degree-of-freedom system. (d) Message, noise and two disturbance signal ensembles in two-degree-of-freedom system.

time invariant linear L so as to minimize the error. If $n(t)$ did not exist, then the loop transmission L could be made very large over a wide frequency band, and the system would respond excellently to $m(t)$ and very poorly to $d(t)$. The presence of $n(t)$ prevents such a solution. However, if a pre-filter could be inserted, as in Fig. 9.15-1b, then G could be chosen to provide the optimum filtering of N, and then L could be assigned as large a gain-bandwidth as needed to attenuate D. It is supposed in this problem that one must, for whatever the reason may be, use precisely the structure of Fig. 9.15-1a, with its one degree of freedom.

As before, there is an ensemble of each of the three inputs, and the ensemble average error is to be minimized. In Fig. 9.15-1a the ideal output (for a follow-up system) is M, while the actual output is $TM + TN + D(1 - T)$, where $T = L/(1 + L)$. Consequently, the transform of a sample error function is $E_\alpha = M_\alpha - TM_\alpha - TN_\alpha - D_\alpha(1 - T) = (M_\alpha - D_\alpha)(1 - T) - TN_\alpha = M_\alpha'(1 - T) - TN_\alpha$, where $M_\alpha' \triangleq M_\alpha - D_\alpha$. E_α has the same form as the error due to a message signal M_α' and a noise signal N_α, with no disturbance signal. Consequently, the problem is in precisely the same form as the one treated in Sections 9.8-9.13. The results there obtained may therefore be used, providing there is no correlation between $m'(t) = m(t) + d(t)$ and $n(t)$.

In using the results of Sections 9.8-9.13, $\Phi_{m'}$ must be properly interpreted. Its value is $\Phi_{m'} = \mathsf{E}\,[M_\alpha(j\omega) - D_\alpha(j\omega)]\,[M_\alpha(-j\omega) - D_\alpha(-j\omega)] = \mathsf{E}\,|M_\alpha|^2 + \mathsf{E}\,|D_\alpha|^2 - \mathsf{E}\,[D_\alpha(j\omega)\,M_a(-j\omega) + M_\alpha(j\omega)\,D_\alpha(-j\omega)]$. If the message and disturbance signals are not correlated, i.e., $\mathsf{E}\int_0^\infty m_\alpha(t)\,d_\alpha(t - \tau)\,dt = 0$ for all τ, then $\Phi_{m'} = \Phi_m + \Phi_d$.

Example. $\Phi_n = 0.25$, $\Phi_m = -1/s^2$, $\Phi_d = 1/(9 - s^2)$. $\Phi_{m'} = \Phi_m + \Phi_d = 2(s^2 - 4.5)/s^2(9 - s^2)$, $\Phi_i = -0.25(14.52 - s^2)(2.48 - s^2)/s^2(9 - s^2)$. $[\Phi_{m'}/\Phi_i^-]^+ = [-4(s^2 - 4.5)/s(s + 3)(s - 3.81)(s - 1.575)]^+ = s^{-1} + 0.193(s + 3)^{-1}$. Therefore $T_{\text{opt}} = [\Phi_{m'}/\Phi_i^-]^+(1/\Phi_i^+) = 2.39(s + 2.52)/(s + 1.575)(s + 3.81)$.

Optimum Time-Invariant Filter for Message Signal and Two Disturbance Signals

In Fig. 9.15-1c there is again the constraint of only one degree of freedom. If the ideal output is $m(t)$, then the transform of the error is $E_\alpha(s) = M_\alpha - M_\alpha T - D_{1\alpha}(1 - T) + D_{2\alpha}T = (M_\alpha - D_{1\alpha})(1 - T) + D_{2\alpha}T$. This is equivalent to the previous problem, with $M_\alpha - D_{1\alpha}$ replacing M_α' and $D_{2\alpha}$ replacing N_α. The procedure for finding the optimum T is therefore exactly the same

Optimization of a Two-Degree-of-Freedom Configuration

In Fig. 9.15-1d if the disturbance $d_2(t)$ did not exist, there would be no problem to determine G and H. $T = G/(1 + GH)$ would be chosen to perform the optimum filtering of $n(t)$ from $m(t)$. Since the output due to $d_1(t)$ is $D_1/(1 + GH)$,

GH would be chosen with as large a gain-bandwidth as required to attenuate the disturbance output. However, when D_2 is introduced, with its output of $-D_2GH/(1+GH)$, large GH tends to increase the output due to D_2. There must then be a compromise between the various factors.

The optimum (in the minimum ensemble-mean square sense) is found by formulating the error squared function in the usual manner. Here it is easily found that $E = M - (M+N)T - D_1S - D_2(1-S) = [M(1-T) - NT] + [D_2(S-1) - D_1S] = E_1 + E_2$, where $T = G/(1+GH)$, and the sensitivity function $S = 1/(1+GH)$. If the message and noise signal ensembles are each not correlated with either of the two disturbance signal ensembles, then the error minimization is achieved by separately minimizing $\mathsf{E}\int_0^\infty e_1^2\,dt$ and $\mathsf{E}\int_0^\infty e_2^2\,dt$. If, furthermore, the message and noise signals are uncorrelated, and the two disturbance signals are uncorrelated, the results of Sections 9.8-9.12 may be used; otherwise Section 9.13 is used.

Optimization of Multiple-Loop Systems

In Fig. 9.15-2, the problem is to choose H_1, H_2 so as to minimize the ensemble average squared output $\mathsf{E}\int_0^\infty c^2(t)dt \triangleq \epsilon^2$, where $C = [D_1P_2 - N_1H_1P - N_2H_2P + D_2(1+P_1H_1)]S$, and $S \triangleq (1+P_1H_1+PH_2)^{-1}$. Let $S_2 \triangleq (1+P_1H_1)S$, and it is readily found that $C = S_2(D_2 - N_1P_2) - N_2(1-S_2) + P_2S(D_1+N_1)$. The problem is now one of choosing S, S_2 to minimize ϵ^2. This constitutes the solution of the feedback problem. A prefilter can then be inserted in front of Fig. 9.15-2, and chosen so as to obtain T_{opt} for the optimum filtering of noise from the useful system input. The resulting structure can then be transformed into any other equivalent form which maintains S, S_2, and T_{opt}, invariant.

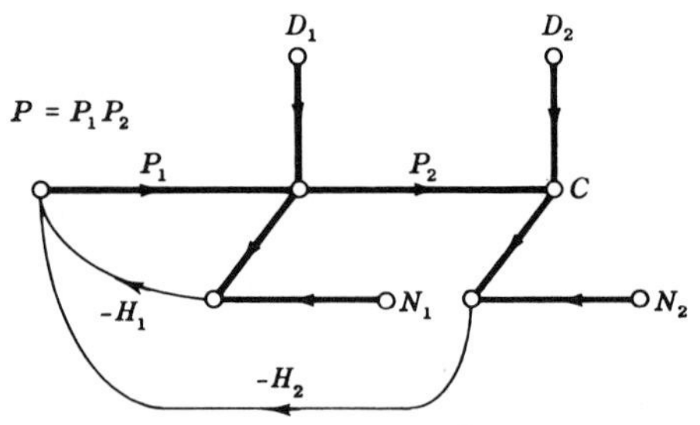

FIG. 9.15-2. Two-loop structure with ensembles of disturbance and feedback transducer noise signals.

In order to find S, S_2, we proceed as in Section 9.8, i.e., write $\epsilon^2 = \mathsf{E}\int_0^\infty c^2(t)dt = \mathsf{E}\int_{-\infty}^\infty C(s)C(-s)df$, and replace C by its equivalent in terms of S, S_2. The result is

$$\epsilon^2 = \int_{-\infty}^{\infty} df\{P_2\bar{P}_2S\bar{S}(\Phi_{D1}+\Phi_{N1}) + S_2\bar{S}_2(P_2\bar{P}_2\Phi_{N1}+\Phi_{D2}) + \Phi_{N2}(1-S_2)(1-\bar{S}_2) - \Phi_{N1}P_2\bar{P}_2(S\bar{S}_2+\bar{S}S_2)\} \tag{9.15,1}$$

where $\Phi_{D1} \stackrel{\Delta}{=} \mathsf{E}\, D_1\bar{D}_1$, etc., and $\bar{D}_1$ represents $D_1(-s)$, etc. In obtaining Eq. (9.15,1), it has been assumed, for the sake of simplicity, that all cross-correlation functions are zero, i.e., $\mathsf{E}\, D_1\bar{D}_2$, $\mathsf{E}\, N_2\bar{N}_1$, ... are zero. However, the same development may be followed when this is not so.

The calculus of variations is next used (cf. Section 9.12). S, S_2 are replaced by $S + k\Delta$, $S_2 + k_2\Delta_2$, respectively, in Eq. (9.15,1), and ϵ^2 becomes a function of k, k_2, i.e., of the form

$$\epsilon^2 = A + k^2B + k(F + \bar{F}) + k_2{}^2G + kk_2J + k_2(Q + \bar{Q}). \tag{9.15,2}$$

A is not needed in the development. F and Q are given in Eqs. (9.15,4a and b). B, J, and G are needed for investigating the signs of $\partial^2\epsilon^2/\partial k^2$, $\partial^2\epsilon^2/\partial k\partial k_2$, and $\partial^2\epsilon^2/\partial k_2{}^2$. They are

$$\begin{aligned} B &= \int_{-\infty}^{\infty} (\Phi_{D1} + \Phi_{N1})P_2\bar{P}_2\Delta\bar{\Delta}\,df = 0.5\partial^2\epsilon^2/\partial k^2 \\ G &= \int_{-\infty}^{\infty} (\Phi_{D2} + P_2\bar{P}_2\Phi_{N1} + \Phi_{N2})\Delta_2\bar{\Delta}_2\,df = 0.5\partial^2\epsilon^2/\partial k_2{}^2 \\ J &= -\int_{-\infty}^{\infty} \Phi_{N1}P_2\bar{P}_2(\Delta\bar{\Delta}_2 + \bar{\Delta}\Delta_2)\,df = \partial^2\epsilon^2/\partial k\partial k_2. \end{aligned} \tag{9.15,3}$$

The conditions for S, S_2 to minimize ϵ^2 are[1]:

(1) $\partial\epsilon^2/\partial k = \partial\epsilon^2/\partial k_2 = 0$ at $k = k_2 = 0$.

(2) $J^2 - 4GB < 0$ and $B + G > 0$ at $k = k_2 = 0$.

These ensure that the resulting S, S_2 determine a minimum ϵ^2 rather than a maximum. It is seen from Eqs. (9.15,3) that B, G, and J are independent of S, S_2. Obviously $B + G > 0$. The other inequality is also satisfied, and the proof is as follows: Let $\Delta = a + jb$, $\Delta_2 = a_2 + jb_2$, $\Phi_{N1}P_2\bar{P}_2 = \beta > 0$; then $J^2 = 4[\int_{-\infty}^{\infty} \beta(aa_2 + bb_2)df]^2$ and $4BG > \gamma = 4\int_{-\infty}^{\infty} \beta(a^2 + b^2)df \int_{-\infty}^{\infty} \beta(a_2{}^2 + b_2{}^2)df$. Consider any small portion of the path of integration. For this portion, $J^2 - \gamma = \beta^2[(aa_2 + bb_2)^2 - (a^2 + b^2)\,(a_2{}^2 + b_2{}^2)]df = -\beta^2 df(ab_2 - ba_2)^2 < 0$. But $4BG > \gamma$. Hence $J^2 - 4BG < 0$.

We therefore seek S, S_2 such that $\partial\epsilon^2/\partial k = \partial\epsilon^2/\partial k_2 = 0$ at $k = k_2 = 0$, which, from Eq. (9.15,2), requires that F (and therefore $\bar{F}$) $= 0$, Q (and therefore $\bar{Q}$) $= 0$, i.e.,

$$F = \int_{-\infty}^{\infty} [(\Phi_{D1} + \Phi_{N1})S - \Phi_{N1}S_2]\bar{\Delta}\,df \stackrel{\Delta}{=} \int_{-\infty}^{\infty} \psi\bar{\Delta}\,df = 0 \tag{9.15,4a}$$

$$Q = \int_{-\infty}^{\infty} \Phi_{N1}P_2\bar{P}_2(S - S_2) + \Phi_{N2}(1 - S_2) - \Phi_{D2}S_2]\,\bar{\Delta}_2\,df \stackrel{\Delta}{=} \int_{-\infty}^{\infty} \psi_2\bar{\Delta}_2\,df = 0. \tag{9.15,4b}$$

[1] W. Kaplan, "Advanced Calculus," p. 126. Addison-Wesley, Reading, Massachusetts, 1952.

It is recalled (Section 9.12) that Δ, Δ_2 are arbitrary except that $S + k\Delta$, $S_2 + k_2\Delta_2$ represent the classes of all suitable S, S_2 functions. Hence $\bar{\Delta}$, $\bar{\Delta}_2$ must have all their poles in the right-half plane.

It follows from Cauchy's residue theorem (Section 7.2) that Eqs. (9.15,4) are satisfied[1] if the integrands $\psi\bar{\Delta}$, $\psi_2\bar{\Delta}_2$ (1) have no poles in either the left-half plane or the right-half plane, and (2) are Os^{-x} at infinity with $x \geqslant 2$ (Os^{-x} is defined in Section 9.12). Since the poles of $\bar{\Delta}$, $\bar{\Delta}_2$ are arbitrarily located in the right-half plane, the first condition is satisfied only if ψ, ψ_2 in Eqs. (9.15,4) have no lhp poles. If we also add the constraint that S, S_2 are nonzero at infinity [because $S = (1 + L)^{-1}$, $S_2 = (1 + P_1H_1)S$, and we certainly do not want infinite L at infinite s], then the second condition requires that each of Φ_{D1}, Φ_{N1}, $\Phi_{N1}P_2\bar{P}_2$, Φ_{D2} is Os^{-x}, $x \geqslant 2$. This also ensures that ϵ^2 [Eq. (9.15,1)] is infinite.

The next step is to solve for S, S_2 in terms of ψ_1, ψ_2, resulting in:

$$S = [\psi(\Phi_{N2} + \Phi_{D2} + P_2\bar{P}_2\Phi_{N1}) - \Phi_{N1}(\psi_2 - \Phi_{N2})]\, K^{-1} \tag{9.15,5a}$$

$$-S_2 = [(\Phi_{D1} + \Phi_{N1})\,(\psi_2 - \Phi_{N2}) - \Phi_{N1}P_2\bar{P}_2\psi]\, K^{-1} \tag{9.15,5b}$$

$$K = \Phi_{D1}(\Phi_{N2} + \Phi_{D2} + P_2\bar{P}_2\Phi_{N1}) + \Phi_{N1}(\Phi_{N2} + \Phi_{D2}). \tag{9.15,6}$$

Since ψ, ψ_2 can have no lhp poles, the poles of S, S_2 must be the zeros of K. Let the number of poles of Φ_{N1}, Φ_{N2}, ... be $2n_1$, $2n_2$, ... and their excess of poles over zeros be $2e_{n1}$, $2e_{n2}$, etc. Then it is readily established that the number of lhp zeros of K (i.e., poles of S, S_2) is equal to

$$x = n_1 + n_2 + d_1 + d_2 + p_2 - y$$
$$y = \text{smallest of } e_{d1} + e_{n2},\quad e_{d1} + e_{d2},\quad e_{d1} + e_{p2},\quad e_{n1} + e_{n2},\quad e_{n1} + e_{d2}. \tag{9.15,7}$$

The denominator of S and of S_2 will be denoted by the polynomial $\lambda_x(s)$, of degree x, whose zeros are the lhp zeros of $K(s)$, and whose leading coefficient is unity.

The numerators of S, S_2 are determined from the condition that ψ, ψ_2 of Eqs. (9.15,4a and b) have no lhp poles. Since $\psi = \Phi_{D1}S + (S - S_2)\Phi_{N1}$, S must have d_1 zeros at the d_1 lhp poles of Φ_{D1}, and $S = S_2$ at the n_1 lhp poles of Φ_{n1}. Also, in order that the x lhp poles of S, S_2 disappear from ψ, it is necessary that at each of these poles (denoted by s_i),

$$\lim_{s\to s_i}\, [\Phi_{D1}(s_i) + \Phi_{N1}(s_i)]\,(s - s_i)S(s_i) = \lim_{s\to s_i}\, \Phi_{N1}(s_i)\,(s - s_i)S_2(s_i). \tag{9.15,8}$$

The $\lim_{s\to s_i}(s - s_i)S(s_i)$ is, of course, the residue of S in the pole at s_i. Equation (9.15,8) also ensures that the poles at s_i are canceled out of ψ_2, in view of the

[1] The derivation presented here is based on S. Darlington, Linear least-squares smoothing and prediction, with applications, Section 3.4. *Bell System Tech. J.* **37** (5), 1221–1294 (1958).

definition of the s_i as zeros of K in Eq. (9.15,6). The following additional conditions must be satisfied in order that ψ_2 [in Eq. (9.15,4b)] has no lhp poles: S_2 must have d_2 zeros at the d_2 lhp poles of Φ_{D2}; $S_2 = 1$ at the n_2 lhp poles of Φ_{N2}; and $S = S_2$ at the p_2 lhp poles of P_2 (we assume here that P_2 has no rhp poles).

Thus a total of $f = d_1 + n_1 + x + d_2 + n_2 + p_2$ conditions are imposed on the numerators of S, S_2, in order that ψ and ψ_2 have no lhp poles. These f conditions require two numerator polynomials with a total of f coefficients, i.e., a total of $f - 2$ zeros. We furthermore insist that S, S_2 each have as many zeros as poles [because $S = (1 + L)^{-1}$, etc., and L is at least not to be infinite at infinite s], and therefore we want $0.5(f - 2) = x$; hence $x = d_1 + n_1 + d_2 + n_2 + p_2 - 2$. It follows from Eq. (9.15,7) that our demand that S, S_2 be finite at infinity can be satisfied only if $y \ngtr 2$ (each of $e_{d1} + e_{n2}$, $e_{d1} + e_{d2}$, $e_{d1} + e_{p2}$, etc., must be at least 2, because of the Os^{-2} condition). Otherwise, at least one of S, S_2 will emerge with more zeros than poles. If this condition is not satisfied in a practical problem, then one or more of the energy spectra should first be suitably modified [by means of $(k^2 - s^2)/k^2$ factors and eventually letting $k \to \infty$].

Summary

The procedure may be summarized as follows [it is assumed that y of Eq. (9.15,7) is 2]:

(1) Let $S = F\lambda_{d1}\lambda_{(x-d1)}/\lambda_x$, $S_2 = F_2\lambda_{d2}\lambda_{(x-d2)}/\lambda_x$; F, F_2 are constants and λ_n is a polynomial of degree n, with leading coefficient unity.

(2) The zeros of λ_x are the x lhp zeros of $K(s)$ of Eq. (9.15,6).

(3) The zeros of λ_{d1} are the d_1 lhp poles of $D_1(s)$.

(4) The zeros of λ_{d2} are the d_2 lhp poles of $D_2(s)$.

(5) The $2x - d_1 - d_2 + 2$ coefficients of $F\lambda_{(x-d1)}$, $F_2\lambda_{(x-d2)}$ are chosen so that:

(a) $S = S_2$ at the poles of P_2 (assumed all left-hand plane) and lhp poles of Φ_{N1}($p_2 + n_1$ equations),

(b) $S_2 = 1$ at the lhp poles of Φ_{N2} (n_2 equations),

(c) $$\lim_{s\to s_i} (s - s_i)\,[\Phi_{D1}(s_i) + \Phi_{N1}(s_i)]\,(s - s_i)S(s_i) = \lim_{s\to s_i} \Phi_{N1}(s_i)\,(s - s_i)S_2(s_i),$$

at the x zeros of λ_x (lhp poles of K).

From Eq. (9.15,7), $p_2 + n_1 + n_2 + x = 2x - d_1 - d_2 + 2$ (when $y = 2$); hence there are precisely as many equations as unknowns, and S, S_2 are uniquely determined.

It would be desirable that S and S_2 emerge to be unity at infinity, because the resulting $P_1H_1 + PH_2 = S^{-1} - 1$, and $PH_2(1 + P_1H_1)^{-1} = S_2^{-1} - 1$ will then be zero at infinity. What conditions must the Φ_N's and Φ_D's satisfy so that the resulting optimum S, S_2 should emerge unity at infinity? We shall find the answer for the case $e_{N1} = e_{N2} = e_{D1} = e_{D2} = e_{P2} = 1$. The procedure is as follows. The leading terms in the expansion of ψ, ψ_2 at infinity are found from their definitions in Eqs. (9.15,4a and b). The results are $K_{D1}s^{-2}$, $-K_{D2}s^{-2}$, where K_{D1} is the multiplying factor of Φ_{D1}, and K_{D2} that of Φ_{D2} (i.e., $K_{D1}s^{-2} = \lim_{s\to\infty} \Phi_{D1}$, etc.). In deriving these results, it was assumed that S, S_2 were unity at infinite s. These limiting values of ψ, ψ_2 are used in examining S, S_2 of Eqs. (9.15,5a and b) at infinite s. As $s \to \infty$, it is found that $S, S_2 \to (K_{N1} + K_{D1})(K_{N2} + K_{D2})$. Hence $(K_{N1} + K_{D1})(K_{N2} + K_{D2}) = 1$ in order that S and S_2 emerge unity at infinity. If the original energy spectra do not satisfy this condition, it is possible to modify them at high frequencies. Alternatively, and more simply, one may not bother to modify them, and emerge with loop transmissions which are nominally finite at infinity. However, they may be allowed to decrease as rapidly as stability permits at any frequency beyond which the frequency response of the disturbance is satisfactorily low, i.e., beyond which the disturbances do not require attenuation. The resulting design will result in an ϵ^2 which is negligibly larger than the optimum.

It is clear that the above technique may be extended to systems with larger numbers of loops.

Example. Suppose $\Phi_{N1} = \Phi_{N2} = 50/(100 - s^2)$, $\Phi_{D1} = \Phi_{D2} = 100/(25 - s^2)$, $P_2 = 5/(s + 1)$, and the cross-correlation spectra are zero.

Design:

(a) Find the lhp zeros of K of Eq. (9.15,6). The result (slide-rule accuracy) is: $\lambda_x = (s + 1.8)(s + 8.8)^2$.

(b) Let $S = (s + 5)(Fs^2 + Es + G)/\lambda_x$, $S_2 = (s + 5)(F_2s^2 + E_2s + G_2)\lambda_x$ (because Φ_{D1}, Φ_{D2} each have a lhp pole at -5).

(c) $S_2 = 1$ at the lhp pole of Φ_{N2}, so
(1) $5(G_2 - 10E_2 + 100) = (1.2)^2 8.2$.

(d) $S = S_2$ at the lhp poles of P_2 and Φ_{N1}, so
(2) $G - E + F = G_2 - E_2 + F_2$
(3) $G - 10E + 100F = G_2 - 10E_2 + 100F_2$.

(e) The poles of S, S_2 do not appear in ψ, hence
(4) $(\Phi_{D1} + \Phi_{N1})(s + 8.8)^2 S = \Phi_{N1}(s + 8.8)^2 S_2$ at $s = -8.8$.
(5) $d/ds[(\Phi_{D1} + \Phi_{N1})(s + 8.8)^2 S] = d/ds[\Phi_{N1}(s + 8.8)^2 S_2]$ at $s = -8.8$.
(6) $(\Phi_{D1} + \Phi_{N1})(s + 1.8)S = \Phi_{N1}(s + 1.8)S_2$ at $s = -1.8$.

These six linear equations are solved for E, F, G, E_2, F_2, G_2.

§ 9.16 Optimization with Nonminimum Phase Plant

In the optimization problems treated so far, the optimization procedure led to the specification of the system transfer function [e.g., $G/(1 + G)$ in Fig. 9.15-1c, $G/(1 + GH)$ in Fig. 9.15-1d]. It was implicitly assumed that G could be completely specified. The latter is not true when G includes within it a plant with rhp zeros. Such zeros cannot be canceled out by rhp poles, because the inevitable inexact cancellation leads to closed-loop rhp poles. Instead, such rhp zeros must be *a priori* admitted as zeros of G. The optimization procedure must be suitably amended to include such constraints.

A convenient starting point is Eq. (9.12,3), but in place of T_{opt} we write $T_{\text{opt}} = FT'_{\text{opt}}$, where F includes the fixed portion that must be included in any solution for T. F can have one or more rhp zeros, a pure delay factor e^{-sT}, or any transform function whatsoever which is constrained to be a part of $T(s)$. It is important to remember that $T = T_{\text{opt}} + k\Delta(s)$ comprises the class of all permissible functions. Therefore, $\Delta(s)$ too must have the factor F; let $\Delta = F\Delta'(s)$. Hence Eq. (9.12,3) becomes

$$\int_{-\infty}^{\infty} [T'_{\text{opt}}F(s)F(-s)\Phi_i - F(-s)T_I(s)\Phi_m]\Delta'(-s)df = 0. \qquad (9.16,1)$$

The rest is easy. Equation (9.16,1) has precisely the same form as Eq. (9.12,3), if $\Phi_{i'} \triangleq F(s)F(-s)\Phi_i$ of Eq. (9.16,1) is used in place of Φ_i of Eq. (9.12,3), and $T_{I'} \triangleq T_I(s)F(-s)$ of Eq. (9.16,1) is used in place of T_I of Eq. (9.12,3). Hence, by analogy with Eq. (9.11,1),

$$T'_{\text{opt}} = \frac{1}{(\Phi_{i'})^+}\left[\frac{T_{I'}\Phi_m}{(\Phi_{i'})^-}\right]^+. \qquad (9.16,2)$$

Example 1. $T_I = 1$, $\Phi_n = 1$, $\Phi_m = 5/(4 - s^2)$, $F = (s - 1)$. Therefore $\Phi_i = \Phi_n + \Phi_m = (9 - s^2)/(4 - s^2)$; $\Phi_{i'} = -\Phi_i(s-1)(s+1) = [(s+1)(s+3)/(s+2)][(1-s)(3-s)/(2-s)] = (\Phi_{i'})^+(\Phi_{i'})^-$; $T_{I'} = -(s+1)$. $T_{i'}\Phi_m/(\Phi_{i'})^- = -5(s+1)/(s+2)(1-s)(3-s) = [\frac{1}{3}/(s+2)] + \ldots$ Hence $[T_{I'}\Phi_m/(\Phi_{i'})^-]^+ = \frac{1}{3}(s+2)$, and $T'_{\text{opt}} = [(s+2)/(s+1)(s+3)][\frac{1}{3}/(s+2)] = \frac{1}{3}/(s+1)(s+3)$. Hence $T_{\text{opt}} = \frac{1}{3}(s-1)/(s+1)(s+3)$.

Example 2. In this problem the plant has in it a pure time lag e^{-sT}, so that $F = e^{-sT}$, $F(-s) = e^{sT}$, and $\Phi_{i'} = \Phi_i e^{sT}e^{-sT} = \Phi_i$, $T_{I'} = e^{sT}$. $[T_{I'}\Phi_m/(\Phi_{i'})^-]^+ = [5e^{sT}/(2+s)(3-s)]^+ = \{e^{sT}[(s+2)^{-1} - (s-3)^{-1}]\}^+ = \mathscr{L}e^{-2(t+T)} = e^{-2T}/(s+2)$. Therefore $T'_{\text{opt}} = e^{-2T}/(s+3)$ and $T_{\text{opt}} = e^{-sT}e^{-2T}/(s+3)$.

Unstable Open-Loop Systems

If the plant has rhp poles at α, β, then the loop transmission $L(s)$ should also have the same rhp poles (see Section 5.16); thus $L = L_1/(s - \alpha)(s - \beta)$. If the single-degree-of-freedom structure must be used, then $T = L/(1 + L) = L_1/[(s - \alpha)(s - \beta) + L_1]$, and $T - 1 = -(s - \alpha)(s - \beta)/[(s - \alpha)(s - \beta) + L_1]$ has zeros at α, β. Hence $T - 1$ may be written in the form $T - 1 = -(s - \alpha)(s - \beta)F(s)$. It is important to note that in the variational technique of Section 9.12 $T = T_{\text{opt}} + k\Delta(s)$ represents the class of suitable response functions. In this specific case this means that $\Delta(s)$ has zeros at α, β, i.e., $\Delta(s) = (s - \alpha)(s - \beta)\Delta_1(s)$, in order that $T - 1$ has zeros at α, β independent of k. Equation (9.12,3) becomes

$$\int_{-\infty}^{\infty} \{[1 - (s - \alpha)(s - \beta)F_{\text{opt}}]\Phi_i - T_I\Phi_m\}(s + \alpha)(s + \beta)\Delta_1(-s)df = 0.$$

The integrand may be written as $\{-F_{\text{opt}}(s^2 - \alpha^2)(s^2 - \beta^2)\Phi_i - (T_I\Phi_m - \Phi_i)(s + \alpha)(s + \beta)\}\Delta_1(-s)$. Let $\Phi_i' \triangleq -(s^2 - \alpha^2)(s^2 - \beta^2)\Phi_i$; $(T_I\Phi_m)' \triangleq (T_I\Phi_m - \Phi_i)(s + \alpha)(s + \beta)$. By analogy with Eqs. (9.12,3) and (9.11,1),

$$F_{\text{opt}} = \frac{1}{(\Phi_i')^+}\left[\frac{(T_I\Phi_m)'}{(\Phi_i')^-}\right]^+. \tag{9.16,4}$$

It is important to note that inequality (9.12,5) must be suitably interpreted and, if necessary, Φ_n modified to ensure that the inequality is satisfied. This is done in the following example.

Example. $T_I = 1$, $\Phi_n = 1$, but instead we take $\Phi_n = a^2/(a^2 - s^2)$ in order to satisfy (9.12,5). $\Phi_m = 5/(4 - s^2)$, and the plant has a pole at $s = 1$. Hence $\Phi_i = [9a^2 - (5 + a^2)s^2]/(4 - s^2)(a^2 - s^2)$. $\Phi_i' = (1 - s^2)\Phi_i$, $(T_I\Phi_m)' = -(s + 1)\Phi_n$.

$$F_{\text{opt}} = \frac{(2 + s)(a + s)}{(1 + s)[3a + (5 + a^2)^{0.5}s]}\left[\frac{-(s + 1)a^2(2 - s)}{(a + s)(1 - s)[3a - (5 + a^2)^{0.5}s]}\right]^+$$

$$= \frac{(2 + s)(a - 1)a(2 + a)}{(1 + s)[3a + (5 + a^2)^{0.5}s](1 + a)[3 + (5 + a^2)^{0.5}]}.$$

As a approaches infinity, Φ_n approaches unity, F_{opt} approaches $(s + 2)/(s + 1)(s + 3)$, and $T = 1 - (s - 1)F(s) = (3s + 5)/(s + 1)(s + 3)$, $L = T/(1 - T) = (3s + 5)/(s - 1)(s + 2)$.

In the two-degree-of-freedom system (Fig. 9.15-1b), $T = GL/(1 + L)$, so that $T(\alpha) = G(\alpha)$. Hence no restrictions are imposed on $T(s)$ by rhp poles of $L(s)$.

§ 9.17 Optimization under Constraints[1]

It is often necessary to optimize under one or more constraints. For example, the stress applied to mechanical parts of a system, or the temperature of some

[1] Optimization under constraints is the principal theme of the book by G. C. Newton, L. A. Gould, and J. F. Kaiser, "Analytical Design of Linear Feedback Controls." Wiley, New York, 1957; see also S. S. Chang, "Synthesis of Optimum Control Systems." McGraw-Hill, New York, 1961.

member, might not be allowed to exceed a specified amount. The optimization procedure can often be extended to such problems by the Lagrange multiplier technique.[1]

We assume that the constraint may be formulated in the same manner as the error minimization problem, i.e., as an ensemble average of the integral of a squared signal. For example, to avoid plant saturation or overheating, $\mathsf{E} \int_0^\infty x_\alpha^2(t)\, dt \leqslant Q$, where $x_\alpha(t)$ is a sample signal input to the plant. This is actually a constraint on the ensemble average squared value, and some individual values of $\int_0^\infty x_\alpha^2(t)\, dt$ may be considerably larger than Q. In order that the above approach be meaningful, it is therefore necessary that the designer knows the relation between the mean and the peak values of $\int_0^\infty x_\alpha^2(t)\, dt$, or, better still, the distribution of the values of $\int_0^\infty x_\alpha^2(t)\, dt$.

The Lagrange multiplier procedure is as follows: If $I = \mathsf{E} \int_0^\infty x_\alpha^2(t)\, dt \leqslant Q$, then the quantity to be minimized is not $\epsilon^2 = \mathsf{E} \int_0^\infty [e(t)]^2\, dt$, but

$$\epsilon^2 + \nu^2 I = \mathsf{E} \left\{ \int_0^\infty [e_\alpha(t)]^2 dt + \nu^2 \int_0^\infty [x_\alpha(t)]^2 dt \right\}. \tag{9.17,1}$$

Next, we relate $x_\alpha(t)$ to the system parameters. For example, if $x(t)$ represents the signal input to a plant as in Fig. 9.17-1a, then, from the equivalent Fig. 9.17-1b,

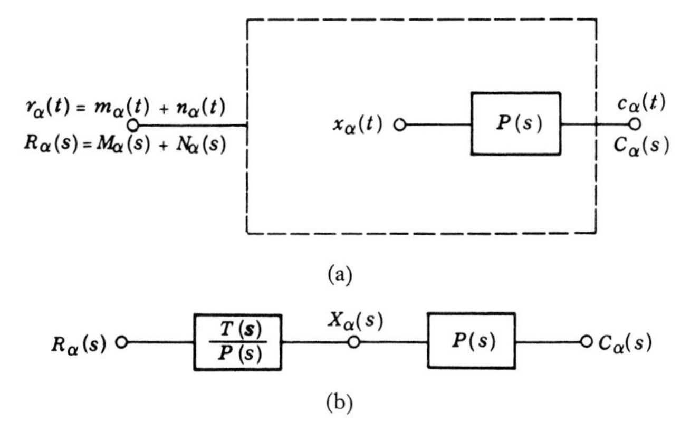

FIG. 9.17-1. (a) Formulation of the problem—limitation on $\mathsf{E} \int_0^\infty x_\alpha^2(t) dt$; $T(s) = C_\alpha(s)/R_\alpha(s)$. (b) Plant input related to T and P.

$X_\alpha = R_\alpha T/P = (M_\alpha + N_\alpha)\, T/P$, and by the Parseval relation, Eq. (9.17,1) becomes

$$\epsilon^2 + \nu^2 I = \mathsf{E} \int_{-\infty}^{\infty} [E_\alpha \bar{E}_\alpha + \nu^2 (M_\alpha + N_\alpha)(\bar{M}_\alpha + \bar{N}_\alpha) T\bar{T}/P\bar{P}]\, df$$

[1] F. B. Hildebrand, "Methods of Applied Mathematics," pp. 141-144. Prentice-Hall, New York, 1952.

where $\bar{E}$, $\bar{M}$, etc., represent $E(-s)$, $M(-s)$, etc. The procedure henceforth is exactly the same as in Section 9.8. We write $e_\alpha(t) = c_{I\alpha}(t) - [m_{o\alpha}(t) + n_{o\alpha}(t)]$. If the message and noise signals are uncorrelated, then ϵ^2 is obtained from Eq. (9.12,1). Also, if the message and noise signals are not correlated,

$$\nu^2 I = \nu^2 \int_{-\infty}^{\infty} \frac{T\bar{T}}{P\bar{P}} \mathsf{E}(M_\alpha \bar{M}_\alpha + N_\alpha \bar{N}_\alpha + N_\alpha \bar{M}_\alpha + M_\alpha \bar{N}_\alpha) df$$

$$= \nu^2 \int_{-\infty}^{\infty} \frac{T\bar{T}}{P\bar{P}} (\Phi_m + \Phi_n) df.$$

Therefore Eq. (9.17,1) may be written

$$\epsilon^2 + \nu^2 I = \int_{-\infty}^{\infty} \left\{ (T_I - T)(\bar{T}_I - \bar{T})\Phi_m + T\bar{T}\Phi_n + \nu^2 \Phi_i \frac{T\bar{T}}{P\bar{P}} \right\} df.$$

The procedure hereafter is precisely the same as before. We replace T by $T + k\Delta$, differentiate with respect to k, and set the derivative equal to zero at $k = 0$. The result is $\int_{-\infty}^{\infty} [T\Phi_i[1 + (\nu^2/P\bar{P})] - T_I \Phi_m] \bar{\Delta}\, df = 0$ for all permissible $\bar{\Delta}$. If the intuitive approach of Section 9.11 is used, then the unrealizable optimum, denoted by T_{opt}^*, is $T_{\text{opt}}^* = T_I \Phi_m / \Phi_i (1 + \nu^2/P\bar{P})$, and the realizable optimum is

$$T_{\text{opt}} = \frac{1}{[\Phi_i(1 + \nu^2/P\bar{P})]^+} \left\{ \frac{T_I \Phi_m}{[\Phi_i(1 + \nu^2/P\bar{P})]^-} \right\}^+. \qquad (9.17,2)$$

Equation (9.17,2) can be rigorously justified by the methods of Section 9.12.

T_{opt} is found as a function of the unknown ν^2, and then $I = \int_{-\infty}^{\infty} (T_{\text{opt}} \bar{T}_{\text{opt}} / P\bar{P}) \Phi_i df$ is evaluated as a function ν^2, and set equal to Q, which is the maximum permitted value of I. The value of ν^2 is thereby obtained and substituted in T_{opt}.

Example. $T_I = 1$, $P = 1/s$, $\Phi_n = 1$, $\Phi_m = 5/(4 - s^2)$; $\Phi_i = (9 - s^2)/(4 - s^2)$ and $(1 + \nu^2/P\bar{P}) = 1 - \nu^2 s^2$. $\Phi_i(1 + \nu^2/P\bar{P}) = [(3 - s)(1 - \nu s)/(2 - s)] [(s+3)(1+\nu s)/(s+2)]$ and $T_{\text{opt}} = 1/(\nu s + 1)(s + 3)(1 + 2\nu)$. Therefore $I = \nu^{-2}(1 + 2\nu)^{-2} \int_{-\infty}^{\infty} df\, s^2/(s^2 - 9)(s^2 - \nu^{-2})$. The integral can be evaluated by means of Cauchy's residue theorem since the integrand is Os^{-2}. The result is $I = [2\nu(3\nu + 1)(2\nu + 1)^2]^{-1} = Q$. As soon as Q is specified by the specific constraint, ν can be found, and T_{opt} is precisely known.

§ 9.18 The Optimum Filter Problem in Sampled-Data Systems

The statistical design techniques of Sections 9.8-9.17 may be applied to sampled-data systems. The terminology and mathematical tools of sampled-data systems (Sections 11.1-11.5) are used in the balance of this chapter. Two different problems will be treated. In the first (Fig. 9.18-1), the system output at only the sampling instants is considered; there is no concern with the inter-sample behavior. In the second (Fig. 9.19-1), the optimum filter is based on the continuous output of the system.

The first problem is formulated in Fig. 9.18-1, where the object is to find

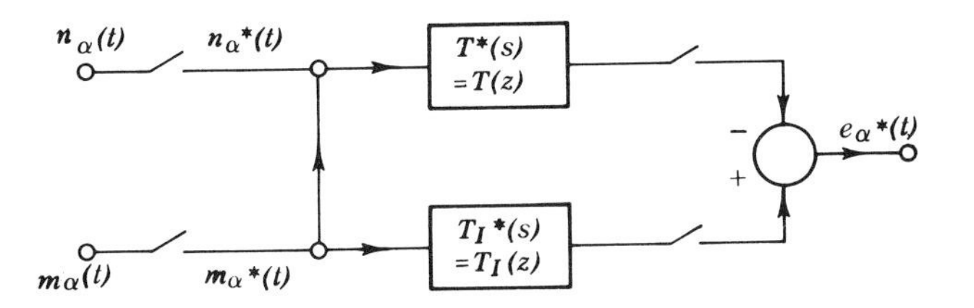

FIG. 9.18-1. The sampled-data filter problem in which the sampled input and error are considered.

that $T(z)$ which minimizes the ensemble average of $\sum_{n=0}^{\infty} e_\alpha^2(nT)$. The problem is very similar to that of Sections 9.8-9.13. We may therefore use the development of these sections by suitably replacing variables and occasionally changing an equation. The reader is asked to return to Section 9.8. There is no change until Fig. 9.8-1, where the sampled values of the functions replace the continuous values. Figure 9.18-1 replaces Fig. 9.8-2. There is a difficulty in using the Parseval relation of Eq. (9.8,1), because it involves the integral of the product of two series of impulses. This problem is avoided as follows.

In the unbounded integral $I = \int_{-\infty}^{\infty} M^*(s)\, M^*(-s)\, df = \int_0^{\infty} [\sum_{n=0}^{\infty} m(nT)\, \delta(t - nT)]^2\, dt$, if we use $z = e^{sT}$ then $I = \int_{C_1} M(z)\, M(z^{-1})\, (j2\pi z T)^{-1}\, dz$, where C_1 consists of an infinite number of excursions around the unit circle in the z plane. However, if only one excursion is made [corresponding to the $\pm\infty$ limits in f being changed to $\omega/2\pi$, $(\omega + \omega_s)/2\pi$], then the resulting modified integral is finite and is equal to $T^{-1} \sum_{n=0}^{\infty} m^2(nT)$. Hence we may use the Parseval relation in the z domain if the limits of integration are changed to 0, $(\omega_s/2\pi)$. (The reader may prove the above, with the aid of the following hints. First prove $m_k \triangleq m(kT) = (2\pi j)^{-1} \int_C M(z)\, z^{k-1}\, dz$ with C the unit circle, by writing $m(t) = (2\pi j)^{-1} \int_{-j\infty}^{j\infty} M(s)\, e^{st}\, ds$, replacing t by kT, dividing the $j\omega$ axis into vertical strips each $j\omega_s$ long, interchanging summation and integration, and using [Eq. (11.3,1)] $M^*(s) = T^{-1} \sum_{-\infty}^{\infty} M(s + jn\omega_s)$. Then write $\sum_{k=0}^{\infty} m_k^2 = \sum_0^{\infty} m_k (2\pi j)^{-1} \int_C M(z)\, z^{k-1}\, dz$, interchange the order of summation and integration, replace z by z^{-1}, and finally obtain $\int_0^{\omega_s} M^*(s)\, M^*(-s)\, d\omega/2\pi = \int_C M(z)\, M(z^{-1})\, z^{-1}\, dz = T^{-1} \sum_{n=0}^{\infty} m^2(nT)$.)

The above permits us to continue with the development which follows Eq. (9.8,1); $e_\alpha^*(t)$, $m_\alpha^*(t)$ replace $e_\alpha(t)$, $m_\alpha(t)$, etc. Equation (9.8,2) becomes $\Phi_{m^*} \overset{\Delta}{=} \mathsf{E}\, M_\alpha^*(s)\, M_\alpha^*(-s) = \mathsf{E}\, M_\alpha(z)\, M_\alpha(z^{-1})$, etc. It is noted that the poles (and zeros) of $\Phi_{m^*}(z)$ outside the unit circle are the images of those inside the unit circle; i.e., if there is a pole at $z = \alpha + j\beta$, then there is another at $(\alpha + j\beta)^{-1}$. Those inside the unit circle are associated with an impulse series existing for positive time only, and those with poles outside the unit circle correspond to an impulse series in negative time only. In Eq. (9.8,3) the integration limits are $0, \omega_s/2\pi$; $T_I(s)$, $T(s)$, Φ_m are replaced by $T_I^*(s)$, $T^*(s)$, Φ_{m^*}, etc. We then go on to Section 9.9 seeking a time domain method of finding Φ_{m^*}. Equation (9.9,1) becomes $\phi_{m_\alpha^*}(\tau) \overset{\Delta}{=} \int_0^\infty m_\alpha^*(t_1)\, m_\alpha^*(t_1 - \tau)\, dt_1$, and is zero except at $\tau = kT$, k an integer. The technique described in connection with Fig. 9.9-3 may be used to find $\phi_{m^*}(\tau)$, which replaces $\phi_m(\tau)$. The samples of $m_\alpha(t)$ in Fig. 9.9-3 are replaced by those of $m_\alpha^*(t)$. It is important to note that generally $\phi_{m^*}(\tau) \neq \phi_m^*(\tau)$, unless $m_\alpha(t)$ is a stationary ensemble. In the latter case, the above sampling technique is in fact often used to find $\phi_m(\tau)$. The cross-correlation impulse series $\phi_{m^*n^*}(\tau)$ is obtained in the same manner. The arguments used in Section 9.10 are not affected by the change in the limits of integration, so the unrealizable optimum pulse transfer functions is $T_I(z)\, \Phi_{m^*}(z)/[\Phi_{i^*}(z)]$ where $\Phi_{i^*}(z) = \Phi_{m^*}(z) + \Phi_{n^*}(z)$. The nonrigorous arguments of Section 9.11 are also not affected by the finite limits of integration. The equivalent of Eq. (9.11,1) is

$$T_{\text{opt}}(z) = \frac{1}{\Phi_{i^*}^+(z)} \left[\frac{T_I(z)\, \Phi_{m^*}(z)}{\Phi_{i^*}^-(z)}\right]^+ \tag{9.18,1}$$

where $\Phi_{i*}^+(z)$ has all its poles and zeros inside the unit circle. The plus sign associated with the brackets indicates that, just as in (9.11,1), the transform of only the positive time portion is to be used. Section 9.12 may be used to provide a rigorous proof of Eq. (9.18,1).

Example. Suppose $T_I(z) = 1$, $\Phi_{m^*} = -z/[(z - 0.5)(z - 2)]$, $\Phi_{n^*} = 0.5$. Hence $\Phi_{i^*} = [(z - 0.233)/(z - 0.5)]\,[(z - 4.29)/(z - 2)] = (\Phi_{i^*}^+)(\Phi_{i^*}^-)$, and $T_{\text{opt}}(z) = (z - 0.5)/(z - 0.233)\,[-z/(z - 0.5)(z - 4.29)]^+ = [(z - 0.5)/(z - 0.233)]\,[0.264z/(z - 0.5)] = 0.264z/(z - 0.233)$.

It should be evident from the above that the problems solved in Sections 9.13 and 9.15-9.17 for continuous systems may be solved in the same manner for sampled-data systems, wherein only the error at sampling instants is considered.

§ 9.19 Optimum Filter for Minimizing the Continuous Squared Error in Sampled-Data Systems

The problem is formulated in Fig. 9.19-1. The reader is again asked to return to Section 9.8. Parseval's relation with its integration limits at $\pm\infty$ may be used,

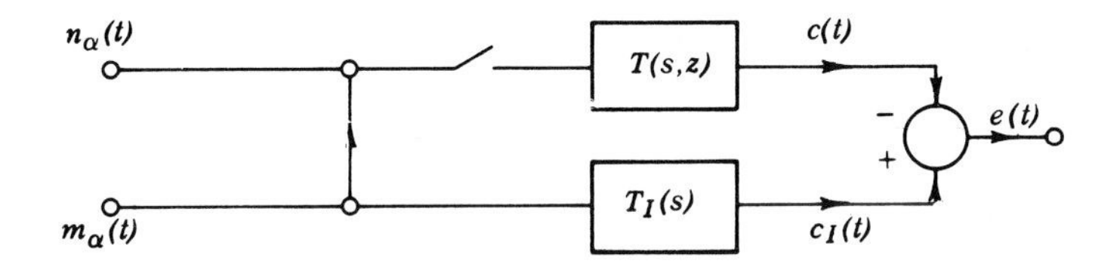

FIG. 9.19-1. Sampled-data filter problem in which continuous ideal output and continuous error are considered.

because at most, the integrand has the form $F_1^*(s)\,F_2(-s)$ or $F_1(s)\,F_2^*(-s)$. The ensemble-averaged error squared integral becomes

$$\epsilon^2 = \int_{-\infty}^{\infty} T_I(s)\,T_I(-s)\,\Phi_m - T_I(s)\,T(-s)\,\mathsf{E}[M_\alpha(s)\,M_\alpha^*(-s) + M_\alpha(s)\,N_\alpha^*(-s)]$$
$$- T(s)\,T_I(-s)\,\mathsf{E}[M_\alpha^*(s)\,M_\alpha(-s) + N_\alpha^*(s)\,M_\alpha(-s)] + T(s)\,T(-s)[\Phi_{m^*}(z)$$
$$+ \Phi_{m^*n^*}(z) + \Phi_{n^*m^*}(z) + \Phi_{n^*}(z)]df,$$

where Φ_m is defined in Eq. (9.8,2); Φ_{m^*} has been defined in Section 9.18, and $\Phi_{m^*n^*} \triangleq \mathsf{E}\,M_\alpha^*(s)\,N_\alpha^*(-s)$, $\Phi_{n^*m^*} \triangleq \mathsf{E}\,N_\alpha^*(s)\,M_\alpha^*(-s)$. We also let

$$\Phi_{mm^*} \triangleq \mathsf{E}\,M_\alpha(s)\,M_\alpha^*(-s), \qquad \Phi_{mn^*} \triangleq \mathsf{E}\,M_\alpha(s)\,N_\alpha^*(-s),$$
$$\Phi_{m^*m} \triangleq \mathsf{E}\,M_\alpha^*(s)\,M_\alpha(-s), \qquad \Phi_{n^*m} \triangleq \mathsf{E}\,N_\alpha^*(s)\,M_\alpha(-s).$$

If we pursue the development in Section 9.9, we are led to the autocorrelation function defined by $\phi_{mm^*}(\tau) \triangleq \mathsf{E}\int_0^\infty m_\alpha(t_1)\,m_\alpha^*(t_1-\tau)\,dt_1$, but it should be noted that while $\phi_{m^*m}(\tau)$ is an even function of τ at $\tau = kT$, it is not necessarily so at other values of τ. However, if $m_\alpha(t)$ has stationary statistics, then $\phi_{m^*m}(\tau)$ [which is then defined as $\mathsf{E}\lim_{A\to\infty} A^{-1}\int_{-A}^{A} m_\alpha(t_1)\,m_\alpha^*(t_1-\tau)\,dt_1$] is identical to $\phi_m(\tau)$. Equation (9.9,3) becomes $\Phi_{mm^*} = (\mathscr{L} + \mathscr{L}^-)\,\phi_{mm^*}(\tau)$. Similarly,

$$\phi_{m^*m}(\tau) = \mathsf{E}\int_0^\infty m_\alpha^*(t_1)\,m_\alpha\,(t_1-\tau)dt_1, \qquad \phi_{m^*m}(\tau) = \phi_{mm^*}(-\tau);$$
$$\Phi_{m^*m} = (\mathscr{L} + \mathscr{L}^-)\,\phi_{m^*m}(\tau),$$
$$\phi_{m^*n} = \mathsf{E}\int_0^\infty m_\alpha^*(t_1)\,n_\alpha\,(t_1-\tau)dt_1, \qquad \Phi_{m^*n} = (\mathscr{L} + \mathscr{L}^-)\phi_{m^*n}(\tau),$$

etc. The procedure described in Fig. 9.9-3 may be used to find the correlation function of a continuous and sampled signal pair.

From Sections 9.10-9.13, it is readily found that

$$T_{\text{opt}}(s, z) = \frac{1}{\Phi_{i*}^{+}(z)} \left\{ \frac{T_I\,[\Phi_{mm^*}(s) + \Phi_{mn^*}(s)]}{\Phi_{i*}^{-}(z)} \right\}^{+} \tag{9.19,1}$$

where $\Phi_{i^*} = \Phi_{m^*} + \Phi_{n^*} + \Phi_{m^*n^*} + \Phi_{n^*m^*}$. The extra complication in (9.19,1) over (9.11,1) or (9.18,1) is that, unlike the latter two, T_{opt} is not a function of s alone or $z = e^{sT}$ alone, but of both. This makes it more difficult to evaluate $\{\,\}^+$ of (9.19,1), which we denote as $K(s, z) = \mathscr{L}k(t)$. To find $k(t)$ it is usually necessary to use the inversion formula[1] $h(t) = (2\pi j)^{-1} \int_{c-j\infty}^{c+j\infty} H(s)\, e^{st}\, ds$, which may be evaluated by means of Cauchy's residue theorem.

Example. *Uncorrelated Inputs with Stationary Statistics.* Suppose $T_I(s) = e^{\alpha s}$, $T = 1$, $\Phi_m = -4/(s^2 - 4) = \Phi_{mm^*}$, so $\phi_m(\tau) = e^{-2|\tau|}$. Since the statistics are stationary, $\phi_{m^*}(\tau) = \phi_m{}^*(\tau)$ and $\Phi_{m^*} = (\mathscr{L} + \mathscr{L}^-)\, \phi_{m^*}(\tau)$. However, the impulse at $t = 0$ must not be counted twice (in continuous systems it does not matter because the integral of a finite number over zero time is zero). Hence $\Phi_{m^*} = (\mathscr{L} + \mathscr{L}^-)\, \phi_m{}^*(\tau) - 1 = z/(z - e^{-2T}) + z^{-1}/(z^{-1} - e^{-2T}) - 1 = 7.265z/(z - 0.135)\,(7.40 - z)$. Also, $\Phi_{n^*} = \Phi_n = 1$, so $\Phi_{i^*} = (z - 0.068)\,(z - 14.7)/(z - 0.135)\,(z - 7.40)$.

$$\left[T_I \frac{\Phi_{mm^*}}{\Phi_{i*}^{-}}\right]^{+} = [K(s, e^{sT})]^{+} = \left\{ \frac{-4(e^{sT} - 7.40)e^{\alpha s}}{(s-2)\,(s+2)\,(e^{sT} - 14.7)} \right\}^{+}.$$

$\mathscr{L}^{-1}(K)^+ = (2\pi j)^{-1} \int_{c-j\infty}^{c+j\infty} K(s, e^{sT})\, e^{st}\, ds$, for $t > 0 = \Sigma$ residues in lhp poles. There is only one lhp pole of $K(s, e^{sT})$ and it is at $s = -2$. (The zeros of $e^{sT} - 14.7$

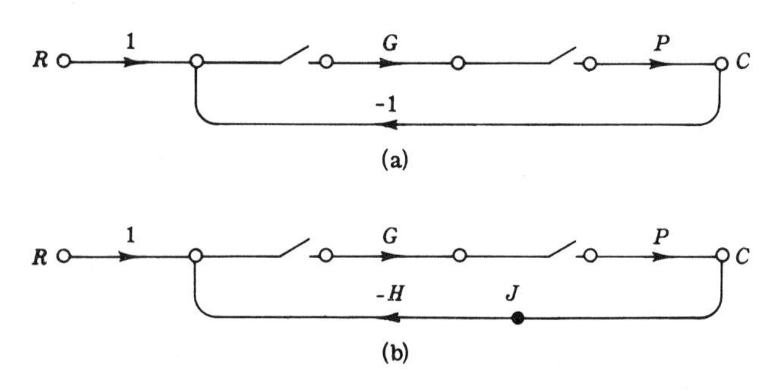

FIG. 9.19-2. Two feedback structures for realizing $T_{\text{opt}}(s, z)$.

[1] B. Van der Pol and H. Bremmer, "Operational Calculus," p. 18. Cambridge Univ. Press, London and New York, 1955.

are all in the rhp.) The residue in the pole at -2 is $(e^{-2T} - 7.40)\, e^{-2(\alpha+t)}/(e^{-2T} - 14.7) = 0.5\, e^{-2(\alpha+t)}\, \mu(t)$. Hence $(K)^+ = 0.5\, e^{-2\alpha}/(s + 2)$, and $T_{\rm opt} = 0.5(z - 0.135)\, e^{-2\alpha}/(s + 2)\,(z - 0.068)$.

In order to independently control the continuous and the sampled portion in $T_{\rm opt}(s, e^{sT})$, with the one-degree-of-freedom structure, a configuration such as that shown in Fig. 9.19-2a is required. Here $T(s, z) = C/R^* = G^*P/(1 + G^*P^*) = T_{\rm opt}$. If $P = B/(s + 2)$, $P^* = Bz/(z - 0.135)$, and $G^*/(1 + G^*P^*) = A(z - 0.135)/(z - 0.068)$, with $AB = 0.5\, e^{-2\alpha}$. Hence $G^* = A(z - 0.135)/[z(1 - AB) - 0.068]$. If Fig. 9.19-2b is used, with $T = G^*P/(1 + G^*PH^*)$, then presumably $L^* = G^*PH^*$ is determined by the desired benefits of feedback, so $T_{\rm opt} = G^*P/(1 + L^*)$ determines $G^*(s)$ and $P(s)$, and then $H(s)$ is deduced such that $G^*PH^* = L^*$. The problem is simpler if there is a sampler at node J, in Fig. 9.19-2b. Similar procedures are used in other configurations. The extension of the above to the problems considered in Sections 9.15-9.17 is straightforward.

Additional References

Paul M. De Russo, Performance Limitations of Digitally Controlled Systems, Rep. 7849-R-5. Servomechanisms Laboratory, M.I.T., Cambridge, Massachusetts, December 1958.

S. S. L. Chang, Optimum Transmission of Continuous Signal Over a Sampled Data Link, *Tech. Rept.* 400-411. Dept. of Electrical Engineering, New York University, New York, June 1960.

R. M. Stewart, Statistical design and evaluation of filters for the restoration of sampled data. *Proc. IRE* 44, 253-257 (1956).

J. T. Tou, Statistical design of digital control systems. *Trans. IRE* **AC-5**, 290-297 (1960).

J. R. Ragazzini and G. F. Franklin, "Sampled-Data Control Systems," Chapter 10, pp. 250-281. McGraw-Hill, New York, 1958.

S. S. L. Chang, Statistical design theory for digital-controlled continuous systems. *Trans. AIEE Part II.* **77**, 191-201 (1958).

CHAPTER TEN

Synthesis of Linear, Multivariable Feedback Control Systems[1]

§ 10.1 Introduction

This chapter is concerned with control systems built around multivariable plants. These are plants with n inputs and m outputs in which $n > 1$ and $m \geqslant 1$. The input and output variables can have any and different units—voltage, current, displacement, torque, rate of flow, temperature, etc. The plant output variables are controlled by the plant input variables and it is presumed that the relations between the two sets of variables are linear ones. Almost any chemical process is a multivariable process. Other examples of multivariable plants are: air frames, missiles, engines of all kinds, paper mills, steel mills, etc. Actually, most control systems are multivariable systems. In the so-called single variable plant (single input and single output), it is assumed that the other variables (e.g., the reference phase of the two phase induction motor) are maintained at a constant value.

The successful design of a multivariable feedback control system is greatly facilitated by the adoption of a synthesis point of view. In the single variable system, it is often possible to design by cut and try, but such an approach is usually hopeless in the multivariable system. The reason is that there are simply too many degrees of freedom to expect that their random adjustment will converge to a satisfactory solution. Even if by good fortune an acceptable design is thus obtained, it is extremely difficult to answer such questions as "What is the effect of plant parameter variation on the various transmission functions, on the system stability; how does the system react to disturbances, etc.?" In the synthesis approach, all such relevant questions are asked at the very beginning, and the desired system behavior is incorporated into the design procedure. It will be found that the intelligent synthesis of a linear, multivariable system, although laborious, is at least logical and straightforward. Of course, the synthesis

[1] This chapter is primarily based on I. M. Horowitz, Synthesis of Linear Multivariable Feedback Control Systems, Research Rept. 164, Hughes Research Labs., Malibu, Calif., Dec. 15, 1960.

approach imposes a burden not present in the cut and try approach. This is the burden of deciding beforehand what is wanted. To answer the question "What do I want?" requires mental exertion that one often prefers to avoid, no matter what problem is being considered.

§ 10.2 Clarification of Design Objectives and System Constraints

Consider a plant with n independent input variables and m dependent output variables. It is assumed that the output variables can be controlled only via the plant, or, in other words, we must work with the given plant, and may not substitute another for it. There are, therefore, mn relations between the plant input and output variables, and these relations can be put in the form of an $m \times n$ (m rows, n columns) plant matrix (symbol P) of transfer functions. The system (not the plant) input variables, the plant input variables, and the plant (also system) output variables (see Fig. 10.2-1) are denoted, respectively, by the column matrices[1] R, X, C.

It is assumed that there are as many system input variables as there are plant input variables, i.e., as many r's as there are x's, because n input variables suffice to control the n plant input variables; nothing is gained by postulating any more. Let it be assumed, in the meantime, that $n = m$, so that P is a square $n \times n$ matrix. The n x's uniquely determine the n c's from the set of n linear simultaneous equations, denoted by the matrix equation $C = PX$. Likewise, any set of values of c's is realizable by assigning a suitable set of values of x's.

[1] An easily interpreted notation is of the utmost importance here. Ordinarily, in the literature, lower case letters are used to represent time functions and their upper case equivalents are used to represent their respective transforms. However, in this chapter, unless specifically otherwise stated, upper case letters are used to represent matrices and the lower case letters to represent the respective elements of the matrix.

Thus the plant transfer and system transfer matrices are, respectively,

$$P = \begin{bmatrix} p_{11} \cdots p_{1n} \\ p_{21} \cdots p_{2n} \\ \cdot \\ \cdot \\ p_{m1} \cdots p_{mn} \end{bmatrix}, \qquad T = \begin{bmatrix} t_{11} \cdots t_{1n} \\ \cdot \\ \cdot \\ \cdot \\ t_{n1} \cdots t_{nn} \end{bmatrix}$$

The matrices of the transforms of the system input signals, the plant input signals, and the system output signals are, respectively,

$$R = \begin{bmatrix} r_1 \\ \cdot \\ \cdot \\ \cdot \\ r_n \end{bmatrix}, \qquad X = \begin{bmatrix} x_1 \\ \cdot \\ \cdot \\ \cdot \\ x_n \end{bmatrix}, \qquad C = \begin{bmatrix} c_1 \\ \cdot \\ \cdot \\ \cdot \\ c_m \end{bmatrix}.$$

Thus $X = P^{-1}C$, and so long as P is not singular,[1] the required x's are available. The relations between the system output variables (the c's) and the system input variables (the r's) are denoted by the matrix equation $C = TR$, where T

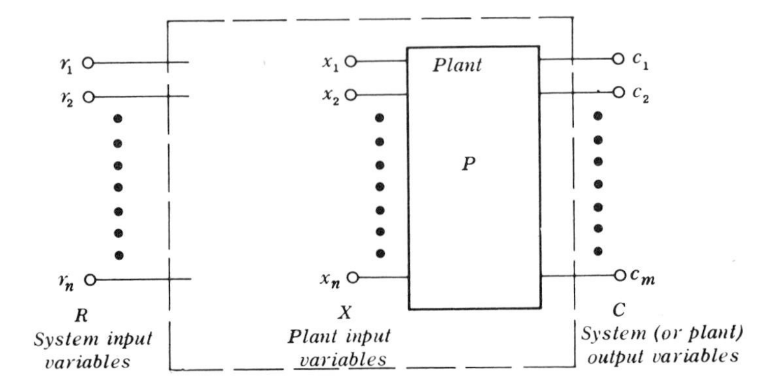

FIG. 10.2-1. The multivariable system with its 3 essential matrices of variables R, X. C.

is an $n \times n$ matrix called the system (transmission) matrix. The n^2 elements of T are independently realizable because $C = TR = PX$ and hence $X = P^{-1}TR$. If no feedback is used, it is necessary to obtain a network with n inputs and n outputs and whose transfer function matrix is $P^{-1}T$. In Fig. 10.2-2 the desired T is realized with a nonfeedback configuration.

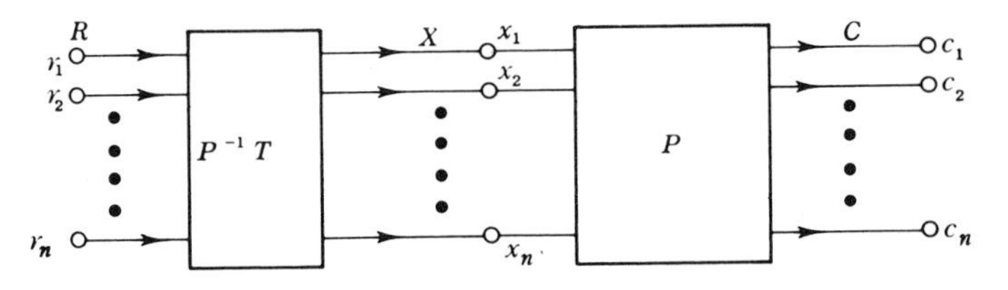

FIG. 10.2-2. Synthesis of desired T matrix without feedback.

The open-loop structure of Fig. 10.2-2 is satisfactory if all the following conditions are satisfied.

(1) P is known within the same tolerance specified for T.

(2) The parameters of P do not vary more than the tolerances specified for T,

---

[1] P is a square matrix which becomes singular when its determinant is zero. In such a case, the elements of C are always related to each other in the same way, no matter what the x's are. Independent values of all the elements of C are then unobtainable. It is assumed here that P is not singular. When the parameters of P vary, it is assumed that P is not singular at any combination of values of the elements of P. If this is not so, i.e., if P is singular for any or all of its element values, then for these values one must specify a suitably restricted T matrix. This degenerate case is not considered here.

or if they do, then their variation is known, and it is possible to realize the matrix of time-varying transfer functions, $P^{-1}T$.

(3) There are no additional inputs to P, or if there are, their values are known as functions of time and $P^{-1}T$ (or the elements of R) can be varied so as to negate their effect.

(4) The elements of P and of P^{-1} do not have any right half-plane (rhp) poles.

It is only when one or more of the above conditions are not satisfied that feedback around the plant is needed.

Consider the case when the number of plant input variables is less than the number of plant output variables, i.e., $n < m$. Then $m - n$ output variables must be dependent functions of the remaining output variables. Therefore these $m - n$ uncontrollable outputs may be omitted and the plant represented by a square $n \times n$ matrix. For example, let $n = 2$, $m = 3$, as depicted in Fig. 10.2-3. The plant is described by the set of equations: $c_1 = p_{11}x_1 + p_{12}x_2$, $c_2 = p_{21}x_1 + p_{22}x_2$, $c_3 = p_{31}x_1 + p_{32}x_2$. It is impossible to realize three independent c's. Any two equations with two given c's determine x_1, x_2 and these in turn determine the third c. Therefore one c may as well be entirely deleted; P becomes a square matrix of order 2, and only four independent T matrix elements are realizable. If c_3 is deleted, then the four independent elements of T are t_{11}, t_{12}, t_{21}, t_{22}.

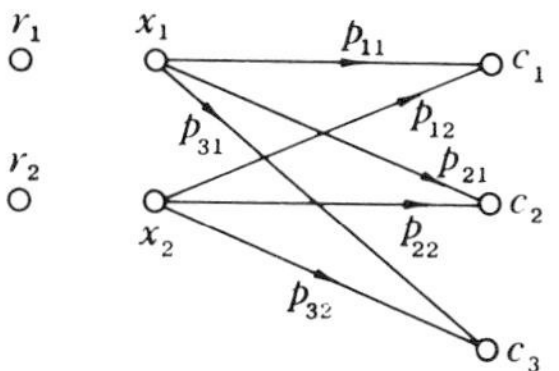

FIG. 10.2-3. Plant with $n = 2$ $m = 3$.

However, it is not always possible to dispose of the problem so easily. For example, consider again the structure of Fig. 10.2-3. It has previously been noted that only four independent transfer functions are realizable. If they are to be t_{11}, t_{12}, t_{21}, t_{22}, then c_3 may be ignored, and p_{31}, p_{32} omitted entirely from the plant representation.[1] But suppose the four independent transmissions to be realized are t_{11}, t_{22}, t_{31}, t_{32}, and one must perforce accept whatever t_{12}, t_{21} that result. It is extremely inconvenient to work with nonsquare matrices. When parameter variations are insignificant, this problem may be overcome by noting that the specification of any four t's willy nilly fixes all the other t's. It is possible to solve for t_{21} (or t_{12}) and use it in place of t_{11} (or t_{22}) and omit c_1 (or c_2) entirely. The t_{21} transfer function is found as follows: Let $r_1 = 1$, $r_2 = 0$. In this special case $c_1 = t_{11}r_1 + t_{12}r_2 = t_{11}$, and $c_3 = t_{31}r_1 + t_{32}r_2 = t_{31}$ are known, because t_{11} and t_{31} are known. Knowing c_1 and c_3, it is possible to find c_2 from the previous three equations for c_1, c_2, and c_3. The first and last of these equations provide two simultaneous equations (c_1, c_3 are known, and of course the p's are known) in

[1] In practice, it might be convenient to use feedback signals from c_3, so c_3, p_{31}, and p_{32} would be retained for this purpose. However, this would fall into the category of a plant with internal variables available for feedback purposes. Only the simplest case of a plant with no accessible internal points is considered here.

the two unknowns x_1, x_2. One solves for x_1 and x_2 and substitutes in the second equation to evaluate c_2, which is precisely equal to t_{21} for these special values of r_1 and r_2.This approach is impractical when the four t_{ij}'s of interest are to be made insensitive to plant parameter variation. The reason is that (as will be seen) there must be a feedback signal from c_i in order that any t_{ij} may be made insensitive to parameter variation. Hence reduced sensitivity of all four t_{11}, t_{22}, t_{31}, t_{32} requires feedback from c_1, c_2, c_3 and therefore these three variables must be retained in the equations. This special case will be treated in Section 10.16.

Hence excluding this special case, it is possible to work with a square plant matrix even though $m > n$. There is no point in postulating more system inputs than there are plant inputs (i.e., more r' sthan x's), just as there is no point in assuming two r's in the single variable plant. If there are in fact two input points (two r's) in the single variable plant, then the system response to each input is a separate distinct problem, and the single variable synthesis technique is separately applied for each problem. A similar approach is taken in the multivariable problem when there are more r's than x's.

On the other hand, if the number of plant inputs exceeds the plant outputs, i.e., $n > m$, then one may invent $n - m$ fictitious outputs, precisely equal to the $n - m$ excess plant inputs. The result is again an $(n \times n)$ P matrix. For examplè, if $n = 4$, $m = 2$, then let the augmented

$$P = \begin{bmatrix} p_{11} & p_{12} & p_{13} & p_{14} \\ p_{21} & p_{22} & p_{23} & p_{24} \\ 0 & 0 & 1 & 0 \\ 0 & 0 & 0 & 1 \end{bmatrix}.$$

We are therefore justified in assuming henceforth that P is a square $n \times n$ matrix.

An important assumption is being made throughout this chapter. This is that for a given $n \times n$ plant, a specific $n \times n$ matrix of system transfer functions is desired. In such a case, the designer has nothing to say on how to utilize the plant. This is clarified by considering the opposite situation. Suppose that in Fig. 10.2-4, the designer's interest is only in control of c_1 by r_1 and r_2, i.e., in the

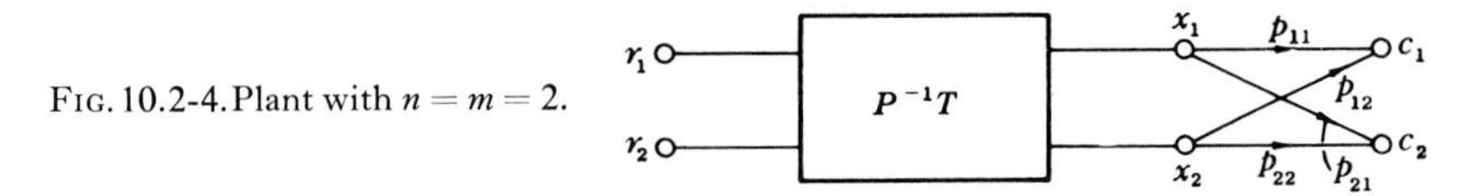

FIG. 10.2-4. Plant with $n = m = 2$.

T elements t_{11}, t_{12}; and the effect on c_2 is disregarded (i.e., t_{21}, t_{22} are of no consequence). For any values of r_1 and r_2, $c_1 = p_{11}x_1 + p_{12}x_2 = t_{11}r_1 + t_{12}r_2$. The inputs and the desired system transfer functions t_{11} and t_{12} determine c_1, but this one equation in the unknowns x_1, x_2 is insufficient to uniquely fix x_1 and x_2 or their ratios. The designer can use this freedom to optimize in some sense—

perhaps to throw most of the burden of achieving c_1 on that part of the plant which is most economical to operate. This freedom exists only because c_2 and the transfer functions contributing to c_2 (t_{21}, t_{22}), are neglected. If, however, all the elements of the matrix are specified, then there is no such freedom. The signal levels that exist in the various parts of the plant, in order to achieve the desired combination of outputs, are completely fixed. This is seen as follows. For any combination of inputs, the combination of desired outputs is completely determined by the matrix equation $C = TR$. However, C is completely determined by $C = PX$. This last matrix equation gives n simultaneous equations in the n unknowns $x_1, x_2, ..., x_n$. So long as P is not a singular matrix, the x's are completely determined, and with them the signals in the various parts of the plant.

It is outside the scope of this chapter to decide how to best utilize the plant, and it is therefore assumed that all the elements in T are specified. This means that if the plant has 5 inputs and 5 outputs, it is assumed that there are 5 input command points (5 r's) and the specifications include a desired 5th order T matrix with each of its 25 elements specified, at least over their significant frequency range. This assumption is also made for the case $m < n$, where $n - m$ fictitious outputs are invented. In other words, the decision has already been made how to best utilize the plant.[1] Thus, this chapter is primarily devoted to the *feedback* problem in multivariable systems.

§ 10.3 Role of System Configuration

The simplest design problem concerns a plant in which the only plant variables that need be considered are the plant input and output variables. It is assumed that no other *internal* plant variables are available, or, if available, they are not used, for whatever the reasons may be. The more general case, when internal plant variables are available for feedback purposes, is considerably more complicated (cf. the multiple-loop design of Chapter 8 with the single-loop design of Chapter 6). It is also assumed that control over any output variables is achieved only via the plant. This is a very reasonable and realistic assumption in the vast majority of control problems. With these constraints, the design capabilities are completely determined. There are then only two matrices of functions that may be independently realized (cf. in the analo-

[1] For a treatment and bibliography of the *control* problem in multivariable systems, see M. D. Mesarovic, "The Control of Multivariable Systems." Wiley, New York, 1960.

For work on the *filter* problem in multivariable systems, see R. C. Amara, The Linear Least Squares Synthesis of Continuous and Sampled Data Multivariable Systems, Tech. Rept. No. 40, Stanford Electronics Labs., Stanford, Calif., July 28, 1958; H. C. Hsieh and C. T. Leondes, Techniques for the optimum synthesis of multipole control systems with random processes and inputs, *Trans. IRE* **AC-4**, 212-231 (1959).

gous single variable system, only two functions may be independently realized—see Section 6.1). This is seen as follows.

There are only three essential matrices of variables, each matrix represented by a node in Fig. 10.3-1a. Since no signal is allowed to reach node C except

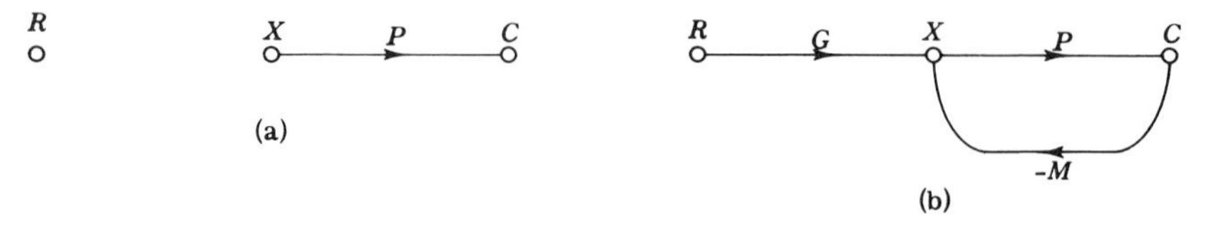

FIG. 10.3-1. A canonical feedback structure for system with three essential matrices of variables.

through P, the system response is entirely determined by the values of X. Also, X is completely determined by the signals received from R and C. There are, accordingly, only two essential matrices of transfer functions, one from R to X, and the other from C to X, with a total of $2n^2$ elements. Therefore, to find the effect of any independent external signals applied at R, X, or C nodes, Fig. 10.3-1b is sufficiently general, and no other configuration has any more generality. Also, as far as the dependence of any of the above effects on P, Fig. 10.3-1b is sufficiently general. The important point (previously noted in Chapter 6 for the single variable system and emphasized here again) is that the plant constraints (which are the accessible points of the plant) determine the degrees of freedom in the system. If these are q in number, then the most general configuration is the one with at least q independent functions, and there is usually a variety of configurations possessing at least q independent functions.

§ 10.4 Principal Steps in the Design of Multivariable Systems

The two principal reasons for using feedback are: (1) to reduce the sensitivity of the system response to plant parameter variations, and (2) to control the effect of disturbances entering the system at any points in the plant. The design problem consists of attaining these two feedback objectives, while at the same time realizing a desired set of overall system transfer functions. Systematic and intelligent design requires a quantitative statement of design objectives, the range and nature of parameter variation and disturbances, and specification of the desired system response. Assuming that these have been ascertained, the detailed quantitative steps for attaining these objectives are presented in this chapter. There are three principal steps in the design:

(1) The benefits of feedback, in the form of reduced system sensitivity to plant parameter variation, or in the form of attenuated plant disturbances, are

attained by means of sufficiently large loop transmissions over sufficiently large bandwidths. The first step then is to ascertain the required magnitudes of loop transmissions over the siginficant frequency ranges, in order to obtain the desired benefits of feedback. For this purpose, relations between the effect of the parameter variations (and effect of disturbances) and the loop transmission functions are developed in subsequent sections. Thus the sensitivity reduction and disturbance attenuation fix the values of the loop transmission functions over the significant frequency ranges.

It is not possible, in the two-degree-of-freedom configuration, to independently control the system sensitivities to plant parameters and the disturbance attenuations. Therefore some loop transmissions may be larger than necessary for one of these, in order to adequately minimize the second. When both parameter variations and disturbances are simultaneously important, it is necessary to ensure that the disturbances are properly attenuated for all possible values of the plant parameters (see Section 10.15).

(2) Beyond the significant frequency ranges, the loop transmissions are, as usual, decreased as fast as possible, consistent with the requirement of stability despite plant parameter variation. This step is considerably more difficult in the multivariable system than in the single variable system. The reason is not so much due to the larger number of varying parameters. It is due to the fact that the function which must be shaped for stability is a complicated combination of the individual loop transmission functions, and each of the latter has varying parameters. In this step the designer is always confronted with a choice between design labor economy and economy in gain-bandwidth requirements of the compensation functions.

(3) The second available degree of freedom is used to obtain the desired set of system response functions. Since these are often specified only up to a few octaves beyond their half-power points, the freedom (if any) that exists in their values at higher frequencies may be used to simplify to some extent the expressions for the compensating functions.

§ 10.5 The System Transfer Matrix

One of the important reasons for using feedback in control is to reduce the sensitivity of the control system to plant parameter variations. In the following sections, relations are developed to enable the designer to specify the appropriate system functions which keep the system sensitivity to plant parameter variations within desired bounds. In view of the discussion in previous sections, it is assumed that P is an $n \times n$ matrix and the only accessible plant variables are the plant inputs (denoted by the matrix X) and the plant outputs. The sensitivity relations are developed in such a manner that they can be applied to any two-degree-of-freedom configuration. Some typical configurations are shown in Fig. 10.5-1. Figure 10.5-1a is shown in detail in Fig. 10.5-2 for $n = 2$.

To attain the desired generality for the system sensitivity functions, let[1] the

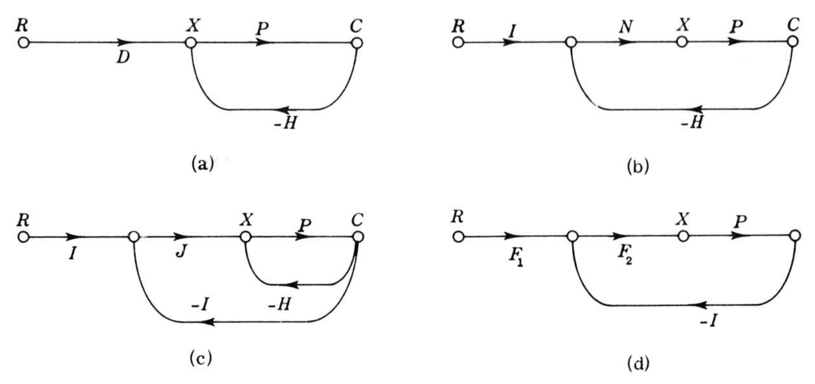

FIG. 10.5-1. Some two-degree-of-freedom structures. (Each element represents a matrix; I is the unit diagonal matrix.)

matrix equation $X = GR - MC$ define the matrices G and M. Thus, in Fig. 10.5-1a, $M = H$, $G = D$; in Fig. 10.5-1b, $M = NH$, $G = N$; in Fig. 10.5-1c, $M = H + J$, $G = J$; in Fig. 10.5-1d, $G = F_2F_1$, $M = F_2$. Also, $C = PX = P(GR - MC)$, and collecting terms in C, $C + PMC = (I + PM)\,C = PGR$, where I is the unit diagonal matrix. Let

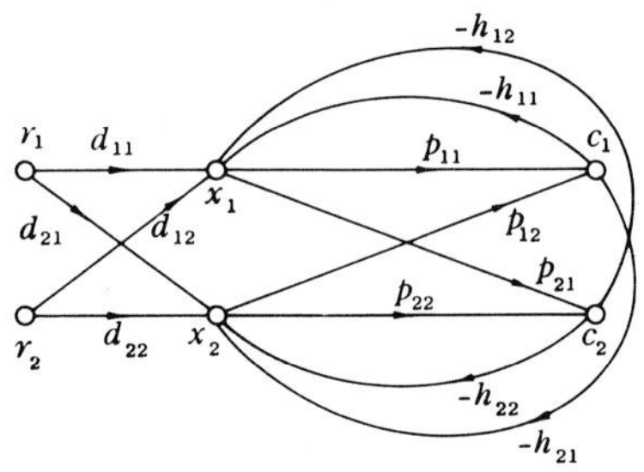

FIG. 10.5-2. Details of Fig. 10.5-1a.

$$L \triangleq PM \qquad (L \text{ is called the loop transmission matrix}). \tag{10.5,1}$$

Therefore, in general, $(I + L)\,C = PGR$, $C = (I + L)^{-1}\,PGR = TR$, which thereby defines

$$T = (I + L)^{-1}\,PG; \tag{10.5,2a}$$

$$T^{-1} = G^{-1}P^{-1}(I + L). \tag{10.5,2b}$$

The physical significance of L is obtained by a method analogous to that used in the single variable system. In Fig. 10.5-3, cuts are made just before the output points. The cuts are made such that the plant is unaffected but all the

[1] It should be noted that in general $NH \neq HN$, $MC \neq CM$, etc., since each upper case letter represents a matrix. Only a very elementary knowledge of matrix theory is needed here. This can be gained by reading pp. 30-48 in E. A. Guillemin, "The Mathematics of Circuit Analysis." Wiley, New York, 1949.

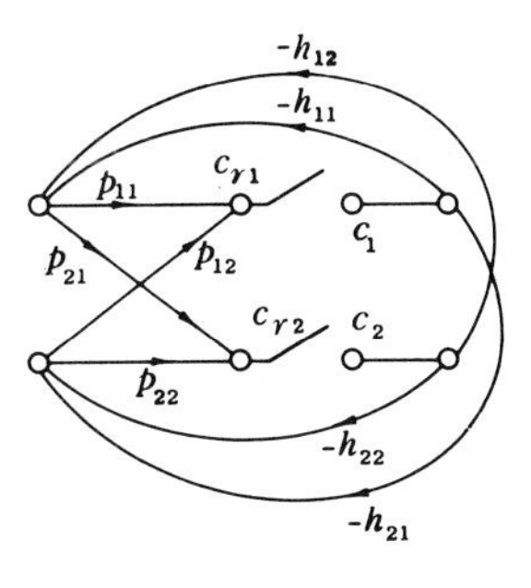

FIG. 10.5-3. Physical interpretation of loop transmission matrix.

feedback loops are opened. Now assume the input $R = 0$ and inject a negative unit signal at c_1, i.e., let $c_1 = -1$, $c_2 = c_3 = ... = c_n = 0$. The resulting values of the returned signals $c_{r1}, c_{r2}, ..., c_{rn}$ of Fig. 10.5-3 comprise the first column of L, i.e., $c_{r1} = l_{11}$, $c_{r2} = l_{21}$, etc. If all c's are made zero except for $c_2 = -1$, then the resulting c_{ri}'s form the second column of L, etc. Thus $l_{\alpha\beta}$ is the returned signal $c_{r\alpha}$ due to a signal -1 at c_β, when the loops are cut in the manner of Fig. 10.5-3. When L is a diagonal matrix, then with unit negative signal at c_α, all c_r's are zero except for $c_{r\alpha}$.

§ 10.6 Effect of Parameter Variations on the System Transmission Matrix

The next step in the development of a synthesis procedure is to obtain simple expressions which enable the designer to choose the elements of L so that the variations in the elements of T are maintained within specified tolerances. A variety of expressions which are useful for this purpose will be derived and their use illustrated with numerical examples. Each has its own advantages and disadvantages and these will be noted.

Let T be the nominal transmission matrix when the plant has the nominal matrix P. When the plant parameters assume a different set of values, let the resulting plant matrix be P' and the resulting different system matrix be T'. Thus let

$$P' = P + \Delta P, \qquad T' = T + \Delta T, \qquad L' = P'M. \tag{10.6,1}$$

Also, from Eqs. (10.5,2), $T' = (I + L')^{-1} P'G$, $T'^{-1} = G^{-1}P'^{-1}(I + L')$. Subtracting (10.5,2b) from T'^{-1}, and replacing L' by $P'M$, L by PM, we get $T'^{-1} - T^{-1} = G^{-1}P'^{-1}(I + P'M) - G^{-1}P^{-1}(I + PM) = G^{-1}[P'^{-1} + M - P^{-1} - M] = G^{-1}(P'^{-1} - P^{-1})$. Premultiply the above by T, giving $TT'^{-1} - I = TG^{-1}(P'^{-1} - P^{-1})$, and then postmultiply the result by T'. The result is $T - T' = -\Delta T = TG^{-1}(P'^{-1} - P^{-1})\, T'$. TG^{-1} may be replaced by its equivalent, $(I + L)^{-1} P$ [see Eq. (10.5,2a)]. Therefore,

$$\Delta T = (I + L)^{-1} (I - PP'^{-1})\, T' \triangleq (I + L)^{-1}\, VT'. \tag{10.6,2}$$

Equation (10.6,2) would be very useful for sensitivity design, but for the unfortunate fact that T', which is unknown, appears on the right-hand side of the equation. But consider the case when the specifications demand that the variation in each element of T is to be kept small over some common frequency range. Then over[1] this frequency range, each $t'_{ij} = t_{ij} + \tau_{ij} \doteq t_{ij}$. For this special case, Eq. (10.6,2) becomes

$$\Delta T \doteq (I + L)^{-1} (I - PP'^{-1}) T = (I + L)^{-1} VT. \tag{10.6,3}$$

Equation (10.6,3) is convenient for choosing L so as to satisfy the specifications on ΔT, because the only unknown on its right-hand side is L. It is especially convenient for this purpose when L is a diagonal matrix. The application of Eq. (10.6,3) is illustrated by several examples.

Example 1.

Specifications: The elements of the plant matrix P are $p_{11} = k_{11}/(s + 1)$, $p_{12} = k_{12}/(s + 1)$, $p_{21} = k_{21}/(s + 1)$, $p_{22} = k_{22}/(s + 2)$, where k_{11}, k_{22} can each vary independently from 1 to 4, while k_{12}, k_{21} can each vary independently from 0.1 to 0.5. Over the significant system frequency range, each t_{ij} is to be approximately equal and close to unity, and its variation, denoted by τ_{ij}, is to be no more than 0.05 in magnitude.

Design: The nominal plant parameter values are taken as $k_{11} = k_{22} = 1$, $k_{12} = k_{21} = 0.1$. The elements of the P' matrix are $p'_{11} = k'_{11}/(s + 1)$, etc. Then

$$P'^{-1} = \frac{(s + 1)}{\Delta_{k'}} \begin{bmatrix} k'_{22} & -k'_{12} \\ -k'_{21} & k'_{11} \end{bmatrix},$$

where $\Delta_{k'}$ is the determinant of the k' matrix, i.e., $\Delta_{k'} = k'_{11}k'_{22} - k'_{12}k'_{21}$. Let the elements of $I - PP'^{-1}$ be denoted by v_{ij}, i.e., v_{ij} is the element in the ith row and jth column of the matrix $I - PP'^{-1}$. Then

$$v_{11} = \frac{k'_{22}(k'_{11} - 1) - k'_{21}(k'_{12} - 0.1)}{\Delta_{k'}}, \qquad v_{12} = \frac{(k'_{12} - 0.1k'_{11})}{\Delta_{k'}}, \tag{10.6,4}$$

and v_{21}, v_{22} may be obtained from v_{12}, v_{11} by interchanging the subscripts 1 and 2. At low frequencies, $T' \doteq T$, whose elements are all 1. Using Eq. (10.6,3) with a diagonal L, it is found that due to the symmetry in T and P, each element of ΔT has the same value, which is $\tau \doteq (v_{11} + v_{12})/(1 + l)$, which is to be $\leqslant 0.05$. The values of the k's which lead to the maximum magnitude of $v_{11} + v_{12}$ are obtained by cut and try. They are $k'_{11} = k'_{22} = 4$, $k'_{12} = k'_{21} = 0.5$, and they lead to $0.75/(1 + l) \leqslant 0.05$; consequently $1 + l \geqslant 15$. Thus the minimum values of the elements of L in the significant frequency range are ascertained.

It has been noted that Eq. (10.6,3) may be used only over the frequency

[1] The elements of ΔT are denoted by τ_{ij}, those of T by t_{ij} and those of T' by t'_{ij}, i.e., $t'_{ij} = t_{ij} + \tau_{ij}$. The elements of L are denoted by l_{ij}.

range in which *each* $t'_{ij} \doteq t_{ij}$. The specifications may, however, permit *some* t'_{ij} to differ appreciably from the nominal t_{ij}, while in the same frequency range, *other* $t'_{\alpha\beta} \doteq t_{\alpha\beta}$. It is then not permissible to replace T' by T over this frequency range, unless one is willing to overdesign and choose the elements of L so that all $t'_{uv} \doteq t_{uv}$. This latter course is obviously uneconomical. For this type of problem it is necessary to use a better approximation of T', as follows. The $t'_{\alpha\beta}$ elements, whose variation is to be kept small, are replaced by $t_{\alpha\beta}$ but the t'_{ij} elements, with the larger permitted variations, are retained and their values estimated. This better approximation of T' is then used in Eq. (10.6,2) to find the required levels of loop transmissions. One should then check the accuracy of the estimate of the t'_{ij}, but generally it is not necessary to repeat the calculations, because the loop gain levels are chosen to ensure that the τ_{ij} are no larger than permitted. This procedure is illustrated by means of a numerical example.

Example 2.

Specifications: The plant is that given in Example 1. But here τ_{11}, τ_{12} must not be more than 0.05 up to 2 rps; however, τ_{12}, τ_{22} must not be more than 0.05 up to $\frac{1}{2}$ rps, but are allowed to increase to 0.25 at 2 rps. Ar 2 rps the nominal design values of the t_{ij} are all $0.7 \angle -60°$.

Design: From zero frequency to $\frac{1}{2}$ rps, the result of Example 1 can be used with $1 + l \geqslant 15$. Next consider the situation at 2 rps. What is the variation in t'_{21} and t'_{22} due to the permitted τ_{21}, τ_{22}? This is obtained from Fig. 10.6-1a. At $s = j2$, $\Delta T \doteq (I + L)^{-1} (I - PP'^{-1}) T'_{\text{approx}}$ becomes, for a diagonal L,

$$\begin{bmatrix} \tau_{11} & \tau_{12} \\ \tau_{21} & \tau_{22} \end{bmatrix} \doteq 0.7 \angle -60° \begin{bmatrix} \dfrac{v_{11}}{(1 + l_1)} & \dfrac{v_{12}}{(1 + l_1)} \\ \dfrac{v_{21}}{(1 + l_2)} & \dfrac{v_{22}}{(1 + l_2)} \end{bmatrix} \begin{bmatrix} 1 & 1 \\ x & x \end{bmatrix},$$

where x is shown in Fig. 10.6-1b. For each τ_{ij} one must find the combination of k's and x's which leads to the maximum $| \tau_{ij} |$. This is done by cut and try,

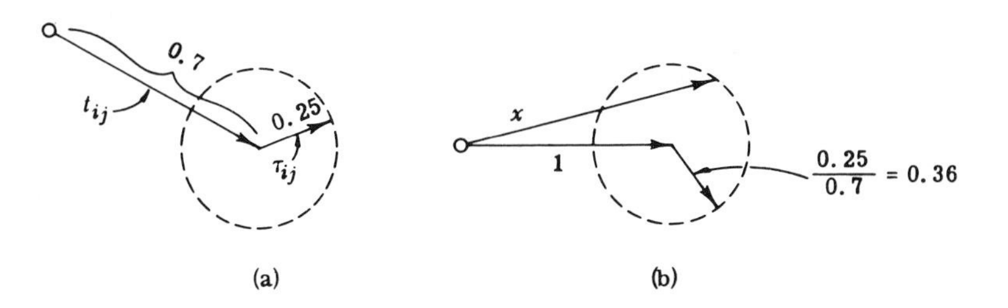

FIG. 10.6-1. (a) Determination of $t'_{ij} = t_{ij} + \tau_{ij}$. (b) Normalization of (a) by dividing by $0.7 \angle -60°$.

although for more complicated systems a systematic procedure for doing this would be required. It is found that the maximum τ_{ij} results when $k'_{11} = k'_{22} = 4$, $k'_{12} = k'_{21} = 0.5$ and $x = 1.36$. This leads to $0.05 \geqslant \tau_{11} = \tau_{12} = 0.7\,(v_{11} + 1.36v_{12})/(1 + l_1) = 0.55/(1 + l_1)$. Therefore $|\, 1 + l_1 \,| \geqslant 11$, so $l_1(j2) = 12$ is chosen, because of the worst possibility that $\angle l_1(j2) = 180°$. Similarly, $0.25 \geqslant \tau_{21} = \tau_{22} = 0.7\,(v_{21} + 1.36\,v_{22})/(1 + l_2) = 0.98/(1 + l_2)$. Therefore $|\, 1 + l_2 \,| \geqslant 4$.

§ 10.7 Comparison of Diagonal and Antidiagonal Loop Transmission Matrices

When a diagonal L matrix is used, then, from Eq. (10.6,2), each element in the xth row of $(I - PP'^{-1})\,T'$ is divided by $1 + l_x$. This appears to be satisfactory if the required amount of loop transmission is the same for each element in the xth row. This was the case in Examples 1 and 2 (Section 10.6), where the expressions for τ_{11} and τ_{12} (the first row of ΔT) were identical, and those for τ_{21} and τ_{22} (the second row of ΔT) were also identical. In addition, the maximum permitted values of τ_{11} and τ_{12} were the same, and that of τ_{21} equaled that of τ_{22}. But suppose in Example 1 it is specified that at zero frequency τ_{11} and τ_{21} must be less than 0.05, while τ_{12} and τ_{22} may be as much as 0.25. Since l_1 controls τ_{11} and τ_{12}, and l_2 controls τ_{21} and τ_{22}, it would be necessary to use the larger value of l in each case, and thus overdesign for τ_{12} and τ_{22}. Would a nondiagonal L be better in such a case? It is readily found that nonzero l_{11}, l_{12}, l_{21}, l_{22} do not lead to smaller required levels of loop transmission.

Can any general conclusion be drawn? Consider the general case for $n = 2$. In Eq. (10.6,3), let

$$Q \triangleq (I - PP'^{-1})\,T = VT \tag{10.7,1}$$

with q_{ij} denoting the elements of the Q matrix. Then Eq. (10.6,3) becomes

$$\Delta T = \begin{bmatrix} \tau_{11} & \tau_{12} \\ \tau_{21} & \tau_{22} \end{bmatrix} \doteq \frac{\begin{bmatrix} [q_{11}(1 + l_{22}) - q_{21}l_{12}] & [q_{12}(1 + l_{22}) - q_{22}l_{12}] \\ [-q_{11}l_{21} + q_{21}(1 + l_{11})] & [-q_{12}l_{21} + q_{22}(1 + l_{11})] \end{bmatrix}}{(1 + l_{11})\,(1 + l_{22}) - l_{12}l_{21}}. \tag{10.7,2}$$

Let the elements δ_{11}, δ_{12}, δ_{21}, δ_{22} denote the maximum permitted magnitudes of τ_{11}, τ_{12}, τ_{21}, τ_{22}, respectively. The four l_{ij}'s must be chosen so as to satisfy the four inequalities, $|\, \tau_{ij} \,| \leqslant \delta_{ij}$.

Consider, for example, the inequality

$$|\, \tau_{11} \,| \doteq \left| \frac{q_{11}(1 + l_{22}) - q_{21}l_{12}}{(1 + l_{11})\,(1 + l_{22}) - l_{12}l_{21}} \right| \leqslant \delta_{11}. \tag{10.7,3}$$

Is there any value in using both l_{12} and l_{22} in the above equation? There would be some value in it, if l_{12} and l_{22} could be chosen so that the two terms in the numerator would wholly or at least partially cancel each other. We could then achieve the inequality with small magnitudes for the l's. It should be recalled, however, that the q's are independently varying quantities because they are functions of the p_{ij}'s, and the latter independently vary over a range of values. Partial cancellation is feasible only if the variations in q_{11} are reasonably well related to the variations in q_{21}. Furthermore, in order that it may permit economizing on the loop transmission magnitudes, the partial cancellation must be achieved not at a single frequency, but over a range of frequencies. Only in special cases, when the q's are reasonably consistently related to each other over the significant frequency range, would it be feasible to achieve this partial cancellation of terms simultaneously over the range of q values and over the required frequency range. The individual q's can vary independently in Examples 1 and 2, so consistent partial cancellation in the sum $q_{11}(1 + l_{22}) - q_{21}l_{12}$ is not possible. There is therefore no value in retaining both l_{22} and l_{12} in Eq. (10.7,2). In fact, it may be worse than if only one was kept, because the extreme case when the terms aid each other would have to be considered. Either l_{12} or l_{22} may therefore be made equal to zero. The question is, which one of these, and which one of l_{21} and l_{11}?

A little investigation indicates that for $n = 2$, and in the general case, when partial cancellation is not feasible due to independent parameter variation, there are two alternatives. One is a diagonal L, the other is an antidiagonal L. In an antidiagonal L, only l_{21}, l_{12} are finite for $n = 2$, only l_{13}, l_{22}, l_{31} are finite if $n = 3$, etc. On the other hand, if we choose (for $n = 2$) $l_{12} = l_{11} = 0$, then $\tau_{11} = q_{11}$, $\tau_{12} = q_{12}$. [The q's are defined in Eq. (10.7,1).] L is not able in this case to reduce τ_{11} and τ_{12} to any extent whatsoever. The diagonal L and the antidiagonal L are two simple loop transmission matrices that are sufficiently general to handle the parameter variation problem for $n = 2$. In general, it is necessary (in order that the system sensitivity to all plant parameter variations may be made small) that L has at least one element in each row and column. If $n = 3$, there are six different matrices which have only one element in each row and column.

It is apparent from the above that there is quite a bit of freedom (which increases with the order of the system) in choosing L so as to obtain the desired system insensitivity to parameter variation. This freedom would not exist if the sensitivities of the n^2 elements of T were precisely defined as analytic functions in s. The n^2 elements of L would thereupon also be precisely determined. But the sensitivity specifications are hardly ever stated in such a manner. They are usually stated as in the preceeding examples. Such specifications only require a sufficient magnitude of return difference at each c_i in order to achieve the required insensitivity. Due to the multiple paths that exist in the plant, it is possible to achieve each return difference in n different ways, leading to $n!$

canonic L matrices. This latitude in the form of L may be useful, as it enables the designer to choose the simplest means of achieving the design objectives. The actual feedback structure and elements that represent the best choice must be independent of the way the terminals are numbered. If a diagonal L is best, this means that in Fig. 10.5-3, $c_{r1} = -l_1c_1$, $c_{r2} = -l_2c_2$. If the output terminal numbers are interchanged, then the old p_{11} becomes the new p_{21}, etc., but a diagonal L should again represent the best choice and the design should emerge with the new l_1, l_2 the same as the old l_2, l_1, respectively. In a third order system if the L matrix has only the elements l_{13}, l_{21}, l_{32} and the output terminals 1, 2, 3 are renumbered 2, 3, 1, respectively, then the new L matrix should also have the elements l_{13}, l_{21}, l_{32} but in the synthesis these should emerge with the same values as the old l_{32}, l_{13}, l_{21}, respectively, i.e., the final structure and element values will be the same.

That an antidiagonal L matrix has sufficient generality can be seen from Eq. (10.6,3), $\Delta T \doteq (I+L)^{-1}(I-PP'^{-1})\,T \triangleq (I+L)^{-1}Q$. If L is antidiagonal, then

$$\Delta T = \begin{bmatrix} \tau_{11} & \tau_{12} \\ \tau_{21} & \tau_{22} \end{bmatrix} \doteq \begin{bmatrix} 1 & l_1 \\ l_2 & 1 \end{bmatrix}^{-1} \begin{bmatrix} q_{11} & q_{12} \\ q_{21} & q_{22} \end{bmatrix} \leqslant \begin{bmatrix} \delta_{11} & \delta_{12} \\ \delta_{21} & \delta_{22} \end{bmatrix}.$$

The result is (assuming $|\,l_1l_2\,| \gg 1$)

$$\begin{aligned} l_1 &\geqslant (q_{11} - q_{21}l_2^{-1})/\delta_{21}, \qquad (q_{12} - q_{22}l_2^{-1})/\delta_{22} \\ l_2 &\geqslant (q_{21} - q_{11}l_1^{-1})/\delta_{11}, \qquad (q_{22} - q_{12}l_1^{-1})/\delta_{12}. \end{aligned} \tag{10.7,4}$$

These inequalities apply to the *magnitudes* of the various quantities and it should be remembered that the q's in general vary independently if the plant parameters vary independently.

For $n = 3$, Eq. (10.6,3) becomes

$$\begin{aligned} l_1 &\geqslant (q_{11} - q_{31}l_3^{-1})/\delta_{31}, \qquad (q_{12} - q_{32}l_3^{-1})/\delta_{32}, \qquad (q_{13} - q_{33}l_3^{-1})/\delta_{33} \\ l_2 &\geqslant q_{21}/\delta_{21}, \qquad q_{22}/\delta_{22}, \qquad q_{23}/\delta_{23} \\ l_3 &\geqslant (q_{31} - q_{11}l_1^{-1})/\delta_{11}, \qquad (q_{32} - q_{12}l_1^{-1})/\delta_{12}, \qquad (q_{33} - q_{13}l_1^{-1})/\delta_{13}. \end{aligned} \tag{10.7,5}$$

(It is recalled that δ_{ij} is the maximum permitted value of τ_{ij}.) It should be clear from the above how to extend the relations for larger n.

If a diagonal L is used in Eq. (10.6, 3), $T \doteq (I+L)^{-1}(I-PP'^{-1})\,T$, with $(I-PP'^{-1})\,T \triangleq Q$, then for $n = 2$, the result is

$$\begin{aligned} &|\,1 + l_1\,| \text{ is the larger of } q_{11}/\delta_{11},\ q_{12}/\delta_{12} \\ &|\,1 + l_2\,| \text{ is the larger of } q_{21}/\delta_{21},\ q_{22}/\delta_{22} \end{aligned} \tag{10.7,6}$$

and, in general, for an nth-order system

$$|\,1 + l_\alpha\,| \geqslant q_{\alpha j}/\delta_{\alpha j}, \qquad \text{for all} \quad j = 1, 2, \ldots, n. \tag{10.7,7}$$

It is possible to fabricate problems in which use of the diagonal L in one case, or the antidiagonal L in the other case, requires smaller magnitudes of loop transmissions. For example, if the maximum magnitudes of q_{11}, q_{22}, q_{12}, q_{21} are, respectively, 0.1, 0.1, 1, 1, and the maximum magnitudes of τ_{11}, τ_{22}, τ_{12}, τ_{21} are, respectively, 0.10, 0.10, 0.01, 0.01, then an antidiagonal L requires $l_{12} = l_{21} = 16$, whereas a diagonal L requires $l_{11} = l_{22} = 100$. If the corresponding quantities are 1, 1, 0.1, 0.1 for the q's and 0.1, 0.1, 0.01, 0.01 for the τ's, then a diagonal L requires $l_{11} = l_{22} = 9$, while an antidiagonal L requires $l_{12} = l_{21} = 100$. However, if all the q's are 1 and the δ's are 0.01, then the l's must be 100 in both cases.

Effect of Plant Parameter Variations on the Signal Levels in the Plant

A question of considerable importance concerns the signal levels which exist in the plant elements, when the plant parameters have varied. Thus we noted in Section 10.2 that the division, among the plant elements, of the burden of achieving the desired output is fixed as soon as one fixes the T matrix and the nominal P matrix. Do different choices of the form of the L matrix lead to significant differences in the division of the burden, for the same parameter variations?

As previously noted in Section 10.2, the plant signal levels are completely determined by the T', P', and R matrices. The sensitivity specifications determine the bounds on T'. Hence the plant signal levels are independent of the specific canonic L matrix which is chosen to achieve the sensitivity specifications, providing of course that the identical sensitivity functions are synthesized in all cases. In practice, different choices of L matrices may give the same *bounds* of performance but result in different outputs within these bounds, for the same plant parameter values. However, it is certain that the plant signal levels (for different choices of L matrices), are closely the same if the regions enclosed by the bounds are small. Thus in Fig. 10.5-1, in general, $C = PX$ and $C' = P'X' = C + \Delta C$. If each element of ΔC is much smaller than the corresponding element of C, the sensitivity is small. Hence $P'X' \doteq PX$, and $X' \doteq P'^{-1}PX$ is therefore relatively independent of L. (The elements of X' completely determine the plant signal levels.) The form of the L matrix will undoubtedly affect the precise values of ΔC, but the resulting effect on X' is small, if the elements of ΔC are small in comparison with those of C. Therefore the type of L matrix that is chosen does not have a major effect on the division of the burden among the plant elements, as their parameters vary, if the elements of T do not vary much over their significant frequency ranges.

The above considerations represent only one factor in the choice of the type of L matrix. Another important factor is based on stability considerations, so we now turn to the stability problem.

§ 10.8 Loop Shaping for Stability with Plant Parameter Variation

Additional relations for the effect of parameter variation are developed in order to increase our understanding of the subject. It is possible to eliminate T' on the right-hand side of Eq. (10.6,2) and thereby obtain an expression which does not contain any quantities which change as the plant parameters vary. To do this, replace T' in Eq. (10.6,2) by $T + \Delta T$ and collect terms in ΔT, leading to $[I - (I + L)^{-1} (I - PP'^{-1})] \Delta T = (I + L)^{-1} (I - PP'^{-1}) T$. Premultiply both sides of the equation by $I + L$, which gives $[(I + L) - (I - PP'^{-1})] \Delta T = (L + PP'^{-1}) \Delta T = (I - PP'^{-1}) T$. Next premultiply by $(L + PP'^{-1})^{-1}$ to get

$$\Delta T = (L + PP'^{-1})^{-1} (I - PP'^{-1}) T. \tag{10.8,1}$$

T' may be similarly obtained by replacing ΔT in Eq. (10.6,2) by $T' - T$, collecting terms in T', then premultiplying by $(I + L)$, and finally by $(L + PP'^{-1})^{-1}$, leading to

$$T' = (L + PP'^{-1})^{-1} (I + L)T \tag{10.8,2}$$

This last equation is the exact counterpart of the very useful equation $t'/t = (1 + l)/(l + pp'^{-1})$ [Eq. (6.7,1)] for the single input-output plant ($n = 1$).

The approximate expressions developed and used in Section 10.6 are very useful in those frequency regions when the elements of ΔT are small and therefore $T' \doteq T$. In these regions the loop transmissions are fairly large. However, sooner or later in frequency, the loop transmissions must decrease in magnitude and approach zero numerically or minus infinity on the decibel scale. At very large frequencies, $L \doteq 0$, $T' \doteq P'P^{-1}T$, and the feedback is ineffective. The greatest design difficulty is in the neighborhood of the crossover region where the magnitudes of one or more loop transmissions are neither large nor small, but near unity. This is the region in which the feedback is positive and there is danger of instability, or at least of large peaking in one or more of the t_{ij}''s, at some combination of plant parameter values.

There are two ways to treat the design problem in the crossover region. One way is to devote attention to the stability problem alone, and handle the large peaking (of the t_{ij}''s) problem by assigning stability margins. This is the easier and less exact way, because the required stability margins are usually only guessed at. To apply this method, it need only be noted that the rhp poles of T', in Eq. (10.8,2), are the rhp zeros of the determinant of $(L + PP'^{-1})$. How else can T' in Eq. (10.8,2) have rhp poles? The nominal T will obviously have only stable matrix elements. If $(I + L)$, i.e., L, has rhp poles, these will be canceled out by similar poles of $(L + PP'^{-1})^{-1}$, as may be noted from Eq. (10.8,3) below. The conclusion is that only the determinant of $(L + PP'^{-1})$ need be examined for rhp poles. For example, if $n = 2$, if L is diagonal, and if the elements of PP'^{-1} are denoted by λ_{ij}, then

$$\Delta \triangleq \det(L + PP'^{-1}) = (l_1 + \lambda_{11})(l_2 + \lambda_{22}) - \lambda_{12}\lambda_{21} = \Delta_\lambda + \lambda_{11}l_2 + \lambda_{22}l_1 + l_1 l_2.$$

The problem is to pick l_1 and l_2 so that the determinant Δ has no rhp zeros for any of the permitted values of the λ's. The shaping of the l's is done so as to secure what one hopes to be reasonable stability margins, in order that there is no excessive peaking in the t'_{ij}'s.

The second way is the exact way, in that attention is paid to the individual elements of T'. It is obviously more complicated. For example, from Eq. (10.8,2), if $n = 2$, if L is diagonal, and if $PP'^{-1} \triangleq \Lambda$ (elements λ_{ij}), then

$$\begin{aligned} t'_{11} &= [t_{11}(1 + l_1)(\lambda_{22} + l_2) - \lambda_{12}t_{21}(1 + l_2)]/\Delta, \\ t'_{12} &= [t_{12}(1 + l_1)(l_2 + \lambda_{22}) - \lambda_{12}t_{22}(1 + l_2)]/\Delta, \end{aligned} \tag{10.8,3}$$

and t'_{22}, t'_{21} are obtained from t'_{11}, t'_{12} by interchanging the subscripts 1, 2. It is recalled that Δ is the determinant of $L + PP'^{-1}$.

Normally, if there is peaking in any of the t'_{ij}'s, it is due to the common denominator Δ of Eq. (10.8,3) and not to the numerator terms. The maximum values of the numerator terms, for example, of t'_{11}, may be written as $t_{11} \mid (1 + l_1) \mid \mid (\lambda_{22} + l_2) \mid + \lambda_{12}t_{21} \mid 1 + l_2 \mid$, and in the crossover region there is normally no reason why these should be large. It is generally a good idea to use for the nominal P that combination of parameter values which leads to the smallest magnitudes of the p_{ij}'s. This was done in the examples of Section 10.6. When this is done, the elements of PP'^{-1}, i.e., the λ's, generally have peak magnitudes not much greater than one. Therefore a brief scrutiny of the numerator elements of t'_{11}, etc., should ordinarily suffice to determine their maximum values. Thereafter, attention is concentrated on the common denominator, to ensure no undesirable peaking in the t_{ij} elements, for the entire range of plant parameter variation.

§ 10.9 Design Example with a Diagonal *L*

The greater difficulty in working with Eq. (10.8,1) or (10.8,2), rather than with Eq. (10.6,3), is due to the inversion of $(L + PP'^{-1})$ that is required in the former two. This inversion mixes up the known and varying PP'^{-1} terms with the unknown L terms, and complicates the problem of choosing the elements of L. It should be noted that the choice of a diagonal L (or any other of the $n!$ canonic L structures) considerably simplifies the stability design problem, since in the resulting nth order matrix, only n loop transmissions need shaping, rather than n^2. There arises the question as to the effect of parameter variations on L. This is of no consequence. Equation (10.8,2) is exact, and L there refers to the nominal value of L.

Example 1.

Specifications: The design begun in Example 1 of Section 10.6 can now be completed. It is recalled that the elements of P were $p_{ij} = k_{ij}/(s + 1)$, where k_{11}, k_{22} can vary independently from 1 to 4, and k_{12}, k_{21} from 0.1 to 0.5. The nominal values of the k's were taken as $k_{11} = k_{22} = 1$, $k_{12} = k_{21} = 0.1$. A diagonal L was used. It was found that we could take $l_1 = l_2 = l$, and that at very low frequencies $| 1 + l | \geqslant 15$, for a diagonal L. Additional design specifications must be stated before the loop transmissions can be completely specified. These are as follows. The t_{ij}'s are not to change more than 5 % at zero frequency, 25 % at 1 rps, and beyond 2 rps the t_{ij}'s must not exceed 0.3 in magnitude, for any values of plant parameters. Also, the nominal t_{ij}'s are all the same, and over their significant frequency range are given by $2/(s^2 + 1.3s + 1)(s + 2)$.

Design: The methods of Section 10.6 [a suitable approximation of Eq. (10.6,2)] are first used to find the required values of l_1, l_2 at 1 rps. Because of the symmetry in P, P' and the specifications, let $l_1 = l_2 = l$ and $t'_{11} = t'_{22}$, $t'_{12} = t'_{21}$. Therefore, from Eqs. (10.6,2) and (10.6,4), $\tau_{11} = \tau_{22} = (t'_{11}v_{11} + t'_{12}v_{12})/(1 + l)$, where it is recalled that v_{ij} are the elements of the matrix $V = I - PP'^{-1}$. It is readily ascertained that the worst possibility is at $k'_{11} = k'_{22} = 4$, $k'_{12} = k'_{21} = 0.5$, and $t'_{11} = t'_{12} = 1.25 \angle \theta$, leading to $\tau_{11} = \tau_{22} = 0.934 \angle \theta/(1 + l)$. For different values of θ, there are different values of the τ_{11} that result in $| t'_{11} | < 1.25$, but the smallest value of τ_{11} is 0.25. Therefore this is used, and it leads to $| 1 + l(j) | \geqslant 3.7$. It is easily seen that this value of $| 1 + l(j) |$ also suffices for $t'_{12} = t'_{21}$.

It was previously found, in Example 1 of Section 10.6, that $| 1 + l(0) | \geqslant 15$, and therefore we choose $l(0) = 14$ [because $\angle l(0) = 0$]. Also take $| l(j) | \approx 3.5$ [because it is guessed that $\angle l(j) \not< -90°$]. The next step is to reduce $| l(j\omega) |$ as fast as possible, subject to satisfying the specifications. For this part of the design, the exact equation (10.8,2) and its expansion in Eq. (10.8,3) are used. The determinant of $L + PP'^{-1}$ (denoted as Δ) is examined, and the elements of L are shaped to achieve conservative stability margins.

In this problem, the determinant of $(L + PP'^{-1})$ may be written as $\Delta = l^2 + \alpha l + \beta$, where $\alpha = (k_{11}k'_{22} + k'_{11}k_{22} - k_{12}k'_{21} - k_{21}k'_{12})/\Delta_{k'}$, $\beta = \Delta_k/\Delta_{k'}$, $\Delta_k = k_{11}k_{22} - k_{12}k_{21}$, etc.

The zeros of Δ are the zeros of $l + \beta l^{-1} + \alpha$. In order that Δ may have no rhp zeros, it is necessary and sufficient that no locus of $f \triangleq l + \beta l^{-1}$, for all possible values of β, should encircle the corresponding value of $-\alpha$ (note β and α are functions of the variable plant parameters). This extension of the Nyquist criterion is fairly obvious. It is readily found that $-\alpha$ lies between -0.5 and -2, and that $1 > \beta > 1/15.75$. One proceeds by cut and try. The first try is $l = l_A = 7/(s + 0.5)(s + 1)$. This choice of l gives $l(0) = 14$, $| l(j) | = 4.4$, so it satisfies the sensitivity specifications at 0 rps and at 1 rps, and thereafter decreases fairly rapidly. The locus of $l_A(j\omega)$ is shown dotted in Fig.

10.9-1. To it must be added β/l_A with $1 > \beta > (15.75)^{-1}$. One extreme value of $f_A = l_A + \beta/l_A$ is shown in the figure.

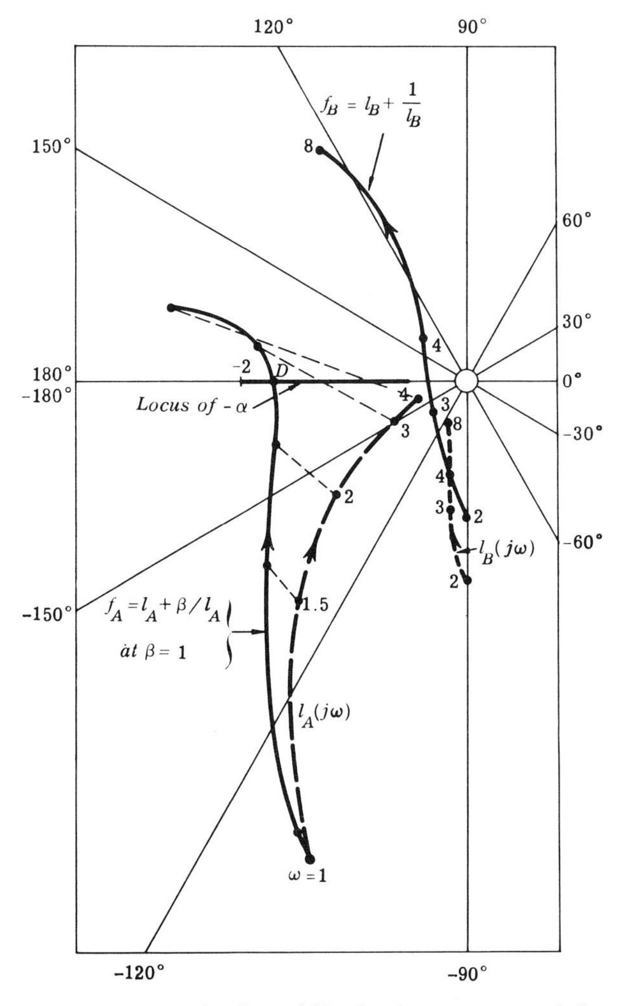

FIG. 10.9-1. Loop shaping for stability despite parameter variations.

Although the extreme value of $f_A = l_A + \beta l_A^{-1}$ cuts the $-\alpha$ line at point D, the design is not necessarily unstable. To check this, we note that this extreme f_A occurs at $k'_{11} = k'_{22} = 1$, $k'_{12} = k'_{21} = 0.1$, at which $-\alpha \doteq -2$, and therefore at this setting the system is stable. To justify this last conclusion, the complete locus of $f_A = l_A + l_A^{-1}$ is derived. We need the locus of $f_A(s)$ as s encircles the right half-complex plane. In Fig. 10.9-2a, as R approaches ∞, $l_A \to 0$ and $1/l_A$ becomes very large, so f_A approaches $1/l_A$ as s becomes large. Since l_A approaches

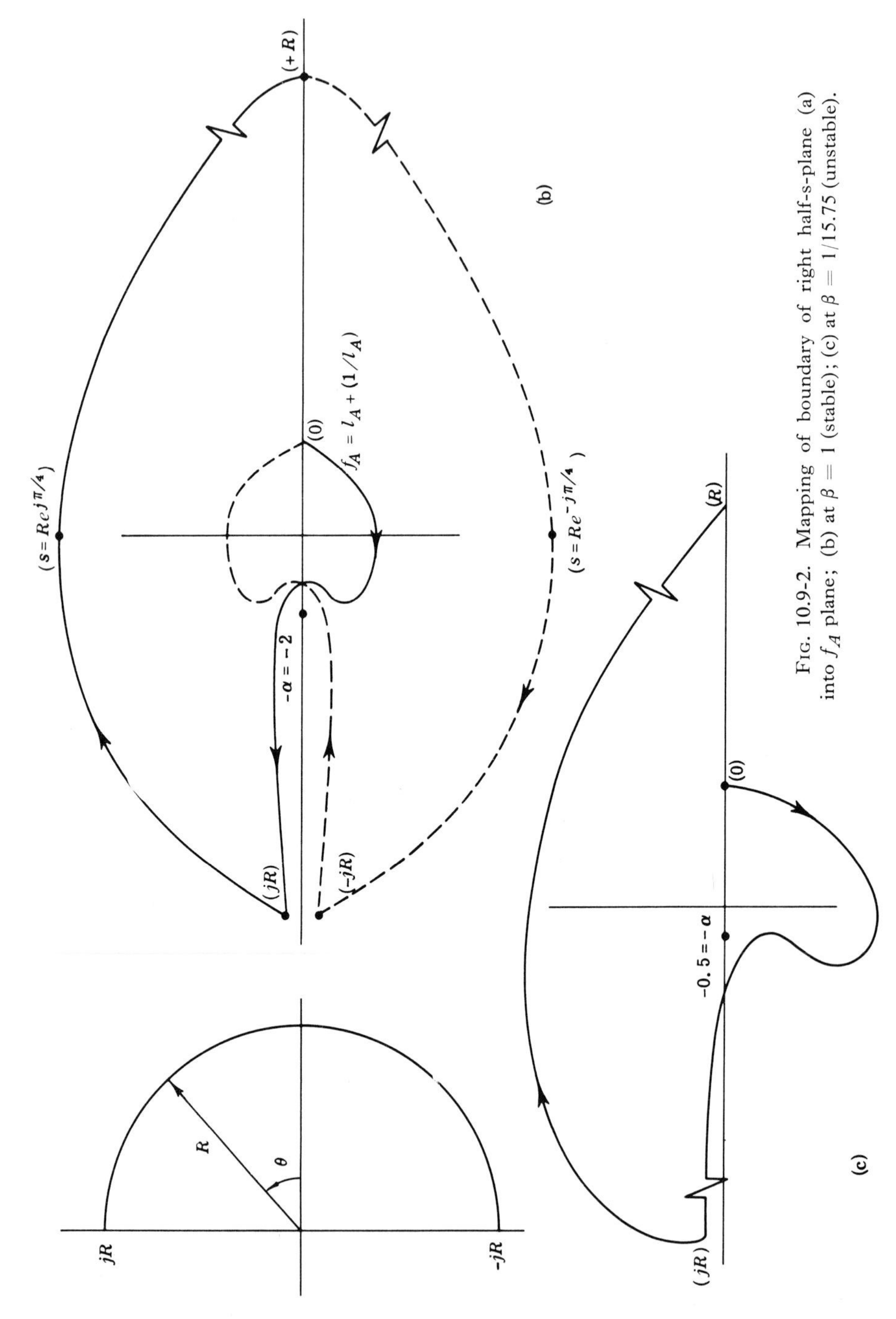

FIG. 10.9-2. Mapping of boundary of right half-s-plane (a) into f_A plane; (b) at $\beta = 1$ (stable); (c) at $\beta = 1/15.75$ (unstable).

$7/s^2$ at large s, f_A approaches $1/l_A$, which approaches $0.14\, R^2 e^{j2\theta}$. Consequently, the complete locus of f_A must be as shown in Fig. 10.9-2b. It is seen that f_A does not encircle the critical point $-\alpha = -2$. However, at small β, the design is unstable with a frequency of oscillation near 6 rps. Thus at $\beta = 1/15.75$, which corresponds to $k'_{11} = k'_{22} = 4$, $k'_{12} = k'_{21} = 0.5$, the critical point that must not be encircled is $-0.5 = -\alpha$. The locus of $f_A = l_A + (1/15.75 l_A)$, for positive frequencies, is qualitatively sketched in Fig. 10.9-2c, and it does encircle $-\alpha = -0.5$. This first unsuccessful attempt does at least teach the designer what modification must be made in l to ensure a stable design.

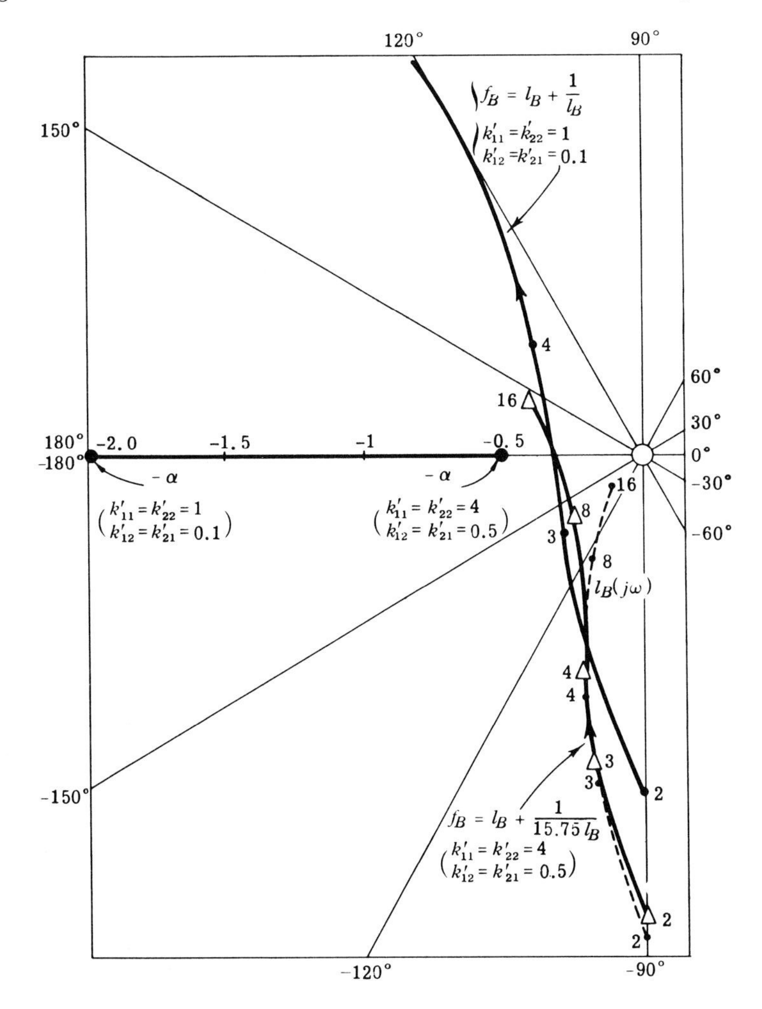

FIG. 10.-9-3a. Nyquist loci for modified system which is stable for entire range of plant parameter values.

It is seen (see Fig. 10.9-1) how the term β/l contributes to the instability. It is also clear that the design for l must be more conservative, with less phase lag for the same decrease in magnitude of l. (This is very much in accord with our single variable system experience.) Such a more conservative l would permit $l + \beta l^{-1}$ to cross the real axis, in Fig. 10.9-1, to the right of the $-\alpha$ locus. This is, in fact, always the way to secure a stable design—by letting the loop gain decrease more slowly, so as to incur less phase lag. The price paid for this is the need to maintain the loop gain at the higher level over a larger frequency range. This price can be decreased by allowing more phase lag at the low frequencies. From Fig. 10.9-1, it is seen that the phase lag of l should not be much more than 90° until l is fairly small (in order for l to cross the real axis on the right-hand side of the locus of $-\alpha$); but one can let the phase lag be 90° even at fairly low frequencies, and so decrease the price that must be paid in the gainbandwidth of the loop transmissions. One pole is therefore chosen as small as the specifications at 0 rps and at $\omega = 1$ permit. The result is a pole of l at -0.25, so that the drop in $|l|$ is from 14 at zero frequency to 3.5 at 1 rps, i.e., -12 db. The next pole is chosen so as to satisfy the stability requirements. The second choice of l, denoted as $l_B = 56/(s + 0.25)(s + 16)$ is sketched in Fig. 10.9-1 and in Fig. 10.9-3a. The locus of $f_B = l_B + \beta l_B^{-1}$ is also sketched in Figs. 10.9-1 and 10.9-3a.

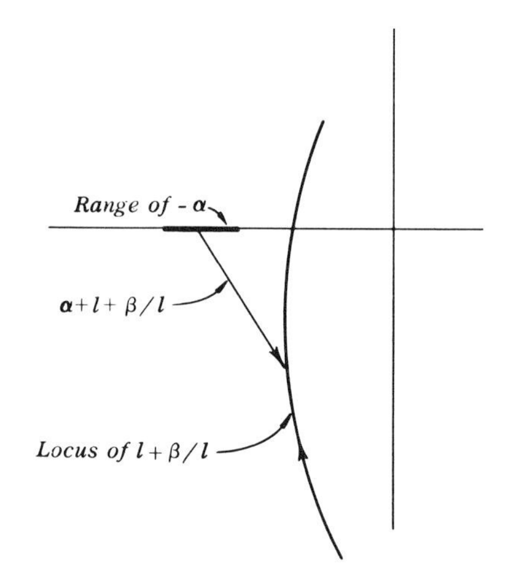

FIG. 10.9-3b. Construction for finding $\alpha + l + \beta l^{-1}$.

It is clear from Fig. 10.9-3a that the system is stable for the entire range of plant parameter values. However, one must check whether the peaking specifications at high frequencies are satisfied. This is done as follows.

The exact expressions for the primed t's is given by Eq. (10.8,3). The denominator is, in this case, $l(l + \alpha + \beta l^{-1})$. The construction in Fig. 10.9-3b shows how to find $\alpha + l + \beta/l$. Using this construction and with the aid of Fig. 10.9-3a, it is found that the minimum value of the denominator is about 0.18 at $\omega \approx 12$ and $l(j12) \approx 0.25 \angle -125°$. This minimum value of the denominator occurs at $\beta = 1/15.75$, at which Eq. (10.8,3) becomes

$$\begin{bmatrix} t'_{11} & t'_{12} \\ t'_{21} & t'_{22} \end{bmatrix} \approx \frac{\sqrt{2}\, t \begin{bmatrix} 1 & 1 \\ 1 & 1 \end{bmatrix}}{(0.18)}.$$

Therefore the maximum increase in each t is by a factor of 7.8. The specifications dictate that $|t'(j12)| \not> 0.3$. Therefore $|7.8t(j12)| \leqslant 0.3$, and the specifications are satisfied if the nominal $|t(j12)|$ is less than 0.032. Actually, the nominal $|t(j12)|$ is considerably less than 0.032. The design therefore satisfies all the sensitivity specifications.

The design cannot be completed until T has been specified and the configuration chosen. Suppose the configuration of Fig. 10.5-1a is used, in which $L = PH$, and therefore $H = P^{-1}L$. In this example, this gives

$$H = \frac{56(s+1)}{(s+0.25)(s+16)} \begin{bmatrix} 1 & -0.1 \\ -0.1 & 1 \end{bmatrix}.$$

It should be noted that a diagonal L does not mean a diagonal H. In fact, H will in general be nondiagonal, and there has to be some cancellation to ensure nominal diagonal L. Thus, in Fig. 10.5-3, unit signal at c_1 must lead to zero signal at c_{r2} for a diagonal L. Therefore h_{21} must have the proper value to cancel out the signal which reaches c_{r2} by way of $-h_{11}$ and p_{21}. This raises the interesting question as to whether a diagonal H would be feasible. If there is no serious disadvantage in a diagonal H in the sensitivity aspects of the design, it would seem that a simpler implementation is possible by using a diagonal H. It is, in fact, possible to use a diagonal H, and relations similar to (10.7,5) and (10.7,7) can be developed in terms of the elements of a diagonal or antidiagonal H[1]. The difference between a diagonal and nondiagonal H need not be great, however—for example, see Fig. 10.9-4. Also, the saving, if any, in elements at

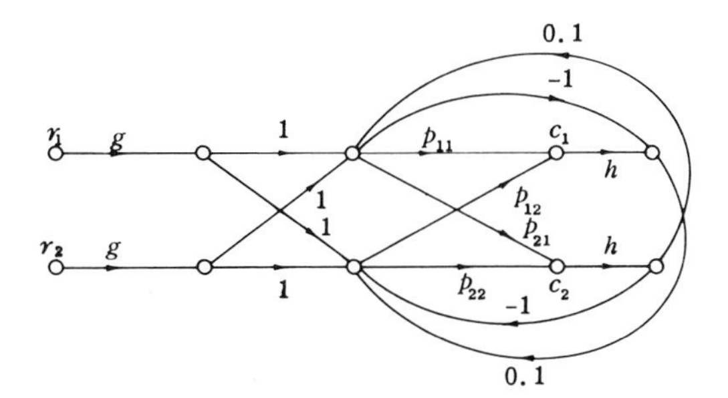

FIG. 10.9-4. The final design.

$$g = \frac{6.75(s+1)}{(s+0.25)(s^2+1.3s+1)(s+2)}; \qquad h = \frac{56(s+1)}{(s+0.25)(s+16)}.$$

best applies only to the structures of Fig. 10.5-1a and d. In the others, e.g., in Fig. 10.5-1b, a diagonal NH does not result in diagonal H and N, because N is in general not diagonal, due to its obligations in realizing T.

[1] I. M. Horowitz, Synthesis of Linear Multivariable Feedback Control Systems, Research Rept. 164, Hughes Research Labs., Malibu, Calif., Dec. 15, 1960, Section 2.5.

G is needed (Fig. 10.5-1a) in order to complete the present design. Using Eq. (10.5,2a), i.e., $T = (I + L)^{-1} PG$, there results

$$G = P^{-1}(I + L)T$$

$$= \frac{(0.9)(s+1)(s^2+16.25s+60)}{(s+0.25)(s+16)} \begin{bmatrix} 1 & 1 \\ 1 & 1 \end{bmatrix} \frac{2\zeta(s)}{(s^2+1.3s+1)(s+2)}.$$

This result was obtained for G by letting

$$T = 2 \begin{bmatrix} 1 & 1 \\ 1 & 1 \end{bmatrix} \frac{\zeta(s)}{(s^2+1.3s+1)(s+2)}.$$

The function $\zeta(s)$ was assigned to each element of T to permit minor (high-frequency) adjustment of T, in order thereby to simplify G. To simplify G, let $\zeta(s) = (s + 16)\, 3.75/(s^2 + 16.25s + 60)$, so that

$$G = \frac{(s+1)6.75}{(s+0.25)(s^2+1.3s+1)(s+2)} \begin{bmatrix} 1 & 1 \\ 1 & 1 \end{bmatrix}.$$

The complete design is shown in Fig. 10.9-4. Note that it is not necessary to construct four filters for G and for H; only two suffice in each case, if isolating stages are used.

A somewhat different procedure may be used to shape l so that $\Delta = l^2 + \alpha l + \beta$ has no rhp zeros, and so that the sensitivity specifications are satisfied. In this alternative procedure, write $\Delta = (l + \lambda_1)(l + \lambda_2)$ and find the regions enclosed by the loci of λ_1, λ_2 as the plant parameters vary. The loop transmission l is then appropriately shaped in the manner of Section 6.7.

§ 10.10 Example—Stability Problem with an Antidiagonal L

It was noted at the end of Section 10.7 that the low frequency magnitudes of the elements of L represent only one factor in choosing between a diagonal or an antidiagonal L. Another factor is the shaping of the elements of L for stability despite plant parameter variation. The importance of this factor will be seen by attempting an antidiagonal L in the problem treated in Section 10.9.

It is recalled that the problem of Section 10.9 is that of Example 1 of Section 10.6. The specifications are repeated here. The plant matrix elements are $p_{ij} = k_{ij}/(s + 1)$; k_{11}, k_{22} vary independently from 1 to 4, and k_{12}, k_{21} from 0.1 to 0.5. At low frequencies, each $t_{ij} \doteq 1$, and each $|\tau_{ij}| \leqslant 0.05$. The nominal values for P were taken as $k_{11} = k_{22} = 1$, $k_{12} = k_{21} = 0.1$. A diagonal L led to $l_1 = l_2 = l \geqslant 14$ at zero frequency. The higher frequency specifications and stability problems were satisfactorily solved by the diagonal design detailed in Section 10.9. We now attempt an antidiagonal L design for the same problem. Equation (10.7,4) is used to find the required magnitudes of the two elements of L. The elements of the $(I - PP'^{-1})\, T \triangleq Q = VT$ matrix are available from Eq. (10.6,4). They are approximately $q_{11} = q_{22} = 0.75$, $q_{12} = q_{21} = 0.03$. From Eq. (10.7,4), we get $l_1 = l_2 = l \geqslant 15$ at 0 rps, which is basically the same value

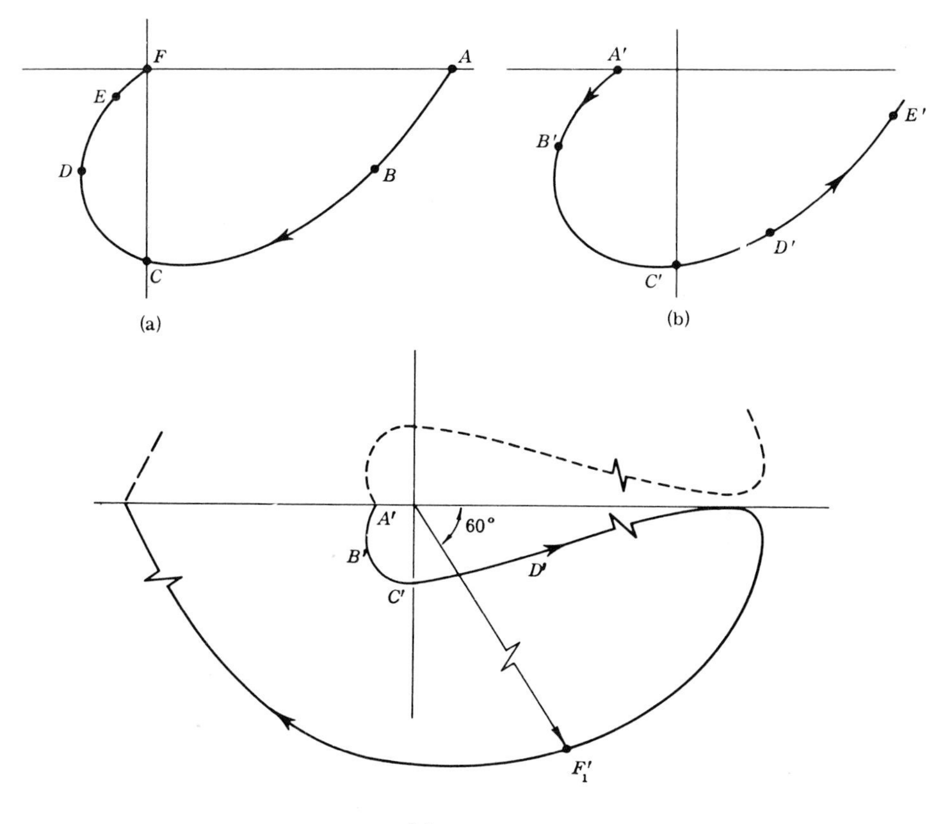

FIG. 10.10-1. Appropriate Nyquist loci for antidiagonal L matrix with $l_1 = l_2 = l$. (a) Locus of $l(j\omega) = k/(s+a)(s+b)$. (b) Initial part of locus of $-\beta/l$. (c) Complete locus of $-\beta/l$.

needed in the diagonal L design. There is therefore no difference between the two designs as regards the low frequency demands on the corresponding elements of L.

Next, consider the stability problem. In the antidiagonal case, $\det(L + PP'^{-1}) = \Delta = -l(l + \gamma - \beta l^{-1})$, where $\beta = 0.99/\Delta_{k'}$, $\gamma = [0.1(k'_{11} + k'_{22}) - (k'_{12} + k'_{21})]/\Delta_{k'}$. Note that β varies from 1 to 1/15.75, and γ varies from -1.07 (when $k'_{11} = k'_{22} = 1$, $k'_{12} = k'_{21} = 0.5$) to 0.05 (when $k'_{11} = k'_{22} = 2$, $k'_{12} = k'_{21} = 0.1$).

If we choose l of the form $l = k/(s+a)(s+b)$ with a locus which is therefore as shown in Fig. 10.10-1a, then $-\beta/l$ (for $\beta > 0$) has the locus shown in Fig. 10.10-1b. In order to complete the locus, let s complete the infinite semicircle around the right half-plane, as shown in Fig. 10.9-2a. At $s = R\exp(j60°)$, $1/l \doteq R^2 \exp(j120°)$, and $-\beta/l \doteq \beta R^2 \exp(-j60°)$. Figure 10.10-1b is

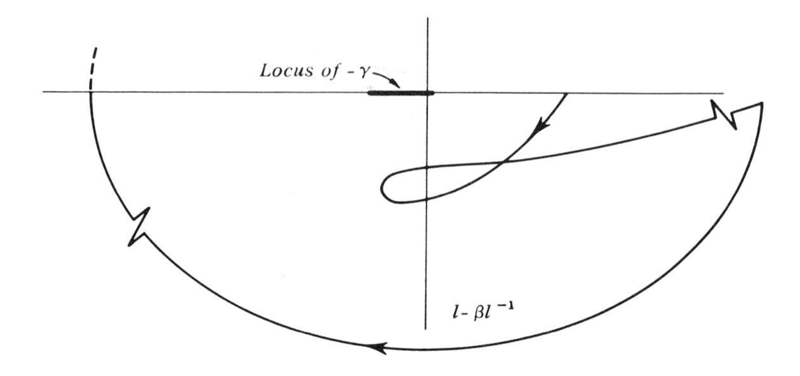

FIG. 10.10-1d. Locus of $l - \beta l^{-1}$.

therefore completed as shown in Fig. 10.10-1c. The locus of $l - \beta l^{-1}$ therefore has the form shown in Fig. 10.10-1d, in which the locus of $-\gamma$ is included. Clearly, the system is unstable for all permitted β, γ. It is similarly unstable if the sign of l is changed. If the sign of only l_1 or l_2 is changed, then a stable design is possible. Thus, suppose $l_1 = l$, $l_2 = -l$, then the determinant Δ becomes $l\{l + \beta l^{-1} - \mu\}$, with $\mu\Delta_{k'} = 0.1(k'_{11} - k'_{22}) - (k'_{12} - k'_{21})$. A stable design is possible, but must be very conservative because the critical region which must not be encircled, i.e., μ, extends along the positive and negative real axis (see Fig. 10.10-2). For a stable design, the locus of $l + \beta l^{-1}$ must pass to the right

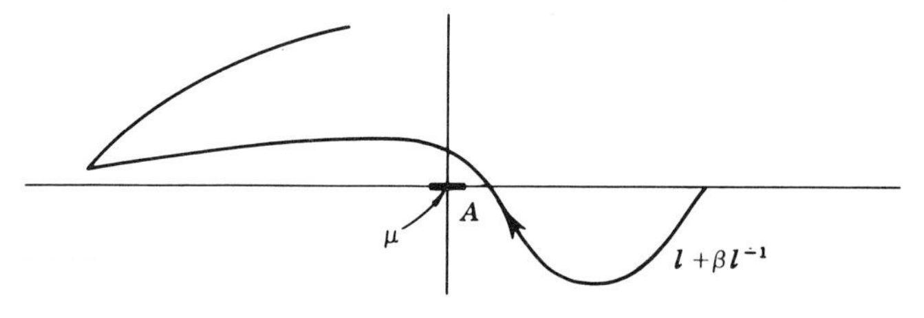

FIG. 10.10-2. Locus for antidiagonal L, case $l_1 = -l_2 = l$.

of A. This means an $l(s)$ must be used which becomes small at a phase lag no more than 90°.

It is perhaps possible that in some problems an antidiagonal L requires smaller low frequency loop gain levels, but a more conservative design at the crossover frequency range. The latter consideration is of greater importance,[1] and the

[1] The reason is that in the frequency range in which the l's > 1, the effect of feedback transducer noise on the plant inputs is independent of the actual values of l's (just as in the single input-output system—see Section 8.2). However, in the range in which the l's < 1, the noise is amplified by functions of l/p. Hence it is important to reduce the l's as rapidly as possible. See also Sec. 10.18.

diagonal L would then be preferable. However, there is no apparent reason for assuming that the antidiagonal L will always lead to such more conservative designs. One should examine each problem. It is perhaps possible that for some problems the diagonal L will require the more conservative design in the crossover region. Thus, in the above example, if μ was negative at all times, while the corresponding α of the diagonal design of Section 10.9 (Fig. 10.9-1) was also negative at all times, then the antidiagonal design would be a better one.[1]

§ 10.11 Design Examples

Example 1.

Specifications: The theoretical work of the previous sections will be illustrated with two design problems. In the first problem, the plant matrix elements are $p_{11} = a/(\tau s + 1)$, $p_{12} = b/(\tau s + 1)$, $p_{21} = 0$, $p_{22} = c$. The quantities, a, b, c, τ can each have values anywhere from 1 to 4. The elements t_{11} and t_{22} are to have bandwidths of $\frac{1}{2}$ and 1 rps, respectively. The values of t_{11} and t_{22} are not to change by more than 10 % from 0 rps up to $\frac{1}{2}$ and 1 rps, respectively. Also, t_{12} and t_{21} must be less than 0.05 from 0 rps up to $\frac{1}{2}$ and 1 rps, respectively. (This corresponds to "noninteraction," which will be discussed in Section 10.12.)

Design: The nominal plant values are taken as $\tau = 4$, $a = b = c = 1$. These values give the smallest magnitudes of the elements of P, and therefore tend to make the PP'^{-1} elements small. This is convenient for the graphical part of the design. The elements of $Q = (I - PP'^{-1})\,T$ are $q_{11} = [1 - (\tau s + 1)/a(4s + 1)]\,t_{11}$, $q_{12} = [-(a - b)/ac(4s + 1)]\,t_{22}$, $q_{21} = 0$, $q_{22} = [1 - c^{-1}]\,t_{22}$. The maximum q's are $q_{11} = 0.75$, $q_{12} = 3$, $q_{21} = 0$, $q_{22} = 0.75$. If a diagonal L is used, we get from Eq. (10.7,7), at 0 rps, $1 + l_1 \geqslant 0.75/0.10$, $3/.05$, and therefore $l_1(0) = 59$; $1 + l_2 \geqslant 0$, $0.75/0.10$, so $l_2(0) = 6.5$. At 0.5 rps, in the same manner, $l_1 = 54$, $l_2 = 7$, and at 1 rps, $l_2 \approx 7$.

The stability problem is considerably simplified when a diagonal L is used here, because $q_{21} = 0$. The determinant of $L + PP'^{-1}$ is therefore simply $[l_1 + (\tau s + 1)/a(4s + 1)]\,[l_2 + c^{-1}]$, so that the loop shaping problem is no more difficult than for the single variable design, except that it must be done separately for l_1 and for l_2. Consider first $l_1 + (\tau s + 1)/a(4s + 1) \triangleq l_1 + f$. At $s = 0$, $f = 1/a$ ranges from 1 to 0.25. At $s = j$, the range of f is shown in Fig. 10.11-1a. Similar loci are sketched in the figure at $s = j2$ and at infinite s. The next step is to choose an l_1 that satisfies the low frequency specifications, which properly avoids the loci of f in the higher frequency range and which goes to zero at infinity at least as $1/s^3$ (one excess pole for the plant and one each for the possible

[1] See Section 10.17 for additional comments on this subject.

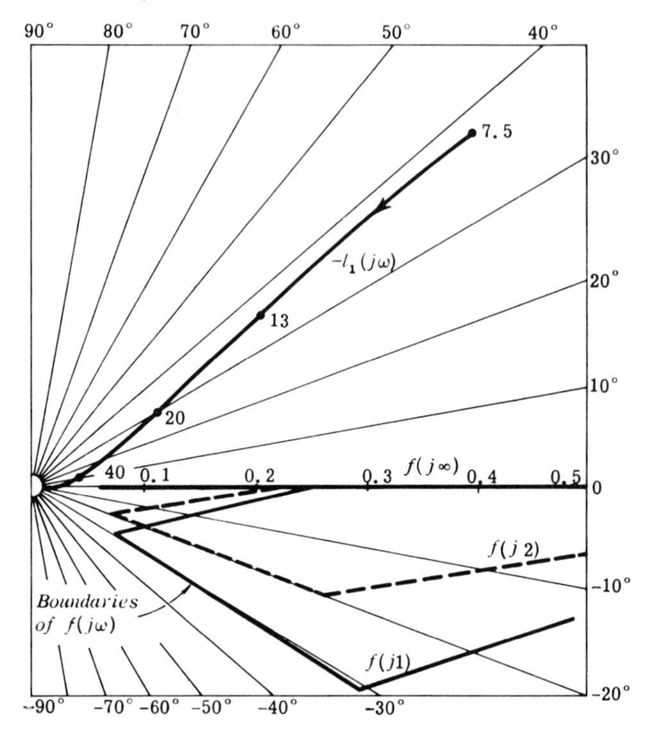

FIG. 10.11-1a. Example 1—range of $f(j)$, $f(j2)$, $f(j\infty)$ due to plant parameter variations.

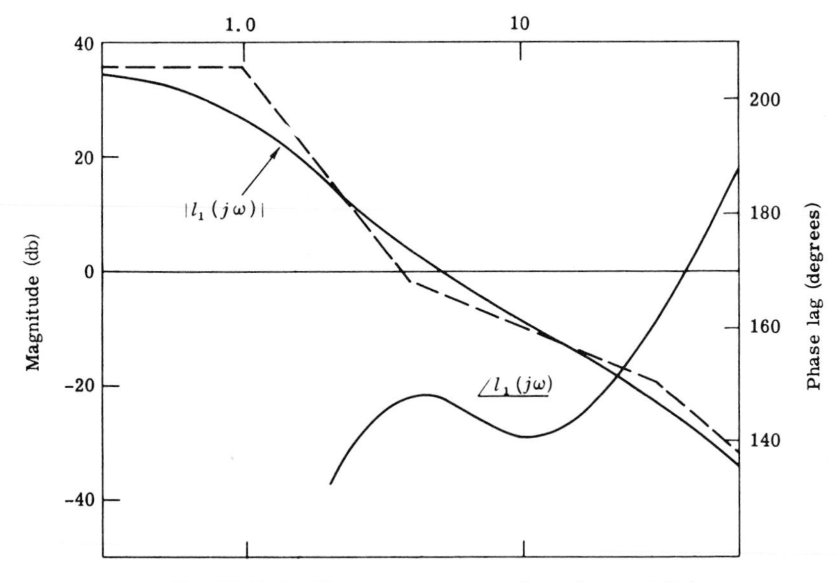

FIG. 10.11-1b. Frequency response of satisfactory $l_1(j\omega)$.

two compensating networks in the loop transmission). With a little cut and try, we choose $l_1 = [11{,}250(s + 4)^2/(s + 1)^3 (s + 30)(s + 100)]$. The Bode sketch of l_1 is drawn in Fig. 10.11-1b, while the polar locus of $-l_1$ is shown in Fig. 10.11-1a.

A satisfactory l_2 is even easier to find in view of the simplicity of the factor $(l_2 + c^{-1})$. We take $l_2 = 180/(s + 3)(s + 8)$. The design is easily completed. Suppose the structure used is that shown in Fig. 10.5-1d, in which I is, as usual, the unit diagonal matrix. Then $L = PF_2$, and $T = (I + L)^{-1} PF_2F_1$. Therefore

$$F_2 = P^{-1}L = \begin{bmatrix} 4s + 1 & -1 \\ 0 & 1 \end{bmatrix} \begin{bmatrix} l_1 & 0 \\ 0 & l_2 \end{bmatrix} = \begin{bmatrix} (4s + 1)l_1 & -l_2 \\ 0 & l_2 \end{bmatrix}.$$

Also, $F_1 = F_2^{-1}P^{-1}(I + L)\,T$ is obtained in a straightforward manner as soon as T is specified analytically. As before, the high frequency values of T can be chosen to simplify the expressions for F_1.

Design with an antidiagonal L matrix:

In this problem, use of Eq. (10.7,4) leads to $|l_1(0)| \geqslant 30$, and $|l_2(0)| \geqslant 16$, as compared to 59 and 6.5 in the diagonal L design. However, there are two disadvantages in using an antidiagonal L. One is the fact that $\Delta = L + PP'^{-1}$ no longer factors into the simple form $(l_1 + f_1)(l_2 + f_2)$; instead, it has the form $l_1 + f_1 l_2 + f_2 l_1 l_2$ so that loop shaping for stability is more difficult. The second disadvantage is that it is found that precisely the phenomenon discussed in Section 10.10 occurs here. Either l_1 or l_2 must be assigned a minus sign in order that it may be at all possible to have a stable system. Even so, the design must be very conservative with phase lag less than 90^0 until the loop transmission magnitude is quite small. The antidiagonal design is therefore not attempted here.

Example 2.

Specifications: A plant with three inputs and two outputs. The plant matrix elements are $p_{11} = a_1/(\tau s + 1)$, $p_{12} = 0.5a_2/(\tau s + 1)$, $p_{13} = 1/(\tau s + 1)$, $p_{21} = 0.5\,p_{11}$, $p_{22} = 2p_{12}$, $p_{23} = 0.5p_{13}$, $p_{31} = p_{32} = 0$, $p_{33} = 1$, from which the determinant of P is $\Delta_p = 0.75a_1a_2/(\tau s + 1)^2$. The parameters τ, a_1, a_2 may each have values anywhere from 1 to 4. The bandwidth of t_{11} and t_{22} is to be each 1 rps, and the values of t_{11} and t_{22} are to vary by not more than 10% in this range due to plant parameter variation. The cross transmissions t_{12}, t_{21}, etc., are to be less than 0.05 over the 1-rps bandwidth.

Design: The nominal values of the plant parameters are taken as $a_1 = a_2 = 1$, $\tau = 4$. Accordingly, the elements of $\Lambda = PP'^{-1}$ are $\lambda_{11} = (a_2 - 0.25a_1)(\tau s + 1)\,d^{-1}$, $\lambda_{12} = 0.5(\tau s + 1)(a_1 - a_2)\,d^{-1}$, $\lambda_{13} = 0.75a_2(a_1 - 1)\,d^{-1}$, $\lambda_{21} = -\lambda_{12}$, $\lambda_{22} = (a_1 - 0.25a_2)(\tau s + 1)\,d^{-1}$, $\lambda_{23} = 2\lambda_{13}$, $\lambda_{31} = \lambda_{32} = 0$, $\lambda_{33} = 1$, where $d = 0.75a_1a_2(4s + 1)$. At 0 rps, $t_{11} = t_{22} =$

$t_{33} = 1$, while $t_{ij} \doteq 0$ for all $i \neq j$. Therefore, from $Q = (I - PP'^{-1})\,T$, $q_{11} = 1 - (a_2 - 0.25a_1/0.75a_1a_2)$ and $(q_{11})_{\max} = 1$ at $a_2 = 1$, $a_1 = 4$. Similarly, the other maximum values of the q's at 0 rps are $q_{11} = q_{22} = 1$, $q_{12} = q_{21} = 0.5$, $q_{13} = 0.75$, $q_{23} = 0.375$, $q_{31} = q_{32} = q_{33} = 0$. From Eq. (10.7,7) for a diagonal L, $1 + l_1 \geqslant 1/0.10$, $0.5/0.05$, $0.75/0.05$, therefore, $l_1 \geqslant 15$; $1 + l_2 \geqslant 0.5/0.05$, $1/0.10$, $0.375/0.05$, therefore, $l_2 \geqslant 10$. Also, $l_3 = 0$ (an antidiagonal L requires $l_1 \geqslant 20$, $l_2 \geqslant 10$). $Q_{\max}$ at 1 rps is similarly obtained and it is found that a magnitude of 10 is needed for both l_1 and l_2 at 1 rps.

For stability analysis the determinant of $L + PP'^{-1} = L + \Lambda$ is needed. It is

$$\Delta = 9a_1a_2\,\{a_1a_2l_1l_2(s + 0.25)^2 + (1/3)\,(\tau s + 1)\,(s + 0.25) \times [l_1(a_1 - 0.25a_2) + l_2(a_2 - 0.25a_1)] + (1/16)\,(\tau s + 1)^2\}.$$

The problem is simplified by letting $l_1 = l_2 = l$, even though a 0-rps level of 10 would suffice for l_2. Using 15 in place of 10 is not too big a price to pay for the simplification of the design work. Accordingly, let $l_1 = l_2 = l$ and manipulate the determinant into the following form:

$$\Delta = 9a_1a_2l\{a_1a_2(s + 0.25)^2\,l + (1/16)\,(\tau s + 1)^2\,l^{-1} + (\tau s + 1)\,(s + 0.25)\,(a_1 + a_2)\,0.25\}.$$

Let $f = l(s + 0.25)/(\tau s + 1)$, and then

$$\Delta = 9a_1^2a_2^2(\tau s + 1)^2 f\{f + (1/16)\,(a_1a_2f)^{-1} + (a_1 + a_2)/4a_1a_2\}.$$

The last factor can be written as $f + \beta f^{-1} + \alpha$ in the manner of Example 1, Section 10.9, and the procedure thereafter is the same as in that example.[1] A trial value of f is chosen such that the low frequency specifications of l are satisfied and such that l has the minimum required excess of poles over zeros. A polar sketch of f is made, from which the range of β/f are deduced and added to f. Next one checks whether the locus of $f + \beta f^{-1}$ encircles the forbidden region $-\alpha$. In this problem $f = l(s + 0.25)/(\tau s + 1)$, so the above must be repeated to cover the range of τ. The procedure is now described in detail.

The designer begins with the value of τ which he thinks will result in the greatest stability problem. Here it is when $\tau = 1$. The first trial value of f is $f = 4800\,(s + 0.25)\,(s + 10)/(s + 1)^3\,(s + 40)\,(s + 80)$. Note $f(0) = 3.75$, so $l(0) = 4f(0) = 15$, and f and l have an excess of three poles over zeros. A Bode sketch of f is shown in Fig. 10.11-2b. Why was f so chosen? First there was noted the range of $-\alpha = -(a_1 + a_2)/4a_1a_2$, and its value closest to the origin, which is

---

[1] It has since been noted that Δ is readily factored into

$$\Delta = 9a_1a_2[a_1(s + 0.25)l + 0.25(\tau s + 1)]\,[a_2(s + 0.25)l + 0.25(\tau s + 1)].$$

Therefore the method of Section 6.7 may be used for each individual factor and the more complicated procedure described here is unnecessary. Nevertheless, this description will be useful to the reader for those cases where the factoring of Δ is much more difficult.

$-\alpha = -0.125$ at $a_1 = a_2 = 4$. From our experience with Example 1 of Section 10.9, it is known that instability results when the lag angle of f is too large with small $|f|$. It is convenient to get an idea of how large the lag angle may be in a stable design when $|f|$ is small. For example, suppose when $|f| = 0.10$, its lag angle is 160°. To $f = 0.1 \angle -160°$, add β/f. The latter is $1/64f = (10/64) \angle -160°$ at $a_1 = a_2 = 4$. The resulting $f + \beta f^{-1} = -0.24 + j0.02$ is shown in Fig. 10.11-2a.

It is known that the form of the frequency locus of $f + \beta f^{-1}$ at low and high frequencies is as shown in Fig. 10.11-2a (if f has three more poles than zeros; compare with Fig. 10.9-2 where the function has two more poles than zeros). From Fig. 10.11-2a, it is clear that the system would be unstable, because $-\alpha$ is encircled. Thus the lag angle of 160° is too great when $|f| = 0.10$. In this way it is readily found that a lag angle of about 145° at $|f| = 0.10$ would be acceptable. The above is sufficient to serve as a guide in the selection of f. In other cases one or two more points may be taken. The trial design of f (Fig. 10.11-2b) was chosen as follows.

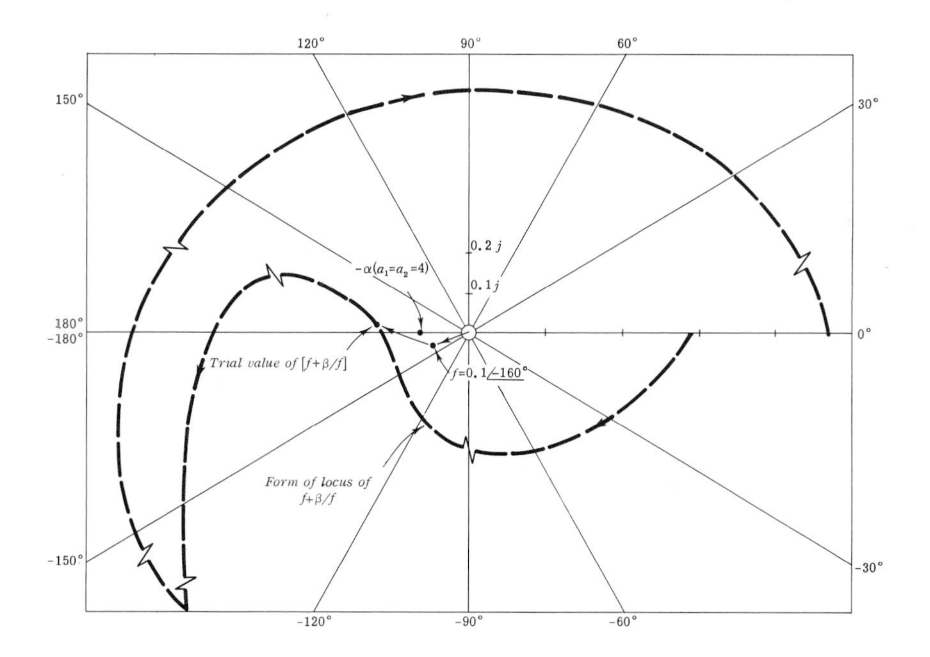

FIG. 10.11-2a. Construction showing that lag angle of 160° (at $|f| = 0.10$) is too large.

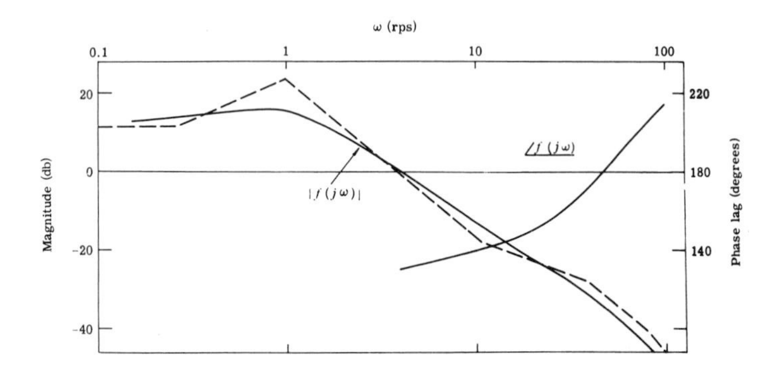

FIG. 10.11-2b. Example 2—tentative choise of $f(j\omega)$.

Begin with $f(0) = 3.75$, which is 11.5 db. The specifications require that $|l(j)| \approx 10$, hence $|f(j)|$ must be 7.2 (17 db) when $\tau = 1$, since $f = l(s + 0.25)/(\tau s + 1)$. Therefore introduce a lead break point at $\omega = 0.25$ but soon thereafter try to reduce $|f|$ as fast as possible, subject to the requirement that when $|\mathrm{f}| = 0.10$, the phase lag is not more than 145°. This fast reduction of f is achieved by letting $|f|$ come down very quickly, but lead break points are introduced as soon as is necessary to prevent the lag angle from being too large.

There is also the matter of complexity; a slope of -12 db per octave is used here, but if a more complex loop transmission is acceptable, it would be possible to decrease $|f|$ faster. If gain-bandwidth is at a high premium, due perhaps to a plant and sensors with inherently low gain-bandwidth, then it would be worthwhile to use a more complex $f(s)$ and to reduce $|f(s)|$ faster. The procedure would be to determine, as was done for $|f| = 0.10$, the maximum phase lag tolerable at several values of $|f|$, and then bring $|f|$ down faster, until these phase lags are achieved. For example, suppose 160° lag is permissible up to $|f| = 0.5$ and 140° is required up to $|f| = 0.05$. At a frequency slightly higher than 1 rps (f cannot be changed for $\omega < 1$ because of the low frequency sensitivity specifications), one could let the magnitude of f fall down faster than in the present design. It is known that this can be done because the present design lag angle is less than the maximum permitted 160°. It is known from the results of Chapter 7 that so long as the lag angle of f is less than its maximum permitted value, it is possible to increase the rate at which $|f|$ falls and thus decrease the gain-bandwidth area of $f(s)$.

In any case, the above approach led to

$$f(s) = 4800\,(s + 0.25)\,(s + 10)/(s + 1)^3\,(s + 40)\,(s + 80)$$

of Fig. 10.11-2b. Next make sure the design is indeed a stable one. This is done by first making a polar sketch of $f(j\omega)$ in Fig. 10.11-3. The loci of $f + \beta f^{-1}$ are next obtained for various values of β. The corresponding values of α are also found (recall that our present concern is with the zeros of $f + \beta f^{-1} + \alpha$). This is all done in Fig. 10.11-3. It is seen that the system is stable, at least for $\tau = 1$.

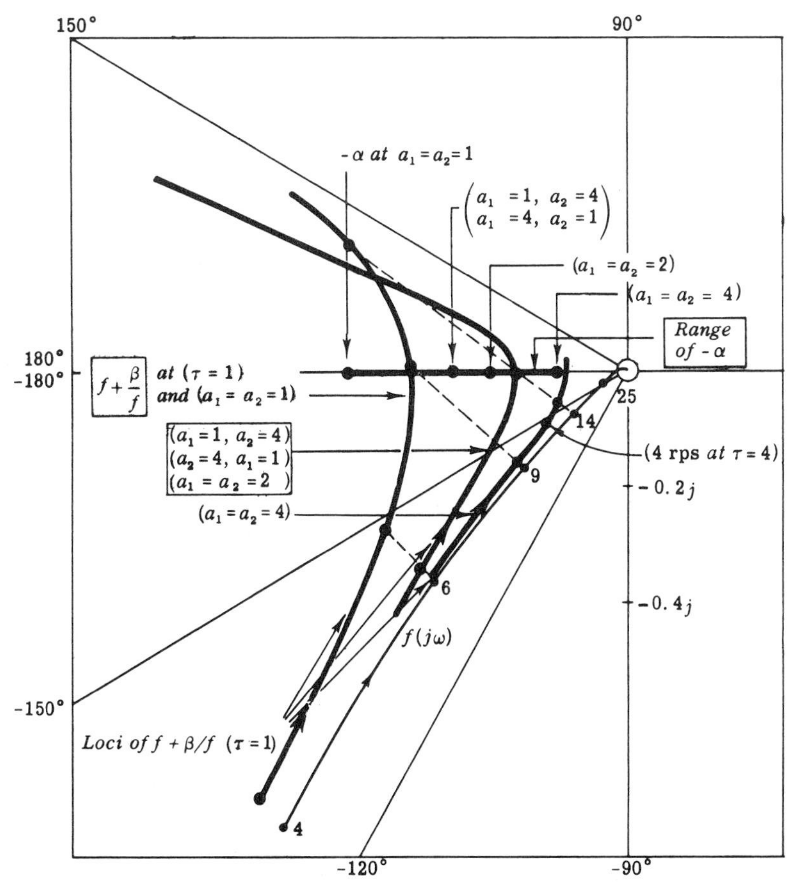

FIG. 10.11-3. Example 2—checking the design for stability.

Next consider the situation at the other extreme value of τ, i.e., at $\tau = 4$. The new value of f is obtained by multiplying the old value by 0.25 $(s + 1)/(s + 0.25)$. The lag angle is increased at the low frequencies but the destabilizing effect is negligible in the critical region. It is also easily seen that there is no need to worry about any other values of τ. The design of the L matrix is therefore complete. The balance of the synthesis is straightforward and requires no comment.

§ 10.12 Noninteraction[1]

A multivariable system is called noninteracting when the desired T matrix is diagonal. In such a case the input r_j elicits a response c_j only and $c_i = 0$ for $i \neq j$. Actually, it is possible to secure this condition at only one combination of plant parameter values. If there is any plant parameter variation there will be some off-diagonal transmissions. However, these may be kept as small as desired by the usual procedure for keeping ΔT small, i.e., by choosing the elements of L sufficiently large. The two examples of Section 10.11 were, in fact, noninteracting systems and there the specifications required the off-diagonal transmissions to be $\leqslant 0.05$ over the significant frequency range.

If there is plant parameter variation, it should therefore be recognized that small t'_{ij} cannot be specified over an indefinite frequency range. The price for the small off-diagonal transmissions must be paid by sufficiently large loop transmissions over the frequency range in which this behavior is desired. The sensitivity problem in a noninteracting system is, therefore, basically the same as in the general system. However, since T is forced to be a diagonal matrix, it is convenient to work, at least at the start, with the exact equation (10.6,2), viz., $\Delta T = (I + L)^{-1} (I - PP'^{-1}) T' = (I + L)^{-1} VT'$, where $V = I - PP'^{-1}$. The diagonal elements of T' are the usual t'_{11}, t'_{22}, etc. The nondiagonal elements, however, are elements of ΔT because T is diagonal, i.e., when $i \neq j$, $t'_{ij} = t_{ij} + \tau_{ij} = \tau_{ij}$. Hence Eq. (10.6,2) becomes (for $n = 3$ and L diagonal)

$$\begin{bmatrix} (1 + l_1)\tau_{11} & (1 + l_1)\tau_{12} & (1 + l_1)\tau_{13} \\ (1 + l_2)\tau_{21} & (1 + l_2)\tau_{22} & (1 + l_2)\tau_{23} \\ (1 + l_3)\tau_{31} & (1 + l_3)\tau_{32} & (1 + l_3)\tau_{33} \end{bmatrix}$$

$$= \begin{bmatrix} v_{11} & v_{12} & v_{13} \\ v_{21} & v_{22} & v_{23} \\ v_{31} & v_{32} & v_{33} \end{bmatrix} \begin{bmatrix} t'_{11} & \tau_{12} & \tau_{13} \\ \tau_{21} & t'_{22} & \tau_{23} \\ \tau_{31} & \tau_{32} & t'_{33} \end{bmatrix}. \quad (10.12,1)$$

This matrix equation is simple enough for the designer to use in his sensitivity design. It is useful for checking the justifiable approximations. For example, $(1 + l_1)\, \tau_{11} = v_{11}t'_{11} + v_{12}\tau_{21} + v_{13}\tau_{31}$, exactly. The v's and the maximum permissible values of t'_{11}, τ_{21}, and τ_{31} are known. Therefore, the designer easily checks whether he may neglect $v_{12}\tau_{21} + v_{13}\tau_{31}$. If so, one minimum value of l_1 is ascertained. Similarly, $(1 + l_1)\, \tau_{12} = v_{11}\tau_{12} + v_{12}t'_{22} + v_{13}\tau_{32}$, exactly, and a similar investigation of the permissible approximations is fairly

[1] A. S. Boksenboom and R. Hood, General Algebraic Method Applied to Control Analysis of Complex Engine Types, Tech. Rept. No. 980. NACA, Washington, D.C., April 1949; R. J. Kavanagh, Noninteraction in linear multivariable systems, *Trans. AIEE Part II*. **76**, 95-100 (1957); Herbert Freeman, Stability and realizability conditions in the synthesis of multipole control systems, *Trans. AIEE, Part II*. **77**, 1-15 (1958).

easily made. When $v_{13}\tau_{32}$ cannot be neglected, it is not impractical to use a cut and try iterative procedure.

Finally, after the diagonal L matrix has been chosen from sensitivity and stability considerations, attention is turned to the G matrix. It is fairly easy to evaluate the elements of G, because, from Eq. (10.5,2a), $G = P^{-1}(I + L)\,T$, and $(I + L)\,T$ is a diagonal matrix.

§ 10.13 Loop Shaping for Stability—The General Second-Order Case

In previous sections, numerical problems were considered in which there was some degeneracy, which somewhat simplified the stability problem. However, there was thereby obtained some feeling for the problem, and this will help us to consider the more difficult general case. The stability problem in multivariable systems is that of shaping the elements of the L matrix such that Δ, which is the determinant of $L + PP'^{-1}$, has no rhp zeros (recall the discussion in Section 10.8). This problem is greatly reduced when L is either a diagonal or antidiagonal matrix (or any canonical matrix). In such a case, if the system is of order n, there are only n transmissions to shape in place of n^2. Furthermore, it was shown in Section 10.7 that if the elements of Q vary independently, there is no sacrifice made in any respect in using a diagonal or antidiagonal L.

For $n = 2$, the most general problem is therefore treated by considering either

$$\begin{aligned}\Delta &= (l_1 + \lambda_{11})(l_2 + \lambda_{22}) - \lambda_{12}\lambda_{21}\\ &= l_1 l_2 + l_1\lambda_{22} + l_2\lambda_{11} + \Delta_\lambda\end{aligned}$$

or

$$\begin{aligned}\Delta &= \lambda_{11}\lambda_{22} - (l_1 + \lambda_{12})(l_2 + \lambda_{21})\\ &= -\,l_1 l_2 - l_1\lambda_{21} - l_2\lambda_{12} + \Delta_\lambda,\end{aligned}$$

according as to whether L is diagonal or antidiagonal. (The λ's are the elements of $\Lambda = PP'^{-1}$ and Δ_λ is the determinant of Λ.) The general form of Δ is the same for both cases, so we will consider the diagonal L only. Three distinct possibilities can be visualized:

(1) In the significant frequency region in which the l's act to decrease the sensitivity, the levels of l_1, l_2 and the frequency ranges over which the levels are specified are substantially the same. In this case, take $l_1 = l_2 = l$, and then $\Delta = l^2 + \alpha l + \beta = l(l + \beta l^{-1} + \alpha)$ and the procedure previously used in Section 10.11, Example 2, is followed.

(2) The level of one l, say l_1, is higher than that of l_2, but the bandwidth of l_2 is larger than that of l_1, such that at the higher frequencies the two are not substantially different. In that case, write $l_1 = l_2 h$ with h known and then the problem is the same as for case (1).

(3) Suppose there are substantial differences in the minimum bandwidths or magnitudes of l_1 and l_2. If the designer forces the two to be equal over all or part of the frequency range, he is overdesigning for the sake of easing the theoretical design labor. This may sometimes be justified, but at other times it is not, and one should therefore consider this more difficult general case. Let l_1 be the loop

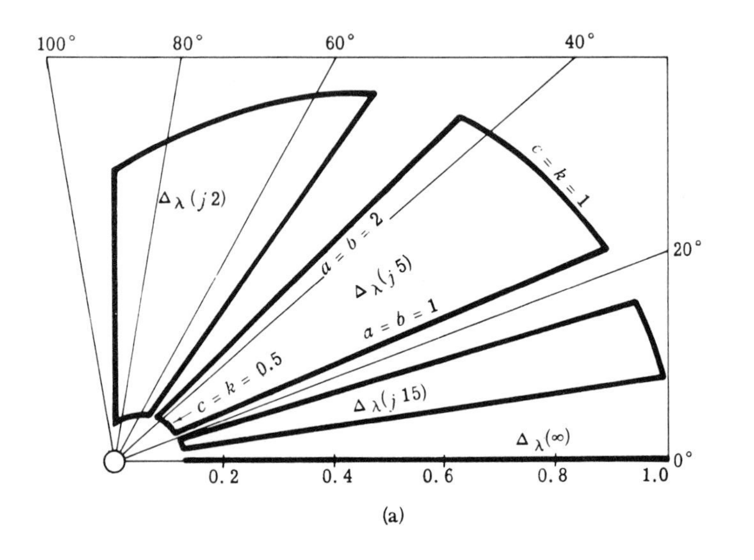

(a)

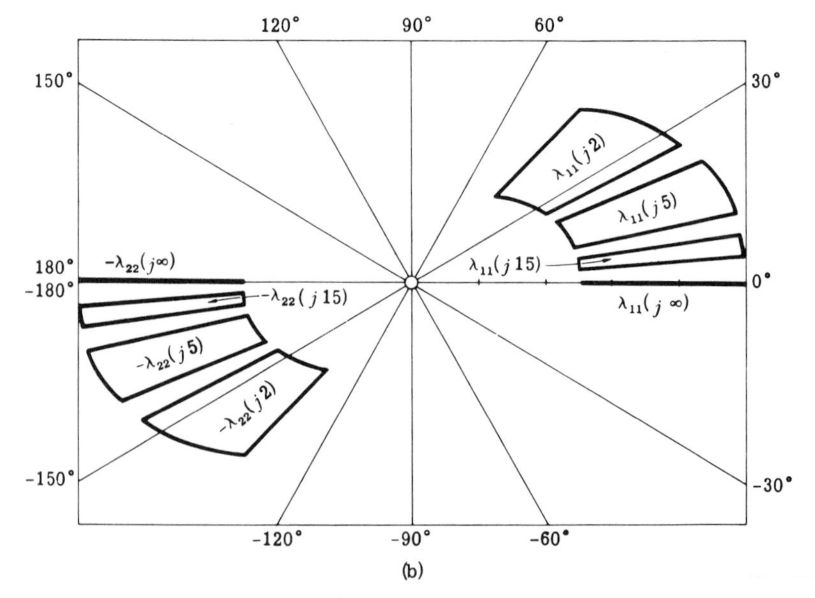

(b)

FIG. 10.13-1. Range of Δ_λ, λ_{11}, λ_{22} due to plant parameter variations.

transmission which is substantially larger than l_2. It is assumed that there is at least a factor of 3 here to justify the greater design labor of different l_1 and l_2.

This third case is treated in detail in this section. This is done by considering a typical numerical example, with sufficient discussion as to method to enable the reader to apply it to other problems.

Example—*Design Problem.*

Specifications: The elements of $\Lambda \triangleq PP'^{-1}$ are given as $\lambda_{11} = ks/(s + a)$, $\lambda_{22} = ks/(s + b)$, and $\Delta_\lambda = ck^2s^2/(s + a)(s + b)$. The values of k and c vary independently from 0.5 to 1 and those of a and b from 1 to 2. The low frequency requirements on l_1 and l_2 are as follows: the value of l_1 must be 100 at 0 rps, 10 at 2 rps, while that of l_2 must be 5 at 0 rps, and 1.5 at 1 rps.

The problem is to shape l_1 and l_2 so that no zeros of Eq. (10.13,1), i.e., of $f + \lambda_{22}$ below, are in the right half-plane, where f is defined by

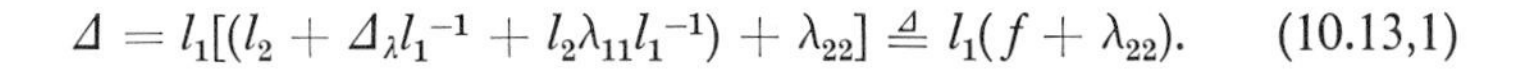

$$\Delta = l_1[(l_2 + \Delta_\lambda l_1^{-1} + l_2\lambda_{11}l_1^{-1}) + \lambda_{22}] \triangleq l_1(f + \lambda_{22}). \qquad (10.13,1)$$

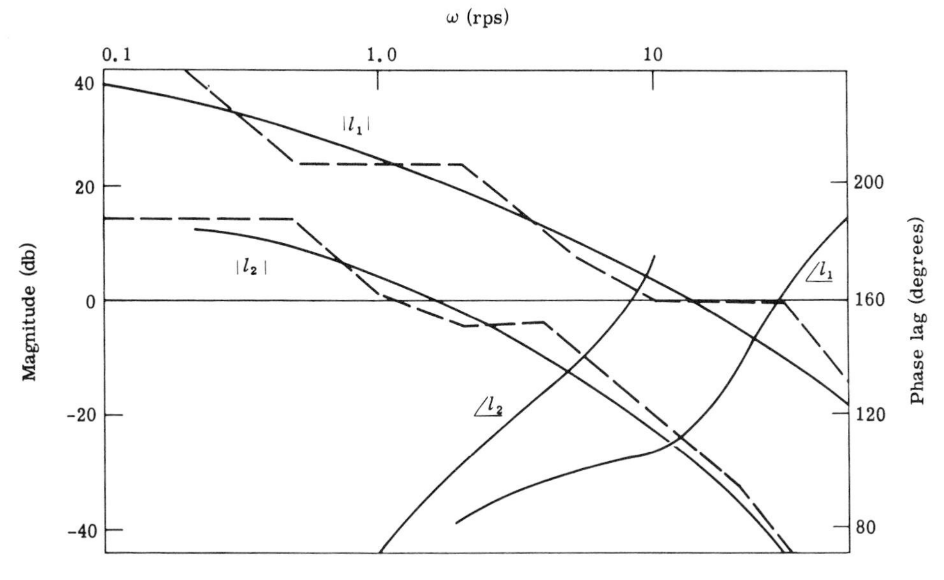

FIG. 10.13-2. Step-by-step derivation of l_1, l_2.

Procedure: (1) Find the boundaries of Δ_λ, λ_{11}, $-\lambda_{22}$ for a few specific frequencies. This is done in Figs. 10.13-1a and b at 2, 5, and 15 rps and at infinity. The λ's are deliberately chosen such that the above quantities have maximum magnitudes near unity. It is always advisable to pick the nominal plant values such that the maximum values of λ_{11}, λ_{22}, and Δ_λ are less than unity or at least as small as possible. Such a choice facilitates an estimation of the contributions of the individual terms in $f = l_2 + \Delta_\lambda l_1^{-1} + \lambda_{11} l_2 l_1^{-1}$. Usually such a contraction of the areas occupied by λ_{11}, etc., is achieved by taking the smallest plant matrix element values as the nominal values. This was done in all previous examples.

(2) The low frequency values of l_1 and l_2 are prescribed by sensitivity requirements. The procedure is next as follows: Sketch the low frequency portions of $| l_1 |$ and $| l_2 |$ on a Bode plot (see Fig. 10.13-2). Guess the crossover frequency (the 0 db point) of the smaller, i.e., of l_2, and the value of l_1 at this frequency. From the given low frequency specifications, a reasonable guess is that $| l_2 | = 1$ at $\omega = 2$ rps, at which $| l_1 | = 10$. From Eq. (10.13,1), in view of the relative magnitudes of the terms at $s = j2$, and the fact that $| \lambda_{11} |$, $| \Delta_\lambda |$ are less than one, $f(j2) \doteq l_2$. Now refer to Fig. 10.13-3, where the boundary of $-\lambda_{22}$ has been qualitatively sketched. Consider the zeros of $f + \lambda_{22}$. Figures 10.13-3a and b

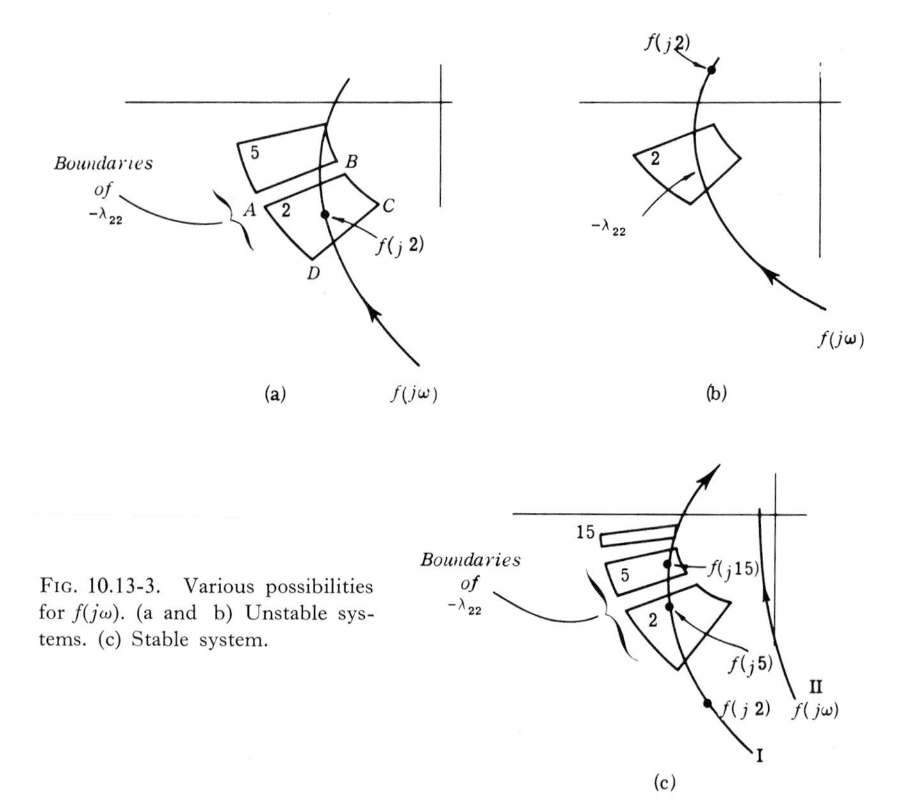

FIG. 10.13-3. Various possibilities for $f(j\omega)$. (a and b) Unstable systems. (c) Stable system.

represent unstable systems, i.e., there are zeros of $f + \lambda_{22}$ in the right half-plane, while Fig. 10.13-3c represents a stable system. These conclusions follow from the Nyquist criterion by examining whether $f + \lambda_{22}$ encircles the origin and noting that $f + \lambda_{22}$ at $s = j2$, for example, is a vector originating anywhere inside $ABCD$ in Fig. 10.13-3a and terminating at $f(j2)$.

In the present problem it was noted that at the crossover frequency of l_2, estimated at 2 rps, $f \doteq l_2$. Therefore the above (Fig. 10.13-3) stability considerations in conjunction with Fig. 10.13-1b immediately determine that the maximum permitted phase lag of $l_2(j2)$ is 135° [because $-135°$ is a boundary of $-\lambda_{22}(j2)$ in Fig. 10.13-1b]. It is preferable to use phase lags as large as possible for l_1 and l_2 (to permit fast reduction of their magnitudes and thereby conserve gain-bandwidth). But it is seen (because of the need for stability margins) that it is not possible to go much beyond 100° lag for l_2 at $\omega = 2$ rps. It is noted that the phase lag of $l_1(j2)$ is of very minor importance [because $|l_1(j2)|$ is large and appears in the denominator in the terms in f, i.e., at $s = j2$, l_2 is easily the dominant term in f]. It is also important to note that if $|l_2(j2)|$ is less than the contemplated value of unity but with the same lag angle, the effect is a stabilizing one, and vice versa.

It will be seen that the following statements are generally true and are in agreement with our single variable system experience: Reduction in $|l_1|$ or $|l_2|$ at the same lag angle has a stabilizing effect and vice versa. Reduction in the lag angle of l_1 or l_2 at the same magnitudes has a stabilizing effect and vice versa. However, the relative importance of l_1 or of l_2 changes appreciably in different frequency regions and *this in fact is the key to the two variable loop shaping problem*—i.e., knowing on which one and when to throw the main burden. It has been noted that there should be a fair difference in the significant frequency requirements of l_1 and l_2 to justify their separate formulation. The first frequency region to consider is the one at which the smaller (here l_2) has 0 db. In this region it is l_2 that dominates in f, and its maximum reasonable lag value is easily obtained. Next, Fig. 10.13-2 is examined. (Figure 10.13-2 shows the final complete l_2—the reader should, however, construct l_1 and l_2 as he goes along, in accordance with the detailed reasoning that is described here.) It is seen (by the usual Bode plot shaping) that a lag of about 100° with crossover at 2 rps for l_2 seems reasonable. The situation about an octave away, at 5 rps, is next examined. At crossover, the slope of a loop transmission is probably about -6 db per octave, so it is guessed that $l_2(j5)$ is no more than $0.5 \angle -130°$.

The designer is now in a region in which *no part of f can be neglected*. Such a region occurs sooner or later, but it will be seen that fortunately there is only one such region. The greatest design labor is in this region because it may be necessary to carefully sketch the boundaries of Δ_λ/l_1 and of $\lambda_{11}l_2/l_1$ in $f = l_2 + l_1^{-1}(\Delta_\lambda + l_2\lambda_{11})$. First the boundaries of $\Delta_\lambda + l_2\lambda_{11}$ (using the above guess value for l_2) are sketched at $s = j5$ in Fig. 10.13-4. As for l_1, it had to be 20 db at 2 rps, so if it is assumed that its slope is about -10 db per octave, then $|l_1(j5)| \approx 2.5$. The effect of multiplying $\Delta_\lambda + l_2\lambda_{11}$ by $1/l_1$ is to contract it by a factor 0.4 and swing it counterclockwise by the lag angle of $l_1(j5)$. Suppose the latter is 90°. The resulting boundaries of $(\Delta_\lambda + l_2\lambda_{11})/l_1$ are shown in Fig.

10.13-4 (these boundaries should be carefully ascertained), and it is then possible to obtain $f(j5) = l_2 + (\Delta_\lambda + l_2\lambda_{11})/l_1$, whose boundaries are also sketched in Fig. 10.13-4. It is seen (refer to the discussion of Fig. 10.13-3) that the guessed at

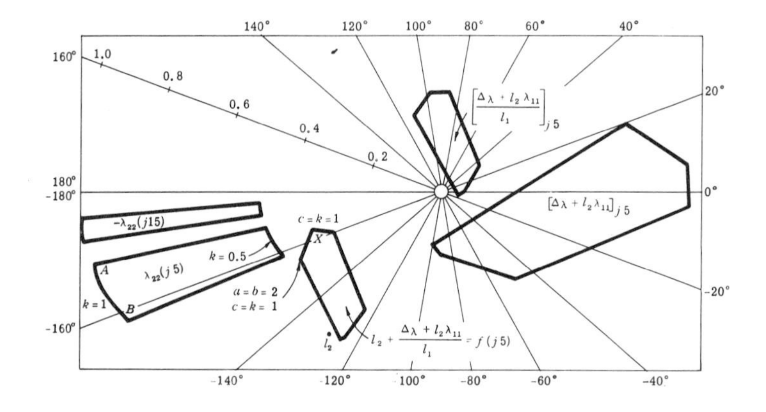

FIG. 10.13-4. Determination of boundaries of $f(j5)$.

values are reasonably good for a stable design. Also, the proximity of part of the boundary (region X in Fig. 10.13-4) to $-\lambda_{22}(j5)$ is not worrisome, because the X region corresponds to $k = 1$, and at $k = 1$, AB is the corresponding boundary of $-\lambda_{22}(j5)$.

It is of even greater importance to see the effect of changes in l_1 and l_2 on stability, because the final design values of l_1 and l_2 will certainly not be precisely the above guessed at values, and it is desirable to be able to intelligently adjust them *without* having to repeat the above. If the lag angle only of l_2 is decreased, the effect on the first term in f is obviously stabilizing. The effect on the $\lambda_{11}l_2/l_1$ term is more difficult to see, but it is also slightly stabilizing. If $|l_2|$ alone is decreased, the effect is also stabilizing but to a lesser extent. It is easier to see that decreasing the phase lag only of l_1 is unambiguously stabilizing. Again the results are in accordance with our single variable system experience.

Next turn to an octave or two higher, say $s = j15$. At this frequency the l_2 terms in Eq. (10.13,1) can certainly be ignored and so can $\lambda_{11}l_2/l_1$ and consequently $f(j15) \approx \Delta_\lambda/l_1$. We prefer that the left-hand boundary of the loci of f is similar to curve II in Fig. 10.13-3c, and not to curve I or to Fig. 10.13-3a. It is possible to have a stable design with the latter, but careful checking would be necessary, while with curve II, the system is obviously stable.

Consider $f(j15) \doteq \Delta_\lambda/l_1(j15)$. From Fig. 10.13-1a, it is seen that $|\Delta_\lambda|$ varies from 0.12 to 1.0, while its phase varies from 8.5° to 16.5°. Also, the crossover frequency for l_1 is not far from 15 rps, so $-\angle l_1$ may be around 140°. Thus $|\Delta_\lambda/l_1|$ will range from about 0.10 to 1.0, while its phase may be from about 150° to 160°. Therefore a reasonable guess is that $\Delta_\lambda/l_1 \doteq f(j15)$ may lie anywhere in the indicated region in Fig. 10.13-5.

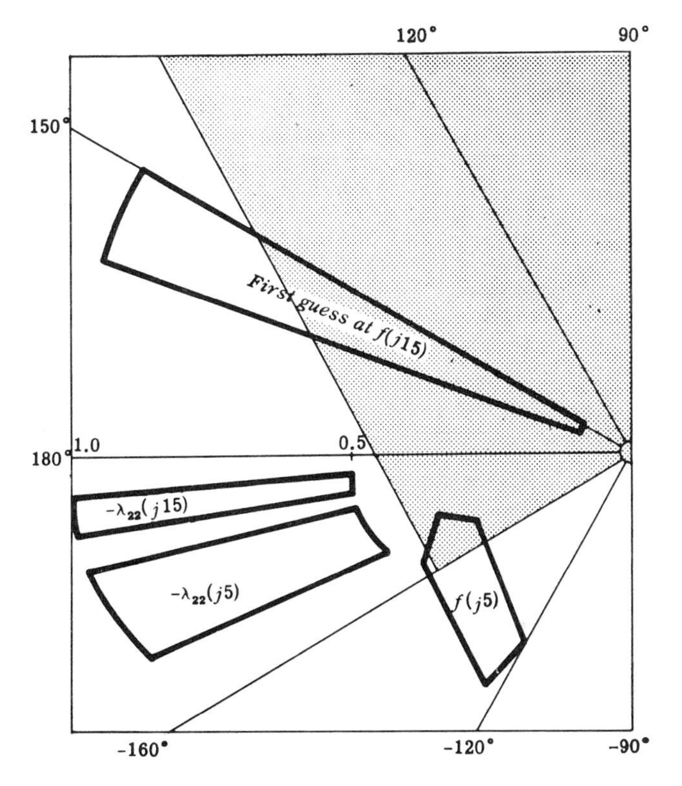

FIG. 10.13-5. Determination of satisfactory range of $f(j15)$.

The question is whether the loci of f are like II in Fig. 10.13-3c, or like I. If $f(j15)$ does indeed range over the region shown in Fig. 10.13-5, then the latter applies. We can make sure that II (Fig. 10.13-3c) applies by insisting that the range of $f(j15)$ is restricted to the shaded region shown in Fig. 10.13-5. What are the corresponding restrictions on $l_1(j15)$? If $|l_1| = 1$, then $|f|_{\max} = 1$, and its angle, from Fig. 10.13-5, should be less than 143°, i.e, $\angle\Delta_\lambda - \angle l_1 < 143°$. Therefore $\angle l_1 > -143° + \angle\Delta_\lambda$, and since $(\angle\Delta_\lambda)_{\max} = 17°$, $\angle l_1 > -126°$. Similarly, if $|l_1| = 0.5$, $\angle l_1 > -155°$, etc. The larger $|l_1(j15)|$, the smaller is its permitted lag angle.

At higher frequencies $f \doteq \Delta_\lambda/l_1$, and since $\angle\Delta_\lambda \doteq 0$, the locus of f therefore lies entirely in the second quadrant (90° to 180°) so long as $90° < \angle -l_1 < 180°$.

There is no need to worry any longer about stability, so long as $| l_1 |$ steadily decreases as its lag angle steadily increases—which is obviously what must happen. One can therefore now turn to the precise Bode plot shaping of l_1 and l_2 in Fig. 10.13-2, using the above results as a guide. In addition, the effects of deviations in l_1 and l_2 from the above guessed at values are known and the designer is therefore in full command of the situation. The trial values of l_1 and l_2 are shown in Fig. 10.13-2. They are $l_1 = 34{,}600(s + 0.5)^2 (s + 5) (s + 10) [(s + 0.2)^2 (s + 2)^2 (s + 30)^3]^{-1}$ and $l_2 = 200(s + 1) (s + 2) [(s + 0.5)^2 (s + 4)^2 (s + 20)]^{-1}$. The value of l_2 was shaped with care (checking the specifications and the above stability considerations) only up to about 5 rps. The value of l_1 was shaped by similar considerations.

We now check our design. At 2 rps, $l_2 = 0.8 \angle -103°$, which is more conservative than the estimated $1 \angle -100°$. At 5 rps, $l_2 = 0.25 \angle -137°$, which is, on the whole, more conservative than the estimated $l_2 = 0.5 \angle -130°$, and $l_1 = 3.5 \angle -100°$, which is more liberal than the estimated $l_1 = 2.5 \angle -90°$, but a quick check reveals that the latter is more than compensated for by the more conservative l_2. At 15 rps, $l_1 \doteq 1 \angle -120°$, which satisfies the requirement that $\angle l_1 > -126°$ if $| l_1(j15) | = 1$. We can now appreciate the great value of our previous examinations of the effects of deviations in l_1 and l_2 from their estimated values. It was found that the more conservative (in the same sense as it is used in single variable systems) the design, the better the stability. This is eminently reasonable for open-loop stable plants.

Although a specific numerical example has been detailed in the above, it is clear that the procedure is fairly general and applicable to most problems. The distinct difference in levels between l_1 and l_2 is certainly a great help, but it is a realistic assumption, because if it did not exist, one might as well pick $l_1 = l_2$. The procedure reveals that there is only one frequency at which a fair amount of labor must be expended to obtain the boundary of f fairly precisely. Even this would not exist if the difference in level between l_1 and l_2 was greater. It is, however, important that this work need be done only once. Thereafter it is known what corrective action, if any, must be taken. The effects of changes in the estimated l_1 and l_2 are checked in advance and this knowledge can be used to quickly achieve the final satisfactory l_1 and l_2.

§ 10.14 The General Loop Shaping Problem for Higher Order Multivariable Systems

Loop shaping for stability for the two variable system turned out to be not too difficult a problem, because only in one frequency range was it necessary to consider both l_1 and l_2. What happens when one goes to higher order systems? Consider $\Delta = | L + PP'^{-1} |$ for $n = 3$ with L diagonal (the form of Δ will be the same for any one of the $n!$ canonic matrices). The result is

$$\Delta = l_1 l_2 l_3 + l_1 l_2 \lambda_{33} + l_1 l_3 \lambda_{22} + l_2 l_3 \lambda_{11} + l_1 \Lambda_{11} + l_2 \Lambda_{22} + l_3 \Lambda_{33} + \Delta_\lambda \quad (10.14,1)$$

in which $\Lambda_{11} = \lambda_{22}\lambda_{33} - \lambda_{23}\lambda_{32}$, etc., and Δ_λ is the determinant of PP'^{-1}. It is assumed:

(1) that l_1 is substantially larger than l_2 which in turn is substantially larger than l_3; otherwise, two or more of the l's can be made identical over all or at least over the higher frequency range (alternative 2 of Section 10.13), and the loop shaping problem is the same as that for $n = 2$;

(2) that it is possible to select the nominal plant values such that all the λ functions have maximum magnitudes of one. We write

$$\Delta / l_1 l_2 = l_3 + \lambda_{22} l_3 l_2^{-1} + \lambda_{11} l_3 l_1^{-1} + \Lambda_{11} l_2^{-1} + \Lambda_{22} l_1^{-1} + \Lambda_{33} l_3 l_1^{-1} l_2^{-1} + \Delta_\lambda l_1^{-1} l_2^{-1} + \lambda_{33} \triangleq f + \lambda_{33}. \quad (10.14,2)$$

The next step is to sketch $-\lambda_{33}$ for a number of frequencies.

Procedure: (1) In view of assumption (2), crossing of the $-\lambda_{33}$ loci by f can first occur only when $|f| \approx 1$. Examination of Eq. (10.14,2) reveals that due to assumption (1), $f \doteq l_3$ in this range. The term $\lambda_{22} l_3 / l_2$ may make a slight contribution. The predominance of l_3, however, makes it easy to state reasonable requirements on $\angle l_3$ after making a reasonable estimate as to the frequency at which $|l_3| \approx 1$. If necessary, suitable requirements on l_2 can also be easily stated. Let this frequency range be denotes by ω_a. This region is similar to $\omega = 2$ rps of Section 10.13.

(2) At ω_b which is an octave or so above ω_a, $|l_3| \approx 0.5$ or less and at worst $f(j\omega_b) \approx l_3 + \Lambda_{11} l_2^{-1} + \lambda_{22} l_3 l_2^{-1}$. The problem in this frequency region is similar to that at 5 rps in Section 10.13 (Figs. 10.13-2 to 4). The restrictions on the magnitude and phase of l_2 and l_3 at ω_b are thereby obtained.

(3) The next frequency region of interest, denoted by ω_c, is when $|l_2|$ is between 0.5 and 1. Here l_3 is so small that it can be neglected. In the worst case (i.e., when the differences in the levels of the l's are barely enough to justify keeping all three distinct), $f \approx \Lambda_{11} l_2^{-1} + \Delta_\lambda l_1^{-1} l_2^{-1} + \Lambda_{22} l_1^{-1}$, and l_2, l_1 are in the same relative situation as were l_3, l_2, respectively, at ω_b. The procedure is the same as at ω_b in (2).

(4) The next frequency region to consider is ω_d, corresponding approximately to the crossover frequency of l_1. In this region $l_3, l_2 \ll l_1$ and $f \approx \Delta_\lambda / l_1 l_2$. This region corresponds to $\omega = 15$ rps of Section 10.13 (Fig. 10.13-5), and the procedure is the same.

(5) At higher frequencies one term predominates, i.e., $f \doteq \Delta_\lambda / l_1 l_2$.

It is seen from the above that thanks to assumptions (1) and (2), the loop shaping of a third-order system with substantial parameter variation is logical and straightforward. In the worst case there are two regions in which f may have to be represented by three terms, whereas in the second-order case there is only one such region. Furthermore, at every step in the process one checks the effect of smaller $|l_i|$ at the same lag angle, and of reduced lag angle at the same magnitude. It is expected that the effect is always stabilizing, but this is easily checked at every step. Knowing these effects, it is possible, in the final precise shaping of the l's, to use values substantially different from the estimated values (just as we did in the example in Section 10.13), without having to redesign. This rational and economical procedure can obviously be applied to higher order systems, where $n > 3$. A three-term approximation of f will suffice even in the most difficult frequency regions. The number of such regions is at most equal to $n - 1$.

§ 10.15 Design for Rejection of Disturbances

Introduction

One of the important reasons for using feedback in a control system is to make the system insensitive to some desired extent to disturbances entering the system at any point other than the command signal entry points. When the disturbances are in the nature of step functions or impulses, there are usually two design requirements. These are: (1) the steady state effects are to be kept below some minimum value, and (2) the effects of the disturbance are to be below some minimum within a specified time.

Thus an abrupt disturbance acting on the plant cannot instantaneously be attenuated by the feedback, unless the loop transmission has infinite bandwidth, which is of course impossible. One must allow some time for the feedback to assert itself. Therefore an important design specification in the attenuation of abrupt disturbances is the time one is willing to wait for the feedback to act. The shorter this time, the larger must be the loop transmission bandwidth.

If the disturbances are of the impulsive type, then only requirement (2) applies, unless there is integration between the disturbance and the output. The design procedure is now developed by means of an example, where the disturbances are in the above categories.

Example.

Specifications: The plant parameter variations are very small and feedback is being used primarily to attenuate disturbances acting on the plant. The matrices of the maximum simultaneously occurring disturbances (see Fig. 10.15-1) and of the plant are, respectively,

$$D_A = \begin{bmatrix} 2/s \\ 1 \\ 0 \end{bmatrix}, \quad D_B = \begin{bmatrix} 3/s \\ 0 \\ 0 \end{bmatrix}, \quad P = \frac{1}{(s+1)} \begin{bmatrix} 1 & 0.5 & 1 \\ 0.5 & 1 & 0.5 \\ 0 & 0 & 1 \end{bmatrix}.$$

It is required that the disturbances yield total steady state output values less than 0.1 and that their transient responses should decrease to less than 0.2 within 0.2 second. It is possible in general to have more complicated individualized specifications such as the effect on c_2 due to d_{A1}, etc., but the above stated specifications in terms of the maximum permitted total effect on the output are certainly realistic.

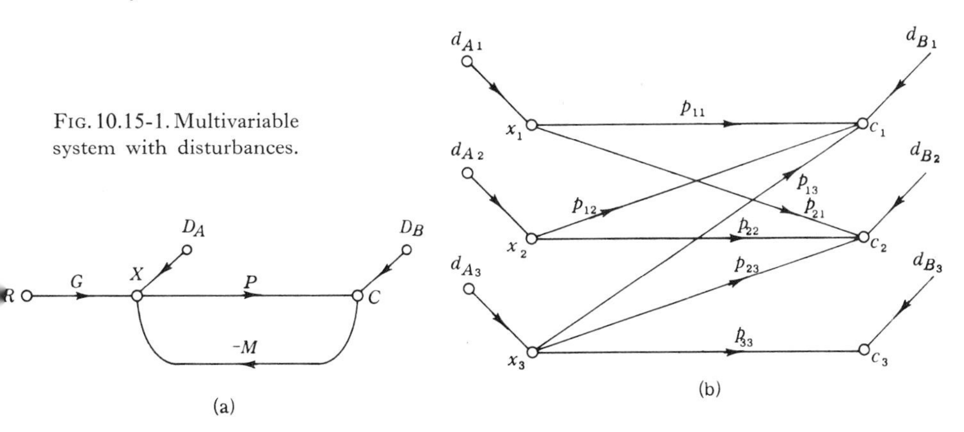

FIG. 10.15-1. Multivariable system with disturbances.

It is readily seen that it is the elements of the loop transmission matrix L that are responsible for the attenuation of the disturbances. Thus in Fig. 10.15-1a when $R = 0$, $C_D = D_B + PX = D_B + P(D_A - MC_D)$. Therefore $(I + PM)\, C_D = (I + L)C_D = D_B + PD_A$ and

$$C_D = (I + L)^{-1}\,(D_B + PD_A). \tag{10.15,1}$$

This equation applies for any configuration so long as L is the loop transmission matrix. With no feedback, $C_D = D_B + PD_A$, so it is clearly the elements of L that are responsible for the attenuation of the disturbances. Let $K = (I + L)^{-1}$ and $D = D_B + PD_A$. Then, from Eq. (10.15,1), $c_i = k_{i1}d_1 + k_{i2}d_2 + \dots + k_{in}d_n$, where d_j is the element in the jth row of $D_B + PD_A$.

If the d_j are independent of each other, then the individual terms in the last equation are also independent. In order to make c_i small, each individual term must be even smaller. Therefore, in such cases it is indeed advantageous to make K, and consequently L, a diagonal matrix (L could be any one of the canonic $n!$ L matrices). Then

$$c_i = k_{ii}d_i = d_i/(1 + l_i), \tag{10.15,2}$$

and the disturbance d_i is attenuated by the factor $1 + l_i$. When the d_j are related to each other, it may be possible to pick the k_{ij} so that the terms in $c_i = \sum k_{ij}\, d_j$ wholly or partially cancel. In such a case large attenuation can be achieved with smaller magnitude loop transmissions. Such special cases should be noted and exploited of course, but we assume that the disturbances are unrelated and therefore choose L diagonal.

Equation (10.15,2) is the basic design equation. It determines the magnitudes of the loop transmissions over the significant frequency ranges. Beyond the latter it is desirable to reduce the $|l_i|$ as fast as possible to their high frequency asymptotic slopes (i.e., with an excess of two, three, or more poles over zeros, as the case may be). This reduction must be done subject to stability requirements.

It is noted in this specific example that the disturbances must be fairly well attenuated within 0.2 second, while the plant element response time is in the order of 1 second. Therefore the elements of M in $L = PM$ will have to be of an anticipatory nature with rising frequency characteristics. The general broad demands on the elements of L are fairly easily obtained by considering

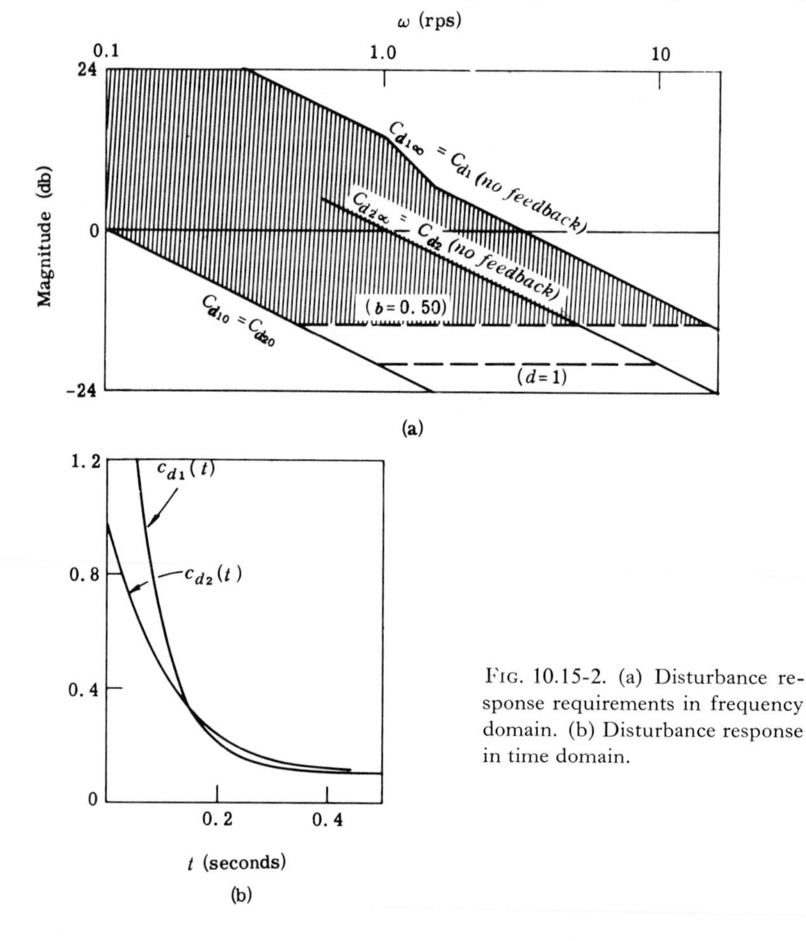

FIG. 10.15-2. (a) Disturbance response requirements in frequency domain. (b) Disturbance response in time domain.

the situation at the two extremes of zero frequency and infinite frequency. Consider Eq. (10.15,1) at infinite frequency. As $s \to \infty$, $L \to 0$ and $C_D \to D_B + PD_A$. This corresponds to the fact that the feedback cannot act instantaneously to reduce the disturbances. The high frequency asymptotes of C_D are therefore

$$D_B + PD_A = \begin{bmatrix} 3/s \\ 0 \\ 0 \end{bmatrix} + \left(\frac{1}{s+1}\right) \begin{bmatrix} 1 & 0.5 & 1 \\ 0.5 & 1 & 0.5 \\ 0 & 0 & s+1 \end{bmatrix} \begin{bmatrix} 2/s \\ 1 \\ 0 \end{bmatrix} = \begin{bmatrix} (3.5s+5)/s(s+1) \\ 1/s \\ 0 \end{bmatrix}.$$

These are sketched in Fig. 10.15-2a, and labeled[1] $C_{d1\infty}$, $C_{d2\infty}$. They are also labeled C_d (no feedback), because these would be the disturbance outputs in an open-loop system.

Next consider the situation at 0 rps. The 0 rps demands on L are directly available from the specifications by applying the final value theorem. It is necessary that $\lim_{t\to\infty} c_{di}(t) < 0.1$, therefore $\lim_{s\to 0} s\, C_{di}(s) < 0.1$. These low frequency asymptotes are sketched in Fig. 10.15-2a and labeled C_{d10}, C_{d20}.

The problem now is to join the two asymptotes in the most economical manner. If there were no feedback at all, the C_d would be $C_{d1\infty}$, $C_{d2\infty}$, which are of course too large. Feedback is being used to decrease the C_d values. At low frequencies, 34 db of reduction is needed for C_{d1}, 20 db for C_{d2}, and therefore 34 db of l_1 and 20 db of l_2 are needed. At very large frequencies the feedback is ineffective and the loop gains go to zero. How soon can we let C_{d1}, C_{d2} (which begin at C_{d10}, C_{d20}) merge with $C_{d1\infty}$, $C_{d2\infty}$? The sooner it is done, the easier it is on the design, i.e., less loop gain-bandwidth is required. This merging of the asymptotes must be done so as to satisfy the specifications that the response be damped to 0.2 within 0.2 second.

The simplest approximation is a horizontal (actually only the asymptote is horizontal) line between the two asymptotes, viz., $C_{d1} = 3.5(s+b)/s(s+35b)$, which results in $c_{d1}(t) = 0.1 + 3.4e^{-35bt}$, and it is readily seen that $b = 0.50$ is required in order to satisfy the specs (see Fig. 10.15-2b). This permits us to join the low and high frequency asymptotes of $C_{d1}(j\omega)$ with the dotted horizontal line labeled $b = 0.50$ in Fig. 10.15-2a. The gain-bandwidth area of l_1 is the difference in area between C_{d1} (no feedback) and the actual C_{d1}. This area is shown shaded in Fig. 10.15-2a. The larger the value of b, the faster the feedback responds, i.e., the faster c_{d1} is damped out, but the lower the horizontal line joining the two asymptotes must be drawn, and therefore the larger is the gain-bandwidth area of l_1. Similarly, for C_{d2}, we take the simplest form, $C_{d2} = (s+d)/s(s+10d)$ and in precisely the same manner as before find that the minimum d is 1, with $c_{d2}(t) = 0.1 + 0.9e^{-10t}$ (see Fig. 10.15-2b).

It is appropriate to point out an important problem relating to time domain synthesis. We assumed a very simple form for C_{d1} and C_{d2} and emerged with

---

[1] For the balance of this section we shall use C_{d1}, C_{d2} for the elements of the C_D matrix, and c_{d1}, c_{d2} to indicate time functions.

the gain-bandwidth requirements of l_1 and l_2. Can the latter be decreased at all? In other words, what shape of C_{d1} satisfies the time domain specifications with a minimum difference in area between C_{d1} and C_{d1} (no feedback) in Fig. 10.15-2a? This same type of problem appears whatever the form of the disturbance and its required attenuation. This problem was considered in Section 9.4.

Returning to the design proper, since we have the matrices C_D, D_A, D_B, P, it is clear that we can use Eq. (10.15,1) to solve for L. However, it is well to anticipate some difficulties. Since L is diagonal, we have $C_{dj} = d_j/(1 + l_j)$ where d_j is the element in the nth row of $D_B + PD_A = D$. Suppose l_j is to have an excess of x poles over zeros. Then

$$1 + l_j = 1 + \frac{K(s^h + a_1 s^{h-1} + \dots)}{s^{h+x} + b_1 s^{h+x-1} + \dots}$$

$$= \frac{s^{h+x} + b_1 s^{h+x-1} + \dots + b_{x-1} s^{h-1} + (K + b_x) s^h + \dots}{s^{h+x} + b_1 s^{h+x-1} + \dots + b_{x-1} s^{h-1} + b_x s^h + \dots}.$$

Thus $1 + l_j$ has its first x numerator coefficients equal, respectively, to its first x denominator coefficients. Conversely, since $1 + l_j = d_j/C_{dj}$, it is necessary that the first x numerator coefficients of d_j/C_{dj} should be equal, respectively, to its first x denominator coefficients, if we want the l_j that is cranked out to emerge with x more poles than zeros. If C_{dj} is not specified in this manner, the resulting l_j will emerge with a smaller or no excess of poles over zeros. In general it is extremely difficult to specify C_{dj} in this manner when $x > 2$. Fortunately, it is possible to get around this by using Eq. (10.15,1) to determine the loop transmissions over their significant frequency ranges only. Beyond the latter, the l's are brought down as fast as possible, consistent with stability requirements as well as the disturbance response requirements.

The above remarks are illustrated by first using Eq. (10.15,1) to solve for L. We have

$$\begin{bmatrix} 3.5\,(s + 0.5)/s(s + 17.5) \\ (s + 1)/s(s + 10) \\ 0 \end{bmatrix}$$

$$= \begin{bmatrix} 1/(1 + l_1) & 0 & 0 \\ 0 & 1/(1 + l_2) & 0 \\ 0 & 0 & 1/(1 + l_3) \end{bmatrix} \begin{bmatrix} (3.5s + 5)/s(s + 1) \\ 1/s \\ 0 \end{bmatrix}.$$

Therefore $l_1 = (17.4s + 25)/(s + 0.5)\,(s + 1)$, $l_2 = 9/s + 1$, $l_3 = 0$. These expressions for the l's are used only as a first approximation to the actual values. In particular, their high-frequency values require modification in order that they may have the proper excess of poles over zeros. For example, if the structure of Fig. 10.15-1a is used, an excess of two poles over zeros may suffice, while if Fig. 10.5-1b is used, there should be at least three poles more than zeros, because a cascade of three structures appears in L and the plant elements have an excess of only one pole over zeros. To find the final suitable l_1, the Bode sketch of the approximate l_1 in Fig. 10.15-2a is used. Note that for $\omega > 15$, $|C_{d1}| \approx |C_{d1}\text{(no feedback)}|$. This corresponds to the fact that for $\omega > 15$, $|l_1| < 1$ and therefore $|1 + l_1| \approx 1$. Hence additional poles may be assigned to l_1 beyond 15 rps with negligible effect on $c_{d1}(t)$. Similarly, additional poles may be assigned to l_2 beyond 10 rps. Of course proper gain and phase margins should be assigned to l_1 and l_2 to prevent undue peaking in the system response. We accordingly take $l_1 = 27{,}400\,(s + 1.4)\,(s + 17)/(s + 0.5)\,(s + 1)\,(s + 30)^3$, $l_2 = 4600\,(s + 8)/(s + 1)\,(s + 16)^3$. These are sketched in Fig. 10.15-3.

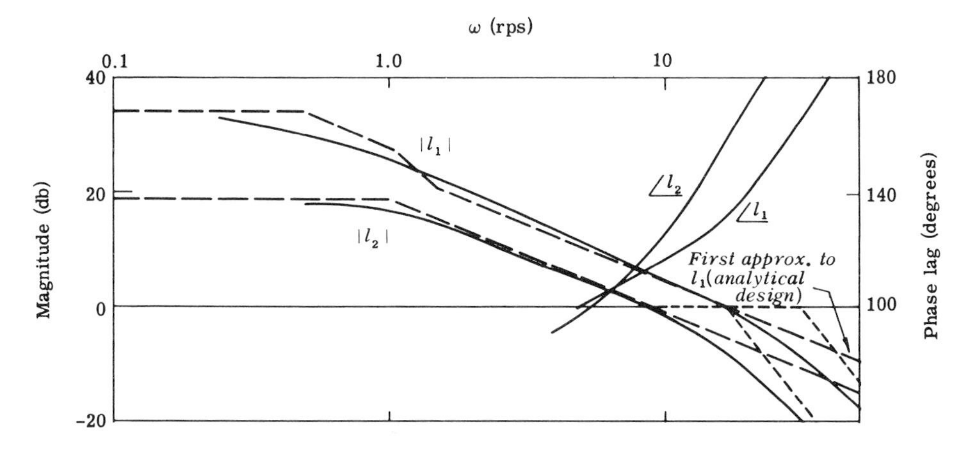

FIG. 10.15-3. Bode plots of l_1, l_2.

With L completely specified and P known, M in $L = PM$ is easily obtained. Also, from Eq. (10.5,2a), $T = (I + L)^{-1}PG$; so the G matrix is chosen to realize the desired T matrix elements. There is no need to go through these calculations since the procedure is straightforward.

Design for Other Disturbance Functions

In the above, the design problem for disturbances consisting of impulses and steps was treated. The same basic design procedure is used for other kinds of inputs. The frequency responses of the c_d's with no feedback are ascertained and sketched on a Bode plot. Next, the maximum permitted attenuated responses are ascertained and their frequency responses sketched. The procedure for finding the latter was considered in Section 9.4. The difference between the two gain-bandwidth areas is made up by the loop transmissions. The more steady state attenuation is desired, the greater are the loop gain levels at 0 rps. The faster the disturbances and the faster they must be attenuated by the feedback loops, the greater the required bandwidths of the loop transmissions. The equation

$C_D = (I + L)^{-1} (D_B + PD_A)$ should be used to obtain any l, only over the frequency range in which $|l| > 1$. Beyond this range one assigns additional poles of l to obtain the required excess of poles over zeros. This is done with adequate stability margins so that there is no excessive peaking in the system frequency response and to obtain the desired damping for the dominant zeros of $1 + l$.

Simultaneous Design for Parameter Variation and Disturbance Attenuation

The design for disturbance attenuation which has been described in this section is based on no parameter variation, because the assumed diagonal L is truly diagonal at only one set of values of the plant parameters. Let P' be the new plant matrix. Equation (10.15,1) may be rewritten as $(I + P'M) C_D = (D_B + P'D_A) = C_D + P'P^{-1}PMC_D = C_D + P'P^{-1}LC_D$. The matrix $P'P^{-1}$ is the inverse of $\Lambda = PP'^{-1}$. For the second-order system (with L diagonal) the above leads to

$$c_{d1}(1 + \gamma_{11}l_1) + \gamma_{12}l_2c_{d2} = d_{b1} + p'_{11}d_{a1} + p'_{12}d_{a2}$$

$$\gamma_{21}l_1c_{d1} + c_{d2}(1 + \gamma_{22}l_2) = d_{b2} + p'_{21}d_{a1} + p'_{22}d_{a2}$$

where the γ's are the elements of Λ^{-1}. The solution for c_{d1} is

$$c_{d1} = \frac{d_{b1} + p'_{11}\,d_{a1} + p'_{12}\,d_{a2} - \gamma_{12}\,l_2c_{d2}}{\gamma_{11}\,(\gamma_{11}^{-1} + l_1)}. \tag{10.15,3}$$

The minimum values of $|c_{d1}(j\omega)|$ as a function of ω are determined by the attenuation specifications. These can always be achieved by making l_1, l_2 sufficiently large. In Eq. (10.15,3) the range of $d_{b1} + p'_{11}d_{a1} + p'_{12}d_{a2}$ at any value of ω is known and may be sketched just as in Fig. 6.12-1. The maximum value of c_{d2} (as given by the specifications) is also known, so the maximum contribution of $\gamma_{12}l_2c_{d2}$ may be estimated. To be on the safe side one may assign to c_{d2} that phase angle which maximizes the numerator. The range of γ_{11}, γ_{11}^{-1} is also sketched. The permissible location of l_1 which satisfies the specification on c_{d1} at that frequency may then be ascertained in exactly the same manner as in Fig. 6.12-1. This procedure is repeated at a few frequencies in the significant (attenuation) range and a satisfactory $l_1(j\omega)$ is then chosen. The process is repeated for $l_2(j\omega)$. The above determines $|l_1|_{\min}(j\omega)$, $|l_2|_{\min}(j\omega)$ for disturbance attenuation. The sensitivity design procedure which has been outlined in Sections 10.6 and 10.7 is unaffected by the presence of disturbances. The procedure is used to ascertain the minimum levels of loop transmission which are required to satisfy the sensitivity specifications and these are compared with those required for disturbance attenuation. The levels that are assigned must of course be the larger of the two.

10.16 The Feedback Problem for Plants with More Outputs Than Inputs

We return to the problem briefly examined in Section 10.2, where the number of plant outputs exceeds the number of plant inputs. In Section 10.2 the plant under consideration had two inputs and three outputs. It was noted that only four out of the six system transmission functions are independently realizable. It was also shown that the feedback problem is the same as that already treated if these four are assigned to only two rows in the T matrix, i.e., if the four chosen are t_{i1}, t_{i2}, t_{j1}, t_{j2}. If the plant has three inputs and five outputs, then nine independent $t_{\alpha\beta}$'s are realizable, and the sensitivity problem is the same as that treated in the preceding sections if the nine are assigned to only three rows in the T matrix, etc. But what if the t's specified are scattered among the rows, e.g., t_{11}, t_{22}, t_{31}, t_{32}, for the plant with two inputs and three outputs?

It is possible to question the wisdom of such a choice, as follows. If the two inputs r_1 and r_2 are applied simultaneously, then $c_1 = t_{11}r_1 + t_{12}r_2$, and the t_{11} component is only a part of c_1. Hence the designer has not achieved control of c_1 by specifying t_{11} without t_{12}. It seems that the choice of t_{11}, t_{22} (without t_{12}, t_{21}) makes sense only when the inputs r_1, r_2 are separately applied. The plant specialist might answer this question as follows. Leave the control problem to us. We have studied the typical system inputs, desired outputs, etc., and under the constraint of only four independently realizable t's, we have made a reasonable compromise. You are the feedback specialist. We would like you to design the feedback loops so that all six t_{ij}'s are insensitive, to a specified extent, to plant parameter variations. If this is not possible, tell us what is possible.

The answer to the above is that it is not possible to decrease the sensitivity of the six $t_{\alpha\beta}$'s or even of three elements in three different rows. We can only decrease the sensitivity of any two rows of transmission functions. This can be formally proven, but the reader might be convinced by the following reasoning. It was shown in Section 8.15 that parameter variations are equivalent to disturbances. In the plant under consideration there are then three disturbances at the plant output points and the problem is whether all three can be attenuated. The attenuation is achieved by means of appropriate feedback signals at the plant inputs. But if there are only two plant inputs then only two plant outputs can be controlled. Hence we are able to directly attenuate only two of these disturbances. This is equivalent to decreasing the sensitivity of the elements in only two rows of the T matrix. (Of course the third disturbance might also thereby be attenuated but we cannot guarantee this.)

We shall also indicate the formal proof of the above. If the plant has two inputs and three outputs, then the feedback network M (in the notation of Section 10.5) has three inputs (from the plant outputs) and two outputs (to the plant inputs). M has two rows and three columns, i.e., six independent elements. $L = PM$ is then a square 3×3 matrix, but L is a singular matrix. The effect of plant paramter variations is given by Eq. (10.6,2), $\Delta T = (I + L)^{-1} VT' = (I + L)^{-1}Q'$, so we must evaluate $(I + L)^{-1}$. When this is done in terms of the elements of P and of M, then it is found that all the elements of $(I + L)^{-1}$ have the common denominator $1 + P_{13}M_{31} + P_{23}M_{32} + P_{33}M_{33} + l_{11} + l_{22} + l_{33}$ (as usual P_{ij} is the cofactor of the ijth element in the P matrix). Each $P_{\alpha 3}M_{3\alpha}$ is proportional to l^2. However, all the matrix numerator elements also have terms proportional to l^2. These are $P_{13}M_{31}$, $P_{23}M_{31}$, $P_{33}M_{31}$ for the first row; $P_{13}M_{32}$, $P_{23}M_{32}$, $P_{33}M_{32}$ for the second row; $P_{13}M_{33}$, $P_{23}M_{33}$, $P_{33}M_{33}$ for the third row. If M_{31}, M_{32}, M_{33} are chosen of the same order of magnitude, then no elements of ΔT can be made small. If we choose M_{31} large but M_{32}, M_{33} small, then only the second and third rows of ΔT can be made small, etc.

One might try choosing $M_{31} = M_{32} = M_{33} = 0$ and examining what remains of $(I + L)^{-1}$ to see if something better is possible. It is readily found that the situation is not thereby improved. Thus for plants in which $m > n$ (m outputs, n inputs), only n rows of ΔT can be made small, or n disturbance outputs can be attenuated. After the choice has been made, the design procedure is thereafter exactly the same as in plants where $n = m$.

10.17 Discussion of the Feedback Mechanism in Multivariable Systems

The treatment in this chapter has thus far been restricted to multivariable plants in which there is access only to the plant input and output points. These are analogous to the two-degree-of-freedom systems of Chapter 6. We now

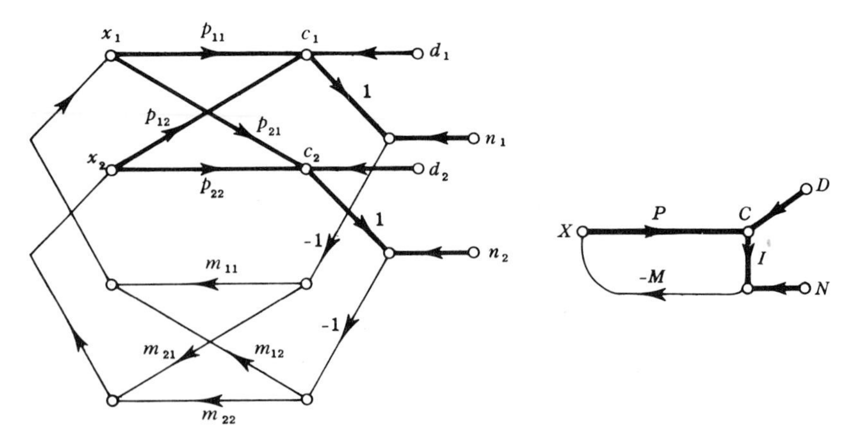

FIG. 10.17-1. Feedback structure for disturbance attenuation. (Heavy lines represent the plant with its disturbance and feedback transducer noise constraints.)

wish to consider the disturbance attenuation problem in plants, some of whose internal variables are available for feedback purposes. The objective is the same as in the analogous single input-output systems of Sections 8.2-8.3, i.e., to use the additional freedom to satisfy the feedback specifications, with decreased system sensitivity to feedback transducer noise. The problem is so much more complicated in multivariable systems that it is useful to first understand some of the important factors in the disturbance attenuation mechanism in such systems.

In Fig. 10.17-1, the output due to the feedback transducer noise is $C_N = -L(N + C_N)$, where $L = PM$, so $C_N = -(I + L)^{-1}LN$. In the significant frequency range where the $| l_{ij} |$'s are large, $I + L \doteq L$, $C_N \doteq -N$ and nothing can be done about it. However, in the higher frequency range where the $| l_{ij} |$'s are small, $C_N \doteq -LN$. Hence it is important to reduce the $| l_{ij} |$ elements as fast as possible. In Fig. 10.17-1 the outputs due to d_1, d_2 are

$$c_1 = [(1 + l_{22})\, d_1 - l_{12}d_2]/\Delta, \qquad c_2 = [-l_{21}d_1 + (1 + l_{11})\, d_2]/\Delta \tag{10.17,1}$$

where $\Delta = 1 + l_{11} + l_{22} + l_{11}l_{22} - l_{12}l_{21}$. Presumably c_1, c_2 can be made small (with small $| l |$'s) by choosing $(1 + l_{22})\, d_1 \doteq l_{12}d_2$, $(1 + l_{11})\, d_2 \doteq l_{21}d_1$. This is feasible only if (1) d_1 and d_2 are related in a known, constant manner, and (2) plant parameter variations, if any, are so interrelated that these last two equations continue to be satisfied. These two necessary conditions are rarely satisfied simultaneously, so we shall assume that small c_1, c_2 are attained by sufficiently large $| l |$'s over the frequency range of interest.

It was noted in Section 10.7 that the canonic L matrices (at least one element in each row and column of L) have the least number of nonzero elements for handling the most general parameter variation problem. It is readily seen (for example, by invoking Section 8.15) that this is also true for disturbance attenuation. If parameter variation is small, then there appear to be definite advantages in choosing the diagonal L out of the $n!$ canonic matrix possibilities. One advantage is that the diagonal L permits faster reduction of the $| l_{ij} |$'s, and we have just noted that this decreases the effect of feedback transducer noise in the higher frequency range. That such faster reduction is so achieved is seen by considering the denominator of c_1, c_2 in Eq. (10.17,1). If L is diagonal, then $\Delta = (1 + l_{11})(1 + l_{22})$ and the stability problem is precisely the same as in two separate single input-output systems. However, if L is antidiagonal, then $\Delta = 1 - l_{12}l_{21}$, and the $l_{12}l_{21}$ product must be shaped for stability. The product $l_{12}l_{21}$ can then decrease at the maximum rate determined by stability considerations, and therefore each of l_{12} and l_{21} must decrease more slowly than their product. The superiority of the diagonal L is even more when the elements of L are nonminimum phase. The parameter variation case was discussed in Sections 10.7 and 10.10, and it is recalled that in the specific examples that were there considered, the diagonal L was found to be better, although there was insufficient evidence that this is always so.

The physical reason for the superiority of diagonal L (in the no parameter variation case) is as follows. Attenuation of d_1 at c_1 (see Fig. 10.17-1) requires a closed loop of signal flow from c_1 back to itself. In a diagonal L this signal flow is from c_1 via m_{11}, m_{21} to x_1, x_2 and then directly through p_{11}, p_{12} to c_1. However, in the antidiagonal L, the signal from c_1 must first reach c_2 by way of $m_{11}p_{21}$ and $m_{21}p_{22}$ (i.e., via l_{21}), and only then can it go on through $m_{12}p_{11}$ and $m_{22}p_{12}$ (i.e., via l_{12}) to c_1. One might be thereby deluded into thinking that the benefits of feedback are also proportional to $l_{12}l_{21}$; for example, if each p_{ij} contains an integration, that the effective K_v is infinite. This is only partly true. It is true for the attenuation of d_1 at c_1. However, it is not true for the effect of d_2 on c_1 (or of d_1 on c_2) because d_2 is first multiplied by l_{12} before it reaches c_1; i.e., at c_1 the effective disturbance is $d_1 + l_{12}d_2$. The above is compounded in higher order systems. For example, for $n = 3$, if only l_{13}, l_{21}, l_{32} are nonzero, then the feedback from c_1 back to itself must first go through c_2 (via l_{21}), then to c_3 (l_{32}), and only then back to c_1 (via l_{13}). The disturbance d_1 is attenuated at c_1 by the product $l_{13}l_{21}l_{32}$, but d_2, d_3, lead to effective disturbances $l_{32}l_{13}d_2$, $l_{13}d_3$ at c_1. If L is diagonal, then $\det 1 + L = (1 + l_{11})(1 + l_{22})(1 + l_{33})$ when $n = 3$.

Another advantage of the diagonal L (in the small parameter variation case) is that it may thereby be possible to use some l_{ii}'s with smaller gains and bandwidts. Thus with L diagonal the disturbance outputs are $c_i = d_i/(1 + l_{ii})$ and l_{ii} need have only the gain and bandwidth required for the attenuation of d_i, which may perhaps, in some problems, be considerably less than that required for d_j. Otherwise (for $n = 2$) $c_1 = [(1 + l_{22})\, d_1 - l_{12}d_2]/\Delta$. In order to have small c_1, it is necessary to consider both d_1 and d_2. If $l_{12} \sim l_{22}$, then each must have the larger of the gains and the bandwidths needed for the attenuation of d_1 and d_2 in the diagonal L case.

An extreme case of the above is when there are disturbances at only some of the plant access points. For example, in Fig. 10.17-1, if $d_2 = 0$, then there is no need to use l_{22} at all (i.e., only one feedback transducer is needed), providing parameter variations are negligible. When the parameters vary, we must use Eq. (10.15,3), etc. For the case $l_2 = d_2 = 0$, $c_1 = d_1/(1 + \gamma_{11}l_1)$, $c_2 = -\gamma_{21}l_1/(1 + \gamma_{11}l_1)$. Hence it is impossible to secure small c_2 unless γ_{21} is small (note that $\gamma_{21} = 0$ if there are no parameter variations) over the entire range of parameter variations. Hence l_{22} is also required in the severe parameter variation case, even though there is only the single disturbance d_1. The physical reason is that feedback from c_1 back to c_1 is through x_1, x_2 and these also affect c_2. It is possible to choose the m_{ij} such that the two different signals $p_{21}x_1$ and $p_{22}x_2$ cancel each other, but variations in p_{21}, p_{22} will generally change this and then the feedback from c_1 leads to an equivalent disturbance at c_2, which can be attenuated only by feedback from c_2 itself. With finite l_{22} we get

$$c_2 = -\gamma_{21}l_{11}d_1/[1 + \gamma_{11}l_1 + \gamma_{22}l_{22} + l_{11}l_{22}\Delta_\gamma] \doteq -\gamma_{21}d_1/\Delta_\gamma l_{22}.$$

Hence the range over which it is necessary that $| l_{22} | > 1$ is of the same order of magnitude as that for l_{11}, the precise relation depending on the specifications for c_2 and the maximum value of $\gamma_{21}/\Delta_\gamma$. The above is true for any order system. The conclusion is that when there are nonnegligible parameter variations, then the existence of a single disturbance d_i at any c_i requires all the l_{jj} (with cross-over frequencies of the same order of magnitude as l_{ii}), in order to control the effect of d_i at all c_j.

10.18 Multivariable Plants with Internal Variables Available for Feedback

We are now ready to consider design for disturbance attenuation in multivariable systems with more than two (matrix) degrees of freedom. The simplest of the latter is of the cascade type, a simple example of which is shown in Fig. 10.18-1 (for $n = 2$). The cascade type may be defined as one in which

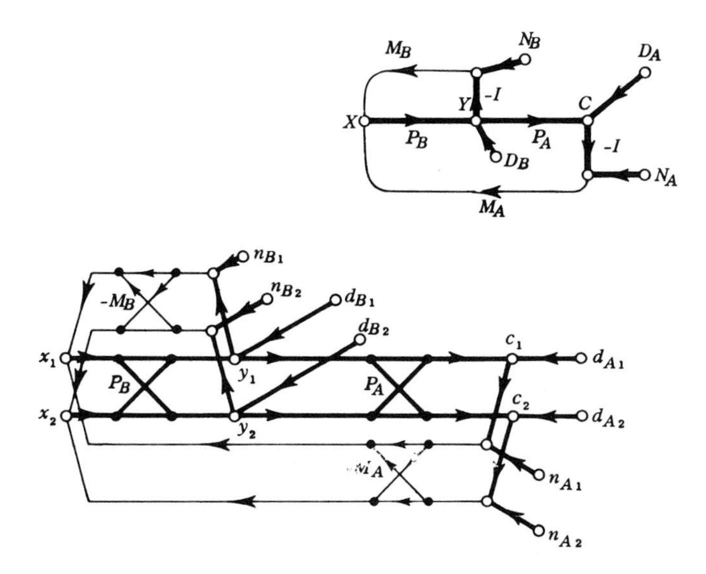

FIG. 10.18-1. Feedback system for a cascade second-order three (matrix)-degree-of-freedom multivariable plant. (Heavy lines represent the plant with its disturbances and feedback transducer noise constraints.)

the complete plant matrix P can be written in the form $P = P_A P_B P_C \dots$ with P_A, P_B, ... representing square matrices all of the same order and in which the C, Y, ... variables are available for feedback purposes. The problem is to assign the feedback burden among the loop transmissions so as to minimize the effect of feedback transducer noise. The following notation is used. The matrices of the variables C, Y, X due to D_A will be denoted by C_A, Y_A, X_A, etc. (those of the matrix elements by c_{Ai}, y_{Ai}, etc.), and their values due to N_A will be denoted by C_{NA}, Y_{NA}, X_{NA}, etc. (elements by c_{NAi}, etc.).

In Fig. 10.18-1, $C_A = D_A + P_A Y_A$, $Y_A = P_B X_A$, $X_A = -M_B Y_A - M_A C_A$. By combining these, there results

$$[I + P_A(I + P_B M_B)^{-1} P_B M_A]\, C_A \triangleq (I + L_A)\, C_A = D_A. \quad (10.18,1)$$

Similarly,

$$C_{NA} = -(I + L_A)^{-1} L_A N_A \quad (10.18,2)$$

$$(I + P_B M_B + P_B M_A P_A)\, Y_B = D_B. \quad (10.18,3)$$

In the significant frequency range, attenuation is achieved by having the elements of $L_A C_A$ larger than the corresponding ones of C_A, so $(I + L_A) \doteq L_A$ and Eq. (10.18,2) becomes $C_{NA} \doteq -N_A$ and nothing can be done about it. Thus, in this range C_{NA} is independent of M_B, so we might as well (ideally) take $M_B = 0$ in order that at least $C_{NB} = 0$. Therefore the outside feedback loop $P_A P_B M_A$ should handle the entire feedback burden for both D_A and D_B up to the crossover frequencies required for D_A (cf. Section 8.2). Let ω_{A1}, ω_{A2}, ... indicate the required minimum crossover frequencies for d_{A1}, d_{A2}, ... attenuation, i.e., d_{A1} need be attenuated only for $\omega < \omega_{A1}$ (see Section 8.2), and similarly ω_{B1}, ω_{B2}, ... are the minimum crossover frequencies for d_{B1}, d_{B2}, ... attenuation. Let us consider the case $\omega_{A1} > \omega_{B1}$, $\omega_{A2} > \omega_{B2}$, etc., and therefore in this special case M_B is not needed for D_B attenuation, but may nevertheless be useful in shaping the system loop transmissions for stability, as follows.

The determinant whose zeros must be examined for stability is that of $I + L_A$ [Eq. (10.18,1)]. However, it is more convenient to put it into the form [by premultiplying by $(I + P_B M_B) P_A^{-1}$ and then premultiplying by P_A], $I + L \triangleq I + P_A(P_B M_B) P_A^{-1} + P_A P_B M_A$. The reason for the greater convenience of $I + L$ over $I + L_A$ is that in the first part of the design ($\omega < \omega_A$) it is the elements of (diagonal) $P_A P_B M_A$ which are specified. If the burden for $\omega > \omega_x$, with $\omega_x > \omega_A$, is passed on to M_B, then preferably (recall Section 10.17) the elements of L are diagonal. $I + L$ can then be written as $(1 + l_1)(1 + l_2)$ ... and the l_i's must have definite minimum magnitudes as functions of frequency (denoted by l_{m1}, l_{m2}, ...) in order that the system be stable. The burden of maintaining any l_{mi} may however, be transferred from the $P_A P_B M_A$ term in L to the second term $P_A(P_B M_B) P_A^{-1}$. Hence the design of each channel may be considered separately. The transition frequency for each channel is obtained in the same manner as in the single input-output system, and is readily found to be at that frequency at which $N_A = P_A N_B$ (see Section 8.2, "Case $\omega_{2m} > \omega_{1m}$"), i.e., for the jth channel it is the frequency at which $n_{Aj} = \Sigma_i\, p_{Aji} n_{Bi}$ (rms addition is required if the noise is statistical).

Of course, in practice the transition must be a gradual one and extend over several octaves (recall Figs. 8.2-2 to 6).

The above is a special case in that it was assumed that each d_A crossover frequency ω_{Ai} was larger than the respective d_B crossover frequency ω_{Bi}. The general case when one or more $\omega_{Bi} > \omega_{Ai}$ is more difficult than the same problem in single input-output systems. The reason is as follows. In the region $\omega < \omega_{A,\min}$ ($\omega_{A,\min}$ is the minimum ω_A crossover frequency), the outside loop matrix (M_A) only is ideally used. In this region L_A becomes closely equal to $P_A P_B M_A$ [see Eq. (10.18,1)] and the latter is preferably diagonal. In the region $\omega > \omega_{A,\min}$, it may be desirable to let the outer M_A loop handle some of the d_B's over part of the frequency range (e.g., as in the example of Section 8.2). But then Eq. (10.18,3) must be used, where preferably (as discussed in Section 10.17) $P_B M_A P_A = P_A^{-1}(P_A P_B M_A) P_A$ is diagonal. These two are both diagonal only if the diagonal $P_A P_B M_A$ has equal elements. One possibility is to modify M_A over a transition frequency range such that below this range $P_A P_B M_A$ is approximately diagonal (the nondiagonal elements are relatively small) and above this range $P_B M_A P_A$ is approximately diagonal. This may possibly be practical only when all of the ω_A crossover frequencies are approximately equal. Another possibility (when the ω_A crossover frequencies are not concentrated in one region, and when it is clearly advisable to shift to M_B at higher frequencies) is to use M_A only for D_A attenuation, design a diagonal $P_A P_B M_A$ for this purpose, and let its individual diagonal elements decrease reasonably rapidly (without worrying about stability) beyond the respective individual crossover frequencies ω_{Ai}. The resulting nondiagonal $P_B P_A M_A$ is then evaluated and inserted in Eq. (10.18,3) for D_B attenuation. No attempt is made to modify M_A to obtain an approximately diagonal $P_B M_A P_A$ in the higher frequencies. Instead the $P_B M_B$ term is used to obtain satisfactory attenuation of D_B. With increasing frequency more of the elements in $P_B M_A P_A$ will hopefully become small, and then $P_B M_B$ dominates in Eq. (10.18,3) and can be diagonal. However, it should be noted that many elements of the nondiagonal $P_B M_A P_A$ may sometimes remain large right up to the largest ω_A crossover frequency. This is not all bad, in that these terms may help in Eq. (10.18,3) to attenuate D_B. However, nondiagonal $P_B M_B + P_B M_A P_A$ makes the disturbance attenuation design and the eventual design for stability much more difficult to perform. There is no problem, of course, when all the elements of the diagonal $P_A P_B M_A$ are equal, for then $P_B M_A P_A = P_A P_B M_A$.

Another way to deal with the problem is to forego the use of a diagonal $P_A P_B M_A$ in designing for D_A attenuation. Instead, one insists on a diagonal $K \triangleq P_B M_A P_A$ and designs the nondiagonal $P_A P_B M_A = P_A K P_A^{-1}$ to satisfactorily attenuate D_A in Eq. (10.18,1) (with $M_B \doteq 0$). It is then easy to work with Eq. (10.18,3) and make smooth transitions from $K = P_B M_A P_A$ to a diagonal $P_B M_B$. The disadvantage of the above is that, as noted in Section 10.17, the design can be inefficient when $P_A P_B M_A$ is nondiagonal.

The above discussion reveals the difficulties encountered in designing the simplest multivariable systems with internal variables available for feedback purposes. The problem is more difficult with other multivariable plants, for example, those of Fig. 10.18-2. As noted in Section 8.3, it is then highly desirable

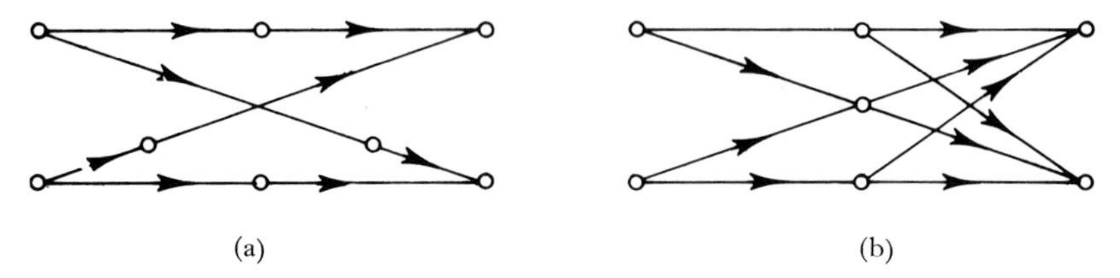

FIG. 10.18-2. Two other plant structures with internal variables available for feedback.

to drop the requirement of "no plant modification," and instead to permit the "feedback" aspects of the design to be considered in setting up the plant.

CHAPTER ELEVEN

Sampled-Data Feedback Control Systems

§ 11.1 Definition of Sampled-Data Systems

A sampled-data system (with or without feedback) is defined here as one in which the signal at one or more points consists of a discrete sequence of pulses. This is clarified in Fig. 11.1-1. In Fig. 11.1-1a the sampler is a switch which closes once every T seconds and remains closed for an interval of τ seconds (with $\tau < T$). Thus, if $f(t)$ is as shown in Fig. 11.1-1b, the sampler output, denoted by $f^*(t)$, is shown in Fig. 11.1-1c. In a sampled-data *feedback* system, one or more samplers is present in at least one closed loop.

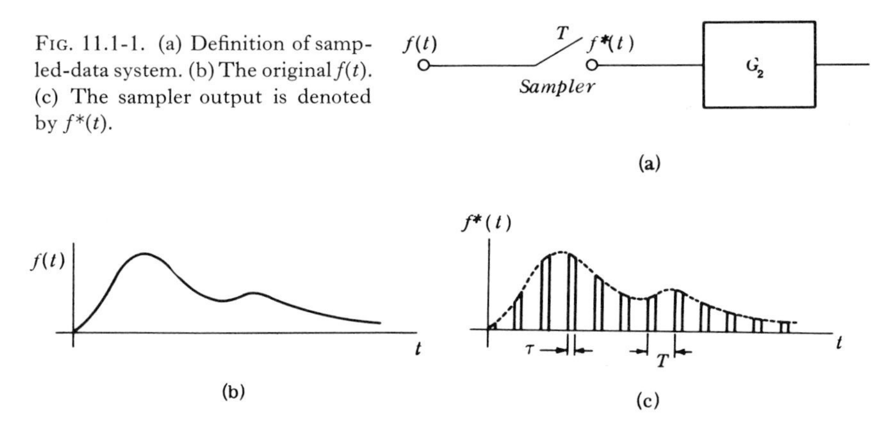

FIG. 11.1-1. (a) Definition of sampled-data system. (b) The original $f(t)$. (c) The sampler output is denoted by $f^*(t)$.

This chapter is concerned only with those systems in which the sampling frequency ω_s is constant, where $\omega_s \triangleq 2\pi/T$. It is also assumed that τ is much less than the effective time constants of the devices or networks into which the sampled signals are applied (e.g., G_2 in Fig. 11.1-1a). Therefore these devices see the input as a sequence of impulses with varying areas. In addition, it is assumed that the change in $f(t)$ over the interval τ is sufficiently small[1] that the

[1] Systems with finite sampling duration are treated in the following texts: E. I. Jury, "Sampled-Data Control Systems." Wiley, New York, 1958; J. T. Tou, "Digital and Sampled-Data Control Systems." McGraw-Hill, New York, 1959.

nth impulse area may be taken as $\tau f(nT)$, in place of $\int_{nT}^{nT+\tau} f(t)dt$. The output of the sampler is therefore $\tau \sum_{n=0}^{\infty} f(nT)\, \delta(t - nT)$, with $\delta(t - nT)$ denoting a unit impulse at $t = nT$. It is convenient to assign to every sampler an amplifier whose gain is $1/\tau$. Consequently, with the inclusion of this amplifier (which is henceforth assumed to be a part of every sampler), the sampler output is

$$f^*(t) = f(0)\delta(t) + f(T)\delta(t - T) + ... + f(nT)\delta(t - nT) + ...$$
$$= \sum_{n=0}^{\infty} f(nT)\delta(t - nT). \tag{11.1,1}$$

This series can also be expressed as the product of two functions, viz., $f^*(t) = f(t)\, m(t) = f(t)\, [m_1(t) + 0.5\delta(t)]$, where $m_1(t)$ is the infinite sequence of unit impulses spaced T seconds apart, as shown in Fig. 11.1-1d. The $m_1(t)$ series will be treated as the limit of a sequence of finite-width unit area pulses which are symmetrical with respect to the origin. The half-unit impulse at $t = 0+$ is therefore needed to make up for the fact that one-half the area of the $m_1(t)$ impulse at $t = 0$ really exists for negative t.

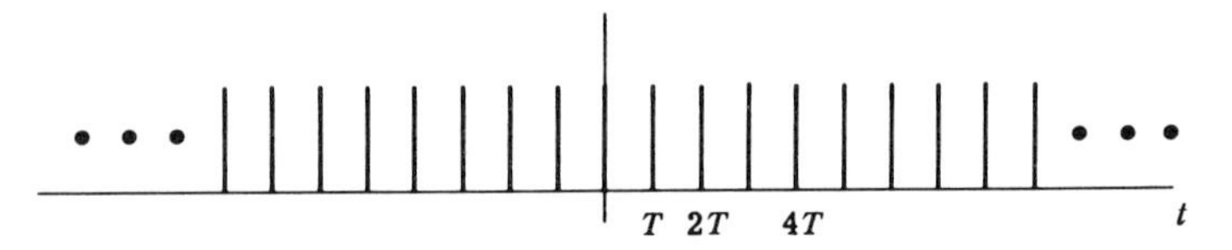

FIG. 11.1-1d. The infinite unit impulse sequence $m_1(t)$.

§ 11.2 The Importance of Sampled-Data Systems

What are the reasons for introducing a sampler into a system? Some of the principal reasons are as follows:

(1) Some systems are inherently of a sampled-data nature as, for example, search and tracking radars, the automatic periodic quality check in a production line, and any system in which the output is of a discrete rather than a continuous nature. Here too are included systems in which there is any time-sharing of a device, as, for example, in the telemetering of a variety of data over the same transmission channel. It will be seen that the feedback properties of sampled-data feedback systems are inferior in some respects to those of continuous-data systems. However, in many control systems, especially of the regulator type (where an output variable must be maintained at a constant value), the plant disturbances and parameter variations are so slow that continuous monitoring is unnecessary. In such a case, feedback transducers, data processing facilities, and possibly long and expensive feedback communication facilities can be shared among a number of control systems.

(2) A vital part of a feedback control system may consist of a very delicate and sensitive element, usually as the error sensing device, since the ultimate accuracy of the system is that of the error-sensing device. Part of the price paid for its high sensitivity may be the requirement that it be loaded only very lightly. Such light loading can be achieved by means of sampling.

(3) Systems which contain a digital computer are inherently sampled-data systems, because the digital computer operates on and yields output signals which are quantized both in magnitude and time. The magnitude quantization is ignored here—it is assumed that the quanta are sufficiently small to justify this.

There are several reasons why there is a tendency toward the use of digital computers in complex feedback control systems.

(1) In many complex control problems, which are usually multivariable in nature, there is a large variety of data received from the various sensors. This is especially true in chemical plants, steel mills, refineries, process controls, and wherever automation is attempted in an industry. It is also the case in the modern jet aircraft, in missile guidance, fire control, etc. This large amount of data must be processed in a variety of ways before it can be used for control purposes. Included among such data processing are calibration operations on the data, determination of maxima and minima, comparisons with stored data, and insertion of the data into equations to determine functions of interest. For example, in one oil refinery, over 100 process variables must be measured and operated upon in order to determine such operating guides as rates of feed of various catalysts and fuels.[1] The digital computer is essential for this type of data processing, and is therefore present in any case. The above data processing usually constitute open-loop operations, but the digital computer may also be used to realize the compensation networks which are needed in the closed-loop feedback portion of the systems. Also, in multivariable systems, the standard type of network synthesis of a matrix of transfer functions may be replaced by an equivalent digital compensator which may be programmed on a digital computer. Furthermore, the program (i.e., the compensation) can easily be adjusted and varied. The digital computer as a tool in the synthesis of compensation networks is not necessarily more economical that the more usual type of network consisting of amplifiers, resistors, capacitors, etc. (see Figs. 11.13-1 and 2). One must consider the specific problem. However, computer programming (which takes the place of network synthesis) is usually simpler to formulate and achieve. It is therefore convenient for experimentation purposes. After a final fixed design has been found satisfactory, it is probably more economical, in most cases, to use the usual network synthesis techniques.

(2) In the more sophisticated type of feedback systems, where one is interested in the optimum operation of a process, the digital computer is, in most cases,

[1] T. M. Stout, Analog or digital computer for process control. *Trans. IRE* **AC-3**, 3 (1957).

needed to perform the operations needed. Thus the contemplated complex optimum systems (of the future) are usually sampled-data systems. While such systems are not considered here, the sampled-data feedback theory that is presented in this chapter is a necessary prerequisite for the study of these advanced feedback systems.

(3) The quantization and coding of data permit a tremendous increase in the signal-to-noise ratio of a signal transmission channel. Basically, this is because the distortion and noise of any one link in the over-all chain is not passed on to the next, so that the signal-to-noise ratios of the individual links need not be many times greater than that desired for the over-all system.[1] In such transmission systems, there are very few elementary signal building blocks. The information content of the signal is in their number and arrangement. Signal reception consists of deciding whether the signal exists at all. If it is concluded that it does exist, the signal shape is regenerated before it is passed on to the next link. Such methods of information transmission, which among other things involve time sampling, are becoming of increasing importance.

The above should indicate the current (and especially future) importance of sampled-data systems. In this chapter the synthesis philosophy of previous chapters is extended to sampled-data systems, i.e., procedures are developed for realizing (1) a desired over-all system transfer function, (2) the required insensitivity of the above function to plant parameter variations, and (3) a desired insensitivity of the system output to disturbances affecting the plant.

It will be seen that the sampling constraint introduces definite limits on what may be achieved in each of the above three design requirements. This is in contrast with the continuous type system where there is no theoretical limit (in minimum phase systems) on what can be achieved in each of the above three, providing only that the plant has the physical capacity to achieve the objectives and that compensating elements with the required gain-bandwidth products can be found. Even if the latter two requirements present no problem in the sampled-data system, there are limits as to what may be achieved. One of the important objectives of sampled-data feedback control theory is to clarify what these limits are. The effects of sampling on the last two design objectives are treated in later sections. The effect on the first is considered in the next section.

It is worthwhile noting, however, that if sampling is inherent only in the filter portion of the system, then it is not necessarily wise to introduce the sampler into the feedback part of the system. Thus, suppose the input command signal is constrained to be of a sampled nature and feedback is used because of plant parameter variations and plant disturbances; but there is no inherent constraint which compels a sampler to be present in the closed loop. In such a case the structure in Fig. 11.16-2 is superior (with respect to the system feedback properties) to the one shown in Fig. 11.15-1.

---

[1] W. R. Bennett, Synthesis of active networks. *In* "Proceedings of the Symposium on Modern Network Synthesis" (J. E. Fox, ed.), pp. 45-61. Interscience, New York.

§ 11.3 Effect of Sampling on Achievable System Accuracy

It is generally desired that the control system output follow the input signal, or some function of the input signal, with a minimum error. One possible source of error may be due to the sheer physical inability of the plant to follow a signal which is changing too quickly. But suppose that the plant does have the physical capacity to do this. In such a case, in the continuous system, the error can theoretically be made as small as the resolution of the error-sensing device (if the input is free of noise). This is not the case in the sampled-data system.

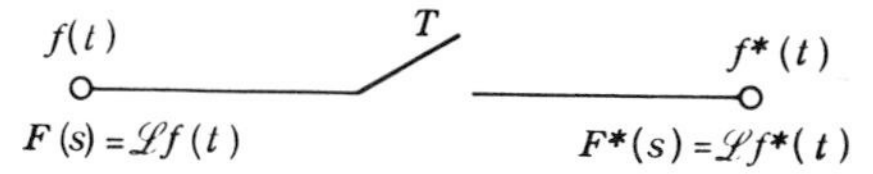

$f(t)$ T $f^*(t)$

$F(s) = \mathcal{L}f(t)$ $F^*(s) = \mathcal{L}f^*(t)$

FIG. 11.3-1. Definition of terms.

Thus, consider the sampled signal $f^*(t)$ in Fig. 11.3-1, where $f^*(t) = f(t)\,m(t)$ and $m(t) - \frac{1}{2}\delta(t) = m_1(t)$ is the impulse sequence of Fig. 11.1-1d. The latter can be represented by its Fourier series, viz., $m_1(t) = \sum_{-\infty}^{\infty} M_n e^{jn\omega_s t}$ with $\omega_s \triangleq 2\pi/T$ and $M_n = T^{-1}\int_0^T m_1(t)\, e^{-jn\omega_s t}\, dt$. It is easily found that $M_n = 1/T$ for all n, so that $f^*(t) = f(t)\; T^{-1}\sum_{-\infty}^{\infty} e^{jn\omega_s t} + 0.5\, f(0+)\; \delta(t)$. Let $F(s)$, $F^*(s)$ denote the Laplace transforms of $f(t)$, $f^*(t)$, respectively. Therefore

$$F^*(s) = \mathcal{L}f^*(t) = T^{-1}\mathcal{L}\{f(t) + f(t)\, e^{j\omega_s t} + f(t)e^{-j\omega_s t} + f(t)e^{j2\omega_s t} + \ldots\} + \tfrac{1}{2}f(0+)$$
$$= T^{-1}\{F(s) + F(s - j\omega_s) + F(s + j\omega_s) + F(s - j2\omega_s) + \ldots\} + \tfrac{1}{2}f(0+).$$

Consequently, if $f(0+) = 0$,

$$\boxed{F^*(s) = \frac{1}{T}\sum_{-\infty}^{\infty} F(s + jn\omega_s)} \tag{11.3,1}$$

The transform of a sampled signal thus consists of an infinite series of terms, each of which is derived from the transform of the original signal. Clearly, $F^*(s)$ is periodic in s with period $j\omega_s$ because, the series being infinite, $F^*(s_1) = F^*(s_1 + j\omega_s)$. The poles of $F^*(s)$ are therefore infinite in number and are those of $F(s)$ repeated over and over again at intervals of $j\omega_s$. Thus, if the poles of $F(s)$ are as shown in Fig. 11.3-2a, those of $F^*(s)$ are shown in Fig. 11.3-2b. The zeros of $F^*(s)$ are not those of $F(s)$, but they too must be infinite in number and periodically spaced, in order that $F^*(s)$ may be periodic.

Equation (11.3,1) for $F^*(s)$ may be used to obtain graphically the value of $F^*(s)$ along the line $s = j\omega$, i.e., the frequency spectrum of the sampled signal. For example, if $F(s) = 1/s(s + 1)$ with the Nyquist sketch shown in Fig. 11.3-3a, then to find $F^*(j\omega)$, one simply performs graphically the summation indicated

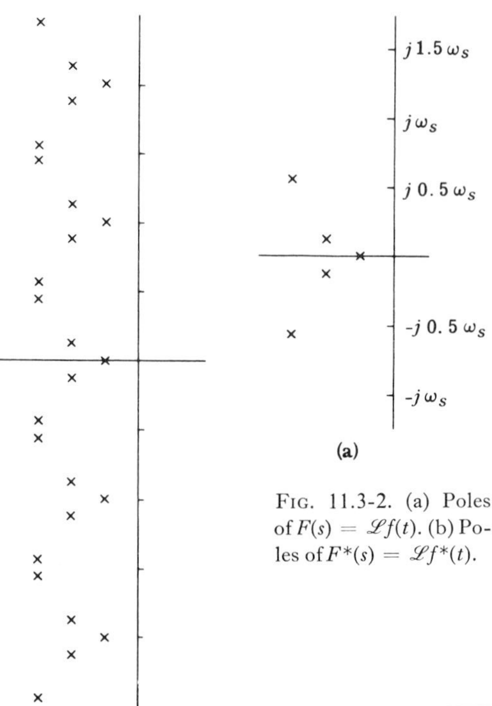

FIG. 11.3-2. (a) Poles of $F(s) = \mathscr{L}f(t)$. (b) Poles of $F^*(s) = \mathscr{L}f^*(t)$.

by Eq. (11.3,1). For example, if $\omega_s = 2\pi/T = 4$ rps, then $TF^*(j1) = F(j1) + F(j5) + F(-j3) + F(j9) + F(-j7)$... and these terms are added graphically as shown in Fig. 11.3-3b. $F^*(j0.5\omega_s)$ and $F^*(0)$ are real, unless $F(s)$ has an odd number of poles or zeros at $s = j0.5n\omega_s$ with n zero or an integer.

The addition of the infinite number of periodic spectra is also shown in Fig. 11.3-4. In Fig. 11.3-4a a typical $|F(j\omega)|$ is displayed. To find $F^*(j\omega)$, an infinite number of these, properly spaced, as in Fig. 11.3-4b, must be added in accordance with Eq. (11.3,1). The result is qualitatively as shown by the curve of Fig. 11.3-4c.

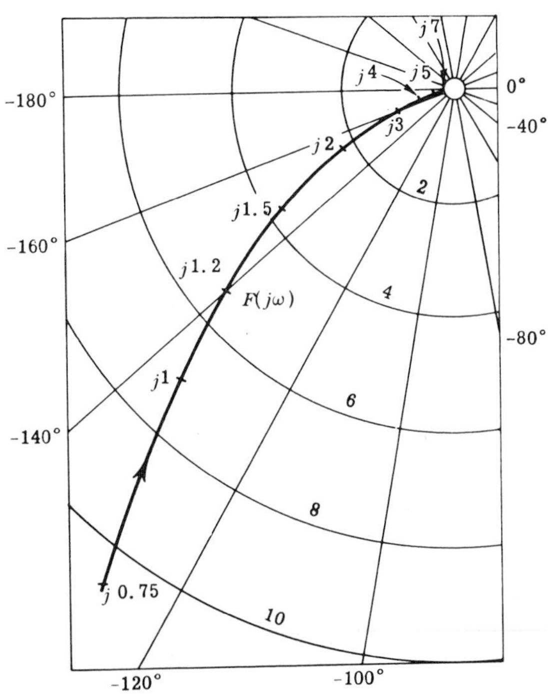

FIG. 11.3-3. (a) Nyquist sketch of $F(j\omega)$.

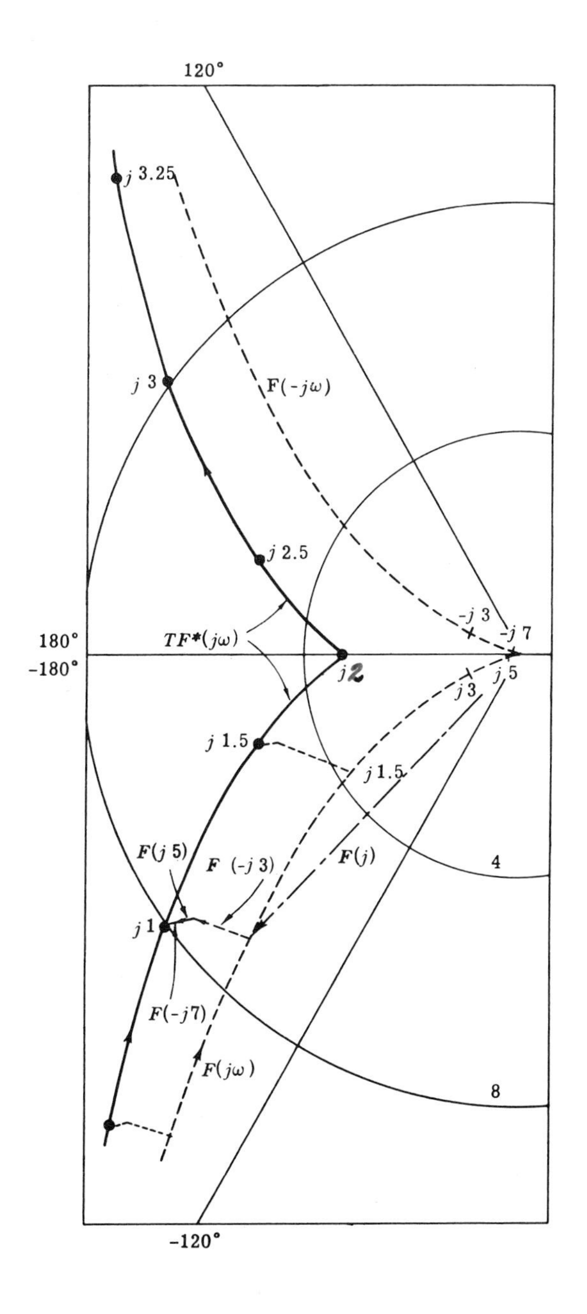

FIG. 11.3-3. (b) Nyquist sketch of $TF^*(j\omega)$.

Clearly, for the case shown in Fig. 11.3-4c, it is not possible to pass the signal represented by $F^*(j\omega)$ through any frequency filter and recover $F(j\omega)$ [unless the original signal $f(t)$ is known *a priori*, which is obviously a trivial case]. The regions of overlap in Fig. 11.3-4b obscure the original value of $F(j\omega)$ in the final sketch of $F^*(j\omega)$ in Fig. 11.3-4c. However, if $F(j\omega)$ is bandwidth limited, so $F(j\omega) = 0$ for $\omega > \omega_o$, as in Fig. 11.3-5a, then, so long as $0.5\omega_s \geqslant \omega_o$, there is no region of overlap in the addition operation of Eq. (11.3,1) (see Fig. 11.3-5b). In such a case, it is possible to recover the original signal $f(t)$ by passing $f^*(t)$ through the ideal, low-pass filter $H(s)$ of Fig. 11.3-6, where $\omega_y > \omega_x > 2\omega_o$ (with ω_y as defined in Fig. 11.3-5b). This filter (if its phase is zero) has the impulse response[1] $h(t) = (\sin 0.5\omega_x t)/0.5\omega_x t$. When the input to the filter is the series $f^*(t) = \sum_0^\infty f(nT)\,\delta(t - nT)$, the filter output is

$$f_o(t) = h(t) * [f^*(t)] = \Big[\sum_0^\infty f(nT)\delta(t - nT)\Big] * \Big[\frac{\sin 0.5\omega_x t}{0.5\omega_x t}\Big]$$

$$= \sum f(nT)\frac{\sin 0.5\omega_x(t - nT)}{0.5\omega_x(t - nT)}. \qquad (11.3,2)$$

The latter must be exactly $f(t)$ because its transform is $H(s)\,F^*(s)$ which is exactly equal to $F(s)$. This constitutes proof of the *sampling theorem*, which states that a bandwidth limited signal (whose spectrum is zero for $\omega > \omega_o$, with ω_o finite) is completely characterized by its samples, providing the sampling frequency $\omega_s \geqslant 2\omega_o$; and the formula for recovering the original signal from its samples is Eq. (11.3,2).

It follows from the above that if the input signal to the control system is not contaminated with noise, if it is bandwidth limited and is sampled, then the ideal frequency response characteristic of the control system is that shown in Fig. 11.3-6. However, it is seen from Eq. (11.3,2) that any bandwidth limited signal must extend over infinite time ($-\infty$ to ∞), because $(\sin t)/t$ extends over infinite time. No signal of finite duration is therefore bandwidth limited, and there must be some loss of frequency content (and consequently of the signal) by sampling and ideal filtering.

An upper bound for the resulting instantaneous error due to sampling and ideal filtering has been derived[2] and is as follows. The maximum error is given in terms of parameters of the Fourier transform of the signal, $f(t)$. If $g(\omega) = (2\pi)^{-0.5}\int_{-\infty}^{\infty} f(t)\,e^{-j\omega t}\,dt$ is the transform of $f(t)$, and if there exist $M_1 > 0$, $M_2 > 0$, $K_1 > 2$, $K_2 > 2$, $\omega_o > 0$, such that $|\,g(\omega)\,| < M_1\,|\,\omega\,|^{-K_1}$

[1] E. A. Guillemin, "Communications Networks," Vol. 2, p. 477. Wiley, New York, 1947.

[2] K. R. Johnson, On sin x/x sampling in the case of non-band-limited functions. *TR* 195. Lincoln Labs., M.I.T., Cambridge, Massachusetts, Feb. 16, 1959.

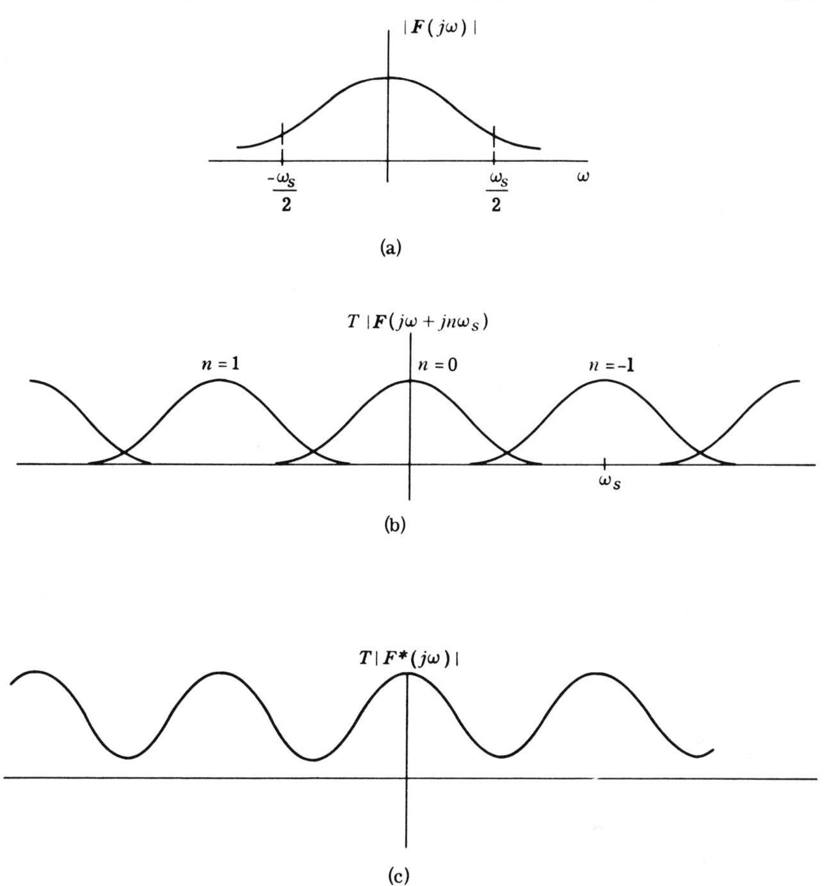

FIG. 11.3-4. Effect of sampling on signal frequency response; case where original signal is not recoverable.

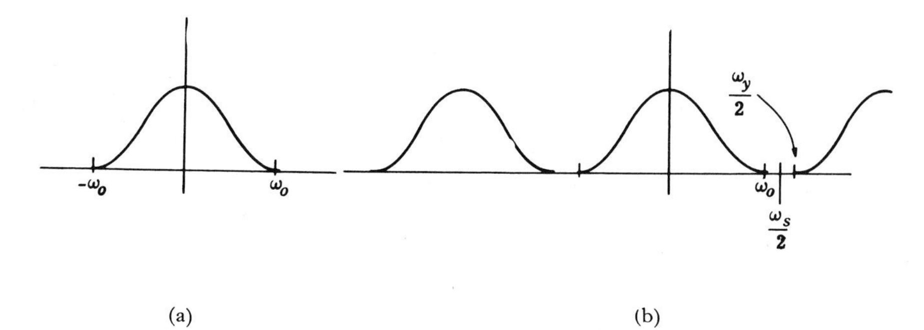

FIG. 11.3-5. Effect of sampling on signal frequency response; case where original signal is recoverable.

and $| g'(\omega) | < M_2 | \omega |^{-K_2}$ for $\omega \geqslant \omega_o$, and if $f(t)$ is sampled at the frequency $\omega_s = 2\pi/T$ rps, with $\omega_s \geqslant 2\omega_o$, then

$$| f(t) - f_o(t) | < 6M_1(0.5\omega_s)^{1-K_1} + M_2(0.5\omega_s)^{2-K_2} \frac{(K_2 + 4)}{(K_2 - 1)} \quad (11.3,3)$$

where $f_o(t)$ has been defined in Eq. (11.3,2). For example, suppose $g(\omega)$ falls off at the rate of $1/\omega^3$ beyond $\omega = 5$ rps, and $| g(j5) | = 0.1$. Clearly, $| g'(\omega) |$ must therefore decrease at the rate of $3/\omega^4$ for $\omega > 5$. Consequently, $K_1 = 3$, $K_2 = 4$, $M_1 = 12.5$, $M_2 = 3$, and an upper bound (the error may never have this value, however) of the approximation error for any t, using Eq. (11.3,3), is $83/\omega_s^2$.

There is another error in addition to the above, and that is due to the unrealizability of the ideal filter of Fig. 11.3-6 (it is not realizable because it has a response before the input is applied). In addition, the effect of noise superimposed on the useful signal must be considered. Finally, if there is sampling in the feedback loop, then there are constraints on the admissible form of the optimum filter. The optimum response function for the minimization of the mean square error can be found by means of statistical techniques, and this is done in Sections 9.18 and 9.19.

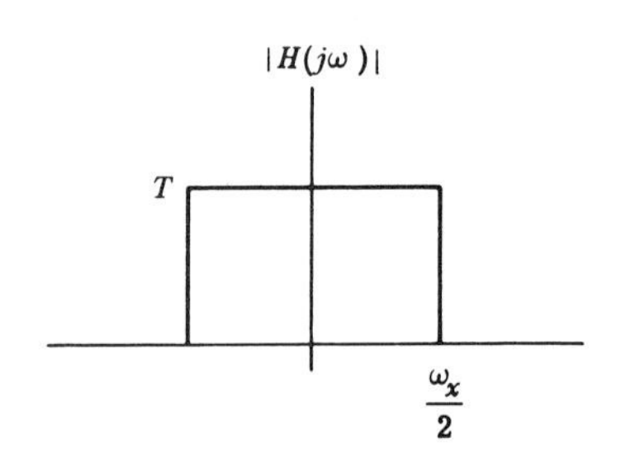

FIG. 11.3-6. Ideal low pass filter characteristic.

This section has treated the limitations due to sampling on the accuracy with which a sampled-data system can follow an input signal, i.e., the effect of sampling on the filter problem. It is impossible, as yet, to consider the limitations due to sampling on the other two design objectives (i.e., the feedback objectives) listed in the latter part of Section 11.2. Before this can be done, a considerable number of theoretical tools must first be developed. These tools are developed in the following sections.

§ 11.4 z Transforms[1]

Consider again $F^*(s)$, the Laplace transform of the sampled signal $f^*(t)$ of Fig. 11.3-1. From Eq. (11.3,1), $F^*(s) = T^{-1} \sum_{n=-\infty}^{\infty} F(s + jn\omega_s)$. This infinite

[1] Sections 11.4 and 11.5 are based entirely on the following references: L. A. MacColl, "Fundamental Theory of Servomechanisms." Van Nostrand, New York, 1945; W. Hurewicz, *in* "Theory of Servomechanisms" (H. James, N. Nichols, and R. Phillips, eds.), Chapter 5, Radiation Laboratory Series, Vol. 25, McGraw-Hill, New York, 1947; D. F. Lawden, A general theory of sampling servosystems. *Proc. IEE* (*London*), *Part IV*. **98**, 31 (1951); R. H. Barker, The pulse transfer function and its application to sampling servo systems. *Proc. IEE* (*London*), *Part IV*, Monograph 43. **99**, 302-317 (1952); J. R. Ragazzini and L. A. Zadeh, The analysis of sampled-data systems, *Trans. AIEE, Part II. Applications & Ind.* **71**, 225-232 (1952).

series is inconvenient to work with. A much more convenient closed form for $F^*(s)$ can be obtained by using Eq. (11.1,1), $f^*(t) = \Sigma_{n=0}^{\infty} f(nT)\, \delta(t - nT)$, and transforming term by term. The result is

$$F^*(s) = \sum_{0}^{\infty} f(nT)e^{-nsT} = \sum_{0}^{\infty} f(nT)z^{-n}, \qquad \text{with } z = e^{sT}. \qquad (11.4,1)$$

There is usually no difficulty in obtaining a closed form for the sum of the infinite series of Eq. (11.4,1). For example, if $f(t) = e^{-at}$, then $f^*(t) = \Sigma_0^{\infty} e^{-anT} \delta(t - nT)$, and $F^*(s) = 1 + e^{-(a+s)T} + e^{-2(a+s)T} + ... + e^{-n(a+s)T} + ... = 1/(1 - e^{-sT} e^{-aT})$. This closed form applies only for $| e^{-(s+a)T} | < 1$, i.e., $\sigma > - a$ (where $s + \sigma + j\omega$), which is precisely the region in the s plane for which $F(s)$ is defined.[1] The transforms of most sampled time functions of interest can be obtained in closed form in the above manner.

One more simplification may be made. Even though $F^*(s)$ is in closed form, it has the inconvenient properties of periodicity, with an infinite number of poles, when the variable is s. One can avoid this by using the new variable z, defined as

$$z \triangleq e^{sT}. \qquad (11.4,2)$$

Now $F^*(s)$ becomes a function of z, and in the literature is denoted by $F(z)$. It is important to note that $F(z) \neq [F(s)]_{s=z}$; rather, $F(z) = [F^*(s)]_{z=e^{sT}}$. For example, if $f(t) = e^{-at}$, with $F^*(s) = (1 - e^{-sT} e^{-aT})^{-1}$, then $F(z) = (1 - z^{-1}e^{-aT})^{-1} = z/(z - e^{-aT})$. *Thus the z transform of a signal is the Laplace transform of the sampled signal with z substituted for e^{sT}.* The z transforms of some commonly used signals are given in Table 11.4-1.

$F(z)$, a rational function in z, has a finite number of poles and zeros. It is obviously much easier to work with than $F^*(s)$. It should be noted that any transform is, in a sense, a code only for the real thing, which is the time function itself. Engineers who have worked with the code known as the Laplace transform, have become so familiar with it that from the properties of the transform they have a good feeling for the time function represented. If the z transform is analytically more convenient for sampled-data systems and is to be used, the engineer should obtain the same familiar feeling for this new code.

Such a familiarity can be established by relating important concepts and regions in the s domain to their equivalent in the z domain. The $s = j\omega$ axis is readily found to correspond to the unit circle $| z | = 1$ in the z domain, and the left half of the s plane maps an infinite number of times into the interior of the

[1] It is seen that the above is identical to Eq. (11.3,1) if $0.5f(0 +) = 0.5$ is added to the latter by expanding the former in partial fractions. The poles of the former are at $s = - a \pm jn\omega_s$, at which the residues are all $1/T$. Also at $s = 0$ the former is $(1 - e^{-aT})^{-1}$ which is equal to $0.5 + T^{-1} \Sigma_{-\infty}^{\infty} (a + jn\omega_s)^{-1}$. Therefore the above, when thus expanded, is precisely $(1/T) \Sigma (s + a + jn\omega_s)^{-1} + 0.5$.

TABLE 11.4-1

z TRANSFORMS

$f(t)$ (zero for $t < 0$)	$F(s)$	$F(z)$
$\delta(t)$	1	1
$u(t)$	$\frac{1}{s}$	$\frac{z}{z-1}$
t	$\frac{1}{s^2}$	$\frac{Tz}{(z-1)^2}$
$\frac{t^2}{2!}$	$\frac{1}{s^3}$	$\frac{T^2 z(z+1)}{2(z-1)^3}$
$\frac{t^3}{3!}$	$\frac{1}{s^4}$	$\frac{T^3 z(z^2+4z+1)}{6(z-1)^4}$
$\frac{t^n}{n!}$	$\frac{1}{s^{n+1}}$	$\lim_{\alpha\to 0} \frac{(-1)^n}{n!} \frac{\partial^n}{\partial \alpha^n}\left[\frac{z}{z-e^{-\alpha T}}\right]$
e^{-at}	$\frac{1}{s+a}$	$\frac{z}{z-e^{-aT}}$
te^{-at}	$\frac{1}{(s+a)^2}$	$\frac{Tze^{-aT}}{(z-e^{-aT})^2}$
$\frac{\partial^{n-1}}{\partial a^{n-1}} e^{-at}$	$\frac{(n-1)!}{(-1)^{n-1}} \frac{1}{(s+a)^n}$	$\frac{\partial^{n-1}}{\partial a^{n-1}} \frac{1}{1-z^{-1}e^{-aT}}$
$\sin \omega_1 t$	$\frac{\omega_1}{s^2+\omega_1^2}$	$\frac{z \sin \omega_1 T}{z^2 - 2z\cos\omega_1 T + 1}$
$\cos \omega_1 t$	$\frac{s}{s^2+\omega_1^2}$	$\frac{z(z-\cos\omega_1 T)}{z^2 - 2z\cos\omega_1 T + 1}$
$e^{-\zeta\omega_n t} \sin \omega_n t\sqrt{1-\zeta^2}$	$\frac{\omega_n\sqrt{1-\zeta^2}}{s^2+2\zeta\omega_n s+\omega_n^2}$	$\frac{ze^{-\zeta\omega_n T}\sin\omega_n T\sqrt{1-\zeta^2}}{z^2-2ze^{-\zeta\omega_n T}\cos\omega_n T\sqrt{1-\zeta^2}+e^{-2\zeta\omega_n T}}$
$e^{-\zeta\omega_n t} \cos \omega_n t\sqrt{1-\zeta^2}$	$\frac{s+\zeta\omega_n}{s^2+2\zeta\omega_n s+\omega_n^2}$	$\frac{z(z-e^{-\zeta\omega_n T}\cos\omega_n T\sqrt{1-\zeta^2})}{z^2-2ze^{-\zeta\omega_n T}\cos\omega_n T\sqrt{1-\zeta^2}+e^{-2\zeta\omega_n T}}$
$e^{-\alpha T}f(t)$	$F(s+\alpha)$	$F(ze^{\alpha T})$
$\frac{\partial}{\partial\alpha} f(t,\alpha)$	$\frac{\partial}{\partial\alpha} F(s,\alpha)$	$\frac{\partial}{\partial\alpha} F(z,\alpha)$
$f(t-nT)$	$e^{-nsT}F(s)$	$z^{-n}F(z)$

unit circle in the z domain (Figs. 11.4-1a and b). Consequently, for a system to be stable, the z transform of its impulse response must have no poles outside the unit circle. The Nyquist criterion applies in the z domain with the critical boundary as the unit circle rather than the imaginary axis.

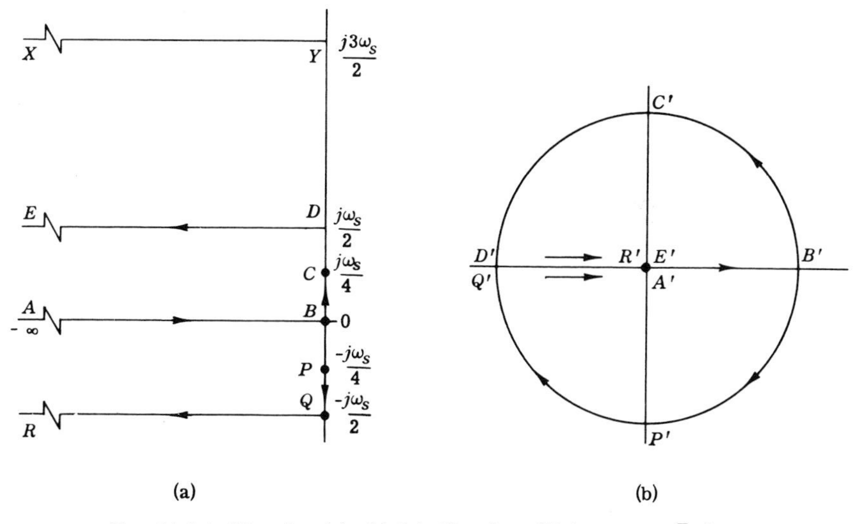

FIG. 11.4-1. Mapping (a) of left half-s-plane (b) into $z = e^{sT}$ plane.

Every one of the left half-plane (lhp) strips of Fig. 11.4-1a, $j\omega_s$ wide, must map into the unit circle in the z domain, in order that the infinite number of poles of $F^*(s)$ in the s plane should give a finite number of poles in the z plane. Thus the strips $XYDE$ and $EDQR$ in the s plane map identically on the unit circle in the z plane. A time function which goes to zero at infinity must have its z transform poles inside the unit circle. Poles on the boundary correspond to signals that are sinusoidal or steps, ramps, etc.

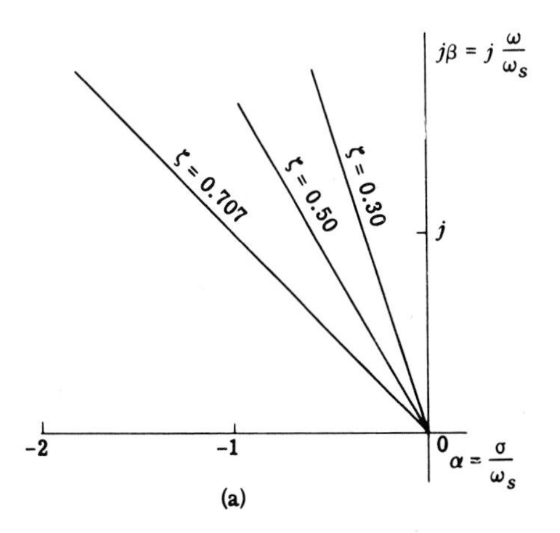

FIG. 11.4-2. Details of mapping from normalized s plane into z plane. (a) Normalized s plane.

Figures 11.4-2a and b are useful for relating pole positions in the z plane to their equivalents in the s plane, and vice versa. In Fig. 11.4-2a, the s plane is normalized with respect to the sampling frequency $\omega_s = 2\pi/T$, so that the

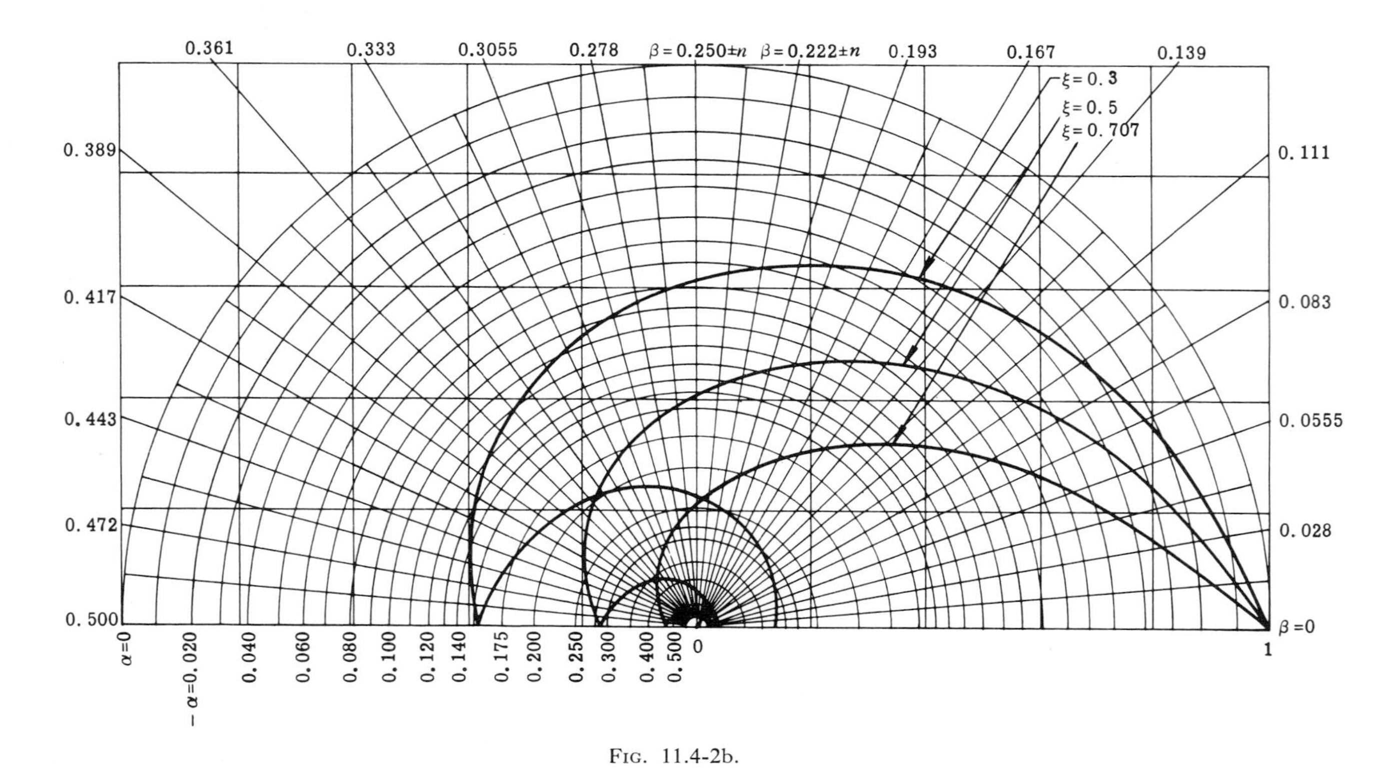

FIG. 11.4-2b.
z plane with loci of constant α, β and ξ.

normalized s plane's axes are $\alpha = \sigma/\omega_s$ and $j\beta = j\omega/\omega_s$. The loci of constant α and β in the s plane map as shown in Fig. 11.4-2b. The loci of constant damping factor in the s plane map as indicated in the z plane. One must be careful in interpreting a pole-zero pattern in the z plane. In the interpretation of a pole-zero pattern in the s plane, one notes the poles and zeros, and mentally performs a partial fraction expansion of the transfer function. The time function is then visualized as a series of terms, each term corresponding to a real pole or a pair of complex poles. In the z domain, for a similar sort of interpretation, one wants the following type of expansion: $\Sigma A_i z/(z - a_i) + \Sigma B_j z/[(z - a_j)^2 + b_j^2]$. Each term leads to a series of impulses (see Table 11.4-1), but the envelope of each series is the analogous continuous time function. Thus, $A_i z/(z - a_i)$ gives an infinite series of impulses whose envelope is $A_i \exp [(t/T) \ln a_i]$. For example, $z/(z - 0.502)$, according to Fig. 11.4-2b, corresponds to $\alpha = -0.11$, $\beta = 0$, and its impulse series is therefore $\Sigma_{n=0}^{\infty} \exp [(-0.11)\, \omega_s nT]\, \delta(t - nT) = \Sigma_{n=0}^{\infty} \exp [(-0.11)\, 2\pi n]\, \delta(t - nT)$.

The z transform $z/[(z - a)^2 + b^2]$ can be written as

$$(z/2jb)\,\{[z - (a + jb)]^{-1} - [z - (a - jb)]^{-1}\};$$

this corresponds in the s domain to

$$(2jb)^{-1}\,\{[s - (\alpha + j\beta)\,\omega_s]^{-1} - [s - (\alpha - j\beta)\omega_s]^{-1}\},$$

whose inverse transform is $f(t) = [e^{\alpha\omega_s t}/b] \sin \beta\omega_s t$. The sampled (often called the pulsed) time function is therefore $f(t)$, evaluated at $t = nT$. For example, if $T = 1$, $a = 0.52$, $b = 0.3$, then, from Fig. 11.4-2b (entering at $z = 0.52 + j0.3$), $-\alpha = \beta = 0.083$, and the inverse transform is $(1/0.3)\, \Sigma_{n=0}^{\infty} \{\exp [(-0.038)\, 2\pi n] \sin [(0.083)\, 2\pi n]\delta(t - nT)\}$.

Note that the mapping of a line $\beta = K$ from the s plane (K being any number) into the z plane is indistinguishable from the mapping of any line $\beta = K \pm \nu$ (ν being any integer). Thus, a pole pair $(-\alpha \pm jK)\, \omega_s$ in the s plane maps into exactly the same points in the z plane as the pole pair $[-\alpha \pm j(K + \nu)]\, \omega_s$. Although the inverse transforms in s result in different continuous time functions, the sampled values of these functions are the same. One may therefore take any integral value of ν. For example, $z/[(z - 0.39)^2 + (0.46)^2]$ can correspond in the s domain to $(0.139\omega_s/0.46)\, [(s + 0.123\omega_s)^2 + (0.139\omega_s)^2]^{-1}$ [the sampled values of $(1/0.46) \exp (-0.123\omega_s t) \sin 0.139\omega_s t$], or it can correspond in the s domain to $(1.139\omega_s/0.46)\, [(s + 0.123\omega_s)^2 + (1.139\omega_s)^2]^{-1}$ [the sampled values of $(1/0.46) \exp (-0.123\omega_s t) \sin 1.139\omega_s t$]. The sampled values of these two time functions are identical. Similarly, $z/(z^2 - z + 0.5)$ corresponds (for $T = 1$) to $F(s) = (\pi/2)/[(s + 0.347)^2 + (\pi/4)^2]$, with envelope $2 \exp (-0.347t) \sin (\pi/4)t$, or to $F(s) = 4.5\pi/[(s + 0.347)^2 + (9\pi/4)]$,[2] with envelope $2 \exp (-0.347t) \sin (9\pi/4)\, t$. The two envelopes are compared in

Fig. 11.4-3. It is seen that their sampled values (at $t = nT$, with $T = 1$) are identical, although their intersampled values radically differ. Thus the z transform by itself can be used only to find the sampled values of a signal. Additional information is needed to evaluate the intersample values. This problem will be discussed in Section 11.7.

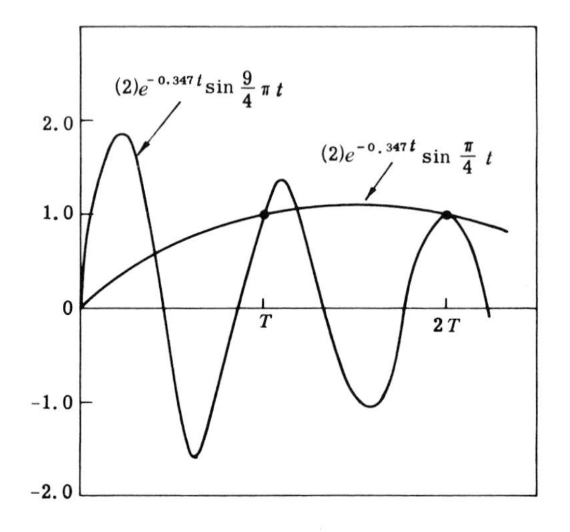

FIG. 11.4-3. Two different signals with identical sampled values and z transforms.

The z transform pole-zero pattern can therefore be mentally expanded in the above manner and the individual terms correlated with the s domain and the corresponding transients. One looks, of course, for dominant poles, which, from Fig. 11.4-2, are clearly those nearest the unit circle. The relative magnitudes of the residues must, of course, be taken into account, and for this the positions of the zeros are important.

Another point to note is that while the poles of $F(s)$ map directly as poles of $F(z)$ by the transformation $z = e^{sT}$, the zeros of $F(s)$ do not map into the z plane as zeros of $F(z)$. Thus, given $f(t)$ with transform $F(s) = (s + a)/(s + p_1)(s + p_2)$, the z transform of $f(t)$ [i.e., the transform of $f^*(t)$ in the z domain] has poles at $z = e^{-p_1 T}$ and $z = e^{-p_2 T}$. However, $F(z)$ does not have a zero at e^{-aT}. To find $F(z)$, write

$$\begin{aligned} F(s) &= (s + a)/(s + p_1)(s + p_2) \\ &= (p_2 - p_1)^{-1} [(a - p_1)(s + p_1)^{-1} - (a - p_2)(s + p_2)^{-1}], \end{aligned}$$

and from Table 11.4-1,

$$F(z) = (p_2 - p_1)^{-1} [z(a - p_1)(z - e^{-p_1 T})^{-1} - z(a - p_2)(z - e^{-p_2 T})^{-1}],$$

with a zero at $[(a - p_1) e^{-p_2 T} - (a - p_2) e^{-p_1 T}]/(p_2 - p_1)$.

§ 11.5 z Transforms and the Pulse Transfer Function

The real usefulness of the z transform appears when one considers the outputs of networks or devices to which are applied pulsed inputs, i.e., inputs consisting of sampled time functions. In Fig. 11.5-1, $C(s) = G(s)\,F^*(s)$. $C(s)$,

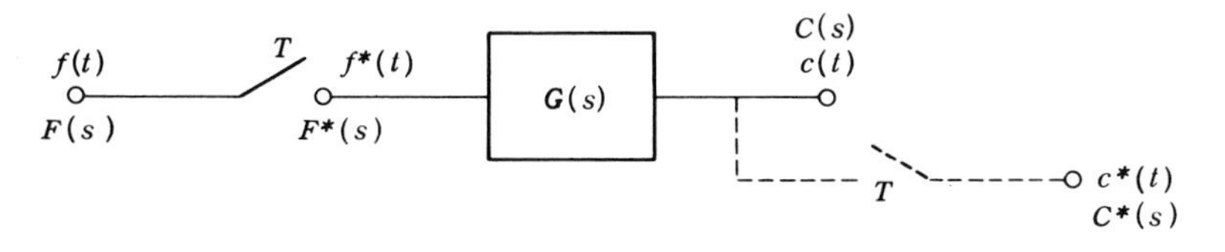

FIG. 11.5-1. Definition of pulse transfer function C^*/F^*

which is the Laplace transform of $c(t)$, is a function of the variables s and e^{sT}. For example, if $f(t) = e^{-aT}$ and $G(s) = 1/(s+b)$, then $C(s) = e^{sT}/(e^{sT} - e^{-aT})\,(s+b)$. Even for this very simple problem, there results an expression for $C(s)$ from which it appears extremely diffcult to extract any useful information. Basically, it is difficult to work with because the output consists of a succession of transients (since the network is driven by a series of impulses), and $C(s)$ contains within it *all* the information about the *complete* function $c(t)$. However, if one is willing to settle for only *part* of the information about $c(t)$, it is possible to obtain an expression from which one may extract fairly easily this *partial* information about $c(t)$.

Suppose that only the values of the output at $t = 0, T, 2T, 3T$, etc., are being sought. For this purpose, a fictitious sampler is attached to the output, as shown by the dashed lines in Fig. 11.5-1. This fictitious sampler is imagined to be in exact synchronism with the real sampler which samples $f(t)$. Now consider $C^*(s)$ which is the transform of $c^*(t)$. Recall Eq. (11.3,1) which is: $F^*(s) = T^{-1}\sum_{-\infty}^{\infty} F(s + jn\omega_s)$; so $C^*(s) = T^{-1}\sum_{-\infty}^{\infty} C(s + jn\omega_s)$, where $C(s) = G(s)\,F^*(s) = T^{-1}\,G(s)\sum_{-\infty}^{\infty} F(s + jn\omega_s)$. Therefore

$$C^*(s) = \frac{1}{T^2}\Big\{G(s)\sum_{-\infty}^{\infty} F(s + jn\omega_s) + G(s + j\omega_s)\sum_{-\infty}^{\infty} F[s + j(n+1)\omega_s] + \dots$$

$$+ G(s + j\nu\omega_s)\sum_{-\infty}^{\infty} F[s + j(n+\nu)\omega_s] + \dots + \dots$$

$$+ G(s - j\nu\omega_s)\sum_{-\infty}^{\infty} F[s + j(n-\nu)\omega_s] + \dots\Big\}.$$

However, $\sum_{-\infty}^{\infty} F(s + jn\omega_s) = \sum_{-\infty}^{\infty} F[s + j(n+1)\,\omega_s] = \sum_{-\infty}^{\infty} F[s + j(n+\nu)\,\omega_s]$, etc., because the summation is over infinite n. Consequently, $C^*(s) = T^{-2} \sum_{-\infty}^{\infty} F(s + jn\omega_s) \sum_{-\infty}^{\infty} G(s + j\nu\omega_s) = F^*(s) G^*(s)$. Thus,

$$\text{if} \quad C(s) = G(s) F^*(s), \quad \text{then} \quad C^*(s) = G^*(s) F^*(s) \quad (11.5,1)$$

or, in terms of z, if $c(t) = \mathscr{L}^{-1} C(s) = \mathscr{L}^{-1} G(s) F^*(s)$, then the transform of $c^*(t)$ is $C(z) = F(z) G(z)$. $G^*(s) = T^{-1} \sum_{-\infty}^{\infty} G(s + jn\omega_s)$ and its equivalent $G(z)$ is called the *pulse transfer function* (P.T.F.) of the network represented by $G(s)$. [Again the reader is cautioned: $G(z) \neq [G(s)]_{s=z}$; rather, $G(z) = [G^*(s)]_{z=e^{sT}}$. $G^*(s)$ is the Laplace transform of the sampled impulse response of the $G(s)$ network.] $G(z)$, which is often denoted as $ZG(s)$, is also sometimes called the z transform of the network, although, strictly, it is the z transform of the impulse response of the network. The P.T.F. thus relates the transform of a pulsed output to that of a pulsed input and to the z transform of the network's impulse response. *It is relevant only when the input is truly pulsed.*

Equation (11.5,1) is very important, for thanks to it z transforms offer a great advantage over Laplace transforms in sampled-data systems. In words, when a sampled (pulsed) time function is applied to a network, the *sampled* values of the output (the pulsed output) are related to the sampled input by the P.T.F. of the network. The P.T.F. is the z transform of the impulse response of the network.

The following relations are easily established:

$$Z[G_1(s) + G_2(s)] = Z[G_1(s)] + Z[G_2(s)] = G_1(z) + G_2(z) = [G_1(s) + G_2(s)]^*,$$

$$Z[G_1(s)\, G_2(s)] = Z[G_2(s)\, G_1(s)] \text{ which is } \textit{not}\text{, in general, equal to } G_1(z)\, G_2(z).$$

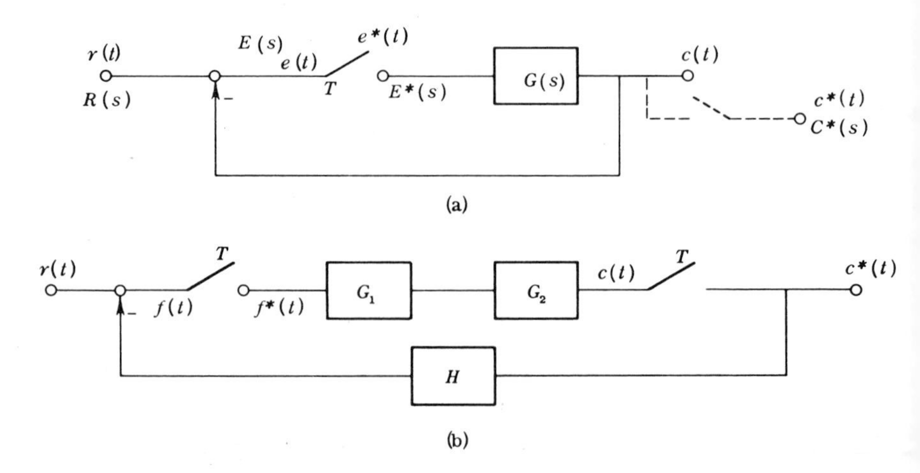

FIG. 11.5-2. Analysis of two sampled-data feedback structures (Examples 1 and 2).

This is important. Also,

$$Z[G(s)\, e^{-nsT}] = Z \text{ transform of } g(t - nT)$$
$$= e^{-nsT}[g(0) + g(T)\, e^{-sT} + g(2T)e^{-2sT} + \ldots] = z^{-n}\, G(z) \quad (11.5,2)$$

where $G(s) = \mathscr{L} g(t)$. With this notion of the pulse transfer function, the analysis of sampled-data feedback systems is straightforward.

Example 1 (Fig. 11.5-2a). In Fig. 11.5-2a, $E^*(s) = R^*(s) - C^*(s)$. But $C(s) = G(s)\, E^*(s)$, so that, using Eq. (11.5,1), $C^*(s) = G^*(s)\, E^*(s)$, and then $E^*(s) = R^*(s) - G^*(s)\, E^*(s)$, leading to $E^*(s) = R^*(s)/[1 + G^*(s)]$. Finally, $C(s) = G(s)\, E^*(s) = G(s)\, R^*(s)/[1 + G^*(s)]$. It is impossible in this case to obtain a transfer function $C(s)/R(s)$ which is independent of input; because the output is the same for all inputs having the same values at $t = nT$, even though their intersample values differ radically. However, one can obtain a P.T.F. by applying Eq. (11.5,1) to the last equation, resulting in $C^*(s) = G^*(s)\, E^*(s) = G^*(s)\, R^*(s)/[1 + G^*(s)]$ or $C(z)/R(z) = T(z) = G(z)/[1 + G(z)]$.

Example 2 (Fig. 11.5-2b). Here,

$$C^*(s) = [G_1G_2]^*\, F^*(s) = [G_1G_2]^*\, [R(s) - H(s)\, C^*(s)]^*$$
$$= [G_1G_2]^*\, [R^*(s) - H^*(s)\, C^*(s)].$$

Therefore $C^*(s) = (G_1G_2)^*R^*/[1 + (G_1G_2)^*H^*]$, and $C(z)/R(z) = T(z) = G_1G_2(z)/[1 + G_1G_2(z)\, H(z)]$. One usually writes $G_1G_2{}^*$ and $G_1G_2(z)$ in place of $[G_1G_2]^*$. It is important to note that $G_1G_2{}^* \neq G_1{}^*G_2{}^*$, i.e., $G_1G_2(z) \neq G_1(z)\, G_2(z)$. The latter is the transform of the impulse response (the P.T.F. of the system) in Fig. 11.5-3a, while the former is the P.T.F. of the system in Fig. 11.5-3b. Rapid methods of deriving the output of a sampled-data system are presented in the next section.

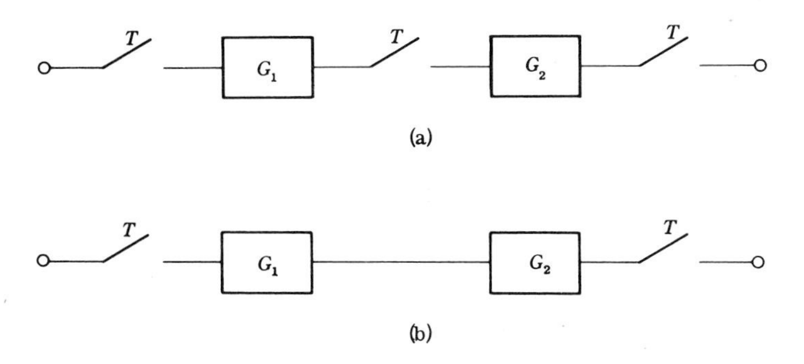

FIG. 11.5-3. Distinction between (a) $G_1{}^*G_2{}^*$ and (b) $G_1G_2{}^*$.

§ 11.6 The Fundamental Feedback Equation for Sampled-Data Systems

Mason's rule (Section 1.6) can sometimes be used in sampled-data systems for simple, rapid evaluation of the system transfer function. It can be used if it is possible to eliminate s-transfer functions and transforms, and replace them by P.T.F.'s and z transforms.

Example 1. Figure 11.5-2a can be redrawn as in Fig. 11.6-1a, which in turn can be replaced by Fig. 11.6-1b. Mason's rule may be applied to Fig. 11.6-1b,

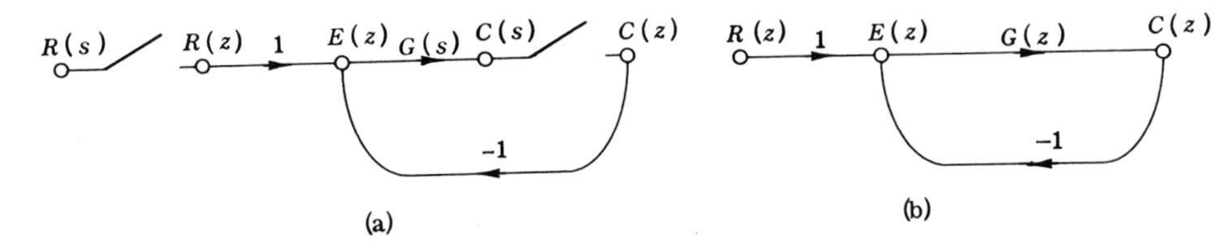

FIG. 11.6-1. Case where Mason's rule is directly applicable.

because in this homogeneous system all operations with respect to z are precisely the same as their continuous analogs are with respect to s. Mason's rule gives $C(z) = G(z)\,R(z)/[1 + G(z)]$. Also in Fig. 11.6-1a, $C(s) = G(s)\,E(z)$, and since $C(z) = G(z)\,E(z)$, it follows that $E(z) = R(z)/[1 + G(z)]$, and therefore $C(s) = G(s)\,R(z)/[1 + G(z)]$.

Example 2 (Fig. 11.6-2a, with $H_1 = 1$). First consider the problem for $H_1 = 1$. Mason's rule can then be applied if Fig. 11.6-2a is replaced by Fig. 11.6-2b, resulting in $C(z)/R(z) = G_1G_2(z)/[1 + G_2H_2(z) + G_1G_2(z)]$. The intermediate step of drawing Fig. 11.6-2b can usually be avoided by mentally noting whether it can be executed. If so, then Mason's rule is applied directly to the original figure. But care must be taken to apply the star at the correct points, i.e., to each collective group of functions between samplers. Thus, in Fig. 11.6-2a, (with $H_1 = 1$), $\Delta = 1 + G_1G_2(z) + G_2H_2(z)$, because in going around the appropriate loop, one must pass through G_1 and G_2 (or H_2 and G_2) before reaching a sampler. Similarly, if the input first passes through a pre-filter, G, in Fig. 11.6-2a, then $R(z)$ in the above must be replaced by $GR(z) = (GR)^*$.

The Fundamental Feedback Equation

It should be clear from the above comments, that Mason's rule cannot be directly applied for a general H_1 in Fig. 11.6-2a. The superposition technique of Chapter 2, which led to the fundamental feedback Eq. (2.2,7), provides a fairly simple technique for evaluating this and more complicated structures. We must re-derive the fundamental feedback equation, because samplers are

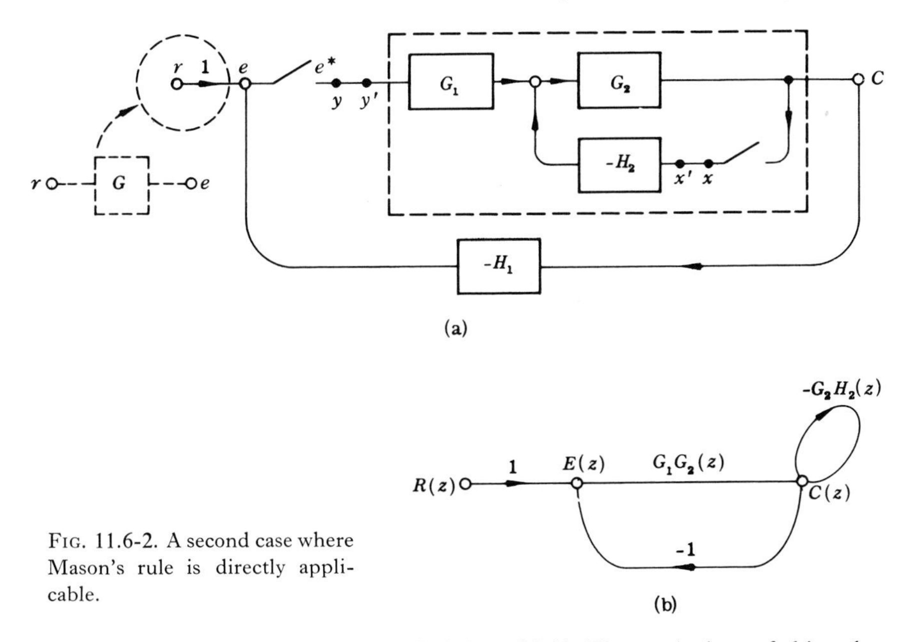

FIG. 11.6-2. A second case where Mason's rule is directly applicable.

present, and it is incorrect to let $(G\mathscr{S})^* = G^*\mathscr{S}$. The equivalent of this substitution is (correctly) used in the derivation in continuous systems. In sampled-data systems (refer to Fig. 11.6-3, which is a duplicate of Fig. 2.2-1b), we write

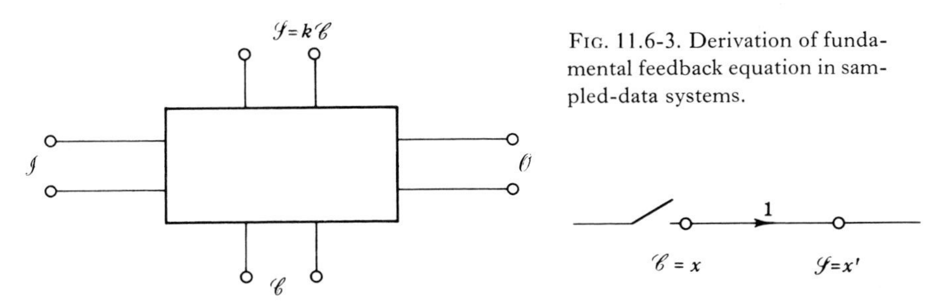

FIG. 11.6-3. Derivation of fundamental feedback equation in sampled-data systems.

$\mathscr{O} = (t_{oi}\mathscr{I}) + (\mathscr{S}t_{os})$. The terms $t_{oi}\mathscr{I}$, $t_{os}\mathscr{S}$ are enclosed in parentheses to indicate that it may not be permissible to pull out any individual factors [e.g., if $t_{oi}\mathscr{I} = (t_1\mathscr{I})^*$]. Also, $\mathscr{S} = k\mathscr{C} = k[(t_{ci}\mathscr{I}) + (t_{cs}\mathscr{S})]$, and we enclose $t_{ci}\mathscr{I}$, $t_{cs}\mathscr{S}$ in parentheses for the same reason as before.

It is impossible to derive the fundamental feedback equation unless we can factor out $\mathscr{S}$ in the above equation, i.e., unless we may write $k[(t_{ci}\mathscr{I}) + (t_{cs}\mathscr{S})] = (kt_{ci}\mathscr{I}) + (kt_{cs})(\mathscr{S})$. We therefore must restrict $\mathscr{S}$ to variables which always permit such factoring. This is done by taking $\mathscr{S} = x'$, $k = 1$, and $\mathscr{C} = x$ in Fig. 11.6-3, i.e., $x' = x$ are nodes *immediately following any sampler*. Since $x' = \mathscr{S}$ consists of impulses, and since $\mathscr{C}$ follows a sampler, $\mathscr{S}^* = \mathscr{S}$ and $(t_{cs}\mathscr{S}) = (t_{cs}\mathscr{S})^* = (t_{cs}\mathscr{S}^*)^* = t_{cs}^*\mathscr{S}^*$. We shall always pick $\mathscr{C}$ and $\mathscr{S}$ in this manner, and may therefore collect terms in $\mathscr{S}$, resulting in $\mathscr{S}[1 - t_{cs}] = (t_{ci}\mathscr{I})$ and $(\mathscr{S}t_{os}) = (t_{os})(\mathscr{S})$, from which it follows that

$$\boxed{\mathscr{O} = (t_{oi}\mathscr{I}) + \frac{(t_{ci}\mathscr{I})\, t_{os}}{1 - t_{cs}}} \tag{11.6,1}$$

In the continuous case, $\mathscr{I}$ may be factored out, leading to Eq. (2.2,7).

Example 3. Equation (11.6,1) is applied to Fig. 11.6-2a, and either xx' or yy' may be used for the $\mathscr{C}\mathscr{S}$ combination. If the former is used, then $t_{oi}R$, $t_{ci}R$ are obtained from Fig. 11.6-4a, with $t_{ci}R = (t_{oi}R)^*$, and $t_{oi}R =$

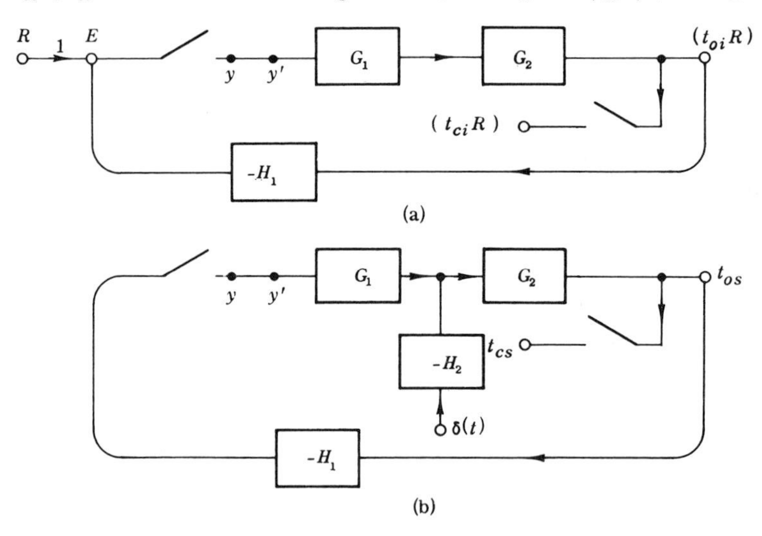

FIG. 11.6-4. (a) Structure for calculating $(t_{ci}R)$, $(t_{oi}R)$ (Example 3). (b) Structure for calculating t_{os}, t_{cs} (Example 3).

$R^*G_1G_2/(1 + G_1G_2H_1^*)$ (this may be obtained by Mason's rule, or from the fundamental feedback equation applied to yy' in Fig. 11.6-4a). Also, $t_{ci}R = (t_{oi}R)^* = R^*G_1G_2^*/(1 + G_1G_2H_1^*)$. The functions t_{os}, t_{cs} are obtained from Fig. 11.6-4b, with $t_{cs} = t_{os}^*$ and

$$t_{os} = -H_2G_2 + (G_2H_1H_2{}^*G_1G_2)/(1 + G_1G_2H_1{}^*).$$

[To find t_{os}, apply Eq. (11.6,1) to Fig. 11.6-4b, with $\mathscr{C}\mathscr{S} = yy'$, and then

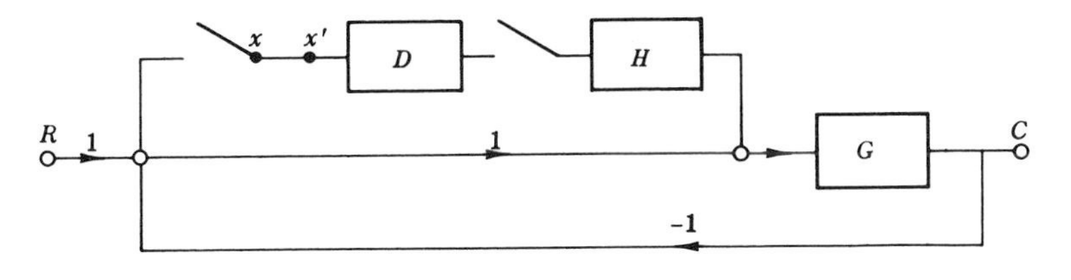

(a)

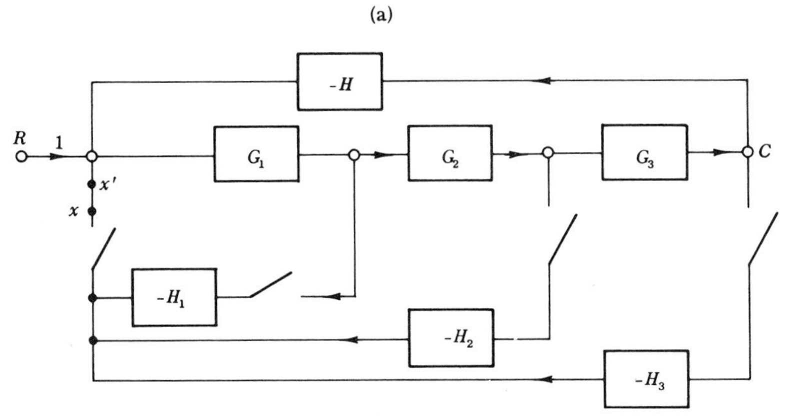

(b)

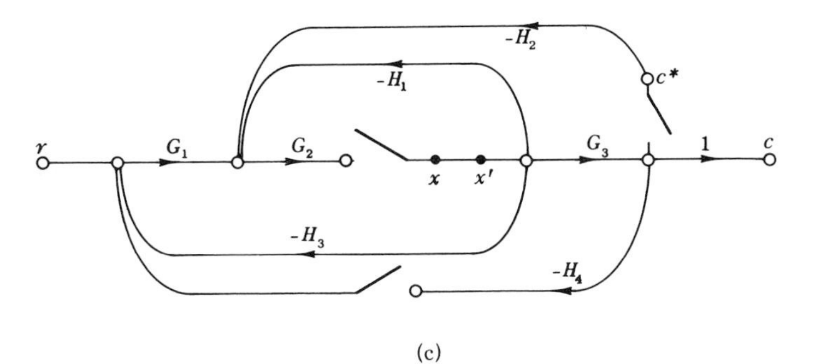

(c)

FIG. 11.6-5a to 5c. Examples for applying fundamental feedback equation.

$t_{oi}\mathscr{I} = -H_2G_2$, $t_{ci}\mathscr{I} = H_2G_2H_1{}^*$, $t_{os} = G_1G_2$, $t_{cs} = -G_1G_2H_1{}^*$.] Combining, as per Eq. (11.6,1), and simplifying,

$$C(s) = \frac{R^*[G_1G_2(1 + H_2G_2{}^*) - G_1G_2{}^*H_2G_2]}{(1 + H_2G_2)^*(1 + H_1G_1G_2{}^*) - G_1G_2{}^*H_1H_2G_2{}^*},$$

from which C^*/R^* is easily obtained, using Eq. (11.5,1).

Example 4 (Fig. 11.6-2a, using yy' for $\mathscr{C}\mathscr{S}$). Let $\mathscr{I} = R$; $Rt_{oi} = 0$, $Rt_{ci} = R^*$. If we accept, in the meantime, that the output at C due to y' in Fig. 11.6-2a is $T = Y'G_1G_2 - Y'G_1G_2{}^*G_2H_2/(1 + H_2G_2{}^*)$, then $t_{os} = T$ (with $Y' = 1$), and $t_{cs} = -(H_1T)^*$, with $Y' = 1$. The final result is the same as in Example 3. In order to evaluate T (the dotted portion of Fig. 11.6-2a), use $\mathscr{I} = Y'$, xx' for $\mathscr{S}$, and then $\mathscr{I}t_{oi} = G_1G_2Y'$, $\mathscr{I}t_{ci} = Y'G_1G_2{}^*$, $t_{os} = -H_2G_2$, $t_{cs} = -H_2G_2{}^*$, so that T is as given above.

Example 5[1] (Fig. 11.6-5a). We take $\mathscr{C}\mathscr{S} = xx'$, and then, by inspection,

$$C = \underbrace{\left(\frac{GR}{1+G}\right)}_{t_{oi}R} + \frac{\overbrace{[R/(1+G)]^*}^{t_{ci}R}\ \overbrace{[D^*H\,G/(1+G)]}^{t_{os}}}{1 + \underbrace{\{D^*\,[HG/(1+G)]^*\}}_{-t_{cs}}}.$$

Example 6 (Fig. 11.6-5b). Using xx' for $\mathscr{C}\mathscr{S}$, and $G = G_1G_2G_3$, and noting that $t_{cs} = t_{ci}$ with $R = 1$, we have, by inspection,

$$C = \underbrace{\left(\frac{GR}{1+GH}\right)}_{t_{oi}R}$$
$$-\frac{\overbrace{\left[\left(\frac{G_1R}{1+GH}\right)^* H_1{}^* + \left(\frac{G_1G_2R}{1+GH}\right)^* H_2{}^* + \left(\frac{GR}{1+GH}\right)^* H_3{}^*\right]}^{t_{ci}R}\ \overbrace{\left[\frac{G}{1+GH}\right]}^{t_{os}}}{1 + (t_{ci})_{R=1}}.$$

Example 7 (Fig. 11.6-5c). Using xx' for $\mathscr{C}\mathscr{S}$, we have, by inspection;

$$C = \frac{\overbrace{(G_1G_2R)^*}^{t_{ci}R}\ \overbrace{G_3}^{t_{os}}}{1 + \underbrace{[G_2(H_1 + G_3{}^*H_2 + G_3H_4{}^*G_1 + H_3G_1)]^*}_{-t_{cs}}}.$$

§ 11.7 Intersample Behavior and Inverse z Transformation

It might seem that to design sampled-data systems, one need only parallel the synthesis techniques of the continuous feedback system, with its s domain methods, by using analogous z domain methods. This is, to a certain extent, true and is attempted in later sections, except that one runs into some difficulties, as will be seen. But before embarking upon this, it is once again emphasized

[1] Figures 11.6-5a and b are taken from R. Ash, W. H. Kim, and G. M. Kranc, A general flow graph technique for the solution of multiloop sampled systems, *J. Basic Eng.* **82**, 360 (1960). A different method is presented in the paper.

that the z transform provides information as to the output values at the sampling instants only. This important point is illustrated in this section by means of a fairly vivid example. Also, this section presents some procedures for finding the intersample behavior.

Consider $T(z) = 1.254z/(z^2 + 0.119z + 0.1353)$, with $\omega_s = 2\pi$. Such a transfer function is obtained from each of Figs. 11.7-1a, b, and c. The pro-

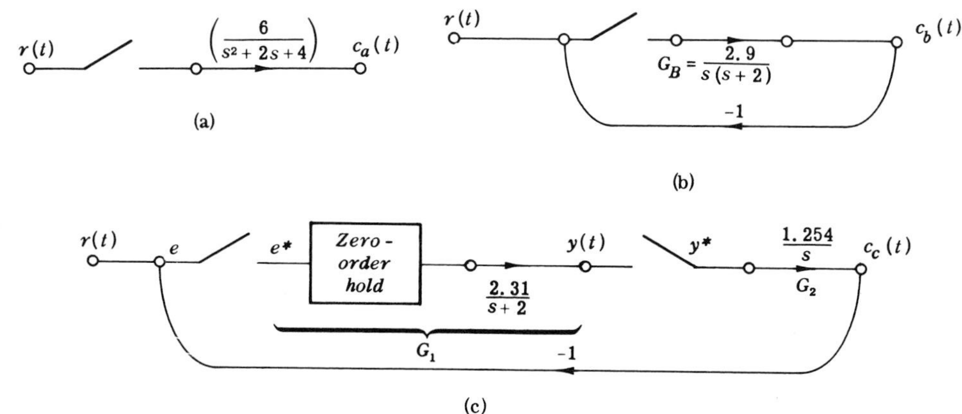

FIG. 11.7-1. Three systems with the same z transform, $T(z) = C(z)/R(z)$, but with different intersample outputs.

cedure for finding $T(z)$ is shown here for each case. In Fig. 11.7-1a, $T(z) = Z\{6/[(s+1)^2 + (\sqrt{3})^2]\}$, which from Table 11.4-1[1] [with $T = 1$, $\omega_o = \sqrt{3}$, $a = 1$, $e^{-2aT} = 0.1353$, $2e^{-aT}\cos\omega_o T = -(2)(0.368)\cos 80.7° = -0.119$] gives the above value for $T(z)$. In Fig. 11.7-1b, $T(z) = G(z)/[1 + G(z)]$, with

$$G(z) = 2.90Z\left(\frac{1}{s(s+2)}\right) = 1.45Z\left(\frac{1}{s} - \frac{1}{s+2}\right) = 1.45\left(\frac{z}{z-1} - \frac{z}{z-e^{-2T}}\right)$$

$$= \frac{1.254z}{(z-1)(z-0.1353)}.$$

Substituting in $G/(1+G)$ gives the indicated result. In Fig. 11.7-1c, the zero-order hold is a device which is used to obtain an approximate restoration of the original signal. Its impulse response is shown in Fig. 11.7-2, and its transfer function is consequently $(1 - e^{-sT})/s$. In Fig. 11.7-1c,

$$C^*/R^* = G_1^*G_2^*/(1 + G_1^*G_2^*); \qquad G_2^* = Z(1.254/s) = 1.254z/(z-1),$$

and

$$G_1^* = Z\,\frac{2.31(1-e^{-sT})}{s^2(s+2)} = Z\,\frac{2.31(1-z^{-1})}{s^2(s+2)} \qquad \text{[from Eq. (11.5,2)]}$$

$$= Z\,[2.31(1-z^{-1})]\left[\frac{0.5}{s^2} - \frac{0.25}{s} + \frac{0.25}{s+2}\right].$$

[1] Fifth entry from bottom of Table, letting $s^2 + 2\zeta\omega_n s + \omega_n^2 = (s+a)^2 + \omega_o^2$

Therefore

$$G_1^* = 2.31(1 - z^{-1})\,[0.5z(z-1)^{-2} - 0.25z(z-1)^{-1} + 0.25z(z-e^{-2})^{-1}].$$

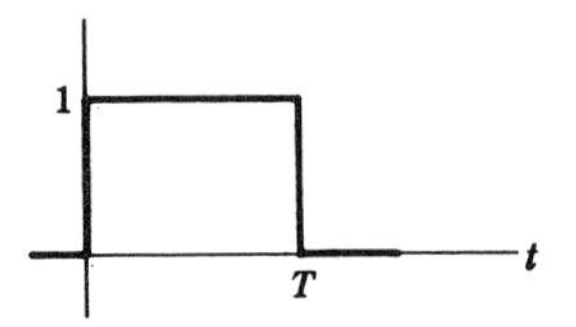

FIG. 11.7-2. Impulse response of zero-order hold.

The balance is straightforward, and results in the indicated $T(z)$. Since $T(z)$ is the same for each of the three structures, it follows that for *any* input, the sampled value of the output is the same for each system. However, there are very significant differences between the intersample values of the outputs of the three systems.

To find a system's sampled output, there are the following two principal methods:

(1) Expand $z^{-1}C(z)$ in partial fractions, and identify the terms of $C(z)$ with those listed in Table 11.4-1.

(2) Expand $C(z)$ in the form $C(z) = \sum_0^\infty a_n/z^n$ (by long division), from which $c(nT) = a_n$, because $z^{-n} = e^{-nsT}$.

These two methods are illustrated by the above example. Here, for a step input

$$C(z) = R(z)T(z) = \frac{z}{z-1}\,\frac{1.254z}{z^2 + 0.119z + 0.135}$$

$$= \frac{z}{z-1} - z\,\frac{(z+0.595) - 0.195}{z^2 + 0.119z + 0.135}.$$

Therefore, from Table 11.4-1,

$$c^*(t) = \sum \delta(t - nT)\,[1 - e^{-nT}\cos 1.73nT + 0.535e^{-nT}\cos 1.73nT].$$

This expression is correct only for $t = nT$, with n any positive integer.

A simpler procedure is to expand $C(z)$ into a series of the form a_n/z^n. This is usually easiest done by dividing. Thus

$$C(z) = 1.254z^2/(z^3 - 0.881z^2 + 0.0163z - 0.135).$$

Division gives $1.254z^{-1} + 1.11z^{-2} + 0.96z^{-3} + 0.99z^{-4} + 1.01z^{-5} + z^{-6} + \cdots$.

The intersample system output is next considered. That of Fig. 11.7-1c [denoted by $c_c(t)$] is easiest to find. Clearly, $c_c(t) = 1.254 \int_0^t y^*(t)\,dt$, and since $y^*(t)$ consists of impulses, $c_c(t)$ must consist of the staircase function shown in Fig. 11.7-3. Next consider Fig. 11.7-1b. Since the input to G_b consists of impulses, the output of G_b in any interval between nT and $(n+1)\,T$ must have the form $A + Be^{-2t}$. The values of A and B change from one interval to the next. Consider any interval $nT \leqslant t \leqslant (n+1)\,T$. Two boundary values are easily obtained. Write $c_b(\tau) = A + Be^{-2\tau}$, where $\tau = t - nT$. At $\tau = 0$,

$c_b(0) = A + B = c^*(nT)$; at $\tau = T$, $c_b(T) = A + Be^{-2T} = c^*[(n + 1)\,T]$, and one can solve for A and B from the known values of c^*. This procedure can be repeated for every interval. The result is shown in Fig. 11.7-3. Figure 11.7-1a can be handled in the same way. In each interval the output has the form $Ae^{-t}\sin(1.732t + \theta)$, and for $nT \leqslant t \leqslant (n + 1)\,T$, $c_a(\tau) = Ae^{-\tau}\sin(1.732\tau + \theta)$, with $\tau = t - nT$; $c^*(nT) = A\sin\theta$, $c^*[(n + 1)\,T] = Ae^{-T}\sin(1.732T + \theta)$, which permits solving for A and θ in each interval. The result is shown in Fig. 11.7-3.

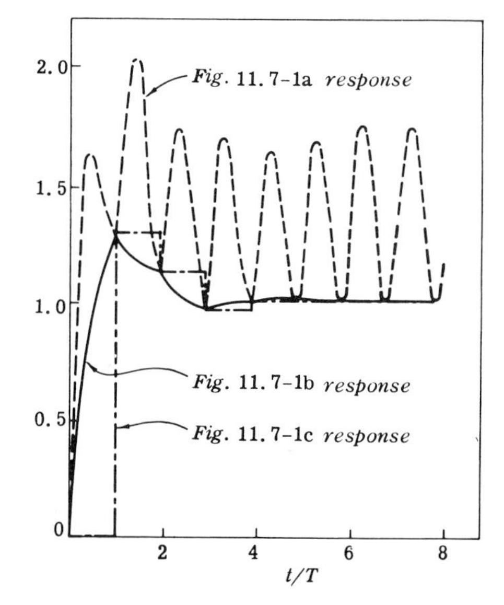

FIG. 11.7-3. Step responses of the three systems of Fig. 11.7-1.

It is clear from Fig. 11.7-3 that the intersample behavior can be violently different for the same sampled output. This is because the intersample behavior is determined by the final stage which is driven by impulses. When this final stage has no more than two poles, the values at the beginning and end of each interval suffice to give the exact behavior in that interval. When the transfer function of the final stage has more than two poles, the above procedure can be generalized, but the amount of calculation tends to obscure one's qualitative picture of the situation. Actually, one can easily get a qualitative estimate of the intersample behavior from the impulse response of the final network which is driven by impulses. If it is highly oscillatory, the intersample behavior will be highly oscillatory, etc. Such a qualitative estimate usually suffices.

The following method[1] can be used to find the exact intersample values. It also has the advantage that the general *form* of the intersample behavior is fairly easily available. One is usually interested in the general form rather than the precise values. Basically, the idea is to break up the final stage into a sum of simple stages, each of whose individual continuous output is easily available. Consider, for example, Fig. 11.7-4 with $G_3 = \Sigma A_i/(s + a_i)$ (a_i can be complex). The continuous output is determined by G_3 and the x' impulse series, which we denote as y^*. The y^* series is available by z-transform methods, and then, in place of G_3, the equivalent representation in Fig. 11.7-4 is used. The individual

[1] J. Sklansky and J. R. Ragazzini, Analysis of errors in sampled-data feedback systems. *Trans. AIEE, Part II*, **74**, 65-71 (1955).

$c_1(t)$, $c_2(t)$, etc., are available almost by inspection. What is important is that one has a good idea of the form of the intersample output without any detailed calculations.[1]

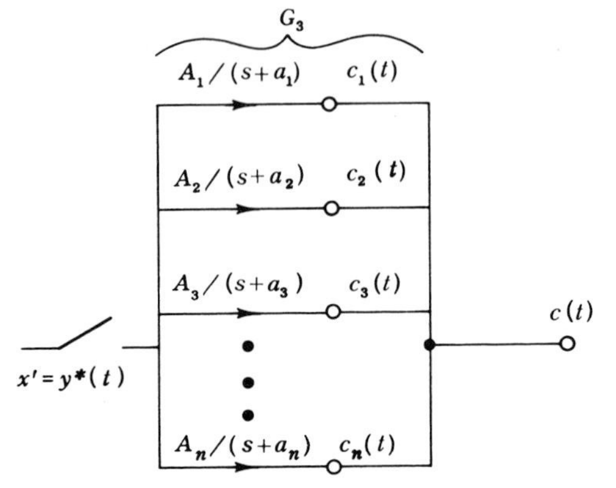

FIG. 11.7-4. A method of finding the intersample output.

§ 11.8 Elementary Design of Sampled-Data Feedback Control Systems

Sufficient tools have been developed to consider the elementary design of sampled-data feedback systems. By elementary design methods are meant those in which (1) the single-degree-of-freedom configuration of Fig. 11.8-1a is

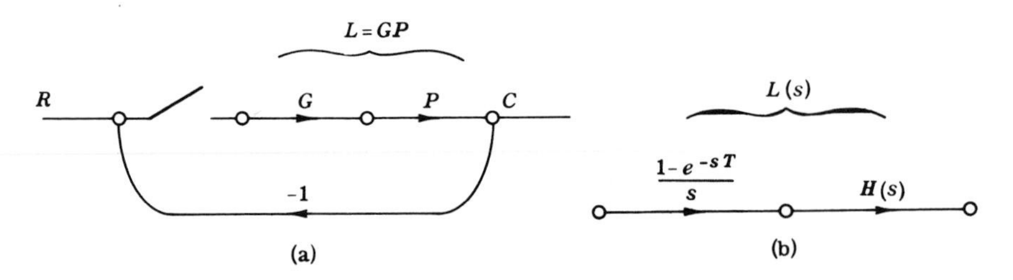

FIG. 11.8-1. (a) Elementary sampled-data feedback system. (b) Form of $L(s)$ for numerical example.

used, (2) primary attention is paid to simple input signals such as the step and ramp functions; and (3) the input is assumed to be noise free. Since only one

[1] Other methods for finding intersample behavior are discussed by E. I. Jury, "Sampled-Data Control Systems." Wiley, New York, 1958; J. T. Tou, "Digital and Sampled-Data Control Systems." McGraw-Hill, New York, 1959.

degree of freedom is available in this structure, only one independent system function may be realized, and traditionally this is the system transfer function. There is also imposed the important restriction that there is only one sampler in the system. Thus $G(s)$ and $P(s)$ are rational functions of s, i.e., they represent the usual continuous type building blocks. It will soon be seen that this restriction to one sampler imposes design difficulties.

A seemingly attractive approach to the design problem is to select a suitable *continuous* output, and take its *sampled* values as the desired sampled-data output. For example, suppose the system is designed on the basis of its step function response, and, to be specific, suppose that the sampled step response of a continuous system whose transfer function is $T(s) = (1 - 0.366s)(s^2 + s + 1)^{-1}$ is considered satisfactory. Such a choice is not unreasonable if the plant has a rhp zero at $1/0.366 = 2.73$. One writes[1]

$$C(z) = Z\frac{1 - 0.366\,s}{s(s^2 + s + 1)} = Z\left(\frac{1}{s} - \frac{s + 1.366}{s^2 + s + 1}\right)$$

$$= z\left(\frac{1}{z - 1} - \frac{z + e^{-0.5T}(\sin 0.866T - \cos 0.866T)}{z^2 - 2ze^{-0.5T}\cos 0.866T + e^{-T}}\right).$$

$$= R(z)T(z) = \frac{z}{z - 1}T(z) = \frac{z}{z - 1}\frac{L(z)}{1 + L(z)},$$

for Fig. 11.8-1a. Solving for $L(z)$, and assuming $T = 1$, the result is $L(z) = (0.144z + 0.436)[(z - 1)(z + 0.0685)]^{-1}$.

The form of $L(z)$ suggests that $L(s)$ includes a zero-order hold. Consequently, $L(s)$ has the form shown in Fig. 11.8-1b, and $L(z) = Z[(1 - e^{-sT})H(s)/s] = [(z - 1)/z]\,Z[H(s)/s]$ [see Eq. (11.5,1)]. Therefore

$$Z\frac{H(s)}{s} = z\frac{0.144z + 0.436}{(z - 1)^2(z + 0.0685)} = z\left(\frac{0.543}{(z - 1)^2} - \frac{0.374}{z - 1} + \frac{0.374}{z + 0.0685}\right).$$

Finally,[2] $H(s)/s = (0.543/s^2) - (0.374/s) + 0.374(s + A)(s^2 + 5.36s + 17.3)^{-1}$. A can have any value. If it is chosen to minimize the degree of the numerator of $H(s)$, then $A = 3.87$, and $H(s) = -3.56(s - 2.64)[s(s^2 + 5.36s + 17.3)]^{-1}$. The rhp zero at 2.64 is sufficiently close to 2.73 to assume that it is contributed by the plant. In any case A can be chosen so that $H(s)$ has a zero precisely at 2.73. In this example the design has turned out fairly satisfactory. The rhp plant zero has been accounted for, and the intersample system output, which is

[1] It would be incorrect to choose $T(z) = ZT(s)$, because the response of such a system is $C(z) = T(z)R(z)$, while for the continuous system $C(s) = T(s)R(s)$. In general, $T(z)R(z) \neq Z[T(s)R(s)]$.

[2] The z transform of $e^{-\alpha t}\cos(\pi t/T)$ is easily seen to be $z/(z + e^{-\alpha T})$. The z transform of $e^{-\alpha t}\sin(\pi t/T)$ is zero because $[\sin(\pi t/T)]_{t=nT}$ is zero. The Laplace transform of $e^{-\alpha t}\{\cos(\pi t/T) + [(A - \alpha)T/\pi]\sin(\pi t/T)\}$ is $(s + A)[(s + \alpha)^2 + \pi^2/T^2]^{-1}$. Thus A can have any value without affecting the z transform.

determined by $L(s) = (1 - e^{-sT}) H(s)/s$, is not any more oscillatory than the function which was originally used as the basis for the design.

Does this approach always lead to a satisfactory design? Let us consider two more examples. In the second example $T = 1$ and the system is designed on the basis of the step response function $C(s) = [s(s^2 + s + 1)]^{-1}$. As in the previous example, the procedure is to expand $C(s)$ in partial fractions and, with the aid of Table 11.4-1, evaluate $C(z) = ZC(s)$, and set it equal to $zT(z)/(z-1) = zL(z)/(z-1)[1 + L(z)]$, since the structure of Fig. 11.8-1a is being used. In this example the results are: $T(z) = (0.341z + 0.242)/(z^2 - 0.785z + 0.368)$; $L(z) = (0.341z + 0.242)/(z-1)(z - 0.126) = (1 - z^{-1}) Z[H(s)/s]$, from which $H(s) = -0.105(s - 13.2)/s(s + 2.07)$. This result is not attractive, partly because of the special effort required to realize a transfer function with a rhp zero. A more important reason is that we usually dread encountering plants with rhp zeros, because of the resulting restrictions on the system feedback properties (recall Sections 7.9 and 7.12), and in this case we are asked to deliberately introduce a rhp zero. We should be aware that the system feedback properties are being ignored when design is solely on the basis of system response. If we nevertheless pursue this course, it must be because we presumably hope that the feedback properties will somehow turn out satisfactory. However we must certainly stop and question the procedure when it requires the deliberate degradation of the system feedback properties.

Difficulties may also arise if the output is chosen by specifying the output at several sample points. There is no reason why this should lead to a more satisfactory $L(s)$. For example, suppose[1] the sampled step response $c(nT)$ is chosen to have the following values: $c(T) = 1$, $c(2T) = 1.2$, $c(3T) = 0.95$, $c(\infty) = 1$. We therefore pick $C(z)$ with four unknowns, viz., $C(z) = [z/(z-1)][(a_1 z + a_o)/(z^2 + b_1 z + b_o)]$. We use the final value theorem (whose proof is left as an exercise for the reader), $\lim_{t\to\infty} c(t) = \lim_{z\to 1} (z-1) C(z)$, and therefore $(a_1 + a_o)/(1 + b_1 + b_o) = 1$. Long division of $C(z)$, and equating the quotient with the pre-assigned $c(nT)$, leads to $a_1 = 1$, $a_o = 0.45$, $b_1 = 0.25$, $b_o = 0.2$. This gives $L(z) = (z + 0.45)[(z-1)(z + 0.25)]^{-1} = [(z-1)/z] Z[H(s)/s]$; and finally $H(s)/s = (1.16/s^2) - (0.128/s) + 0.128(s + A)(s^2 + 2.772s + 11.82)^{-1}$, with any value for A. If A is chosen to minimize the degree of the numerator of $H(s)$, the result is $H(s) = 1.70(s + 8.1)/s(s^2 + 2.77s + 11.8)$, whose complex pole pair has a damping factor of 0.4. The intersample output may be unsatisfactory despite the smoothing effect of the zero-order hold. The basic difficulty with the approach being used is that from the specifications of the sampled output values, no matter how they are chosen, it is very difficult to see beforehand the nature of the required continuous $L(s)$ transfer function, and thus the nature of the intersampled system

[1] This numerical example is taken from J. G. Truxal, "Control Systems Synthesis," pp. 539-540. McGraw-Hill, New York, 1955.

behavior. What is even worse is that after one or more unsatisfactory attempts have been made, there is no indication what changes in $T(z)$ are required in order to obtain a satisfactory $L(s)$.

The poles of $L(z)$ which lead to the low damped complex pole pairs of $L(s)$ can be avoided as follows. Suppose a continuous output with system transfer function $T_c(s)$ is used as the model for the sampled system, i.e., $C(s) = s^{-1}T_c(s) = s^{-1} - [(1 - T_c(s)]\, s^{-1}$, and therefore $Z[C(s)] = z/(z-1) - zN(z)/D(z) = z[D(z) - (z-1)\,N(z)]/[(z-1)\,D(z)]$, where $zN(z)/D(z) = Z\{[1 - T_c(s)]/s\}$. $Z[C(s)]$ is set equal to $T(z)\,R(z) = T(z)z/(z-1)$, where $T(z)$ is the P.T.F. of the sampled-data system. Therefore $T(z) = 1 - [(z-1)\,N(z)/D(z)] = L(z)/[1 + L(z)] = n(z)/[d(z) + n(z)]$, where $L(z) = n(z)/d(z)$. Consequently, $n(z) = D(z) - (z-1)\,N(z)$, $d(z) = (z-1)\,N(z)$, and one must restrict the zeros of $N(z)$ if those of $d(z)$ are to be restricted. Thus, in the previous example, $C(s) = \omega_n{}^2/s(s^2 + 2\zeta\omega_n s + \omega_n{}^2) = s^{-1} - (s + 2\zeta\omega_n)\,(s^2 + 2\zeta\omega_n s + \omega_n{}^2)^{-1}$. It is readily found that

$$N(z) = z - e^{-\zeta\omega_n'}[\cos \omega_n'(1-\zeta^2)^{0.5} - \zeta(1-\zeta^2)^{-0.5} \sin \omega_n'(1-\zeta^2)^{0.5}]$$

where $\omega_n' = \omega_n T$. If $L(z)$ is to have no pole on the negative real axis [because such poles lead to low damped complex poles of $L(s)$], it is necessary that $\omega_n'(1-\zeta^2)^{0.5}$ rad must lie in the sector shown in Fig. 11.8-2a. The resulting permissible values of ω_n' as functions of ζ are plotted in Fig. 11.8-2b. [Note $\tan \psi = (1-\zeta^2)^{0.5}/\zeta$.]

For more complicated $T_c(z)$, it becomes more difficult to ascertain the restrictions on $C(s)$ which correspond to restrictions on $L(z)$. Also, there is the problem of the rhp zeros of $L(s)$. Thus the basic difficulty in the above design procedure is that it is extremely difficult to specify the sampled output values so that the resulting $L(s)$ turns out reasonable. In addition, after one unsuccessful trial, no insight is thereby gained to assure any better chances of success in additional trials.

What can be done in view of the above difficulties? There are several possibilities, none of which, unfortunately, are completely satisfactory.

(1) Assume a convenient practical form for $L(s)$ of Fig. 11.8-1. Solve for the resulting $C(z)$ and relate it to the sampled values of a continuous $C(s)$. If possible, classify the results in charts, graphs, etc., which would permit one to go from a desired sampled $C(s)$ to the required (practical) $L(s)$. This is done in Section 11.9 for only two rather simple forms of $L(s)$. But it could, with the help of a computer, be extended to more complicated $L(s)$ and thus make available a dictionary of required (practical and convenient) $L(s)$ for desired sampled $C(s)$.

(2) One may proceed from the desired sampled $C(s)$ to the resulting $L(z)$, just as in the previous examples. The inverse transform of $L(z)$ is then obtained by dividing the denominator of $L(z)$ into its numerator. Then one seeks an approximation to $L(s)$, denoted by $L_a(s)$, whose continuous inverse transform

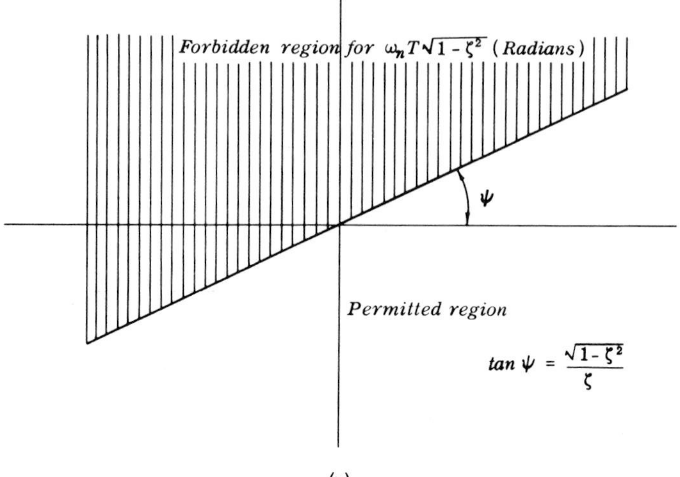

(a)

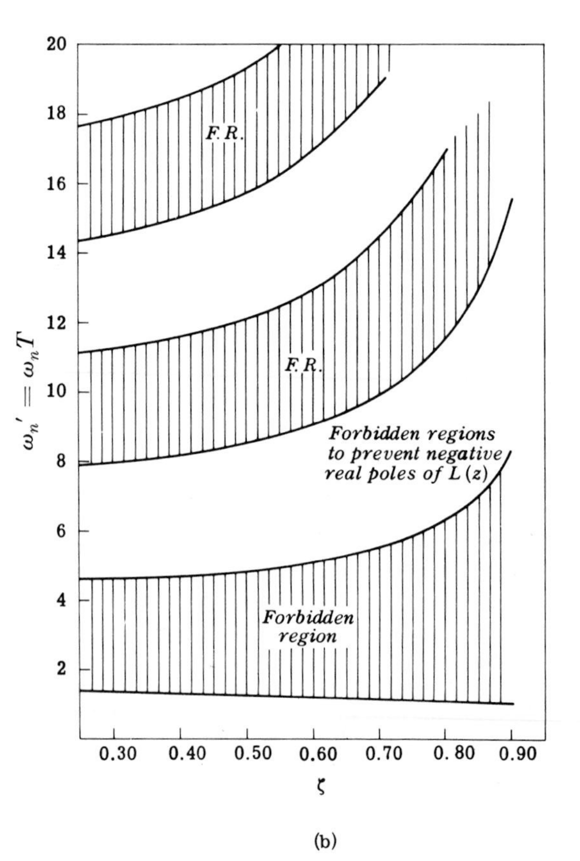

(b)

FIG. 11.8-2. Restrictions on z-transform parameters to avoid low damped poles of $L(s)$. (a) Permissible region. (b) Forbidden regions.

has approximately the same values at $t = nT$ as the inverse transform of $L(z)$. The approximation may be done in the time domain. If possible, $L_a(s)$ is chosen minimum phase, and with no low damped poles. This method might be called the Approximation method, and its usefulness depends completely on how well and easily $L_a(s)$ can be obtained from $L(z)$. The design is basically complete as soon as a satisfactory $L_a(s)$ is obtained. Some approximation techniques are presented in Chapter 12.

(3) One may abandon the z-transform formulation and return to the s plane. Since it is practically impossible to proceed analytically with the resulting transcendental functions, one must perforce work with the open-loop frequency response, and shape it so as to achieve the design specifications and a satisfactory closed-loop response.[1] The shaping is not as simple as in the continuous-data system, because $L^*(j\omega) = (1/T) \sum_{-\infty}^{\infty} L[j(\omega + n\omega_s)]$ and vector addition (Fig. 11.3-3) is required. One must also re-examine the correlation between the frequency and time response. The advantage of this method is that in working directly with the continuous-data $L(s)$, one can avoid an undesirable form for $L(s)$. The procedure is discussed in Section 11.10.

(4) As previously noted, it is extremely difficult to foresee the character of $L(s)$ of Fig. 11.8-1 that results from a specified $C(z)$. It is easier to consider the relationship between a specific $L(z)$ and the $L(s)$ that it leads to. Based upon experience and rough qualitative reasoning, some rules for choosing $L(z)$ to avoid undesirable forms of $L(s)$ are given in Section 11.11. One could then work with the open-loop $L(z)$ and shape it, by root locus, to achieve a satisfactory design. It is also possible to transform from the z variable to still another variable, $w = (z - 1)/(z + 1)$. In the w domain, the Bode chart can be used in exactly the same simple elegant way in which it was used in continuous-data systems. Just as in continuous-data systems, the relation between the open-loop and the closed-loop functions is available from the constant $|T|$ and $\arg T$ loci or the Nichols chart. There only remains the necessity to correlate time response with the w-domain system function representation. This is done by working out a variety of examples. This method can therefore be considered the counterpart of the Bode plot shaping method for continuous-data systems. It is presented in Section 11.11.

§ 11.9 Design from Selected Form of *L*(*s*)

We pick *a priori* the form of $L(s)$ and find the resulting response to a step function. The details are worked out for two cases only. The first is shown in Fig. 11.9-1.

[1] W. K. Linvill, Sampled-data control systems studies through comparison with amplitude modulation. *Trans. AIEE Part II*, **70**, 1779-1788 (1951); W. K. Linvill and R. W. Sittler, Extension of conventional techniques to the design of sampled-data systems, *Convention Record, IRE Part I*, pp. 99-104 (1953).

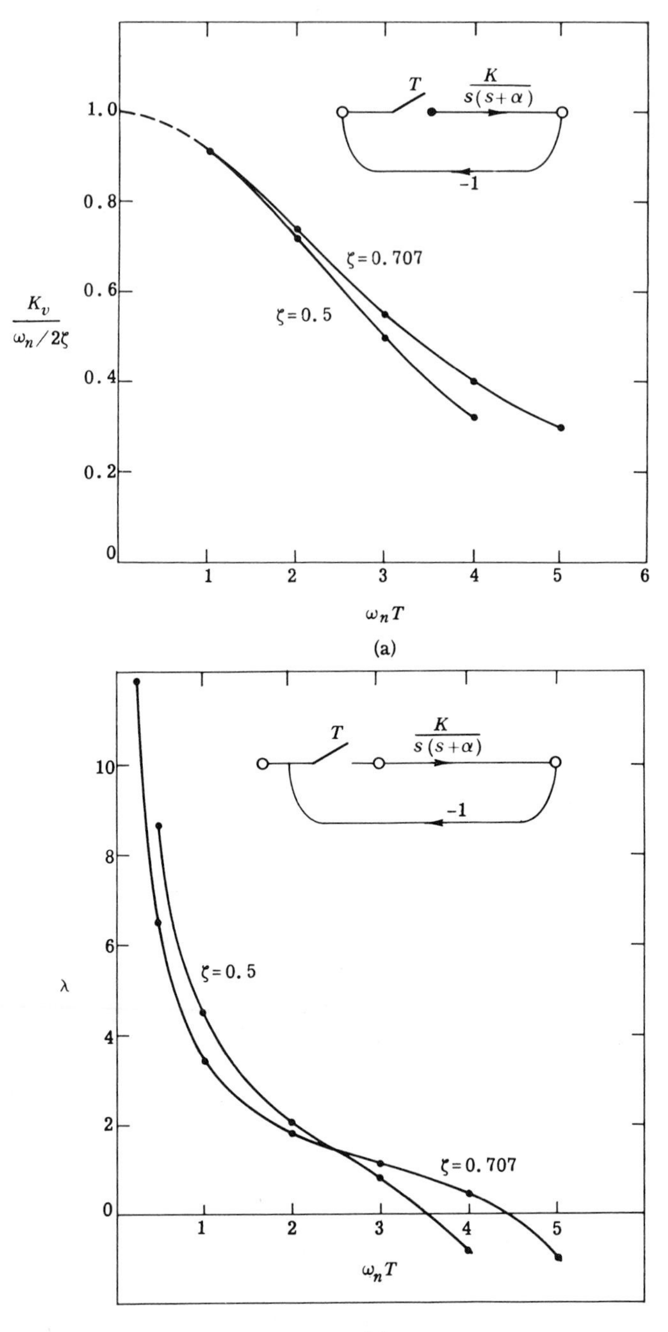

(b)

FIG. 11.9-1. Relations between open- and closed-loop parameters for $L(s) = K/s(s + \alpha)$. In (b), $K = 2\zeta\omega_n T K_v$; $\alpha = 2\zeta\omega_n$.

Case 1. In Fig. 11.9-1,

$$T(z) = [K\alpha^{-1}z(1 - e^{-\alpha T})]\{z^2 - z[1 + e^{-\alpha T} - K\alpha^{-1}(1 - e^{-\alpha T})] + e^{-\alpha T}\}^{-1}.$$

It is easy to see that the position constant $K_p = \lim_{z\to 1} L(z)$ is infinite, and the velocity constant $K_v = \lim_{z\to 1} T^{-1}(z - 1) L(z) = K/\alpha T$. (The derivation of these relations for K_p and K_v are left as an exercise for the reader—see Section 11.10.) For a step function input, $C(z) = z(z - 1)^{-1} T(z) = z(z - 1)^{-1} - z(z - e^{-\alpha T})\{z^2 - z[1 + e^{-\alpha T} - K\alpha^{-1}(1 - e^{-\alpha T})] + e^{-\alpha T}\}^{-1}$. This corresponds to the sampling of a continuous output with transform and parameter relations

$$C(s) = \omega_n(s + \lambda\zeta\omega_n)/\lambda\zeta s(s^2 + 2\zeta\omega_n s + \omega_n^2) \tag{11.9,1a}$$

$$\left(\zeta - \frac{1}{\lambda\zeta}\right) = \frac{\sqrt{1 - \zeta^2}\,[e^{-\zeta\omega_n T}\cos\omega_n T\sqrt{1 - \zeta^2} - e^{-2\zeta\omega_n T}]}{e^{-\zeta\omega_n T}\sin\omega_n T\sqrt{1 - \zeta^2}} \tag{11.9,1b}$$

$$\alpha = 2\zeta\omega_n \tag{11.9,1c}$$

$$2e^{-\zeta\omega_n T}\cos\omega_n T\sqrt{1 - \zeta^2} = 1 + e^{-\alpha T} - K\alpha^{-1}(1 - e^{-\alpha T}). \tag{11.9,1d}$$

Thus the sampling tends to increase the overshoot and, as will be seen, to decrease the velocity constant.

Suppose one selects the desired ζ, ω_n, and λ of Eq. (11.9,1a). What are the corresponding required values of K, α, T? Of course, if T is fixed, then only two of ζ, ω_n, and λ may be independently chosen. Equation (11.9,1b) determines T, and α, K are obtained from Eqs. (11.9,1c, d). The resulting $K_v = K/\alpha T$. To facilitate design, the values of λ and K_v for given $\omega_n T$ and ζ have been calculated, and are presented in normalized form in Figs. 11.9-1a and b. It is seen that K_v decreases as the sampling period is increased. As the latter becomes very small, K_v approaches the value for a continuous system, i.e., $(K_v)_{\text{cont.}} = \omega_n/2\zeta$. For example, suppose one wants $\lambda = 2$, $\zeta = 0.707$ for the analogous $T(s)$ of Eq. (11.9,1a). From Fig. 11.9-1b, $\omega_n T = 1.85$ and $K_v = 0.77\omega_n/1.414 = 1.01/T$.

The second case that is worked out is shown in Fig. 11.9-2a, where $L(s) = (1 - e^{-sT})K/s^2(s + \alpha)$. It turns out that Eq. (11.9,1b) and therefore Fig. 11.9-1b are correct for this case too. However, the relations between ζ, $\omega_n T$ and K, α are different from before. They are now

$$\begin{aligned} e^{-2\zeta\omega_n T} &= K\alpha^{-2} + e^{-\alpha T}[1 - K\alpha^{-2}(1 + \alpha T)], \\ 2e^{-\zeta\omega_n T}\cos\omega_n T\sqrt{1 - \zeta^2} &= 1 + e^{-\alpha T} - K\alpha^{-2}(T\alpha - 1 + e^{-\alpha T}). \end{aligned} \tag{11.9,2}$$

Consequently $K_v = K\alpha^{-1}$ is different here. The value of $2\zeta K_v/\omega_n$ vs. $\omega_n T$ is plotted in Fig. 11.9-2a for 2 values of ζ. Graphs (Fig. 11.9-2b) are also included to facilitate finding the parameters K and α for given $\omega_n T$ and ζ.

The velocity constant can be increased by means of a lag network with trans-

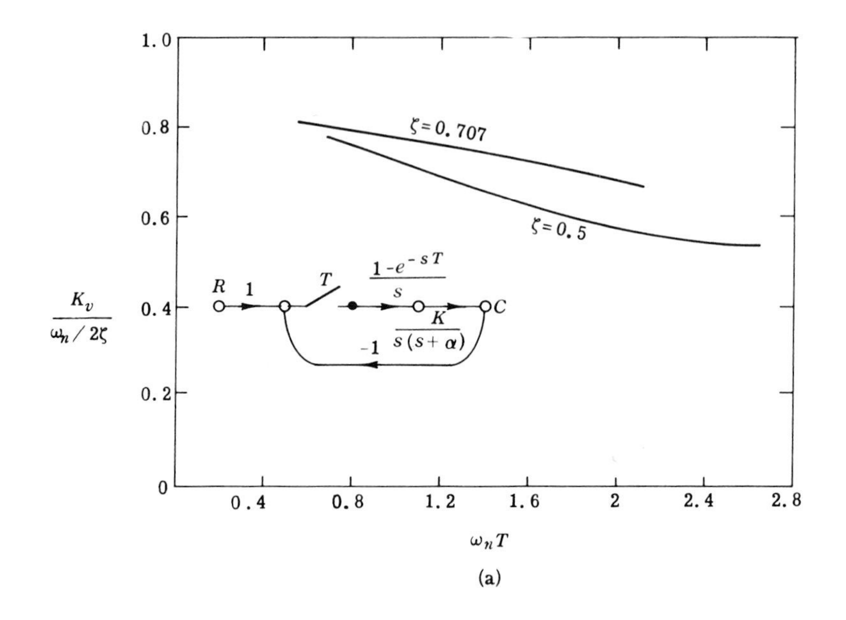

(a)

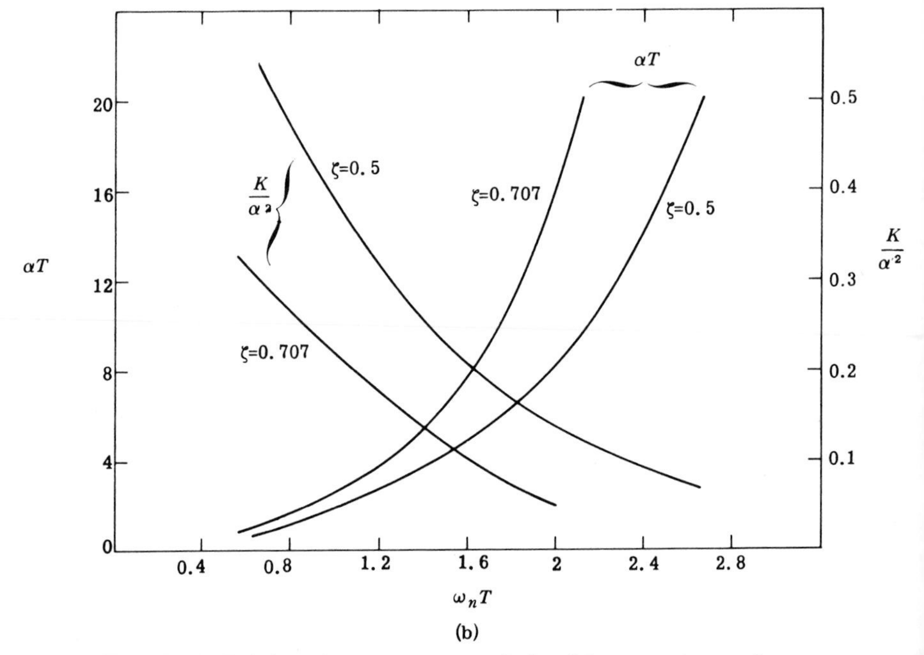

(b)

FIG. 11.9-2. Relations between open- and closed-loop parameters for $L(s) = K(1 - e^{-sT})/s^2(s + \alpha)$.

fer function $F(s) = (s + mb)/m(s + b)$. In lag compensation, the corner frequencies b and mb are much less than ω_s, so in view of Eq. (11.3,1), $F^*(s) \doteq F(s)$ very closely. The lag compensation can therefore be added at the very end of the design just as if the system were of the continuous-data type. The extension of the above approach to more complicated $L(s)$ is straightforward but tedious, and would require the use of a computer.

§ 11.10 Elementary Design by the Frequency Response Method

In this section, we apply the frequency response method of Chapter 5 to the elementary design of sampled-data feedback control systems. As in Chapter 5, it is assumed that the specifications prescribe the error constant, the crossover frequency, and the stability margins. The problem is to shape $L^*(j\omega)$ to achieve these specifications. The material is presented in the following order:

(1) The validity of the error constants for the evaluation of steady state error is re-examined.

(2) The constraints (due to sampling) on the loop transmission crossover frequency and its stability margins are noted.

(3) The design procedure is presented by means of a typical numerical example.

(4) Some feeling for the relationship between the time and frequency domains is obtained by presenting some examples.

(1) Error Constants

Do the error constants have the same meaning in sampled-data systems as they do in continuous systems? In Fig. 11.8-1, $C(s) = R^*(s)\,L(s)/[1 + L^*(s)]$. If the input is a step, then $R^*(s) = e^{sT}/(e^{sT} - 1)$, $E = R - C$, $\lim_{t\to\infty} c(t) = \lim_{s\to 0} sC(s) = L(0)/T[1 + L^*(0)]$, and $e(\infty) = 1 - c(\infty) = [T + TL^*(0) - L(0)]/T[1 + L^*(0)]$. Usually $|L(0)| \gg L(jn\omega_s)$, so, from Eq. (11.3,1), $L^*(0) \doteq T^{-1}L(0)$, and $e(\infty) \doteq [1 + L^*(0)]^{-1}$. Therefore the sampled-data position constant is $K_p^* \doteq L^*(0)$. If $L(s)$ has a pole at the origin, then $L^*(0) = T^{-1}L(0)$ exactly, and K_p is infinite, with zero final position error. When the same reasoning is applied to a ramp input, finite K_p leads to infinite position error, but infinite K_p and finite K_v lead to an indeterminate position error, because the latter varies during the intersample intervals. However, the position error at the sampling instants can be obtained from the final value theorem in the z domain, which is: Given $F(z) = Zf^*(t)$, then $\lim_{n\to\infty} f(nT) = \lim_{z\to 1}(z - 1)\,F(z)$. (The proof is left to the reader, and it should be noted that just as in the s-transform final value theorem, it is valid only when it gives a finite result.) The above leads to $\lim_{n\to\infty} e(nT) = K_v^*$, where $K_v^* \triangleq$

$\lim_{s \to 0} sL^*(s) = T^{-1} \lim_{z \to 1} (z - 1) L(z)$. Similar results are obtained for the higher order error constants.

(2) Constraints on Crossover Frequency and Stability Margins

Are there any constraints on the value of the crossover frequency? $L^*(j\omega)$ is periodic with period $j\omega_s$ and $L^*(j\omega) = \bar{L}^*(-j\omega)$, so that $L^*(j\omega)$ need be evaluated for only $0 \leqslant \omega \leqslant 0.5\omega_s$. Also, from Eq. (11.3,1), $L^*(j0.5\omega_s)$ is real [unless $L(s)$ has a pole at $s = j0.5n\omega_s$]. A graphical method of deriving $L^*(j\omega)$ from $L(j\omega)$ was described in connection with Figs. 11.3-3a and b. If $L(s)$ has at least two more poles than zeros, and $L(0)$ is finite, then $L^*(j\omega)$ has the general form of curve A or B in Fig. 11.10-1a; i.e., curves C and D are impossible (proved in Section 11.11). Curve B in Fig. 11.10-1a results from an $L(j\omega)$ of the form[1]

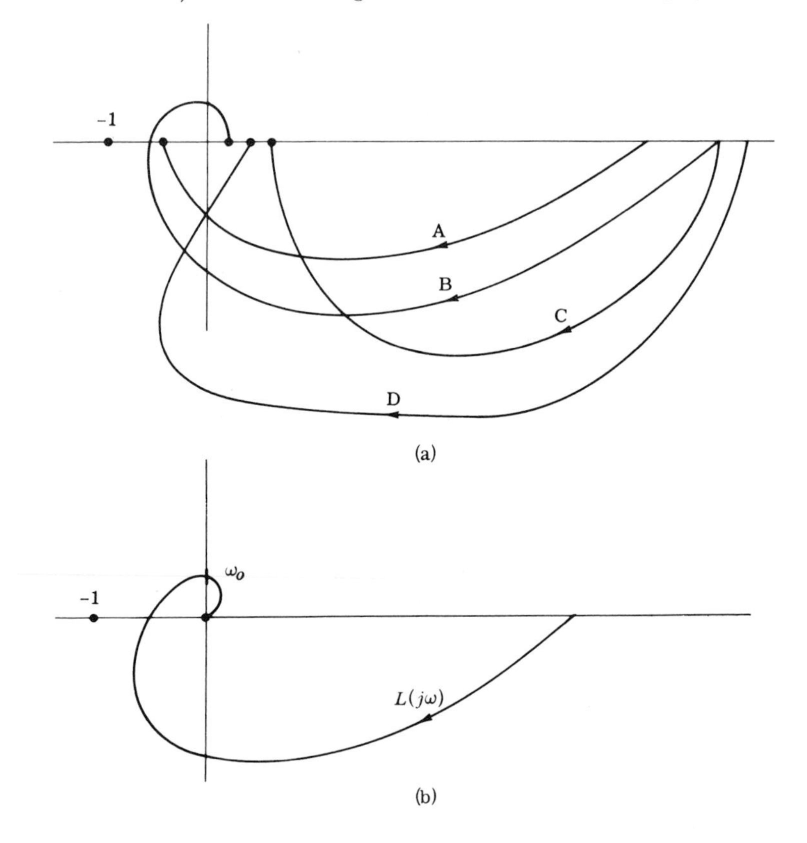

FIG. 11.10-1. Some tentative loci of $L^*(j\omega)$.

[1] Y. Z. Tsypkin, Frequency method of analyzing intermittent regulating systems, *in* "Frequency Response" (R. Oldenburger, ed.), pp. 309-324. Am. Soc. Mech. Engrs., New York, 1956.

shown in Fig. 11.10-1b, where the critical factor is the requirement that $\omega_o \leqslant 0.5\omega_s$. It is easy to see (from Figs. 11.3-3a and b) that, for curve A in Fig. 11.10-1a, $L^*(j\omega)$ tends to have larger lag angles than $L(j\omega)$; thus the sampling has a destabilizing influence. For curves of the B type, $L^*(j\omega)$ tends to have smaller lag angle than $L(j\omega)$, but this stabilizing influence is more than

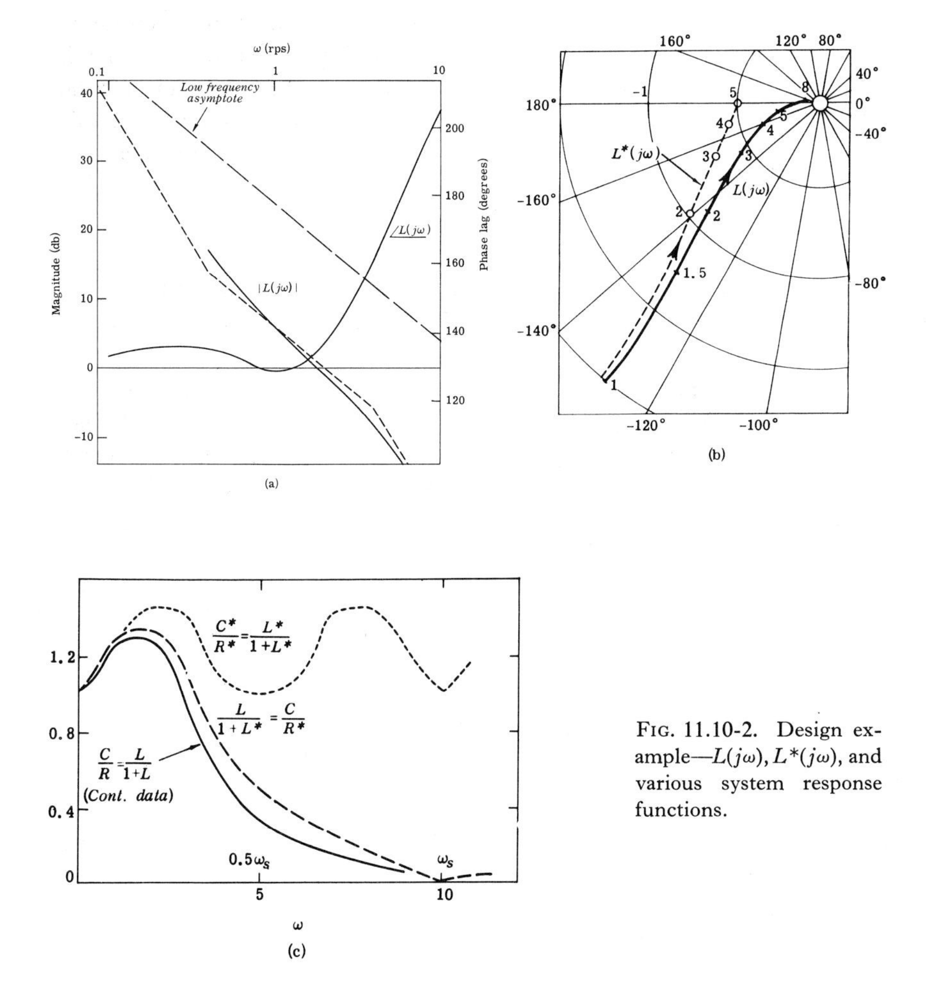

FIG. 11.10-2. Design example—$L(j\omega)$, $L^*(j\omega)$, and various system response functions.

canceled by the *a priori* need in curve B for very large lag angles. If $L(s)$ has a pole at the origin, then only the low frequency regions of the curves in Figs. 11.10-1 are modified. The shapes of the curves in the higher frequency region are substantially the same.

Stable designs demand that the -1 point be located as shown in Fig. 11.10-1a. Consequently, the crossover frequency ω_c must be less than $0.5\omega_s$. Furthermore, the closer ω_c is to $0.5\omega_s$, the more difficult it is to simultaneously achieve large phase and gain margins. For example, if $\omega_c = 0.25\omega_s$, then only one octave is available for reducing $| L^*(j\omega) |$. Fast reduction is necessary in order to achieve a large gain margin, and such fast reduction is accompanied by large phase lags, which works against a large phase margin. The only recourse is to have a horizontal step for $m\omega_c < \omega < \omega_c$ (with m about 0.5 or so), so that most of the lag at ω_c is due to the fast decrease of L^* for $\omega > \omega_c$. But then the crossover frequency (in the sense of the frequency up to which the benefits of feedback are available) is really $m\omega_c$ and not ω_c. This point is rather important, for the benefits of feedback are determined by $L^*(j\omega)$, and the above discussion reveals the significant limitation on the effective loop transmission, due to sampling.

(3) Design Procedure

The design procedure is illustrated by a numerical example in which $\omega_s = 10$ and it is desired that $\omega_c = 2$ rps, $K_v{}^* = 25$, and the phase margin is to be about 35°.

The design is initiated by drawing the low frequency asymptote of $L(j\omega)$. Since $K_v{}^*/j\omega \doteq L^*(j\omega) \doteq T^{-1}L(j\omega)$ for small ω, we take $L(j\omega) \doteq TK_v{}^*/j\omega = 5\pi/j\omega$, for small ω. (See Fig. 11.10-2a.) In the first trial, we try to achieve approximate crossover at $\omega = 2$, with a phase margin of about 40° (an extra 5° is allowed for the destabilizing effect of the sampling). Clearly, low frequency lag compensation is needed, with a frequency ratio of three octaves in order to reduce $L(j2)$ from 18 db to 0 db. We allow about 10° for the effect of the lag compensation at 2 rps, and therefore have $90° - 50° = 40°$ for the inevitable additional poles. Assuming three excess poles over zeros are required, one is assigned to $s = -4$ and the other at -10. The resulting $L(j\omega)$ is sketched in Figs. 11.10-2a and b. $L^*(j\omega)$ is obtained by means of the construction described in Fig. 11.3-3 and is shown in Fig. 11.10-2b. The specifications have been fairly closely achieved, but the gain margin is only 6 db approximately, and cannot be substantially increased without affecting the other design specifications.

Some frequency response magnitudes (for the above design) are shown in Fig. 11.10-2c. That of $C/R = L/(1 + L)$ is of a continuous-data system with the $L(s)$ of Fig. 11.10-2a. The exact time response to the sampled input is available from C/R^*, but it is difficult to calculate, because it consists of a succession of transients. We shall therefore be content in attempting to establish a correlation between the sampled step response and C^*/R^*.

(4) Time-Frequency Correlation

Curves of C^*/R^* are presented in Figs. 11.10-3a and b. The former were obtained from Case 1 of Section 11.9 [Eq. (11.9,1)] for different values of αT

and K/α. The latter were obtained from the z transforms of a complex pole pair. The corresponding sampled step-response curves are sketched in Figs. 11.10-4a, b, and c. (The sampled values are joined by straight lines, but obviously these lines do not represent the intersample output.) The effect of bandwidth, peaking, etc., is fairly obvious.

Comments on the Frequency Response Method

The singular advantage of the frequency response method is that one is working directly with $L(j\omega)$, which is the function which must eventually be realized. Constraints imposed on $L(s)$ due to the nature of the plant (e.g., a nonminimum phase plant) are therefore easily taken into account. The other design methods are quite inadequate in this last respect. Also, $L^*(j\omega)$ is very closely related to $L(j\omega)$, and the former determines the benefits of feedback in a sampled-data system. Therefore, just as in the continuous-data system, the system filter and feedback properties are apparent, and can be kept in hand. The designer is not sacrificing knowledge of the system feedback properties, as he must in some of the other methods. One disadvantage of the method is the need to work both with the Bode plot and the Nyquist plot, in order to derive $L^*(j\omega)$ from $L(j\omega)$. Another is in the indirect correlation between frequency and time, in contrast with some of the other methods which are essentially time domain synthesis procedures. However, one does not often want a very precise time response. Therefore this last disadvantage will disappear as more curves like those in Figs. 11.10-3 and 4 are prepared. Also, the first disadvantage tends to disappear as one gains design experience.

§ 11.11 Loop Transmission Function in the *w* Plane

The limitations on the achievable gain and bandwidth of $L^*(j\omega)$ were noted in the last section. The benefits of feedback are of course determined by $L^*(j\omega)$. In a nonelementary design requiring a two-degree-of-freedom structure, L^* is chosen not to achieve the desired system response (filter properties), but to achieve the desired feedback properties. System feedback design may of course be done by shaping $L^*(j\omega)$, as in the last section. But due to the importance of the problem, another method is presented in this section, one which has some advantage over that of Section 11.10.

In the z plane, the boundary which is analogous to $s = j\omega$ is $z = e^{j\theta}$. Along this boundary

$$|z - a|^2 = a^2 + 1 - 2a\cos\theta = (a-1)^2 + 4a\sin^2(0.5\theta)$$
$$= (a-1)^2\,[1 + (4a\sin^2 0.5\theta)/(a-1)^2].$$

Therefore, if log sin 0.5θ is used as abscissa, and magnitude is plotted in decibels, one get the same 6 db per octave asymptotic slope in sketching $|z - a|$,

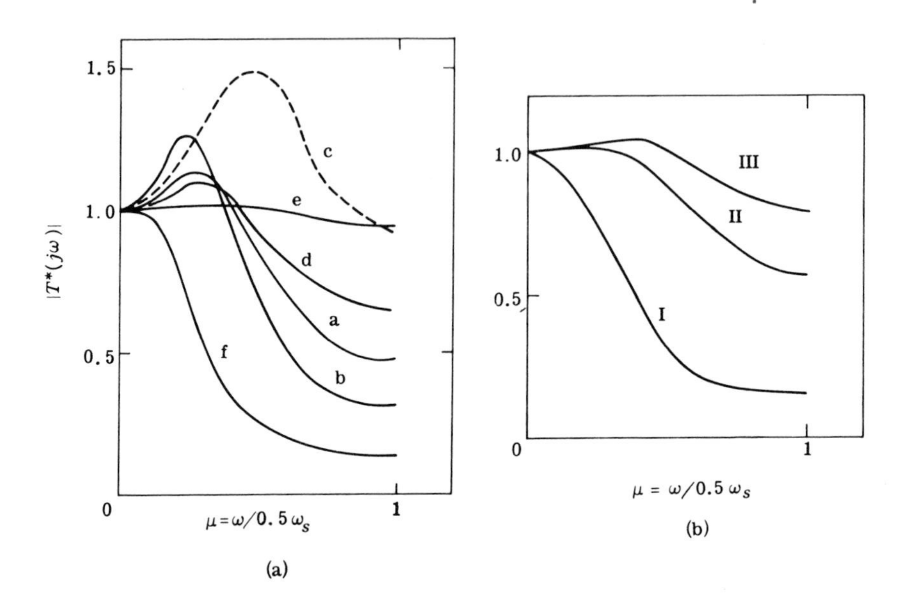

FIG. 11.10-3. Some frequency response curves of $C^*/R^*(j\omega) = T^*(j\omega)$.

with corner or break point at $0.5(a - 1)\, a^{-0.5}$. Unfortunately, the phase of $(z - a)$ is not symmetrical about any point, and its shape is a function of the value of a. Consequently, rather than work in the z domain, it is better to turn to a new variable[1] which results in Bode loop function shaping techniques, which are just as elegant and simple as the Bode plot techniques in continuous-data systems. It will also be seen that the feedback properties of sampled-data feedback systems (sensitivity to parameter variation and disturbance attenuation) are nicely clarified by working in this new domain.

This new variable is defined as

$$w = u + jv = (z - 1)/(z + 1) \tag{11.11,1}$$

with u, v real. It is readily found that at $s = j\omega$, $u = 0$, $v = \tan 0.5\omega T = \tan \pi\omega/\omega_s$, i.e., the $j\omega$ axis, which maps into the unit circle in the z plane, maps into the jv axis in the w plane. Also, the interior of the unit circle in the z plane maps into the left half-w plane. Each semi-infinite strip of $j\omega_s$ width, in the s plane, maps into the entire left half-w plane. Details of the mapping are shown in Fig. 11.11-1.

[1] C. W. Johnson, D. P. Nordling, and D. P. Lindorff, Extension of continuous-data system design techniques to sampled-data control system. *Trans. AIEE Part II*, **74**, 252-263 (1955).

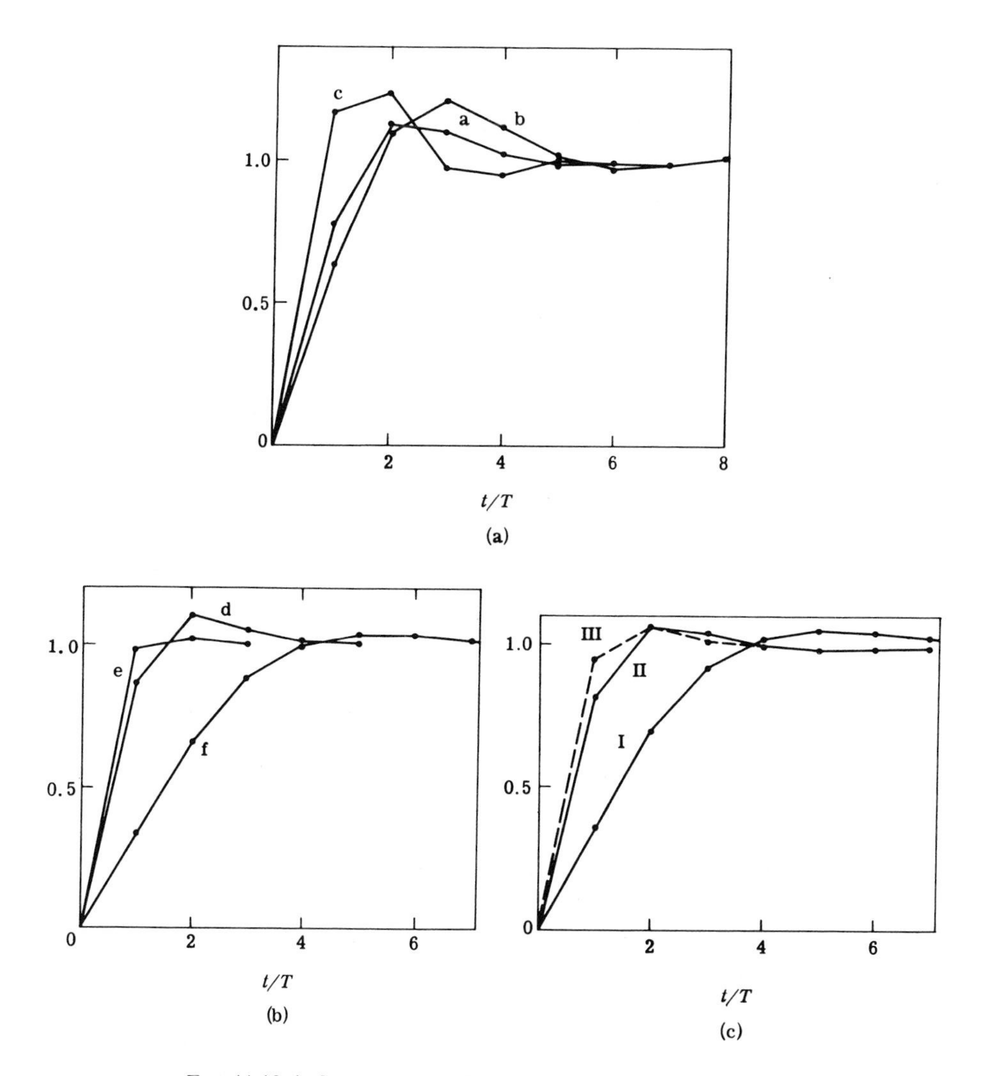

FIG. 11.10-4. Step responses for curves of Fig. 11.10-3.

Curve	αT	K/α
a	1.5	1.0
b	1.0	1.0
c	1.5	1.5
d	2.0	1.0
e	4.0	1.0
f	1.0	0.5

Curve ($\zeta = 0.707$)	$\omega_n/0.5\omega_s$
I	0.25
II	0.5
III	0.60

The transfer function $1/(s + a)$ becomes the pulsed transfer function $z/(z - e^{-aT})$, and in the w domain, it becomes $(w + 1)\{(1 + e^{-aT}) \times [w + (1 - e^{-aT})/(1 + e^{-aT})]\}^{-1}$. The equivalent Bode chart in the jv domain has $\log v$ as its abscissa, and the magnitude and phase of functions of $w = jv$

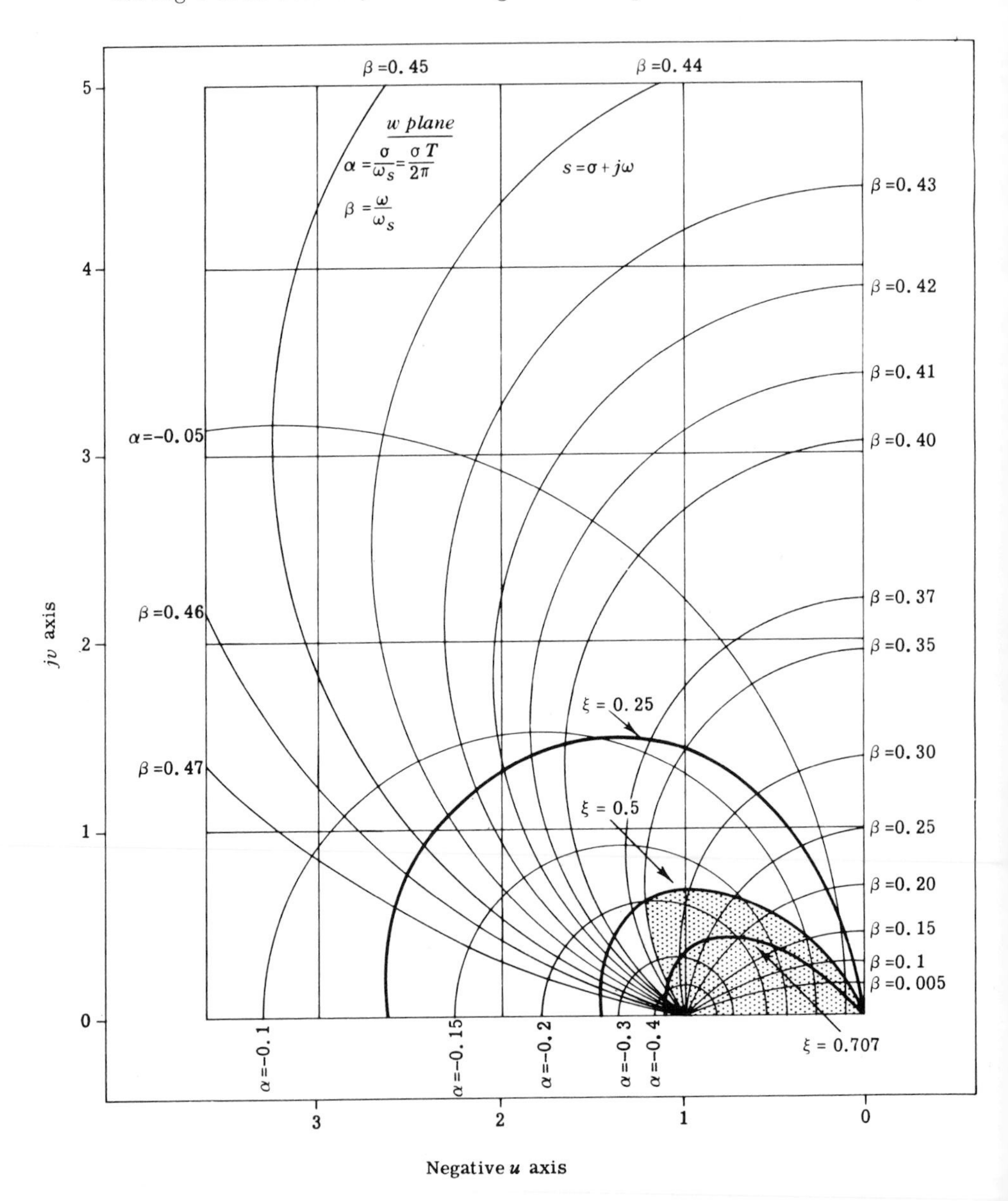

FIG. 11.11-1. Mapping of $s = \sigma + j\omega = (\alpha + j\beta)\omega_s$ plane into $w = u + jv$ plane.

are sketched in exactly the same manner as are functions of ω in the $j\omega$ domain. Thus the variable v becomes the equivalent frequency variable.

It will be seen that the feedback properties of a sampled-data system are well clarified in the w domain. It is therefore important to understand what are the significant regions in the w domain. The $j\omega$ axis maps on the jv axis, and the entire negative-real σ axis ($s = \sigma + j\omega$) maps in the w plane on the line $-1 \leqslant u \leqslant 0$. The line $\sigma < 0$, $j\omega = j0.5\omega_s$ maps on the line $u < -1$. Bandwidth in the w domain is not completely independent of damping factor, as it is in the s domain. In Fig. 11.11-1, if a straight line is drawn through the origin, and one locates a pair of poles on this line, the damping factor decreases as one moves away from the origin (i.e., the equivalent s-plane poles, which characterize the associated envelope of the sampled time function, have decreasing damping factors). Some lines of constant s-plane damping factor ($\zeta = 0.707$, 0.50, 0.25) are shown in Fig. 11.11-1. It is therefore seen that the poles and zeros of $L(w)$ will be concentrated in an area not too far from the origin, probably within the $\zeta = 0.50$ region, which forms one of the boundaries of the lightly shaded region in Fig. 11.11-1.

The notation $L(w)$ is used to denote $L^*(s)$ with $s = (1/T) \ln [(1 + w)/(1 - w)]$, or $L(z)$, $z = (1 + w)/(1 - w)$. Are there any limitations on the loop transmission $L(w)$, or can its poles and zeros be selected freely? It is known that $L(z)$ must have more poles than zeros in order that $L(s)$ go to zero at infinity. If there is to be no pure time delay in $L(s)$, $L(z)$ should have an excess of only one pole over zeros. Now a factor $(w + a)$ in $L(w)$ becomes $[z(1 + a) - (1 - a)]/(z + 1)$ in $L(z)$, with a zero at $z = (1 - a)/(1 + a)$. Consequently, one zero of $L(w)$ must be at 1 (i.e., at infinite z) in order that $L(z)$ have one more pole than zeros. Thus, to ensure that $L(z)$ has one more pole than zeros, $L(w)$ must have in its numerator the factor $-(w - 1)$. Hence $L(w)$ is nonminimum phase. The limitations thereby imposed on the achievable loop transmission gain and bandwidth were noted in Sections 7.9 and 7.12. Suppose a pole at -1 is assigned to $L(w)$ to cancel the effect of the $-(w - 1)$ factor on $|L(w)|$. This combination of pole and zero is an all-pass function which contributes appreciable phase lag, as shown in Fig. 11.11-2. The severe restrictions that this phase lag imposes on the permitted gain and bandwidth of $L(w)$ can be seen immediately. For example, suppose $L(w)$, excluding the all-pass $-(w - 1)(w + 1)^{-1}$ part, consists of only one pole at the origin. If a 30° phase margin is desired, then the "crossover" frequency must be at $v = 0.58$, because in Fig. 11.11-2 the all-pass network phase lag at $v = 0.58$ is 60°, which, added to the 90° lag due to the pole at the origin, gives a total of 150° lag; for 40° phase margin it is 0.45. The crossover frequency can be increased by letting $L(w)$ have as many zeros as poles. Ordinarily, $L(w)$ will have as many zeros as poles, unless $L(z)$ is assigned a zero at $z = -1$. This is obvious, by noting that $z = -1$ corresponds to infinite w, so that a zero at infinite w requires a zero of $L(z)$ at $z = -1$. In order to maximize the crossover frequency of $L(w)$, it is better to let $L(w)$ have as many zeros as poles. In such a case, the phase lag of $L(jv)$ approaches 180° as v approaches infinity. Otherwise, if $L(z)$ has a zero at $z = -1$, the phase lag approaches 270°. [This proves the statement made in Section 11.10 that curves C and D for $L^*(j\omega)$ in Fig. 11.10-1a are impossible if $L(s)$ has at least two more poles than zeros.]

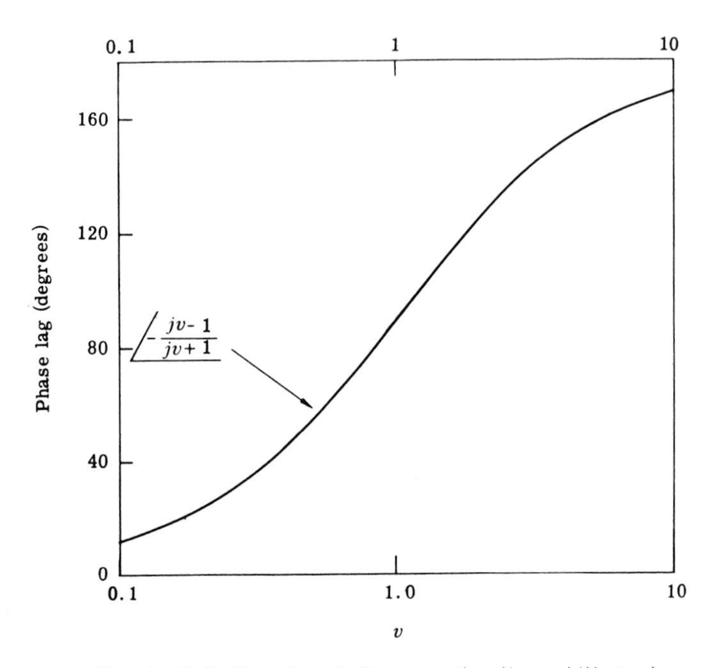

FIG. 11.11-2. Phase lag of all-pass section $(1 - w)/(1 + w)$.

Suppose all the component parts of $L(s)$, i.e., $G(s)$ and $P(s)$ in Fig. 11.8-1, are to consist of the usual continuous type elements as opposed to pulsed circuits or digital controllers (discussed in Section 11.13). In realizing the corresponding network, it is advantageous to expand $L(z)$ in a series of the form $\sum A_i z/(z - a_i)$, and to obtain from Table 11.4-1 the corresponding continuous transfer function. For the configuration of Fig. 11.8-1 then, $L(z)$ should have a zero at the origin, and, consequently $L(w)$ must have a zero at $w = -1$. On the other hand, if the plant is to be preceded by a hold circuit, no zero at $w = -1$ is required.

If the poles of $L(s)$ are to be confined to the negative real axis, then the poles of $L(w)$ are restricted to lie on the negative real w axis between the origin and -1 [this is seen from Eq. (11.11,1)]. In the configuration of Fig. 11.12-1b there is no need to avoid poles of $L(w)$ which lead to s-plane highly underdamped poles, because they can be assigned to $G(s)$, and therefore will not adversely affect intersample behavior. In the structure of Fig. 11.8-1, they do affect the intersample behavior.

In designing for the system feedback properties, the advantage of working with $L(jv)$ in place of $L^*(j\omega)$ is that there is no need to graphically add the higher harmonics; i.e., the Bode plot alone suffices in place of the Bode-Nyquist combination. The disadvantage is in the need to translate all the old familiar concepts into the new formulation. In this connection, it may be noted that in working with $L(jv)$ there is no control of the excess poles over zeros of the final $L(s)$ which must be constructed. It is therefore necessary to add to $L(s)$ additional poles sufficiently far off to have only a minor effect on the resulting new $L(z)$. How far off must these additional poles be placed?

Consider a nominal $L_o(s)$ derived from a v-axis design which emerges with an insufficient number of excess poles. It is only the sampled values that are used in the feedback loop. Therefore additional poles can be added so long as these negligibly affect the sampled values. Let $L(s) = L_o(s)\, X(s)$, with $X(s)$ representing the additional poles, i.e., $X(s) = abc.../(s + a)(s + b)(s + c)\ldots$. Clearly, if $x(t) = \mathscr{L}^{-1}X(s)$ is substantially over in a fraction of the sampling period, the sampled values of $l(t) = \mathscr{L}^{-1}L(s)$ will not differ by much from those of $l_o(t)$. Thus $\tau \approx 0.1T$ [where τ is the effective time constant of $x(t)$] should be satisfactory. If $X(s) = a/(s + a)$, this would require $a \approx 2\omega_s$. Thus the far-off additional poles need not actually be so very far off. However it may be a good idea to check their effect on the stability margins of $L(jv)$.

§ 11.12 Use of Pulsed Networks as Compensators

Some of the difficulties associated with the configuration of Fig. 11.12-1a disappear if more samplers are allowed in the system. Consider, for example, the structure in Fig. 11.12-1b. Since the poles of $G(z)$ do not directly affect the

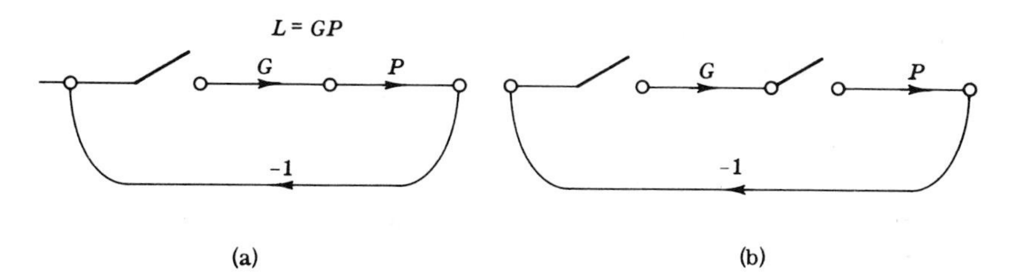

FIG. 11.12-1. (a) Elementary feedback structure (b) Modification which removes some of the difficulties.

intersample system output, they can be lightly damped. Consequently $L(w)$ may have poles beyond $w = -1$ without an oscillatory intersample response, because these poles may be assigned to G. This permits $L(w)$ to extend over a wider band. Also, it is no longer inconvenient to specify $T(z)$ directly, either on the basis of sampling a continuous output or from any other considerations.

It will be recalled that one difficulty (Section 11.8) with the configuration of Fig. 11.12-1a was that of picking $T(z)$ such that the resulting $L(s)$ did not have oscillatory poles. Now we do not care if $L(s)$ does have such poles.

The lightly damped poles of $L(z)$ can be assigned to $G(z)$ and not to $P(z)$. The intersample behavior is determined primarily by the $P(s)$ network, so the higher damped poles are assigned to $P(z)$. This may require $P(s)$ to consist of compensation networks in cascade with the plant, in order to cancel out the plant poles. In this regard, it is important to recall that, just as in continuous systems, rhp poles of the plant transfer function cannot be canceled out exactly. They must appear as poles of the loop transfer function. On the other hand, rhp plant zeros are more difficult to accommodate in sampled-data systems than in continuous systems. In the latter, $T(s) = C(s)/R(s)$ must incorporate such zeros. But in sampled-data systems, this can be done only when $P(z)$ consists *only* of the plant and then the rhp zero is known. If $P(s)$ includes compensation, the zero in the z domain is no longer the same, and is unknown until the compensation is known. The constraint of the rhp plant zero can be incorporated directly only if the compensation included in the $P(s)$ block is known.

There are three principal methods for realizing the $G(z)$ transfer function. One method is to use lumped continuous networks by transforming in the usual manner from $G(z)$ to $G(s)$. This has been previously discussed and requires no further comment. It may be inconvenient to realize low frequency complex poles or rhp zeros by means of networks consisting of inductors, resistors, and capacitors. In such a case, active RC techniques may be used (see Appendix). A second method is Sklansky's[1] pulsed RC networks, and the third is the digital controller.

Design of Pulsed RC Networks

In this method, the most general structure for realizing $G(z)$ of Fig. 11.12-1b is that shown in Fig. 11.12-2. Let the desired z transfer function be $G(z) = N(z)/D(z) = [z^m + a_1 z^{m-1} + \dots + a_m]\,[z^n + b_1 z^{n-1} + \dots + b_n]^{-1}$. In order that

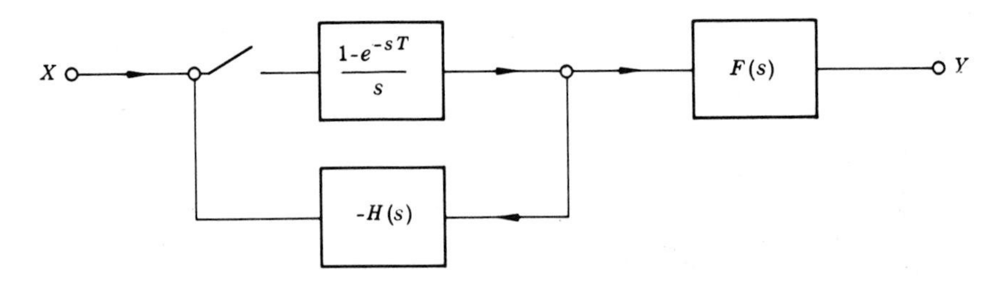

FIG. 11.12-2. General pulsed RC structure for realizing $G(z) \triangleq Y(z)/X(z)$.

[1] J. Sklansky, Pulsed RC networks for sampled-data systems, *Convention Record, IRE Part II.* Circuit Theory, pp. 81-99 (1956).

$G(z)$ be realizable, $n \geqslant m$; otherwise $G(z)$ must produce an output before there is an input. The object is to realize $G(z)$ by means of the structure of Fig. 11.12-2, such that $F(s)$ and $H(s)$ have negative real poles only, so that they are realizable by RC networks. Actually, it would also be desirable to have the zeros of $F(s)$ and $H(s)$ restricted to the left half-plane, but this is not guaranteed by the design. From Fig. 11.12-2,

$$G(z) = \frac{\frac{z-1}{z} Z\left[\frac{F(s)}{s}\right]}{1 + \frac{z-1}{z} Z\left[\frac{H(s)}{s}\right]} = \frac{N(z)}{D(z)} = \frac{z^m + a_1 z^{m-1} + \dots + a_m}{z^n + b_1 z^{n-1} + \dots + b_n}. \tag{11.12,1}$$

Divide numerator and denominator of the right-hand side of Eq. (11.12,1) by $d(z) = (z-1)(z-\alpha_1)\dots(z-\alpha_n)$ with $0 < \alpha_i < 1$, so that

$$\frac{(z-1)z^{-1} Z\, F(s)s^{-1}}{1 + z^{-1}(z-1)ZH(s)/s} = \frac{N(z)/d(z)}{D(z)/d(z)} = \frac{(z^m + a_1 z^{m-1} + \dots + a_m)/(z-\alpha_1)\dots(z-\alpha_n)}{(z^n + b_1 z^{n-1} + \dots + b_n)/(z-\alpha_1)\dots(z-\alpha_n)}. \tag{11.12,2}$$

From Eq. (11.12,2), the following identifications are made:

$$Z\frac{F(s)}{s} = \frac{zN(z)}{(z-1)\,d(z)} = \frac{z}{z-1}\left[\frac{z^m + a_1 z^{m-1} + \dots + a_m}{(z-\alpha_1)\dots(z-\alpha_n)}\right] \tag{11.12,3a}$$

$$Z\frac{H(s)}{s} = \frac{z}{z-1}\left[\frac{D(z)}{d(z)} - 1\right] = \frac{z}{z-1}\left[\frac{z^n + b_1 z^{n-1} + \dots + b_n}{(z-\alpha_1)\dots(z-\alpha_n)} - 1\right]. \tag{11.12,3b}$$

To find $s^{-1}F(s)$, expand $N(z)/(z-1)\,d(z)$ in partial fractions, so that

$$Z\frac{F(s)}{s} = z\left[\frac{z^m + a_1 z^{m-1} + \dots + a_m}{(z-1)(z-\alpha_1)\dots(z-\alpha_n)}\right] = z\left[\frac{A_o}{z-1} + \sum \frac{A_i}{z-\alpha_i}\right]. \tag{11.12,4}$$

Consequently, $s^{-1}F(s) = A_o s^{-1} + \sum A_i(s+\beta_i)^{-1}$, with $\alpha_i = e^{-\beta_i T}$, and the poles of $F(s)$ are restricted to the negative real axis, and $F(s)$ has no poles at infinity. As previously noted, it would be desirable to restrict the zeros of $F(s)$ to the left half-plane, but the above procedure does not guarantee it. Similar remarks will apply to the zeros of $H(s)$. The α_i in $d(z)$ are best chosen so that $\beta_i = -T^{-1}\ln\alpha_i$, are not too closely spaced on the negative real axis. Similarly,

$$Z\frac{H(s)}{s} = z\left[\frac{D(z) - d(z)}{(z-1)d(z)}\right] = z\left[\frac{B_o}{z-1} + \sum \frac{B_i}{z-\alpha_i}\right] \tag{11.12,5}$$

by means of a partial fraction expansion of the right-hand side of Eq. (11.12, 3b).

Therefore $s^{-1}H(s) = B_0 s^{-1} + \sum B_i(s + \beta_i)^{-1}$. In the above, $d(s)$ could have been chosen with more than n negative real zeros, but there is no point in doing this.

It has been assumed in the above that $G(z)$ has no poles or zeros on $0 < z < 1$. If $G(z)$ has some (say x) zeros there, let $n(z)$ contain these zeros and write

$$G(z) = \frac{N(z)}{D(z)} = \frac{N_1(z)n(z)}{D(z)} = \frac{N_1(z)}{D(z)/n(z)}.$$

Proceed as before, dividing through by $d(z)$, etc., but $d(z)$ need have only n-x zeros, i.e., $(1 - z^{-1})\, Zs^{-1} F(s) = N_1(z)/d(z)$, and $1 + (1 - z^{-1})\, Zs^{-1} H(s) = D(z)/n(z)\, d(z)$. Thus $H(s)$ will have x poles which are not canceled out by poles of $F(s)$.

If only $D(z)$ has zeros (y in number) on $0 < z < 1$, then write

$$G(z) = \frac{N(z)}{D(z)} = \frac{N(z)}{D_1(z)q(z)} = \frac{N(z)/q(z)d(z)}{D_1(z)/d(z)}$$

with $d(z)$ of degree $n - y$. Now $F(s)$ will have y poles which are not canceled out by poles of $H(s)$.

Finally, if $N(z)$ has x zeros [contained in $n(z)$] and $D(z)$ has y zeros [contained in $q(z)$] on $0 < z < 1$, then write

$$G(z) = \frac{N(z)}{D(z)} = \frac{N_1(z)n(z)}{D_1(z)q(z)} = \frac{N_1(z)/q(z)d(z)}{D_1(z)/n(z)d(z)}$$

with $d(z)$ of degree n-x-y.

Example. $G(z) = (z^2 + 0.4z + 0.5)/z(z^2 - z + 0.8) = N(z)/D(z)$. Let $d(z) = (z - 0.2)(z - 0.4)(z - 0.6)$. Therefore

$$Z\frac{1}{s}F(s) = z\frac{z^2 + 0.4z + 0.5}{(z-1)(z-0.2)(z-0.4)(z-0.6)}$$
$$= z\left(\frac{9.9}{z-1} - \frac{9.7}{z-0.2} + \frac{34.2}{z-0.4} - \frac{34.4}{z-0.6}\right).$$

Consequently,

$s^{-1}F(s) = 9.9s^{-1} - 9.7(s + 1.61)^{-1} + 34.2(s + 0.92)^{-1} - 34.4(s + 0.51)^{-1}$, if $T = 1$, and

$$F(s) = [1.8(s^2 + 0.10s + 4.1)]/[(s + 0.51)(s + 0.92)(s + 1.61)].$$

From Eq. (11.12,5),

$Zs^{-1}H(s) = z[3.17(z-1)^{-1} - 2(z-0.2)^{-1} + 9.34(z-0.4)^{-1} - 10.5(z-0.6)^{-1}]$.

Therefore

$s^{-1}H(s) = 3.17s^{-1} - 2(s + 1.61)^{-1} + 9.34(s + 0.92)^{-1} - 10.5(s + 0.51)^{-1}$, and $H(s) = 0.83(s + 2.87)\,[(s + 1.61)(s + 0.92)(s + 0.51)]^{-1}$.

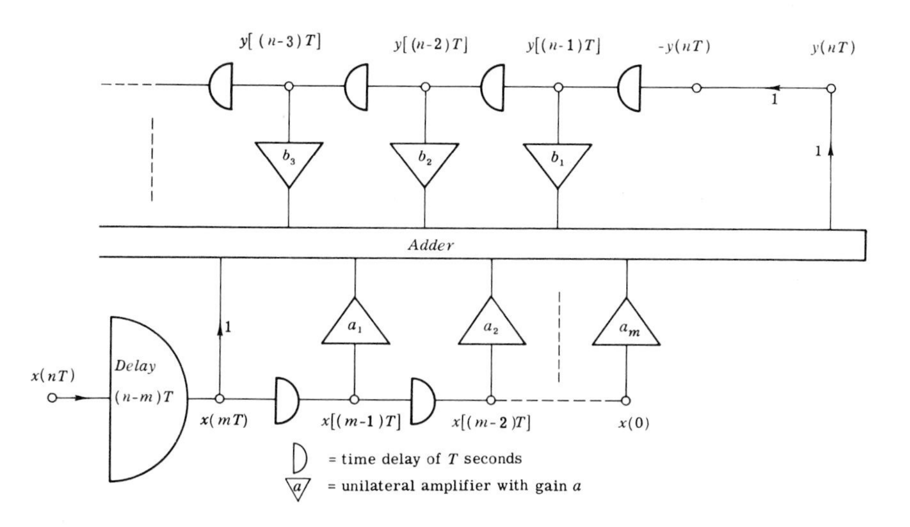

FIG. 11.13-1. Digital controller realization of P.T.F.

§ 11.13 The Digital Controller

The pulse transfer function can be realized by means of a suitable program on a digital computer. Let

$$G(z) = Y(z)/X(z) = [z^m + a_1 z^{m-1} + \ldots + a_m]/[z^n + b_1 z^{n-1} + \ldots + b_n],$$

$(n \geqslant m)$, represent the desired P.T.F. The above can be written in the form (by dividing by z^{-n} and multiplying through):

$$(1 + b_1 z^{-1} + b_2 z^{-2} + \ldots + b_n z^{-n})\, Y(z) = (z^{m-n} + a_1 z^{m-n-1} + \ldots + a_m z^{-n})\, X(z).$$

This equation has the following time domain interpretation:

$$y(nT) = -\,[b_1 y(n-1)\,T + b_2 y(n-2)\,T + \ldots + b_n y(0)] + [x(mT) + a_1 x(m-1)T + \ldots + a_m x(0)].$$

The latter is implemented in Fig. 11.13-1, in which $n + m$ delay and storage elements are required. The number can be reduced[1] to n by writing $Y(z) = [z^{m-n} + a_1 z^{m-n-1} + \ldots + a_m z^{-n}]\, Y_1(z)$, where $Y_1(z) = X(z)[1 + b_1 z^{-1} + b_2 z^{-2} + \ldots + b_n z^{-n}]^{-1}$, and $Y_1(z)$ is implemented in Fig. 11.13-2a. Next, $Y(z) = [z^{m-n} + a_1 z^{m-n-1} + \ldots + a_m z^{-n}]\, Y_1(z)$ is implemented in Fig. 11.13-2b. The two (Figs. 11.13-2a and b) are combined, and only n storage and delay elements are required.

[1] R. H. Barker, The pulse transfer function and its application to sampling servo systems. *Proc. IEE, Monograph No. 43, Part IV* **99**, 302-317 (1952).

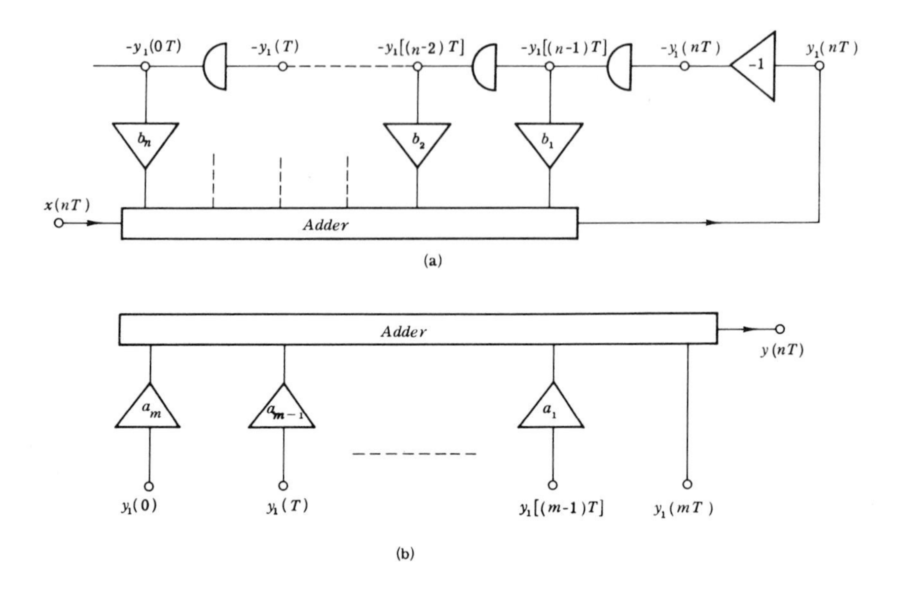

FIG. 11.13-2. More economical digital controller synthesis.

§ 11.14 Sensitivity in the Single-Degree-of-Freedom Configuration

It was noted in Section 5.22 that in the continuous one-degree-of-freedom system, the system sensitivity to plant parameter variations in a significant frequency range is greater than if no feedback were used at all. This property of the one-degree-of-freedom system is accentuated in sampled-data systems. It is recalled that the poor sensitivity properties of the one-degree-of-freedom configuration are due to the fact that in such systems the sensitivity function is completely determined by the system transfer functions. Thus, both in continuous systems [Eq. (5.22, 1)] and in sampled-data systems (in the s, z, or w domains), $S_P{}^T \triangleq (\Delta C/C')/(\Delta L/L') = 1 - T$, where C', L' are the appropriate functions at the new parameter values. Figure 5.22-1 is redrawn here as Fig. 11.14-1.

A typical $T(jv)$ is sketched in Fig. 11.14-1. Infinite v corresponds to $\omega = 0.5\omega_s$ and at this point T is finite[2] (see Figs. 11.10-2). Since $S_P{}^T = 1 - T$, the origin

[2] The situation is made worse if $T(jv)$ is deliberately made zero at infinity [i.e. by letting $L(w)$ have one more pole than zeros]. For in such a case the lag of $L(jv)$ as $v \to \infty$, is 270° in place of 180°. The frequency range for which $|S_P{}^T| > 1$ is thereby increased.

for the sensitivity function is at the point 1, indicated by A in Fig. 11.14-1. A circle of radius 1 centered at A is drawn. For all points inside this circle, $|S_P{}^T| = |1 - T| < 1$, so that sensitivity decrease is realized. For all points

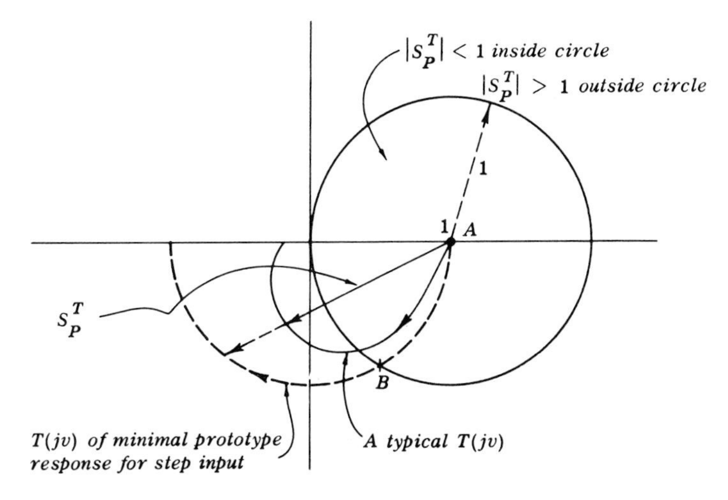

FIG. 11.14-1. Inherent limitations in sensitivity properties of elementary sampled-data feedback systems.

outside this circle, $|S_P{}^T| = |1 - T| > 1$, so that for such points the sensitivity is greater than if no feedback were used at all. A significant portion of our typical $T(jv)$ is outside the circle. If the phase lag of T increases quickly as a function of v, then the smaller the value of "frequency" at which T lies outside the critical circle, and the large the "frequency" range of T for which the sensitivity is greater than if no feedback were used at all. In sampled-data systems, $L(w)$ and $T(w)$ are always nonminimum phase, with a consequent tendency for more phase lag in the low frequency range. In addition, the tendency is for $T^* = L^*/(1 + L^*)$ to be not small at $v = \infty$ ($\omega = 0.5\omega_s$). For example, a value of -0.2 for L^* at $0.5\omega_s$ is quite small (in magnitude) and it results in $T^* = -0.25$. A more typical value of $L^* = -0.4$ results in $T^* = -0.67$. Consequently the poor sensitivity properties of the one-degree-of-freedom configuration are accentuated in sampled-data systems.

The Sensitivity of Minimal Prototype Response Functions

A minimal prototype response function[1] $T_m(z)$, is one (1) whose sampled transient response is as fast as possible, and (2) whose sampled steady state error

[1] J. R. Ragazzini and G. F. Franklin, "Sampled-Data Control Systems," Chapter 7. McGraw-Hill, New York, 1958.

is zero, for a specific class of inputs. It is also recalled that (3) $T(z)$ must have one more pole than zeros for a realistic plant which cannot respond instantaneously and which has no pure time delays in it. The error series is $E(z) = R(z)[1 - T(z)]$, and condition (2) states that the series must be finite. Condition (1) requires that the series have the minimum number of terms. Condition (3) requires that $1 - T(z)$ must have as many zeros as poles. These three conditions suffice for finding $T_m(z)$ for any input.

For example, if the input is a step function, then $E(z) = [1 - T(z)]z/(z-1)$. A finite series for $E(z)$ requires that $z - 1$ be canceled out, i.e., $1 - T(z) = A(z)(z-1)$, and then $E(z) = zA(z)$. $A(z)$ with the smallest number of terms [condition (1)], which satisfies condition (3), is K/z. Also, $K = 1$, because $e(0) = r(0) = 1$. Therefore $E(z) = 1$ and $T(z) = 1 - (E/R) = 1 - R^{-1} = z^{-1}$. If the input is a ramp, then $E(z) = [1 - T(z)]\,Tz/(z-1)^2$. From condition (2), $1 - T(z) = (z-1)^2 A(z)$, and therefore $E(z) = TzA(z)$. $A(z)$ with the smallest number of terms which satisfies condition (3) is K/z^2, giving $E(z) = TK/z$. Hence $T(z) = 1 - (E/R) = [z^2 - K(z-1)^2]/z^2$. $K = 1$ in order to satisfy condition (3), leading to $T(z) = 2z^{-1} - z^{-2}$.

It is easy to see that the minimal prototype response functions are very sensitive to parameter variations. For example, that for a step input has $T(z) = z^{-1}$ or $T(w) = (1-w)/(1+w)$, an all-pass function. Hence, in Fig. 11.14-1, most of $T(jv)$ lies in the region $|S_P^T| > 1$ (actually for all $\omega > \omega_s/6$, because B in Fig. 11.14-1 corresponds to $v = \tan 90°$, at which $\omega = \omega_s/6$). For a ramp input $T(w) = (1-w)(1+3w)/(1+w)^2$, which has even larger excursions outside the region $|S_P^T| < 1$, i.e., the feedback is positive over a very large portion of the frequency range. It is noted that these response functions have all their poles at $z = 0$, i.e., at $w = -1.0$. The large sensitivity of such functions is also seen from Fig. 11.11-1—the characteristics are very crowded at $w = -1.0$. A very slight change in pole position moves the operating point to a radically different equivalent s-plane value.

Use of z Transforms to Calculate the Effect of Parameter Variations

The z-transform notation is very convenient for calculating the effect of parameter variations on the system response. Let $T(z)$, $L(z)$ represent the original functions, and T', L', etc., the final values due to parameter changes. In the one-degree-of-freedom system $T = L/(1+L)$, and it is easily found that $C' - C = \Delta C = C'(1-T)\,\Delta L/L'$. Let $\Delta L/L' = \sum_0^\infty \beta_n z^{-n}$, $\Delta T = T' - T = \tau_n z^{-n}$, $\Delta C = \sum \delta_n z^{-n}$, $C = \sum c_n z^{-n}$, $C' = \sum (c_n + \delta_n) z^{-n}$. Substituting and equating coefficients in the expression for ΔC results in $\delta_1 = \beta_o c_1/(1-\beta_o)$, $\delta_2 = (\beta_o c_2 + \beta_1 c_1' - \beta_o c_1' \tau_1)/(1-\beta_o)$, and, in general,

$$\delta_n(1-\beta_o) = \beta_o c_n + \sum_1^{n-1} \beta_i c'_{n-i} - \tau_1 \sum_0^{n-2} \beta_i c'_{n-i-1} - \tau_2 \sum_0^{n-3} \beta_i c'_{n-i-2} - \dots$$
$$- \tau_{n-1}\beta_o c_1'.$$

Another way to calculate the effect of parameter changes is to manipulate $\Delta C = (\Delta L/L')C'(1 - T)$ into $C'[1 - (L/L')(1 - T)] = C$, and then $C/C' = 1 - (1 - T)\Delta L/L'$. The series for C/C' is inverted in order to find C'.

The above is illustrated with an example in which $T(z) = (z + 0.45)/(z^2 + 0.25z + 0.2)$. The nominal step-function response, sketched in Fig. 11.14-2, is $C_o(z) = z^{-1} + 1.2z^{-2} + 0.95z^{-3} + 0.973z^{-4} + 1.017z^{-5} + 1.001z^{-6} + \ldots$, and $T(z) = z^{-1} + 0.2z^{-2} - 0.25z^{-3} + 0.023z^{-4} + 0.44z^{-5} - 0.016z^{-6} + \ldots$. Suppose the plant gain constant increases by 25% so that $\Delta P/P' = 0.20$. Then from either of the two relations $\Delta C(z) = 0.25z^{-1} - 0.0125z^{-2} - 0.122z^{-3} + 0.055z^{-4} + 0.023z^{-5} - 0.076z^{-6} + \ldots$. The resulting $C'(z) = C_o(z) + \Delta C(z)$ is also sketched in Fig. 11.14-2. It is seen that the system step response is quite sensitive to parameter changes.

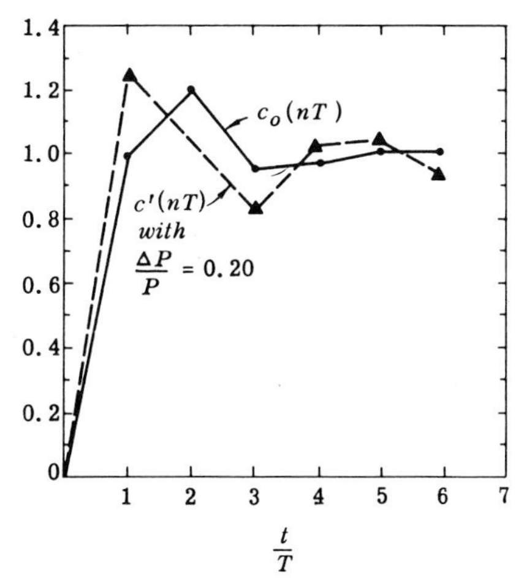

FIG. 11.14-2. Effect of parameter change on step response.

In the structure of Fig. 11.12-1a, it is more difficult to ascertain the effect of plant time constant variations than it is for Fig. 11.12-1b. Thus, in continuous systems, when $L(s) = G(s)\,P(s)$, $\Delta L/L' = \Delta P/P'$, but this is not generally true in sampled-data systems. For example, if $P(s) = 1/(s + d)$ and $G(s) = (s + d)/(s + b)$, then $L(s) = 1/(s + b)$ and $L(z) = z/(z - e^{-bT})$. However, if d becomes d', then $L'(s) = (s + d)/(s + b)\,(s + d')$,

$$L'(z) = z(d' - b)^{-1}\,[(d - b)\,(z - e^{-bT})^{-1} + (d' - d)\,(z - e^{-d'T})^{-1}],$$

and

$$\Delta L/L' = (d' - d)\,(e^{-d'T} - e^{-bT})/[z(d' - b) - e^{-d'T}(d - b) - e^{-b'T}(d' - d)],$$

which is not equal to $\Delta P/P' = (e^{-d'T} - e^{-dT})/(z - e^{-dT})$. If pulsed compensation is used as in Fig. 11.12-1b, where $L(z) = G(z)\,P(z)$, then $\Delta L/L'(z) = \Delta P/P'(z)$. However, in Fig. 11.12-1a, if the $G(s)$ impulse response is much faster than that of $P(s)$ and of $P'(s)$, then $L(z) \approx G(z)\,P(z)$, and we can take $\Delta L/L'(z) \approx \Delta P/P'(z)$.

§ 11.15 Sensitivity Limitations in the Two-Degree-of-Freedom Structure

It is not surprising that the one-degree-of-freedom system does not permit independent control of the system transfer function and the system sensitivity to plant parameter variations. In the analogous continuous case, the problem was solved (Chapter 6) with two-degree-of-freedom structures. This is also formally the situation in sampled-data systems, but there is one significant difference. The constraint on the achievable loop gain-bandwidth is not in any way affected by the fact that there are now two degrees of freedom. The second degree of freedom can, however, be used to limit the effective bandwidth of the transfer function. If it is sufficiently narrow, then small sensitivity over the significant band of $T^*(j\omega)$ is achievable. Design for simultaneous realization of the system transfer function and its insensitivity to parameter variations can be carried out either in the $j\omega$ frequency domain or in terms of the equivalent jv domain. The former is similar to the detailed design procedure given in Chapter 6, with the modification noted in Section 11.10; it is therefore not repeated here. The second method is quite similar to the first, but since it involves working in the less familiar jv domain, it will be described in this section.[1]

Consider the two-degree-of-freedom structure of Fig. 11.15-1, and let

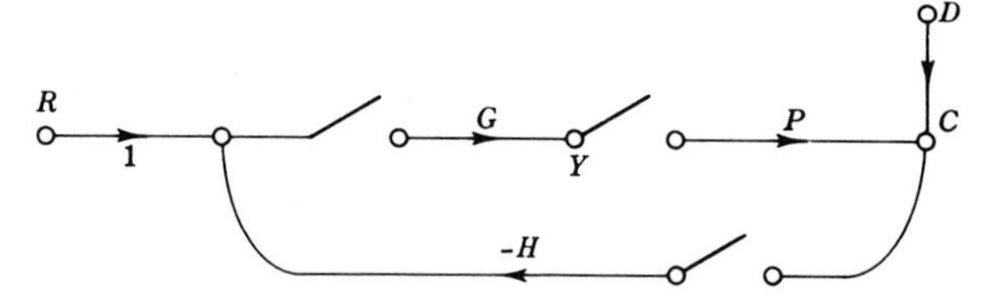

FIG. 11.15-1. Two-degree-of-freedom sampled-data feedback system.

T_o, P_o, L_o represent the usual functions when the plant transfer function has its nominal value P_o; and let T, P, L represent the corresponding functions when the plant transfer function has a different value P. Then $T_o(z) = G(z) P_o(z)/[1 + G(z) P_o(z) H(z)]$, $T(z) = G(z) P(z)/[1 + G(z) P(z) H(z)]$, and $T_o(z)/T(z) = [(P_o/P)(z) + L_o(z)] [1 + L_o(z)]^{-1}$, where $L_o(z) = G(z) P_o(z) H(z)$, $L(z) = G(z) P(z) H(z)$. The above is identical to Eq. (6.7,1) except that in the former the variable is $z = e^{sT}$ and not s. This is, of course, due to the fact that the relations between the discrete sampled output and the discrete sampled input are being studied here. It is, therefore, possible to parallel here the work of Chapter 6, except for the change in variable. However, for the reasons discussed in Section 11.11, it is preferable to work in the $w = u + jv$ domain in place of the z domain. The basic design equation is then

$$\frac{T_o(jv)}{T(jv)} = \frac{(P_o/P)(jv) + L_o(jv)}{1 + L_o(jv)}. \tag{11.15,1}$$

The design procedure is henceforth precisely the same as in Section 6.7 for the analogous continuous system design problem. The specifications are given in terms of the maximum allowable variation in $T_o/T(jv)$ instead of $T_o/T(j\omega)$.

[1] For a time-domain approach to the sensitivity problem see J. B. Cruz, Jr., Sensitivity Considerations for Time-Varying Sampled-Data Feedback Systems, *Trans. IRE*, AC-6, No. 2, pp. 228-236, May, 1961.

The boundaries of $P_o/P(jv)$ are found at a selected number of values of v. From these boundaries and from the specifications, the boundaries of the permissible range of $L_o(jv)$ are found, etc. This procedure has been given in detail in Section 6.7, and is therefore not repeated here. Instead, the reader's memory will be refreshed by working out a detailed numerical design problem.

Example.

Design specifications: The structure in Fig. 11.15-1 is being used, where the plant, which includes a zero-order hold, has the transfer function $P(s) = aA(1 - e^{-sT})/s^2(s + a)$. The sampling period is $T = 2$ seconds, and A and a vary independently over the ranges 1 to 2 and 0.5 to 1, respectively. The bandwidth of $T(jv)$ is to extend approximately to $v = 0.3$. It is believed that sufficient insensitivity will be attained if, despite the above parameter variations, $T(jv)$ does not vary more than zero at $v = 0$, 10% at 0.1, and about 20% at 0.3. At higher frequencies, there is some concern about the maximum peaking in $T(jv)$. Any high frequency peaking due to parameter variation should be kept to a few decibels.

Design procedure: The sensitivity design procedure will now be briefly reviewed. Equation (11.15,1) is used for picking the loop transmission $L_o(w)$ to satisfy the sensitivity specifications. Suppose, for example, that at $v = v_1$, the entire range of variations of $P_o(jv_1)/P(jv_1)$ (which is, in general, a complex function) is contained within the area indicated in Fig. 11.15-2, and that at v_1, $-L_o(jv_1)$ has the value indicated by the point A. Then, from Eq. (11.15,1), $T_o(jv_1)/T(jv_1) = \mathbf{AB}/\mathbf{AC}$, where C is fixed and B may lie anywhere in the shaded area. The extreme values of $T_o/T(jv_1)$ are obtained by letting B assume its extreme values.

Conversely, given the complete area of variation of $(P_o/P)\,(jv_1)$ and the extreme permissible values of $T_o/T(jv_1)$, the range of permissible $L_o(jv_1)$ can easily be found. Similar boundaries of permissible locations of $L_o(jv)$ may be sketched at other values of "frequency," the resulting boundaries depending on the range of $P_o/P(jv)$ and the specified limits on $T_o/T(jv)$. With these boundaries, it is a relatively simple matter to obtain a satisfactory $L_o(jv)$, providing, of course, that the requirements do not exceed the basic capabilities, in view of the previously discussed (Section 11.11) limitations on $L_o(jv)$.

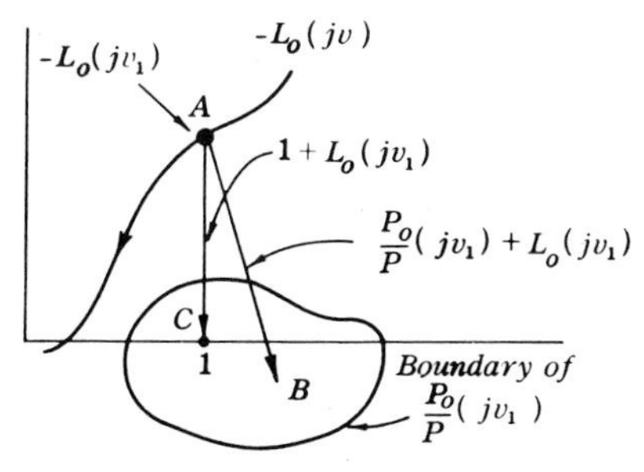

FIG. 11.15-2. $T_o/T(jv_1) = \mathbf{AB}/\mathbf{AC}$.

In applying the above procedure to our specific numerical problem, the region of variation of $P_o/P(jv)$ for a number of values of v is first obtained. Here, the nominal $P(s) = (1 - e^{-sT})\, Aa/s^2(s + a)$, and if $a_o = 0.5$, $A_o = 1$, the resulting $P_o/P(w)$ is $P_o/P(w) = [M(w)\,(-0.078w^2 - 0.389w + 0.463)]\{A(w + 0.463)\,[w^2N(w) + Qw + J]\}^{-1}$, where $M = w + (1 - e^{-2a})/(1 + e^{-2a})$, $N = -1 - a^{-1} + 2/a(1 + e^{-2a})$, $Q = [e^{-2a}(2 + a^{-1}) - a^{-1}]/(1 + e^{-2a})$,

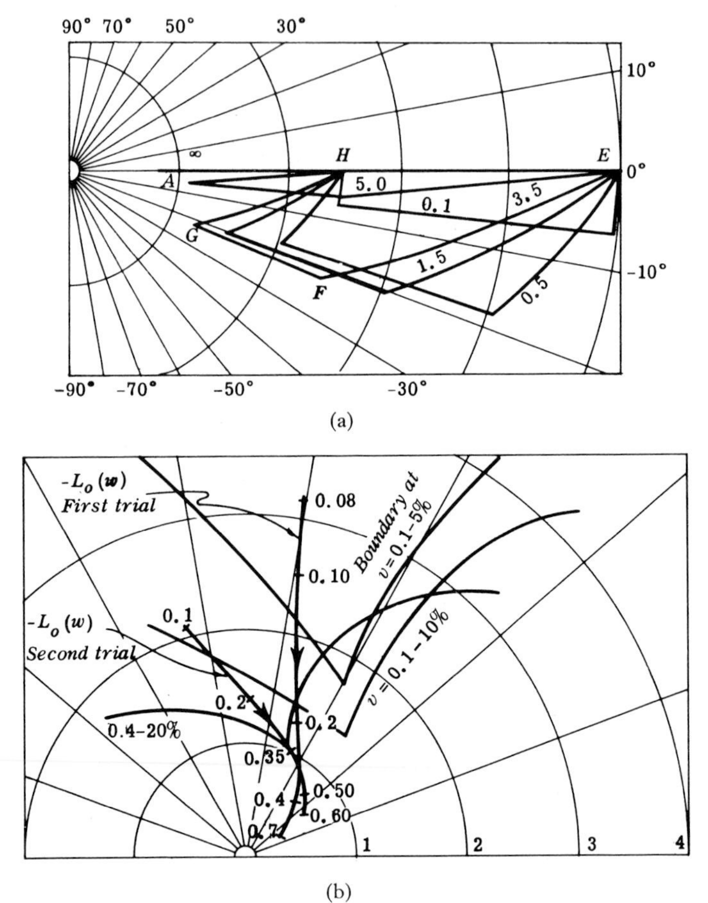

FIG. 11.15-3a to 3e. Steps in choice of $L_o(w)$ and its shaping for stability despite parameter variations. (a) Range of $P_o/P(jv)$ for several values of v. (b) Boundaries of permissible $-L_o(jv)$ at several values of v. (c) Bode plot of $L_o(jv)$. (d) Polar plot of $L_o(jv)$ at high frequencies.

$J = (1 - e^{-2a})/(1 + e^{-2a})$. It is far easier to find the boundaries of $P_o/P(j\omega)$ than of $P_o/P(jv)$. This is the price paid for the advantage that is gained in shaping $L_o(jv)$ instead of $L^*(j\omega)$. At a fixed value of A and $w = jv$, the locus of

$P_o/P(w)$, as a function of a, is obtained by calculating P_o/P for a few values of a. The effect on this locus of varying A is obvious, since A appears only as a multiplier. Thus, in Fig. 11.15-3a, E, F, G, H mark the region of variation of $P_o/P(w)$ for $w = j3.5$. The calculations and sketches are repeated for several values of jv, and the results are shown in Fig. 11.15-3a; EHA is the locus of $P_o/P(jv)_{v\to\infty}$. Next, the boundaries of the permissible values of $L_o(jv)$ are

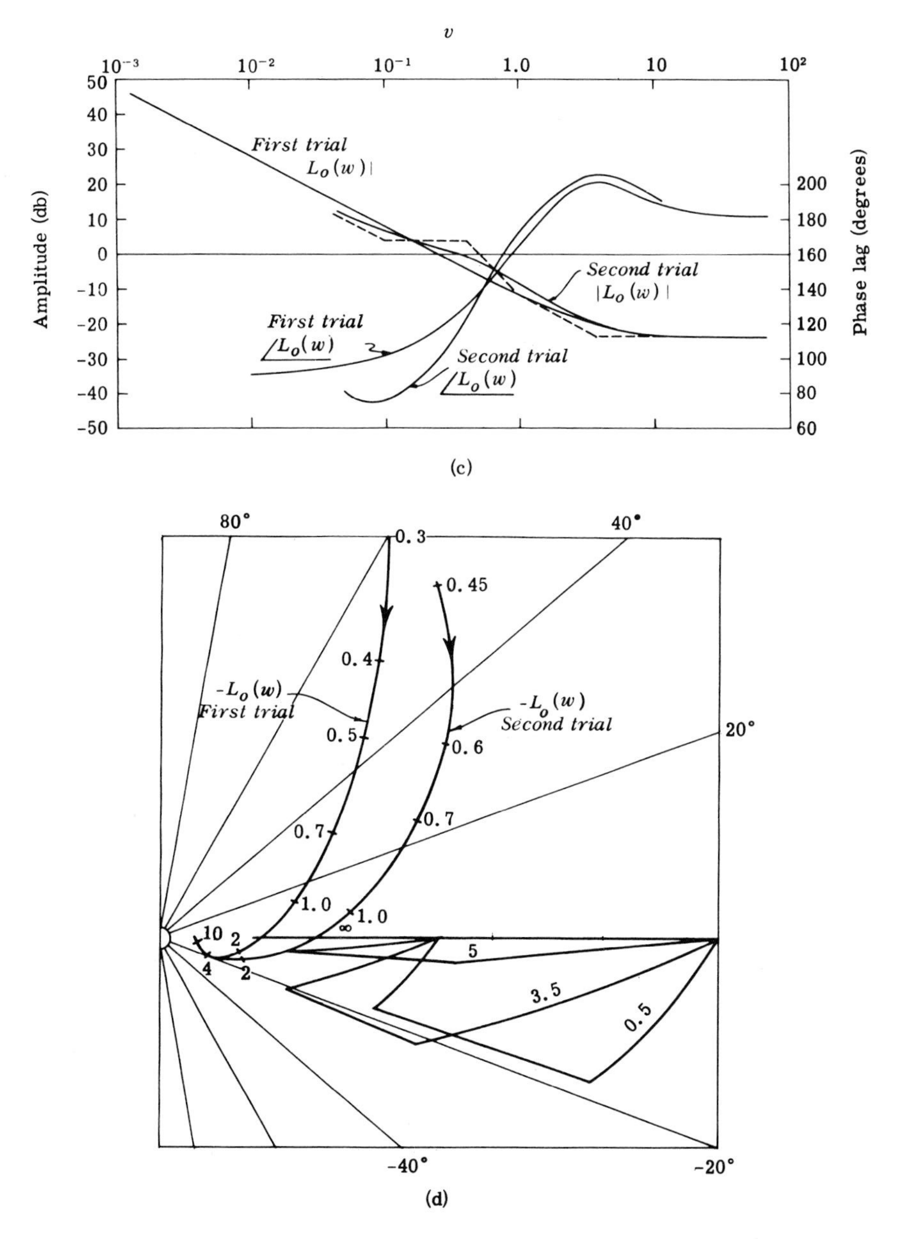

obtained—see Fig. 11.15-3b. The requirements on $L_o(w)$ are then apparent.

A simple $L_o(w)$ that is used as first trial is $L_o(w) = -0.0625(w-1)(w+4)/w(w+1)$. Its Bode sketch is shown in Fig. 11.15-3c and polar sketches at low and high frequencies are shown in Figs. 11.15-3b and d, respectively. While this $L_o(w)$ is more than satisfactory around $v = 0.10$, it is quite poor at higher frequencies (30 % maximum increase at $v = 0.35$, 90 % at $v = 0.50$). In order to obtain a $L_o(w)$ with somewhat more gain at $v \approx 0.3$, some sacrifice in the lower frequency gain is made with $L_o(w) = -0.063(w+0.1)(w-1)(w+4)/w(w+0.4)^2$. This too is sketched in Figs. 11.15-3b and d. At $v = 0.35$, the variation in T is 9 %, which is higher than that for the first trial, but its performance at $v \sim 0.4$ (20 % maximum at $v = 0.35$, 35 % at $v = 0.5$) is satisfactory. It is therefore used in the design.

To complete the design, the system transfer $T_o(w)$ is needed. The problem of the choice of the system transfer function in sampled-data systems has been treated in previous sections. This section is not concerned with the above problem but rather with that of obtaining the desired insensitivity of the chosen function to plant parameter variations. In this specific example, it is assumed that the function in its significant frequency region is given by $K(w-1)/(w^2+0.45w+0.10)$. Therefore, let $T_o(w) = -K(w-1)M(w)/(w^2+0.45w+0.10)$, with $M(w)$ to be chosen later. For the configuration of Fig. 11.15-1,

$$L_o(w) = GP_oH = \frac{-0.063(w+0.1)(w-1)(w+4)}{w(w+0.4)^2},$$

$$G = T_o\frac{1+L_o}{P_o} = \frac{12.5K(w+0.071)(w+0.463)(w^2+0.57w+0.379)M(w)}{(w^2+0.45w+0.10)(w+0.4)^2(w+6.0)},$$

and

$$H = \frac{L_o}{GP_o} = \frac{L_o}{T_o(1+L_o)} = \frac{(0.067/K)(w+0.1)(w^2+0.45w+0.10)(w+4)}{M(w)(w+0.071)(w^2+0.574w+0.379)}.$$

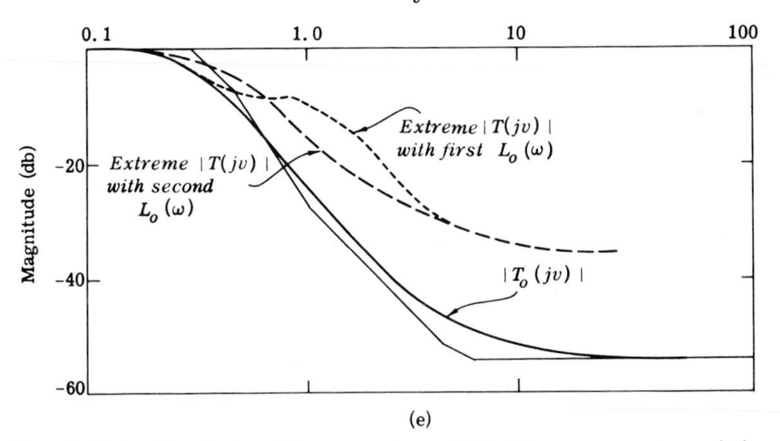

FIG. 11.15-3e. Nominal and Extreme values of $T(jv)$ due to parameter variation.

$M(w)$ is selected so that the maximum peaking of $T(w)$ at higher frequencies is kept within the desired bounds. Thus at $v = 1.5$, it is found from Fig. 11.15-3d that there is a maximum increase of 18 db in $T(j1.5)$. If $T(j1.5)$ is never to be more than -15 db, an additional 10-db attenuation is needed. From such considerations, $M(w)$ is selected to be $(w + 4)(w + 6)/(w + 0.463)$, and $K = 0.0463/24$ in order that $T_o(0) = 1$. The resulting extreme peak values of $T(jv)$ for the two trial values of $L(jv)$ are shown in Fig. 11.15-3e.

The balance of the design procedure is straightforward. Previous sections have described techniques for realizing transfer functions which are functions of z, either by means of continuous type networks or by digital computers. Therefore $G(w)$ and $H(w)$ are first converted into the z notation and the resulting transfer functions are then realized.

§ 11.16 Limitations on Achievable Sensitivity Reduction—Use of Multirate Sampling

It is clear from Section 11.15 that the only limitation on the achievable sensitivity reduction is the limitation on the achievable loop gain-bandwidth. The latter is limited because of the rhp zero that $L(w)$ must have at $w = 1$. It was shown in Section 11.11 how difficult it is to achieve a crossover frequency as high as $v = 1.0$. If there are substantial parameter variations, one must settle for considerably less, as in the last example. The value of $v = 0.4$ may be taken as a reasonable value for the maximum frequency at which sensitivity reduction is achievable. Therefore the bandwidth of $T_o(jv)$ should not be more than 0.4 if

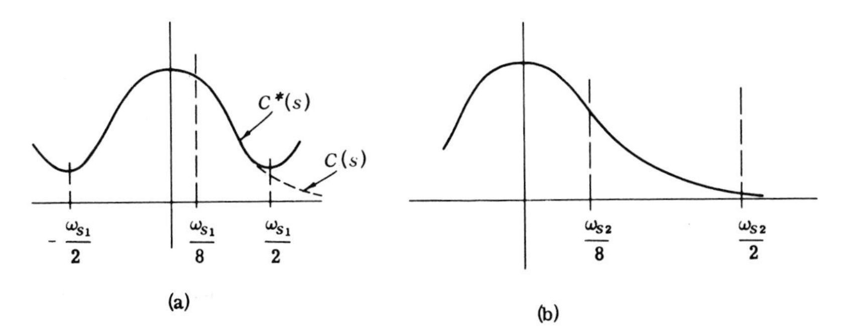

FIG. 11.16-1. Need for larger ω_s to achieve small sensitivity over significant frequency range.

the system transfer function is to be insensitive to plant parameter variation over its significant range. This may require the sampling frequency to be increased. Thus suppose it is found that a sampling period of 1 second, with a bandwidth of 0.6, is sufficient for reproducing the input signal with the desired accuracy.

It is impossible to achieve sensitivity reduction over a bandwidth of 0.6. The bandwidth should therefore be reduced to approximately 0.40. However, the ability of the system to follow the input is thereby degraded. In order to restore it to its previous value, the sampling period must be decreased.

The price that must be paid in increase of sampling frequency can be seen by studying Figs. 11.16-1a and b. In Fig. 11.16-1a the sampling frequency ω_{s1} has been chosen so that the spectrum of $C^*(s)$ is not too greatly distorted by the sampling. However, sensitivity reduction is achievable approximately over $|\omega| < \omega_{s1}/8$ only (since $v = \tan \pi\omega/\omega_s$ and $v_{\max} \approx 0.4$, $\omega_{\max}/\omega_s \approx 0.12$). To remedy this, it is necessary that the significant frequency range of the output lie within $\omega_s/8$. If the significant range is considered to be $\omega_{s1}/4$, then $\omega_{s2}/8 = \omega_{s1}/4$ and therefore $\omega_{s2} = 2\omega_{s1}$, i.e., the sampling frequency must be doubled. The result is shown in Fig. 11.16-1b. From the viewpoint of signal reproduction, the system sampling period is better than need be.

It is usually considered desirable to choose the sampling period T as large as possible. This permits greater time sharing of equipment, less loading of the sensitive elements, and more efficient utilization of the digital computer, if the latter is used for realizing the required compensation networks G and H of Fig. 11.15-1. It is therefore a good idea to use different sampling periods for the filtering (system transfer function realization) part of the system and for the sensitivity reduction part (loop transmission) of the system. Thus the structure of Fig. 11.16-2 is superior to that of Fig. 11.15-1, because the major burden of

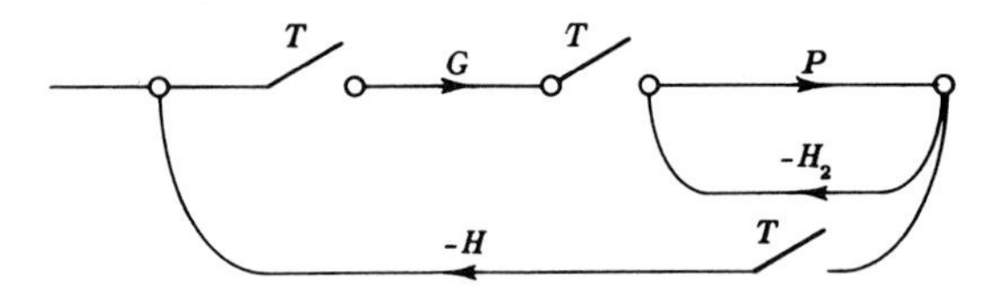

FIG. 11.16-2. Structure with better feedback properties.

sensitivity reduction can be borne by the continuous PH_2 loop. Let $P_e(s) = P/(1 + PH_2)$, and then the sampled-data synthesis procedure of Section 11.15 is used with $P_e(s)$ replacing $P(s)$ of Fig. 11.15-1. The difference is that the variations in P_e are considerably less than those in $P(s)$. There is then no need to increase the sampling frequency beyond that needed from filter considerations.

Figure 11.16-2 cannot be used if one of the reasons for sampling is to lightly load a sensitive feedback transducer. In such a case, two (or more) transducers can be used—one for gross accuracy, which can be loaded continuously or at a sampling frequency $\omega_{s2} > \omega_{s1}$, and the other is the sensitive transducer with ω_{s1} as its sampling frequency. This is shown in Fig. 11.16-3, where it is assumed that n is an integer.

In order to fit Fig. 11.16-3 into our synthesis procedures, it is necessary to find the pulse transfer function (in terms of the sampling frequency $\omega_{s1} = 2\pi/T_1$) of the dashed portion of Fig. 11.16-3. For this purpose, the pulse transfer functions of the systems in Figs. 11.16-4a and b are needed. Consider Fig.

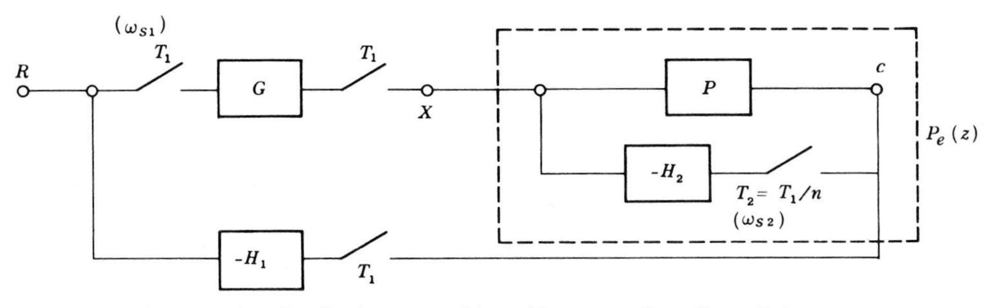

FIG. 11.16-3. Feedback system with multirate sampling. $T_2 = T_1/n$

11.16-4a, where the higher rate sampler follows the slower sampler. From Eqs. (11.3,1) and (11.5,1), $C(s) = F(s)\, R^*(s) = F(s)\, T^{-1} \sum_{k=-\infty}^{\infty} R(s + jk\omega_s)$. The Laplace transform of the sampled output (at T/n) is

$$C^*_{(n)}(s) = nT^{-1} \sum_{\nu=-\infty}^{\infty} F[s + j\nu(n\omega_s)]\, T^{-1} \sum_{k=-\infty}^{\infty} R(s + j\nu n\omega_s + jk\omega_s).$$

(a)

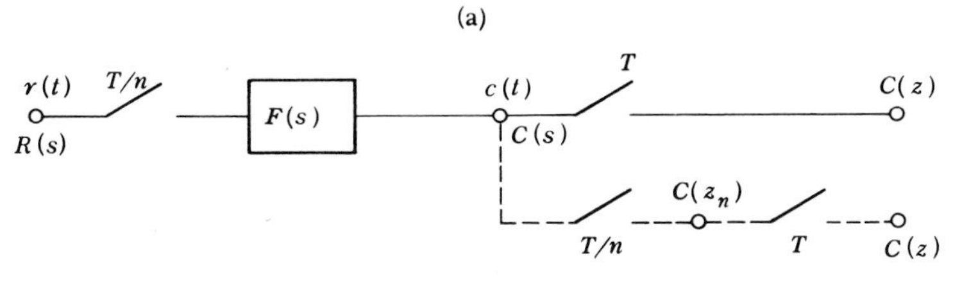

(b)

FIG. 11.16-4a to 4c. Some multirate sampling configurations. In (a), $z = e^{sT} = (z_n)^n$; $z_n = e^{sT/n}$.

However, $T^{-1} \sum_{k=-\infty}^{\infty} R(s + j\nu n\omega_s + jk\omega_s) = T^{-1} \sum_{k=-\infty}^{\infty} R(s + jk\omega_s) = R^*(s)$, for any integral value of ν. Hence $C^*_{(n)}(s) = R^*(s)\, nT^{-1} \sum F[s + j\nu(n\omega_s)] = R^*(s)\, F^*_{(n)}(s)$, i.e.,

$$C(z_n) = R(z)\, F(z_n), \tag{11.16,1a}$$

with

$$z_n \triangleq e^{sT/n} = z^{1/n}. \tag{11.16,1b}$$

Thus, to find the output [which consists of $c(t)$ sampled at the rate of T/n], evaluate $R(z)$ and the P.T.F. $F(z)$, replace z by $z^{1/n}$ in $F(z)$ only (or in evaluating the P.T.F. use T/n in place of T); then take the product of the two and expand in a series in $1/z^{1/n}$.

In Fig. 11.16-4b, the slower sampler follows the higher rate sampler. Here $C(z_n) = R(z_n)\,F(z_n)$ is found in the usual manner, i.e., the sampling period is T/n. $C(z)$ is found by sampling $C(z_n)$ with the sampling period T. This can be done by finding the pulse series for $c(z_n)$ and rejecting (if $n = 4$) all but the 1st, 5th, 9th, etc., terms of the series. Another way is to first recreate the continuous signal $c(t)$, and then sample the latter at intervals of T seconds. For example, if $r(t) = u(t)$ and $F(s) = 1/(s+1)$, then

$$\begin{aligned} C(z_n) &= R(z_n)\,F(z_n) = [z_n/(z_n - 1)]\,[z_n/(z_n - e^{-T/n})] \\ &= [z_n/(1 - e^{-T/n})]\,[(z_n - 1)^{-1} - e^{-T/n}(z_n - e^{-T/n})^{-1}] \end{aligned}$$

whose continuous counterpart is $c(t) = (1 - e^{-T/n})^{-1}\,[1 - e^{-T/n}e^{-t}]$. The latter is sampled with sampling period T. Therefore $C(z) = ZC(z_n) = (1 - e^{-T/n})^{-1}\,[z(z-1)^{-1} - ze^{-T/n}(z - e^{-T})^{-1}]$. The above procedure can always be followed to find $C(z)$ in Fig. 11.16-4b.

For later use, it is noted that it is possible to write $C(z) = ZC(z_n) = Z[R(z_n)\,F(z_n)]$, as

$$Z[F(s)R(z_n)] = Z\Big[F(s)\frac{n}{T}\sum_{\nu=-\infty}^{\infty} R(s + j\nu n\omega_s)\Big]$$

$$= \frac{1}{T}\sum_{\alpha=-\infty}^{\infty}\Big\{F(s + j\alpha\omega_s)\frac{n}{T}\sum_{\nu=-\infty}^{\infty} R(s + j\alpha\omega_s + j\nu n\omega_s)\Big\}. \qquad (11.16,2)$$

Conversely, whenever the latter appears, it can be written in the form $Z[R(z_n)\,F(z_n)]$ and evaluated in the manner previously shown.

This result is used to solve for $C(z)$ in Fig. 11.16-4c. From Eq. (11.16,1a),

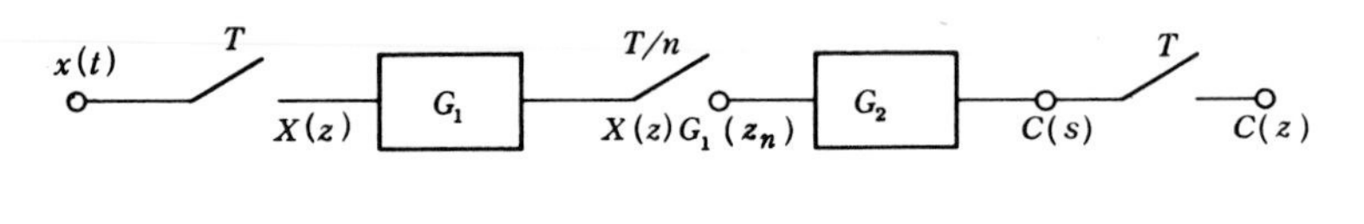

(c)

FIG. 11.16-4c.

$C(z) = Z[X(z)\,G_1(z_n)\,G_2(s)]$. To find $C(z)$, write

$$\begin{aligned} C(s) &= G_2(s)G_1(z_n)X(z) \\ &= G_2(s)\frac{n}{T}\sum_{\beta} G_1(s + j\beta n\omega_s)\frac{1}{T}\sum_{\gamma} X(s + j\gamma\omega_s). \end{aligned}$$

Therefore

$$C(z) = ZC(s)$$

$$= \frac{1}{T}\sum_k \left\{ G_2(s + jk\omega_s) \frac{n}{T}\sum_\beta G_1(s + jk\omega_s + j\beta n\omega_s) \frac{1}{T}\sum_\gamma X(s + jk\omega_s + j\gamma\omega_s) \right\}.$$

The last can be factored out, leaving

$$C(z) = \frac{1}{T}\sum_\gamma X(s + j\gamma\omega_s) \frac{1}{T}\sum_k \left\{ G_2(s + jk\omega_s) \frac{n}{T}\sum_\beta G_1[s + j\omega_s(k + \beta n)] \right\}$$

$$= X(z) \frac{1}{T}\sum_k \left\{ G_2(s + jk\omega_s) \frac{n}{T}\sum_\beta G_1[s + j\omega_s(k + \beta n)] \right\}.$$

From Eq. (11.16,2), the latter part can be written as $Z[G_2(z_n)\, G_1(z_n)]$. Therefore, in Fig. 11.16-4c,

$$C(z) = ZC(s) = X(z)\, Z[G_2(z_n)\, G_1(z_n)]. \tag{11.16,3}$$

This result makes it possible to solve Fig. 11.16-3. Consider first the dashed portion of Fig. 11.16-3, (i.e., imagine $H_1 = 0$). From Eq. (11.16,1a), $C(z_n) = P(z_n)\, X(z) - H_2 P(z_n)\, C(z_n)$. Therefore $C(z_n) = P(z_n)\, X(z)/[1 + H_2 P(z_n)]$ and $C(z) = X(z)\, ZP(z_n)/[1 + H_2 P(z_n)]$. Consequently $C(z)/X(z) = ZP(z_n)/[1 + H_2 P(z_n)] = P_e(z)$ is regarded as the new plant whose parameter variations are less than those of $P(z)$, because of the local feedback loop with the faster sampling rate. Consequently, in Fig. 11.16-3, $X(z) = G(z)[R(z) - H_1(z)\, C(z)] = G(z)[R(z) - H_1(z)\, P_e(z)\, X(z)]$. Hence $X(z) = G(z)\, R(z)/[1 + G(z)\, H_1(z)\, P_e(z)]$. Finally, $C(z) = P_e(z)\, X(z)$, and $T(z) = C(z)/R(z) = P_e(z)\, G(z)/[1 + G(z)\, H_1(z)\, P_e(z)] = P_e(z)\, G(z)/[1 + L(z)]$. The procedure henceforth is exactly the same as in Section 11.15. $L(z)$ is chosen to handle the parameter variations of $P_e(z)$, or to attenuate disturbances, and $G(z)$ is chosen so as to obtain the desired $T(z)$.

§ 11.17 Design for Attenuation of Disturbances

The fundamental problem in designing for the attenuation of disturbances is to pick a loop transmission function so as to obtain the desired attenuation. In Fig. 11.17-1, $C_D(z) = D(z)/[1 + L(z)]$, so that disturbance attenuation is achieved by making $|L(z)|$ sufficiently large over the required frequency range. The design details can be executed either in the z, jv, or $j\omega$ domain. In either case the available procedures are similar to those described for continuous-data systems. It is noted, however, that the system is open-loop with no feedback during the intersample intervals. Another important difference between the two is in the limitations on the achievable loop gain-bandwidth of $L^*(j\omega)$, or of the equivalent $L(jv)$. It may therefore be impossible to satisfy the specifications unless a larger sampling frequency, or multirate sampling (Section 11.16),

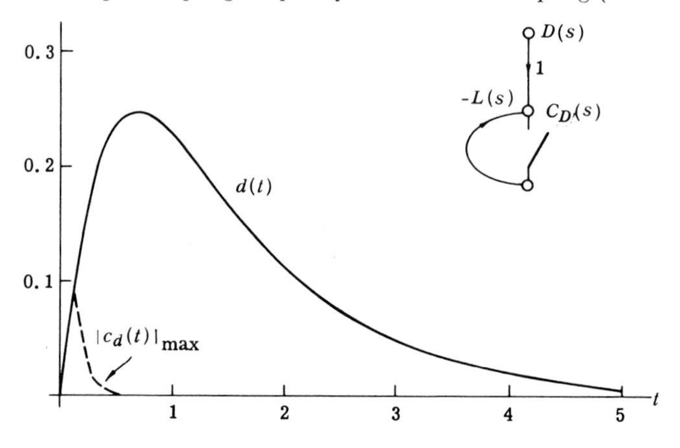

FIG. 11.17-1. Sampled-data feedback loop subject to disturbances.

or continuous-data feedback around the plant is used (Fig. 11.16-2). Or else a multiple-loop design (Chapter 8) may be possible, where continuous-data feedback may profitably be used around some portions of the plant. The latter will not be considered here.

Design in the z Domain

The principal features of a z-domain design procedure paralleling the s-domain technique of Sections 9.4-9.6 are described by means of a numerical example. It is assumed that the design may be realistically based on a "typical" continuous disturbance signal, whose z transform is then readily obtained.

Example. Suppose $D(s) = 1/(s+1)(s+2)$, and the maximum tolerable effect of the disturbance is that shown in Fig. 11.17-2 (for $t > 0.7$, $|c_d(t)| < 10^{-4}$). The maximum value of sampling period which satisfied the specifications is to be used.

The worst possibility is when the origin for $d(t)$ is at the first sampling instant, for then the system is open-loop until $t = T$, at which $d(t) = e^{-T} - e^{-2T}$. T must be chosen so that $e^{-T} - e^{-2T} < 0.10$, i.e., $T < 0.12$ second. Suppose we round it off to $T = 0.10$, i.e., $\omega_s = 20\pi$ rps, and then $D(z) = ZD(s) = 0.087z/(z-0.818)(z-0.905)$. Following the procedure described in Section 9.4, we assign to $T_D(z) = [1 + L(z)]^{-1}$ zeros at $z = 0.818, 0.905$. Also if, as is here assumed, the plant includes a motor whose pole at $s = 0$ (i.e., at $z = 1$) is retained, for the zero dc sensitivity thereby available, then $T_D(z)$ is assigned a zero at $z = 1$. Hence $T_D(z) = (z-0.818)(z-0.905)(z-1)/(z-a)(z^2 - 2bz + c^2)$. Note

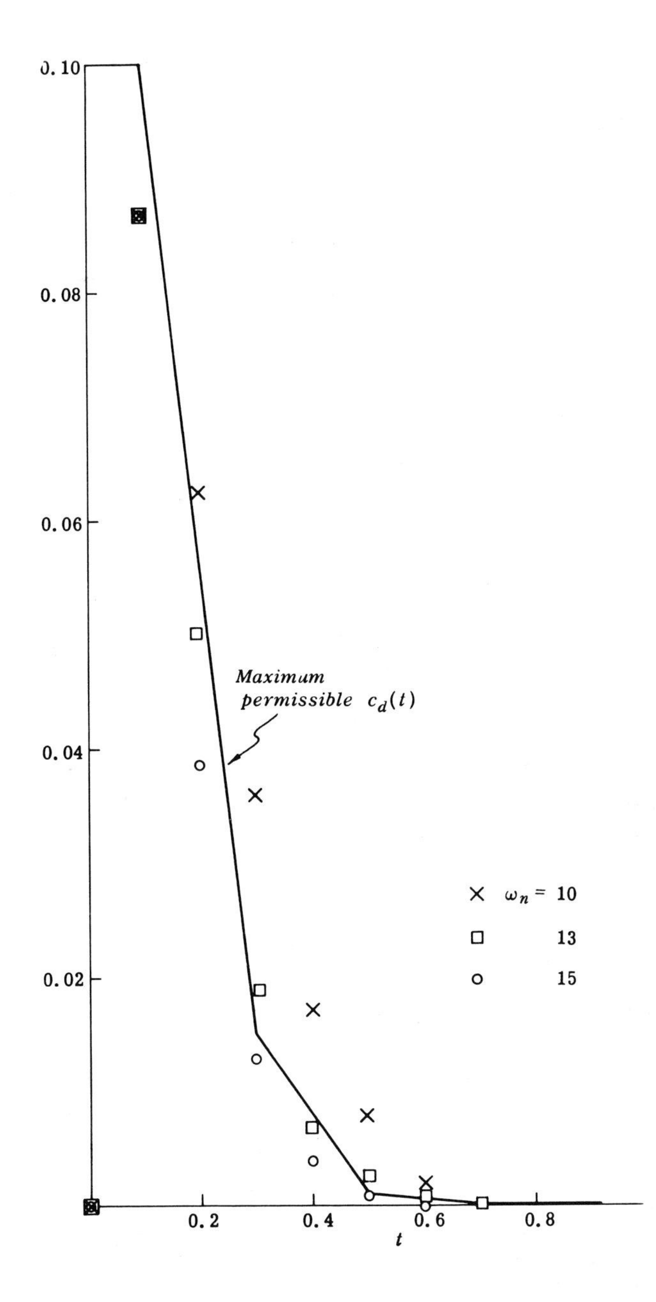

FIG. 11.17-2. Design for attenuation of disturbances.

that we have chosen $T_D(z)$ such that $T_D(\infty) = 1$, because $T_D(z) = 1/[1 + L(z)]$ and any practical $L(z)$ is zero at infinity.

As a result of the above, $C_D(z) = D(z)\, T_D(z) = 0.087z(z - 1)/(z - a) \times (z^2 - 2bz + c^2)$, and the problem is to choose a, b, c such that the specifications on $c_d(t)$ are satisfied. To simplify the problem (and simultaneously economize on the loop gain and bandwidth by using the maximum tolerable output), we may choose $a \doteq 1$, corresponding to lag compensation. In the expansion of $C_D(z)$ in the form $\Sigma A_i z/(z - \alpha_i)$, there is then a term $A_1 z/(z - a)$, associated with $A_1 e^{-\alpha t}$, with $a = e^{-\alpha T}$. If $a \doteq 1$, then α is very small, i.e., the term contributes a very long tail to $c_d(t)$. It is therefore necessary that A_1 should be less than 10^{-4}. The balance of $C_D(z)$ may be written as $C_{D_1}(z)$ and it is very closely equal to $0.087z/(z^2 - 2bz + c^2)$, which corresponds in the time domain to the sampled impulse response of a second-order system. Examination of Fig. 5.12-1 indicates that $\zeta = 1$ should be used for the equivalent continuous $c_{d_1}(t)$. We therefore take $b = c$, and then $C_{D_1}(z) = 0.087z/(z - b)^2$. From Table 11.4-1, $Z(s + \omega_n)^{-2} = Tze^{-\omega_n T}/(z - e^{-\omega_n T})^2$; hence $C_{D_1}(z) = Z[0.087/bT(s + \omega_n)^2] = Z[0.087(bT)^{-1}\, te^{-\omega_n t}]$, with $b = e^{-\omega_n T}$ and $T = 0.1$ second. We try $\omega_n = 10, 13$, and 15, and the resulting $c_{d_1}(t)$ are shown in Fig. 11.17-2. It is seen that $\omega_n = 13$ almost exactly satisfies the specifications. If this value is used, then $b = e^{-\omega_n T} = 0.273$.

The magnitude of the residue in the pole of $C_D(z)$ at $a \doteq 1$, is very closely equal to $0.87(1 - a)\,(1 - 0.273)^{-2} = 0.165(1 - a)$ and it must be less than 10^{-4}. Hence $1 - a = 10^{-4}/(0.165) \approx (6)\, 10^{-4}$, and $T_D(z) = (z - 0.818)\,(z - 0.905)\,(z - 1)/(z - 0.9994)\,(z - 0.273)^2$. The next step is to find $L(z)$ from $T_D = 1/[1 + L(z)]$, leading to $L(z) = 1.18(z - 0.57) \times (z - 0.9994)\,[(z - 1)\,(z - 0.905)\,(z - 0.818)]^{-1}$. [Note that it is easier to first find $L_1(z) = T_{D_1}^{-1} - 1$, and then add a dipole whose zero may be taken to be at 0.9994.] $L(s)$ is found from $L(z)$. It is noted that $L(s)$ must contain a zero-order hold circuit because $L(z)$ is not zero at $z = 0$. [The hold circuit can be avoided by letting $T_D(z)$ be one at $z = 0$, for then $L(z) = T_D^{-1} - 1$ is zero at $z = 0$. This adjustment of T_D can be made after a satisfactory $T_D(z)$ has been found, by inserting a dipole $(z - \alpha)/(z - \beta)$ near the origin. Such a dipole corresponds to a pole-zero pair far out on the negative real axis in the analogous continuous system, i.e., they have negligible effect on the system response. The α/β ratio is chosen so that the new $T_D(0) = 1$.] In this problem, if a zero-order hold is used, then $L(s) = (1 - e^{-sT})\, X(s)/s$ and $L(z) = Z(1 - e^{-sT})\, X(s)/s = (1 - z^{-1})\, ZX(s)/s$. Therefore,

$$
\begin{aligned}
ZX(s)/s &= [z/(z - 1)]\, L(z) \\
&= 1.18z(z - 0.57)\,(z - 0.9994)\,[(z - 1)^2\,(z - 0.905)\,(z - 0.818)]^{-1} \\
&= 1.18z\,\{0.015(z - 1)^{-2} + 25.4(z - 1)^{-1} \\
&\qquad - 40.5(z - 0.905)^{-1} + 15.1(z - 0.818)^{-1}\}.
\end{aligned}
$$

From Table 11.4-1, $X(s)/s = 1.18\,\{0.15s^{-2} + 25.4s^{-1} - 40.5(s+1)^{-1} + 15.1(s+2)^{-1}\}$, and $X(s) = 12.3(s+4.93)\,(s+0.0058)\,[s(s+1)\,(s+2)]^{-1}$. [Note that the zero of the dipole maps very closely in accordance with the relation $z = e^{sT} \doteq 1 + sT$, for very small sT. The dipole could have been neglected in calculating $X_1(s)$, and only then the dipole inserted.]

It is important to note that $X(s)$ has only one more pole than zeros. If the plant has an excess of two poles over zeros and two more are allowed for the compensating networks, then three more poles must be assigned to $X(s)$. These poles should be inserted sufficiently far away so as to negligibly effect the resulting $1 + L(z)$. The discussion of Section 11.11 indicates how far off is sufficient. In addition, the effect of these additional poles on stability should be checked. Finally, in a two-degree-of-freedom system, the second degree of freedom is used to obtain a satisfactory system response function $T(z)$.

Intersample Output

The intersample disturbance output may be obtained by considering the last stage which is driven by impulses. For example, in Fig. 11.15-1, $C_D = D - Y^*P$. The designer is usually only interested in the nature of the intersample output, not in its precise value. If the impulse response of P has only moderate overshoot, then $\mathscr{L}^{-1}Y^*P$ in between samples will not be excessively oscillatory. P should be modified if necessary for this purpose. If also $d(t)$ is not oscillatory in between its sampled values, then the intersample values of $c_d(t)$ will be satisfactory. However, if $d(t)$ is highly oscillatory in between samples, then the intersample output will obviously be unsatisfactory, because ignorance of the precise disturbance function and the plant parameter variation render it impossible to assign the precise required oscillatory response to $P(s)$ such as to cancel out $d(t)$ in between samples.

It is fairly easy to recognize when the intersample response will be unsatisfactory. It can often be done by simple visual comparison of the sampling period with the intersample character of the typical $d(t)$. Another way is to first make a rough continuous design, i.e., assume that $C_D(s) = D(s)/[1 + L_c(s)]$, and find $L_c(s)$ that satisfies the specifications. The frequency response of $L_c(j\omega)$ is then examined. It is known (Section 11.10) that $|L^*(j\omega)|$ can be more than one only for $\omega < 0.3\omega_s$ approximately. Therefore, if $|L_c(j\omega)|$ must be significantly more than one for $\omega > 0.3\omega_s$, then the sampling frequency must be increased, or the specifications relaxed. The above is useful as a guide in selecting ω_s.

Generality of the Synthesis Procedure

The synthesis procedure which was illustrated for a specific example is valid in general, providing the typical disturbance approach is satisfactory. A satisfactory ω_s is first chosen. $T_D(z)$ is assigned zeros at the poles of $D(z)$. Additional poles and zeros of $T_D(z)$ are assigned such that $C_D(z)$ satisfies the specifications.

Formally, the procedure is one of finding the z transform of the extreme $c_d(t)$, or of a reasonable approximation, and setting it equal to $C_D(z) = D(z)\, T_D(z)$. In the design example, a second-order approximation of $c_d(t)$ was satisfactory; but if such a simple approximation is unsatisfactory, then the techniques of Chapter 12 may be used to find $C_D(s)$ and from the latter, $C_D(z)$. $T_D(z)$ is then determined. If necessary, $T_D(z)$ is assigned additional poles and/or zeros, for example, so that $T_D(z) = 1$ at infinite z [because $T_D = (1 + L)^{-1}$ and $L(\infty) = 0$], or if a hold circuit is to be avoided. Such additional poles and/or zeros can always be assigned sufficiently close to the origin to negligibly affect the response. The resulting $T_D(z)$ is used to determine $L(z)$ and $L(s)$. Additional far-off poles are assigned to $L(s)$ for satisfactory asymptotic behavior, in accordance with the guide suggested in Section 11.11. The final stage driven by impulses is modified, if necessary, in order to obtain satisfactory intersample output. G of Fig. 11.15-1 must then also be modified so that the design values of $L(z)$ and $T(z)$ are not altered. When the disturbances, etc. are of a statistical character then the techniques of Sec. 9.15 may be modified for application to sampled-data systems, in the same manner as those of Sections 9.8-9.13 were modified in Sections 9.18-9.19.

Finally, it would appear that the topics treated for continuous systems in Chapters 8 and 10 (multiple-loop design, parallel plants, multivariable systems etc.) could be applied to sampled-data feedback theory.

CHAPTER TWELVE

Approximation Methods in Feedback System Design

§ 12.1 Introduction

In previous chapters, it was generally taken for granted that the necessary design data and specifications were available in the form most useful for the purpose at hand. For example, the plant characteristics and the desired system response function were assumed to be available as rational functions $P(s)$, $T(s)$, respectively. In actual practice, it is often necessary for the feedback engineer to derive these transfer functions from other data. The representation of data originally in one form into another more convenient form is nearly always accompanied by some errors—hence the designation "approximation" problems. The reasons for the errors will become apparent when we consider the specific problems.

The following approximation problems are treated in this chapter.

(1) Time Domain Synthesis (Sections 12.2-12.6)

In Fig. 12.1-1, the input $x(t)$ and the output $y(t)$ are presumed to be available graphically. It is necessary to find a rational function approximant to $H(s)$,

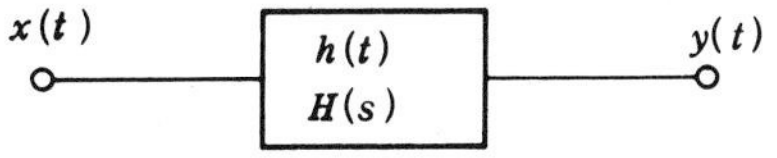

FIG. 12.1-1. Problem: to find $H(s)$ when $x(t)$, $y(t)$ are available graphically.

i.e., $H_a(s) = \mathscr{L}h_a(t)$ where $h_a(t) \approx h(t) = \mathscr{L}^{-1}H(s)$, and $H_a(s)$ is a rational function. There are several origins of this problem.

(1) One is in plant identification. One of the initial tasks in feedback system design is to determine the plant transfer function. Sometimes the latter is derivable from theoretical considerations. More often it must be experimentally determined. One method is that shown in Fig. 12.1-1. A known input signal $x(t)$ is applied and the output signal $y(t)$ is recorded. If the system is linear, then $Y(s) = H(s)\,X(s)$, where $Y(s) = \mathscr{L}y(t)$, $X(s) = \mathscr{L}x(t)$. If $y(t)$ and $x(t)$ are available analytically, then the exact $Y(s)$ and $X(s)$ can be formally derived.

However, in the experimental case, $y(t)$ is not available analytically. It is therefore necessary to find the transform of a function which is available graphically.

(2) Another way in which the time domain synthesis problem arises is as follows. It may be convenient to prescribe a desired performance in the time domain. This is especially true for ideal performances, some of which were mentioned in Chapter 9. In such cases, the desired time domain performance is available graphically, and it is necessary to specify the corresponding $T(s)$. Straightforward transformation is not usually convenient, because the transforms of idealized performances are often transcendental functions.

(3) The typical system inputs and the typical external disturbances acting on a system may be known as time functions, and their transform representation is desirable for proceeding with the synthesis procedure which was described in Sections 9.4-9.6.

(2) Inverse Transformations (Sections 12.11-12.13)

Feedback system design is almost always done in terms of the transform domain, but the real test of the system is its behavior in real time. The designer must make sure of his theoretical transform domain design, by finding the theoretical time domain system performance. Straightforward inverse transformation, using factoring to locate the system poles, followed by partial fraction expansion and term by term inversion, plotting, etc., may be quite long and tedious. There is, therefore, much interest in faster procedures, even if they are not so exact.

(3) Numerical Convolution (Sections 12.8-12.10)

The convolution of two time functions is an operation which often confronts the feedback engineer. A numerical convolution procedure which is easily derived from z-transform theory is presented in Sections 12.8-12.10. It is shown how it may also be used for solving linear constant-coefficient and linear time-varying differential equations.

(4) Other Problems

Other approximation problems which are considered are: rational fraction approximation of frequency response data (Section 12.7), and numerical Fourier analysis (Section 12.14).

§ 12.2 Time Domain Synthesis—The Method of Moments[1]

It is assumed that a time function, $h(t)$, is available either graphically or analytically. In the latter case, $H(s) = \mathscr{L}h(t)$ is not a rational function; otherwise there is no approximation problem. The problem is to find $H_a(s)$, which is a rational fraction approximating the true $H(s) = \mathscr{L}h(t)$. In this method, $H(s)$ is expanded in a Taylor series, as follows:

$$\begin{aligned} H(s) = \mathscr{L}h(t) &= \int_0^\infty h(t)e^{-st}dt = \int_0^\infty h(t)\Big[1 - st + \frac{s^2t^2}{2!} - \frac{s^3t^3}{3!} + \dots\Big]dt \\ &= \int_0^\infty h(t)dt - s\int_0^\infty th(t)dt + \frac{s^2}{2!}\int_0^\infty t^2h(t)dt - \frac{s^3}{3!}\int_0^\infty t^3h(t)dt + \dots \\ &= m_o + sm_1 + s^2m_2 + \dots + s^nm_n + \dots \end{aligned} \tag{12.2,1}$$

where $m_i \triangleq [(-1)^i/i!]\int_0^\infty t^ih(t)\,dt$ is called the ith moment of the function $h(t)$. The moments of $h(t)$ are the coefficients of the Taylor series expansion of $H(s) = \mathscr{L}h(t)$. In the method of moments, the approximant $h_a(t) = \mathscr{L}^{-1}H_a(s)$ is chosen so that as many as possible of its moments are equal to those of $h(t) = \mathscr{L}^{-1}H(s)$. This is equivalent to choosing $H_a(s)$, such that as many as possible of the Taylor series coefficients of $H_a(s)$ are equal to the respective coefficients of the Taylor series expansion of $H(s)$. Thus, suppose we *a priori* pick $H_a(s)$ with a single-degree numerator and a second-degree denominator, i.e., $H_a(s) = (a_o + a_1s)/(1 + b_1s + b_2s^2)$. The four unknowns are chosen so that the first four coefficients of the Taylor series of $H(s)$ and of $H_a(s)$ are, respectively, equal to each other. The Taylor series expansion of $H_a(s) = (a_o + a_1s)(1 + b_1s + b_2s^2)^{-1}$ is

$$\begin{aligned} H_a(s) = a_o &+ (a_1 - a_ob_1)\,s + (a_ob_1^2 - a_ob_2 - a_1b_1)\,s^2 \\ &+ (2a_ob_1b_2 - a_ob_1^3 + a_1b_1^2 - a_1b_2)\,s^3 + \dots. \end{aligned}$$

We therefore set $a_o = m_o$, $a_1 - a_ob_1 = m_1$, $a_ob_1^2 - a_ob_2 - a_1b_1 = m_2$, $2a_ob_1b_2 - a_ob_1^3 + a_1b_1^2 - a_1b_2 = m_3$, and solve for a_o, a_1, b_1, and b_2.

In actual practice, it is easier to write

$$m_o + sm_1 + s^2m_2 + s^3m_3 + \dots \approx (a_o + a_1s)/(1 + b_1s + b_2s^2),$$

cross-multiply mentally, and equate coefficients to obtain the linear equations

$$m_o = a_o,\ b_1m_o + m_1 = a_1,\ b_2m_o + b_1m_1 + m_2 = 0,\ b_2m_1 + b_1m_2 + m_3 = 0.$$

[1] W. H. Huggins, Network Approximation in the Time Domain, Air Force Cambridge Research Labs., Cambridge, Massachusetts, Oct., 1949, Report E5048; F. Ba-Hli, A general method for time domain network synthesis. *Trans. IRE* **CT-1**, 21-28 (1954).

Example 1. A given step response $f(t)$ is shown in Fig. 12.2-1a. It represents an idealized response (Fig. 9.3-1d) rather than any realistic response, but this is of no matter here. We could use the method of moments to approximate

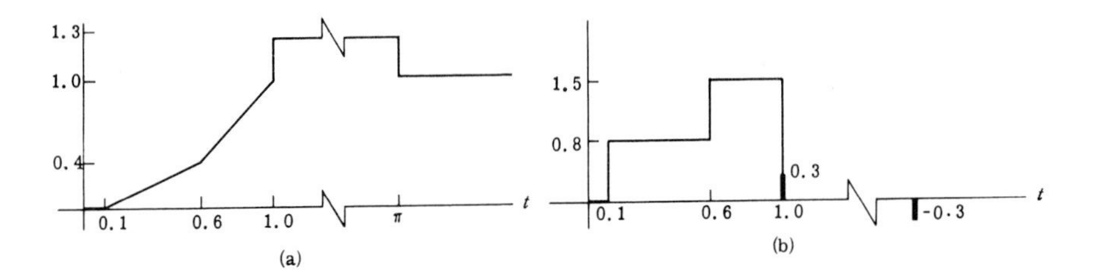

FIG. 12.2-1. (a) Idealized step-response function. (b) Derivative of (a).

$F(s) = \mathscr{L}f(t)$ by working with $f(t) - 1$, in order to obtain finite moments; or we can differentiate $f(t)$ and approximate $H(s) = \mathscr{L}h(t)$ with $h(t) = f'(t)$. It can be shown that, in general, the two methods lead to identical results. (This is left as an exercise for the student.)

The derivative of $f(t)$ is shown in Fig. 12.2-1b, and its transform is easily found:

$$H(s) = 0.8s^{-1}(e^{-0.1s} - e^{-0.6s}) + 1.5s^{-1}(e^{-0.6s} - e^{-s}) + 0.3(e^{-s} - e^{-\pi s}).$$

The Taylor series of $H(s)$ may be obtained by expanding the exponential functions, multiplying through, and collecting terms. This leads to: $H(s) =$

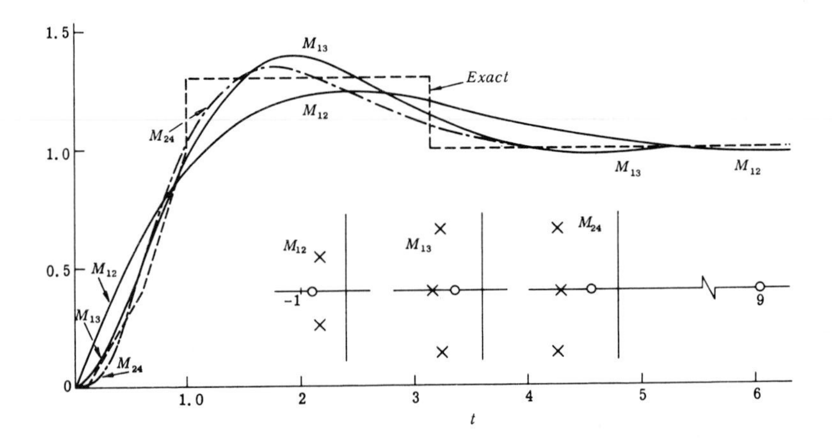

FIG. 12.2-1c. Comparison of original function and approximants obtained by the method of moments.

$1 + 0.022s - 1.105s^2 + 1.44s^3 - 1.19s^4 + 0.76s^5 \ldots$. We try the simple approximant $H_a(s) = (1 + a_1 s)(1 + b_1 s + b_2 s^2)^{-1}$, and in accordance with the previous discussion, $b_1 + 0.022 = a_1$; $b_2 + 0.022b_1 - 1.105 = 0$; $1.44 - 1.105b_1 + 0.022b_2 = 0$. The result is $H_a(s) = 1.22(s + 0.764)(s^2 + 1.2s + 0.928)^{-1}$. The resulting step response, $1 + 1.29e^{-0.6t} \sin(0.754t - 50.6°)$, is labeled M_{12} in Fig. 12.2-1c. It may be considered a fairly good approximation, in view of the simplicity of $H_a(s)$.

An attempt is made to obtain a better approximation by taking $H_a(s) = (1 + a_1 s)(1 + b_1 s + b_2 s^2 + b_3 s^3)^{-1} \approx 1 + 0.022s - 1.105s^2 + 1.44s^3 - 1.19s^4 + \ldots$, cross-multiplying, and equating coefficients of s^i to obtain the equations: $0.022 + b_1 = a_1$; $-1.105 + 0.022b_1 + b_2 = 0$; $1.44 - 1.105b_1 + 0.022b_2 + b_3 = 0$; $-1.19 + 1.44b_1 - 1.105b_2 + 0.022b_3 = 0$. These four equations are solved, leading to $H_a(s) = 4.65(s + 0.60)[(s + 1.1)(s^2 + 1.8s + 2.53)]^{-1}$. The resulting step response is labeled M_{13} in Fig. 12.2-1c. It appears that an excess of two poles over zeros suffices for good approximation at small values of time. We therefore try next $H_a(s) = (1 + a_1 s + a_2 s^2)(1 + b_1 s + b_2 s^2 + b_3 s^3 + b_4 s^4)^{-1}$, and find $b_4 \approx 0$ (using the slide rule), and $H_a(s) = -0.82(s + 0.61)(s - 9.08)[(s + 1.3)(s^2 + 2.75s + 3.51)]^{-1}$, whose step response is $1 + 2.77e^{-1.3t} - 3.77e^{-1.375t}(\cos 1.27t + 0.5 \sin 1.27t)$, and is labeled M_{24} in Fig. 12.2-1c. Although M_{24} appears to be more desirable than M_{13}, because it has less overshoot and smaller settling time, the difference between the two, in terms of approximation to the exact curve, is not significant.

The idealized exact step response of Fig. 12.2-1a was obtained from theoretical considerations noted in Section 9.3. The time domain synthesis problem also arises in the problem of identification. For example, suppose that a plant impulse

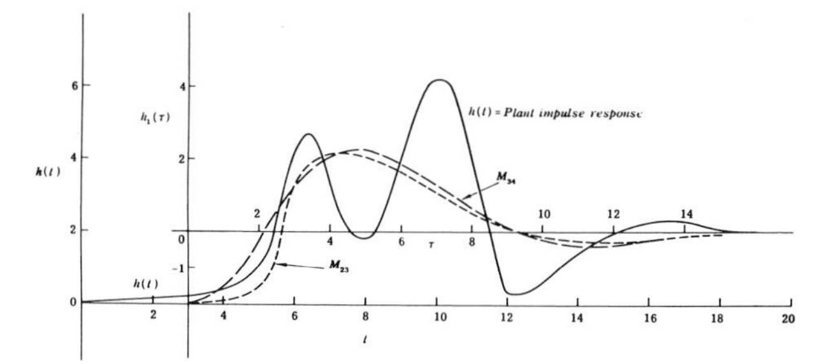

FIG. 12.2-2. Example for which method of moments is inherently very poor.

response $h(t)$ is that shown in Fig. 12.2-2, and the problem is to determine the plant transfer function, which is the Laplace transform of the time function $h(t)$ of Fig. 12.2-2. In order to have finite moments, the ordinate is shifted by two units. The time function has a very long delay, so we first approximate an appreciable portion of the delay by a pure time delay, i.e., $h_1(\tau) = [h(t) - 2]\,\mu(t-3)$, $\tau = t - 3$. $\mathscr{L}h_1(\tau) = H_1(s)$, and we seek a rational fraction approximation of $H_1(s)$ by means of the method of moments.

The first step is to calculate the moments, m_i of Eq. (12.2,1). The integrals $\int_0^\infty t^i h_1(t)\,dt$ are evaluated numerically. The result is $H_1(s) \approx 5.2 - 34.3s + 67.7s^2 - 28.4s^3 + 188s^4 + 576s^5 - 877s^6 + 561s^7 \ldots$. In the first trial we take H_{1a} with a second-degree numerator and a third-degree denominator, leading to: $5.2 - 34.3s + 67.7s^2 - 28.4s^3 + 188s^4 \approx (-2s^2 + a_1 s + a_o)(s^3 + b_2 s^2 + b_1 s + b_o)^{-1}$. In this case, one degree of freedom has been used to ensure that $h_{1a}(0) = -2$. Cross-multiplying, equating coefficients, etc., results in $H_{1a}(s) = -2(s + 3.95)(s - 0.25)[(s + 1.5)(s^2 + 0.58s + 0.27)]^{-1}$, with the corresponding $h_{1a}(t)$ labeled M_{23} in Fig. 12.2-2. The poor approximation should not be surprising. It would be unreasonable to expect a fair approximation of such a complex $h_1(\tau)$ with an $H_a(s)$ consisting of only three poles and two zeros.

We next try an $H_a(s)$ with four poles and three zeros, and while we still do not expect even a fair approximation to the original $h_1(t)$, it seems reasonable to expect a definite improvement over the first approximation. Equating coefficients in the usual manner, etc., leads to $H_{2a}(s) = -2(s + 1.82)(s + 0.52)(s - 0.26)[(s^2 + 0.57s + 0.322)(s^2 + s + 0.284)]^{-1}$. The corresponding $h_{2a}(t)$ is labeled M_{34} in Fig. 12.2-2. There is little, if any, improvement in the approximation; certainly insufficient to justify the labor involved in attempting higher order approximations. We suspect that the accuracy of approximation will not increase, no matter how complex the approximant. The reasons for this are discussed in the following section.

§ 12.3 Comments on the Method of Moments

(1) In this method, the approximant $H_a(s)$ is chosen to match $H(s)$ at $s = 0$. For example, if five moments are matched, then $H(0) = H_a(0)$, $H'(0) = H_a'(0)$, ..., up to $H^{\mathrm{IV}}(0) = H_a^{\mathrm{IV}}(0)$. However, the higher derivatives of $H_a(s)$ may differ greatly from those of $H(s)$. All the available freedom is used to achieve an excellent fit at $s = 0$. The point $s = 0$ in the s domain is directly related to infinite time in the time domain. Consequently, excellent accuracy can be expected at large values of time. Since the time functions must go to zero at very large time (in order that the moments be finite), the value of the good accuracy at large time is questionable. To offset this, it is possible to choose the excess of poles over zeros of $H_a(s)$ to determine (to a limited extent) the behavior of $h_a(t)$ at $t = 0$. Thus, if $H_a(s)$ has two more poles than zeros, then

$h_a(0) = 0$, $h_a'(0) \neq 0$. If the excess is three, then $h_a(0) = h_a'(0) = 0$, but $h_a''(0) \neq 0$. However, the above is not guaranteed, because one or more coefficients may turn out to be zero (as in the case of M_{24} in the first example of Section 12.2).

A better way to achieve an improved approximation is to split the available freedom between large time and small time approximation. Consider the Taylor series approximation of $h(t)$: $h(t) = h(0) + th'(0) + (t^2/2!)\, h''(0) + \ldots.$ The values of $h(0)$, $h'(0)$, etc., are very simply related to the parameters of $H(s) = (\alpha_1 s^{n-1} + \alpha_2 s^{n-2} + \ldots + \alpha_n)\,(s^n + \beta_1 s^{n-1} + \ldots + \beta_n)^{-1}$. The relations are obtained by expanding $H(s)$ in an infinite series in powers of s^{-1}. This can be done by dividing $H(s)$ into 1, and leads to $H(s) = \alpha_1/s + (\alpha_2 - \alpha_1\beta_1)/s^2 + (\alpha_3 - \alpha_1\beta_2 + \alpha_1\beta_1^2 - \alpha_2\beta_1)/s^3 + \ldots.$ The latter corresponds to $h(t) = \alpha_1 + (\alpha_2 - \alpha_1\beta_1)\, t + \ldots.$ As many degrees of freedom as may be desired can be used to secure the desired behavior at $t = 0$, leaving that number less for approximating $H(s)$ at $s = 0$.

(2) A serious shortcoming of the method of moments is that there is no guarantee that the approximant $h_a(t)$ will converge to the original $h(t)$, no matter how complex $H_a(s)$ may be. This is because the power series representation of a function is valid only for the region in which the power series converges. At best, the approximant $H_a(s)$ can be made very similar to the original $H(s)$ in this circle of convergence centered at $s = 0$, and whose circumference reaches to the nearest singularity of the exact $H(s)$. If the most significant part of $H(j\omega)$ lies in this circle of convergence, then we can expect a good fit (but again, not necessarily one that converges to an exact fit). However, if the situation is that shown in Fig. 12.3-1, where an important portion of $H(j\omega)$ is in the region $\omega_b > \omega_a$, then it is impossible to obtain a good approximation, no matter how many moments are matched.

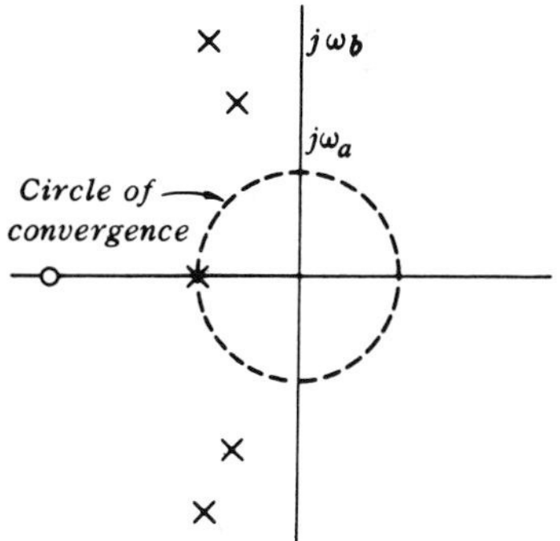

FIG. 12.3-1. Function for which method of moments is not suitable.

We might attempt to circumvent the above by using some or all of the variables to match $H(s)$ at some point other than $s = 0$, viz.,

$$H(s) = H(s_o) + (s - s_o)H'(s_o) + [(s - s_o)^2/2!]H''(s_o) + \ldots$$

$$= \int_0^\infty h(t)e^{-st}dt = \int_0^\infty h(t)e^{-s_o t}\left[1 - (s - s_o)t + \frac{(s - s_o)^2}{2!}\,t^2 - \ldots\right]dt$$

$$= \int_0^\infty h(t)e^{-s_o t}dt - (s - s_o)\int_0^\infty th(t)e^{-s_o t}dt + 0.5(s - s_o)^2\int_0^\infty t^2h(t)e^{-s_o t}dt + \ldots.$$

The disadvantage of the above is that there is more work involved in evaluating the integrals $\int_0^\infty t^i h(t)\, e^{-s_o t}\, dt$. Furthermore, some initial work must be done to estimate the location of s_o, i.e., to find the frequency region where $H(s)$ is significant.

(3) Another shortcoming of the method of moments is that it is basically an approximation in the s domain—$H_a(s)$ approximates the true $H(s)$. In order to evaluate the worth of the approximation, it is necessary to find the inverse transform of $H_a(s)$ and compare it to $h(t)$. This involves additional time and labor.

(4) The chief merit of the method of moments is its extreme simplicity. The shortcoming listed in (2) may even be construed as a merit, in that after we have tried a relatively simple form of $H_a(s)$ (say a third-degree denominator), we know pretty well the best that can be done with the method of moments, and need not go any further. The method of moments should therefore not be attempted on time functions whose transforms might be expected to have poles scattered over the s plane [such that the circle of convergence does not encompass the part of the s plane in which $H(s)$ is significant].

§ 12.4 The z-Transform Method

In the z-transform method, the approximant is chosen such that its sampled output is precisely equal to the sampled values of the specified time function.

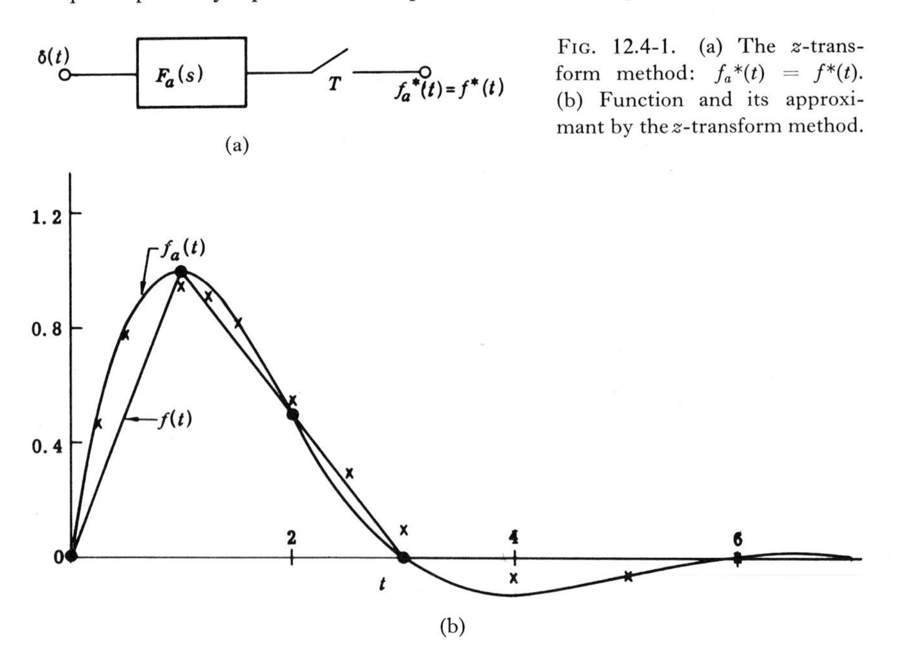

FIG. 12.4-1. (a) The z-transform method: $f_a{}^*(t) = f^*(t)$. (b) Function and its approximant by the z-transform method.

The problem may be posed as one of finding the approximant $F_a(s)$ such that its sampled output, $f_a^*(t)$ in Fig. 12.4-1a corresponds to a given desired sampled function. For example, consider the triangular $f(t)$ of Fig. 12.4-1b. We must pick the desired regularly spaced sampled values. Suppose we pick $T = 1$, and $f^*(t)$ is to correspond to the dotted points (0, 1, 0.5, 0) shown in Fig. 12.4-1b. In this first trial, only four points are used. The simplest $F_a(z)$ is $F_a(z) = a_1z/(z^2 + b_1z + b_2)$. Why is this the simplest form? The reason is that since $F_a(z)$ has one more pole than zero, $f_a^*(0) = 0$. Therefore only three points need be matched, and only three degrees of freedom are needed.

We write $z^{-1} + 0.5z^{-2} + 0z^{-3} \approx a_1z/(z^2 + b_1z + b_2) = F_a(z)$ and choose the a's and b's so that the first three terms of the expansion of $F_a(z)$ in z^{-1} matches the left-hand side of the equation. The result is $a_1 = 1$, $b_1 = -0.5$, $b_2 = 0.25$. The behavior of $f_a^*(t)$ for larger time is easily obtained by dividing $z^2 + b_1z + b_2$ into a_1z, leading to $f_a^*(t) = z^{-1} + 0.5z^{-2} - (1/8)\,z^{-4} - (1/16)\,z^{-5} + (1/64)\,z^{-7} + \ldots$. To find $F_a(s)$, we write $F_a(z) = z/(z^2 - 0.5z + 0.25)$, and, from Table 11.4-1, $F_a(s) = Bb/[(s + a)^2 + b^2]$, with $0.25 = e^{-2aT}$, $0.5 = 2e^{-aT}\cos bT$, $Be^{-aT}\sin bT = 1$, leading to $a = 0.693$, $b = 1.047$, $B = 2.31$. The resulting $f_a(t) = 2.31e^{-0.693t}\sin 1.047t$ is sketched in Fig. 12.4-1b.

It is not surprising that there is a large positive error in Fig. 12.4-1b at $t < 1$, and a negative error for $t > 3$. The synthesis procedure guarantees that $h_a(t)$ will have precisely the assigned values at 0, 1, 2, 3. It is clearly impossible to draw a smooth curve [with the smoothness demanded by the simple $F_a(s)$ attempted] without the above error characteristics. Realizing this, we could improve matters by modifying the original choice of the desired sampled values. Thus the negative error could be reduced by choosing $f_a^*(3) = 0.1$, and $f_a^*(4) = 0$; the error at $t < 1$ can be reduced by taking $f_a^*(1) = 0.95$, $f_a^*(2) = 0.55$. We now write (letting $w = z^{-1}$) $0.95w + 0.55w^2 + 0.1w^3 \approx a_1w(b_2w^2 + b_1w + 1)^{-1}$. Cross-multiplying and equating coefficients, there results

$$F_a(z) = 0.95z/(z^2 - 0.58z + 0.23) = 0.95w + 0.55w^2 + 0.10w^3 - 0.068w^4 - 0.063w^5 - 0.013w^6 + 0.006w^7 + \ldots;$$

$F_a(s) = (2.48)\,(0.923)/[(s + 0.735)^2 + (0.923)^2]$, and $f_a(t) = 2.48\,e^{-0.735t}\sin 0.923t$. The latter is indicated by the crosses in Fig. 12.4-1b.

Clearly, not much improvement is possible with $T = 1$. To improve the approximation, a smaller value of T must be used. We could take $T = 0.5$ and pick (with $w = z^{-1}$)

$$f_a^*(t) = 0.5w + 0.95w^2 + 0.75w^3 + 0.5w^4 + 0.22w^5 + 0.05w^6 + 0.01w^7;$$

$F_a(z) = (0.5z^3 + a_2z^2 + a_1z)/(z^4 + b_3z^3 + b_2z^2 + b_1z + b_o)$. The six available variables suffice for the given $f_a^*(t)$.

Comments on the z-Transform Method

The z-transform method is a very simple and straightforward method. It guarantees that the final approximant passes precisely through the specified sample points. There is, however, no guarantee that the transfer function is a stable one. For example, in Fig. 12.4-2, the approximant passes through the

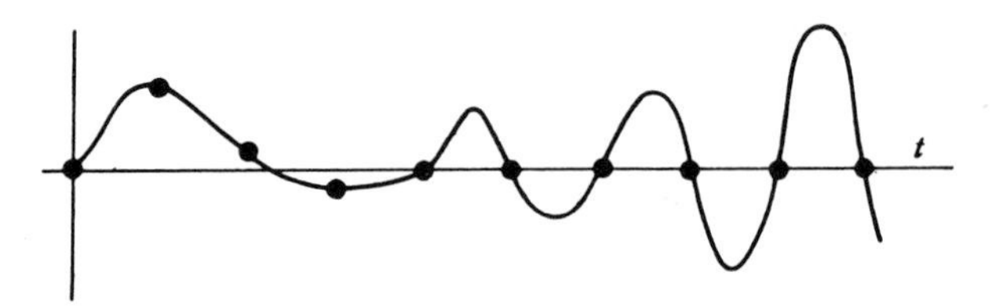

FIG. 12.4-2. Unstable approximant by the z-transform method.

specified points, but it is an exponentially increasing oscillatory time function. Such a phenomenon may occur when one tries to force a relatively simple form of $F_a(s)$ [an $f_a(t)$ consisting of only a few terms] to achieve an output which is beyond its ability. For example, consider $f(t)$ of Fig. 12.4-3, where we try $1 + w + 0(w^2) \approx (1 + b_1 w + b_2 w^2)^{-1}$. The result is $b_1 = -1$, $b_2 = 1$, so that $F_a(z) = z^2/(z^2 - z + 1)$, corresponding to poles of $F_a(s)$ on the $j\omega$ axis, i.e., a steady state sinusoidal oscillation. No difficulty arises, however, if we take (more reasonably) $1 + w + 0.1w^2 + 0w^3 \approx (1 + a_1 w)(1 + b_1 w + b_2 w^2)^{-1}$. The result is

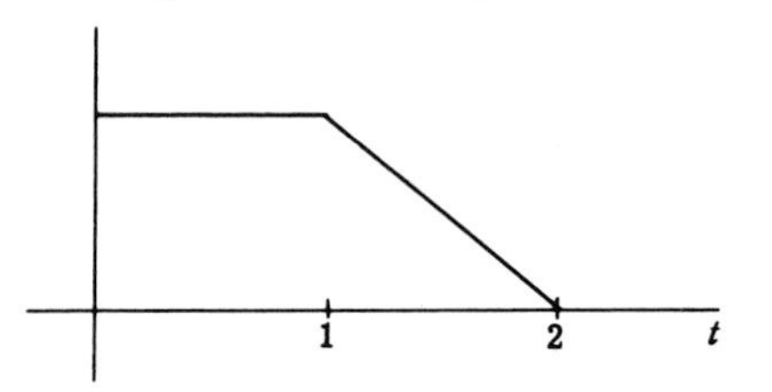

FIG. 12.4-3. Function to be approximated by the z-transform method.

$$F_a(z) = (z^2 + 8z/9)/(z^2 - z/9 + 1/90)$$
$$= 1 + 1/z + 0.1/z^2 - 1/900z^4 - 1/8100z^5 \dots$$

Nevertheless, the major shortcoming of the z-transform method is its inability to guarantee in advance that the final approximant has all its poles in the left half-plane. The seriousness of this shortcoming was forcefully brought to our attention when we applied the z-transform method to $h_1(\tau)$ of Fig. 12.2-2. Sampling at intervals of $\tau = 1$ second was first attempted, and it resulted in an approximant with some rhp poles. The sampling interval was halved, but the resulting approximant also had rhp poles. A great deal of straightforward but tedious labor was involved in obtaining the approximants. The labor would not be all wasted, if the final results at least gave the designer some feeling for what was wrong. The values at the sampling points could then be modified to ensure a final stable function. But this is not the case, in general. The

z-transform method would be a very useful time domain synthesis technique, but for this serious weakness.

There is a way[1] of overcoming the stability problem, but it requires the sacrifice of one-half of the available degrees of freedom. It is to pick the poles of the z-transform function $F_a(z)$ in advance. Only the numerator of $F_a(z)$ is then available to the designer, so he can only ensure that $f_a(t)$ passes through half the number of sampled points. If n poles are assigned to $F_a(z)$, $f_a(t)$ is guaranteed to pass only through n equally spaced assigned values. The accuracy of this modified z-transform method (at values other than the sampling points) depends a great deal upon a judicious selection of the poles. The latter problem arises in the method of orthogonalized exponentials of Section 12.6, so that comments on the selection of the poles are postponed to Section 12.6.

§ 12.5 Guillemin's Fourier Series Method

The method of moments has the disadvantage that it is generally not possible to obtain any desired accuracy by using higher order approximants. Furthermore, the actual approximating time function is not known until $F_a(s)$ has been obtained and its inverse transform calculated. The latter objection also applies to some extent to the z-transform method. There, the values of $f_a(t)$ are known at $t = nT$. However, the nature of the intersample behavior, and of the asymptotic behavior at large t, are not known until $F_a(z)$ has been found. It is very desirable to have methods (1) wherein $f_a(t)$ is known precisely before $F_a(s)$ is evaluated, (2) which are guaranteed to converge to the exact prescribed $f(t)$ as closely as desired, if a sufficiently high-order $F_a(s)$ is used. Several methods which belong in this category are described in the following sections. One of these is Guillemin's[2] Fourier series method.

Fourier analysis is a well-established powerful method for approximating a periodic function by a sum of very simple functions. Guillemin has devised a method for applying Fourier decomposition to time domain synthesis. Consider a typical aperiodic $f(t)$ such as that in Fig. 12.5-1a. The first step is to select a suitable point (P in the figure) at which $f(t)$ is virtually over, and consider the interval OP to be half a period: $OP = T/2$. Next, the semiperiodic function $f_p(t)$ (Fig. 12.5-1b) is constructed, and a Fourier analysis of $f_p(t)$ is performed. A finite number of terms of the expansion of $f_p(t)$ will be used. Let this finite series be denoted by $f_p{}^* = f_1 + f_2$. Since $f_p{}^* = f_1 + f_2$ consists of a finite sum of sinusoids, it will only approximate $f_p(t)$. Suppose $f_p{}^* = f_1 + f_2$ (with transform $F_1 + F_2$) has the value shown by the dotted semiperiodic curve in

[1] L. E. Franks, On the Use of Delay Lines as Network Elements, Tech. Rept. No. 18. Stanford Electronics Lab., Stanford, California, July 29, 1957.

[2] E. A. Guillemin, "Synthesis of Passive Networks," pp. 707-726. Wiley, New York, 1957.

Fig. 12.5-1b. Let f_1 consist of the odd harmonics, and f_2 consist of the even harmonics of $f_p{}^* = f_1 + f_2$. Then $f_1(t) = -f_1(t + T/2)$, and $f_2(t) = f_2(t + T/2)$, and will appear as shown in Figs. 12.5-1c and d. If the function $f_2 - f_1$ is also constructed (Fig. 12.5-1e), it is seen that $f_2 - f_1$ is just $f_1 + f_2$ with a phase lead of 180°.

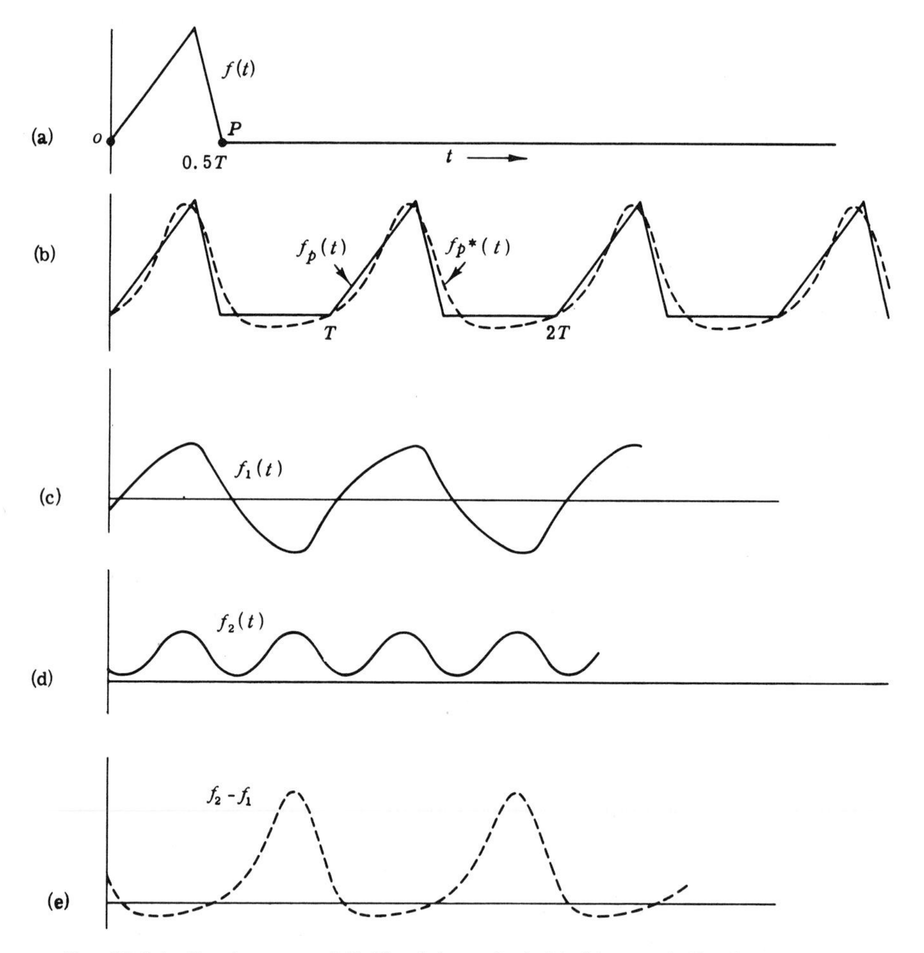

FIG. 12.5-1. Development of Guillemin's method. (a) $f(t)$—aperiodic. (b) Solid curve: $f_p(t)$—periodic continuation of $f(t)$; dashed curve: $f_p{}^*(t) = f_1 + f_2$—a finite series approximation of $f_p(t)$. (c) $f_1(t)$—odd harmonics of $f_p{}^*(t)$. (d) $f_2(t)$—even harmonics of $f_p{}^*(t)$. (e) $f_2 - f_1 = f_2(t + 0.5T) + f_1(t + 0.5T) = f_p{}^*(t + 0.5T)$.

One way to extract an aperiodic approximant out of the periodic $f_1 + f_2$ is to subtract from the latter a delayed $f_1 + f_2$. Thus, let $(f_1 + f_2)^*$ be $f_1 + f_2$, delayed by one period. Subtract $(f_1 + f_2)^*$ from $f_1 + f_2$ (see Fig. 12.5-2). The result is an aperiodic approximant with transform $F_d = (F_1 + F_2)(1 - e^{-sT})$.

By taking a sufficient number of harmonics of the Fourier series, the accuracy can be made as great as desired. The difficulty with the above is the need to realize the pure delay function e^{-sT}. Guillemin removes this difficulty by the following extremely ingenious scheme.

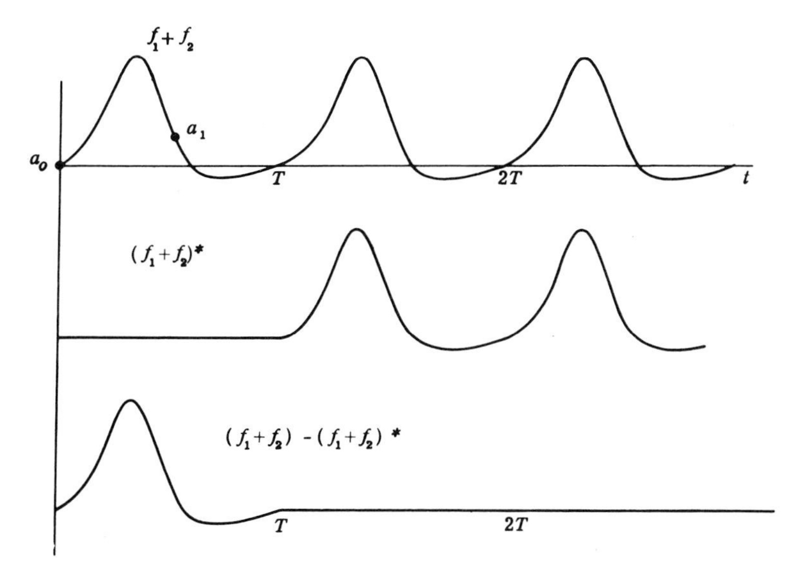

FIG. 12.5-2. Development of Guillemin's method (continued).

Let $A^2(s)$ be an approximation to the required delay e^{-sT}. We define $A(s)$ as follows:

$$(F_2 + F_1)\, A(s) \triangleq F_2 - F_1. \tag{12.5,1}$$

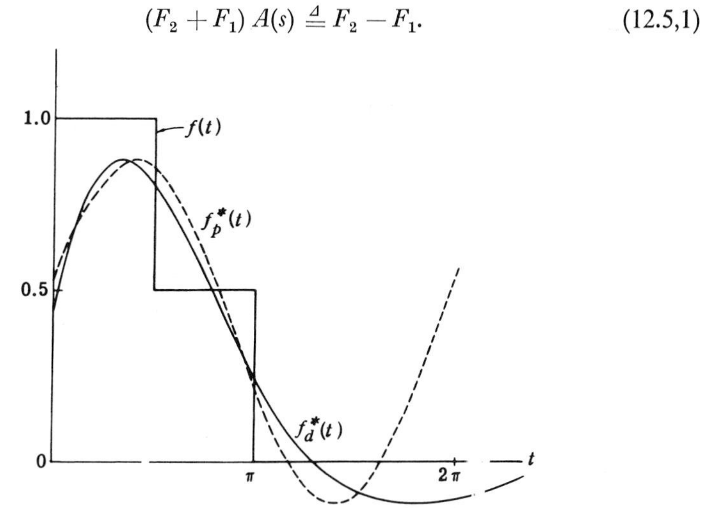

FIG. 12.5-3. Example of Guillemin's method (one harmonic).

Thus $a(t) \triangleq \mathscr{L}^{-1}A(s)$ is a function which, convolved with $f_1 + f_2$ (of Fig. 12.5-1b), results in the same function, with a phase lead of 180° (Fig. 12.5-1e). $A(s)$ is thus almost equivalent to $e^{-sT/2}$. We therefore take as our final approximant:

$$F_d{}^* = (F_1 + F_2)\,[1 - A^2(s)] = (F_1 + F_2)\left[1 - \frac{(F_2 - F_1)^2}{(F_2 + F_1)^2}\right] = \frac{4F_1(s)F_2(s)}{F_1(s) + F_2(s)}. \tag{12.5,2}$$

Example 1. The function $f(t)$ which is to be approximated by one with a rational transform is shown in Fig. 12.5-3. The first step is to find the Fourier series expansion of a periodic function $f_p(t)$ of period 2π, which has the value of $f(t)$ for $0 < t < \pi$, and is zero from π to 2π. Let $f_p(t) = A_o + A_1 \cos t + A_2 \cos 2t + \dots + B_1 \sin t + B_2 \sin 2t + \dots$, with

$$A_o = (1/2\pi) \int_0^{\pi} f(t)dt = (1/2\pi)\,(\pi/2 + 0.5\pi/2) = 0.375,$$

$$A_n = (1/\pi) \int_0^{\pi} f(t) \cos nt\; dt, \qquad B_n = (1/\pi) \int_0^{\pi} f(t) \sin nt\; dt.$$

As a first trial, we take a very simple approximation, using $f_p{}^*(t) = 0.375 + 0.159 \cos t + 0.477 \sin t$, which is sketched (dotted lines) in Fig. 12.5-3. Here $f_1 = 0.159 \cos t + 0.477 \sin t$, so $F_1 = (0.159s + 0.477)/(s^2 + 1)$; $f_2 = 0.375$, $F_2 = 0.375/s$. Therefore

$$F_d{}^* = \frac{4F_2F_1}{F_1 + F_2} = \frac{0.446(s+3)}{(s + 0.447)^2 + (0.71)^2} = \frac{0.446[s + 0.447 + (0.71)\,(3.6)]}{(s + 0.447)^2 + (0.71)^2},$$

and its inverse transform is $f_d{}^* = 0.446\, e^{-0.447t}(\cos 0.71t + 3.6 \sin 0.71t) = 1.66\, e^{-0.447t} \sin(0.71t + 15.5°)$, which is sketched in Fig. 12.5-3. The approximation is not bad, considering the simplicity of $F_d{}^*$.

The above example is really an unfair test of Guillemin's method (because of the small number of terms used) but it nevertheless does reveal the significant feature of the method. This is that the accuracy of the approximation in the region $0 < t < T/2$ is available by comparing the finite harmonic series $f_p{}^*(t)$ with the original $f(t)$ in that interval, i.e., there is comparatively little difference between $f_d{}^*$ and $f_p{}^*$, in the region $0 < t < T/2$. Therefore this method is almost truly a time domain method, in that the approximation is made in the time domain and not in the frequency domain. There is no need to find the inverse transform of $F_d{}^*$, in order to check on the quality of the approximation (as there was in the method of moments, and to a certain extent, in the z-transform method). We can presumably therefore use as many harmonics in the series as accuracy demands, and be assured that the final $f_d{}^*$ will be very close to $f_p{}^*(t)$ for $0 < t < T/2$. What of the error for $t > T/2$? Experience and theory suggest that the total error in this interval is of the same order of magnitude as the *total* error in the first interval.

Examination of the Approximating Error

We have $F_d^* = (F_2 + F_1)(1 - A^2) = (F_2 + F_1) - A(F_2 + F_1)A$. But $A = (F_2 - F_1)/(F_2 + F_1)$ [from Eq. (12.5,1)], so that $F_d^* = (F_2 + F_1) - A(F_2 - F_1)$; and in the time domain,

$$f_d^* = (f_2 + f_1) - a * (f_2 - f_1). \tag{12.5,3}$$

It is convenient to write $f_2 - f_1$ as the sum of two components. It is recalled that $f_2 - f_1$ is $f_2 + f_1$ with a phase lead of 180°. (Note Figs. 12.5-1b and e.) Therefore (see Fig. 12.5-4) $f_2 - f_1 = (f_2 + f_1)_{0.5T} + x(t)$, where $(f_2 + f_1)_{0.5T}$ is $(f_2 + f_1)$ delayed by $T/2$. Consequently [from Eq. (12.5,3)],

$$f_d^* = [(f_2 + f_1) - a * (f_2 + f_1)_{0.5T}] - a * x(t).$$

However, it is recalled that, by definition, $a * (f_2 + f_1)$ is equal to $f_2 - f_1$. Therefore $a * (f_2 + f_1)_{0.5T} = (f_2 - f_1)_{0.5T}$, i.e., it results in $f_2 - f_1$, delayed by $T/2$. Hence

$$\begin{aligned}(f_2 + f_1) - a * (f_2 + f_1)_{0.5T} &= f_2 + f_1 \qquad (\text{for } 0 < t < T/2)\\ &= 0 \text{ for } t > T/2.\end{aligned}$$

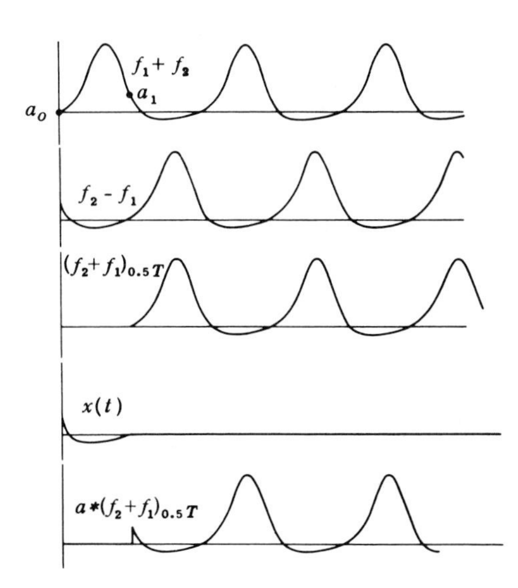

FIG. 12.5-4. Evaluation of the approximation error.

Thus the error $e(t)$ in using A^2 in place of e^{-sT} is precisely

$$e(t) = a(t) * x(t). \tag{12.5,4}$$

The calculation of $e(t)$ is simplified by writing $a(t) \triangleq \delta(t - T/2) + a_2(t)$. Now it is recalled [from Eq. (12.5,1)] that

$$a(t) * (f_2 + f_1) = [\delta(t - T/2) + a_2] * (f_2 + f_1) = f_2 - f_1 = x(t) + (f_2 + f_1)_{0.5T}$$

and $\delta(t - T/2) * (f_2 + f_1)$ is simply $(f_2 + f_1)_{0.5T}$. Accordingly, $a_2(t) * (f_2 + f_1) = x(t)$, and $e(t) = a(t) * x(t) = [\delta(t - T/2) + a_2(t)] * x(t) = x(t - T/2) + a_2(t) * x(t) = e_1 + e_2$. The value of $e_1 = x(t - T/2)$ is available from $x(t)$, which is $f - f_p^*$ in the region $T/2 \leqslant t \leqslant T$; and $e_2 = a_2(t) * x(t)$, with $a_2 * (f_2 + f_1) = x(t)$. We can easily evaluate a_2 by the method of Sections 12.8 and 12.9, and then evaluate e_2. In this way, it is possible to determine the error, and take as many harmonics as may be required in order to achieve any desired accuracy.

Value of f_d^* at $t = 0$

From the initial value theorem,

$$f_d^*(0) = \lim_{s\to\infty} \frac{4sF_1(s)sF_2(s)}{sF_1(s) + sF_2(s)} = \frac{4f_1(0)f_2(0)}{f_1(0) + f_2(0)}.$$

Let $f_2(0) + f_1(0) = f_{12}(0)$, and, from Fig. 12.5-4, $f_2(0) - f_1(0) = f_2(T/2) + f_1(T/2) \triangleq f_{12}(T/2)$. Therefore $2f_1(0) = f_{12}(0) - f_{12}(T/2) = a_o - a_1$, $2f_2(0) = f_{12}(0) + f_{12}(T/2) = a_o + a_1$, and finally $f_d^*(0) = (a_o^2 - a_1^2)/a_o$. Hence, if we want $f_d^*(0) = 0$, we must ensure that $(a_o^2 - a_1^2)/a_o = 0$, either by choosing $a_o = - a_1 \neq 0$, or by choosing $a_1 = a_o = 0$. We may sometimes want $(d/dt) f_d^*(t) |_{(t=0)} = 0$. From the initial value theorem,

$$f_d'^*(0) = \lim_{s\to\infty} \frac{4s^2F_1F_2}{F_1 + F_2} = \lim_{s\to\infty} \frac{4s^2F_1s^2F_2}{s^2F_1 + s^2F_2} = \frac{4f_1'(0)f_2'(0)}{f_1'(0) + f_2'(0)}.$$

In precisely the same manner as before, it is found that

$$f_d'^*(0) = [f_{12}'^2(0) - f_{12}'^2(0.5T)]/f_{12}'(0).$$

Example 2. The function $f(t)$ to be approximated is shown in Fig. 12.5-5. We

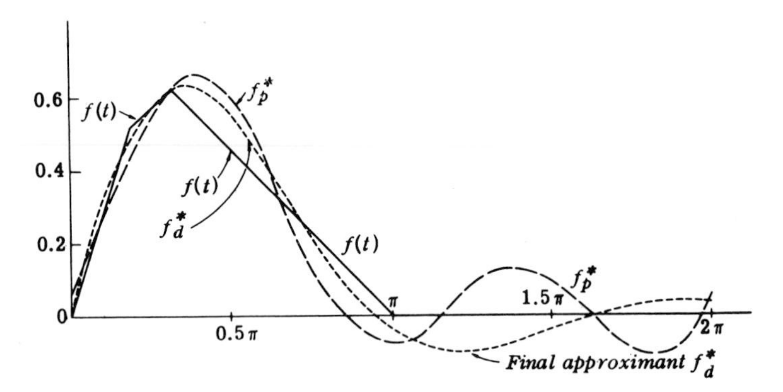

FIG. 12.5-5. Example of Guillemin's method (two harmonics).

try approximating with two harmonics, and find $f_p^* = 0.168 + 0.071 \cos t + 0.256 \sin t - 0.177 \cos 2t + 0.146 \sin 2t$, which is sketched in Fig. 12.5-5. We want $f_d^*(0)$ to be zero and note that at present $a_o = 0.062$, $a_1 = - 0.080$.

We can force $f_d^*(0)$ to be zero by adjusting it so that $a_0 = -a_1$. This can be done in a variety of ways. For example, we could change the constant term from 0.168 to $0.168 + 0.009 = 0.177$, since this would make the new $a_0 = 0.062 + 0.009 = 0.071$, and the new $a_1 = -0.080 + 0.009 = -0.071$; or we could change $-0.177 \cos 2t$ to $-0.168 \cos 2t$ for the same effect. If the former is chosen, then

$$F_1 = \frac{0.071s + 0.256}{s^2 + 1}, \qquad F_2 = \frac{0.177}{s} + \frac{(0.146)(2) - 0.177s}{s^2 + 4};$$

$$F_d^* = \frac{4F_1F_2}{F_1 + F_2} = \frac{1.168(s + 2.42)(s + 3.61)}{(s + 0.85)(s + 5.82)(s^2 + 1.05s + 2.03)}$$

$$= 1.168 \left\{ \frac{0.482}{s + 0.85} - \frac{0.0515}{s + 5.82} - \frac{0.43[(s + 0.525) - (1.56)1.325]}{(s + 0.525)^2 + (1.325)^2} \right\}$$

whose inverse transform is

$$f_d^* = 1.168[0.482\, e^{-0.85t} - 0.052\, e^{-5.82t} - 0.43\, e^{-0.525t} (\cos 1.325t - 1.56 \sin 1.325t)].$$

The latter is sketched in Fig. 12.5-5.

Comments on Guillemin's Fourier Series Method

(1) The most tedious parts of the method are: (a) finding the Fourier coefficients, (b) sketching f_p^* in order to check on the accuracy of the approximation. Simplified procedures for doing the above are presented in Section 12.14 and Sections 12.13 and 12.14, respectively.[1]

(2) The method does not guarantee that the poles of the approximant (the zeros of $F_1 + F_2$) will all lie in the left half-plane, and occasionally rhp poles do result. [Note that the zeros of $F_1 + F_2$ are the poles of $A(s) = \mathscr{L}a(t)$, and recall that $a * (f_2 + f_1) = f_2 - f_1$.]

It has been shown[2] that, in general, rhp poles must have very small residues. They may therefore be neglected, and their neglect will have little effect on the accuracy of the approximation. To neglect them, we expand F_d^* in partial fractions, and remove the offending terms. Or, if there is a zero near the rhp pole, then both the zero and the pole can be removed, the argument being that they practically cancel each other out.

[1] Manually operated mechanical devices for Fourier analysis are available. Gerber Scientific Instrument Company, Hartford, Connecticut.

[2] M. Strieby, Time Domain Synthesis by Means of Trigonometric Polynomial Approximations. Tech. Rept. 308, Section 3.3. Research Lab. of Electronics, M.I.T., Cambridge, Massachusetts, Jan. 5, 1956.

§ 12.6 Expansion in Orthogonalized Exponentials[1]

The time domain synthesis problem is to find $F_a(s)$, a rational function in s, whose inverse transform approximates a given time function $f(t)$. The inverse transform of a rational $F_a(s)$ is a sum of exponentials. Therefore, a direct method of solving the problem is to first approximate the given time function $f(t)$ by a sum of exponentials, viz.,

$$f(t) \approx f_a(t) = \Sigma\, e^{-\alpha_i t}\,[A_i \sin \omega_i t + B_i \cos \omega_i t] \tag{12.6,1}$$

and then $F_a(s) = \Sigma\,[B_i(s + \alpha_i) + A_i\omega_i]\,[(s + \alpha_i)^2 + \omega_i^2]^{-1}$. This is true time domain synthesis in that the approximation is done entirely in the time domain; the exact approximation error is $e(t) = f(t) - f_a(t)$. There is no additional error in going from the time domain to the s domain.

The problem is thus one of approximating the given $f(t)$ by a sum of exponentials, as in Eq. (12.6,1). A reasonable procedure is to pick an error criterion, and then choose the unknowns in Eq. (12.6,1) (A_i, B_i, α_i, and ω_i) so as to minimize the error for the given number of terms used in the approximation. The error criterion most suitable for analytic work is the weighted integral squared-error, i.e., $\mathscr{E} = \int_0^\infty w^2(t)\, e^2(t)\, dt$, where $e(t) \triangleq f(t) - f_a(t)$, and the weighting function $w^2(t)$ is included, so as to permit the designer to obtain maximum accuracy wherever he might want it. A little thought will indicate how enormously complicated it would be to use the above (partial differentiation of $\mathscr{E}$ with respect to each unknown) to choose *all* the unknowns in Eq. (12.6,1).

The problem becomes more tractable, however, if some of the unknowns, specifically the α_i and ω_i [the poles of $F_a(s)$], are *a priori* specified; then the A_i and B_i are chosen so as to minimize $\mathscr{E}$. In such a case, setting to zero the partial derivative of $\mathscr{E}$ with respect to each A_j and B_j leads to exactly $2n$ equations, which may be solved for the $2n$ unknowns. A typical equation is obtained as follows:

$$\frac{\partial \mathscr{E}}{\partial A_k} = \frac{\partial}{\partial A_k}\int_0^\infty w^2(t)\,(f - f_a)^2 dt = 2\int_0^\infty (f_a - f)\,\frac{\partial f_a}{\partial A_k}\, w^2(t) dt = 0.$$

[1] The time domain synthesis technique described in this section is only occasionally useful to the control engineer for numerical design work. The concepts are, however, rather important, and are frequently used in the literature. The material of this section has been adapted principally from the work of Kautz and of Huggins: W. H. Kautz, Transient synthesis in the time domain, *Trans. IRE* **CT-1** (3), 29-39 (1954); Network Synthesis for Specified Transient Response, Tech. Rept. No. 209, Research Lab. of Electronics, M.I.T., Cambridge, Massachusetts, April 23, 1952; W. H. Huggins, Representation and Analysis of Signals, Part I. The Use of Orthogonalized Exponentials. AFCRC Rept. No. TR-57-357, Astia No. AD133741, John Hopkins University, Electrical Engineering Dept., Baltimore, Maryland, Sept. 30, 1957.

This leads to

$$\int_0^\infty w^2(t)e^{-\alpha_k t}\sin\omega_k t \sum e^{-\alpha_i t}(A_i \sin\omega_i t + B_i\cos\omega_i t)\,dt$$
$$= \int_0^\infty w^2(t)f(t)e^{-\alpha_k t}\sin\omega_k t\,dt. \tag{12.6,2}$$

The amount of labor still appears to be enormous. In addition, the A's and B's which result are functions of n, i.e., of the number of terms used in the approximation. After a specific n has been tried, and the work done finding the A's and B's, the designer might well decide that more terms are needed. If so, he must recalculate *all* the A's and B's.

There is a way to reduce the labor somewhat and at the same time eliminate the need to recalculate all the A's and B's if more terms are added. Consider Eq. (12.6,1), and let us rewrite it as $f_a(t) = \sum_{j=1}^{2n} D_j\varphi_j(t)$, where the φ_j are fixed and known linear combinations of the exponentials. The error function is differentiated with respect to the D_j. Equation (12.6,2) becomes

$$\int_0^\infty w^2(t)f(t)\varphi_k(t)dt = \int_0^\infty w^2(t)\varphi_k(t)[D_1\varphi_1(t) + D_2\varphi_2(t) + \ldots + D_{2n}\varphi_{2n}(t)]dt. \tag{12.6,3}$$

Suppose the φ's are deliberately chosen so that $\int_0^\infty w^2(t)\,\varphi_k(t)\,\varphi_j(t)dt = 0$ for all $j \neq k$, and equals $M_j \neq 0$ for $j = k$. The φ functions are then said to be an *orthogonal* set with respect to the weighting function $w^2(t)$, over the interval 0 to ∞. If $M_j = 1$ for all j, they are called an *orthonormal* set. Equation (12.6,3) then becomes the very much simpler $\int_0^\infty w^2(t)f(t)\varphi_k(t)\,dt = D_k M_k$. There is now no need to solve a set of equations—each equation determines an unknown D. Furthermore, each D_k is independent of n, the number of terms used in the approximation. Hence, if the designer should later decide to use more terms, he need only calculate the coefficients of the added terms. The coefficients previously calculated remain at their former values. The expansion is said to have the property of *finality*.

An important consideration is whether the orthogonal series φ_k is finite or infinite. If the former, then it is certainly impossible, in general, for the series $f_a(t) = \sum D_k\varphi_k(t)$ to approximate $f(t)$ with arbitrarily small error. If there are an infinite number of φ_k, it is still not necessarily true that $\mathscr{E}$ may be made arbitrarily small. Any orthogonal set of functions which does have this property is called a *complete* set. It has been shown[1] that the set is complete if the poles used to generate the set [in Eq. (12.6,1)] are such that $\sum_{i=1}^\infty \alpha_i\{1 + [(\alpha_i - 0.5)^2 + \omega_i^2]\}^{-1}$ approaches infinity. If $f(t)$ is square integrable [$\int_0^\infty |f(t)|^2\,dt$ is finite], and it is expanded in an orthogonal series which is complete [$f(t) \approx \sum^n D_j\varphi_j$], then $\lim_{n\to\infty}\int_0^\infty |f(t) - \sum^n D_j\varphi_j|^2\,dt = 0$. If $f(t)$ is also continuous, then

[1] R. Paly and N. Wiener, Fourier transforms in the complex domain. *Am. Math. Soc. Colloquium Publs.* **19**, p. 34 (Szasz's theorem) (1934).

$f(t) = \lim_{n\to\infty} \Sigma^n D_j\varphi_j$. It is conceptually important to know that any time function which is square integrable and continuous can be approximated to any desired accuracy by a finite orthogonal exponential set. For engineering purposes, all functions of interest are continuous. All the results in this book which were derived for rational transforms and transfer functions (specifically in Chapters 4, 7, and 9) are therefore completely general.

Conversion of Fourier Set to Exponential Set

The well-known Fourier series represents an orthogonal set (over the interval 0 to 2π, with unit weight) whose poles all lie on the $j\omega$ axis. It therefore cannot be directly used for time domain synthesis. Guillemin's method of Section 12.5 represents one technique for applying Fourier series to time domain synthesis. Another way is to predistort $f(t)$ by first multiplying it by e^{at} and then proceed with the Fourier expansion. This procedure is perhaps best described by means of a numerical example.

Example 1. The original $f(t)$ is drawn in Fig. 12.6-1. The function $f_1(t) \triangleq f(t)\,e^{2t}$

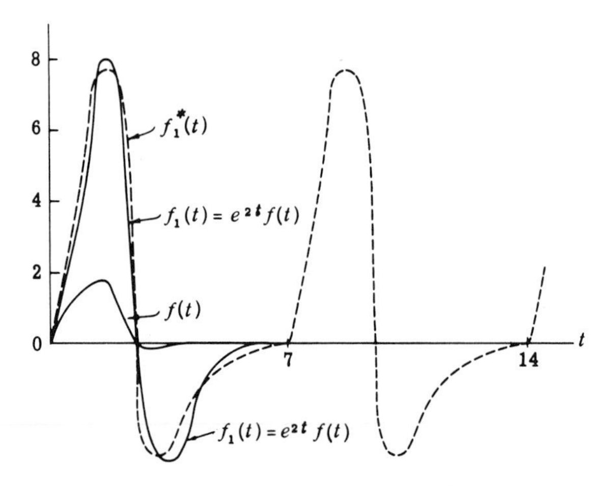

FIG. 12.6-1. Conversion of Fourier series to exponential set.

is obtained by point-by-point multiplication (see figure). Next, a Fourier series expansion of $f_1(t)$ is made with $T = 7$ seconds. A finite number of terms is used, and let us suppose that with two harmonics, the approximation obtained [denoted by $f_1{}^*(t)$] is that shown in the figure, i.e.,

$$f_1{}^*(t) = B_o + B_1 \cos \frac{2\pi t}{7} + B_2 \cos \frac{4\pi}{7} + A_1 \sin \frac{2\pi t}{7} + A_2 \sin \frac{4\pi t}{7}.$$

Then

$$F_1{}^*(s) = \mathscr{L}[f_1{}^*(t)] = \frac{B_o}{s} + \frac{B_1 s + A_1\omega_1}{s^2 + \omega_1{}^2} + \frac{B_2 s + A_2\omega_2}{s^2 + 4\omega_1{}^2}$$

where $\omega_1 = 2\pi/7$. Next, an aperiodic function is extracted from $f_1{}^*(t)$, by taking $f_a(t) = e^{-2t}f_1{}^*(t)$. The transform of $f_a(t) = e^{-2t}f_1{}^*(t)$ is

$$F_a(s) = F_1{}^*(s+2) = \frac{B_o}{(s+2)} + \frac{B_1(s+2) + A_1\omega_1}{(s+2)^2 + \omega_1{}^2} + \frac{B_2(s+2) + A_2\omega_2}{(s+2)^2 + 4\omega_1{}^2}.$$

It is noted that the above operations in effect apply a weighting function of e^{2t} to the error; the actual approximation error in the interval $0 \leqslant t \leqslant T$ is $e^{-2t}[f_1(t) - f_1{}^*(t)]$. This procedure leads to transform poles located on a line parallel to the $j\omega$ axis. There is some freedom in the choice of the fundamental period T, and of the exponent in $f_1(t) = e^{-at}f(t)$. However, it is desirable that $e^{-at}f(t) \to f(0)$ as $t \to T$, for then $f_1{}^*(t)$ is continuous at $t = T$, which leads to faster convergence of the Fourier series.[1]

Conversion of Orthogonal Polynomials into Orthogonal Exponentials

The well-known polynomial sets which are orthogonal over a finite interval are convertible into exponential sets, orthogonal over the semi-infinite interval $0 \leqslant t < \infty$. For example, the Legendre orthonormal set of polynomials can be regarded as generated by orthonormalizing, with unit weight, the functions $x^0, x, x^2, x^3, \ldots$ over the interval $-1 \leqslant x \leqslant 1$. The result is $\varphi_k = [(2k+1)/2]^{0.5} P_k(x)$, $k = 0, 1, 2, \ldots$, and $P_k(x)$ is the well-known[2] Legendre polynomial of the nth degree. If we let $x = 1 - 2e^{-2at}$, then the interval $-1 \leqslant x \leqslant 1$ is transformed into the interval $0 \leqslant t \leqslant \infty$. The effective weighting of the error is by the factor $dx/dt = 4ae^{-2at}$. The procedure therefore is to transform $f(t)$ into a function of x, and expand the latter function in terms of the normalized Legendre polynomials as functions of x. At this point, x is replaced by $1 - 2e^{-2at}$, and the Laplace transform of the series is obtained term by term. The resulting transform has its poles at $s = -2a, -4a, -6a$, etc. Other well-known orthogonal polynomial sets can be similarly transformed into exponential sets.[3]

Generation of Orthonormal Sets

There is no limit to the number of orthonormal sets that may be generated. Suppose we pick arbitrarily a set of pole positions, $s_\alpha, s_\beta, \ldots, s_1, \bar{s}_1, s_2, \bar{s}_2, s_3, \bar{s}_3, \ldots$ (the s_α are the real poles). Then we can take the first member of the set as $\varphi_1 = A_1 \exp(s_1 t) + \bar{A}_1 \exp(\bar{s}_1 t)$, with A_1 chosen so that $\int_0^\infty \varphi_1{}^2\, dt = 1$; the second member as $\varphi_2 = A_2\varphi_1 + A_3 \exp(s_2 t) + \bar{A}_3 \exp(\bar{s}_2 t)$; and choose A_2 so

[1] Cornelius Lanczos, "Applied Analysis," Chapter 4. Prentice-Hall, Englewood Cliffs, New Jersey, 1956.

[2] As tabulated, for example, in E. Jahnke and E. Emde, "Tables of Functions," 4th ed., p. 107. Dover Publ., New York, 1945.

[3] W. H. Kautz, Network synthesis for specified transient response. Tech. Rept. No. 209, Appendix D. Research Lab. of Electronics, M.I.T., Cambridge, Massachusetts, April 23, 1952.

that $\int_0^\infty \varphi_1\varphi_2\, dt = 0$, A_3 so that $\int_0^\infty \varphi_2^2\, dt = 1$. As φ_3 we take $\varphi_3 = B_1\varphi_1 + B_2\varphi_2 + C_3 \exp(s_3 t) + \bar{C}_3 \exp(\bar{s}_3 t)$, and choose the three available variables so that the following three conditions are satisfied: $\int_0^\infty \varphi_3\varphi_1\, dt = 0 = \int_0^\infty \varphi_3\varphi_2\, dt$, and $\int_0^\infty \varphi_3^2\, dt = 1$. This process can be continued indefinitely. However, rather than pick the set as above in the time domain, it is much easier to generate the set in the s domain, by the Kautz method, which is next described.

Kautz Method for s-Domain Generation of Exponential Orthonormal Sets

The method is based upon the Parseval relation (Section 9.8),

$$2\pi j \int_0^\infty f_1(t) f_2(t) dt = \int_{-j\infty}^{j\infty} F_1(s) F_2(-s) ds.$$

Therefore, if two time functions are orthogonal, their Fourier transforms are also orthogonal. If the $F(s)$ are rational functions and $F_j(s)\, F_k(s)$ goes to infinity as $1/s^2$, then the right-hand integral is zero if the integrand has no poles in either half-plane. If there are any poles, the value of the integral is available from Cauchy's residue theorem. These facts are the key to generating the desired set. For example, suppose we pick the set of simple poles at $-1, -2, -3$, etc. Then $\varphi_1 = Ae^{-t}$, $\Phi_1 \triangleq \mathscr{L}\varphi_1 = A/(s+1)$;

$$\int_0^\infty \varphi_1^2(t) dt = \frac{A^2}{2\pi j} \oint \frac{ds}{(s+1)(1-s)} = A^2\{\text{residue at} -1 \text{ of } [(s+1)(1-s)]^{-1}\}$$

$$= -A^2\{\text{residue at} +1 \text{ of } [(s+1)(1-s)]^{-1}\} = 0.5A^2.$$

Therefore, for orthonormality, $A = (2)^{0.5}$. To determine the form of φ_2, note that we want $\int_0^\infty \varphi_1\varphi_2\, dt = A(2\pi j)^{-1} \oint (s+1)^{-1}\, \Phi_2(-s)\, ds = 0$. It is sufficient that the integrand have no poles in one of the two half-planes. Since the poles of all Φ_i must lie in the left half-plane, $\Phi_2(-s)$ must have its poles in the right half-plane. We therefore choose $\Phi_2(-s) = B(s+1)/[(s-2)(s-1)]$, in order that $\Phi_1(s)\, \Phi_2(-s)$ has no lhp poles. One pole of $\Phi_2(-s)$ is at $+2$, in order that φ_2 is independent of φ_1; another pole is necessary in order that $\Phi_1(s)\, \Phi_2(s)$ goes to zero at large s as $1/s^2$. The latter is taken at $+1$ because the poles must be chosen from the *a priori* specified list, and each new member of the orthogonal set must introduce only one new pole. The factor B is chosen so that $\int_0^\infty \varphi_2^2\, dt = 1$.

In general, it can be shown, in the above manner, that the orthonormal set generated from the set of real poles at $-s_1, -s_2, -s_3, \ldots$ has the form

$$\Phi_k(s) = (2s_k)^{0.5} \frac{(s-s_1)(s-s_2)\ldots(s-s_{k-1})}{(s+s_1)(s+s_2)\ldots(s+s_{k-1})(s+s_k)}.$$

If the *a priori* specified set of poles has any or all complex pairs, the Kautz procedure is as follows. The first complex pole pair, $(s^2 + \alpha_1 s + \omega_1^2)^{-1}$, that is introduced, say into φ_k, leads to two poles instead of the customary one, so one additional zero is needed. This zero factor is $(s + \omega_1)$. For φ_{k+1} the new factors are $(s - \omega_1)/(s^2 + \alpha_1 s + \omega_1^2)$. Also, the normalizing factor is $(2\alpha_k)^{0.5}$. For example, consider the set of poles $-1, -1 \pm j, -2 \pm 2j, -3, -4, -5 \pm 5j$. The orthonormal set is

$$\Phi_1 = (2)^{0.5}(s + 1)^{-1};$$
$$\Phi_2 = (2)^{0.5}(s - 1)(s + \sqrt{2})/(s + 1)(s^2 + 2s + 2);$$
$$\Phi_3 = (2)^{0.5}(s - 1)(s - \sqrt{2})/(s + 1)(s^2 + 2s + 2);$$
$$\Phi_4 = (4)^{0.5}(s - 1)(s^2 - 2s + 2)(s + \sqrt{8})/(s + 1)(s^2 + 2s + 2)(s^2 + 4s + 8);$$
$$\Phi_5 = (4)^{0.5}(s - 1)(s^2 - 2s + 2)(s - \sqrt{8})/(s + 1)(s^2 + 2s + 2)(s^2 + 4s + 8);$$
$$\Phi_6 = (6)^{0.5}(s - 1)(s^2 - 2s + 2)(s^2 - 4s + 8)/(s + 1)(s^2 + 2s + 2)(s^2 + 4s + 8)(s + 3);$$
$$\Phi_7 = (4/3)^{0.5}\,\Phi_6(s - 3)/(s + 4);$$
$$\Phi_8 = (1.25)^{0.5}\,\Phi_7(s + \sqrt{50})/(s^2 + 10s + 50);$$
$$\Phi_9 = (1.25)^{0.5}\,\Phi_7(s - \sqrt{50})/(s^2 + 10s + 50).$$

While on this subject of the relation between the orthogonal term and its transform, it is worth noting that the expansion of a given function can sometimes be performed much easier in the s domain than in the time domain. Thus $f(t) \approx D_1\varphi_1 + D_2\varphi_2 + D_3\varphi_3 + \ldots$ with $D_k = \int_0^\infty f(t)\,\varphi_k dt = (2\pi j)^{-1}\int_{-\infty}^{\infty} F(s)\,\Phi_k(-s)\,ds$. Cauchy's residue theorem is used to evaluate the integral. This method can be used only if $F(s)$ is known analytically, and is appropriate, for example, when $f(t)$ consists of piecewise linear segments or when $f(t)$ can be reasonably approximated in this manner. When $F(s)$ is transcendental, care must be taken to find all the lhp poles of $F(s)$, and to avoid the branch points.

Choosing the Set of Poles

What pole set should be used as a basis for the expansion of a specified time function in an orthonormal exponential set? This is extremely important in relation to economy in the number of terms needed to obtain a specific accuracy of approximation. For example, $f(t) = e^{-\alpha_1 t}\sin(\omega_1 t + \theta_1)$ can be approximated to any degree of accuracy by the exponential Legendre set, but it is obviously much more economical to use a set whose first member is $[(s + \alpha_1)^2 + \omega_1^2]^{-1}$. There is, unfortunately, as yet no reasonably simple procedure for finding the set of poles which is "natural" for a given time function.[1] The simplest proce-

[1] The methods that have been suggested require nearly as much labor as that involved in the final expansion after the poles are known, and so almost double the work. See W. H. Kautz, Network synthesis for specified transient response, Tech. Rept. No. 209, p. 27. Research Lab. of Electronics, M.I.T., Cambridge, Massachusetts, April 23, 1952. Also J. E. Storer, "Passive Network Synthesis," pp. 307-315. McGraw-Hill, New York. 1957.

dure is to use experience as a guide—note similar time functions and the poles of their transforms. Fortunately, the economy in the number of terms is not usually critically sensitive to the precise location of the poles. Those cases where it is critically sensitive (e.g., low-damped exponential sinusoids) are easily recognized by inspection.

Accuracy Requirements of Approximants

It is clear that in the methods described in Sections 12.5 and 12.6 the accuracy of approximation is closely related to the amount of computational labor the analyst is willing to invest. It is therefore important to have some appreciation for the kind of accuracy that is needed in the problem under consideration. In feedback control problems the given function may be the transient response of the plant or transducer, it may represent a typical input signal or disturbance, or the autocorrelation function of an ensemble of signals. In the latter cases the approximate transform that is obtained is used to find appropriate disturbance or error response functions (Sections 9.4-9.6) or to derive the mean square $T_{opt}(s)$ for filtering noise from the useful system input (Sections 9.8-9.17). If the relation between the time signal and its transform is not a sensitive one, then minor deviations from the transform will have only a minor effect on the

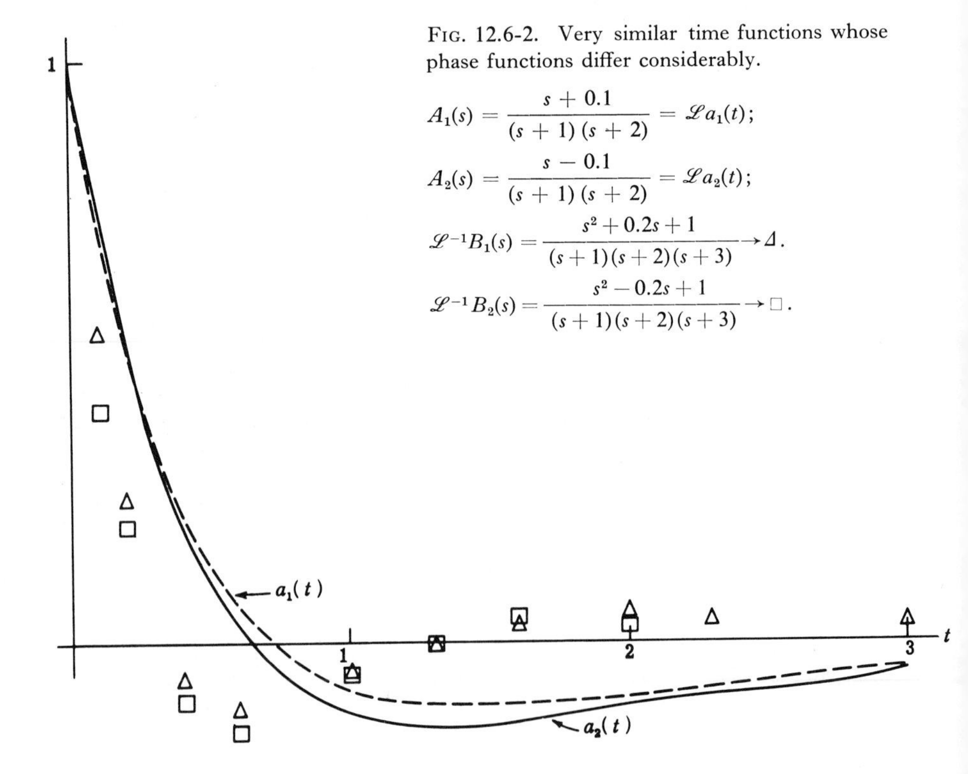

FIG. 12.6-2. Very similar time functions whose phase functions differ considerably.

$$A_1(s) = \frac{s+0.1}{(s+1)(s+2)} = \mathscr{L}a_1(t);$$

$$A_2(s) = \frac{s-0.1}{(s+1)(s+2)} = \mathscr{L}a_2(t);$$

$$\mathscr{L}^{-1}B_1(s) = \frac{s^2+0.2s+1}{(s+1)(s+2)(s+3)} \rightarrow \Delta.$$

$$\mathscr{L}^{-1}B_2(s) = \frac{s^2-0.2s+1}{(s+1)(s+2)(s+3)} \rightarrow \square.$$

resulting disturbance response function, $T_{\text{opt}}(s)$, etc., because these functions are basically derived from time considerations (error, disturbance output vs. time, average squared error, etc.). In deriving these functions the properties of the frequency response that are important are the bandwidth, amplitude peaking, accumulated phase lag, etc. (recall Section 5.10). If these rather gross properties of the transform are not much affected by accuracy of approximation, then there is no need for great accuracy. Obviously the amplitude peaking and the time response are very sensitive to the position of poles which are close to the $j\omega$ axis. If there are no such underdamped poles, then the sensitivity of the time function to its transform is small and there is no need for great accuracy.

The problem must be examined more carefully when the approximant represents the transfer function of the plant or of any element in a feedback loop. It is possible to have two elements with impulse responses which are very similar, but their effect in a feedback loop can be extraordinarily different. For example, compare $a_1(t)$, $a_2(t)$ in Fig. 12.6-2. The time responses are very similar, but they differ considerably in their frequency responses (in their phase characteristics). It does not matter if A_1 is used to represent a_2 (or vice versa) in the optimum filter problem, but it does make a great deal of difference if a_2 is the impulse response of a plant in a feedback loop. In the latter case, a feedback design based on $A_1(s)$ in place of the correct $A_2(s)$ could easily result in an unstable system. Similar conclusions hold for B_1 and B_2 of Fig. 12.6-2. These curves illustrate the tremendous usefulness of the transform method with its stability criteria and how difficult and hopeless feedback system design would be if it had to be executed completely in the time domain. The conclusion is that in feedback applications one must be suspicious of elements whose impulse responses have approximants with zeros close to the $j\omega$ axis. Plant identification by means of a frequency response run is more suitable for such plants. There is less danger of error if the approximant has a zero further off from the $j\omega$ axis—for example, see Fig. 12.6-3.

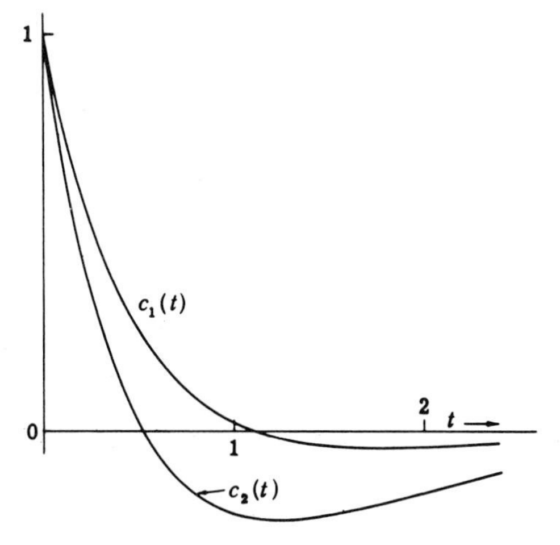

FIG. 12.6-3. Larger difference in time functions when zeros are further from $j\omega$ axis.

$$C_1(s) = \frac{s + 0.5}{[(s + 1)(s + 2)]};$$

$$C_2(s) = \frac{s - 0.5}{[(s + 1)(s + 2)]}.$$

§ 12.7 Rational Function Approximation of Frequency Response Data

The control engineer often obtains the characteristics of the plant in the form of frequency response data, and he may wish to translate this data into a rational function $P(s)$. It should be noted that it is, in many cases, possible to get along without such a rational function representation. The shaping of the loop transmission functions in order to obtain the desired benefits of feedback can be carried out without the rational function representation of the plant. In fact, as noted throughout the book, the feedback part of the design is usually best done in terms of frequency response.

Frequency response methods can also be used to obtain a satisfactory response function. In a specific problem, after the design specifications have been determined, the loop transmission functions may be chosen, in the frequency domain, in accordance with the desired feedback properties of the system. The system response may, in general, be written in the form $T = GP/(1 + L)$. The broad correlation between frequency response and time response (Section 5.10) often suffices for determining the desired $T(j\omega)$. The desired T and the designed $P/(1 + L)$ are accordingly available in terms of frequency response, and they consequently determine the frequency response of G. At this point, a rational function representation of G may be very useful for the actual network realization of G. This postponement of the rational function approximation problem to G may be justified in that there is then no need to calculate and factor $1 + L$.

We have previously noted, in Section 8.14, that there exist elegant techniques for complex curve fitting, which the engineer is rarely advised to use. Cut and try shaping on the Bode plots suffice for most problems that confront the feedback control engineer. A semigraphical method such as Linvill's (see Section 8.14) may be used as the final step in the approximation.

§ 12.8 Numerical Convolution

This section considers the problem of convolving two time functions $[f(t) = \mathscr{L}^{-1}F(s)$ and $g(t) = \mathscr{L}^{-1}G(s)]$ which are available graphically. In most problems, it is not convenient to find $F(s)$ and $G(s)$, and then evaluate the inverse transform of $F(s)\,G(s)$. It is usually easier to perform the operation $f(t) * g(t) = \int_0^t f(t - \tau)\,g(\tau)\,d\tau$ by numerical methods. Of the many methods presented in the literature, we have chosen one[1] which is easily derivable from the z-transform theory of Chapter 11.

The following results from z-transform theory are used in the development: Consider a staircase function $h(t)$, such as that shown in Fig. 12.8-1a. We

[1] The material of this section has been adapted from the paper by J. G. Truxal, Numerical analysis for network design. *Trans. IRE* **CT-1** (3), 49-60 (1954).

take $h(0) = h(0-) = 0$, but $h'(0) = a_o \delta(t)$. The Laplace transform of the sampled $h(t)$ (with sampling period T) is: $\bar{H}(z) = a_1 z^{-1} + a_2 z^{-2} + \ldots$, where $z = e^{sT}$. We use $\bar{H}(z)$ to distinguish this series which does not include a_o,

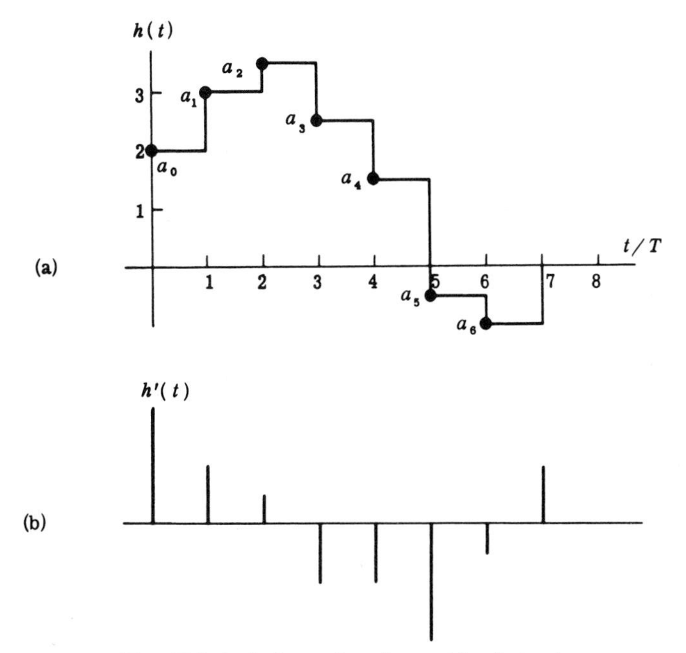

FIG. 12.8-1. Staircase function and its derivative.

from the customarily used series $H(z) = \bar{H}(z) + a_o$. The derivative of $h(t)$ is the impulse series of Fig. 12.8-1b, and the Laplace transform of $h'(t)$ is $Zh'(t) = a_o + (a_1 - a_o) z^{-1} + (a_2 - a_1) z^{-2} + \ldots$. It is easy to see that

$$Zh'(t) = (1 - z^{-1}) [\bar{H}(z) + a_o] = (1 - z^{-1}) H(z). \qquad (12.8,1)$$

Equation (12.8,1) is not true, in general. It is true only if $h(t)$ is a staircase function. It is important to note that we have taken $h(0) = \lim_{t \to 0-} h(t) = 0$ and therefore $h'(0)$ is the impulse a_o.

Similarly, consider a polygon function, such as the one shown in Fig. 12.8-2a [note $p(0) = 0$]. Then $p'(t)$ is the staircase function of Fig. 12.8-2b and $p''(t)$ is the impulse series of Fig. 12.8-2c. The condition $p(0) = 0$ ensures that $p'(t)$ has no impulse at $t = 0$. If $P(z) = Zp(t) = a_1 z^{-1} + a_2 z^{-2} + \ldots$ (i.e., the Laplace transform of the sampled function), then $Zp'(t) = T^{-1} [a_1 + (a_2 - a_1) z^{-1} + (a_3 - a_2) z^{-2} + \ldots] = T^{-1}(z - 1) P(z)$. From Eq. (12.8,1), $Zp''(t) = [(z - 1)/z] Zp'(t) = T^{-1} z^{-1}(z - 1)^2 P(z)$. Therefore,

$$Zp''(t) = P(z) (z - 1)^2/Tz = (Tz)^{-1} (z - 1)^2 Zp(t) \qquad (12.8,2)$$

[if $p(t)$ is a strictly polygon function]. A strictly polygonal function is one whose derivative is a staircase function (no impulses).

We can continue in this manner with higher order functions. Suppose $m(t)$ consists of confluent parabolic segments of period T [$m(t)$ and $m'(t)$ are con-

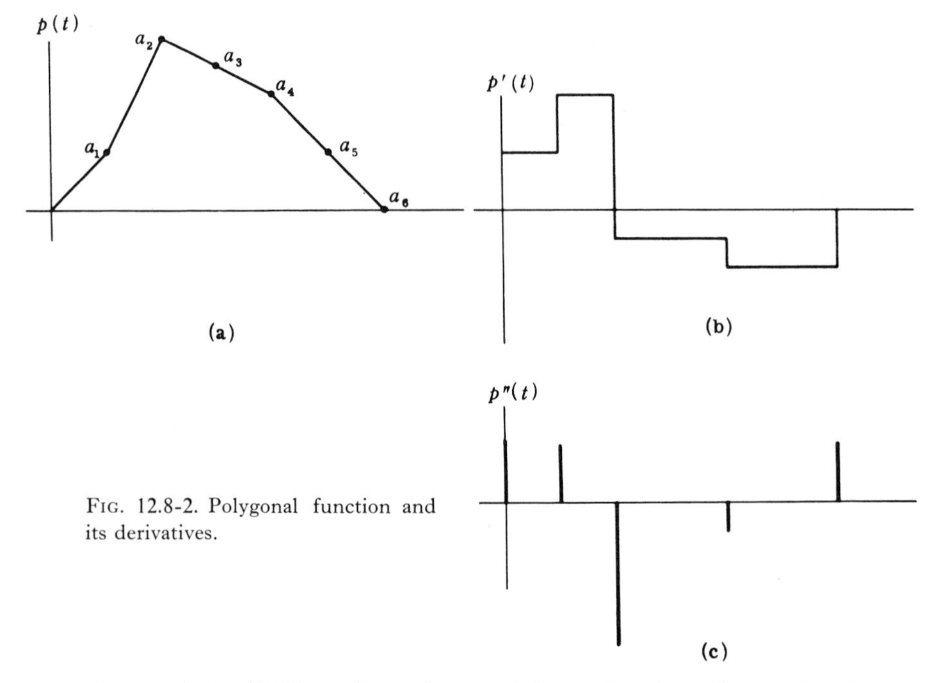

FIG. 12.8-2. Polygonal function and its derivatives.

tinuous, but $m''(t)$ is a discontinuous staircase function with no impulses—this requires $m(0) = m'(0) = 0$]. It can be shown in the same manner as in the above that

$$Zm'''(t) = \frac{2(z-1)^3}{T^2 z(z+1)} M(z) = \frac{2(z-1)^3}{T^2(z+1)z} Zm(t) \qquad (12.8,3)$$

[only if $m(t)$ is strictly parabolic, as previously described]. It should be noted that $z/(z-1)$ in Eq. (12.8,1) is $Z(1/s)$; $Tz/(z-1)^2$ in Eq. (12.8,2) is $Z(1/s^2)$; $T^2 z(z+1)/2(z-1)^3$ in Eq. (12.8,3) is $Z(1/s^3)$ (see Table 11.4-1). In general, $Zf^k(t) = Zf(t)/Z(s^{-k})$, $k = 0, 1, 2, ..., n$, providing $f(t)$ is an nth-order function. An nth-order function is here defined as one in which $f(t), f'(t), ..., f^{(n-1)}(t)$ are continuous, and $f^n(t)$ is a staircase function. An nth-order function must therefore have $f(0) = f'(0) = ... = f^{n-1}(0)$. The staircase function is zero order, the polygon is first order, etc. Conversely, if $g(t)$ consists of a series of impulses located at $t = nT$, with transform $G(z)$, a single integration yields a staircase function $g^{-1}(t)$, such that

$$Zg^{-1}(t) = Zs^{-1}\, Zg(t) = zG(z)/(z-1). \qquad (12.8,4a)$$

A second integration gives $g^{-2}(t)$, a polygon function (first order), whose sampled values are available from

$$Zg^{-2}(t) = Zs^{-2}\, Zg(t) = TzG(z)/(z-1)^2. \tag{12.8,4b}$$

In general, n integrations results in a function of order $n-1$, whose sampled values are available from

$$Zg^{-n}(t) = Zs^{-n}\, Zg(t) = G(z)\, Zt^{n-1}/(n-1)!. \tag{12.8,5}$$

The Convolution of Two Impulse Series

The problem is to evaluate $c(t) = f(t) * g(t) = \int_0^t f(t-\tau)\, g(\tau)\, d\tau$, where $f(t)$ and $g(t)$ are only available graphically. Consider the special case where f and g are impulse series equally spaced at intervals of T seconds; viz.,

$$f(t) = a_o\delta(t) + a_1\delta(t-T) + a_2\delta(t-2T) + \dots$$
$$g(t) = b_o\delta(t) + b_1\delta(t-T) + b_2\delta(t-2T) + \dots .$$

Then $F(s) = a_o + a_1 z^{-1} + a_2 z^{-2} + \dots$, $G(s) = b_o + b_1 z^{-1} + b_2 z^{-2} + \dots$, with $z = e^{sT}$. Consequently,

$$\begin{aligned} c(t) &= \mathscr{L}^{-1} F(s)\, G(s) \\ &= \mathscr{L}^{-1}[a_o b_o + (a_o b_1 + a_1 b_o)\, z^{-1} + (a_o b_2 + a_1 b_1 + a_2 b_o)\, z^{-2} + \dots] \\ &= a_o b_o \delta(t) + (a_o b_1 + a_1 b_o)\delta(t-T) + (a_o b_2 + a_1 b_1 + a_2 b_o)\delta(t-2T) + \dots . \end{aligned}$$

The convolution of two series of impulse functions is clearly a very simple operation. Summarizing, $c(t)$ is an impulse series and

$$C(z) = Zc(t) = Z[f(t) * g(t)] = F(z)\, G(z), \tag{12.8,6}$$

if $f(t)$ and $g(t)$ are impulse series.

Impulse Approximation of Continuous Functions

We have seen how easy it is to find the convolution of two series of equally spaced impulses. An approximation of $c(t) = f(t) * g(t)$ is therefore readily obtained by approximating $f(t)$ and $g(t)$ by two impulse series, i.e., $f_a(t) \approx f(t)$, $g_a(t) \approx g(t)$ and $c(t) \approx c_a(t) = f_a(t) * g_a(t)$. The impulse approximation of $f(t)$ is shown in Fig. 12.8-3a. In Fig. 12.8-3b each impulse has the area $a_k T$, where $a_k = f(kT)$. It is usually more accurate to position the impulses at $t = (k+0.5)T$, $k = 0, 1, 2, \dots$, and let each impulse have the area $f[(k+0.5)T]$ as in Fig. 12.8-3c.

In the latter case, the origin is shifted to the right by $T/2$, and is shifted back after all the calculations are completed. In any case, the operation $c_a(t) = f_a(t) * g_a(t)$ is easily performed.

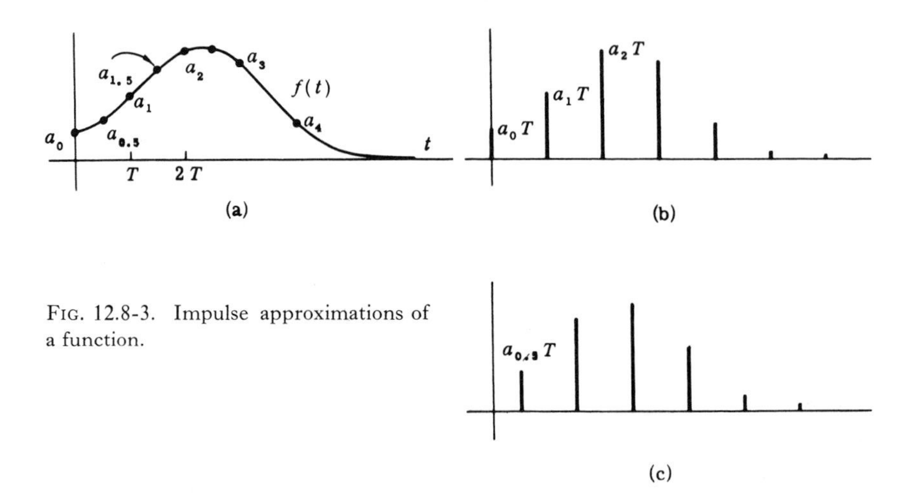

FIG. 12.8-3. Impulse approximations of a function.

Mixed Staircase and Impulse Approximations

The accuracy obtained in the impulse approximation is not great, unless very small T is used, or unless the functions really do consist of equally spaced impulses. Better accuracy may be obtained by approximating one of the curves by a staircase function, as in Fig. 12.8-4. The heights of the stairs at kT may be taken as $f(kT)$, but this leads to the rather poor approximation of Fig. 12.8-4a.

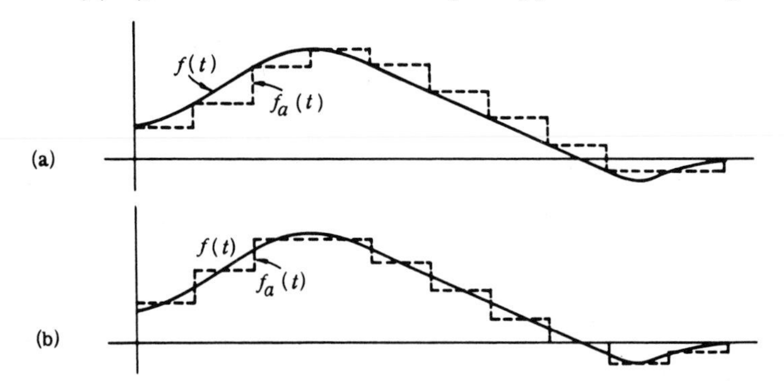

FIG. 12.8-4. (a) Staircase approximation of a function. (b) A better staircase approximation of the function.

In Fig. 12.8-4b a better approximation is obtained by taking the height of the stair at kT to be $f(k + 0.5) T$. The analyst can use his own judgment in choosing the values of $f_a(t)$. Suppose, in the meantime, that we retain the impulse approximation (of the type shown in Fig. 12.8-3b) for $g_a(t)$. Then $c_a(t) = \int_0^t f_a(t - \tau)\, g_a(\tau)\, d\tau$ is a staircase function, because the convolution of an impulse series with a staircase function leads to a staircase function. Hence $c_a'(t)$ is an impulse series, and is obtained as follows:

$$c_a'(t) = \frac{d}{dt}\left[\int_0^t f_a(t - \tau) g_a(\tau) d\tau\right] = \int_0^t f_a'(t - \tau) g_a(\tau) d\tau + f_a(0) g_a(t)$$

$$= \int_0^t f_a'(\tau) g_a(t - \tau) d\tau + f_a(0) g_a(t) \tag{12.8,7}$$

because, in general,

$$\frac{d}{dx}\left[\int_{f_1(x)}^{f_2(x)} F(x, y) dy\right] = \int_{f_1(x)}^{f_2(x)} \frac{\partial F(x, y)}{\partial x} dy + f_2'(x) F[x, f_2(x)] - f_1'(x) F[x, f_1(x)]. \tag{12.8,8}$$

The function $f_a'(t)$ is an impulse series, because it is the derivative of a staircase function, and, from Eq. (12.8,1), $Zf_a'(t) = [(z - 1)/z]\, F_a(z)$. [Note that $f_a(0) = 0$, as noted in the comment following Eq. (12.8,1). If we took $f_a(0) = a_o$, then $f_a'(0) = 0$, and the term $f_a(0)\, g_a(t)$ in Eq. (12.8,7) would exactly make up for the missing $a_o \delta(t)$ impulse in $F_a(z)$.] From Eq. (12.8,6) and the fact that $f_a(0) = 0$, $Zc_a'(t) = Zf_a'(t)\, Zg_a(t) = [(z - 1)/z]\, F_a(z)\, G_a(z)$. We want the staircase function $c_a(t)$, and use Eq. (12.8,4a), i.e.,

$$Zc_a(t) = [z/(z - 1)]\, Zc_a'(t) = F_a(z)\, G_a(z). \tag{12.8,9}$$

The points obtained from Eq. (12.8,9) are the exact sampled values of the staircase function $c_a(t)$.

Staircase Approximations

For increased accuracy, both $f(t)$ and $g(t)$ may be approximated by staircase functions. The resulting $c_a(t) = f_a(t) * g_a(t)$ is therefore a polygonal function, and $c_a''(t)$ is an impulse series. Differentiation of $c_a'(t) = \int_0^t f_a'(\tau)\, g_a(t - \tau)\, d\tau + f_a(0)\, g_a(t)$, of Eq. (12.8,7) [recall $f_a(0) = 0 = g_a(0)$], leads to the impulse series

$$c_a''(t) = \int_0^t f_a'(\tau)\, g_a'(t - \tau)\, d\tau + f_a'(t)\, g_a(0) + f_a(0)\, g_a'(t). \tag{12.8,10}$$

In the present case, $c_a''(t) = \int_0^t f_a'(\tau)\, g_a'(t - \tau)\, d\tau = \int_0^t f_a'(t - \tau)\, g_a'(\tau)\, d\tau$. Each of $f_a'(t)$ and $g_a'(t)$ consists of an impulse series and, from Eq. (12.8,1), $Zf_a'(t) = [(z - 1)/z]\, F_a(z)$, $Zg_a'(t) = [(z - 1)/z]\, G_a(z)$, so that $Zc_a''(t) = [(z - 1)^2/z^2]\, F_a(z)\, G_a(z)$. To find $C_a(z)$, we use Eq. (12.8,4b), with

$$C_a(z) = Zc_a(t) = [Tz/(z - 1)^2]\, Zc_a''(t) = Tz^{-1} F_a(z)\, G_a(z). \tag{12.8,11}$$

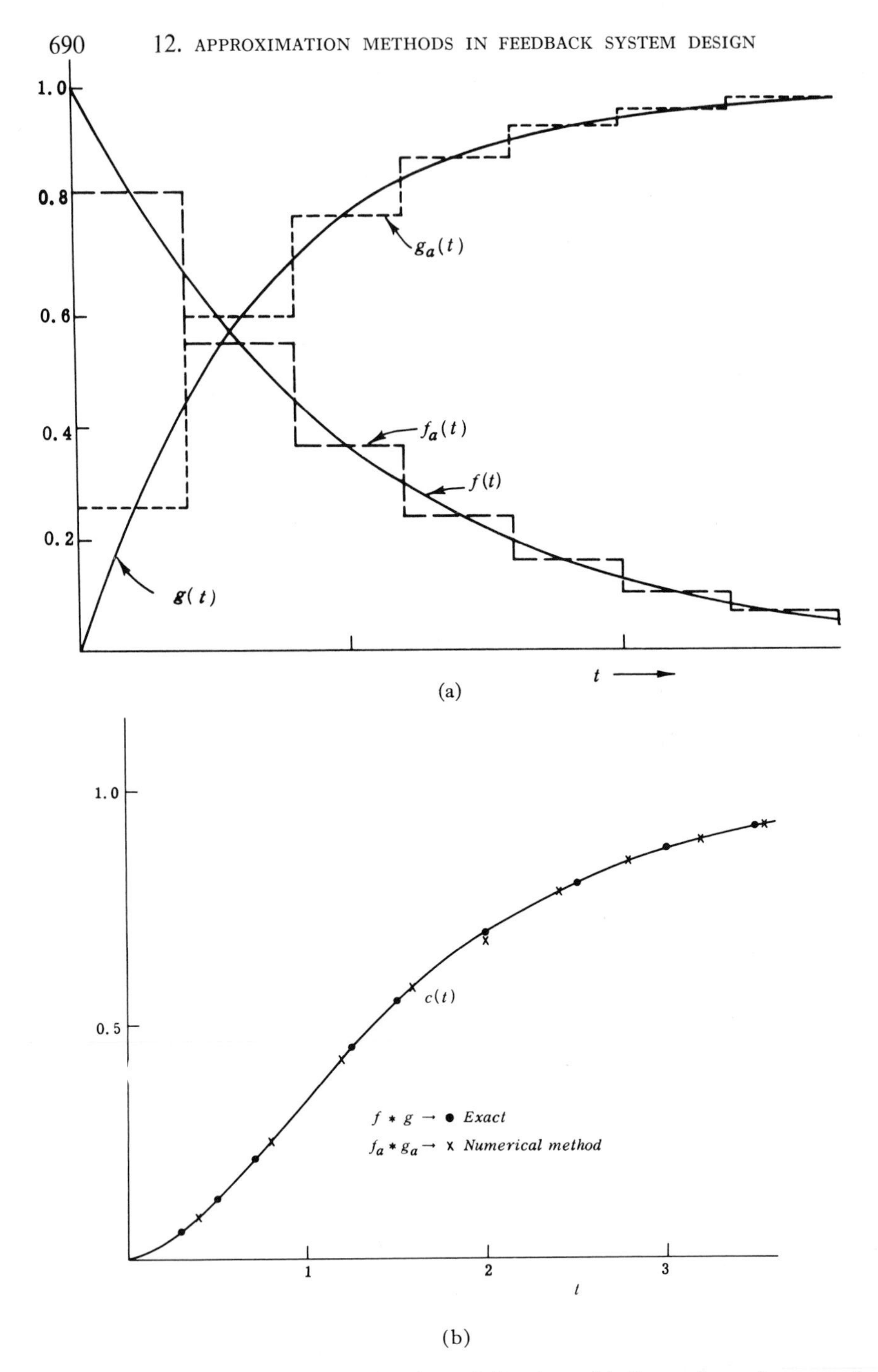

FIG. 12.8-5. (a) Staircase approximations of functions. (b) Comparison of exact and approximate convolution results.

The points obtained from Eq. (12.8,11) are the exact sampled values of the polygonal function $c_a(t)$, which, in turn, is an approximation of $c(t)$. The results are exact if $f(t)$ and $g(t)$ are staircase functions.

Example. We shall use Eq. (12.8,11) to approximately calculate $f(t) * g(t)$ of Fig. 12.8-5a. We choose $T = 0.4$ and use the staircase approximations $f_a(t)$ and $g_a(t)$ shown in the figure, i.e., $F_a(z)$: 0.819, 0.550, 0.370, 0.246, 0.167, 0.110, 0.074, 0.050, 0.036, 0.020, ...; $G_a(z)$: 0.260, 0.595, 0.775, 0.878, 0.935, 0.966, 0.982, 0.990, 0.995, ...; and $C_a(z) = (0.4/z) F_a(z) G_a(z)$. The exact $c(t)$ is compared to $c_a(nT)$ in Fig. 12.8-5b.

Mixed Polygonal Approximations

For functions of the type shown in Fig. 12.8-5, the accuracy is improved by using polygonal approximations. [If $f(t)$ is like that shown in Fig. 12.8-6, then

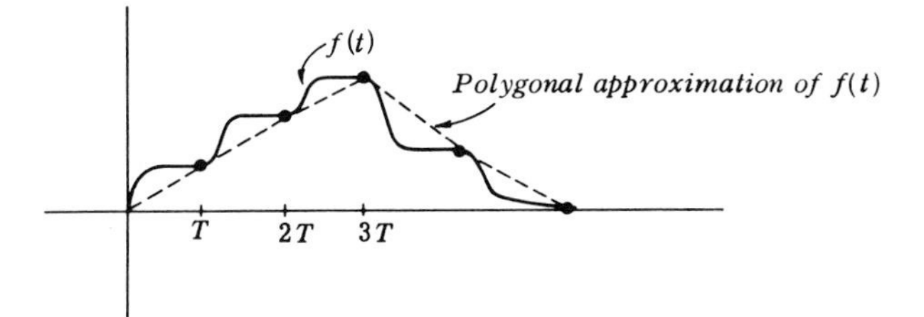

FIG. 12.8-6. A function for which staircase approximation is better than polygonal approximation.

a staircase approximation is better than a polygonal one.] Suppose $f(t)$ is approximated by a polygonal (first-order) curve, while the staircase (zero-order) approximation is retained for $g(t)$. Then $c_a(t) = f_a(t) * g_a(t)$ consists of confluent parabolic segments (second-order curve) and $c_a'''(t)$ consists of an impulse series. Differentiating Eq. (12.8,10),

$$c_a'''(t) = \int_0^t f_a''(t-\tau) g_a'(\tau) d\tau + f_a'(0) g_a'(t) + f_a''(t) g_a(0) + f_a(0) g_a''(t). \qquad (12.8,12)$$

As usual, we take $g_a(0) = g_a(0-) = 0$. However, there is a difficulty if $f_a(0+) \neq 0$, because if we take $f_a(0) = 0$, then $f_a'(0)$ is an impulse, and $f_a''(0)$ is a doublet. There are two ways to overcome this difficulty (both lead to the same result, of course). The first is to write $f_a = f_{a1} + f_{a2}$, with $f_{a2} = f_a(0+)$ a constant, and $f_{a1}(t) = f_a(t) - f_a(0+)$ a true polygonal function which is zero at the origin. We then have $c_a(t) = f_{a1}(t) * g_a(t) + f_a(0+) * g_a(t) = c_{a1}(t) + c_{a2}(t)$. The first part, $c_{a1}(t)$, consists of a second-order curve, and the second, $c_{a2}(t) = f_a(0+) * g_a(t)$, consists of a first-order curve (polygonal). Hence $c_a(t)$ is perforce a mixed curve. To find $c_{a1}(t)$, we use Eq. (12.8,12), with $f_a'(0) =$

$f_a(0) = g_a(0) = 0$, and $c_{a1}'''(t) = \int_0^t f_{a1}''(t - \tau)\, g_a'(\tau)\, d\tau$. Consequently, from Eqs. (12.8,1) and (12.8,2),

$$Zc_{a1}'''(t) = Zf_{a1}''(t)\, Zg_a'(t) = [(z-1)^3/Tz^2]\, F_{a1}(z)\, G_a(z)$$

and from Eq. (12.8,5) (for $n = 3$)

$$\begin{aligned} C_{a1}(z) = Zc_{a1}(t) &= Zs^{-3}Zc_{a1}'''(t) \\ &= [0.5T^2z(z+1)/(z-1)^3]\,[(z-1)^3/Tz^2]\,[F_{a1}(z)\, G_a(z)] \\ &= 0.5T(1+z^{-1})\, F_{a1}(z)\, G_a(z). \end{aligned} \tag{12.8,13}$$

The above points, determined from Eq. (12.8,13), are joined by confluent parabolic segments, and give $c_{a1}(t)$. To find $c_{a2}(t)$, note that

$$c_{a2}(t) = f_a(0+) * g_a(t) = f_a(0+) \int_0^t g_a(t) dt,$$

and since $g_a(t)$ is a staircase function, $C_{a2}(z) = f_a(0+)\, Z(1/s^2)/Z(1/s)$, $G_a(z) = [T/(z-1)]\, f_a(0+)\, G_a(z)$, and the points thereby obtained are joined by straight line segments.

The second way to obtain these same results is to let $f_a(0) = f_a(0+)$, which is finite, and then $f_a'(0) = f_a'(0+)$, $f_a''(0) = f_a''(0+)$ are zero. However, in Eq. (12.8,12) the terms $f_a'(0)\, g_a'(t)$ and $f_a(0)\, g_a''(t)$ can no longer be disregarded. The former is a constant times an impulse series, so its third integration is a second-order curve which makes up for the first impulse missing in $f_a''(0)$. The second term is a constant times a series of doublets, whose third integration is exactly the polygonal curve $c_{a2}(t)$.

Polygonal Approximations

Polygonal approximations of both $f(t)$ and $g(t)$ are clearly better for the usual functions encountered. In this case, $c_a(t)$ is a third-order curve (its first and second derivatives are continuous, while its third derivative is discontinuous at $t = kT$). Also, $c_a(t)$ must be differentiated four times to yield an impulse series given by $Zc_a^{\mathrm{IV}}(t) = [(z-1)^4/T^2z^2]\, F_a(z)\, G_a(z)$, and from Eq. (12.8,5), with $n = 4$,

$$\begin{aligned} Zc_a(t) &= Zs^{-4}\, Zc_a^{\mathrm{IV}}(t) = [T^3z(z^2+4z+1)/6(z-1)^4]\, Zc_a^{\mathrm{IV}}(t) \\ &= [T(z^2+4z+1)/6z]\, F_a(z)\, G_a(z). \end{aligned} \tag{12.8,14}$$

It is essential here that $f_a(0+) = g_a(0+) = 0$. If this is not the case, then we may write $f_a = f_{a1} + f_a(0+)$, $g_a = g_{a1} + g_a(0+)$, and

$$\begin{aligned} c_a = [f_{a1} + f_a(0+)] * [g_{a1} + g_a(0+)] &= f_{a1} * g_{a1} + f_{a1} * g_a(0+) \\ &+ f_a(0+) * g_{a1} + f_a(0+) * g_a(0+) = c_{a1} + c_{a2} + c_{a3} + c_{a4}. \end{aligned}$$

Equation (12.8,14) gives $C_{a1}(z) = (T/6)\,[(z^2 + 4z + 1)/z]\,F_{a1}(z)\,G_{a1}(z)$. The second and third terms, $c_{a2}(t)$ and $c_{a3}(t)$, are the convolutions of a constant with a polygonal curve, so they are second-order curves. Hence

$$C_{a2}(z) = g_a(0+)\,F_{a1}(z)\,Z(s^{-3})/Z(s^{-2}) = 0.5T(z+1)\,g_a(0+)\,F_{a1}(z)/(z-1);$$

$C_{a3}(z) = 0.5T(z+1)\,f_a(0+)\,G_{a1}(z)/(z-1)$. The last term, c_{a4}, is the convolution of two constants, so it is a first-order curve (linear); i.e., $c_{a4}(kT) = f_a(0+)\,g_a(0+)\,kT$ for $k = 0, 1, 2, \ldots$.

Higher Order Approximations

It must not be assumed that the higher the order of the approximation, the more accurate is the final result. For the curve shown in Fig. 12.8-6, a staircase approximation is much better than a polygonal. For the curve shown in Fig. 12.8-7,

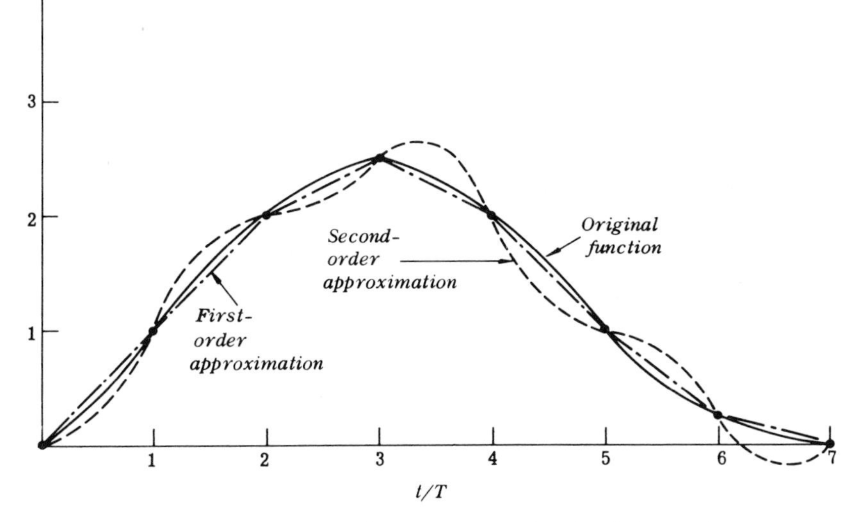

FIG. 12.8-7. A function for which first-order approximation is better than zeroth- or second-order approximation.

a first-order polygonal approximation is clearly much better than the second-order (confluent parabolic) approximation. The parabolic approximation is obtained by writing $y = At^2$ for $0 \leqslant t \leqslant T$, and finding A so that $y(T) = AT^2$. For $T \leqslant t \leqslant 2T$, we use $y = A_1\tau_1^{\,2} + B_1\tau_1 + C_1$, with $\tau_1 = t - T$, and choose A_1, B_1, C_1 so that $y(\tau_1 = 0) = y(T) = C_1$, $y'(\tau_1 = 0) = B_1 = y'(t = T) = 2AT$, and $y(\tau_1 = T) = A_1T^2 + B_1T + C_1 = y(t = 2T)$. The process is continued for each interval. In general, therefore, that approximation which is best suited for the given curve should be chosen.

§ 12.9 Discussion of Technique and Application to Inverse Convolution

The technique described in Section 12.8 can give fairly accurate values for $c(t) = f(t) * g(t)$ even when rather poor approximations of $f(t)$ and $g(t)$ are used. The reason is that one of the principal operations on $f(t)$, $g(t)$ for finding $c(nT)$ is an integration operation. The approximations of $f(\tau)$, $g(t - \tau)$ may be very poor, but if they are such that the integral of their product is the same as that of the exact functions, then $c_a(nT) = c(nT)$ exactly. For example, in Fig. 12.9-1, if in the interval $0 < t < T$, f_a and g_a are chosen as shown, then $c_a(T) \doteq c(T)$, despite the extremely poor approximations of $f(t)$, $g(t)$ by f_a, g_a. For this reason the analyst should appropriately distort the values of $F(z)$, $G(z)$ in accordance with the type of approximation that is being used. He should take for $f_a(nT)$ a value such that $Tf_a(nT)$ is closely equal to $\int_{(n-1)T}^{nT} f(t)\,dt$, as was done, for example, in Fig. 12.8-5a. The analyst has $f(t)$ and $g(t)$ before him. He can readily see what kind of distortion of $f(nT)$, $g(nT)$ are advisable. The second-order quadrate approximation is usually poorer than the first-order (polygonal), firstly because of the oscillation it produces in $f_a(t)$ (see Fig. 12.8-7), and secondly it is much more difficult to see what distortion of the data is needed to preserve the area of the products.

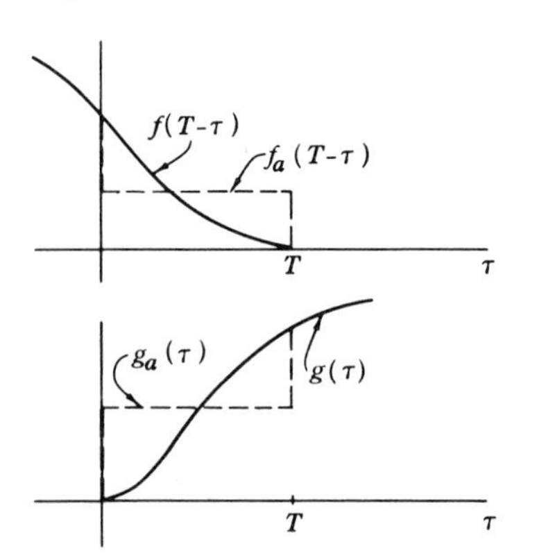

FIG. 12.9-1. Very poor approximations (f_a, g_a) may be satisfactory for convolution operation.

It is precisely for the above reason that numerical inverse convolution of $c(t) = f(t) * g(t)$ is much more difficult. Suppose $c(t)$, $f(t)$ are known and the problem is to find $g(t)$. It is possible for two violently different $g_1(t)$, $g_2(t)$ to give the same values of $c(nT)$ (see, for example, Fig. 12.9-1). Hence the results of the inverse convolution depend enormously on the assumed nature of $g(t)$. Conversely, in the convolution problem one knows the nature of the functions which are being operated on and very accurate values of the unknown $c(t)$ can be obtained without *a priori* knowledge of the nature of $c(t)$. In the inverse convolution problem it is advisable to get some idea of the nature of $g(t)$ by using the physical interpretation of convolution as that of folding, multiplication and integration (see Fig. 9.4-4). Then sufficiently small sampling intervals (smaller than those used in convolution) must be used such that the approximation used appears to be a reasonable approximation of what the actual $g(t)$ probably looks like. It is also apparent from the physical interpretation of convolution that the calculated value of $g_a(nT)$ is quite sensitive to the value of $g_a[(n-1)T]$. It is therefore a good idea to first calculate only a few values of $g_a(nT)$ and check (by direct convolution) whether the resulting $f(t) * g_a(t)$ agrees reasonably well with $c(0.5nT)$. This is necessary because $g_a(t)$ must be a good approximation of $g(t)$ for all t if the results at $g_a(nT)$ are to be reliable. If these is poor agreement then a smaller value of T must be used. Obviously for the methods of Section 12.8, polygonal approximations of $f(t)$ and $g(t)$ will usually be best.

§ 12.10 Numerical Solution of Differential Equations

Let us first consider the numerical evaluation of $y(t) = \int_0^t f(t)\,dt$. Suppose $f(t)$ is like that shown in Fig. 12.8-6, so it is well approximated by a staircase function. We may write $y_a(t) = \int_0^t [\int_0^t f_a'(t)\,dt]\,dt$. [Note that $\int_0^t f'(t)\,dt = f(t) - f(0) = f(t)$, if, as noted after Eq. (12.8,1), $f(0) = f(0-)$.] From Eq. (12.8,4b), $Zy_a(t) = [Zf_a'(t)]\,[Tz/(z-1)^2]$. But from Eq. (12.8,1), $Zf_a'(t) = (1-z^{-1})F_a(z)$, so $Zy_a(t) = TF_a(z)\,(z-1)^{-1}$ and $y_a(t)$ is a first-order curve. Similarly, if $f(t)$ is approximated by a first-order curve, we then write $y_a(t) = \int_0^t f_a(t)\,dt = \int_0^t \int\int f_a''(t)\,(dt)^3$, and, from Eq. (12.8,5), $Zy_a(t) = [Z(1/s^3)]Zf_a''(t)$; and from Eq. (12.8,2), $Zf_a''(t) = Zf_a(t)/[Z(1/s^2)]$. Consequently, $Zy_a(t) = Zf_a(t)\,Z(1/s^3)/Z(1/s^2)$. This relation is correct, however, only if $f(0) = 0$; otherwise $f(t)$ should be expressed as the sum of a polygonal function and a staircase function.

The above is the essential tool in the numerical solution of differential equations. Consider, for example, the following linear, constant coefficient, differential equation: $y^{IV} + a_1 y''' + a_2 y'' + a_3 y' + a_4 y = f(t)$. The above numerical technique may be applied by integrating four times in order to eliminate all derivatives:

$$y + a_1 \int_0^t y dt + a_2 \int_0^t \int y(dt)^2 + a_3 \int_0^t \int\int y(dt)^3 + a_4 \int_0^t \int\int\int y(dt)^4$$

$$= \int_0^t \int\int\int f(t)\,(dt)^4 = g(t).$$

Suppose it is *a priori* assumed that a polygonal approximation of y is warranted. Then, by the technique previously discussed,

$$Zy(t)\left[1 + a_1 \frac{Z(s^{-3})}{Z(s^{-2})} + a_2 \frac{Z(s^{-4})}{Z(s^{-2})} + a_3 \frac{Z(s^{-5})}{Z(s^{-2})} + a_4 \frac{Z(s^{-6})}{Z(s^{-2})}\right] = Zg(t).$$

The series for $Y(z)$ is then readily obtained. In the above, it is implicitly assumed that $y(0+) = 0$. The value of $y(0+)$ is easily available from the differential equation, and if it is not zero, then it should be written as $y = y_1 + y_2$, etc. The above procedure (of eliminating all the derivatives) leads to considerably more accurate results than is available from the application of the z-transform technique directly to the differential equation. Integrations tend to reduce the approximation errors, while differentiation tends to increase their effect.

The same procedure may be applied to linear differential equations with time varying coefficients. The equation must first be prepared for the operation. For example, consider $f_o(t)\ddot{y} + f_1(t)\dot{y} + f_2(t)y = g(t)$. Since $f_o\ddot{y} = (d^2/dt^2)f_oy - 2\dot{f}_o\dot{y} - y\ddot{f}_o = (d^2/dt^2)f_oy - (2d/dt)\dot{f}_oy + \ddot{f}_oy$, and $f_1\dot{y} = (d/dt)f_1y - \dot{f}_1y$, we may replace the original equation by $(d^2/dt^2)f_oy + (d/dt)[(f_1 - 2\dot{f}_o)y] + [f_2 + \ddot{f}_o - \dot{f}_1]y = g(t)$, or $(d^2/dt^2)f_oy + (d/dt)(h_1y) + h_2y = g(t)$. Integrating the above twice leads to

$$f_oy + \int_0^t h_1y\,dt + \int_0^t \int h_2y\,(dt)^2 = \int_0^t \int g(t)\,(dt)^2.$$

An estimate must be made of the nature of a suitable approximation of each of f_oy, h_1y, h_2y, and, if necessary, $g(t)$. It is permissible to use different-order curves for each of these, if f_o, h_1, etc., are such as to warrant them. Let us assume that polygonal approximations are considered satisfactory. Then $Zf_oy + Zh_1yZs^{-3}/Zs^{-2} + Zh_2yZs^{-4}/Zs^{-2} = Zg(t)\,Zs^{-4}/Zs^{-2}$. The series for each of Zf_oy, Zh_1y, Zh_2y are easily written in terms of the series y, which becomes available by equating coefficients.

Error Bounds

The weakness of the above and many other numerical methods is in the absence of good error estimates. Unfortunately, even in those techniques where expressions for error bounds are available, they are often so poor as to be useless. In the ideal technique, the value of the interval T and the type of approximation would be *a priori* chosen on the basis of the maximum allowable error. In the absence of such a theory of errors, the designer must depend on intuition and experience. In convolving two known functions, a spot check of the answer can be made by performing a fairly accurate graphical convolution at one value of time. An inverse convolution problem can be grossly checked simply by inspection; also, one or more spot checks can be made with little work. In the case of a differential equation, the answer may be substituted into the equation and one or more spot checks made.

§ 12.11 Approximate Inverse Transformation

The problem of finding the time function corresponding to a given transform is one that continually confronts the feedback system designer. This is because the design is nearly always carried out in the transform domain, and the designer is often not completely satisfied with his design until he knows the time domain system response to typical inputs and disturbances. He must therefore evaluate the appropriate inverse transform. If the system itself is available for testing purposes, or its analog (usually on the analog computer) is available, then the time domain behavior is readily found. In other cases the inverse transform must be found by numerical methods.

There is no single numerical procedure which can be singled out as best. There is the matter of individual taste; for example, one might prefer to spend 2 hours with a method that does not require intense concentration, rather than 1 hour with a different method that does require such concentration and is therefore more subject to error. There is also the significant matter of the information that is available. For example, partial fraction expansion and term-by-term inversion may sometimes be preferred if the system poles are already known, but another method may be faster if the transfer function denominator is not available in factored form. On the other hand, if the frequency response is only available graphically, then a third method may be preferable. We will present here a variety of procedures, each with its own area of suitability.

(1) Partial Fraction Expansion

An exact analytical method of inverse transformation of rational transforms is partial fraction expansion and term-by-term inversion of the transform. The time function is then available as a sum of damped sinusoids. This method is usually considered as very long and tedious. However, it can be speeded up by the use of standard curves and adjustable scales. The method may then be quite practical, especially if the transform denominator is already in factored form. This is seen as follows.

The two most troublesome terms in partial fraction expansions have the forms $D/(s+d)$, $[A(s+\zeta\omega_n)+B\omega_n(1-\zeta^2)^{0.5}]\,[(s+\zeta\omega_n)^2+\omega_n{}^2(1-\zeta^2)]^{-1}$ and their inverse transforms are, respectively, $f_1(t)=De^{-dt}$, $f_2(t)=Ke^{-\zeta\omega_n t}\sin[\omega_n t(1-\zeta^2)^{0.5}+\theta]$ with $K=(A^2+B^2)^{0.5}$, $\tan\,\theta=A/B$. Adjustable time[1] and vertical scales are very useful here. Only one curve is then needed to characterize $f_1(t)=De^{-dt}$. Also, $f_2(t)$ becomes a function of only two parameters, ζ and m, i.e., we may write $f_2(t)=e^{-\zeta t}\sin[t(1-\zeta^2)^{0.5}+m]$. One family of curves suffices for evaluating $f_2(t)$ for any ζ and m. For example, suppose $f_2(t)$ (denoted as f_{2a}) is available for $m=0$, $\zeta=0.707$, and $f_{2b}(t)$ for $m=1$, $\zeta=0.707$ is required. We have $f_{2a}=e^{-\zeta t}\sin t\sqrt{1-\zeta^2}$ and $f_{2b}=e^{-\zeta\tau}\sin[\tau\sqrt{1-\zeta^2}+m]$. Let $\tau=t-t_1$, where $m=t_1\sqrt{1-\zeta^2}$ and then $f_{2b}=f_{2a}\,e^{\zeta t_1}$. Hence f_{2b} may be obtained from the curve of f_{2a} vs. t by shifting

[1] Gerber Scientific Instrument Company, Hartford, Connecticut.

the origin of the time axis in the latter to t_1 and multiplying the readings by $e^{\zeta t_1}$. Similarly, the abscissa scale must be stretched or contracted in order to accommodate any ω_n in $f_2(t)$. If flexible vertical and time scales are available, then there is no need to multiply the abscissa and ordinate readings of f_{2a}; the scales are stretched or contracted and f_{2a} is copied (with t_1 as origin) at the new scale settings. This process is repeated (the scales being reset) for each term in the series and the individual curves are finally added. It is easy to see in advance which of the individual curves may be neglected at different ranges of t. The family of curves of Fig. 5.12-1b may be used, or the reader may prepare his own set from the data of Table 5.13-1.

(2) Use of Predistortion and Convolution

Transforms which often occur and which involve a limited number of parameters may be evaluated once and for all and tabulated or put into graphical form. This was done in Section 5.12 for the impulse and step responses of a complex pole pair. It was also shown in Section 5.13 how these two sets of data (or the two families of curves) could be used to readily obtain the inverse transforms of functions of the form $A(s + \lambda\zeta\omega_n)\,[(s)\,(s^2 + 2\zeta\omega_n s + \omega_n^2)]^{-1}$. Families of the response could thus be obtained (one family of curves for a fixed ζ, using λ as parameter).

The inverse of $F(s) = (s + m)\,(s^2 + 2\zeta\omega_n s + \omega_n^2)^{-1}$ may be considered easy to obtain, since it has the form $Ae^{-at}\sin(bt + \theta)$, which can be standardized into one family of curves, as previously indicated. To evaluate $[s(s + \alpha) \times (s^2 + 2\zeta s + 1)]^{-1}$, the numerical convolution of $f_1(t) = \mathscr{L}^{-1}[1/(s + \alpha)]$ with $f_2(t) = \mathscr{L}^{-1}[1/s(s^2 + 2\zeta s + 1)]$ is convenient, using the method of Section 12.8. The reason is that $f_1(t)$ affects only the early portion of the $f_2(t)$ time function. From the very process of numerical convolution, it is known when to stop the computation. Also, from the *a priori* known $f_1(t)$, $f_2(t)$, it is simple to choose a reasonable value of the sampling interval. The smaller the ζ as compared to α, the sooner may the computation be terminated.

To evaluate $F(s) = [(s + \alpha)\,(s^2 + 2\zeta s + 1)]^{-1}$, with $\zeta > \alpha$, it is convenient to let $s + \alpha = p$. The transform then has the form $[p(p^2 + 2\zeta_o\omega_o p + \omega_o^2)]^{-1}$, whose inverse, denoted by $g(t)$, has been sketched (Fig. 5.12-1a) and tabulated (Table 5.13-1). $\mathscr{L}^{-1}F(s) = f(t)$ is then easily obtained from the relation $f(t) = e^{-\alpha t}g(t)$. This cannot be done if $\alpha > \zeta$. The numerical convolution of $\mathscr{L}^{-1}(s + \alpha)^{-1}$ with $\mathscr{L}^{-1}(s^2 + 2\zeta s + 1)^{-1}$ is convenient when $\alpha > \zeta$ by a reasonable factor; the stronger the inequality, the sooner may the computation be terminated. The evaluation of $(s + a)\,[s(s^2 + 2\zeta s + 1)\,(s + b)\,(s + c)]^{-1}$ can be performed by convolving $\mathscr{L}^{-1}(s + a)\,[(s + b)\,(s + c)]^{-1}$ with $\mathscr{L}^{-1}[s(s^2 + 2\zeta s + 1)]^{-1}$. The latter is available and the former is easily evaluated.

The numerical convolution technique is, in general, very helpful in simplifying inverse transformation, because the engineer is rarely interested in precise inversion. It is often possible to split up a complicated function into two simpler ones, where one of the latter predominates for small t, and the other at large t. For example, the five poles and single zero in Fig. 12.11-1a are conveniently

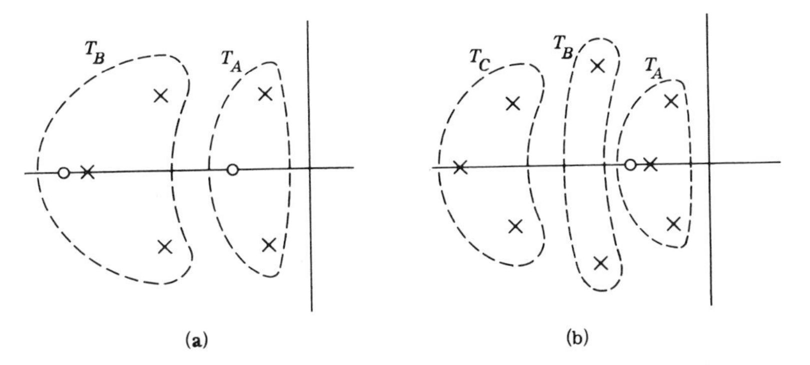

FIG. 12.11-1. Simplifying inverse transformation by means of convolution.

split up as shown. T_B is important at small t; the numerical convolution of its inverse with that of T_A need be carried out for only small and intermediate values of time. What is also important is that the engineer retains a feeling for the physical meaning of his functions—he does not get lost in the algebra. One can extend this to three divisions of the original transform as in Fig. 12.11-1b.

Partial Fraction Expansion

Partial fraction expansion is facilitated by means of synthetic division and the remainder theorem. These are illustrated by means of the following example.

Example.

$$F(s) = \frac{N(s)}{D(s)} = \frac{s^4 + 5s^3 + 2s^2 + 3s + 10}{(s + 1.4)^2(s^2 + s + 1)\,(s^2 + 2s + 10)}$$

$$= \frac{A}{(s + 1.4)^2} + \frac{B}{s + 1.4} + \frac{Cs + D}{s^2 + s + 1} + \frac{Es + F}{s^2 + 2s + 10}.$$

To find A, note that $A = \lim_{s \to -1.4} (s + 1.4)^2 F(s) = N(-1.4)/D_1(-1.4)$, where $D(s) = (s + 1.4)^2 D_1(s)$. Let $N(s) = Q_1(s + 1.4) + R$, where R is therefore the remainder that is obtained when $N(s)$ is divided by $s + 1.4$. R is readily obtained by synthetic division, viz.,

1	1.4	1	5	2	3	10
		1	1.4	5.04	− 4.256	10.1584
			3.6	− 3.04	7.256	− 0.1584

Hence $Q_1 = s^3 + 3.6s^2 - 3.04s + 7.256$, and $R = N(-1.4) = -0.1584$. Similarly, $d_2(-1.4) = (s^2 + s + 1)_{-1.4} = 1.56$, $d_3(-1.4) = (s^2 + 2s + 10)_{-1.4} = 9.16$, and $A = -0.1584/(1.56)(9.16)$. To find B, note that $B = \{(d/ds)(s + 1.4)^2 F(s)\}_{s=-1.4} = \{(N'D_1 - ND_1')/D_1^2\}_{s=-1.4}$. $N(-1.4)$ and $D_1(-1.4)$ are already known; $N' = 4s^3 + 15s^2 + 4s + 3$. Note $N'(-1.4) = Q_1(-1.4)$. By synthetic division,

1	1.4	4	15	4	3
		4	5.6	13.16	− 12.824
			9.4	− 9.16	15.824

$N'(-1.4) = 15.824$. To find $D_1'(-1.4)$, we use $D_1' = (2s + 1)\, d_2 + d_1(2s+2)$, and with d_2, d_3 known, D_1' is easily evaluated.

To find C, D, we use $[F(s)(s^2 + s + 1)]_{s=-0.5+j.866} = [Cs + D]_{s=-0.5+j.866}$. To evaluate $N(s)$ at $s = s_o = -0.5 + j0.866$, we again use the remainder theorem, i.e., let $N(s) = Q_2(s)(s^2 + s + 1) + R(s)$, and if s_o is a zero of s^2+s+1, then $N(s_o) = R(s_o)$. By synthetic division, $R(s) = 2s + 13$, viz.,

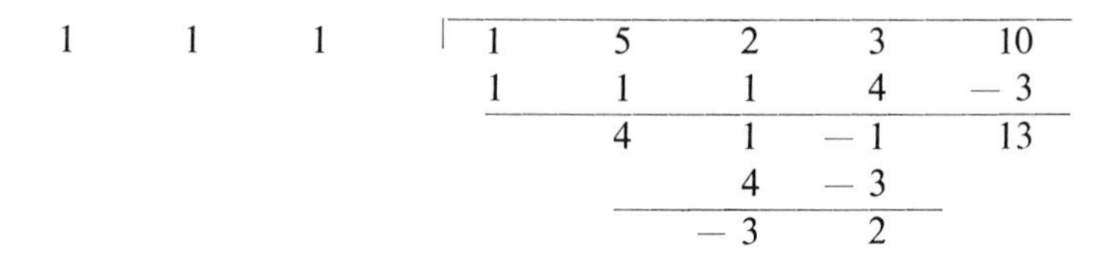

1	1	1	1	5	2	3	10
			1	1	1	4	− 3
				4	1	− 1	13
					4	− 3	
					− 3	2	

Therefore $N(s_o) = (2s + 13)_{so} = 12 + j1.732$; $d_1(s_o)$ and $d_3(s_o)$ are individually easy to obtain. Finally, let $[F(s)(s^2 + s + 1)]_{so} = e + jf$; then $C(-0.5 + j0.866) + D = e + jf$; so $0.866C = f$ and $-0.5C + D = e$. E and F are similarly obtained.

It is noted that there is a difference of two in the degrees of the numerator and denominator of $F(s)$. Therefore, as a check on the partial fraction expansion, it is easy to use: 5th degree coefficients, $B + C + E = 0$; 4th degree coefficients, $A + B(1.4 + 1 + 2) + C(2.8 + 2) + D + E(2.8 + 1) + F = 0$.

§ 12.12 Inverse Transformation with Unfactored Transforms

Many numerical methods[1] have been suggested by engineers for the approximate inversion of Laplace transforms. There have been some attempts on the

[1] R. Boxer and S. Thaler, A simplified method of solving linear and nonlinear systems, *Proc. IRE* **44** (1), 89-101 (1956); A. Tustin, A method of analyzing the behavior of linear systems in terms of time series. *J. Inst. Elec. Engrs.* (*London*) **94**, Part II-A (1947); A. Madwed, Number Series Method of Solving Linear and Nonlinear Differential Equations. M.I.T. Instr. Lab., Rept. 6445-T-26. Cambridge, Massachusetts, April, 1950.

part of mathematicians to establish these methods on a proper mathematical basis. The results due to Wasow[1] are presented here without proof.

$F(s)$ is the transform which is to be evaluated approximately.

Rule 1. Express $F(s)$ in the form $F(s) = P(s)/[1 + Q(s)]$, where $P(s)$ and $Q(s)$ are zero and differentiable at $s = \infty$. Expand $P(s)$ and $Q(s)$ in powers of s^{-1}. Replace every s^{-r} by $[T^r/(r-1)!]\, R_{r-1}(z)$, except for s^{-1}, which is replaced by $TR_0(z) - T/2$ [the $R_{r-1}(z)$ are later defined]. Divide by T and expand the resulting function of z in powers of z^{-1}. Then the coefficient of z^{-n} is approximately $f(nT)$. The error (for $n > 1$) is proportional to T^2; so the smaller the T, the smaller the error.

When $p(t) = \mathcal{L}^{-1}P(s)$ and $q(t) = \mathcal{L}^{-1}Q(s)$ have their first k ($k > 0$) derivatives equal to zero at $t = 0$, then a better approximation (error proportional to T^{2+k}) is possible by the following modification of the procedure.

Rule 2. After expanding $P(s)$ and $Q(s)$ in powers of s^{-1}, replace every s^{-r} by $[T^r/(r-1)!]\, R_{r-1}(z) - T^r B_r/r!$, where B_r is the rth Bernoulli number (later defined), and $R_k(z)$ is later defined. Divide by T and expand the resulting function in powers of z^{-1}. The coefficient of z^{-n} is approximately equal to $f(nT)$.

Definition of $R_n(z)$.

$$R_n(z) = \sum_{r=0}^{\infty} r^n s^{-n} = \sum_{k=1}^{n} k!\, S_n^k \frac{z}{(z-1)^{k+1}}, \quad \text{for } n > 0.$$

S_n^k are positive integers known as Stirling numbers of the second kind. They are available from the relation $m^q = S_q^1 m + S_q^2 m(m-1) + \dots + S_q^q m(m-1)\dots(m-k+1)$. Some values of S_n^k are listed[2] in Table 12.12-1.

TABLE 12.12-1

STIRLING NUMBERS OF THE SECOND KIND (S_n^k)

	k							
n	1	2	3	4	5	6	7	8
1	1							
2	1	1						
3	1	3	1					
4	1	7	6	1				
5	1	15	25	10	1			
6	1	31	90	65	15	1		
7	1	63	301	350	140	21	1	
8	1	127	966	1701	1050	266	28	1

[1] W. Wasow, Discrete approximations to the Laplace transformation, *Z. angew. Math. u. Phys.* **8**, 401-417 (1957).

[2] Ch. Jordan, "Calculus of Finite Differences," 2nd ed., p. 170. Chelsea Publ., New York, 1947.

Some of the $R_n(z)$ are as follows:

$R_o = z/(z - 1), \qquad R_1 = z/(z - 1)^2,$
$R_2 = z(z + 1)/(z - 1)^3, \qquad R_3 = z(z^2 + 4z + 1)/(z - 1)^4,$
$R_4 = z(z^3 + 11z^2 + 11z + 1)/(z - 1)^5,$
$R_5 = z(z^4 + 26z^3 + 66z^2 + 26z + 1)/(z - 1)^6.$

Some Bernoulli numbers[1] are herewith listed:

$$B_3 = B_5 = B_{2n+1} = 0 \text{ for all nonzero integral } n,$$

$B_o = 1, \quad B_1 = -0.5, \quad B_2 = 1/6, \quad B_4 = -1/30, \quad B_6 = 1/42,$
$B_8 = -1/30, \quad B_{10} = 5/66, \quad B_{12} = -691/2730, \quad B_{14} = 7/16.$

Example.

$$F(s) = \frac{1}{s(s+1)} = \frac{s^{-2}}{1 + s^{-1}} \approx \frac{TR_1(z)}{1 + TR_o(z) - T/2} = \frac{Tz(z-1)^{-2}}{1 + Tz(z-1)^{-1} - T/2}$$

(using Rule 1). We take $T = 0.5$. Therefore $F(z) = 0.5z/(1.25z^2 - 2z + 0.75)$, and division gives 0, 0.4, 0.64, 0.784, 0.866, 0.917, ... as compared to the exact 0, 0.394, 0.638, 0.777, 0.865, 0.918, ... at $t = kT$.

§ 12.13 Inverse Transformation—A Graphical Method[2]

Suppose $H(j\omega) = A(\omega)\, e^{j\theta(\omega)}$ is available graphically (amplitude and phase). If $|H(j\omega)| = A(\omega)$ only is available, and if $H(s)$ is minimum phase, then its phase may be obtained by the methods of Chapter 7. If it is not minimum phase, it is impossible to derive its phase unless additional information is supplied. In Fig. 12.13-1, imagine the input $r(t)$ is a periodic square wave of period T,

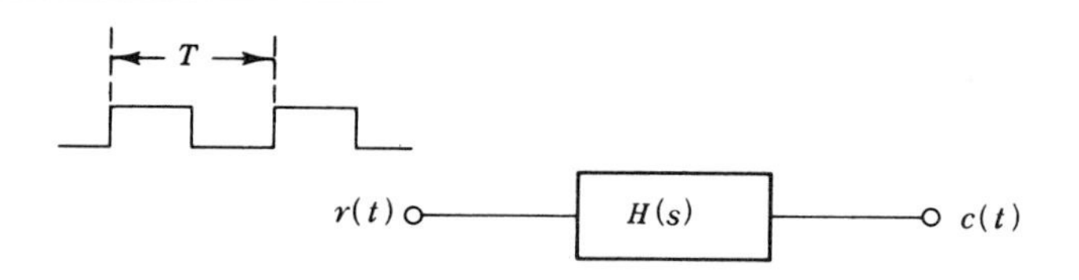

FIG. 12.13-1. Use of periodic square wave input to evaluate step response.

[1] E. Jahnke, E. Emde, and F. Losche, "Tables of Functions," 6th ed. McGraw-Hill, New York, 1960.

[2] L. V. Hamos, B. Jansson, and Th. Persson, Simplified methods for the evaluation of transients in linear systems, *Acta Polytech. NR 112, Phys. Appl. Math.* **2** (3), 5-35 (1952).

which is chosen such that $0.5T$ is much larger than the system step response settling time. Since $r(t)$ is a periodic square wave, it may be written as

$$r(t) = 0.5 + \sum_{n=1,3,5,\ldots}^{\infty} \frac{2}{n\pi} \sin n\omega_1 t,$$

with $\omega_1 = 2\pi/T$. The system steady state output consists of an infinite number of terms,

$$c(t) = 0.5H(0) + \frac{2}{\pi} \sum_{1,3,\ldots} \frac{A(n\omega_1)}{n} \sin [n\omega_1 t + \theta(n\omega_1)]. \qquad (12.13,1$$

Equation (12.13,1) may be evaluated graphically or analytically by using a finite number of terms. The resulting $c_a(t)$ is used as an approximation of the inverse transform of $H(s)/s$ [i.e., one half-period portion of $c_a(t)$].

Let us now consider the errors in this approach, in order thereby to choose an appropriate value of $\omega_1 = 2\pi/T$ and the proper number of terms required in the series of Eq. (12.13,1). There are three sources of error: (1) use of a finite number of terms in Eq. (12.13,1), (2) use of a finite T, (3) computational or graphical errors. Suppose we use up to $n = 25$, and pick ω_1 such that $A(25\omega_1) = k = 0.1$. The rms error[1] (square root of the average value of the square of the error, over the interval T) is then [assuming that for $\omega > 25\omega_1$, $A(\omega) \to K/\omega^3$ at least, with $K/(25\omega_1)^3 = k$],

$$\epsilon = T^{-0.5}[2k\pi^{-1}(25)^3] \{27^{-8} + 29^{-8} + \ldots\}^{0.5} \approx k10^{-2}(\omega_1)^{0.5}.$$

Next consider the error due to a finite T. If we use the approximate 0.1% settling time of Eq. (5.10,6b), $t_{s0.1} = (1/2f_x) [(9\omega_b/\omega_x) M - 3] \approx 10\pi/\omega_x$, and estimate that $\omega_x/\omega_{0.1} = 1/3$, then $t_{s0.1} \approx 30\pi/\omega_{0.1}$, where $\omega_{0.1} = 25\omega_1$. Therefore $t_{s0.1} \approx 30\pi/25\omega_1 \approx 0.5T$. This is very good, for it indicates that the choice of $k = 0.1$, $n = 25$ leads to 99.9% of the transient being over after $0.5T$ second. Also $\epsilon \approx 10^{-3}(\omega_1)^{0.5}$, or $\epsilon(T)^{0.5} \approx \frac{1}{4} 10^{-2}$ [we have implicitly assumed $| H(0) | = 1$]. Therefore the value of the error squared area over the interval $\frac{1}{2} T$ is $\approx$ (3) 10^{-6} second. If the error is constant, then its magnitude is $(1.7)10^{-3}$. Clearly the computational or graphical errors predominate if $n = 25$, $k = 0.1$, and $A(\omega)$ decreases at least as fast as K/ω^3 for $\omega > \omega_1$, where $A(\omega_1) = 0.1$. The analyst may then use this method with confidence.

The authors[2] have prepared a set of charts to facilitate graphical sketching of $c(t)$. With these charts, the work is greatly simplified. The charts consist of families of sinusoids, one family for each odd harmonic in Eq. (12.13,1).

---

[1] C. Lanczos, "Applied Analysis," Chapter 4, Section 3. Prentice-Hall, Englewood Cliffs, New Jersey, 1956.

[2] L. V. Hamos, B. Jansson, and Th. Persson, Simplified methods for the evaluation of transients in linear systems. *Acta Polytech. NR112, Phys. Appl. Math.* **2** (3), 5-35 (1952).

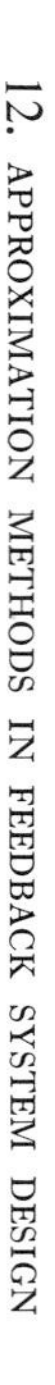

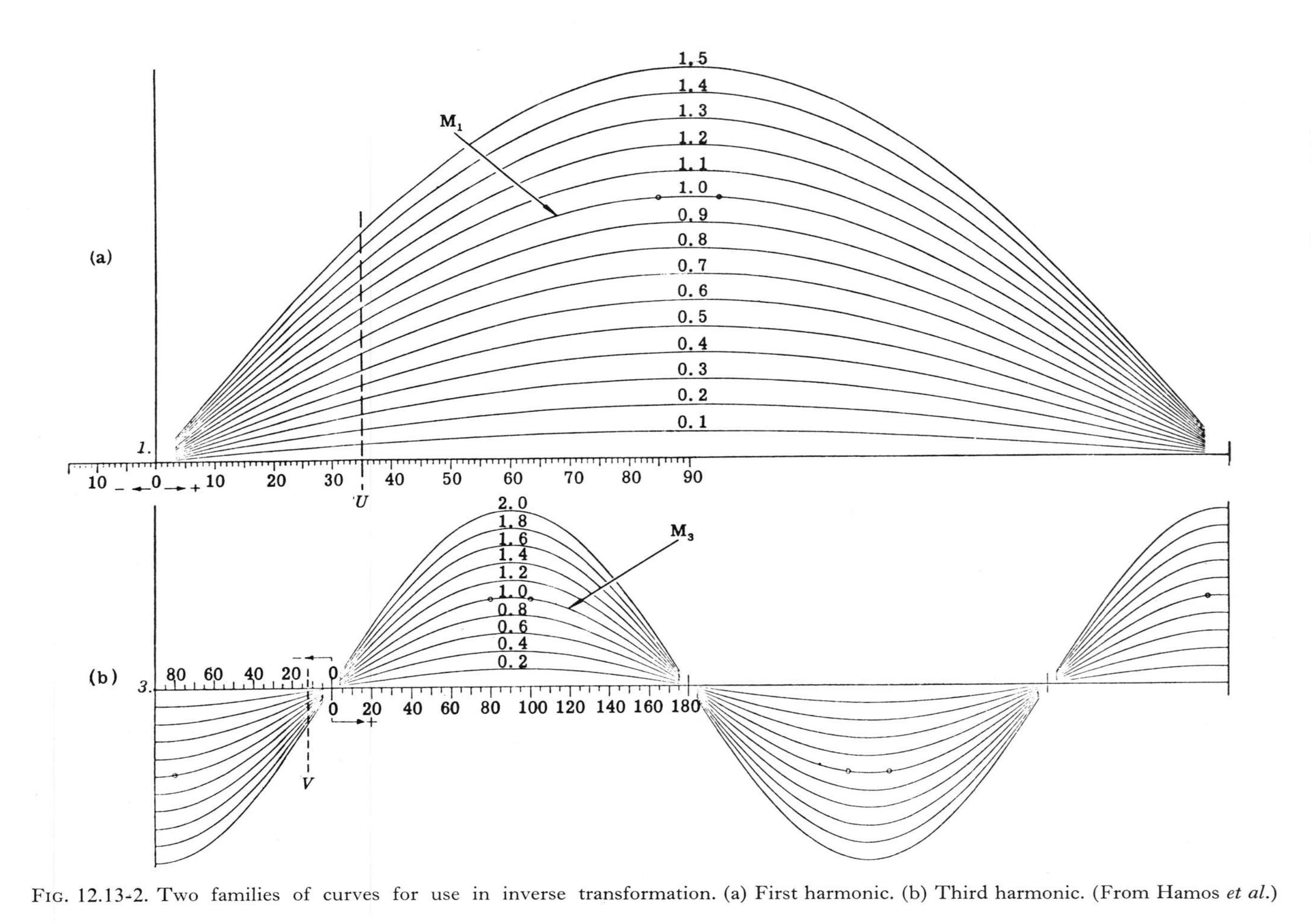

FIG. 12.13-2. Two families of curves for use in inverse transformation. (a) First harmonic. (b) Third harmonic. (From Hamos *et al.*)

Portions of two such families for the 1st and 3rd harmonics are sketched in Fig. 12.13-2. The ratio of M_3/M_1 is chosen to be 1/3 because of Eq. (12.13,1).[1] In the family of the 5th harmonic curves, $M_5/M_1 = 1/5$, etc. These ratios are precisely the ratios of the amplitudes of the harmonics of $r(t)$ of Fig. 12.13-1. Suppose $A(\omega_1) = 1.1$, $A(3\omega_1) = 1.2$, $A(5\omega_1) = 0.95, \ldots$. Then the curve labeled 1.1 is picked in Fig. 12.13-2a, the one labeled 1.2 is picked in Fig. 12.13-2b, etc. This accounts for the amplitudes of the output harmonics. There is a degree scale associated with each family of curves in order to account for the phase. A sheet of transparent paper is used. It is placed in turn, in accordance with the above, over the appropriate member of each family of harmonics, and the sinusoid is traced. As many are used as the resolution and error of this graphical method permit. For example, suppose $\theta(\omega_1) = 35°$, $A(\omega_1) = 1.1$. The origin ($t = 0$) of $c(t)$ on the transparent paper is lined up with U in Fig. 12.13-2a and the curve labeled 1.1 is traced. Suppose $\theta(3\omega_1) = -12°$, $A(3\omega_1) = 1.2$, then the origin of $c(t)$ is lined up with V in Fig. 12.13-2b and the curve labeled 1.2 is traced. The individual sinusoids are then added to give the step response.

§ 12.14 Numerical Fourier Analysis

Numerical methods of Fourier analysis are very well known, and have been described in the literature. The method described here[2] is not well known, and has the advantage that the time function data need not be multiplied by the cosine and sine functions, i.e., the formulae $2T^{-1}\int_0^T f(t) \,{\cos \atop \sin}\, 2\pi n T^{-1}\, t\, dt$ are not used. Instead, summations of the form $\Sigma \pm f(t_k)$ are used, but the k's are irregularly spaced. The general appropriate formulae can be derived in a rigorous manner. We shall show that the formulae are correct for some simple cases and state the general formulae.

Given $h(t)$, a function which exists for $-0.5T \leqslant t \leqslant 0.5T$ only, then we can write a Fourier series for $h(t)$, which is valid for this time interval. For simplicity, let $T = 1$, and then

$$h(t) = a_o + a_1 \cos 2\pi t + a_2 \cos 4\pi t + \ldots$$
$$+ b_1 \sin 2\pi t + b_2 \sin 4\pi t + \ldots = h_e(t) + h_o(t),$$

where $h_e(t) = 0.5[h(t) + h(-t)]$ is the even part of h, and $h_o(t) = 0.5[h(t) - h(-t)]$ is the odd part of h.

At $t = 1/12$,

$$h_e(1/12) = a_o + a_1(\sqrt{3}/2) + a_2(1/2) - a_4(1/2) - a_5(\sqrt{3}/2) - a_6 - a_7(\sqrt{3}/2)\ldots .$$

At $t = 5/12$,

$$h_e(5/12) = a_o - a_1(\sqrt{3}/2) + a_2(1/2) - a_4(1/2) + a_5(\sqrt{3}/2) - a_6 + a_7(\sqrt{3}/2)\ldots .$$

---

[1] M_i is the peak value of the curve labeled 1.0 in the ith family.

[2] L. Hyvarinen, Fourier analysis, a new numerical method *Valtion Tek. Tutkimuslaitos, Julkaisu No.* **40** (1958).

Hence $(1/\sqrt{3})\,[h_e(1/12) - h_e(5/12)] = a_1 - a_5 - a_7 + a_{11} + a_{13} - a_{17} - a_{19} + \ldots$. At

$$t = 1/24, \quad h_e(1/24) = a_o + a_1 \cos 15^\circ + a_2 \cos 30^\circ + a_3 \cos 45^\circ + a_4 \cos 60^\circ + \ldots,$$
$$t = 5/24, \quad h_e(5/24) = a_o + a_1 \cos 75^\circ - a_2 \cos 30^\circ - a_3 \cos 45^\circ + a_4 \cos 60^\circ + \ldots,$$
$$t = 7/24, \quad h_e(7/24) = a_o - a_1 \cos 75^\circ - a_2 \cos 30^\circ + a_3 \cos 45^\circ + a_4 \cos 60^\circ - \ldots,$$
$$t = 11/24, h_e(11/24) = a_o - a_1 \cos 15^\circ + a_2 \cos 30^\circ - a_3 \cos 45^\circ + a_4 \cos 60^\circ - \ldots.$$

Therefore $(1/2\sqrt{3})\,\{h_e(1/24) - h_e(5/24) - h_e(7/24) + h_e(11/24)\} = a_2 - a_{10} - a_{14} + a_{22} + a_{26} - \ldots$.

In general,

$$\frac{1}{n\sqrt{3}} \sum_{\nu=-\infty}^{\infty} (-1)^\nu h_e\left(\frac{1-6\nu}{12n}\right) = a_n - a_{5n} - a_{7n} + a_{11n} + a_{13n} - \ldots$$

$$\frac{1}{n\sqrt{3}} \sum_{\nu=-\infty}^{\infty} (-1)^\nu h_o\left(\frac{1-3\nu}{6n}\right) = b_n - b_{5n} + b_{7n} - b_{11n} + b_{13n} - \ldots.$$

Although the summation on the left-hand side is formally from $\nu = -\infty$ to $\nu = +\infty$, only a finite number of terms exist because $h_e(t) = 0$ for $|t| > 0.5$. If $h(-t) = 0$ then $h_e(t) = h(t)/2 = h_o(t)$, and

$$(1/2\sqrt{3})\,[h(1/12) - h(5/12)] = a_1 - a_5 - a_7 + a_{11} + a_{13} \ldots,$$
$$(1/4\sqrt{3})\,[h(1/24) - h(5/24) - h(7/24) + h(11/24)] = a_2 - a_{10} - a_{14} + a_{22} + a_{26} \ldots,$$
$$(1/6\sqrt{3})\,[h(1/36) - h(5/36) - h(7/36) + h(11/36) + h(13/36) - h(17/36)] = a_3 - a_{15} - a_{21} + a_{33} + \ldots,$$
$$(1/8\sqrt{3})\,[h(1/48) - h(5/48) - h(7/48) + h(11/48) + h(13/48) - h(17/48) - h(19/48) + h(23/48)] = a_4 - a_{20} - a_{28} + \ldots.$$

The analysis is greatly facilitated by normalizing all time functions to be analyzed, so that their period always corresponds to a fixed interval. A template can then be prepared with markings corresponding to the points $(1 - 6\nu)/12n$ for a_n and $(1 - 3\nu)/6n$ for b_n. The markings for $n = 1$ to 7 are shown in Fig. 12.14-1. The templates have been prepared on the assumption that only the first term on the right-hand side of each of the above equations need be retained. This is almost always justified for all $n \geqslant 2$. If $|a_5|$, $|a_7|$ is not each very much less than $|a_1|$ then the values of a_5, a_7 can be used in the first equation to solve for a_1.

The above analysis can be used with hardly any modification to evaluate the Fourier transform of an aperiodic time function. It is only necessary to use the relation between the transform of an aperiodic function $h(t)$, of duration T, to the Fourier series of the periodic function $h_p(t) = \sum_{\nu=-\infty}^{\infty} h(t - \nu T)$. We have $h_p(t) = \sum_{-\infty}^{\infty} c_n \exp(jn2\pi t/T)$, where $c_n = (1/T)\int_{-T/2}^{T/2} h(t) \exp(-jn2\pi t/T)$. However, $H(j\omega) = \mathscr{F}h(t) = \int_{-T/2}^{T/2} f(t)\, e^{-j\omega t}\, dt$. Therefore $c_n = (1/T)\, H(jn2\pi/T)$, and $H(j\omega)$ is easily available from $c_n = 0.5(a_n - jb_n)$ of the previous.

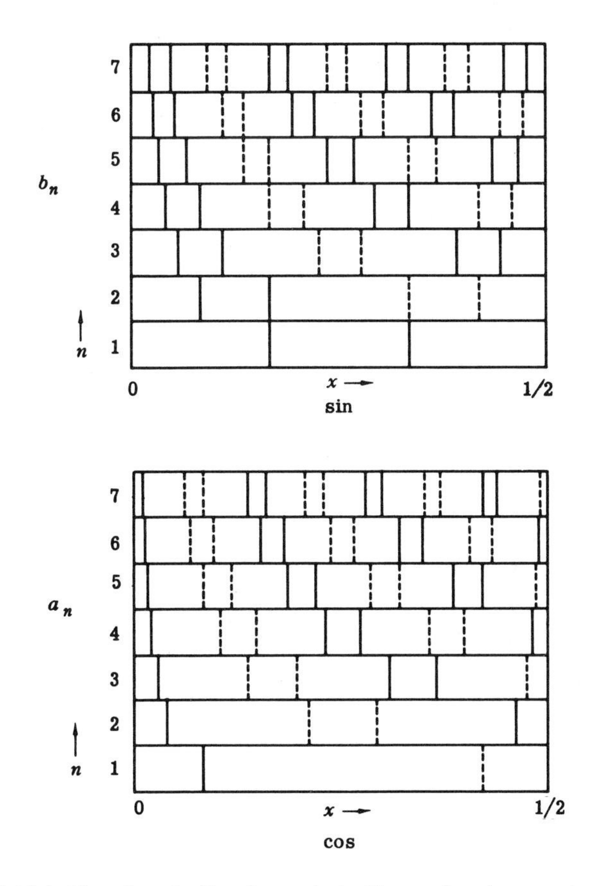

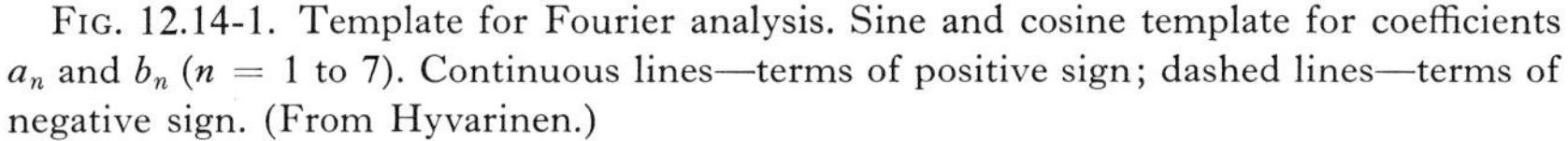
FIG. 12.14-1. Template for Fourier analysis. Sine and cosine template for coefficients a_n and b_n ($n = 1$ to 7). Continuous lines—terms of positive sign; dashed lines—terms of negative sign. (From Hyvarinen.)

Application to Inverse Transform Computation

Exactly the same procedure and the identical templates may be used to find the time function which is the inverse transform of a given Fourier transform function.

Thus, given a transform function $H = R(f) + jX(f)$ extending over the interval $-f_o/2$ to $f_o/2$, and where $R(f)$ is an even function of f, and $X(f)$ is an odd function of f, then let

$$H_p(f) = R_p(f) + jX_p(f) = \sum_{-\infty}^{\infty} H(f - \nu f_o)$$

$$= A_o + A_1 \cos 2\pi f/f_o + A_2 \cos 4\pi f/f_o + \ldots$$

$$+ j[B_1 \sin 2\pi f/f_o + B_2 \sin 4\pi f/f_o + \ldots].$$

We identify $R_p(f)$ with $h_e(t)$, and $X_p(f)$ with $h_o(t)$. We can follow through with the previous development, and obtain the equivalent relations, i.e. (normalizing with $f_o = 1$),

$$\frac{1}{n\sqrt{3}} \sum_{\nu=-\infty}^{\infty} (-1)^\nu R\left(\frac{1-6\nu}{12n}\right) = A_n - A_{5n} - A_{7n} + \ldots$$

$$\frac{1}{n\sqrt{3}} \sum_{\nu=-\infty}^{\infty} (-1)^\nu X\left(\frac{1-3\nu}{6n}\right) = B_n - B_{5n} + B_{7n} - \ldots$$

All that is left to do is to identify the A_i's and B_i's with $h(t_i) = [\mathcal{F}^{-1}H(f)]_{t_i}$. This is done as follows:

$$A_n = \frac{2}{f_o} \int_{-\infty}^{\infty} R(f) \cos \frac{2\pi nf}{f_o} df$$

$$B_n = \frac{2}{f_o} \int_{-\infty}^{\infty} X(f) \sin \frac{2\pi nf}{f_o} df.$$

Also

$$h(t) = \int_{-\infty}^{\infty} H(f) e^{j2\pi ft} df$$

$$= \int_{-\infty}^{\infty} (R + jX) (\cos 2\pi ft + j \sin 2\pi ft) df$$

$$= \int_{-\infty}^{\infty} R \cos 2\pi ft df - \int_{-\infty}^{\infty} X \sin 2\pi ft df$$

[because $h(t)$ is a real function of time]

$$= h_e(t) + h_o(t).$$

If furthermore, $h(t) = 0$ for $t < 0$, then $h_e = h(t)/2$ so $h(t) = 2 \int_{-\infty}^{\infty} R \cos 2\pi ft\, df$. Therefore $A_n = f_o^{-1} h[n/f_o]$, and, for $f_o = 1$, $A_n = h(n)$. Therefore the templates are used to find A_n (exactly as they are used to find a_n) and thereby $h(t = n)$ is obtained.

Appendix

Active *RC* Realization of Transfer Functions with Complex Poles[1]

A simple circuit for the active *RC* realization of a transfer function with a pair of complex poles is shown in Fig. A1. In this circuit $T(s) = E_o/I_i =$

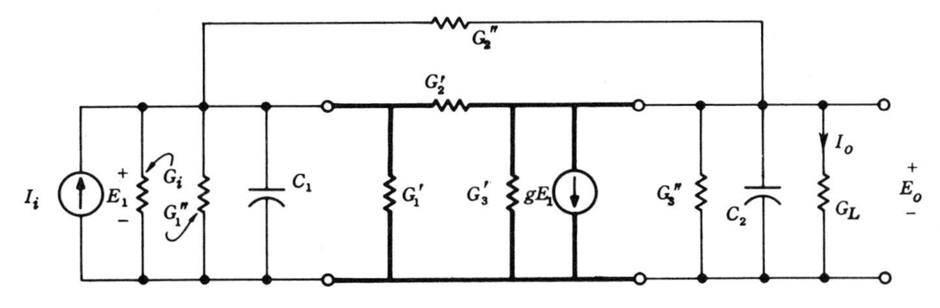

FIG. A1. Active *RC* circuit for realizing a pair of complex transfer poles (model of active element in dark lines).

$E_i = I_i G_i^{-1}$; $G_1 = G_i + G_1' + G_1''$; $G_2 = G_2' + G_2''$; $G_3 = G_3' + G_3'' + G_L$

$- g(C_1C_2)^{-1}/(s^2 + 2\zeta\omega_n s + \omega_n^2)$, and ζ, ω_n are functions of the element values. The heavy line portion of Fig. A1 represents the model of any active element in the low frequency range. This range is satisfactory for most feedback control applications. G_i, G_L represent the source and load resistances, respectively. Of course, all or part of C_1, C_2 can be similarly absorbed. G_1'', G_2'', G_3'' represent any additional conductances which may perhaps be added for purpose of sensitivity reduction. It is assumed that G_i, G_L, ζ, ω_n are *a priori* given. The problem is to pick the remaining element values to (1) secure the desired ζ, ω_n; (2) have ζ, ω_n insensitive to a specified extent to variations of the four active parameters G_1', G_2', G_3', g; (3) use whatever freedom there remains to secure a design which is optimum in some sense.

[1] I. M. Horowitz, Optimum design of single-stage gryrator-*RC* filters with prescribed sensitivity. *Trans. IRE* **CT-8** (2), 88-92 (1961).

In this design procedure, we consider optimum that design which realizes the first two specifications with a "least active" active element. The heart of the active element is the controlled source with its transconductance g; G_1', G_2', G_3' are parasitics. In the ideal active element, $G_1' = G_2' = G_3' = 0$, and the ratios $\nu_1' = g/G_1'$, $\nu_2' = g/G_2'$, $\nu_3' = g/G_3'$ are infinite. The smaller these ratios, the poorer (less active) is the active element. The ratio ν_1' is the short-circuit current gain if $G_2' = 0$; ν_2' is the open-circuit voltage gain if $G_2' = 0$, and ν_3' is a measure of the ratio of forward to reverse transmission. In the practical circuit of Fig. A1, we let $\nu_1 = g/G_1$, $\nu_2 = g/G_2$, $\nu_3 = g/G_3$ with $\nu_i < \nu_i'$. The optimum design is therefore that which realizes the specifications with the smallest possible ν's. Some design parameters tend to lower the values of some of the ν's, but increase the values of the others. However, it is shown in the reference that one relation which lowers the values of all the ν's is

$$\zeta^{-2} = \nu_2\nu_3(\nu_2 + \nu_3)^{-1}[\nu_3^{-1} + (1 + \nu_1)(\nu_1 + \nu_2)^{-1}]. \tag{A1}$$

In a typical design, $\nu_2 \gg \nu_1$, $\nu_1 \gg 1$ so that Eq. (A1) is approximately $\nu_2 \doteq [\zeta^2\nu_1\nu_3 - (1 - \zeta^2)(\nu_1 + \nu_3)]/(1 - \zeta^2)$. It is also shown that

$$\begin{aligned} g &= \zeta\omega_n\nu_2\nu_3C_2/(\nu_2 + \nu_3), \\ C_1 &= \zeta^2\nu_2\nu_3^2(\nu_2 - 1)\, C_2/\nu_1(\nu_2 + \nu_3)\,[\zeta^2\nu_2(1 + \nu_3) - (\nu_2 + \nu_3)]. \end{aligned} \tag{A2}$$

Sensitivity to Parameter Variations

The circuit sensitivity to parameter variations is an important factor in the choice of the ν's. It is convenient to use the pole sensitivity function defined in Section 4.3, i.e., $S_k^{s_i} = ds_i/(dk/k)$, where s_i is the system pole at $-\zeta\omega_n + j\omega_n(1 - \zeta^2)^{0.5}$, and ds_i is its shift due to the change dk in the parameter k. The rule given in Section 4.3 is used to find the effect of parameter changes, and the results are given in Table A1. (See the reference for the details.)

The required sensitivity reduction can be achieved in two ways. One way is to use the shunting effect of G_1'', G_2'', G_3''. For example, $G_1 = G_1' + G_1'' + G_i$; hence $S_{G_1'}^{G_1} = G_1'/G_1 = \nu_1/\nu_1'$. If $G_i + G_1'' = 4G_1'$, then $S_{G_1'}^{G_1} = 0.2$, and the system sensitivity to variations in G_1' is reduced by a factor of 5.

Another way to reduce the sensitivity to the active parameters is by adjusting the impedance level of the active structure. For example, in the common emitter transistor connection $G_1' = (1 - \alpha)/r_e'$, $G_2' = r_e/r_e'r_c$, $G_3' = r_b/r_e'r_c$, $g = \alpha/r_e'$, $r_e' = r_e + r_b(1 - \alpha)$. However, if an external resistor R is added in series with the emitter, then we may incorporate R into the transistor, providing $G_1' = (1 - \alpha)/R_e'$, $G_2' = R_e/R_e'r_c$, $G_3' = r_b/R_e'r_c$, $g = \alpha/R_e'$, $R_e = R + r_e$, $R_e' = R_e + r_b(1 - \alpha)$. The system sensitivity is reduced in two ways. One is due to the fact that the variation in each of g, G_1', G_2', G_3' is less than before. The drift in $g = \alpha/(r_e' + R_e)$, due to the drift in r_e', is reduced by the factor

TABLE A1

k = element	Real part of $S_k^{s_i}$	Imaginary part of $S_k^{s_i}$	Zero frequency Gain sensitivity
G_3	$0.5\lambda_3\zeta\omega_n$	Zero	Zero
C_2	$0.5\zeta\omega_n$	$-0.5\omega_n(1-\zeta^2)^{0.5}$	Zero
g	Zero	$0.5\omega_n[1-\nu_2^{-1}(1-\zeta^2)^{-0.5}]$	$\nu_2/(\nu_2-1)$
R_2	$0.5\zeta\omega_n\left[1-\lambda_3+\frac{1-\lambda_3\zeta^2}{\lambda-\zeta^2(1-\lambda_3)}\right]$	$-0.5\omega_n(1-\zeta^2)^{0.5}$	$-1/(\nu_2-1)$
C_1	$0.5\zeta\omega_n$	$-0.5\omega_n(1-\zeta^2)^{0.5}$	Zero
G_1	$-0.5\zeta\omega_n(\lambda-1+\zeta^2\lambda_3)/\lambda$	Zero	Zero

$\lambda_3 \triangleq \nu_2/(\nu_2+\nu_3)$, $\lambda \triangleq \zeta^2\nu_2\nu_3/(\nu_2+\nu_3)$.

$S_k^{s_i} = ds_i/(dk/k)$ = sensitivity of pole at $s_i = -\zeta\omega_n + j\omega_n(1-\zeta^2)^{0.5}$ to variations in element values.

R_e'/r_e'. S_α^g is unaffected, but fortunately the drift in α is negligible. Similarly, R reduces the drift in $G_1' = (1-\alpha)/(r_e'+R)$, due to the drift in r_e'. We shall, however, ignore this reduction in the drift of G_1', G_2', G_3', because of the difficulty of taking it into account in the design. This leads to an extra safety factor in the final design. The second way in which a finite R decreases the sensitivity is because g, G_1', G_3' are thereby decreased by the factor $m = R_e'/r_e'$. The source and load conductances G_i, G_L are, however, fixed. Therefore ν_1', ν_3' do not change, but ν_1, ν_3 are decreased. Hence $S_{G_1'}^{G_1} = G_1'/G_1 = \nu_1/\nu_1'$ is decreased, as is $S_{G_3'}^{G_3} = \nu_3/\nu_3'$. However, the effect on $S_{G_2'}^{G_2}$ is different, because $G_2' = R_e/r_cR_e'$; for example, if $r_e' = 40\Omega$, $r_e = 30\Omega$, $r_c = 10^6\Omega$, then for $R = 0$, $G_2' = (0.75)\,10^{-6}$, and even if R is very large, $G_2' = 10^{-6}$. Hence $\nu_2' = g/G_2'$ decreases as R is increased, in contrast to ν_1', ν_3' whose values are not affected by R.

To what extent should each of these two methods be used for sensitivity reduction? It is reasonable to make the choice so as to maximize the gain, which at $s = j\omega_n$ is

$$\left|\frac{E_o}{E_i}\right| = \frac{\nu_2^2(1-\zeta^2)(\nu_1^{-1}-\nu_1'^{-1})}{(\nu_2-1)2\zeta} \tag{A3}$$

where $E_i = I_iG_i^{-1}$.

It is difficult to indicate in a general manner how the sensitivity reduction burden is divided between the two methods. The procedure will therefore be explained by means of a detailed design example.

Example.

Specifications: $\zeta = 0.707$, $\omega_n = 100$ rps, $R_i = 5000\Omega$, $R_L = 30{,}000\Omega$. The drift in the real and imaginary pole coordinates is to be less than 4 rps. Design is attempted by means of a single transistor in the common emitter (CE) connection with the typical values: $\alpha = 0.98$, $r_e = 30\Omega$, $r_b = 500\Omega$, $r_c = 10^6\Omega$, so that $r_e' = 40$,

$$G_1' = (1 - \alpha)/r_e' = (5)\,10^{-4}, \qquad g' = \alpha/r_e' = (24.5)\,10^{-3},$$
$$G_2' = r_e/r_c r_e' = (0.75)\,10^{-6}, \qquad G_3' = r_b/r_c r_e' = (12.5)\,10^{-6},$$
$$\nu_1' = g'/G_1' = 49, \qquad \nu_2' = g'/G_2' = 33{,}000, \qquad \nu_3' = g'/G_3' = 2000.$$

Suppose that due to environmental conditions the maximum changes in g', G_1', G_2', G_3' are each 40%, It is assumed in what follows that these drifts are such that their effects all add. This is the worst possibility, because in practice they may partially cancel.

Design procedure: The design is initiated by using Table A1 to find the maximum drift in the transfer poles, due to the variation in g'. We do this because in this procedure $S_{g'}^{g}$ can be reduced only by taking $m \triangleq R_e'/r_e'$ sufficiently large. If we assume $\nu_2 \gg 1$, then $|S_{g'}^{s_i}| \doteq 50$. Since g varies at most by 40%, it alone leads to $\mathscr{I}m\,\Delta s_i$ equal approximately to (0.40) (50) = 20. The specifications require $\Delta s_i < 4$, so a minimum value of $m = R_e'/r_e' = 5$ is required in order that $S_{g'}^{g} = 0.25$. Actually, the drift due to G_2' may add to that due to g' (because the parameter variations are assumed to be independent), so we try $m = 7$, i.e., $R + r_e' = 7r_e'$, $R = 240\Omega$, $R_e = 270\Omega$. This leads to G_1', G_3' decreasing by a factor of 7, and $G_2' = R_e/r_c R_e' = (0.97)\,10^{-6}$. Also, $\nu_1 = \nu_1'/[1 + (G_i/G_1')] = 12.9$; similarly, $\nu_3 = 100$, and from Eq. (A1), $\nu_2 = 1170$. Table A1 is then used to find the maximum drifts in the complex poles (taking the shunting effects of G_i, G_L, G_2'' into account). The results are listed in Table A2. For example, $S_{G_3}^{s_i} = (0.5)\,(0.92)\,70.7 = 32.6$ ($\lambda_3 = 0.92$, $\zeta\omega_n = 70.7$). This is multiplied by 0.4 (because of 40% maximum drift in G_3'), and is further reduced by the factor $S_{G_3'}^{G_3} = \nu_3/\nu_3' = 0.05$, leading to a net $\Delta s_i = 0.65$. For G_2', we find the new $G_2' = R_e'/r_c R_e$ and the new $\nu_2' = g'_{\text{old}}/7G_2' = 4080$. Hence $S_{G_2'}^{O_2} = \nu_2/\nu_2' = 1170/4080 = 0.29$.

TABLE A2

	$\mathscr{R}e\,\Delta s_i$	$\mathscr{I}m\,\Delta s_i$
g	0	2.8
G_3	0.65	0
G_2	0.43	4.0
G_1	3.7	0

The design is unsatisfactory, because $\mathscr{I}m\, \Delta s_i > 4$. It is seen from Table A2 that the effect of G_2' must be decreased. In order to reduce the effect of G_2', we demand that $\nu_2'/\nu_2 = 12$, so as to decrease the G_2' component of $\mathscr{I}m\, \Delta s_i$ from (0.4) (35.3) = 14.1 at $\nu_2 = \nu_2'$ to 1/12 of the latter at $\nu_2' = 12\nu_2$. At $m = 7$, $R_e' = 280$, $R_e = 270$, $\nu_2' = g/G_2' = \alpha r_c/R_e = 3630$, and therefore $\nu_2 = 3630/12 = 303$. From the approximation to Eq. (A1), $\nu_1\nu_3 - (\nu_1 + \nu_3) \doteq 303$. How shall we shoose ν_1 and ν_3? If we take G_1''' of Fig. A1 to be zero, then $\nu_1 = \nu_1'/[1 + mG_i/G_1'] = 12.9$, which forces ν_3 to be 26.6. This requires nonzero G_3''' in Fig. A1, i.e., $\nu_3 = \nu_3'[1 + m(G_L + G_3'')/G_3'] = 2000$ $[1 + m(G_L + G_3'')/G_3']^{-1} = 26.6$, so $G_L + G_3'' = 74G_3'/m$, with $G_3'/m = (12.5)\,10^{-6}/7$, $G_L = 1/30{,}000$. From Eq. (A3), the voltage gain is 6.1. This modification of the output load is best if the voltage gain is of primary interest. If current gain is of primary interest, then it is better to have $G_3''' = 0$, and modify the source resistance. The equation for current gain (at $s = j\omega_n$) is $|I_o/I_i| = \nu_2^2(1 - \zeta^2)(\nu_3^{-1} - \nu_3'^{-1})/2\zeta(\nu_2 - 1)$. The contributions to Δs_i are listed in Table A3.

TABLE A3

Element	$\mathscr{R}e\, \Delta s_i$	$\mathscr{I}m\, \Delta s_i$
g	0	2.8
G_3	0.17	0
G_2	0.14	1.2
G_1	3.5	0
	3.8	4.0

We next experiment with other values of m to see if more gain may be attained. For example, $m = 10$ leads to 2.0 for the g contribution to $\mathscr{I}m\, \Delta s_i$. Therefore $\nu_2'/\nu_2 = 7$ is required in order that the G_2 contribution to $\mathscr{I}m\, \Delta s_i$ is only 2.0. This leads to $R_e' = 400$, $\nu_2' = 2510$, $\nu_2 = 359$. If $G_1''' = 0$, then $\nu_1 = 9.8$, $\nu_3 = 41.9$, and the voltage gain is 10.3. Similarly, $m = 12$ leads to a gain of 12, and $m = 15$ apparently leads to a gain of 13.5. However, when $m = 15$, the required ν_3 is 55, which requires the load resistance to be increased, because $\nu_3 = 2000[1 + (15)\,(80{,}000)/(30{,}000)]^{-1} = 48.7$, when $G_1'' = 0$. Finite G_1'' decreases ν_3, so it is necessary to add 3600Ω in series with the load resistance. Since $R_L = 30{,}000$, this means that only 30/33.6 of the gain is across the load; so at $m = 15$, the net gain is only (13.5) (30)/33.6 = 12. We take $m = 12$, which requires $\nu_2'/\nu_2 = 6$, and since $\nu_2' = (0.98)\,10^6/470 = 2080$, $\nu_2 = \nu_2'/6 = \nu 347$. Also, $\nu_1 = 8.5$, and we solve Eq. (A1) and find $\nu_3 = 47.2$. $G_3''' + G_L = [(2000/47.2) - 1]\,(12.5)\,10^{-6}/12 = (43)\,10^{-6}$, so $R_3'' = 103\text{K}\Omega$. $R_2' = r_c R_e'/R_e = (1.02)\,10^6$, so $R_2'' = R_2'/(6-1) = 204\text{K}\Omega$. The value of C_2 is obtained from Eq. (A2), since it is known that $g = (24.5)10^{-3}/12$; the result is

$C_2 = 0.695\mu f$. From the second equation in (A2), $C_1 = 3.48\mu f$. The final circuit is shown in Fig. A2.

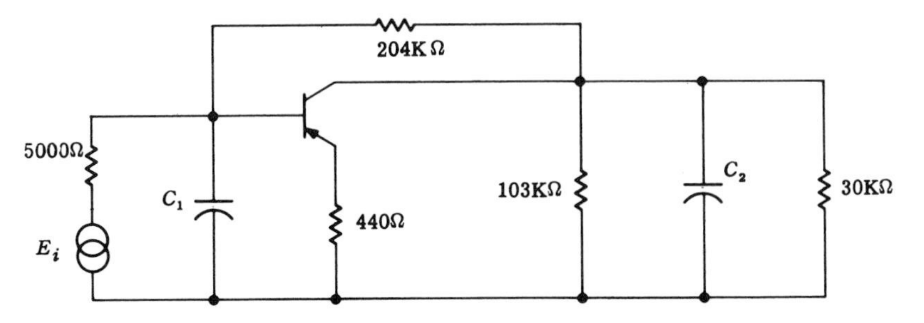

FIG. A2. Design example.

Use of Incremental Sensitivity Function for Large Parameter Variations

$S_k^{s_o} \triangleq ds_o/(dk/k)$ is defined for small parameter variations. For large variations it is really necessary to write $\Delta s_o = \int_{k_1}^{k_2} Sdk/k$, where we let S represent $S_k^{s_o}$. It is recalled that S is the residue in the pole in question of a ratio of polynomials. In the above design problem the denominator polynomial is always $s^2 + 2\zeta\omega_n s + \omega_n^2$, whose coefficients are not being allowed to vary much. Although the coefficients of the numerator polynomials (different for different network elements) may vary to a greater extent, they too do not vary very much, because they are functions of ζ, ω_n and the ν_i's. If we may assume that S does not vary much, then $\Delta s_o \approx S \int_{k_1}^{k_2} dk/k = S \ln k_2/k_1 = S \ln (1 + \beta)$, if $k_2/k_1 = 1 + \beta$, $\beta > 0$. We have been using the more approximate relation $\Delta s_o \approx S\Delta k/k_1 = \beta S$ which is of course larger than the more accurate $S(1 + \beta)$. Hence we have erred on the safe side by using the incremental relation for large element variations.

Index